D0805501

MACHINE TOOL PRACTICES

MACHINE TOOL PRACTICES

FOURTH EDITION

Richard R. Kibbe

Oxnard Community College
Oxnard, California

John E. Neely

Lane Community College (ret.)
Eugene, Oregon

Roland O. Meyer

Lane Community College
Eugene, Oregon

Warren T. White

San Jose State University
San Jose, California

REGENTS/PRENTICE HALL
Englewood Cliffs, New Jersey 07632

Library of Congress Cataloging-in-Publication Data

Machine tool practices / Richard R. Kibbe . . . [et. al.].—4th ed.
p. cm.
Includes index.
ISBN 0-13-541848-8
1. Machine-tools. 2. Machine-shop practice. I. Kibbe, Richard R.
TJ1185.M224 1991
621.9'02—dc20 90-44140
CIP

Editorial/production and supervision: **Marcia Krefetz**
Cover design: **Linda Rosa**
Manufacturing buyers: **Mary McCartney and Ed O'Dougherty**

Printed in the United States of America
10 9 8 7 6 5 4 3

ISBN 0-13-541848-8

Prentice-Hall International (UK) Limited, *London*
Prentice-Hall of Australia Pty. Limited, *Sydney*
Prentice-Hall Canada Inc., *Toronto*
Prentice-Hall Hispanoamericana, S.A., *Mexico*
Prentice-Hall of India Private Limited, *New Delhi*
Prentice-Hall of Japan, Inc., *Tokyo*
Simon & Schuster Asia Pte. Ltd., *Singapore*
Editora Prentice-Hall do Brasil, Ltda., *Rio de Janeior*

CONTENTS

PREFACE

The fourth edition of this text is designed for students training to become machinists, either through apprenticeship training, vocational schools, or community college programs. The content deals with topics usually presented in a combined lecture and laboratory program.

To better meet the needs of users of this book, the authors made a careful study of the entire contents. Many users of previous editions were consulted and their comments solicited so that the fourth edition could be updated to meet present-day needs of students, instructors, and current industry training standards. The following are some of the special features included in this textbook:

- Each section begins with an introductory overview, followed by instructional units with clearly stated objectives. Instructional units in each section contain easy-to-read information and instructions that accurately reflect the state-of-the art in industrial settings.
- The book is extensively illustrated with photographs of actual machining operations. Graphic explanations are used to highlight important concepts and common errors and difficulties encountered by machinist.
- Many units are designed around specific projects that provide much of the performance experience for the student. The structure of the book allows instructors to insert easily projects that are more applicable to their specific individual programs.
- Self-tests at the end of most units enable students to evaluate their own progress and understanding of the text material. Self-test answers are given in Appendix 1.

Additions to the fourth edition include:

- Updated information and illustrations for cutting tools and materials
- New and updated illustrations throughout the book
- New computer numerical control material and nontraditional machining information added to Section O

Although we have updated the book to reflect current machining technology, we have preserved the classical shop practice while deleting that which is truly not used or no longer relevant to this subject area. We feel that the standard machine shop practice is still very much relevant to machining technology even in the high-technology computer age. Students of modern industrial and manufacturing technology will still require solid backgrounds in standard shop practice if they are to thoroughly understand and appreciate computer-controlled machining as well as other high-technology manufacturing processes.

Instructors will find the following materials available to supplement the textbook:

- An *Instructor's Manual* containing suggestions on how to use the textbook for conventional and competency-based education, post-tests, and answer keys. The post-tests can be freely reproduced by the users of this book.
- A workbook titled *Student Workbook for Machine Tool Practices*. We feel that this adjunct

publication plays an extremely important part in maximizing the use of the book. The workbook contains process worksheets with projects, alternative projects, and additional tables. These features are keyed to the text material and thus greatly enhance the use of the book as a complete instructional system. Use of the student workbook is highly recommended.

Richard R. Kibbe
John E. Neely
Roland O. Meyer
Warren T. White

ACKNOWLEDGMENTS

The authors all wish to express their sincere thanks and appreciation to Doris Neely for her considerable efforts in organizing and detailing this edition of *Machine Tool Practices*, and for the difficult task of coordinating the work of the four authors, who live in different locations. It is only because of her untiring devotion to the preparation of the manuscript that it was completed in time for publication.

The authors wish to thank the following for their contributions to this book.

American Society for Metals, Metals Park, Ohio

American Society of Mechanical Engineers, New York, New York

California State University at Fresno, Department of Industrial Arts and Technology, Fresno, California

DeAnza College, Engineering and Technology, Cupertino, California

Epsilon Pi Tau Fraternity, Alpha Lambda Chapter, California State University at Fresno, Fresno, California

Foothill College district, Office of Technical Education, Individualized Machinist's Curriculum Project, Los Altos Hills, California

Lane Community College, Mechanics Department, Eugene, Oregon

National Machine Tool Builders Association, McLean, Virginia

National Screw Machine Products Association, Cleveland, Ohio

North County Technical School, St. Louis County, Missouri

Yuba College, Applied Arts Department, Marysville, California

Our special thanks go to the following firms and their employees with whom we corrsponded and who supplied us with invaluable technical information and illustrations.

Accurate Diamond Tool Corporation, Hackensack, New Jersey

Aloris Tool Company, Inc., Clifton, New Jersey

American Chain & Cable Company, Inc., Wilson Instrument Division, Bridgeport, Connecticut

American Iron & Steel Institute, Washington, D.C.

American Reishauer Corporation, Elgin, Illinois

American SIP Corporation, Elmsford, New York

Ameropean Industries, Inc., Hamden, Connecticut

Ames Research Center, National Aeronautics and Space Administration (NASA), Mountain View, California

Barber-Colman Company Rockford, Illinois

Barnes Drill Company, Rockford, Illinois

Barret Centrifugals, Worcester, Massachusetts

Bay State Abrasives, Division of Dresser Industries, Westborough, Massachusetts

Bendix Corporation, South Bend, Indiana

Bethlehem Steel Corporation, Bethlehem, Pennsylvania

Boyer-Schultz Corporation, Broadview, Illinois

Bridgeport Milling Machine Division, Textron, Inc., Bridgeport, Connecticut

Brown & Sharpe Manufacturing Company, North Kingstown, Rhode Island

Bryant Grinder Corporation, Springfield, Vermont

Buck Tool, Kalamazoo, Michigan

The Carborundum Company, Niagara Falls, New York

Cincinnati Incorporated, Cincinnati, Ohio

Cincinnati Milacron, Inc., Cincinnati, Ohio

Clausing Corp., Kalamazoo, Michigan

Cleveland Twist Drill Company, Cleveland Ohio

Cone-Blanchard Machine Company, Windsor, Vermont

Dake Corporation, Grand Haven, Michigan

Desmond-Stephan Manufacturing Companby, Urbana, Ohio

Diamond Abrasive Corporation, New York, New York

DoALL Company, Des Plains, Illinois

Dover Publications, Inc., New York, New York

The du Mont Corporation, Greenfield, Massachusetts

El-Jay Inc., Eugene, Oregon

Elm Systems, Inc., Arlington Heights, Illinois

Enco Manufacturing Company, Chicago, Illinois

Engis Corporation, Morton Grove, Illinois

Erix Tool AB, Sweden

Ex-Cell-O Corporation, Troy, Michigan

Exolon Company, Tonawanda, New York

Federal Products Corporation, Providence, Rhode Island

Fellows Corporation, Springfield, Vermont

Floturn, Inc., Division of Lodge & Shipley Company, Cincinnati, Ohio

Fred V. Fowler Co., Inc., Newton, Massachusetts

Gaertner Scientific Corporation, Chicago, Illinois

General Electric Company, Detroit, Michigan, and Specialty Materials Department, Worthington, Ohio

Geometric Tool, New Haven, Connecticut

Giddings & Lewis, Fond Du Lac, Wisconsin

Gleason Works, Rochester, New York

Great Lakes Screw, Chicago, Illinois

Hammond Machinery Builders, Kalamazoo, Michigan

Hardinge Brothers, Inc., Elmira, New York

Harig Products, Inc., Elgin, Illinois

Heald Machine Division, Cincinnati Milacron Company, Worcester, Massachusetts

Hewlett-Packard Company, Palo Alto and Santa Clara, California

Hitachi Magna-Lock Corporation, Big Rapids, Michigan

Illinois/Eclipse, Division of Illinois Tool Works, Inc., Chicago, Illinois

Industrial Information Controls, Inc., Eden Prairie, Minnesota

Industrial Plastics Products, Inc., Forest Grove, Oregon

Industrial Press, New York, New York

Ingersoll Milling Machine Company, Rockford, Illinois

Japax Inc., Kawasaki City, Japan

Jarvis Products Corporation, Middletown, Connecticut

K & M Tool, Inc., Eugene, Oregon

Kastro-Racine, Inc., Monroeville, Pennsylvania

Kearney & Trecker Corporation, Milwaukee, Wisconsin

Kennametal, Inc., Latrobe, Pennsylvania

K. O. Lee Company, Aberdeen, South Dakota

Landis Tool Company, Division of Litton Industries, Waynesboro, Pennsylvania

Lapmaster Division, Crane Packing Company, Morton Grove, Illinois

LeBlond, Inc., Cincinnati, Ohio

Louis Levin & Son, Inc., Culver City, California

Lodge & Shipley Company, Cincinnati, Ohio

M & M Tool Manufacturing Company, Dayton, Ohio

Madison Industries, Division of Amtel, Inc., Providence, Rhode Island

Mahr Gage Company, New York, New York

Mattison Machine Works, Rockford, Illinois

Mazak Corporation, Florence, Kentucky

Megadiamond Industries, New York, New York

Minnesota Mining and Manufacturing Company (3M), St. Paul, Minnesota

The MIT Press, Cambridge, Massachusetts

Monarch Machine Tool Company, Sidney, Ohio

Moog, Inc., Hydra Point Division, Buffalo, New York

Moore Special Tool Company, Bridgeport, Connecticut

MTI Corporation, New York, New York

National Broach & Machine Division, Lear Siegler, Inc., Detroit, Michigan

National Twist Drill & Tool Division, Lear Siegler, Inc., Rochester, Michigan

Newcomer Products, Inc., Latrobe, Pennsylvania

Norton Company, Worcester, Massachusetts

PMC Industries, Wickliffe, Ohio

Pratt & Whitney Machine Tool Division of The Colt Industries Operating Corporation, West Hartford, Connecticut

Precision Diamond Tool Company, Elgin, Illinois

Ralmike's Tool-a-Rama, South Plainfield, New Jersey

Rank Scherr-Tumico Inc., Des Plaines, Illinois

Rockford Machine Tool Company, Rockford, Illinois

Sipco Machine Company, Marion, Massachusetts

Harry M. Smith & Associates, Santa Clara, California

Snap-On Tools Corporation, Kenosha, Wisconsin

Southwestern Industries, Inc., Los Angeles, California

Speedfam Corporation, Des Plaines, Illinois

Standard Gage Company, Poughkeepsie, New York

L.S. Starrett Company, Athol, Massachusetts

Sunnen Products Company, St. Louis, Missouri

Superior Electric Company, Bristol, Connecticut

Surface Finishes, Inc., Addison, Illinois

Syclone Products, Inc., Ranchita, California

Taft-Pierce Manufacturing Company, Woonsocket, Rhode Island

Taper Micrometer Corporation, Worcester, Massachusetts

Tinius Olsen Testing Machine Company, Inc., Willow Grove, Pennsylvania

TRW, Inc., Greenfield, Massachusetts; Plymouth, Michigan; Buffalo, New York

Ultramatic Equipment Company, Addison, Illinois

Unison Corporation, Madison Heights, Michigan

Waldes Kohinoor, Inc., Long Island City, New York

Walton Company, Hartford, Connecticut

Warner & Swasey Company, Cleveland, Ohio, and King of Prussia, Pennsylvania

Warren Tool Corporation, Hiram, Ohio

WCI Machine Tools & Systems Company, Cincinnati, Ohio

Weldon Tool Company, Cleveland, Ohio

Wells Manufacturing Corporation, Three Rivers, Michigan

Whitnon Spindle Division, Mite Corporation, Farmington, Connecticut

Wilton Tool Division, Wilton Corporation, Des Plaines, Illinois

R. R. K.
J. E. N.
R. O. M.
W. T. W.

SECTION **A**

INTRODUCTION

Of all the manufacturing processes that can be applied to the shaping and forming of raw materials into useful products, machining processes will always remain among the most important. The fundamental cutting processes in machining, those of bringing the work into contact with a cutting tool, are still very much in evidence and will always remain mainstays of the industry. What has changed is the way in which these processes are applied, the cutting tool materials, material used for products, and the methods of material removal. These new methods are quite different from the classical chip-producing processes. They include the use of lasers, electrical energy, electrochemical processes, ultrasound, high-pressure water jets, and high-temperature plasma arcs as material removal tools.

Computers have enhanced automated manufacturing; machining is no exception. Numerical control (NC) of machine tools has been available for many years. Today, typical machine tools of all types are equipped with their own computer numerical controls (CNCs) and almost every machining process can now be efficiently automated with an exceptional degree of accuracy, reliability, and repeatability. In fact, machining processes have become so sophisticated and reliable that the human operator of a machine tool may now be replaced with a robot (Figure A-1). This application of computer-driven automated equipment will profoundly affect the employment of machinists and machine operators in future years.

The computer has found its way into almost every other phase of manufacturing as well. One

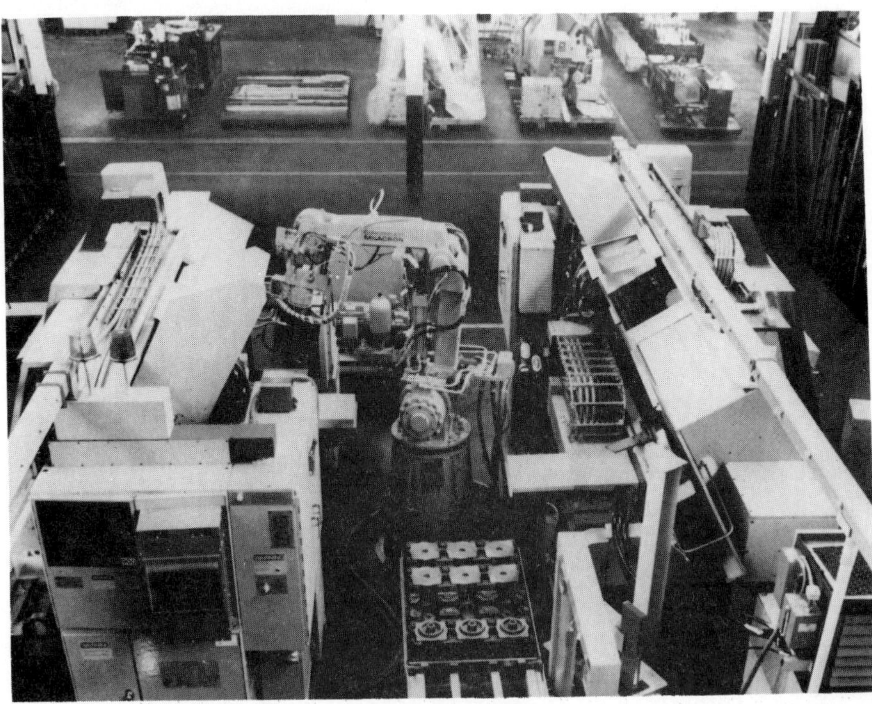

FIGURE A-1 The machine operator of tomorrow's automated manufacturing industry. A computer-controlled industrial robot loads and unloads parts on a CNC turning center (Cincinnati Milacron).

FIGURE A-2 Computer-aided design (CAD) is fast replacing the drawing board in engineering design applications (Cincinnati Milacron).

FIGURE A-3 A specialty CNC machining center forms a work cell in a flexible manufacturing system (FMS). Robot transporters move parts from cell to cell all under computer control (Kearney and Trecker Corporation).

FIGURE A-4 Parts to be machined are mounted on pallets so that they may be moved from machining center to machining center in different work cells in the flexible manufacturing system (Kearney and Trecker Corporation).

important area is computer-aided design (CAD) (Figure A-2). The age of drawing board design is fast drawing to a close. Design is now done on computer terminals and manufacturing equipment control programs are generated directly in computer-aided manufacturing (CAM), or fully computer-aided design and manufacturing (CAD/CAM), or computer-integrated manufacturing (CIM).

The ultimate outcome of manufacturing automation will be an entirely automated factory in which material will be automatically transported between manufacturing cells that will perform similar manufacturing or machining operations. Such a system, called a flexible manufacturing system (FMS), will be fully computer controlled and already has become an industry standard (Figures A-3 and A-4).

■ ■ ■

Career Opportunities in Machining and Related Areas

The influence of high technology in machining manufacturing has had, and will continue to have a significant effect on the types and numbers of jobs available in this field. Many exciting career opportunities are available for those willing to prepare themselves. However, like all technology, machining manufacturing has become a very specialized business and this trend is likely to continue.

TRADE LEVEL OPPORTUNITIES

MACHINE OPERATOR The machine operator will be widely employed for some time to come. The machine operator's responsibilities will be to operate computer-driven (CNC) machine tools such as turning or machining centers. The operator will observe machine functions and tool performance, change and inspect parts, and perhaps have some limited duties in setting up and adjusting machine programming.

Preparation for machine operator will consist of familiarization with conventional machining processes, tooling selection and application, machine control unit operations, reading drawings, related math, limited machine setup, and quality control inspection measurement functions. The machine operators can receive training through trade schools, community college programs, or industrial training programs, or they may learn on the job. The CNC operator will generally work in a fast-paced production environment in companies ranging in size from very small to very large.

SETUP The setup person is responsible for setting up the machine tool and assisting the operator in establishing a first article acceptable part at the start of production. Setup people will generally have considerable CNC experience at the operator level and will be very familiar with jigs, fixtures, cutting tools, and CNC program operations. The setup person is most likely to learn the work through on-the-job training. Setup people will also work in fast-paced production shops, in medium-sized to large companies.

GENERAL MACHINIST The general machinist will have the capability to set up and operate all of the common conventional machine shop equipment. This person may receive training through an industrial apprenticeship lasting about four years, through broad-scope community college and trade school education often tied to a local machining industry, or through on-the-job experience of several years. The general machinist is not exactly disappearing, but much of the production work heretofore done by this individual can now be routinely accomplished on CNC machine tools by machine operators.

The general machinist may work in a job shop environment where many different types of work are performed. In a large company where modeling and prototyping are accomplished, the general machinist will find varied and extremely interesting work.

AUTOMOTIVE MACHINIST The automotive machinist will work in an engine rebuilding shop where engines are overhauled. This individual's responsibilities will be somewhat like those of the general machinist, with specialization in engine work including boring, milling, and some types of grinding applications. Training for this job may be obtained on the job or through college or trade school programs.

MAINTENANCE MACHINIST The maintenance machinist has broad responsibilities. This individual may be involved in plant equipment maintenance, machine tool rebuilding, or general mechanical repairs including welding and electrical. The maintenance machinist is often involved with general machine shop work as well as with general industrial mechanical work. The general maintenance machinist is often a vital member of the manufacturing support team in industries of all sizes. This individual may receive training through college and trade school programs or on the job.

TOOL, DIE, AND MOLD MAKER The tool, die and mold maker is often considered to be at the upper end of the machinist trade occupations. These individuals are essential in almost every machining manufacturing industry. The tool and die maker will usually be an experienced general machinist with superior talents developed over a number of years of shop apprenticeship and more years of experience. Tool and die makers may receive training through industrial apprenticeships or college and trade school programs. They may also be selected for industrial training in companies large enough to have an in-house tool and die shop. Although these individuals are often chosen only after several years of on-the-job experience, it is possible to start out in tool and die work through an apprenticeship program. Tool and die makers often receive premium pay for their work and will be involved with many high-precision machining applications, tool design, material selection, metallurgy, and general manufacturing processes.

INSPECTOR Inspectors will be necessary to handle the dimensional measurement quality control functions. Many of these individuals will become inspectors after some time as machine operators, general machinists, or possibly toolmakers. The machined parts inspector of today will require a knowledge of computerized measurement equipment (Figure A-3). Inspectors may be trained on the job or through college and trade school programs.

PRODUCTION TECHNICIANS Production technicians will be involved in many different tasks in machining manufacturing industries. Some of these responsibilities relative to maintenance may be similar to those of the maintenance machinist. The production technician will be responsible for installation of production equipment. Both electrical/electronic and mechanical technicians will be needed as manufacturing industries shift toward more computer-integrated manufacturing. The technician, especially in the electrical/electronic areas, will need excellent diagnostic skills so that expensive equipment can be quickly serviced and returned to production. Technicians may obtain training through college and trade school programs or on the job. Areas of knowledge will include pneumatics, hydraulics, electrical/electronic systems, computers, and the interrelationships of all these systems in complex manufacturing systems.

MARINE MACHINIST The marine machinist is generally employed in the shipbuilding industry. This trade has many different aspects, including installation, testing, and repairs of all types of shipboard mechanical systems, including hydraulics and pneumatics as well as conventional and nuclear steam systems. The marine machinist may also use portable machine tools that are brought to the job, including milling, boring, and drilling machines. Marine machinists often use optical instruments to align and locate mechanical components and machine tools.

APPRENTICE MACHINIST The apprentice machinist learns the trade by entering a formal training program sponsored by private industrial, trade union, or government entities. The period of training is typically four years long and is a combination of on-the-job experiences and formal classroom education. Apprenticeship curriculum standards are often quite universal, representing the collective inputs of all levels of the trade from production through management. Serving an apprenticeship represents one of the best and well-established methods of learning a skilled trade.

HELPERS AND LIMITED MACHINISTS Many manufacturing industries use limited machinists or machinists' helpers. These individuals assist the journeyperson by providing general help. These trades are often fairly low skill since the person does not have full responsibility for the work at hand. However, helpers and limiteds may be able to advance to journeyperson status after a suitable training period.

PROFESSIONAL CAREER OPPORTUNITIES

At the professional level, many exciting career opportunities are also available. These careers will require college preparation and include industrial technology (IT); industrial engineering (IE); manufacturing engineering; materials engineering; mechanical, electrical, electronic, and computer systems engineering; and CNC programming.

The industrial engineer, industrial technologist, and manufacturing engineer are often involved with the applications of manufacturing technology. These individuals design tooling, set up manufacturing systems, apply computers to manufacturing requirements, and write CNC programs.

Design engineers, often using computers, design products and manufacturing equipment and apply new materials in product design.

Machining and You

Whether a career in machining or a related area is for you will depend on your personal goals and how much effort you are willing to expend in preparation. No matter which area you might like to pursue, whether it be at trade or professional level, a working knowledge at the shop level of the machining processes and the related subjects described in this text will provide an excellent basis on which to build an exciting career in industrial manufacturing technology.

UNITS IN THIS SECTION

The units in this section deal with the important areas of safety in the machine shop, mechanical hardware, and reading shop drawings. These fundamental technical foundation areas are necessary for anyone involved in any phase of mechanical technology. To start off your study in the proper way, take the time now to familiarize yourself with these important fundamentals.

UNIT 1

Shop Safety

CAUTION **Safety is not often thought about as you proceed through your daily tasks. Often you expose yourself to needless risk because you have experienced no harmful effects in the past. Unsafe habits become almost automatic. You may drive your automobile without wearing a seat belt. You know this to be unsafe, but you have done it before and so far no harm has resulted. None of us really likes to think about the possible consequences of an unsafe act. However, safety can and does have an important effect on anyone who makes his or her living in a potentially dangerous environment such as a machine shop. An accident can reduce or end your career as a machinist. You may spend several years learning the trade and more years gaining experience. Experience is a particularly valuable asset. It can only be gained through time spent on the job.**

This becomes economically valuable to you and to your employer. Years spent in training and gaining experience can be wasted in an instant if you should have an accident, not to mention a possible permanent physical handicap for you and hardship on your family. Safety is an attitude that should extend far beyond the machine shop and into every facet of your life. You must constantly think about safety in everything you do.

Objectives

After completing this unit, you should be able to:

1. Identify common shop hazards.
2. Identify and use common shop safety equipment

Personal Safety

EYE PROTECTION

Eye protection is a primary safety consideration around the machine shop. Machine tools produce metal chips, and there is always a possibility that these may be ejected from a machine at high velocity. Sometimes they can fly many feet. Furthermore, most cutting tools are made from hard materials. They can occasionally break or shatter from the stress applied to them during a cut. The result can be more flying metal particles.

Eye protection must be worn at all times in the machine shop. There are several types of eye protection available. Plain safety glasses are all that are required in most shops. These have shatterproof lenses that may be replaced if they become

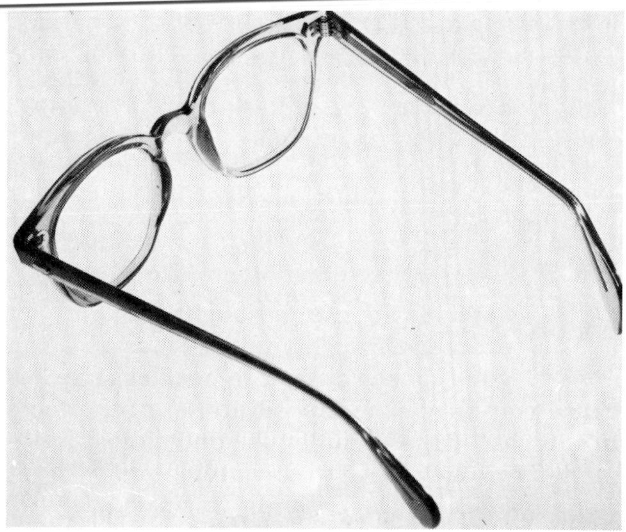

FIGURE A-5 Common fixed-bow safety glasses (Epsilon Pi Tau Fraternity, CSU, Fresno).

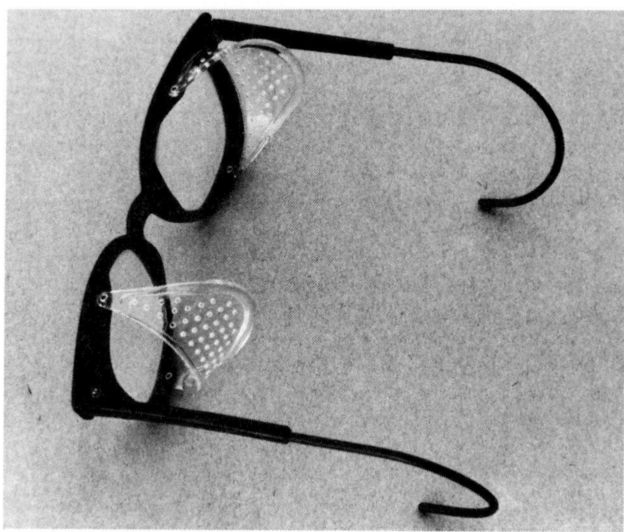

FIGURE A-6 Perforated side shield safety glasses (Epsilon Pi Tau Fraternity, CSU, Fresno).

scratched. The lenses have a high resistance to impact. Common types include fixed-bow safety glasses (Figure A-5) and flexible bow safety glasses. The flexible bows may be adjusted to the most comfortable position for the wearer.

Side shield safety glasses must be worn around any grinding operation. The side shield protects the side of the eye from flying particles. Side shield safety glasses may be of the solid or perforated type (Figure A-6). The perforated side shield fits closer to the eye. Bows may wrap around the ear. This prevents the safety glasses from falling off.

If you wear prescription glasses, you may want to cover them with a safety goggle. The full face shield may also be used (Figure A-7). Prescription glasses can be made as safety glasses. In industry, prescription safety glasses are sometimes provided free to employees.

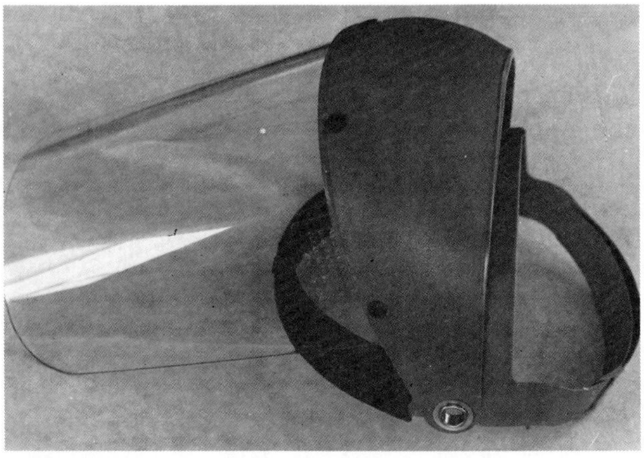

FIGURE A-7 Safety face shield.

FOOT PROTECTION

Generally, the machine shop does not present too great a hazard to the feet. However, there is always a possibility that you could drop something on your foot. A safety shoe is available, with a steel toe shield designed to resist impacts. Some safety shoes also have an instep guard. Shoes must be worn at all times in the machine shop. A solid leather shoe is recommended. Tennis shoes and sandals should not be worn. You must never even enter a machine shop with bare feet. Remember that the floor is often covered with razor-sharp metal chips.

EAR PROTECTION

The instructional machine shop usually does not present a noise problem. However, an industrial machine shop may be adjacent to a fabrication or punch press facility. New safety regulations are quite strict regarding exposure to noise. Several types of sound suppressors and noise-reducing ear plugs may be worn. Excess noise can cause a permanent hearing loss. Usually this occurs over a period of time, depending on the intensity of the exposure. Noise is considered an industrial hazard if it is continuously above 85 **decibels**, the units used in measuring sound waves. If it is over 115 decibels for short periods of time ear protection must be worn (Figure A-8). Earmuffs or earplugs should be used wherever high intensity noise occurs. A considerate worker will not create excessive noise when it is not necessary. Table A-1 shows the decibel level of various sounds; sudden sharp or high-intensity noises are the most harmful to your eardrums.

FIGURE A-8 Sound suppressors are designed to protect the ears from damage caused by loud noises.

TABLE A-1 Decibel Level of Various Sounds

130—Painful sounds; jet engine on ground
120—Airplane on ground: reciprocating engine
110—Boiler factory
—Pneumatic riveter
100—
—Maximum street noise
—Roaring lion
90—
—Loud shout
80—Diesel truck
—Piano practice
—Average city street
70—
—Dog barking
—Average conversation
60—
—Average city office
50—
—Average city residence
40—One typewriter
—Average country residence
30—Turning page of newspaper
—Purring cat
20—
—Rustle of leaves in breeze
—Human heartbeat
10—
0—Faintest audible sound

GRINDING DUST AND HAZARDOUS FUMES

Grinding dust is produced by abrasive wheels and consists of extremely fine metal particles and abrasive wheel particles. These should not be inhaled. In the machine shop, most grinding machines have a vacuum dust collector (Figure A-9). Grinding may be done with coolants that aid in dust control. A machinist may be involved in portable grinding operations. This is common in such industries as shipbuilding. You should wear an approved respirator if you are exposed to grinding dust. Change the respirator filter at regular intervals. Grinding dust can present a great danger to health. Examples include the dust of such metals as beryllium, or the presence of radioactivity in nuclear systems. In these situations, the spread of grinding dust must be carefully controlled.

Some metals such as zinc give off toxic fumes when heated above their boiling point. Some of these fumes when inhaled cause only temporary sickness, but other fumes can be severe or even fatal. The fumes of mercury and lead are especially dangerous, as their effect is cumulative in your body and can cause irreversible damage. Cadmium and beryllium compounds are also very poisonous. Therefore, when welding, burning, or heat treating metals, adequate ventilation is an absolute necessity. This is also true when parts are being carburized with compounds containing potassium cyanide. These **cyanogen compounds** are deadly poisonous and every precaution should be taken when using them. Kasenite, a trade name for a carburizing compound that is not toxic, is often found in school shops and in machine shops. Uranium salts are toxic, and all radioactive materials are extremely dangerous.

FIGURE A-9 Vacuum dust collector on grinders.

CLOTHING, HAIR, AND JEWELRY

Wear a short-sleeved shirt or roll up long sleeves above the elbow. Keep your shirt tucked in and remove your necktie. It is recommended that you wear a shop apron. If you do, keep it tied behind you. If apron strings become entangled in the machine, you may be reeled in as well. A shop coat may be worn as long as you roll up long sleeves. Do not wear fuzzy sweaters around machine tools.

If you have long hair, keep it secured properly. In industry, you may be required to wear a hair net so that your hair cannot become entangled in a moving machine. The result of this can be disastrous (Figure A-10).

Remove your wristwatch and rings before operating any machine tool. These can cause serious injury if they should be caught in a moving machine part.

FIGURE A-10 Long hair may be caught and reeled into the machine.

HAND PROTECTION

There is reallly no device that will totally protect your hands from injury. Next to your eyes, your hands are the most important tools you have. It is up to you to keep them out of danger. Use a brush to remove chips from a machine (Figure A-11). Do not use your hands. Chips are not only razor sharp, they are often extremely hot. Resist the temptation to grab chips as they come from a cut. Long chips are extremely dangerous. These can often be eliminated by properly sharpening your cutting tools. Chips should *not* be removed with a rag. The metal particles become imbedded in the cloth and they may cut you. Furthermore, the rag may be caught in a moving machine. Gloves must not be worn around most machine tools, although they are acceptable when working with a band saw blade. If a glove should be caught in a moving part, it will be pulled in, along with the hand inside it.

FIGURE A-11 Use a brush to clear chips.

Various cutting oils, coolants, and solvents may affect your skin. The result may be a rash or an infection. Avoid direct contact with these products as much as possible and wash your hands as soon as possible after contact.

LIFTING

Improper lifting (Figure A-12) can result in a permanent back injury that can limit or even end your career. Back injury can be avoided if you lift properly at all times. If you must lift a large or heavy object, get some help or use a hoist or forklift. Don't try to be a "superman" and lift something that you know is too heavy. It is not worth the risk.

Objects within your lifting capability can be lifted safely by using the following procedure (Figure A-13):

1. Keep your back straight.

2. Squat down, bending your knees.

3. Lift smoothly, using the muscles in your legs to do the work. Keep your back straight. Bending over the load puts an excessive stress on your spine.

4. Position the load so that it is comfortable to carry. Watch where you are walking when carrying a load.

5. If you are placing the load back at floor level, lower it in the same manner you picked it up.

SCUFFLING AND HORSEPLAY

The machine shop is no place for scuffling and horseplay. This activity can result in a serious injury to you, a fellow student, or worker. Practical

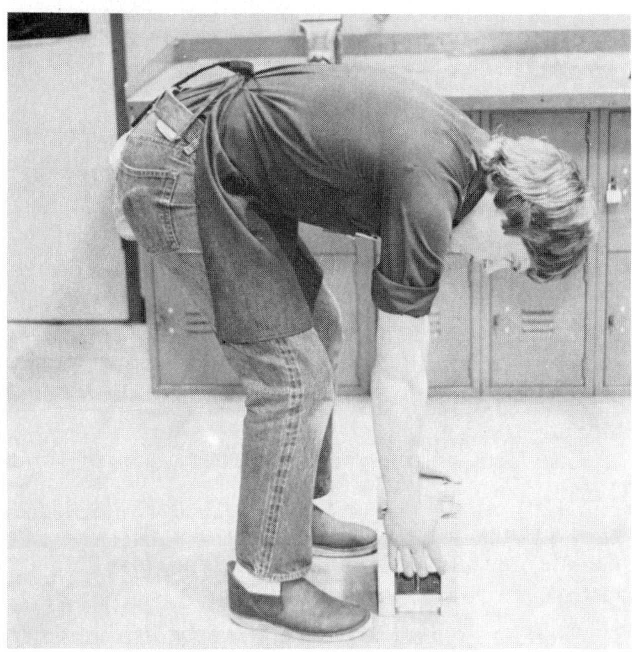

FIGURE A-12 The wrong way to lift, placing excessive strain on the back (Lane Community College).

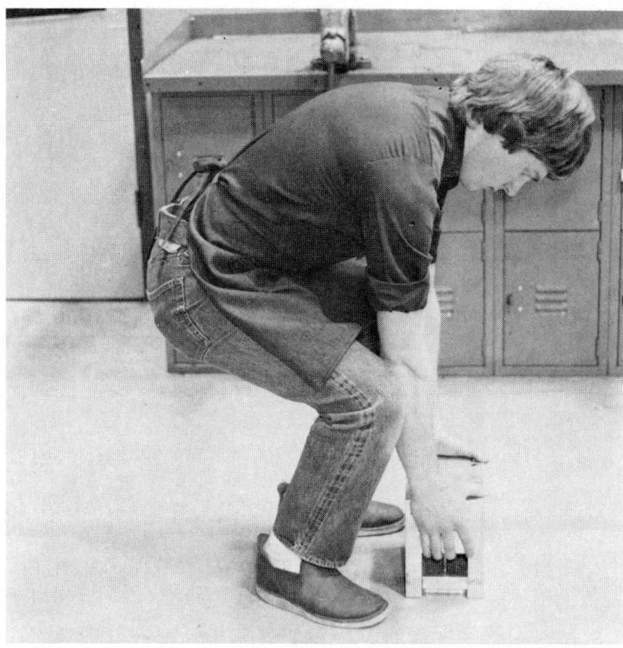

FIGURE A-13 The right way to lift with knees bent, using leg muscles to do the work (Lane Community College).

joking is also very hazardous. What might appear to be a comical situation to you could result in a disastrous accident to someone else. In industry, horseplay and practical joking are often grounds for dismissal of an employee.

INJURIES

If you should be injured, report it immediately to your instructor.

Identifying Shop Hazards

A machine shop is not so much a dangerous place as a potentially dangerous place. One of the best ways to be safe is to be able to identify shop hazards before they can involve you in an accident. By being aware of potential danger, you can better make safety part of your work in the machine shop.

COMPRESSED AIR

Most machine shops have compressed air. This is needed to operate certain machine tools. Often flexible air hoses are hanging about the shop. Few people realize the large amount of energy that can be stored in a compressed gas such as air. When this energy is released, extreme danger may be present. You may be tempted to blow chips from a machine tool using compressed air. This is not good practice. The air will propel metal particles at high velocity. They can injure you or someone on the other side of the shop. Use a brush to clean chips from the machine. Do not blow compressed air on your clothing or skin. The air can be dirty and the force can implant dirt and germs into your skin. Air can be a hazard to ears as well. An eardrum can be ruptured.

Should an air hose break, or the nozzle on the end come unscrewed, the hose will whip about wildly. This can result in an injury if you happen to be standing nearby. When an air hose is not in use, it is good practice to shut off the supply valve. The air trapped in the hose should be vented. When removing an air hose from its supply valve, be sure that the supply is turned off and the hose has been vented. Removing a charged air hose will result in a sudden venting of air. This can surprise you and an accident might result.

HOUSEKEEPING

Keep floor and aisles clear of stock and tools. This will insure that all exits are clear if the building should have to be evacuated. Material on the floor, especially round bars, can cause falls. Clean oils or coolants that spill on the floor. Several preparations designed to absorb oil are available. These may be used from time to time in the shop. Keep oily rags in an approved safety can (Figure A-14).

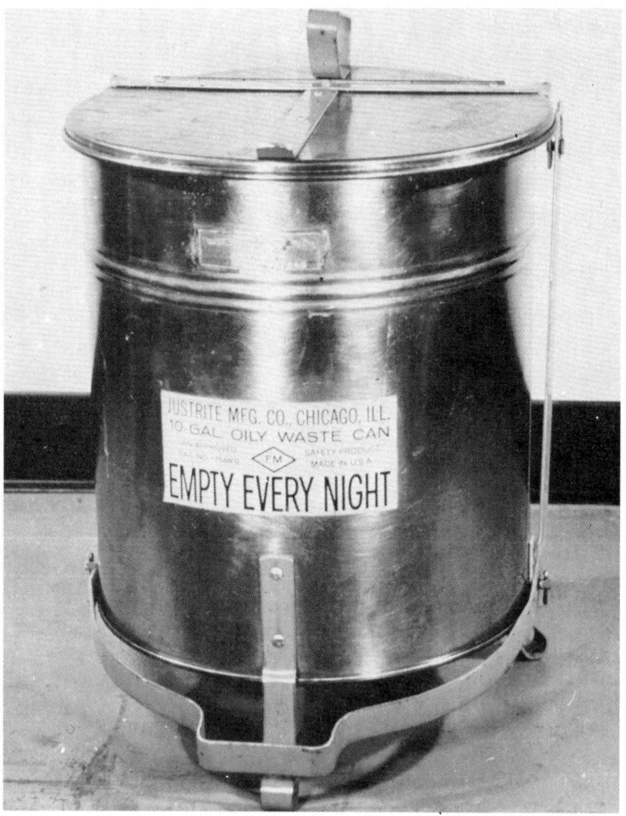

FIGURE A-14 Store oil soaked rags in an approved safety can.

This will prevent possible fire from spontaneous combustion.

FIRE EXTINGUISHERS

It is an important safety consideration to know the correct fire extinguisher to use for a particular fire. For example, if you should use a water-based extinguisher on an electrical fire, you could receive a severe or fatal electrical shock. Fires are classed according to types as given in Table A-2.

There are four basic types of fire extinguishers used other than the use of tap water.

1. The dry chemical type is effective on classes B and C fires.
2. The pressurized water and loaded stream types are safe only on class A fires. These types may actually spread an oil or gasoline fire.
3. The dry chemical multi-purpose extinguisher may be safely used on classes A, B, and C fires.
4. Pressurized carbon dioxide (CO_2) can be used on classes B and C fires.

You should always make yourself aware of the locations of fire extinguishers in your working area. Take time to look at them closely and note their types and capabilities. This way, if there should ever be an oil-based or electrical fire in your area, you will know how to put it out safely.

ELECTRICAL

Electricity is another potential danger in a machine shop. Your exposure to electrical hazard will

TABLE A-2 Types of Extinguishers Used on the Classes of Fire (Brodhead-Garrett Company)

	Pressurized Water	Loaded Stream	Co_2	Regular Dry Chemical	All Use Dry Chemical
Class A Fires Paper, wood, cloth, etc., where quenching by water or insulating by general-purpose dry chemical is effective.	Yes Excellent	Yes Excellent	Small surface fires only	Small surface fires only	Yes Excellent Forms smothering film, prevents reflash
Class B Fires Burning liquids (gasoline, oils, cooking fats, etc.), where smothering action is required	No Water will spread fire	Yes Has limited capability	Yes Carbon dioxide has no residual effects on food or equipment	Yes Excellent Chemical smothers fire	Yes Excellent Smothers fire, prevents reflash
Class C Fires Fires in live electrical equipment (motors, switches, appliances, etc.), where a nonconductive extinguishing agent is required	No Water is a conductor of electricity	No Water is a conductor electricity	Yes Excellent CO_2 is a nonconductor, leaves no residue	Yes Excellent Nonconducting smothering film; screens operator from heat	Excellent Nonconducting smothering film; screens operator from heat

be minimal unless you become involved with machine maintenance. A machinist is mainly concerned with the on and off switch on a machine tool. However, if you are adjusting the machine or accomplishing maintenance, you should unplug it from the electrical service. If it is permanently wired, the circuit breaker may be switched off and tagged with an appropriate warning. In industry, this procedure often means that the operator must sign a clearance stating that electrical service has been secured. Service cannot be restored until the operator signs a restoration order. Normally you will not disconnect the electrical service for routine adjustments such as changing speeds. However, when a speed change involves a belt change, you must insure that no other person is likely to turn on the machine while your hands are in contact with belts and pulleys.

CARRYING OBJECTS

If material is over 6 feet long it should be carried in the horizontal position. If it must be carried in the vertical position, be careful of light fixtures and ceilings. If the material is both long and over 40 pounds in weight, it should be carried by two people, one at each end. Heavy stock, even if it is short, should be carried by two people.

Machine Hazards

There are many machine hazards. Each section of this book will discuss the specific dangers applicable to that type of machine tool. Remember that a machine cannot distinguish between cutting metal and cutting fingers. Do not think that you are strong enough to stop a machine should you become tangled in moving parts. You are not. When operating a machine, think about what you are going to do before you do it. Go over a safety checklist.

1. Do I know how to operate this machine?
2. What are the potential hazards involved?
3. Are all guards in place?
4. Are my procedures safe?
5. Am I doing something that I probably should not do?
6. Have I made all the proper adjustments and tightened all locking bolts and clamps?
7. Is the workpiece secured properly?
8. Do I have proper safety equipment?
9. Do I know where the stop switch is?
10. Do I think about safety in everything that I do?

Industrial Safety and Federal Law

In 1970, Congress passed the Williams-Steiger Occupational Safety and Health Act. This act took effect on April 28, 1971. The purpose and policy of the act is "to assure so far as possible every working man and woman in the Nation safe and healthful working conditions and to preserve our human resources."

The Occupational Safety and Health Act is commonly known as OSHA. Prior to its passage, industrial safety was the individual responsibility of each state. The establishment of OSHA added a degree of standardization to industrial safety throughout the nation. OSHA encourages states to assume full responsibility in administration and enforcement of federal occupational safety and health regulations.

DUTIES OF EMPLOYERS AND EMPLOYEES

Each employer under OSHA has the general duty to furnish employment and places of employment free from recognized hazards causing or likely to cause death or serious physical harm. The employer has the specific duty of complying with safety and health standards as defined under OSHA. Each employee has the duty to comply with safety and health standards and all rules and regulations established by OSHA.

OCCUPATIONAL SAFETY AND HEALTH STANDARDS

Job safety and health standards consist of rules for avoiding hazards that have been proven by research and experience to be harmful to personal safety and health. These rules may apply to all employees, as in the case of fire protection standards. Many standards apply only to workers engaged in specific types of work. A typical standard might state that aisles and passageways shall be kept clear and in good repair, with no obstruction across or in aisles that could create a hazard.

COMPLAINTS OF VIOLATIONS

Any employee who believes that a violation of job safety or health standards exists may request an inspection by sending a signed written notice to OSHA. This includes anything that threatens physical harm or represents an imminent danger. A copy must also be provided to the employer. However, the name of the person complaining need not be revealed to the employer.

ENFORCEMENT OF OSHA STANDARDS

OSHA inspectors may enter a plant or school at any reasonable time and conduct an inspection. They are not permitted to give prior notice for this. They may question any employer, owner, operator, agent, or employee in regard to any safety violation. The employer and a representative of the employees have the right to accompany the inspector during the inspection.

If a violation is discovered, a written citation is issued to the employer. A reasonable time is permitted to correct the condition. The citation must be posted at or near the place of the violation. If, after a reasonable time, the condition has not been corrected, a fine may be imposed on the employer. If the employer is making an attempt to correct the unsafe condition but has exceeded the time limit, a hearing may be held to determine progress.

PENALTIES

Willful or repeated violations may incur penalties up to $10,000. Citations issued for serious violations incur mandatory penalties of $1000. A serious violation where extreme danger exists may be penalized up to $1000 for each day the violation exists.

OSHA EDUCATION AND TRAINING PROGRAMS

The Occupational Safety and Health Act provides for programs to be conducted by the Department of Labor. These programs provide for education and training of employers and employees in recognizing, avoiding, and preventing unsafe and unhealthful working conditions. The act also provides for training an adequate supply of qualified personnel to carry out OSHA's purpose.

SELF-TEST

1. What is the primary piece of safety equipment in the machine shop?

2. What can you do if you wear prescription glasses?

3. Describe proper dress for the machine shop.

4. What can be done to control grinding dust?

5. What hazards exist from coolants, oils, and solvents?

6. Describe proper lifting procedure.

7. Describe at least two compressed air hazards.

8. Describe good housekeeping procedures.

9. How should long pieces of material be carried?

10. List at least five points from the safety checklist for a machine tool.

UNIT 2

Mechanical Hardware

Many precision-machined products produced in the machine shop are useless until assembled into a machine, tool, or other mechanism. This assembly requires many types of fasteners and other mechanical hardware. In this unit, you will be introduced to many of these important hardware items.

Objectives

After completing this unit, you should be able to:

1. Identify threads and threaded fasteners.
2. Identify thread nomenclature on drawings.
3. Discuss standard series of threads.
4. Identify and describe applications of common mechanical hardware found in the machine shop.

Threads

The thread is an extremely important mechanical device. It derives its usefulness from the inclined plane, one of the six simple machines. Almost every mechanical device is assembled with threaded fasteners. A thread is a helical groove that is formed on the outside or inside diameter of a cylinder (Figure A-15). These helical grooves take several forms. Furthermore, they have specific and even spacing. One of the fundamental tasks of a machinist is to produce both external and internal threads using several machine tools and hand tools. The majority of threads appear on threaded fasteners. These include many types of bolts, screws, and nuts. However, threads are used or a number of other applications aside from fasteners. These include threads for adjustment purposes, measuring tool applications, and the transmission of power. A close relative to the thread, the helical auger, is used to transport material.

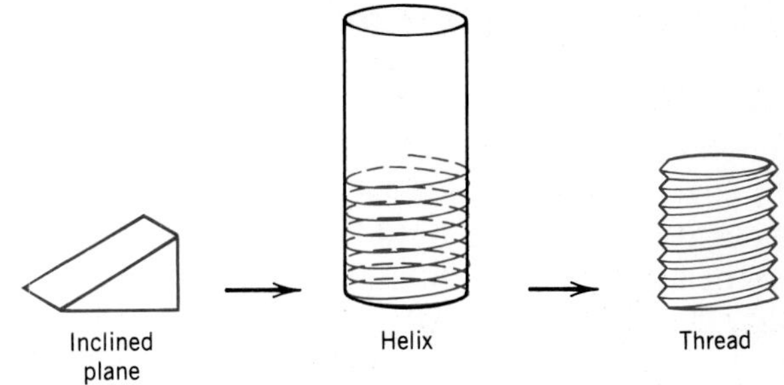

FIGURE A-15 Thread helix.

Thread Forms

There are a number of thread forms. In later units, you will examine these in detail, and you will have the opportunity to make several of them on a machine tool. As far as the study of machined hardware is concerned, you will be most concerned with the unified thread form (Figure A-16). The unified thread form is an outgrowth of the American National Standard form. In order to help standardize manufacturing in the United States, Canada, and Great Britain, the unified form was developed. Unified threads are a combination of the American National and the British Standard Whitworth forms. Unified threads are divided into the following series:

UNC	National Coarse
UNF	National Fine
UNS	National Special

Identifying Threaded Fasteners

Unified coarse and *unified fine* refer to the number of threads per inch of length on standard threaded fasteners. A specific diameter of bolt or nut will have a specific number of threads per inch of length. For example, a $\frac{1}{2}$-in.-diameter Unified National Coarse bolt will have 13 threads per inch of length. This bolt will be identified by the following marking:

$\frac{1}{2}$ in.–13 UNC

The $\frac{1}{2}$ in. is the **major diameter** and **13** is the **number of threads per inch of length**. A $\frac{1}{2}$-in.-diameter Unified National Fine bolt will be identified by the following marking:

$\frac{1}{2}$ in.–20 UNF

The $\frac{1}{2}$ in. is the major diameter and 20 is the number of threads per inch.

The Unified National Special Threads are identified in the same manner. A $\frac{1}{2}$-in.-diameter UNS bolt may have 12, 14, or 18 threads per inch. These are less common than the standard UNC and UNF. However, you may see them in machining technology. There are many other series of threads used for different applications. Information and data on these can be found in machinists' handbooks. You might wonder why there needs to be a UNC and UNF series. This has to do with thread applications. For example, an adjusting screw might require a fine thread, while a common bolt may require only a coarse thread.

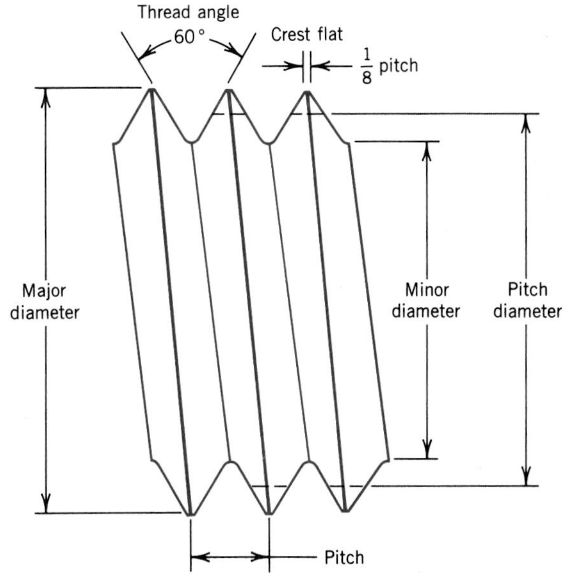

FIGURE A-16 Unified thread form.

Classes of Thread Fits

The preceding information was necessary for an understanding of **thread fit classes**. Some thread applications can tolerate loose threads, while other applications require tight threads. For example, the head of your car's engine is held down by a threaded fastener called a stud bolt, or simply a stud. A stud is threaded on both ends. One end is threaded into the engine block. The other end receives a nut that bears against the cylinder head. When the head is removed, it is desirable to have the stud remain screwed into the engine block. This end requires a tighter thread fit than the end of the stud accepting the nut. If the fit on the nut end is too tight, the stud may unscrew as the nut is removed.

Unified **thread fits** are classified as **1A, 2A, 3A**, or **1B, 2B, 3B**. The **A symbol** indicates an **external** thread. The **B symbol** indicates an **internal** thread. This notation is added to the thread size and number of threads per inch. Let us consider the $\frac{1}{2}$-in.-diameter bolt discussed previously. The complete notation reads

$\frac{1}{2}$–13 UNC 2A

On this particular bolt, the class of fit is 2. The symbol A indicates an external thread. If the notation had read

$\frac{1}{2}$–13 UNC 3B

this would indicate an internal thread with a class 3 fit. This could be a nut or a hole threaded with a tap. Taps are a very common tool for producing an internal thread.

Classes 1A and 1B have the greatest manufacturing tolerance. They are used where ease of assembly is desired and a loose thread is not objectionable. Class 2 fits are used on the largest percentage of threaded fasteners. Class 3 fits will be tight when assembled. Each class of fit has a specific tolerance on major diameter and pitch diameter. These data may be found in machinists' handbooks and are required for the manufacture of threaded fasteners.

Standard Series of Threaded Fasteners

Threaded fasteners, including all common bolts and nuts, range from quite small machine screws through quite large bolts. Below a diameter of $\frac{1}{4}$ in., threaded fasteners are given a number. Common UNC and UNF series threaded fasteners are listed in Table A-3. Above size 12, the major diameter is expressed in fractional form. Both series continue up to about 4 inches.

All of the sizes listed in the table are very common fasteners in all types of machines, automobiles, and other mechanisms. Your contact with

TABLE A-3 UNC and UNF Threaded Fasteners

UNC			UNF		
Size	Major Diameter (in.)	Threads/Inch	Size	Major Diameter (in.)	Threads/Inch
			0	.059	80
1	.072	64	1	.072	72
2	.085	56	2	.085	64
3	.098	48	3	.098	56
4	.111	40	4	.111	48
5	.124	40	5	.124	44
6	.137	32	6	.137	40
8	.163	32	8	.163	36
10	.189	24	10	.189	32
12	.215	24	12	.215	28
$\frac{1}{4}$ in.	.248	20	$\frac{1}{4}$ in.	.249	28
$\frac{5}{16}$ in.	.311	18	$\frac{5}{16}$ in.	.311	24
$\frac{3}{8}$ in.	.373	16	$\frac{3}{8}$ in.	.373	24
$\frac{7}{16}$ in.	.436	14	$\frac{7}{16}$ in.	.436	20
$\frac{1}{2}$ in.	.498	13	$\frac{1}{2}$ in.	.498	20
$\frac{9}{16}$ in.	.560	12	$\frac{9}{16}$ in.	.561	18
$\frac{5}{8}$ in.	.623	11	$\frac{5}{8}$ in.	.623	18
$\frac{3}{4}$ in.	.748	10	$\frac{3}{4}$ in.	.748	16
$\frac{7}{8}$ in.	.873	9	$\frac{7}{8}$ in.	.873	14
1 in.	.998	8	1 in.	.998	12

these common sizes will be so frequent that you will soon begin to recall them from memory.

METRIC THREADS

With the importation of foreign manufactured hardware in recent years, especially in the automotive and machine tool areas, metric threads have become the prevalent thread type on many kinds of equipment.

The metric thread form is similar to the unified and based on an equilateral triangle. The root may be rounded and the depth somewhat greater. An attempt has been made through international efforts (ISO) to standardize metric threads. The ISO metric thread series now has 25 thread sizes with major diameters ranging from 1.6 millimeters (mm) to 100 mm.

Metric thread notations take the following form:

$$M\ 10 \times 1.5$$

where M is the major diameter and 1.5 is the thread pitch in millimeters. This thread would have a major diameter of 10 mm and a pitch (or lead) of 1.5 mm. ISO metric thread major diameters and respective pitches are shown in Table A-4.

TABLE A-4 ISO Metric Threads

Diameter (mm)	Pitch (mm)	Diameter (mm)	Pitch (mm)
1.6	.35	20.0	2.5
2.0	.40	24.0	3.0
2.5	.45	30.0	3.5
3.0	.50	36.0	4.0
3.5	.60	42.0	4.5
4.0	.70	48.0	5.0
5.0	.80	56.0	5.5
6.3	1.00	64.0	6.0
8.0	1.25	72.0	6.0
10.0	1.50	80.0	6.0
12.0	1.75	90.0	6.0
14.0	2.00	100.0	6.0
16.0	2.00		

Common Externally Threaded Fasteners

Common mechanical hardware includes threaded fasteners such as bolts, screws, nuts, and thread inserts. All of these are used in a variety of ways to hold parts and assemblies together. Complex assemblies such as airplanes, ships, or automobiles may have many thousands of fasteners taking many forms.

BOLTS AND SCREWS

A general definition of a bolt is "an externally threaded fastener that is inserted through holes in an assembly." A bolt is tightened with a nut (Figure A-17 right). A screw is an externally threaded fastener that is inserted into a threaded hole and tightened or released by turning the head (Figure A-17, left). From these definitions, it is apparent that a bolt can become a screw or the reverse can be true. This depends on the application of the hardware. Bolts and screws are the most common of the threaded fasteners. These fasteners are used to assemble parts quickly and they make disassembly possible.

The strength of an assembly of parts depends to a large extent on the diameter of the screws or bolts used. In the case of screws, strength depends on the amount of thread engagement. Thread engagement is the distance that a screw extends into a threaded hole. The minimum thread engagement should be a distance equal to the diameter of the screw used; preferably it would be $1\frac{1}{2}$ times the screw diameter. Should an assembly fail, it is better that the screw break than to have the internal thread stripped from the hole. It is generally easier to remove a broken screw than to drill and tap for a larger screw size. With a screw engagement of $1\frac{1}{2}$ times its diameter, the screw will usually break rather than strip the thread in the hole.

Machine bolts (Figure A-18) are made with hexagonal or square heads. These bolts are often used in the assembly of parts that do not require a precision bolt. The body diameter of machine bolts is usually slightly larger than the nominal or standard size of the bolt. Body diameter is the diameter

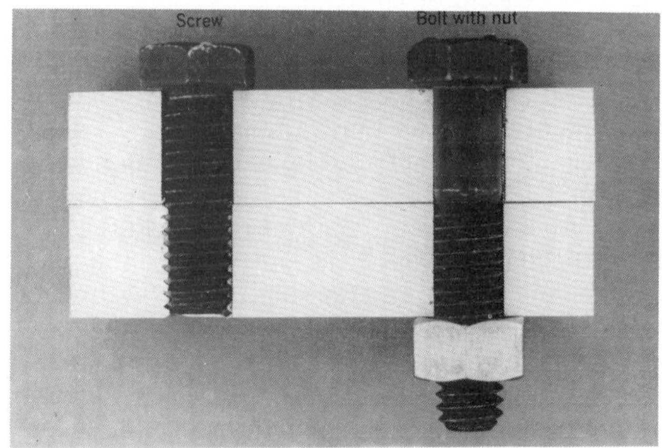

FIGURE A-17 Screw and bolt with nut.

FIGURE A-18 Square head bolt and hex head bolt. Square nut and hex nut.

FIGURE A-19 Stud bolt.

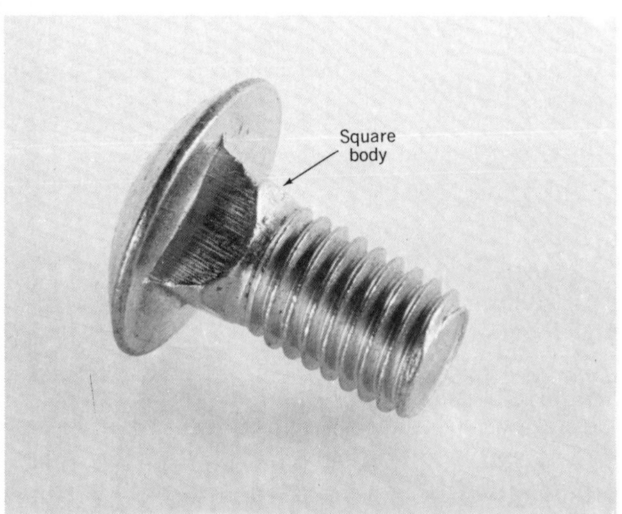

FIGURE A-20 Carriage bolt.

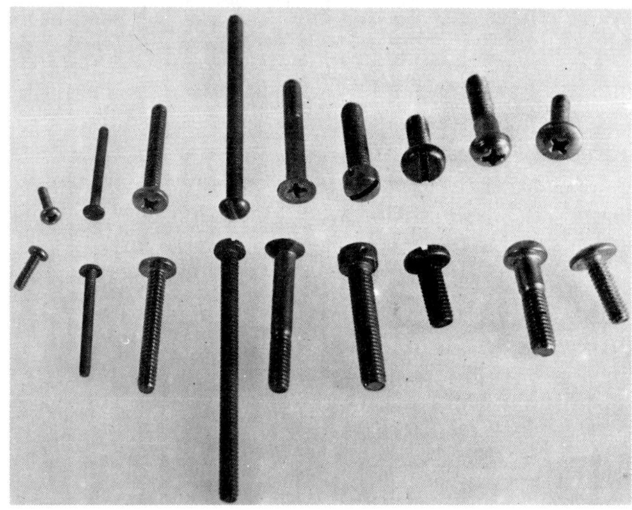

FIGURE A-21 Machine screws.

of the unthreaded portion of a bolt below the head. A hole that is to accept a common bolt must be slightly larger than the body diameter. When machine bolts are purchased, nuts are frequently included. Common bolts are made with a class 2A unified thread and come in both UNC and UNF series. Sizes in hexagonal head machine bolts range from $\frac{1}{4}$ in. in diameter to 4 in. Square head machine bolts are standard to a $1\frac{1}{2}$-in. diameter.

Stud bolts (Figure A-19) have threads on both ends. Stud bolts are used where one end is semipermanently screwed into a threaded hole. A good example of the use of stud bolts is an automobile engine. The stud bolts are tightly held in the cylinder block and easily changed nuts hold the cylinder heads in place. The end of the stud bolt screwed into the tapped hole has a class 3A thread, while the nut end is a class 2A thread.

Carriage bolts (Figure A-20) are used to fasten wood and metal parts together. Carriage bolts have round heads with a square body under the head. The square part of the carriage bolt, when pulled into the wood, keeps the bolt from turning while the nut is being tightened. Carriage bolts are manufactured with class 2A coarse threads.

Machine screws are made with either coarse or fine thread and are used for general assembly work. The heads of most machine screws are slotted to be driven by screwdrivers. Machine screws are available in many sizes and lengths (Figure A-21). Several head styles are also available (Figure A-22). Machine screw sizes fall into two categories. Fraction sizes range from diameters of $\frac{1}{4}$ in. to $\frac{3}{4}$ in. Below $\frac{1}{4}$ in. diameter, screws are identified by numbers from 0 to 12. A No. 0 machine screw has a diameter of .060 in. (60 thousandths of an inch). For each number above zero add .013 in. to the diameter.

EXAMPLE

Find the diameter of a No. 6 machine screw:

No. 0 diameter = .060 in.
No. 6 diameter = .060 in. + (6 × .013 in.)
= .060 in. + .078 in.
= .138 in.

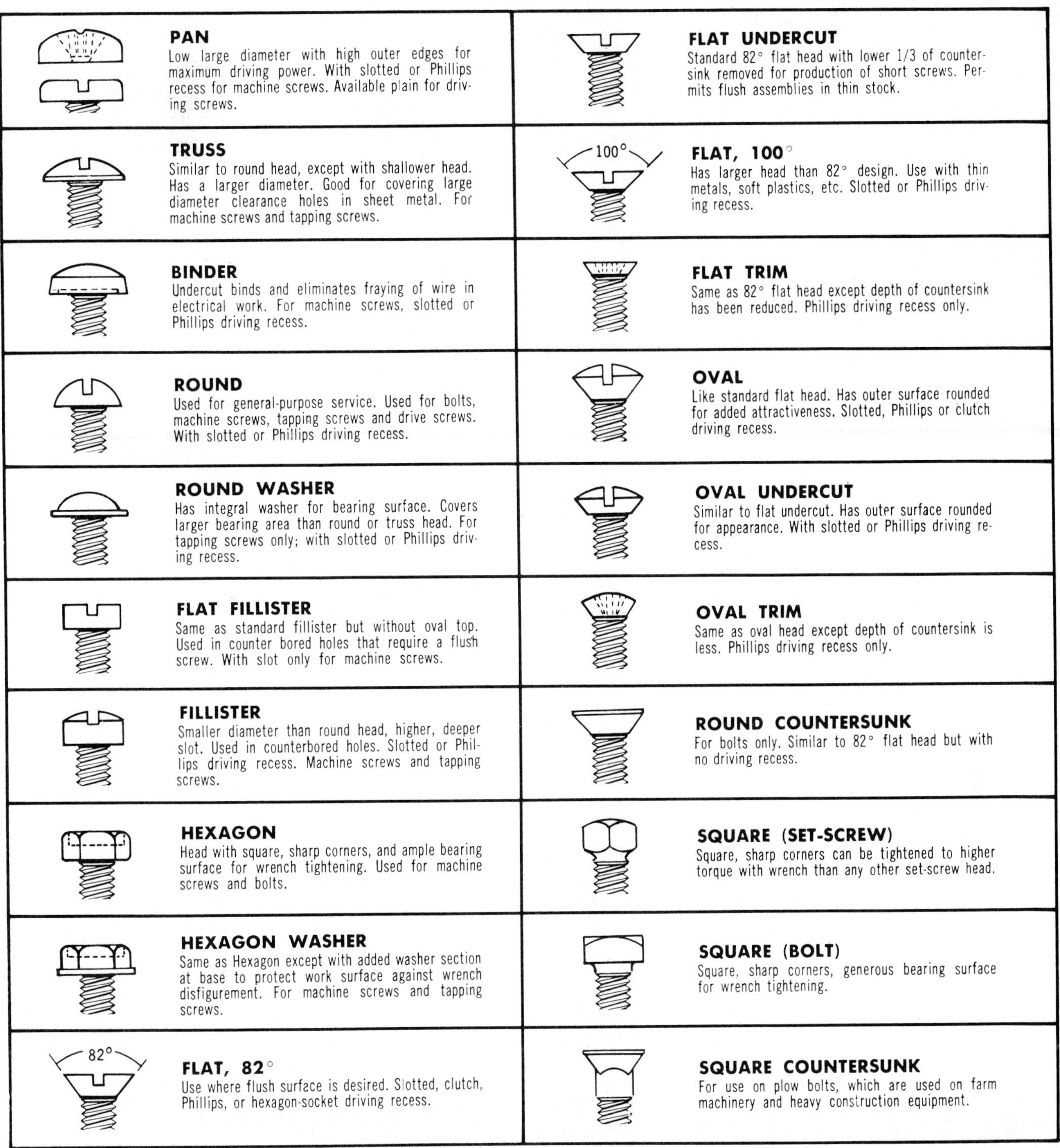

PAN
Low large diameter with high outer edges for maximum driving power. With slotted or Phillips recess for machine screws. Available plain for driving screws.
TRUSS
Similar to round head, except with shallower head. Has a larger diameter. Good for covering large diameter clearance holes in sheet metal. For machine screws and tapping screws.
BINDER
Undercut binds and eliminates fraying of wire in electrical work. For machine screws, slotted or Phillips driving recess.
ROUND
Used for general-purpose service. Used for bolts, machine screws, tapping screws and drive screws. With slotted or Phillips driving recess.
ROUND WASHER
Has integral washer for bearing surface. Covers larger bearing area than round or truss head. For tapping screws only; with slotted or Phillips driving recess.
FLAT FILLISTER
Same as standard fillister but without oval top. Used in counter bored holes that require a flush screw. With slot only for machine screws.
FILLISTER
Smaller diameter than round head, higher, deeper slot. Used in counterbored holes. Slotted or Phillips driving recess. Machine screws and tapping screws.
HEXAGON
Head with square, sharp corners, and ample bearing surface for wrench tightening. Used for machine screws and bolts.
HEXAGON WASHER
Same as Hexagon except with added washer section at base to protect work surface against wrench disfigurement. For machine screws and tapping screws.
82°
FLAT, 82°
Use where flush surface is desired. Slotted, clutch, Phillips, or hexagon-socket driving recess.
FLAT UNDERCUT
Standard 82° flat head with lower 1/3 of countersink removed for production of short screws. Permits flush assemblies in thin stock.
100°
FLAT, 100°
Has larger head than 82° design. Use with thin metals, soft plastics, etc. Slotted or Phillips driving recess.
FLAT TRIM
Same as 82° flat head except depth of countersink has been reduced. Phillips driving recess only.
OVAL
Like standard flat head. Has outer surface rounded for added attractiveness. Slotted, Phillips or clutch driving recess.
OVAL UNDERCUT
Similar to flat undercut. Has outer surface rounded for appearance. With slotted or Phillips driving recess.
OVAL TRIM
Same as oval head except depth of countersink is less. Phillips driving recess only.
ROUND COUNTERSUNK
For bolts only. Similar to 82° flat head but with no driving recess.
SQUARE (SET-SCREW)
Square, sharp corners can be tightened to higher torque with wrench than any other set-screw head.
SQUARE (BOLT)
Square, sharp corners, generous bearing surface for wrench tightening.
SQUARE COUNTERSUNK
For use on plow bolts, which are used on farm machinery and heavy construction equipment.

FIGURE A-22 Machine screw head styles (Great Lakes Screw).

Machine screws 2 in. long or shorter have threads extending all the way to the head. Longer machine screws have a $1\frac{3}{4}$-in. thread length.

Capscrews (Figure A-23) are made with a variety of different head shapes and are used where precision bolts or screws are needed. Capscrews are manufactured with close tolerances and have a finished appearance. Capscrews are made with coarse, fine, or special threads. Capscrews with a 1-in. diameter have a class 3A thread. Those greater than a 1-in. diameter have a class 2A thread. The strength of screws depends mainly on the kind of material used to make the screw. Different screw materials are aluminum, brass,

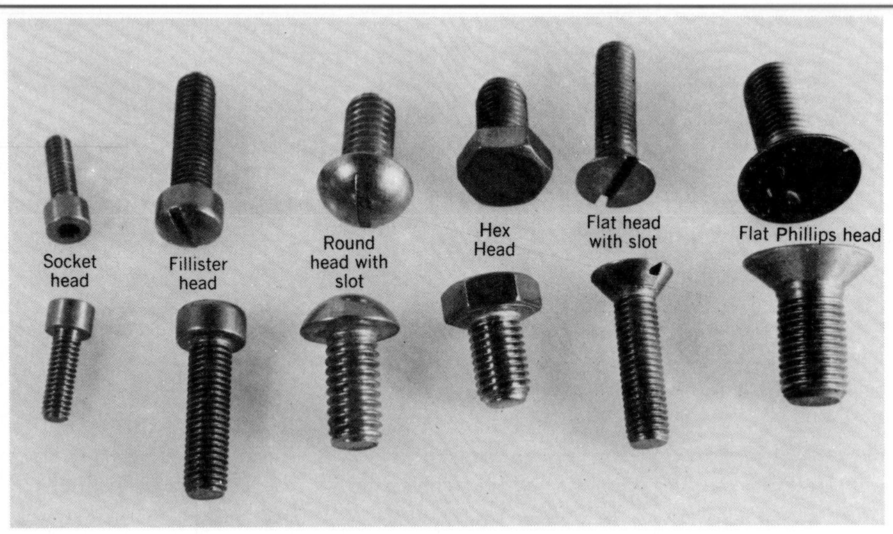

FIGURE A-23 Cap screws.

Bolt head marking	SAE – Society of Automotive Engineers ASTM – American Society for Testing and Materials SAE – ASTM Definitions	Material	Minimum tensile strength in pounds per square inch (PSI)
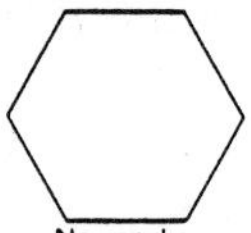 No marks	SAE grade 1 SAE grade 2 Indeterminate quality	Low carbon steel Low carbon steel	65,000 PSI
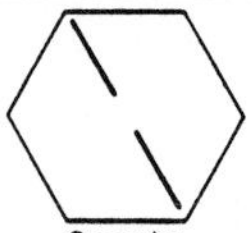 2 marks	SAE grade 3	Medium carbon steel, cold worked	110,000 PSI
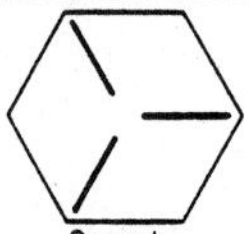 3 marks	SAE grade 5 ASTM – A 325 Common commercial quality	Medium carbon steel, quenched and tempered	120,000 PSI
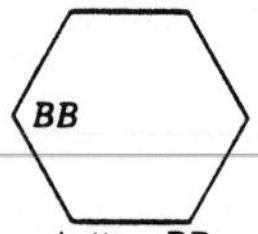 *BB* Letters *BB*	ASTM – A 354	Low alloy steel or medium carbon steel, quenched and tempered	105,000 PSI
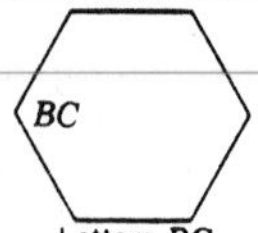 *BC* Letters *BC*	ASTM – A 354	Low alloy steel or medium carbon steel, quenched and tempered	125,000 PSI
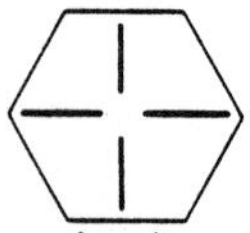 4 marks	SAE grade 6 Better commercial quality	Medium carbon steel, quenched and tempered	140,000 PSI
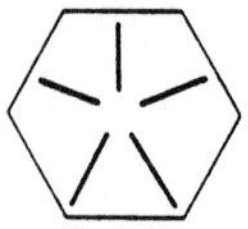 5 marks	SAE grade 7	Medium carbon alloy steel, quenched and tempered, roll threaded after heat treatment	133,000 PSI
 6 marks	SAE grade 8 ASTM – A 345 Best commercial quality	Medium carbon alloy steel, quenched and tempered	150,000 PSI

FIGURE A-24 Grade markings for bolts.

bronze, low carbon steel, medium carbon steel, alloy steel, stainless steel, and titanium. Steel hex head capscrews come in diameters from $\frac{1}{4}$ to 3 in., and their strength is indicated by symbols on the hex head (Figure A-24). Slotted head capscrews can have flat heads, round heads, or fillister heads. Socket head capscrews are also made with socket flat heads and socket button heads.

Setscrews (Figure A-25) are used to lock pulleys or collars or shafts. Setscrews can have square heads with the head extending above the surface; more often, the setscrews are slotted or have socket heads. Slotted or socket head setscrews usually disappear below the surface of the part to be fastened. A pulley or collar with the setscrews below the surface is much safer for persons working around it. Socket head setscrews may have hex socket heads or spline socket heads. Setscrews are manufactured in number sizes from 0 to 10 and in fractional sizes from $\frac{1}{4}$ to 2 in. Setscrews are usually made from carbon or alloy steel and hardened.

Square head setscrews are often used on toolholders (Figure A-26) or as jackscrews in leveling machine tools (Figure A-27). Setscrews have several different points (Figure A-28). The flat point setscrew will make the least amount of indentation on a shaft and is used where frequent adjustments are made. A flat point setscrew is also used to provide a jam screw action when a second setscrew is tightened on another setscrew to prevent its release through vibration. The oval point setscrew will make a slight indentation as compared with the cone point. With a half dog or full dog point setscrew holding a collar to a shaft, alignment between shaft and collar will be maintained even when the parts are disassembled and reassembled. This is because the shaft is drilled with a hole of the same diameter as the dog point. Cup-pointed setscrews will make a ring-shaped depression in the shaft and will give a very slip-resistant connection. Square head setscrews have a class 2A thread and are usually supplied with a coarse thread. Slotted and socket head setscrews have a class 3A UNC or UNF thread.

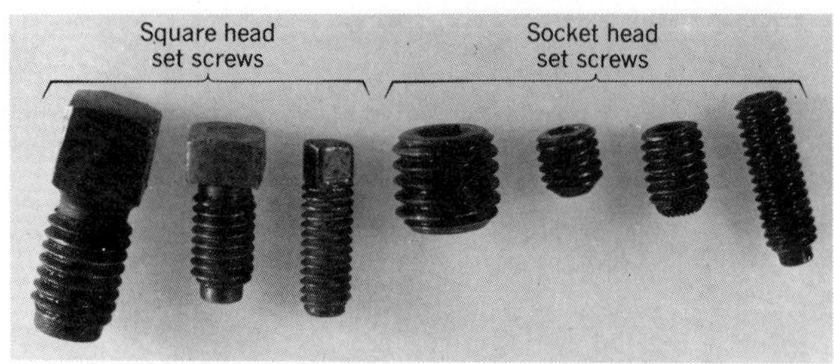

FIGURE A-25 Socket and square head setscrews.

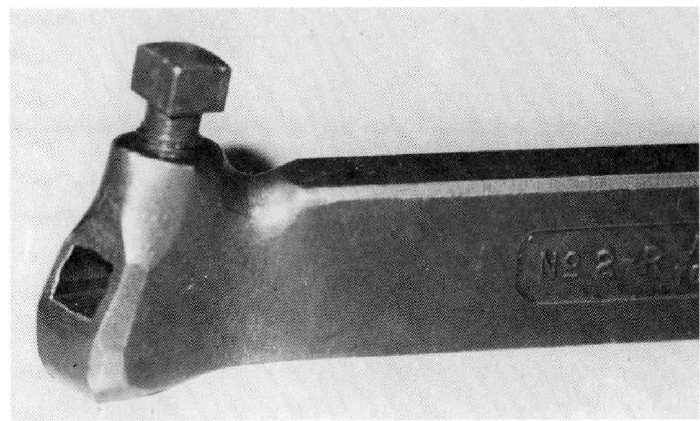

FIGURE A-26 Square head setscrews are found in tool holders (Lane Community College).

FIGURE A-27 Square head jack screw (Lane Community College).

FIGURE A-28 Setscrew points.

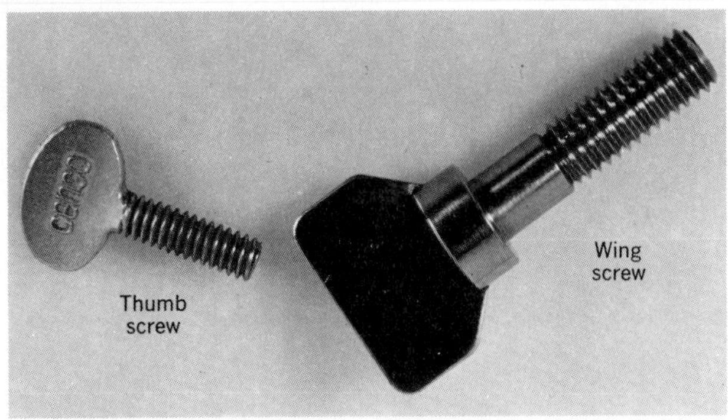

FIGURE A-29 Thumbscrew and wing screw.

Thumbscrews and wing screws (Figure A-29) are used where parts are to be fastened or adjusted rapidly without the use of tools.

Thread-forming screws (Figure A-30) form their own threads and eliminate the need for tapping. These screws are used in the assembly of sheet metal parts, plastics, and nonferrous material. Thread-forming screws form threads by displacing material with no cutting action. These screws require an existing hole of the correct size.

Thread-cutting screws (Figure A-31) make threads by cutting and producing chips. Because of the cutting action these screws need less driving torque than thread-forming screws. Applications are similar to those for thread-forming screws. These include fastening sheet metal, aluminum brass, die castings, and plastics.

Drive screws (Figure A-32) are forced into the correct size hole by hammering or with a press. Drive screws make permanent connections and are often used to fasten name plates or identification plates on machine tools.

Common Internally Threaded Fasteners

NUTS

Common nuts (Figure A-33) are manufactured in as many sizes as there are bolts. Most nuts are either hex (hexagonal) or square in shape. Nuts are identified by the size of the bolt they fit, not by their outside size. Common hex nuts are made in different thicknesses. A thin hex nut is called a jam nut. It is used where space is limited or where the strength of a regular nut is not required. Jam nuts are often used to lock other nuts (Figure A-34). Regular hex nuts are slightly thinner than their size designation. A $\frac{1}{2}$-in. regular hex nut is $\frac{7}{16}$ in. thick. A $\frac{1}{2}$-in. heavy hex nut is $\frac{31}{64}$ in. thick. A $\frac{1}{2}$-in.-high hex nut measures $\frac{11}{16}$ in. thick. Other common nuts include various stop nuts or locknuts. Two common types are the elastic stop nut and the compression stop nut. They are used in applications where the nut might vibrate off the bolt. Wing nuts and thumb nuts are used where quick assembly or disassembly by hand is desired. Other hex nuts are slotted and castle nuts. These nuts have slots cut into them. When the slots are aligned with holes in a bolt, a cotter pin may be used to prevent the nut from turning. Axles and spindles on vehicles have slotted nuts to prevent wheel bearing adjustments from slipping.

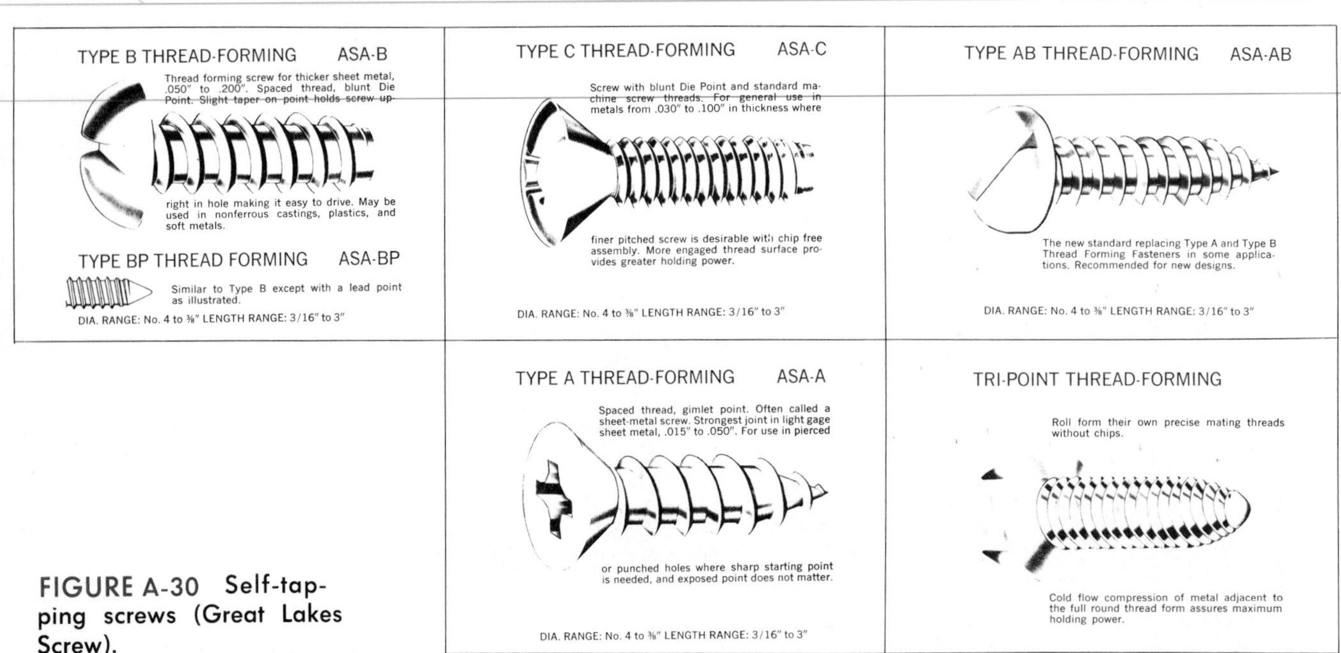

TYPE B THREAD-FORMING ASA-B
Thread forming screw for thicker sheet metal, .050" to .200". Spaced thread, blunt Die Point. Slight taper on point holds screw upright in hole making it easy to drive. May be used in nonferrous castings, plastics, and soft metals.
TYPE BP THREAD FORMING ASA-BP
Similar to Type B except with a lead point as illustrated.
DIA. RANGE: No. 4 to ⅜" LENGTH RANGE: 3/16" to 3"
TYPE C THREAD-FORMING ASA-C
Screw with blunt Die Point and standard machine screw threads. For general use in metals from .030" to .100" in thickness where finer pitched screw is desirable with chip free assembly. More engaged thread surface provides greater holding power.
DIA. RANGE: No. 4 to ⅜" LENGTH RANGE: 3/16" to 3"
TYPE AB THREAD-FORMING ASA-AB
The new standard replacing Type A and Type B Thread Forming Fasteners in some applications. Recommended for new designs.
DIA. RANGE: No. 4 to ⅜" LENGTH RANGE: 3/16" to 3"
TYPE A THREAD-FORMING ASA-A
Spaced thread, gimlet point. Often called a sheet-metal screw. Strongest joint in light gage sheet metal, .015" to .050". For use in pierced or punched holes where sharp starting point is needed, and exposed point does not matter.
DIA. RANGE: No. 4 to ⅜" LENGTH RANGE: 3/16" to 3"
TRI-POINT THREAD-FORMING
Roll form their own precise mating threads without chips.
Cold flow compression of metal adjacent to the full round thread form assures maximum holding power.

FIGURE A-30 Self-tapping screws (Great Lakes Screw).

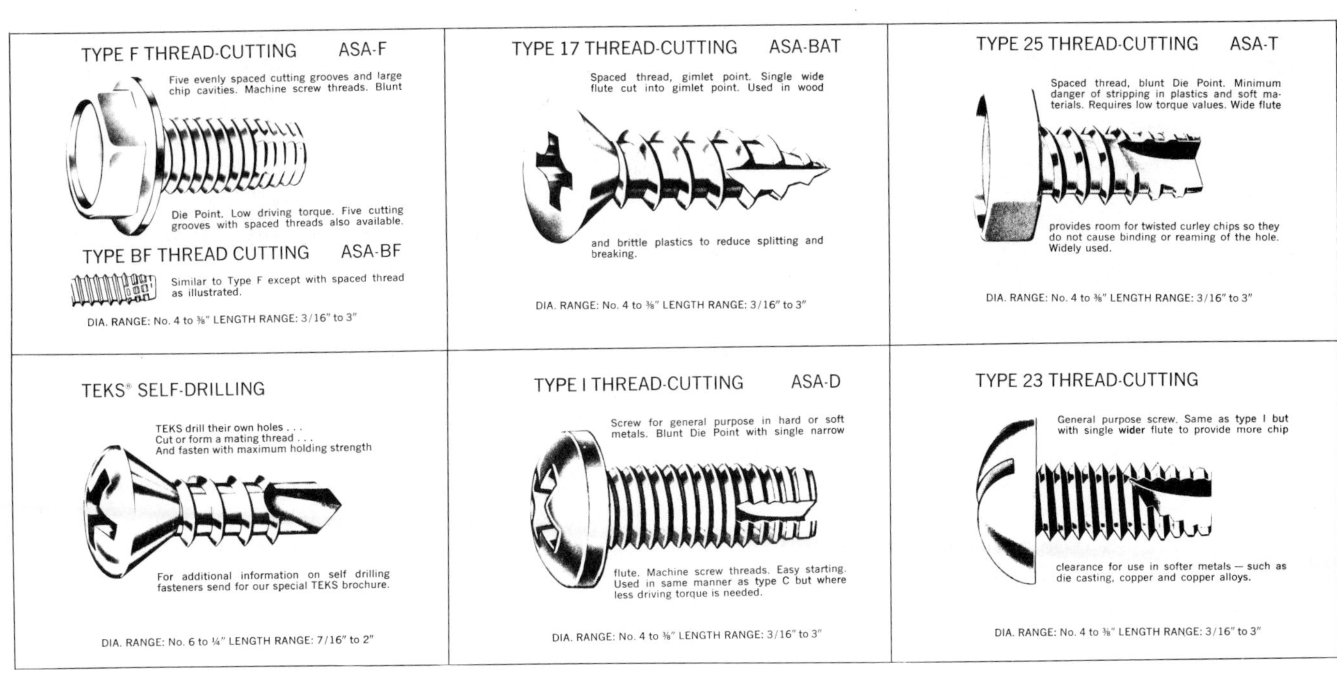

FIGURE A-31 Thread-cutting screws (Great Lakes Screw).

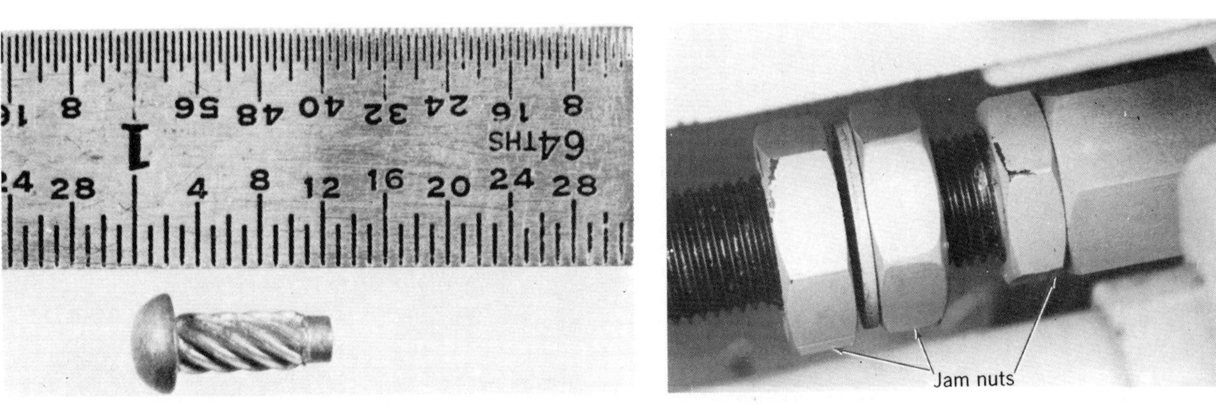

FIGURE A-32 Drive screw.

FIGURE A-34 Jam nuts.

FIGURE A-33 Common nuts.

Cap or acorn nuts are often used where decorative nuts are needed. These nuts also protect projecting threads from accidental damage. Nuts are made from many different materials, depending on their application and strength requirements.

Internal Thread Inserts

Internal thread inserts may be used when an internal thread is damaged or stripped and it is not possible to drill and tap for a larger size. A thread insert retains the original thread size. However, it is necessary to drill and tap a somewhat larger hole to accept the thread insert.

One common type of internal thread insert is the wedge type. The thread insert has both external and internal threads. This type of thread insert is screwed into a hole tapped to the same size as the thread on the outside of the insert. The four-wedges are driven in using a special driver (Figure A-35). This holds the insert in place. The internal thread in the insert is the same as the original hole.

A second type of internal thread insert is also used in repair applications as well as in new installations. Threaded holes are often required in products made from soft metals such as aluminum. If bolts, screws, or studs were to be screwed directly into the softer material, excessive wear could result, especially if the bolt is taken in and out a number of times. To overcome this problem, a thread insert made from a more durable material may be used. Stainless steel inserts are frequently used in aluminum (Figure A-36). This type of thread insert requires an insert tap, an insert driver, and a thread insert (Figure A-37). After the hole for the thread insert is tapped, the insert driver is used to screw the insert into the hole (Figure A-38). The end of the insert coil must be broken off and removed after the insert is screwed into place. The insert in the illustration is used to repair spark plug threads in engine blocks.

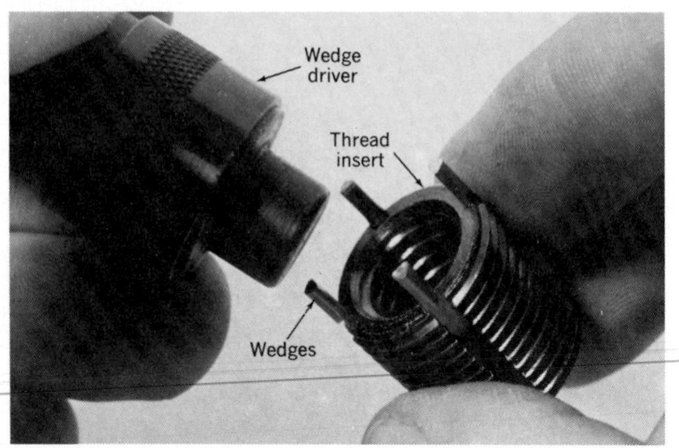

FIGURE A-35 Wedge-type internal thread insert.

FIGURE A-36 Stainless steel thread insert used in an aluminum valve housing.

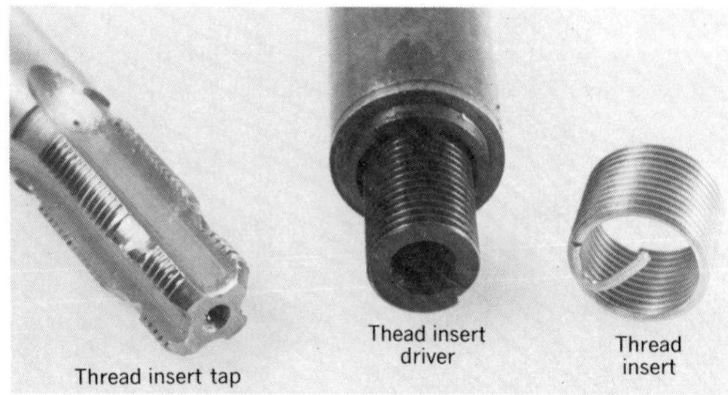

FIGURE A-37 Thread insert tap, driver, and repair insert for spark plug holes.

FIGURE A-38 Thread insert driver.

Washers, Pins, Retaining Rings, and Keys

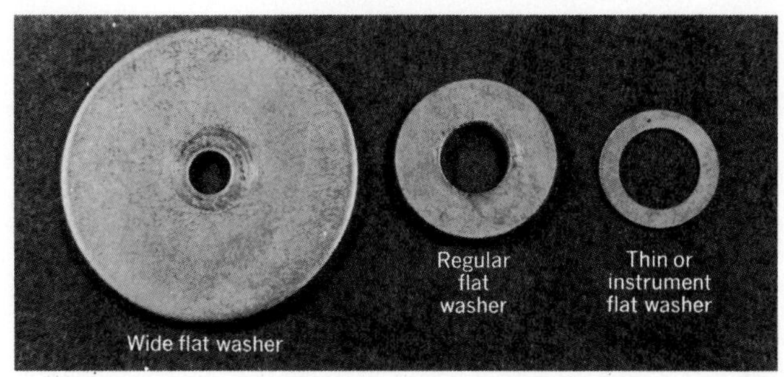

FIGURE A-39 Wide, regular, and thin (or instrument) flat washers.

WASHERS

Flat washers (Figure A-39) are used under nuts and bolt heads to distribute the pressure over a larger area. Washers also prevent the marring of a finished surface when nuts or screws are tightened. Washers can be manufactured from many different materials. The nominal size of a washer is intended to be used with the same nominal-size bolt or screw. Standard series of washers are narrow, regular, and wide. For example, the outside diameter of a $\frac{1}{4}$-in. narrow washer is $\frac{1}{2}$ in., the outside diameter of a $\frac{1}{4}$-in. regular washer is almost $\frac{3}{4}$ in., and the diameter of a wide $\frac{1}{4}$-in. washer measures 1 in.

Lock washers (Figure A-40) are manufactured in many styles. The helical spring lock washer provides hardened bearing surfaces between a nut or bolt head and the components of an assembly. The spring-type construction of this lock washer will hold the tension between a nut and bolt assembly even if a small amount of looseness should develop. Helical spring lock washers are manufactured in series: light, regular, heavy, extra duty, and hi-collar. The hi-collar lock washer has an outside diameter equal to the same nominal-size socket head

EXTERNAL TYPE
External type lock washers provide greater torsional resistance due to teeth being on largest radius. Screw heads should be large enough to cover washer teeth. Available with left hand or alternate twisted teeth.

INTERNAL TYPE
For use with small screw heads or in applications where it is necessary to hide washer teeth for appearance or snag prevention.

EXTERNAL-INTERNAL TYPE
For use where a larger bearing surface is needed such as extra large screw heads or between two large surfaces. More biting teeth for greater locking power. Excellent for oversize or elongated screw holes.

HEAVY DUTY INTERNAL TYPE
Recommend for use with larger screws and bolts on heavy machinery and equipment.

DOME TYPE PLAIN PERIPHERY
For use with soft or thin materials to distribute holding force over larger area. Used also for oversize or elongated holes. Plain periphery is recommended to prevent surface marring.

DOME TYPE TOOTHED PERIPHERY
For use with soft or thin materials to distribute holding force over larger area. Used also for oversize or elongated holes. Toothed periphery should be used where additional protection against shifting is required.

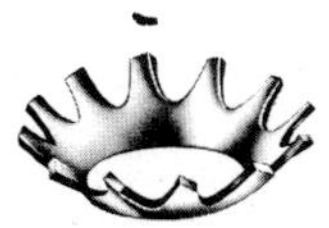

COUNTERSUNK TYPE
Countersunk washers are used with either flat or oval head screws in recessed countersunk applications. Available for 82° and 100° heads and also internal or external teeth.

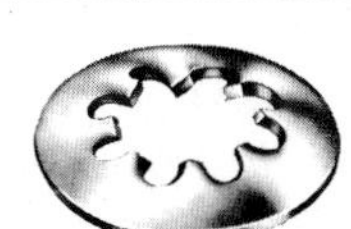

DISHED TYPE PLAIN PERIPHERY
Recommended for the same general applications as the dome type washers but should be used where more flexibility rather than rigidity is desired. Plain periphery for reduced marring action on surfaces.

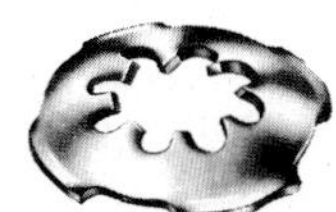

DISHED TYPE TOOTHED PERIPHERY
Recommended for the same general applications as the dome type washers but should be used where more flexibility rather than rigidity is desired. Toothed periphery offers additional protection against shifting.

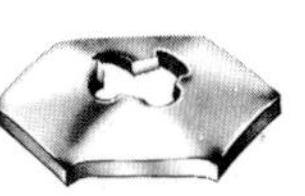

PYRAMIDAL TYPE
Specially designed for situations requiring very high tightening torque. The pyramidal washer offers bolt locking teeth and rigidity yet is flexible under heavy loads Available in both square and hexagonal design.

FINISH TYPE
Recommended where marring or tearing of surface material by turning screw head must be prevented and for decorative use.

HELICAL SPRING LOCK TYPE
Spring lock washers may be used to eliminate annoying rattles and provide tension at fastening points.

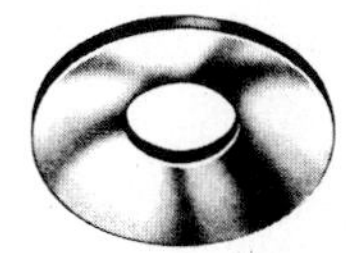

CONE SPRING TYPE

CONE SPRING TYPE SERRATED PERIPHERY
Same general usage as the cone type with plain periphery but with the added locking action of a serrated periphery. Takes high tightening torque.

FLAT TYPE
For use with oversize and elongated screw holes. Spreads holding force over a larger area. Used also as a spacer. Available in all metals.

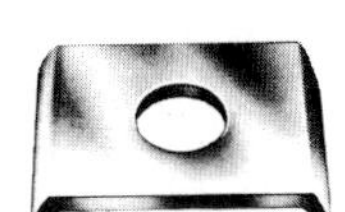

SPECIAL TYPES
Special washers with irregular holes, cup types, plate types with multiple holes or tab types may be supplied upon request. Consult our engineering department for any of your special needs.

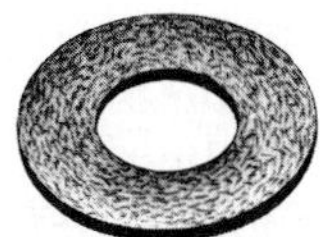

FIBER AND ASBESTOS
In cases where insulation or corrosion resistance is more important than strength, fiber or asbestos washers are available.

DOUBLE SEMS
Two washers securely held from slipping off, yet free to spin and lock. Prevents gouging of soft metals.

FIGURE A-40 Lock washers (Great Lakes Screw).

capscrew. This makes the use of these lock washers in a counterbored bolt hole possible. Counterbored holes have the end enlarged to accept the bolt head. A variety of standard tooth lock washers are produced, the external type providing the greatest amount of friction or locking effect between fastener and assembly. For use with small head screws and where a smooth appearance is desired, an internal tooth lock washer is used. When large bearing area is desired or where the assembly holes are oversized, an internal–external tooth lock washer is available. A countersunk tooth lock washer is used for a locking action with flat head screws.

PINS

Pins (Figure A-41) find many applications in the assembly of parts. Dowel pins are heat treated and precision ground. Their diameter varies from the nominal dimension by only plus or minus .0001 in. ($\frac{1}{10000}$ of an inch). Dowel pins are used where very accurate alignments must be maintained between two or more parts. Holes for dowel pins are reamed to provide a slight press fit. Reaming is a machining process during which a drilled hole is slightly enlarged to provide a smooth finish and accurate diameter. Dowel pins only locate. Clamping pressure is supplied by the screws. Dowel pins may be driven into a blind hole. A blind hole is closed at one end. When this kind of hole is used, provision must be made to let the air that is displaced by the pin escape. This can be done by drilling a small through hole or by grinding a narrow flat the full length of the pin. Always use the correct lubricant when making screw and pin assemblies.

One disadvantage of dowel pins is that they tend to enlarge the hole in an unhardened workpiece if they are driven in and out several times. When parts are intended to be disassembled frequently, **taper pins** will give accurate alignment. Taper pins have a taper of $\frac{1}{4}$ in. per foot of length and are fitted into reamed taper holes. If a taper pin hole wears larger because of frequent disassembly, the hole can be reamed larger to receive the next larger size of taper pin. Diameters of taper pins range in size from $\frac{1}{16}$ in. to $\frac{11}{16}$ in. measured at the large end. Taper pins are identified by a number from 7/0 (small diameter) to number 10 (large diameter) as well as by their length. The large end diameter is constant for a given size pin, but the small diameter changes with the length of the pin. Some taper pins have a threaded portion on the large end. A nut can be threaded on the pin and used to pull the pin from the hole much like a screw jack. This facilitates removal of the pin.

A **grooved pin** is either a cylindrical or a tapered pin with longitudinal grooves pressed into

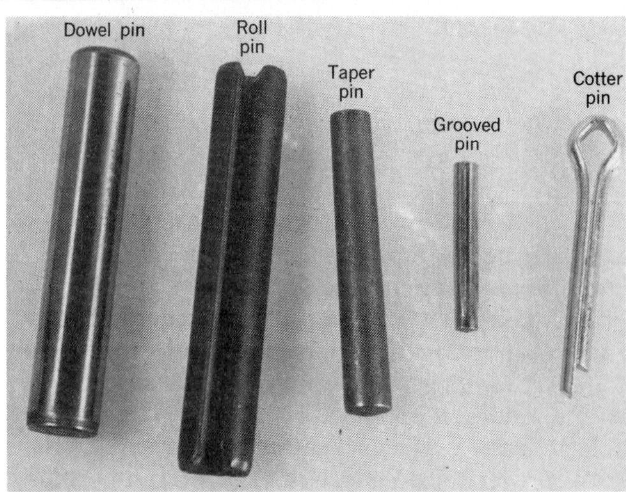

FIGURE A-41 Pins.

the pin body. This causes the pin to deform. A groove pin will hold securely in a drilled hole even after repeated removal.

Roll pins can also be used in drilled holes with no reaming required. These pins are manufactured from flat steel bands and rolled into cylindrical shape. Roll pins, because of their spring action, will stay tight in a hole even after repeated disassemblies.

Cotter pins are used to retain parts on a shaft or to lock a nut or bolt as a safety precaution. Cotter pins make a quick assembly and disassembly possible.

RETAINING RINGS

Retaining rings are fasteners used in many assemblies. Retaining rings can be easily installed in machined grooves, internally in housings, or externally on shafts or pins (Figure A-42). Some types of retaining rings do not require grooves but have a self-locking spring-type action. The most

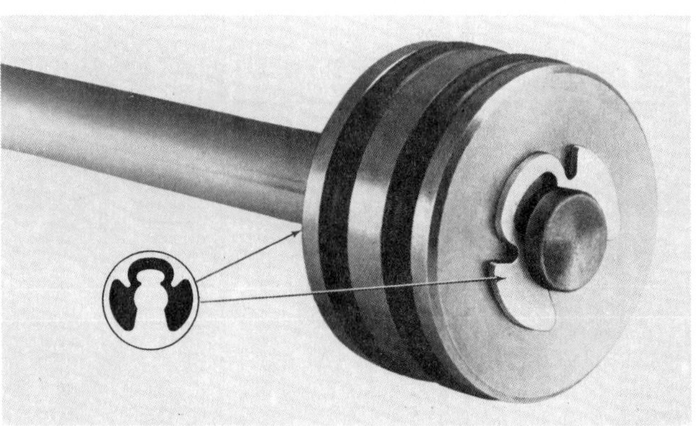

FIGURE A-42 External retaining ring used on a shaft (Waldes Kohinoor, Inc.).

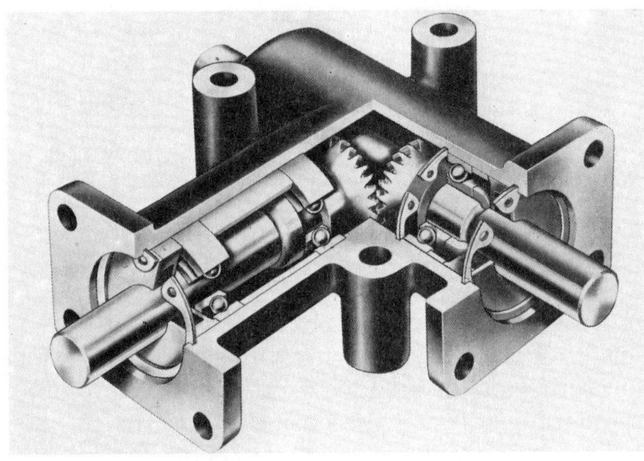

FIGURE A-43 Internal retaining rings used to retain bearings (Waldes Kohinoor, Inc.).

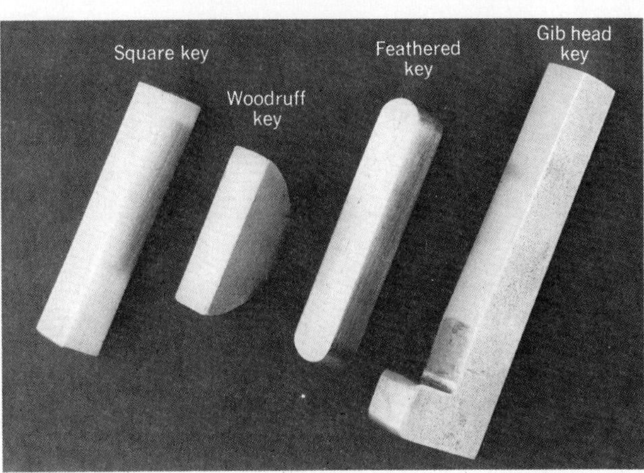

FIGURE A-44 Keys (Lane Community College).

common application of a retaining ring is to provide a shoulder to hold and retain a bearing or other part on an otherwise smooth shaft. They may also be used in a bearing housing (Figure A-43). Special pliers are used to install and remove retaining rings.

KEYS

Keys (Figure A-44) are used to prevent the rotation of gears or pulleys on a shaft. Keys are fitted into key seats in both the shaft and the external part. Keys should fit the key seats rather snugly. **Square keys**, where the width and the height are equal, are preferred on shaft sizes up to a $6\frac{1}{2}$-in. diameter. Above a $6\frac{1}{2}$-in. diameter, rectangular keys are recommended. **Woodruff keys**, which are almost in the shape of a half circle, are used where relatively light loads are transmitted. One advantage of woodruff keys is that they cannot change their axial location on a shaft because they are retained in a pocket. A key fitted into an end milled pocket will also retain its axial position on the shaft. Most of these keys are held under tension with one or more set screws threaded through the hub of the pulley or gear. Where extremely heavy shock loads or high torques are encountered, a **taper key** is used. Taper keys have a taper of $\frac{1}{8}$ in. per foot. Where a tapered key is used, the key seat in the shaft is parallel to the shaft axis and a taper to match the key is in the hub. Where only one side of an assembly is accessible, a **gib head taper key** is used instead of a plain taper key. When a gib head taper key is driven into the key seat as far as possible, a gap remains between the gib and the hub of the pulley or gear. The key is removed for disassembly by driving a wedge into the gap to push the key out. A **feathered key** is a key that is secured in a key seat with screws. A feathered key is often a part of a sliding gear or sliding pulley.

SELF-TEST

1. What is the difference between a bolt and a screw?
2. How much thread engagement is recommended when a screw is used in an assembly?
3. When are class 3 threads used?
4. What is the difference between a machine bolt and a capscrew?
5. What is the outside diameter of a No. 8 machine screw?
6. Where are setscrews used?
7. When are stud bolts used?
8. Explain the difference between thread-forming and thread-cutting screws.
9. Where are castle nuts used?
10. Where are cap nuts used?
11. Explain two reasons why flat washers are used.
12. What is the purpose of a helical spring-lock washer?
13. When is an internal–external tooth lock washer used?
14. When are dowel pins used?
15. When are taper pins used?
16. When are roll pins used?
17. What are retaining rings?
18. What is the purpose of a key?
19. When is a woodruff key used?
20. When is a gib head key used?

UNIT 3

Reading Drawings

From earliest times, people have communicated their thoughts through drawings. The pictorial representation of an idea is a vital line of communication between the designer and the people who produce the final product. Technological design would be impossible were it not for the several different ways an idea may be represented by a drawing. The drawing also provides an important testing phase for an idea. Many times an idea may be rejected at the drawing board stage before a large investment is made to equip a manufacturing facility and risk production of an item that does not meet the design requirements.

This does not mean that all design problems can be solved in the drafting room. Almost anything can be represented by a drawing, even those designs that would be quite impossible to manufacture. It is important that the designer be aware of the problems that confront the machinist. On the other hand, you must fully understand all of the symbols and terminology on the designer's drawing. You must then interpret these terms and symbols in order to transform the ideas of the designer into useful products.

Objective

After completing this unit, you should be able to:

Read and interpret common detail drawings found in the machine shop.

Isometric Drawing

An isometric drawing (Figure A-45) is one method used to represent an object in three dimensions. In the isometric format, the lines of the object remain parallel and the object is drawn about the three isometric axes that are 120 degrees apart.

Oblique Drawing

Object lines in the oblique drawing (Figure A-46) also remain parallel. The oblique differs from the isometric in that one axis of the object is parallel to the plane of the drawing. Isometric and oblique are not generally used as working drawings for the

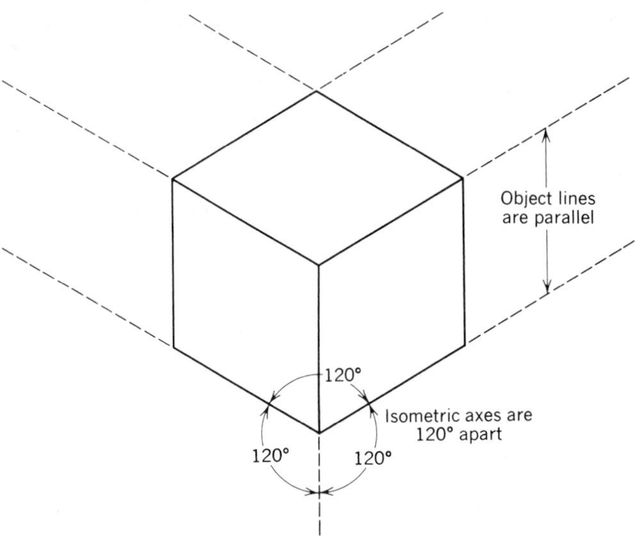

FIGURE A-45 Isometric drawing.

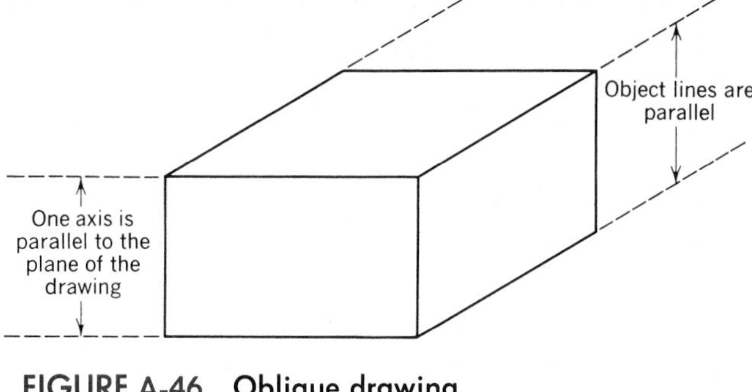

FIGURE A-46 Oblique drawing.

machinist. However, you may occasionally see them in the machine shop.

Exploded Drawings

The exploded drawing (Figure A-47) is a type of pictorial drawing designed to show several parts in their proper location prior to assembly. Although the exploded view is not used as the working drawing for the machinist, it has an important place in mechanical technology. Exploded views appear extensively in manuals and handbooks that are used for repair and assembly of machines and other mechanisms.

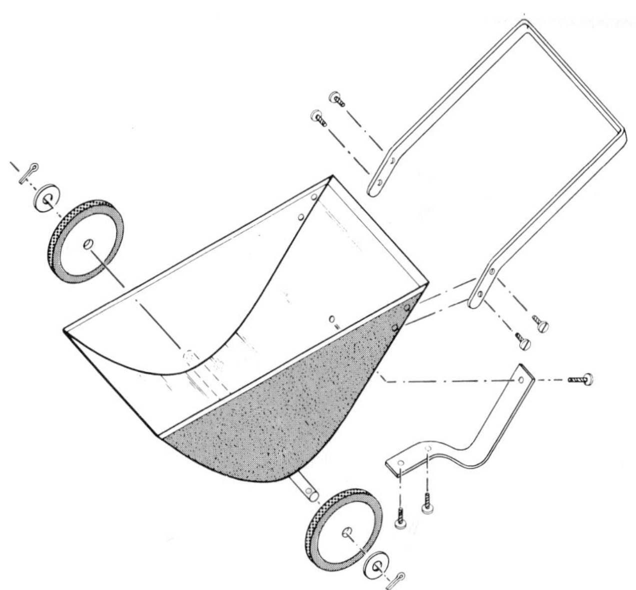

FIGURE A-47 Exploded drawing.

Orthographic Drawings

THE ORTHOGRAPHIC PROJECTION DRAWING

In almost every case, the working drawing for the machinist will be in the form of the **three-view** or **orthographic drawing**. The typical orthographic format always shows an object in the three-view combination of side, end, and top (Figure A-48). In some cases, an object can be completely shown by a combination of only two orthographic views. However, any orthographic drawing must have a minimum of two views in order to show an object completely. The top view may be referred to as the plan view. The front or side views may be referred to as the elevation views. The terms *plan* and *elevation* may appear on some drawings, especially those of large complex parts or assemblies.

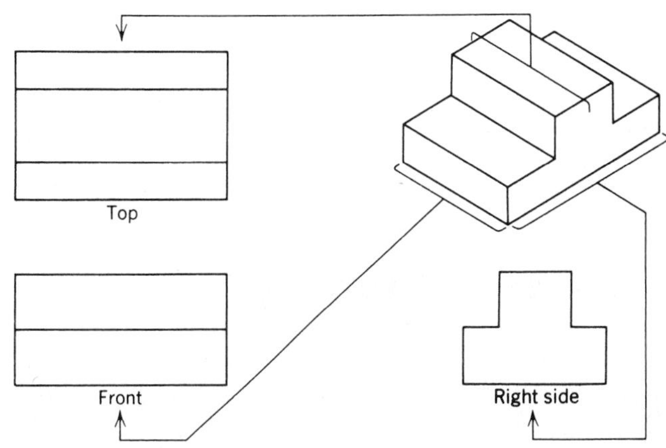

FIGURE A-48 Standard orthographic drawings.

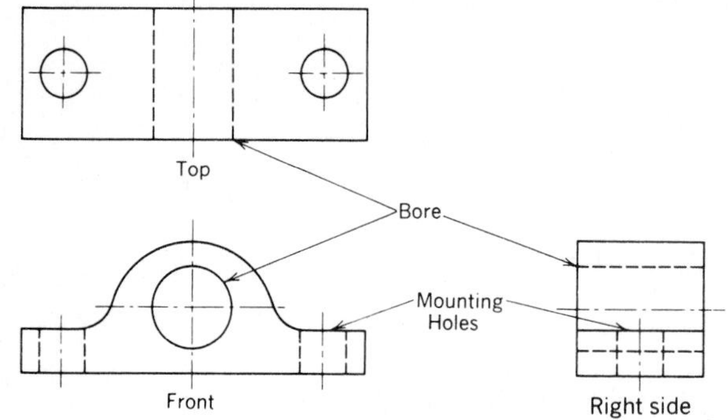

FIGURE A-49 Hidden lines for part features not visible.

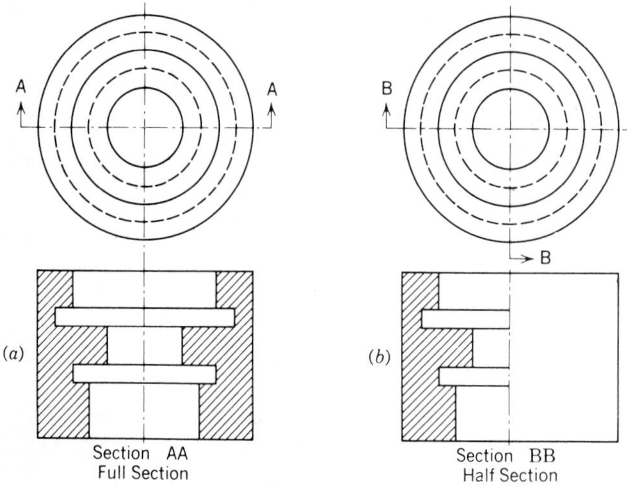

FIGURE A-50 Sectioned drawings.

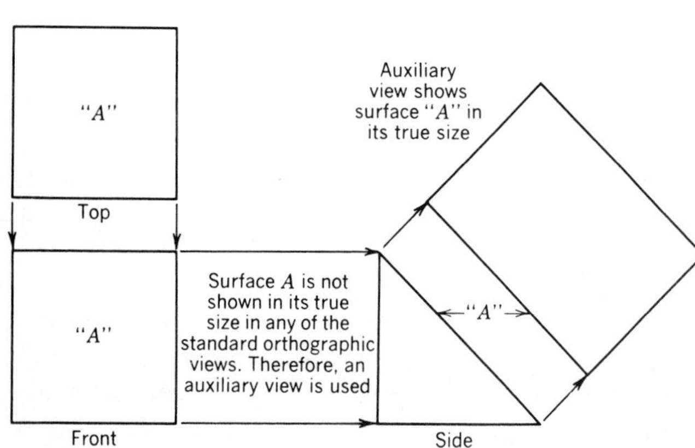

FIGURE A-51 Auxiliary view.

HIDDEN LINES FOR PART FEATURES NOT VISIBLE

Features that are not visible are indicated by dashed lines. These are called hidden lines, as they indicate the locations of part features hidden from view. The plain bearing (Figure A-49) is shown in a typical orthographic drawing. The front view is the only one in which the hole through the bearing, or bore, can be observed. In the side and top views, the bore is not visible. Therefore, it is indicated by dashed or hidden lines. The mounting holes through the base are visible only in the top view. They appear as hidden lines in the front and side views.

SECTIONED VIEWS

When internal features are complex to the extent that indicating them as hidden lines would be confusing, a sectioned drawing may be employed. Two common styles of sections are used. In the full section (Figure A-50*a*), the object has been cut completely through. In the half section (Figure A-50*b*), one-quarter of the object is removed. The section indicator line shows the plane at which the section is taken. For example (Figure A-50*a*), the end view of the object shows the section line marked "AA." Section line BB (Figure A-50*b*) indicates the portion removed in the half section. An object may be sectioned at any plane as long as the section plane is indicated on the drawing.

AUXILIARY VIEWS

One of the reasons for adopting the orthographic drawing is to represent an object in its true size and shape. This is not possible with the pictorial drawings discussed earlier. Generally, the orthographic drawing meets this requirement. However, the shape of certain objects is such that their actual size and shape are not truly represented. An **auxiliary view** may be required (Figure A-51). On the object shown in the figure, surface A does not appear in its true size in any of the standard orthographic views. Therefore, surface A is projected to the auxiliary view, thus revealing its true size.

Reading and Interpreting Drawings

SCALE

In some cases, an object may be represented by a drawing that is the same size as the object. In other cases, an object may be too large to draw full size, or a very small part may be better represented by a drawing that is larger. Therefore, all drawings are drawn to a specific **scale**. For example, when the drawing is the same size as the object, the scale is said to be full, or 1 = 1. If the drawing is one-half size, the scale is one-half, or $\frac{1}{2} = 1$. A drawing twice actual size would be double scale, or 2 = 1. The scale used is generally indicated on the drawing.

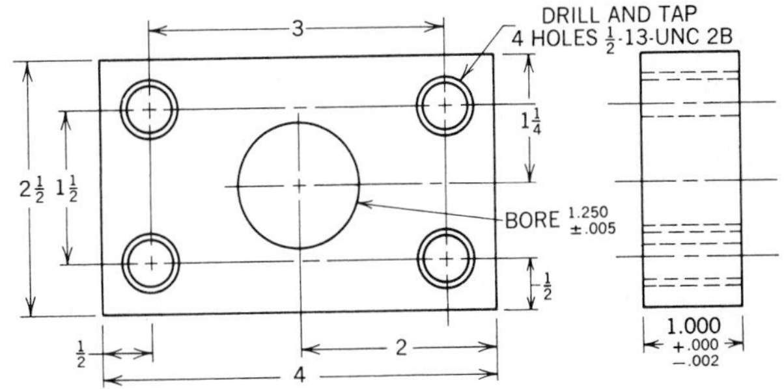

FIGURE A-52 Fractional inch dimensioning.

DIMENSIONING OF DETAIL AND ASSEMBLY DRAWINGS

You will primarily come in contact with the **detail drawing**. This is a drawing of an individual part and, in almost all cases, will appear in orthographic form. Depending on the type of work you are doing, you may also see an **assembly drawing**. The assembly drawing is a drawing of subassemblies or several individual parts assembled into a complete unit. For example, a drawing of a complete automobile engine would be an assembly drawing. In addition, a detail drawing of each engine component would also exist.

A detail drawing contains all the essential information you need in order to make the part. Most important are the **dimensions**. Dimensioning refers generally to the sizes specified for the part and the locations of its features. Furthermore, dimensions reflect many design considerations, such as the fit of mating parts, that will affect the operation characteristics of all machines. Much of the effort you expend performing the various machining operations will be directed toward controlling the dimensions specified on the drawing.

Several styles of dimensioning appear on drawings. The most common of these is the standard **fractional inch** notation (Figure A-52). The outline of the part along with several holes are dimensioned according to size and location. Generally, the units of the dimensions are not shown. Note that certain dimensions are specified to come within certain ranges. This is known as **tolerance**.

Tolerance refers to an acceptable range of part size or feature location and is generally expressed in the form of a minimum and maximum limit. The bore (Figure A-52) is shown to be 1.250 in. ± .005 in. This notation is called a **bilateral tolerance** because the acceptable size range is both above and below the nominal (normal) size of 1.250 in. The bored hole could be any size from 1.245 to 1.255 in diameter. The thickness of the part is specified as 1.000 + .000 and − .002. This tolerance is **unilateral** as all the range is on one side of nominal. Thus the thickness could range from .998 to 1.000 in.

No tolerance is specified for the outside dimensions of the part or the locations of the various features. Since these dimensions are indicated in standard fractional form, the tolerance is taken to be plus or minus $\frac{1}{64}$ of an inch unless otherwise specified on the drawing. This range is known as **standard tolerance** and applies only to dimensions expressed in standard fractional form.

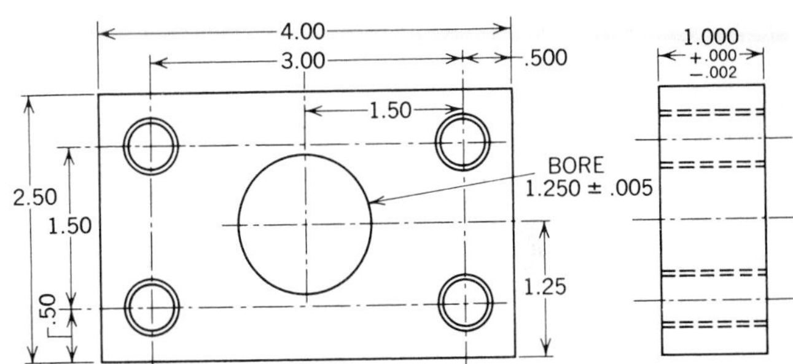

FIGURE A-53 Decimal inch dimensioning.

Another system of dimensioning used in certain industries is that of **decimal fraction** notation (Figure A-53). In this case, tolerance is determined by the number of places indicated in the decimal notation:

2 places	.00 tolerance is ± .010
3 places	.000 tolerance is ± .005
4 places	.0000 tolerance is ± .0005

Always remember that standard tolerances apply only when no other tolerance is specified on the drawing.

The **coordinate** or **absolute** system of dimensioning (Figure A-54) may be found in special applications such as numerically controlled machining. In this system, all dimensions are specified from the same zero point. The figure shows the dimensions expressed in decimal form. Standard fracation notation may also be used.

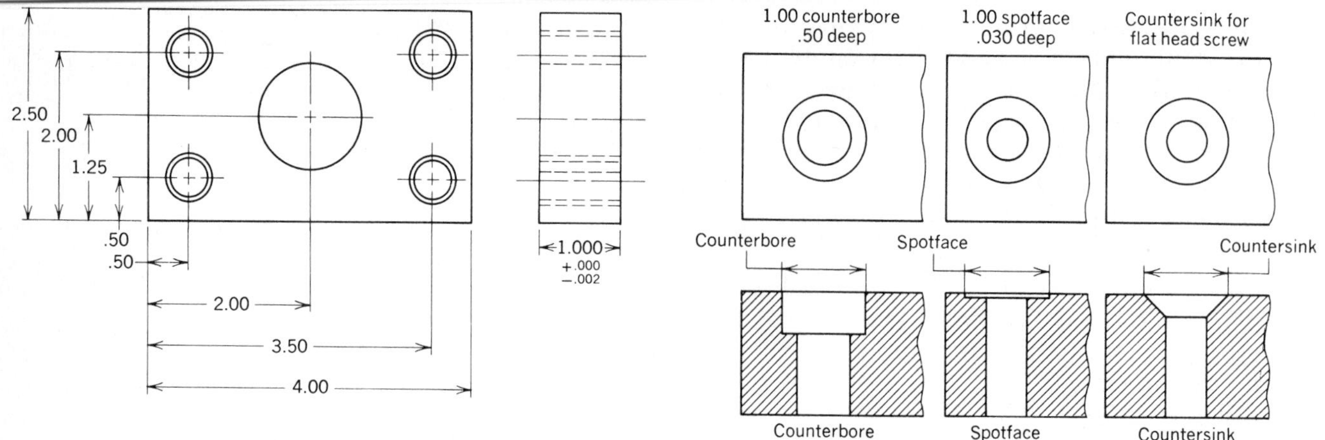

FIGURE A-54 Absolute or coordinate dimensioning.

FIGURE A-56 Countersinking, counterboring, and spotfacing symbols.

Standard tolerances apply unless otherwise specified.

With the increase in metrification in recent years, some industries have adopted a system of dual dimensioning of drawings with both metric and inch notation (Figure A-55). Dual dimensioning has, in some cases, created a degree of confusion for the machinist. Hence industry is constantly devising improved methods by which to differentiate metric and inch drawing dimensions. You must use caution when reading a dual-dimensioned drawing to ensure that you are conforming your work to the proper system of measurement for your tools. In the figure, metric dimensions appear above the line and inch dimensions appear below the line.

ABBREVIATIONS FOR MACHINE OPERATIONS (FIGURE A-56)

Working drawings contain several symbols and abbreviations that convey important information to the machinist. For example, certain machining operations may be abbreviated. **Countersinking** is a machining operation in which the end of a hole is

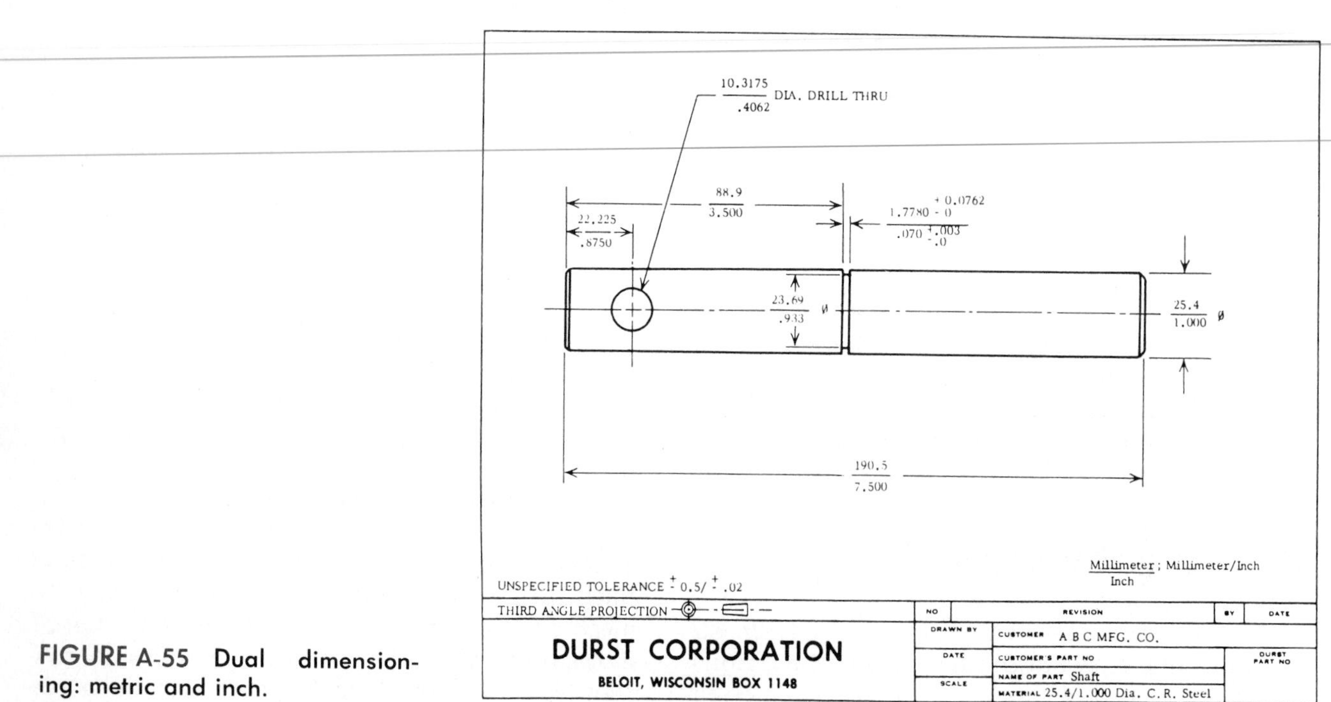

FIGURE A-55 Dual dimensioning: metric and inch.

shaped to accept a flat head screw. On a drawing, countersinking may be abbreviated as CS or CSK. The desired angle will also be specified. In **counterboring**, the end of a hole is enlarged in diameter so that a bolt head may be recessed. Counterboring may be abbreviated C'BORE. **Spotfacing** is usually spelled out. This operation is similar to counterboring except that the spotfacing depth is only sufficient to provide a smooth and flat surface around a hole.

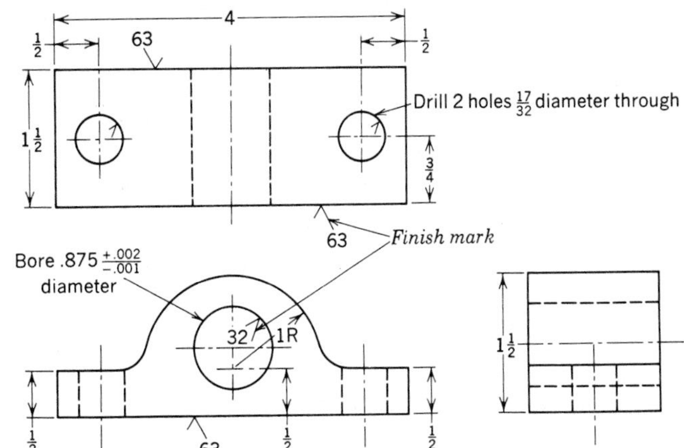

FIGURE A-57 Finish marks.

FINISH MARKS (FIGURE A-57)

Very often you will perform work on a part that has already been partially shaped. An example of this might be a casting or forging. The **finish mark** is used to indicate which surfaces are to be machined. Furthermore, the finish mark may also indicate a required degree of surface finish. For example, a finish mark notation of 4, 32, or 64 refers to a specific surface finish. The numerical data with the surface finish symbol indicates the average height of the surface deviations measured in **microinches** (millionths of inches). A microinch equals one millionth of an inch (.000001 in.).

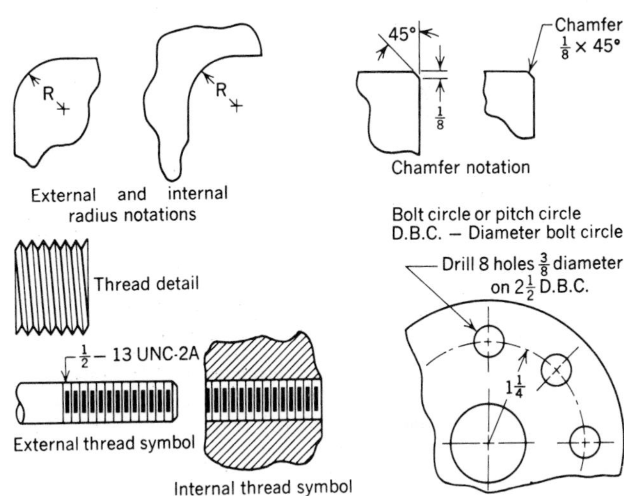

FIGURE A-58 Other symbols and abbreviations.

OTHER COMMON SYMBOLS AND ABBREVIATIONS (FIGURE A-58)

External and internal radii are generally indicated by the symbol *R* and the specified size. **Chamfers** may be indicated by size and angle as shown in the figure. Threads are generally represented by symbols, but they may be drawn in detail. Threads will also have a notation that indicates type, size, and fit. Consider the notation $\frac{1}{2}$–13 UNC 2A. This thread notation indicates the following:

$\frac{1}{2}$	Major thread diameter, in inches
13	Number of threads per inch
UNC	Form and series of thread
2	Class of fit
A	External thread (internal is denoted B)

A specific bolt circle or pitch circle is often indicated by the abbreviation **D.B.C.** or **B.C.** meaning **diameter of bolt circle**. The size of the diameter is indicated by normal dimensioning or with an abbreviation such as $1\frac{1}{2}$ D.B.C. or $1\frac{1}{2}$ B.C.

Drawing Formats

A designer's idea may at first appear as a freehand sketch in one of the pictorial forms discussed previously. After further discussion and examination, the decision may be made to have a part or an assembly manufactured. This necessitates suitable orthographic drawings that can be supplied to the machine shop. The original drawings produced by the drafting department are not used directly by the machine shop. These original drawings must be carefully preserved, as a great deal of time and money has been invested in them. Were they to be sent directly to the machine shop, they would soon be destroyed by constant handling. Therefore, a copy of the original drawing is made.

Several methods are employed to obtain copies of original drawings. One of the most common is **blueprinting**. Any number of drawings may be made and distributed to the various departments of a manufacturing facility. For example, assembly drawings are needed in the assembly area while

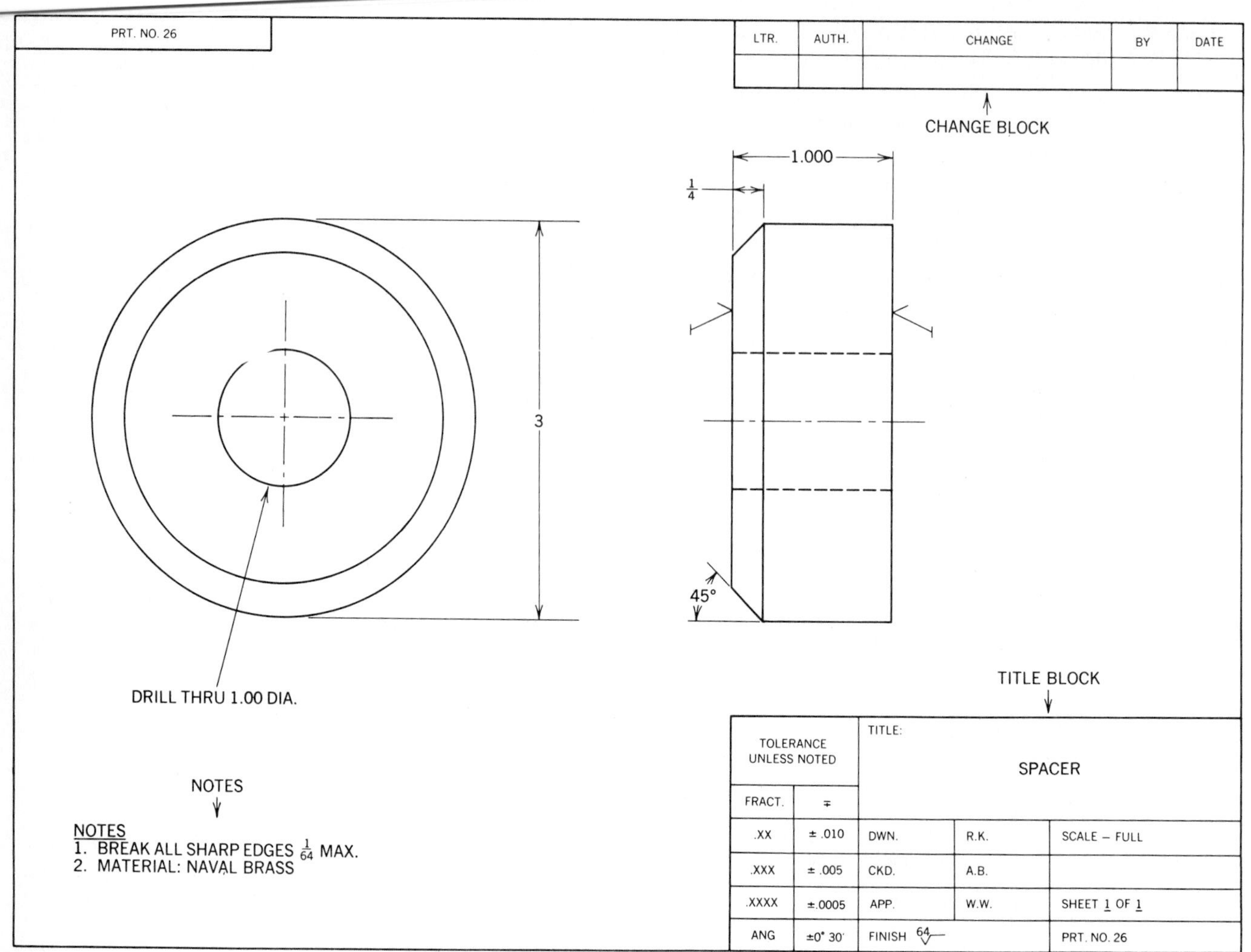

FIGURE A-59 Detail drawing.

drawings of individual parts are required at the machine tool stations and in the inspection department.

The typical detail drawing format (Figure A-59) contains a suitable title block. In many cases, the name of the firm appears in the title block (Figure A-55). The block also contains the name of the part, specified tolerance, scale, and the initials of the draftsman. A finish mark notation may also appear. The drawing may also contain a change block. Often designs may be modified after an original drawing is made. Subsequent drawings will reflect any changes. A drawing may also contain one or more general notes. The notes contain important information for the machinist. Therefore, you should always find and read any general notes appearing on a blueprint.

A typical assembly drawing format contains esentially the same information as found on the detail plan (Figure A-60). However, assembly drawings generally show only those dimensions that pertain to the assembly. Dimensions of the individual parts are found on the detail plans. In addition to the normal information, a **bill of materials** appears on the assembly plan. This bill contains the part number, description, size, material, and required quantity of each piece in the assembly. Often, the source of a specific item not manufactured by the assembler wil be specified in the bill of material. An assembly drawing may also include a list of references to detail drawings of the parts in the assembly. Any general notes containing information regarding the assembly will also be included.

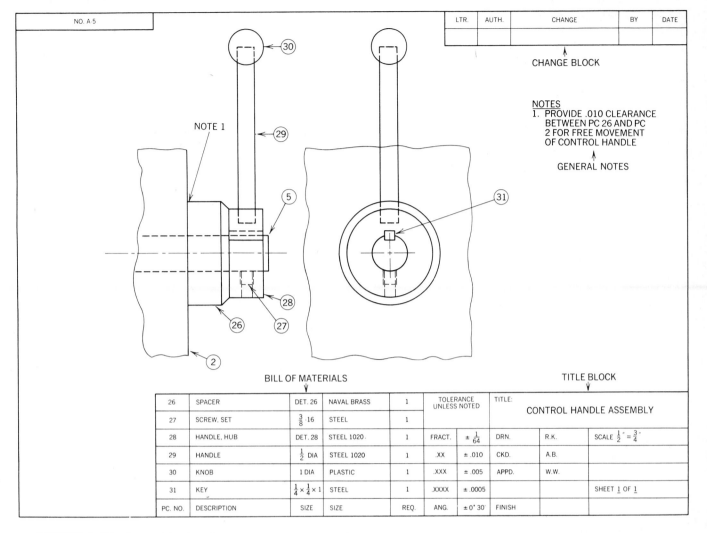

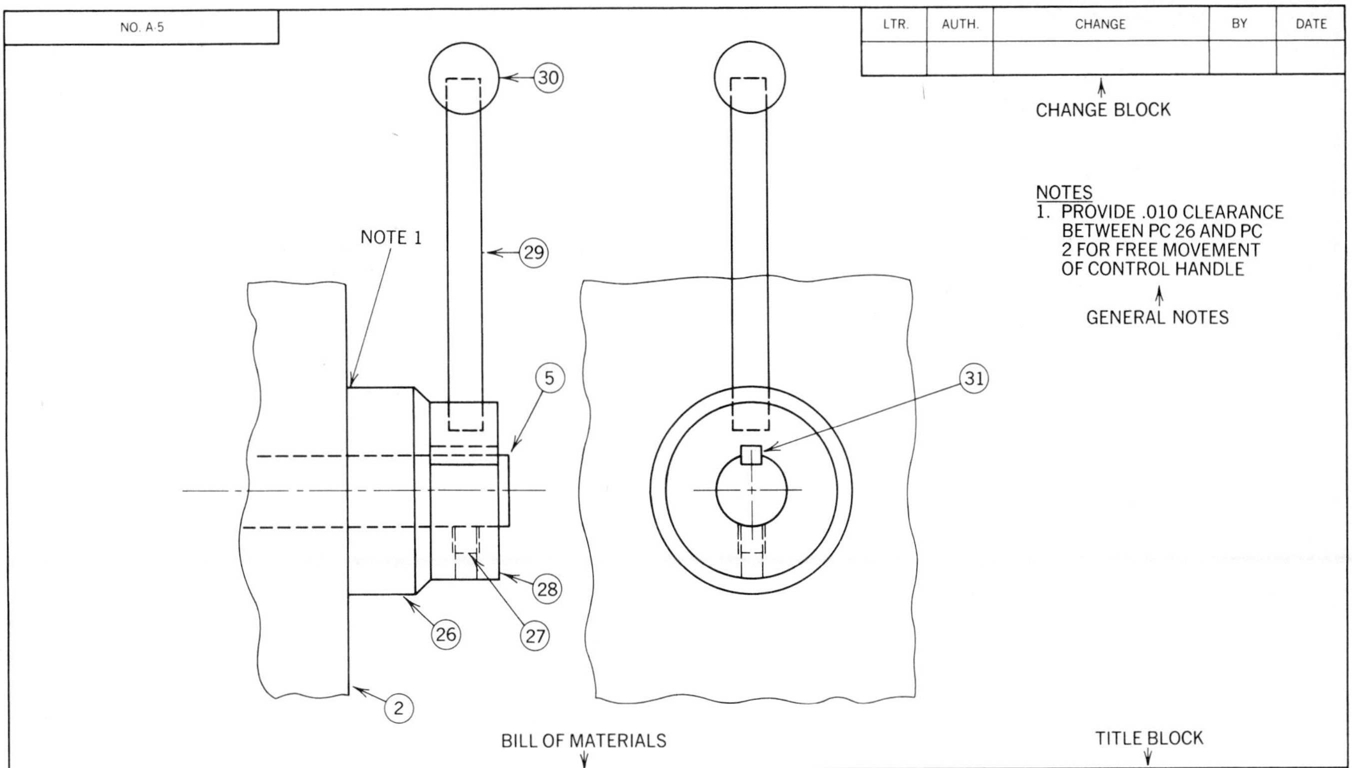

NO. A-5

LTR.	AUTH.	CHANGE	BY	DATE

CHANGE BLOCK

NOTES
1. PROVIDE .010 CLEARANCE BETWEEN PC 26 AND PC 2 FOR FREE MOVEMENT OF CONTROL HANDLE

GENERAL NOTES

NOTE 1

30 29 5 28 26 27 2 31

BILL OF MATERIALS

TITLE BLOCK

26	SPACER	DET. 26	NAVAL BRASS	1	TOLERANCE UNLESS NOTED		TITLE: CONTROL HANDLE ASSEMBLY		
27	SCREW, SET	$\frac{3}{8}$-16	STEEL	1					
28	HANDLE, HUB	DET. 28	STEEL 1020	1	FRACT.	$\pm\frac{1}{64}$	DRN.	R.K.	SCALE $\frac{1}{2}" = \frac{3}{4}"$
29	HANDLE	$\frac{1}{2}$ DIA	STEEL 1020	1	.XX	±.010	CKD.	A.B.	
30	KNOB	1 DIA	PLASTIC	1	.XXX	±.005	APPD.	W.W.	
31	KEY	$\frac{1}{4}\times\frac{1}{4}\times 1$	STEEL	1	.XXXX	±.0005			SHEET 1 OF 1
PC. NO.	DESCRIPTION	SIZE	SIZE	REQ.	ANG.	±0° 30′	FINISH		

FIGURE A-60 Assembly drawing.

SELF-TEST

1. Sketch the object in Figure A-61 as it would appear in correct orthographic form.

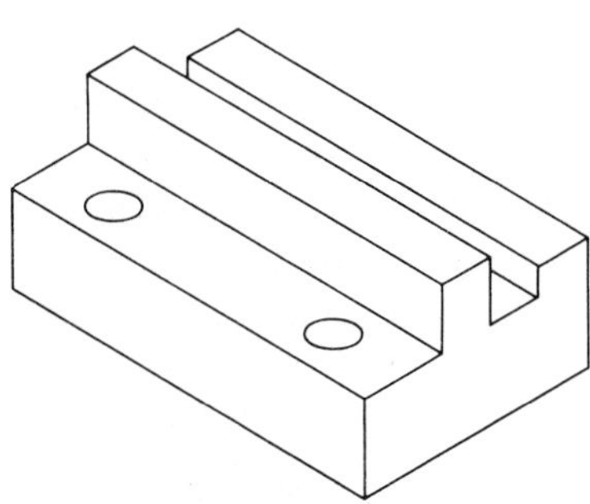

FIGURE A-61

For problems 2 to 10, refer to Figure A-62.

2. What is the minimum size of the hole through the clevis head?

3. What length of thread is indicated on the drawing?

4. What is the tolerance of the slot in the clevis head?

5. What radius is specified where the shank and clevis head meet?

6. What is the total length of the part?

7. What is the width of the slot in the clevis head?

8. Name two machining operations specified on the drawing.

9. What is the size and angle of the chamfer on the thread end?

10. What does note 1 mean?

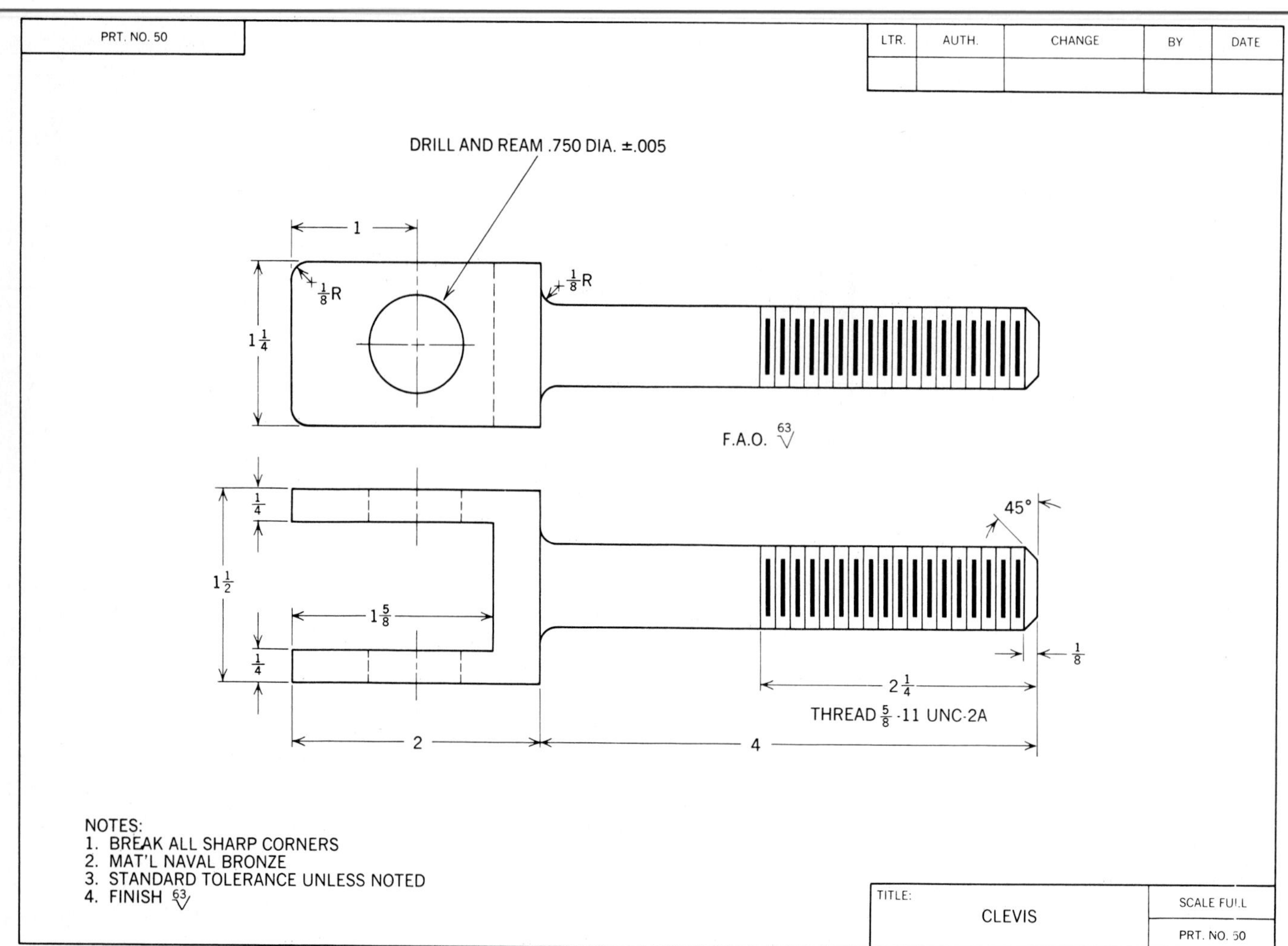
PRT. NO. 50
LTR.
AUTH.
CHANGE
BY
DATE
DRILL AND REAM .750 DIA. ±.005
1
1/8 R
1/8 R
1 1/4
F.A.O. 63
1/4
1 1/2
1 5/8
1/4
45°
1/8
2 1/4
THREAD 5/8 -11 UNC-2A
2
4
NOTES:
1. BREAK ALL SHARP CORNERS
2. MAT'L NAVAL BRONZE
3. STANDARD TOLERANCE UNLESS NOTED
4. FINISH 63
TITLE:
CLEVIS
SCALE FULL
PRT. NO. 50

FIGURE A-62

SECTION B

HAND TOOLS

Our ability to make and use tools has been directly responsible for all technical advance. Prior to the development of advanced metalworking, natural materials such as stone, flint, and wood provided the only tool materials. When metals and metalworking techniques became better established, tool development was advanced greatly, which led to the many fine tools of today. A study of tools must logically begin with those used by hand for hand operations. In this section you will be introduced to the basic complement of hand tools used in all branches of mechanical technology.

Work-holding devices were not developed in early times. Artisans in many Middle East and Asian countries still preferred to use their feet instead of a vise to hold the workpiece. Machinists today tend to take the bench vise for granted, seldom realizing that they could hardly get along without it.

Arbor presses and hydraulic shop presses are very useful and powerful shop tools. If they are used incorrectly, however, they can be very hazardous to the operator, and workpieces can be ruined.

Noncutting tools such as screwdrivers, pliers, and wrenches should be properly identified. It is impossible to request a particular tool from the toolroom without knowing its correct name.

Cutting hand tools such as hacksaws, files, hand reamers, taps, and dies are very important to a machinist. In this section you will also be introduced to the pedestal grinder and its important functions in the machine shop.

The units that follow in this section will instruct you in the identification, selection, use, and safety of these important hand tools and hand-operated machines.

Hand Tool Safety

Tools described in this section are quite safe if they are used as they were designed to be used. For example, a screwdriver is not meant to be a chisel and a file is not meant to be a pry bar. Wrenches should be the correct size for the nut or bolt head so they will not slip. Inch measure wrenches should not be used on metric fasteners. When a wrench slips, skinned knuckles are often the result. Hacksaws should be held by the handle, not the frame. Fingers wrapped around the frame tend to get mashed. Files should never be used without a handle because the tang can severely damage the hand or wrist. Safety precautions are noted throughout this section for using other equipment, such as presses and pedestal grinders.

UNIT 1

Arbor and Shop Presses

The arbor press and the small shop press are very common sights in most machine shops. It would be difficult indeed to get along without them. You will find these tools extremely useful when you know how to use them, but if you are not instructed in their use, they can be dangerous to you and destructive to the workpiece.

Objectives

After completing this unit, you should be able to:

1. Install and remove a bronze bushing using an arbor press.
2. Press on and remove a ball bearing from a shaft on an arbor press using the correct tools.
3. Press on and remove a ball bearing from a housing using an arbor press and correct tooling.
4. Install and remove a mandrel using an arbor press.
5. Install and remove a shaft with key in a hub using the arbor press.

Types

The arbor press is an essential piece of equipment in the small machine shop. Without it a machinist would be forced to resort to the use of a hammer or sledge to make any forced fit, a process that could easily damage the part.

Two basic types of hand powered arbor presses are manufactured and used: the hydraulic (Figure B-1) and the mechanical (Figure B-2). The lever gives a "feel" or a sense of pressure applied, which is not possible with the power-driven presses. This pressure sensitivity is needed when small delicate parts are being pressed so that a worker will know when to stop before collapsing the piece.

Uses

The major uses of the arbor press are bushing installation and removal, ball and roller bearing installation and removal (Figure B-3), pressing shafts into hubs (Figure B-4), pressing mandrels into workpieces, straightening and bending, and

FIGURE B-1 Fifty-ton capacity hydraulic shop presses (Lane Community College).

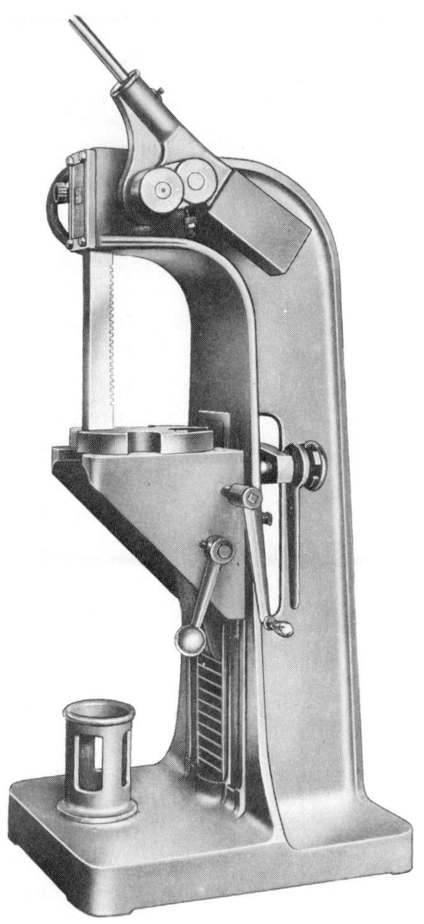

FIGURE B-2 Simple rachet floor-type arbor press (Dake Corporation).

FIGURE B-3 Roller bearing being removed from axle (Lane Community College).

broaching keyseats. A keyseat is an axially located rectangular groove in a shaft or hub. Keyseats in shafts are cut in milling machines, but keyseats in hubs of gears, sprockets, or other driven members must be either broached or cut in a keyseater, a special type of vertical shaper.

Procedures

INSTALLING BUSHINGS

A bushing is a short metal tube, machined inside and out to precision dimensions, and usually made to fit into a bore, or accurately machined hole. Many kinds of bushings are used for various purposes and are usually installed with an **interference fit** or press fit. This means that the bushing is slightly larger than the hole into which it is pressed. The amount of interference will be considered in greater detail in a later unit. There are many bushings made of many materials, including bronze and hardened steel, but they all have one thing in common: they must be lubricated with high-pressure lube before they are pressed into the bore. Oil is not used as it will simply wipe off and cause the bushing to seize the bore. Seizing is the condition where two unlubricated metals tend to weld together under pressure. In this case it may cause the bushing to be damaged beyond repair.

FIGURE B-4 Shaft being pressed into hub (Lane Community College).

The bore should always have a strong chamfer, that is, an angled or beveled edge, since a sharp

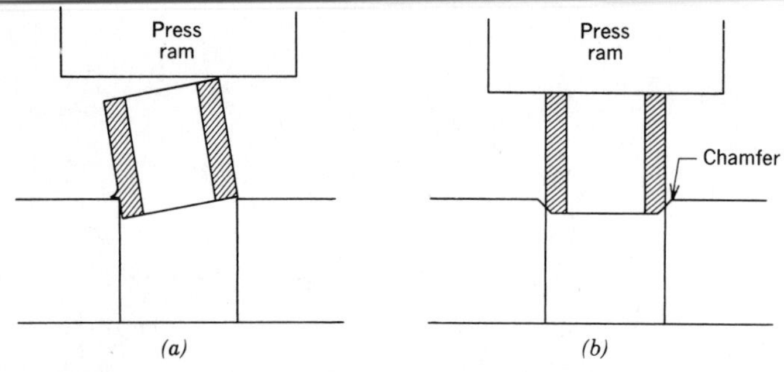

FIGURE B-5(*a*) Bushing being pressed where bore is not chamfered and bushing is misaligned; (*b*) bushing being pressed into correctly chamfered hole in correct alignment.

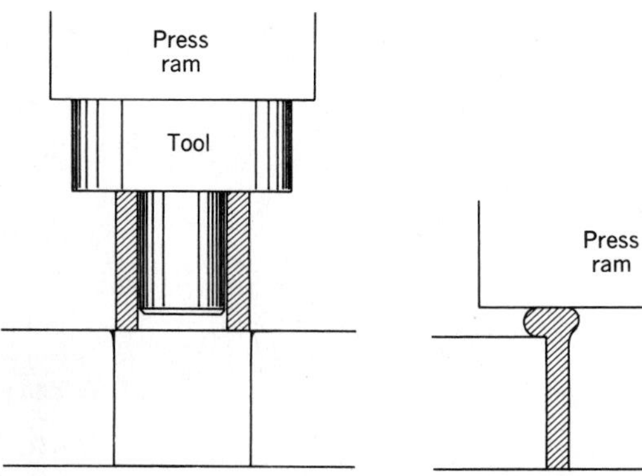

FIGURE B-6 Special tool to keep bushing square to press ram.

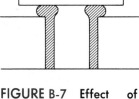

FIGURE B-7 Effect of excessive pressure on bushing that exceeds bore length.

edge would cut into the bushing and damage it (Figure B-5). The bushing should also have a long tapered chamfer or start so it will not dig in and enter misaligned. Bushings are prone to go in crooked if there is a sharp edge, especially hardened steel bushings. Care should be taken to see that the bushing is straight entering the bore, and that it continue into the bore in proper alignment. This should not be a problem if the tooling is right; that is, if the end of the press ram is square and if it is not loose and worn. The proper bolster plate should also be used under the part so that it cannot tilt out of alignment. Sometimes special tooling is used to guide the bushing (Figure B-6). When the workpiece is resting on a solid bolster plate, only the pressure needed to force the bushing into place should be applied, especially if the bushing is longer than the bore length. Excessive pressure might distort and collapse the bushing and cause it to be undersized (Figure B-7). However, some bolster plates provide various size holes so a bushing or part can extend through. In that case, the press ram can be brought to the point where it contacts the workpiece, and there is no danger of upsetting the bushing.

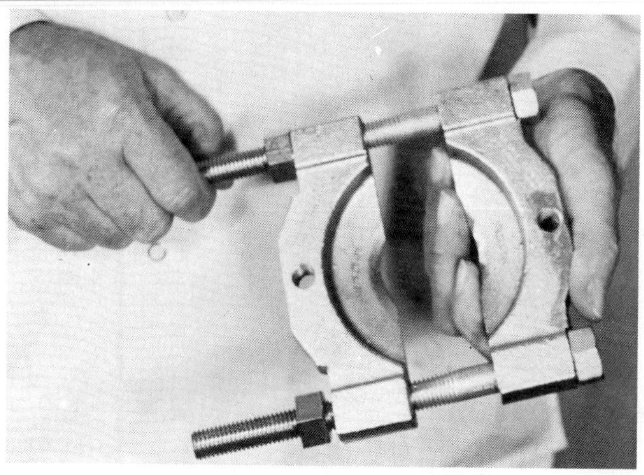

FIGURE B-8 Bearing puller, a special tool for supporting inner bearing race (see Figure B-3) (Lane Community College).

BALL AND ROLLER BEARINGS

Ball and roller bearings pose special problems when they are installed and removed by pressing. This is because the pressure must be applied directly against the race and not through the balls or rollers since this could destroy the bearing. Frequently, when removing ball bearings from a shaft, the inner race is hidden by a shoulder and cannot be supported in the normal way. In this case, a special tool called a bearing puller is used (Figure B-8). On the inner and outer races, bearings may be installed by pressing on the race with a steel tube of the proper diameter. As with bushings, high-pressure lubricant should be used.

Sometimes there is no other way to remove an old ball bearing except by exerting pressure through the balls. When this is done, there is a real danger that the race may be violently shattered. In this case a scatter shield must be used. A scatter shield is a heavy steel tube about 8 to 12 in. long and is set up to cover the work. The shield is placed around the bearing during pressing to keep shattered parts from injuring the operator. It is a good safety practice to always use a scatter shield when ball bearings are removed from a shaft by pressing. Safety glasses should be worn during all pressing operations.

BORES AND SHAFTS

Holes in the hubs of gears, sprockets, and other machine parts are also frequently designed for a force fit. In these instances, there is usually a key-

FIGURE B-9 Chamfer on key helps in alignment of parts being pressed together (Lane Community College).

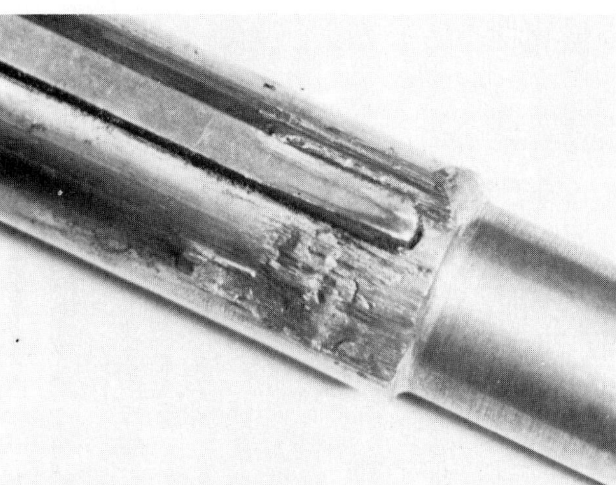

FIGURE B-10 This shaft had just been made by a machinist and was forced into an interference fit bore for a press fit. No lubrication was used and it immediately seized and welded to the bore, which was also ruined (Lane Community College).

seat that needs to be aligned. A keyseat is a groove in which a key is placed. This key, in turn, also fits into a slot in the hub of a gear or pulley, and secures the part against the shaft, keeping it from rotating. When pressing shafts with keys into hubs with keyseats, it is sometimes helpful to chamfer the leading edge of the key so that it will align itself properly (Figure B-9). Seizing will occur in this operation, as with the installation of bushings, if high-pressure lubricant is not used (Figure B-10).

MANDRELS

Mandrels, cylindrical pieces of steel with a slight taper, are pressed into bores in much the same way that shafts are pressed into hubs. There is one important difference, however; since the mandrel is tapered about .006 of an inch per foot, it can be installed only with the small end in first.

The large end of the mandrel may have a flat where the lathe dog screw can rest. The large end may also be determined by measuring with a micrometer, or by trying the mandrel in the bore. The small end should start into the hole, but the large end should not. Apply lubricant and press the mandrel in until definite resistance is felt (Figure B-11).

FIGURE B-11 Mandrel being lubricated and pressed into part for further machining (Lane Community College).

KEYSEAT (KEYWAY) BROACHING

The process of broaching is just one of the machining processes. Broaching can be done on both internal and external surfaces. In keyseat broaching, a slot or groove is cut inside the bore through a hub or pulley so that a key can be retained.

Although many types of keyseating machines are in use in many machine shops, keyseat broaching is often done on arbor presses. Broaching is the process of cutting out shapes on the interior of a metal part. Keyseats are only one type of cutting that can be done by the push-type procedure. Such internal shapes as a square or hexagon can

FIGURE B-12 Hexagonal shape being broached. (The person in the photo is only posing and not operating the machine; safety glasses should always be worn when operating machinery) (The duMont Corporation).

FIGURE B-13 A typical set of keyseat (keyway) broaches (The duMont Corporation).

also be cut by this method (Figure B-12). All that is needed for these procedures is the proper size of arbor press and a set of keyseat broaches (Figure B-13), which are hardened cutters with stepped teeth so that each tooth cuts only a definite amount when pushed or pulled through a part. These are available in inch and metric dimensions.

Broaching keyseats (multiple-pass method) is done as follows:

STEP 1. Choose the bushing that fits the bore and the broach, and put it in place in the bore.

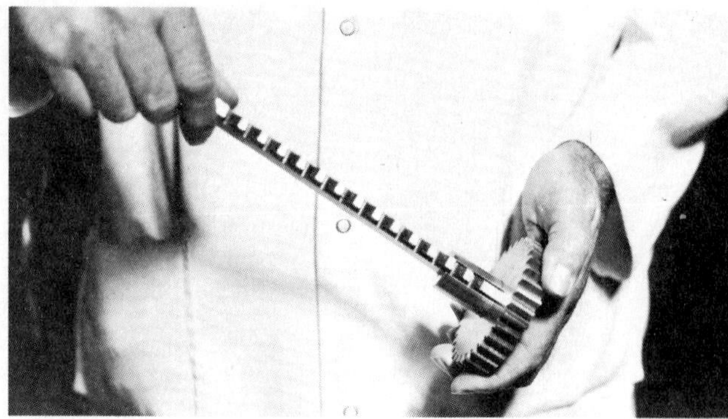

FIGURE B-14 Broach with guide bushing inserted into gear (Lane Community College).

FIGURE B-15 Broach with guide bushing placed in arbor press that is ready to lubricate and to perform first pass (Lane Community College).

STEP 2. Insert the correct-size broach into the bushing slot (Figure B-14).

STEP 3. Place this assembly in the arbor press (Figure B-15).

STEP 4. Lubricate.

STEP 5. Push the broach through.

STEP 6. Clean the broach.

STEP 7. Place second-pass shim in place.

FIGURE B-16 Shims in place behind broach that is ready to lubricate and make final cut on part (Lane Community College).

STEP 8. Insert broach.

STEP 9. Lubricate.

STEP 10. Push the broach through.

STEP 11. If more than one shim is needed to obtain the correct depth, repeat the procedure (Figure B-16).

The tools should be cleaned and returned to their box and the finished keyseat should be deburred and cleaned.

Production or single-pass broaching requires no shims or second-pass cuts, and with some types no bushings need be used (Figure B-17).

Two important things to remember when push broaching are alignment and lubrication. Misalignment, caused by a worn or loose ram, can cause the broach to hog (dig in) or break. Sometimes this can be avoided by facing the teeth of the broach toward the back of the press and permitting the bushing to protrude above the work to provide more support for the broach. After starting the cut, relieve the pressure to allow the broach to center itself. Repeat this procedure during each cut.

FIGURE B-17 Production push broaching without bushing on shims (The duMont Corporation).

At least two or three teeth should be in contact with the work. If needed, stack two or more workpieces to lengthen the cut. The cut should never exceed the length of the standard bushing used with the broach. Never use a broach on material harder than Rockwell C35. You will study hardness testing later in this book. If it is suspected that a part is harder than mild steel, its hardness should be determined before any broaching is attempted.

Use a good high-pressure lubricant. Also apply a sulfur base cutting oil to the teeth of the broach. Always lubricate the back of the keyseat broach to reduce friction, regardless of the material to be cut. Brass is usually broached dry, but bronzes cut better with oil or soluble oil. Cast iron is broached dry, and kerosene (solvent) or cutting oil is recommended for aluminum.

BENDING AND STRAIGHTENING

Bending and straightening are frequently done on hydraulic shop presses. Mechanical arbor presses are not usually used for this purpose. There is a definite safety hazard in this type of operation as a poor setup can allow pieces under pressure to suddenly fly out of the press. Brittle materials such as cast iron or hardened steel bearing races can suddenly break under pressure and explode into fragments.

A shaft to be straightened is placed between two nonprecision vee blocks—steel blocks with a vee-shaped groove running the length of the blocks that support a round workpiece. In the vee blocks, the shaft is rotated to detect runout, or the amount of bend in the shaft. The rotation is measured on a dial indicator, which is a device capable of detecting very small mechanical movements, and read from a calibrated dial. The high point is found and marked on the shaft (Figure B-18). After removing the indicator, a soft metal pad such as copper is placed between the shaft and the ram and pressure is applied (Figure B-19). The shaft should be bent back to a straight position and then very slightly beyond that point. The pressure is then removed and the dial indicator is again put in position. The shaft is rotated as before, and the position of the mark noted, as well as the amount of runout. If improvement has been found, continue the process; but if the first mark is opposite the new high point, too much pressure has been applied. Repeat the same steps, applying less pressure on opposite side.

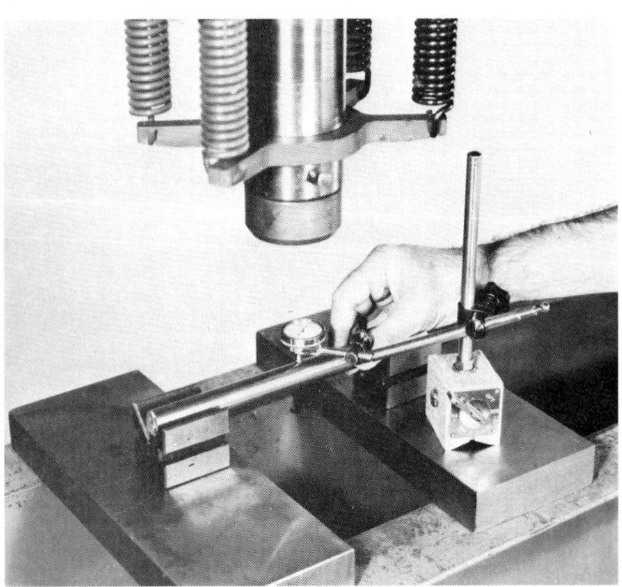

FIGURE B-18 Part being indicated for runout prior to straightening (Lane Community College).

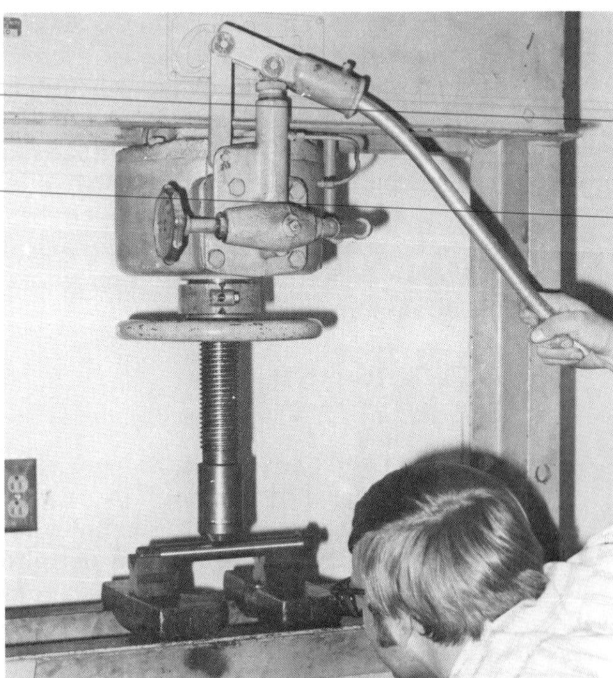

FIGURE B-19 Pressure being applied to straighten shaft (Lane Community College).

Other straightening jobs on flat stock and other shapes are done in a similar fashion. Frequently, two or more bends will be found that may be opposite or are not in the same direction. This condition is best corrected by straightening one bend at a time and checking with a straightedge and feeler gage. Special shop press tooling is sometimes used for simple bending jobs in the shop.

SELF-TEST

1. Why is it important to know how to use the arbor press properly and how to set up pressing operations correctly?

2. What kinds of arbor presses are made? What makes them different from large commercial presses?

3. List several uses of the arbor press.

4. A newly machined steel shaft with an interference fit is pressed into the bore of a steel gear. The result is a shaft ruined beyond repair; the bore of the gear is also badly damaged. What has happened? What caused this failure?

5. The ram of an arbor press is loose in its guide and the pushing end is rounded off. What kind of problems could be caused by this?

6. When a bushing is pushed into a bore that is located over a hole in the bolster plate of a press, how much pressure should you apply to install the bushing: 30 tons, 10 tons, or just enough to seat the bushing into the bore?

7. When pressing a shaft from the innner race of a ball bearing, where should the bearing be supported on the bolster plate of the press?

8. What difference is there in the way a press fit is obtained between mandrels and ordinary shafts?

9. Prior to installing a bushing with the arbor press, what two important steps must be taken?

10. Name five ways to avoid tool breakage and other problems when using push broaches for making keyseats in the arbor press.

UNIT 2

Work-Holding and Hand Tools

The bench vise is a basic but very necessary tool in the shop. With proper care and use, this work-holding tool will give many years of faithful service. Hand tools are essential in all of the mechanical trades. This unit will help you learn the names and uses of most of the noncutting tools used by machinists.

Objectives

After completing this unit, you should be able to:

1. Identify various types of vises, their uses, and maintenance.
2. Identify the proper tool for a given job.
3. Determine the correct use of a selected tool.

Types of Vises

Vises of various types are used by machinists when doing hand or bench work. They should be mounted in such a way that a long workpiece can be held in a vertical position extending alongside the bench (Figure B-20). Some bench vises have a solid base, and others have a swivel base (Figure B-21). The machinist's bench vise is measured by the width of the jaws (Figure B-22).

Toolmakers often use small vises that pivot on a ball and socket for holding delicate work. Hand-held vises, called pin vises, are made for holding very small or delicate parts.

Most bench vises have hardened insert jaws that are serrated for greater gripping power. These crisscross serrations are sharp and will dig into finished workpieces enough to mar them beyond

FIGURE B-20 When long work is clamped in the vise vertically, it should clear the workbench (Lane Community College).

FIGURE B-21 Swivel-base bench vise (Lane Community College).

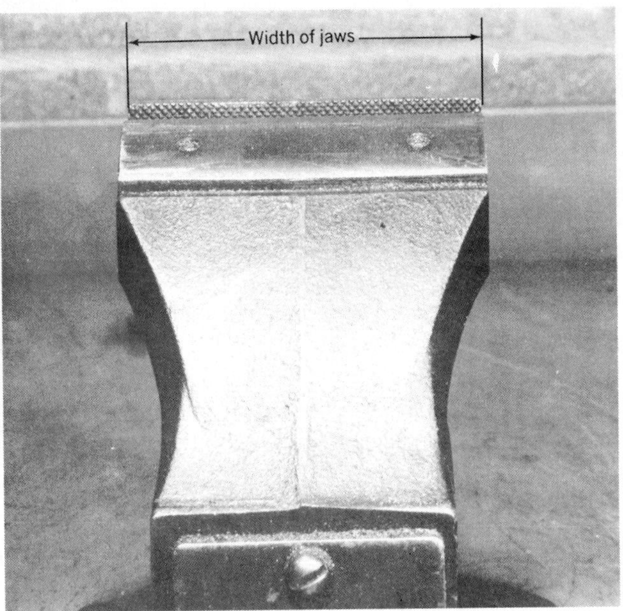

FIGURE B-22 How to measure a vise (Lane Community College).

FIGURE B-23 View of the soft jaws placed on the vise (Lane Community College).

FIGURE B-24 Never hammer on the slide bar of a vise. This may crack or distort it (Lane Community College).

repair. Soft jaws (Figure B-23) made of copper, other soft metals, or wood, are used to protect a finished surface on a workpiece. These soft jaws are made to slip over the vise jaws. Some vises used for sheet metal work have smooth, deep jaws.

Uses of Vises

Vises are used to hold work for filing, hacksawing, chiseling, and bending light metal. They are also used for holding work when assembling and disassembling parts.

Vises should be placed on the workbench at the correct working height for the individual. The top of the vise jaws should be at elbow height. Poor work is produced when the vise is mounted too high or too low. A variety of vise heights should be provided in the shop or skids made available to stand on.

Care of Vises

Like any other tool, vises have limitations. "Cheater" bars or pipes should not be used on the handle to tighten the vise. Heat from a torch should not be applied to work held in the jaws as the hardened insert jaws will then become softened. There is usually one vise in a shop reserved for heating and bending.

Heavy hammering should not be done on a bench vise. The force of bending should be against the fixed jaw rather than the movable jaw of the vise. Bending light, flat stock or small round stock in the jaws is permissible if a light hammer is used. The movable jaw slide bar (Figure B-24) should never be hammered upon as it is usually made of thin cast iron and can be cracked quite easily. An anvil is often provided behind the solid jaw for the purpose of light hammering.

Bench vises should occasionally be taken apart so that the screw, nut, and thrust collars can be cleaned and lubricated (Figure B-25). The screw and nut should be cleaned in solvent. A heavy grease should be packed on the screw and thrust collars before reassembly.

Clamps

C-clamps are used to hold workpieces on machines such as drill presses, as well as to clamp parts together. The size of the clamp is determined by the largest opening of its jaws. Heavy duty C-clamps (Figure B-26) are used by machinists to hold heavy parts such as steel plates together for drilling or other machining operations. The clamp shown in

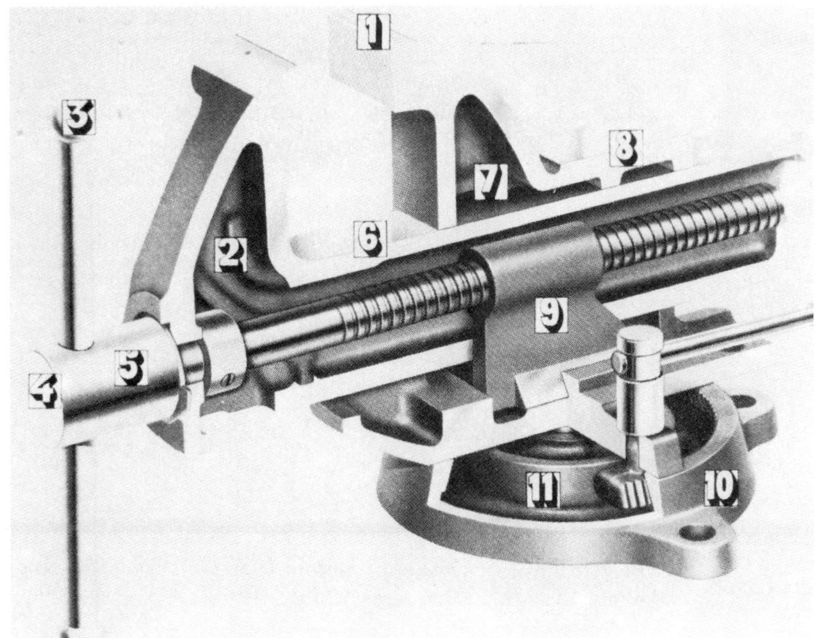

FIGURE B-25 Cutaway view of a vise: (1) replaceable hardened tool steel faces pinned to jaw; (2) malleable iron front jaw; (3) steel handle with ball ends; (4) cold-rolled steel screw; (5) bronze thrust bearing; (6) front jaw beam; (7) malleable iron back jaw body; (8) anvil; (9) nut, mounted in back jaw keyseat for precise alignment; (10) malleable iron swivel base; (11) steel tapered gear and lock bolt (The Warren Group, Division of Warren Tool Corporation).

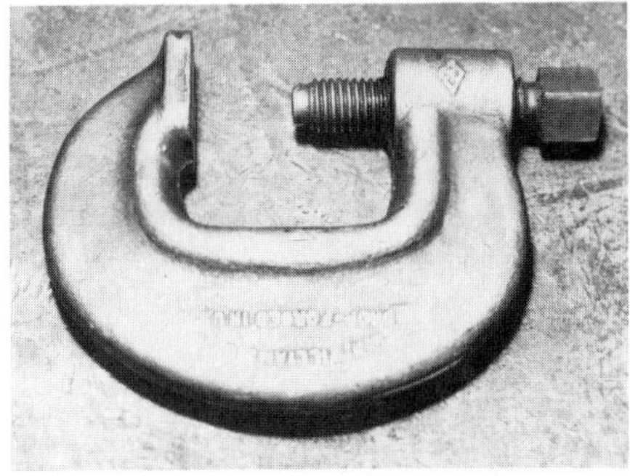

FIGURE B-26 Heavy-duty C-clamp (Lane Community College).

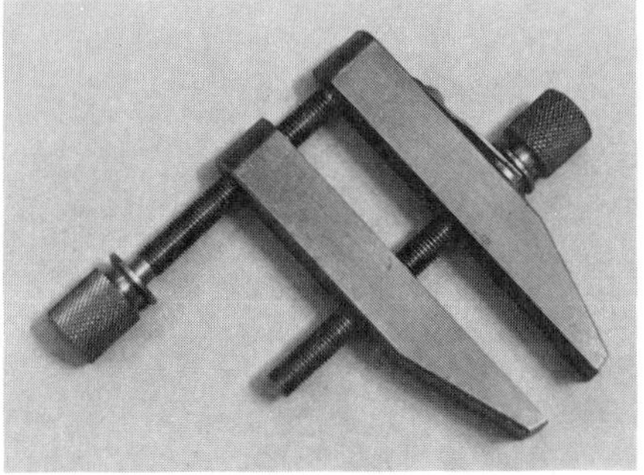

FIGURE B-28 Single-size parallel clamps.

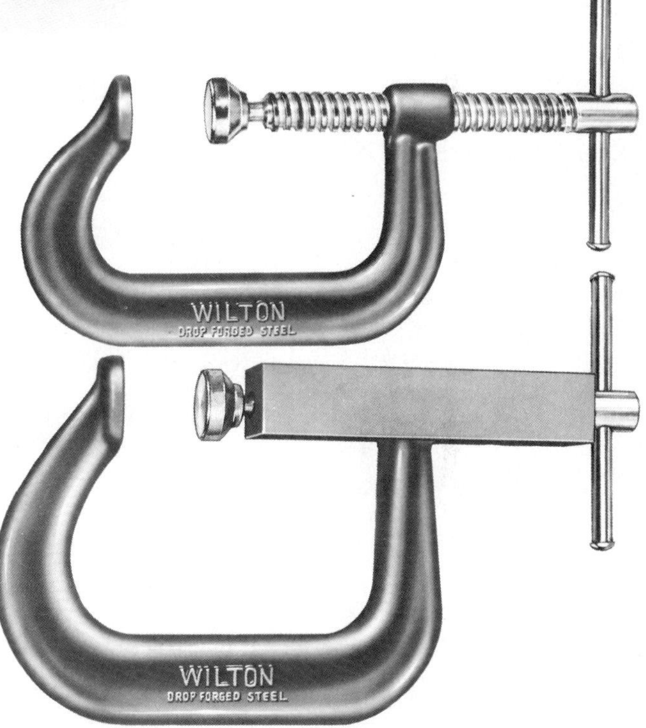

FIGURE B-27 Two types of C-clamps (Wilton Corporation).

the bottom view (Figure B-27) has a shielded screw. The clamp screw is protected by a sheet metal cover. Thus, the screw is protected from dirt and damage. Parallel clamps (Figure B-28) are used to hold small parts. Since they do not have as much holding power as C-clamps, this usually limits the use of parallel clamps to delicate work. Precision measuring setups are usually held in place with parallel clamps.

Pliers

Pliers come in several shapes and with several types of jaw action. Simple combination or slip joint pliers (Figure B-29) will do most jobs for

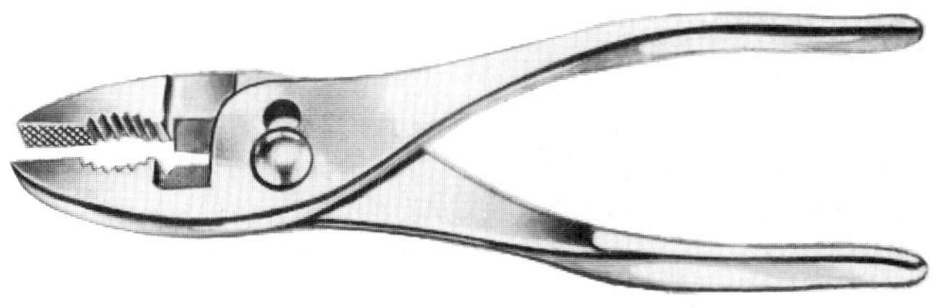

FIGURE B-29 Slip joint or combination pliers (Snap-on Tools Corporation).

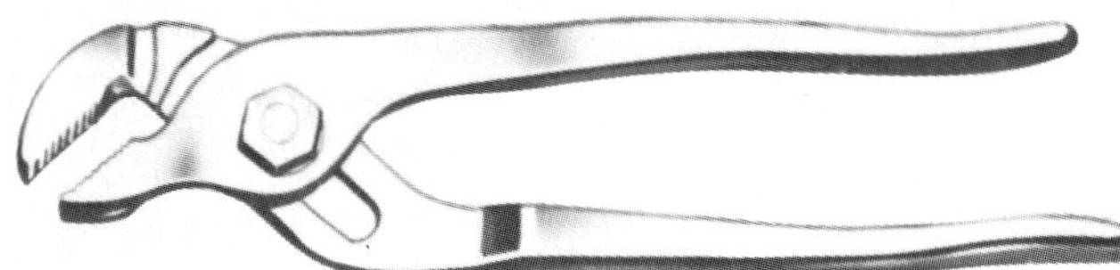

FIGURE B-30 Interlocking joint or water pump pliers (Snap-on Tools Corporation).

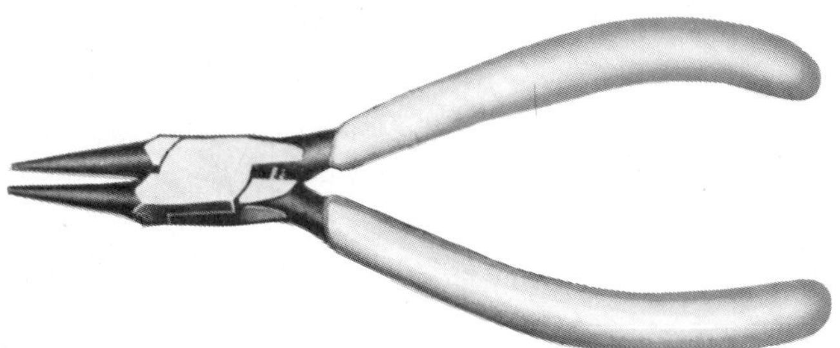

FIGURE B-31 Round nose or wire looper pliers (Snap-on Tools Corporation).

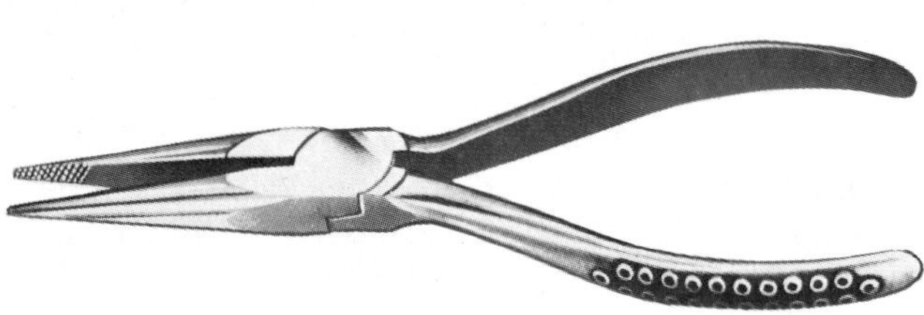

FIGURE B-32 Needlenose pliers, straight (Snap-on Tools Corporation).

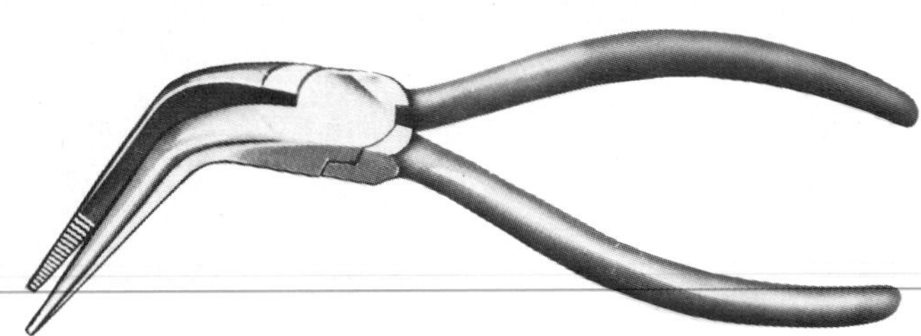

FIGURE B-33 Needlenose pliers, bent (Snap-on Tools Corporation).

FIGURE B-34 Side cutting pliers (Snap-on Tools Corporation).

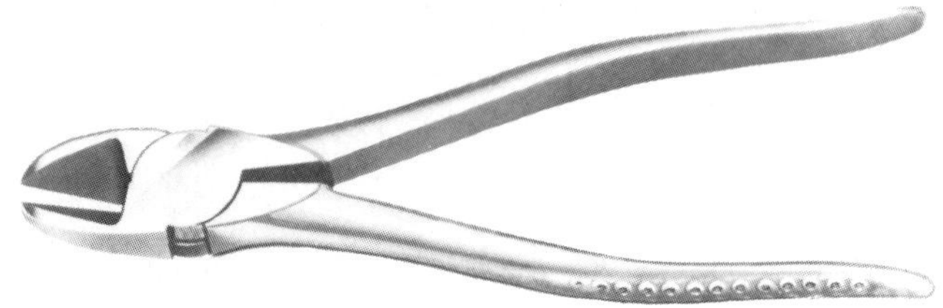

FIGURE B-35 Diagonal cutters (Snap-on Tools Corporation).

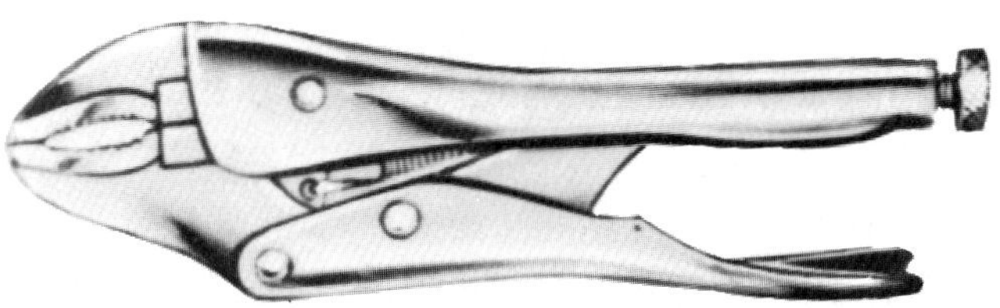

FIGURE B-36 Vise grip wrench (Snap-on Tools Corporation).

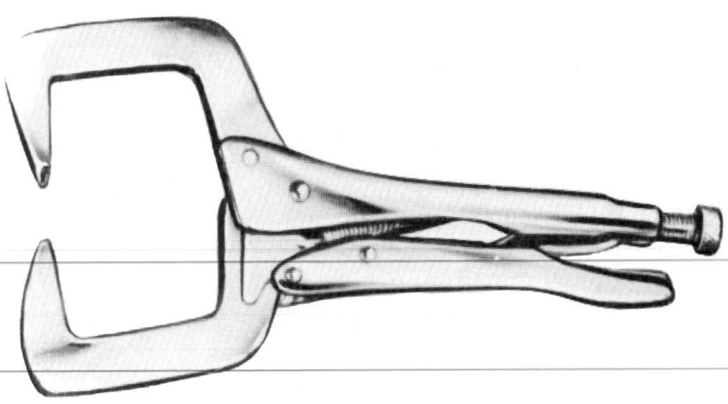

FIGURE B-37 Vise grip C-clamp (Snap-on Tools Corporation).

which you need pliers. The slip joint allows the jaws to expand to grasp a larger size workpiece. They are measured by overall length and are made in 5-, 6-, 8-, and 10-in. sizes.

Interlocking joint pliers (Figure B-30), or water pump pliers, were made to tighten packing gland nuts on water pumps on cars and trucks, but are useful for a variety of jobs. Pliers should never be used as a substitute for a wrench, as the nut or bolt head will be permanently deformed by the serrations in the plier jaws and the wrench will no longer fit properly. Round nose pliers (Figure B-31) are used to make loops in wire and shape light metal. Needlenose pliers are used for holding small delicate workpieces in tight spots. They are available in both straight (Figure B-32) and bent nose (Figure B-33) types. Linemen's pliers (Figure B-34) can be used for wire cutting and bending. Some types have wire stripping grooves and insulated handles. Diagonal cutters (Figure B-35) are only used for wire cutting.

The lever-jawed locking wrench has an unusually high gripping power. The screw in the handle adjusts the lever action to the work size (Figure B-36). They are made with special jaws for various uses such as the C-clamp types used in welding (Figure B-37).

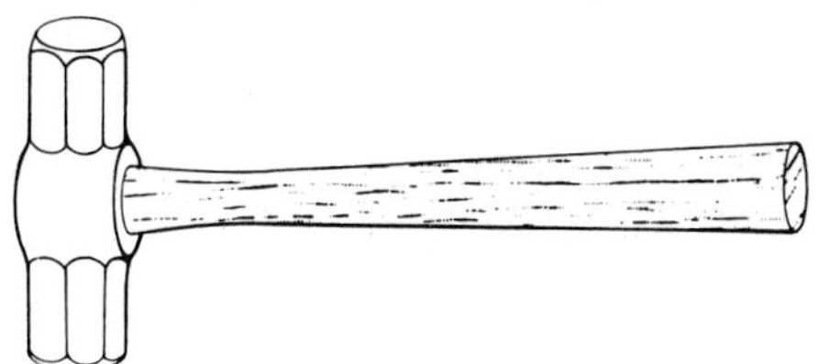

FIGURE B-38 Maul.

FIGURE B-39 Ball peen hammer.

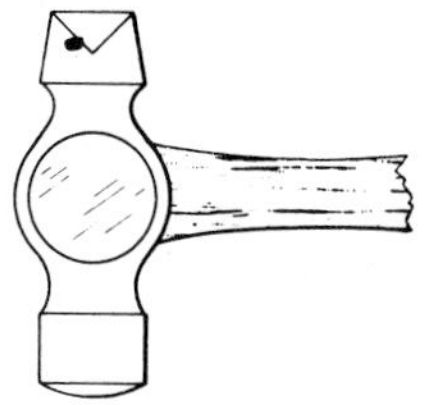

FIGURE B-40 Straight peen hammer.

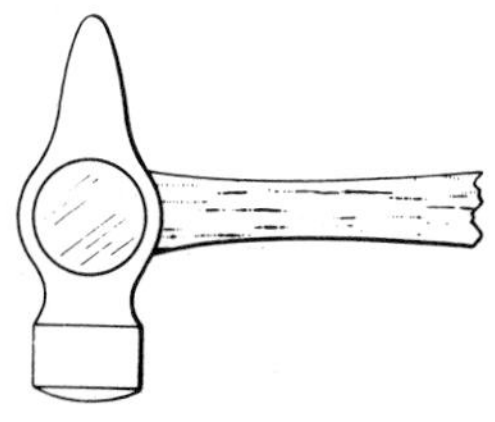

FIGURE B-41 Cross peen hammer.

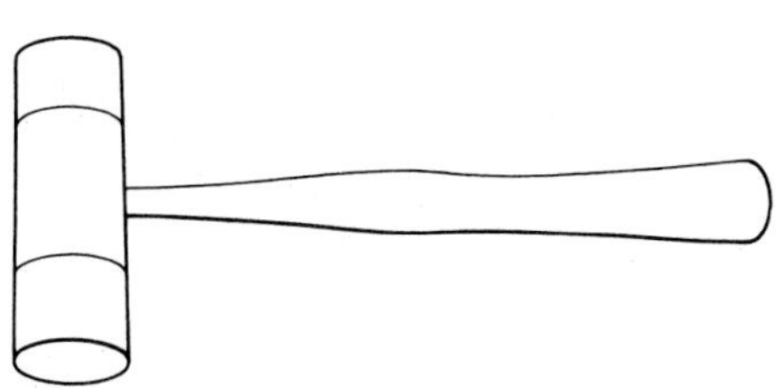

FIGURE B-42 Plastic hammer.

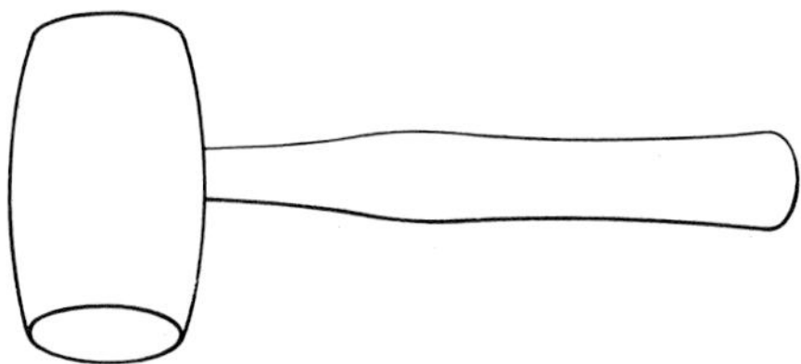

FIGURE B-43 Lead hammer.

FIGURE B-44 Adjustable wrench showing the correct direction of pull. The movable jaw should always face the direction of rotation (Snap-on Tools Corporation).

FIGURE B-45 Open end wrench (Snap-on Tools Corporation).

Hammers

Hammers are classified as either hard or soft. Hard hammers have steel heads such as blacksmith types or mauls made for heavy hammering (Figure B-38). The ball peen hammer (Figure B-39) is the one most frequently used by machinists. It has a rounded surface on one end of the head, which is used for upsetting or riveting metal, and a hardened striking surface on the other. Two hammers should never be struck together on the face, as pieces could break off. Hammers are specified according to the weight of the head. Ball peen hammers range from 2 oz. to 3 lb. Those under 10 oz. are used for layout work. Two other shop hammers are the straight peen (Figure B-40) and the cross peen (Figure B-41).

Soft hammers are made of plastic (Figure B-42), brass, copper, lead (Figure B-43), or rawhide and are used to position workpieces that have finishes that would be damaged by a hard hammer. A dead blow hammer is sometimes used in place of a lead hammer because, like the lead hammer, the dead blow hammer does not have a tendency to rebound. When a hammer bounces away from a workpiece, the work will not remain in place, but will move slightly. The movable jaw on most machine tool vises tends to move slightly upward when tightened against the workpiece. Thus, the workpiece is moved upward and out of position. The machinist must then use a dead blow hammer or lead hammer to reposition it.

Wrenches

A large variety of wrenches is made for different uses such as turning capscrews, bolts, and nuts. The adjustable wrench, commonly called a crescent wrench (Figure B-44), is a general-purpose tool and will not suit every job, especially those requiring work in close quarters. The wrench should be rotated toward the movable jaw and should fit the nut or bolt tightly. The size of the wrench is determined by its overall length in inches.

Open end wrenches (Figure B-45) are best suited to square-headed bolts, and usually fit two

FIGURE B-46 Box wrench (Snap-on Tools Corporation).

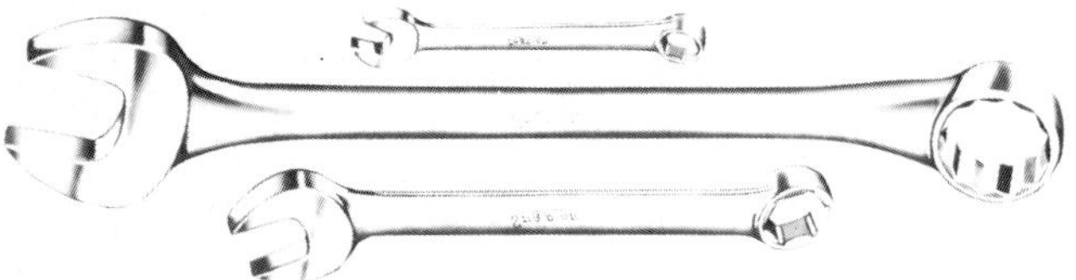

FIGURE B-47 Combination wrench (Snap-on Tools Corporation).

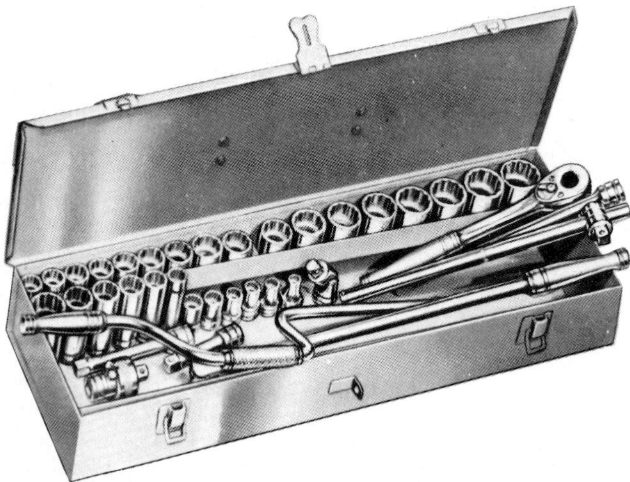

FIGURE B-48 Socket wrench set (Snap-on Tools Corporation).

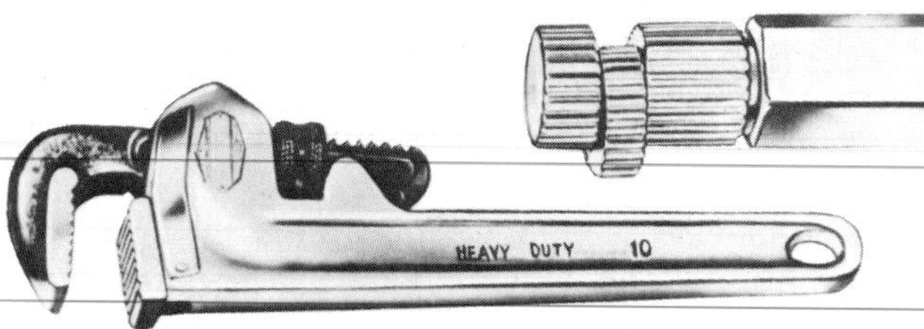

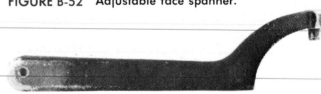

FIGURE B-49 Pipe wrenches, external and internal (Snap-on Tools Corporation).

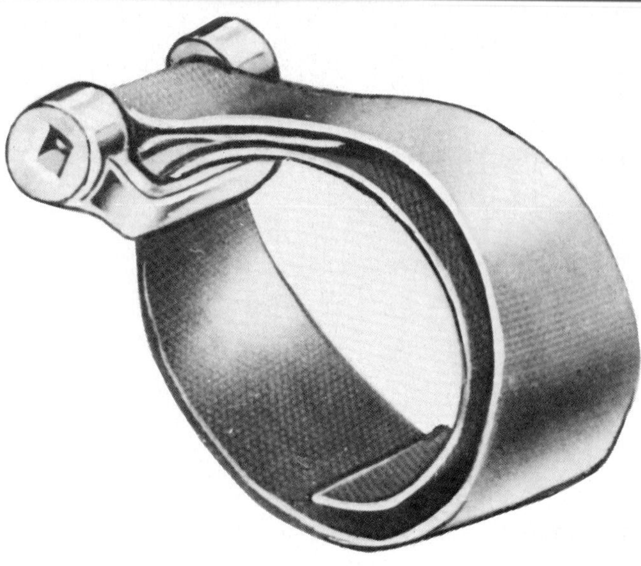

FIGURE B-50 Strap wrench (Snap-on Tools Corporation).

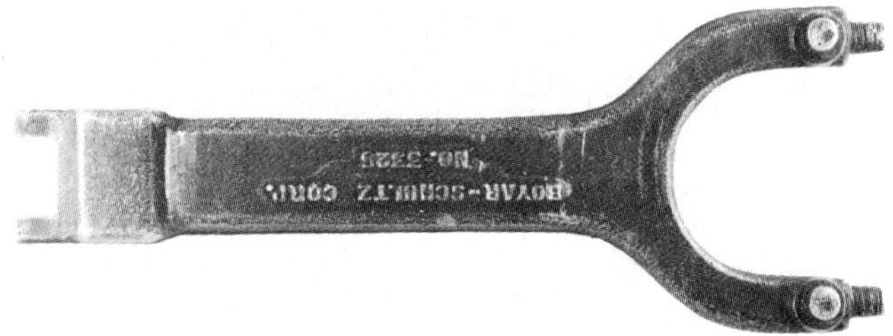

FIGURE B-51 Fixed face spanner.

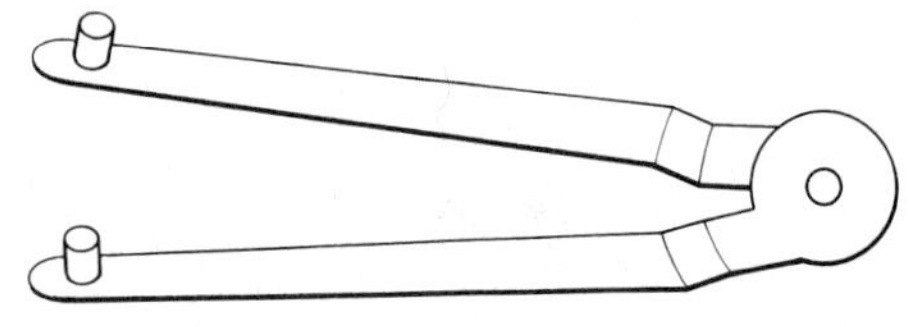

FIGURE B-52 Adjustable face spanner.

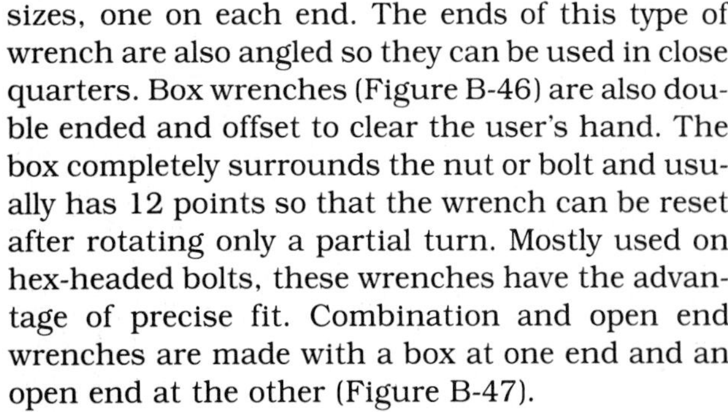

FIGURE B-53 Hook spanner.

sizes, one on each end. The ends of this type of wrench are also angled so they can be used in close quarters. Box wrenches (Figure B-46) are also double ended and offset to clear the user's hand. The box completely surrounds the nut or bolt and usually has 12 points so that the wrench can be reset after rotating only a partial turn. Mostly used on hex-headed bolts, these wrenches have the advantage of precise fit. Combination and open end wrenches are made with a box at one end and an open end at the other (Figure B-47).

Socket wrenches are similar to box wrenches in that they also surround the bolt or nut and usually are made with 12 points contacting the six-sided nut. Sockets are made to be detached from various types of drive handles (Figure B-48).

Pipe wrenches (Figure B-49), as the name implies, are used for holding and turning pipe. These wrenches have sharp serrated teeth and will damage any finished part on which they are used. Strap wrenches (Figure B-50) are used for extremely large parts or to avoid marring the surface of tubular parts.

Spanner wrenches come in several basic types, including face and hook. Face types are sometimes called pin spanners (Figure B-51). Spanners are

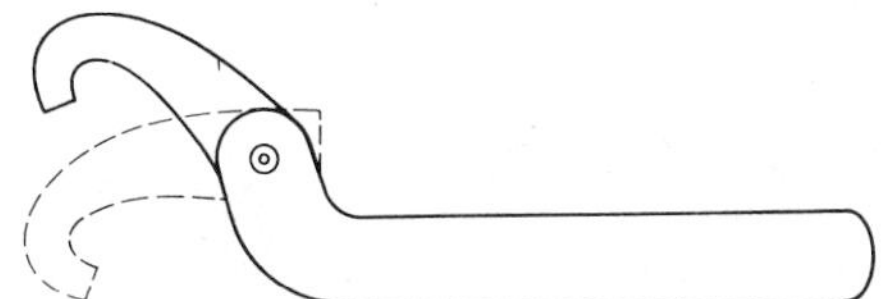

FIGURE B-54 Adjustable hook spanner.

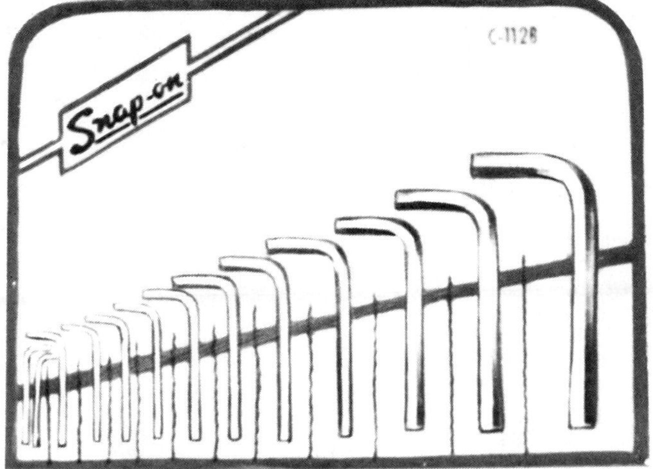

FIGURE B-55 Socket head wrench set (Snap-on Tools Corporation).

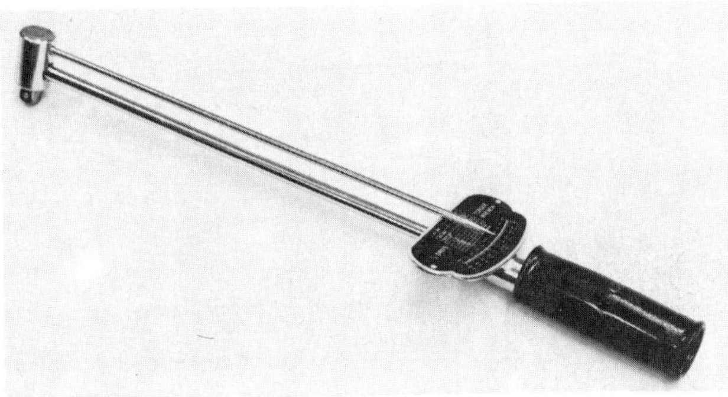

FIGURE B-56 Beam-type torque wrench (Lane Community College).

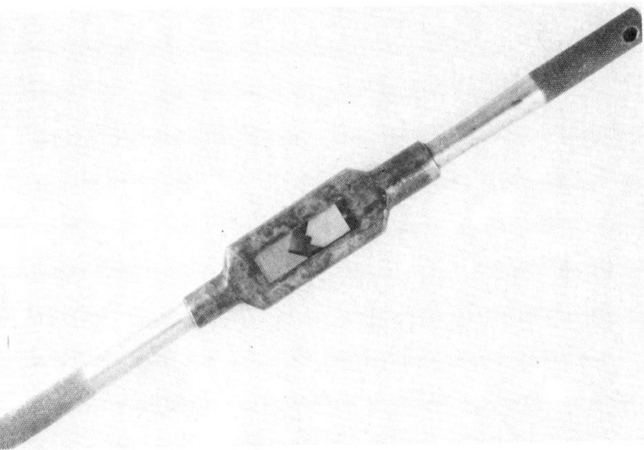

FIGURE B-57 Hand tap wrench.

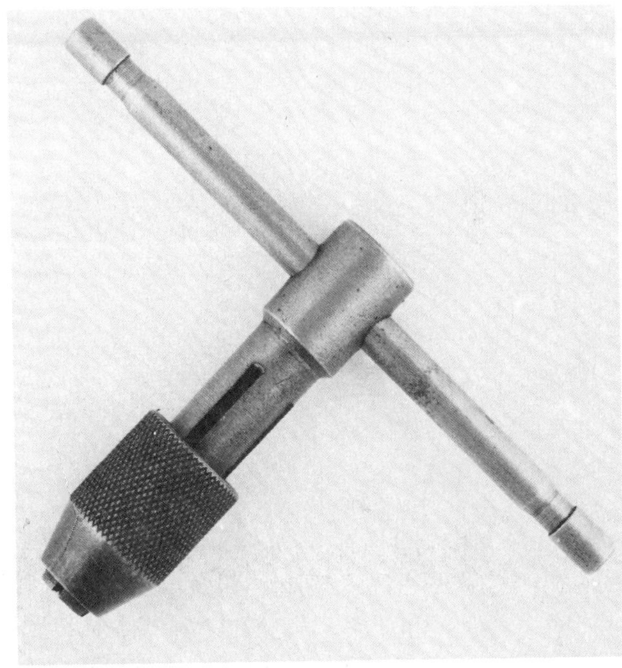

FIGURE B-58 T-handle tap wrench.

made in fixed sizes or adjustable types (Figures B-52 to B-54).

Socket head wrenches (Figure B-55) are six-sided bars having a 90-degree bend near one end. They are used with socket head capscrews and socket setscrews.

Torque wrenches (Figure B-56) are widely used by machinists and mechanics to provide the correct amount of tighening torque on a screw or nut. A dial reads in English measure (inch-pounds and foot-pounds) or in metric measure (kilogram-centimeters and newton-meters).

The hand tap wrench (Figure B-57) is used for medium-sized and large taps. The T-handle tap wrench (Figure B-58) is used for small taps $\frac{1}{4}$-in. and under, as its more sensitive "feel" results in less tap breakage.

CAUTION **Here are safety hints for using wrenches:**

1. Make sure that the wrench you select fits properly. If it is a loose fit, it may round off the corners of the nut or bolt head.

2. Pull on a wrench instead of pushing to avoid injury.

3. Never use a wrench on moving machinery.

4. Do not hammer on a wrench or extend the handle for additional leverage. Use a larger wrench.

Screwdrivers

The two types of screwdrivers that are most used are the standard (Figure B-59) and Phillips (Figure B-60). Both types are made in various sizes and in several styles, straight, shank, and offset (Figures B-61 and B-62). It is important to use the right-width blade when installing or removing screws (Figure B-63). The shape of the tip is important also. If the tip is badly worn or incorrectly ground, it will tend to jump out of the slot. Never use a screwdriver as a chisel or pry bar. Keep a screwdriver in proper shape by using it only on the screws for which it was meant.

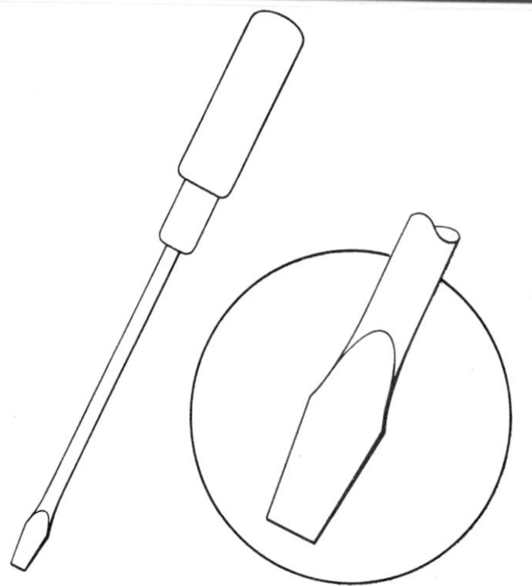

FIGURE B-59 Screwdriver, standard.

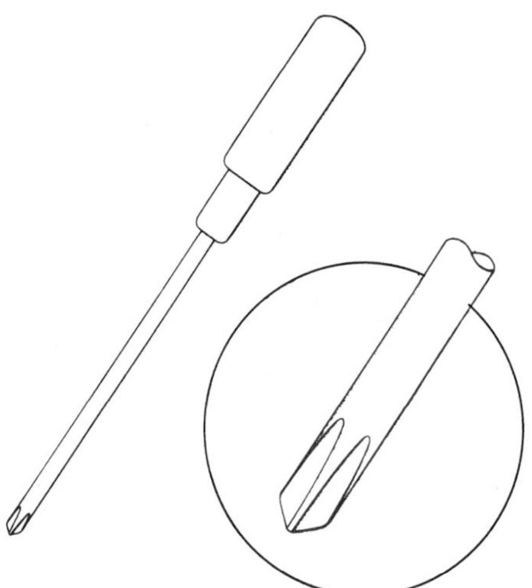

FIGURE B-60 Screwdriver, Phillips.

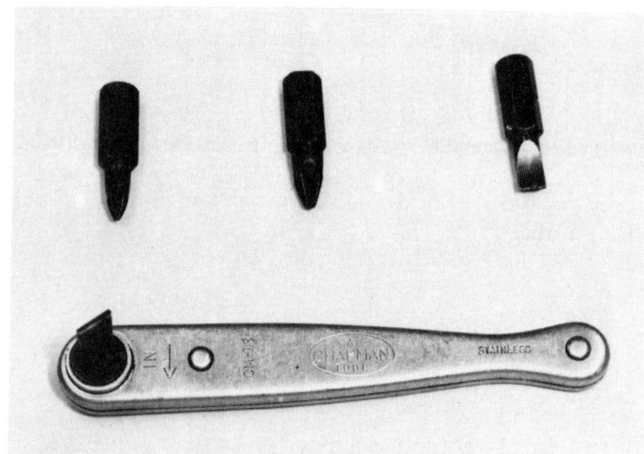

FIGURE B-62 Ratchet offset screwdriver with interchangeable points (Lane Community College).

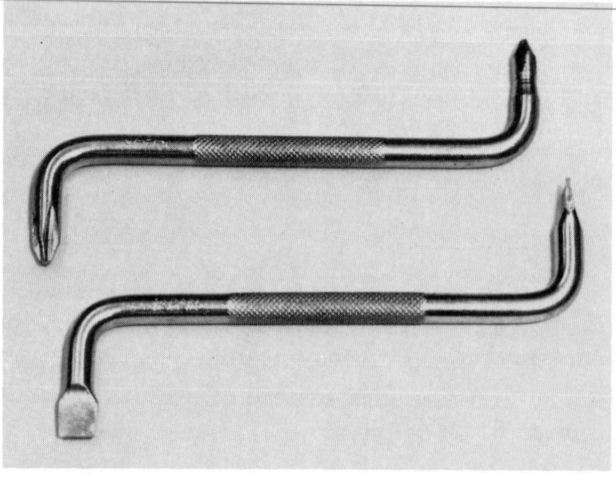

FIGURE B-61 Standard and Phillips offset screwdrivers (Lane Community College).

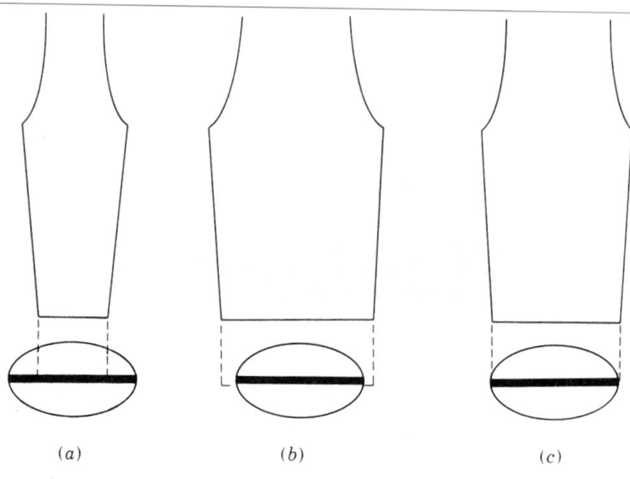

FIGURE B-63 The width of a screwdriver blade: (*a*) too narrow; (*b*) too wide; (*c*) correct width.

Chisels and Punches

Chisels and punches (Figure B-64) are very useful tools for machinists. The tool at the top of the illustration is a pin punch, used to drive out straight, taper, and roll pins. The drift punch below it is used as a starting punch for driving out pins. In the middle is a center punch that makes a starting point for drilling. The two bottom tools are cold chisels. Cold chisels are made in many shapes and are useful for cutting off rivet heads and welds.

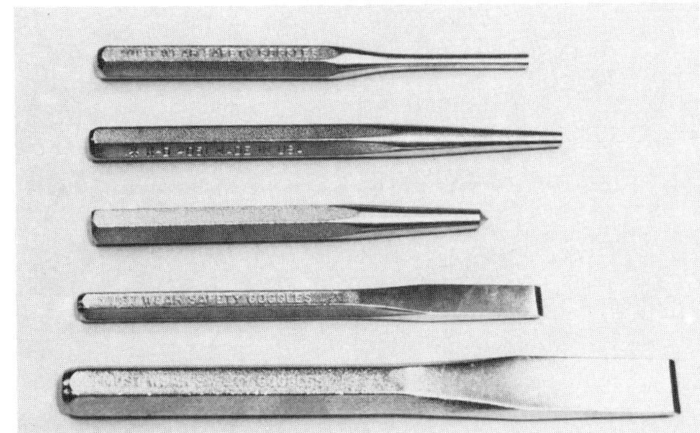

FIGURE B-64 Common chisels and punches that are used by machinists. Top to bottom: pin punch, drift punch, center punch, small flat chisel, and large flat chisel (Lane Community College).

SELF-TEST

1. What clamping position should be considered when mounting a vise on a workbench?

2. Name two types of bench vises.

3. How is the machinist's bench vise measured for size?

4. Explain two characteristics of the insert jaws on vises.

5. How can a finished surface be protected in a vise?

6. Name three things that should never be done to a vise.

7. How should a vise be lubricated?

8. A four-inch machinist bench vise has jaws 4 in. wide. True or false?

9. What is the purpose of soft jaws?

10. Parallel clamps are used for heavy duty clamping work, and C-clamps are used for holding precision setups. True or false?

11. In order to remove a nut or bolt, slip joint or water pump pliers make a good substitute for a wrench when a wrench is not handy. True or false?

12. What advantage does the lever-jawed wrench offer over other similar tools such as pliers?

13. Would you use a 3-lb. ball peen hammer for layout work? If not, what size do you think is right?

14. Some objects should never be struck with a hard hammer—a finished machine surface or the end of a shaft, for instance. What could you use to avoid damage?

15. A machine has a capscrew that needs to be tightened and released quite often. Which wrench would be best to use in this case: the adjustable or box-type wrench? Why?

16. Why should pipe wrenches never be used on bolts, nuts, or shafts?

17. What are two important things to remember about standard screwdrivers that will help you avoid problems in their use?

UNIT 3

Hacksaws

The hacksaw is one of the more frequently used hand tools. The hand hacksaw is a relatively simple tool to use, but the facts and rules presented in this unit will help you improve your use of the hacksaw.

Objective

After completing this unit, you should be able to:

Identify, select, and use hand hacksaws.

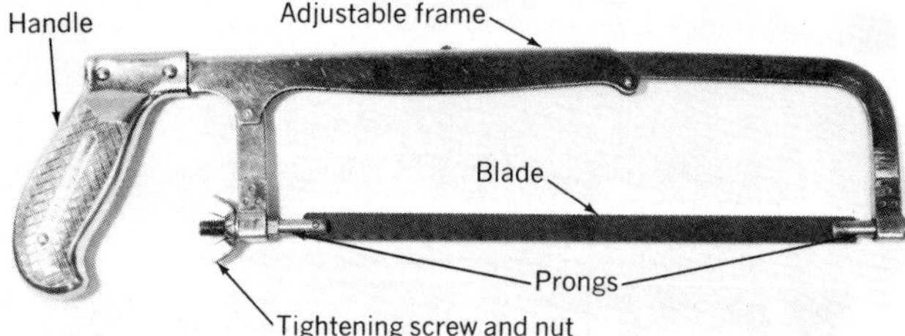

FIGURE B-65 The parts of a hacksaw.

FIGURE B-66 Straight sawing with a hacksaw (Lane Community College).

The hacksaw consists of three parts: the frame, the handle, and the saw blade (Figure B-65). Frames are either the solid or adjustable type. The solid frame can only be used with one length of saw blade. The adjustable frame can be used with hacksaw blades from 8 to 12 in. in length. The blade can be mounted to cut in line with the frame or at a right angle to the frame (Figures B-66 and B-67). By turning the blade at right angles to the frame, you can continue a cut that is deeper than the capacity of the frame. If the blade is left in line with the frame, the frame will eventually hit the workpiece and limit the depth of cut.

Most hacksaw blades are made from high speed steel, and in standard lengths of 8, 10, and 12 in. Blade length is the distance between the centers of the holes at each end. Hand hacksaw blades are generally $\frac{1}{2}$ in. wide and .025 in. thick. The kerf, or cut, produced by the hacksaw is wider than the .025-in. thickness of the blade because of the set of the teeth (Figure B-68).

The *set* refers to the bending of teeth outward from the blade itself. Two kinds of set are found on hand hacksaw blades. The first is the straight or alternate set (Figure B-69), where one tooth is bent to the right and the next tooth to the left for the length of the blade. The second kind of set is

FIGURE B-67 Sawing with the blade set at 90 degrees to the frame (Lane Community College).

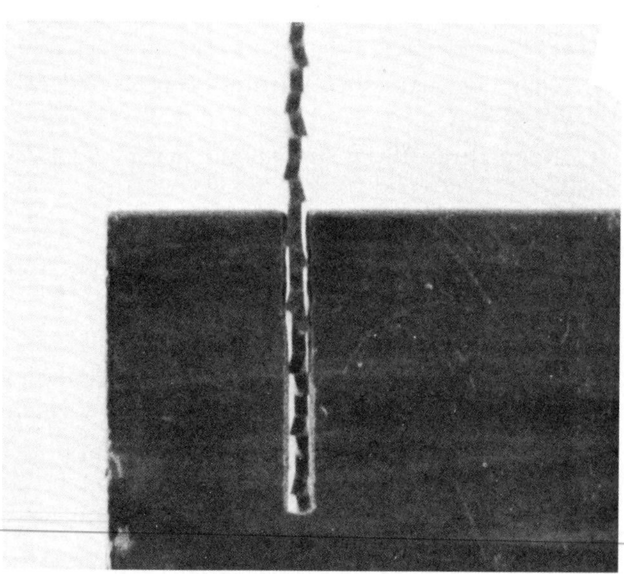

FIGURE B-68 The kerf is wider than the blade because of the set of the teeth (Lane Community College).

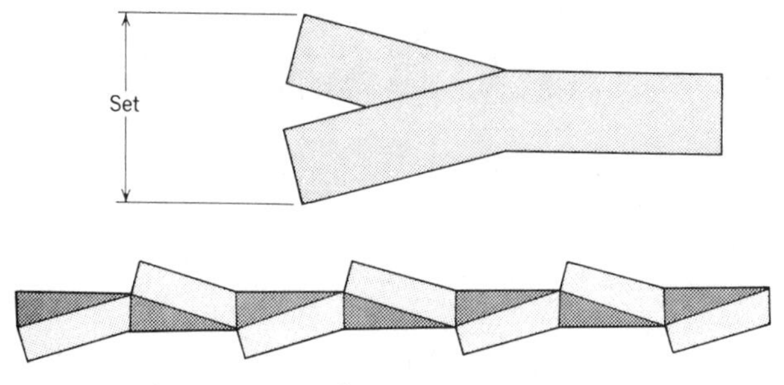

FIGURE B-69 The straight (alternate) set.

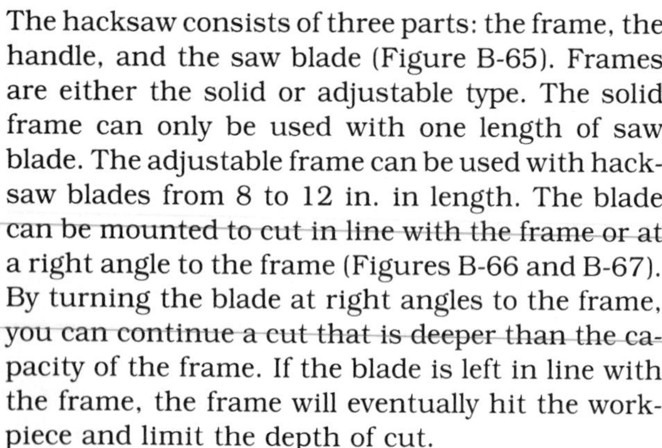

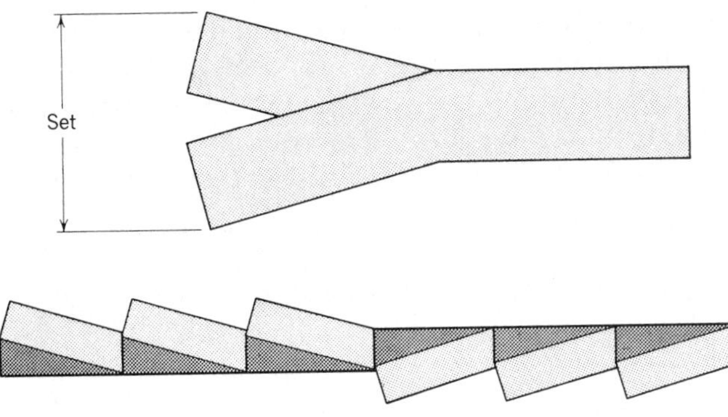

FIGURE B-70 The wavy set.

FIGURE B-71 The pitch of the blade is expressed as the number of teeth per inch (Lane Community College).

FIGURE B-72 A new blade must be started on the opposite side of the work, not in the same kerf as the old blade (Lane Community College).

the wavy set, in which a number of teeth are gradually bent to the right and then to the left (Figure B-70). A wavy set is found on most fine-tooth hacksaw blades.

The spacing of the teeth on a hand hacksaw blade is called the pitch and is expressed in teeth per inch of length (Figure B-71). Standard pitches are 14, 18, 24, and 32 teeth per inch, with the 18-pitch blade used as a general-purpose blade.

The hardness and size of thickness of a workpiece determine to a great extent which pitch blade to use. As a rule, you should use a coarse-tooth blade on soft materials, to have sufficient clearance for the chips, and a fine-tooth blade on harder materials. But you also should have at least three teeth cutting at any time, which may require a fine-tooth blade on soft materials with thin cross sections.

Hand hacksaw blades fall into two categories: soft-backed or flexible blades and all-hard blades. On the flexible blades only the teeth are hardened, the back being tough and flexible. The flexible blade is less likely to break when used in places difficult to get at such as in cutting off bolts on machinery. The all-hard blade is, as the name implies, hard and very brittle and should be used only where the workpiece can be rigidly supported, as in a vise. On an all-hard blade even a slight twisting motion may break the blade. All-hard blades, in the hands of a skilled person, will cut true straight lines and give long service.

The blades are mounted in the frame with the teeth pointing away from the handle so that the hacksaw cuts only on the forward stroke. No cutting pressure should be applied to the blade on the return stroke as this tends to dull the teeth. The sawing speed with the hacksaw should be from 40 to 60 strokes per minute. To get the maximum performance from a blade, make long, slow, steady strokes using the full length of the blade. Sufficient pressure should be maintained on the foward stroke to keep the teeth cutting. Teeth on a saw blade will dull rapidly if too little or too much pressure is put on the saw. The teeth will dull also if too fast a cutting stroke is used; a speed in excess of 60 strokes a minute will dull the blade because friction will overheat the teeth.

The saw blade may break if it is too loose in the frame or if the workpiece slips in the vise while sawing. Too much pressure may also cause the blade to break. A badly worn blade, where the set has been worn down, will cut a too narrow kerf, which will cause binding and perhaps breakage of the blade. When this happens and a new blade is used to finish the cut, turn the workpiece over and start with the new blade from the opposite side and make a cut to meet the first one (Figure B-72). The set on the new blade is wider than the old kerf. Forcing the new blade into an old cut will immediately ruin it by wearing the set down.

A cut on a workpiece should be started with only light cutting pressure, with the thumb or fingers on one hand acting as a guide for the blade. Sometimes it helps to start a blade when a small vee-notch is filed in the workpiece. When a workpiece is supported in a vise, make sure that the

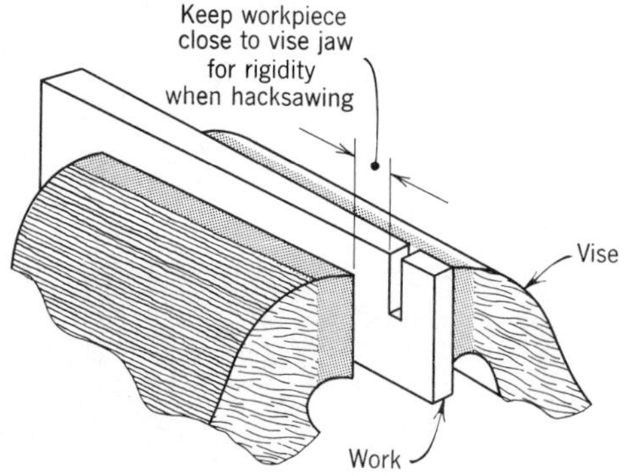
Keep workpiece
close to vise jaw
for rigidity
when hacksawing
Vise
Work

FIGURE B-73 The workpiece is being sawed close to the vise to avoid vibration and chatter (Lane Community College).

cutting is done close to the vise jaws for a rigid setup free of chatter (Figure B-73). Work should be positioned in a vise so that the saw cut is vertical. This makes it easier for the saw to follow a straight line. At the end of a saw cut, just before the pieces are completely parted, reduce the cutting pressure or you may be caught off balance when the pieces come apart and cut your hands on the sharp edges of the workpiece. To saw thin material, sandwich it between two pieces of wood for a straight cut. Avoid bending the saw blades because they are likely to break, and when they do, they usually shatter in all directions and could injure you or others nearby.

SELF-TEST

1. What is the kerf?

2. What is the set on a saw blade?

3. What is the pitch of the hacksaw blade?

4. What determines the selection of a saw blade for a job?

5. Hand hacksaw blades fall into two basic categories. What are they?

6. What speed should be used in hand hacksawing?

7. Give four causes that make saw blades dull.

8. Give two reasons why hacksaw blades break.

9. A new hacksaw blade should not be used in a cut started with a blade that has been used. Why?

10. What dangers exist when a hacksaw blade breaks while it is being used?

UNIT 4

Files

Files are often used to put the finishing touches on a machined workpiece, either to remove burrs or sharp edges or as a final fitting operation. Intricate parts or shapes are often entirely produced by skilled workers using files. In this unit you are introduced to the types and uses of files in metalworking.

Objective

After completing this unit, you should be able to:

Identify eight common files and some of their uses.

Types of Files

Files are tools that anyone in metalwork will use. Often, through lack of knowledge, these tools are misused. Files are made in many different lengths ranging from 4 to 18 in. (Figure B-74). Files are manufactured in many different shapes and are used for many specific purposes. Figure B-75 shows the parts of a file. When a file is measured, the length is taken from the heel to the point, with the tang excluded. Most files are made from high-carbon steel and are heat-treated to the correct hardness range. They are manufactured in four different cuts: single, double, curved tooth, and rasp. The single cut, double cut, and curved tooth are commonly encountered in machine shops. Rasps are usually used with wood. Curved tooth files will give excellent results with soft materials such as aluminum, brass, plastic, or lead.

Files also vary in their coarseness: rough, coarse, bastard, second cut, smooth, and dead smooth. The files most often used are the bastard, second cut, and smooth grades. Different sizes of files within the same coarseness designation will have varying sizes of teeth (Figure B-76): the longer the file, the coarser the teeth. For maximum metal removal a double cut file is used. If the emphasis is on a smooth finish, a single cut file is recommended.

The face of most files is slightly convex because they are made thicker in the middle than on the ends. Because of this curvature only some of the teeth are cutting at any one time, which makes them penetrate better. If the face were flat, it would be difficult to obtain an even surface because of the tendency to rock a file while filing. Some of this curvature is also offset by the pressure applied to make the file cut. New files do not cut as well as slightly used ones, since on new files some teeth are longer than most of the others and leave scratches on a workpiece.

Files are either blunt or tapered (Figure B-77). A blunt file has the same cross-sectional area from heel to point, whereas a tapered file narrows toward the point.

Files fall into five basic categories: mill and saw files, machinists' files, Swiss pattern files, curved tooth files, and rasps. Machinists', mill, and saw files are classified as American pattern files. Mill files (Figure B-78) were originally designed to sharpen large saws in lumber mills, but now they

FIGURE B-74. Files are made in several different lengths (Lane Community College).

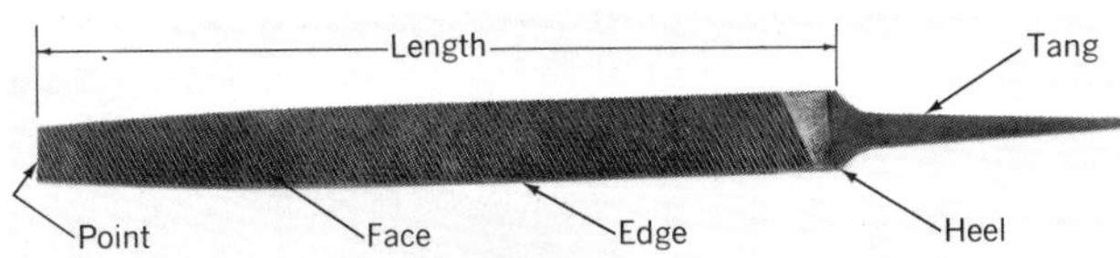

FIGURE B-75 The parts of a file.

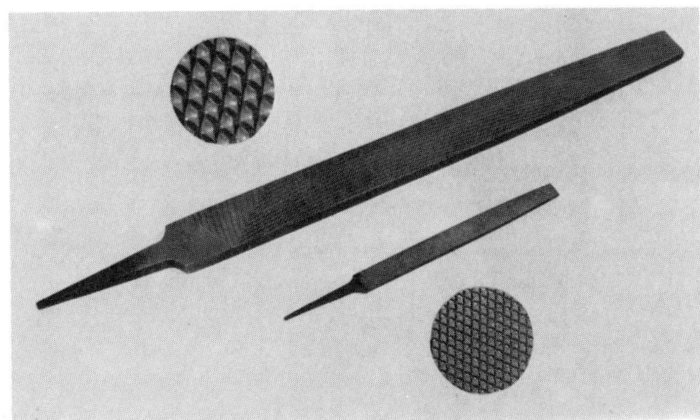

FIGURE B-76 These two files are both bastard cut, but since they are of different lengths, they have different coarsenesses (Lane Community College).

FIGURE B-77 Blunt and tapered file shapes.

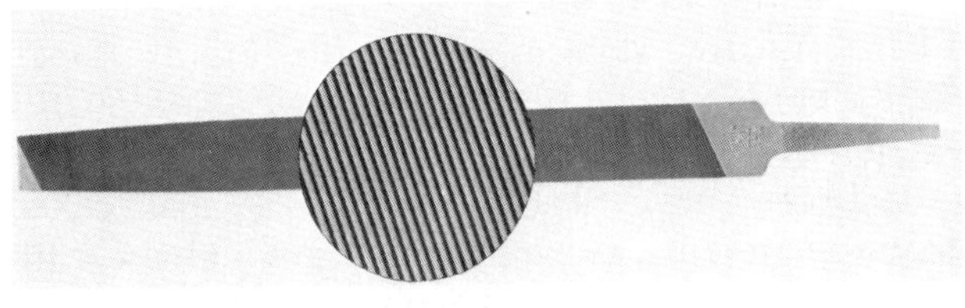

FIGURE B-78 A mill file.

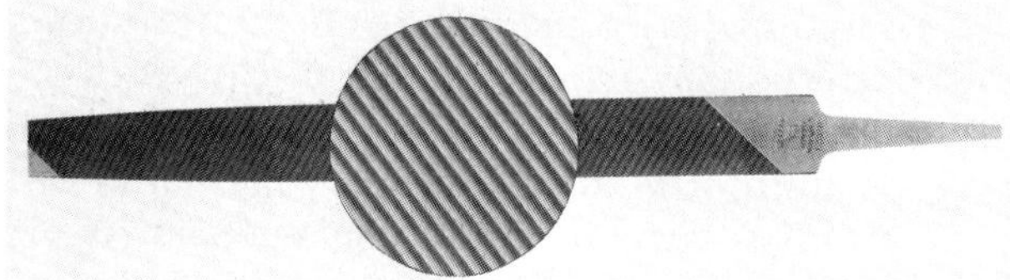

FIGURE B-79 The lathe file has a longer angle on the teeth to clear the chips when filing on the lathe.

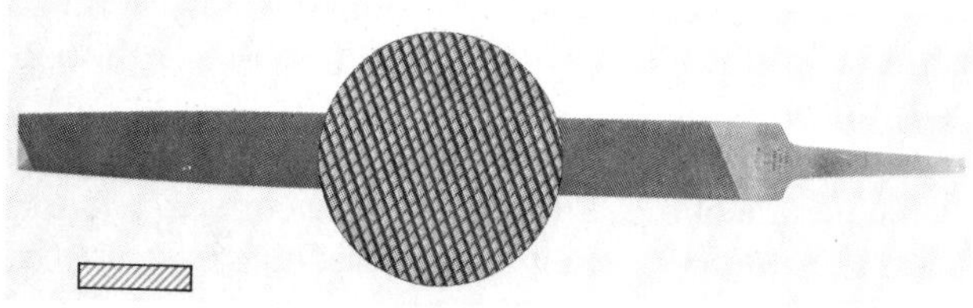

FIGURE B-80 The flat file is usually a double cut file.

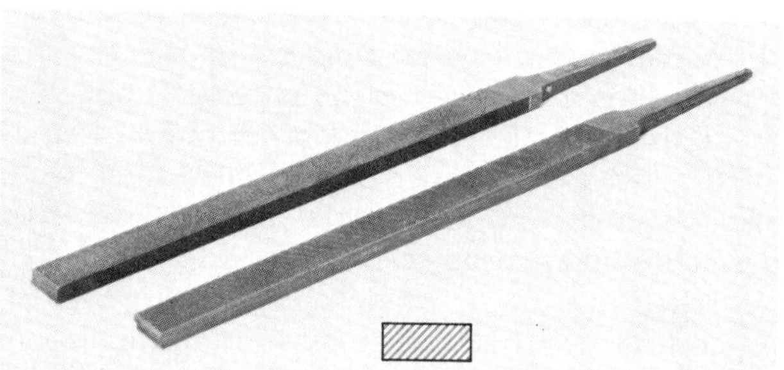

FIGURE B-81 Two pillar files.

are used for draw filing, filing on a lathe (Figure B-79), or filing a finish on a workpiece. Mill files are single cut and work well on brass and bronze. Mill files are slightly thinner than an equal-sized flat file, a machinist's file (Figure B-80) that is usually double cut. Double cut files are used when fast cutting is needed. The finish produced is relatively rough.

Pillar files (Figure B-81) have a narrower but thicker cross section than flat files. Pillar files are parallel in width and taper slightly in thickness. They also have one or two safe edges that allow filing into a corner without damaging the shoulder. Square files (Figure B-82) usually are double cut and are used to file in keyseats, slots, or holes.

If a very thin file is needed with a rectangular cross section, a warding file (Figure B-83) is used. This file is often used by locksmiths when filing notches into locks and keys. Another file that will fit into narrow slots is a knife file (Figure B-84). The included angle between the two faces of this file is approximately 10 degrees.

Three-square files (Figure B-85), also called three-cornered files, are triangular in shape with the faces at 60-degree angles to each other. These files are used for filing internal angles between 60

FIGURE B-82 Square file (Lane Community College).

FIGURE B-83 Warding file (Lane Community College).

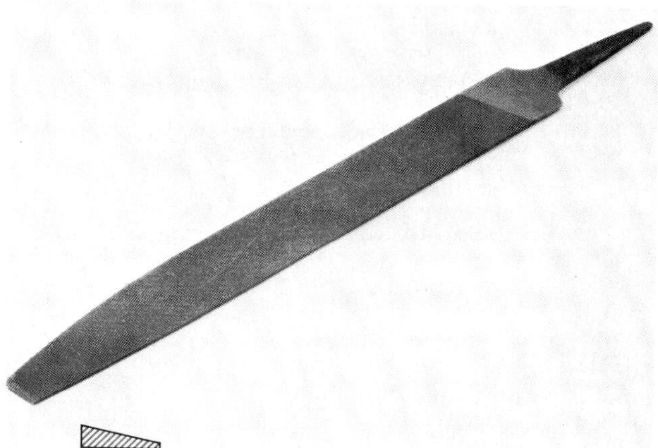

FIGURE B-84 Knife file (Lane Community College).

FIGURE B-85 Three-square files are used for filing angles between 60 and 90 degrees (Lane Community College).

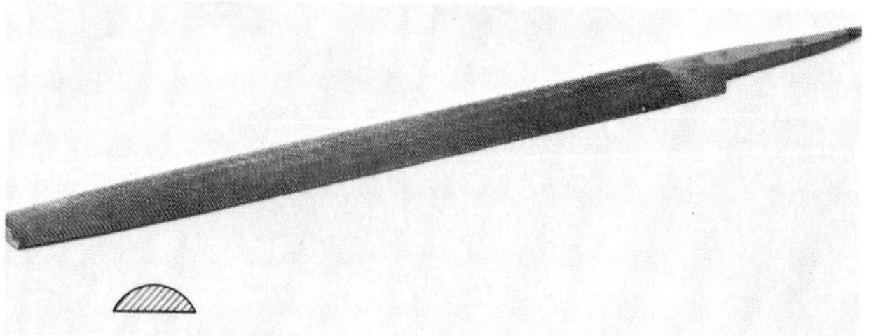

FIGURE B-86 Half-round files are used for internal curves (Lane Community College).

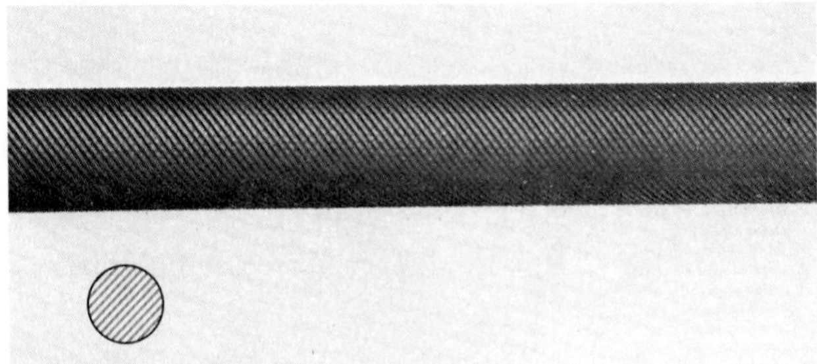

FIGURE B-87 Round files are used to file a small radius or to enlarge a hole (DeAnza College).

FIGURE B-88 Set of Swiss pattern files. Since these small files are very delicate and can be broken quite easily, great care must be exercised in their use (DeAnza College).

FIGURE B-89 Die sinker's rifflers.

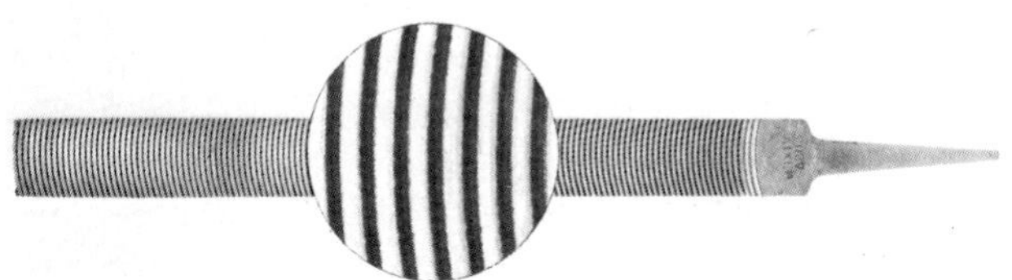

FIGURE B-90 Curved tooth files are used on soft metals (DeAnza College).

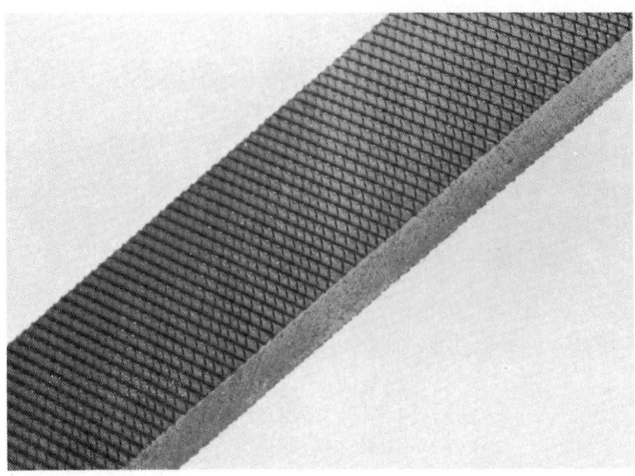

FIGURE B-91 A file with a safe edge will not cut into shoulders or corners when filing is being done (DeAnza College).

and 90 degrees as well as to make sharp corners in square holes. Half-round files (Figure B-86) are available to file large internal curves. Half-round files, because of their tapered construction, can be used to file many different radii. Round files (Figure B-87) are used to file small radii or to enlage holes. These files are available in many diameter sizes.

Swiss pattern files (Figure B-88) are manufactured to much closer tolerances than American pattern files, but are made in the same shapes. Swiss pattern files are more slender, as they taper to finer points and their teeth extend to the extreme edges. Swiss pattern files range in length from 3 to 10 in. and their coarseness is indicated by numbers from 00 (coarse) to 6 (fine). Swiss pattern files are made with tangs to be used with file handles or as needle files with round or square handles that are part of the files. Another type of Swiss pattern files are die sinkers' rifflers (Figure B-89). These files are double-ended with cutting surfaces on either end. Swiss pattern files are used primarily by tool and die makers, mold makers, and other workers engaged in precision filing on delicate instruments.

Curved tooth files (Figure B-90) cut very freely and remove material rapidly. The teeth on curved tooth files are all of equal height and the gullets or valleys between teeth are deep and provide sufficient room for the filings to curl and drop free. Curved tooth files are manufactured in three grades of cut—standard, fine, and smooth—and in lengths from 8 to 14 in. These files are made as rigid tang types for use with a file handle, or as rigid or flexible blade types used with special handles. Curved tooth file shapes are flat, half-round, pillar, and square.

The bastard cut file (Figure B-91) has a safe edge that is smooth. Flat filing may be done up to

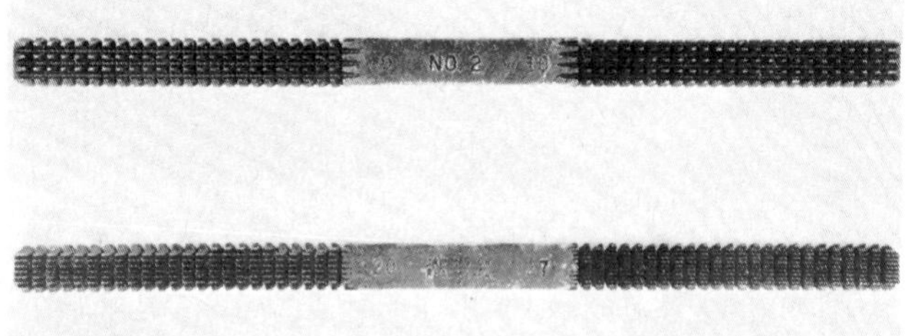

FIGURE B-92 Thread files (Lane Community College).

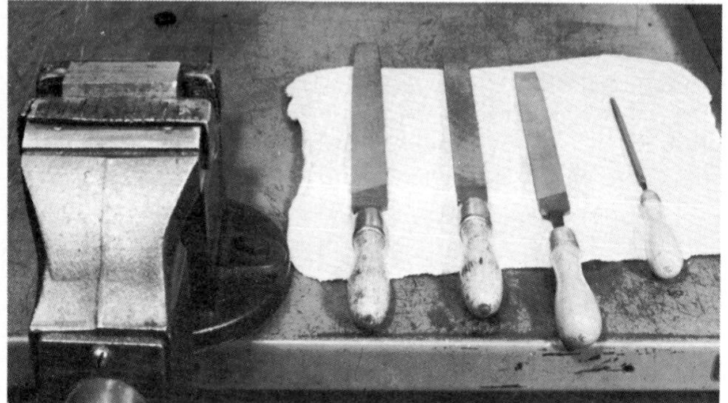

FIGURE B-93 Files should be kept neatly arranged so that they will not strike each other and damage the cutting edges (Lane Community College).

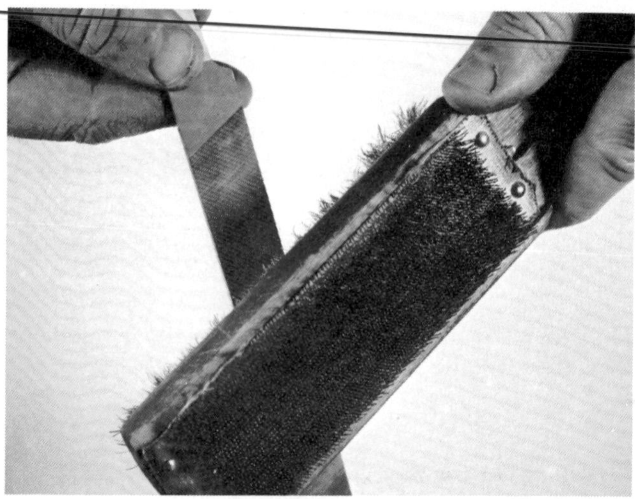

FIGURE B-94 Using a file card to clean a file (DeAnza College).

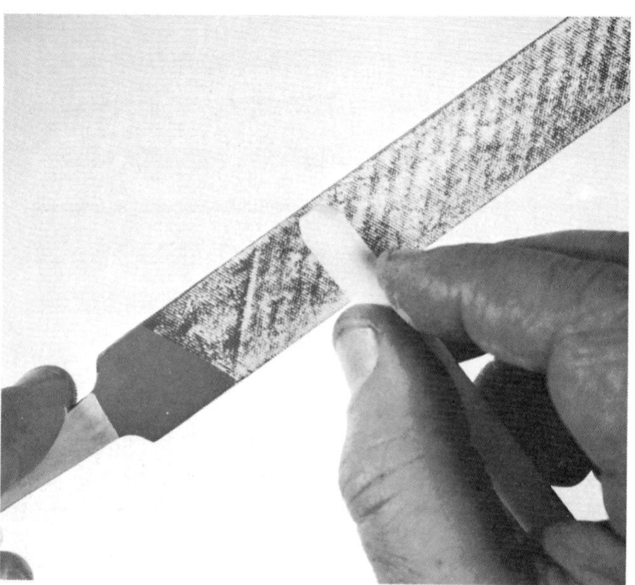

FIGURE B-95 Using chalk on the file to help reduce pinning.

the shoulders of the workpiece without fear of damage. Files of other cuts and coarseness are also available with safe edges on one or both sides.

Thread files (Figure B-92) are used to clean up and reshape damaged threads. They are square in cross section and have eight different thread pitches on each file. The thread file of the correct pitch is most effectively used when held or stroked against the thread while it is rotating in a lathe. A thread can be repaired, however, even when it cannot be turned in a lathe.

Care and Use of Files

Files do an efficient job of cutting only while they are sharp. Files and their teeth are very hard and brittle. Do not use a file as a hammer or as a pry bar. When a file breaks, particles will fly quite a distance at high speed and may cause an injury. Files should be stored so that they are not in contact with any other file. The same applies to files on a workbench. Do not let files lie on top of each other because one file will break teeth on the other (Figure B-93). Teeth on files will also break if too much pressure is put on them while filing. On the other hand, if not enough pressure is applied while filing, the file only rubs the workpiece and dulls the teeth. A dull file can be identified by its shiny, smooth teeth and by the way it slides over the work without cutting. Dulling of teeth is also caused by the filing of hard materials or by filing too fast. A good filing speed is 40 to 50 strokes per minute, but remember that the harder the material, the slower the strokes should be; the softer the material, the coarser the file should be.

Too much pressure on a new file may cause **pinning**, that is, filings wedged in the teeth; the result is deep scratches on the work surface. If the pins cannot be removed with a file card (Figure B-94), try a piece of brass, copper, or mild steel and push it through the teeth. Do not use a scriber or other hard object for this operation. A file will not pin as much if some blackboard chalk is applied to the face (Figure B-95). *Never* use a file without

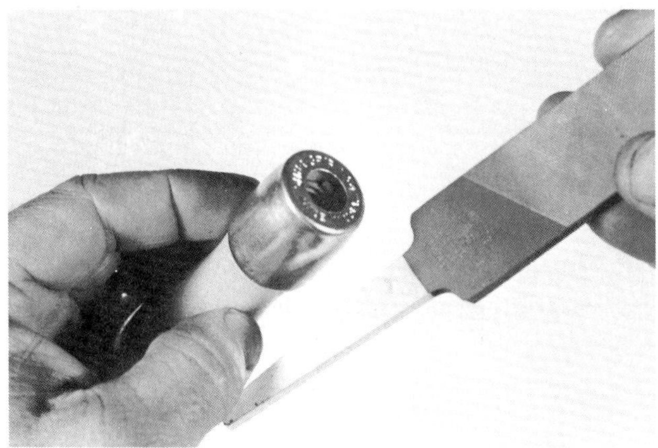

FIGURE B-96 A file should never be used without a file handle. This style of handle is designed to screw on rather than be driven on the tang.

FIGURE B-98 Proper filing position (Lane Community College).

FIGURE B-97 The crosshatch pattern shows that this piece has been filed from two directions, thus producing a flatter surface (Lane Community College).

FIGURE B-99 Draw filing (Lane Community College).

a file handle, or the pointed tang may cause serious hand or wrist injury (Figure B-96).

Many filing operations are performed with the workpiece held in a vise. Clamp the workpiece securely, but remember to protect it from the serrated vise jaws with a soft piece of material such as copper, brass, wood, or paper. The workpiece should extend out of the vise so that the file clears the vise jaws by $\frac{1}{8}$ to $\frac{1}{4}$ in. Since a file cuts only on the forward stroke, no pressure should be applied on the return stroke. Letting the file drag over the workpiece on the return stroke helps release the small chips so that they can fall from the file. However, this can also dull the file and scratch the part, so do it cautiously.

Use a stroke as long as possible; this will make the file wear out evenly instead of just in the middle. To file a flat surface, change the direction of the strokes frequently to produce a crosshatch pattern (Figure B-97). By using a straightedge steel rule to test for flatness, we can easily determine where the high spots are that have to be filed away. It is best to make flatness checks often because, if any part is filed below a given layout line, the rest of the workpiece may have to be brought down just as far.

Figure B-98 shows how a file should be held to file a flat surface. A smooth finish is usually obtained by draw filing (Figure B-99), whereby a single cut file is held with both hands and drawn back and forth on a workpiece. The file should not be pushed over the ends of the workpiece as this would leave rounded edges. To get a smooth finish it sometimes helps to hold the file as shown in

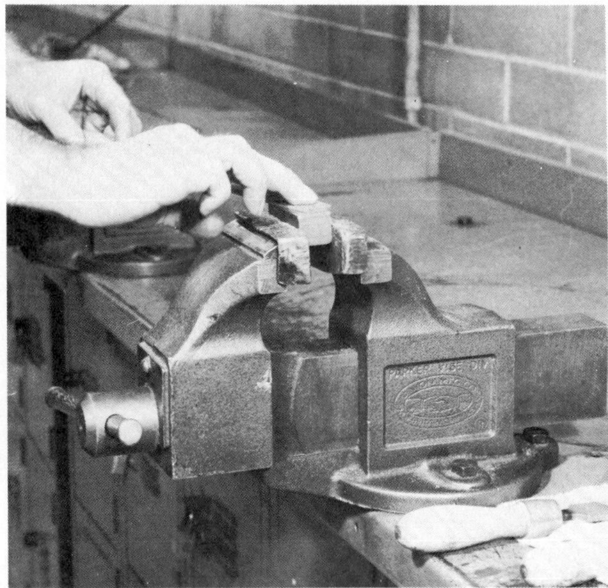

FIGURE B-100 Use this procedure to correct high spots on curvatures on the workpiece. Apply pressure with short strokes only where cutting is needed (Lane Community College).

Figure B-100, making only short strokes. The pressure is applied by a few fingers and does not extend over the ends of the workpiece. When a round file or half-round file is used, the forward stroke should also include a clockwise rotation for deeper cuts and a smoother finish. A tendency of people who are filing is to run their hands or fingers over a newly filed surface. This deposits a thin coat of skin oil on the surface. When filing is resumed, the file will not cut for several strokes, but will only slip over the surface causing the file to dull more quickly.

SELF-TEST

1. How is a file identified?
2. What are the four different cuts found on files?
3. Name four coarseness designations for files.
4. Which of the two kinds of files—single cut or double cut—is designed to remove more material?
5. Why are the faces of most files slightly convex?
6. What difference is there between a blunt and a tapered file?
7. What difference exists between a mill file and an equal-sized flat file?
8. What is a warding file?
9. An American pattern file differs in what way from a Swiss pattern file?
10. What are the coarseness designations for needle files?
11. Why should files be stored so they do not touch each other?
12. What happens if too much pressure is applied when filing?
13. What causes a file to get dull?
14. Why should a handle be used on a file?
15. Why should workpieces be measured often?
16. What happens when a surface being filed is touched with the hand or fingers?
17. How does the hardness of a workpiece affect the selection of a file?
18. How can rounded edges be avoided when a workpiece is draw filed?
19. Should pressure be applied to a file on the return stroke?
20. Why is a round file rotated while it is being used?

UNIT 5

Hand Reamers

Holes produced by drilling are seldom accurate in size and often have rough surfaces. A reamer is used to finish a hole to an exact dimension with a smooth finish. Hand reamers are often used to finish a previously drilled hole to an exact dimension and a smooth surface. When parts of machine tools are aligned and fastened with capscrews or bolts, the final operation is often the hand reaming of a hole in which a dowel pin is placed to maintain the alignment. Hand reamers are designed to remove only a small amount of material from a hole—usually from .001 to .005 in. These tools are made from high-carbon or high-speed steel. This unit will describe some commonly used hand reamers and how they are used.

Objectives

After completing this unit, you should be able to:

1. Identify at least five types of hand reamers.
2. Hand ream a hole to a specified size.

Features of Hand Reamers

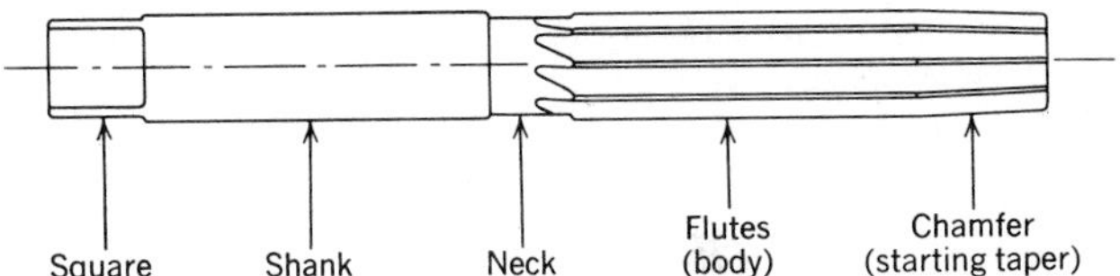

FIGURE B-101 Major features of the hand reamer.

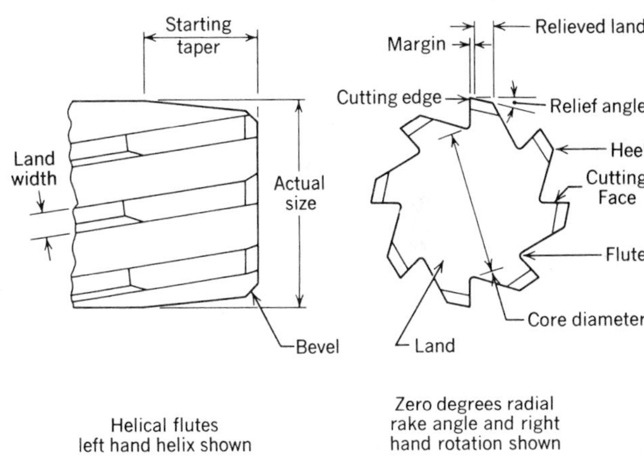

FIGURE B-102 Functional details of the hand reamer (Bendix Industrial Tools Division).

Figure B-101 shows the major features of the most common design of hand reamer. Another design is available with a pilot ahead of the starting taper (see *Machinery's Handbook* for details). The square on the end of the shank permits the clamping of a tap wrench or T-handle wrench to provide the driving torque for reaming. The diameter of this square is between .004 and .008 in. smaller than the reamer size, and the shank of the reamer is between .001 and .006 in. smaller, to guide the reamer and permit it to pass through a reamed hole without marring it. It is very important that these tools not be put into a drill chuck, because a burred shank can ruin a reamed hole as the shank is passed through it.

Hand reamers have a long starting taper that is usually as long as the diameter of the reamer, but may be as long as one-third of the fluted body. This starting taper is usually very slight and may not be apparent at a casual glance. Hand reamers do their cutting on this tapered portion. The gentle taper and length of the taper help to start the reamer straight and keep it aligned in the hole.

Details of the cutting end of the hand reamer are shown in Figure B-102. The full diameter or actual size of the hand reamer is measured where the starting taper ends and the margin of the land appears. The diameter of the reamer should be measured only at this junction, as the hand reamer is generally back tapered or reduced in outside diameter by about .0005 to .001 in. per inch of length toward the shank. This back tapering is done to reduce tool contact with the workpiece. When hand reamers become dull, they are resharpened at the starting taper, using a tool and cutter grinder.

The function of the hand reamer is like that of a scraper, rather than an aggressive cutting tool like most drills and machine reamers. For this reason hand reamers typically have zero or negative radial rake on the cutting face, rather than the positive radial rake characteristic of most machine reamers. (See Section H, Unit 6, "Reaming in the Drill Press.") The right-hand cut with a left-hand helix is considered standard for hand reamers. The left-hand helix produces a negative axial rake for the tool, which contributes to a smooth cutting action.

Most reamers, hand or machine types, have staggered spacing on teeth, which means that the

FIGURE B-103 Straight flute hand reamer (TRW, Inc.).

FIGURE B-104 Helical flute hand reamer (TRW, Inc.).

FIGURE B-105 Straight flute expansion hand reamer (TRW, Inc.).

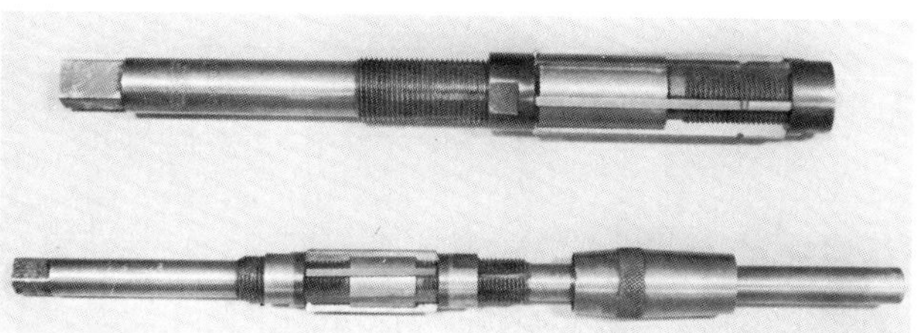

FIGURE B-106 Adjustable hand reamers. The lower reamer is equipped with a pilot and tapered guide bushing for reaming in alignment with a second hole (Lane Community College).

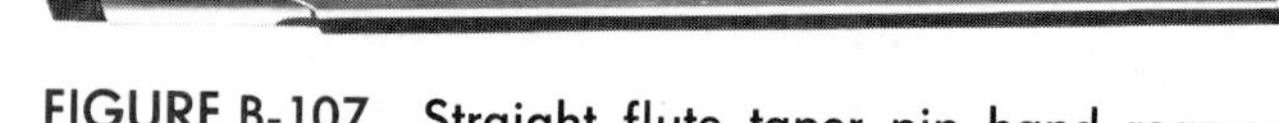

FIGURE B-107 Straight flute taper pin hand reamer (TRW, Inc.).

FIGURE B-108 Spiral flute taper pin hand reamer (TRW, Inc.).

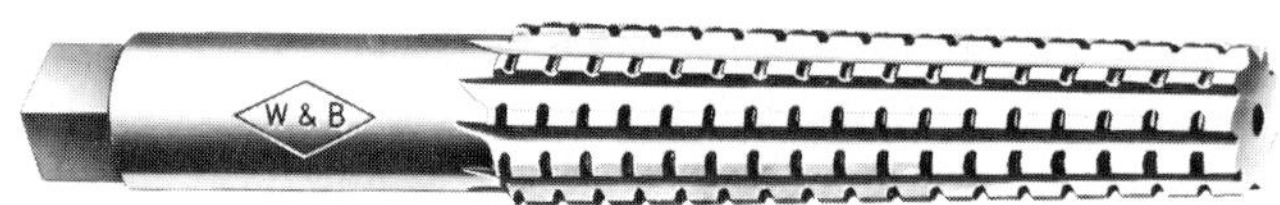

FIGURE B-109 Morse taper socket roughing reamer (TRW, Inc.).

FIGURE B-110 Morse taper socket finishing reamer (TRW, Inc.).

flutes or body channels are not precisely of uniform spacing. The difference is very small, only a degree or two, but it tends to reduce chatter by reducing harmonic effects between cutting edges. Harmonic chatter is especially a problem with adjustable hand reamers which often leave a tooth pattern in the work.

Hand reamers are made with straight flutes (Figure B-103) or with helical flutes (Figure B-104). Most hand reamers are manufactured with a right-hand cut, which means they will cut when rotated in a clockwise direction. Helical or spiral fluted reamers are available with a right-hand helix or a left-hand helix. Helical flute reamers are especially useful when reaming a hole having keyseats or grooves cut into it, as the helical flutes tend to bridge the gaps and reduce binding or chattering.

Hand reamers for cylindrical holes are made as solid (Figures B103 and B104) or as expansion types (Figure B-105). Expansion reamers are designed for use where it is necessary to enlarge a hole slightly for proper fit, such as in maintenance applications. These reamers have an adjusting screw that allows limited expansion to an exact size. The maximum expansion of these reamers is approximately .006 in. for diameters up to $\frac{1}{2}$ in., .010 in. for diameters between $\frac{1}{2}$ and 1 in., and .012 in. for diameters between 1 and $1\frac{1}{2}$ in. These tools are frequently broken by attempts to expand them beyond these limits.

Helical flute expansion reamers are especially adapted for the reaming of bushings or holes having a keyseat or straight grooves because of their bridging and shearing cutting action. Expansion reamers have a slightly undersized pilot on the end that guides the reamer and helps to keep it in alignment.

The adjustable hand reamer (Figure B-106) is different from the expansion reamer in that it has inserted blades. These cutting blades fit into tapered slots in the body of the reamer and are held in place by two locking nuts. The blades have a taper corresponding to the taper of the slots that keeps them parallel at any setting. Adjustments in reamer size are made by loosening one nut while tightening the other. Adjustable hand reamers are available in diameters from $\frac{1}{4}$ to 3 in. The adjustment range varies from $\frac{1}{32}$ in. on the smaller diameter reamers to $\frac{5}{16}$ in. on the larger size reamers. Only a small amount of material should be removed at one time, as too large a cut will usually cause chatter.

Taper pin reamers (Figures B-107 and B-108) are used for reaming holes for standard taper pins used in the assembly of machine tools and other parts. Taper pin reamers have a taper of $\frac{1}{4}$ in. per foot of length, and are manufactured in 18 different sizes numbered from 8/0 to 0 and on up to size 10. The smallest size, number 8/0, has a large end diameter of .0514 in. and the largest reamer, a number 10, has a large end diameter of .7216 in. The sizes of these reamers are designed to allow the small end of each reamer to enter a hole reamed by the next smaller size reamer. As with other hand reamers, the helical flute reamer will cut with more shearing action and less chattering, especially on interrupted cuts.

Morse taper socket reamers are designed to produce holes for American Standard Morse taper shank tools. These reamers are available as roughing reamers (Figure B-109) and as finishing reamers (Figure B-110). The roughing reamer has notches ground at intervals along the cutting edges. These notches act as chip breakers and make the tool more efficient at the expense of fine finish. The finishing reamer is used to impart the final size and finish to the socket. Morse taper socket reamers are made in sizes from No. 0, with a large-end diameter of .356 in., to No. 5, with a large-end diameter of 1.8005 in. There are two larger Morse tapers, but they are typically sized by boring rather than reaming.

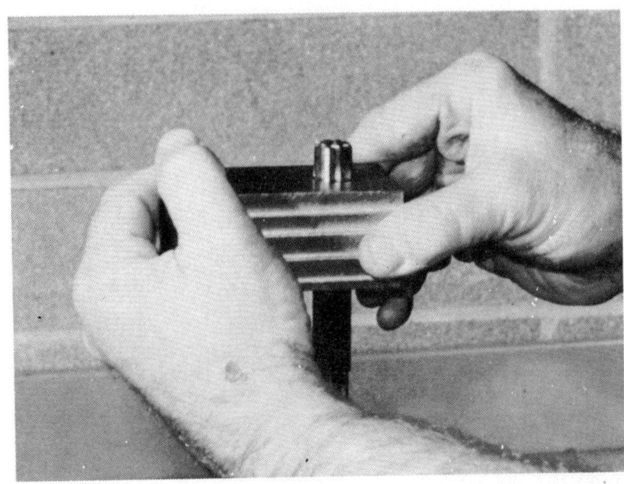

FIGURE B-111 Hand reaming a small workpiece with the reamer held in a vise (Lane Community College).

Using Hand Reamers

A hand reamer should be turned with a tap wrench or T-handle wrench rather than with an adjustable wrench. The use of a single end wrench makes it almost impossible to apply torque without disturbing the alignment of the reamer with the hole. A hand reamer should be rotated slowly and evenly, allowing the reamer to align itself with the hole to be reamed. Use a tap wrench large enough to give a steady torque and to prevent vibration and chatter. Use a steady and large feed; feeds up to one-quarter of the reamer diameter per revolution can be used. Small and lightweight workpieces can be reamed by fastening the reamer vertically in a bench vise and rotating the work over the reamer by hand (Figure B-111).

In all hand reaming with solid, expansion, or adjustable reamers, never rotate the reamer backwards to remove it from the hole, as this will dull it rapidly. If possible, pass the reamer through the hole and remove it from the far side without stopping the forward rotation. If this is not possible, it should be withdrawn while maintaining the forward rotation.

The preferred stock allowance for hand reaming is between .001 and .005 in. Reaming more material than this would make it very difficult to force the reamer through the workpiece. Reaming too little, on the other hand, results in excessive tool wear because it forces the reamer to work in the zone of material work-hardened during the drilling operation. This stock allowance does not apply to taper reamers, for which a hole has to be drilled at least as large as the small diameter of the reamer. The hole size for a taper pin is determined by the taper pin number and its length. These data can be found in machinist's handbooks.

Since cylindrical hand reaming is restricted to small stock allowances, it is most important that you be able to drill a hole of predictable size and of a surface finish that will assure a finished cleanup cut by the reamer. It is a good idea to drill a test hole in a piece of scrap of similar composition and carefully measure both for size and for an enlarged or bellmouth entrance. You may find it necessary to drill a slightly smaller hole before drilling the correct reaming size to assure a more accurate hole size. Carefully spot drill the location before drilling the hole in your actual workpiece. The hole should than be lightly chamfered with a countersinking tool to remove burrs and to promote better reamer alignment.

The use of a cutting fluid also improves the cutting action and the surface finish when reaming most metals. Exceptions are cast iron and brass, which should be reamed dry.

When a hand reamer is started it should be checked for squareness on two sides of the reamer, 90 degrees apart. Another way to assure alignment of the reamer with the drilled holes is to use the drill press as a reaming fixture. Put a piece of cylindrical stock with a 60-degree center in the drill

FIGURE B-112 Using the drill press as a reaming fixture (Lane Community College).

chuck (Figure B-112) and use it to guide and follow the squared end of the reamer as you turn the tool with the tap wrench. Be sure to plan ahead so that you can drill, countersink, and ream the hole without moving the table or head of the drill press between operations.

On deep holes, or especially on holes reamed with taper reamers, it becomes necessary to remove the chips frequently from the reamer flutes to prevent clogging. Remove these chips with a brush to avoid cutting your hands.

Reamers should be stored so they do not contact one another to avoid burrs on the tools that can damage a hole being reamed. They should be kept in their original shipping tubes or set up in a tool stand. Always check reamers for burrs or for pickup of previous material before you use them. Otherwise, the reamed hole can be oversized or marred with a rough finish.

SELF-TEST

1. How is a hand reamer identified?
2. What is the purpose of a starting taper on a reamer?
3. What is the advantage of a spiral flute reamer over a straight flute reamer?
4. How does the shank diameter of a hand reamer compare with the diameter measured over the margins?
5. When are expansion reamers used?
6. What is the difference between an expansion and an adjustable reamer?
7. What is the purpose of cutting fluid used while reaming?
8. Why should reamers not be rotated backwards?
9. How much reaming allowance is left for hand reaming?

UNIT 6

Identification and Uses of Taps

Most internal threads produced today are made with taps. These taps are available in a variety of styles, each one designed to perform a specific type of tapping operation efficiently. This unit will help you identify and select taps for threading operations.

Objectives

After completing this unit, you should be able to:

1. Identify common taps.
2. Select taps for specific applications.

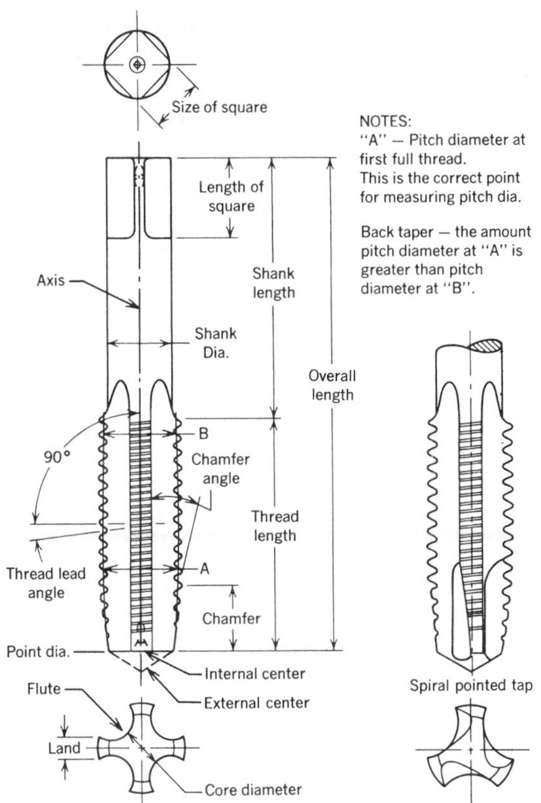

FIGURE B-113 General tap terms (Bendix Industrial Tools Division).

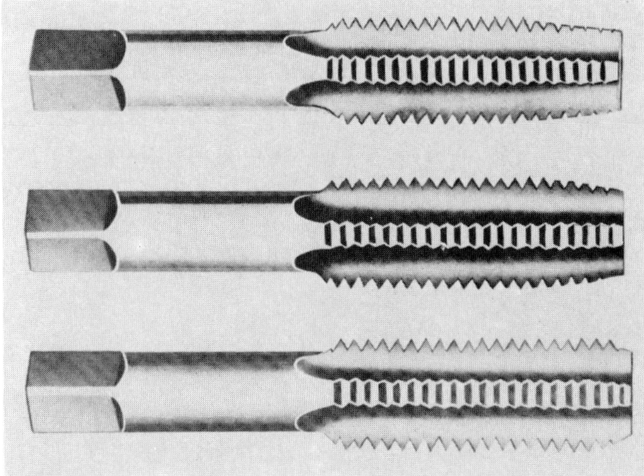

FIGURE B-114 Chamfer designations for cutting taps. Top to bottom: starting tap, plug tap, and bottoming tap.

FIGURE B-115 Interrupted thread tap (DeAnza College).

FIGURE B-116 Identifying marking on a tap.

Identifying Common Tap Features

Taps are used to cut internal threads in holes. This process is called tapping. Tap features are illustrated in Figures B-113 and B-114. The active cutting part of the tap is the chamfer, which is produced by grinding away the tooth form at an angle, with relief back of the cutting edge, so that the cutting action is distributed progressively over a number of teeth. The fluted portion of the tap provides space for chips to accumulate and for the passage of cutting fluids. Two-, three-, and four-flute taps are common.

The major diameter (Figure B-113) is the outside diameter of the tool as measured over the thread crests at the first full thread behind the chamfer. This is the largest diameter of the cutting portion of the tap, as most taps are back tapered or reduced slightly in thread diameter toward the shank. This back taper reduces the amount of tool contact with the thread during the tapping process, hence, making the tap easier to turn.

Taps are made from either high carbon steel or high speed steel and have a hardness of about Rockwell C63. High-speed steel taps are far more common in manufacturing plants than carbon steel taps. High-speed steel taps typically are ground after heat treatment to ensure accurate thread geometry.

Another identifying characteristic of taps is the amount of chamfer at the cutting end of a tap (Figure B-114). A set consists of three taps, taper, plug, and bottoming taps, which are identical except for the number of chamfered threads. The taper tap is useful in starting a tapped thread square with the part. The most commonly used tap, both in hand and machine tapping, is a plug tap. Bottoming taps are used to produce threads that extend almost to the bottom of a blind hole. A blind hole is one that is not drilled clear through a part.

Serial taps are also made in sets of three taps for any given size of tap. Each of these taps has one, two, or three rings cut on the shank near the square. The No. 1 tap has smaller major and pitch diameters and is used for rough cutting the thread. The No. 2 tap cuts the thread slightly deeper, and the No. 3 tap finishes it to size. Serial taps are used when tough metals are to be tapped by hand. Another tap used for tough metal such as stainless steel is the interrupted thread tap (Figure B-115). This tap has alternate teeth removed to reduce tapping friction.

Figure B-116 shows the identifying markings of a tap, where $\frac{5}{8}$ in. is the nominal size, 11 is the number of threads per inch, and NC refers to the standardized National Coarse thread series. G is the symbol used for ground taps. H3 identifies the tolerance range of the tap. HS means that the tap material is high-speed steel. Left-handed taps will also be identified by an LH or left-hand marking on the shank. More information on taps may be found in *Machinery's Handbook*.

Other Kinds and Uses of Taps

Spiral pointed taps (Figure B-117), often called gun taps, are especially useful for machine tapping of through holes or blind holes with sufficient chip room below the threads. When turning the spiral point, the chips are forced ahead of the tap (Figure B-118). Since the chips are pushed ahead of the tap, the problems caused by clogged flutes, especially breakage of taps, are eliminated if it is a through hole. If a spiral pointed tap is used to tap a blind hole, sufficient hole depth is necessary to accommodate the chips that are pushed ahead of the tap. Also, since they are not needed for chip disposal, the flutes of spiral pointed taps can be made shallower, thus increasing the strength of the tap.

Spiral pointed taps can be operated at higher speeds and require less torque to drive than ordinary hand taps. Figure B-119 shows the design of the cutting edges. The cutting edges (*A*) at the point of the tap are ground at an angle (*B*) to the axis. Fluteless spiral pointed taps (Figure B-120) are recommended for production tapping of through holes in sections no thicker than the tap diameter. This type of tap is very strong and rigid, which reduces tap breakage caused by misalignment. Fluteless spiral point taps give excellent results when tapping soft materials or sheet metal.

Spiral fluted taps are made with helical instead of straight flutes (Figure B-121), which draw the chips out of the hole. This kind of tap is also used when tapping a hole that has a keyseat or spline, as the helical lands of the tap will bridge the interruptions. Spiral fluted taps are recommended for tapping deep blind holes in ductile materials such as aluminum, magnesium, brass, copper, and die-cast metals. Fast spiral fluted taps (Figure B-122) are similar to regular spiral fluted taps, but the faster spiral flutes increase the chip lifting action and permit the spanning of comparably wider spaces.

Thread-forming taps (Figure B-123) are fluteless and do not cut threads in the same manner as conventional taps. They are forming tools and their action can be compared with external thread rolling. On ductile materials such as aluminum, brass, copper, die castings, lead, and leaded steels these taps give excellent results. Thread-forming taps are held and driven just as are conventional

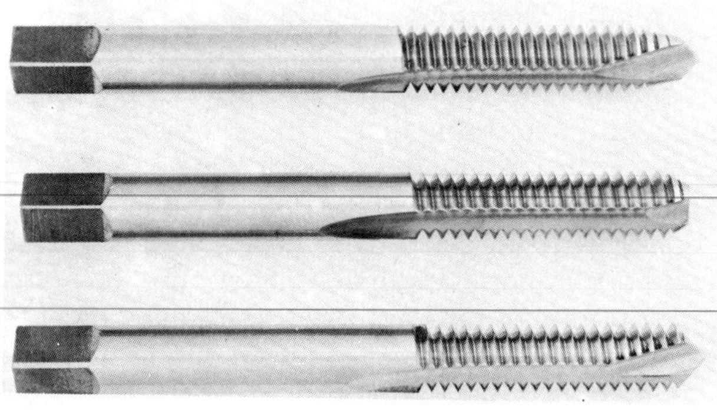

FIGURE B-117 Set of spiral pointed (or gun) taps (DeAnza College).

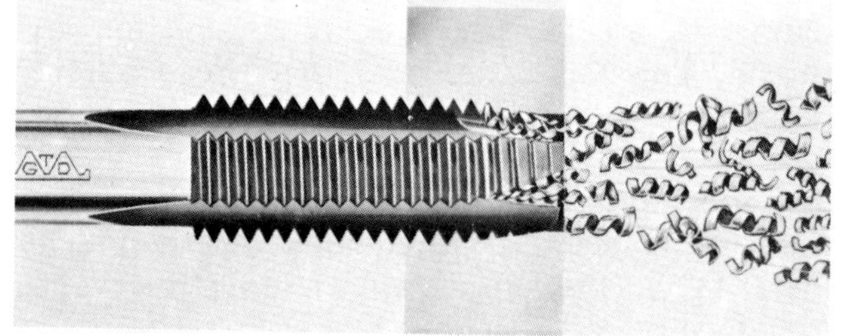

FIGURE B-118 Cutting action of spiral pointed taps (TRW, Inc.).

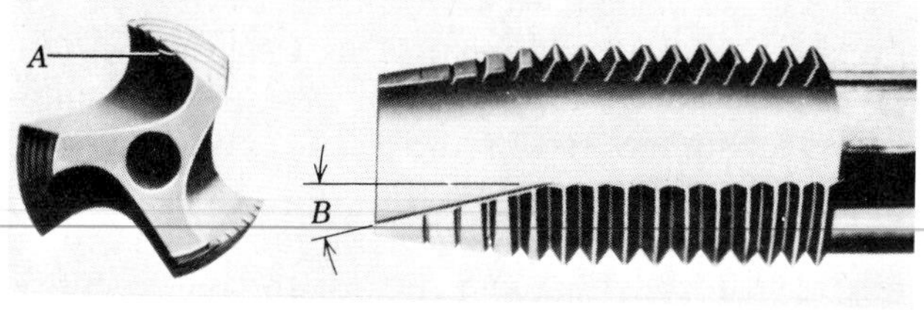

FIGURE B-119 Detail of spiral pointed tap (TRW, Inc.).

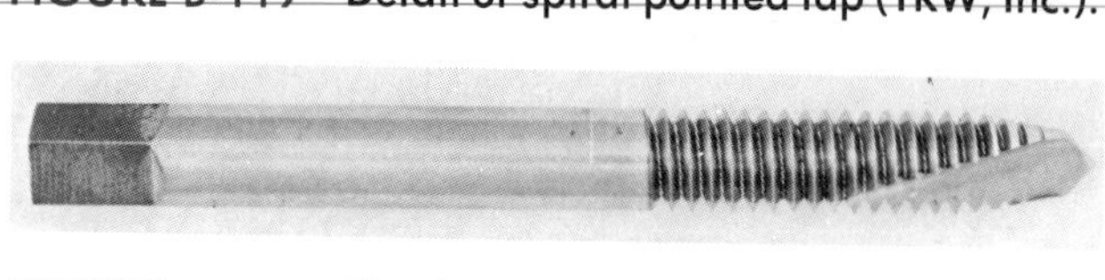

FIGURE B-120 Fluteless spiral pointed tap for thin materials (DeAnza College).

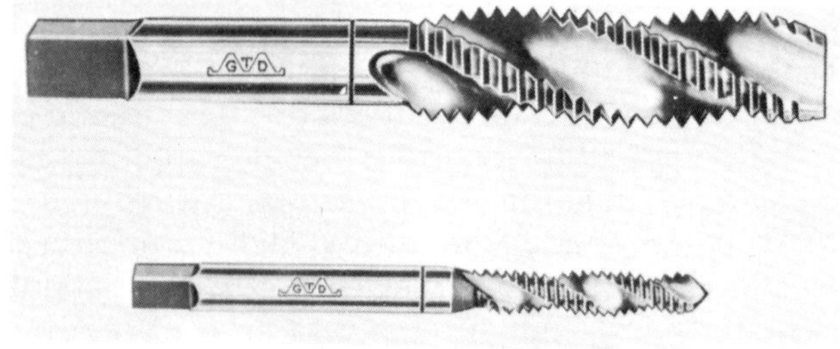

FIGURE B-121 Spiral fluted taps—regular spiral (TRW, Inc.).

taps, but because they do not cut the threads no chips are produced. Problems of chip congestion and removal often associated with the tapping of blind holes are eliminated. Figure B-124 shows how the forming tap displaces metal. The crests of the thread that are at the minor diameter may not be flat but will be slightly concave because of the flow of the displaced metal. Threads produced in this manner have improved surface finish and increased strength because of the cold working of the metal. The size of the hole to be tapped must be closely controlled, since too large a hole will result in a poor thread form and too small a hole will result in the breaking of the tap.

A tapered pipe tap (Figure B-125) is used to tap holes with a taper of $\frac{3}{4}$ in. per foot for pipes with a matching thread and to produce a leakproof fit. The nominal size of a pipe tap is that of the pipe fitting and not the actual size of the tap. When tapping taper pipe threads, every tooth of the tap engaged with the work is cutting until the rotation is stopped. This takes much more torque than does the tapping of a straight thread in which only the chamfered end and the first full thread are actually cutting. Straight pipe taps (Figure B-126) are used for tapping holes or couplings to fit taper-threaded pipe and to secure a tight joint when a sealer is used.

A pulley tap (Figure B-127) is used to tap set screw and oilcup holes in the hubs of pulleys. The long shank also permits tapping in places that might be inaccessible for regular hand taps. When used for tapping pulleys, these taps are inserted through holes in the rims, which are slightly larger than the shanks of the taps. These holes serve to guide the taps and assure proper alignment with the holes to be tapped.

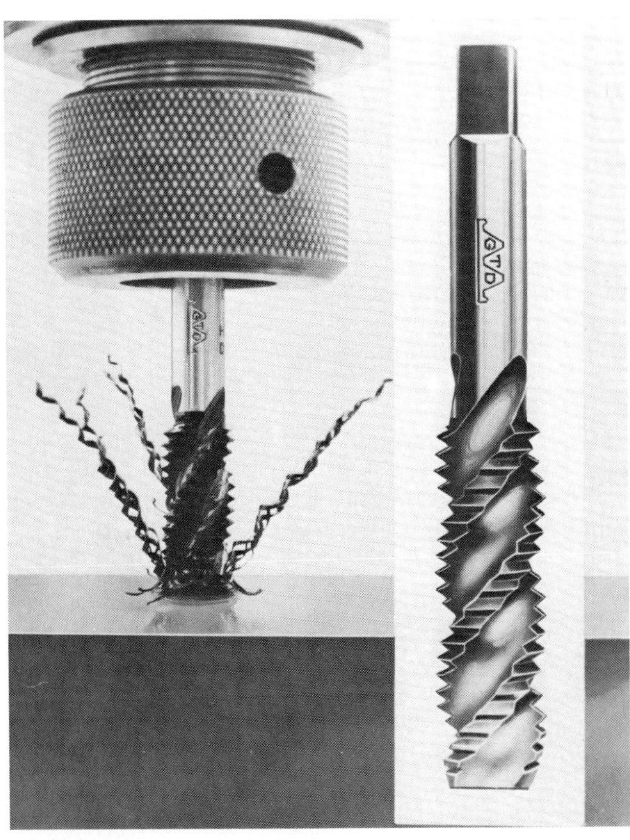

FIGURE B-122 Spiral fluted tap—fast spiral. The action of the tap lifts the chips out of hole to prevent binding (TRW, Inc.).

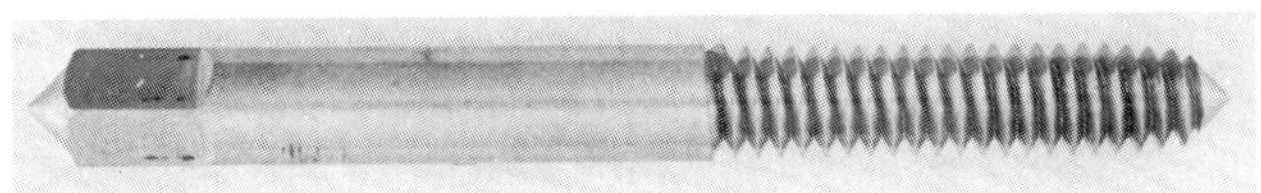

FIGURE B-123 Fluteless thread-forming tap (DeAnza College).

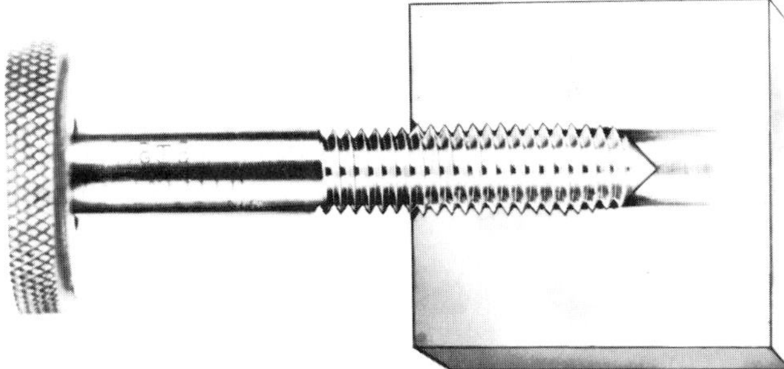

FIGURE B-124 The thread-forming action of a fluteless thread-forming tap (TRW, Inc.).

FIGURE B-125 Taper pipe tap (TRW, Inc.).

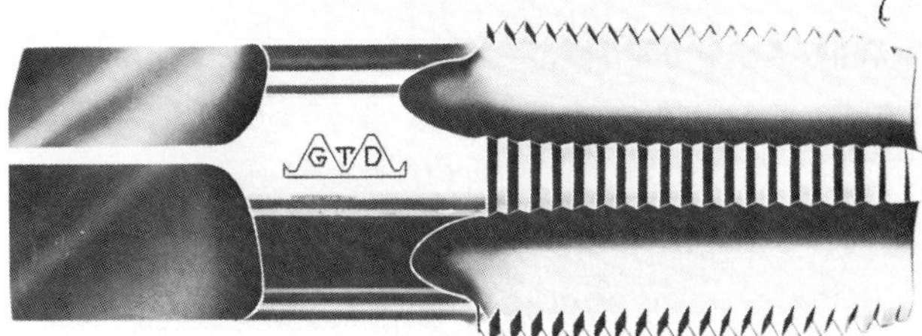

FIGURE B-126 Straight pipe tap (TRW, Inc.).

FIGURE B-127 Pulley tap (TRW, Inc.).

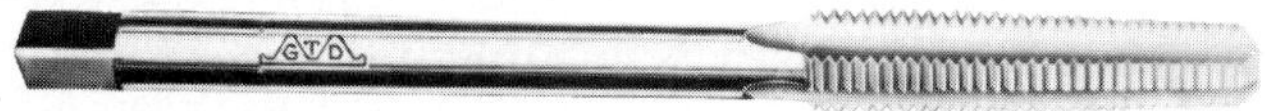

FIGURE B-128 Nut tap (TRW, Inc.).

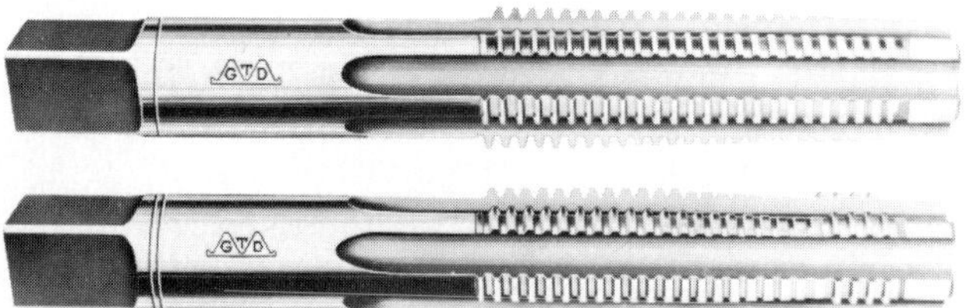

FIGURE B-129 Set of Acme thread taps. The upper tap is used for roughing, the lower tap for finishing (TRW, Inc.).

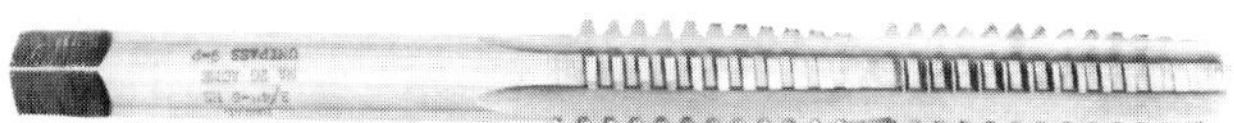

FIGURE B-130 A tandem Acme tap designed to rough and finish cut the thread in one pass (Lane Community College).

Nut taps (Figure B-128) differ from pulley taps in that their shank diameters are smaller than the root diameter of the thread. The smaller shank diameter makes the tapping of deep holes possible. Nut taps are used when small quantities of nuts are made or when nuts have to be made from tough materials such as some stainless steels or similar alloys.

Figure B-129 shows Acme taps for roughing and finishing. Acme threads are used to provide accurate movement—for example, in lead screws on machine tools—and for applying pressure in various mechanisms. On some Acme taps the roughing and finishing operation is performed with one tap (Figure B-130). The length of this tap usually requires a through hole.

Rake and Hook Angles on Cutting Edges

When selecting a tap for the most efficient cutting, the cutting face geometry will be an important factor. It should vary depending on the material to be tapped. Cutting face geometry is expressed in terms of rake and hook (Figure B-131). The rake of a tap is the angle between a line through the flat cutting face and a radial line from the center of the tool to the tooth tip. The rake can be negative, neutral, or positive. Hook angle, on the other hand, relates to the concavity of the cutting face. It is defined by the intersection of the radial line with the tangent line through the tooth tip (tangential hook) or by the average angle of the tooth face from crest to root (chordal). Unlike the rake angle, hook

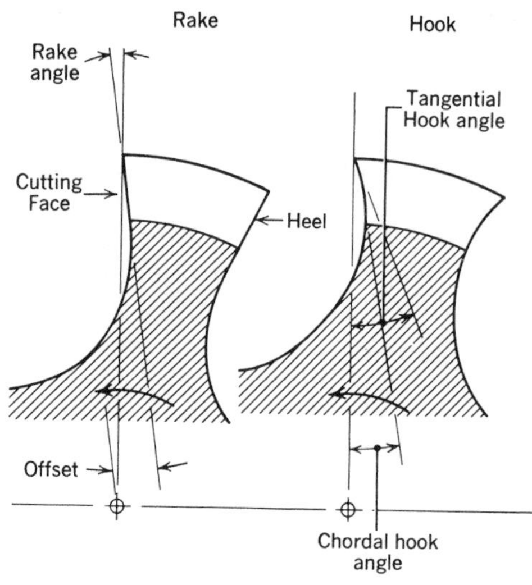

FIGURE B-131 Rake and hook angles on cutting taps (Bendix Industrial Tool Division).

TABLE B-1 Recommended Tap Rake Angles

0–5 Degrees	8–12 Degrees	16–20 Degrees
Bakelite	Bronze	Aluminum and alloys
Plastics	Hard rubber	Zinc die castings
Cast iron	Cast steel	Copper
Brass	Carbon steel	Magnesium
Hard rubber	Alloy steel	
	Stainless steel	

angle cannot be negative. Table B-1 gives the rake angle recommendations for some workpiece materials. In general, the softer or more ductile the material, the greater the rake angle. Harder and more brittle materials call for reduced rake angles.

Reducing Friction in Tapping

As discussed earlier in this unit, a tap is usually back tapered along the thread to relieve the friction between the tool and the workpiece. There is another form of relief often applied to taps with the same results. When the fully threaded portion of the tap is cylindrical (other than back taper), it is called a concentric thread (Figure B-132). If the pitch diameter of the fully threaded portion of the tap is brought uniformly closer to the axis of the tap as measured from face to back (heel), it has eccentric relief. This means less tool contact with the workpiece and less friction. A third form of friction relief combines the concentric thread and the eccentric thread relief, and is termed con-eccentric. The concentric margin gives substantial guidance and the relief following the margin reduces

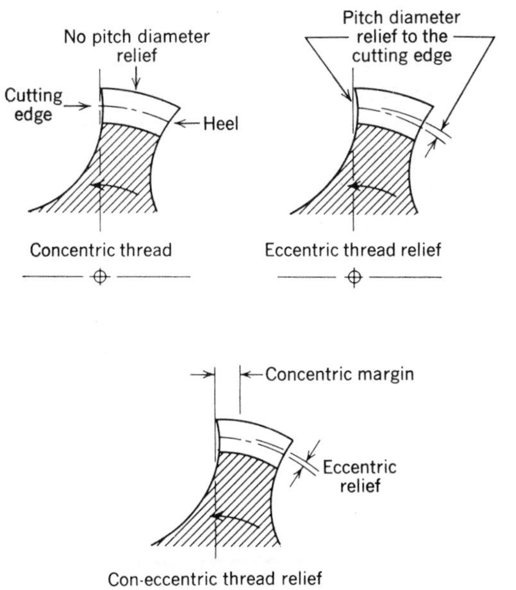

FIGURE B-132 Pitch diameter relief forms on taps (Bendix Industrial Tools Division).

friction. Relief is also provided behind the chamfer of the tap to provide radial clearance for the cutting edge. Relief may also be provided in the form of a channel that runs lengthwise down the center of the land (Figure B-133), termed a concave groove land relief.

Other steps may also be taken to reduce friction and to increase tap life. Surface treatment of taps is often an answer if poor thread forming or tap breakage is caused by chips adhering to the flutes or welding to the cutting faces. These treatments generally improve the wear life of taps by increasing their abrasion resistance.

Different kinds of surface treatments are used by tap manufacturers. Liquid nitride produces a very hard shallow surface on high speed steel tools when these tools are immersed in cyanide salts at

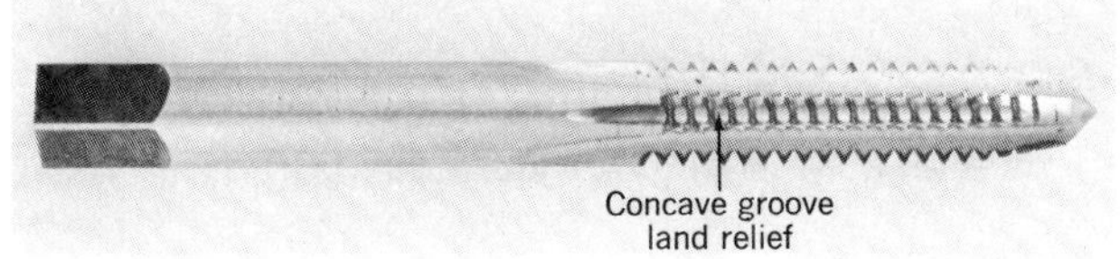

FIGURE B-133 Tap with concave groove land relief (DeAnza College).

closely controlled temperatures. Oxide finishes are usually applied in steam tempering furnaces and can be identified by their bluish-black color. The oxide acts as a solid lubricant. It also holds liquid lubricant at the cutting edges during a tapping operation. Oxide treatments prevent welding of the chips to the tool and reduce friction between the tool and the work. Chrome plating is a very effective treatment for taps used on nonferrous metals and some soft steels. The chromium deposit is very shallow and often referred to as flash chrome plating. Titanium nitride is another effective coating to extend the working life of taps.

SELF-TEST

1. What difference exists between a set of taps and serial taps?
2. Where is a spiral pointed tap used?
3. When is a fluteless spiral pointed tap used?
4. When is a spiral fluted tap used?
5. How are thread-forming taps different from conventional taps?
6. How are taper pipe taps identified?
7. Why are finishing and roughing Acme taps used?
8. Why are the rake angles varied on taps for different materials?
9. Name at least three methods used by tap manufacturers to reduce friction between the tap and the workpiece material.

UNIT 7

Tapping Procedures

Today's mass production of consumer goods depends to a large extent on the efficient and secure assembly of parts using threaded fasteners. It takes skill to produce usable tapped holes, so a worker in the metal trades must have an understanding of the factors that affect the tapping of a hole, such as the work material and its cutting speed, the proper cutting fluid, and the size and condition of the hole. A good machinist can analyze a tapping operation, determine whether or not it is satisfactory, and usually find a solution if it is not. In this unit, you will learn about common tapping procedures.

Objectives

After completing this unit, you should be able to:

1. Select the correct tap drill for a specific percentage of thread.
2. Determine the cutting speed for a given work material–tool combination.
3. Select the correct cutting fluid for tapping.
4. Tap holes by hand or with a drill press.
5. Identify and correct common tapping problems.

Taps are used to cut internal threads in holes. The actual cutting process is called tapping and can be performed by hand or with a machine. A tap wrench (Figure B-134) or a T-handle tap wrench (Figure B-135) attached to the tap is used to provide driving torque while hand tapping. To obtain a greater accuracy in hand tapping, a hand tapper (Figure B-136) is used. This fixture acts as a guide for the tap to insure that it stays in alignment and cuts concentric threads.

Holes can also be tapped in a drill press that has a spindle reverse switch, which is often foot operated for convenience. Drill presses without reversing switches can be used for tapping with a tapping attachment (Figure B-137). Some of these tapping attachments have an internal friction clutch where downward pressure on the tap turns

FIGURE B-134 Tap wrench

FIGURE B-135 T-handle tap wrench.

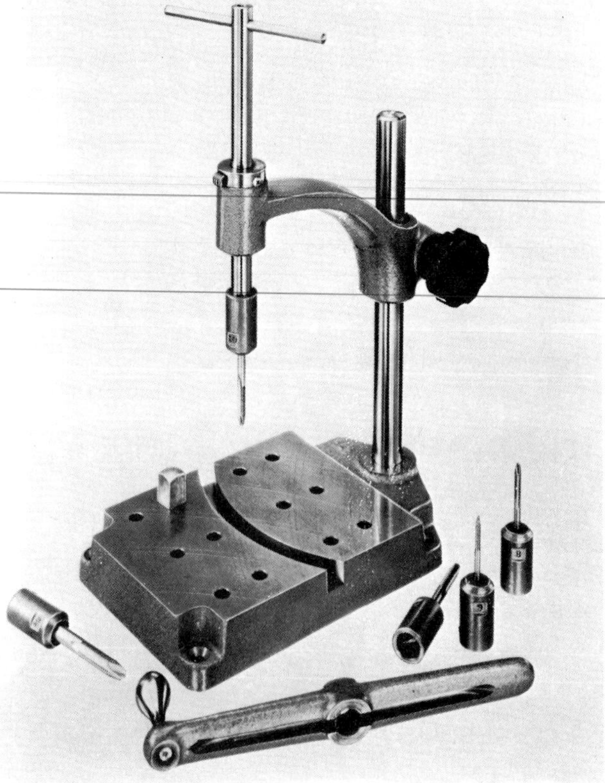

FIGURE B-136 Hand tapper (Ralmike's Tool-A-Rama).

the tap forward and feeds it into the work. Releasing downward pressure will automatically reverse the tap and back it out of the workpiece. Some tapping attachments have lead screws that provide tap feed rates equal to the lead of the tap. Most of these attachments also have an adjustment to limit the torque to match the size of tap, which eliminates most tap breakage.

Thread Percentage and Hole Strength

The strength of a tapped hole depends largely on the workpiece material, the percentage of full thread used, and the length of the thread. The workpiece material is usually selected by the designer, but the machinist can often control the percentage of thread produced and the depth of the thread. The percentage of thread produced is dependent on the diameter of the drilled hole. Tap drill charts generally give tap drill sizes to produce 75 percent thread. (See Appendix Table 10 for a tap drill chart.)

An example will illustrate the relationships between the percentage of thread, torque required to drive the tap, and resulting thread strength. An increase in thread depth from 60 to 72 percent in AISI 1020 steel required twice the torque to drive the tap, but it increased the strength of the thread by only 5 percent. The practical limit seems to be 75 percent of full thread, since a greater percentage of thread does not increase the strength of the threaded hole in most materials.

In some difficult-to-machine materials such as titanium alloys, high tensile steels, and some stainless steels, 50 to 60 percent thread depth will give sufficient strength to the tapped hole. Threaded assemblies are usually designed so that the bolt breaks before the threaded hole strips. Common practice is to have a bolt engage a tapped hole by 1 to $1\frac{1}{2}$ times its diameter.

Drilling the Right Hole Size

The condition of the drilled hole affects the quality of the thread produced, as an out-of-round hole leads to an out-of-round thread. Bellmouthed holes will produce bellmouth threads. When an exact hole size is needed, the hole should be reamed before tapping. This is especially important for large diameter taps and when fine pitch threads are used. The size of the hole to be drilled is usually obtained from tap drill charts, which usually show a 75 percent thread depth. If a thread depth other than 75 percent is wanted, use the following formula to determine the proper hole size:

$$\text{outside diameter of thread} - \frac{.01299 \times \text{percentage of thread}}{\text{number of threads per inch}} = \text{hole size}$$

For example, to calculate the hole size for a 1 in.–12 thread fastener with a 70 percent thread depth:

$$1 - \frac{.01299 \times 70}{12} = .924 \text{ in.}$$

FIGURE B-137 Drill press tapping attachment (Lane Community College).

Speeds for Tapping

When tapping a thread by hand with a tap wrench, speed is not a consideration at all but, when using a tapping machine or attachment, speed is very important. The quality of the thread produced also depends on the speed at which a tap is operated. The selection of the best speed for tapping is limited, unlike the varying speeds and feeds possible with other cutting tools, because the feed per revolution is fixed by the lead of the thread. Excessive speed develops high temperatures that cause rapid wear of the tap's cutting edge Dull taps produce rough or torn and off-size threads. High cutting speeds prevent adequate lubrication at the cutting edges and often create a problem of chip disposal.

When selecting the best speed for tapping, you should consider not only the material being tapped, but also the size of the hole, the kind of tap holder being used, and the lubricant being

TABLE B-2 Recommended Cutting Speeds and Lubricants for Machine Tapping

Material	Speeds (ft/min)	Lubricant
Aluminum	90–100	Kerosene and light base oil
Brass	90–100	Soluble oil or light base oil
Cast iron	70–80	Dry or soluble oil
Magnesium	20–50	Light base oil diluted with kerosene
Phosphor bronze	30–60	Mineral oil or light base oil
Plastics	50–70	Dry or air jet
Steels		
Low carbon	40–60	Sulfur-base oil
High carbon	25–35	Sulfur-base oil
Free machining	60–80	Soluble oil
Molybdenum	10–35	Sulfur-base oil
Stainless	10–35	Sulfur-base oil

used. Table B-2 gives some guidelines in selecting a speed and a lubricant for some materials when using high-speed steel taps.

These cutting speeds in feet per minute have to be translated into rpm to be useful. For example, calculate the rpm when tapping a $\frac{3}{8}$–24 UNF hole in free-machining steel. The cutting speed chart gives a cutting speed between 60 and 80 feet per minute. Use the lower figure; you can increase the speed once you see how the material taps. The formula for calculating rpm is:

$$\frac{\text{cutting speed (CS)} \times 4}{\text{diameter } (D)} \text{ or } \frac{60 \times 4}{3/8} = 640 \text{ rpm}$$

Lubrication is one of the most important factors in a tapping operation. Cutting fluids used when tapping serve as coolants, but are more important as lubricants. It is important to select the correct lubricant because the use of a wrong lubricant may give results that are worse than if no lubricant was used. For lubricants to be effective, they should be applied in sufficient quantity to the actual cutting area in the hole. (See Section F for more information on cutting speeds and cutting fluids.)

Solving Tap Problems

In Table B-3, common tapping problems are presented with some possible solutions. Occasionally, it becomes necessary to remove a broken tap from a hole. If a part of the broken tap extends out of the workpiece, removal is relatively easy with a pair of pliers. If the tap breaks flush with or below the surface of the workpiece, a tap extractor can be used (Figure B-138). Before trying to remove a broken tap, the chips in the flutes should be removed. A jet of compressed air or cutting fluid can be used for this.

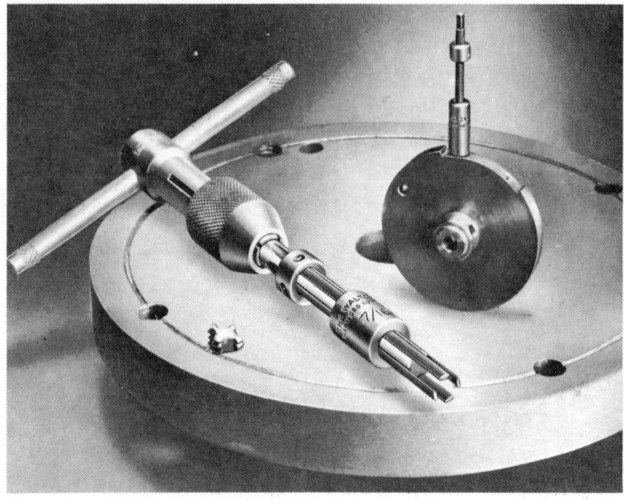

FIGURE B-138 Tap extractor (Walton Co.).

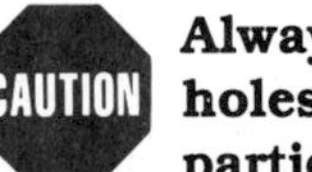

Always stand aside when cleaning out holes with compressed air as chips and particles tend to fly out at high velocity.

When the chips are packed so tightly in the flutes or the tap is jammed in the work so that a tap extractor cannot be used, the tap may be broken up with a pin punch and removed piece by piece. If the tap is made from carbon steel and cannot be pin punched, the tap can be annealed so it becomes possible to drill out.

On high-speed steel taps it may be necessary to use an electrical discharge machine (EDM), sometimes called a tap disintegrator, to remove the broken tap. These machines erode away material from extremely hard workpieces while they are immersed in a fluid. The shape of the hole conforms precisely to that of the electrode.

Tapping Procedure, Hand Tapping

STEP 1. Determine the size of the thread to be tapped and select the tap.

STEP 2. Select the proper tap drill with the aid of a tap drill chart. A taper tap should be selected for hand tapping; or if a drill press or tapping machine is to be used for alignment, use a plug tap.

STEP 3. Fasten the workpiece securely in a drill press vise. Calculate the correct rpm for the drill used:

$$\text{rpm} = \frac{\text{CS} \times 4}{D}$$

Drill the hole using the recommended coolant. Check the hole size.

STEP 4. Countersink the hole entrance to a diameter slightly larger than the major diameter of the threads (Figure B-139). This allows the tap to be started more easily, and it protects the start of the threads from damage.

STEP 5. Mount the workpiece in a bench vise so that the hole is in a vertical position.

STEP 6. Tighten the tap in the tap wrench.

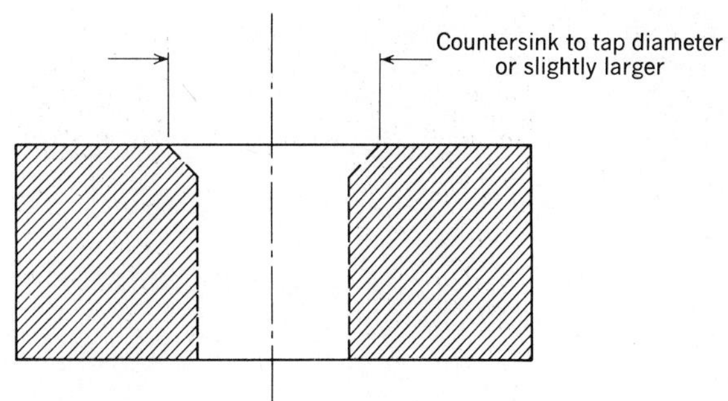

FIGURE B-139 Preparing the workpiece.

TABLE B-3 Common Tapping Problems and Possible Solutions

Causes of Tap Breakage	Solutions
Tap hitting bottom of hole or bottoming on packed chips	Drill hole deeper. Eject chips with air pressure (*Caution:* Stand aside when you do this and always wear safety glasses.) Use spiral fluted taps to pull chips out of hole. Use a thread-forming tap.
Chips are packing in flutes	Use tap style with more flute space. Tap to a lesser depth or use a smaller percentage of threads. Select a tap that will eject chips forward (spiral point) or backward (spiral fluted).
Hard materials or hard spots	Anneal the workpiece. Reduce cutting speed. Use longer chamfers on tap. Use taps with more flutes.
Inadequate lubricant	Use the correct lubricant and apply a sufficient amount of it under pressure at the cutting zone.
Tapping too fast	Reduce cutting speed.
Excessive wear	
Abrasive materials	Improve lubrication. Use surface-treated taps. Check the alignment of tap and hole to be tapped.
Chips clogging flutes	
Insufficient lubrication	Use better lubricant and apply it with pressure at the cutting zone.
Excessive speed	Reduce cutting speed.
Wrong-style tap	Use a more free-cutting tap such as spiral pointed tap, spiral fluted tap, interrupted thread tap, or surface-treated taps.
Torn or rough threads	
Dull tap	Resharpen.
Chip congestion	Use tap with more chip room. Use lesser percentage of thread. Drill deeper hole. Use a tap that will eject chips.
Inadequate lubrication and chips clogging flutes	Correct as suggested previously.
Hole improperly prepared	Torn areas on the surface of the drilled, bored, or cast hole will be shown in the minor diameter of the tapped thread.
Undersize threads	
Pitch diameter of tap too small	Use tap with a large pitch diameter.
Excessive speed	Reduce tapping speed.
Thin wall material	Use a tap that cuts as freely as possible. Improve lubrication. Hold the workpiece so that it cannot expand while it is being tapped. Use an oversized tap.
Dull tap	Resharpen.
Oversize or bellmouth threads	
Loose spindle or worn holder	Replace or repair spindle or holder.
Misalignment	Align spindle, fixture, and work.
Tap oversize	Use smaller pitch diameter tap.
Dull tap	Resharpen.
Chips packed in flutes	Use tap with deeper flutes, spiral flutes, or spiral points.
Buildup on cutting edges of tap	Use correct lubricant and tapping speed.

FIGURE B-140 Starting the tap (Lane Community College).

FIGURE B-141 Tapping a thread by hand (Lane Community College).

FIGURE B-142 Checking the tap for squareness (Lane Community College).

FIGURE B-143 Using the drill press as a tapping fixture (Lane Community College).

STEP 7. Cup your hand over the center of the wrench (Figure B-140) and place the tap in the hole in a vertical position. Start the tap by turning two or three turns in a clockwise direction for a right-hand thread. At the same time keep a steady pressure downward on the tap. When the tap is started, it may be turned as shown in Figure B-141.

STEP 8. After the tap is started for several turns, remove the tap wrench without disturbing the tap. Place the blade of a square against the solid shank of the tap to check for squareness (Figure B-142). Check from two positions 90 degrees apart. If the tap is not square with the work, it will ruin the thread and possibly break in the hole if you continue tapping. Back the tap out of the hole and restart.

STEP 9. Use the correct cutting oil on the tap when cutting threads.

STEP 10. Turn the tap clockwise one-quarter to one-half turn and then turn it back a three-quarter turn to break the chip. This is done with a steady motion to avoid breaking the tap.

STEP 11. When tapping a blind hole, use the taps in the order of starting, plug, and then bottoming. Remove the chips from the hole before using the bottoming tap and be careful not to hit the bottom of the hole with the tap.

STEP 12. Figure B-143 shows a 60-degree point center chucked in a drill press to align a tap squarely with the previously drilled hole. Only very slight follow-up pressure should be applied to the tap. Too much downward pressure will cut a loose, oversized thread.

SELF-TEST

1. What kind of tools are used to drive taps when hand tapping?

2. What is a hand tapper?

3. What is a tapping attachment?

4. Which three factors affect the strength of a tapped hole?

5. How deep should the usable threads be in a tapped hole?

6. When should tap drill holes be reamed?

7. What causes taps to break while tapping?

8. What causes rough and torn threads?

9. What causes oversized threads in a hole?

10. Give three methods of removing broken taps from holes.

UNIT 8

Thread-Cutting Dies and Their Uses

A die is used to cut external threads on the surface of a bolt or rod. Many machine parts and mechanical assemblies are held together with threaded fasteners, most of which are mass-produced. If necessary, the threaded portion of a bolt may be extended with a die toward the head, but this should be done with unhardened bolts, as cutting heat-treated bolts will dull the die. In this unit you will be introduced to some thread-cutting dies and their uses.

Objectives

After completing this unit, you should be able to:

1. Identify dies used for hand threading.

2. Select and prepare a rod for threading.

3. Cut threads with a die.

Dies are used to cut external threads on round materials. Some dies are made from carbon steel, but most are made from high speed steel. Dies are identified by the markings on the face as to the size of thread, number of threads per inch, and form of thread, such as NC, UNF, or other standard designations (Figure B-144).

Common Types of Hand Threading Dies

The die shown in Figure B-144 is an example of a round split adjustable die, also called a button die. These dies are made in all standardized thread sizes up to $1\frac{1}{2}$-in. thread diameters and $\frac{1}{2}$-in. pipe threads. The outside diameters of these dies vary from $\frac{5}{8}$ to 3 in.

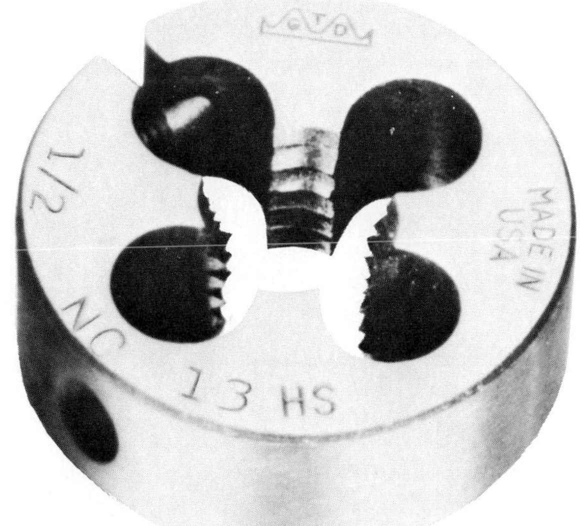

FIGURE B-144 Markings on a die. (Example shown is a round split adjustable die.)

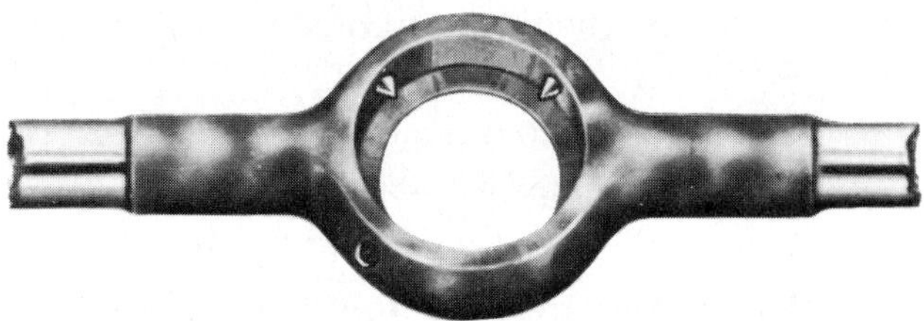

FIGURE B-145 Diestock for round split adjustable dies (TRW, Inc.).

FIGURE B-146 Die halves for two-piece die (TRW, Inc.).

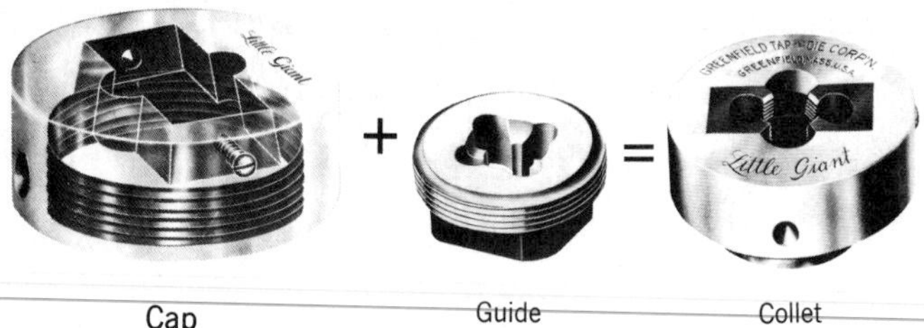

FIGURE B-147 Components of a split adjustable die collet (TRW, Inc.).

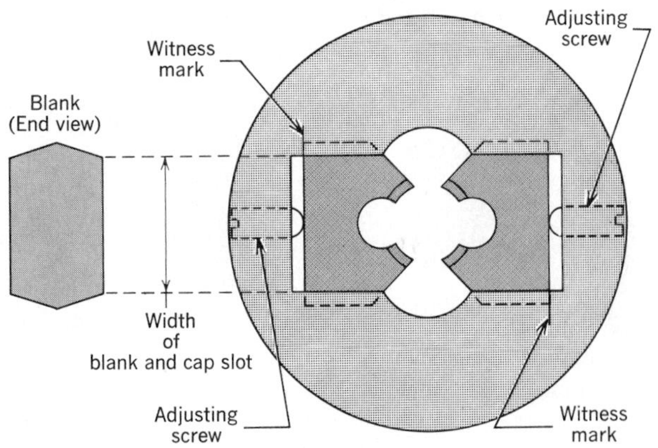

FIGURE B-148 Setting the die position to the witness marks on the die and collet assembly (Courtesy of TRW, Inc.).

Adjustments on these dies are made by turning a fine pitch screw that forces the sides of the die apart or allows them to spring together. The range of adjustment of round split adjustable dies is very small, allowing only for a loose or tight fit on a threaded part. Adjustments made to obtain threads several thousandths of an inch oversize will result in poor die performance because the heel of the cutting edge will drag on the threads. Excessive expansion may cause the die to break.

Some round split adjustable dies do not have the built-in adjusting screw. Adjustments are then made with the three screws in the die stock (Figure B-145). Two of these screws on opposite sides of the die stock hold the die in the die stock and also provide closing pressure. The third screw engages the split in the die and provides opening pressure. These dies are used in a die stock for hand threading or in a machine holder for machine threading.

Another type of threading die is the two-piece die, whose halves (Figure B-146) are called blanks. These blanks are assembled in a collet consisting of a cap and the guide (Figure B-147). The normal position of the blanks in the collet is indicated by witness marks (Figure B-148). The adjusting screws allow for precise control of the cut thread size. The blanks are inserted in the cap with the tapered threads toward the guide. Each of the two die halves is stamped with a serial number. Make sure the halves you select have the same numbers. The guide used in the collet serves as an aid in starting and holding the dies square with the work being threaded. Each thread size uses a guide of the same nominal or indicated size. Collets are held securely in die stocks (Figure B-149) by a knurled setscrew that seats in a dimple in the cap.

Hexagon rethreading dies (Figure B-150) are used to recut slightly damaged or rusty threads. Rethreading dies are driven with a wrench large enough to fit the die. Solid square dies (Figure B-151) have the same uses as hexagon rethreading dies. All of the die types discussed previously are also available in pipe thread sizes. Square dies are used to cut new threads and have sufficient chip clearance for this purpose.

Hand Threading Procedures

Threading of a rod should always be started with the leading or throat side of the die. This side is identified by the chamfer on the first two or three threads and also by the size markings. The chamfer distributes the cutting load over a number of threads, which produces better threads and less chance of chipping the cutting edges of the die. Cutting oil or other threading fluids are very important in obtaining quality threads and main-

FIGURE B-149 Diestock for adjustable die and collet assembly (Courtesy of TRW, Inc.).

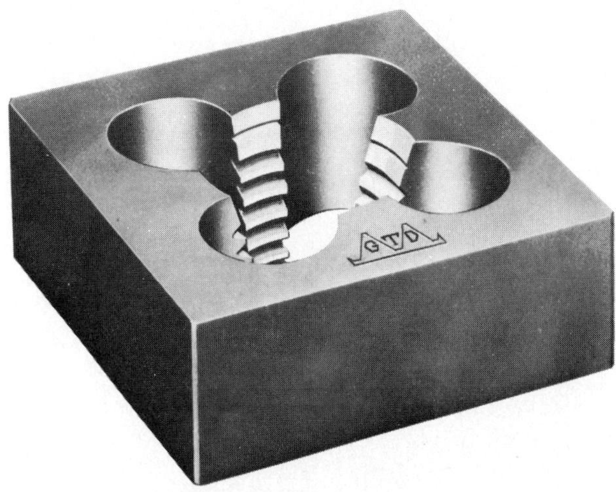

FIGURE B-151 Solid square die (TRW, Inc.).

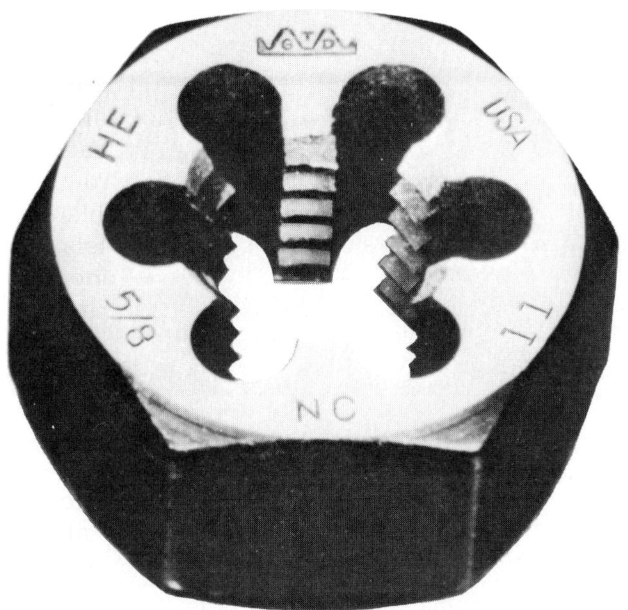

FIGURE B-150 Hexagon rethreading die (Courtesy of TRW, Inc.).

FIGURE B-152 Threading a rod with a hand die in a lathe (Lane Community College).

taining long die life. Once a cut is started with a die, it will tend to follow its own lead, but uneven pressure on the die stock will make the die cut variable helix angle or "drunken" threads.

Threads cut by hand often show a considerable accumulated lead error. The lead of a screw thread is the distance a nut will move on the screw if it is turned one full revolution. This problem is caused by the dies being relatively thin compared to the diameter of thread that they cut. Only a few threads in the die can act as a guide on the already cut threads. This error usually does not cause problems when standard or thin nuts are used on the threaded part. However, when an item with a long internal thread is assembled with a threaded rod, it usually gets tight and then locks, not because the thread depth is insufficient, but because there is a lead error. This lead error can be as much as one-fourth of a thread in one inch of length.

The outside diameter of the material to be threaded should not be over the nominal size of the thread and preferably a few thousandths of an inch (.002 to .005 in.) undersized. After a few full threads are cut, the die should be removed so that the thread can be tested with a nut or thread ring gage. A thread ring gage set usually consists of two gages, a go and a no go gage. As the names imply, a go gage should screw on the thread, while the no go gage will not go more than $1\frac{1}{2}$ turns on a thread of the correct size. Do not assume that the die will cut the correct-size thread; always check by gaging or assembling. Adjustable dies should be spread open for the first cut and set progressively smaller for each pass after checking the thread size.

It is very important that a die is started squarely on the rod to be threaded. A lathe can be used as a fixture for cutting threads with a die (Figure B-152). The rod is fastened in a lathe chuck for rotation, while the die is held square because it is supported by the face of the tailstock spindle. The carriage or the compound rest prevents the die stock from turning while the chuck is rotated

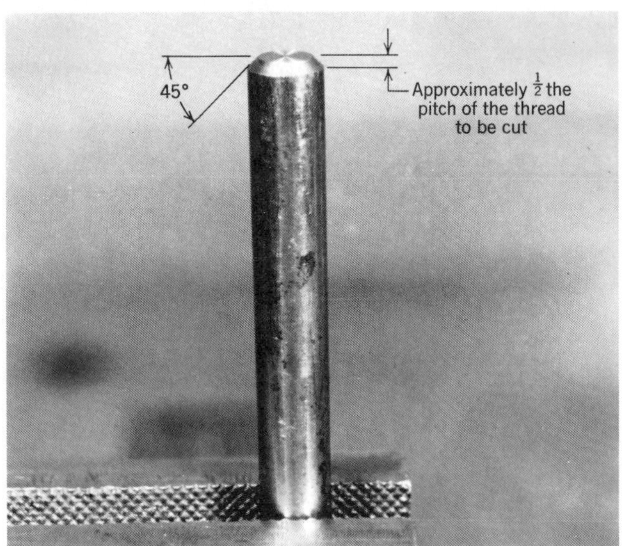

FIGURE B-153 Chamfer workpiece before using die (Lane Community College).

FIGURE B-154 Start the die with one hand (Lane Community College).

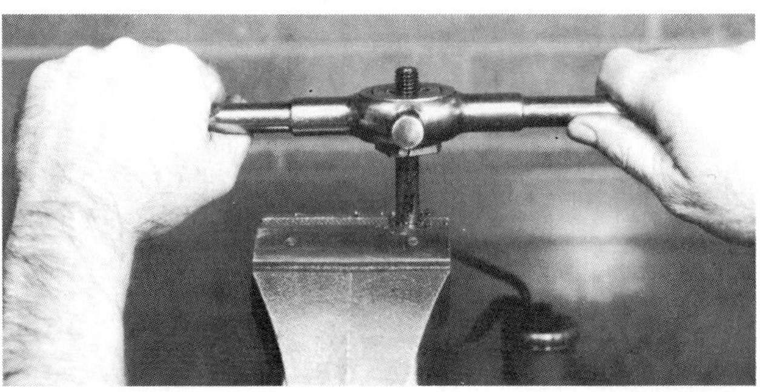

FIGURE B-155 Use both hands to turn the threading die (Lane Community College).

by hand. As the die advances, the tailstock spindle is also advanced to stay in contact with the die. Do not force the die with the tailstock spindle, or a loose thread may result. A die may be used to finish to size a long thread that has been rough threaded on the lathe.

It is always good practice to chamfer the end of a workpiece before starting a die (Figure B-153). The chamfer on the end of a rod can be made by grinding on a pedestal grinder, by filing, or with a lathe. This will help in starting the cut and it will also leave a finished thread end. While cutting threads with a hand die, the die rotation should be reversed after each full turn forward to break the chips into short pieces that will fall out of the die. Chips jammed in the clearance holes will tear the thread.

Threading Procedure, Threading Dies

STEP 1. Select the workpiece to be threaded and measure its diameter. Then chamfer the end. This may be done on a grinder or with a file. The chamfer should be at least as deep as the thread to be cut.

STEP 2. Select the correct die and mount it in a die stock.

STEP 3. Mount the workpiece in a bench vise. Short workpieces are mounted vertically and long pieces usually are held horizontally.

STEP 4. To start the thread, place the die over the workpiece. Holding the die stock with one hand (Figure B-154), apply downward pressure and turn the die.

STEP 5. When the cut has started, apply cutting fluid to the workpiece and die and start turning the die stock with both hands (Figure B-155). After each complete revolution forward, reverse the die one-half turn to break the chips.

STEP 6. Check to see that the thread is started square, using a machinist's square. Corrections can be made by applying slight downward pressure on the high side while turning.

STEP 7. When several turns of the thread have been completed, you should check the fit of the thread with a nut, thread ring gage, thread micrometer, or the mating part. If the thread fit is incorrect, adjust the die with the adjustment screws and take another cut with the adjusted die. Continue making adjustments until the proper fit is achieved.

STEP 8. Continue threading to the required thread length. To cut threads close to a shoulder, invert the die after the normal threading operation and cut the last two or three threads with the side of the die that has less chamfer.

SELF-TEST

1. What is a die?

2. What tool is used to drive a die?

3. How much adjustment is possible with a round split adjustable die?

4. What is the purpose of the guide in a two-piece adjustable die collet?

5. What are important points to watch when assembling two-piece dies in a collet?

6. Where are hexagon rethreading dies used?

7. Why do dies have a chamfer on the cutting end?

8. Why are cutting fluids used?

9. What diameter should a rod be before being threaded?

10. Why should a rod be chamfered before being threaded?

UNIT 9

Off-Hand Grinding

Although a machine tool, the pedestal grinder is used for many hand grinding operations, especially sharpening and shaping drills and tool bits. In this unit you will study the setup, use, and safety aspects of this important machine.

Objective

After completing this unit, you should be able to:

Describe setup, use, and safety on the pedestal grinder.

Off-Hand Grinding on Pedestal Grinders

The pedestal grinder is really a machine tool. However, since the workpiece is handheld, it is more logical to discuss this machine in conjunction with hand tools. Furthermore, you must be familiar with the pedestal grinder, as you will be using it very early in your study of machine tool practices.

The pedestal grinder gets its name from the floor stand or pedestal that supports the motor and abrasive wheels. The pedestal grinder is a common machine tool that you will use almost daily in the machine shop. This grinding machine is used for general purpose, off-hand grinding where the workpiece is handheld and applied to the rapidly rotating abrasive wheel. One of the primary functions of the pedestal grinder is the shaping and sharpening of tool bits and drills in machine shop work. Pedestal grinders are often modified for use with rotary wire brushes or buffing wheels.

Large, heavy-duty pedestal grinders are sometimes found in machine shops. These grinders are used for rough grinding (snagging) welds, castings, and general rough work. These machines are generally set up in a separate location from the tool grinders. Rough grinding metal parts should never be done on tool grinders because it causes the wheels to become rounded, grooved, uneven, and out-of-round. In that condition, they are useless for tool grinding.

SETUP OF THE PEDESTAL GRINDER

The pedestal grinder in your shop stands ready for use most of the time. If it becomes necessary to replace a worn wheel, the side of the guard must be removed and the tool rest moved out of the way. A piece of wood may be used to prevent the wheel from rotating so that the spindle nut can be turned and removed (Figure B-156). Remember that the left side of the spindle has left-handed threads, while those on the right side are right-handed.

A new wheel should be ring tested to determine if there are any cracks or imperfections (Figure B-157). Gently tap the wheel near its rim with a screwdriver handle or a piece of wood and listen for a clear ringing sound like a bell. A clear ring indicates a sound wheel that is safe to use; if a dull thud is heard, the wheel may be cracked and should not be used. The flanges and the spindle should be clean before mounting the wheel. Be sure that the center hole in the wheel is the correct size for the grinder spindle. If a bushing must be used, be sure that it is the correct size and properly installed (Figure B-158). Place a clean, undamaged blotter on each side between wheel and flanges. The spindle nut should be tightened just enough to hold the wheel firmly. Excessive tightening will break the wheel.

After the guard and cover plate have been replaced, the tool rest should be brought up to the wheel so that between $\frac{1}{16}$- and $\frac{1}{8}$-in. clearance exists between the rest and wheel (Figure B-159). If there is excessive space between the tool rest and the wheel, a small workpiece, such as a tool bit that is being ground, may flip up and catch between the wheel and tool rest.

Your finger may be caught between workpiece and grinding wheel resulting in a serious injury.

The clearance between the tool rest and wheel should never exceed $\frac{1}{8}$ in. The spark guard, located on the upper side of the wheel guard (Figure B-160), should be adjusted to within $\frac{1}{16}$ in. of the wheel. This protects the operator if the wheel should shatter.

FIGURE B-156 Using a piece of wood to hold the wheel while removing the spindle nut (Lane Community College).

FIGURE B-158 The wheel is mounted with the proper bushing in place (Lane Community College).

FIGURE B-157 The ring test is made before mounting the wheel (Lane Community College).

FIGURE B-159 The tool rest is adjusted (Lane Community College).

DRESSING THE GRINDING WHEEL

Stand aside out of line with the rotation of the grinding wheel and turn on the grinder. Let the wheel run idle for one full minute. A new wheel does not always run exactly true and therefore must be dressed. A Desmond dresser (Figure B-161) may be used to sharpen and to some extent true the face of the wheel. Pedestal grinder wheels often become grooved, out-of-round, glazed, or misshapen, and therefore must be frequently dressed to obtain proper grinding results.

The grinding wheel dresser should be used so that the notch on the lower side is hooked behind the work rest. However, the dresser is often used in the manner shown in Figure B-161. When the tool rest extends along the side of the wheel, the proper method of using the dresser is impossible.

FIGURE B-160 The spark guard is adjusted (Lane Community College).

FIGURE B-161 The wheel is being dressed (Lane Community College).

USING THE PEDESTAL GRINDER

Bring the workpiece into contact with the wheel gently without bumping. Grind only on the face of the wheel. The workpiece will heat from friction during the grinding operation. It may become too hot to hold in just a few seconds. To prevent this, frequently cool the workpiece in the water pot attached to the grinder. Be especially careful when grinding drills and tool bits so that they do not become overheated. Excessive heat may permanently affect tool steel metallurgical properties.

Screwdrivers are probably the most misused of all tools, so they are often twisted or misshapen so much that they will no longer fit a screw slot or will damage the slot if they are used (Figure B-162). Screwdrivers can be ground flat on their sides, but a better method is to hollow grind them on a pedestal grinder (Figure B-163). The end of the tool

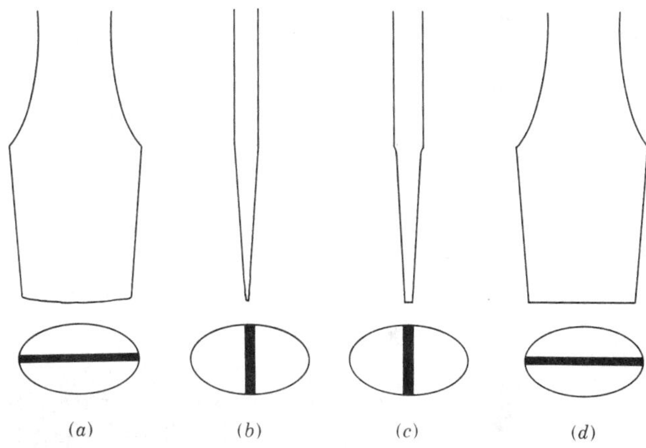

FIGURE B-162 The proper shape of the end of screwdriver blade. Blades (*a*) and (*b*) are badly worn; blades (*c*) and (*d*) are ground correctly.

FIGURE B-163 Hollow grind on a screwdriver being formed on the periphery of the wheel (Lane Community College).

FIGURE B-164 The end of the blade being squared on the wheel (Lane Community College).

should then be squared (Figure B-164) and given the proper thickness to fit the screw slot.

Center punches should also be hollow ground (Figure B-165) when they are sharpened. They should be evenly rotated while grinding the point to the correct angle. Flat cold chisels should also be hollow ground (Figure B-166) when they are sharpened.

Layout tools, such as spring dividers and scribers, should be kept sharp by honing on a fine, flat stone in the same way a knife is sharpened. However, when it becomes necessary to reshape a layout tool, extreme care must be exercised to avoid overheating the thin point. A fine grit wheel should be used for sharpening most cutting tools and they should be frequently cooled in water. Woodcutting tools are almost always made of plain carbon steel, which loses its hardness when overheated on the grinder. If one of these tools becomes blue-colored from the heat of grinding, it has become too soft for cutting purposes and must either be carefully ground back past the softened edge or rehardened and tempered.

FIGURE B-165 This punch is being correctly sharpened to produce a 90-degree angle for use as a center punch. It must be rotated while being ground. A sharper angle of 60 degrees may be ground to produce a prick punch (Lane Community College).

FIGURE B-166 Flat chisel being ground to produce a 60-degree angle (Lane Community College).

SAFETY CHECKPOINTS ON THE PEDESTAL GRINDER

CAUTION **Always wear appropriate eye protection when dressing wheels or grinding on the pedestal grinder. Be sure that grinding wheels are rated at the proper speed for the grinder that you are using. The safety shields, wheel guards, and spark guard must be kept in place at all times while grinding. The tool rest must be adjusted and the setting corrected as the diameter of the wheel decreases from use. Grinding wheels and rotary wire brushes may catch loose clothing or long hair (Figure B-167). Long hair should be contained in an industrial-type hairnet. Wire wheels often throw out small pieces of wire at high velocities.**

Nonferrous metals such as aluminum and brass should never be ground on the aluminum oxide wheels found on most pedestal grinders. These metals fill the voids or spaces between the abrasive particles in the grinding wheel so that more pressure is needed to accomplish the desired grinding. This additional pressure sometimes

FIGURE B-167 The force and speed of this action was such that the operator's head was jerked suddenly into the guard. Note that the cast aluminum guard was shattered as a result of the impact.

causes the wheel to break or shatter. Pieces of grinding wheel may be thrown out of the machine at extreme velocities. Always use silicon carbide abrasive wheels for grinding nonferrous metals. Excessive pressure should never be used in any grinding operation. If this seems to be necessary, it means that the improper grit or grade of abrasive is being used or the wheel is glazed and needs to be dressed. Always use the correct abrasive grit and grade for the particular grinding that you are doing. (For grinding wheel selection, see Section N.)

SELF-TEST

1. What is the primary function of the pedestal grinder in a machine shop?

2. Why should a tool grinder never be used for rough grinding metal?

3. When a wheel needs to be reshaped, sharpened, and to some extent trued, what tool is usually used on a pedestal grinder?

4. When sharpening layout tools and reshaping screwdrivers, what is the most important concern?

5. Name at least three safety factors to remember when using the pedestal grinder.

SECTION C

DIMENSIONAL MEASUREMENT

All of us, no matter what we may be doing, are totally surrounded by measurement. Measurement can be generally defined as the assignment of a value to time, length, and mass. We cannot escape measurement. Our daily lives are greatly influenced by the clock, a device that measures time. Mass or weight is measured in almost every product we buy, and the measure of length is incorporated in every creation of humans, ranging from the minute components of a watch to many thousands of miles of superhighways extending across a continent.

Measurement, in the modern age, has been developed to an exact science known as **metrology.** As the hardware of technology has become more complex, a machinist is ever more concerned with that branch of the science called dimensional metrology. Furthermore, mass production of goods has made necessary very complex systems of metrology to check and control the critical dimensions that control standardization and interchangeability of parts. Components of an automobile, for example, may be manufactured at locations far removed from each other and then brought to a central assembly point, with the assurance that all parts will fit as intended by the designer. In addition, the development and maintenance of a vast system of carefully controlled measurement has permitted manufacturers to locate their factories close to raw materials and available labor. Because of the standardization of measurement, industry has been able to diversify its products. Thus, manufacturers can do what they do best, and manufacturing effort can be directed toward the quality of a product and its production at a competitive price. As a result, metrology affects not only the technical aspects of production but also the economic aspects. Metrology is a common thread woven through the entire fabric of manufacturing from the drafting room to the shipping dock.

Measurement Needs of the Machinist

A machinist is mainly concerned with the measurement of **length;** that is, the distance along a line between two points (Figure C-1). It is length that defines the **size** of most objects. **Width** and **depth** are simply other names for length. A machinist measures length in the basic units of linear measure such as **inches, millimeters,** and, in advanced metrology, wavelengths of light. In addition, the machinist sometimes needs to measure the relationship of one surface to another, which is commonly called **angularity** (Figure C-2). **Squareness,** which is closely related to angularity, is the measure of deviation from true perpendicularity. A machinist will measure angularity in the basic units of angular measure, **degrees, minutes,** and **seconds of arc.**

In addition to the measure of length and angularity, a machinist also needs to measure such things as **surface finish, concentricity, straightness,** and **flatness.** He or she also occasionally comes in contact with measurements that involve circularity, sphericity, and alignment (Figure C-3). However, many of these more specialized measurement techniques are in the realm of the inspector or laboratory metrologist and appear infrequently in general machine shop work.

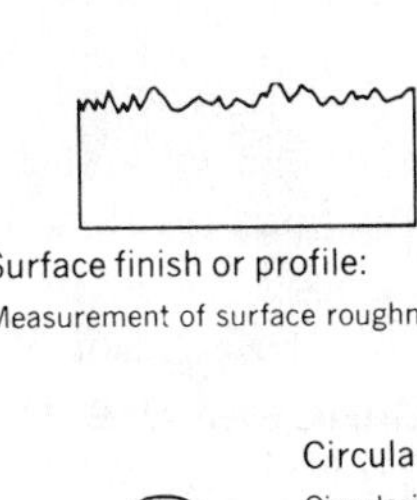

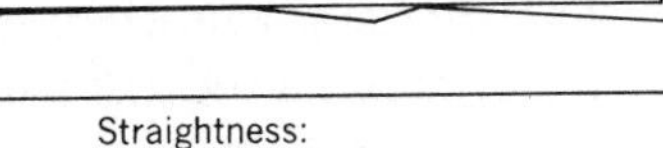

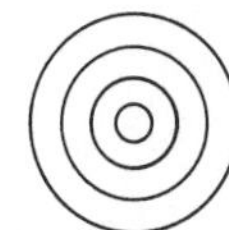

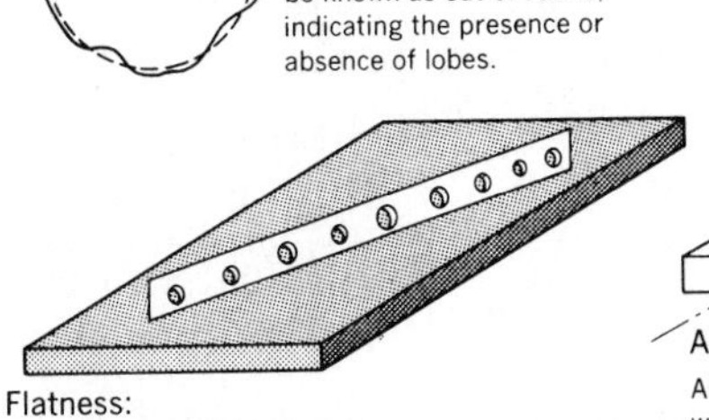

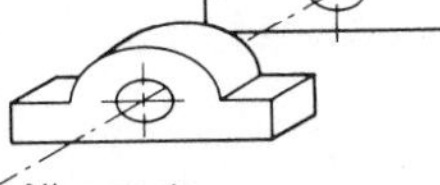

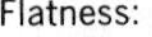

Surface finish or profile:
Measurement of surface roughness
Straightness:
Straightness refers to the deviation of a surface from a true line. Straightness is generally a single axis measurement
Concentricity:
Concentricity refers to two more circles with the same center. The measurement of deviation from true concentricity may be called indicated runout or eccentricity
Circularity or roundness:
Circularity refers to the conformity to the true circle. Measured deviation from true circularity may be known as out of round, indicating the presence or absence of lobes.
D
D
Sphericity:
Sphericity is the measurement of diameter and circularity in all planes.
Alignment:
Alignment determines the degree to which two or more components are colinear (along the same line). Alignment may refer to separate components or features of the same component (distortion)
Flatness:
Flatness is the deviation of a surface from a true plane. Flatness is generally a multiaxis measurement

FIGURE C-1 The measurement of length may appear under several different names.

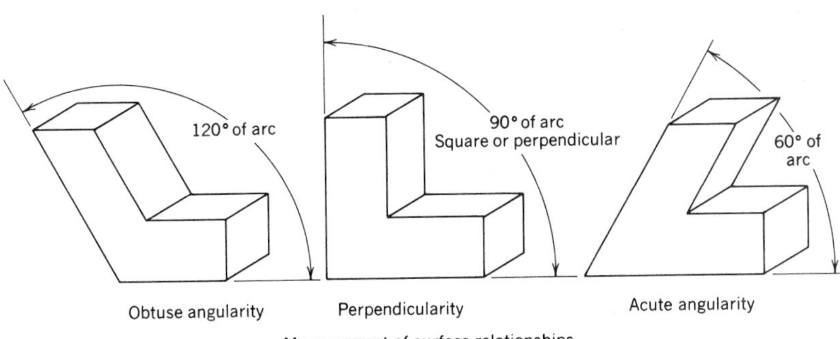

FIGURE C-2 Measurement of surface relationships or angularity.

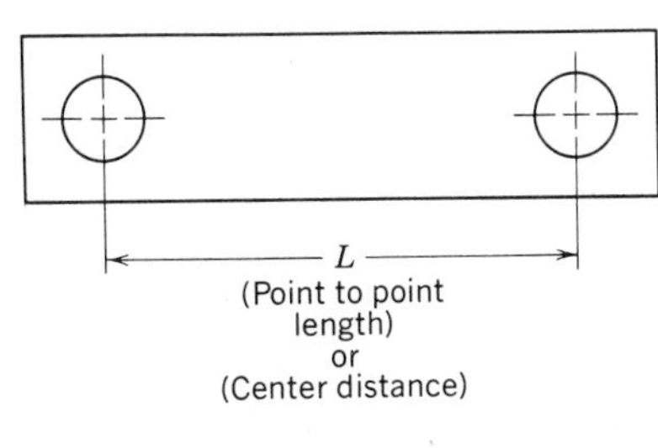

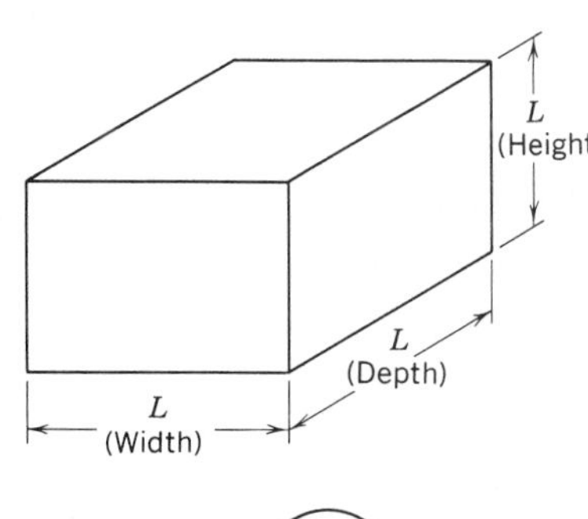

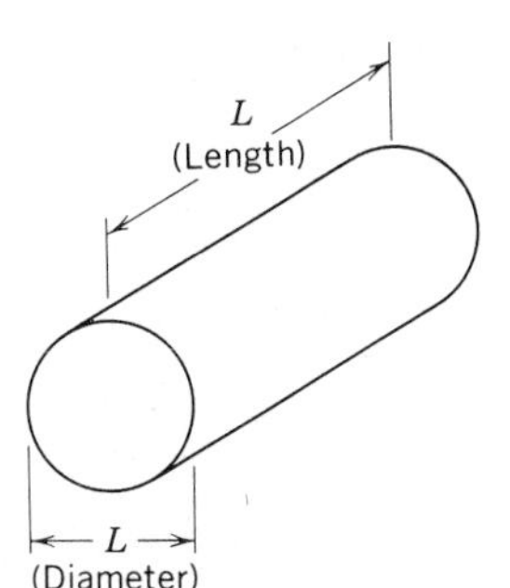

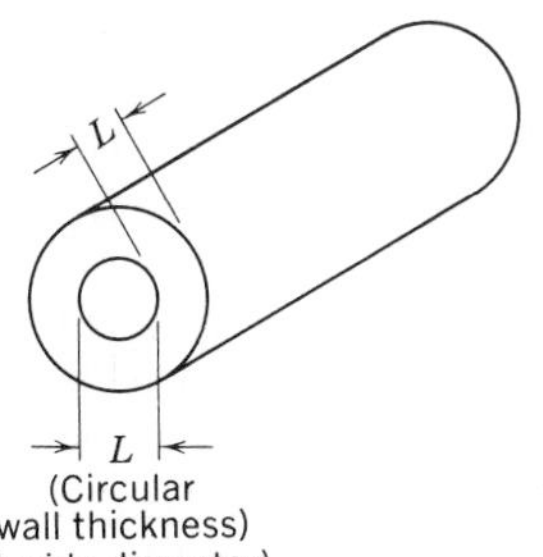

FIGURE C-3 Other measurements encountered by the machinist.

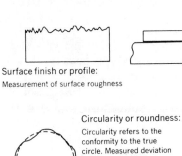

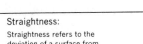

General Principles of Metrology

A machinist has available a large number of measuring tools designed for use in many different applications. However, not ever tool is equally suited for a specific measurement. As with all the tools of a machinist, selecting the proper measuring tool for the specific application at hand is a primary skill. The successful outcome of a machinist's work may indeed depend upon the choice of measuring tools. In this regard, a machinist must be familiar with several important terms and principles of metrology.

ACCURACY

Accuracy in metrology has a twofold meaning. First, accuracy can refer to whether or not a specific measurement is actually within its stated size. For example, a certain drill has its size stamped on its shank. A doubtful machinist decides to verify the drill size using a properly adjusted micrometer. The size is found to be as stated. Therefore, the size stamped on the drill is accurate. Second, accuracy refers to the act of measurement itself with regard to whether or not the specific measurement taken is within the capability of the measuring tool selected. A machinist obtains a drill with the size marked on the shank and decides to verify it using a steel rule. The edge of this rule with the finest graduations is selected; the machinist then lays the drill over the marks. Sighting along the drill, he discovers that it is really three graduations, on his rule, smaller than the size stamped on the shank. He then reasons that the size marked on the drill must be in error. In this example, the act of measurement is not accurate because the inappropriate measuring tool was selected and the improper procedure was used. **User accuracy** is also an important consideration. If, when the machinist measured the drill with his micrometer, as described in the first example, he did not bother to confirm the accuracy of the instrument prior to making the measurement, an inaccuracy that can be attributed to the user may have resulted.

PRECISION

The term **precision** is relative to the specific measurement being made, with regard to the degree of exactness required. For example, the distance from the earth to the moon, measured to within one mile, would indeed be a precise measurement. Likewise, a clearance of five-thousandths of an inch between a certain bearing and journal might be precise for that specific application. However, five-thousandths of an inch clearance between ball and race on a ball bearing would not be considered precise, as this clearance would be only a very few millionths of an inch. **There are many degrees of precision dependent on application and design requirements.** For a machinist, any measurement made to a degree finer than one sixty-fourth of an inch or one-half millimeter can be considered a **precision measurement** and must be made with the appropriate precision measuring instrument.

RELIABILITY

Reliability in measurement refers to the ability to obtain the desired result to the degree of precision required. Reliability is most important in the selection of the proper measuring tool. A certain tool may be reliable for a certain measurement, but totally unreliable in another application. For example, if it were desired to measure the distance to the next town, the odometer on an automobile speedometer would yield quite a reliable result, provided a degree of precision of less than one-tenth mile is not required. On the other hand, to measure the length of a city lot with an odometer is much less likely to yield a reliable result. This is explained by examining another important principle of metrology, that of the discrimination of a measuring instrument.

DISCRIMINATION

Discrimination refers to the degree to which a measuring instrument divides the basic unit of length it is using for measurement. The automobile odometer divides the mile into 10 parts; therefore, it discriminates to the nearest tenth of a mile. A micrometer, one of the most common measuring instruments of a machinist, subdivides an inch into 1000 or, in some cases, 10,000 parts. Therefore, the micrometer discriminates to .001 or .0001 of an inch. If a measuring instrument is used beyond its discrimination, a loss of reliability will result. Consider the example cited previously regarding the measurement of a city lot. Most lots are less than one-tenth of a mile in length; therefore, the discrimination of the auto odometer for this measurement is not sufficient for reliability.

THE 10:1 RATIO FOR DISCRIMINATION

In general, a measuring instrument should **discriminate 10 times finer** than the smallest unit that it will be asked to measure. The odometer, which discriminates to a tenth of a mile, is most reliable for measuring whole miles. To measure the length of a city lot in feet requires an instrument that discriminates at least to one-tenth of a foot. Since most surveyors' measuring tapes used for this application discriminate to tenths and in some cases to hundredths of a foot, they are an appropriate tool for the measurement.

THE POSITION OF A LINEAR MEASURING INSTRUMENT WITH REGARD TO THE AXIS OF MEASUREMENT

A large portion of the measurements made by a machinist is linear in nature. These measurements attempt to determine the shortest distance between two points. In order to obtain an accurate and reliable linear measurement, **the measuring instrument must be exactly in line with the axis of that measurement.** If this condition is not met, reliability will be in question. The alignment of the measuring instrument with the axis of measurement applies to all linear measurements (Figure C-4). The figure illustrates the alignment of the instrument with the axis of measurement using a simple graduated measuring device. Only under the reliable condition can the measurement approach accuracy. Misalignment of the instrument, as illustrated in the unreliable situation, will result in inaccurate measurements.

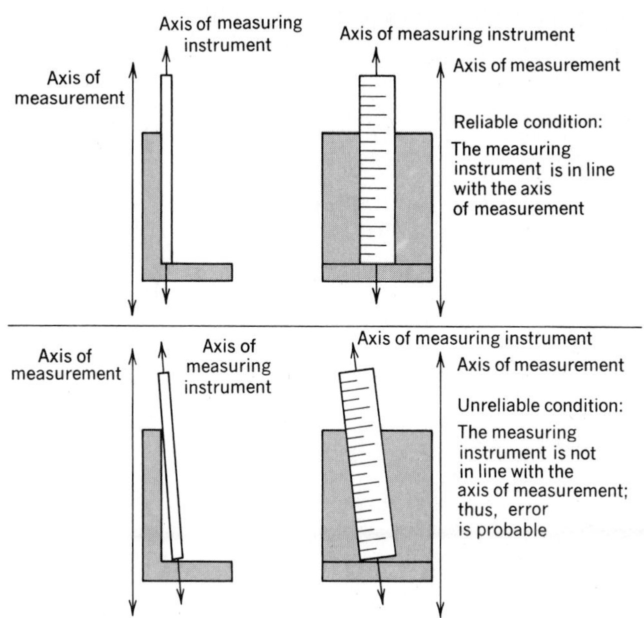

FIGURE C-4 The axis of a linear measuring instrument must be in line with the axis of measurement.

RESPONSIBILITY OF THE MACHINIST IN MEASUREMENT

The units in this section discuss most of the common measuring tools available to a machinist. The capabilities, discrimination, and reliability as well as procedures for use are examined. It is, of course, the responsibility of a machinist to select the proper measuring instrument for the job at hand. When faced with a need to measure, a machinist should ask the following questions:

1. What degree of accuracy and precision must this measurement meet?
2. What degree of measuring tool discrimination does this required accuracy and precision demand?
3. What is the most reliable tool for this application?

CALIBRATION Accurate and reliable measurement places a considerable amount of responsibility on a machinist. He or she is responsible for the conformity of his or her measuring tools to the appropriate standards. This is the process known as **calibration.** In a large industrial facility, all measuring tools are periodically cycled through the metrology laboratory where they are calibrated against appropriate standards. Any adjustments are then made to bring the tools into conformity to the standards. Only through this process can standardized measure within an individual plant or within an entire industrial nation be maintained. Most of the common measuring instruments provide a factory standard. Even though calibration cannot be carried out under laboratory conditions, the instruments should at least be checked periodically against available standards.

VARIABLES IN MEASURING A machinist should be further aware that any measurement is **relative to the conditions under which it is taken.** A common expression that is often used around the machine shop is: "the measurement is right on." There is, of course, little probability of obtaining a measurement that is truly exact. Each measurement has a certain degree of deviation from the theoretical exact size. This degree of error is dependent on many variables including the measuring tool selected, the procedure used, the temperature of the part, the temperature of the room, the cleanliness of the room air, and the cleanliness of the part at the time of measurement. The deviation of a measurement from exact size is taken into consideration by the designer. Every measurement has a tolerance, meaning that the measurement is acceptable within a specific range. Tolerance can be quite small depending on design requirements. When this condition exists, reliable measurement becomes more difficult because it is more heavily influenced by the many variables present. Therefore, before a machinist makes any measurement, he or she should stop for a moment and consider the possible variables involved. He or she should then consider what might be done to control as many of these variables as possible.

If you understand these basic principles of dimensional metrology and you assume the proper responsibility in the selection, calibration, and application of measuring instruments, you will experience little difficulty in performing the many measurements encountered in the science of machine tools and machining practices.

Tools for Dimensional Measurement

There are several hundred measuring instruments available to a machinist. In this modern age there is a measuring instrument that can be applied to almost any conceivable measurement. Many instruments are simply variations and combinations of a few common precision measuring tools. As you begin your study of machine tool practices, you will be initially concerned with the use, care, and applications of the common measuring instruments found in the machine shop. These will be discussed in detail within this section.

In addition to these, there is a large variety of instruments that are designed for many specialized uses. Some of these are rarely seen in the school or general purpose machine shop. Others are intended for use in the tool room or metrology laboratory where they are used in the calibration process. Your contact with these instruments will depend on the particular path you take while learning the trade.

Many measuring instruments have undergone modernization in recent years. Even though the function of these tools is basically the same, many have been redesigned and equipped with mechanical or electronic digital displays. These features make the instruments easier to read and improve accuracy. As a machinist, you must be skilled in the use of all the common measuring instruments. In addition, you should be familiar with the many important instruments used in production machining, inspection, and calibration. In the following pages many of these tools will be briefly described so that you may become familiar with the wide selection of measuring instruments available to the machinist.

FIXED GAGES AND AIR GAGES

FIXED GAGES In production machining, where large numbers of duplicate parts are produced, it may only be necessary to determine if the part is within acceptable tolerance. Many types of fixed gages are used. The adjustable limit snap gage (Figure C-5) is used to check outside diameter. One anvil is set to the minimum limit of the tolerance to be measured. The other anvil is set to the maximum limit of the tolerance. If both anvils slip over the part, an undersized condition is indicated. If neither anvil slips over the part, an oversized condition is indicated. The gage is set initially to a known standard such as gage blocks.

Threaded products are often checked with fixed gages. The **thread plug gage** (Figure C-6) is

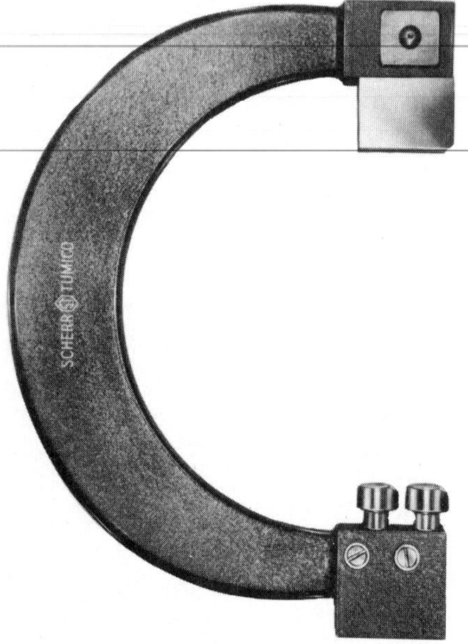

FIGURE C-5 Adjustable limit snap gage (Rank Scherr-Tumico, Inc.).

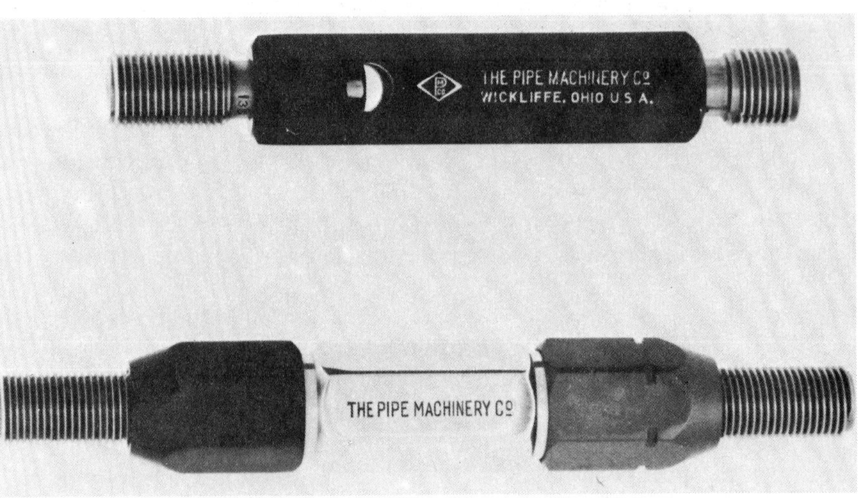

FIGURE C-6 Thread plug gage (PMC Industries).

used to check internal threads. The **thread ring gage** (Figure C-7) is used to check external threads. These are frequently called **go** and **no go** or **not go** gages. One end of the plug gage is at the low limit of the tolerance, while the other end is at the high limit of the tolerance. The thread gage functions in the same manner. Thread gages appear in many different forms (Figure C-8).

Fixed gages are also used to check internal and external tapers (Figure C-9). Plug gages are used for internal holes (Figure C-10). A ring gage is used for external diameters (Figure C-11).

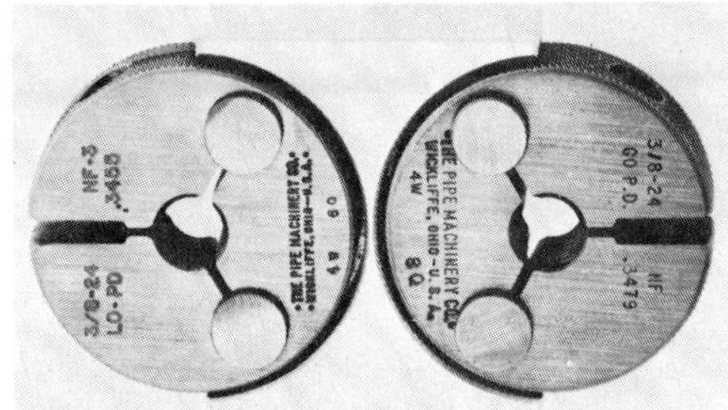

FIGURE C-7 Thread ring gage (PMC Industries).

FIGURE C-8 Fixed thread gages appear in many different forms (PMC Industries).

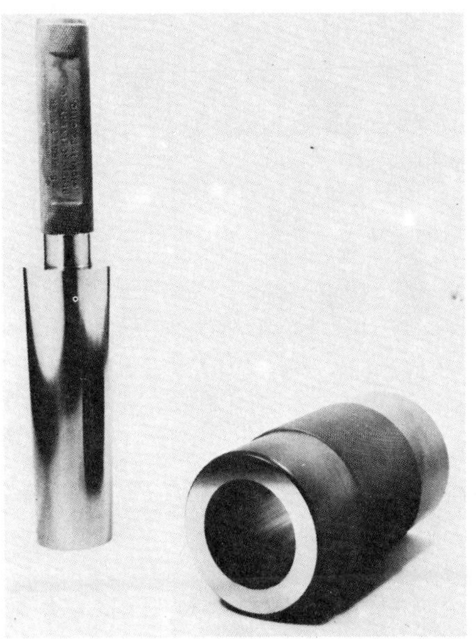

FIGURE C-9 Taper plug and taper ring gage (PMC Industries).

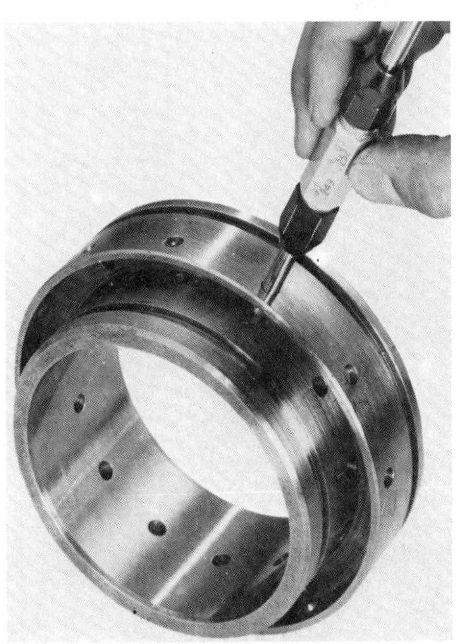

FIGURE C-10 Using the cylindrical plug gage (PMC Industries).

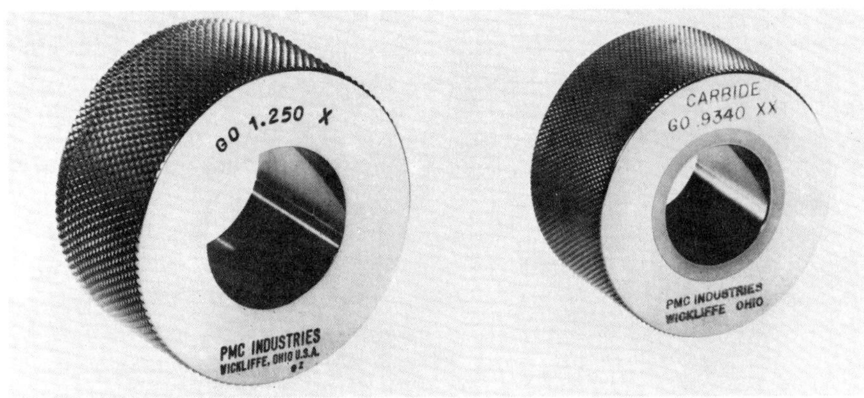

FIGURE C-11 Cylindrical ring gages (PMC Industries).

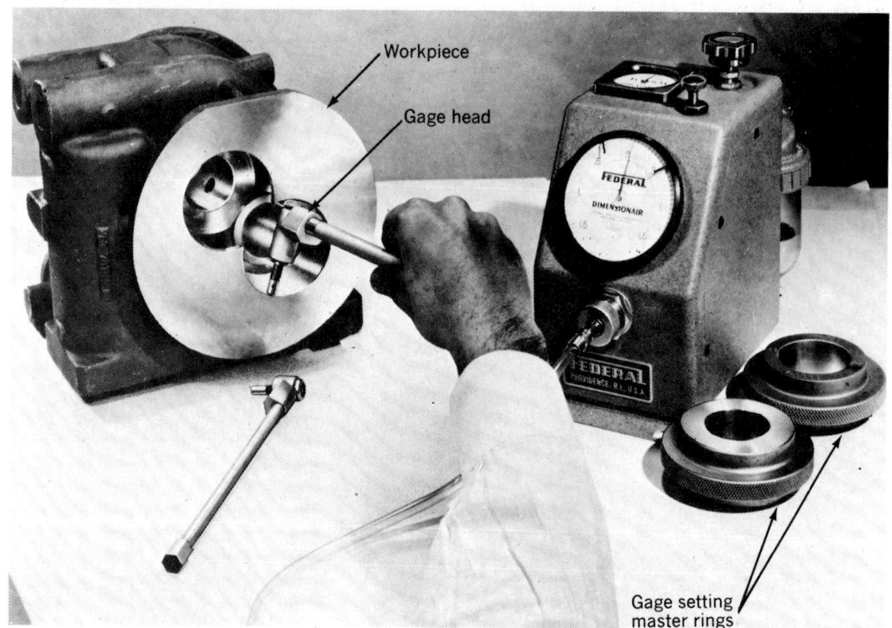

FIGURE C-12 Pressure-type air gage. (Federal Products Corporation).

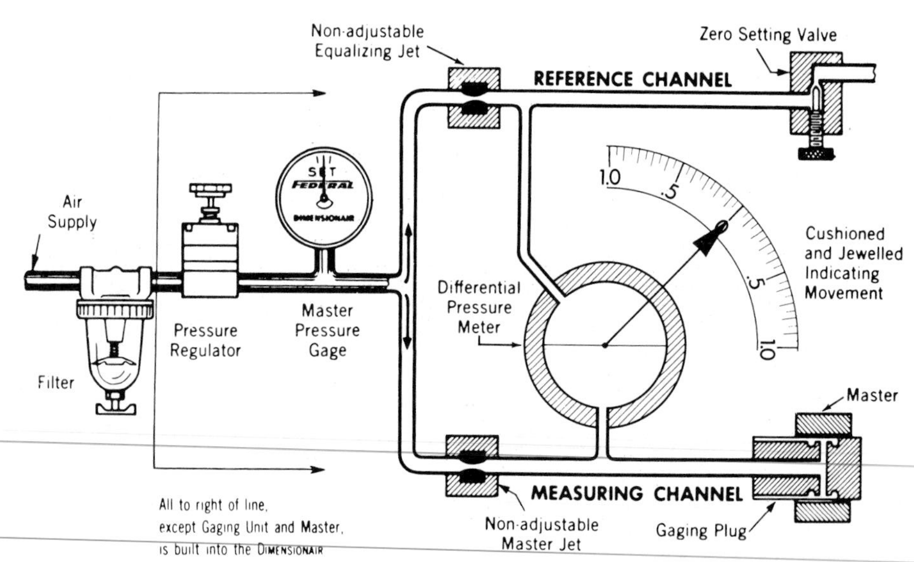

FIGURE C-13 Pressure type air gage system (Federal Products Corporation).

AIR GAGES Air gages are also known as **pneumatic comparators.** Two types of air gages are used in comparison measuring applications. In the pressure-type air gage (Figure C-12), filtered air flows through a reference and measuring channel. A sensitive differential pressure meter is connected across the channels (Figure C-13). The gage head is adjusted to a master setting gage. Air gage heads may be ring, snap (Figure C-14), or plug types (Figure C-15). Air flowing through the reference and measuring channel is adjusted until the differential pressure meter reads zero with the setting master in place. A difference in workpiece size above or below the master size will cause more or less air to escape from the gage head. This, in turn, will change the pressure on the reference channel. The pressure change will be indicated on the differential pressure meter. The meter scale is graduated in suitable linear units. Thus workpiece size above or below the master can be directly determined.

In the column or flow-type air gage (Figure C-16), airflow from the gage head is indicated on a flow meter or rotameter (Figure C-17). This type of air gage is also set to master gages. In the case of the plug gage shown, if the workpiece is oversized,

FIGURE C-14 Pressure-type air snap gage (Federal Products Corporation).

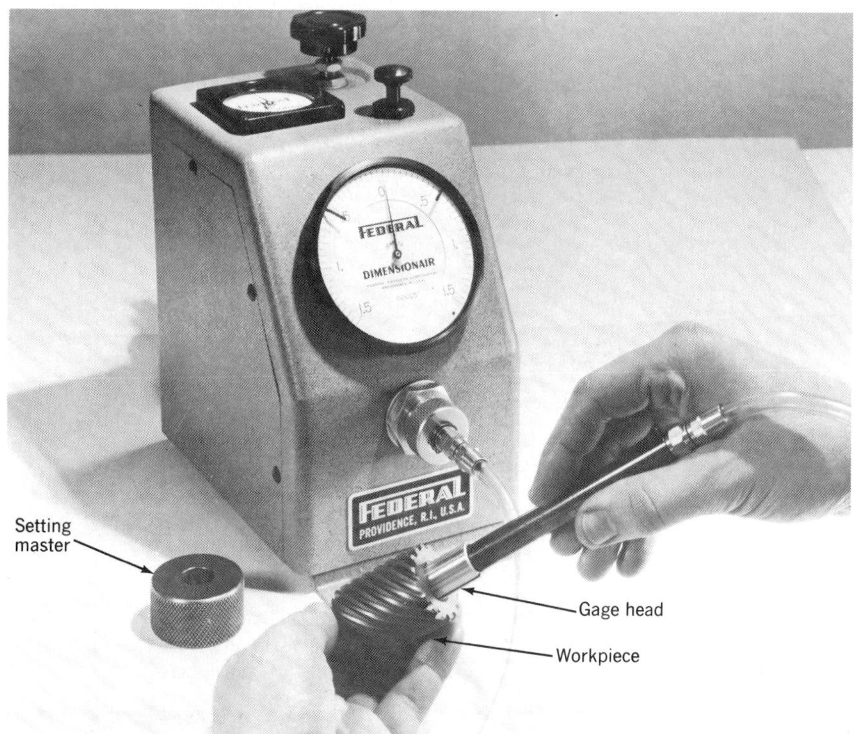

FIGURE C-15 Air plug gage (Federal Products Corporation).

more air will flow from the gage head. An undersized condition will permit less air to flow. Differences in flow are indicated on a suitably graduated flowmeter scale. Workpiece size deviation can be read directly.

Air gages have several advantages. The gage head does not touch the workpiece. Consequently there is no wear on the gage head and no damage to the finish of the workpiece. Variations in workpiece geometry that would be difficult to measure by mechanical means can be detected by air gaging (Figure C-18).

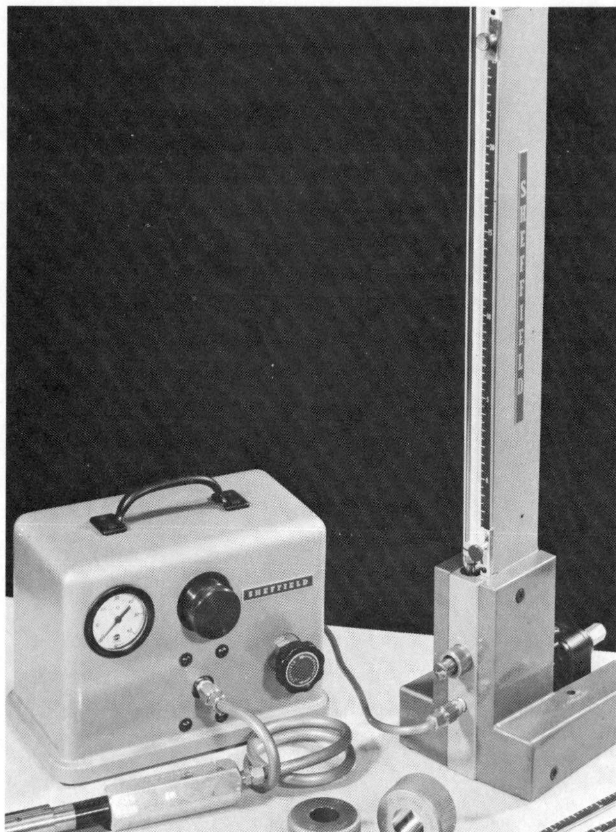

FIGURE C-16 Column or flow-type air gage (Automation and Measurement Division—Bendix Corporation).

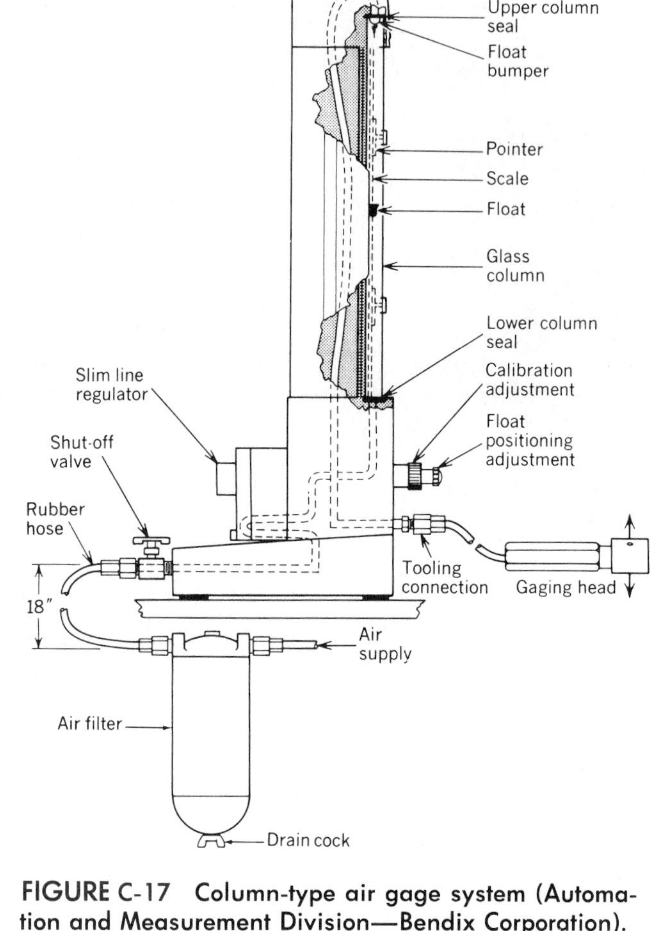

FIGURE C-17 Column-type air gage system (Automation and Measurement Division—Bendix Corporation).

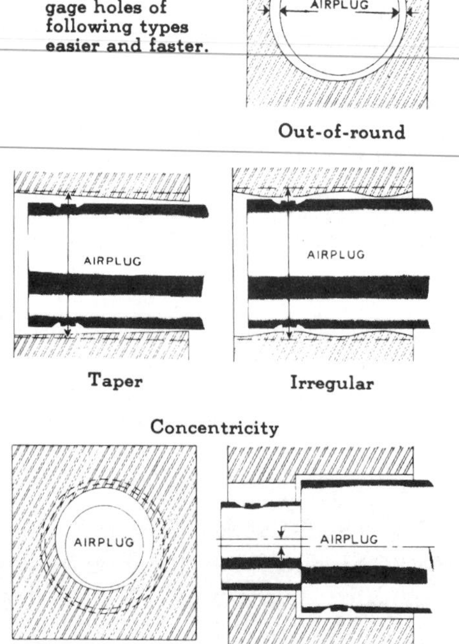

FIGURE C-18 Hole geometry detectable by air gaging (Federal Products Corporation).

MECHANICAL DIAL MEASURING INSTRUMENTS

Measuring instruments that show a measurement on a dial have become very popular in recent years. Several of the common dial instruments are outgrowths from vernier instruments of the same type. Dial instruments have an advantage over their vernier counterparts in that they are easier to read. Dial measuring equipment is frequently found in the inspection department where many types of measurements must be made quickly and accurately.

DIAL THICKNESS GAGE The **dial thickness gage** (Figure C-19) is used to measure the thickness of paper, leather, sheet metal, and rubber. Discrimination is .0005 in.

DIAL INDICATING SNAP GAGES **Dial indicating snap gages** (Figure C-20) are used for determining whether workpieces are within acceptable limits. They are first set to a gage block

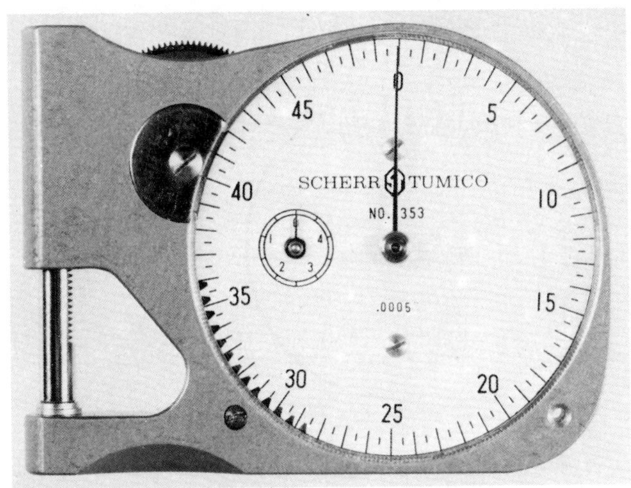

FIGURE C-19 Dial thickness gage (Rank Scherr-Tumico, Inc.).

FIGURE C-20 Using the indicating snap gage (Federal Products Corporation).

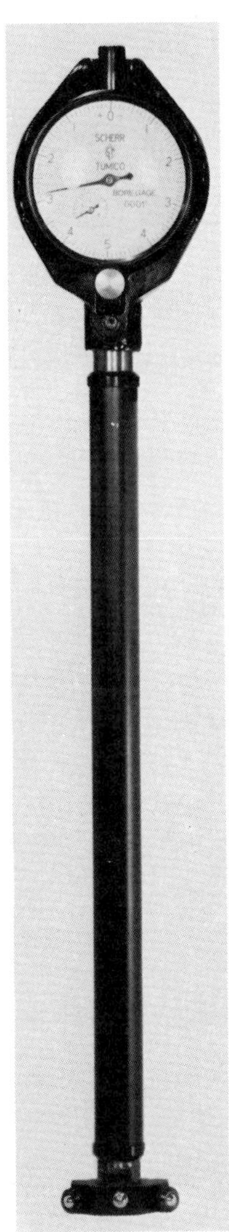

FIGURE C-21 Dial bore gage (Rank Scherr-Tumico, Inc.).

FIGURE C-22 Using the dial bore gage (L.S. Starrett Company).

standard. Part size deviation is noted on the dial indicator.

DIAL BORE GAGE The **dial bore gage** (Figure C-21) uses a three-point measuring contact. This more accurately measures the true shape of a bore (Figure C-22). The dial bore gage is useful for checking engine block cylinders for size, taper, bellmouth, ovality, barrel shape, and hourglass

FIGURE C-23 Hole geometry detectable with the dial bore gage.

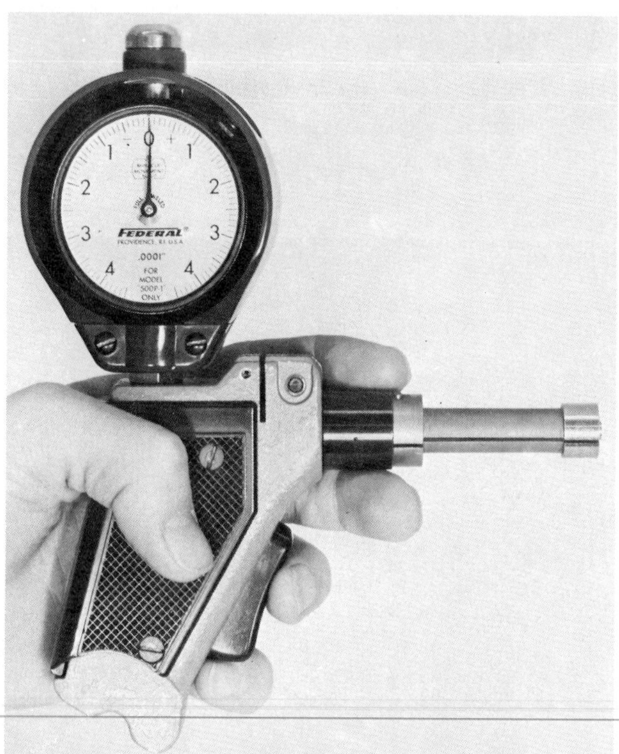

FIGURE C-24 Dial indicating expansion plug bore gage (Federal Products Corporation).

FIGURE C-25 Using the expansion plug bore gage (Federal Products Corporation).

shape (Figure C-23). Dial bore gages are set to a master ring and then compared to a bore diameter. Discrimination ranges from .001 to .0001 in.

DIAL INDICATING EXPANSION PLUG BORE GAGE The **indicating expansion plug gage** (Figure C-24) is used to measure the inside diameter of a hole or bore. This type of gage is built to check a single dimension. It can detect ovality bellmouth, barrel shape, and taper. The expanding plug is retracted and the instrument inserted into the hole to be measured (Figure C-25).

DIAL INDICATING THREAD PLUG GAGE The **indicating thread plug gage** (Figure C-26) is used to measure internal threads. This type of gage need not be screwed into the thread. The measuring anvils retract so that the gage may be inserted into a threaded hole.

DIAL INDICATING SCREW THREAD SNAP GAGE The **dial indicating screw thread snap gage** (Figure C-27) is used to measure an external thread. The instrument may be fitted with suitable anvils for measuring the major, minor, or pitch diameter of screw threads. Discrimination is .0005 or .00005 in., depending on the dial indicator used.

DIAL INDICATING INPROCESS GRINDING GAGE Gages can be built into machining processes. The **indicating inprocess grinding gage** is used to measure the workpiece while it is still running in the machine tool (Figure C-28). The instrument swings down over the part to be measured (Figure C-29). The machine can remain running. These instruments are used in such applications as cylindrical grinding. Discrimination

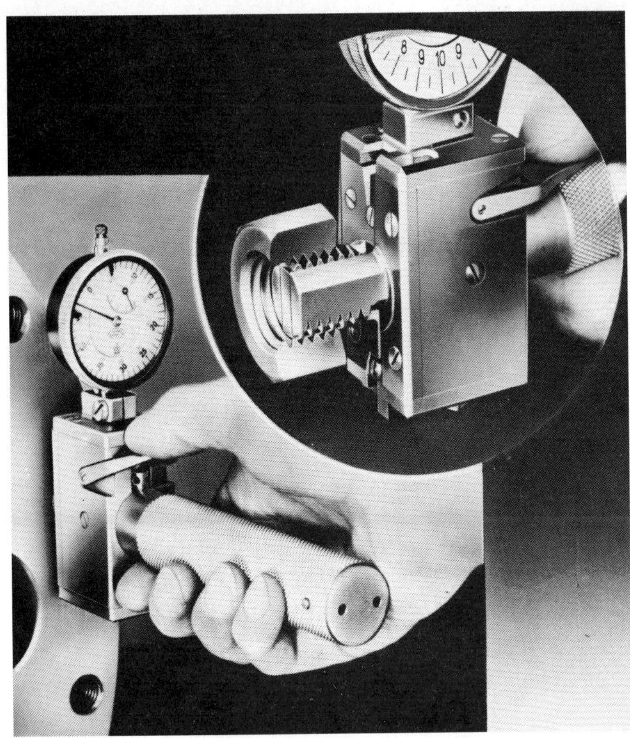

FIGURE C-26 Using the indicating thread plug gage (Mahr Gage Company).

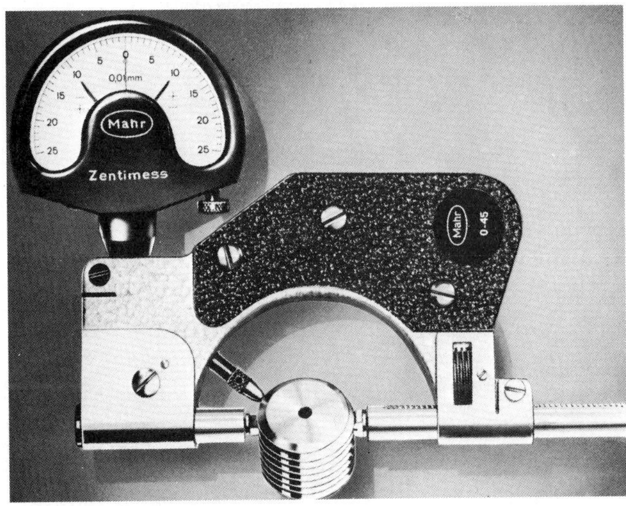

FIGURE C-27 Dial indicating thread snap gage (Mahr Gage Company).

FIGURE C-28 Dial indicator inprocess grinding gage, in retracted position (Federal Products Corporation).

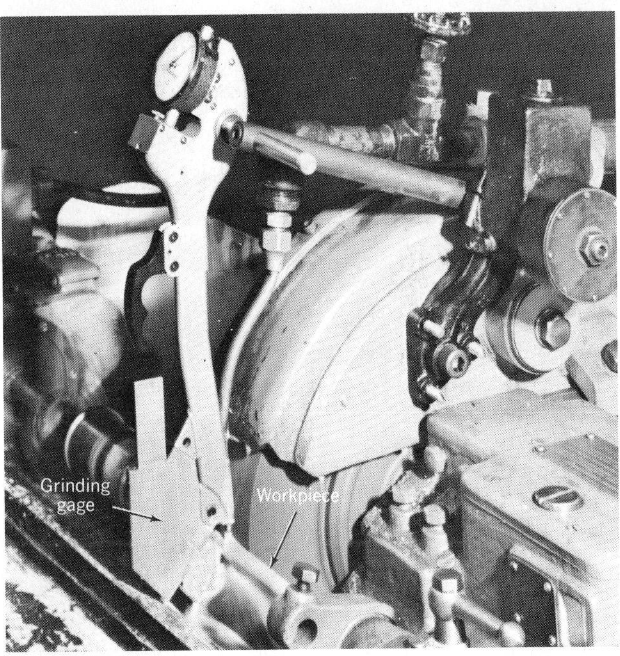

FIGURE C-29 Inprocess grinding gage measuring the workpiece (Federal Products Corporation).

can be .0005 or .00005 in., depending on the dial indicator used. This instrument also has its electronic counterpoint.

MECHANICAL DIAL INDICATING TRAVEL INDICATORS **Mechanical dial indicators** can be used to indicate the travel of machine tool components. This is very valuable to the machinist in controlling machine movement that in turn controls the dimensions of the parts produced. Mechanical dial travel indicators are used in many applications such as indicating table and saddle travel on a milling machine (Figure C-30). They may also be used to indicate vertical travel of quills and spindles. Mechanical dial travel indicators are also useful for indicating travel in metric dimensions. Discrimination is .001 in., .005 in., and .01 mm.

INSPECTION AND CALIBRATION THROUGH MECHANICAL MEASUREMENT

All measuring instruments must be periodically checked against accepted standards if the control that permits interchangeability of parts is to be maintained. Without control of measurement there could be no diverse mass production of parts that will later fit together to form the many products that we now enjoy.

Parts produced on a machine tool must be inspected to determine if their size meets design requirements. Parts produced out of tolerance can greatly increase the dollar cost of production. These must be kept to a minimum, and this is the purpose of part inspection and calibration of measuring instruments.

INDICATING BENCH MICROMETER The **indicating bench micrometer,** commonly called a **supermicrometer** (Figure C-31), is used to inspect tools, parts, and gages. This instrument has a discrimination of .00002 in. (20 millionths). This instrument also is available in an electronic digital model (Figure C-32).

SURFACE FINISH VISUAL COMPARATOR Surface finish may be approximated by visual inspection using the **surface roughness gage** (Figure C-33). Samples of finishes produced by various machining operations are indicated on the gage. These can be visually compared to a machined surface to determine the approximate degree of surface finish.

COORDINATE MEASURING MACHINES The **coordinate measuring machine** (Figure C-34) is an extremely accurate instrument that can measure the workpiece in three dimensions. Coordi-

FIGURE C-30 Mechanical dial travel indicators installed on a milling machine (Southwestern Industries, Inc., Trav-A-Dial®).

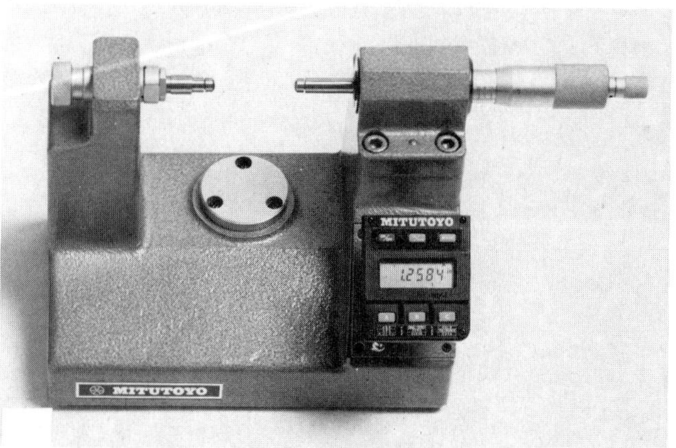

FIGURE C-32 Digital electronic bench micrometer (MTI Corporation).

FIGURE C-31 Indicating bench micrometer or supermicrometer (Colt Industries, Pratt and Whitney Cutting Tool and Gage Division).

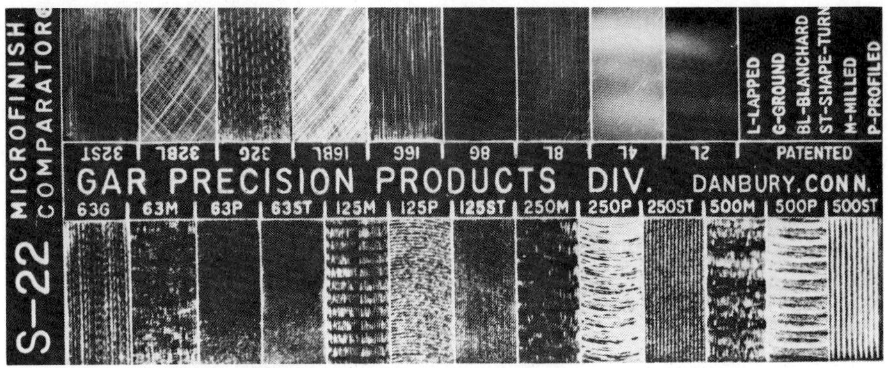

FIGURE C-33 Visual surface roughness comparator gage (DoALL Company).

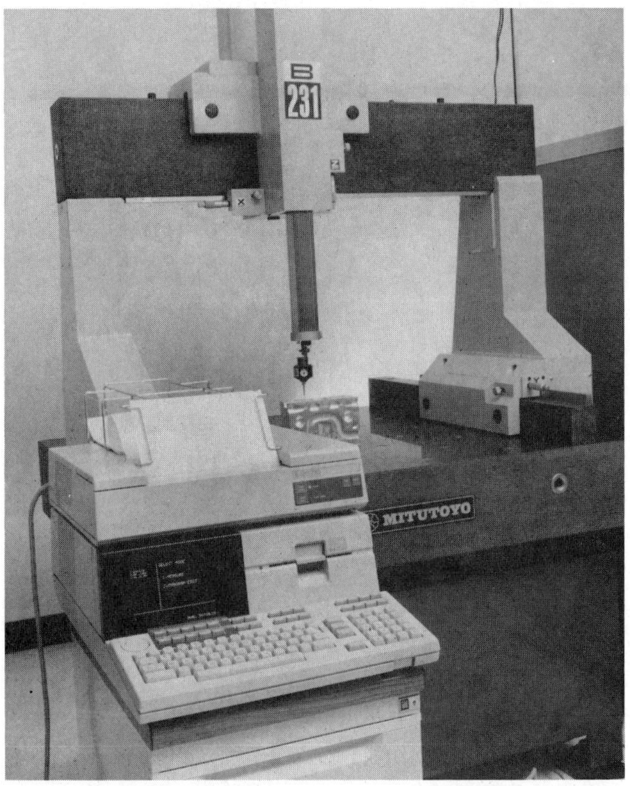

FIGURE C-34 Three-axis coordinate measuring machine with computer and printer (MTI Corporation).

nate measuring machines are very useful for determining the location of a part feature relative to a reference plane, line, or point.

The electronic coordinate measuring machine is an indispensable tool for the inspector and gage laboratory. Many of these instruments are computer equipped, allowing calculations to be made relative to the measurements being taken. Computer printouts may be easily obtained indicating graphs of measurement, as well as graphics illustrating the parts being measured. The computerized coordinate measuring machine is another example of the integration of digital readout electronics and computers into a precision mechanical system for modern high precision measurement applications.

MEASUREMENT WITH ELECTRONICS

REMOTE GAGING Electronic technology has come into wide use in measurement. Electronic equipment can be designed with greater sensitivity than mechanical equipment. Thus, higher discrimination can be achieved. Electronics can be applied in **remote gaging** applications (Figure C-35). In this application, there is no direct connec-

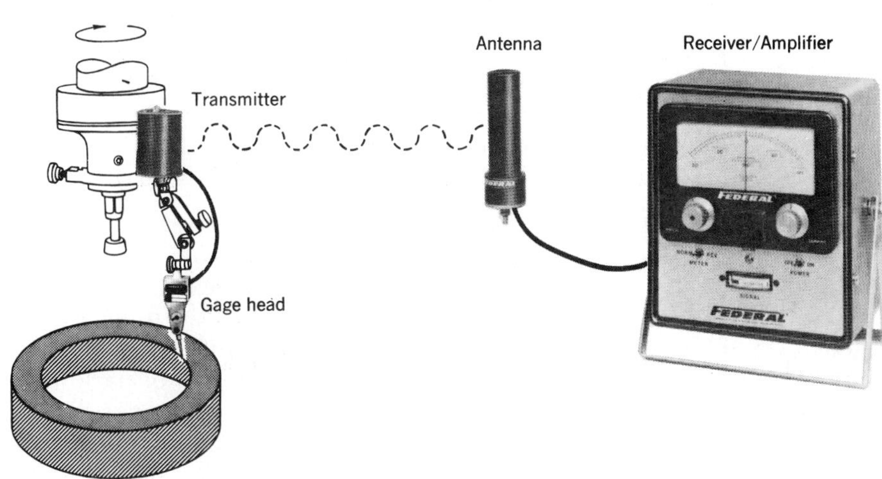

FIGURE C-35 Remote electronic gaging system (Federal Products Corporation).

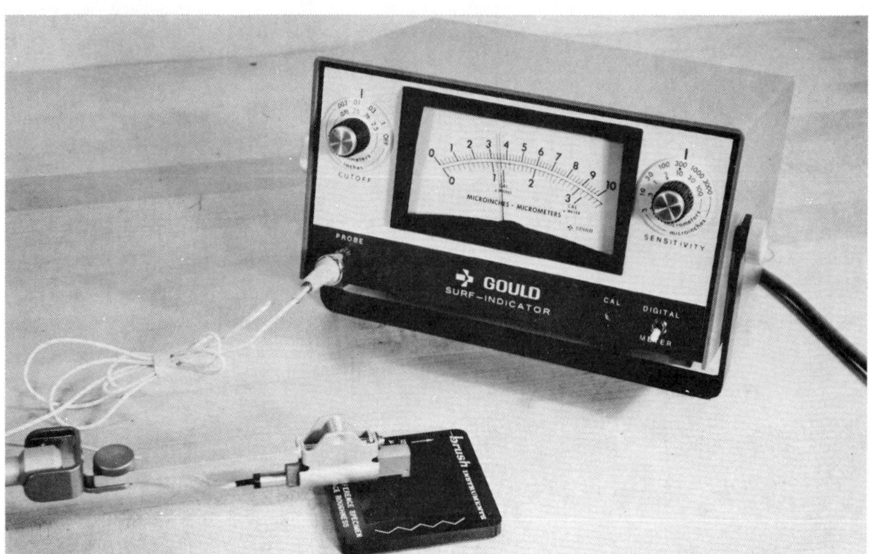

FIGURE C-36 Surface finish indicator being calibrated.

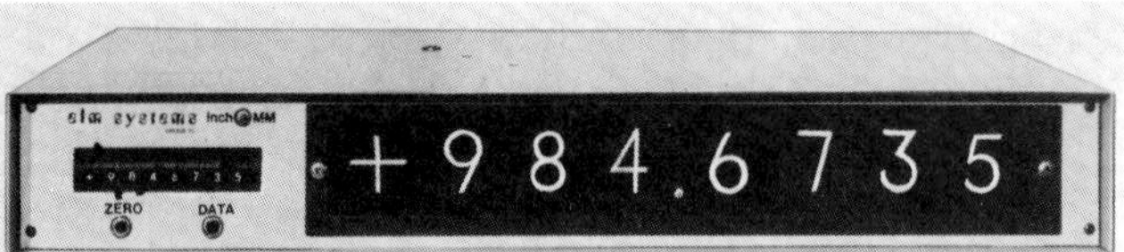

FIGURE C-37 Electronic digital travel indicator display (Elm Systems, Inc.)

tion to the gage head. The head is free to move with the machine tool since there are no attached wires. This facilitates use of the gaging instruments.

SURFACE FINISH INDICATORS Surface finish is critical on many parts such as bearings, gears, and hydraulic cylinders. Surface finish is a measure of **surface roughness** or **profile.** The measurement is in **microinches.** A **microinch** is **one millionth of an inch.** A surface finish indicator (Figure C-36) consists of a diamond stylus connected to a suitably graduated dial (meter). The stylus records surface deviations, which are indicated on the dial.

ELECTRONIC DIGITAL TRAVEL INDICATORS **Electronic digital travel indicators** use a sensor attached to the machine tool. These systems will discriminate to .0001 in. and can be switched to read in metric dimensions. The travel of the machine component is indicated on a digital display (Figure C-37). They are very useful for the accurate positioning of machine tables on such tools as milling machines and jig borers (Figure C-38). A sensor on the machine tool detects movement of the machine components. The amount of travel is displayed on an electronic digital display.

FIGURE C-38 Electronic digital travel indicator system installed on a jig borer (Elm Systems, Inc.).

ELECTRONIC COMPARATORS **Electronic comparators** (Figure C-39) take advantage of the sensitivity of electronic equipment. They are used to make comparison measurements of parts and other measuring tools. For example, gage blocks may be calibrated using a suitable electronic comparator.

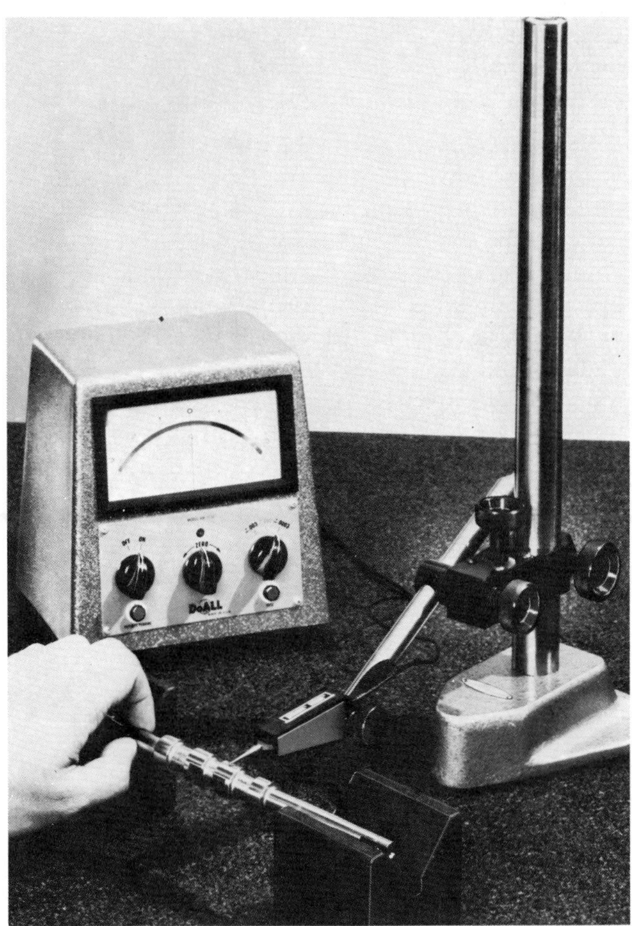

FIGURE C-39 Electronic comparator (DoALL Company).

MEASUREMENT WITH LIGHT

TOOLMAKER'S MICROSCOPE The **toolmaker's microscope** (Figure C-40) is used to inspect parts, cutting tools, and measuring tools. The microscope has a stage that can be precisely rotated and moved in two perpendicular axes. The instrument may be equipped with an electronic accessory measuring system that discriminates to .0001 in. Thus stage movement can be recorded permitting measurements of a workpiece to be made.

The **optical comparator** (Figure C-41) is used in the inspection of parts, cutting tools, and other measuring instruments. Optical comparators project a greatly magnified shadow of the object on a screen. The surface of the workpiece may also be illuminated. Shape patterns or graduated patterns can be placed on the screen and used to make measurements on the workpiece projection.

Optical flats are used in the inspection of other measuring instruments and for the measurement of flatness. They can be used, for example, to reveal the surface geometry of a gage block or measuring faces of a micrometer (Figure C-42). Optical flats take advantage of the principles of light interferometry to make extremely small measurements in millionths of an inch.

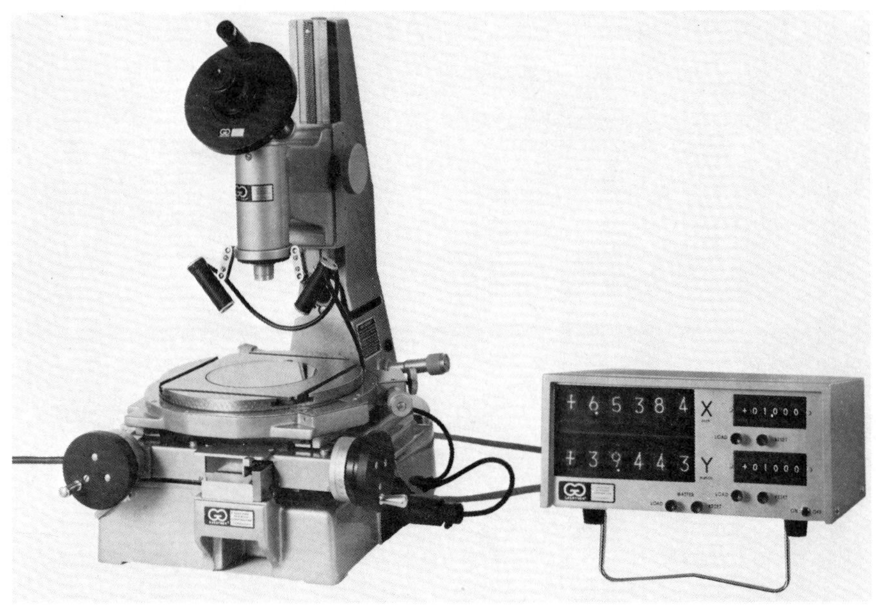

FIGURE C-40 Toolmaker's microscope (Gaertner Scientific Corporation).

AUTOCOLLIMATORS The **autocollimator** in Figure C-43 is being used to check the flatness of a surface plate. The mirror on the left is moved along the straightedge in small increments. Deviations from flatness are shown by angular changes of the mirror. This change is recorded by the instrument.

ALIGNMENT TELESCOPE A machinist may accomplish alignment tasks by optical means. Optical alignment may be used on such applications as ship propeller shaft bearings. Portable machine tools such as boring bars may be positioned by optical alignment. The **dual micrometer alignment telescope** (Figure C-44) is a very useful alignment instrument. The micrometers permit the deviation of the workpiece from the line of sight to be determined.

LASER INTERFEROMETER The term *laser* is an acronym for **light amplification by stimulated emission of radiation.** A laser light beam is a coherent beam. This means that each ray of light follows the same path. Thus, it does not disperse over long distances. This property makes the laser beam very useful in many measurement and alignment applications. For example, the laser beam may be used to determine how straight a machine tool table travels (Figure C-45). Other uses include the checking of machine tool measuring systems (Figure C-46).

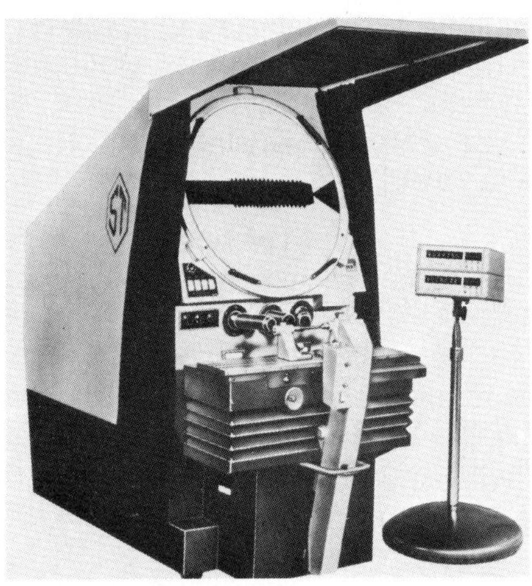

FIGURE C-41 Optical comparator (Rank Scherr-Tumico, Inc.).

FIGURE C-42 Using optical flats to check micrometer measuring faces (DoALL Company).

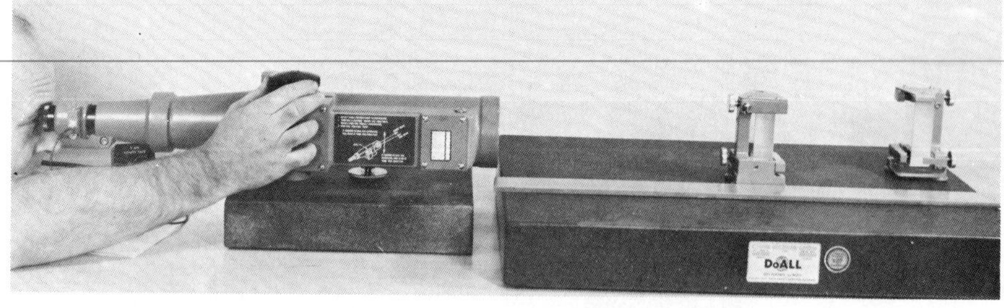

FIGURE C-43 Autocollimator checking a surface plate for flatness.

FIGURE C-44 Dual micrometer alignment telescope (DoALL Company).

FIGURE C-45 Laser interferometer being used for straightness determination (Hewlett-Packard Company).

FIGURE C-46 Laser interferometer checking the measuring system on a jig boring machine.

UNIT 1

Systems of Measurement

Throughout history there have been many systems of measurement. Prior to the era of national and international industrial operations, an individual was often responsible for the manufacture of a complete product. Since the same person made all the necessary parts and did the required assembly, he or she needed to conform only to his or her particular system of measurement. However, as machines replaced people and diversified mass production was established on a national and international basis, the need for standardization of measurement became readily apparent. Total standardization of measurement throughout the world still has not been fully realized. Most measurement in the modern world does, however, conform to either the English (inch-pound-second) or the metric (meter-kilogram-second) system. Metric measurement is now predominant in most of the industrialized nations of the world. The English system is still used to a great extent in U.S. manufacturing. The importation of manufactured goods built to metric specifications has, in recent years, made the use of metric tools and measurements very common in the United States.

Today's machinists must now begin to think in terms of metric measurement. During their careers they may come in contact with metric specifications. However, for the present they will be using inch measurement. Since you are primarily concerned with length measurement, this unit will review the basic length standards of both systems, examine mathematical and other methods of converting from system to system, and look at techniques by which a machine tool can be converted to work in metrics.

Objectives

After completing this unit, you should be able to:

1. Identify common methods of measurement conversion.
2. Convert inch dimensions to metric equivalents and convert metric dimensions to inch equivalents.

The English System of Measurement

The English system of measurement uses the units of inches, pounds, and seconds to represent the measurement of time, length, and mass. Since we are primarily concerned with the measurement of length in the machine shop, we will simply refer to the English system as the inch system. Most of us are thoroughly familiar with inch measurement.

SUBDIVISIONS AND MULTIPLES OF THE INCH

The following table shows the common subdivisions and multiples of the inch that are used by the machinist.

Common Subdivisions

.000001	millionth
.00001	hundred thousandth
.0001	ten thousandth
.001	thousandth
.01	hundredth
.1	tenth
1.00	Unit inch

Common Multiples

12.00	1 foot
36.00	1 yard

Other common subdivisions of the inch are:

$\frac{1}{128}$	.007810 (decimal equivalent)
$\frac{1}{64}$	.015625
$\frac{1}{32}$	.031250
$\frac{1}{20}$	.050000
$\frac{1}{16}$	.062500
$\frac{1}{8}$	.125000
$\frac{1}{4}$	.250000
$\frac{1}{2}$	.500000

Multiples of Feet

3 feet = 1 yard

5280 feet = 1 mile

Multiples of Yards

1760 yards = 1 mile

The Metric System and the International System of Units—SI

The basic unit of length in the metric system is the meter. Originally the length of the meter was defined by a natural standard, specifically, a portion of the earth's circumference. Later, more convenient metal standards were constructed. In 1886, the metric system was legalized in the United States, but its use was not made mandatory. Since 1893 the yard has been defined in terms of the metric meter by the ratio

$$1 \text{ yard} = \frac{3600}{3937} \text{ meter}$$

Although the metric system had been in use for many years in many different countries, it still lacked complete standardization among its users. Therefore, an attempt was made to modernize and standardize the metric system. From this effort has come the **Système International d'Unités,** known as **SI** or the **International Metric System.**

The basic unit of length in SI is the meter, or metre (in the common international spelling). The SI meter is defined by a physical standard that can be reproduced anywhere with unvarying accuracy.

1 meter = 1,650,763.73 wavelengths in a vacuum of the orange-red light spectrum of the krypton-86 atom

Probably the primary advantage of the metric system is that of convenience in computation. All subdivisions and multiples use 10 as a divisor or multiplier. This can be seen in the following table.

.000001	(one-millionth meter or micrometer)
.001	(one-thousandth meter or millimeter)
.01	(one-hundredth meter or centimer)
.1	(one-tenth meter or decimeter)
1.00	*Unit meter*
10	(10 meters or 1 dekameter)
100	(100 meters or 1 hectometer)
1000	(1000 meters or 1 kilometer)
1,000,000	(1 million meters or 1 megameter)

METRIC SYSTEM EXAMPLES

1. One meter (m) = ______ millimeter (mm).
Since a mm is 1/1000 part of an m, there are 1000 mm in a meter.

2. 50 mm = ________ centimeters (cm).
Since 1 cm = 10 mm, 50/10 = 5 cm in 50 mm.

3. Four kilometers (km) = ________ m.
Since 1 km = 1000 m, then 4 km = 4000 m.

4. 582 mm = ________ cm.
Since 10 mm = 1 cm, 582/10 = 58.2 cm.

Conversion Between Systems

Much of the difficulty with working in a two-system environment is experienced in converting from one system to the other. This can be of particular concern to the machinist as he or she must exercise due caution in making conversions. Arithmetic errors can be easily made. Therefore, the use of a calculator is recommended.

CONVERSION FACTORS AND MATHEMATICAL CONVERSION

Since the historical evolution of the inch and metric systems is quite different, there are no obvious relationships between length units of the two systems. You simply have to memorize the basic conversion factors. We know from the preceding discussion that the yard has been defined in terms of the meter. Knowing this relationship, you can derive mathematically any length unit in either system. However, the conversion factor

$$1 \text{ yard} = \frac{3600}{3937} \text{ meter}$$

is a less common factor for the machinist. A more common factor can be determined by the following:

$$1 \text{ yard} = \frac{3600}{3937} \text{ meter}$$

Therefore,

$$1 \text{ yard} = .91440 \text{ meter}$$

$$\left(\frac{3600}{3937} \text{ expressed in decimal form}\right)$$

Then

$$1 \text{ inch} = \frac{1}{36} \text{ of } .91440 \text{ meter}$$

So

$$\frac{.91440}{36} = .025400$$

We know that

$$1 \text{ m} = 1000 \text{ mm}$$

Therefore,

$$1 \text{ inch} = .025400 \times 1000$$

or

1 inch = 25.4000 mm

The conversion factor 1 in. = 25.4 mm is very common and should be memorized. From the example shown it should be clear that in order to find inches knowing millimeters, you must divide inches by 25.4.

1000 mm = ______ inches

$$\frac{1000}{25.4} = 39.37 \text{ inches}$$

In order to simplify the arithmetic, any conversion can always take the form of a multiplication problem.

EXAMPLE

Instead of 1000/25.4, multiply by the reciprocal of 25.4, which is 1/25.4 or .03937. Therefore,

$$1000 \times .03937 = 39.37 \text{ inches}$$

EXAMPLES OF CONVERSIONS [INCH TO METRIC]

1. 17 in. = ______ cm.
Knowing inches, to find centimeters multiply inches by 2.54: 2.54 × 17 in. = 43.18 cm.

2. .807 in. = ______ mm.
Knowing inches, to find millimeters multiply inches by 25.4: 25.4 × .807 in. = 20.49 mm

EXAMPLES OF CONVERSIONS [METRIC TO INCH]

1. .05 mm = ______ in.
Knowing millimeters, to find inches multiply millimeters by .03937: .05 × .03937 = .00196 inches.

2. 1.63 m = ______ in.
Knowing meters, to find inches, multiply meters by 39.37: 1.63 × 39.37 m = 64.173 in.

CONVERSION FACTORS TO MEMORIZE

1 in. = 25.4 or 2.54 cm
1 mm = .03937 in.

OTHER METHODS OF CONVERSION

The conversion chart is a popular device for making conversions between systems. Conversion charts are readily available from many manufacturers. However, most conversion charts give equivalents for whole millimeters or standard fractional inches. If you must find an equivalent for a factor that does not appear on the chart, you must interpolate. In this instance, knowing the common conversion factors and determining the equivalent mathematically is more efficient.

Several electronic calculators designed to convert directly from system to system are available. Of course, any calculator can and should be used to do a conversion problem. The direct converting calculator does not require that any conversion constant be remembered. These constants are permanently programmed into the calculator memory.

CONVERTING MACHINE TOOLS

With the increase in metric measurement in industry, which predominantly uses the inch system, several devices have been developed that permit a machine tool to function in either system. These conversion devices eliminate the need to convert all dimensions prior to beginning a job.

Conversion equipment includes conversion dials (Figure C-47) that can be attached to lathe cross slide screws as well as milling machine saddle and table screws. The dials are equipped with

FIGURE C-47 Inch/metric conversion dials for machine tools (Sipco Machine Co.).

gear ratios that permit a direct metric reading to appear on the dial.

Metric mechanical and electronic travel indicators can also be used. The mechanical dial travel indicator (Figure C-48) uses a roller that contacts a moving part of a machine tool. Travel of the machine component is indicated on the dial. This type of travel indicator discriminates to .01 mm. Whole millimeters are counted on the 1 mm. counting wheel. Mechanical dial travel indicators are used in many applications such as reading the travel of a milling machine saddle and table (Figure C-49).

The electronic travel indicator (Figure C-50) uses a sensor that is attached to the machine tool. Machine tool component travel is indicated on an electronic digital display. The equipment can be switched to read travel in inch or metric dimensions.

FIGURE C-48 Metric mechanical dial travel indicator (Southwestern Industries, Inc., Trav-A-Dial®).

FIGURE C-49 Metric mechanical dial travel indicators reading milling machine saddle and table movement (Southwestern Industries, Inc., Trav-A-Dial®).

Metric conversion devices can be fitted to existing machine tools for a moderate expense. Many new machine tools, especially those built abroad, have dual system capability built into them.

SELF-TEST

Perform the following conversions:

1. 35 mm = ________ in.

2. 125 in. = ________ mm.

3. 6.273 in. = ________ mm.

4. Express the tolerance ±.050 in metric terms to the nearest mm.

5. To find cm knowing mm, (multiply/divide) by 10.

6. Express the tolerance ±.02 mm in terms of inches to the nearest 1/10,000 in.

7. What is meant by SI?

8. Describe methods by which conversions between metric and inch measurement systems may be accomplished.

9. How is the yard presently defined?

10. Can an inch machine tool be converted to work in metric units?

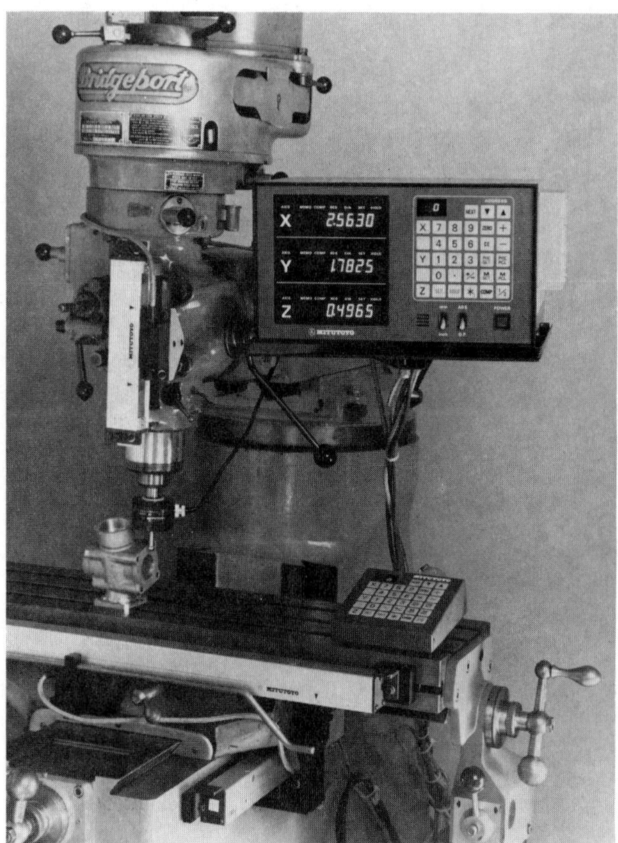

FIGURE C-50 High discrimination digital electronic readout measurement system (DRO) installed on a milling machine (MTI Corporation).

UNIT 2

Using Steel Rules

One of the most practical and common measuring tools available in the machining and inspection of parts is the steel rule. It is a tool that the machinist uses daily in different ways. It is important that anyone engaged in machining be able to select and use steel rules.

Objectives

After completing this unit, you should be able to:

1. Identify various kinds of rules and their applications.
2. Apply rules in typical machine shop measurements.

Scales and Rules

The terms **scale** and **rule** are often used interchangeably and often incorrectly. A rule is a linear measuring instrument whose graduations represent **real units** of lengths and their subdivisions. In contrast, a **scale** is graduated into **imaginary** units that are either smaller or larger than the real units they represent. This is done for convenience where proportional measurements are needed. For example, an architect uses a scale that has graduations representing feet and inches. However, the actual length of the graduations on the architect's scale are quite different from full-sized dimensions.

Discrimination of Steel Rules

The general concept of **discrimination** was discussed in the introduction to this section. Discrimination refers to the extent to which a unit of length has been divided. If the smallest graduation on a specific rule is $\frac{1}{32}$ in., then the rule has a discrimination of, or discriminates to, $\frac{1}{32}$ in. Likewise, if the smallest graduation of the rule is $\frac{1}{64}$ in., then this rule discriminates to $\frac{1}{64}$ in.

The maximum discrimination of a steel rule is generally $\frac{1}{64}$ in., or, in the case of the decimal inch rule, $\frac{1}{100}$ in. The metric rule has a discrimination of .05 mm. Remembering that a measuring tool should never be used beyond its discrimination, the steel rule will not be reliable in trying to ascertain a measurement increment smaller than $\frac{1}{64}$ or $\frac{1}{100}$ in. If a specific measurement falls between the markings on the rule, only this can be said of this reading: it is more or less than the amount of the nearest mark. No further data as to how much more or less can be reliably determined. It is not recommended practice to attempt to read between the graduations on a steel rule with the intent of obtaining reliable readings.

Reliability and Expectation of Accuracy in Steel Rules

For reliability, great care must be taken if the steel rule is to be used at its maximum discrimination. Remember that the markings on the rule occupy a certain width. A good quality steel rule has engraved graduations. This means that the markings are actually cuts in the metal from which the rule is made. Of all types of graduations, engraved ones occupy the least width along the rule. Other rules, graduated by other processes, may have markings that occupy greater width. These rules are not necessarily any less accurate, but they may require more care in reading. Generally, the reliability of the rule will diminish as its maximum discrimination is approached. The smaller graduations are more difficult to see without the aid of a magnifier. Of particular importance is the point from which the measurement is taken. This is the **reference point** and must be carefully aligned at the point where the length being measured begins.

From a practical standpoint, the steel rule finds widest application for measurements no smaller than $\frac{1}{32}$ in. on a fractional rule or $\frac{1}{50}$ in. on a decimal rule. This does not mean that the rule cannot measure to its maximum discrimination, because under the proper conditions it certainly can. However, at or very near maximum discrimination, the time consumed to insure reliable measurement is really not justified. You will be more productive if you make use of a type of measuring instrument with considerably finer discrimination for measurements below the nearest $\frac{1}{32}$ or $\frac{1}{50}$ in. It is good practice to take more than one reading when using a steel rule. After determining the desired measurement, apply the rule once again to see if the same result is obtained. By this procedure, the reliability factor is increased.

Types of Rules

Rules may be selected in many different shapes and sizes, depending on the need. The common **rigid steel rule** is 6 in. long, $\frac{3}{4}$ in. wide, and $\frac{3}{64}$ in. thick. It is engraved with No. 4 standard rule graduations. A No. 4 graduation consists of $\frac{1}{8}$ and $\frac{1}{16}$ in. on one side (Figure C-51) and $\frac{1}{32}$- and $\frac{1}{64}$-in. divisions on the reverse side (Figure C-52). Other common graduations are summarized in the following table.

Graduation Number	Front Side	Back Side
3	32nds	10ths
	64th	50ths
16	50ths	32nds
	100ths	64ths

The No. 16 graduated rule is often found in the aircraft industry where dimensions are specified in decimal fraction notations, based on 10 or a multiple of 10 divisions of an inch rather than 32 or 64 divisions as found on common rules. Many rigid rules are one inch wide.

Another common rule is the **flexible type** (Figure C-53). This rule is 6 in. long, $\frac{1}{2}$ in. wide, and $\frac{1}{64}$ in. thick. Flexible rules are made from hardened and tempered spring steel. One advantage of a flexible rule is that it will bend, permitting measurements to be made in a space shorter than the length of the rule. Most flexible rules are 6 or 12 in. long.

The **narrow rule** (Figure C-54) is very convenient when measuring in small openings, slots, or holes. Most narrow rules have only one set of graduations on each side. These can be number 10, which is 32nds and 64ths, or number 11, which is 64ths and 100ths.

FIGURE C-51 Six-inch rigid steel rule (front side).

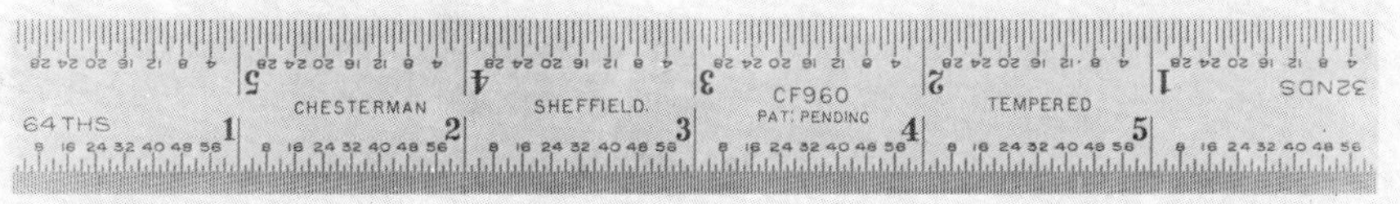

FIGURE C-52 Six-inch rigid steel rule (back side).

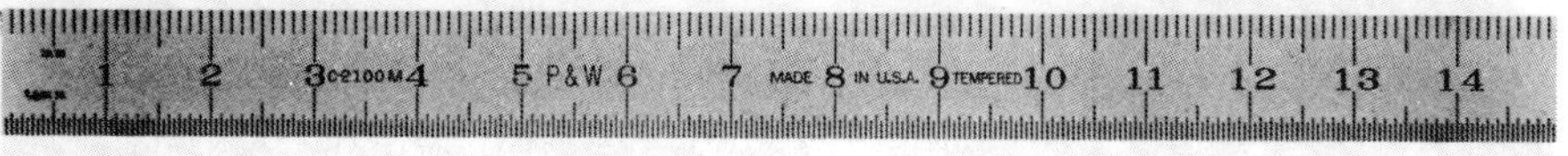

FIGURE C-53 Flexible steel rule (metric).

FIGURE C-54 Narrow rule (decimal inch).

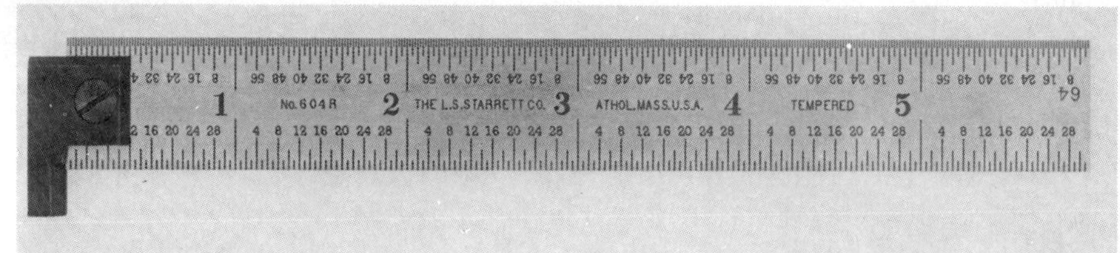

FIGURE C-55 Standard hook rule.

FIGURE C-56 Standard hook rule in use.

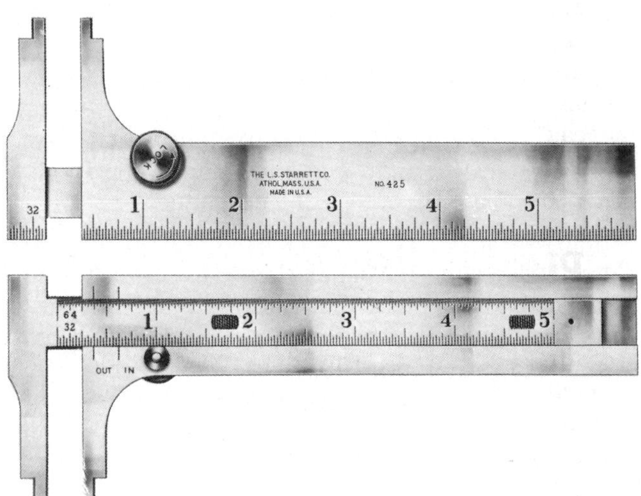

FIGURE C-58 Slide caliper rule (L.S. Starrett Co.).

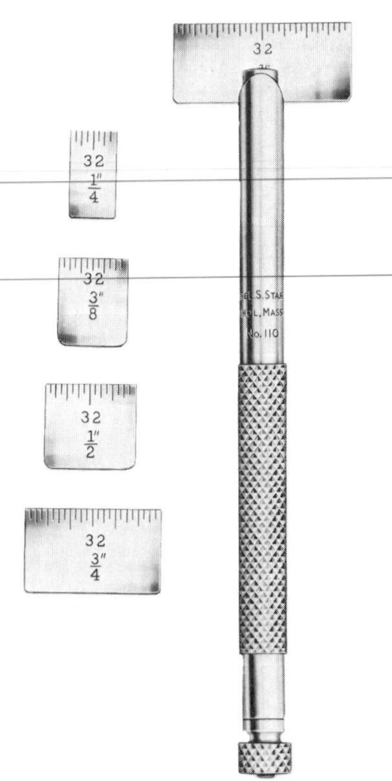

FIGURE C-57 Short rule set with holder (L.S. Starrett Co.).

The **standard hook rule** (Figure C-55) makes it possible to reach through an opening; the rule is hooked on the far side in order to measure a thickness or the depth of a slot (Figure C-56). When a workpiece has a chamfered edge, a hook rule will be advantageous over a common rule. If the hook is not loose or excessively worn, it will provide an easy-to-locate reference point.

The **short rule set** (Figure C-57) consists of a set of rules with a holder. Short rule sets have a range of $\frac{1}{4}$ or 1 in. They can be used to measure shoulders in holes or steps in slots, where space is extremely limited. The holder will attach to the rules at any angle, making these very versatile tools.

The **slide caliper rule** (Figure C-58) is a versatile tool used to measure round bars, tubing, and other objects where it is difficult to measure at the ends and difficult to estimate the diameter with a rigid steel rule. The small slide caliper rule can also be used to measure internal dimensions from $\frac{1}{4}$ in. up to the capacity of the tool.

The **rule depth gage** (Figure C-59) consists of a slotted steel head in which a narrow rule slides. For depth measurements the head is held securely against the surface with the rule extended into the cavity or hole to be measured (Figure C-60). The locking nut is tightened and the rule depth gage can then be removed and the dimension determined.

Care of Rules

Rules are precision tools, and only those that are properly cared for will provide the kind of service they are designed to give. A rule should not be used as a screwdriver. Rules should be kept separate from hammers, wrenches, files, and other hand tools to protect them from possible damage. An occasional wiping of a rule with a lightly oiled shop towel will keep it clean and free from rust.

Applying Steel Rules

When using a steel rule in close proximity to a machine tool, always keep safety in mind. Stop the machine before attempting to make any measurements of the workpiece. Attempting to measure with the machine running may result in the rule being caught by a moving part. This may damage the rule, but worse, may result in serious injury to the operator.

One of the problems associated with the use of rules is that of **parallax error.** Parallax error results when the observer making the measurement is not in line with the workpiece and the rule. You may see the graduation either too far left or too far right of its real position (Figure C-61). Parallax error occurs when the rule is read from a point other than one directly above the point of measurement. The point of measurement is the point at which the measurement is read. It may or may not be the true reading of the size depending on what location was used as the reference point on the rule. Parallax can be controlled by always observing the point of measurement from directly above. Furthermore, the graduations on a rule should be placed as close as possible to the surface being measured. In this regard, a thin rule is preferred over a thick rule.

As a rule is used it becomes worn, usually on the ends. The outside inch markings on a worn rule are less than one inch from the end. This has to be considered when measurements are made. A

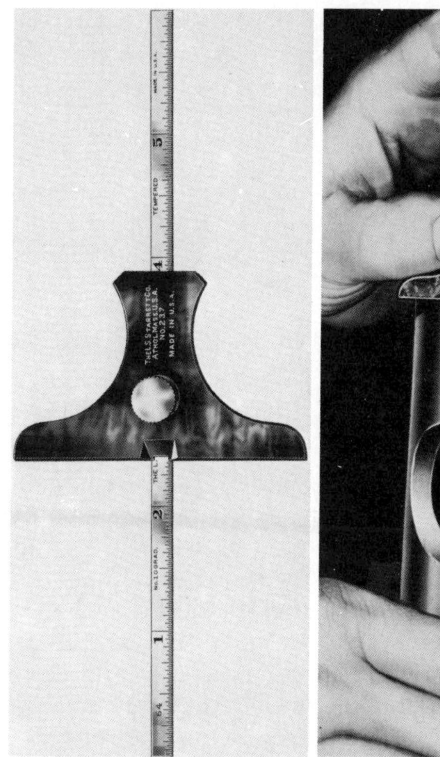

FIGURE C-59 Rule depth gage (L.S. Starrett Co.).

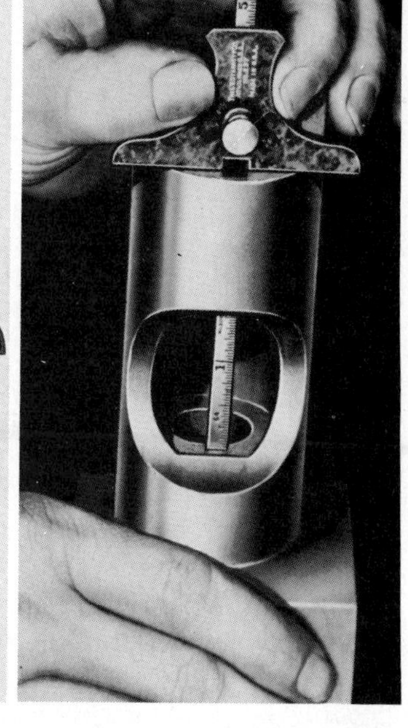

FIGURE C-60 Rule depth gage in use (LS. Starrett Co.).

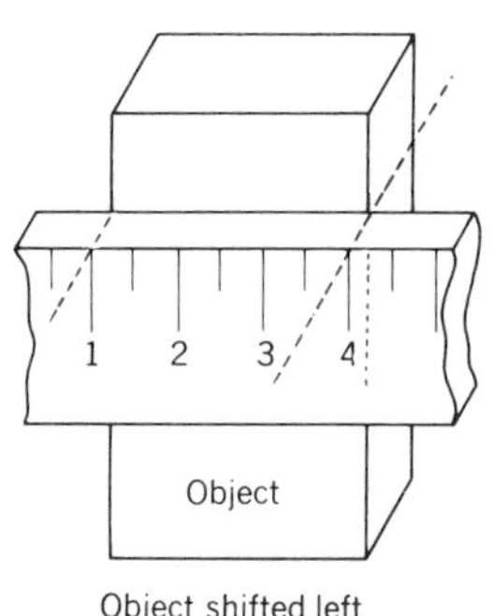

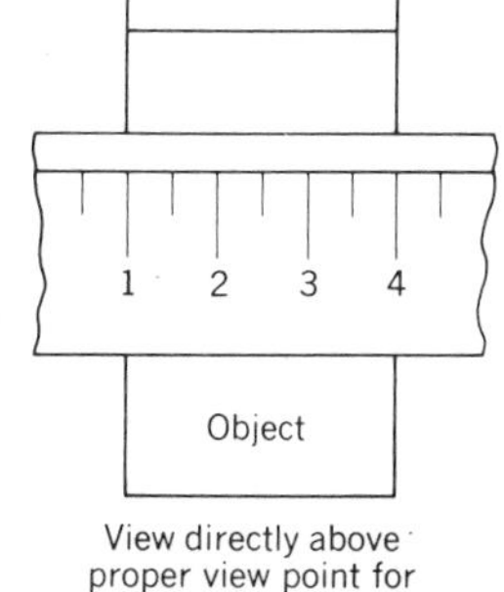

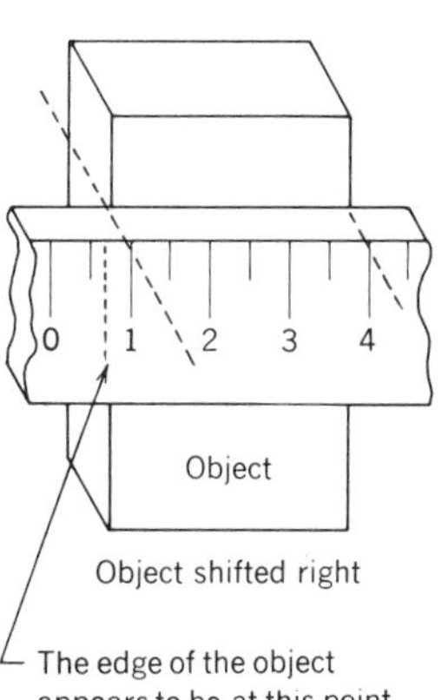

When viewed from directly above, the rule graduations are exactly in line with the edge of the object being measured. However, when the object is shifted right or left of a point directly above the point of measurement, the alignment of the object edge and the rule graduations appears to no longer coincide

FIGURE C-61 Parallax error.

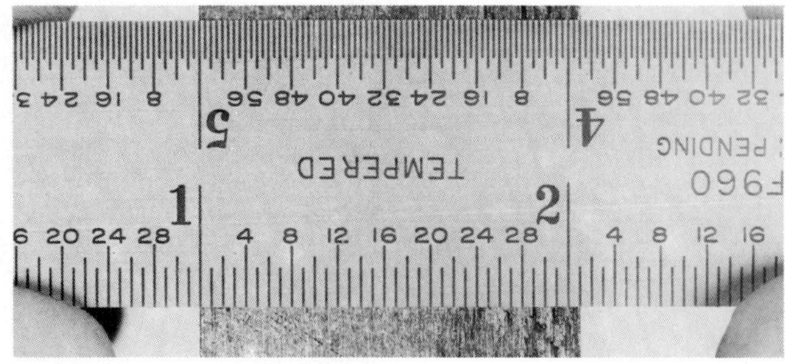

FIGURE C-62 Using the 1-in. mark as the reference point.

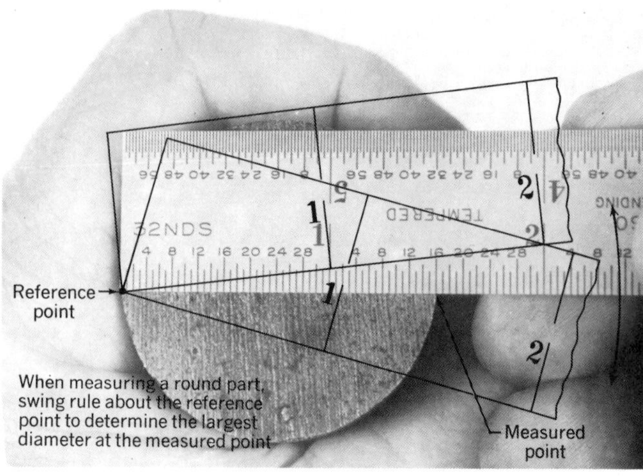

Reference point
When measuring a round part, swing rule about the reference point to determine the largest diameter at the measured point
Measured point

FIGURE C-63 Measuring round objects.

reliable way to measure (Figure C-62) is to use the one inch mark on the rule as the reference point. In the figure, the measured point is at $2\frac{1}{32}$. Subtracting 1 in. results in a size of $1\frac{1}{32}$ for the part.

Round bars and tubing should be measured with the rule applied on the end of the tube or bar (Figure C-63). Select a reference point and set it carefully at a point on the circumference of the round part to be measured. Using the reference point as a pivot, move the rule back and forth slightly to find the largest distance across the diameter. When the largest distance is determined, read the measurement at that point.

Reading Fractional Inch Rules

Most dimensions are expressed in inches and fractions of inches. These dimensions are measured with fractional inch rules. The typical machinist's rule is broken down into 1-, $\frac{1}{2}$-, $\frac{1}{4}$-, $\frac{1}{8}$-, $\frac{1}{16}$-, $\frac{1}{32}$-, and $\frac{1}{64}$-in. graduations. In order to facilitate reading, the 1-, $\frac{1}{2}$-, $\frac{1}{4}$-, $\frac{1}{8}$-, and $\frac{1}{16}$-in. graduations appear on one side of the rule (Figure C-64). The reverse side of the rule has one edge graduated in $\frac{1}{32}$-in. increments and the other edge graduated in $\frac{1}{64}$-in. increments. On the $\frac{1}{32}$-in. side, every fourth mark is numbered and on the $\frac{1}{64}$-in. side, every eighth mark is numbered (Figure C-65). This eliminates the need to count graduations from the nearest whole inch mark. On these rules, the length of the graduation line varies, with the one inch line being the longest, the $\frac{1}{2}$-in. line being next in length, and the $\frac{1}{4}$-, $\frac{1}{8}$-, and $\frac{1}{16}$-in. lines each being consecutively shorter. The difference in line lengths is an important aid in reading a rule. The smallest grad-

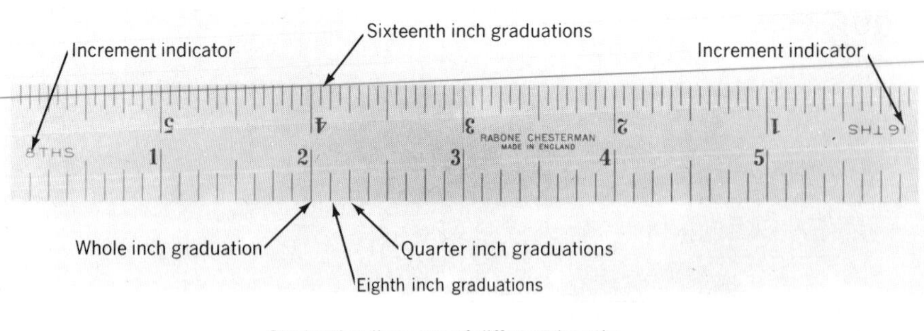

FIGURE C-64 Front-side graduations of the typical machinist's rule.

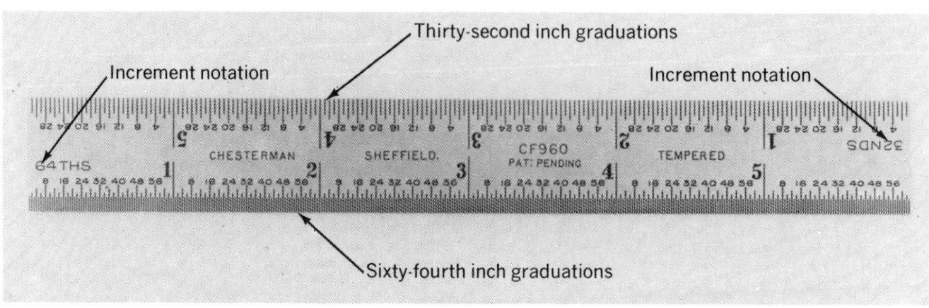

FIGURE C-65 Back-side graduations of the typical machinist's rule.

uation on any edge of a rule is marked by small numbers on the end. Note that the words 8THS and 16THS appear at the ends of the rule. The numbers 32NDS and 64THS appear on the reverse side of the rule, thus indicating thirty-seconds and sixty-fourths of an inch.

EXAMPLES OF FRACTIONAL INCH READINGS

FIGURE C-66 Distance *A* falls on the third $\frac{1}{8}$-in. graduation. This reading would be $\frac{3}{8}$ in.

Distance *B* falls on the longest graduation between the end of the rule and the first full inch mark. The reading is $\frac{1}{2}$ in.

Distance *C* falls on the sixth $\frac{1}{8}$-in. graduation, making it $\frac{6}{8}$ or $\frac{3}{4}$ in.

Distance *D* falls at the fifth $\frac{1}{8}$-in. mark beyond the 2-in. graduation. The reading is $2\frac{5}{8}$ in.

FIGURE C-67 Distance *A* falls at the thirteenth $\frac{1}{16}$-in. mark, making the reading $\frac{13}{16}$ in.

Distance *B* falls at the first $\frac{1}{16}$-in. mark past the 1-in. graduation. The reading is $1\frac{1}{16}$ in.

Distance *C* falls at the seventh $\frac{1}{16}$-in. mark past the 1-in. graduation. The reading is $1\frac{7}{16}$ in.

Distance *D* falls at the third $\frac{1}{16}$-in. mark past the 2-in. graduation. The reading is $2\frac{3}{16}$ in.

FIGURE C-68 Distance *A* falls at the third $\frac{1}{32}$-in. mark. The reading is $\frac{3}{32}$ in.

Distance *B* falls at the ninth $\frac{1}{32}$-in. mark. The reading is $\frac{9}{32}$ in.

Distance *C* falls at the eleventh $\frac{1}{32}$-in. mark past the 1-in. graduation. The reading is $1\frac{11}{32}$ in.

Distance *D* falls at the fourth $\frac{1}{32}$-in. mark past the 2-in. graduation. The reading is $2\frac{4}{32}$ in., which reduced to lowest terms becomes $2\frac{1}{8}$ in.

FIGURE C-69 Distance *A* falls at the ninth $\frac{1}{64}$-in. mark, making the reading $\frac{9}{64}$ in.

Distance *B* falls at the fifty-seventh $\frac{1}{64}$-in. mark, making the reading $\frac{57}{64}$ in.

Distance *C* falls at the thirty-third $\frac{1}{64}$-in. mark past the 1-in. graduation. The reading is $1\frac{33}{64}$ in.

Distance *D* falls at the first $\frac{1}{64}$-in. mark past the 2-in. graduation, making the reading $2\frac{1}{64}$ in.

Reading Decimal Inch Rules

Many dimensions in the auto, aircraft, and missile industries are specified in **decimal notation,** which refers to the division of the inch into 10 parts or a multiple of 10 parts, such as 50 or 100 parts. In this case, a **decimal rule** would be used. Decimal inch dimensions are specified and read as thousandths of an inch. Decimal rules, however, do not discriminate to the individual thousandth because the width of an engraved or etched divi-

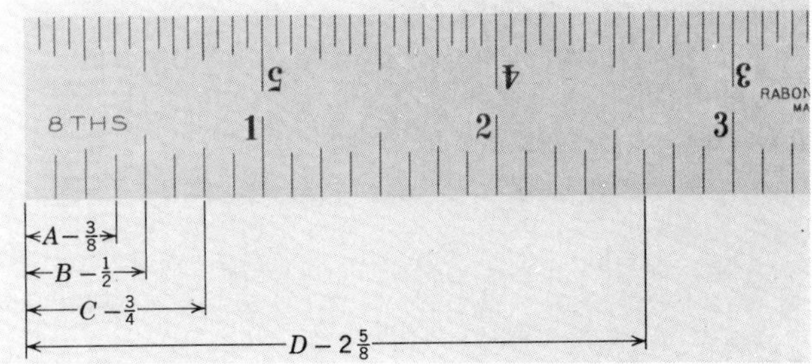

FIGURE C-66 Examples of readings on the $\frac{1}{8}$-in. discrimination edge.

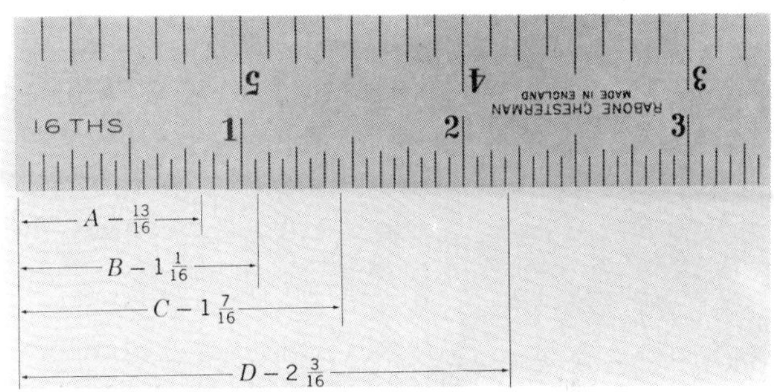

FIGURE C-67 Examples of readings on the $\frac{1}{16}$-in. discrimination edge.

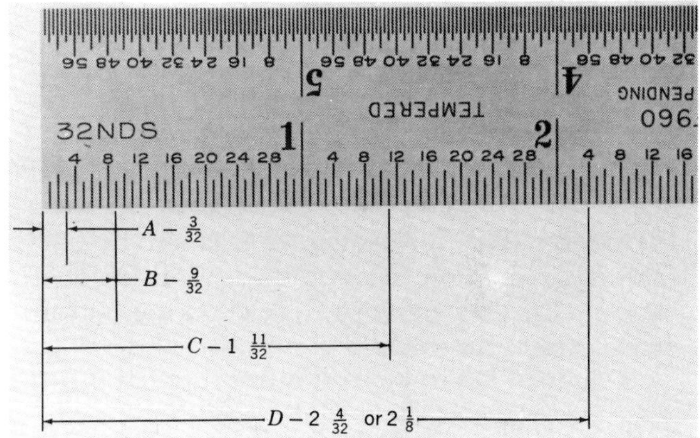

FIGURE C-68 Examples of readings on the $\frac{1}{32}$-in. discrimination edge.

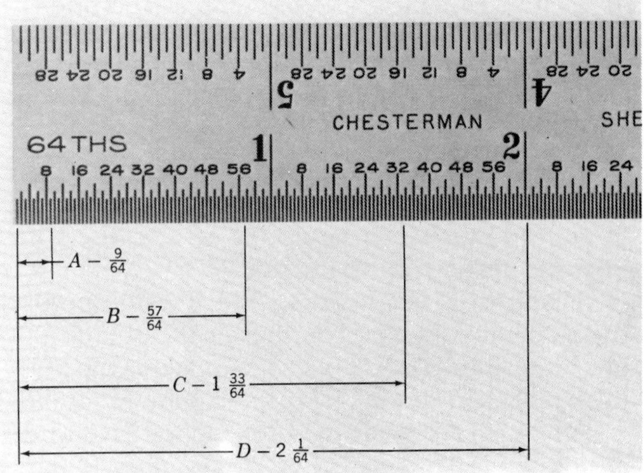

FIGURE C-69 Examples of readings on the $\frac{1}{64}$-in. discrimination edge.

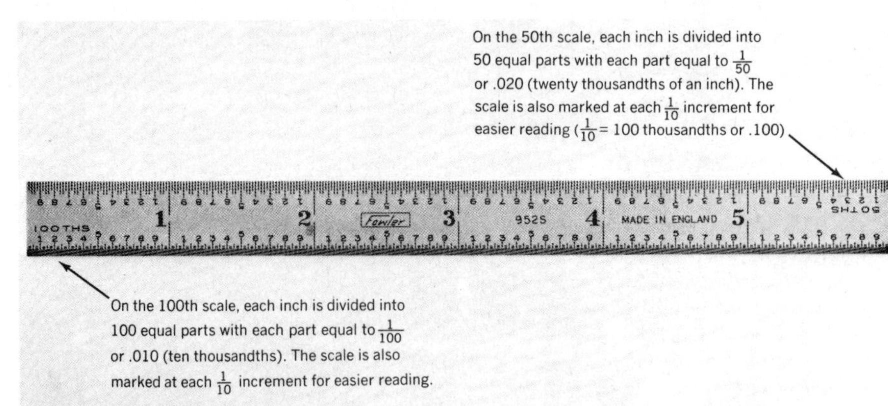

On the 50th scale, each inch is divided into 50 equal parts with each part equal to $\frac{1}{50}$ or .020 (twenty thousandths of an inch). The scale is also marked at each $\frac{1}{10}$ increment for easier reading ($\frac{1}{10}$ = 100 thousandths or .100).
On the 100th scale, each inch is divided into 100 equal parts with each part equal to $\frac{1}{100}$ or .010 (ten thousandths). The scale is also marked at each $\frac{1}{10}$ increment for easier reading.
100THS
50THS
MADE IN ENGLAND
9525

FIGURE C-70 Six-inch decimal rule.

sion on the rule is approximately .003 in. (3 thousandths of an inch). Decimal rules are commonly graduated in increments of $\frac{1}{10}$, $\frac{1}{50}$, or $\frac{1}{100}$ in.

A typical decimal rule may have $\frac{1}{50}$-in. divisions on the top edge and $\frac{1}{100}$-in. divisions on the bottom edge (Figure C-70). The inch is divided into 10 equal parts, making each numbered division $\frac{1}{10}$ in. or .100 in. (100 thousandths of an inch). On the top scale each $\frac{1}{10}$ increment is further subdivided into five equal parts, which makes the value of each of these divisions .020 in. (20 thousandths of an inch).

EXAMPLES OF DECIMAL INCH READINGS

FIGURE C-71 Distance *A* falls on the first marked graduation. The reading is $\frac{1}{10}$ or .100 thousandths in. This can also be read on the 50th-in. scale, as seen in the figure.

Distance *B* can be read only on the 100th-in. scale, as it falls at the seventh graduation beyond the .10 in. mark. The reading is .100 in. plus .070 in., or .170 in. This distance cannot be read on the 50th in. scale because discrimination of the 50th in. scale is not sufficient.

Distance *C* falls at the second mark beyond the .400 in. line. This reading is .400 in. plus .020 in., or .420 in. Since .020 in. is equal to $\frac{1}{50}$ in., this can also be read on the 50th-in. scale, as shown in the figure.

Distance *D* falls at the sixth increment beyond the .400 in. line. The reading is .400 in. plus .060 in., or .460 in. This can also be read on the 50th-in. scale, as seen in the figure.

Distance *E* falls at the sixth division beyond the .700 in. mark. The reading is .700 in. plus .060 in., or .760 in. This can also be read on the 50th-in. scale.

Distance *F* falls at the ninth mark beyond the .700 in. line. The reading is .700 in. plus .090 in., or .790 in. This cannot be read on the 50th in. scale.

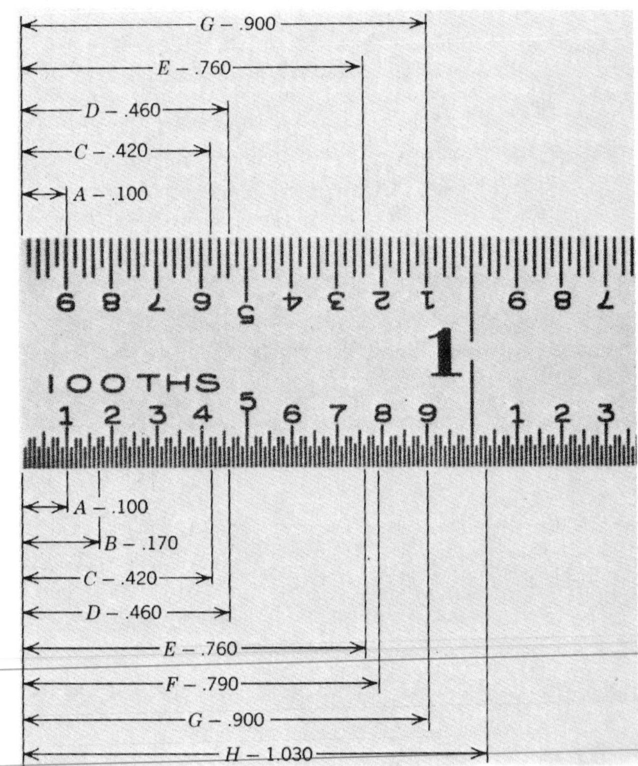

FIGURE C-71 Examples of decimal rule readings.

Distance *G* falls at the .100 graduation on top and at the .900 graduation on the bottom. The reading is .900 or $\frac{9}{10}$ of an inch.

Distance *H* falls three marks past the first full inch mark. The reading is 1.00 in. plus .030 in., or 1.030 in. This cannot be read on the 50th-in. scale.

Reading Metric Rules

Many products are made in metric dimensions requiring a machinist to use a **metric rule.** The typical metric rule has millimeter (mm) and half millimeter graduations (Figure C-72).

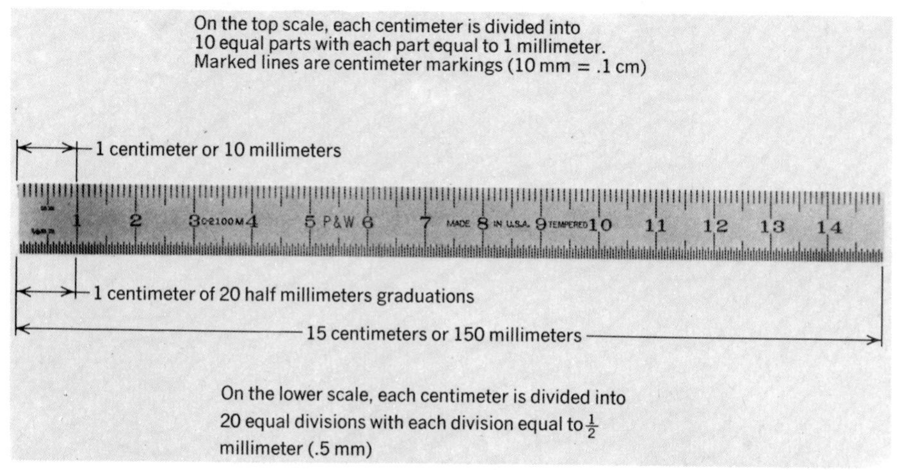

On the top scale, each centimeter is divided into
10 equal parts with each part equal to 1 millimeter.
Marked lines are centimeter markings (10 mm = .1 cm)
1 centimeter or 10 millimeters
1 centimeter of 20 half millimeters graduations
15 centimeters or 150 millimeters
On the lower scale, each centimeter is divided into
20 equal divisions with each division equal to $\frac{1}{2}$
millimeter (.5 mm)

FIGURE C-72 150-mm metric rule.

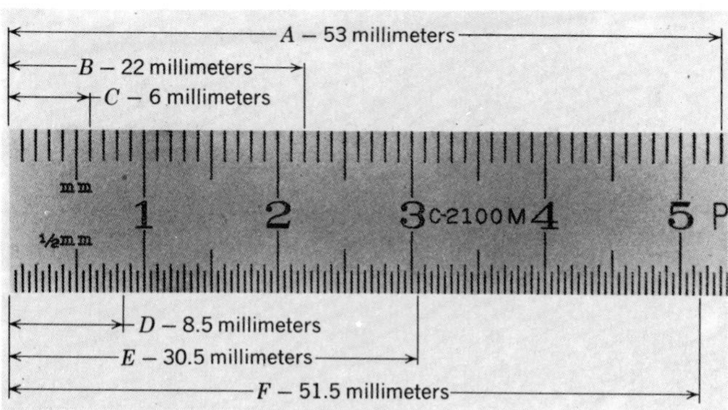

FIGURE C-73 Examples of metric rule readings.

EXAMPLES OF READING METRIC RULES

FIGURE C-73 Distance *A* falls at the fifty-third graduation on the mm scale. The reading is 53 mm.

Distance *B* falls at the twenty-second graduation on the mm scale. The reading is 22 mm.

Distance *C* falls at the sixth graduation on the mm scale. The reading is 6 mm.

Distance *D* falls at the eighteenth $\frac{1}{2}$-mm mark. The reading is 8 mm plus an additional $\frac{1}{2}$ mm, giving a total of 8.5 mm.

Distance *E* falls $\frac{1}{2}$ mm beyond the 3-centimeter (cm) graduation. Since 3 cm is equal to 30 mm, the reading is 30.5 mm.

Distance *F* falls $\frac{1}{2}$ mm beyond the 51-mm graduation. The reading is 51.5 mm. In machine design, all dimensions are specified in mm. Hence 1.5 meters (m) would be 1500 mm.

SELF-TEST READING INCH RULES

Read and record the dimensions indicated by the letter *A* to *H* in Figures C-74*a* to C-74*d*.

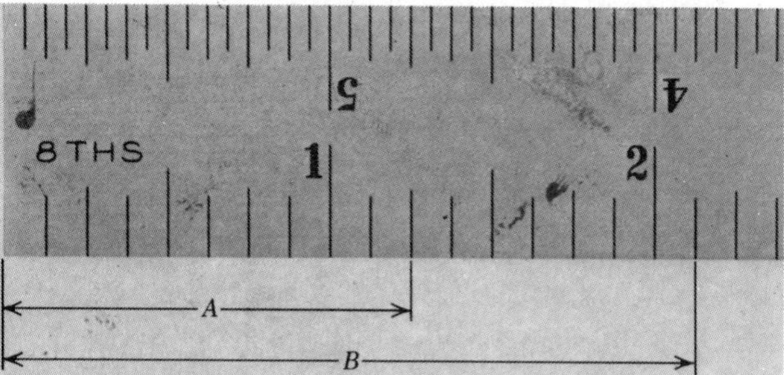

FIGURE C-74*a*

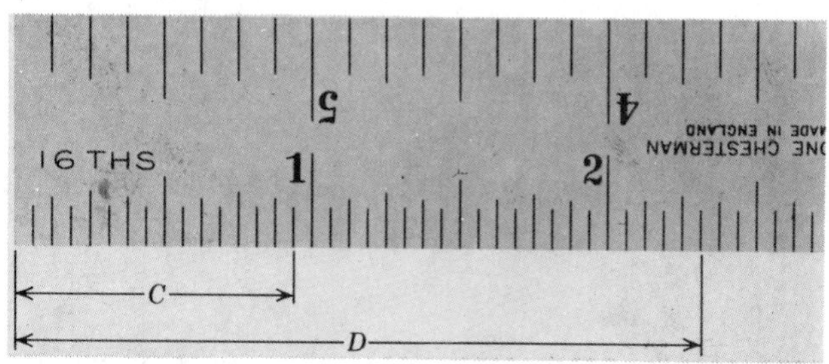

FIGURE C-74*b*

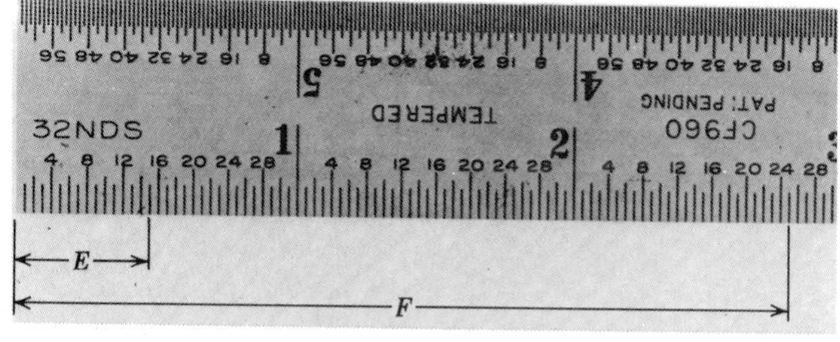

FIGURE C-74*c*

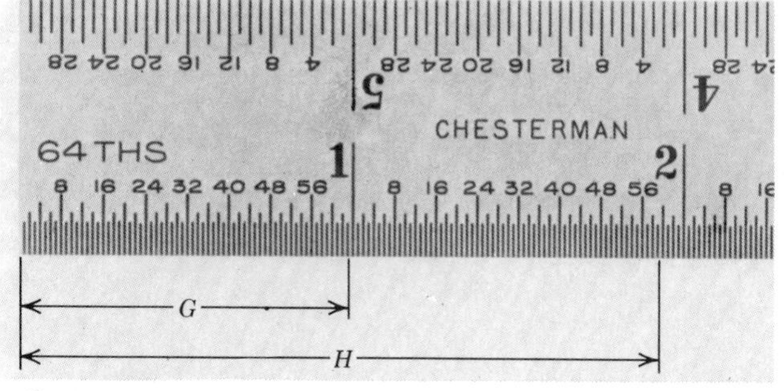

FIGURE C-74*d*

SELF-TEST READING DECIMAL RULES

Read and record the dimensions indicated by the letters *A* to *E* in Figure C-75.

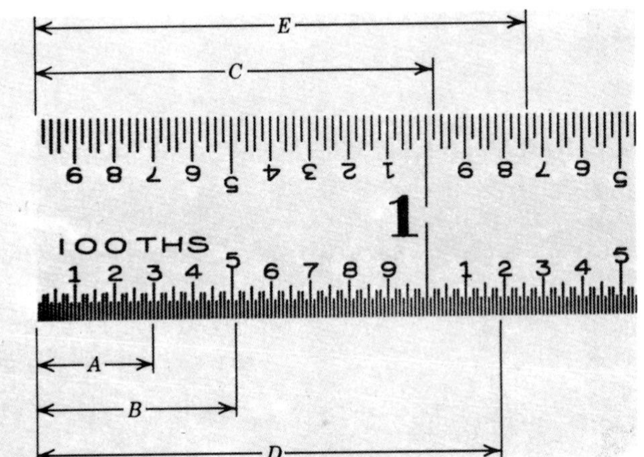

FIGURE C-75 Decimal inch rule.

SELF-TEST READING METRIC RULES

Read and record the dimensions indicated by the letters *A* to *F* in Figure C-76.

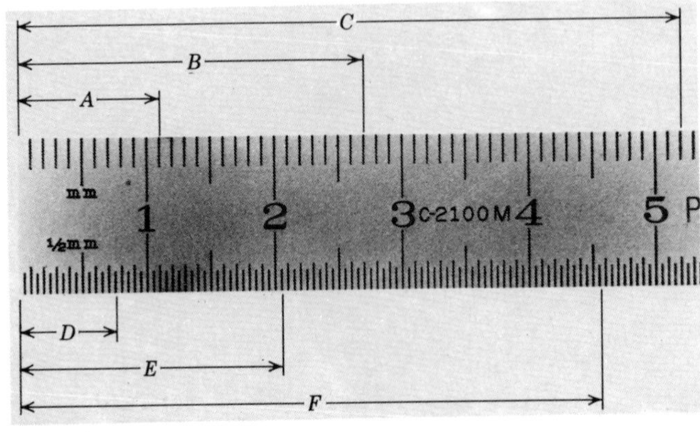

FIGURE C-76

UNIT 3

Using Vernier, Dial, and Digital Instruments for Direct Measurements

The inspection and measurement of machined parts requires various kinds of measuring tools. Often the discrimination of a rule is sufficient, but, in many cases the discrimination of a rule with a vernier scale is required. This unit explains the types, use, and applications of common vernier instruments.

Objectives

After completing this unit, you should be able to:

1. Measure and record dimensions to an accuracy of plus or minus .001 in. with a vernier caliper.
2. Measure and record dimensions to an accuracy of plus or minus .02 mm using a metric vernier caliper.
3. Measure and record dimensions using a vernier depth gage.

Use of the Vernier

For many years the vernier has been used as a device to divide the units in which a measuring tool measures into smaller increments permitting high discrimination measurement. Although the vernier is highly reliable and accurate, the nature of its function makes it somewhat difficult to read, thus requiring more time and skill on the part of the machinist. Modern precision manufacturing techniques and the wide application of digital microelectronics developed in recent years have yielded many types of high discrimination measuring instruments that demonstrate both ease of use and excellent reliability. Thus the use of the mechanical vernier has declined. As digital electronics replace more mechanical measurement, the use of the vernier may in time disappear completely. However, vernier measuring equipment is still very much in use in both school and industrial shops, and it is likely that vernier applications will be around for some time to come. The theory and applications of the vernier as a device for subdividing increments of measurement is discussed in this unit.

Principle of the Vernier

The principle of the **vernier** may be used to increase the discrimination of all graduated scale measuring tools used by a machinist. A vernier system consists of a **main scale** and a **vernier scale.** The vernier scale is placed adjacent to the main scale so that graduations on both scales can be observed together. The spacing of the vernier scale graduations is shorter than the spacing of the main scale graduations. For example, consider a main scale divided as shown (Figure C-77*a*). It is desired to further subdivide each main scale division into 10 parts with the use of a vernier. The spacing of each vernier scale division is made $\frac{1}{10}$ of a main scale division shorter than the spacing of a main scale division. This may sound confusing, but, think of it as 10 vernier scale divisions corresponding to nine main scale divisions (Figure C-77*a*). The vernier now permits the main scale to discriminate to $\frac{1}{10}$ of its major divisions. Therefore, $\frac{1}{10}$ is known as the **least count** of the vernier.

The vernier functions in the following manner. Assume that the zero line on the vernier scale is placed as shown (Figure C-77*b*). The reading on the main scale is two, plus a fraction of a division. It is desired to know the amount of the fraction over two, to the nearest tenth or least count of the vernier. As you inspect the alignment of the vernier scale and the main scale lines, you will note that they move closer together until one line on the vernier scale **coincides** with a line on the main scale. This is the **coincident line** of the vernier and indicates the fraction in tenths that must be added to the main scale reading. The vernier is coincident at the sixth line. Since the least count of the vernier is $\frac{1}{10}$, the zero vernier line is six tenths past two on the main scale. Therefore, the main scale reading is 2.6 (Figure C-77*b*).

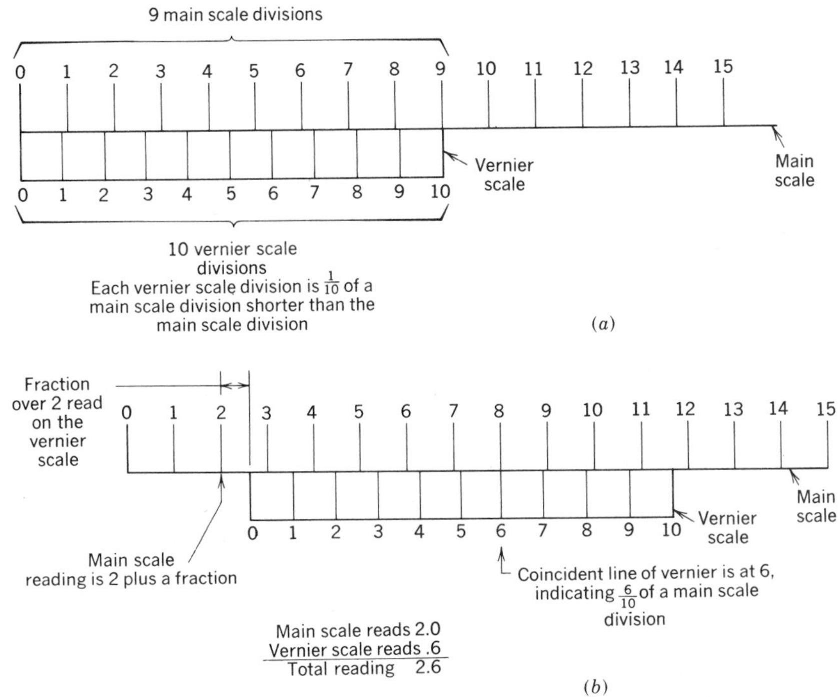

FIGURE C-77 Principle of the vernier.

Discrimination and Applications of Vernier Instruments

Vernier instruments used for linear measure in the inch system discriminate to .001 in. ($\frac{1}{1000}$). Metric verniers generally discriminate to .02 ($\frac{1}{50}$) of a millimeter.

The most common vernier instruments include several styles of **calipers.** The common vernier caliper is used for outside and inside linear measurement. Another style of vernier caliper has the capability of depth measurement in addition to outside and inside capacity. The vernier also appears on a variety of depth gages.

Beyond its most common applications, the vernier also appears on a height gage, which is an extremely important layout tool for a machinist. The vernier is also used on the gear tooth caliper, a special vernier caliper used in gear measurement. As the principle of the vernier can be used to subdivide a unit of angular measure as well as linear measure, it appears on various types of protractors used for angular measurement.

Reliability and Expectation of Accuracy in Vernier Instruments

Reliability in vernier calipers and depth gages is highly dependent on proper use of the tool. The simple fact that the caliper or depth gage has increased discrimination over a rule does not necessarily provide increased reliability. The improved degree of discrimination in vernier instruments requires more than the mere visual alignment of a rule graduation against the edge of the object to be measured. The zero reference point of a vernier caliper is the positively placed contact of the solid saw with the part to be measured. On the depth gage, the base is the zero reference point. Positive contact of the zero reference is an important consideration in vernier reliability.

The vernier scale must be read carefully if a reliable measurement is to be determined. On many vernier instruments the vernier scale should be read with the aid of a magnifier. Without this aid, the coincident line of the vernier is difficult to determine. Therefore, the reliability of the vernier readings can be in question. The typical vernier caliper has very narrow jaws and thus must be carefully aligned with the axis of measurement. On the plain slide vernier caliper, no provision is made for the "feel" of the measuring pressure. Some calipers and the depth gage are equipped with a screw thread fine adjustment that gives them a slight advantage in determining the pressure applied during the measurement.

Generally, the overall reliability of vernier instruments for measurement at maximum discrimination of .001 is fairly low. The vernier should never be used in an attempt to discriminate below .001. The instrument does not have that capability. Vernier instruments are a popular tool on the inspection bench, and they can serve very well for measurement in the range of plus or minus .005 of an inch. With proper use and an understanding of the limitations of a vernier instrument, this tool can be a valuable addition to the many measuring tools available to you.

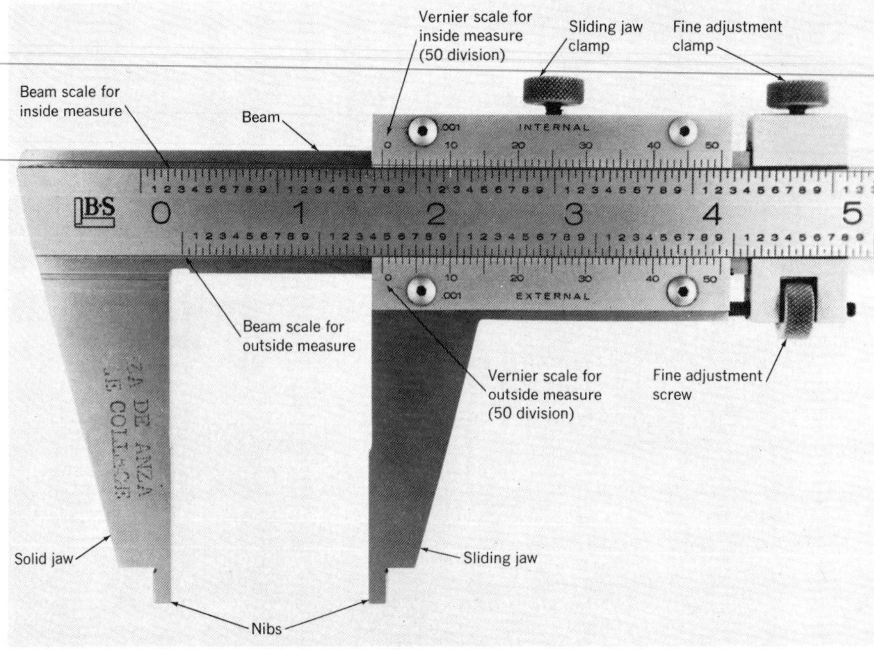

FIGURE C-78 Typical inside–outside, 50-division vernier caliper.

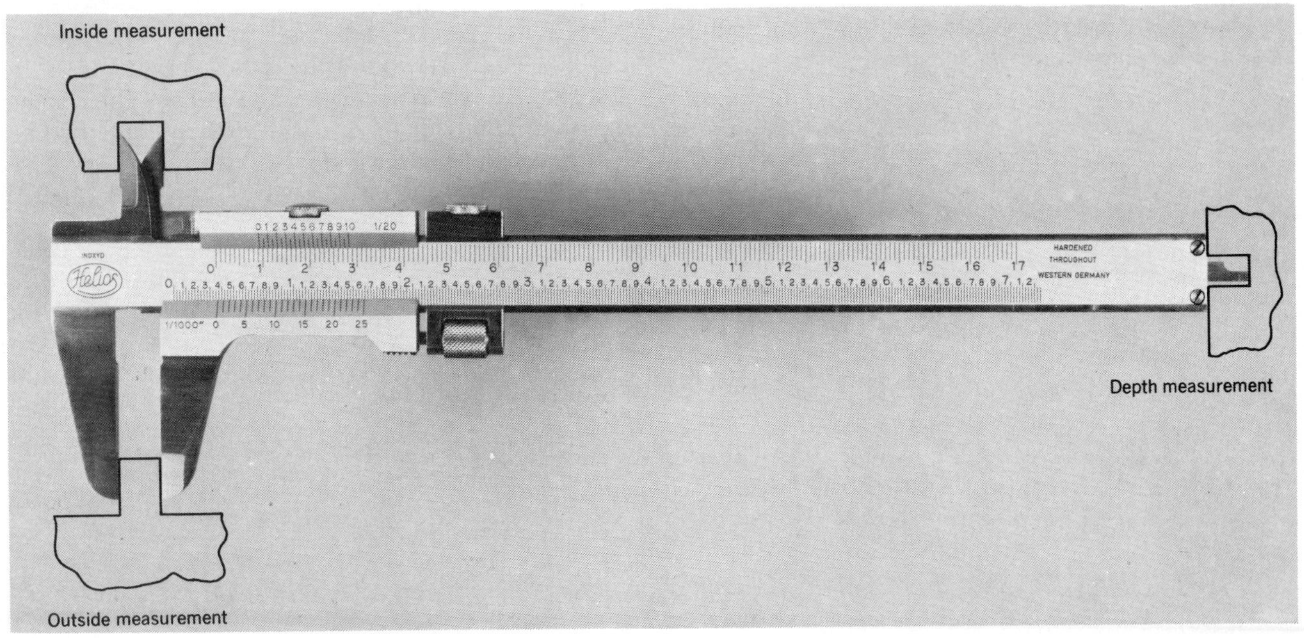

FIGURE C-79 Common vernier caliper.

Vernier Calipers

With a rule, measurements can be made to the nearest $\frac{1}{64}$ or $\frac{1}{100}$ in., but often this is not sufficiently accurate. A measuring tool based on a rule but with much greater discrimination is the **vernier caliper**. Vernier calipers have a discrimination of .001 in. The **beam** or **bar** is engraved with the **main scale.** This is also called the **true scale,** as each inch marking is exactly one inch apart. The beam and the solid jaw are square, or at 90 degrees to each other.

The movable jaw contains the **vernier scale.** This scale is located on the sliding jaw of a vernier caliper or it is part of the base on the vernier depth gage. The function of the vernier scale is to subdivide the minor divisions on the beam scale into the smallest increments that the vernier instrument is capable of measuring. For example, a 25-division vernier subdivides the minor divisions of the beam scale into 25 parts. Since the minor divisions are equal to .025 thousandths of an inch, the vernier divides them into increments of .001 of an inch. This is the finest discrimination of the instrument.

Most of the longer vernier calipers have a fine adjustment clamp for precise adjustments of the movable jaw. Inside measurements are made over the nibs on the jaw and are read on the top scale of the vernier caliper (Figure C-78). The top scale is a duplicate of the lower scale, with the exception that is is offset to compensate for the size of the nibs.

The standard vernier caliper is very common (Figure C-79). This is a versatile tool because of its capacity to make outside, inside, and depth measurements. Many different measuring applications are made with this particular design of vernier caliper (Figure C-80).

Vernier Caliper Procedures

To test a vernier caliper for accuracy, clean the contact surfaces of the two jaws. Bring the movable jaw with normal gaging pressure into contact with the solid jaw. Hold the caliper against a light source and examine the alignment of the solid and

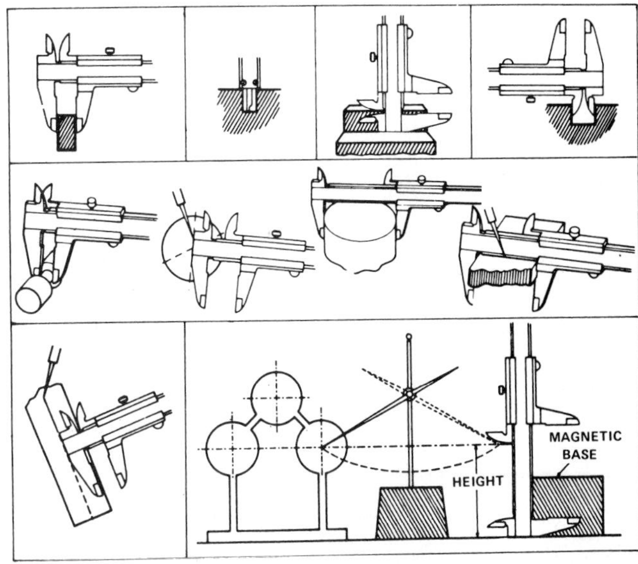

FIGURE C-80 The design of vernier caliper has many applications (M.T.I. Corporation).

movable jaws. If wear exists, a line of light will be visible between the jaw faces. A gap as small as .0001 ($\frac{1}{10,000}$) of an inch can be seen against a light. If the contact between the jaws is satisfactory, check the vernier scale alignment. The vernier scale zero mark should be in alignment with the zero on the main scale. Realignment of the vernier scale to adjust it to zero can be accomplished on some vernier calipers.

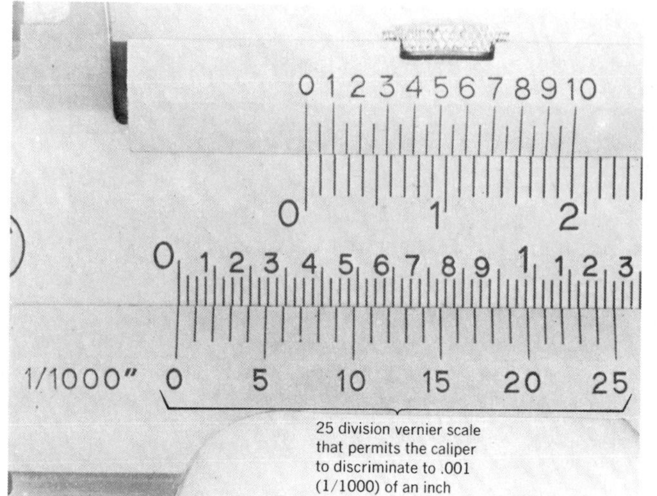

FIGURE C-81 Lower scale is a 25-division vernier.

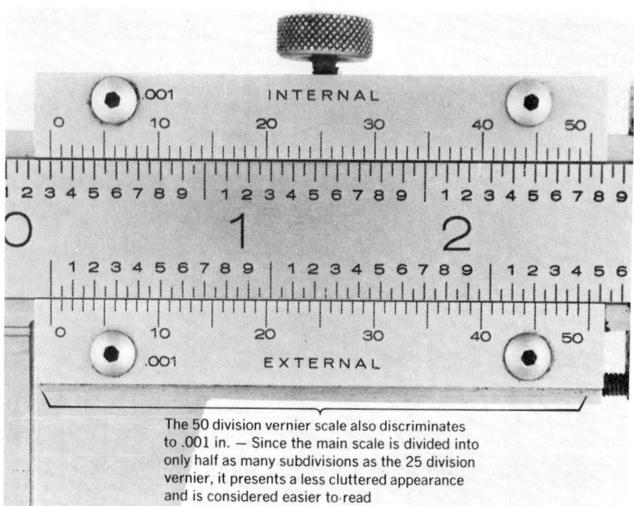

INTERNAL
.001
EXTERNAL
.001
The 50 division vernier scale also discriminates to .001 in. – Since the main scale is divided into only half as many subdivisions as the 25 division vernier, it presents a less cluttered appearance and is considered easier to read

FIGURE C-82 Fifty-division vernier caliper.

Point of coincidence of vernier and main scale graduations

1 full inch

.100

.100

.025

1/20

1/1000"

13 vernier divisions corresponding to .013 thousandths of an inch

25 division vernier scale

1.000	Full inch read on the main scale
.100	Thousandths read on the main scale (1 major division)
.100	Thousandths read on the main scale (1 major division)
.025	Thousandths read on the main scale (1 minor division)
.013	Thousandths read on the vernier scale (13 vernier divisions)
1.238	Total vernier caliper reading

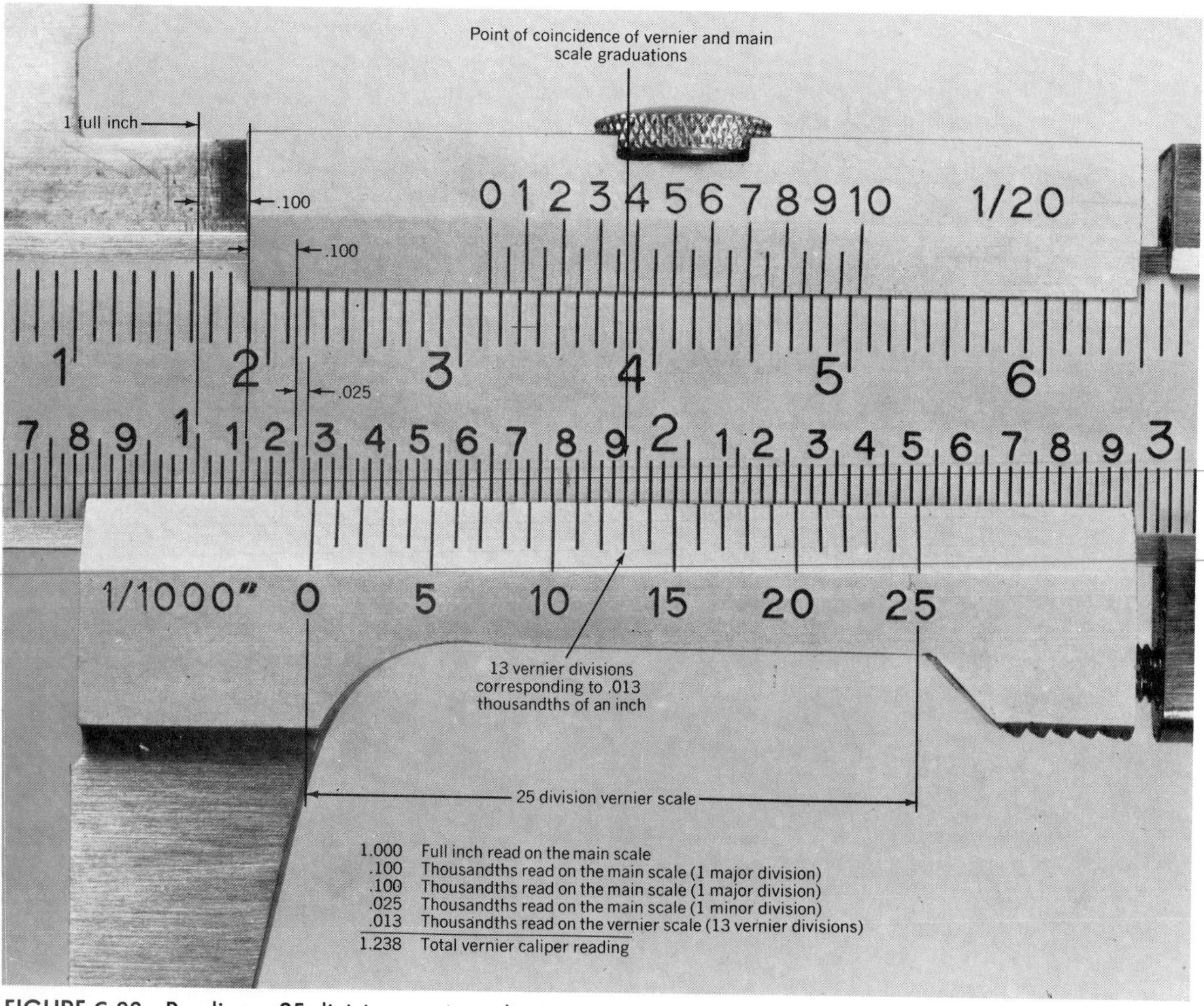

FIGURE C-83 Reading a 25-division vernier caliper.

A vernier caliper is a delicate precision tool and should be treated as such. It is very important that the correct amount of pressure or feel is developed while taking a measurement. The measuring jaws should contact the workpiece firmly. However, excessive pressure will spring the jaws and give inaccurate readings. When measuring an object, use the solid jaw as the reference point. Then move the sliding jaw until contact is made. When measuring with the vernier caliper make certain that the beam of the caliper is in line with the surfaces being measured. Whenever possible, read the vernier caliper while it is still in contact with the workpiece. Moving the instrument may change the reading. Any measurement should be taken at least twice to assure reliability.

Reading Inch Vernier Calipers

Vernier scales are engraved with 25 or 50 divisions (Figures C-81 and C-82). On a 25-division vernier caliper, each inch on the main scale is divided into 10 major divisions numbered from 1 to 9. Each major division is .100 (one hundred-thousandth). Each major division has four subdivisions with a spacing of .025 (twenty-five thousandths). The vernier scale has 25 divisions with the zero line being the index.

To read the vernier caliper, count all of the graduation to the left of the index line. This would be 1 whole inch plus $\frac{2}{10}$ or .200, plus 1 subdivision valued at .025, plus part of one subdivision (Figure C-83). The value of this partial subdivision is determined by the coincidence of one line on the vernier scale with one line of the true scale. For this example, the coincidence is on line 13 of the vernier scale. This is the value in thousandths of an inch that has to be added to the value read on the beam. Therefore, 1 + .100 + .100 + .025 + .013 equals the total reading of 1.238. An aid in determining the coincidental line is that the **lines adjacent to the coincidental line fall inside the lines on the true scale** (Figure C-84).

The 50-division vernier caliper is read as follows (Figure C-85). The true scale has each inch divided into 10 major divisions of .100 in. each, with each major division subdivided in half, thus

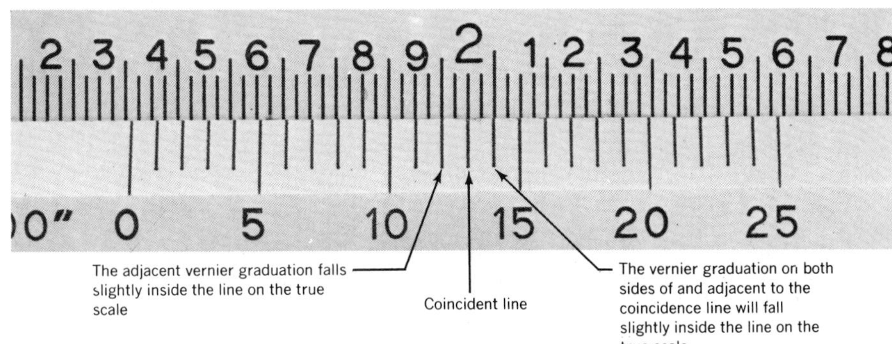

FIGURE C-84 Determining the coincident line on a vernier.

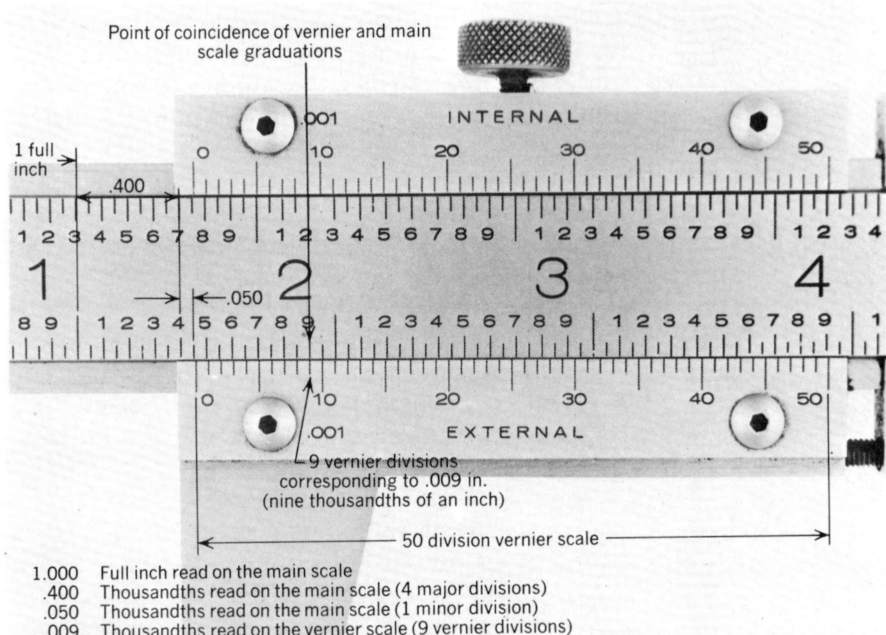

FIGURE C-85 Reading a 50-division vernier.

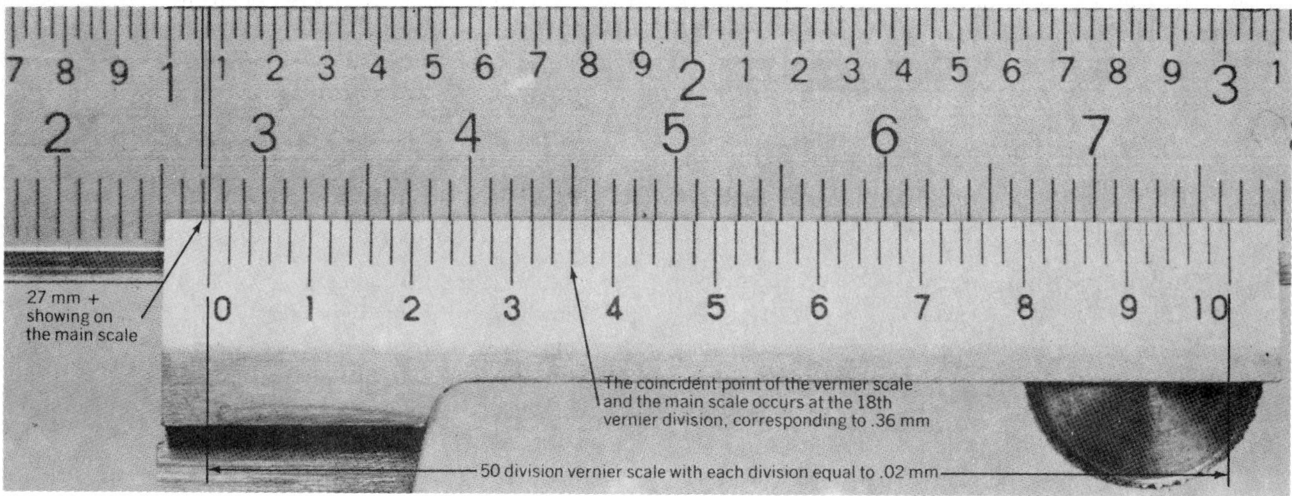

FIGURE C-86 Reading a metric vernier caliper with .02-mm discrimination.

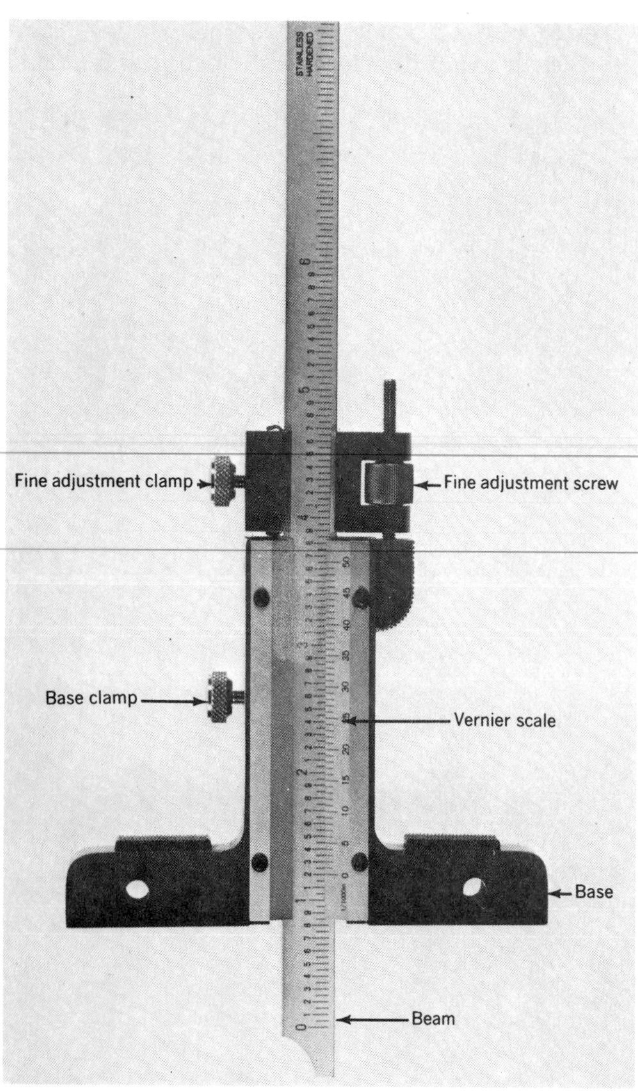

FIGURE C-87 Vernier depth gage with .001-in. discrimination.

being .050 in. The vernier scale has 50 divisions. The 50 division vernier caliper reading shown is read as follows:

Beam whole inch reading	1.000
Additional major divisions	.400
Additional minor divisions	.050
Vernier scale reading	.009
Total caliper reading	1.459

Reading Metric Vernier Calipers

The applications for a metric vernier caliper are exactly the same as those described for an inch system vernier caliper. The discrimination of metric vernier caliper models varies from .02 mm, .05 mm, or .1 mm. The most commonly used type discriminates to .02 mm. The main scale on a metric vernier caliper is divided into millimeters with every tenth millimeter mark numbered. The 10-mm line is numbered 1, the 20-mm line is numbered 2, and so on, up to the capacity of the tool (Figure C-86). The vernier scale on the sliding jaw is divided into 50 equal spaces with every fifth space numbered. Each numbered division on the vernier represents one tenth of a millimeter. The five smaller divisions between the numbered lines represent two hundredths (.02 mm) of a millimeter.

To determine the caliper reading, read, on the main scale, whole millimeters to the left of the zero or the index line of the sliding jaw. The example (Figure C-86) shows 27 mm plus part of an additional millimeter. The vernier scale coincides with the main scale at the eighteenth vernier division.

Since each vernier scale spacing is equal to .02 mm, the reading on the vernier scale is equal to 18 times .02, or .36 mm. Therefore, .36 mm must be added to the amount showing on the main scale to obtain the final reading. The result is equal to 27 mm + .36 mm, or 27.36 mm (Figure C-86).

Reading Vernier Depth Gages

These measuring tools are designed to measure the depth of holes, recesses, steps, and slots. Basic parts of a vernier depth gage include the base or anvil with the vernier scale and the fine adjustment screw (Figure C-87). Also shown is the graduated beam or bar that contains the true scale. To make accurate measurements the reference surface must be flat and free from nicks and burrs. The base should be held firmly against the reference surface while the beam is brought in contact with the surface being measured. The measuring pressure should approximately equal the pressure exerted when making a light dot on a piece of paper with a pencil. On a vernier depth gage, dimensions are read in the same manner as on a vernier caliper.

SELF-TEST—READING INCH VERNIER CALIPERS

Determine the dimensions in the vernier caliper illustrations (Figures C-88*a* and C-88*b*).

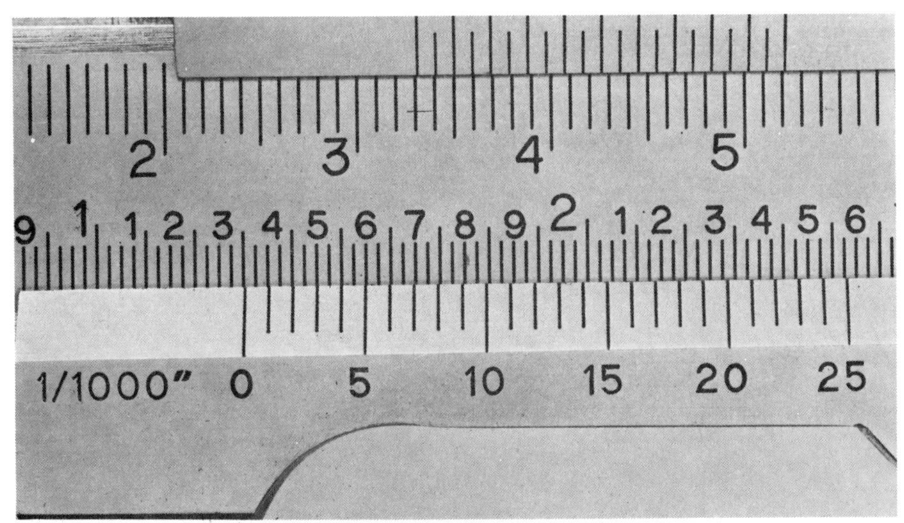

FIGURE C-88*a*

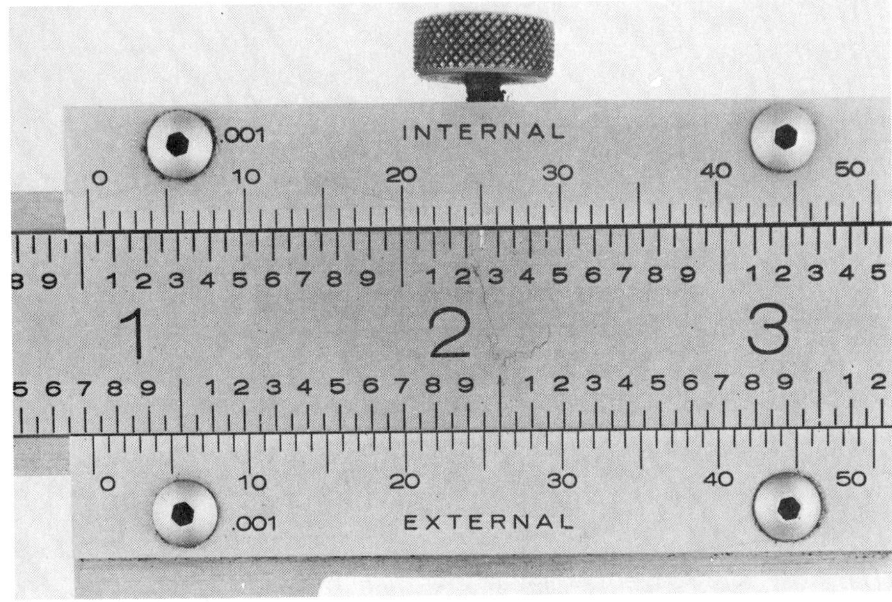

FIGURE C-88*b*

SELF-TEST—READING METRIC VERNIER CALIPERS

Determine the metric vernier caliper dimensions illustrated in Figures C-89*a* and C-89*b*.

SELF-TEST—READING VERNIER DEPTH GAGES

Determine the depth measurement illustrated in Figures C-90*a* and C-90*b*.

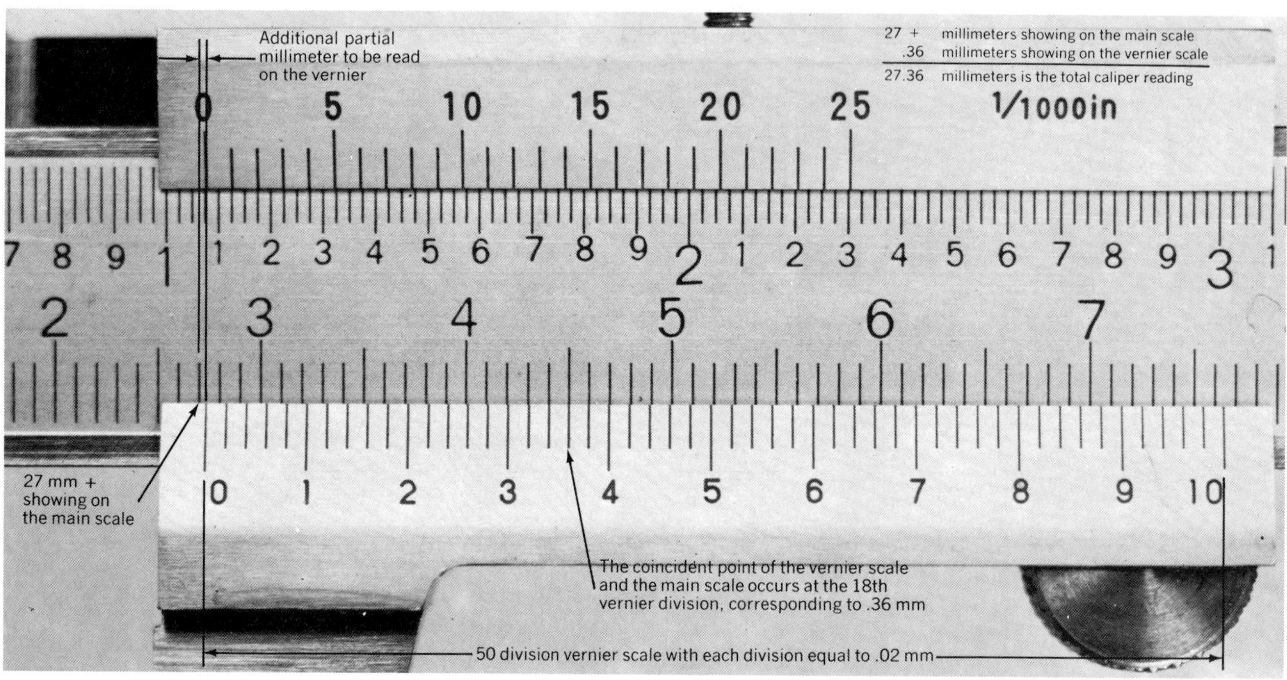

FIGURE C-89*a*

FIGURE C-89*b*

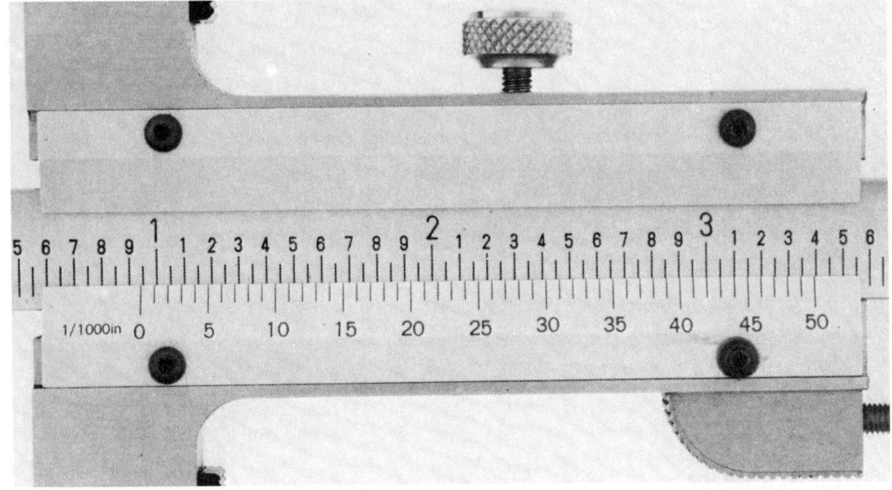

FIGURE C-90*a*

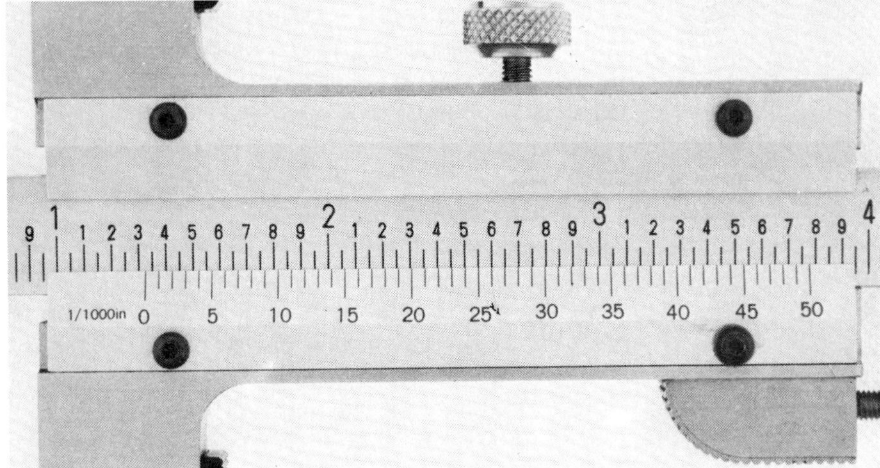

FIGURE C-90*b*

Mechanical Dial Instruments for Direct Measurement

In recent years, many types of mechanical dial measuring instruments have come into wide use. These instruments are direct reading and do not require any interpretation of a vernier. They are extremely reliable and easy to read. Mechanical dial instruments and their electronic digital counterparts have become the industry standard and have replaced vernier instruments in many measurement applications. Common examples of these instruments include the dial caliper and the dial depth gage.

FIGURE C-91 Dial caliper (Harry Smith & Associates).

DIAL CALIPER

An outgrowth from the vernier caliper is the **dial caliper** (Figure C-91). However, this instrument does not employ the principle of the vernier. The beam scale on the dial caliper is graduated only into .10-in. increments. The caliper dial is graduated into either 100 or 200 divisions. The dial hand is operated by a pinion gear that engages a rack on the caliper beam. On the 100-division dial, the hand makes one complete revolution for each .10-in. movement of the sliding jaw along the

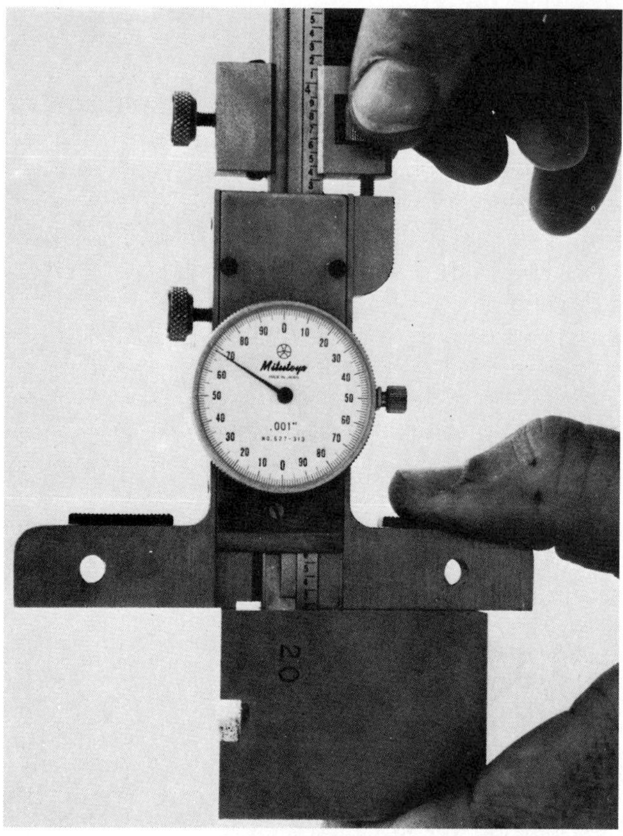

FIGURE C-92 Dial depth gage (Harry Smith & Associates).

beam. Therefore, each dial graduation represents $\frac{1}{100}$ of .10 in., or .001 in. maximum discrimination. On the 200-division dial the hand makes only one-half a revolution for each .10 in. of movement along the beam. Discrimination is also .001 in.

Since the dial caliper is direct reading, the need to determine the coincident line of a vernier scale is eliminated. This greatly facilitates reading of the instruments and, for this reason, the dial caliper has all but replaced its vernier counterpart in many applications. When using the dial caliper, remember what you have learned about the expectation of accuracy in caliper instruments.

DIAL DEPTH GAGES

As with vernier calipers, vernier depth gages have their dial counterparts (Figure C-92). The dial depth gage functions in the same manner as the dial caliper. Readings are direct without the need to use a vernier scale. The dial depth gage has the capacity to measure over several inches of range, depending on the length of the beam. Discrimination is .001 in.

Another type of dial depth gage uses a dial indicator (Figure C-93). However, the capacity and discrimination of this instrument is dependent on the range and discrimination of the dial indicator used. The tool is used primarily in comparison measuring applications.

FIGURE C-93 Dial indicator depth gage (L.S. Starrett Company).

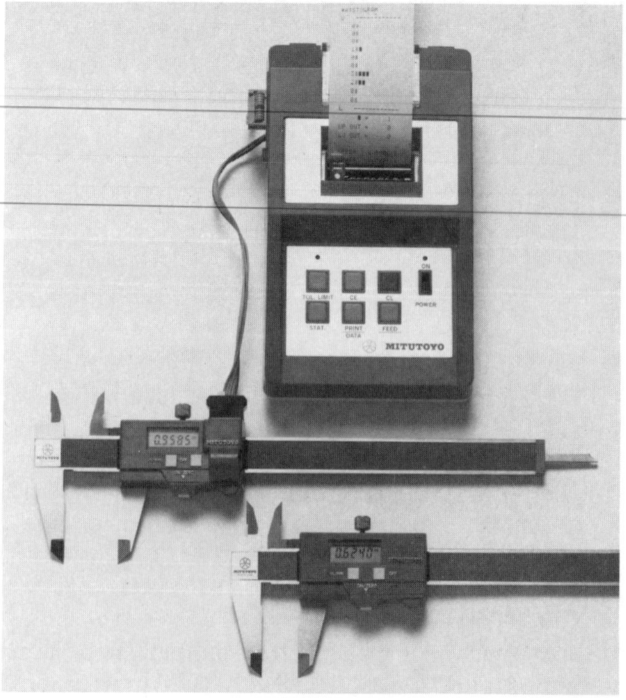

FIGURE C-94 Electronic calipers with computer and printer for listing and graphing measurements (MTI Corporation).

DIGITAL ELECTRONIC CALIPERS

Integrated circuit microelectronics has been revolutionary in the latest generation of precision measuring instruments. An example is the electronic digital readout caliper (Figure C-94). When this instrument is coupled to a computer, printouts may be obtained and a graph (histogram) showing how the measurements are distributed may be generated. This makes the instruments versatile for inspection purposes. As the sliding jaw moves along the beam, the position is shown on the digital display. The display may be set to zero at any point and may also be switched for inch and metric measurement.

UNIT 4

Using Micrometer Instruments

Micrometer measuring instruments are the most commonly used precision measuring tools found in industry. Correct use of them is essential to anyone engaged in making or inspecting machined parts.

Objectives

After completing this unit, with the use of appropriate measuring kits, you should be able to:

1. Measure and record dimensions using outside micrometers to an accuracy of plus or minus .001 of an inch.
2. Measure and record diameters to an accuracy of plus or minus .001 in. using an inside micrometer.
3. Measure and record depth measurements using a depth micrometer to an accuracy of plus or minus .001 in.
4. Measure and record dimensions using a metric micrometer to an accuracy of plus or minus .01 mm.
5. Measure and record dimensions using a vernier micrometer to an accuracy of plus or minus .0001 in. (assuming proper measuring conditions).

Types of Micrometer Instruments

The common types of micrometer instruments, **outside, inside,** and **depth,** are discussed in detail within this unit. The micrometer appears in many other forms in addition to these common types.

BLADE MICROMETER

The blade micrometer (Figure C-95), so-called because of its thin spindle and anvil, is used to mea-

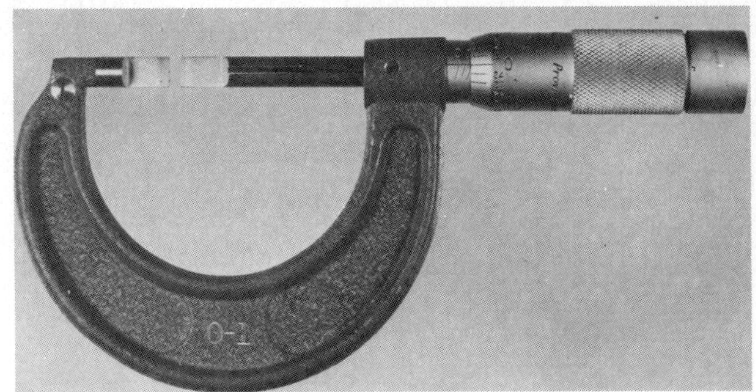

FIGURE C-95 Blade micrometer.

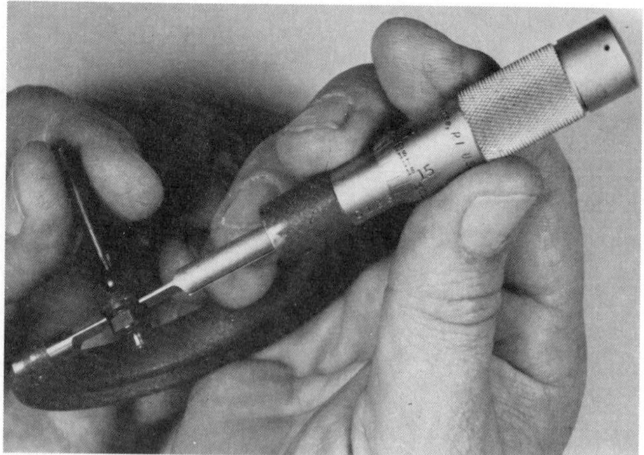
FIGURE C-96 Blade micrometer measuring a groove.

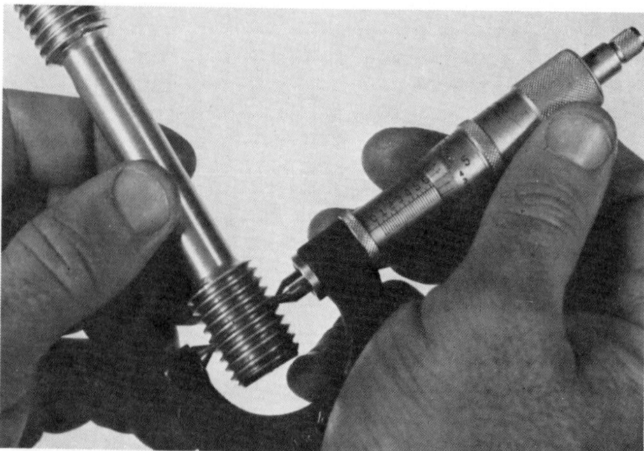
FIGURE C-100 Screw thread comparison micrometer measuring a screw thread.

sure narrow slots and grooves (Figure C-96) where the standard micrometer spindle and anvil could not be accommodated because of their diameter.

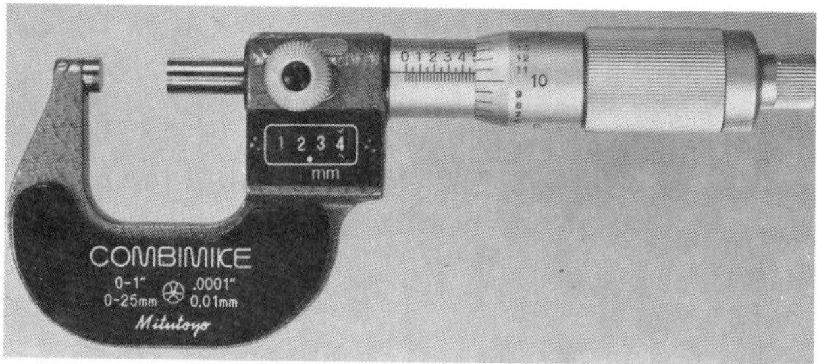

FIGURE C-97 Combination inch/metric micrometer.

COMBINATION METRIC/INCH OR INCH/METRIC MICROMETER

The combination micrometer (Figure C-97) is designed for dual system use in metric and inch measurement. The tool has a digital reading scale for one system while the other system is read from the sleeve and thimble.

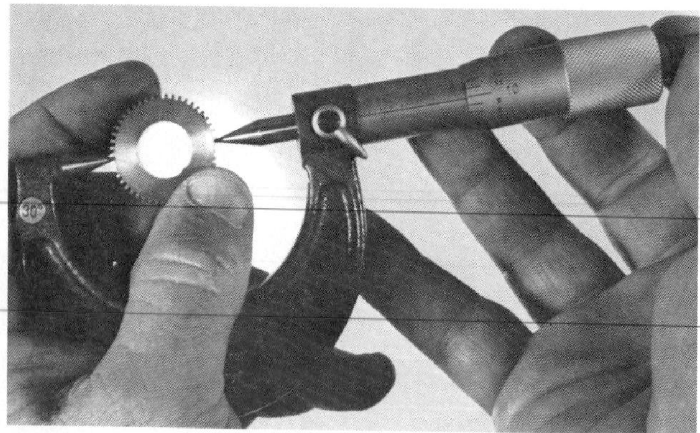

FIGURE C-98 Thirty-degree point comparator micrometer.

POINT MICROMETER AND COMPARATOR MICROMETER

The point micrometer (Figure C-98) is used in applications where limited space is available or where it might be desired to take a measurement at an exact location. Several point angles are available. The 60-degree comparator micrometer (Figure C-99) is usually called a screw thread comparator micrometer. It is most often used to compare screw threads to some known standard like a thread plug gage (Figure C-100).

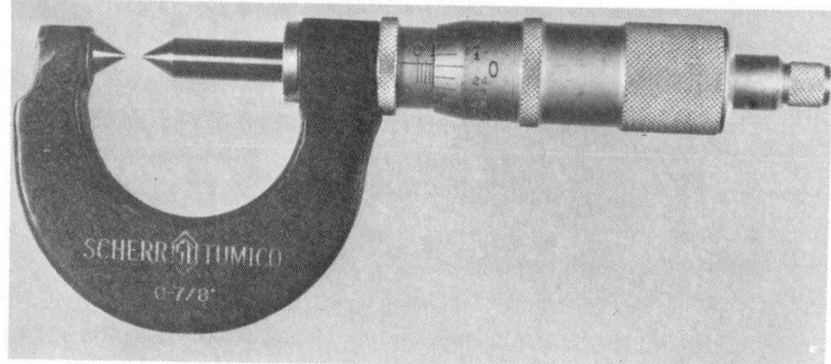

FIGURE C-99 Screw thread comparison micrometer.

DISC MICROMETER

The disc micrometer (Figure C-101) finds application in measuring thin materials such as paper where a measuring face with a large area is needed. It is also useful for such measurements as the one shown in the figure where the distance from the slot to the edge is to be determined.

DIRECT-READING MICROMETER

The direct-reading micrometer, which is also known as a high-precision micrometer, reads directly to .0001 ($\frac{1}{10,000}$) of an inch (Figure C-102).

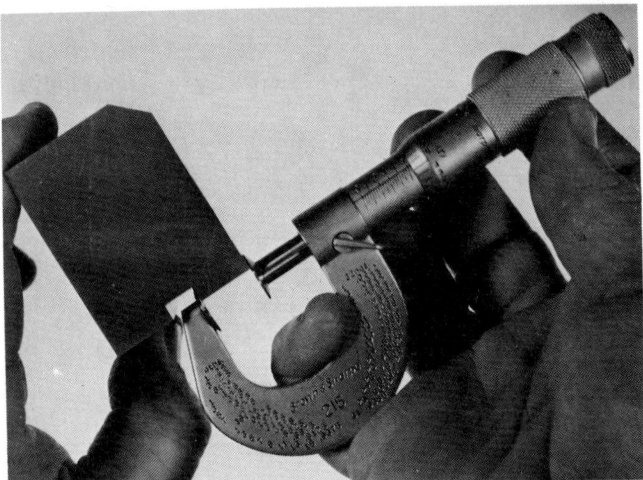

FIGURE C-101 Disc micrometer measuring slot to edge distance.

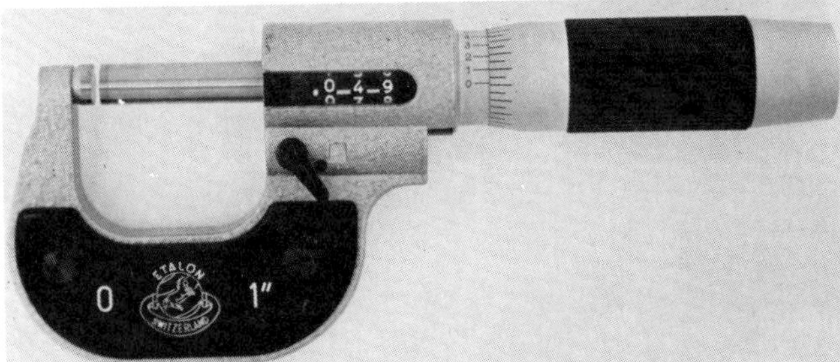

FIGURE C-102 Direct-reading digit micrometer (Harry Smith & Associates).

HUB MICROMETER

The frame of the hub micrometer (Figure C-103) is designed so that the instrument may be put through a hole or bore in order to measure the hub thickness of a gear or sprocket (Figure C-104).

INDICATING MICROMETER

The indicating micrometer (Figure C-105) is useful in inspection applications where a determination of acceptable tolerance is to be made. The instrument has an indicating mechanism built into the frame that permits a dial reading discriminating to .0001 of an inch. When an object is measured, the size deviation above or below the micrometer setting will be indicated on the dial. The indicating dial usually has a range of plus or minus .001 of an inch.

INSIDE MICROMETER CALIPER

The inside micrometer caliper (Figure C-106) has jaws that resemble those on a vernier caliper. This instrument is designed for inside measurement. Thus, the versatility of the caliper and the reliability of the micrometer are combined.

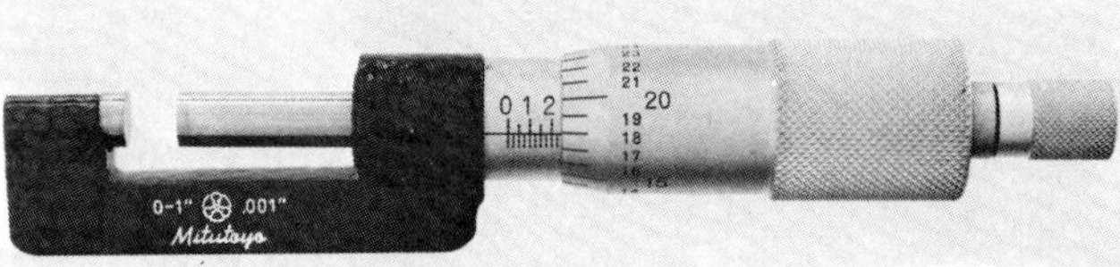

FIGURE C-103 Hub micrometer.

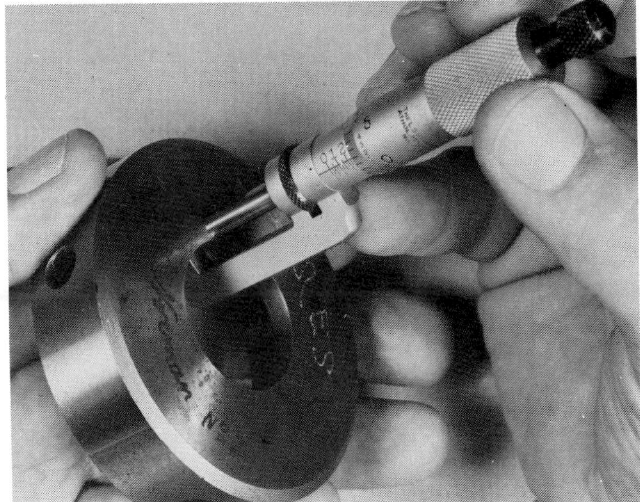

FIGURE C-104 Hub micrometer measuring through a bore.

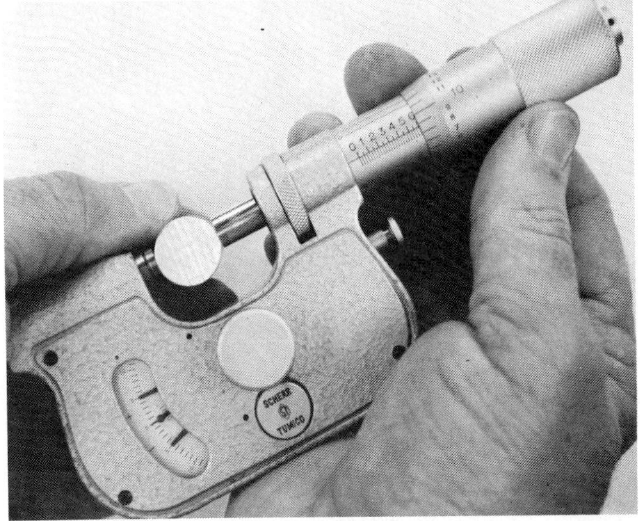

FIGURE C-105 Indicating micrometer (Harry Smith & Associates).

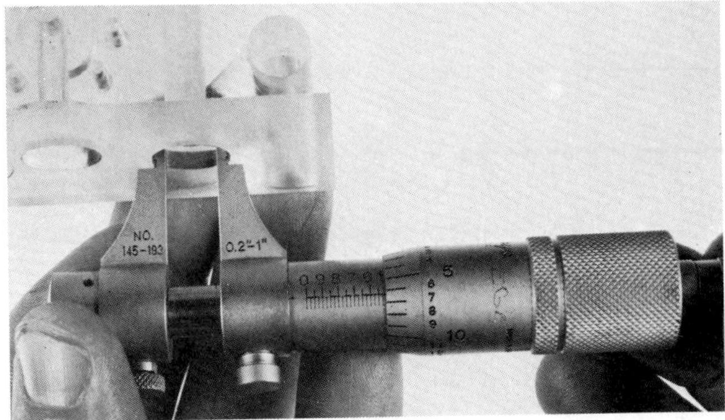

FIGURE C-106 Inside micrometer caliper.

FIGURE C-107 Internal micrometer (Harry Smith & Associates).

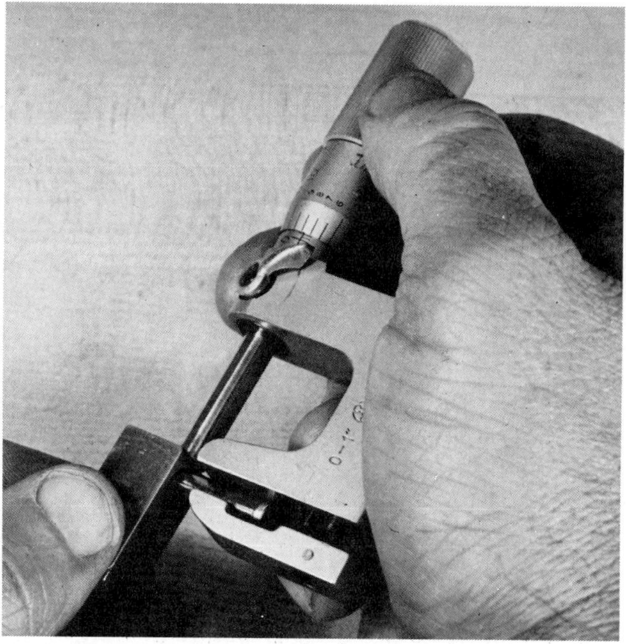

FIGURE C-109 Interchangeable anvil micrometer with pin anvil.

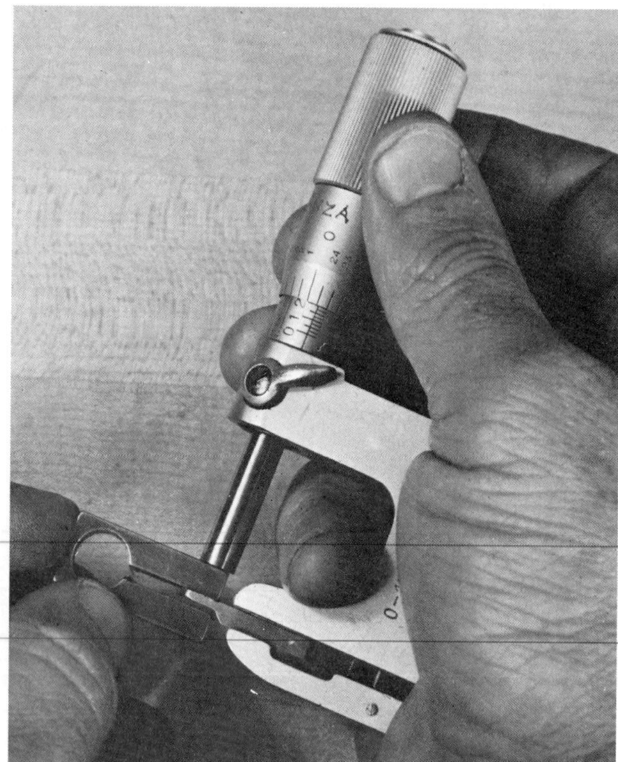

FIGURE C-108 Interchangeable anvil micrometer with flat anvil.

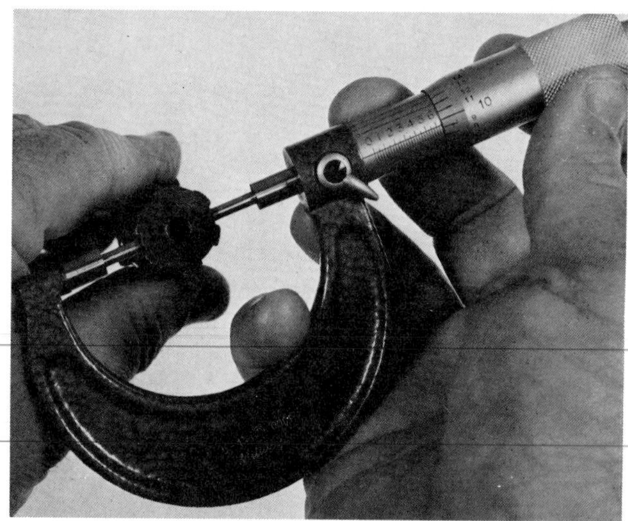

FIGURE C-110 Spline micrometer.

INTERNAL MICROMETER

The internal micrometer (Figure C-107) uses a three-point measuring contact system to determine the size of a bore or hole. The instrument is direct reading and is more likely to yield a reliable reading because its three-point measuring contacts make the instrument self-centering as compared to a tool making use of only two contacts.

INTERCHANGEABLE ANVIL-TYPE MICROMETER

The interchangeable anvil-type micrometer is often called a multi-anvil micrometer. It can be used in a variety of applications. A straight anvil is used to measure into a slot (Figure C-108). A cylindrical anvil may be used for measuring into a hole (Figure C-109). Variously shaped anvils may be clamped into position to meet special measuring requirements.

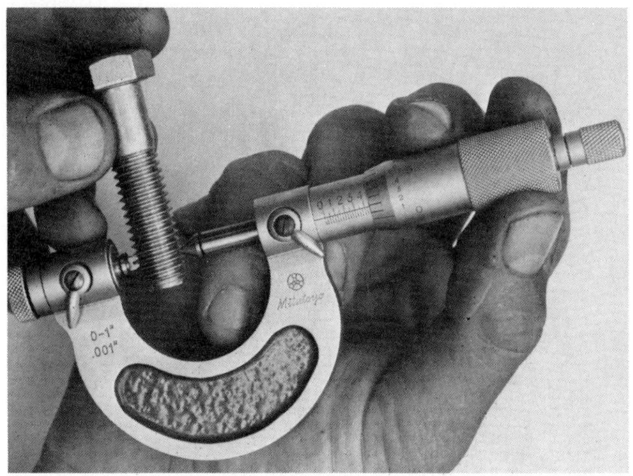

FIGURE C-111 Screw thread micrometer.

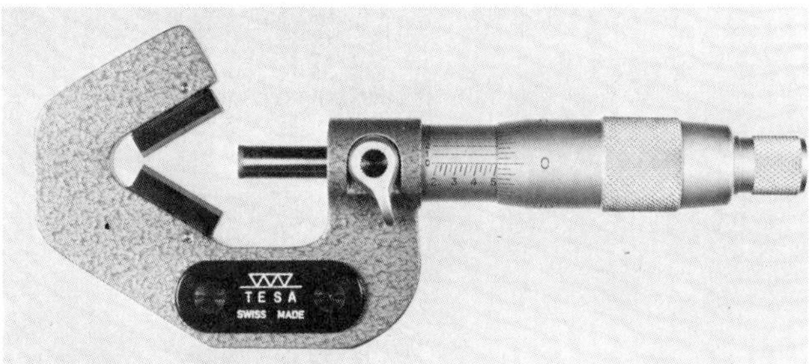

FIGURE C-112 V-anvil micrometer.

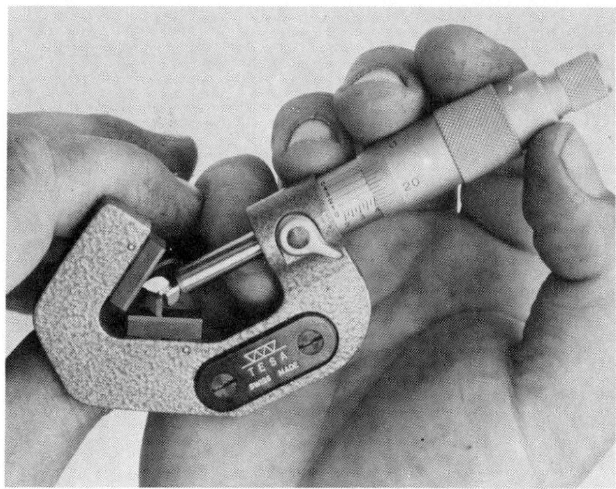

FIGURE C-113 V-anvil micrometer measuring a three-fluted end mill.

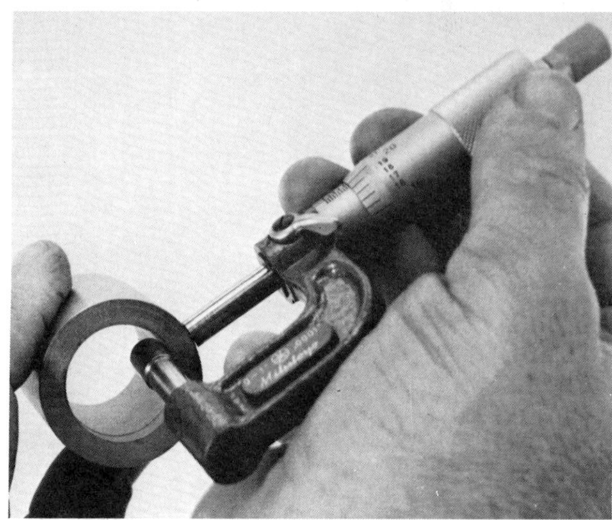

FIGURE C-114 Tubing micrometer measuring a tube wall.

SPLINE MICROMETER

The spline micrometer (Figure C-110) has a small diameter spindle and anvil. The length of the anvil is considerably longer than that of the standard micrometer, and the frame of the instrument is also larger. This type of micrometer is well suited to measuring the minor diameter of a spline.

SCREW THREAD MICROMETER

The screw thread micrometer (Figure C-111) is specifically designed to measure the pitch diameter of a screw thread. The anvil and spindle tips are shaped to match the form of the thread to be measured.

V-ANVIL MICROMETER

The V-anvil micrometer (Figure C-112) is used to measure the diameter of an object with odd-numbered symmetrical or evenly spaced features. They are designed for specific numbers of these features. The type shown is for three-sided objects like the three-fluted end mill being measured (Figure C-113). This design is also very useful in checking out-of-round conditions in centerless grinding that cannot be determined with a conventional outside micrometer caliper. The next most common type of V-anvil micrometer is for five-fluted tools.

TUBING MICROMETER

One type of tubing micrometer has a vertical anvil with a cylindrically shaped tip. Another design is like the ordinary micrometer caliper except that the anvil is a half sphere instead of a flat surface. This instrument is designed to measure the wall thickness of tubing (Figure C-114). The tubing micrometer can also be applied in other applications

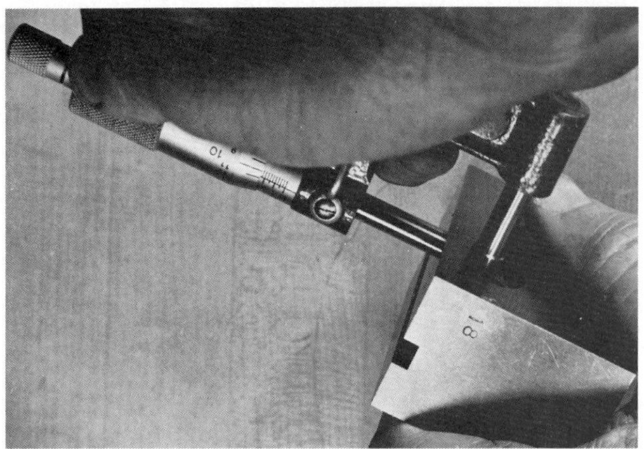

FIGURE C-115 Tubing micrometer measuring hole to edge distance.

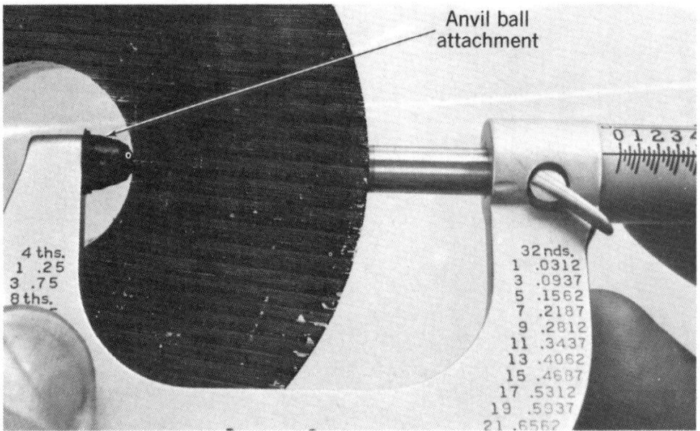

FIGURE C-116 Ball attachment for tubing measurement (Harry Smith & Associates).

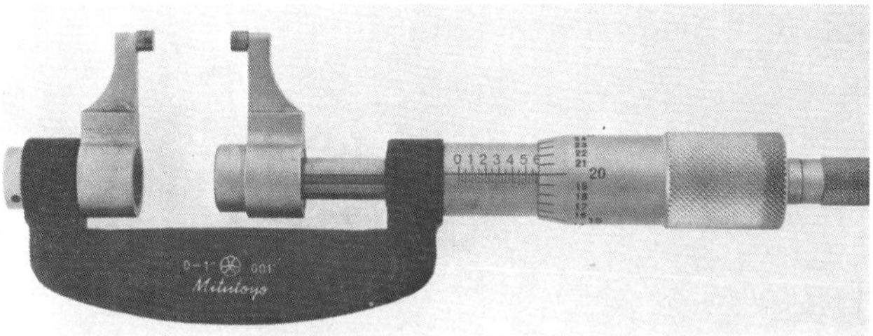

FIGURE C-117 Caliper-type outside micrometer (Harry Smith & Associates).

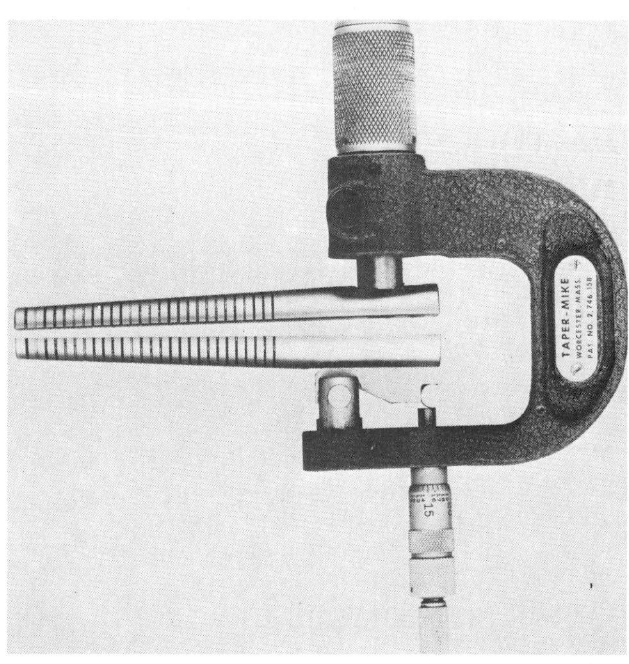

FIGURE C-118 Inside taper micrometer (Taper Micrometer Corp.).

FIGURE C-119 Outside taper micrometer in use (Taper Micrometer Corp.).

such as determining the distance of a hole from an edge (Figure C-115).

A standard outside micrometer may also be used to determine the wall thickness of tube or pipe (Figure C-116). In this application, a ball adapter is placed on the anvil. The diameter of the ball must be subtracted from the micrometer reading in order to determine the actual reading.

CALIPER-TYPE OUTSIDE MICROMETER

The caliper-type outside micrometer is used where measurements to be taken are inaccessible to a regular micrometer (Figure C-117).

TAPER MICROMETER

The taper micrometer can measure inside tapers (Figure C-118) or outside tapers (Figure C-119).

GROOVE MICROMETER

The groove micrometer (Figure C-120) is well suited to measuring grooves and slots, especially in inaccessible places, such as bores.

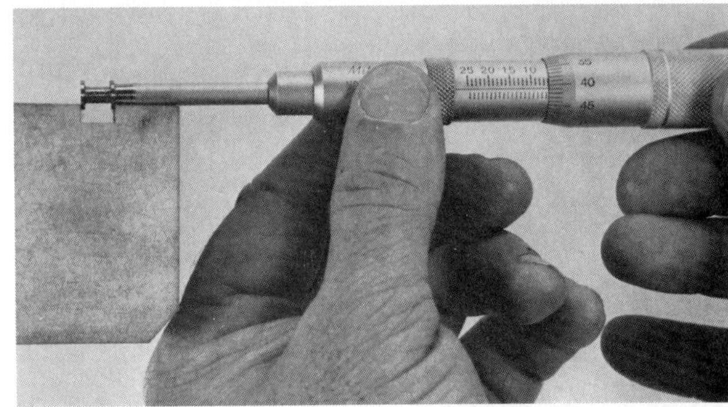

FIGURE C-120 Groove micrometer (Harry Smith & Associates).

DIGITAL ELECTRONIC MICROMETERS

As with calipers and many other common measuring tools, the micrometer is also available in an electronic digital readout model (Figure C-121). The instrument is easy to read, highly accurate, and available in all the common styles of its mechanical counterpart.

Discrimination of Micrometer Instruments

The standard micrometer will discriminate to .001 ($\frac{1}{1000}$) of an inch. In its vernier form, the discrimination is increased to .0001 ($\frac{1}{10,000}$) of an inch. The common metric micrometer discriminates to .01 ($\frac{1}{100}$) of an millimeter. The same rules apply to micrometers as apply to all measuring instruments. The tool should not be used beyond its discrimination. A standard micrometer with .001 discrimination should not be used in an attempt to ascertain measurements beyond that point. In order to measure to a discrimination of .0001 with the vernier micrometer, certain special conditions must be met. These will be discussed in more detail within this unit.

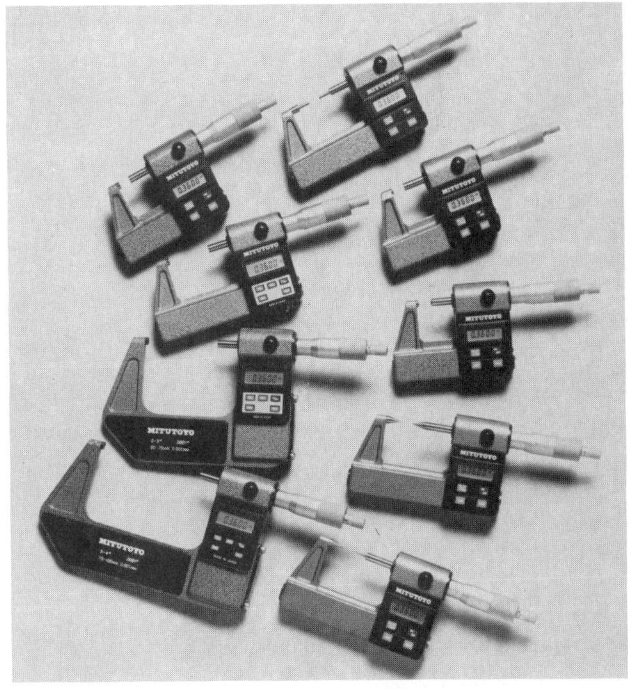

FIGURE C-121 Digital readout micrometers are available in a wide variety of styles (MTI Corporation).

Reliability and Expectation of Accuracy in Micrometer Instruments

The micrometer has increased reliability over the vernier. One reason for this is readability of the instruments. The .001 graduations that dictate the maximum discrimination of the micrometer are placed on the circumference of the thimble. The distance between the marks is therefore increased, making them easier to see.

The micrometer will yield very reliable results to .001 discrimination if the instrument is properly cared for an properly calibrated, and if correct procedure for use is followed. Care and procedure will be discussed in detail within this unit. **Calibration** is the process by which any measuring instrument is compared to a known standard. If the tool deviates from the standard, it may then be adjusted to conformity. This is an additional advantage of the micrometer over the vernier. The micrometer must be periodically calibrated if reliable results are to be obtained.

Can a micrometer measure reliably to within .001? The answer is "no" for the standard micrometer, as this violates the 10 to 1 rule for discrimination. The answer is "yes" for the vernier micrometer, but only under controlled conditions. What then, is an acceptable expectation of accuracy that will yield maximum reliability? This is dependent to some degree on the tolerance specified and can be summarized in the following table.

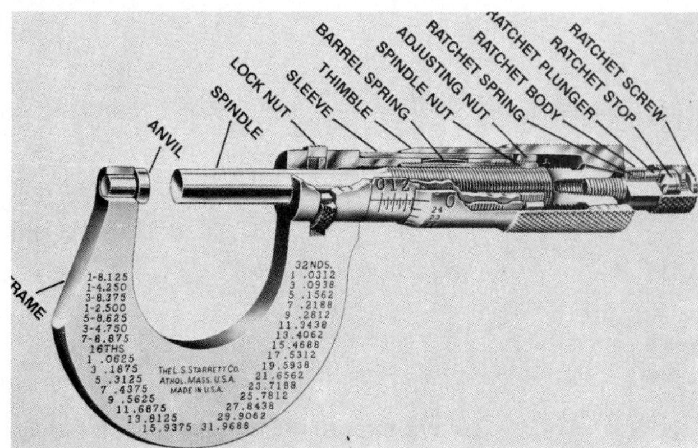

FIGURE C-122 Parts of the outside micrometer (L.S. Starrett Co.).

FIGURE C-123 Micrometers should always be kept on a tool board when used near a machine tool (Lane Community College).

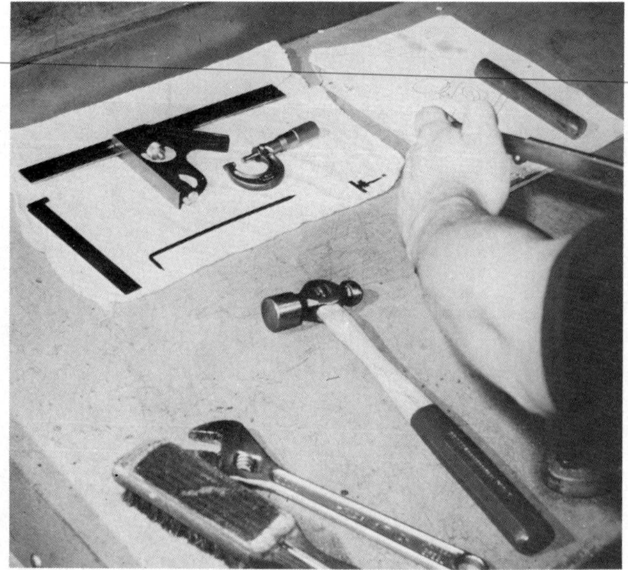

FIGURE C-124 Micrometers should be kept on a clean shop towel when used on the bench (Lane Community College).

Tolerance Specified	Acceptability of the Standard Micrometer	Acceptability of the Vernier Micrometer
+.000 −.001 *or* +.001 −.000	No	Yes (under controlled conditions)
±.001	Yes	Yes (vernier will not be required)

For a specified tolerance within .001 in., the vernier micrometer should be used. Plus or minus .001 in. is a total range of .002 in. or within the capability of the standard micrometers.

The micrometer is indeed a marvelous example of precision manufacturing. These rugged tools are produced in quantity with each one conforming to equally high standards. Micrometer instruments, in all their many forms, constitute one of the fundamental measuring instruments for the machinist.

Care of Outside Micrometers

You should be familiar with the names of the major parts of the typical outside micrometer (Figure C-122). The micrometer uses the movement of a precisely threaded rod turning in a nut for precision measurements. The accuracy of micrometer measurements is dependent on the quality of its construction, the care the tool receives and the skill of the user. A micrometer should be wiped clean of dust and oil before and after it is used. A micrometer should not be opened or closed by holding it by the thimble and spinning the frame around the axis of the spindle. Make sure that the micrometer is not dropped. Even a fall of a short distance can spring the frame. This will cause misalignment between the anvil and spindle faces and destroy the accuracy of this precision tool. A micrometer should be kept away from chips on a machine tool. The instrument should be placed on a clean tool board (Figure C-123) or on a clean shop towel (Figure C-124) close to where it is needed.

Always remember that the machinist is responsible for any measurements that he or she may make. To excuse an inaccurate measurement on the grounds that a micrometer was not properly adjusted or cared for would be less than professional. When a micrometer is stored after use, make sure that the spindle face does not touch the anvil. Perspiration, moisture from the air, or even oils promote corrosion between the measuring faces with a corresponding reduction in accuracy.

Prior to using a micrometer, clean the measuring faces. The measuring faces of many newer micrometers are made from an extremely hard metal called tungsten carbide. These instruments are often known as carbide-tipped micrometers. If you examine the measuring faces of a carbide-tipped micrometer, you will see where the carbide has been attached to the face of the anvil and spindle. Carbide-tipped micrometers have very durable and long-wearing measuring faces. Screw the spindle down lightly against a piece of paper held between it and the anvil (Figure C-125). Slide the paper out from between the measuring faces and blow away any fuzz that clings to the spindle or anvil. At this time, you should test the zero reading of the micrometer by bringing the spindle slowly into contact with the anvil (Figure C-126). Use the ratchet stop or friction thimble to perform this operation. The ratchet stop or friction thimble found on most micrometers is designed to equalize the gaging force. When the spindle and anvil contact the workpiece, the ratchet stop or friction thimble will slip as a predetermined amount of torque is applied to the micrometer thimble. If the micrometer does not have a ratchet device, use your thumb and index finger to provide a slip clutch effect on the thimble. Never use more pressure when checking the zero reading than when making actual measurements on the workpiece. If there is a small error, it may be corrected by adjusting the index line to the zero point (Figure C-127). The manufacturer's instructions provided with the micrometer should be followed when making this adjustment. Also, follow the manufacturer's instructions for correcting a loose thimble-to-spindle connection or incorrect friction thimble or ratchet stop action. One drop of instrument oil applied to the micrometer thread at monthly intervals will help enable it to provide many years of reliable service. Machinists are often judged by their associates on the way they handle and care for their tools. Therefore, if you care for their tools properly, they will more likely be held in higher professional regard.

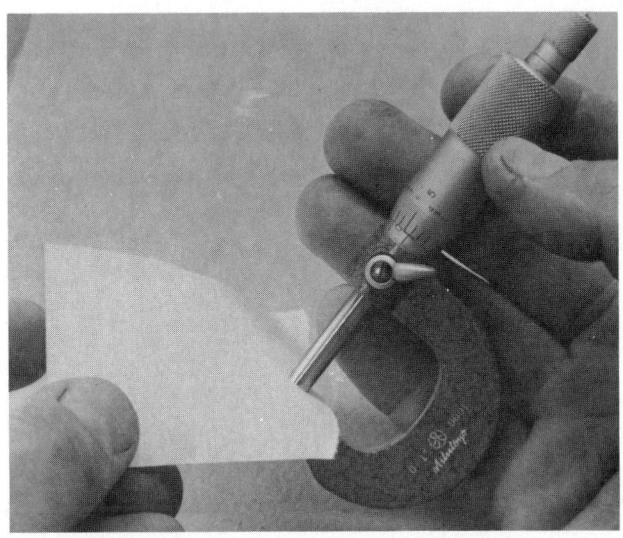

FIGURE C-125 Cleaning the measuring faces (Lane Community College).

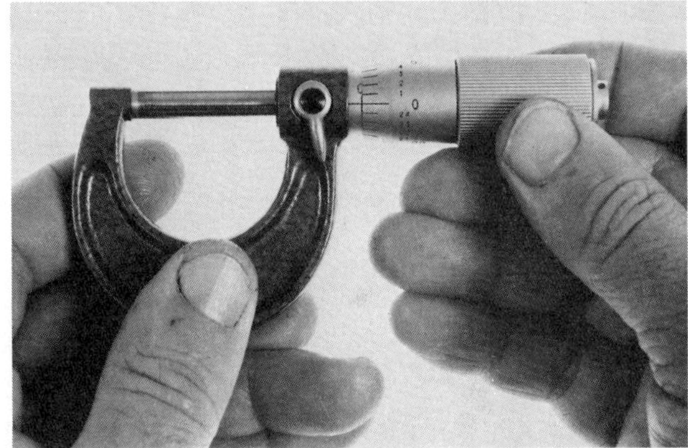

FIGURE C-126 Checking the zero reading.

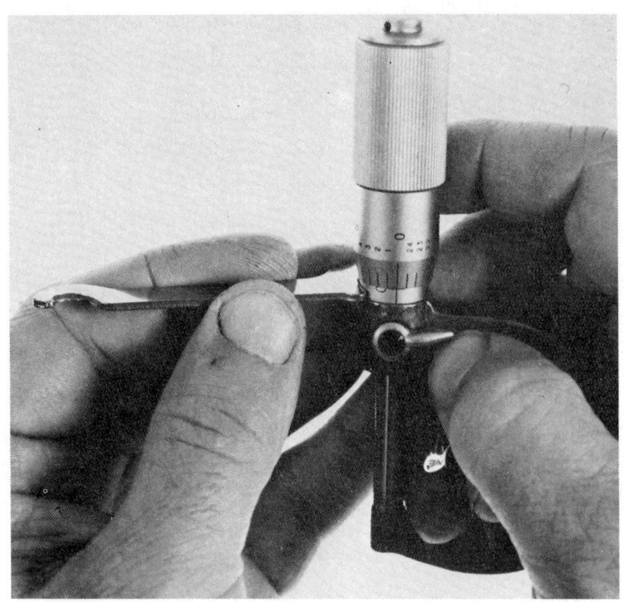

FIGURE C-127 Adjusting the index line to zero.

Reading Inch Micrometers

Dimensions requiring the use of micrometers will generally be expressed in decimal form to three decimal places. In the case of an inch instrument, this would be the thousandths place. You should think in terms of thousandths whenever reading decimal fractions. For example, the decimal .156 of an inch would be read as one hundred and fifty six thousandths of an inch. Likewise, .062 would be read as 62 thousandths.

On the **sleeve** of the micrometer is a graduated scale with 10 numbered divisions, each one being $\frac{1}{10}$ of one inch or .100 (100 thousandths) apart.

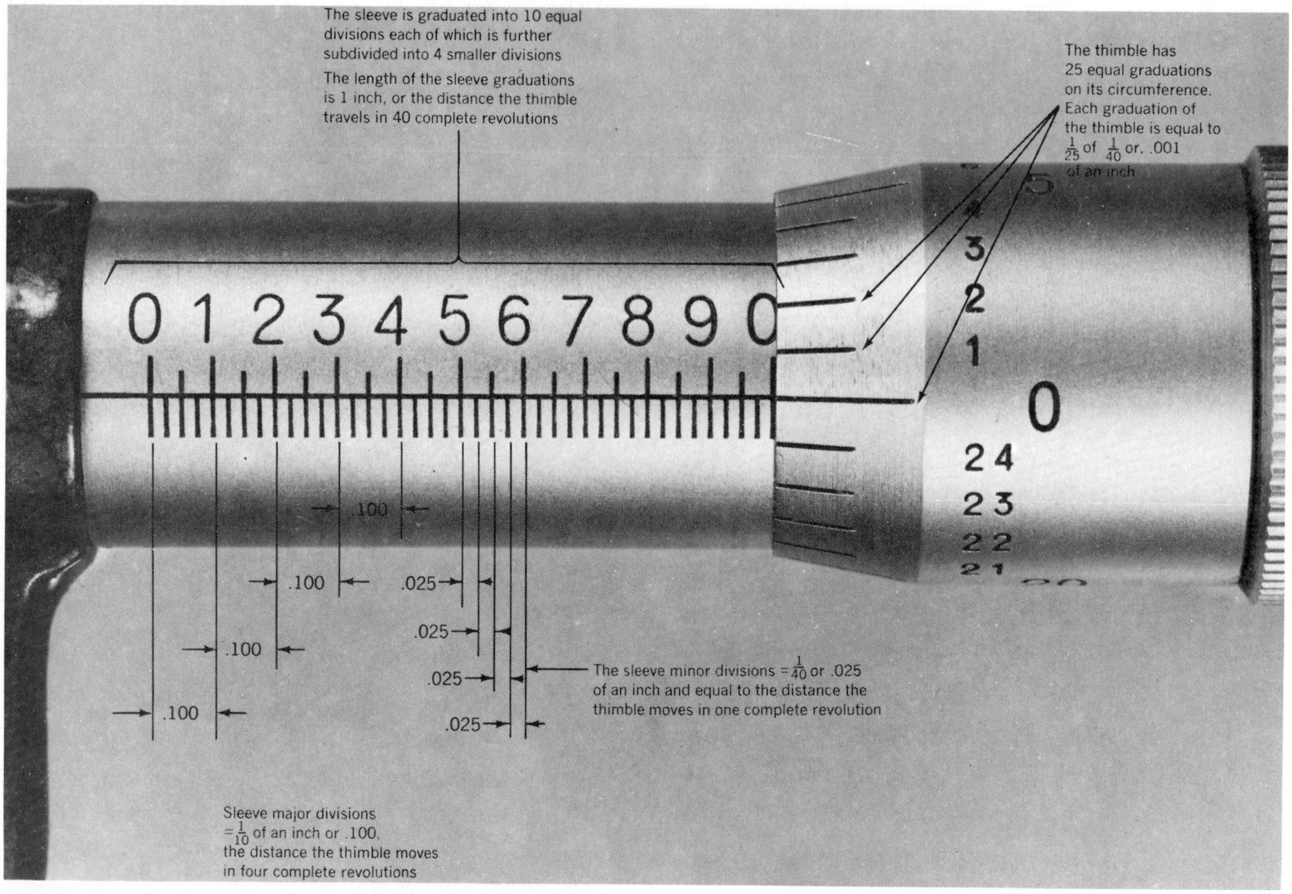

The sleeve is graduated into 10 equal divisions each of which is further subdivided into 4 smaller divisions
The length of the sleeve graduations is 1 inch, or the distance the thimble travels in 40 complete revolutions
The thimble has 25 equal graduations on its circumference. Each graduation of the thimble is equal to $\frac{1}{25}$ of $\frac{1}{40}$ or, .001 of an inch
0 1 2 3 4 5 6 7 8 9 0
.100
.100
.100
.100
.025
.025
.025
.025
The sleeve minor divisions = $\frac{1}{40}$ or .025 of an inch and equal to the distance the thimble moves in one complete revolution
Sleeve major divisions = $\frac{1}{10}$ of an inch or .100, the distance the thimble moves in four complete revolutions

FIGURE C-128 Graduations on the inch micrometer.

0 1 2 3
9 thimble divisions = .009 thousandths
1 minor sleeve division = .025 thousandths
1 minor sleeve division = .025 thousandths
3 major sleeve divisions = .300 thousandths
Total reading = .359 thousandths

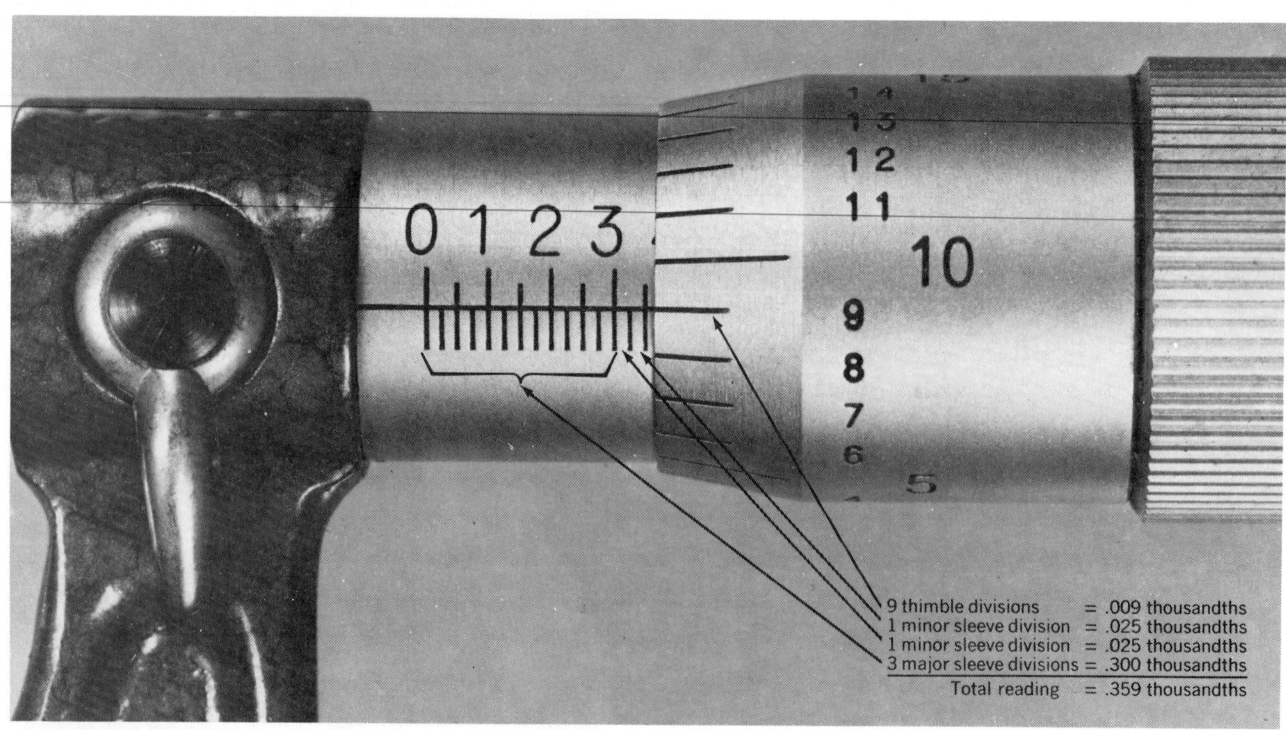

FIGURE C-129 Inch micrometer reading of .359 or three hundred fifty-nine thousandths.

Each of these major divisions is further subdivided into four equal parts, which makes the distance between these graduations $\frac{1}{4}$ of .100 or .025 (25 thousandths) (Figure C-128). The **spindle screw** of a micrometer has 40 threads per inch. When the spindle is turned one complete revolution, it has moved $\frac{1}{40}$ of an inch, or, expressed as a decimal, .025 (25 thousandths).

When you examine the **thimble,** you will find 25 evenly spaced divisions around its circumference (Figure C-128). Because each complete revolution of the thimble causes it to move a distance of .025 in., each thimble graduation must be equal to $\frac{1}{25}$ of .025, or .001 in. (one thousandth). On most micrometers, each thimble graduation is numbered to facilitate reading the instrument. On older micrometers only every fifth line may be numbered.

When reading the micrometer (Figure C-129), first determine the value indicated by the lines exposed on the sleeve. The edge of the thimble exposes three major divisions. This represents .300 in. (300 thousandths). However, there are also two minor divisions showing on the sleeve. The value of these is .025, for a total of .050 in. (50 thousandths). The reading on the thimble is 9, which indicates .009 in. (9 thousandths). The final micrometer reading is determined by adding the total of the sleeve and thimble readings. In the example shown (Figure C-129), the sleeve shows a total of .350 in. Adding this to the thimble, the final reading becomes .350 in. + .009 in., or .359 in.

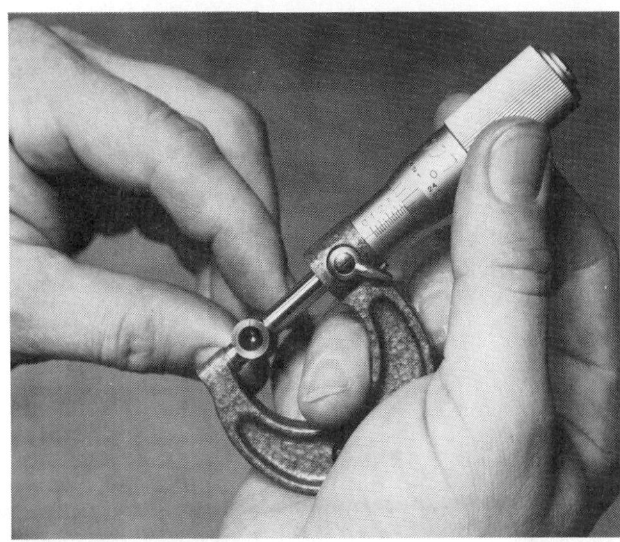

FIGURE C-130 Proper way to hold a micrometer (Lane Community College).

FIGURE C-131 Read a micrometer while still in contact with the workpiece (Lane Community College).

Using the Micrometer

A micrometer should be gripped by the **frame** (Figure C-130), leaving the thumb and forefinger free to operate the thimble. When possible, take micrometer readings while the instrument is in contact with the workpiece (Figure C-131). Use only enough pressure on the **spindle** and **anvil** to yield a reliable result. This is what the machinist refers to as **feel.** The proper feel of a micrometer will come only from experience. Obviously, excessive pressure will not only result in an inaccurate measurement, it will also distort the frame of the micrometer and possibly damage it permanently. You should also remember that too light a pressure on the part by the measuring faces can yield an unreliable result.

The micrometer should be held in both hands whenever possible. This is especially true when measuring cylindrical workpieces (Figure C-132). Holding the instrument in one hand does not permit sufficient control for reliable readings. Furthermore, cylindrical workpieces should be

FIGURE C-132 Hold a micrometer in both hands when measuring a round part (Lane Community College).

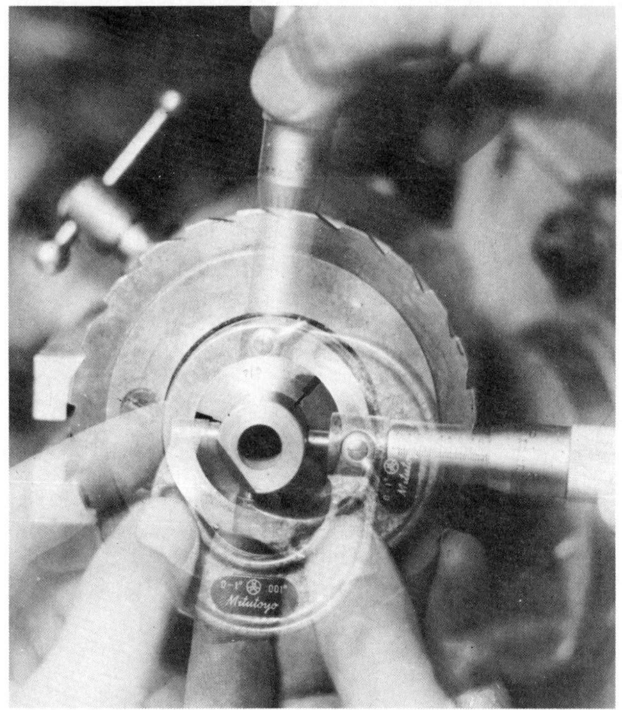

FIGURE C-133 When measuring round parts, take two readings 90 degrees apart (Lane Community College).

checked at least twice with **measurements made 90 degrees apart.** This is to check for an out-of-round condition (Figure C-133). When critical dimensions are measured, that is, any dimension where a very small amount of tolerance is acceptable, make at least two consecutive measurements. Both readings should indicate identical results. If two identical readings cannot be determined, then the actual size of the part cannot be stated reliably. **All critical measurements should be made at a temperature of 68° Fahrenheit (20° Celsius).** A workpiece warmer than this temperature will be larger because of heat expansion.

Outside micrometers usually have a measuring range of 1 in. They are identified by size as to the largest dimensions they measure. A 2-in. micrometer will measure from 1 to 2 in. A 3-in. micrometer will measure from 2 to 3 in. The capacity of the tool is increased by increasing the size of the frame. Typical outside micrometers range in capacity from 0 to 168 in. It requires a great deal more skill to get consistent measurements with large-capacity micrometers.

SELF-TEST

1. Why should a micrometer be kept clean and protected?

2. Why should a micrometer be stored with the spindle out of contact with the anvil?

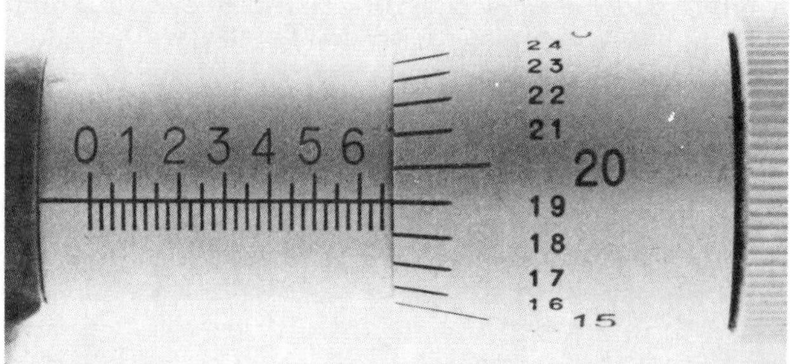

FIGURE C-134*a*

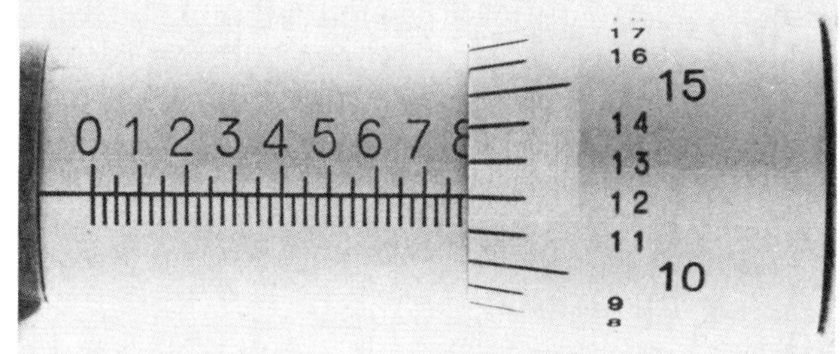

FIGURE C-134*b*

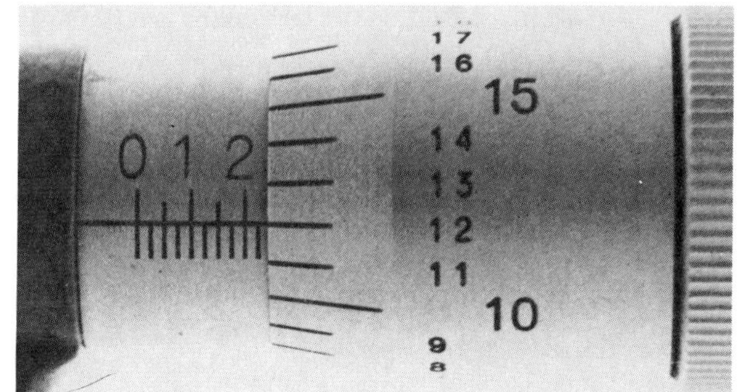

FIGURE C-134*c*

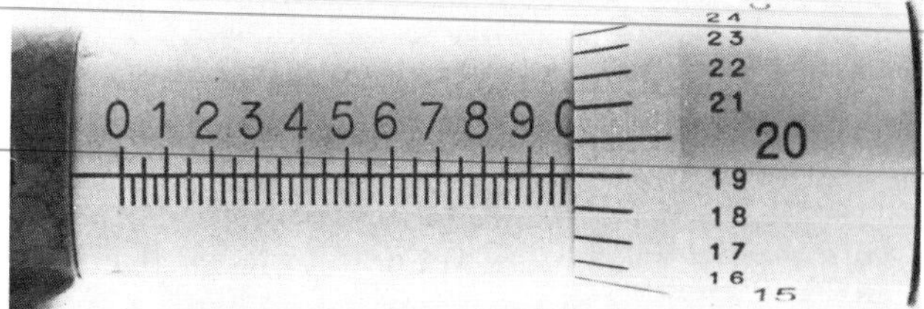

FIGURE C-134*d*

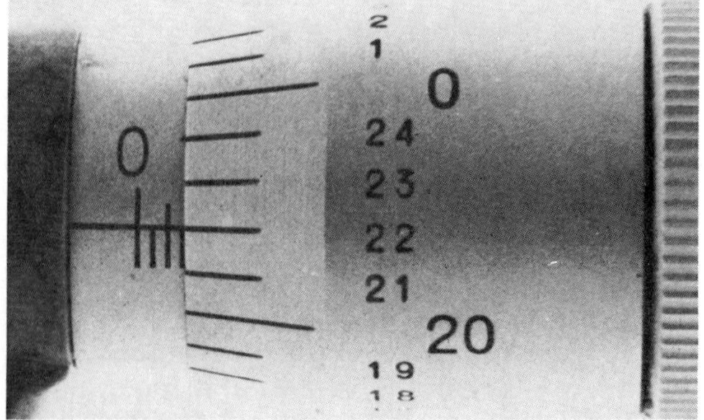

FIGURE C-134*e*

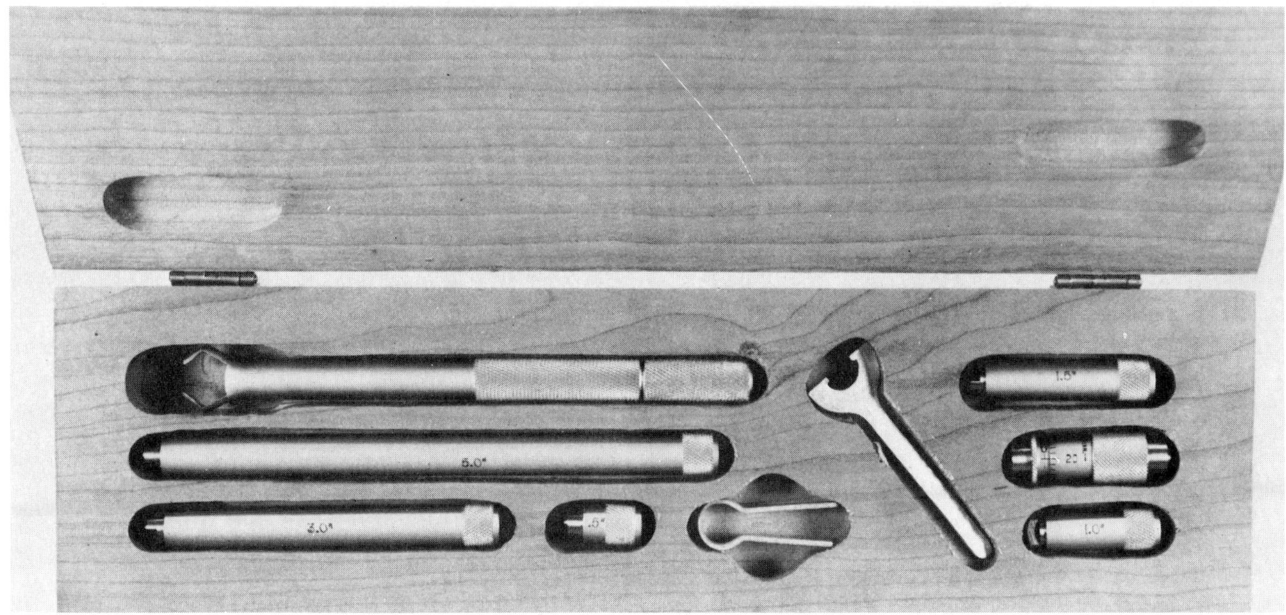

FIGURE C-135 Tubular type inside micrometer set (Harry Smith & Associates).

3. Why are the measuring faces of the micrometer cleaned before measuring?

4. How precise is the standard micrometer?

5. What affects the accuracy of a micrometer?

6. What is the difference between the sleeve and thimble?

7. Why should a micrometer be read while it is still in contact with the object to be measured?

8. How often should an object be measured to verify its actual size?

9. What effect has an increase in temperature on the size of a part?

10. What is the purpose of the friction thimble or ratchet stop on the micrometer?

11. Read and record the five outside micrometer readings in Figures C-134*a* to C-134*e*.

Using Inside Micrometers

Inside micrometers are equipped with the same graduations as outside micrometers. Inside micrometers discriminate to .001 in. and have a measuring capacity ranging from 1.5 to 2.0 in. or more. A typical **tubular type** inside micrometer set (Figure C-135) consists of the **micrometer head** with detachable **hardened anvils** and several **tubular measuring rods** with **hardened contact tips.** The lengths of these rods differ in increments of .5 in. to match the measuring capacity of the micrometer head, which in this case is .5 in. A handle is provided to hold the instrument in places where holding the instrument directly would be difficult.

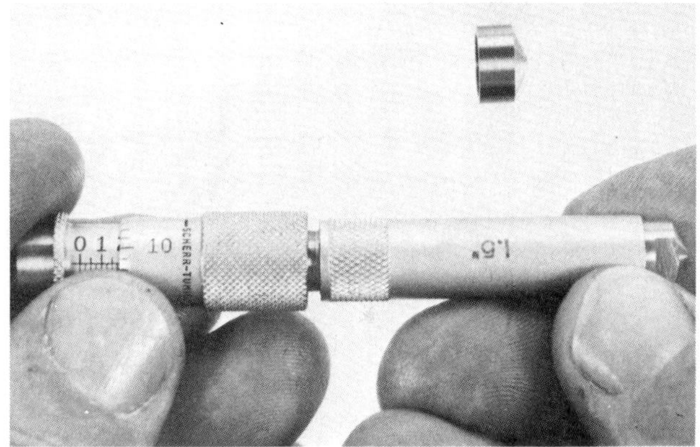

FIGURE C-136 Attaching a 1.5-in. extension rod to an inside micrometer head (Harry Smith & Associates).

Another common type of inside micrometer comes equipped with relatively small diameter solid rods that differ in inch increments, even though the head movement is .5 in. In this case, a .5-in. spacing collar is provided. This can be slipped over the base of the rod before it is inserted into the measuring head.

Inside micrometer heads have a range of .250, .500, 1.000 or 2.000 in. depending on the total capacity of the set. For example, an inside micrometer set with a head range of .500 in. will be able to measure from 1.500 to 12.500 in.

The measuring range of the inside micrometer is changed by attaching the extension rods. Extension rods may be solid or tubular. Tubular rods are lighter in weight and are often found in large range inside micrometer sets. Tubular rods are also more rigid. It is very important that all parts be **extremely clean** when changing extension rods (Figure C-136). Even small dust particles can affect the accuracy of the instrument.

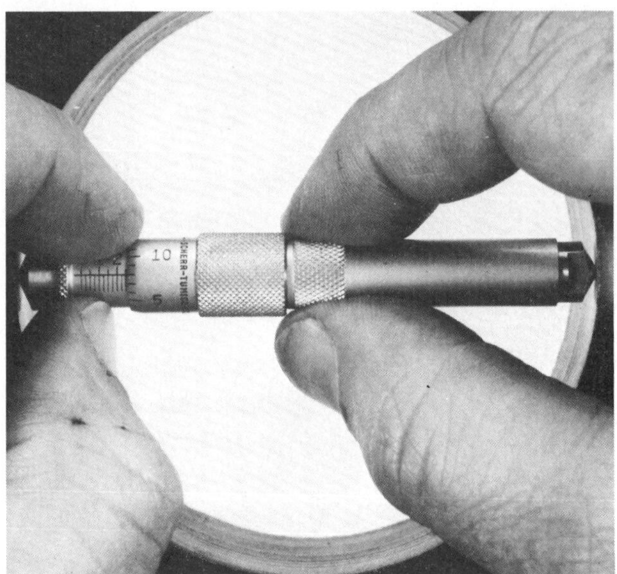

FIGURE C-137 Placing the inside micrometer in the bore to be measured (Harry Smith & Associates).

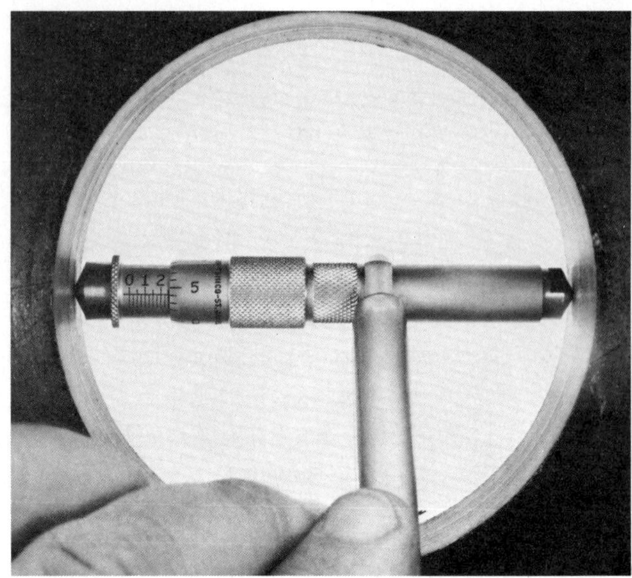

FIGURE C-138 The inside micrometer head used with a handle (Harry Smith & Associates).

FIGURE C-139 Confirming inside micrometer range using a rule (Harry Smith & Associates).

When making internal measurements, set one end of the inside micrometer against one side of the hole to be measured (Figure C-137). An inside micrometer should not be held in the hands for extended periods, as the resultant heat may affect the accuracy of the instrument. A handle is usually provided, which eliminates the need to hold the instrument and also facilitates insertion of the micrometer into a bore or hole (Figure C-138). One end of the micrometer will become the center of the arcing movement used when finding the centerline of the hole to be measured. The micrometer should then be adjusted to the size of the hole. When the correct hole size is reached, there should be a very light drag between the measuring tip and the work when the tip is moved through the centerline of the hole. The size of the hole is determined by adding the reading of the micrometer head, the length of the extension rod, and the length of the spacing collar, if one was used. Read the micrometer **while it is still in place if possible.** If the instrument must be removed to be read, the correct range can be determined by checking with a rule (Figure C-139). A skilled worker will usually use an **accurate outside micrometer to verify** a reading taken with an inside micrometer. In this case, the inside micrometer becomes an easily adjustable transfer measuring tool (Figure C-140). Take at least two readings 90 degrees apart to obtain the size of a hole or bore. The readings should be identical. Inside micrometers do not have a spindle lock. Therefore, to prevent the spindle from turning while establishing the correct feel, the adjusting nut should be maintained slightly tighter than normal.

SELF-TEST

Read and record the five inside micrometer readings shown in Figures C-141*a* to C-141*e*. Micrometer head is 1.500 in. when zeroed.

Obtain an inside micrometer set from your instructor and practice using the instrument on objects around your laboratory. Measure examples such as lathe spindle holes, bushings, bores of roller bearings, hydraulic cylinders, and tubing.

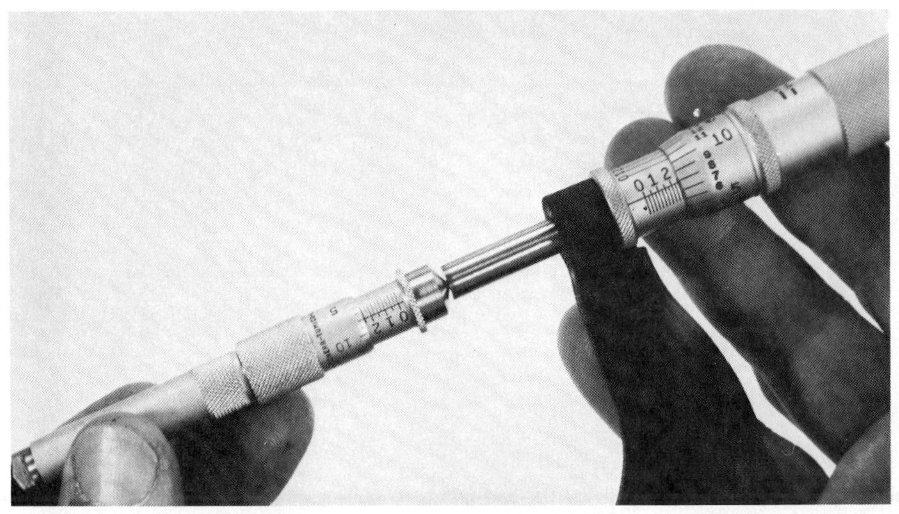

FIGURE C-140 Checking the inside micrometer with an outside micrometer (Harry Smith & Associates).

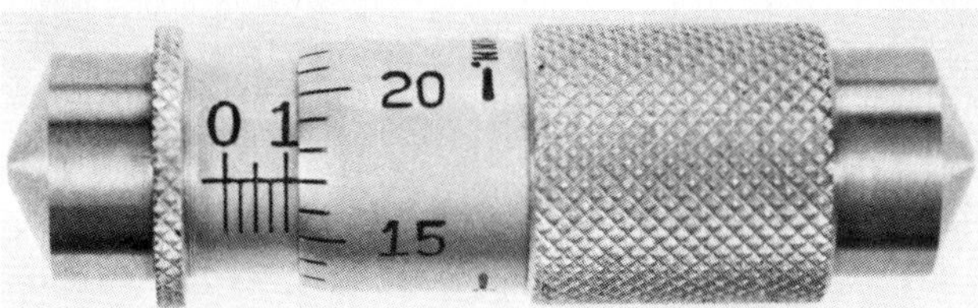

FIGURE C-141*a*

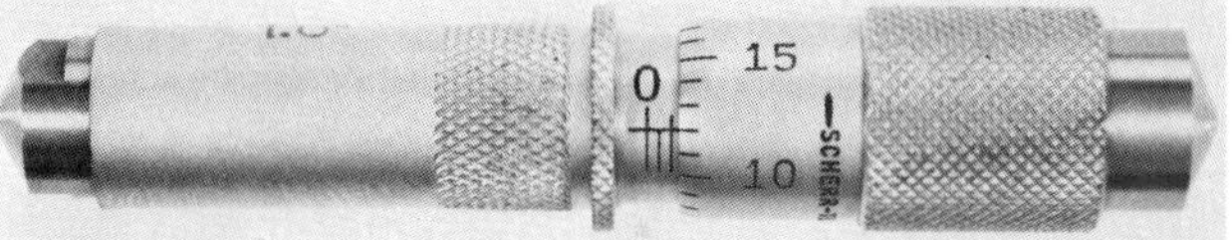

FIGURE C-141*d*

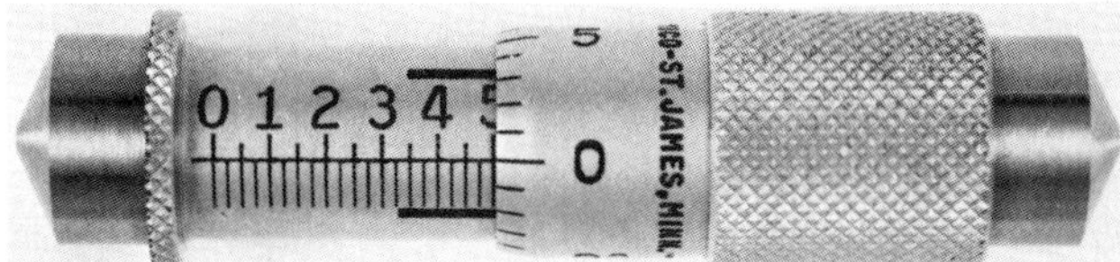

FIGURE C-141*b*

FIGURE C-141*e*

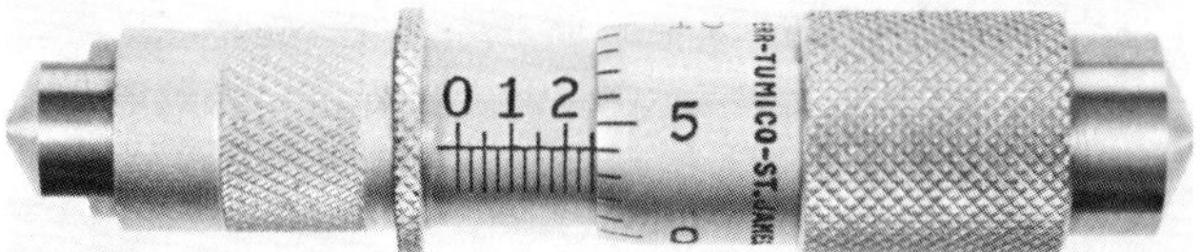

FIGURE C-141*c*

Using Depth Micrometers

A **depth micrometer** is a tool that is used to measure precisely depths of holes, grooves, shoulders, and recesses. Like other micrometer instruments, it will discriminate to .001 in. Depth micrometers usually come as a set with interchangeable rods to accommodate different depth measurements (Figure C-142). The basic parts of the depth micrometer are the **base, sleeve, thimble, extension rod, thimble cap,** and, frequently, a **rachet stop.** The bases of a depth micrometer can be of various widths. Generally the wider bases are more stable, but in many instances, space limitations dictate the use of narrower bases. Some depth microme-

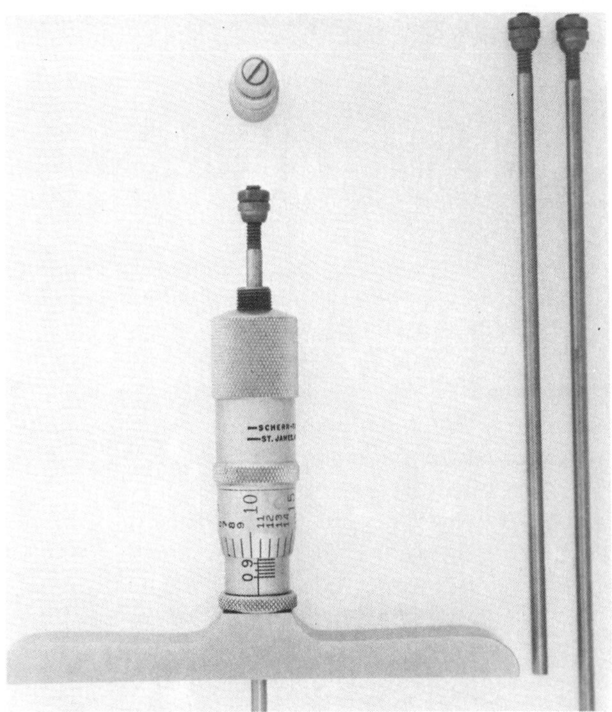

FIGURE C-142 Depth micrometer set.

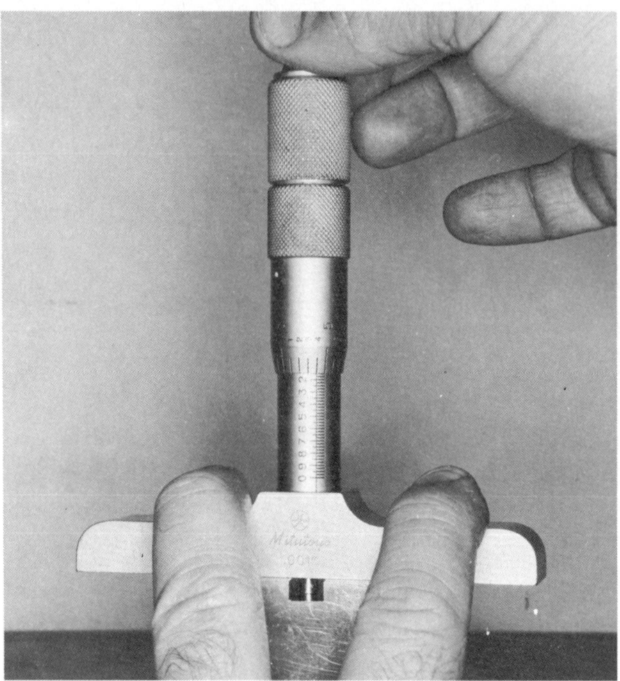

FIGURE C-143 Proper way to hold the depth micrometer.

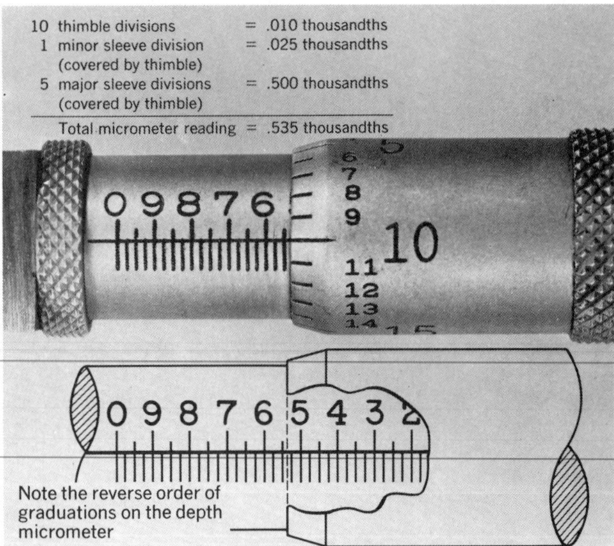

FIGURE C-144 Sleeve graduations on the depth micrometer are numbered in the opposite direction as compared to the outside micrometer.

ters are made with only a half base for measurements in confined spaces.

The extention rods are installed or removed by holding the thimble and unscrewing the thimble cap. Make sure that the seat between the thimble cap and rod adjusting nuts is clean before reassembling the micrometer. Do not overtighten when replacing the thimble cap. Furthermore, **do not attempt to adjust the rod length by turning the adjusting nuts.** These rods are factory adjusted and matched as a set. **The measuring rods from**

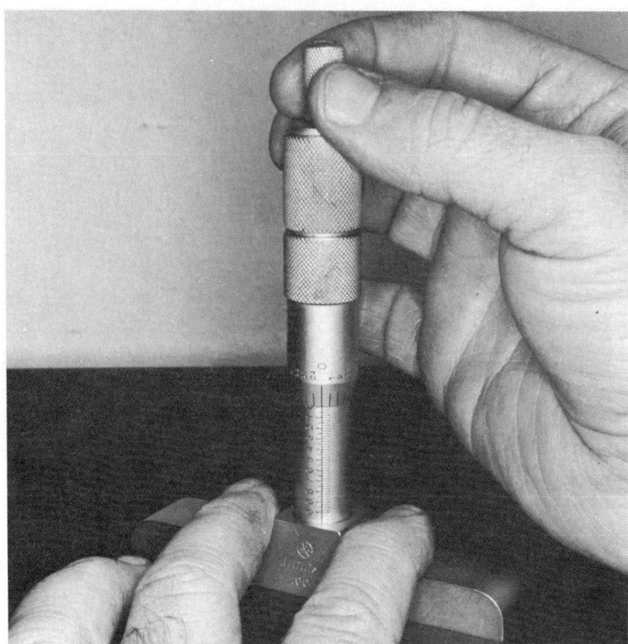

FIGURE C-145 Checking a depth micrometer for zero adjustment using the surface plate as a reference surface.

a specific depth micrometer set should always be kept with that set. Since these rods are factory adjusted and matched to a specific instrument, **transposing measuring rods** from set to set **will** usually **result** in **incorrect measurements.**

When making depth measurements, it is very important that the micrometer base has a smooth and flat surface on which to rest. Furthermore, sufficient pressure must be applied to keep the base in contact with the reference surface. When a depth micrometer is used without a ratchet, a slip clutch effect can be produced by letting the thimble slip while turning it between the thumb and index finger (Figure C-143).

Reading Inch Depth Micrometers

When a comparison is made between the sleeve of an outside micrometer and the sleeve of a depth micrometer, note that the graduations are numbered in the opposite direction (Figure C-144). When reading a depth micrometer, the distance to be measured is the value covered by the thimble. Consider the reading shown (Figure C-144). The thimble edge is between the number 5 and 6. This indicates a value of at least .500 in. on the sleeve major divisions. The thimble also covers the first minor division on the sleeve. This has a value of .025 in. The value on the thimble circumference indicates .010 in. Adding these three values results in a total of .535 in., or the amount of extension of the rod from the base.

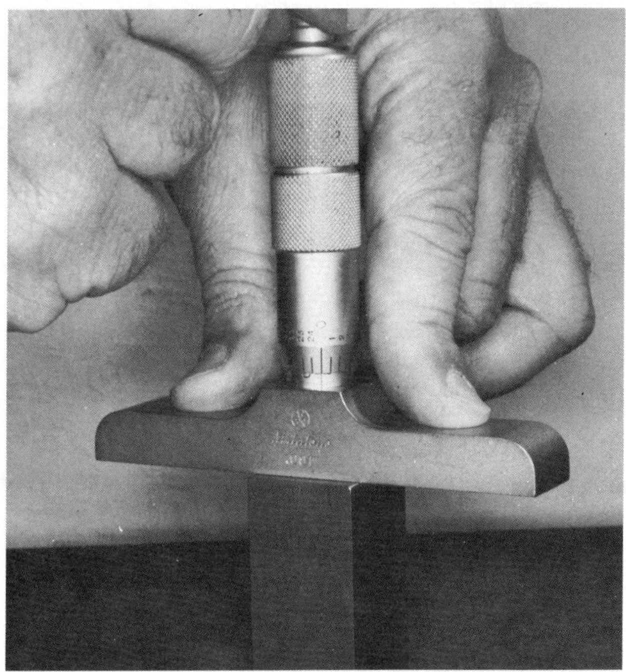

FIGURE C-146 Checking the depth micrometer calibration at the 1.000-in. position in the 0–1 in. rod and a 1-in. square or Hoke-type gage block.

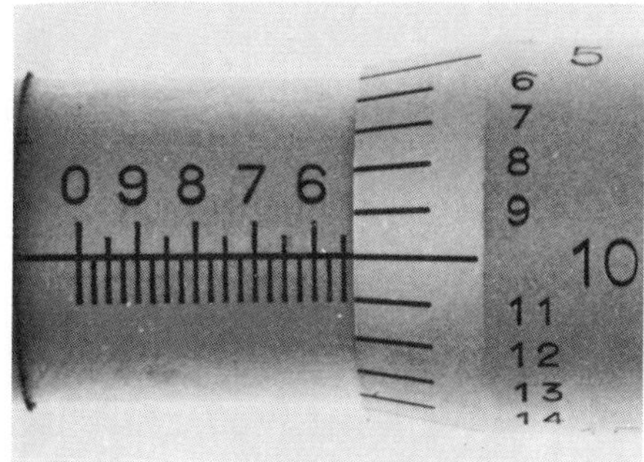

FIGURE C-147*a*

A depth micrometer **should be tested for accuracy** before it is used. When the 0 to 1 in. rod is used, retract the measuring rod into the base. Clean the base and contact surface of the rod. Hold the micrometer base firmly against a flat highly finished surface, such as a surface plate, and advance the rod until it contacts the reference surface (Figure C-145). If the micrometer is properly adjusted, it should read zero. When **testing for accuracy** with the **one-inch extension rod,** set the base of the micrometer on a **one-inch gage block** and measure to the reference surface (Figure C-146). Other extension rods can be tested in a like manner.

SELF-TEST

Read and record the five depth micrometer readings in Figures C-147*a* to C-147*e*.

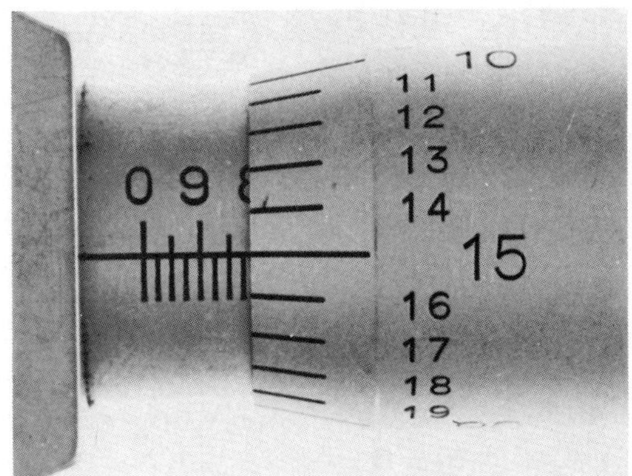

FIGURE C-147*b*

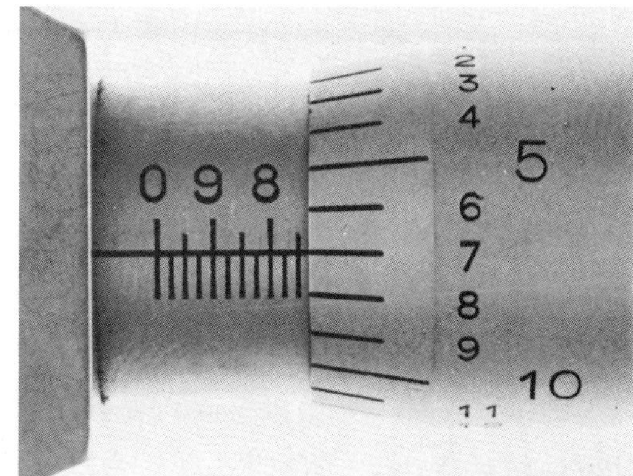

FIGURE C-147*c*

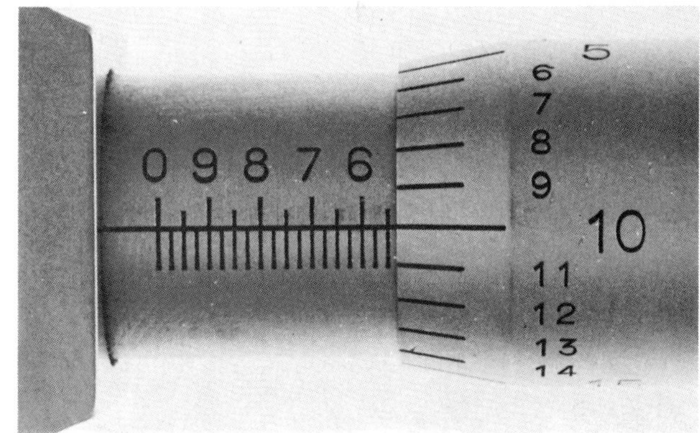

FIGURE C-147*d*

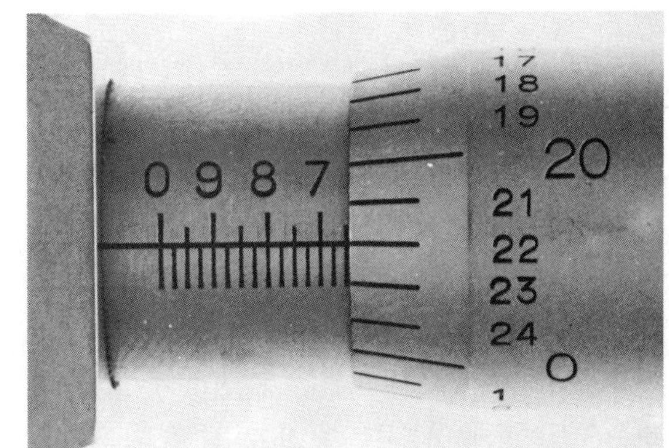

FIGURE C-147*e*

Reading Metric Micrometers

The **metric micrometer** (Figure C-148) has a spindle thread with a .5-mm lead. This means that the spindle will move .5 mm when the thimble is turned one complete revolution. Two revolutions of the thimble will advance the spindle 1 mm. In precision machining, metric dimensions are usually expressed in terms of .01 ($\frac{1}{100}$) of a millimeter. On the metric micrometer the thimble is graduated into 50 equal divisions with every fifth division numbered (Figure C-149). If one revolution of the thimble is .5 mm, then each division on the thimble is equal to .5 mm divided by 50 or .01 mm. The sleeve of the metric micrometer is divided into 25 main divisions above the index line with every fifth division numbered. These are whole millimeter graduations. Below the index line are graduations that fall halfway between the divisions above the line. The lower graduations represent half or .5-mm values. The thimble edge (Figure C-150) leaves the 12-mm line exposed with no. 5-mm line showing. The thimble reading is 32, which is .32 mm. Adding the two figures results in a total of 12.32 mm.

The 15-mm mark (Figure C-151) is exposed on

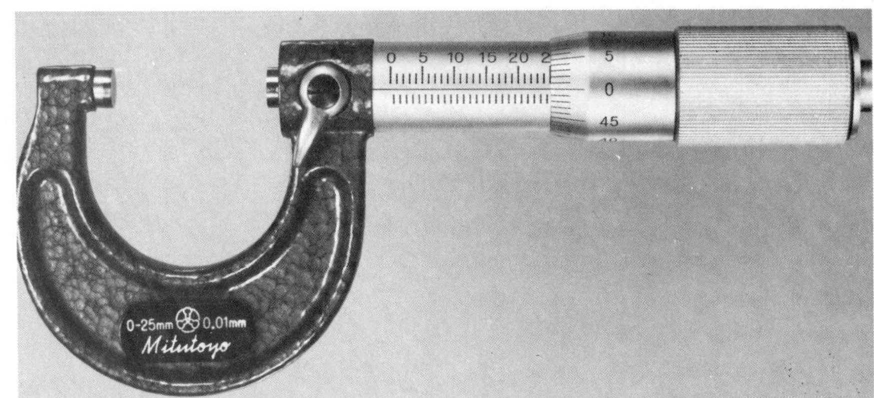
0 5 10 15 20
5
0
45
0-25mm 0.01mm
Mitutoyo

FIGURE C-148 Metric micrometer (Lane Community College).

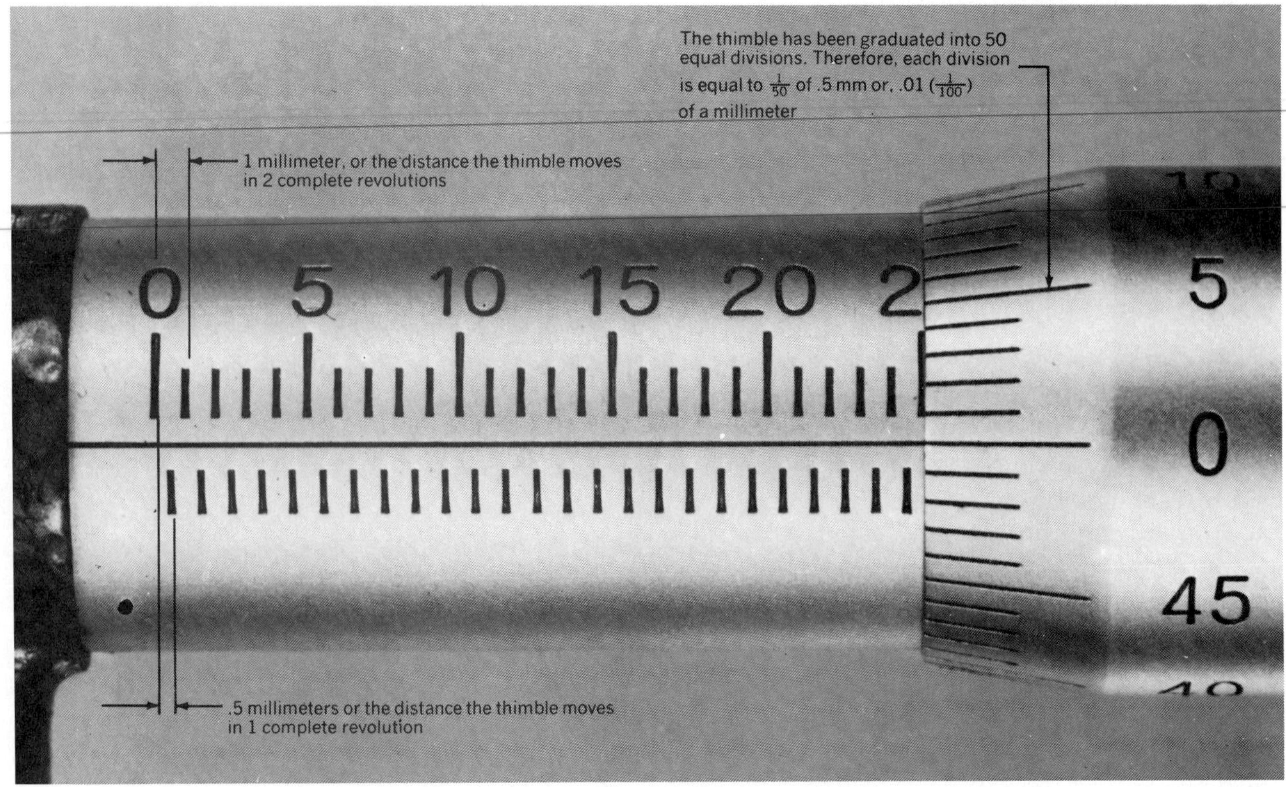
The thimble has been graduated into 50 equal divisions. Therefore, each division is equal to $\frac{1}{50}$ of .5 mm or, .01 ($\frac{1}{100}$) of a millimeter
1 millimeter, or the distance the thimble moves in 2 complete revolutions
0 5 10 15 20 2
5
0
45
.5 millimeters or the distance the thimble moves in 1 complete revolution

FIGURE C-149 Graduations on the metric micrometer.

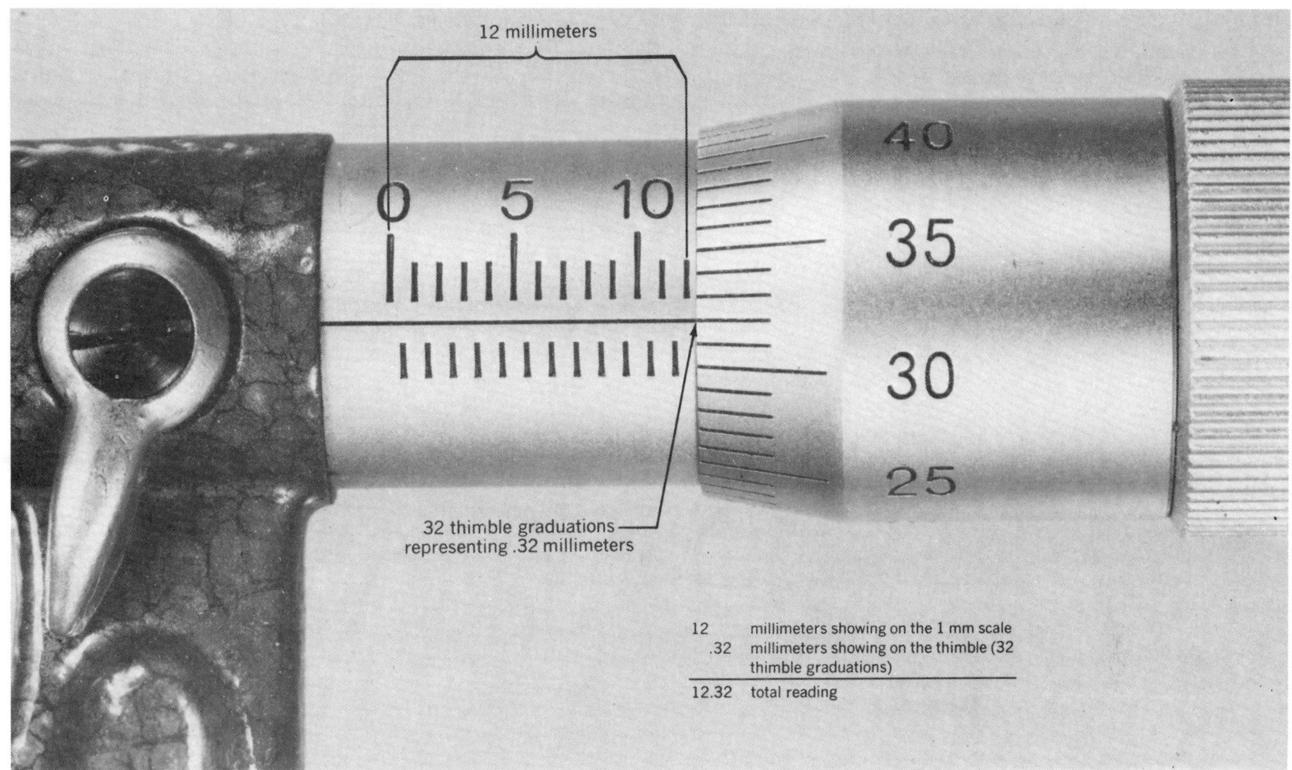

FIGURE C-150 Metric micrometer reading of 12.32 mm.

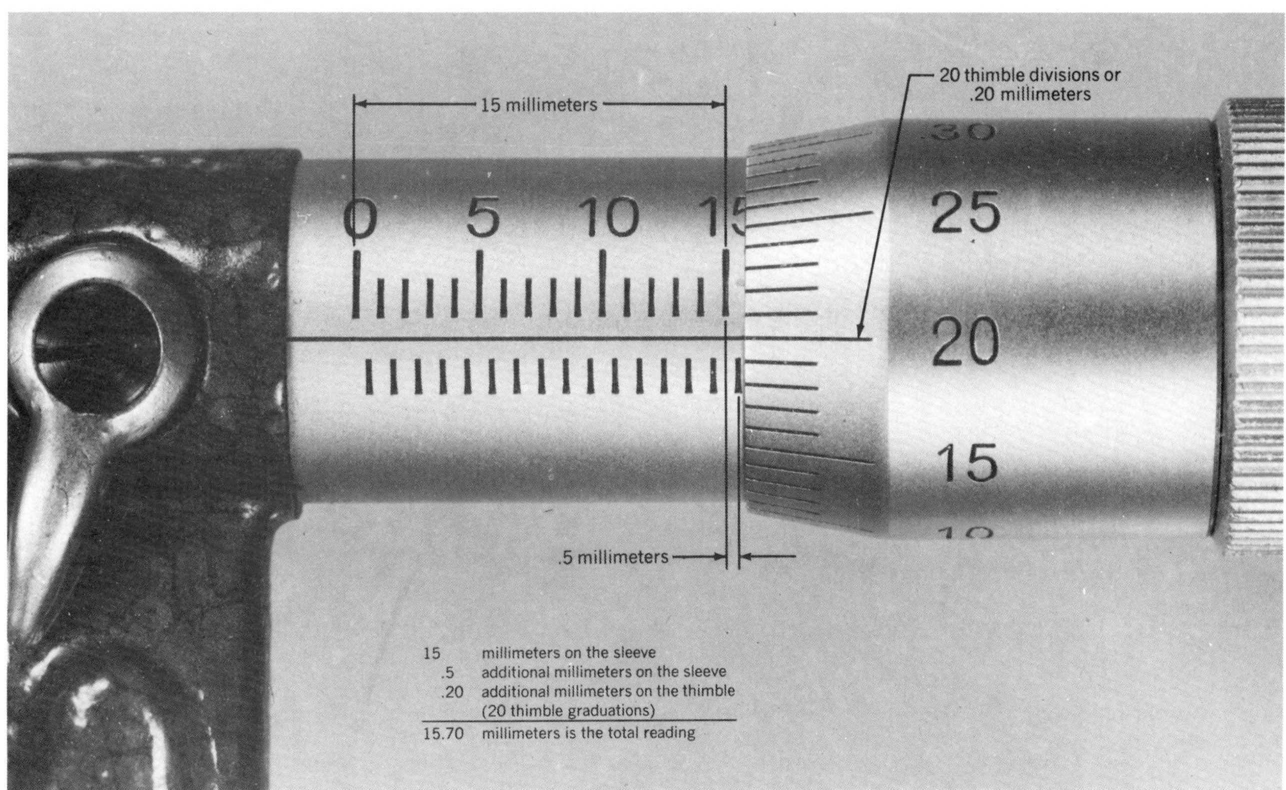

FIGURE C-151 Metric micrometer reading of 15.70 mm.

the sleeve plus a .5-mm graduation below the index line. The thimble reads 20 or .20 mm. Adding these three values, 15.00 + .50 + .20, results in a total of 15.70 mm.

Any metric micrometer should receive the same care discussed in the section on outside micrometers.

FIGURE C-152*a*

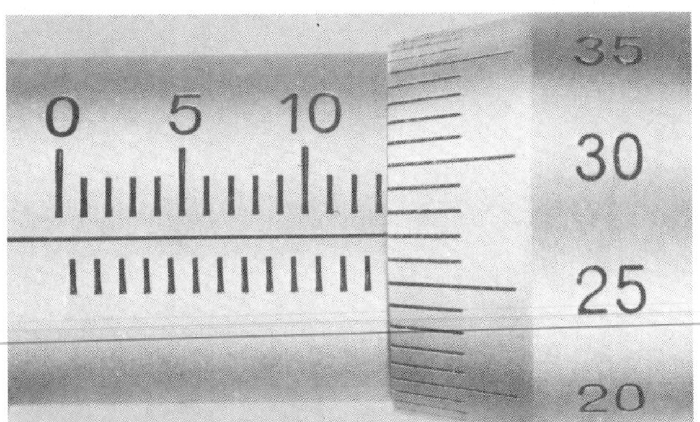

FIGURE C-152*b*

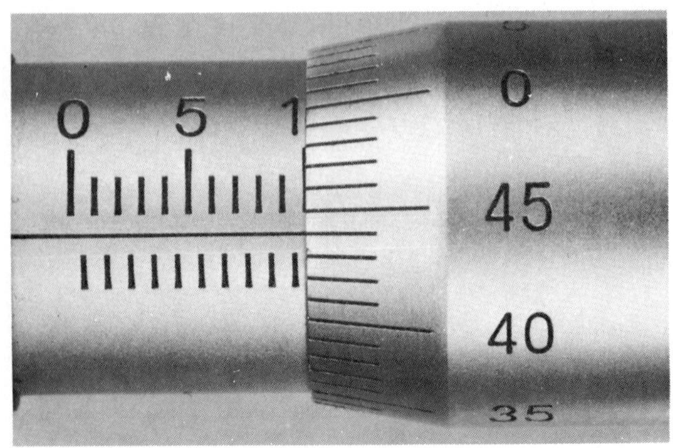

FIGURE C-152*c*

SELF-TEST

Read and record the five metric micrometer readings in Figures C-152*a* to C-152*e*.

Reading Vernier Micrometers

When measurements must be made to a discrimination greater than .001 in., a standard micrometer is not sufficient. With a **vernier micrometer,** readings can be made to a **ten-thousandth part of an inch** (.0001 in.).This kind of micrometer is commonly known as a "tenth mike." A vernier scale is part of the sleeve graduations. The vernier scale consists of 10 lines parallel to the index line and located above it (Figure C-153).

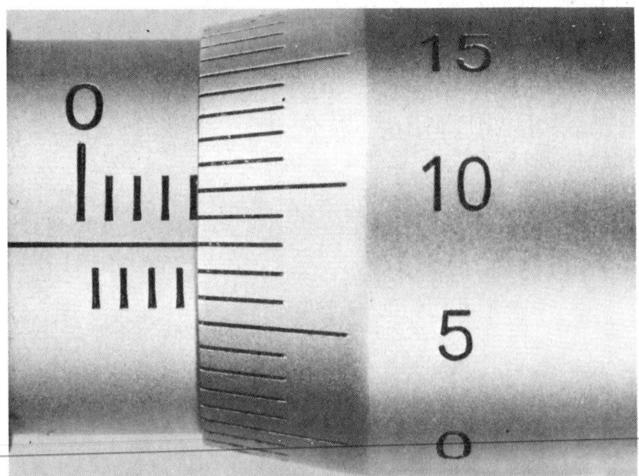

FIGURE C-152*d*

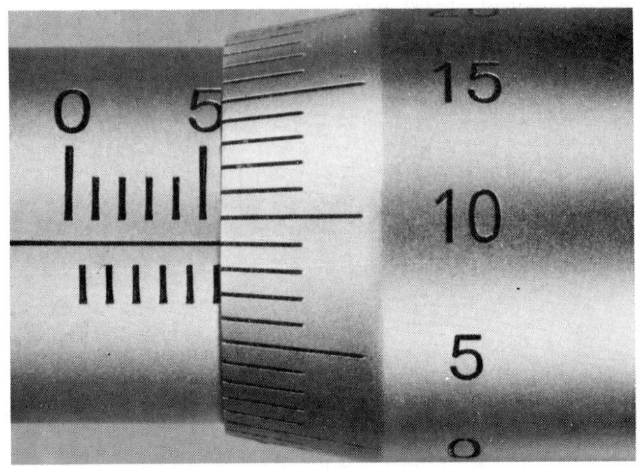

FIGURE C-152*e*

If the 10 spaces on the vernier scale were compared to the spacing of the thimble graduations, the 10 vernier spacings would correspond to 9 spacings on the thimble. Therefore, the vernier scale spacing must be smaller than the thimble spacing. That is, in fact, precisely the case. Since 10 vernier spacings compare to 9 thimble spacings, the vernier spacing is $\frac{1}{10}$ smaller than the thimble space. We know that the thimble graduations correspond to .001 in. (one thousandth). Each vernier spacing must then be equal to $\frac{1}{10}$ of .001 in., or .0001 in. (one ten-thousandth). Thus, according to the principle of the vernier, each thousandth of the thimble is subdivided into 10 parts. This permits the vernier micrometer to discriminate to .0001 in.

To read a vernier micrometer, first read to the nearest thousandth as on a standard micrometer. Then, find the line on the vernier scale that coincides with a graduation on the thimble. The value of this coincident vernier scale line is the value in ten thousandths, which must be added to the thousandths reading, thus making up the total reading. **Remember to add the value of the vernier scale line and not the number of the matching thimble line.**

In the lower view (Figure C-153), a micrometer reading of slightly more than .216 in. is indicated. In the top view, on the vernier scale, the line numbered 3 is in alignment with the line on the thimble. This indicated that .0003 (three ten-thousandths) must be added to the .216 in. for a total reading of .2163. This number is read "two hundred sixteen thousandths and three tenths."

You must exercise cautious judgment when attempting to measure to a tenth of a thousandth using a vernier micrometer. There are many conditions that can influence the reliability of such measurements. The 10 to 1 rule discussed in the section introduction states that for maximum reliability, a measuring instrument must be able to discriminate 10 times finer than the smallest mea-

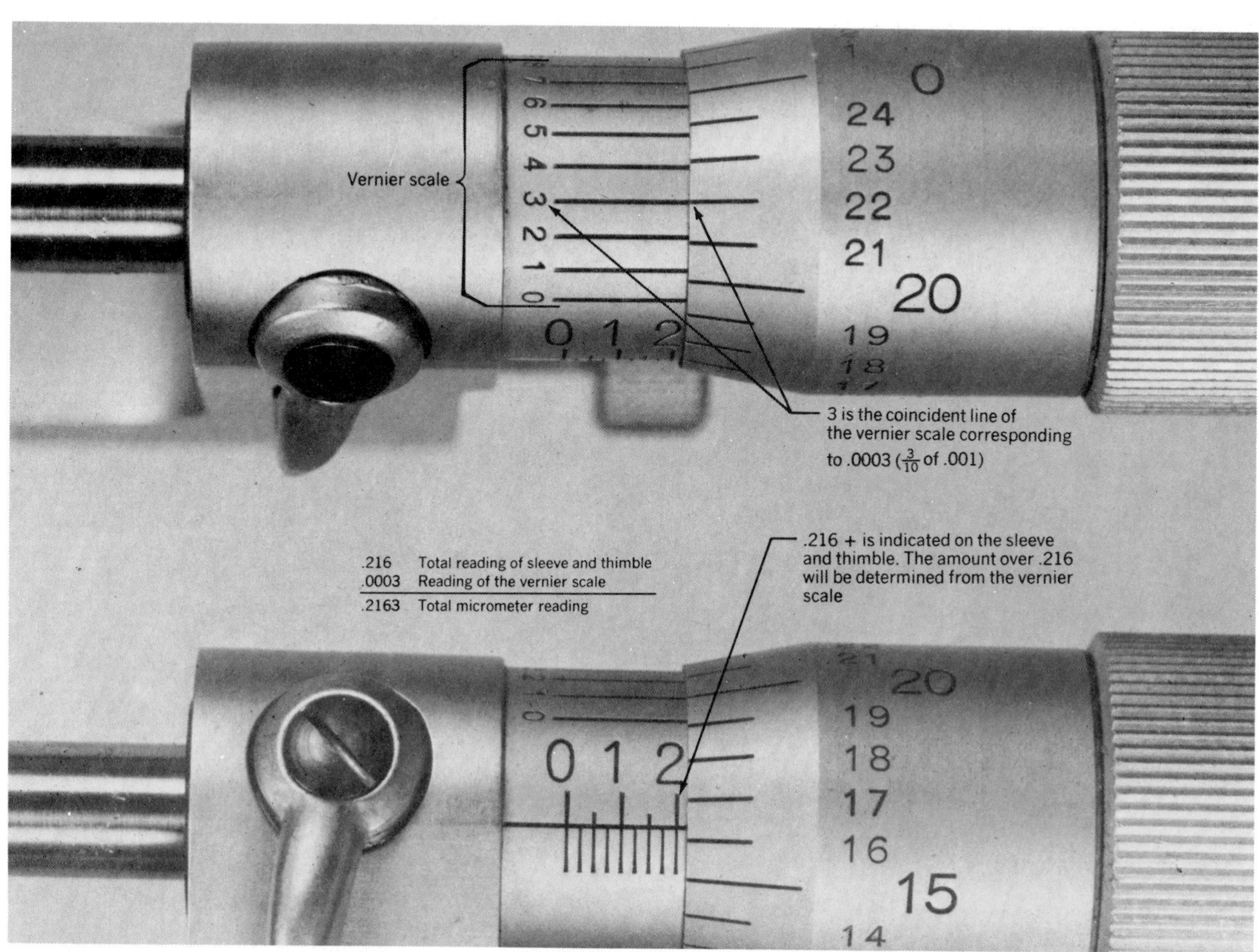

FIGURE C-153 Inch vernier micrometer reading of .2163 in.

surement that it will be asked to make. A vernier micrometer meets this requirement for measurement to the nearest thousandth. However, the instrument does not have the capability to discriminate to a one-hundred thousandth, which it should have if it is to be applied in a tenth of a thousandth measurement. This does not mean that a vernier micrometer should not be used for tenth measurement. The modern micrometer is manufactured with this potential in mind. It does mean that tenth measure should be carried out under **controlled conditions** if truly reliable results are to be obtained. The finish of the workpiece must be extremely smooth. Contact pressure of the measuring faces must be very consistent. The workpiece and instrument must be temperature stabilized. Heat transferred to the micrometer by handling can cause it to deviate considerably. Furthermore, the micrometer must be carefully calibrated against a known standard. Only under these conditions can true reliability be realized.

SELF-TEST

Read and record the five vernier micrometer readings in Figure C-154*a* to C-154*e*.

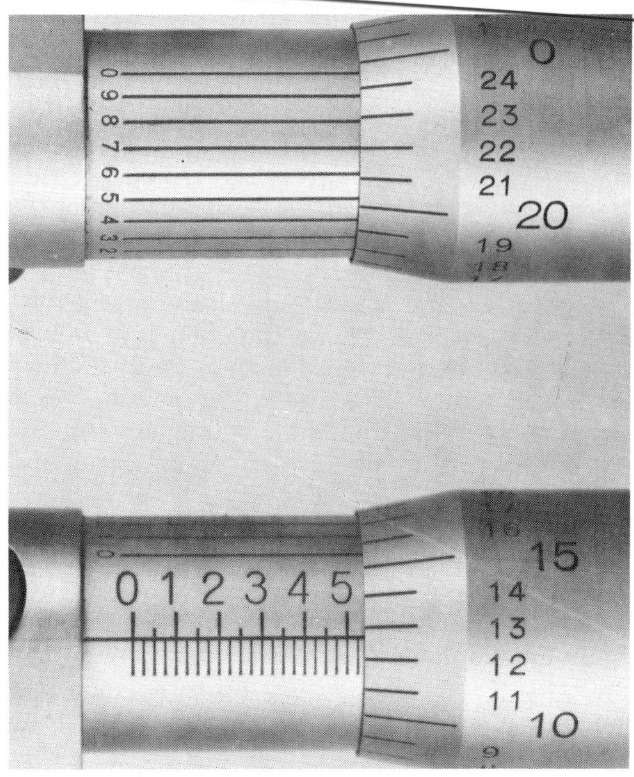

FIGURE C-154*b*

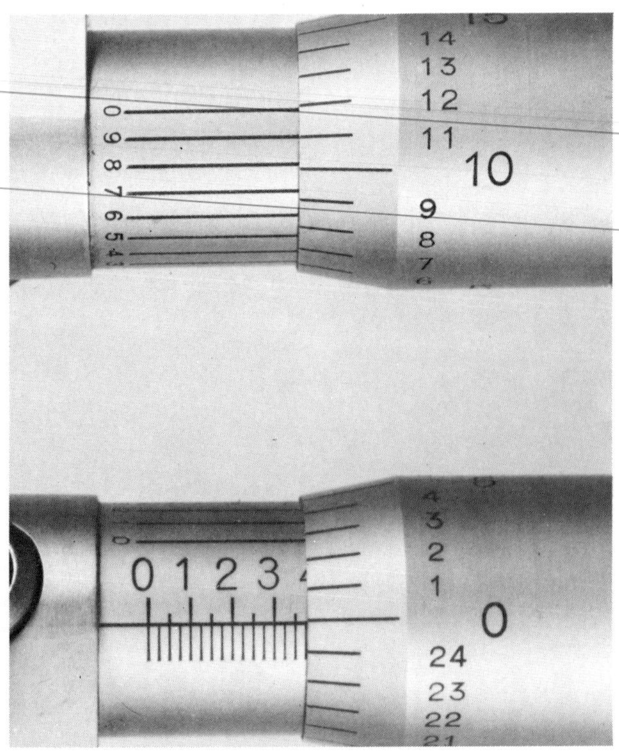

FIGURE C-154*a*

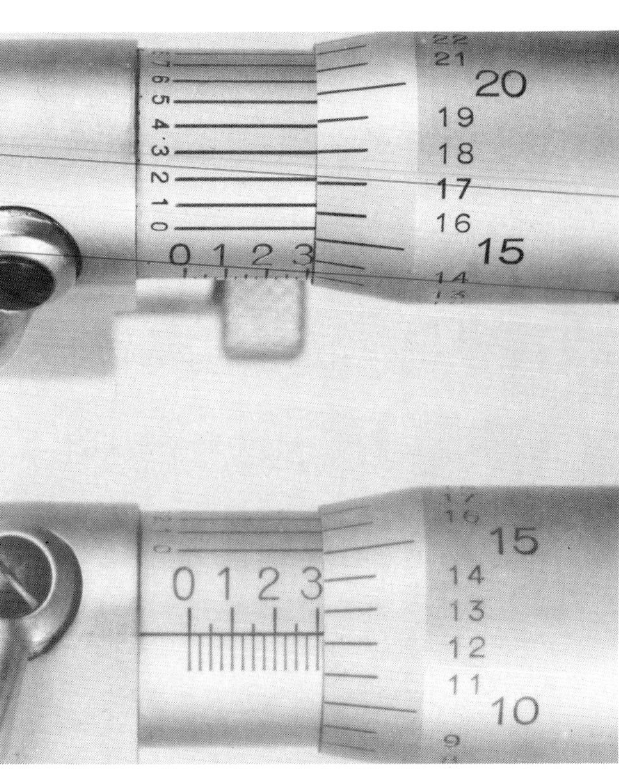

FIGURE C-154*c*

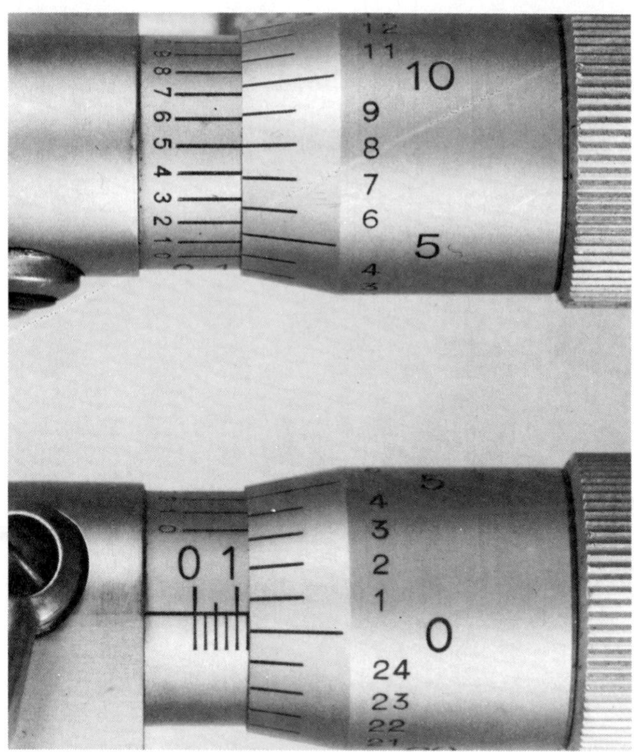

FIGURE C-154*d*

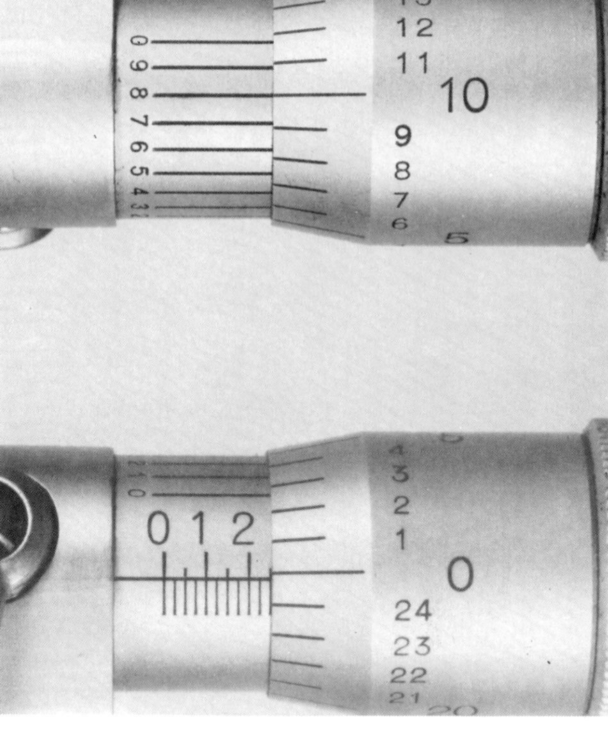

FIGURE C-154*e*

UNIT 5

Using Comparison Measuring Instruments

As a machinist, you will use a large number of measuring instruments that have no capacity within themselves to show a measurement. These tools will be used in comparison measurement applications where they are compared to a known standard, or used in conjunction with an instrument that has the capability of showing a measurement. In this unit you are introduced to the principles of comparison measurement, the common tools of comparison measurement, and their applications.

Objectives

After completing this unit, you should be able to:

1. Define comparison measurement.

2. Identify common comparison measuring tools.

3. Given a measuring situation, select the proper comparison tool for the measuring requirement.

Measurement by Comparison

All of us, at some time, have probably been involved in constructing something in which we used no measuring instruments of any kind. For example, suppose that you had to build some wooden shelves. You have the required lumber available with all boards longer than the shelf spaces. You hold a board to the shelf space and mark the required length for cutting. By this procedure, you have **compared** the length of the board **(the unknown length)** to the shelf space **(the known length or standard).** After cutting the first board to the marked length, it is then used to determine the lengths of the remaining shelves. The board, in itself, has no capacity to show a measurement. However, in this case, it became a measuring instrument.

A great deal of comparison measurement often involves the following steps:

1. A device that has no capacity to show measurement is used to establish and represent an unknown distance.

2. This representation of the unknown is then **transferred** to an instrument that has the capability to show a measurement.

This is commonly known as **transfer measurement.** In the example of cutting shelf boards, the shelf space was transferred to the first board and then the length of the first board was transferred to the remaining boards.

Transfer of measurements may involve some reduction in reliability. This factor must be kept in mind when using comparison tools requiring that a transfer be made. Remember that an instrument with the capability to show measurement directly is always best. **Direct reading instruments should be used whenever possible** in any situation. Measurements requiring a transfer must be accomplished with proper caution if reliability is to be maintained.

Common Comparison Measuring Tools and Their Applications

SPRING CALIPERS

The spring caliper is a very common comparison measuring tool for rough measurements of inside and outside dimensions. To use a spring caliper, set one jaw on the workpiece (Figure C-155). use this point as a pivot and swing the other caliper leg back and forth over the largest point on the diameter. At the same time, adjust the leg spacing. When the correct feel is obtained, remove the caliper and compare it to a steel rule to determine the

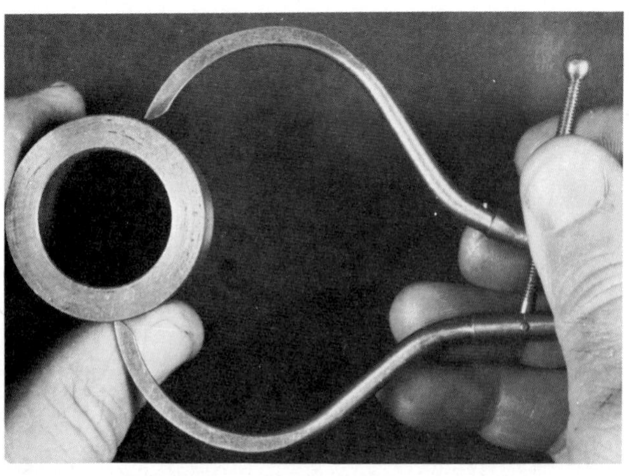

FIGURE C-155 Set one leg of the caliper against the workpiece.

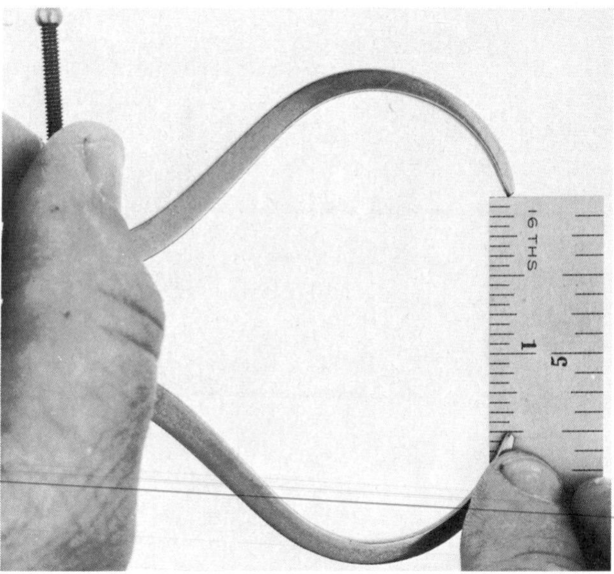

FIGURE C-156 Comparing the spring caliper to a steel rule.

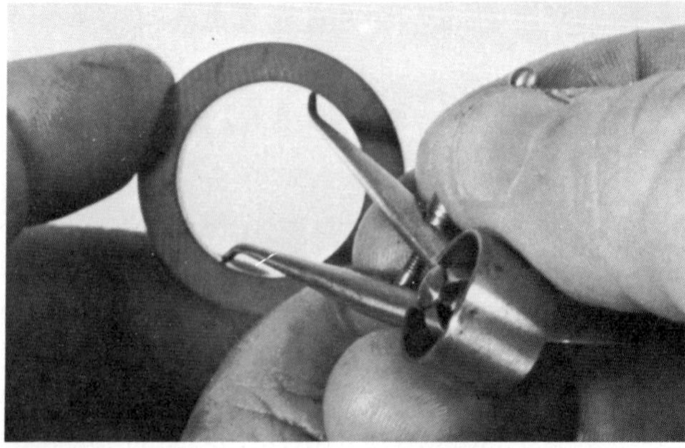

FIGURE C-157 Using an inside spring caliper.

reading (Figure C-156). The inside spring caliper can be used in a similar manner (Figure C-157). The use of the spring caliper is fading, and it has been replaced by measuring instruments of much higher reliability. The spring caliper should be used only for the roughest of measurements.

TELESCOPING GAGE

The telescoping gage is also a very common comparison measuring instrument. Telescoping gages are widely used in the machine shop, and they can accomplish a variety of measuring requirements. The telescoping gage is sometimes called a snap gage. This, however, is incorrect. A snap gage is another type of comparison measuring tool, discussed in the introduction to this section.

Telescoping gages generally come in a set of six gages (Figure C-158). The range of the set is usually $\frac{5}{16}$ to 6 in. (8 to 150 mm). The gage consists of one or two telescoping plungers with a handle and locking screw. The gage is inserted into a bore or slot, and the plungers are permitted to extend, thus conforming to the size of the feature. The gage is then removed and transferred to a micrometer where the reading is determined. The telescoping gage can be a reliable and versatile tool if proper procedure is used in its application.

PROCEDURE FOR USING THE TELESCOPING GAGE

STEP 1. Select the proper gage for the desired measurement range.

STEP 2. Insert the gage into the bore to be measured and release the handle lock screw (Figure C-159). Rock the gage sideways to ensure that you are measuring at the full diameter (Figure C-160). This is especially important in large-diameter bores.

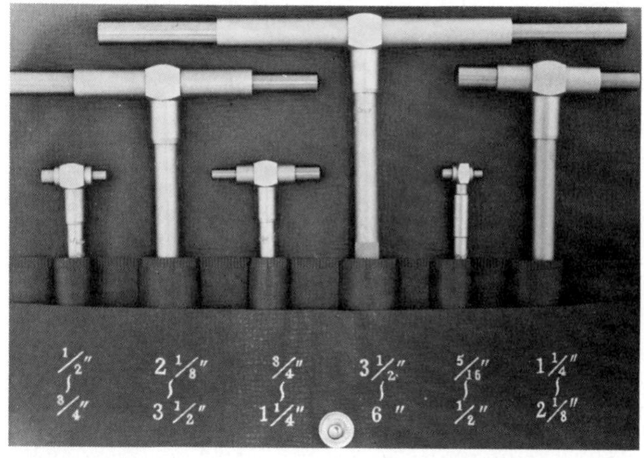

FIGURE C-158 Set of telescoping gages (Harry Smith & Associates).

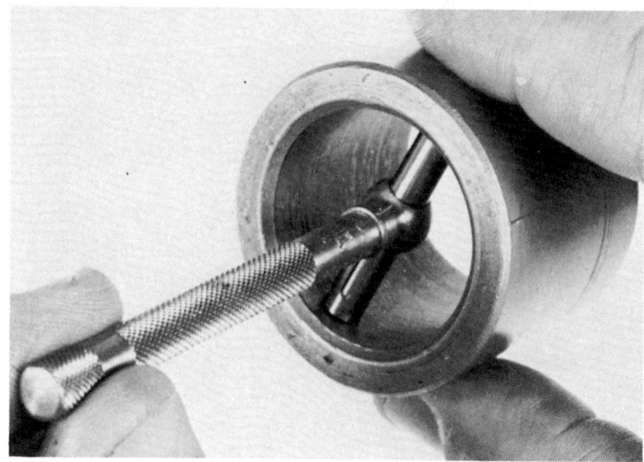

FIGURE C-159 Inserting the telescoping gage into the bore.

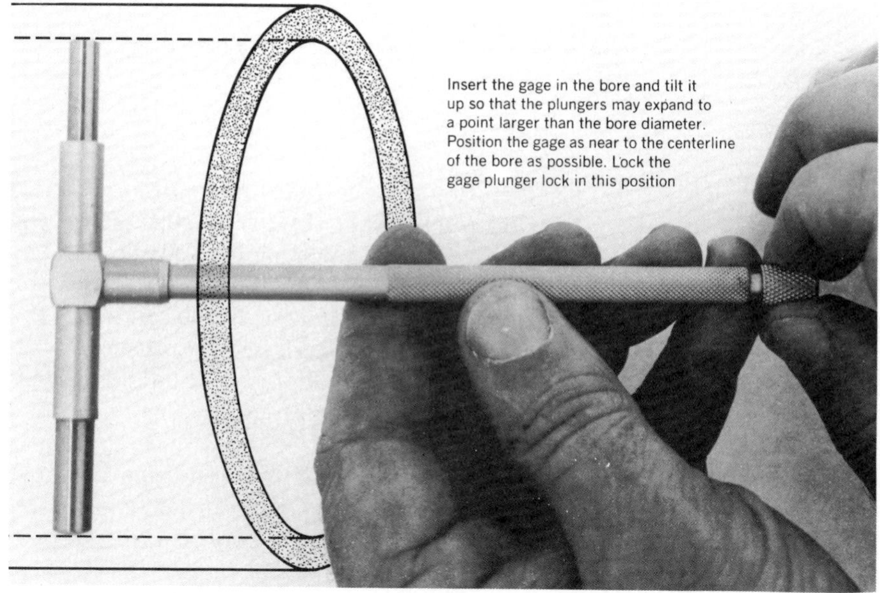

Insert the gage in the bore and tilt it up so that the plungers may expand to a point larger than the bore diameter. Position the gage as near to the centerline of the bore as possible. Lock the gage plunger lock in this position

FIGURE C-160 Release the lock and let the plungers expand larger than the bore.

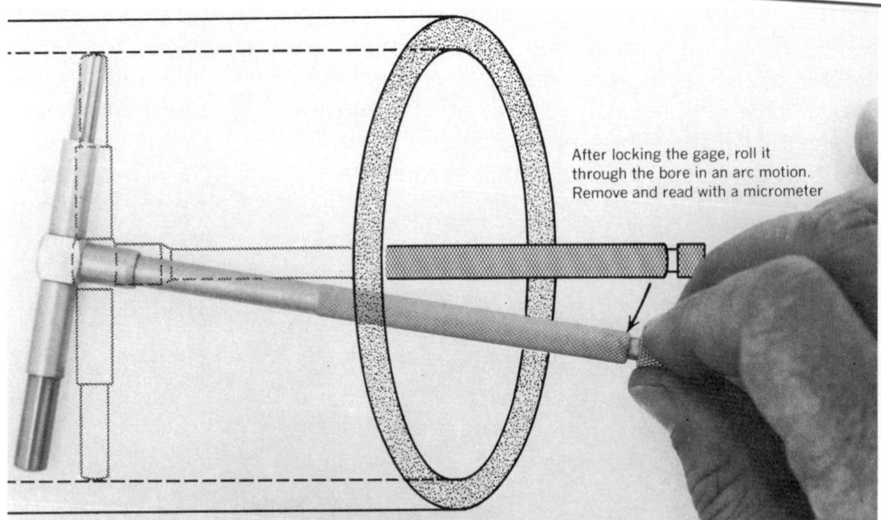

After locking the gage, roll it
through the bore in an arc motion.
Remove and read with a micrometer

FIGURE C-161 Tighten the lock and roll the gage through the bore.

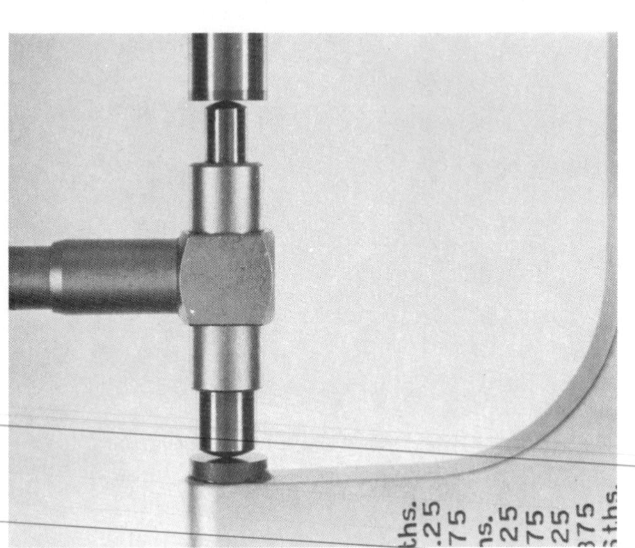

FIGURE C-162 Checking the telescoping gage with an outside micrometer (Harry Smith & Associates).

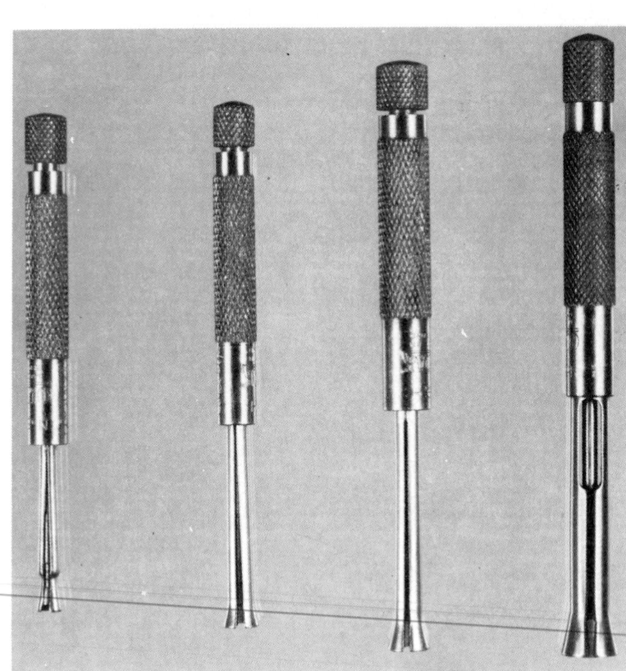

FIGURE C-163 Set of small hole gages (Harry Smith & Associates).

STEP 3. Lightly tighten the locking screw.

STEP 4. Use a downward or upward motion and roll the gage through the bore. The plungers will be pushed in, thus conforming to the bore diameter (Figure C-161). This part of the procedure should be done only once, as rolling the gage back through the bore may cause the plungers to be pushed in further, resulting in an inaccurate setting. If you feel that the gage is not centered properly, release the locking screw and repeat the procedure from the beginning.

STEP 5. Remove the gage and measure with an outside micrometer (Figure C-162). Place the gage between the micrometer spindle and the anvil. Try to determine the same feel on the gage with the micrometer as you felt while the gage was in the bore. Excessive pressure with the micrometer will depress the gage plungers and cause an incorrect reading.

STEP 6. Take at least two readings or more with the telescoping gage in order to verify reliability. If the readings do not agree, repeat steps 2 to 6.

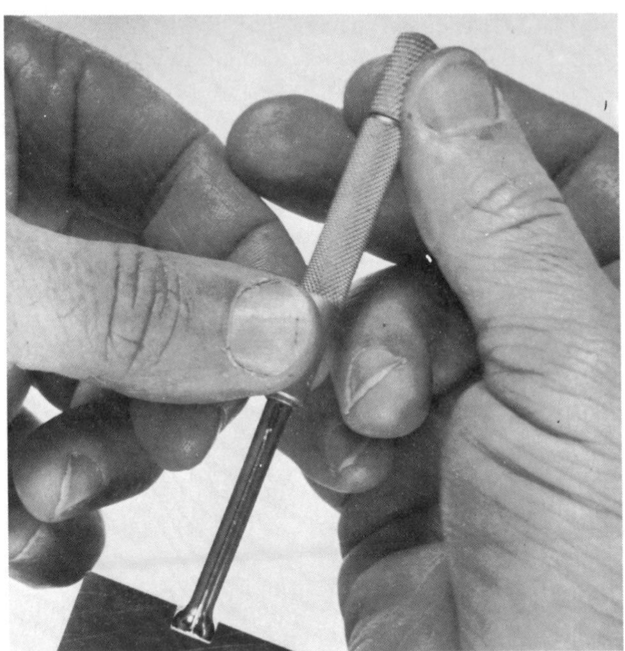

FIGURE C-164 Insert the small hole gage in the slot to be measured (Harry Smith & Associates).

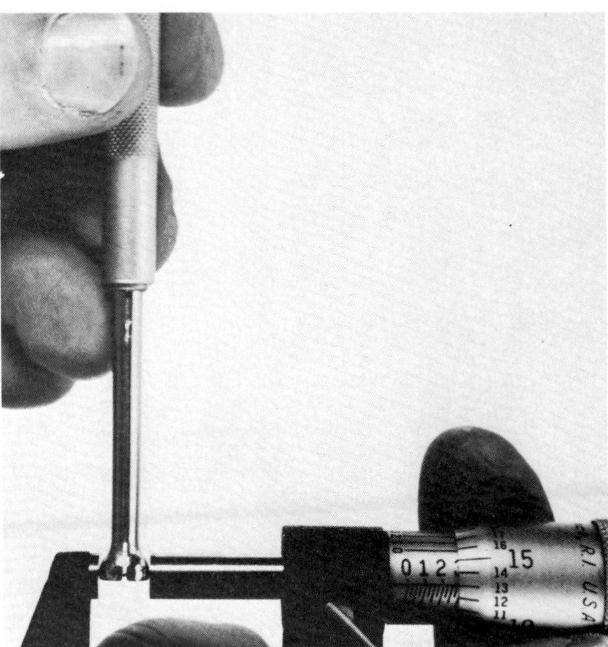

FIGURE C-165 Withdraw the gage and measure with an outside micrometer (Harry Smith & Associates).

SMALL HOLE GAGES

Small hole gages, like telescoping gages, come in sets with a range of $\frac{1}{8}$ to $\frac{1}{2}$ in. (4 to 12 mm). One type of small hole gage consists of a split ball that is connected to a handle (Figure C-163). A tapered rod is drawn between the split ball halves causing them to extend and contact the surface to be measured (Figure C-164). The split ball small hole gage has a flattened end so that a shallow hole or slot may be measured. After the gage has been expanded in the feature to be measured, it should be moved back and forth to determine the proper feel. The gage is then removed and measured with an outside micrometer (Figure C-165).

A second type of small hole gage consists of two small balls that can be moved out to contact the surface to be measured. This type of gage is available in a set ranging from $\frac{1}{16}$ to $\frac{1}{2}$ in. (1.5 to 12 mm). Once again, the proper feel must be obtained when using this type of small hole gage (Figure C-166). After the gage is set, it is removed and measured with a micrometer.

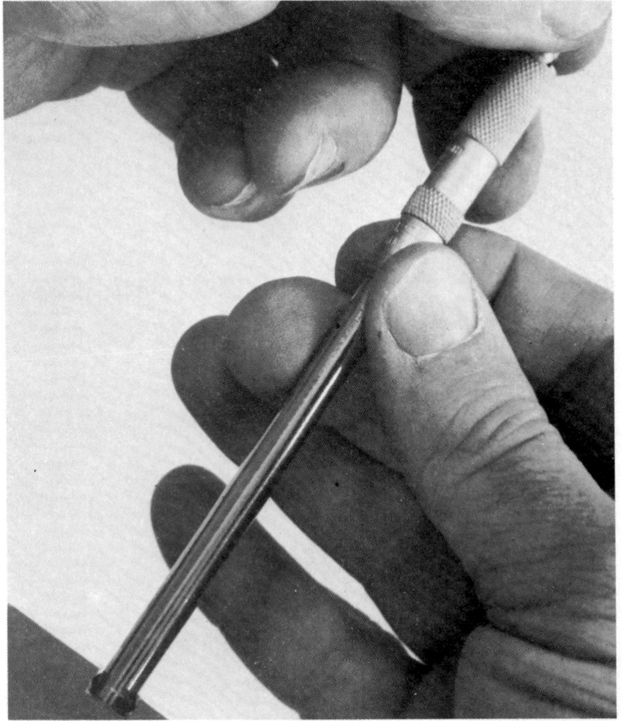

FIGURE C-166 Using the twin ball small hole gage (Harry Smith & Associates).

ADJUSTABLE PARALLELS

For the purpose of measuring slots, grooves, and keyways, the adjustable parallel may be used. Adjustable parallels are available in sets ranging from about $\frac{3}{8}$ to $2\frac{1}{4}$ in. (10 to 60 mm). They are precision ground for accuracy. The typical adjustable parallel consists of two parts that slide together on an angle. Adjusting screws are provided so that clearance in the slide may be adjusted or the parallel locked after setting for a measurement. As the halves of the parallel slide, the width increases or decreases depending on direction. The parallel is placed in the groove or slot to be measured and expanded until the parallel edges conform to the width to be measured. The parallel is then locked with a small screwdriver and measured with a micrometer (Figure C-167). If possible, an adjustable parallel should be left in place while being measured.

RADIUS GAGES

The typical radius gage set ranges in size from $\frac{1}{32}$ to $\frac{1}{2}$ in. (.8 to 12 mm). Larger radius gages are also available. The gage can be used to measure the radii of grooves and external or internal fillets (rounded corners). Radius gages may be separate (Figure C-168) or the full set may be contained in a convenient holder (Figure C-169).

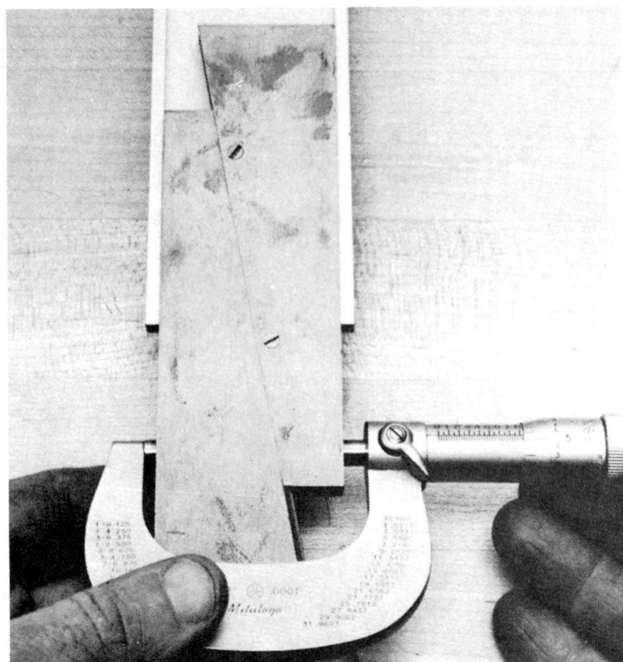

FIGURE C-167 Using adjustable parallels.

FIGURE C-168 Using an individual radius gage.

FIGURE C-169 Radius gage set.

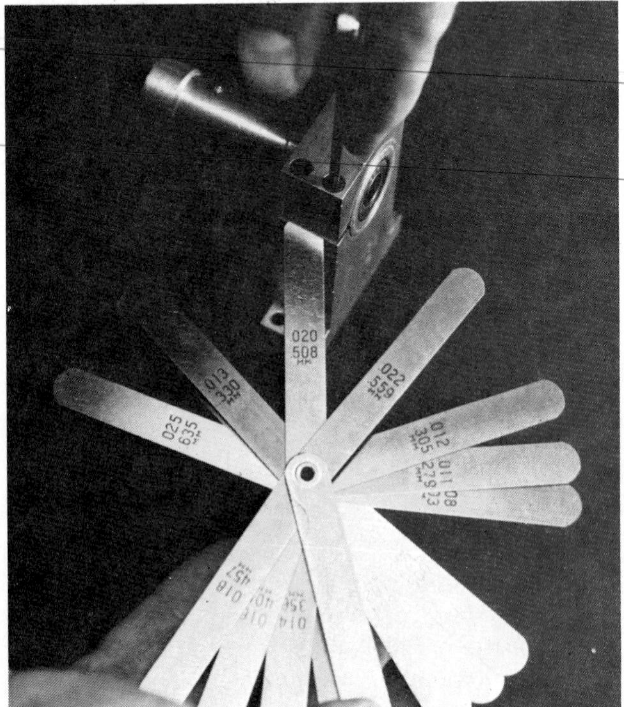

FIGURE C-170 Using a feeler or thickness gage.

FIGURE C-171 Setting the planer gage with an outside micrometer.

THICKNESS GAGES

The thickness gage (Figure C-170) is often called a feeler gage. It is probably best known for its various automotive applications. However, a machinist may use a thickness gage for such measurements as the thickness of a shim, setting a grinding wheel above a workpiece, or determining the height difference of two parts. The thickness gage is not a true comparison measuring instrument, as each leaf is marked as to size. However, it is good practice to check a thickness gage with a micrometer, especially when a number of leaves are stacked together.

PLANER GAGE

The planer gage functions much like an adjustable parallel. Planer gages were originally used to set tool heights on shapers and planers. They can also be used as a comparison measuring tool.

The planer gage may be equipped with a scriber and used in layout. The gage may be set with a micrometer (Figure C-171) or in combination with a dial test indicator and gage blocks. In this application, the planer gage is set by using a test indicator set to zero on a gage block (Figure C-172). This dimension is then transferred to the planer gage (Figure C-173). After the gage has been set,

FIGURE C-172 Setting the dial test indicator to a gage block.

FIGURE C-173 Transferring the measurement to the planer gage.

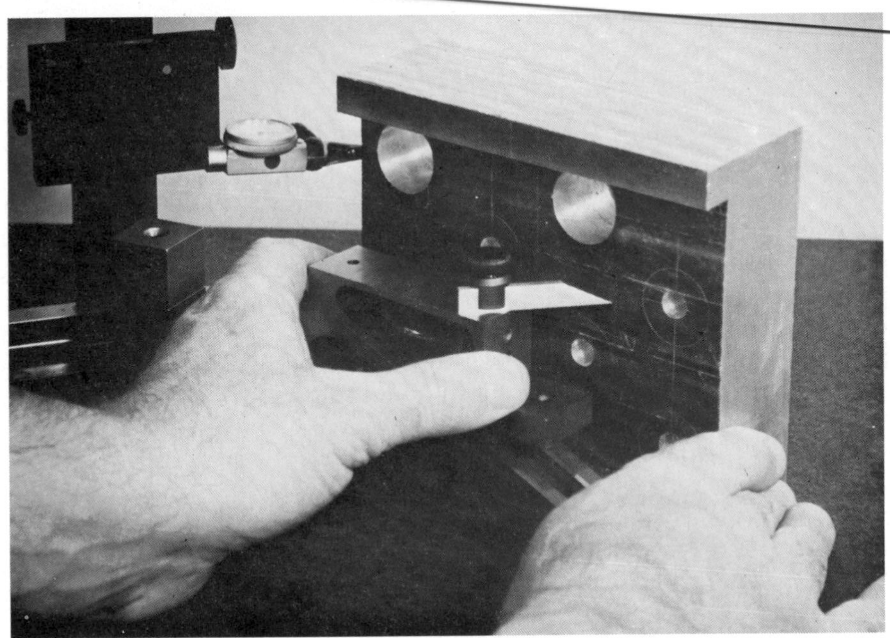

FIGURE C-174 Using the planer gage in layout.

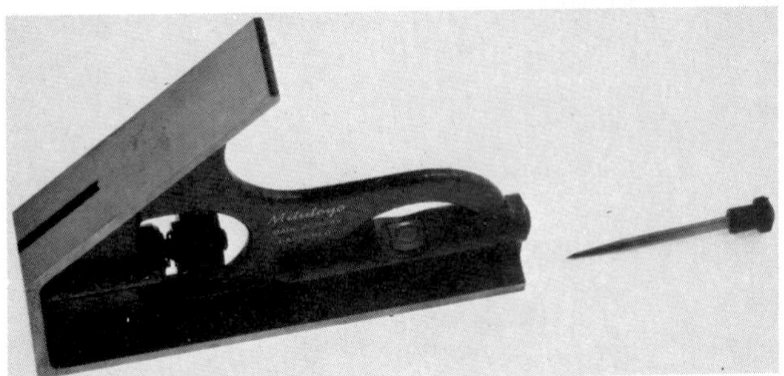

FIGURE C-175 Combination square head with scriber.

the scriber is attached and the instrument used in a layout application (Figure C-174).

SQUARES

The square is an important and useful tool for the machinist. A square is a comparative measuring instrument in that it compares its own degree of perpendicularity with an unknown degree of perpendicularity on the workpiece. You will use several common types of squares.

FIGURE C-176 Machinist's combination set.

FIGURE C-177 Using the combination square.

FIGURE C-178 Solid beam square.

MACHINIST'S COMBINATION SQUARE The combination square (Figure C-175) is part of the combination set (Figure C-176). The combination set consists of a graduated rule, square head, bevel protractor, and center head. The square head slides on the graduated rule and can be locked at any position (Figure C-177). This feature makes the tool useful for layout as the square head can be set according to the rule graduations. The combination square head also has a 45-degree angle along with a spirit level and layout scriber. The combination set is one of the most versatile tools of the machinist.

SOLID BEAM SQUARE On the solid beam square, the beam and blade are fixed. Solid beam squares range in size from 2 to 72 in. (Figure C-178).

PRECISION BEVELED EDGE SQUARE The precision beveled edge square is an extremely accurate square used in the toolroom and in inspection applications. The beveled edge permits a single line of contact with the part to be checked. Precision squares range in size from 2 to 14 in. (Figure C-179).

The squares discussed up to this point do not have any capacity to directly indicate the amount of deviation from perpendicularity. The only determination that can be made is that the workpiece is as perpendicular as or not as perpendicular as the square. The actual amount of deviation from perpendicularity on the workpiece must be determined by other measurements. With the following group of squares, the deviation from perpendicularity can be measured directly. In this respect, the following instruments are not true comparison tools, since they have capacity to show a measurement directly.

FIGURE C-179 Precision beveled edge square (Brown and Sharpe Manufacturing Company).

CYLINDRICAL SQUARE The direct reading cylindrical square (Figure C-180) consists of an accurate cylinder with one end square to the axis of the cylinder. The other end is made slightly out of square with the cylindrical axis. When the non-square end is placed on a clean surface plate, the instrument is actually tilted slightly. As the square is rotated (Figure C-181), one point on the circumference of the cylinder will eventually come into true perpendicularity with the surface plate. On a cylindrical square, this point is marked by a vertical line running the full length of the tool. The cylindrical square has a set of curved lines marked on the cylinder that permits deviation from squareness of the workpiece to be determined. Each curved line represents a deviation of .0002 of an inch over the length of the instruments.

Cylindrical squares are applied in the following manner. The square is placed on a clean surface plate and brought into contact with the part to be checked. The square is then rotated until contact is made over the entire length of the instrument. The deviation from squareness is determined by reading the amount corresponding to the line on the square that is in contact with the workpiece. Cylindrical squares are often used to check the accuracy of another square (Figure C-180). When the instrument is used on its square end, it may be applied as a plain square. Cylindrical squares range in size from 4 to 12 in.

FIGURE C-180 Cylindrical square.

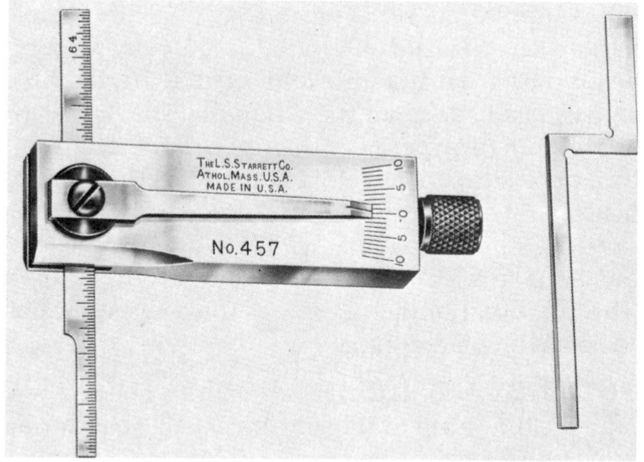

FIGURE C-182 Diemaker's square (L.S. Starrett Company).

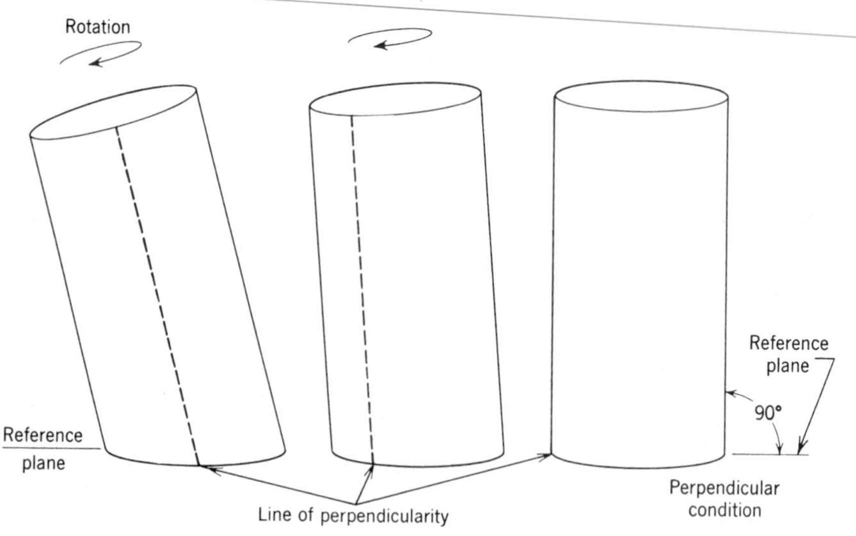

FIGURE C-181 Principle of the cylindrical square.

DIEMAKER'S SQUARE The diemaker's square (Figure C-182) is used in such applications as checking clearance angle on a die (Figure C-183). The instrument can be used with a straight or offset blade. A diemaker's square can be used to measure a deviation of 10 degrees on either side of the perpendicular.

MICROMETER SQUARE The micrometer square (Figure C-184) is another type of adjustable square. The blade is tilted by means of a micrometer adjustment to determine the deviation of the part being checked.

The square is one of the few tools used in measurement that is essentially self-checking. If you have a workpiece with accurately parallel sides, as measured with a micrometer, one end can be observed under the beam of the square and the error observed. Now the part can be rotated under the beam 180 degrees and rechecked. If the error is identical but reversed, the square is accurate. If there is a difference, except for simple reversal, the square should be considered inaccurate. It should be checked against a standard, such as a cylindrical square.

INDICATORS

The many types of indicators are some of the most valuable and useful tools for the machinist. There are two general types of indicators in general use. These are **dial indicators** and **dial test indicators.** Both types generally take the form of a spring loaded spindle that, when depressed, actuates the hand of an indicating dial. At the initial examination of a dial or test indicator, you will note that the dial face is usually graduated in thousandths of an inch or subdivisions of thousandths. This might lead you to the conclusion that the indicator spindle movement corresponds directly to the amount shown on the indicator face. However, this conclusion is to be arrived at only with the most cautious judgment. **Dial test indicators should not be used to make direct linear measurements.** Reasons for this will be developed in the information to follow. Dial indicators can be used to make linear measurements, but only if they are specifically designed to do so and under proper conditions.

As a machinist, you will use dial and test indicators almost daily in the machine shop. Indicators are very essential to the accurate completion of your job. However, do not ask that an indicator do what it was not designed to do. Dial and test indicators when properly used are an invaluable member of the machinist's many tools.

DIAL INDICATORS Dial indicators have discriminations that typically range from .00005 to .001 of an inch. In metric dial indicators, the discriminations typically range from .002 to .01 mm. Indicator ranges or the total reading capacity of the instrument may commonly range from .003 to 2.000 in., or .2 to 50 mm for metric instruments.

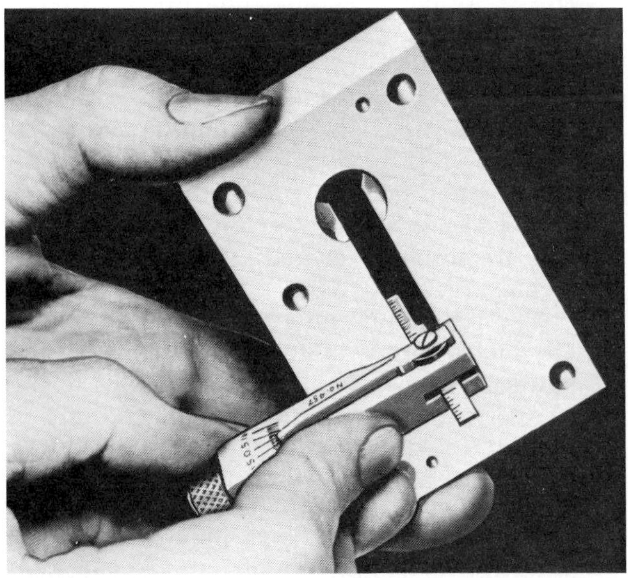

FIGURE C-183 Using the diemaker's square to check die clearance angle (L.S. Starrett Company).

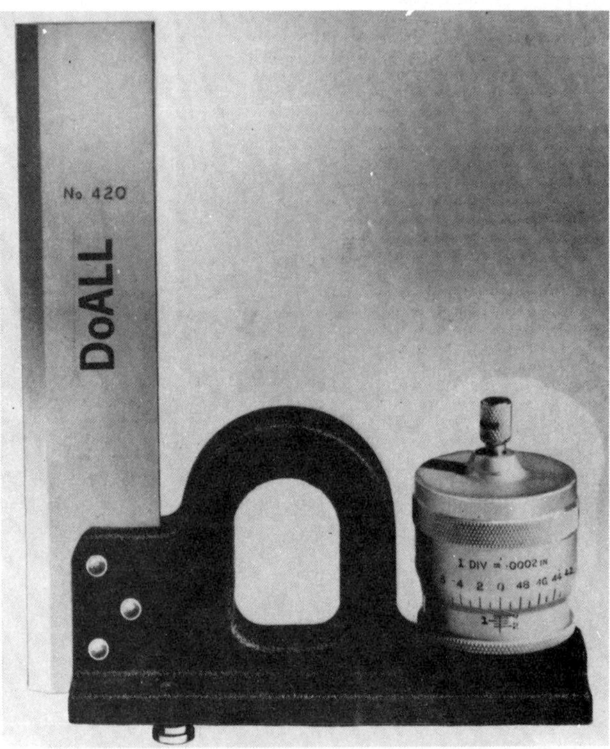

FIGURE C-184 Micrometer square (DoALL Company).

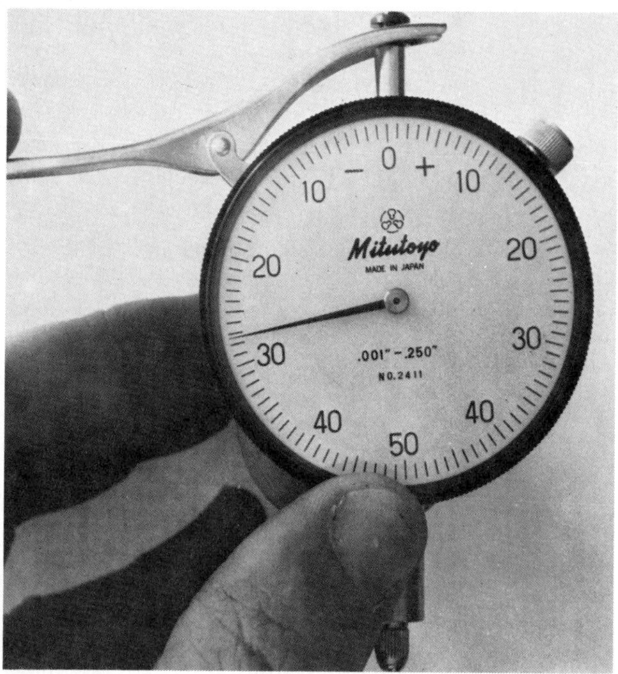

FIGURE C-185 Balanced dial indicator (Harry Smith & Associates).

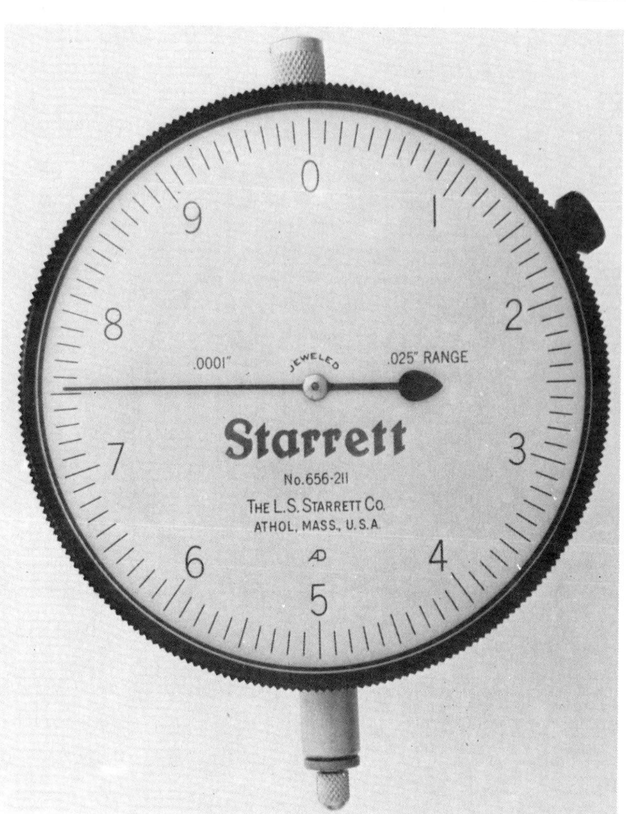

FIGURE C-187 Dial indicator with .025 in. range and .0001 in. discrimination (Harry Smith & Associates).

FIGURE C-186 Dial indicator with 1 in. of travel (Harry Smith & Associates).

On the **balanced** indicator (Figure C-185), the face numbering goes both clockwise and counterclockwise from zero. This is convenient for comparator applications where readings above and below zero need to be indicated. The indicator shown has a lever actuated stem. This permits the stem to be retracted away from the workpiece if desired.

The continuous reading indicator (Figure C-186) is numbered from zero in one direction. This indicator has a discrimination of .0005 and a total range of 1 in. The small center hand counts revolutions of the large hand. Note that the center dial counts each .100 in. of spindle travel. This indicator is also equipped with **tolerance hands** that can be set to mark a desired limit. Many dial indicators are designed for high discrimination and short range (Figure C-187). This indicator has a .0001 discrimination and a range of .025 in.

The **back plunger** indicator (Figure C-188) has the spindle in the back or at right angles to the face. This type of indicator usually has a range of about .200 in. with .001-in. discrimination. It is a very popular model for use on a machine tool. The indicator usually comes with a number of mounting accessories (Figure C-188).

Indicators are equipped with a **rotating face or bezel.** This feature permits the instrument to be set to zero at any desired place. Many indicators

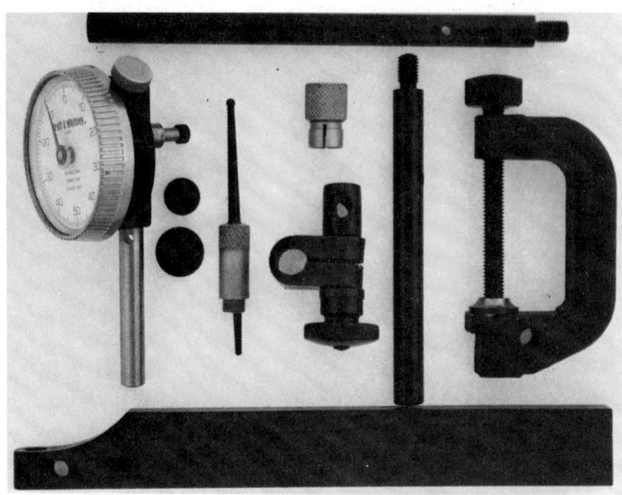

FIGURE C-188 Back plunger indicator with mounting accessories (Harry Smith & Associates).

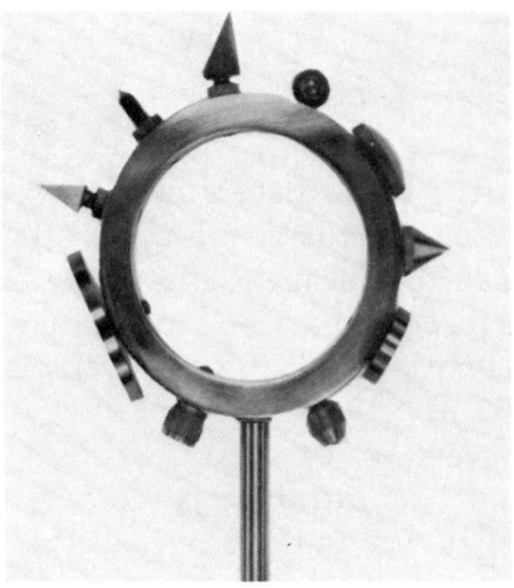

FIGURE C-189 Dial indicator tips with holder (Rank Scherr-Tumico, Inc.).

also have a **bezel lock.** Dial indicators may have removable spindle tips, thus permitting use of different shaped tips as required by the specific application (Figure C-189).

CARE AND USE OF INDICATORS Dial indicators are precision instruments and should be treated accordingly. They **must not be dropped** and should **not be exposed to severe shocks.** Dropping an indicator may bend the spindle and render the instrument useless. Shocks, such as hammering on a workpiece while an indicator is still in contact, may damage the delicate operating mechanism. The spindle should be kept free from dirt and grit, which can cause binding that results in damage and false readings. It is important to **check** indicators **for free travel** before using. When an indicator is not in use, it should be stored carefully with a protective device around the spindle.

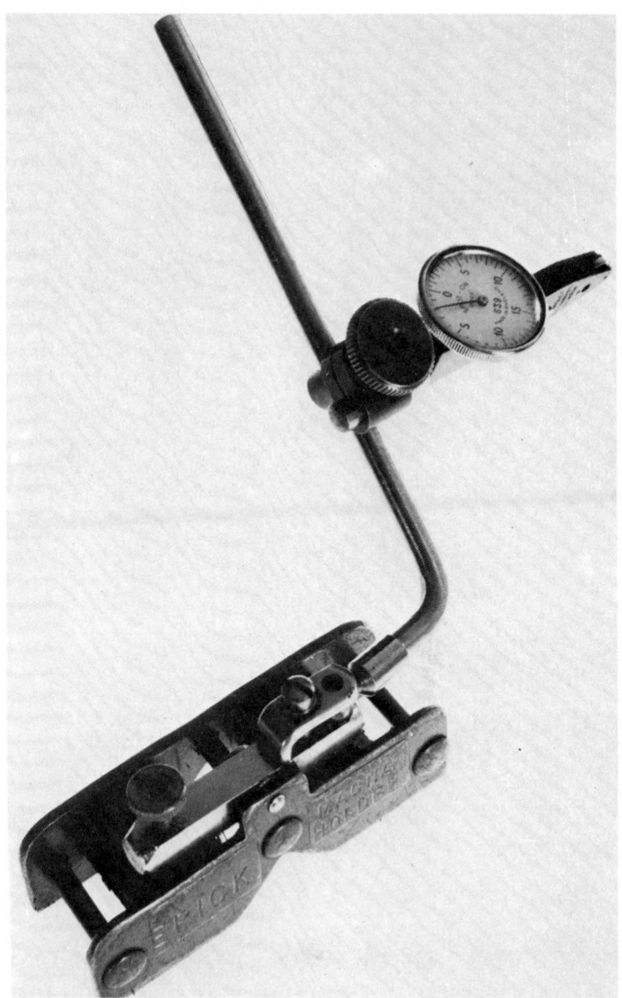

FIGURE C-190 Permanent magnetic indicator base.

One of the problems encountered by indicator users is **indicator mounting.** All indicators must be **mounted solidly** if they are to be reliable. Indicators must be clamped or mounted securely when used on a machine tool. A number of mounting devices are in common use. Some of these have magnetic bases that permit an indicator to be attached at any convenient place on a machine tool. The permanent-magnet indicator base (Figure C-190) is a useful accessory. This type of indicator base is equipped with an adjusting screw that can be used to set the instrument to zero. Another useful magnetic base has a provision for turning off the magnet by mechanical means (Figure C-191). This feature makes for easy locating of the base prior to turning on the magnet. A number of bases making use of flexible link indicator holding arms are also in general use. Often they are not adequately rigid for reliability. In addition to holding an indicator on a magnetic base, it may be clamped to a machine setup by the use of any suitable clamps.

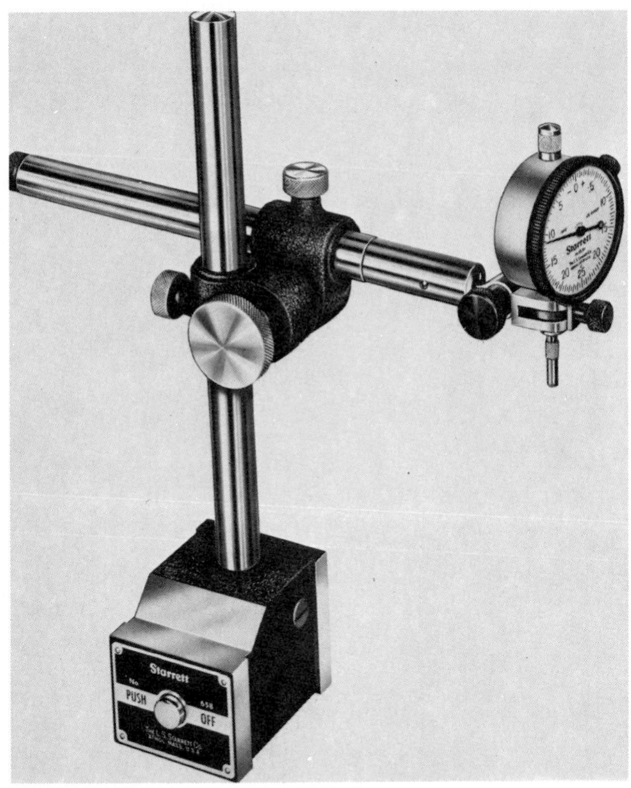

FIGURE C-191 Magnetic-base indicator holder with on/off magnet (L.S. Starrett Company).

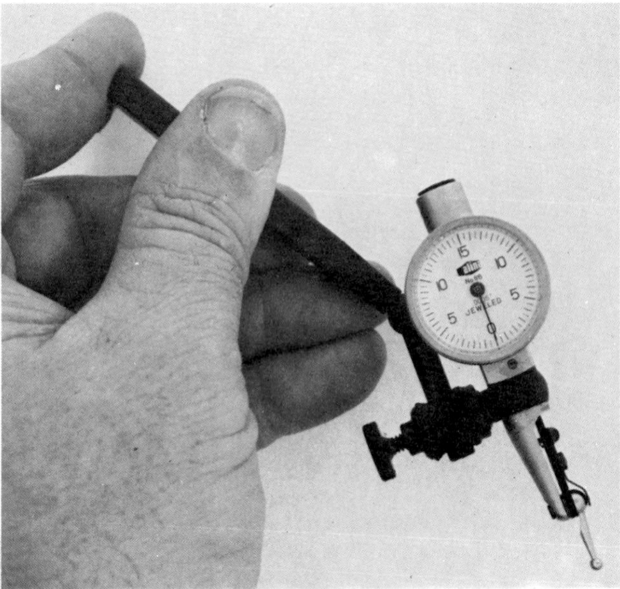

FIGURE C-192 Dial test indicator (Harry Smith & Associates).

DIAL TEST INDICATORS Dial test indicators frequently have a discrimination of .0005 in. and a range of about .030 in. The test indicator is frequently quite small (Figure C-192) so that it can be used to indicate in locations inaccessible to other indicators. The spindle or tip of the test indicator can be swiveled to any desired position. Test indicators are usually equipped with a **movement reversing lever.** This means that the indicator can be actuated by pressure from either side of the tip. The instrument need not be turned around. Test indicators, like dial indicators, have a rotating bezel for zero setting. Dial faces are generally of the balanced design. The same care given to dial indicators should be extended to test indicators.

POTENTIAL FOR ERROR IN USING DIAL INDICATORS Indicators must be used with appropriate caution if reliable results are to be obtained. The spindle of a dial indicator usually consists of a gear rack that engages a pinion and other gears that drive the indicating hand. In any mechanical device, there is always some clearance between the moving parts. There are also minute errors in the machining of the indicator parts. Because of these, small errors may creep into an indicator reading. This is especially true in long travel indicators. For example, if a 1-in. travel indicator with a .001 discrimination had plus or minus 1 percent error at full travel, the following condition could exist if the instrument were to be used for a direct measurement: You wish to determine if a certain part is within the tolerance of .750 ± .003 in. The 1-in. travel indicator has the capacity for this, but remember that it is only accurate to plus or minus 1 percent of full travel. Therefore, .01 × 1.000 in. is equal to ±.010 in., or the total possible error. To calculate the error per thousandth of indicator travel, divide .010 in. by 1000. This is equal to .00001 in., which is the average error per thousandth of indicator travel. This means that at a travel amount of .750 in., the indicator error could be as much as .00001 in. × 750, or ±.0075 in. In a direct measurement of the part, the indicator could read anywhere from .7425 to .7575. As you can see, this is well outside the part tolerance and would hardly be reliable (Figure C-193).

The indicator should be used as a comparison measuring instrument by the following procedure (Figure C-193). The indicator is set to zero on a .750-in. gage block. The part to be measured is then placed under the indicator spindle. In this case, the error caused by a large amount of indicator travel is greatly reduced, because the travel is never greater than the greatest deviation of a part from the basic size. The total part tolerance is .006 in. (±.003 in.) Therefore, 6 × .00001 in. error per thousandth is equal to only ± .00006 in. This is well within the part tolerance and, in fact, cannot even be read on a .001-in. discrimination indicator.

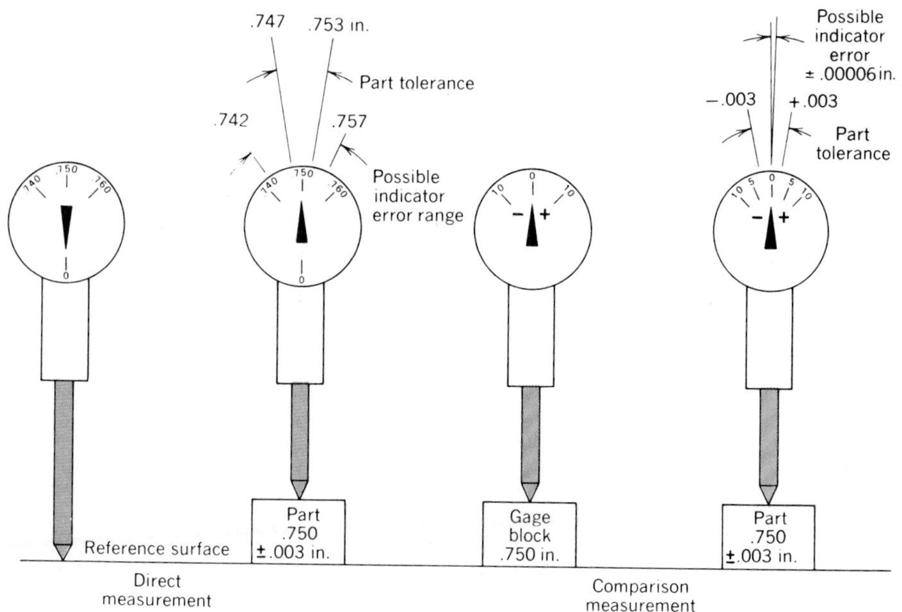

FIGURE C-193 Potential for errors in indicator travel.

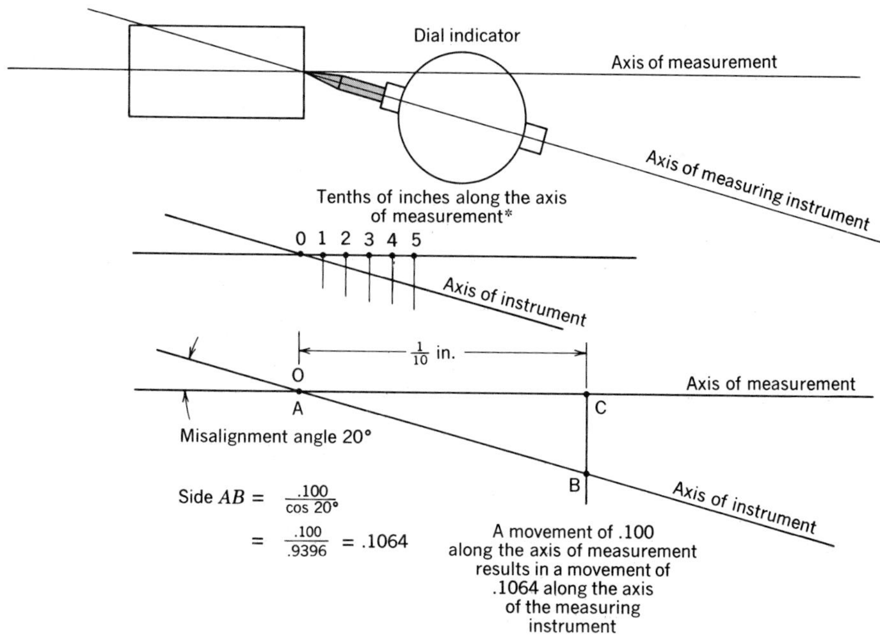

FIGURE C-194 Cosine error.

Of course, you will not know what the error amounts to on any specific indicator. This can only be determined by a calibration procedure. Furthermore, you would probably not use a long travel indicator in this particular application. A moderate to short travel indicator would be more appropriate. Keep in mind that any indicator may contain some **travel error** and that by using a fraction of that travel, this error can be reduced considerably.

In the introduction to this section you learned that the axis of a linear measurement instrument must be in line with the axis of measurement. If a dial indicator is misaligned with the axis of measurement, the following condition will exist: line AC represents the axis of measurement, while line AB represents the axis of the dial indicator (Figure C-194). If the distance from A to C is .100 in., then the distance from A to B is obviously larger, since it is the hypotenuse of triangle ABC. The angle of misalignment, angle A, is equal to 20 degrees. The distance from A to B can then be calculated by the following:

$$AB = \frac{.100}{\cos A}$$

$$AB = \frac{.100}{.9396}$$

$$AB = .1064 \text{ in.}$$

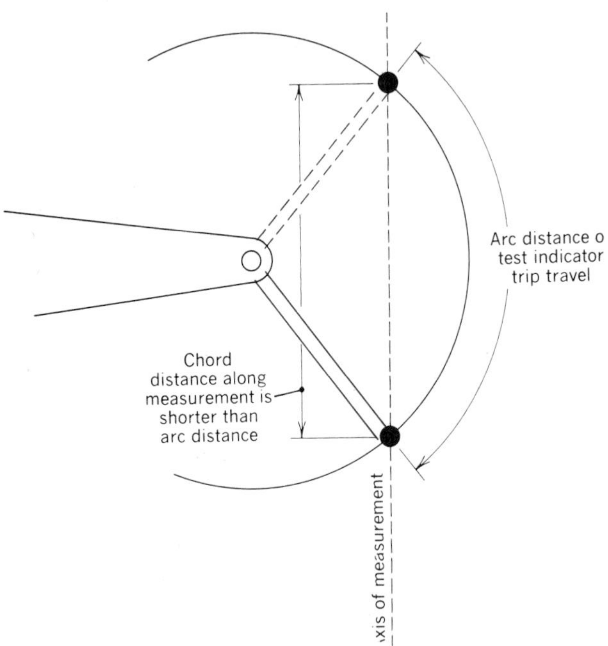

FIGURE C-195 Potential for error in dial test indicator tip movement.

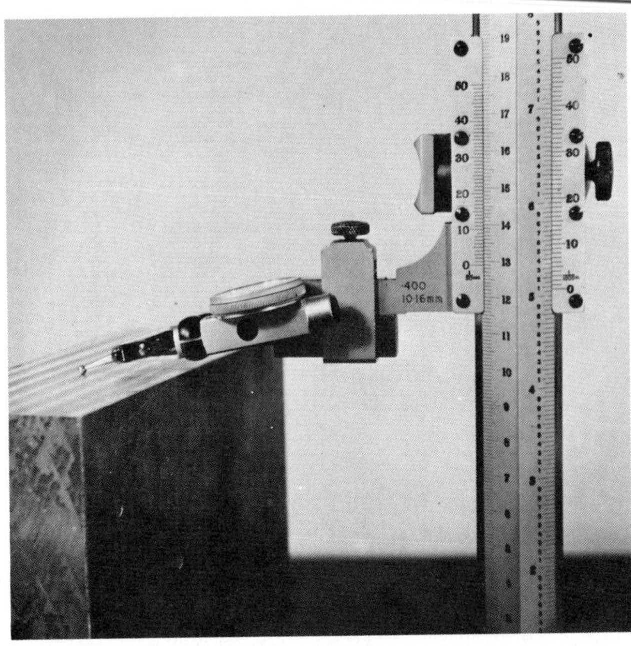
FIGURE C-197 Using the dial test indicator and vernier height gage to measure the workpiece.

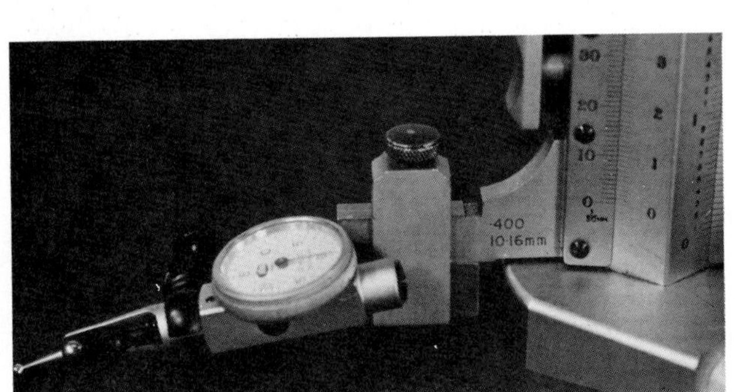
FIGURE C-196 Setting the dial test indicator to zero on the reference surface.

This shows that a movement along the axis of measurement results in a much larger movement along the instrument axis. This **error** is known as **cosine error** and must be kept in mind when using dial indicators. Cosine error is increased as the angle of misalignment is increased.

When using dial test indicators, watch for **arc versus chord length errors** (Figure C-195). The tip of the test indicator moves through an arc. This distance may be considerably greater than the chord distance of the measurement axis. Dial test indicators should **not** be used to make direct measurements. They should only be applied in comparison applications.

USING DIAL TEST INDICATORS IN COMPARISON MEASUREMENTS The dial test indicator is very useful in making comparison measurements in conjunction with the height gage and height transfer micrometer. Comparison measurement using a vernier height gage is accomplished by the following procedure.

STEP 1. Set the height gage to zero and adjust the test indicator until it also reads zero when in contact with the surface plate (Figure C-196). It is very important to use a test indicator for this procedure. Using a scriber tip on a height gage is an inferior way to attempt measurements and can lead to substantial error.

STEP 2. Raise the indicator and adjust the height gage vernier until the indicator reads zero on the workpiece (Figure C-197). Read the dimension from the height scale.

Comparison measurement can also be accomplished using a precision height gage (Figure C-198). The precision height gage shown consists of a series of rings that are moved by the micrometer spindle. The ring spacing is a very accurate 1 in. and the micrometer head has a 1-in. travel. Other designs use projecting gage blocks at inch intervals, or other types of measuring steps. Discrimination of the typical height micrometer is .0001 in. (Figure C-199). Precision height gages come in various height capacities and sometimes have riser blocks as accessories. In Figure C-201, a planer gage is being set using the test indicator and height micrometer. The following procedure is used.

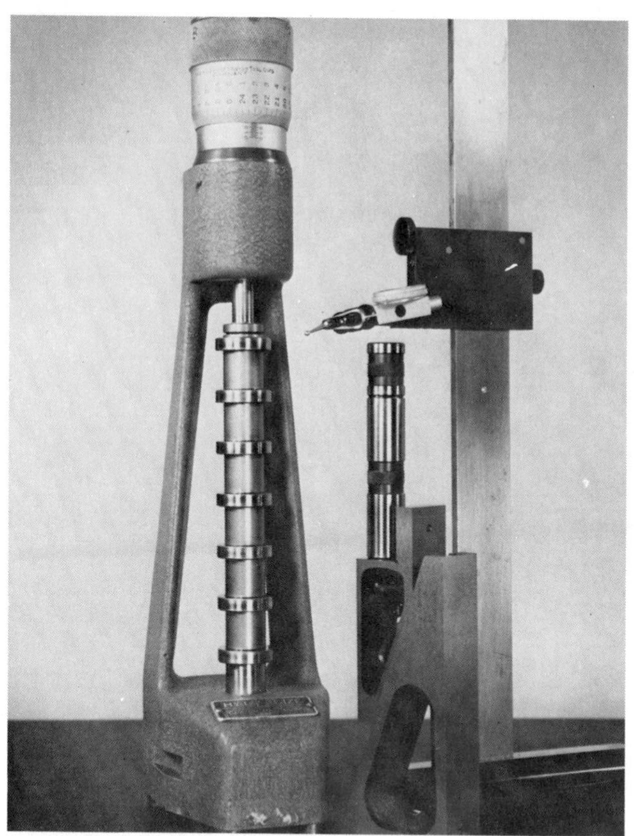

FIGURE C-198 Precision height gage.

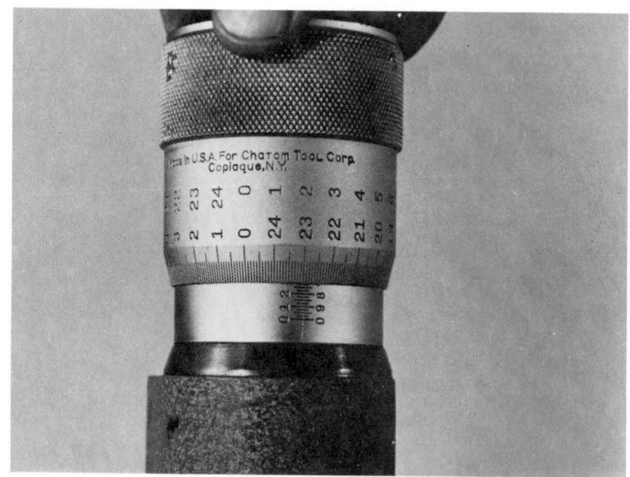

FIGURE C-199 Reading the precision height gage.

STEP 1. The height transfer micrometer is adjusted to the desired height setting. The test indicator is zeroed on the appropriate ring (Figure C-200). A height transfer gage, vernier height gage, or other suitable means can be used to hold the indicator.

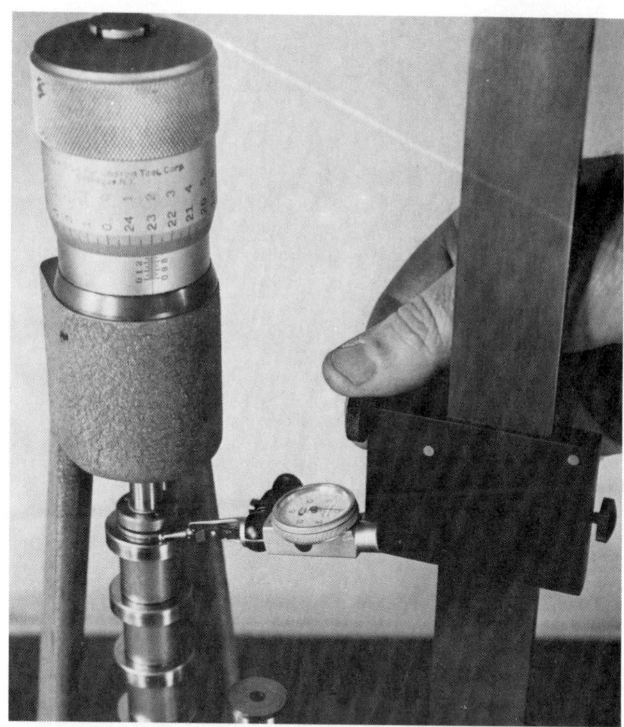

FIGURE C-200 Setting the dial test indicator to the precision height gage.

FIGURE C-201 Transferring the measurement to the planer gage.

STEP 2. The indicator is then moved over to the planer gage. The planer gage is then adjusted until the test indicator reads zero (Figure C-201).

FIGURE C-202 Setting the dial test indicator using the optical height gage.

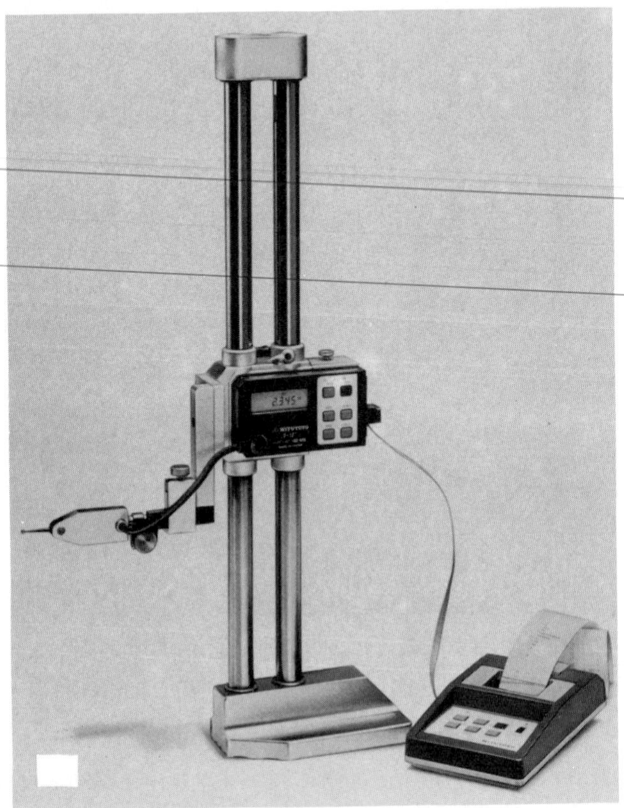

FIGURE C-203 Electronic digital height gage (MTI Corporation).

FIGURE C-204 Dial indicator comparator.

Test indicators can also be used to accomplish comparison measurement in conjunction with an optical height gage (Figure C-202). Digital electronic height gages are also available (Figure C-203).

COMPARATORS

Comparators are exactly what their name implies. They are instruments that are used to compare the size or shape of the workpiece to a known standard. Types include dial indicator, optical, electrical, and electronic comparators. Comparators are used where parts must be checked to determine acceptable tolerance. They may also be used to check the geometry of such things as threads, gears, and formed machine tool cutters. The electronic comparator may be found in the inspection area, toolroom, or gage laboratory and used in routine inspection and calibration of measuring tools and gages.

DIAL INDICATOR COMPARATORS The dial indicator comparator is no more than a dial indicator attached to a rigid stand (Figure C-204). These are dial indicator instruments such as the ones previously discussed. However, in their application as comparator instruments, as many errors as possible have been eliminated by the fixed design of the instrument's components. The indicator is set to zero at the desired dimension by use of gage blocks (Figure C-205). When using a dial indicator comparator, keep in mind the **potential for error** in indicator travel and instrument alignment along the axis of measurement. Once the indicator has been set to zero, parts can be checked for acceptable tolerance (Figure C-206). A

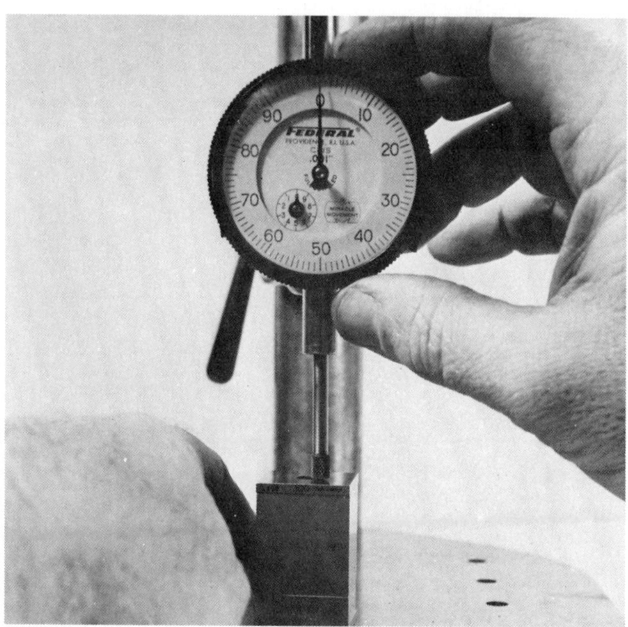

FIGURE C-205 Setting the dial indicator to zero using gage blocks.

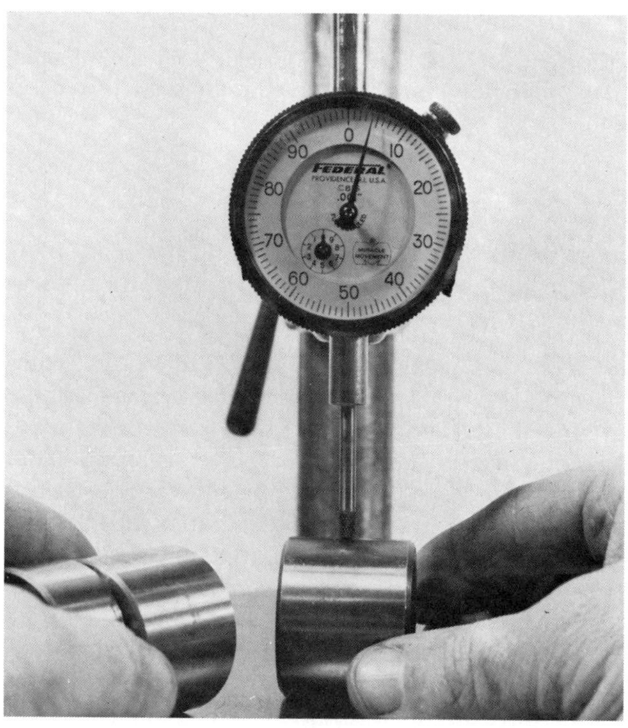

FIGURE C-206 Using the dial indicator comparator.

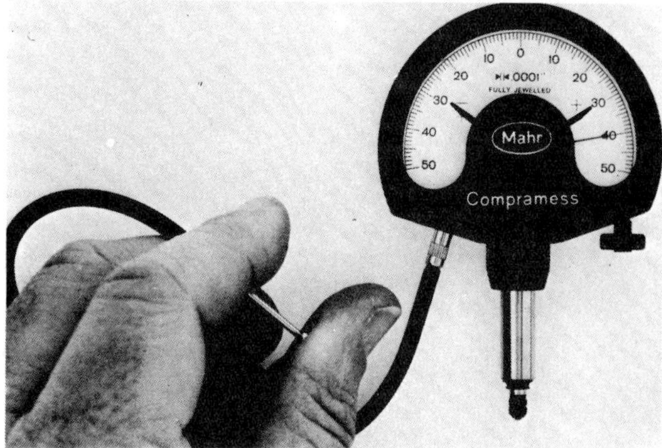

FIGURE C-207 Dial comparator indicator with cable lift (Harry Smith & Associates).

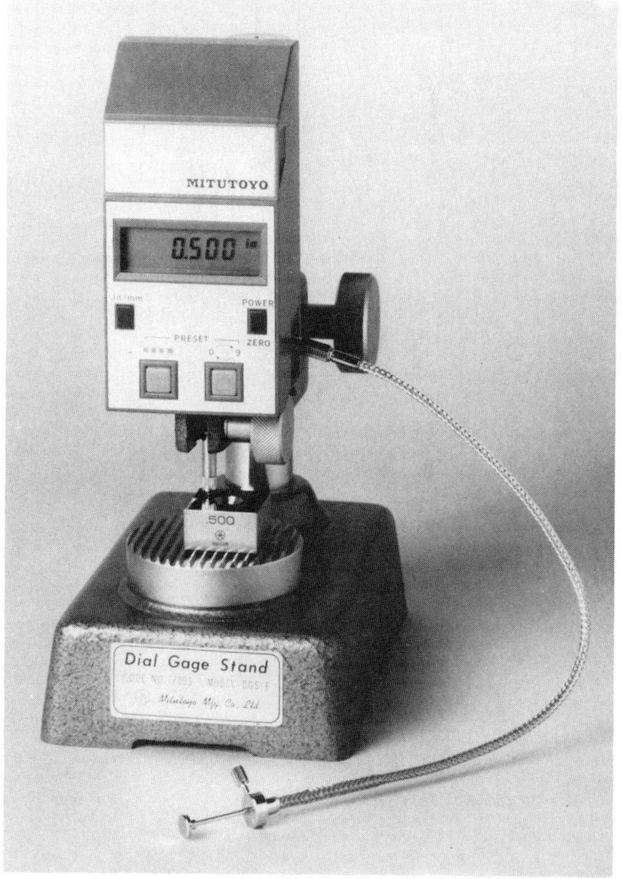

FIGURE C-208 Digital electonic comparator (MTI Corporation).

particularly useful comparator indicator for this is one equipped with tolerance hands (Figure C-207). The tolerance hands can be set to establish an upper and lower limit for part size. On this type of comparator indicator, the spindle can be lifted clear of the workpiece by using the cable mechanism. This permits the indicator to always travel downward as it comes into contact with the work. This is an additional compensation for any mechanical error in the indicator mechanism.

ELECTRONIC DIGITAL INDICATOR COMPARATORS Microelectronic technology has been adapted to indicator comparators as well as many other instruments. With digital readouts, these instruments are easy to read and calibrate, and they demonstrate high reliability and high discrimination. One example is the digital indicator comparator (Figure C-208).

This instrument may be coupled to a microcomputer that will print and graph measurements

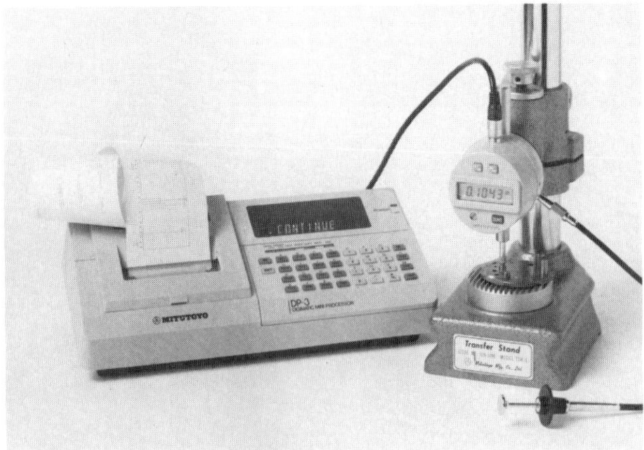

FIGURE C-209 Digital electronic comparator with computer and printer (MTI Corporation).

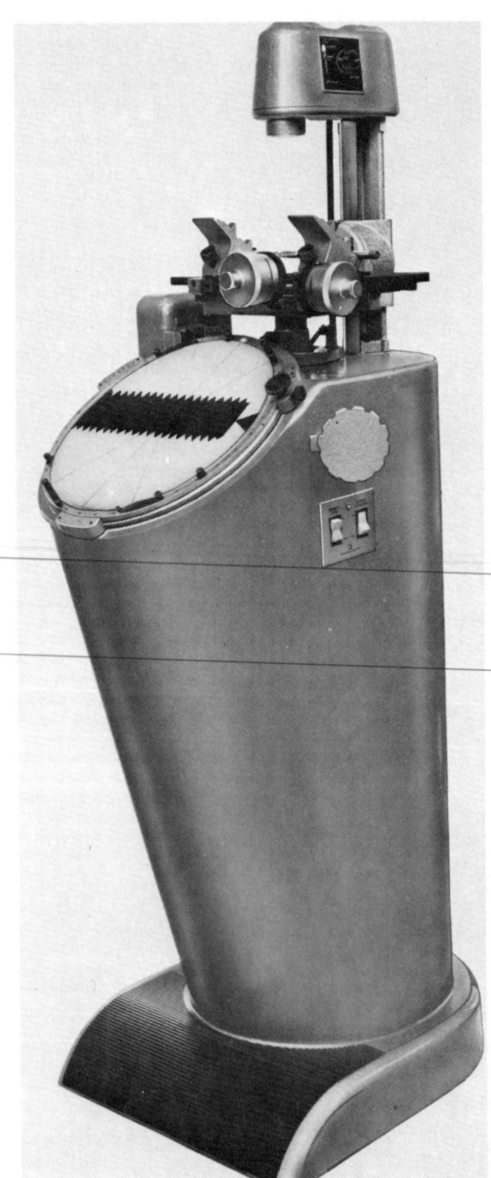

FIGURE C-210 Optical comparator checking a screw thread (Rank Scherr-Tumico, Inc.).

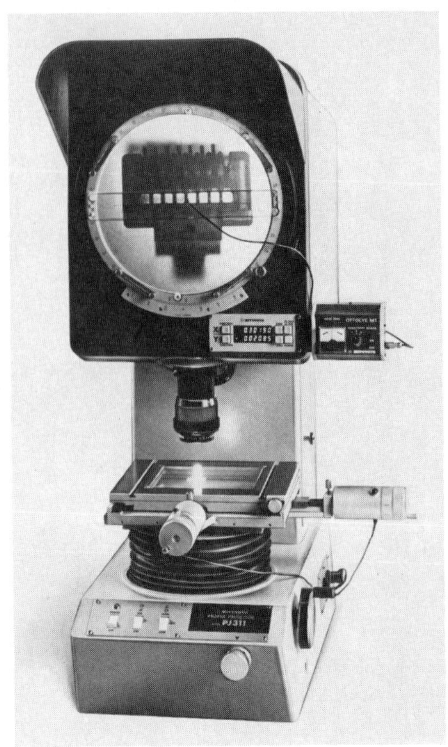

FIGURE C-211 Optical comparator with digital electronic readouts in inch and metric units (MTI Corporation).

as they are taken. The system (Figure C-209) is highly suited to quality control inspection measurement in the machine shop.

OPTICAL COMPARATORS The optical comparator (Figure C-210) projects onto a screen a greatly magnified profile of the object being measured. Various templates or patterns in addition to graduated scales can be placed on the screen and compared to the projected shadow of the part. The optical comparator is particularly useful for inspecting the geometry of screw threads, gears, and formed cutting tools.

Electronic and digital readouts also appear on the optical comparator. These features increase this instrument's reliability, ease of operation, metric/inch selection, high discrimination, and high sensitivity (Figure C-211).

ELECTROMECHANICAL AND ELECTRONIC COMPARATORS Electromechanical and electronic comparators convert dimensional change into changes of electric current or voltage. These changes are read on a suitably graduated scale (meter). Economical mass production of high precision parts requires that fast and reliable measurements be made so that over- and undersized parts can be sorted from those within tolerance. Electromechanical comparators can be used in

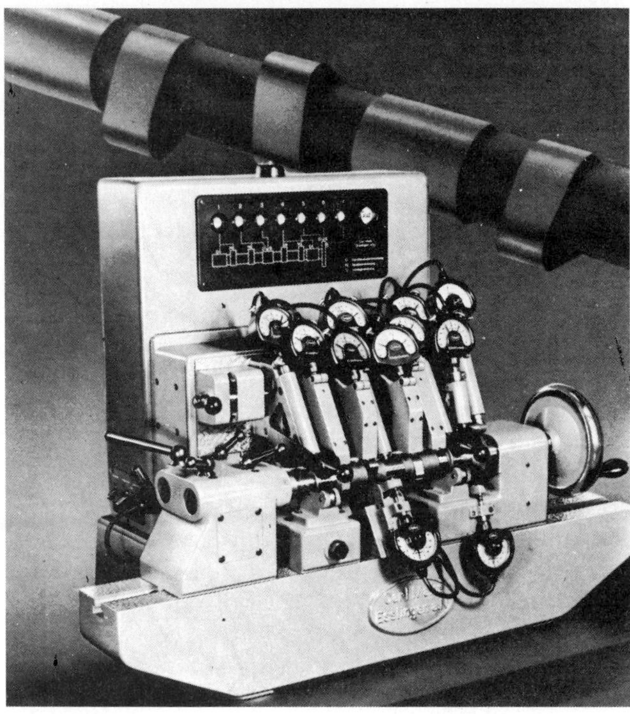

FIGURE C-212 Electromechanical comparator inspecting a camshaft (Mahr Gage Company).

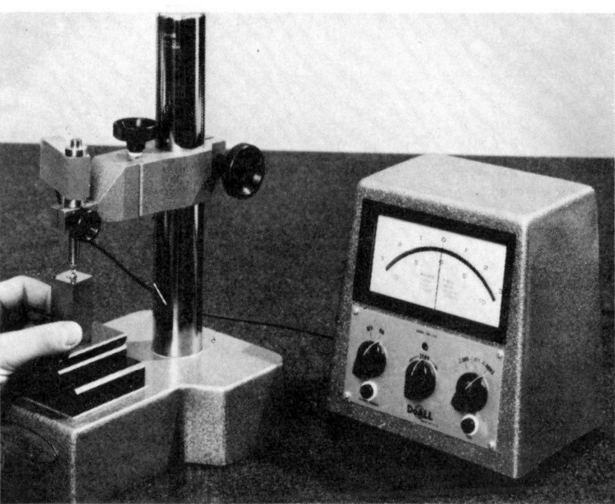

FIGURE C-213 Electronic comparator with maximum discrimination of .00001 in. (DoALL Company).

this application (Figure C-212). The comparator shown is used to check a camshaft.

The electronic comparator (Figure C-213) is a very sensitive instrument. It is used in a variety of comparison measuring applications in inspection and calibration. The comparator is set to a gage block by first adjusting the coarse adjustment. This mechanically moves the measuring probe. Final adjustment to zero is accomplished electronically. This is one of the unique advantages of such instruments. The electronic comparator shown has three scales. The first scale reads ±.003 in. at full range, with a discrimination of .0001 in. The second scale reads ±.001 in. at full range, with a discrimination of .00005 in. The third scale reads ±.0003 at full range, with a discrimination of .00001 in.

SELF-TEST

1. Define comparison measurement.

2. What can be said of most comparison measuring instruments?

3. Define cosine error.

4. How can cosine error be reduced?

Match the following measuring situations with the list of comparison measuring tools. Answers may be used more than once.

5. A milled slot 2 in. wide with a tolerance of ±.002 in.

6. A height transfer measurement.

7. The shape of a form lathe cutter.

8. Checking a combination square to determine its accuracy.

9. The diameter of a $1\frac{1}{2}$-in. hole.

10. Measuring a shim under a piece of machinery.

a. Spring caliper
b. Telescope gage
c. Adjustable parallel
d. Radius gage
e. Thickness gage
f. Planer gage
g. Combination square
h. Solid beam square
i. Beveled edge square
j. Cylindrical square
k. Diemaker's square
l. Micrometer square
m. Dial indicator
n. Dial test indicator
o. Dial indicator comparator
p. Optical comparator
q. Electronic comparator

UNIT 6

Using Gage Blocks

In the introduction to this section, we discussed the need for standardization of measurement. Today's widespread manufacturing can function only if machinists everywhere are able to check and adjust their measuring instruments to the same standards. Gage blocks permit a comparison between the working measurement instruments of manufacturing and recognized international standards of measurement. They are one of the most important measuring tools you will encounter. The practical uses of gage blocks in the metrology laboratory, toolroom, and machine shop include the calibration of precision measuring instruments, the establishment of precise angles, and often measurements involved in the positioning of machine tool components and cutting tools.

Objectives

After completing this unit, you should be able to:

1. Describe the care required to maintain gage block accuracy.
2. Wring gage blocks together correctly.
3. Disassemble gage block combinations and properly prepare the blocks for storage.
4. Calculate combinations of gage block stacks with and without wear blocks.
5. Describe gage blocks applications.

Gage Block Types and Grades

Gage blocks are commonly available individually or in sets. A common gage block set will contain 81 to 88 blocks ranging in thickness from .050 to 4.000 in. The total measuring range of the set is over 25 in. (Figure C-214). Also available are 121-block sets that permit measurement from .010 to 18 in. Sets with 4, 6, 9, 12, and 34 blocks are also used depending on measuring requirements. Sets of extra long blocks are available permitting measurements to 84 in. Metric gage block sets contain blocks ranging from .5 to 100 mm. Angular gage blocks can measure from 0 to 30 degrees. Gage blocks for linear measurement are either rectangular or square.

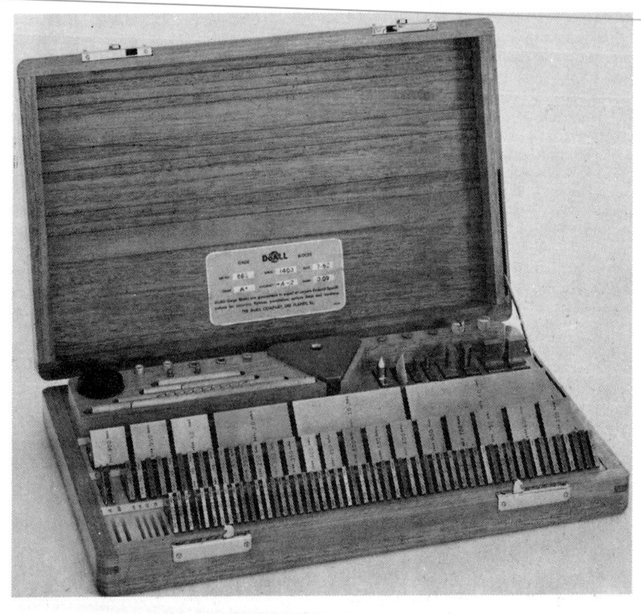

FIGURE C-214 Gage block set with accessories (DoALL Company).

TABLE C-1 Gage Block Tolerances

	Tolerances in Microinches (.000001 in.) for Gage Block Grade:		
Size	Grade 1 (Formerly AA)	Grade 2 (Formerly A+)	Grade 3 (Between A and B)
1 in. and less	+2 −2	+4 −2	+8 −2
2 in.	+4 −4	+8 −4	+16 −8
3 in.	+5 −5	+10 −5	+20 −10
4 in.	+6 −6	+12 −6	+24 −12

The three grades of gage blocks are grade 1 (laboratory), grade 2 (inspection), and grade 3 (shop) (Table C-1). Grades 1 and 2 are manufactured grades. Grade 3 blocks are compromises between grades 1 and 2. Sets of grade 3 can be purchased or they may be created by assembling out-of-tolerance grade 2 and 3 gage blocks that do not meet the tolerances for grade 1. Grade 3 blocks are not used by the inspection or gage laboratory, but are acceptable in the shop for many typical measurement applications.

The Value of Gage Blocks

As you know, a truly exact size cannot be obtained. However, it can be quite closely approached. Gage blocks are one of the physical standards that can closely approach exact dimensions. This makes them useful as measuring instruments with which to check other measuring tools. From the table on gage block tolerances, you can see that the length tolerance on a grade 1 block is ±.000002 in. This is only four millionths of an inch total tolerance. Such a small amount is hard to visualize. Consider that the thickness of a page of this book is about .003 in. Compare this amount with total gage block tolerance and you will note that the page is 750 times thicker than the tolerance. This should indicate that a gage block would be very useful for checking a measuring instrument with .001- or even .0001-in. discrimination.

As a further demonstration of gage block value, consider the following example. It is desired to establish a distance of 20 in. as accurately as possible. Using a typical gage block set, imagine a hypothetical situation where each block has been made to the plus tolerance of .000002 in. over the actual size. This situation would not exist in an actual gage block set, as the tolerance of each block is most likely bilateral. If it required 30 blocks to make up a 20-in. stack, the cumulative tolerance would amount to .000060 in. (60 millionths). As you can see, the 20-in. length is still extremely close to actual size. In a real situation, because of the bilateral tolerance of the gage blocks, the 20-in. stack will actually be much closer to 20.000000 than 20.000060 in. Because the gage block is so close to actual size, cumulative tolerance has little effect even over a long distance.

Preparing Gage Blocks for Use

Gage blocks are, at the same time, rugged and delicate. During their manufacture, they are put through many heating and cooling cycles that stabilize their dimensions. In order for a gage block to function, its **surface** must be **extremely smooth** and **flat.**

Gage blocks are almost always used in combination with each other. This is known as the gage block stack. The secret of gage block use lies in the ability to place two or more blocks together in such a way that most of the air between them is displaced. The space or interface between wrung gage blocks is known as the wringing interval. This is the process of **wringing.** Once this is accomplished, atmospheric pressure will hold the stack together. Properly wrung gage block stacks are essential if cumulative error is to be avoided. Two gage blocks simply placed against each other will have an air layer between them. The thickness of the air layer will greatly affect the accuracy of the stack.

Before gage blocks can be wrung, they must be properly prepared. Burrs, foreign material, lint, grit, and even dust from the air can prevent proper wringing and permanently damage a gage block. The main cause of gage block wear is the wringing of poorly cleaned blocks. Preparation of gage blocks should conform to the following procedure.

STEP 1. Remove the desired blocks from the box and place them on a lint-free tissue. The gage blocks should be handled as little as possible so that heat from fingers will not temporarily affect size.

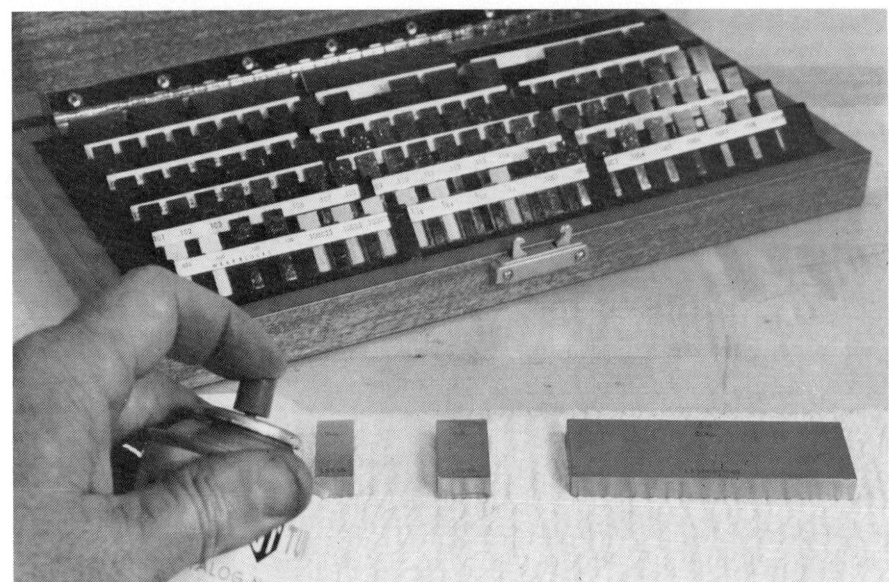

FIGURE C-215 Applying gage block cleaner.

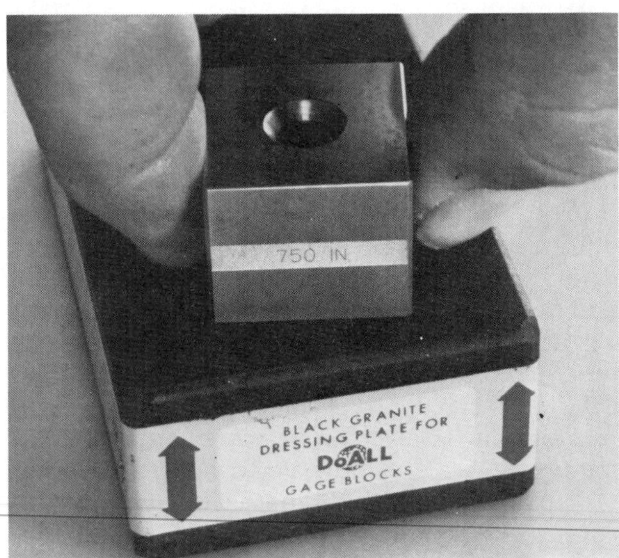

FIGURE C-216 Using the conditioning stone.

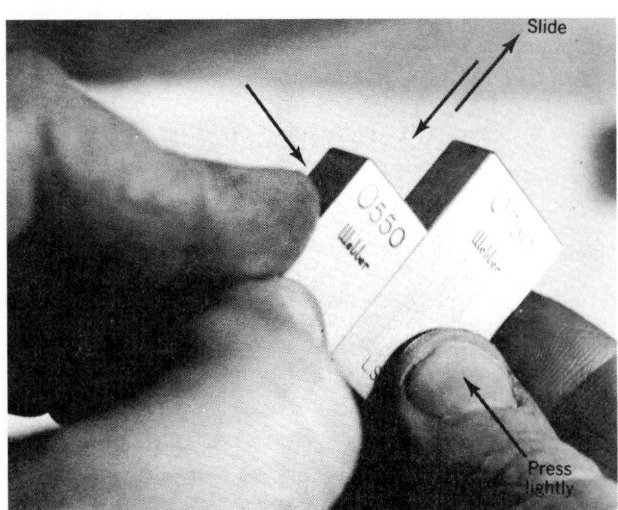

FIGURE C-217 Overlapping gage blocks prior to wringing.

STEP 2. The gage block must be cleaned thoroughly before wringing. This can be done with an appropriate cleaning solvent or commercial gage block cleaner (Figure C-215). Use the solvent sparingly, especially if an aerosol is applied. The evaporation of a volatile solvent can cool the block and cause it to temporarily shrink out of tolerance.

STEP 3. Dry the block immediately with a lint-free tissue.

STEP 4. Any burrs on a gage block can prevent a proper wring and possibly damage the highly polished surface. Deburring is accomplished with a special deburring stone or dressing plate (Figure C-216). The block should be lightly moved over the stone using a single back and forth motion. After deburring, the block must be cleaned again.

Wringing Gage Blocks

Gage blocks should be wrung immediately after cleaning. If more than a few seconds elapses, dust from the air will settle on the wringing surface. The block may require dusting with a camel's hair brush. To wring rectangular gage blocks, the following procedure should be used.

STEP 1. Place the freshly cleaned and deburred mating surfaces together and overlap them about $\frac{1}{8}$ in. (Figure C-217).

STEP 2. Slide the blocks together while lightly pressing together. During the sliding process, you should feel an increasing resistance. This resistance should then level off.

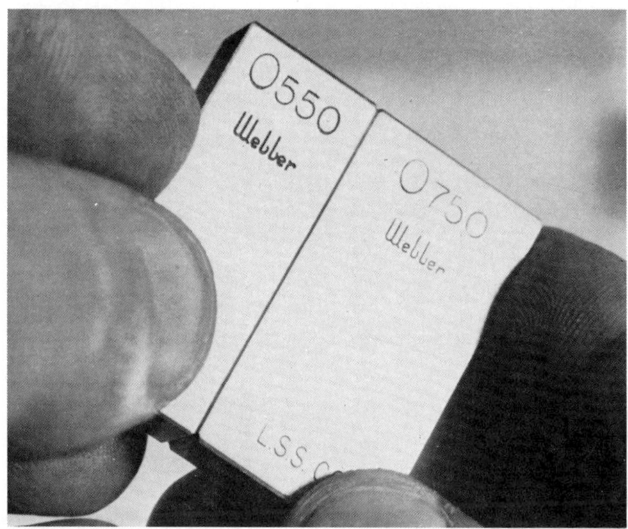

FIGURE C-218 Wrung gage blocks in line.

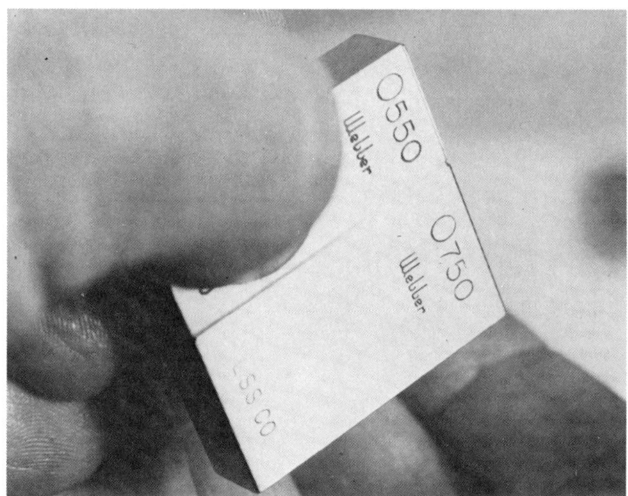

FIGURE C-219 Making sure of a proper wring.

STEP 3. Position the blocks so that they are in line (Figure C-218).

STEP 4. Make sure that the blocks are wrung by holding one block and releasing the other. Hold your hand under the stack in case the block should fall (Figure C-219).

Square gage blocks require the same cleaning and deburring as rectangular blocks. Square gage blocks are wrung by a slightly different technique. Since they are square, they should be placed together at a 45-degree angle. The upper block is then slid over the lower block while at the same time twisting the blocks and applying a light pressure.

During the wringing process, heat from the hands may cause the block stack to expand often well out of tolerance. The stack should be placed on a heat sink in order to normalize the temperature. Generally, gage blocks should be handled as little as possible to minimize heat problems.

If, during the wring, the blocks tend to slide freely, slip them apart immediately and recheck cleaning and deburring. If the blocks fail to wring after proper preparation has been followed, they may be warped or have a surface imperfection. A gage block may be inspected for these conditions by the use of an optical flat.

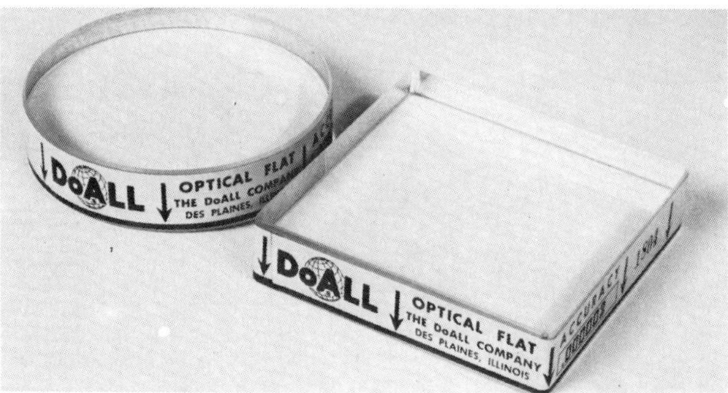

FIGURE C-220 Round and square optical flats (DoALL Company).

FIGURE C-221 Inspecting a gage block under monochromatic light.

Checking Gage Blocks with Optical Flats

An **optical flat** is an extremely flat piece of quartz (Figure C-220). Like gage blocks, there are various grades of flats. First grade or reference optical flats are within .000001 in. (one millionth). Round optical flats range from 1 to 10 in. in diameter. Square flats range from 1 × 1 in. to 4 × 4 in.

The optical flat uses the principles of **light interferometry** to make measurements and reveal surface geometry that could not be detected by other means. The working surface of the flat is placed on the gage block (Figure C-221). The block

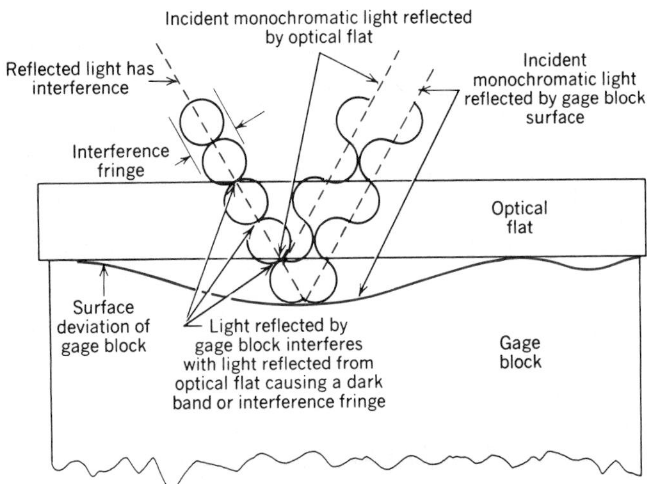

FIGURE C-222 The optical flat in light interference.

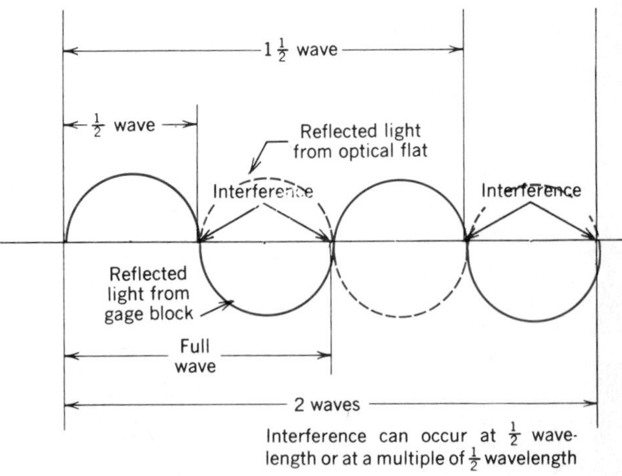

FIGURE C-224 Points of light interference.

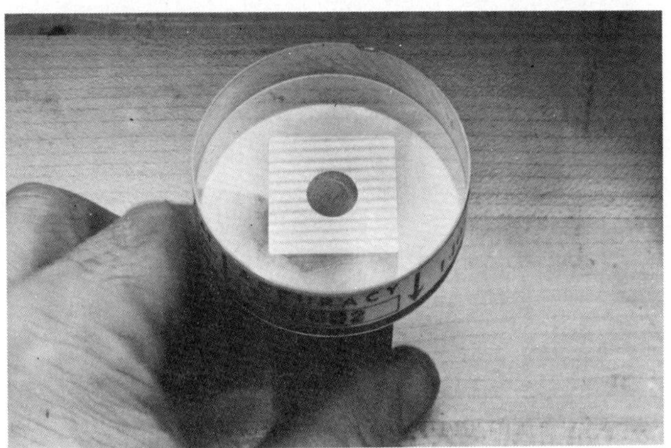

FIGURE C-223 Interference fringe patterns.

and flat are then placed under a single color or **monochromatic** light source. Since it is not possible to produce a truly flat surface, some surface deviation will exist even on the most perfect of gage blocks. Therefore, a portion of the block will be in direct contact with the optical flat. Other portions will not be in contact.

In the areas of no contact, a small space exists between the optical flat and the gage block. Monochromatic light passing through the optical flat is reflected by both the lower surface of the flat and the surface of the gage block. Under certain conditions, dependent on the distance existing between block and flat, light reflected from the block surface will cancel light reflected from the lower surface of the optical flat. This cancellation effect, called **interference**, is directly related to the distance between block and flat. If this distance is the same as or proportional to the wavelength of the light used, interference can occur (Figure C-222). The result of interference produces a dark band or **interference fringe** (Figure C-223). Interference can occur only at $\frac{1}{2}$ wavelength or a multiple of $\frac{1}{2}$ wavelength (Figure C-224). Since the wavelength of the monochromatic light is known, by measuring the spacing of the fringe patterns the actual amount of surface deviation can be determined.

Preparing Gage Blocks for Storage

Gage block stacks should not be left wrung for extended periods of time. The surface finish can be damaged in as little as a few hours time, especially if the blocks were not exceedingly clean at the time of wringing. After use, the stack should be unwrung and the blocks cleaned once again. Blocks should be handled with tissue (Figure C-225). Spray each block with a suitable gage block preservative and replace them in the box. The entire set should then be lightly sprayed with gage block preservative (Figure C-226).

Calculating Gage Block Combinations

In making a gage block stack, a minimum number of blocks should be used, as each surface or wringing interval between blocks can increase the opportunity for error. Poor wringing can make this error relatively large. In order to check your wringing ability, assemble a combination of blocks totaling 4.000 in. Compare this to the 4.000-in. block under a sensitive comparator. Make the comparison immediately after wringing to observe the effect of heat on the stack length. Place the wrung

FIGURE C-225 Handle gage blocks only with tissue.

FIGURE C-226 Applying gage block preservative.

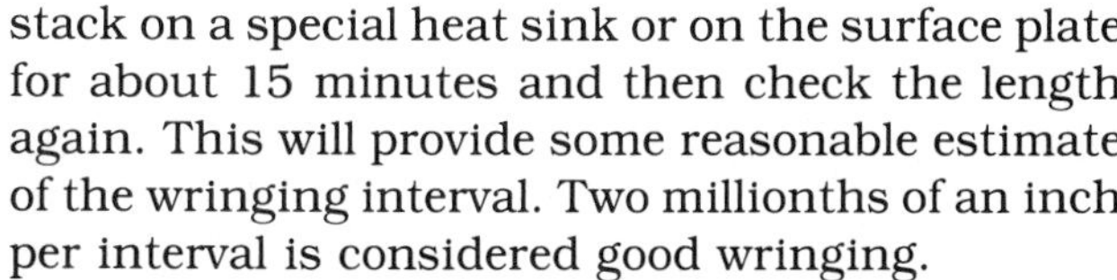

stack on a special heat sink or on the surface plate for about 15 minutes and then check the length again. This will provide some reasonable estimate of the wringing interval. Two millionths of an inch per interval is considered good wringing.

Table C-2 gives the specifications of a typical set of 83 gage blocks. Note that they are in four series.

In the following example, it is desired to construct a gage block stack to a dimension of 3.5752 in. Wear blocks will be used on each end of the stack.

3.5762	
	First, eliminate two .050-in.
.100	wear blocks
3.4762	
	Then eliminate the last figure right,
.1002	by subtracting the .1002-in. block
3.3760	
	Once again, eliminate the last figure
.126	right, by subtracting the .126-in. block
3.250	
	Eliminate the last figure right, using
.250	the .250-in. block
3.000	
	Eliminate the 3.000 in. with the
3.000	3.000-in. block
0.000	

Therefore, the blocks required to construct this stack are:

Quantity	Size
2	.050-in. wear blocks
1	.1002-in. block
1	.126-in. block
1	.350-in. block
1	3.000-in. block

TABLE C-2 Typical 83-piece Gage Block Set

First: .0001 Series—9 blocks

.1001	.1002	.1003	.1004	.1005	.1006	.1007	.1008	.1009

Second: .001 Series—49 Blocks

.101	.102	.103	.104	.105	.106	.107	.108	.109
.110	.111	.112	.113	.114	.115	.116	.117	.118
.119	.120	.121	.122	.123	.124	.125	.126	.127
.128	.129	.130	.131	.132	.133	.134	.135	.136
.137	.138	.139	.140	.141	.142	.143	.144	.145
.146	.147	.148	.149					

Third: .050 Series—19 Blocks

.050	.100	.150	.200	.250	.300	.350	.400	.450	.500	.550
.600	.650	.700	.750	.800	.850	.900	.950			

Fourth: 1.000 Series—4 Blocks

1.000 2.000 3.000 4.000

2 .050 wear blocks

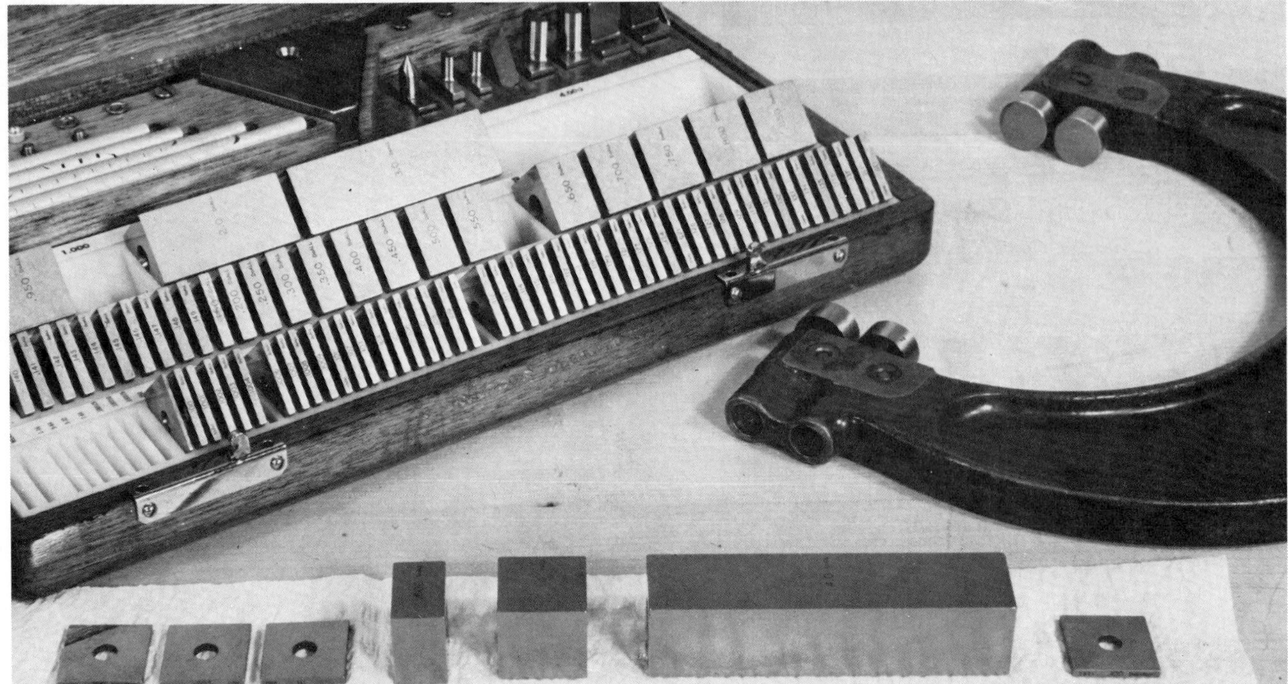

FIGURE C-227 Gage and wear blocks for setting a snap gage.

As a second example, we shall construct a gage block stack of 4.2125 without wear blocks.

$$
\begin{array}{r}
4.2125 \\
\underline{.1005} \\
4.1120 \\
\underline{.112} \\
4.0000 \\
\underline{4.0000} \\
0.0000
\end{array}
$$

Blocks for this stack are:

Quantity	Size
1	.1005
1	.112
1	4.000

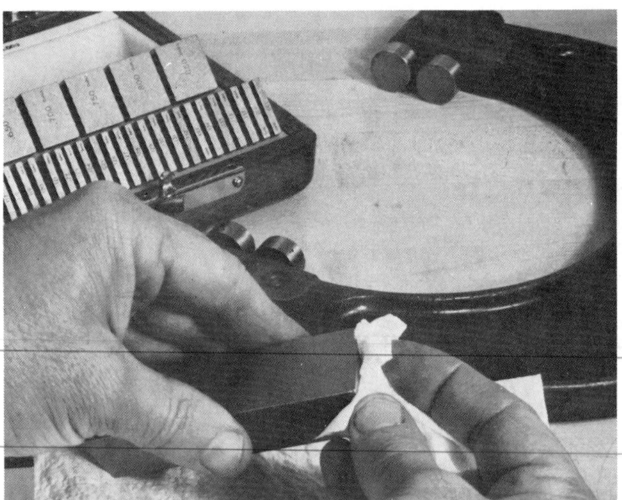

FIGURE C-228 Cleaning blocks prior to wringing.

Using Wear Blocks

When gage blocks are used in applications where direct contact is made, it is advisable to use **wear blocks.** For example, if you were using a gage block stack to calibrate a large number of micrometers, wear blocks would be recommended to reduce the wear on the gage block. Wear blocks are usually included in typical gage blocks sets. They are made from a particularly hard material known as tungsten carbide. A wear block is placed on one or both ends of a gage block stack to protect it from possible damage by direct contact. Wear blocks are usually .050 or .100 in. thick.

Gage Block Applications

Gage blocks are used in setting sine bars for establishing precise angles. The use of the sine bar is discussed in the unit on angular measure. Gage blocks are used to set other measuring instruments such as a snap gage (Figure C-227). The proper blocks are selected for the desired dimension and the stack assembled (Figure C-228).

FIGURE C-229 Setting a snap gage using gage blocks.

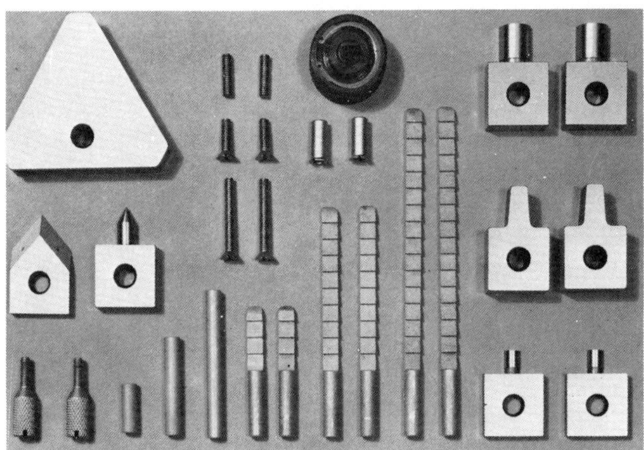

FIGURE C-230 Gage block accessories.

Since this is a direct contact application, wear blocks should be used. The stack is then used to set the gage (Figure C-229).

Gage block measurement is facilitated by various accessories (Figure C-230). Accessories include scribers, bases, gage pins, and screw sets for holding the stack together. In any application where screws are employed to secure gage block stacks, a torque screwdriver must be used. This will apply the correct amount of pressure on the gage block stack. Gage block and accessories can be assembled into precision height gages for layout (Figure C-231). With gage pins (Figure C-232), gage blocks may be used for direct gaging or for checking other measuring instruments. Machine tool applications include the use of gage blocks as auxiliary measuring systems on milling machines, setting cutter heights, and spacing straddle milling cutters.

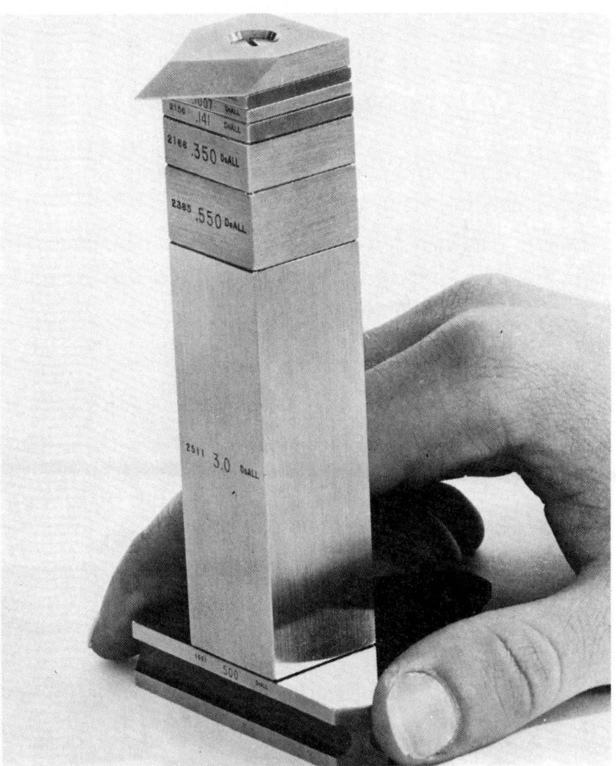

FIGURE C-231 Precision height gage assembled from gage blocks (DoALL Company).

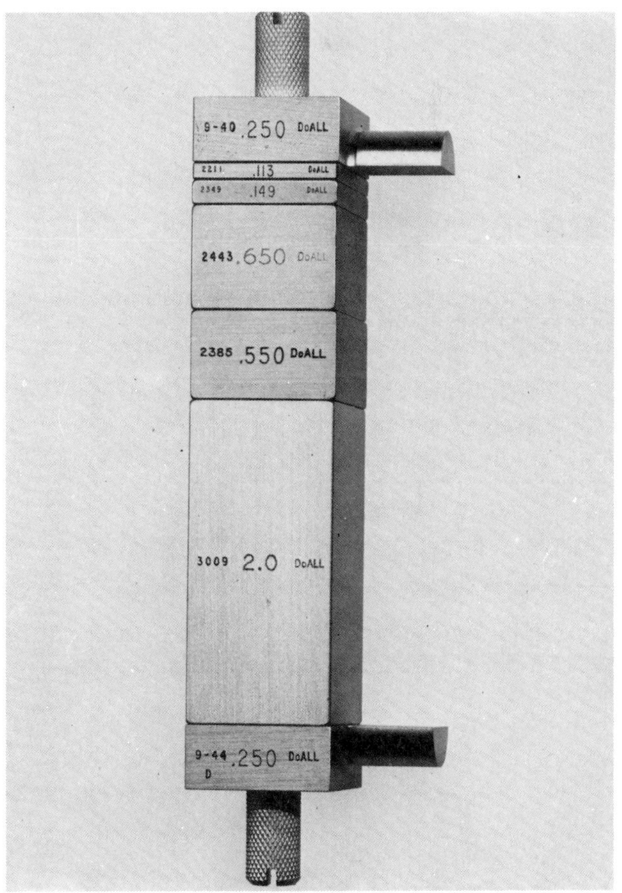

FIGURE C-232 Gage block stack with accessory gage pins (DoALL Company).

SELF-TEST

1. What is a wringing interval?
2. Why are wear blocks frequently used in combination with gage blocks?
3. As related to gage blocks usage, what is meant by the term *normalize*?
4. What length tolerances are allowed for the following grades of gage blocks (under 2.000-in. sizes)? Grade 1; grade 2; grade 3.
5. What is a conditioning stone and how is it used?
6. What does the term *microinch* regarding surface finish of a gage block mean?
7. Describe the handling precautions necessary for the preservation of gage block accuracy.
8. What gage blocks are necessary in order to assemble a stack equal to 3.0213, without using wear blocks?
9. List gage blocks necessary for a stack equal to 1.9643 with wear blocks.
10. Describe at least two gage block applications.

UNIT 7

Using Angular Measuring Instruments

Angular measurement is as important as linear measurement. The same principles of metrology apply to angular measure as to linear measure. Angular measuring instruments have various degrees of discrimination. They must not be used beyond their discrimination. Angular measuring instruments require the same care and handling as any of your precision tools.

Objectives

After completing this unit, you should be able to:

1. Identify common angular measuring tools.
2. Read and record angular measurements using a vernier protractor.
3. Calculate sine bar elevations and measure angles using a sine bar and adjustable parallels.
4. Calculate sine bar elevations and establish angles using a sine bar and gage blocks.

As a machinist, you will find the need to measure **acute angles, right angles,** and **obtuse angles** (Figure C-233). Acute angles are less than 90 degrees. Obtuse angles are more than 90 degrees but less than 180 degrees. Ninety degree or right angles are generally measured with squares. However, the amount of angular deviation from perpendicularity may have to be determined. This requires that an angular measuring instrument be used. Straight angles, or those containing 180 degrees, generally fall into the category of straightness or flatness and are measured by other types of instruments.

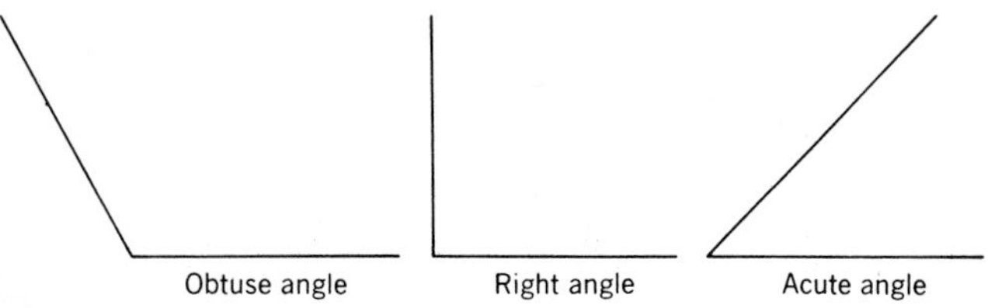

FIGURE C-233 **Acute, right, and obtuse angles.**

Units of Angular Measure

In the inch system, the unit of angular measure is the **degree.**

Full circle = 360 degrees

1 degree = 60 minutes of arc (1° = 60′)

1 minute = 60 seconds of arc (1′ = 60″)

In the metric system, the unit of angular measure is the **radian.** A radian is the length of an arc on the circle circumference that is equal in length to the radius of the circle (Figure C-234). Since the circumference of a circle is equal to $2\pi r$ (radius), there are 2π radians in a circle. Converting radians to degrees gives the equivalent:

$$1 \text{ radian} = \frac{360}{2\pi r}$$

Assuming a radius of 1 unit:

$$1 \text{ radian} = \frac{360}{2\pi}$$

$$= 57°17'44'' \text{ (approximately)}$$

It is unlikely that you will come in contact with much radian measure. All of the common comparison measuring tools you will use read in degrees and fractions of degrees. Metric angles expressed in radian measure can be converted to degrees by the equivalent shown.

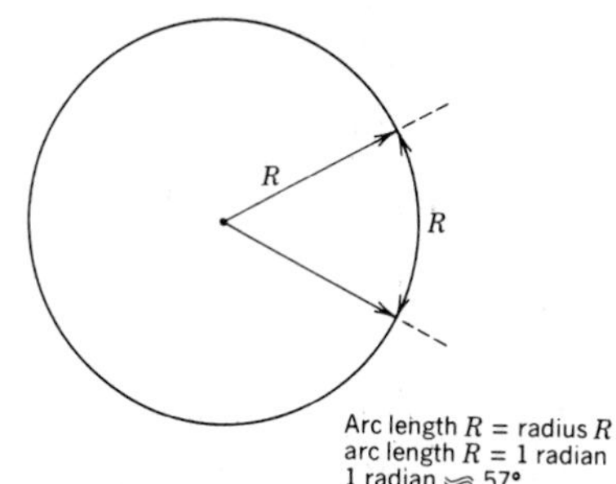

FIGURE C-234 Radian measure.

Reviewing Angle Arithmetic

You may find it necessary to perform angle arithmetic. Use your calculator, if you have one available.

ADDING ANGLES

Angles are added just like any other quantity. One degree contains 60 minutes. One minute contains 60 seconds. Any minute total of 60 or larger must be converted to degrees. Any second total of 60 or larger must be converted to minutes.

EXAMPLES

35° + 27° = 62°

3°15′ + 7°49′ = 10°64′

Since 64′ = 1°4′, the final result is 11°4′.

265°15′52″ + 10°55′17″ = 275°70′69′

Since 69″ = 1°9″ and 70′ = 1°10′, the final result is 276°11′9″.

SUBTRACTING ANGLES

When subtracting angles where borrowing is necessary, degrees must be converted to minutes and minutes must be converted to seconds.

EXAMPLE

15° − 8° = 7°

EXAMPLE

15°3′ − 6°8′ becomes

14°63′ − 6°8′ = 8°55′

EXAMPLE

39°18′13″ − 17°27′52″ becomes

38°77′73″ − 17°27′52″ = 21″50′51″

DECIMAL ANGLES

When using digital readout measuring and dividing equipment for angular measurement, angle fractions may be in decimal form rather than minutes and seconds. For example, 30 degrees 30 minutes would be 30½ degrees of 30.5 degrees. You should familiarize yourself with angle decimal fractions and the methods of converting minutes and seconds to their decimal equivalents.

Decimal fractions of angles are calculated by the following procedures. Since there are 60 minutes in each degree, a fractional portion of a degree becomes a fraction of 60.

EXAMPLE

32 minutes is $\frac{32}{60}$ of 1 degree. Converting to a decimal fraction:

$\frac{32}{60}$ = .5333 degree

23 seconds is $\frac{23}{60}$ of 1 minute

$\frac{23}{60}$ = .3833 minutes

To convert angle decimal fractions to minutes and seconds, multiply the decimal portion times 60.

EXAMPLE

45.5 degrees converted to minutes: decimal portion is .5 × 60 = 30 minutes or 45.5 degrees = 45 degrees and 30 minutes

32.75 degrees converted to minutes: decimal portion is .75 degree × 60 = 45 minutes or 32 degrees and 45 minutes

Some calculators have the capability to convert decimal fractions of angles to minutes and seconds. On other calculators the above computations must be made to accomplish conversions.

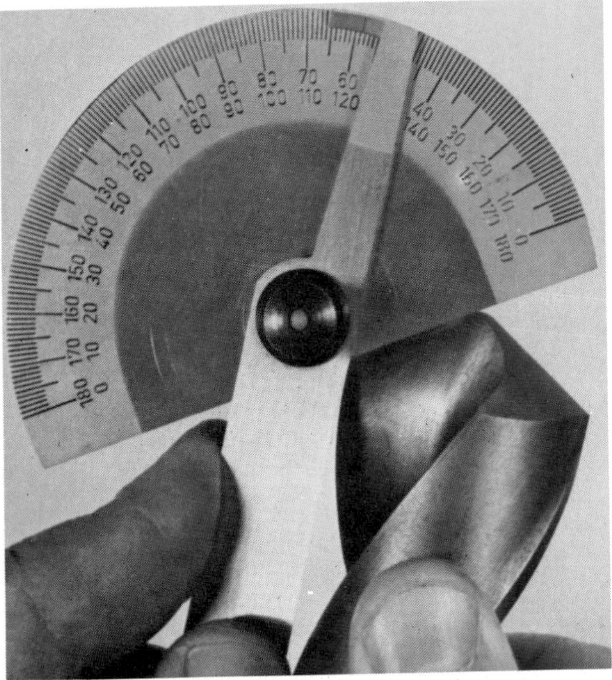

FIGURE C-235 Plate protractor measuring a drill point angle.

Angular Measuring Instrument

PLATE PROTRACTORS

Plate protractors have a discrimination of one degree and are useful in such applications as layout and checking the point angle of a drilll (Figure C-235).

BEVEL PROTRACTORS

The bevel protractor is part of the machinist's combination set. This protractor can be moved along the rule and locked in any position. The protractor has a flat base permitting it to rest squarely on the workpiece (Figure C-236). The combination set protractor has a discrimination of one degree.

DIAL INDICATING SINOMETER ANGLE GAGE

The indicating sinometer (Figure C-237) permits fast and accurate measurement of all angles. Discrimination is 30 seconds of arc per dial division.

UNIVERSAL BEVEL VERNIER PROTRACTOR

The universal bevel vernier protractor (Figure C-238) is equipped with a vernier that permits dis-

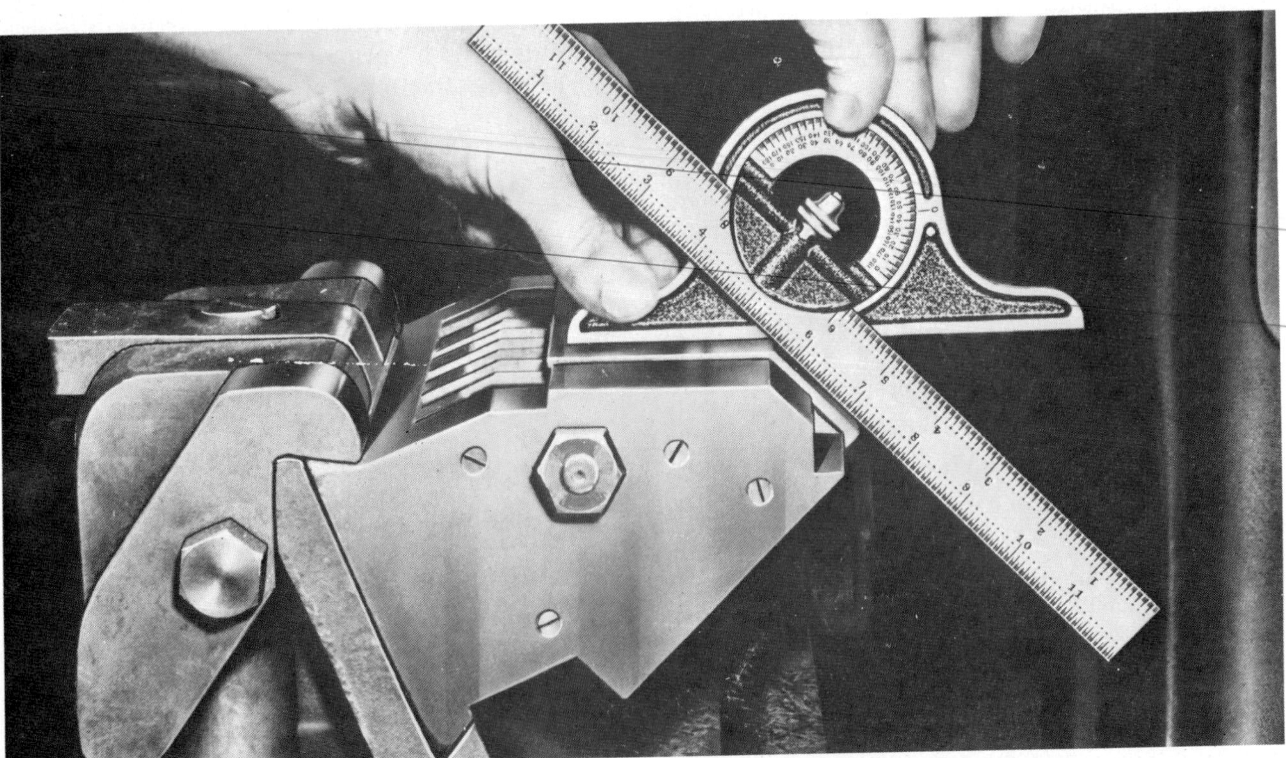

FIGURE C-236 Using the combination set bevel protractor (L.S. Starrett Company).

crimination to $\frac{1}{12}$ of a degree or 5 minutes of arc.

The instrument can measure an obtuse angle (Figure C-239). The acute attachment facilitates the measurement of angles less than 90 degrees (Figure C-240). When used in conjunction with a vernier height gage, angle measurements can be made that would be difficult by other means (Figure C-241).

Vernier protractors are read like any other instrument employing the vernier. The main scale is divided into whole degrees. These are marked in four quarters each 0 to 90 degrees. The vernier divides each degree into 12 parts each equal to 5 minutes of arc.

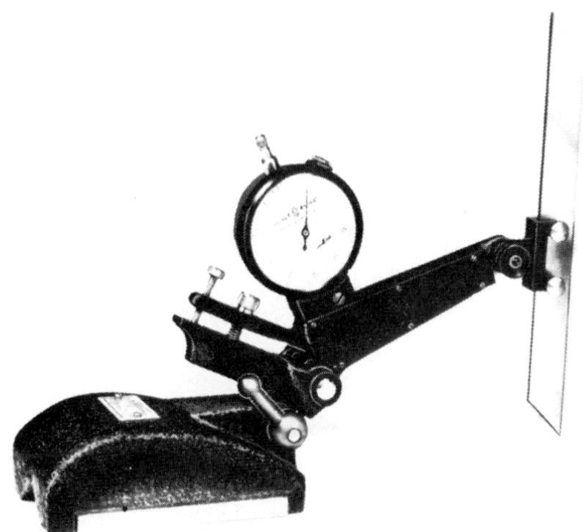

FIGURE C-237 Dial indicating sinometer angle gage (Rank Scherr-Tumico, Inc.).

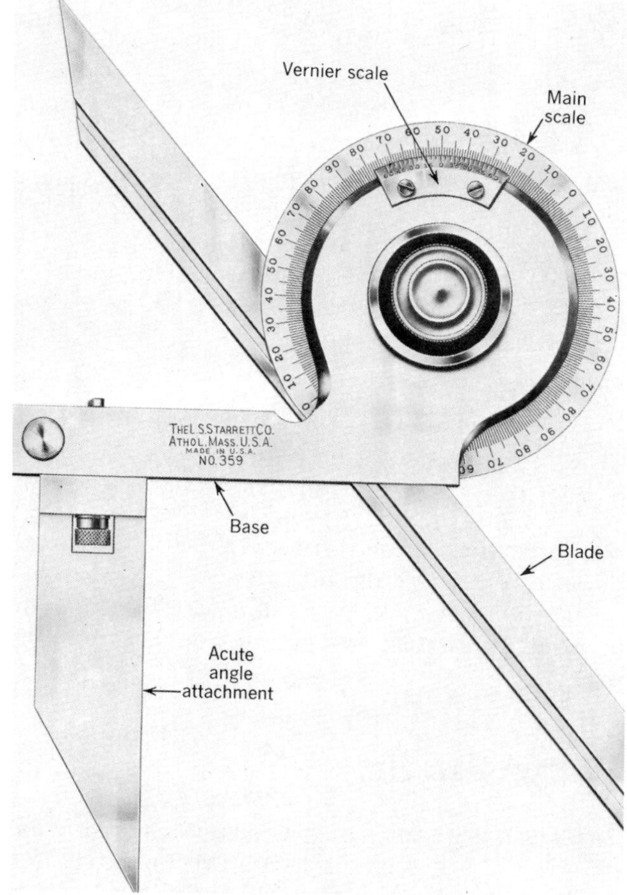

FIGURE C-238 Parts of the universal bevel vernier protractor (L.S. Starrett Company).

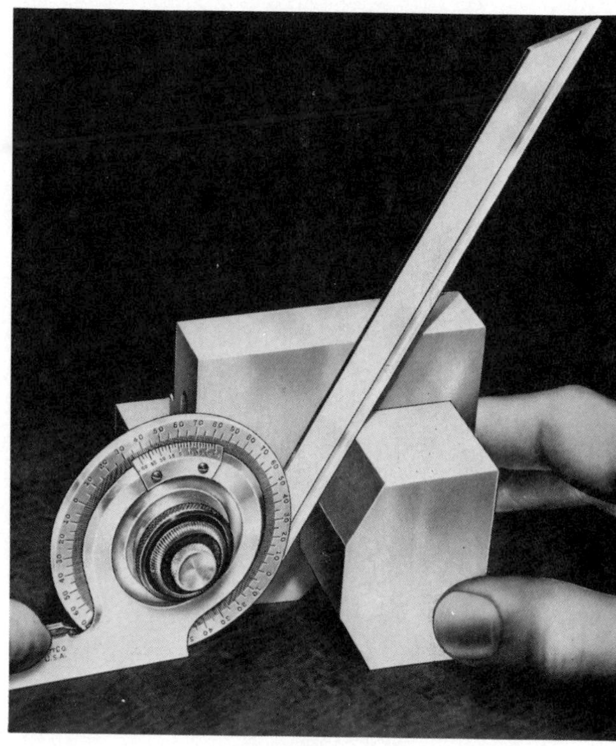

FIGURE C-239 Measuring an obtuse angle with the vernier protractor (L.S. Starrett Company).

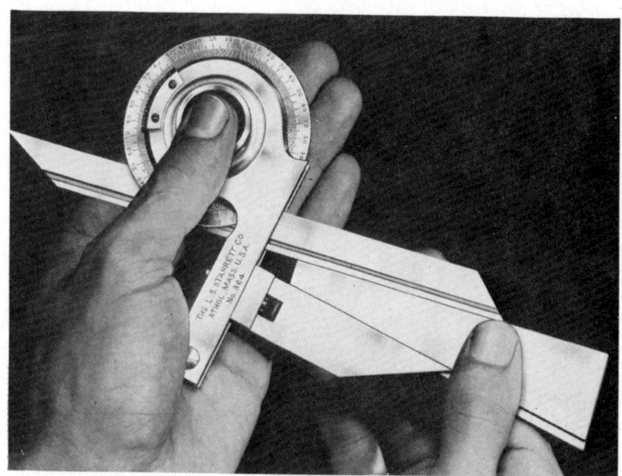

FIGURE C-240 Using the acute angle attachment (L.S. Starrett Company).

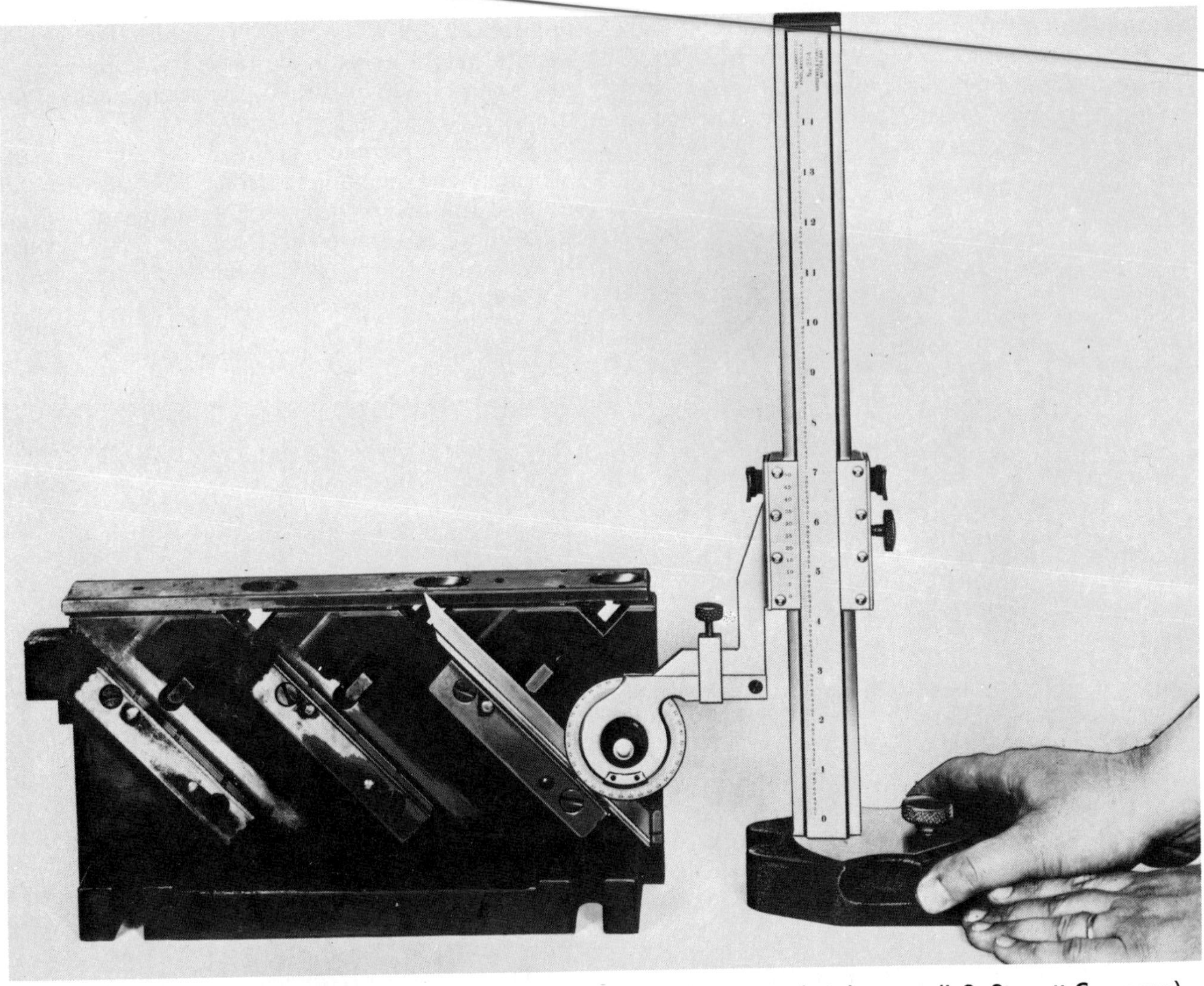

FIGURE C-241 Using the vernier protractor in conjunction with the vernier height gage (L.S. Starrett Company).

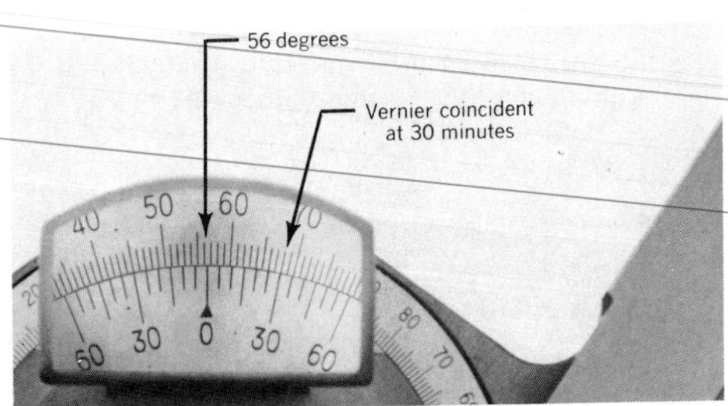

FIGURE C-242 Vernier protractor reading of 56 degrees and 30 minutes.

To read the protractor, determine the nearest full degree mark between zero on the main scale and zero on the vernier scale. **Always read the vernier in the same direction as you read the main scale.** Determine the number of the vernier coincident line. Since each vernier line is equal to 5 minutes, multiply the number of the coincident line by 5. Add this to the main scale reading.

EXAMPLE READING The protractor shown in Figure C-242 has a magnifier so that the vernier may be seen more easily.

Main scale 56°
Vernier coincident at line 6
6 × 5 minutes = 30 minutes
Total reading is 56°30′

For convenience, the vernier scale is marked at 0, 30, and 60, indicating minutes.

The vernier bevel protractor can be applied in a variety of angular measuring applications (Figure C-243).

Using the Sine Bar

Precise angles can be measured using the **sine bar** (Figure C-244). A sine bar is a precision bar that has been hardened and then ground and lapped to very precise dimensions. The sine bar has a precise cylinder attached to each end. The center

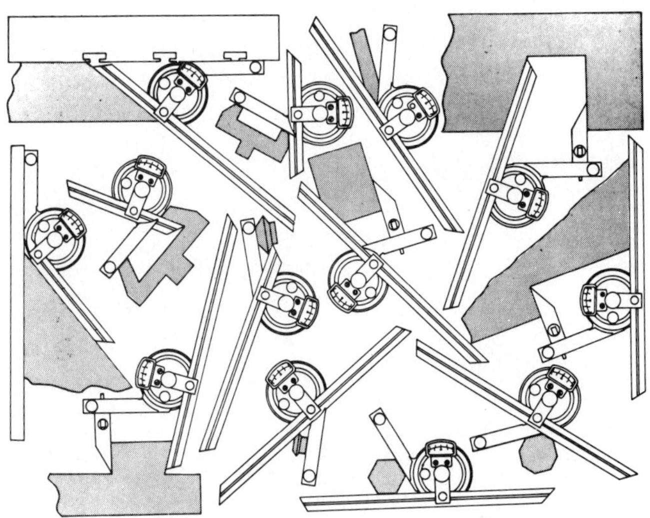

FIGURE C-243 Applications of the vernier bevel protractor (MTI Corporation).

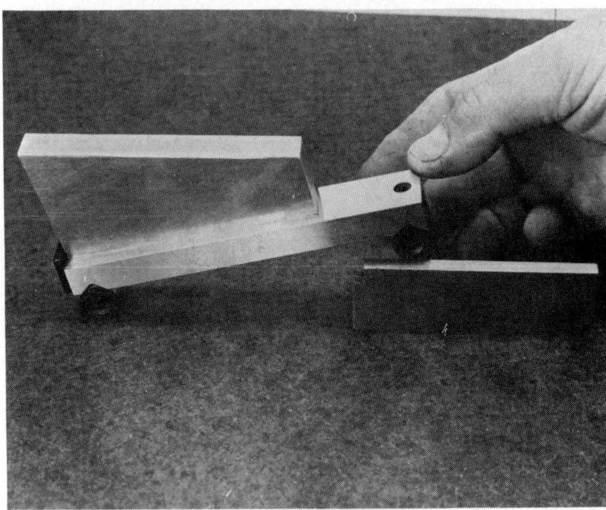

FIGURE C-245 Placing the adjustable parallel under the sine bar.

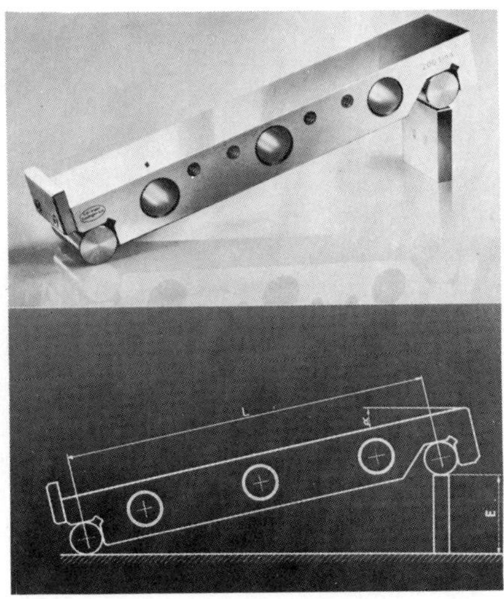

FIGURE C-244 Sine bar (Mahr Gage Company).

FIGURE C-246 Setting the test indicator to zero at the end of the workpiece.

spacing of the cylinders is either 5 or 10 in. and is precisely established.

When in use, the sine bar becomes the hypotenuse of a right triangle. Angles are measured or established by elevating one end of the bar a specified amount. The amount of sine bar elevation for any desired angle is determined by the following formula:

bar elevation = bar length × sine of the desired angle

EXAMPLE

Determine the elevation for 30 degrees using a 5-in. sine bar.

Bar elevation = 5 in. × sin 30°
= 5 × .5
= 2.500 in.

This means that if the bar were elevated 2.500 in., and angle of 30 degrees would be established.

EXAMPLE

Determine the elevation for 42 degrees using a 5-in. sine bar.

Bar elevation = 5 in. × sin 42°
= 5 × .6691
= 3.3456 in.

DETERMINING WORKPIECE ANGLE USING THE SINE BAR AND MEASURING WORKPIECE ANGLE USING THE SINE BAR AND ADJUSTABLE PARALLEL

An angle may be measured using the sine bar and adjustable parallel. The adjustable parallel is used to elevate the sine bar (Figure C-245). The workpiece is placed on the sine bar and a dial test indicator is set to zero on one end of the part (Figure

C-246). The parallel is adjusted until the dial indicator reads zero at each end of the workpiece (Figure C-247). The parallel is then removed and measured with a micrometer (Figure C-248). To determine the angle of the workpiece, simply transpose the sine bar elevation formula and solve for the angle.

Bar elevation = bar length × sine of the angle desired
Sine of the angle desired = elevation/bar length
Sine of angle = 1.9935 (micrometer/reading/5)
Sine = .3987
Angle = 23°29′48″

ESTABLISHING ANGLES USING THE SINE BAR AND GAGE BLOCKS

Extremely precise angles can be measured or established by using gage blocks to elevate the sine bar. Bar elevation is calculated in the same manner. The required gage blocks are properly prepared and the stack is wrung (Figure C-249). The gage block stack totaling 1.9940 in. is placed under the bar (Figure C-250). This will establish an angle of 23°30′11″ using a 5-in. sine bar. The angle of the workpiece is checked using a dial test indicator (Figure C-251).

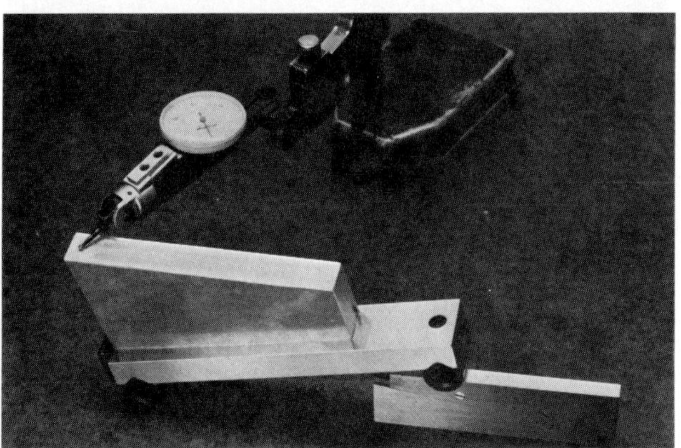
FIGURE C-247 Checking the zero reading at the opposite end of the workpiece.

FIGURE C-249 Wringing the gage block stack for the sine bar elevation.

FIGURE C-250 Placing the gage block stack under the sine bar.

FIGURE C-248 Measuring the adjustable parallel with an outside micrometer.

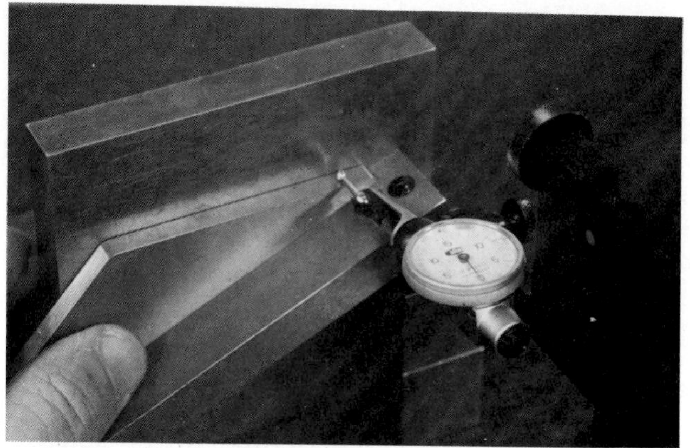

FIGURE C-251 Checking the workpiece using the dial test indicator.

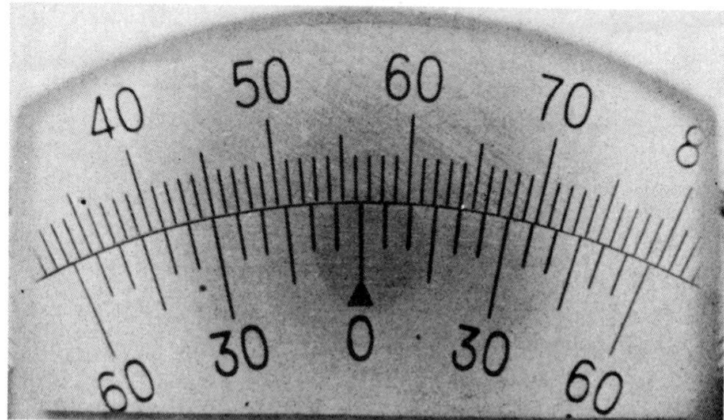

FIGURE C-252*a*

SINE BAR CONSTANT TABLES

The elevations for angles up to about 55 degrees can be obtained directly from a **table of sine bar constants.** Such tables can be found in the Appendix of this text and in machinists' handbooks. The sine bar constant table eliminates the need to perform a trigonometric calculation. The sine bar table may only discriminate to minutes of arc. If discrimination to seconds of arc is required, it is better to calculate the amount of sine bar elevation required.

SELF-TEST

1. Name two angular measuring instruments with one degree of discrimination.

2. What is the discrimination of the universal bevel protractor?

3. Describe the use of the sine bar.

4. Read and record the vernier protractor readings in Figures C-252*a* to C-252*e*.

5. Calculate the required sine bar elevation for an angle of 37 degrees. (Assume a 5-in. sine bar.)

6. A 10-in. sine bar is elevated 2.750 in. Calculate the angle established to the nearest minute.

FIGURE C-252*b*

FIGURE C-252*d*

FIGURE C-252*c*

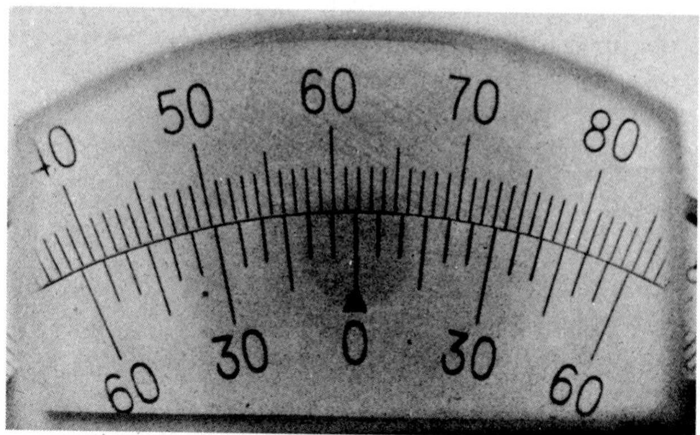

FIGURE C-252*e*

UNIT 8

Tolerances and Fits

Almost no product today is totally manufactured by a single maker. Although you might think that a complex product like an automobile or aircraft is made by a single manufacturer, if you look behind the scenes you would discover that the manufacturer uses many suppliers that make components for the final assembly.

In order to make all the component parts fit together, or **interface,** to form a complex assembly, you have already learned that a standardized system of measurement is essential. You have further learned that the measurement instruments must be compared to known standards in the process of **calibration** in order to maintain their accuracy.

Although standardized measurement is essential to modern industry, perhaps even more important are the design specifications that indicate the **dimensions** of a part. These dimensions control the **size** of a part, its features, and/or their location on the part or relative to other parts. Through this, parts are able to be interchanged and mated to each other to form complete assemblies. The purpose of this unit is to introduce the basic terminology of tolerances and fits.

Objectives

After completing this unit, you should be able to:

1. Describe basic reasons for tolerance specifications.
2. Recognize common geometric dimension and tolerance call outs on drawings.
3. Describe the reasons for press fits and know where to find press fit allowance information.

Limit and Tolerance

Since it is impossible to machine a part to an exact size, a designer must specify an acceptable range of sizes that will still permit the part to fit and function as intended. The maximum and minimum sizes in part dimensions that are acceptable are **limits** between which the actual part dimension must fall. The difference between the maximum and minimum limits is **tolerance,** or the total amount by which a part dimension may vary. Tolerances on drawings are often indicated by specifying a limit, or by **plus** and **minus** notations (Figure C-253). With plus and minus tolerancing, when the tolerance is both above and below the nominal (true theoretical) size, it is said to be **bilateral** (two sides). When the tolerance is indicated all on one side of nominal, it is said to be **unilateral** (one sided).

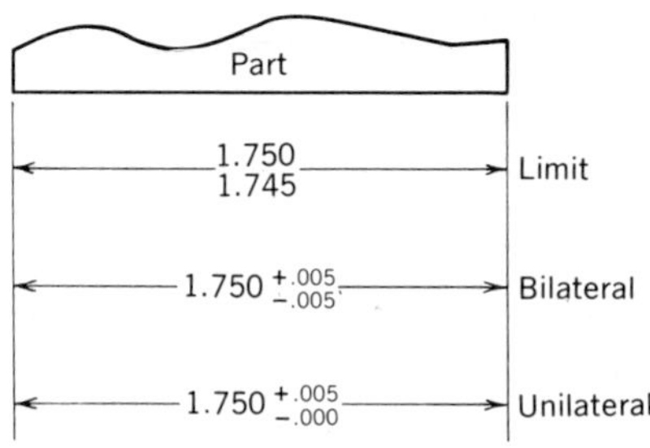

FIGURE C-253 Tolerance notations.

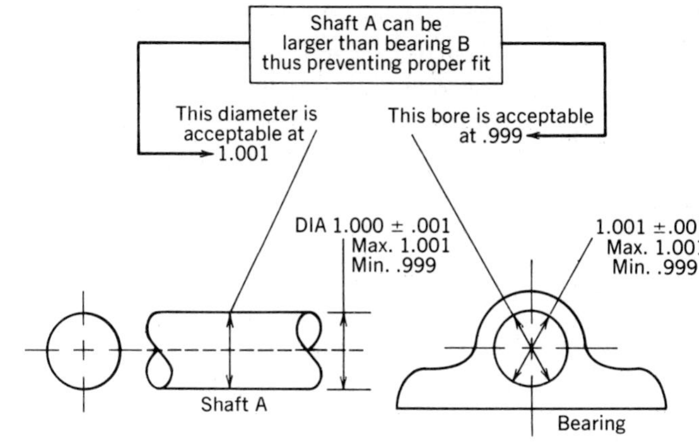

FIGURE C-254 Tolerance overlap can prevent proper fit of mating parts.

How Tolerance Affects Mating Parts

When two parts mate or are interchanged in an assembly, tolerance becomes vitally important. Consider the following example (Figure C-254): The shaft must fit the bearing and be able to turn freely. The diameter of the shaft is specified as 1.000 ± .001. This means that the maximum limit of the shaft is 1.001 and the minimum is .999. The tolerance is then .002 and bilateral.

The maximum limit of the bearing bore is also 1.001 and the minimum limit is .999. The tolerance is once again .002. Will the shaft made by one machine shop fit the bearing made by another machine shop using the tolerances specified? If the shaft is turned to the maximum limit of 1.001 and the bearing is bored to its minimum limit of .999, both parts would be within acceptable tolerance, but would not fit to each other since the shaft is .002 larger than the bearing. However, if the bearing bore was specified in limit form or unilateral tolerance of 1.002 $^{+.002}_{-.000}$, the parts would fit as intended. Even if the shaft was turned to the high limit of 1.001, it would still fit the bearing even though the bore was machined to the low limit of 1.002. Although a machinist is not usually concerned with establishing tolerance and limit specifications, you can easily see how fit problems can be created by overlapping tolerances discussed in this example.

Standard Tolerances

On many drawings you will use in the machine shop, tolerances will be specified at the dimensions. If no particular tolerances are specified at the dimension, accepted standard tolerances may be applied. These are often listed in part of the title block on the drawing and generally conform to the following:

Fractional dimensions $\pm\frac{1}{64}$

Two-place decimal fractions ±.010

Three-place decimal fractions ±.005

Four-place decimal fractions ±.0005

Angles $\pm\frac{1}{2}$ degree

Always check any drawing carefully to determine if standard tolerances apply and what they might be for the particular job you are doing.

Fits

Fit refers to the amount or lack of clearance between two mating parts. Fits can range from free running or sliding, where a certain amount of clearance exists between mating parts, to **press** or **interference** fits where parts are forced together under pressure. Clearance fits can range from a few millionths of an inch, such as would be the case in the component parts of a ball or roller bearing, to a clearance of several thousandths of an inch, for a very low speed drive or control lever application.

Many times a machinist is concerned with press or interference fits. In this case two parts are forced together usually by mechanical or hydraulic pressing. The frictional forces involved then hold the parts together without any additional hardware such as keys or set screws. Tolerances for press fits can become very critical because parts can be easily damaged by attempts to press fit them if there is an excessive difference in their mating dimensions. In addition, press fitting physically deforms the parts to some extent. This can result in damage, mechanical binding, or the need for a secondary resizing operation such as hand reaming or honing after the parts are pressed together.

A very typical example of press fit is when a ball bearing inner race is pressed onto a shaft or an outer race is pressed into a bore. Thus, the bearing is retained by friction and the free running feature is obtained within the bearing itself. Ball to race clearance is only a few millionths of an inch in precision bearings. If a bearing is pressed into a bore or onto a shaft with excessive force because pressing allowances are incorrect, the bearing may be physically deformed to the extent that mechanical binding is present. This will often cause excess friction and heat while in operation resulting in rapid failure of the part. On the other hand, insufficient frictional retention of the part resulting from a press fit that is not sufficient can result in the wrong part turning under load or some of the mechanism falling apart while operating.

PRESS FIT ALLOWANCES

Press fit allowances depend on a number of factors including length of engagements, diameter, material, particular components being pressed, and need for later disassembly of parts.

Soft materials such as aluminum can be pressed very successfully. However, soft materials may experience considerable deformation and these parts may not stand up to repeated pressings. Like metal parts pressed without the benefits of lubrication may gall, making them very difficult if not impossible to press apart. Very thin parts such as tubing may bend or deform to such a degree that the press retention is not sufficient to hold the parts together under design loads. The following general rule can be applied when determining the press allowance for cylindrical parts.

Allowance = .0015 × diameter of part in inches

EXAMPLES

Determine the press allowance for a pin with a .250-in. diameter.

.0015 × .250 = .000375 (slightly more than $\frac{3}{10{,}000}$ of an inch)

Determine the press allowance for a 4.250-in. diameter.

.0015 × 4.250 = .00637 (slightly more than $\frac{6}{1000}$ of an inch)

Generally, pressing tolerances range from a few tenths to a few thousandths of an inch depending on the diameter of the parts and the other factors previously discussed. Proper measurement tools and techniques must be employed to make accurate determinations of the dimensions involved. For further specific dimensions on pressing allowances, consult a machinist's handbook.

PRESS FITS AND SURFACE FINISHES

The surface finish (texture) of parts being press fitted can also play an important part. Smooth finished parts will press fit more readily than rough finished parts. If the roughness height of the surface texture is large (64 μ inch and higher), more frictional forces will be generated in the pressing operation and the chances for misalignment, galling, and seizing will be increased, especially if no lubricant is used. Lubrication will improve this situation to some extent. However, lubrication can be detrimental to press fit retention in some cases. A few molecules of lubricant between fit surfaces, especially if they are quite smooth, can result in the parts slipping apart when subjected to certain pull or push forces.

SHRINK AND EXPANSION FITS

Parts can be fitted by making use of the natural tendency of metals to expand or contract when heated and cooled. By heating a part, it will expand and can be then slipped on a mating part. Upon cooling, the heated part will contract and grip the mating part often with tremendous force. Parts may also be mated by cooling one or the other so that it contracts, thus making it smaller. Upon warming to ambient temperature, it will expand to meet the mating part.

Shrink and expansion fits can have superior holding power over press fits, although special heating and cooling equipment may be necessary. Like press fits, however, allowances are extremely important. Consult a machinist's handbook for proper allowance specifications.

Geometric Dimensioning and Tolerancing

Equally important—and in many cases more important than controlling the size of a particular individual part—is controlling the **form** and **position** of a part or assembly feature. This relates directly to the ability to interchange individualized parts and assemblies. For example, you have undoubtedly purchased standard replacement parts for your auto from many different sources. In many cases, these may be made by manufacturers other than the original maker of your auto. However, they fit and function exactly as the original equipment. To make this kind of interface possible, the manufacturing and engineering community has developed a system of geometric dimensioning and tolerancing that helps a manufacturer control

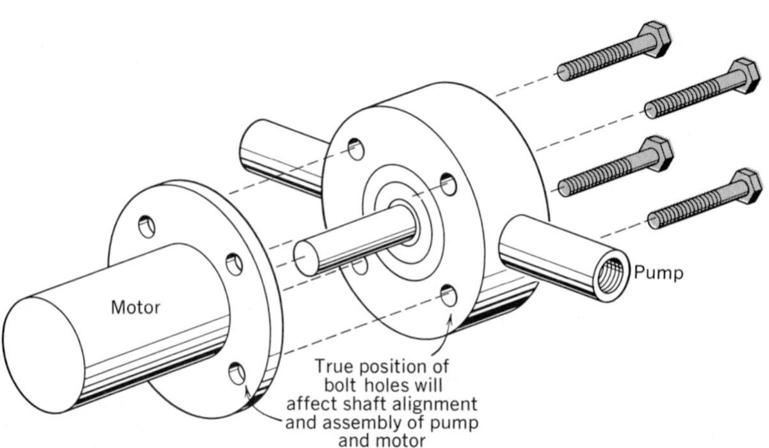
Motor
Pump
True position of
bolt holes will
affect shaft alignment
and assembly of pump
and motor

FIGURE C-255 **Pump and motor bolt hole patterns must match in position in order to accomplish assembly.**

form and position of parts and assemblies. **Geometric dimensioning** and **tolerancing** is a complex subject and would require a great deal of time and space to cover completely. You will learn more about this as you go further into your training. For the present, the following discussion is intended to cover the basic concepts only.

CONTROLLING FEATURE LOCATION

You can see that to bolt the pump to the motor (Figure C-255) it obviously is necessary to ensure that the pattern of bolt holes in the pump matches the pattern of bolt holes in the motor. Also, the bore in the pump housing must match the boss on the motor so that the shaft will engage the motor with proper alignment. If the respective assemblies are made by different manufacturers, you can see that if either bolt pattern position deviates very far from the specified dimensions, the assemblies would be difficult or impossible to interface. As long as the two manufacturers work closely together, the assemblies will interface. However, if the pump manufacturer wants to start using a motor made by another manufacturer, the bolt hole pattern on the new pump motor will also have to interface with the pattern on the pump. This is an example where the **location** or the **true position** of the holes could be more critical than the size of the holes themselves. On drawings the following symbols are used to indicate location control:

⌖ True position

◎ Concentricity

⌯ Symmetry

CONTROLLING FORM

Controlling **form** is equally important. Consider the pump and motor assembly in the previous example. The pump drive shaft must be perpendicular to the impeller case so that it can engage the motor without mechanical binding. Therefore, **perpendicularity** is one example of form that must be controlled during manufacturing. On drawings, the following symbols are used to indicate form control:

⊥ Perpendicularity (squareness)

— Straightness

⏥ Flatness

∠ Angularity

// Parallelism

○ Roundness

⌭ Cylindricity

⌒ Profile of any line

⌓ Profile of any surface

↗ Runout (circular or total)

DATUMS AND BASIC DIMENSIONS

Datums are reference points, lines, and planes taken to be exact for the purpose of calculations and measurements. An initially machined surface on a casting, or example, may be selected as a datum surface and used as a reference from which to measure and locate other part features. Datums are usually not changed by subsequent machining operations and are identified by single or sometimes double letters except (I, O, and Q) inside a rectangular frame. For example,

-A-	-B-

The term **basic** on a drawing represents a true theoretically exact dimension describing location or shape of a part feature. Basic dimensions in theory have no tolerance. They are taken to be exact. Basic dimensions are shown on drawings by the following notations:

1.375 Basic

1.375 BSC

1.375
BSC

1.375

DRAWING FORMATS

The following formats are used to express some of the common geometric dimensions and tolerances on working drawings.

MMC AND RFS

Two other symbols you will encounter on working drawings are **MMC,** or **maximum material condition,** and **RFS,** or **regardless of feature size.** These specifications are identified by the following symbols:

M or **MMC for maximum material condition**
S or **RFS for regardless of feature size**

MMC refers to the maximum amount of material remaining. On an external cylindrical feature this would be the **high limit** of the feature tolerance. For example, a shaft with a diameter of .750 ± .010 would have an MMC diameter of .760 since this would leave maximum material remaining on the part. For an internal cylindrical feature such as a hole, the MMC diameter would be the **low limit** of the tolerance since this would leave maximum material remaining.

RFS, or regardless of feature size, means that the form or position tolerance of a feature must be met no matter what the feature size is. An example of RFS would be a hole located to a true position tolerance call out, where the size of hole itself is not important or not as important as the location.

SELF-TEST

1. Why are tolerances important in manufacturing?
2. What are typical standard tolerances?
3. Name three geometric specifications called out on drawings.
4. What is the general rule for press fit allowances?
5. Describe shrink and expansion fits.

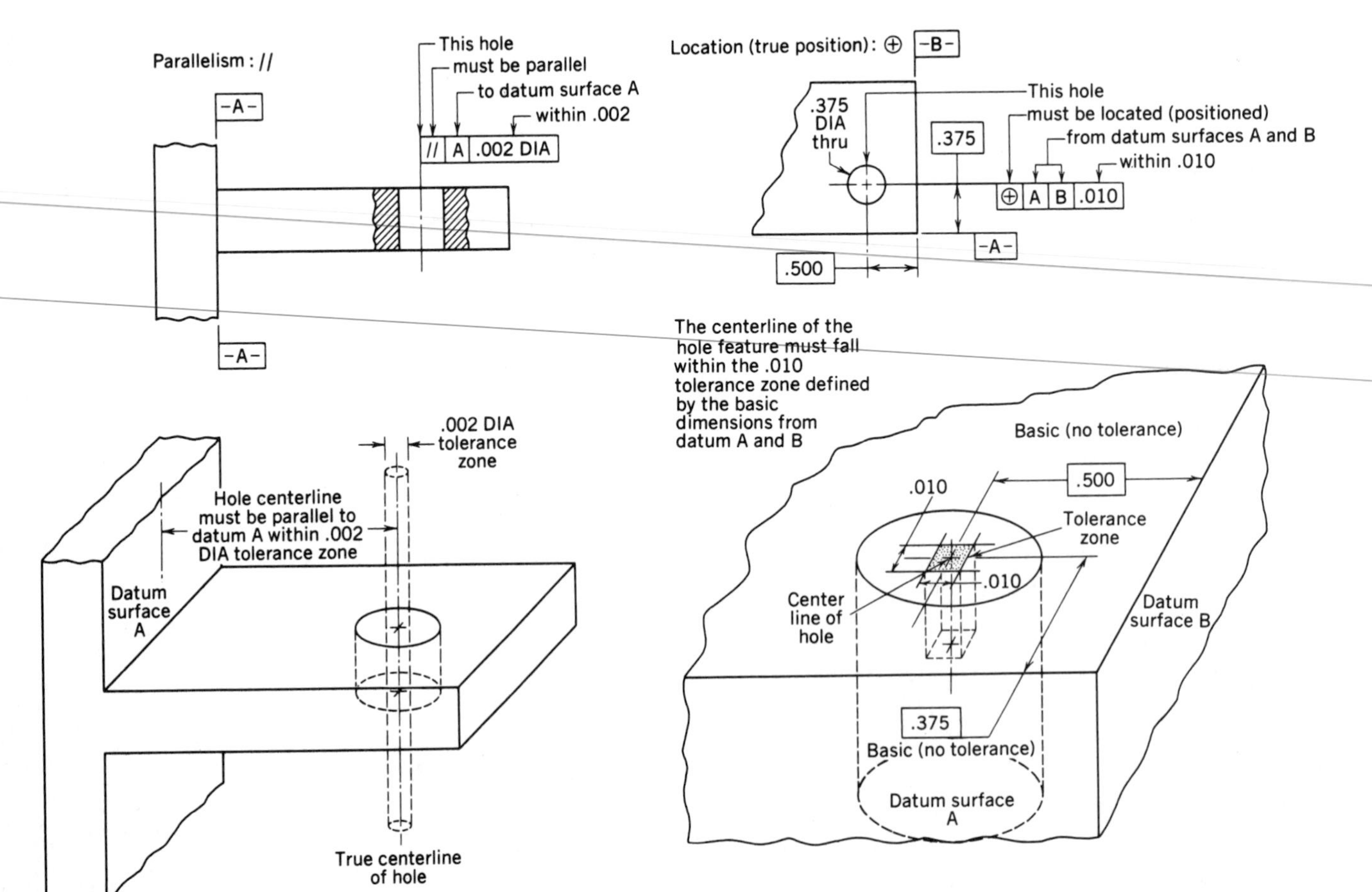

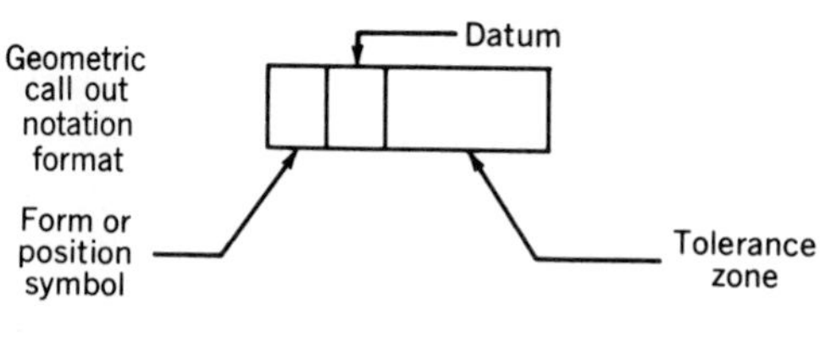

Examples:
Parallelism II

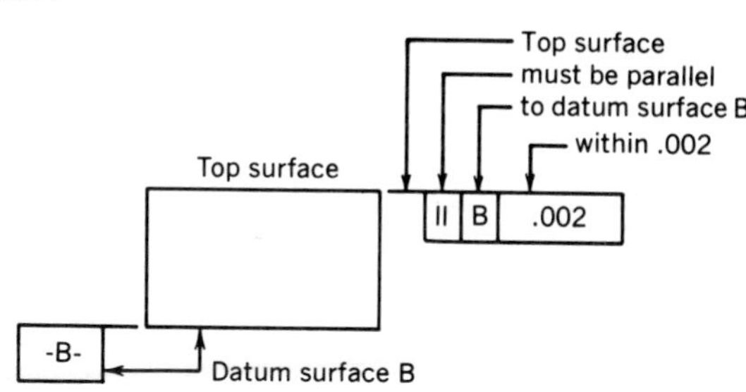

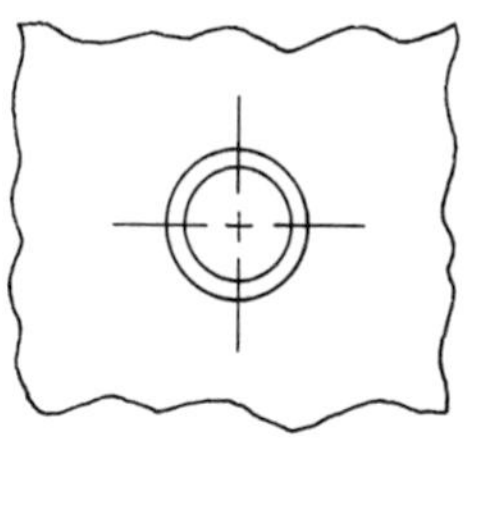

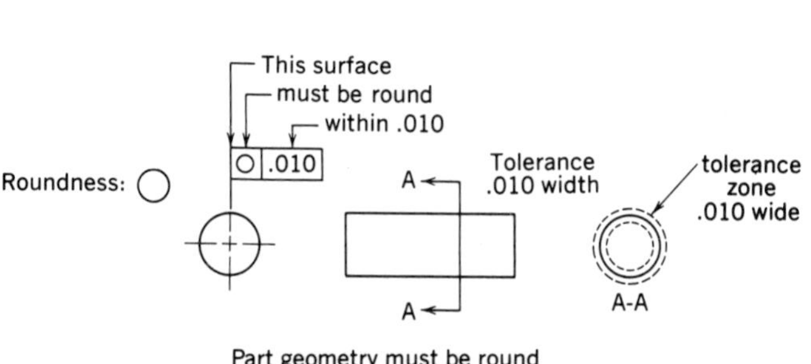

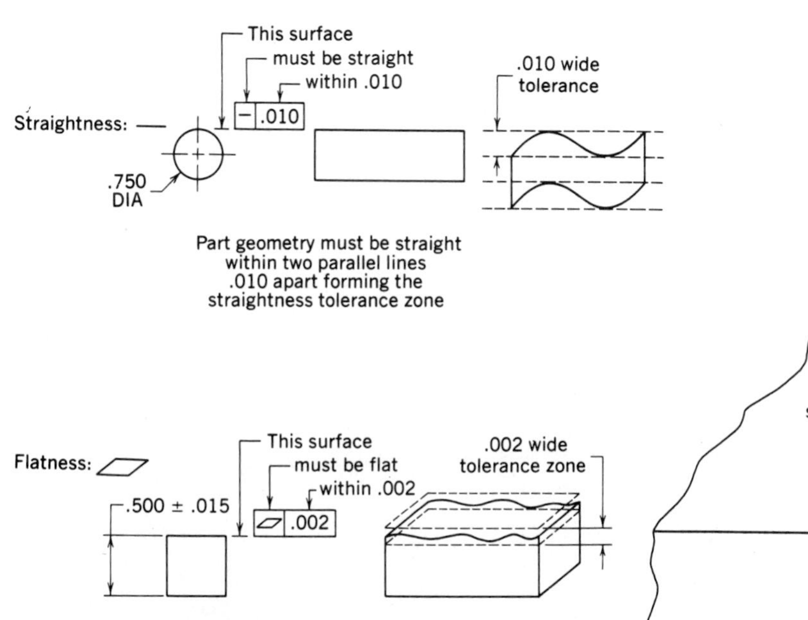

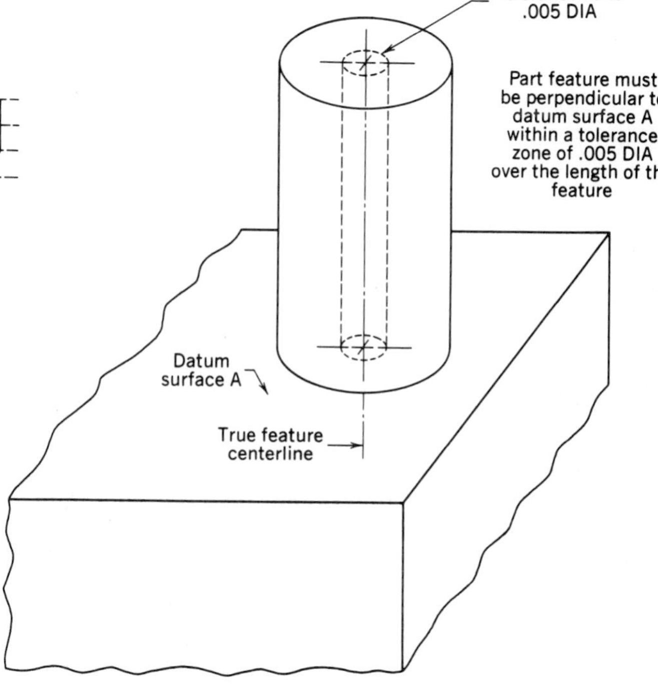

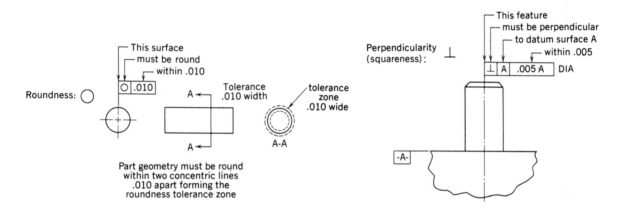

SECTION D

MATERIALS

If there were a single thing to which we could attribute the progress of civilization, it would have to be our ability to make and use tools. The discovery and use of metals would follow close behind, for without metal we would still be fashioning tools of bone and stone.

Nearly everything we need for our present civilization depends upon metals. Vast amounts of iron and steel are used for automobiles, ships, bridges, buildings, machines, and a host of other products (Figure D-1). Almost everything that uses electricity depends on copper and many other metals. Some metals that were impossible to smelt or extract from ores a few years ago are now being used in large quantities. These are usually called space-age metals. There are also hundreds of combinations of metals, called alloys.

We have come a long way since the first iron was smelted, as some believe, by the Hittites about 3500 years ago. Their iron tools, however, were not much better than those of the softer metals, copper and bronze, that were already in use at the time. This was because iron, as iron wire or wrought iron bars, would bend and not hold an edge. The steel making process, which uses iron to make a strong and hard material by heat treatment, was still a long way from being discovered.

Metallic ores are first smelted into metals, and these metals are then formed into the many different products needed in our society. Many metallic ores exist in nature as oxides, in which metals are chemically combined with oxygen. Most iron is removed from the ore by a process called oxidation reduction. Metallic ores are also found as carbonates and silicates.

Modern metallurgy stems from the ancient desire to fully understand the behavior of metals. Long ago, the art of the metalworker was shrouded

FIGURE D-1 Large-scale production of metal products is performed in modern steel mills (Bethlehem Steel Corporation).

FIGURE D-2 Just prior to the industrial Revolution, ironworking had become a highly skilled craft (Dover Publications, Inc.).

in mystery and folklore. Crude methods of making and heat treating small amounts of steel were discovered by trial and error only to be lost and rediscovered later by others (Figure D-2). We have come a long way, indeed, from those early open forges that produced the soft wrought iron in amounts of 20 or 30 lb a day to our modern production marvels that produce more than 100 million tons yearly in the United States.

The modern story of iron and steel begins with the raw materials: iron ore, coal, and limestone. From these ingredients pig iron is produced. Pig iron is the source of almost all our ferrous metals. The steel mill refines it in furnaces, after which it is cast into ingot molds to solidify. The ingot is then formed in various ways into the many steel products that are so familiar to us.

In this section you will investigate the many materials you will work with in the machine shop. You should learn the characteristics of many metals and become familiar with hardening and tempering processes that strengthen and harden by heating and cooling. Different metals often have different properties; for example, cast iron is brittle while soft iron is easily bent; and the difference is due to how they are made. Some steels can be made as hard as needed by heat treatment. You will learn several tests by which you can measure the hardness and resistance of a metal to penetration.

All metals are classified for industrial use by their specific working qualities; you will have to be able to select and identify materials using tables in your handbook or by testing processes in the shop. Some systems are numbers used to classify metals, and others use color codes, which consist of a brand painted on the end of a piece of the material. Spark testing is one popular shop method for identifying metals that you will meet in this book.

It is important to be able to recognize and identify these materials of the trade in order to do a job according to its specifications. This section has been developed to help you achieve this goal.

■ ■ ■

Safety in Material Handling

Safety must be observed when handling material just as it is when using hand and machine tools.

LIFTING AND HOISTING

Machinists were once expected to lift pieces of steel weighing a hundred pounds or more into awkward positions. This is a dangerous practice, however, that results in too many injuries. Hoists and cranes are used to lift all but the smaller parts. Steel weighs about 487 lb per cubic foot; water weighs 62.5 lb per cubic foot, so it is evident that steel is a very heavy material for its size. You can easily be misled into thinking that a small piece of steel does not weigh much. Follow these two rules in all lifting that you do: don't lift more than you can easily handle, and bend your knees and keep your back straight. If a material is too heavy or awkward for you to position it on a machine such as a lathe, use a hoist. Once the workpiece has been hoisted to the required level, it can hang in that position until the clamps or chuck jaws on the machine have been secured.

When lifting heavy metal parts with a mechanical or electric hoist, always stand in a safe posi-

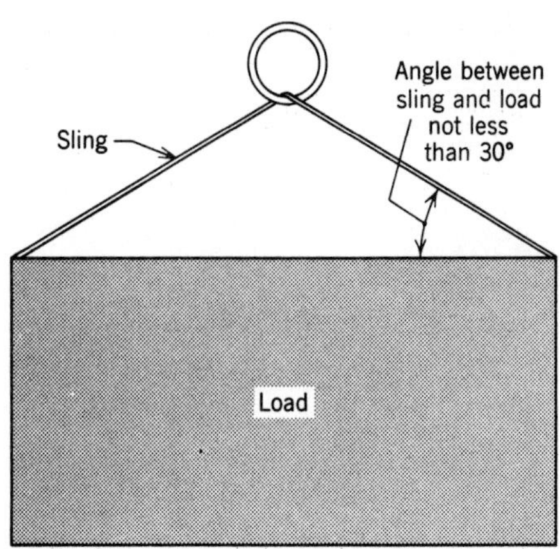

FIGURE D-3 Load sling.

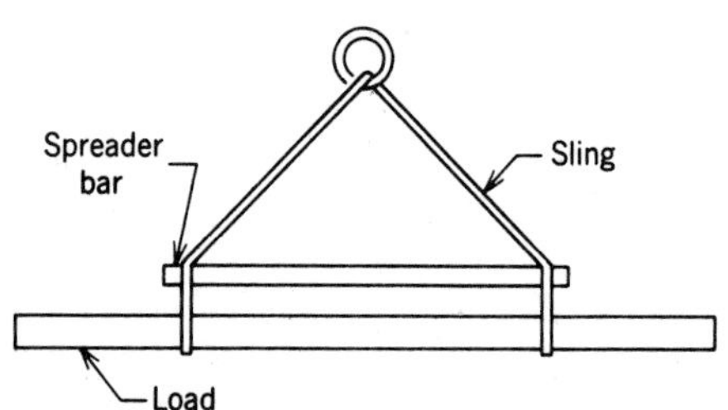

FIGURE D-4 Sling for lifting long bars.

tion, no matter how secure the slings and hooks seem to be. They don't often break, but it can and does happen, and if your foot is under the edge, a painful or crippling experience is sure to follow. Slings should not have less than a 30-degree angle with the load (Figure D-3). When hoisting long bars or shafts, a spreader bar (Figure D-4) should be used so the slings cannot slide together and unbalance the load. When operating a crane, be careful that someone else is not standing in the way of the load or hook. If you are using a block, chain hoist, or electric winch, be sure that the lift capacity rating of the equipment and its support structure is proper for the load.

CARRYING OBJECTS

Carry long stock in the horizontal position. If you must carry it in the vertical position, be careful of light fixtures and ceilings. A better way is to have someone carry each end of a long piece of material. Do not carry sharp tools in your pockets. They can injure you or someone else.

HOT METAL SAFETY

Oxyacetylene torches are often used for cutting shapes, circles, and plates in machine shops. Safety when burning them requires proper clothing, gloves, and eye protection. It is also very important that any metal that has been heated by burning or welding be plainly marked, especially if it is left unattended. The common practice is to write the word *HOT* with soapstone on such items. Whenever arc welding is performed in a shop, the arc flash should be shielded from the other workers. *Never* look toward the arc because if the arc light enters your eye even from the side, the eye can be burned.

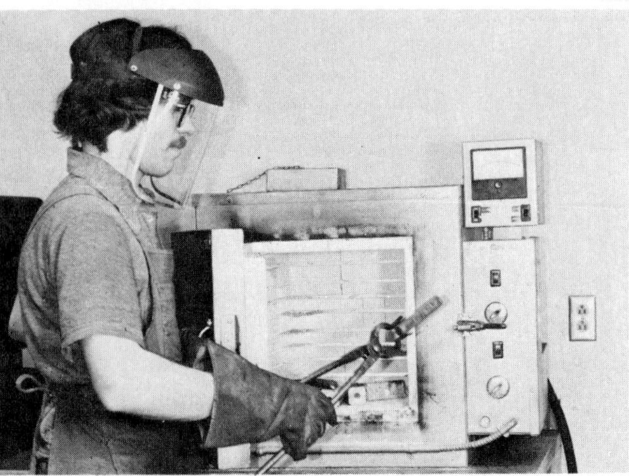

FIGURE D-5 Face shield and gloves are worn for protection while heat treating and grinding (Lane Community College).

When handling and pouring molten metals such as babbitt, aluminum, or bronze, wear a face shield and gloves. Do not pour molten metals where there is a concrete floor, unless it is covered with sand. Concrete contains small amounts of water which instantly become steam when the intense heat of molten metal comes in contact with it. Since concrete is a brittle material, the explosion of the steam causes pieces of the concrete floor to break off the surface and fly like shrapnel.

When heat treating, always wear a face shield and heavy gloves (Figure D-5). There is a definite hazard to the face and eyes when cooling tool steel by oil quenching, that is, submerging it in oil. The oil, hot from the steel, tends to fly upward, so you should stand to one side of the oil tank.

Certain metals, when finely divided as a powder or even as coarse as machining chips, can ignite with a spark or just by the heat of machining. Magnesium and zirconium are two such metals. The fire, once started, is difficult to extinguish, and if water or a water-based fire extinguisher is used, the fire will only increase in intensity. Chloride-based power fire extinguishers are commercially available. These are effective for such fires as they prevent water absorption and form an air-excluding crust over the burning metal. Sand is also used to smother fires in magnesium.

UNIT 1

Selection and Identification of Steels

When the village smithy plied his trade, there were only wrought iron and carbon steel for making tools, implements, and horseshoes, so the task of separating metals was relatively simple. As industry began to need more alloy steels and special metals, they were gradually developed, so today there are many hundreds of these metals in use. Without some means of reference or identification, work in the machine shop would be confusing. Therefore, this unit introduces you to several systems used for marking steels and some ways to choose between them.

Objective

After completing this unit, you should be able to:

Identify different types of metals by various means of shop testing.

Steel Identification Systems

Color coding is used as one means of identifying a particular type of steel. Its main disadvantage is that there is no universal color coding system. Each manufacturer has his own system. The two identification systems most used in the United States are numerical: Society of Automotive Engineers (SAE) and American Iron and Steel Institute (AISI). See Table D-1.

The first two numbers denote the alloy. Carbon, for instance, is denoted by the number 10. The third and fourth digits, represented by x, always denote the percentage of carbon in hundredths of 1 percent. For carbon steel, it could be anywhere from .08 to 1.70 percent. For alloys the second digit designates the approximate percentage of the major alloying element. Steels having over 1 percent carbon require a five-digit number; certain corrosion and heat resisting alloys also use a five-digit number to identify the approximate alloy composition of the metal.

The AISI numerical system is basically the same as the SAE system with certain capital letter prefixes. These prefixes designate the process used to make the steel. The lowercase letters from a to i as a suffix denote special conditions in the steel.

The AISI prefixes are:

B Acid Bessemer carbon steel
C Basic open hearth carbon steel

TABLE D-1 SAE-AISI Numerical Designation of Alloy Steels (x Represents Percent of Carbon in Hundredths)

Carbon steels	
Plain carbon	10xx
Free-cutting, resulfurized	11xx
Manganese steels	13xx
Nickel steels	
.50% nickel	20xx
1.50% nickel	21xx
3.50% nickel	23xx
5.00% nickel	25xx
Nickel–chromium steels	
1.25% nickel, .65% chromium	31xx
1.75% nickel, 1.00% chromium	32xx
3.50% nickel, 1.57% chromium	33xx
3.00% nickel, .80% chromium	34xx
Corrosion and heat-resisting steels	303xx
Molybdenum steels	
Chromium	41xx
Chromium–nickel	43xx
Nickel	46xx and 48xx
Chromium steels	
Low-chromium	50xx
Medium-chromium	511xx
High-chromium	521xx
Chromium–vanadium steels	6xxx
Tungsten steels	7xxx and 7xxxx
Triple-alloy steels	8xxx
Silicon–manganese steels	9xxx
Leaded steels	11Lxx (example)

CB Either acid Bessemer or basic open hearth carbon steel at the option of the manufacturer
D Acid open hearth carbon steel
E Electric furnace alloy steel

Stainless Steel

It is the element chromium (Cr) that makes stainless steels stainless. Steel must contain a minimum of about 11 percent chromium in order to gain resistance to atmospheric corrosion. Higher percentages of chromium make steel even more resistant to corrosion and high temperatures. Nickel is added to improve ductility, corrosion resistance, and other properties.

Excluding the precipitation hardening types that harden over a period of time after solution heat treatment, there are three basic types of stainless steels: the martensitic and ferritic types of the 400 series, and the austenitic types of the 300 series.

The martensitic, hardenable type has carbon content up to 1 percent or more, so it can be hardened by heating to a high temperature, and then quenching (cooling) in oil or air. The cutlery grades of stainless are to be found in this group. The ferritic type contains little or no carbon. It is essentially soft iron that has 11 percent or more chromium content. It is the least expensive of the stainless steels and is used for such things as building trim, pots, and pans. Both ferritic and martensitic types are magnetic.

Austenitic stainless steel contains chromium and nickel, little or no carbon, and cannot be hardened by quenching, but it readily work hardens while retaining much of its ductility. For this reason it can be work hardened until it is almost as hard as a hardened martensitic steel. Austenitic stainless steel is somewhat magnetic in its work hardened condition, but nonmagnetic when annealed or soft.

Table D-2 illustrates the method of classifying the stainless steels. Only a very few of the basic types are given here. You should consult a manufacturer's catalog for further information.

Tool Steels

Special carbon and alloy steels called tool steels have their own classification. There are six major tool steels for which one or more letter symbols have been assigned:

1. Water-hardening tool steels
   W—high-carbon steels
2. Shock-resisting tool steels
   S—Medium carbon, low alloy
3. Cold-worked tool steels
   O—Oil-hardening types
   A—Medium-alloy air-hardening types
   D—High-carbon, high-chromium types
4. Hot-worked tool steels
   H—H1 to H19, chromium-based types
   H20 to H39, tungsten-based types
   H40 to H59, molybdenum-based types
5. High-speed tool steels
   T—Tungsten-based types
   M—Molybdenum-based types

TABLE D-2 Classification of Stainless Steels

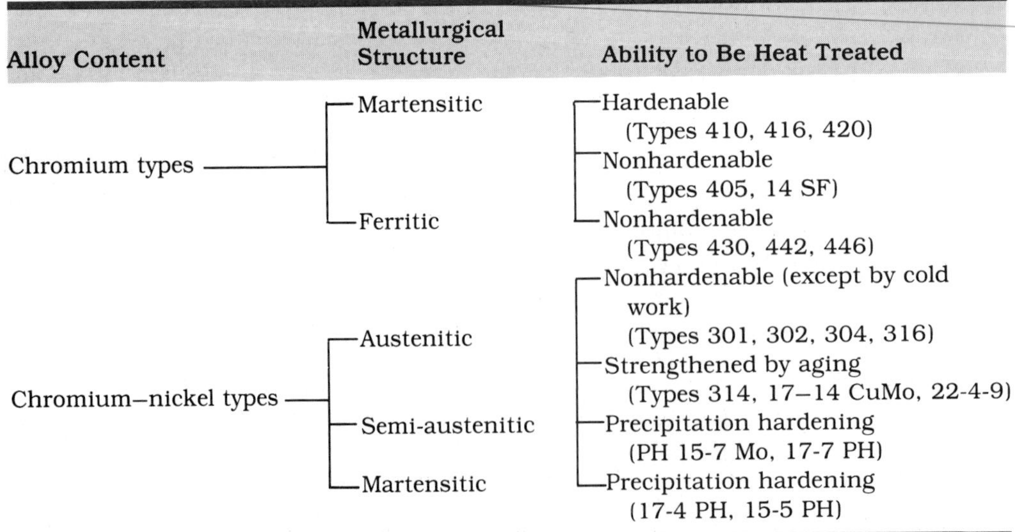

Alloy Content	Metallurgical Structure	Ability to Be Heat Treated
Chromium types	Martensitic	Hardenable (Types 410, 416, 420)
		Nonhardenable (Types 405, 14 SF)
	Ferritic	Nonhardenable (Types 430, 442, 446)
Chromium–nickel types	Austenitic	Nonhardenable (except by cold work) (Types 301, 302, 304, 316)
		Strengthened by aging (Types 314, 17–14 CuMo, 22-4-9)
	Semi-austenitic	Precipitation hardening (PH 15-7 Mo, 17-7 PH)
	Martensitic	Precipitation hardening (17-4 PH, 15-5 PH)

SOURCE: Armco Steel Corporation, Middletown, Ohio, *Armco Stainless Steels*, 1966. The following are registered trademarks of Armco Steel Corporation: 17-4 PH, 15-5 PH, 17-7 PH, and PH 15-7 Mo.

6. Special-purpose tool steels
   L—Low-alloy types
   F—Carbon tungsten types
   P—Mold steels P1 to P19, low-carbon types
   P20 to P39, other types

Several metals can be classified under each group, so that an individual type of tool steel will also have a suffix number that follows the letter symbol of its alloy group. The carbon content is given only in those cases where it is considered an identifying element of that steel.

Type of Steel	Examples
Water hardening: straight carbon tool steel	W1, W2, W4
Manganese, chromium, tungsten: oil-hardening tool steel	01, 02, 06
Chromium (5.0%): air-hardening die steel	A2, A5, A10
Silicon, manganese, molybdenum: punch steel	S1, S5
High-speed tool steel	M2, M3, M30 T1, T5, T15

Shop Tests for Identifying Steels

One of the disadvantages of steel identification systems is that the marking is often lost. The end of a shaft is usually marked. If the marking is obliterated or cut off and the piece is separated from its proper storage rack, it is very difficult to ascertain its carbon content and alloy group. This shows the necessity of returning stock material to its proper rack. It is also good practice always to cut off the unmarked end of the stock material.

Unfortunately, there are always some short ends and otherwise useful pieces in most shops that have become unidentified. Also, when repairing or replacing parts for old or nonstandard machinery, there is usually no record available for material selection. There are many shop methods a machinist may use to identify the basic type of steel in an unknown sample. By process of elimination, the machinist can then determine which of the several steels of that type in the shop is most comparable to the sample. The following are several methods of shop testing that you can use.

VISUAL

Some metals can be identified by visual observation of their surface finishes. Heat scale or black mill scale is found on all hot-rolled (HR) steels. These can be either low carbon (.05 to .30 percent), medium carbon (.30 to .60 percent), high carbon (.60 to 1.70 percent), or alloy steels. Other surface coatings that might be detected are the sherardized, plated, case hardened, or nitrided surfaces. Sherardizing is a process in which zinc vapor is inoculated into the surface of iron or steel.

Cold finish (CF) steel usually has a metallic luster. Ground and polished (G and P) steel has a bright, shiny finish with closer dimensional tolerances than CF. Also cold drawn ebonized, or black, finishes are sometimes found on alloy and resulfurized (free machining) shafting.

Chromium–nickel stainless steel, which is austenitic and nonmagnetic, usually has a white appearance. Straight 12 to 13 percent chromium is ferritic and magnetic with a bluish-white color. Manganese steel is blue when polished, but copper-colored when oxidized. White cast iron fractures will appear silvery or white. Gray cast iron fractures appear dark gray and will smear a finger with a gray graphite smudge when touched.

MAGNET TEST

All ferrous metals such as iron and steel are magnetic; that is, they are attracted to a magnet. Nickel, which is nonferrous (metals other than iron or steel), is also magnetic. U.S. "nickel" coins contain about 25 percent nickel and 75 percent copper, so they do not respond to the magnet test, but Canadian "nickel" coins are attracted to a magnet. Ferritic and martensitic (400 series) stainless steels are also attracted to a magnet and so cannot be separated from other steels by this method. Austenitic (300 series) stainless steel is not magnetic unless it is work hardened.

HARDNESS TEST

Wrought iron is very soft since it contains almost no carbon or any other alloying element. Generally speaking, the more carbon (up to 2 percent) and other elements that steel contains, the harder, stronger, and less ductile it becomes, even if in an annealed state. Thus, the hardness of a sample can help us to separate low carbon steel from an alloy steel or a high-carbon steel. Of course, the best way to check for hardness is with a hardness tester. Rockwell, Brinell, and other types of hardness testing will be studied in another unit. Not all machine shops have hardness testers available, in which case the following shop methods can prove useful.

SCRATCH TEST

Geologists and "rock hounds" scratch rocks against items of known hardness for identification purposes. The same method can be used to check metals for relative hardness. Simply scratch one sample with another and the softer sample will be marked. Be sure all scale or other surface impurities have been removed before scratch testing. A variation of this method is to strike two similar edges of two samples together. The one receiving the deepest indentation is the softer of the two.

FILE TESTS

Files can be used to establish the relative hardness between two samples, as in the scratch test, or they can determine an approximate hardness of a piece on a scale of many steels. Table D-3 gives the Rockwell and Brinell hardness numbers for this file test when using new files. This method, however, can only be as accurate as the skill that the user has acquired through practice.

Care must be taken not to damage the file, since filing on hard materials may ruin the file. Testing should be done on the tip or near the edge.

SPARK TESTING

Spark testing is a useful way to test for carbon content in many steels. When held against a grinding wheel, the metal tested will display a particular spark pattern depending on its content. Spark testing provides a convenient means of distinguishing between tool steel (of medium or high carbon) and low-carbon steel. High-carbon steel (Figure D-6) shows many more bursts than low carbon steel (Figure D-7). It must be noted here that spark testing is by no means an exact method of identifying a particular metal. Other methods of testing should be used if there is any question of the type of metal, especially if it is vitally important that the correct metal be used.

Almost all tool steel contains some alloying elements besides the carbon, which affects the carbon burst. Chromium, molybdenum, silicon, aluminum, and tungsten suppress the carbon burst. For this reason spark testing is not very useful in determining the content of an unknown sample of steel. It is useful, however, as a comparison test. Comparing the spark of a known sample to that of an unknown sample can be an effective method of identification for the trained observer. Cast iron may be distinguished from steel by the characteristic spark stream (Figure D-8). High-speed steel can also be readily identified by spark testing (Figure D-9).

When spark testing, always wear safety glasses or a face shield. Adjust the wheel guard so the spark will fly outward and downward, and away from you. A coarse grit wheel that has been freshly dressed to remove contaminants should be used.

MACHINABILITY TEST

Machinability can be used in a simple comparison test to determine a specific type of steel. For example, two unknown samples identical in appearance and size can be test cut in a machine tool, using the same speed and feed for both. The ease of cutting should be compared, and chips observed for heating color and curl. See Section F for machinability ratings.

TABLE D-3 File Test and Hardness Table

	Rockwell			
Type of Steel	B	C	Brinell	File Reaction
Mild steel	65		100	File bites easily into metal. (Machines well but makes built-up edge on tool.)
Medium-carbon steel		16	212	File bites into metal with pressure. (Easily machined with high-speed tools.)
High-alloy steel High-carbon steel		31	294	File does not bite into metal except with difficulty. (Readily machinable with carbide tools.)
Tool steel		42	390	Metal can only be filed with extreme pressure. (Difficult to machine even with carbide tools.)
Hardened tool steel		50	481	File will mark metal but metal is nearly as hard as the file, and machining is impractical; should be ground.
Case-hardened parts and hardened tool steel		64	739	Metal is as hard as the file; should be ground.

NOTE: Rockwell and Brinell hardness numbers are only approximations since file testing is not an accurate method of hardness testing.
SOURCE: J. E. Neely, *Practical Metallurgy and Materials of Industry*, John Wiley & Sons, New York, 1979.

FIGURE D-6 High-carbon steel. Short, very white or light yellow carrier lines with considerable forking, having many star-like bursts. Many of the sparks follow around the wheel (Lane Community College).

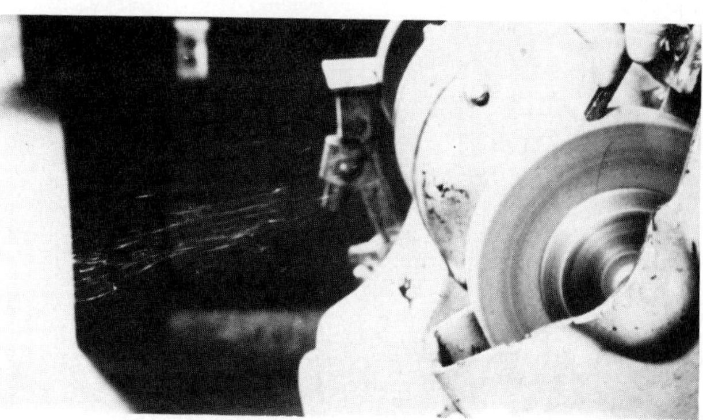

FIGURE D-8 Cast iron. Short carrier lines with many bursts, which are red near the grinder and orange-yellow farther out. Considerable pressure is required on cast iron to produce sparks (Lane Community College).

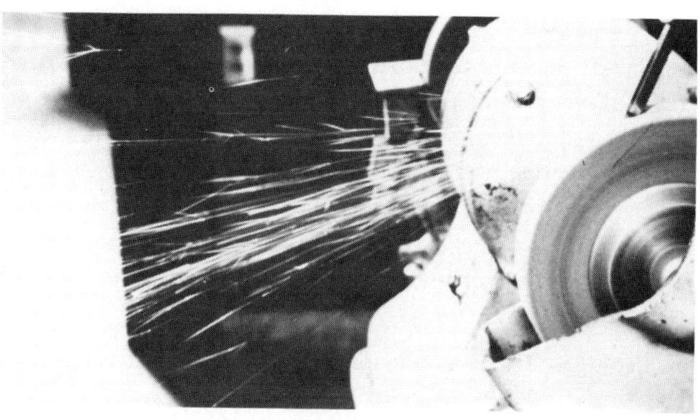

FIGURE D-7 Low-carbon steel. Straight carrier lines having yellowish color with very small amount of branching and very few carbon bursts (Lane Community College).

FIGURE D-9 High-speed steel. Carrier lines are orange, ending in pear-shaped globules with very little branching or carbon sparks. High-speed steel requires moderate pressure to produce sparks (Lane Community College).

OTHER TESTS

Metals can often be identified by their reaction to certain chemicals, usually acids or alkaline substances. Commercial spot testers, available in kits, can be used to identify some metals. Only certain metals can be identified with the chemical test, and the specific amount of an alloying element in a metal is not revealed.

Spectrographic analysis of an unknown metal or alloy can be made in a laboratory. However, a portable spectroscope is now used to sort metals for such purposes as scrap evaluation. This is especially useful for selecting scrap for electric furnaces that produce high-quality steels. Since each element produces characteristic emission lines, it is possible to identify the presence or absence of each element and measure its quantity. These mobile testers are calibrated to a known standard and match signal. A computer compares the test data and a readout on the handheld tester names the metal or alloy.

X-ray analyzers are among the best methods of identifying the content of metals and they are now available as portable units. Although X-ray and spectrographic analyzers are by far the better methods of identifying metals, they are very expensive devices and are probably practical only when large amounts of material must be sorted and identified.

MATERIAL SELECTION

Several properties should be considered when selecting a piece of steel for a job: strength, machinability, hardenability, weldability, fatigue resistance, and corrosion resistance.

Manufacturers' catalogs and handy reference books are available for selection of standard struc-

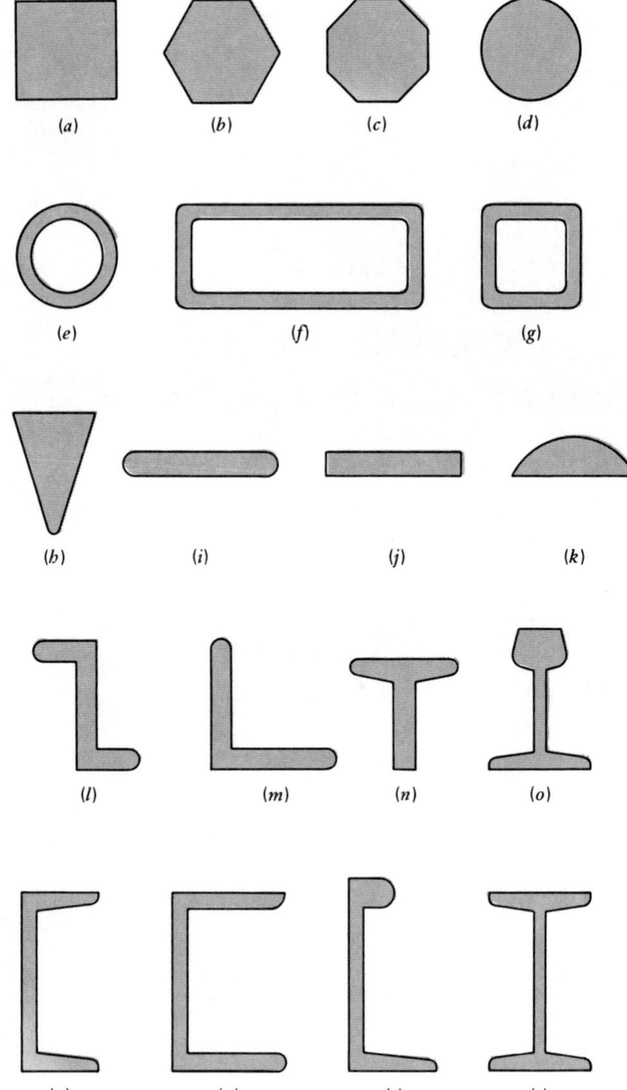

FIGURE D-10 Steel shapes used in manufacturing (J. E. Neely, *Practical Metallurgy and Materials of Industry,* John Wiley & Sons, New York, 1979).

(*a*) Square HR or CR
(*b*) Hexagonal
(*c*) Octagon
(*d*) Round
(*e*) Tubing and pipe (round)
(*f*) HREW (hot rolled electric welded) rectangular steel tubing
(*g*) HREW square steel tubing
(*h*) Wedge
(*i*) HR flat bar (round edge spring steel flats)
(*j*) Flat bar (CR and HR)
(*k*) Half round
(*l*) Zee
(*m*) Angle
(*n*) Tee
(*o*) Rail
(*p*) Channel
(*q*) Car and ship channel
(*r*) Bulb angle
(*s*) Beams—I, H, and wide flange

tural shapes, bars, and other steel products (Figure D-10). Others are available for the stainless steels, tool steels, finished carbon steel, and alloy shafting. Many of these steels are known by a trade name.

A machinist is often called upon to select a shaft material from which to machine finish a part. Shafting is manufactured with two kinds of surface finish: cold finished (CF), found on low-carbon steel, and ground and polished (G and P), found mostly on alloy steel shafts. Tolerances are kept much closer on ground and polished shafts. The following are some common steels used for shafting:

1. SAE 4140 is a chromium–molybdenum alloy with .40 percent carbon. It lends itself readily to heat treating, forging, and welding. It provides a high resistance to torsional and reversing stresses, such as those to which drive shafts are exposed.

2. SAE 1140 is a resulfurized, drawn, free machining bar stock. This material has good resistance to bending stresses because of its fibrous qualities, and it has a high tensile strength. It is best used on shafts where the rpm (revolutions per minute) are high and the torque is low. SAE 1140 is also useful where stiffness is a requirement. It should not be heat treated or welded.

3. Leaded steels have all the free-machining qualities and finishes of resulfurized steels. Leaded alloy steels such as SAE 41L40 have the superior strength of 4140, but are much easier to machine.

4. SAE 1040 is a medium-carbon steel that has a normalized tensile strength of about 85,000 psi. It can be heat treated, but large sections will be hardened only on the surface and the core will still be in a soft condition. Its main advantage is that it is a less expensive way to obtain a higher strength part.

5. SAE 1020 is a low-carbon steel that has good machining characteristics. It normally comes as CF shafting. It is very commonly used for shafting in industrial applications. It has a lower tensile strength than the alloy steels or higher-carbon steels.

Costs of Steel

Steel prices, as prices of most other products, change constantly, so costs can be shown only in example. Steel is usually priced by its weight. A cubic foot of mild steel weighs 489.60 lb, so a square foot 1 in. thick weighs 40.80 lb. From this, you can easily compute the weights for flat materials such as plate. For hexagonal and rounds, it would be much easier to consult a table in a catalog or handbook. Given a price per pound, you should then be able to figure the cost of a desired steel product.

EXAMPLE

A 1 by 6 in. mild steel bar is 48 in. long. If current steel prices are 30 cents per pound, how much does the bar cost?

$$\frac{6 \times 48}{144} = 2\ \text{ft}^2,\ 1\ \text{in. thick}$$

$$2 \times 40.80 = 81.6\ \text{lb}$$

$$81.6 \times \$0.30 = \$24.48$$

SELF-TEST

1. By what universal coding system is carbon and alloy steel designated?

2. What are three basic types of stainless steels and what is the number series assigned to them? What are their basic differences?

3. If your shop stocked the following steel shafting, how would you determine the content of an unmarked piece of each, using shop tests as given in this unit?

a. AISI C1020 CF
b. AISI B1140 (G and P)
c. AISI C4140 (G and P)
d. AISI 8620 HR
e. AISI B1140 (Ebony)
f. AISI C1040

4. A small part has obviously been made by a casting process. Using an inexpensive testing method, how can you determine whether it is a ferrous or a nonferrous metal, or if it is steel or white or gray cast iron?

5. What is the meaning of the symbols O1 and W1 when applied to tool steels?

6. A $2\frac{7}{16}$-in.-diameter steel shaft weighs 1.322 lb per linear inch, as taken from a table of weights of steel bars. A 40-in. length is needed for a job. At 30 cents per pound, what would the shaft cost?

7. When checking the hardness of a piece of steel with the file test, the file slides over the surface without cutting.

a. Is the steel piece readily machinable?
b. What type of steel is it most likely to be?

8. Steel that is nonmagnetic is called ________.

9. What nonferrous metal is magnetic?

10. List at least four properties of steel that should be kept in mind when you select the material for a job.

UNIT 2

Selection and Identification of Nonferrous Metals

Metals are designated as either ferrous or nonferrous. Iron and steel are ferrous metals, and any metal other than iron or steel is called nonferrous. Nonferrous metals such as gold; silver, copper, and tin were in use hundreds of years before the smelting of iron, and yet some nonferrous metals have appeared relatively recently in common industrial use. For example, aluminum was first commercially extracted from ore in 1886 by the Hall–Heroult process, and titanium is a space age metal produced in commercial quantities only since World War II.

In general, nonferrous metals are more costly than ferrous metals. It isn't always easy to distinguish a nonferrous metal from a ferrous metal, nor to separate one from another. This unit should help you to identify, select, and properly use many of these metals.

Objectives

After completing this unit, you should be able to:

1. Identify and classify nonferrous metals by a numerical system.
2. List the general appearance and use of various nonferrous metals.

FIGURE D-11 Structural aluminum shapes used for building trim provides a pleasing appearance.

Aluminum

Aluminum is white or white-gray in color and can have any surface finish from dull to shiny and polished. An anodized surface is frequently found on aluminum products. Aluminum weighs 168.5 pounds per cubic foot (lb/ft^3) as compared to 487 lb/ft^3 for steel, and it has a melting point of 1220°F (660°C) when pure. It is readily machinable and can be manufactured into almost any shape or form (Figure D-11).

TABLE D-4 Aluminum and Aluminum Alloys

Code Number	Major Alloying Elements
1xxx	None
2xxx	Copper
3xxx	Manganese
4xxx	Silicon
5xxx	Magnesium
6xxx	Magnesium and silicon
7xxx	Zinc
8xxx	Other elements
9xxx	Unused (not yet assigned)

Magnesium is also a much lighter metal than steel, as it weighs 108.6 lb/ft^3, and looks much like aluminum. In order to distinguish between the two metals, it is sometimes necessary to make a chemical test. A zinc chloride solution in water, or a copper sulfate solution, will blacken magnesium immediately, but will not change aluminum.

There are several numerical systems used to identify aluminums, such as federal, military, the American Society for Testing and Materials (ASTM), and SAE specifications. The system most used by manufacturers, however, is one adopted by the Aluminum Association in 1954.

From Table D-4 you can see that the first digit of a number in the aluminum alloy series indicates the alloy type. The second digit, represented by an x in the table, indicates any modifications that were made to the original alloy. The last two digits indicate the numbers of similar aluminum alloys of an older marking system, except in the 1100 series, where the last two digits indicate the amount of pure aluminum above 99 percent contained in the metal.

EXAMPLES

An aluminum alloy numbered 5056 is an aluminum–magnesium alloy, where the first 5 represents the alloy magnesium, the 0 represents modifications to the alloy, and 56 are numbers of a similar aluminum of an older marking system. An aluminum numbered 1120 contains no major alloy and has .20 percent pure aluminum above 99 percent.

Aluminum and its alloys are produced as castings or as wrought (cold worked) shapes such as sheets, bars, and tubing. Aluminum alloys are harder than pure aluminum and will scratch the softer (1100 series) aluminums. Pure aluminum and some of its alloys cannot be heat treated so their tempering is done by other methods. The temper designations are made by a letter that follows the four-digit alloy series number:

—F as fabricated. No special control over strain hardening or temper designation is noted.

- —O Annealed, recrystallized wrought products only. Softest temper.
- —H Strain hardened, wrought products only. Strength is increased by work hardening.

The letter —H is always followed by two or more digits. The first digit, 1, 2, or 3, denotes the final degree of strain hardening:

- —H1 Strain hardened only
- —H2 Strain hardened and partially annealed
- —H3 Strain hardened and stabilized

and the second digit denotes higher strength tempers obtained by heat treatment:

- 2 ¼ hard
- 4 ½ hard
- 6 ¾ hard
- 8 full hard

EXAMPLE

5056-H18 is an aluminum–magnesium alloy, strain hardened to a full hard temper.

Some aluminum alloys can be hardened to a great extent by a process called solution heat treatment and precipitation or aging. This process involves heating the aluminum and its alloying elements until it is a solid solution. The aluminum is then quenched in water and allowed to age or is artificially aged by heating slightly. The aging produces an internal strain that hardens and strengthens the aluminum. Some other nonferrous metals are also hardened by this process. For these aluminum alloys the letter —T follows the four-digit series number. Numbers 2 to 10 follow this letter to indicate the sequence of treatment.

- —T2 Annealed (cast products only)
- —T3 Solution heat treated and cold worked
- —T4 Solution heat treated, but naturally aged
- —T6 Solution heat treated and artificially aged
- —T8 Solution heat treated, cold worked, and artificially aged
- —T9 Solution heat treated, artificially aged, and cold worked
- —T10 Artificially aged and then cold worked

EXAMPLE

2024-T6 is an aluminum–copper alloy, solution heat treated and artificially aged.

Cast aluminum alloys generally have lower tensile strength than wrought alloys. Sand castings, permanent mold, and die casting alloys are of this group. They owe their mechanical properties to solution heat treatment and precipitation or to the addition of alloys. A classification system similar to that of wrought aluminum alloys is used (Table D-5).

TABLE D-5 Cast Aluminum Alloy Designations

Code Number	Major Alloy Element
1xx.x	None, 99 percent aluminum
2xx.x	Copper
3xx.x	Silicon with Cu and/or Mg
4xx.x	Silicon
5xx.x	Magnesium
6xx.x	Zinc
7xx.x	Tin
8xx.x	Unused series
9xx.x	Other major alloys

SOURCE: J. E. Neely, *Practical Metallurgy and Materials of Industry*, John Wiley & Sons, New York, 1979.

The cast aluminum 108F, for example, has an ultimate tensile strength of 24,000 psi in the as fabricated condition and contains no alloy. The 220.T4 copper aluminum alloy has a tensile strength of 48,000 psi.

Other Nonferrous Metals

CADMIUM

Cadmium has a blue-white color and is commonly used as a protective plating on parts such as screws, bolts, and washers. It is also used as an alloying element to make metal alloys that melt at low temperature, such as bearing metals, solder, type casting metals, and storage batteries. Cadmium compounds such as cadmium oxide are toxic and can cause illness when breathed. Toxic fumes can be produced by welding, cutting, or machining on cadmium plated parts. Breathing the fumes should be avoided by using adequate ventilation systems. The melting point of cadmium is 610°F (321°C). Its weight is 539.6 lb/ft^3.

COPPER AND COPPER ALLOYS

Copper is a soft, heavy metal that has a reddish color. It has high electrical and thermal conductivity when pure, but loses these properties to a certain extent when alloyed. It must be strain hardened when used for electric wire. Copper is very ductile and can be easily drawn into wire or tubular products. It is so soft that it is difficult to machine, and it has a tendency to adhere to tools. Copper can be work hardened or hardened by solution heat treatment when alloyed with beryllium. The melting point of copper is 1981°F (1083°C). Its weight is 554.7 lb/ft^3.

BERYLLIUM COPPER

Beryllium copper is an alloy of copper and beryllium that can be hardened by heat treating for making nonsparking tools and other products. Machining of this metal should be done after solution heat treatment and aging, not when it is in the annealed state. Machining or welding beryllium copper can be very hazardous if safety precautions are not followed. Machining dust or welding fumes should be removed by a heavy coolant flow or by a vacuum exhaust system. A respirator type of face mask should be worn when around these two hazards. The melting point of beryllium is 2435°F (1285°C), and its weight is 115 lb/ft^3.

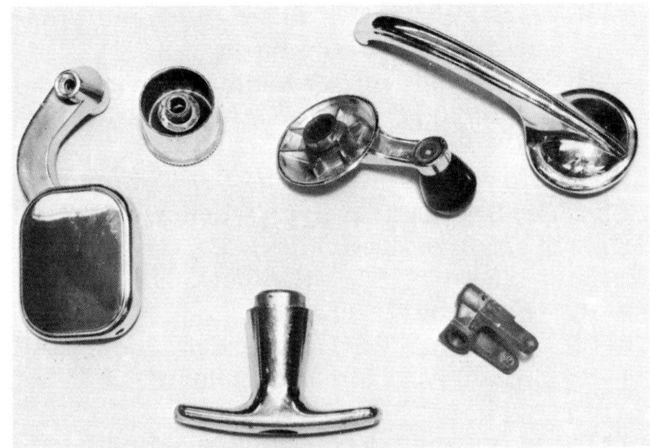

FIGURE D-12 Die-cast parts.

BRASS

Brass is an alloy of zinc and copper. Brass colors usually range from white to yellow, and in some alloys, red to yellow. Brasses range from gilding metal used for jewelry (95 percent copper, 5 percent zinc) to Muntz metal (60 percent copper, 40 percent zinc) used for bronzing rod and sheet stock. Brasses are easily machined. Brass is usually tougher than bronze and produces a stringy chip when machined. The melting point of brasses ranges from 1616 to 1820°F (880 to 993°C), and their weights range from 512 to 536 lb/ft^3.

BRONZE

Bronze is found in many combinations of copper and other metals, but copper and tin are its original elements. Bronze colors usually range from red to yellow. Phosphor bronze contains 91.95 percent copper, 0.05 percent phosphorus, and 8 percent tin. Aluminum bronze is often used in the shop for making bushings or bearings that support heavy loads. (Brass is not normally used for making antifriction bushings.) The melting point of bronze is about 1841°F (1005°C) and its weight is about 548 lb/ft^3. Bronzes are usually harder than brasses, but are easily machined with sharp tools. The chip produced is often granular. Some bronze alloys are used as brazing rods.

CHROMIUM

Chromium is a slightly gray metal that can take a high polish. It has a high resistance to corrosion by most reagents; exceptions are dilute hydrochloric and sulfuric acids. Chromium is widely used as a decorative plating on automobile parts and other products.

Chromium is not a very ductile or malleable metal and its brittleness limits its use as an unalloyed metal. It is commonly alloyed with steel to increase hardness and corrosion resistance. Chrome-nickel and chrome-molybdenum are two very common chromium alloys. Chromium is also used in electrical heating elements such as chromel or nichrome wire. The melting point of chromium is 2939°F (1615°C) and its weight is 432.4 lb/ft^3.

DIE-CAST METALS

Finished castings are produced with various metal alloys by the process of die casting. Die casting is a method of casting molten metal by forcing it into a mold. After the metal has solidified, the mold opens and the casting is ejected. Carburetors, door handles, and many small precision parts are manufactured using this process (Figure D-12). Die-cast alloys, often called "pot metals," are classified in six groups:

1. Tin-based alloys
2. Lead-based alloys
3. Zinc-based alloys
4. Aluminum-based alloys
5. Copper, bronze, or brass alloys
6. Magnesium-based alloys

The specific content of the alloying elements in each of the many die-cast alloys can be found in handbooks or other references on die casting.

LEAD AND LEAD ALLOYS

Lead is a heavy metal that is silvery when newly cut and gray when oxidized. It has a high density, low tensile strength, low ductility (cannot be easily drawn into wire), and high malleability (can be easily compressed into a thin sheet).

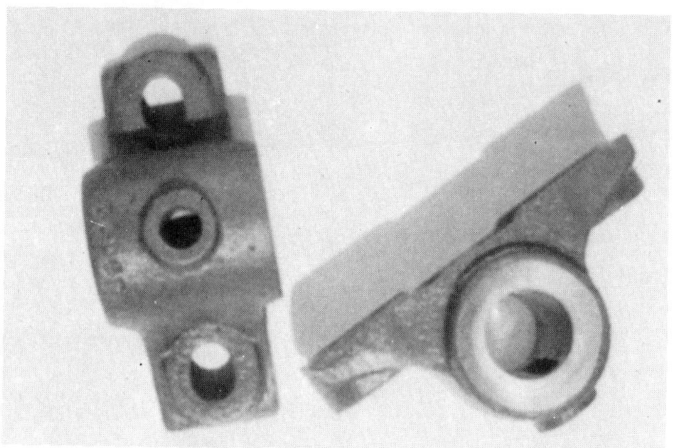

FIGURE D-13 Babbitted pillow block bearings.

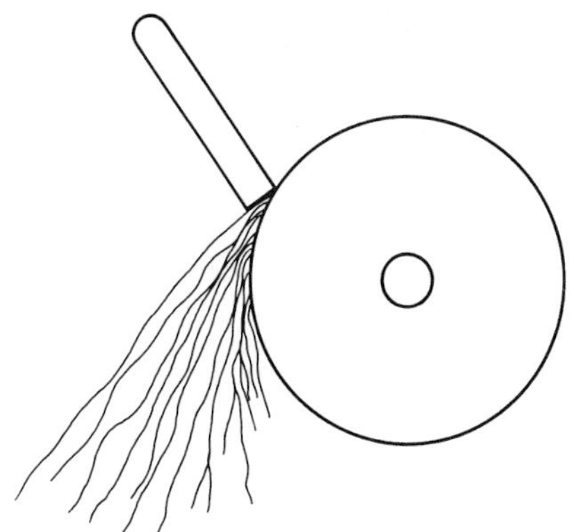

FIGURE D-14 Spark test for nickel.

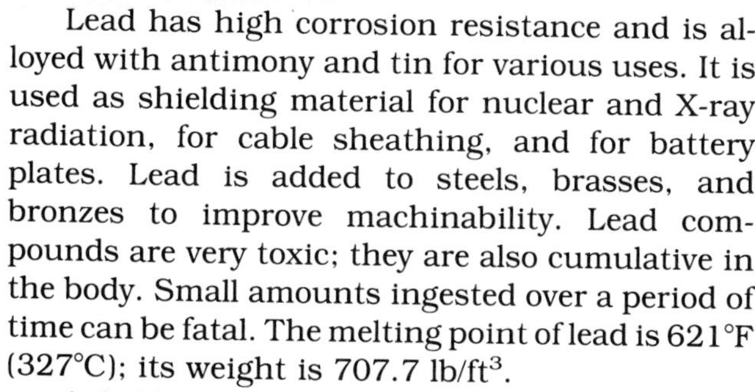

Lead has high corrosion resistance and is alloyed with antimony and tin for various uses. It is used as shielding material for nuclear and X-ray radiation, for cable sheathing, and for battery plates. Lead is added to steels, brasses, and bronzes to improve machinability. Lead compounds are very toxic; they are also cumulative in the body. Small amounts ingested over a period of time can be fatal. The melting point of lead is 621°F (327°C); its weight is 707.7 lb/ft^3.

A babbitt metal is a soft, antifriction alloy metal often used for bearings and is usually tin or lead based (Figure D-13). Tin babbitts usually contain from 65 to 90 percent tin with antimony, lead, and a small percentage of copper added. These are the higher grade and generally the more expensive of the two types. Lead babbitts contain up to 75 percent lead with antimony, tin, and some arsenic making up the difference.

Cadmium-based babbitts resist higher temperatures than other tin- and lead-based types. These alloys contain from 1 to 15 percent nickel or a small percentage of copper and up to 2 percent silver. The melting point of babbitt is about 480°F (249°C).

MAGNESIUM

When pure, magnesium is a soft, silver-white metal that closely resembles aluminum, but weighs less. In contrast to aluminum, magnesium will readily burn with a brilliant white light; thus magnesium presents a fire hazard when machined. Magnesium, which is similar to aluminum in density and appearance, presents some quite different machining problems. Although magnesium chips can burn in air, applying water will only cause the chips to burn more fiercely. Sand or special compounds should be used to extinguish these fires. Thus, when working with magnesium, a water-based coolant should never be used. Magnesium can be machined dry when light cuts are taken and the heat is dissipated. Compressed air is sometimes used as a coolant. Anhydrous (containing no water) oils having a high flash point and low viscosity are used in most production work. Magnesium is machined with very high surface speeds and with tool angles similar to those used for aluminum.

Cast and wrought magnesium alloys are designated by SAE and ASTM numbers, which may be found in metals reference handbooks such as *Machinery's Handbook*. The melting point of magnesium is 1204°F (651°C), and its weight is 108.6 lb/ft^3.

MOLYBDENUM

As a pure metal, molybdenum is used for high-temperature applications and, when machined, it chips like gray cast iron. It is used as an alloying element in steel to promote deep hardening and to increase its tensile strength and toughness. Pure molybdenum is used for filament supports in lamps and in electron tubes. The melting point of molybdenum is 4748°F (2620°C); its weight is 636.5 lb/ft^3.

NICKEL

Nickel is noted for its resistance to corrosion and oxidation. It is a whitish metal used for electroplating and as an alloying element in steel and other metals to increase ductility and corrosion resistance. It resembles pure iron in some ways but has greater corrosion resistance. Electroplating is the coating or covering of another material with a thin layer of metal, using electricity to deposit the layer.

When spark tested, nickel throws short orange carrier lines with no sparks or sprigs (Figure D-14). Nickel is attracted to a magnet, but becomes nonmagnetic near 680°F (360°C). The melting point of nickel is 2646°F (1452°C), and its weight is 549.1 lb/ft^3.

NICKEL-BASED ALLOYS

Monel® is an alloy of 67 percent nickel and 28 percent copper, plus impurities such as iron, cobalt, and manganese. It is a tough but machinable, ductile, and corrosion resistant alloy. Its tensile strength (resistance of a metal to a force tending to tear it apart) is 70,000 to 85,000 lb/in^2. Monel metal is used to make marine equipment such as pumps, steam valves, and turbine blades. On a spark test, monel shoots orange-colored, straight sparks about 10 in. long, similar to those of nickel. K-Monel contains 3 to 5 percent aluminum and can be hardened by heat treatment.

Chromel and nichrome are two nickel–chromium–iron alloys used as resistance wire for electric heaters and toasters. Nickel–silver contains nickel and copper in similar proportions to Monel, but also contains 17 percent zinc. Other nickel alloys, such as Inconel, are used for parts exposed to high temperatures for extended periods.

Inconel, a high-temperature and corrosion-resistant metal consisting of nickel, iron, and chromium, is often used for aircraft exhaust manifolds because of its resistance to high temperature oxidation (scaling). The nickel alloys' melting point range is 2425 to 2950°F (1329 to 1621°C).

FIGURE D-15 The most familiar tin plate product is the steel based tin can (American Iron & Steel Institute).

PRECIOUS METALS

Gold has a limited industrial value and is used in dentistry, electronic and chemical industries, and jewelry. In the past, gold has been used mostly for coinage. Gold coinage is usually hardened by alloying with about 10 percent copper. Silver is alloyed with 8 to 10 percent copper for coinage and jewelry. Sterling silver is 92.5 percent silver in English coinage and has been 90 percent silver for American coinage. Silver has many commercial uses, such as an alloying element for mirrors, in photographic compounds, and electrical equipment. It has a very high electrical conductivity. Silver is used in silver solders that are stronger and have a higher melting point than lead-tin solders.

Platinum, palladium, and iridium, as well as other rare metals, are even more rare than gold. These metals are used commercially because of their special properties such as extremely high resistance to corrosion, high melting points, and high hardness. The melting points of some precious metals are: gold, 1945°F (1063°C); iridium, 4430°F (2443°C); platinum, 3224°F (1773°C); and silver, 1761°F (961°C). Gold has a weight of 1204.3 lb/ft^3. The weight of silver is about 654 lb/ft^3. Platinum is one of the heaviest of metals with a weight of 1333.5 lb/ft^3. Iridium is also a heavy metal, at 1397 lb/ft^3.

TANTALUM

Tantalum is a bluish-gray metal that is difficult to machine because it is quite soft and ductile and the chip clings to the tool. It is immune to attack from all corrosive acids except hydrofluoric and fuming sulfuric acids. It is used for high-temperature operations above 2000°F (1093°C). It is also used for surgical implants and in electronics. Tantalum carbides are combined with tungsten carbides for cutting tools that have high abrasive resistance. The melting point of tantalum is 5162°F (2850°C); its weight is 1035.8 lb/ft^3.

TIN

Tin has a white color with slightly bluish tinge. It is whiter than silver or zinc. Since tin has a good corrosion resistance, it is used to plate steel, especially for the food processing industry (Figure D-15). Tin is used as an alloying element for solder, babbitt, and pewter. A popular solder is an alloy of 50 percent tin and 50 percent lead. Tin is alloyed with copper to make bronze. The melting point of tin is 449°F (232°C). Its weight is 454.9 lb/ft^3.

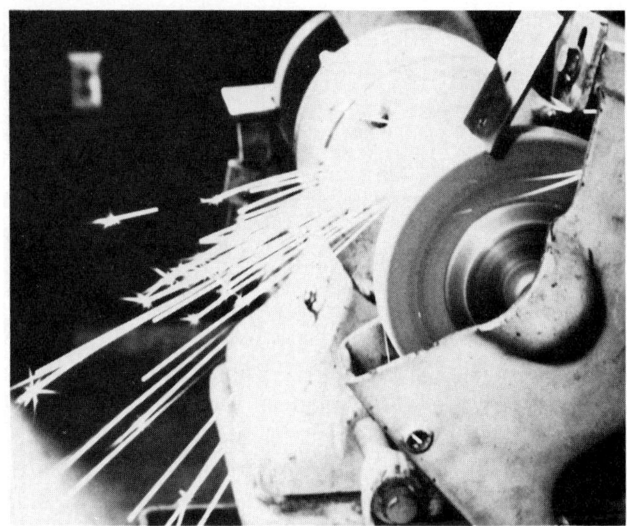

FIGURE D-16 Titanium spark testing (Lane Community College).

TITANIUM

The strength and light weight of this silver-gray metal make it very useful in the aerospace industries for jet engine components, heat shrouds, and rocket parts; however, pure finely divided titanium can ignite and burn when heated to high temperatures. Pure titanium has a tensile strength of 60,000 to 110,000 psi, similar to that of steel; by alloying titanium, its tensile strength can be increased considerably. Titanium weighs about half as much as steel and, like stainless steel, is a relatively difficult metal to machine. Machining can be accomplished with rigid setups, sharp tools, slower surface speed, and proper coolants. When spark tested, titanium throws a brilliant white spark with a single burst on the end of each carrier (Figure D-16). The melting point of titanium is 3272°F (1800°C), and its weight is 280.1 lb/ft^3.

TUNGSTEN

Typically, tungsten has been used for incandescent light filaments. It has the highest known melting point (6098°F, 3370°C) of any metal, but is not resistant to oxidation at high temperatures. Tungsten is used for rocket engine nozzles and welding electrodes and as an alloying element with other metals. Machining pure tungsten is very difficult with single point tools, and grinding is preferred for finishing operations. Tungsten carbide compounds are used to make extremely hard and heat resistant lathe tools and milling cutters by compressing the tungsten carbide powder into a briquette and sintering it in a furnace. Tungsten weighs about 1180 lb/ft^3.

ZINC

The familiar galvanized steel is actually steel plated with zinc and is used mainly for its high corrosion resistance. Zinc alloys are widely used as die-casting metals. Zinc and zinc-based die-cast metals conduct heat much more slowly than aluminum. The rate of heat transfer on similar shapes of aluminum and zinc is a means of distinguishing between them. The melting point of zinc is 787°F (419°C), and it weighs about 440 lb/ft^3.

ZIRCONIUM

Zirconium is similar to titanium in both appearance and physical properties. It was once used as an explosive primer and is a flashlight powder for photography since, like magnesium, it readily combines with oxygen and rapidly burns when finely divided. Machining zirconium, like titanium, requires rigid setups and slow surface speeds. Zirconium has an extremely high resistance to corrosion from acids and sea water. Zirconium alloys are used in nuclear reactors, flash bulbs, and surgical implants such as screws, pegs, and skull plates. When spark tested, it produces a spark similar to that of titanium. The melting point of zirconium is 3182°F (1750°C). Its weight is 339 lb/ft^3.

SELF-TEST

1. What advantages do aluminum and its alloys have over steel alloys? What disadvantages?

2. Describe the meaning of the letter—H when it follows the four digit number that designates an aluminum alloy? The meaning of the letter—T?

3. Name two ways in which magnesium differs from aluminum.

4. What is the major use of copper? How can copper be hardened?

5. What is the basic difference between brass and bronze?

6. Name two uses for nickel.

7. Lead, tin, and zinc all have one useful property in common. What is it?

8. Molybdenum and tungsten are both used in __________ steels.

9. Babbit metals, used for bearings, are made in what major basic types?

10. What type of metal can be injected under pressure into a permanent mold?

UNIT 3

Hardening, Case Hardening, and Tempering

Probably the most important property of carbon steels is their ability to be hardened through the process of heat treatments. Steels must be made hard if they are to be used as tools that have the ability to cut many materials, including other soft steels. Various degrees of hardness are also desirable depending on the application of the tool steels. Heat treating carbon steels involves some very critical furnace operations. The proper steps must be carried out precisely, or a failure of the hardened steel will almost surely result. In some cases only surface hardening of a steel is required. This is accomplished through the process of surface and case hardening.

It is often desirable to slightly reduce the hardness of a steel tool in order to enhance other properties. For example, a chisel must have a hard cutting edge. It must also have a somewhat less hard but tough shank that will withstand hammering. The property of toughness is acquired through the process of tempering.

In this unit you will study the important processes and procedures for hardening, case hardening, and tempering of plain carbon steels.

Objectives

After completing this unit, you should be able to:

1. Correctly harden a piece of tool steel and evaluate your work.
2. Correctly temper the hardened piece of tool steel and evaluate your work.
3. Describe the proper heat treating procedures for other tool steels.

Hardening Metals

Most metals (except copper used for electric wire) are not used commercially in their pure states because they are too soft and ductile and have low tensile strength. When they are alloyed with other elements, such as other metals, they become harder and stronger as well as more useful. A small amount (1 percent) of carbon greatly affects pure iron when alloyed with it. The alloy metal becomes a familiar tool steel used for cutting tools, files, and punches. Iron with 2 to $4\frac{1}{2}$ percent carbon content yields cast iron.

Iron, steel, and other metals are composed of tiny grain structures that can be seen under a microscope when the specimen is polished and etched. This grain structure, which determines the strength and hardness of a steel, can be seen with the naked eye as small crystals in the rough broken section of a piece (Figure D-17). These crystals or grain structures grow from a nucleus as the molten metal solidifies until the grain boundaries are formed. Grain structures differ according to the allotropic form of the iron or steel. An allotropic element is one able to exist in two or more forms with various properties without a change of chemical composition. Carbon exists in three allotropic forms: amorphous (charcoal, soot, coal), graphite, and diamond. Iron also exists in three allotropic forms (Figure D-18): ferrite (at room temperature), austenite (above 1670°F, 911°C) and delta (between 2550 and 2800°F, or 1498 and 1371°C). The points where one phase changes to another are called *critical points* by heat treaters and *transformation points* by metallurgists. The critical points of water are the boiling point (212°F, or 100°C) and the freezing point (32°F, or 0°C). Figure D-19 shows a critical temperature diagram for a carbon steel. The lower critical point is always about 1330°F (721°C) in equilibrium or very slow cooling, but the upper critical point changes as the carbon content changes (Figure D-20).

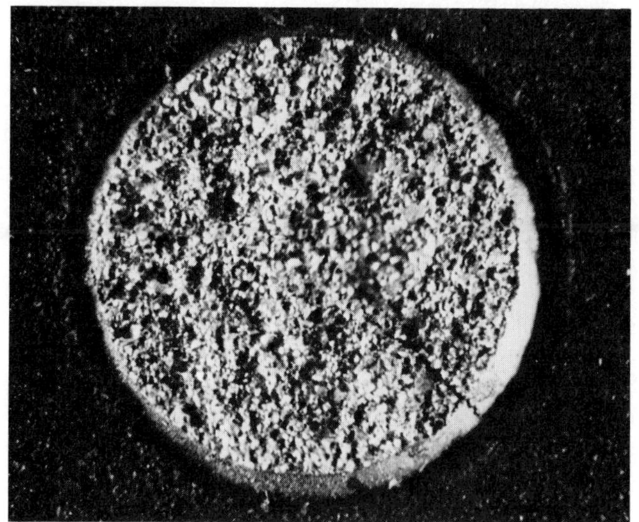

FIGURE D-17 Single fracture of steel.

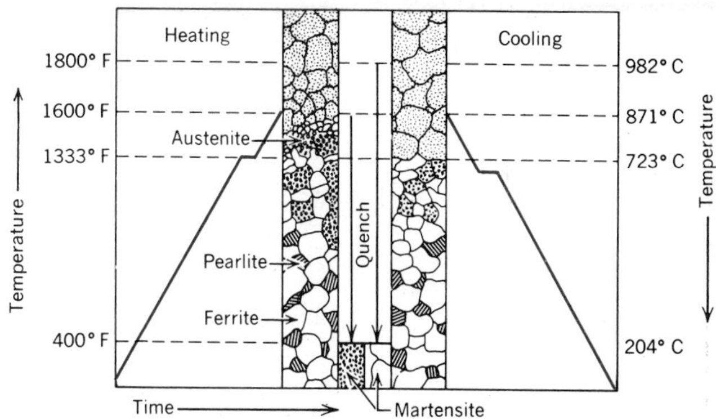

FIGURE D-19 Critical temperature diagram of .83 percent steel showing grain structures in heating and cooling cycles. Center section shows quenching from different temperatures and the resultant grain structure.

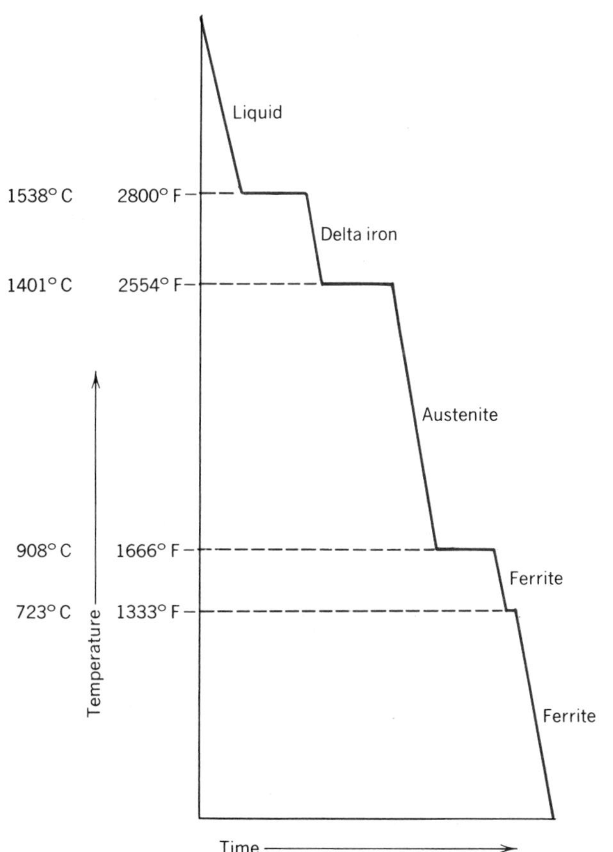

FIGURE D-18 Cooling curve of iron.

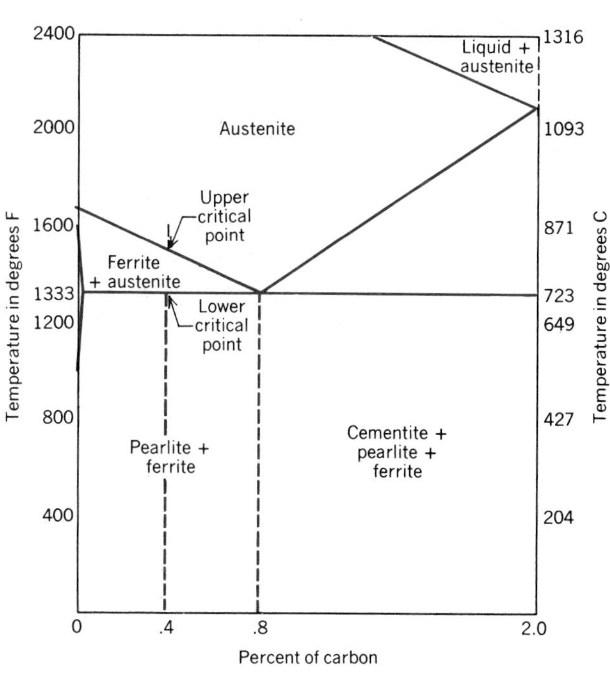

FIGURE D-20 Simplified phase diagram for carbon steels.

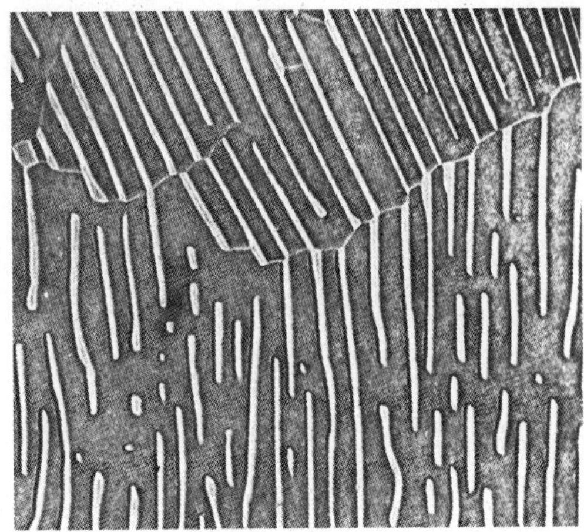

FIGURE D-21 A replica electron micrograph. Structure consists of lamellar pearlite (11,000 ×). (By permission, from *Metals Handbook,* Volume 7, Copyright © American Society for Metals, 1972.)

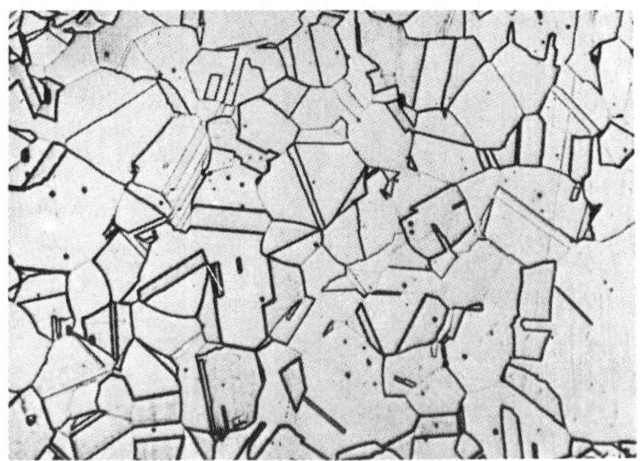

FIGURE D-22 A microstructure of annealed 304 stainless steel that is austenitic at ordinary temperatures (250 ×). (By permission, from *Metals Handbook,* Volume 7, Copyright © American Society for Metals, 1972.)

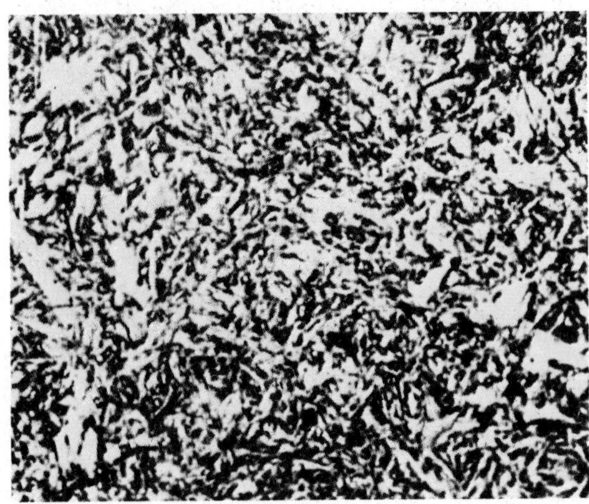

FIGURE D-23 1095 steel, water quenched from 1500°F (816°C) (1000 ×). The needlelike structure shows a pattern of fine untempered martensite. (By permission, from *Metals Handbook,* Volume 7, Copyright © American Society for Metals, 1972.)

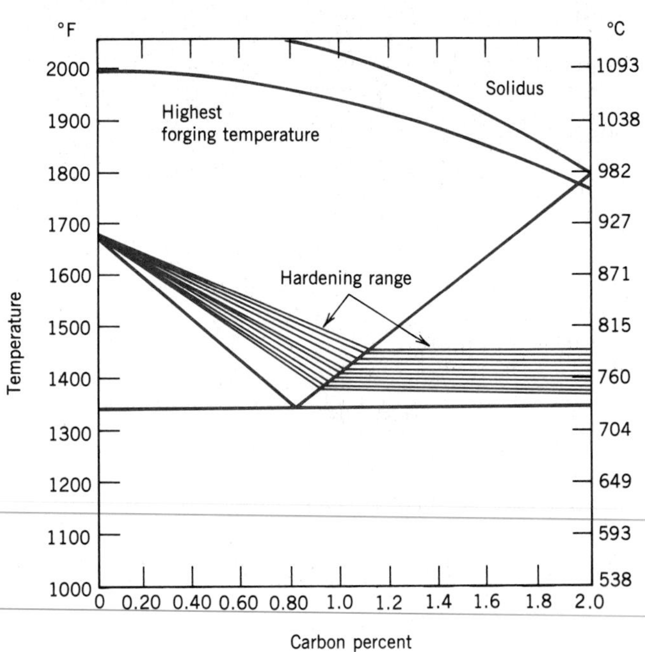

FIGURE D-24 Temperature ranges used for hardening carbon steel (J. E. Neely, *Practical Metallurgy and Materials of Industry,* John Wiley & Sons, New York, 1979).

Heat Treating Steel

When hardening various tool steels, a manufacturer's catalog should be consulted for the correct temperature, time periods, and quenching media. As steel is heated above the critical temperature of 1330°F (721°C), the carbon that was in the form of layers of iron carbide or pearlite (Figure D-21) begins to dissolve in the iron and forms a solid solution called austenite (Figure D-22). When this solution of iron and carbon is suddenly cooled or quenched, a new microstructure is formed. This is called martensite (Figure D-23). Martensite is very hard and brittle, having a much higher tensile strength than the steel with a pearlite microstructure. It is quite unstable, however, and must be tempered to relieve internal stresses in order to have the toughness needed to be useful. AISI-C1095, commonly known as water-hardening (W1) steel, will begin to show hardness when quenched from a temperature just over 1330°F (721°C) but will not harden at all if quenched from a temperature lower than 1330°F (721°C). The steel will become as hard as it can get when heated to 1450°F (788°C) and quenched in water. This quenching temperature changes as the carbon content of the steel changes. It should be 50°F (10°C) above the upper critical limit into the hardening range (Figure D-24).

Low-carbon steels such as AISI 1020 will not, for all practical purposes, harden when they are heated and quenched. Oil- and air-hardening steels harden over a longer period of time and, con-

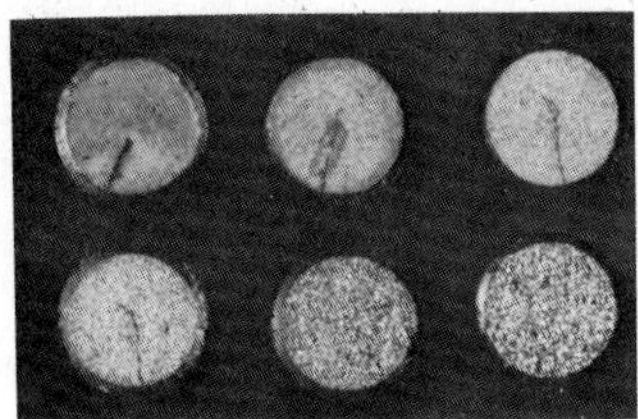

FIGURE D-25 Fractured ends of $\frac{3}{8}$-in.-diameter 1095 water-quenched tool steel ranging from fine grain, quenched from 1475°F (802°C), to coarse grain, quenched from 1800°F (982°C).

sequently, are deeper hardening than water-hardening types, which must be cooled to 200°F (93°C) within 1 or 2 seconds. As you can see, it is quite important to know the carbon content and alloying element so that the correct temperature and quenching medium can be used. Fine-grained tool steels are much stronger than coarse-grained tool steels (Figure D-25). If a piece of tool steel is heated above the correct temperature for its specific carbon content, a phenomenon called grain growth will occur, and a coarse, weak grain structure develops. The grain growth will remain when the part is quenched and, if used for a tool such as a punch or chisel, the end may simply drop off when the first hammer blow is struck. Tempering will not remove the coarse grain structure. If the part has been overheated, simply cooling back to the quenching temperature will not help, as the coarse grain persists well down into the hardening range as shown in Figure D-19. The part should be cooled slowly to room temperature and then reheated to the correct quenching temperature. Steels containing .83 percent carbon can get as hard (RC 67) as any carbon steel containing more carbon.

AISI-C1095, or water-hardening tool steel (W1), can be quenched in oil, depending on the size of the part. For example, for a piece of AISI-C1095 drill rod, if the section is thin or the diameter is small, oil should be used as a quenching medium. Oil is not as severe as water because it conducts heat less rapidly than water, and thus avoids quench cracking. Larger sections or parts would not be fully transformed into martensite if they were oil quenched, but would instead contain some softer transformation structures. Water quench should be used, but remember that W1 is shallow hardening and will only harden about $\frac{1}{8}$ in. deep in small parts.

When using a furnace to heat for quenching, the temperature control should be set for 1450°F (788°C). If the part is small, a preheat is necessary, but if it is thick, it should be brought up to heat slowly. If the part is left in a furnace without a controlled atmosphere for any length of time, the metal will form an oxide scale and carbon will leave

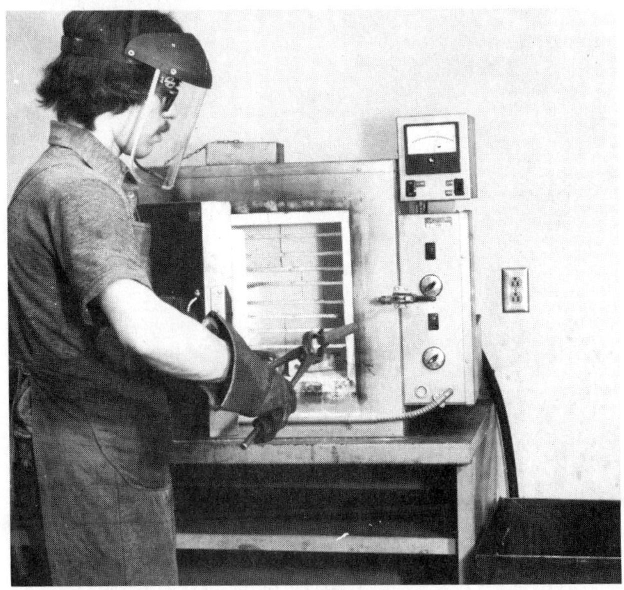

FIGURE D-26 Electric heat-treating furnace. Part is being placed in furnace by heat treater wearing correct attire and using tongs (Lane Community College).

the surface. This decarburization of the surface will cause it to be soft, while the metal directly under the surface will remain hard after the part is quenched. Decarburization takes time. If the part remains at a hardening temperature for only a short time (up to 15 minutes) the loss of carbon is insignificant for most purposes, and it should not be a concern. It would take a high temperature soaking period of several hours in an oxidizing atmosphere to remove the carbon to a depth of .010 in. When heat treating small parts, the oxidation scale on the surface may be more objectionable than decarburization. This oxidation can be avoided by painting the part with a solution of boric acid and water before heating, or wrapping it in stainless steel foil. Place the part in the furnace with tongs, and wear gloves and face shields for protection. When the part has become the same color as the furnace bricks, remove it by grasping one end with the tongs and *immediately* plunge it into the quenching bath. If the part is long like a chisel or punch, it should be inserted into the quench vertically (straight up-and-down), not at a slant. Quenching at an angle can cause unequal cooling rates and bending of the part. Also, agitate the part in an up-and-down or a figure-8 motion to remove any gases or bubbles that might cause uneven quenching.

Furnaces

Electric-, gas-, or oil-fueled furnaces are used for heat treating steels (Figure D-26). They use various types of controls for temperature adjustment.

FIGURE D-27 Thermocouple (Lane Community College).

FIGURE D-28 Input controls on furnace (Lane Community College).

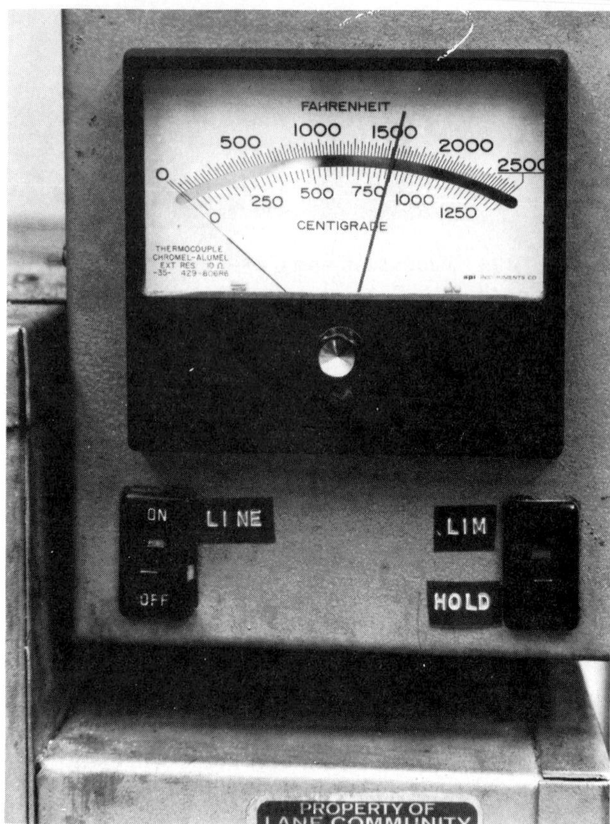

FIGURE D-29 Temperature control. The furnace is cold and the temperature limit control has been set at 1550°F (843°C). When the furnace is turned on and the needle reaches the preset limit control, the furnace will stop heating (Lane Community College).

These controls make use of the principle of the thermocouple (Figure D-27). When two dissimilar metal wires are joined and twisted together and raised to a given temperature, a weak electric current is generated in the wires. This voltage, which varies with the temperature, is amplified and registered on a dial. When a preselected temperature is reached, a switch is tripped that cuts off the furnace. Temperatures generally range up to 2500°F (1371°C). High-temperature salt baths are also used for heating metals for hardening or annealing. One of the disadvantages of most electric furnaces is that they allow the atmosphere to enter the furnace, and the oxygen causes oxides to form on the heated metal. This causes scale and decarburization of the surface of the metal. A decarburized surface will not harden. One way to control this loss of surface carbon is to keep a slightly carbonizing atmosphere or an inert gas in the furnace.

One of the most important factors when heating steel is the rate at which heat is applied. When steel is first heated, it expands. If cold steel is placed in a hot furnace, the surface expands more rapidly than the still-cool core. The surface will then have a tendency to pull away from the center, inducing internal stress. This can cause cracking and distortion in the part. Most furnaces can be adjusted for the proper rate of heat input (Figure D-28) when bringing the part up to the soaking temperature (Figure D-29). This allows a slow, gradual increase of heat that maintains an even internal and external temperature in the part. Only previously hardened or highly stressed parts and those having a large mass need a slow rate of heating. Small parts such as chisels and punches made of plain carbon steel can be put directly into a hot furnace.

Soaking means holding the part for a given length of time at a specified temperature. Another factor is the time of soaking required for a certain size piece of steel. An old rule of thumb allows the steel to soak in the furnace for one hour for each inch of thickness, but there are considerable variations to this rule, since some steels require much more soaking time than others. The correct soaking period for any specific tool steel may be found in tool steel reference books.

FIGURE D-30 Beginning of quench. At this stage heat treater could be burned by hot oil if not adequately protected with gloves and face shield (Lane Community College).

FIGURE D-31 Heat treater is agitating part during quench (Lane Community College).

Quenching Media

In general, seven media are used to quench metals. They are listed here in their order of severity or speed of quenching.

1. Water and salt; that is, sodium chloride or sodium hydroxide (also called brine)
2. Tap water
3. Fused or liquid salts
4. Molten lead
5. Soluble oil and water
6. Oil
7. Air

Most hardening is done with plain (tap) water, quenching oil, or air. The mass (size) of the part has a great influence on cooling rates in quenching and therefore the amount of hardening in many cases. For example, a razor blade made of plain carbon steel would have a severe (sudden) quench in oil and would get very hard; but the same metal as a 2-in.-square block would not have a sufficiently rapid quench in cold water to cause it to harden very much. The reason for this difference is that the hot interior of the larger mass keeps the surface hot enough that the quenching medium cannot remove the heat rapidly enough. The result is a slow rate of cooling, which does not allow hardening to occur. Of course, some tool steels, unlike plain carbon steels, do not need a fast quench in order to harden, so large masses can be quenched in oil or air. These are called oil- and air-hardening steels.

Liquid quenching media goes through three stages. The vapor-blanket stage occurs first because the metal is so hot it vaporizes the media. This envelops the metal with vapor, which insulates it from the cold liquid bath. This causes the cooling rate to be relatively slow during this stage. The vapor transport cooling stage begins when the vapor blanket collapses, allowing the liquid medium to contact the surface of the metal. The cooling rate is much higher during this stage. The liquid cooling stage begins when the metal surface reaches the boiling point of the quenching medium. There is no more boiling at this stage, so heat must be removed by conduction and convection. This is the slowest stage of cooling.

Gloves and face protection must be used in this operation for safety (Figure D-30). Hot oil could splash up and burn the heat treater's face if a face shield were not worn.

It is important in liquid quenching baths that either the quenching medium or the steel being quenched should be agitated (Figure D-31). The

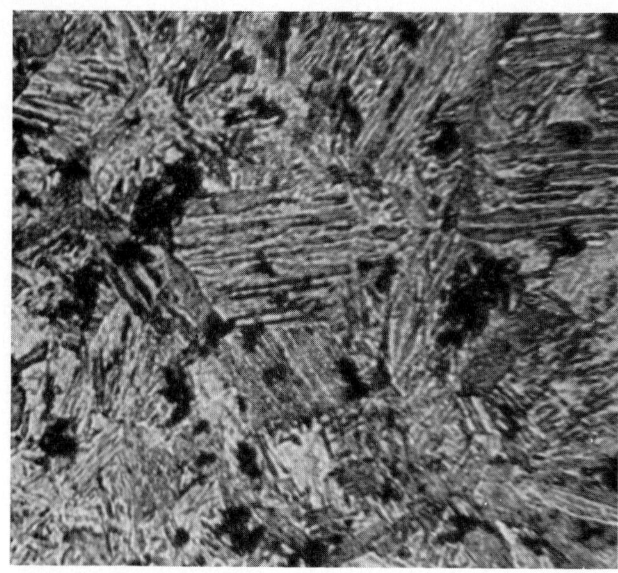

FIGURE D-32 Lower bainite microstructure.

vapor that forms around the part being quenched acts as an insulator and slows down the cooling rate. This can result in incomplete or spotty hardening of the part. Agitating the part breaks up the vapor barrier. An up-and-down motion works best for long, slender parts held vertically in the quench. A figure-8 motion is sometimes used for heavier parts.

Molten salt or lead is often used for isothermal quenching. This is the method of quenching used for austempering. Austempered parts are superior in strength and quality to those produced by the two-stage process of quenching and tempering. The final austempered part is essentially a fine, lower bainite microstructure (Figure D-32). As a rule, only parts such as lawn mower blades and hand shovels that are thin in cross section are austempered.

Another form of isothermal quenching is called martempering; the part is quenched in a lead or salt bath at about 400°F (204°C) until the outer and inner parts of the material are brought to the same uniform temperature. The part is next quenched below 200°F (93°C) to transform all of the austenite to martensite. Tempering is then carried out in the conventional manner. (See the section "Tempering" later in this unit.)

Steels are often classified by the type of quenching medium used to meet the requirements of the critical cooling rate. For example, water quenched steels, which are the plain carbon steels, must have a rapid quench. Oil-quench steels are alloy steels, and they must be hardened in oil. The air-cooled steels are alloy steels that will harden when allowed to cool from the austenitizing temperature in still air. Air is the slowest quenching medium; however, its cooling rate may be increased by movement (by use of fans, for example).

Step or multiple quenching is sometimes used when the part consists of both thick and thin sections. A severe quench will harden the thin section before the thick section has had a chance to cool. The resulting uneven contraction often results in cracking. With this method the part is quenched for a few seconds in a rapid quenching medium, such as water, followed by a slower quench in oil. The surface is first hardened uniformly in the water quench, and time is provided by the slower quench to relieve stresses.

Case Hardening

Low carbon steels (.08 to .30 percent carbon) do not harden to any great extent even when combined with other alloying elements. Therefore, when a soft, tough core and an extremely hard outside surface is needed, one of several case hardening techniques is used. It should be noted that surface hardening is not necessarily case hardening. Flame hardening and induction hardening on the surfaces of gears, lathe ways, and many products depend on the carbon that is already contained in the ferrous metal. On the other hand, case hardening causes carbon from an outside source to penetrate the surface of the steel, carburizing it. Because it raises the carbon content, it also raises the hardenability of the steel.

Carburizing for case hardening can be done by either of two methods. If only a shallow hardened case is needed, roll carburizing may be used. This consists of heating the part to 1650°F (899°C) and rolling it in a carburizing compound, reheating, and quenching in water. In roll carburizing use only a nontoxic compound such as Kasenite unless special ventilation systems are used. Roll carburizing produces a maximum case of about .003 in. but pack carburizing (Figure D-33) can produce a case of $\frac{1}{16}$ in. in 8 hours at 1700°F (926.6°C). The part is packed in a carburizing compound in a metal box and placed in a furnace long enough to cause the carbon to diffuse sufficiently deep into the surface of the material to produce a case of the required depth. The process should consist of heating the part in carburizing compound, cooling to room temperature, and reheating to the correct hardening temperature (usually about 1450°F or 788°C). The part is then removed and quenched in water. After case hardening, tempering is not usually necessary since the core is still soft and tough. Therefore, unlike a hardened piece that is softened by tempering, the surface of a case-hardened piece remains hard, usually RC 60 (as hard as a file) or above.

CAUTION **Some alloys and special heat treatments can be used to case harden steel parts. Liquid carburizing is an industrial method in which the parts are bathed in cyanide carbonate and chloride salts and held at a temperature between 1500 and 1700°F (815.5 and 926.6°C). Cyanide salts are extremely poisonous and adequate worker protection is essential.**

Gas carburizing is a method in which the carbon is supplied from a carburizing gas atmosphere in a special furnace where the part is heated. A gas flame that has insufficient oxygen produces carbon in the form of carbon monoxide (CO). This excess carbon diffuses into the heated metal surface. The same principle is involved when steel is improperly cut with an oxyacetylene torch in which the preheat flame has insufficient oxygen. This carbonizing flame allows carbon to diffuse into the molten metal and the cut edge becomes very hard when it is suddenly cooled (quenched) by the mass of steel adjacent to the cut. This hardening of torch-cut steel workpieces often causes difficulties for a machinist when cutting on the torch-cut edge.

Nitriding is a method of case hardening in which the part is heated in a special container into which ammonia gas is released. Since the temperature used is only 950 to 1000°F (510 to 538°C) and the part is not quenched, warpage is kept to a minimum. The iron nitrides thus formed are even harder than the iron carbides formed by conventional carburizing methods.

Tempering

Tempering, or drawing, is the process of reheating a steel part that has been previously hardened to transform some of the hard martensite into softer structures. The higher the tempering temperature used, the more martensite is transformed, and the softer and tougher (less brittle) the piece becomes. Therefore, tempering temperatures are specified according to the strength and ductility desired. Mechanical properties charts, which can be found in steel manufacturers' handbooks and catalogs, give this data for each type of alloy steel. Table D-6 provides heat-treating information for plain carbon steel (AISI 1095) and some other carbon and alloy steels. Expected hardnesses in Rockwell numbers refer to steel that has attained its full quenched hardness. The hardness readings after

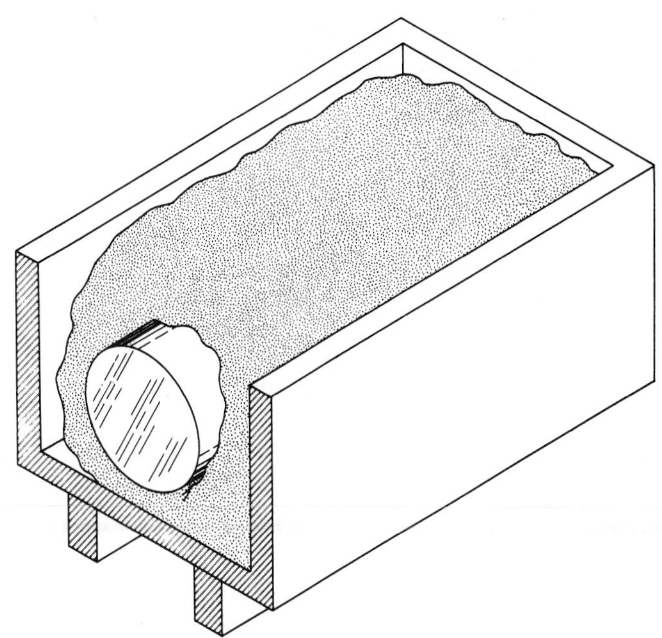

FIGURE D-33 Pack carburizing. The workpiece to be pack carburized should be completely covered with carburizing compound. The metal box should have a close-fitting lid.

TABLE D-6 Typical Heat Treating Information for Direct-Hardening Carbon and Low-Alloy Steels

Grade	Hardening Temperature	Full Hardness	Expected Hardness After Tempering 2 Hours at: 400	500	600	700	800	900	1000	1100	1200
8620	1650/1750	37/43	40	39	37	36	35	32	27	24	20
4130	1550/1625	49/56	47	45	43	42	38	34	32	26	22
1040	1525/1600	53/60	51	48	46	42	37	30	27	22	(14)
4140/4142	1525/1575	53/62	55	52	50	47	45	41	36	33	29
4340	1475/1550	53/60	55	52	50	48	45	42	39	34	30
1144	1475/1550	55/60	55	50	47	45	39	32	29	25	(19)
1045	1475/1575	55/62	55	52	49	45	41	34	30	26	20
4150	1500/1550	59/65	56	55	53	51	47	45	42	38	34
5160	1475/1550	60/65	58	55	53	51	48	44	40	36	31
1060/1070	1475/1550	58/63	56	55	50	43	39	38	36	35	32
1095	1450/1525	63/66	62	58	55	51	47	44	35	30	26

NOTE: Temperatures listed in °F and hardness in Rockwell C. Values were obtained from various recognized industrial and technical publications. All values should be considered approximations.

SOURCE: Pacific Machinery & Tool Steel Co.

tempering would be lower, if for some reason a part were not fully hard in the first place. See Unit 5 in this section for Rockwell and Brinell hardness comparisons.

FIGURE D-34 Tempering a punch on a hot plate.

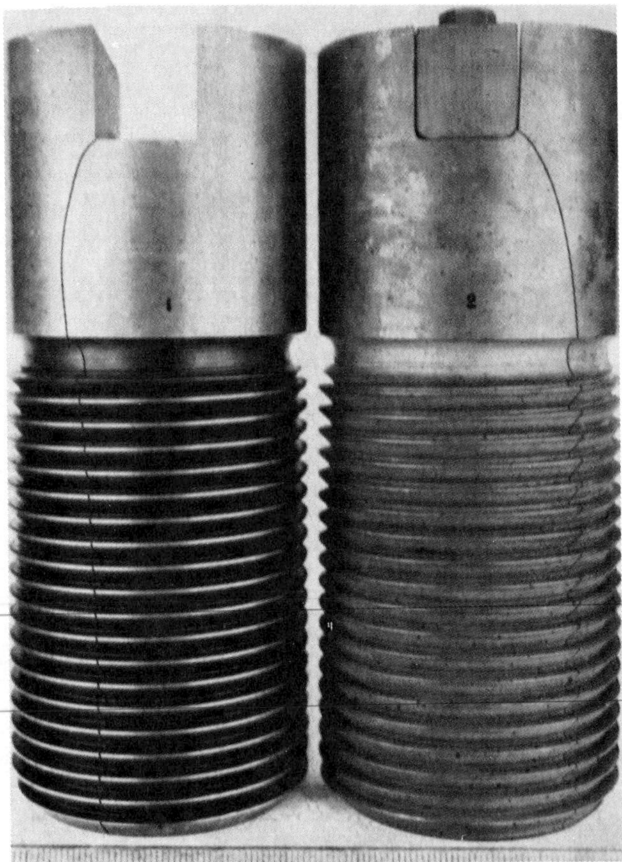

FIGURE D-35 These two breech plugs were made of type L6 tool steel. Plug 1 cracked in the quench through a sharp corner and was therefore not tempered. Plug 2 was redesigned to incorporate a radius in the corners of the slot, and a soft steel plug was inserted in the slot to protect it from the quenching oil. Plug 2 was oil quenched and checked for hardness (Rockwell C 62) and, after tempering at 900°F (482°C), it was found to be cracked. The fact that the as-quenched hardness was measured proves that there was a delay between the quench and the temper that was responsible for the cracking. The proper practice would be to temper immediately at a low temper, check hardness, and retemper to desired hardness (Bethlehem Steel Corporation).

A part can be tempered in a furnace or oven by bringing it to the proper temperature and holding it there for a length of time, then cooling it in air or water.

Some tool steels should be cooled rapidly after tempering to avoid temper brittleness. Small parts are often tempered in liquid baths such as oil, salt, or metals. Specially prepared oils that do not ignite easily can be heated to the tempering temperature. Lead and various salts are used for tempering since they have a low melting temperature.

When there are no facilities to harden and temper a tool by controlled temperatures, tempering by color is done. The oxide color used as a guide in such tempering will form correctly on steel only if it is polished to the bare metal and is free from any oil or fingerprints. An oxyacetylene torch, steel hot plate, or electric hot plate can be used. If the part is quite small, a steel plate is heated from the underside, and the part is placed on top. Larger parts such as chisels and punches can be heated on an electric hot plate (Figure D-34) until the needed color shows, then cooled in water. With this system the tempering process must cease when the part has come to the correct temperature, so the part must be quickly cooled in water to stop further heating of the critical areas. There is no possible soaking time when this method is used.

When grinding carbon steel tools, if the edge is heated enough to produce a color, you have in effect retempered the edge. If the temperature reached is above that of the original temper, the tool has become softer than it was before you began sharpening it. Table D-7 gives the hardnesses of various tools as related to their oxide colors and the temperature at which they form.

Tempering should be done as soon as possible after hardening. A part should not be allowed to cool completely, since untempered, it contains very high internal stresses and tends to split or crack. Tempering will relieve the internal stresses. A hardened part left overnight without tempering may develop cracks by itself. Furnace tempering is one of the best methods of controlling the final condition of the martensite to produce a tempered martensite of the correct hardness and toughness that the part requires. The still-warm part should be put into the furnace immediately. If it is left at room temperature for even a few minutes, it may develop a quench crack (Figure D-35).

A soaking time should also be used when tempering as it is in hardening procedures, and the length of time is related to the type of tool steel used. A cold furnace should be brought up to the correct temperature for tempering. The residual heat in the bricks of a previously heated furnace may overheat the part, even though the furnace has been cooled down.

TABLE D-7 Temper Color Chart

Degrees				
C°	F°	Oxide Color	Suggested Uses for Carbon Tool Steels	
220	425	Light straw	Steel cutting tools, files, and paper cutters	Harder
240	462	Dark straw	Punches and dies	
258	490	Gold	Shear blades, hammer faces, center punches and cold chisels	
260	500	Purple	Axes, wood cutting tools, and striking faces of tools	
282	540	Violet	Spring and screwdrivers	
304	580	Pale blue	Springs	
327	620	Steel gray	Cannot be used for cutting tools	Softer

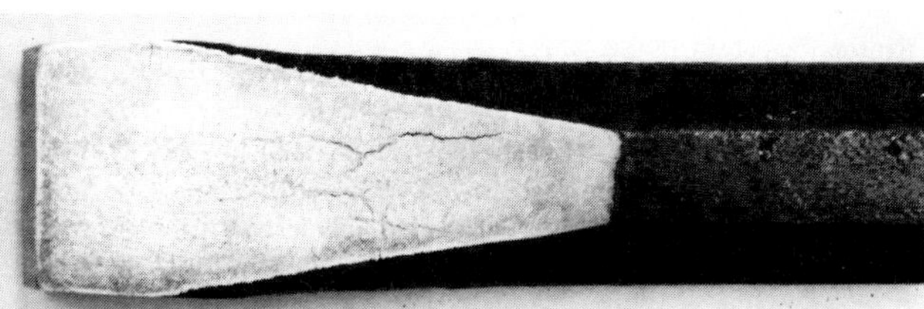

FIGURE D-36 This tool has been overheated and the typical "chicken wire" surface markings are evident. The tool must be discarded (Lane Community College).

Double tempering is used for some alloy steels such as high-speed steels that have incomplete transformation of the austenite when they are tempered for the first time. The second time they are tempered, the austenite transforms completely into the martensite structure.

FIGURE D-37 Drawing die made of type W1 tool steel shows characteristic cracking when water quenching is done without packing the bolt holes (Bethlehem Steel Corporation).

Problems in Heat Treating

Overheating of steels should always be avoided, and you have seen that if the furnace is set too high with a particular type of steel, a coarse grain can develop. The result is often a poor-quality tool, quench cracking, or failure of the tool in use. Extreme overheating causes burning of the steel and damage to the grain boundaries, which cannot be repaired by heat treatment (Figure D-36); the part must be scrapped. The shape of the part itself can be a contributing factor to quench failure and quench cracking. If there is a hole, sharp shoulder, or small extension from a larger cross section (unequal mass), a crack can develop in these areas (Figure D-37). Holes can be filled with steel wool to avoid problems. A part of the tool being held by tongs may be cooled to the extent that it may not harden. The tongs should therefore be heated prior to grasping the part for quenching (Figure D-38).

FIGURE D-38 Heating the tongs prior to quenching a part (Lane Community College).

FIGURE D-39 (Top) Form tool made of type S5 tool steel, which cracked in hardening through the stamped O. The other two form tools, made of type T1 high-speed steel, cracked in heat treatment through deeply stamped + marks. Stress raisers such as these deep stamp marks should be avoided. Although characters with straight lines are most likely to crack, even those with rounded lines are susceptible (Bethlehem Steel Corporation).

As mentioned before, decarburization is a problem in furnaces that do not have controlled atmosphere. This can be avoided in other ways, such as wrapping the part in stainless steel foil, covering it with cast iron chips, or covering it with a commercial compound.

A proper selection of tool steels is necessary to avoid failures in a particular application. If there is shock load on the tool being used, shock resisting tool steel must be selected. If there is to be heat applied in the use of the tool, a hot-work type of tool steel is selected. If distortion must be kept to a minimum, an air hardening steel should be used.

Quench cracks have several characteristics that are easily recognized.

1. In general, the fractures run from the surface toward the center in a relatively straight line. The crack tends to spread open.
2. Since quench cracking occurs at relatively low temperatures, the crack will not show any decarburization. That is, a black scale would form on the cracked surface if it happened at a high temperature, but no scale will be seen on the surface of a quench crack. However, tempering will darken a quench crack.
3. The fracture surfaces will exhibit a fine crystalline structure when tempered after quenching. The fractured surfaces may be blackened by tempering scale.

Some of the most common causes for quench cracks are:

1. Overheating during the austenitizing cycle, causing the normally fine grained steel to become coarse.
2. Improper selection of the quenching medium; for example, the use of water or brine instead of oil for an oil-hardening steel.
3. Improper selection of steel.
4. Time delays between quenching and tempering.
5. Improper design. Sharp changes of section such as holes and keyways (Figure D-39).
6. Improper angle of the work into the quenching bath with respect to the shape of the part, causing nonuniform cooling.
7. Failure to specify the correct size material to allow for cleaning up the outside decarburized surface of the bar before the final part is made. Hot-rolled tool steel bars need $\frac{1}{16}$ in. removed from all surfaces before any heat treating is done.

It is sometimes desirable to normalize the part before hardening it. This is particularly appropriate for parts and tools that have been highly stressed by heavy machining or by prior heat treatment. If they are left unrelieved, the residual stresses from such operations may add to the thermal stress produced in the heating cycle and cause the part to crack even before it has reached the quenching temperature.

There is a definite relationship between grinding and heat treating. Development of surface temperatures ranging from 2000 to 3000°F (1093 to 1649°C) are generated during grinding. This can cause two undesirable effects on hardened tool steels: development of high internal stresses causing surface cracks to be formed, and changes in the hardness and metallurgical structure of the surface area.

One of the most common effects of grinding on hardened and tempered tool steels in that of reducing the hardness of the surface by gradual tempering from the heat of grinding, where the hardness is lowest at the extreme surface but increases with distance below the surface. The depth of this tempering varies with the amount or depth of cut, the use of coolants, and the type of grinding wheel. On the other hand, if extremely high temperatures are produced locally by the grinding wheel and the surface is immediately quenched by the coolant, a martensite having a Rockwell hardness of C 65 to 70 can be formed. This gradiant hardness, being much greater than that beneath the surface of the tempered part, sometimes causes very high stresses that contribute greatly to grinding cracks. Sometimes grinding cracks are visible in oblique or angling light, but they can be easily detected when present by the use of magnetic particles of fluorescent particle testing.

When a part is hardened but not tempered before it is ground, it is extremely liable to stress cracking (Figure D-40). Faulty grinding procedures can also cause grinding cracks. Improper grinding operations can cause tools that have been properly hardened to fail. Sufficient stock should be allowed for a part to be heat treated so that grinding will remove any decarburized surface on all sides to a depth of .010 to .015 in.

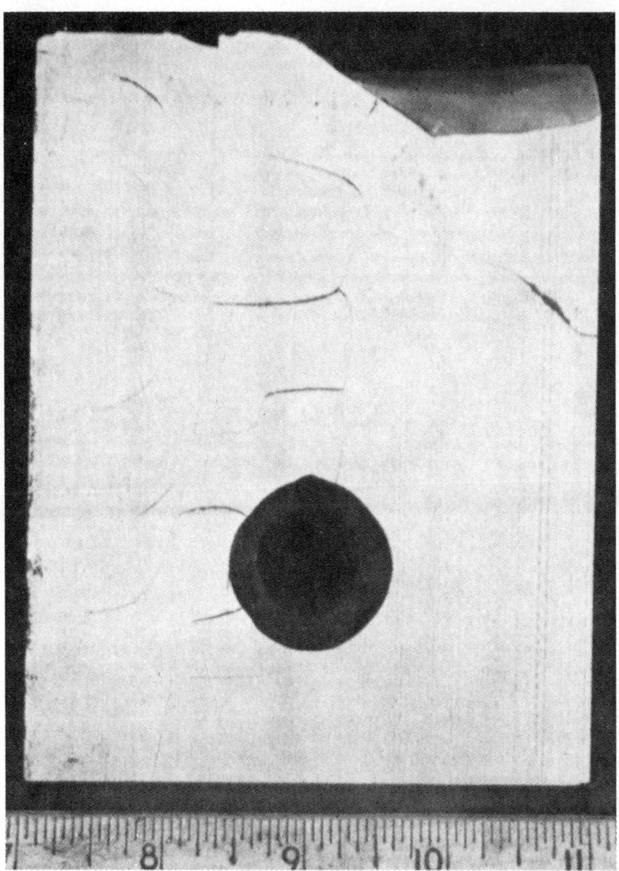

FIGURE D-40 Severe grinding cracks in a shear blade made of type A4 tool steel developed because the part was not tempered after quenching. Hardness was Rockwell C 64, and the cracks were exaggerated by magnetic particle test. Note the geometric scorch pattern on the surface and the fracture that developed from enlargement of the grinding cracks (Bethlehem Steel Corporation).

SELF-TEST

1. If you heated AISI C1080 steel to 1200°F (649°C) and quenched it in water, what would be the result?

2. If you heated AISI C1020 steel to 1500°F (815°C) and quenched it in water, what would happen?

3. List as many problems encountered with water-hardening steels as you can think of.

4. Name some advantages in using air and oil hardening tool steels.

5. What is the correct temperature for quenching AISI C1095 tool steel? For any carbon steel?

6. Why is steel tempered after it is hardened?

7. What factors should you consider when you choose the tempering temperature for a tool?

8. The approximate temperature for tempering a center punch should be ________. The oxide color would be ________.

9. If a cold chisel became blue when the edge was ground on an abrasive wheel, to approximately what temperature was it raised? How would this affect the tool?

10. How soon after hardening should you temper a part?

11. What is the advantage of using low-carbon steel for parts that are to be case hardened?

12. How can a deep case be made?

13. Are parts that are surface hardened always case hardened by carburizing?

14. Name three methods by which carbon may be introduced into the surface of heated steel.

15. What method of case hardening uses ammonia gas?

16. Name three kinds of furnaces used for heat treating steels.

17. What can happen to a carbon steel when it is heated to high temperatures in the presence of air (oxygen)?

18. Why is it absolutely necessary to allow a soaking period for a length of time (that varies for different kinds of steels) before quenching the piece of steel?

19. Why should the part or the quenching medium be agitated when you are hardening steel?

20. Which method of tempering gives the heat treater the most control of the final product: by color or by furnace?

21. Describe two characteristics of quench cracking that would enable you to recognize them.

22. Name four or more causes of quench cracks.

23. In what ways can decarburization of a part be avoided when it is heated in a furnace?

24. Describe two types of surface failures of hardened steel when it is being ground.

25. When distortion must be kept to a minimum, which type of tool steel should be used?

UNIT 4

Annealing, Normalizing, and Stress Relieving

All metals form into tiny grains when cooled, unless they are extremely rapidly cooled. Some grains or crytals are large enough to be seen on a section of broken metal. In general, small grains are best for tools and larger grains for extensive cold working. Small-grain metals are stronger than large-grain ones. Slow cooling promotes the formation of large grains, rapid cooling forms small grains. Since the machinability of steel is so greatly affected by heat treatments, the processes of annealing, normalizing, and stress relieving are important to a machinist. You will learn about these processes in this unit.

Objective

After completing this unit, you should be able to:

Explain the principles of and differences between the various kinds of annealing processes.

Annealing

The heat treatment for iron and steel that is generally called annealing can be divided into several different processes: full anneal, normalizing, spheroidize anneal, stress relief (anneal), and process anneal.

FULL ANNEAL

The full anneal is used to completely soften hardened steel, usually for easier machining of tool steels that have more than .8 percent carbon content. Lower carbon steels are full annealed for other purposes. For instance, when welding has been done on a medium to high carbon steel that must be machined, a full anneal is needed. Full annealing is done by heating the part in a furnace to 50°F (10°C) above the upper critical temperature (Figure D-41), and then cooling very slowly in the

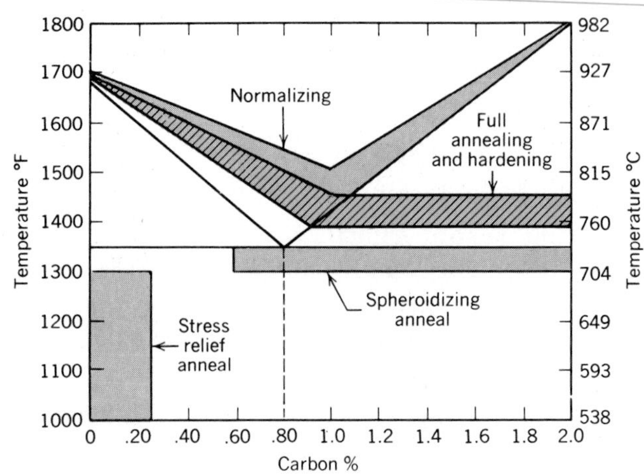

FIGURE D-41 Temperature ranges used for heat treating carbon steels.

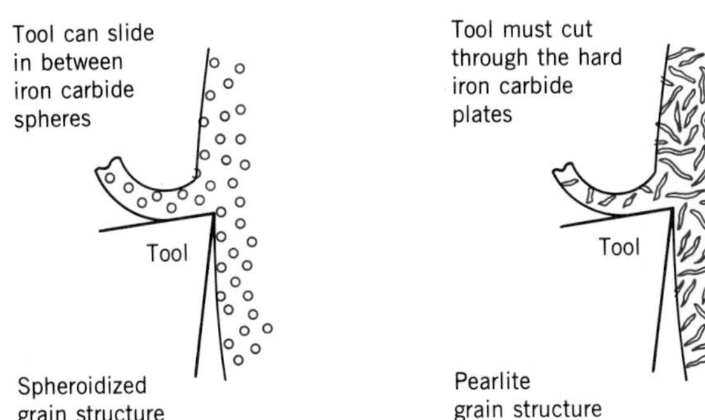

FIGURE D-42 Comparison of cutting action between spheroidized and normal carbon steels.

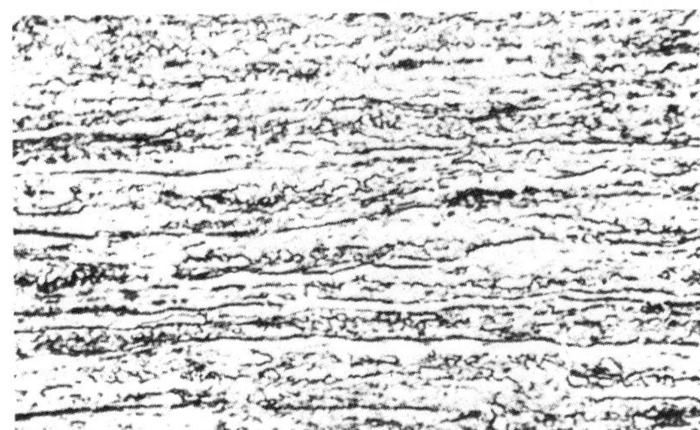

FIGURE D-43 Microstructure of flattened grains of .10 percent carbon steel, cold rolled (1000×). (By permission, from *Metals Handbook,* Volume 7, Copyright © American Society for Metals, 1972.)

furnace or in an insulating material. In the process, the microstructure becomes coarse pearlite and ferrite or pearlite and cementite, depending on the carbon content. It is necessary to heat above the critical temperature for grains containing iron carbides (pearlite) in order to recrystallize them and to reform new soft whole grains from the old hard distorted ones.

NORMALIZING

Normalizing is somewhat similar to annealing, but it is done for different purposes. Medium carbon steels are often normalized to give them better machining qualities. Medium (.3 to .6 percent) carbon steel may be "gummy" when machined after a full anneal, but can be made sufficiently soft for machining by normalizing. The finer, but harder, microstructure produced by normalizing gives the piece a better surface finish. The piece is heated to 100°F (38°C) above the upper critical line, and cooled in still air. When the carbon content is above or below .8 percent, higher temperatures are required (Figure D-41).

Forgings and castings that have unusually large and mixed grain structures are corrected by using a normalizing heat treatment. Stresses are removed, but the metal is not as soft as with full annealing. The resultant microstructure is a uniform fine-grained pearlite and ferrite, including other microstructures depending on the alloy and carbon content. Normalizing also is used to prepare steel for other forms of heat treatment such as hardening and tempering. Weldments are sometimes normalized to remove welding stresses that develop in the structure as well as in the weld.

SPHEROIDIZING

Spheroidizing is used to improve the machinability of high-carbon steels (.8 to 1.7 percent). The cementite or iron carbide normally found in pearlite as flat plates alternating with plates of ferrite

FIGURE D-44 The same .10 percent carbon steel as in Figure 3, but annealed at 1025°F (552°C) (1000×). Ferrite grains (white) are mostly reformed to their original state, but the pearlite grains (dark) are still distorted. (By permission, from *Metals Handbook,* Volume 7, Copyright © American Society for Metals, 1972.)

(iron) is changed into a spherical or globular form by spheroidization (Figure D-42). Low-carbon steels (.08 to .3 percent) can be spheroidized, but their machinability gets poorer since they become gummy and soft, causing tool edge buildup and poor finish. The spheroidization temperature is close to 1300°F (704°C). The steel is held at this temperature for about four hours. Carbides which are very hard that develop from welding on medium to high carbon steels can be made even softer by this process than by the full anneal. This is because the iron carbide spheroids allow the iron to be more ductile than the flat plates of iron carbide.

STRESS RELIEF ANNEAL

Stress relief annealing is a process of reheating low carbon steels to 950°F (510°C). Stresses in the ferrite (mostly pure iron) grains caused by cold working steel such as rolling, pressing, forming, or drawing are relieved by this process. The distorted grains reform or recrystallize into new softer ones (Figures D-43 and D-44).

The pearlite grains and some other forms of iron carbide remain unaffected by this treatment, unless done at the spheroidizing temperature and held long enough to effect spheroidization. Stress relief is often used on weldments as the lower temperature limits the amount of distortion caused by heating (see Figure D-41). Full anneal or normalizing can cause considerable distortion in steel. (See the discussion on recrystallization later in this unit.)

PROCESS ANNEAL

Process annealing is essentially the same as stress relief annealing. It is done at the same temperatures and with low- and medium-carbon steels. In the wire and sheet steel industry, the term is used for the annealing or softening that is necessary during the cold-rolling or wire drawing processes and for removal of final residual stresses in the material. Wire and other metal products that must be continuously formed and reformed would become too brittle to continue after a certain amount of forming. The anneal, between a series of cold-working operations, reforms the grain to the original soft, ductile condition so that cold working can continue. Process anneal is sometimes referred to as bright annealing, and it is usually carried out in a closed container with inert gas to prevent oxidation of the surface.

Recovery, Recrystallization, and Grain Growth

When metals are heated to temperatures less than the recrystallization temperature, a reduction in internal stress takes place. This is done by relieving elastic stresses in the lattice planes, not by reforming the distorted grains. Recovery in annealing processes used on cold-worked metals is usually not sufficient stress relief for further extensive cold working (Figure D-45), yet it is used for some purposes and is called stress relief anneal. Most often, recrystallization is required to reform the distorted grains sufficiently for further cold work.

Recovery is a low-temperature effect in which there is little or no visible change in the microstructure. Electrical conductivity is increased and often a decrease in hardness is noted. It is difficult to make a sharp distinction between recovery and recrystallization. Recrystallization releases much larger amounts of energy than does recovery. The flattened, distorted grains are sometimes reformed to some extent during recovery into polygonal grains, while some rearrangement of defects such as dislocations takes place.

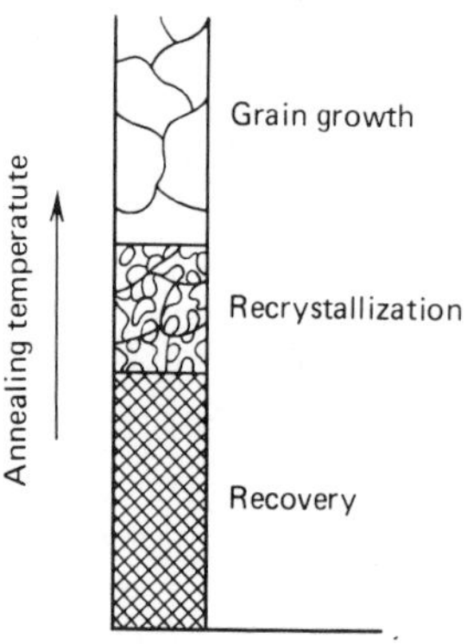

FIGURE D-45 Changes in metal structures that take place during the annealing process (J. E. Neely, *Practical Metallurgy and Materials of Industry*, 2nd ed., copyright © 1984, John Wiley & Sons, Inc.).

Not only does recrystallization release much larger amounts of stored energy, but new, larger grains are formed by the nucleation of stressed grains and the joining of several grains to form larger ones. To accomplish this joining of adjacent grains, grain boundaries migrate to new positions, which changes the orientation of the crystal structure. This is called grain growth.

The following factors affect recrystallization.

1. A minimum amount of deformation is necessary for recrystallization to occur.
2. The larger the original grain size is, the greater will be the amount of cold deformation required to given an equal amount of recrystallization with the same temperature and time.
3. Increasing the time of anneal decreases the temperature necessary for recrystallization.
4. The recrystallized grain size depends mostly on the degree of deformation and, to some extent, on the annealing temperature.
5. Continued heating, after recrystallization (reformed grains) is complete, increases the grain size.
6. The higher the cold working temperature is, the greater will be the amount of cold work required to give equivalent deformation.

Metals that are subjected to cold working become hardened and further cold working cannot be done without danger of splitting or breaking the metal. Various degrees of softening are possible by controlling the recrystallization process. The recrystallization temperatures for some metals are given in Table D-8. If you need to soften work hardened brass, for example, you need to heat it to 660°F (349°C) and cool it in water. Low-carbon steel should not be cooled in water when annealing.

TABLE D-8 Recrystallization Temperatures of Some Metals

Metal	Recrystallization Temperature (°F)
99.999% Alumimun	175
Aluminum bronze	660
Beryllium copper	900
Cartridge brass	660
99.999% Copper	250
Lead	25
99.999% Magnesium	150
Magnesium alloys	350
Monel	100
99.999% Nickel	700
Low-carbon steel	1000
Tin	25
Zinc	50

SOURCE: J. E. Neely, *Practical Metallurgy and Materials of Industry*, 2nd ed., John Wiley & Sons, New York, Copyright © 1984.

SELF-TEST

1. When might normalizing be necessary?
2. At what approximate temperature should you normalize .4 percent carbon steel?
3. What is the spheroidizing temperature of .8 percent carbon steel?
4. What is the essential difference between the full anneal and stress relieving?
5. When should you use stress relieving?
6. What kind of carbon steels would need to be spheroidized to give them free machining qualities?
7. Explain process annealing.
8. How should the piece be cooled for a normalizing heat treatment?
9. How should the piece be cooled for the full anneal?
10. What happens to machinability in low-carbon steels that are spheroidized?

UNIT 5

Rockwell and Brinell Hardness Testers

The Rockwell hardness tester and the Brinell hardness tester are the most commonly used types of hardness testers for industrial and metallurgical purposes. Heat treaters, inspectors, and many others in industry often use these machines. Although only the bench-mounted types of hardness testers are shown in this unit, testers are made in various sizes, including handheld portable units. Two other types of hardness testers are the Vickers diamond pyramid and the Knoop. All of these testers use different scales which can be compared in hardness conversion tables. This unit will direct you into a proper understanding and use of both Rockwell and Brinell hardness testers.

Objectives

After completing this unit, you should be able to:

1. Make a Rockwell test using the correct penetrator, major load, and scale.
2. Make a Rockwell superficial test using the correct penetrator, major load, and scale.
3. Make a Brinell test, read the impression with a Brinell microscope, and determine the hardness number from a table.

The hardness of a metal is its ability to resist being permanently deformed. There are three ways that hardness is measured: resistance to penetration, elastic hardness, and resistance to abrasion. In this unit you will study the hardness of metals by their resistance to penetration.

Hardness varies considerably between different materials. This variation can be illustrated by making an indentation in a soft metal such as aluminum and in a hard metal such as alloy tool steel. The indentation could be made with an ordinary center punch and a hammer, giving a light blow of equal force on each of the two specimens (Figure D-46). Just by visual observation you can tell which specimen is hardest in this case. Of course, this is not a reliable method of hardness testing, but it does show one of the principles of both the Rockwell and Brinell hardness testers: measuring penetration of the specimen by an indenter or penetrator, such as a steel ball or diamond cone.

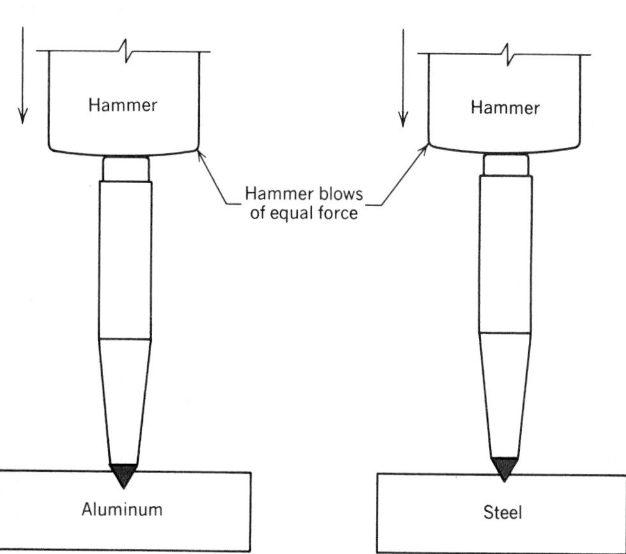

FIGURE D-46 Indentations made by a punch in aluminum and alloy steel.

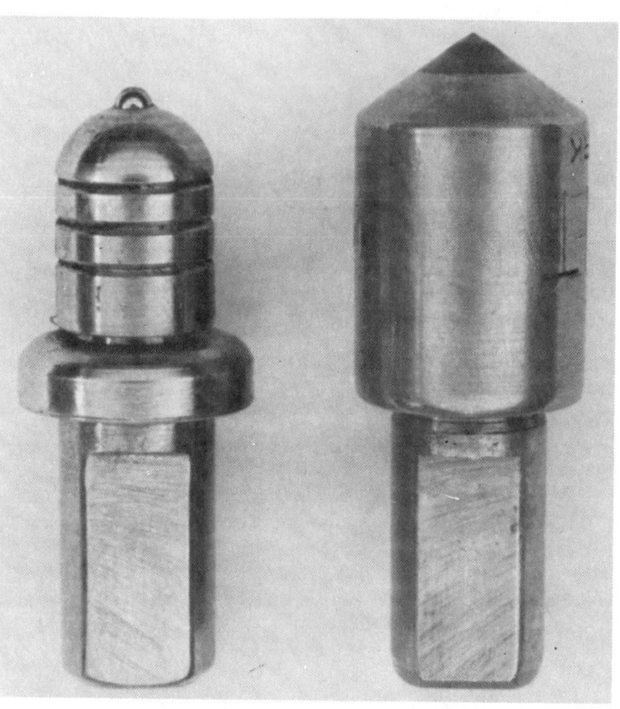

FIGURE D-48 Brale and ball. These two penetrators are the basic types used on the Rockwell Hardness Tester. (*Note:* Brale is a registered trademark of American Chain & Cable Company, Inc., for sphero-conical diamond penetrators.)

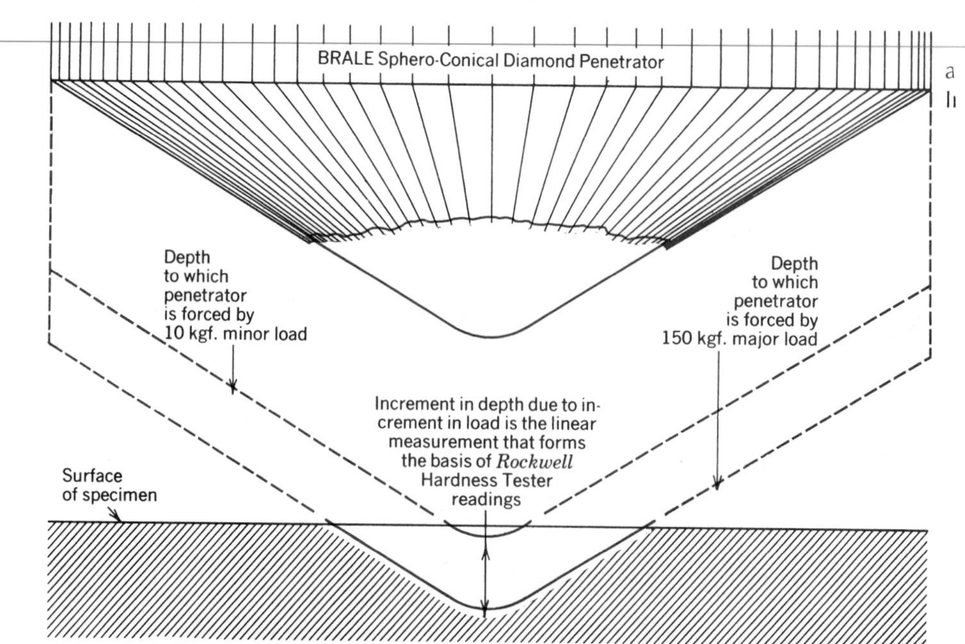

FIGURE D-47 Schematic showing minor and major loads being applied (Wilson Instrument Division of Acco).

Using the Rockwell Hardness Tester

The Rockwell hardness test is made by applying two loads to a specimen and measuring the depth of penetration in the specimen between the first, or minor, load and the major load. The depth of penetration is indicated on the dial when the major load is removed (Figure D-47). The amount of penetration decreases as the hardness of the specimen increases. Generally, the harder the material, the greater its tensile strength, or the ability to resist deformation and rupture when a load is applied. Table D-9 compares hardness by Brinell and Rockwell testers to tensile strength.

There are two basic types of penetrators used on the Rockwell tester (Figure D-48). One is a sphero-conical diamond called a Brale that is used only for hard materials; that is, for materials over B-100, such as hardened steel, nitrided steel, and hard cast irons. When the "C" Brale diamond penetrator is used, the recorded readings should be prefixed by the letter C. The major load used is 150 kgf (kilograms of force). The C scale is *not* used to test extremely hard materials such as cemented

TABLE D-9 Hardness and Tensile Strength Comparison Table

Brinell		Rockwell			Brinell		Rockwell		
Indentation Diameter (mm)	No.	B	C	Tensile Strength (1000 psi, approximately)	Indentation Diameter (mm)	No.	B	C	Tensile Strength (1000 psi, approximately)
2.25	745		65.3		3.75	262	(103.0)	26.6	127
2.30	712		—		3.80	255	(102.0)	25.4	123
2.35	682		61.7		3.85	248	(101.0)	24.2	120
2.40	653		60.0		3.90	241	100.0	22.8	116
2.45	627		58.7		3.95	235	99.0	21.7	114
2.50	601		57.3		4.00	229	98.2	20.5	111
2.55	578		56.0		4.05	223	97.3	(18.8)	—
2.60	555		54.7	298	4.10	217	96.4	(17.5)	105
2.65	534		53.5	288	4.15	212	95.5	(16.0)	102
2.70	514		52.1	274	4.20	207	94.6	(15.2)	100
2.75	495		51.6	269	4.25	201	93.8	(13.8)	98
2.80	477		50.3	258	4.30	197	92.8	(12.7)	95
2.85	461		48.8	244	4.35	192	91.9	(11.5)	93
2.90	444		47.2	231	4.40	187	90.7	(10.0)	90
2.95	429		45.7	219	4.45	183	90.0	(9.0)	89
3.00	415		44.5	212	4.50	179	89.0	(8.0)	87
3.05	401		43.1	202	4.55	174	87.8	(6.4)	85
3.10	388		41.8	193	4.60	170	86.8	(5.4)	83
3.15	375		40.4	184	4.65	167	86.0	(4.4)	81
3.20	363		39.1	177	4.70	163	85.0	(3.3)	79
3.25	352	(110.0)	37.9	171	4.80	156	82.9	(0.9)	76
3.30	341	(109.0)	36.6	164	4.90	149	80.8		73
3.35	331	(108.5)	35.5	159	5.00	143	78.7		71
3.40	321	(108.0)	34.3	154	5.10	137	76.4		67
3.45	311	(107.5)	33.1	149	5.20	131	74.0		65
3.50	302	(107.0)	32.1	146	5.30	126	72.0		63
3.55	293	(106.0)	30.9	141	5.40	121	69.8		60
3.60	285	(105.5)	29.9	138	5.50	116	67.6		58
3.65	277	(104.5)	28.8	134	5.60	111	65.7		56
3.70	269	(104.0)	27.6	130					

Values above 500 are for tungsten carbide ball; below 500 for standard ball.

NOTE 1: This is a condensation of Table 2, *Report J417b, SAE 1971 Handbook*. Values in parentheses are beyond the normal range and are presented for information only.

NOTE 2: The following is a formula to approximate tensile strength when the Brinell hardness is known:

$$\text{tensile strength} = \text{BHN} \times 500$$

SOURCE: Bethlehem Steel Corporation, *Modern Steels and Their Properties*, 7th ed., *Handbook 2757*, 1972.

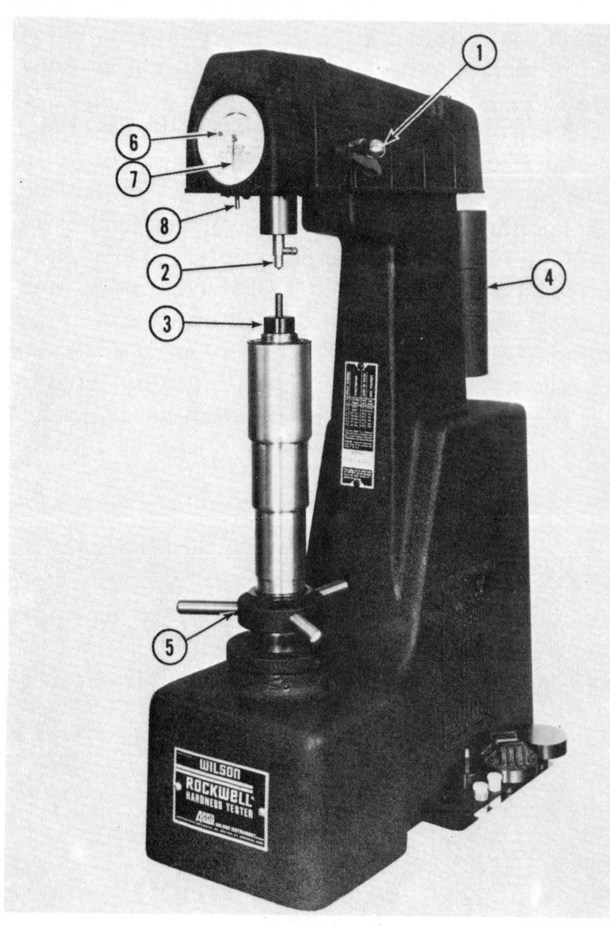

carbides or shallow case hardened steels and thin steel. An A Brale penetrator is used in these cases and the A scale used with a 60-kgf major load.

The second type of penetrator is a $\frac{1}{16}$-in.-diameter ball that is used for testing material in the range of B-100 to B-0, including such relatively soft materials as brass, bronze, and soft steel. If the ball penetrator is used on materials harder than B-100, there is a danger of flattening the ball. Ball penetrators for use on very soft bearing metals are available in sizes of $\frac{1}{2}$, $\frac{1}{4}$, and $\frac{1}{8}$ in. (Table D-10).

Figure D-49 points out the parts used in the testing operation on the Rockwell hardness tester. You should learn the names of these parts before continuing with this unit.

FIGURE D-49 Rockwell hardness tester, listing the names of parts used in the testing operations (Wilson Instrument Division of Acco).

1. Crank handle
2. Penetrator
3. Anvil
4. Weights
5. Capstan handwheel
6. Small pointer
7. Large pointer
8. Lever for setting the bezel.

(*Note:* Rockwell is a registered trademark of American Chain & Cable Company, Inc., for hardness testers and test blocks.)

TABLE D-10 Penetrator and Load Selection

Scale Symbol	Penetrator	Major Load (kgf)	Dial Figures	Typical Applications of Scales
B	$\frac{1}{16}$-in. ball	100	Red	Copper alloys, soft steels, aluminum alloys, malleable iron, etc.
C	Brale	150	Black	Steel, hard cast irons, pearlitic malleable iron, titaneium, deep case-hardened steel, and other materials harder than B-100.
A	Brale	60	Black	Cemented carbides, thin steel, and shallow case-hardened steel.
D	Brale	100	Black	Thin steel and medium case-hardened steel and pearlitic malleable iron.
E	$\frac{1}{8}$-in. ball	100	Red	Cast iron and aluminum and magnesium alloys and bearing metals.
F	$\frac{1}{16}$-in. ball	60	Red	Annealed copper alloys and thin soft sheet metals.
G	$\frac{1}{16}$-in. ball	150	Red	Phosphor bronze, beryllium copper, and malleable irons. Upper limit G-92 to avoid possible flattening of ball.
H	$\frac{1}{8}$-in. ball	60	Red	Aluminum, zinc, and lead.
K	$\frac{1}{8}$-in. ball	150	Red	
L	$\frac{1}{4}$-in. ball	60	Red	Bearing metals and other very soft or thin materials. Use the smallest ball and heaviest load that does not give anvil effect.
M	$\frac{1}{4}$-in. ball	100	Red	
P	$\frac{1}{4}$-in. ball	150	Red	
R	$\frac{1}{2}$-in. ball	60	Red	
S	$\frac{1}{2}$-in. ball	100	Red	
V	$\frac{1}{2}$-in. ball	150	Red	

SOURCE: *Wilson Instruction Manual*, "Rockwell Hardness Tester Models OUR-a and OUS-a," American Chain & Cable Company, Inc., 1973.

FIGURE D-50 Selecting and installing the correct weight (Lane Community College).

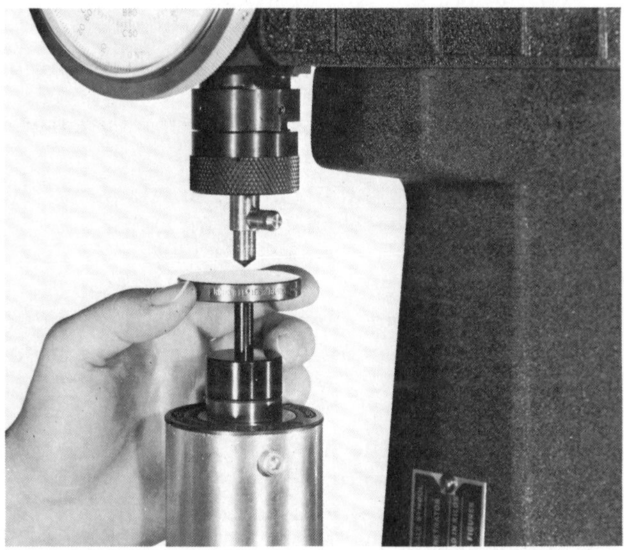

FIGURE D-52 Placing the test block in the machine (Wilson Instrument Division of Acco).

When setting up the Rockwell hardness tester and making the test, follow these steps:

STEP 1. Using Table D-10, select the proper weight (Figure D-50) and penetrator. Make sure that the crank handle is pulled completely forward.

FIGURE D-51 Basic anvils used with Rockwell hardness testers: (*a*) plane; (*b*) shallow V; (*c*) spot; (*d*) Cylindron Jr. (Wilson Instrument Division of Acco).

STEP 2. Place the proper anvil (Figure D-51) on the elevating screw, taking care not to bump the penetrator with the anvil. Make sure that the specimen to be tested is free from dirt, scale, or heavy oil on the underside.

STEP 3. Place the specimen to be tested on the anvil (Figure D-52). Then, by turning the handwheel, gently raise the specimen until it comes

FIGURE D-53 Specimen being brought into contact with the penetrator. This establishes the minor load (Wilson Instrument Division of Acco).

FIGURE D-55 Applying the major load by tripping the crank handle clockwise (Wilson Instrument Division of Acco).

FIGURE D-54 Setting the bezel (Wilson Instrument Division of Acco).

in contact with the penetrator (Figure D-53). Continue turning the handwheel slowly until the small pointer on the dial gage is nearly vertical (near the dot). Now watch the long pointer on the gage and continue raising the work until it is approximately vertical. It should not vary from the vertical position by more than five divisions on the dial. Set the dial to zero on the pointer by moving the bezel until the line marked zero set is in line with the pointer (Figure D-54). You have now applied the minor load. This is the actual starting point for all conditions of testing.

STEP 4. Apply the major load by tripping the crank handle clockwise (Figure D-55).

STEP 5. Wait 2 seconds after the pointer has stopped moving, then remove the major load by pulling the crank handle forward or counterclockwise.

STEP 6. Read the hardness number in Rockwell units on the dial (Figure D-56). The black numbers are for the A and C scales and the red numbers are for the B scale. The specimen should be tested in several places and an average of the test results taken, since many materials vary in hardness even on the same surface.

SUPERFICIAL TESTING

After testing sheet metal, examine the underside of the sheet. If the impression of the penetrator can be seen, then the reading is in error and the superficial test should be used. If the impression can still be seen after the superficial test, then a

FIGURE D-56 Dial face with reading in Rockwell units after completion of the test. The reading is RC 55 (Lane Community College).

FIGURE D-57 Diamond spot anvil (Wilson Instrument Division of Acco).

lighter load should be used. A load 30 kgf is recommended for superficial testing. Superficial testing is also used for case hardened and nitrided steel having a very thin case.

A Brale marked N is needed for superficial testing, as A and C Brales are not suitable. Recorded readings should be prefixed by the major load and the letter N, when using the Brale for superficial testing; for example, 30N78. When using the $\frac{1}{16}$-in. ball penetrator, the same as that used for the B, F, and G hardness scales, the readings should always be prefixed by the major load and the letter T; for example, 30T85. The $\frac{1}{16}$-in. ball penetrator, however, should not be used on material harder than 30T82. Other superficial scales, such as W, X, and Y, should also be prefixed with the major load when recording hardness. See Table D-11 for superficial test penetrator selection.

A spot anvil, as shown in Figure D-51, is used when the tester is being checked on a Rockwell test block. The spot anvil should not be used for checking cylindrical surfaces. The diamond spot anvil (Figure D-57) is similar to the spot anvil, but it has a diamond set into the spot. The diamond is ground and polished to a flat surface. This anvil is used only with the superficial tester, and then only in conjunction with the steel ball penetrator for testing soft metal.

TABLE D-11 Superficial Tester Load and Penetrator Selection

Scale Symbol	Penetrator	Load (kgf)
15N	Brale	15
30N	Brale	30
45N	Brale	45
15T	$\frac{1}{16}$-in. ball	15
30T	$\frac{1}{16}$-in. ball	30
45T	$\frac{1}{16}$-in. ball	45
15W	$\frac{1}{8}$-in. ball	15
30W	$\frac{1}{8}$-in. ball	30
45W	$\frac{1}{8}$-in. ball	45
15X	$\frac{1}{4}$-in. ball	15
30X	$\frac{1}{4}$-in. ball	30
45X	$\frac{1}{4}$-in. ball	45
15Y	$\frac{1}{2}$-in. ball	15
30Y	$\frac{1}{2}$-in. ball	30
45Y	$\frac{1}{2}$-in. ball	45

SOURCE: *Wilson Instruction Manual*. "Rockwell Hardness Tester Models OUR-a and OUS-a," American Chain & Cable Company, Inc., 1973.

SURFACE PREPARATION AND PROPER USE

When testing hardness, surface condition is important for accuracy. A rough or ridged surface caused by coarse grinding will not produce as reliable results as a smoother surface. Any rough scale caused by hardening must be removed before testing. Likewise, if the workpiece has been decarburized by heat treatment, the test area should have this softer "skin" ground off.

Error can also result from testing curved surfaces. This effect may be eliminated by grinding a small flat spot on the specimen. Cylindrical workpieces must always be supported in a V-type centering anvil, and the surface to be tested should

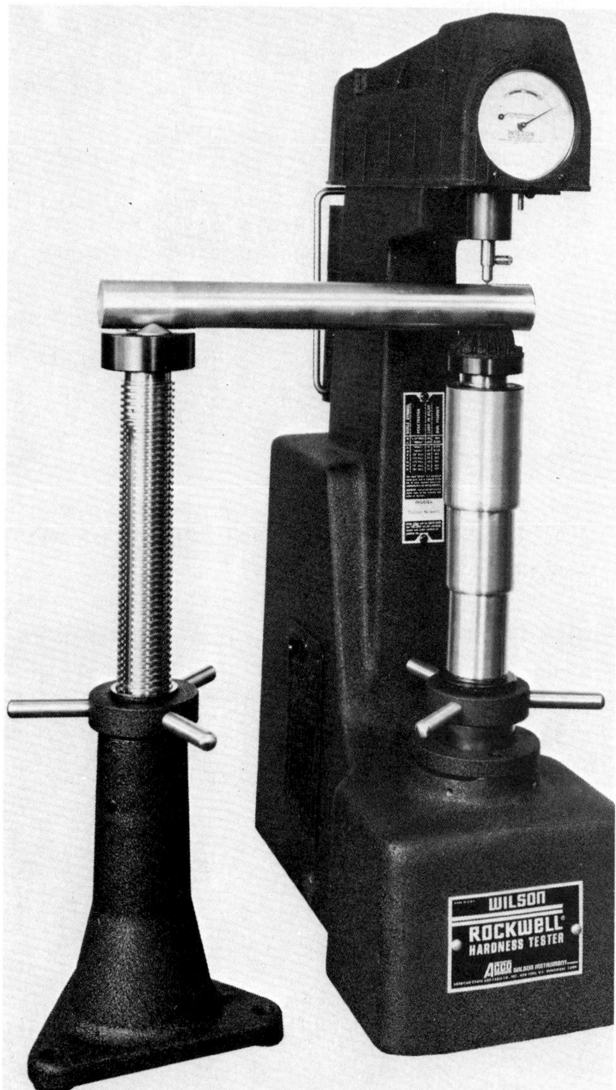

FIGURE D-58 Correct method of testing long, heavy work requires the use of a jack rest (Wilson Instrument Division of Acco).

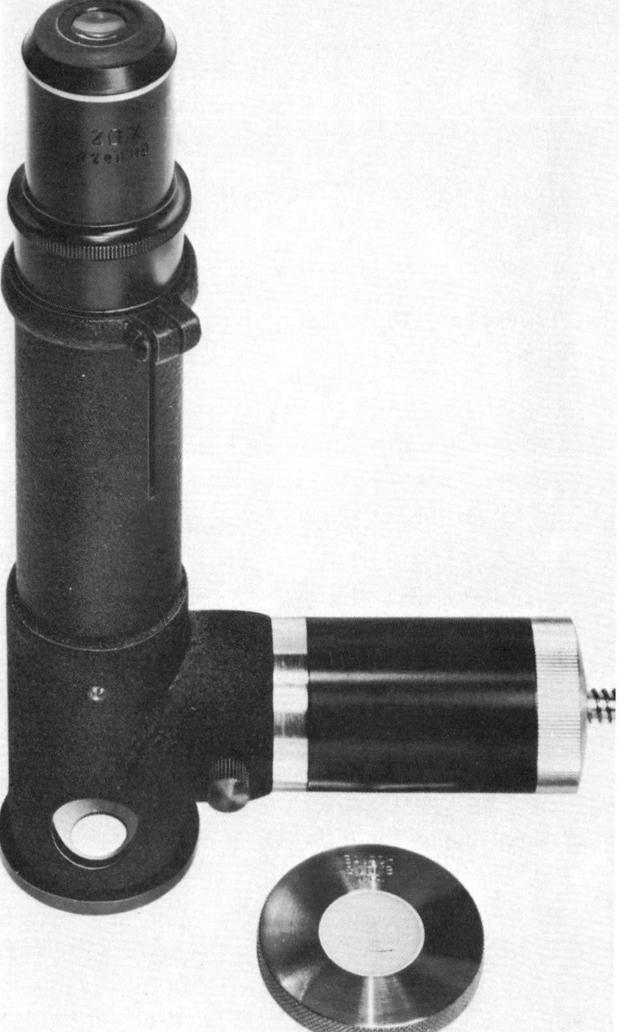

FIGURE D-59 The Olson Brinell microscope provides a fast, accurate means for measuring the diameter of the impression for determining the Brinell hardness number (Tinius Olsen Testing Machine Co., Inc.).

not deviate from the horizontal by more than 5 degrees. Tubing is often so thin that it will deform when tested. It should be supported on the inside by a mandrel or gooseneck anvil to avoid this problem.

Several devices are made available for the Rockwell hardness tester to support overhanging or large specimens. One type, called a jack rest (Figure D-58), is used for supporting long, heavy parts such as shafts. It consists of a separate elevating screw and anvil support similar to that on the tester. Without adequate support, overhanging work can damage the penetrator rod and cause inaccurate readings.

No test should be made near an edge of a specimen. Keep the penetrator at least $\frac{1}{8}$ in. away from the edge. The test block, as shown in Figure D-52, should be used every day to check the calibration of the tester, if it is in constant use.

Using the Brinell Hardness Tester

The Brinell hardness test is made by forcing a steel ball, usually 10 mm in diameter, into the test specimen by using a known load weight and measuring the diameter of the resulting impression. The Brinell hardness value is the load divided by the area of the impression, expressed as follows:

$$\text{BHN} = \frac{P}{(\pi D/2)(D - \sqrt{D^2 - d^2})}$$

where BHN = Brinell hardness number, in kilograms per square millimeter

D = diameter of the steel ball, in millimeters

P = applied load, in kilograms

d = diameter of the impression, in millimeters

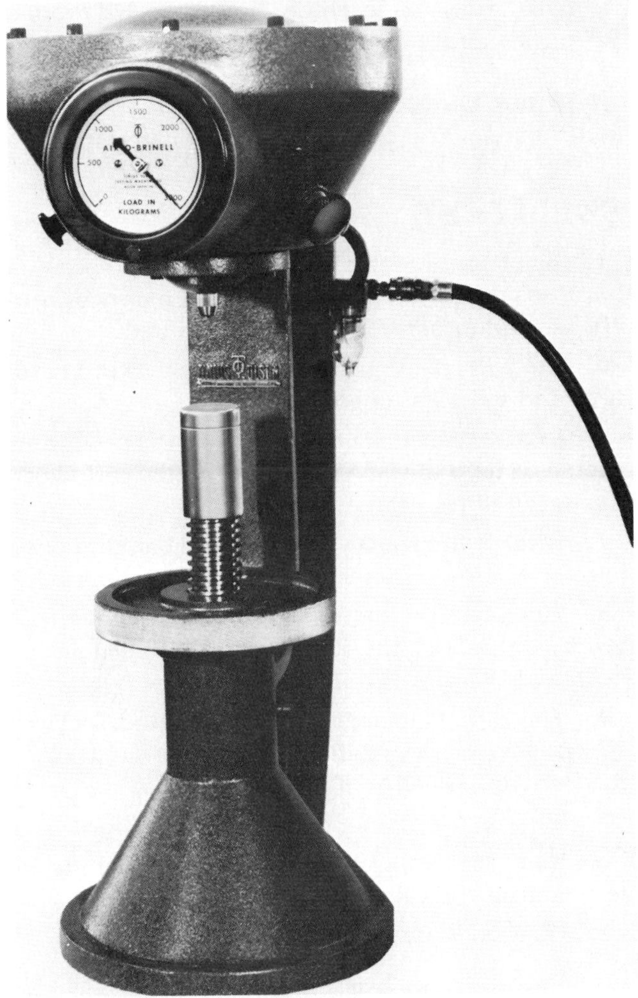

FIGURE D-60 Air-O-Brinell air-operated metal hardness tester (Tinius Olsen Testing Machine Co., Inc.).

A small microscope is used to measure the diameter of the impressions (Figure D-59). Various loads are used for testing different materials: 500 kg for soft materials such as copper and aluminum, and 3000 kg for steels and cast irons. For convenience, Table D-9 gives the Brinell hardness number and corresponding diameters of impression for a 10-mm ball and a load of 3000 kg. The related Rockwell hardness numbers and tensile strengths are also shown. Just as for the Rockwell tests, the impression of the steel ball must not show on the underside of the specimen. Tests should not be made too near the edge of a specimen.

Figure D-60 shows an air-operated Brinell hardness tester. The testing sequence is as follows.

STEP 1. The desired load in kilograms is selected on the dial by adjusting the air regulator (Figure D-61).

FIGURE D-61 Select load. Operator adjusts the air regulator as shown until the desired Brinell load in kilograms is indicated (Tinius Olsen Testing Machine Co., Inc.).

FIGURE D-62 Apply load. Operator pulls out plunger-type control to apply load to the specimen smoothly (Tinius Olsen Testing Machine Co., Inc.).

STEP 2. The specimen is placed on the anvil. Make sure the specimen is clean and free from burrs. It should be smooth enough so that an accurate measurement can be taken of the impression.

STEP 3. The specimen is raised to with $\frac{5}{8}$ in. of the Brinell ball by turning the handwheel.

STEP 4. The load is then applied by pulling out the plunger control (Figure D-62). Maintain

FIGURE D-63 Release load. As soon as the plunger is depressed, the Brinell ball retracts in readiness for the next test (Tinius Olsen Testing Machine Co., Inc.).

the load for 30 seconds for nonferrous metals and 15 seconds for steel. Release load (Figure D-63).

STEP 5. Remove the specimen from the tester and measure the diameter of the impression.

STEP 6. Determine the Brinell hardness number (BHN) by calculation or by using the table. Soft copper should have a BHN of about 40, soft steel from 150 to 200, and hardened tools from 500 to 600. Fully hardened high-carbon steel would have a BHN of 750. A Brinell test ball of tungsten carbide should be used for materials above 600 BHN.

Brinell hardness testers work best for testing softer metals and medium-hard tool steels.

SELF-TEST

1. What one specific category of the property of hardness do the Rockwell and Brinell hardness testers use and measure? How is it measured?

2. State the relationship that exists between hardness and tensile strength.

3. Explain which scale, major load, and penetrator should be used to test a block of tungsten carbide on the Rockwell tester.

4. What is the reason that the steel ball cannot be used on the Rockwell tester to test the harder steels?

5. When testing with the Rockwell superficial tester, is the Brale the same one that is used on the A, C, and D scales? Explain.

6. The $\frac{1}{16}$-in. ball penetrator used for the Rockwell superficial tester is a different one from that used for the B, F, and G scales. True or false?

7. What is the diamond spot anvil used for?

8. How does roughness on the specimen to be tested affect the test results?

9. How does decarburization affect the test results?

10. What does a curved surface do to the test results?

11. On the Brinell tester what load should be used for testing steel?

12. What size ball penetrator is generally used on a Brinell tester?

SECTION **E**

LAYOUT

Layout is the process of placing reference marks on the workpiece. These marks may indicate the shape and size of a part or its features. Layout marks often indicate where machining will take place. Machinists may use layout marks as a guide for machining while checking their work by actual measurement. They may also cut to a layout mark. One of your first jobs after you have obtained material from stock will be to measure and lay out where the material will be cut. This kind of layout may be a simple pencil or chalk mark and is one of the basic tasks of semiprecision layout.

Precision layout can be a complex and involved operation making use of sophisticated tools. In the aircraft and shipbuilding industries, reference points, lines, and planes may be laid out using optical and laser instruments. In the machine shop, you will be concerned with layout for stock cutoff, filing and offhand grinding, drilling, milling, and occasionally in connection with lathe work.

Layout Classifications

The process of layout can be generally classified as **semiprecision** and **precision**. Semiprecision layout is usually done by scale measurement to a tolerance of $\pm\frac{1}{64}$ in. Precision layout is done with tools that discriminate to .001 in. or finer, to a tolerance of $\pm$.001 in. if possible.

Tools of Layout

SURFACE PLATES

The surface plate is an essential tool for many layout applications. A surface plate provides an accurate reference plane from which measurements for both layout and inspection may be made. In many machine shops, where a large amount of layout work is accomplished, a large area surface plate, perhaps 4 by 8 ft may be used. These are often known as layout tables.

Any surface plate or layout table is a precision tool and should be treated as such. It should be covered when not in use and kept clean when being used. No surface plate should be hammered upon, since this will impair the accuracy of the reference surface. As you study machine tool practices, measurement, and layout, the surface plate will play an important part in many of your tasks.

CAST IRON AND SEMISTEEL SURFACE PLATES Cast iron and semisteel surface plates (Figure E-1) are made from good-quality castings that have been allowed to age, thus relieving in-

FIGURE E-1 Cast iron surface plate.

ternal stresses. Aging of the casting reduces distortion after its working surface has been finished to the desired degree of flatness. The cast iron or steel plate will also have several ribs on the underside to provide structural rigidity. Cast plates vary in size from small bench models, a few square inches in area, to larger sizes that may be 4 to 8 ft or larger. The large cast plates are usually a foot or more in thickness with appropriate ribs on the underside to provide for sufficient rigidity. The large iron plate is generally mounted on a heavy stand or legs with provision for leveling. The plate is leveled periodically to insure that its working surface remains flat.

GRANITE SURFACE PLATES The cast iron and semisteel surface plate has all but given way to the granite plate (Figure E-2). Granite is superior to metal because it is harder, denser, impervious to water, and if it is chipped, the surrounding flat surface is not affected. Furthermore, granite, because it is a natural material, has aged in the earth for a great deal of time. Therefore, it has little internal stress. Granite surface plates possess a greater temperature stability than their metal counterparts.

Granite plates range in size from about 12 by 18 in. to 4 by 12 ft. A large granite plate may be from 10 to 20 in. thick and weigh as much as 5 to 10 tons. Some granite plates are finished on two sides, thus permitting them to be turned over and their use extended.

FIGURE E-2 Granite surface plate (DoALL Company).

GRADES OF GRANITE SURFACE PLATES The granite surface plate is available in three grades. Surface plate grade specifications are an indication of the plus and minus deviation of the working surface from an average plane.

Grade	Type	Tolerance
AA	Laboratory grade	± 25 millionths inch
A	Inspection grade	± 50 millionths inch
B	Shop grade	± 100 millionths inch

The tolerances are proportional to the size of the plate. As the size increases above 18 by 18 in., the tolerance widens.

LAYOUT DYES

To make layout marks visible on the surface of the workpiece, a layout dye is used. Layout dyes are available in several colors. Among these are red, blue, and white. The blue dyes are very common. Depending on the surface color of the workpiece material, different dye colors may make layout marks more visible. Layout dye should be applied sparingly in an even coat (Figure E-3).

FIGURE E-3 Applying layout dye to the workpiece (L.S. Starrett Company).

SCRIBERS AND DIVIDERS

Several types of **scribers** are in common use. The pocket scriber (Figure E-4) has a removable tip that can be stored in the handle. This permits the scriber to be carried safely in the pocket. The engineer's scriber (Figure E-5) has one straight and one hooked end. The hook permits easier access to the line to be scribed. The machinist's scriber (Figure E-6) has only one end with a fixed point. **Scribers must be kept sharp.** If they become dull, they must be reground or stoned to restore their points. Scriber materials include hardened steel and tungsten carbide.

When scribing against a rule, hold the rule firmly. Tilt the scriber so that the tip marks as close to the rule as possible. This will insure accuracy. An excellent scriber can be made by grinding a shallow angle on a piece of tool steel (Figure E-7). This type of scriber is particularly well suited to scribing along a rule. The flat side permits the scriber to mark very close to the rule, thus obtaining maximum accuracy.

Spring dividers (Figure E-8) range in size from 2 to 12 in. The spacing of the divider legs is set by turning the adjusting screw. Dividers are usually set to rules. Engraved rules are best as the divider tips can be set in the engraved rule graduations (Figure E-9). Like scribers, divider tips must be kept sharp and at nearly the same length. Spring dividers, as their name implies, are used to divide lines or the circumference of circles. Divisions are marked by using the tip of the divider leg as a scriber.

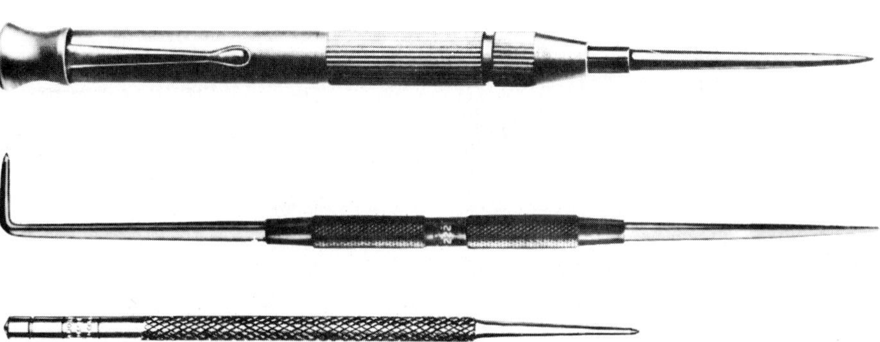

FIGURE E-4 Pocket scriber (Rank Scherr-Tumico, Inc.).

FIGURE E-5 Engineer's scriber (Rank Scherr-Tumico, Inc.).

FIGURE E-6 Machinist's scriber (Rank Scherr-Tumico, Inc.).

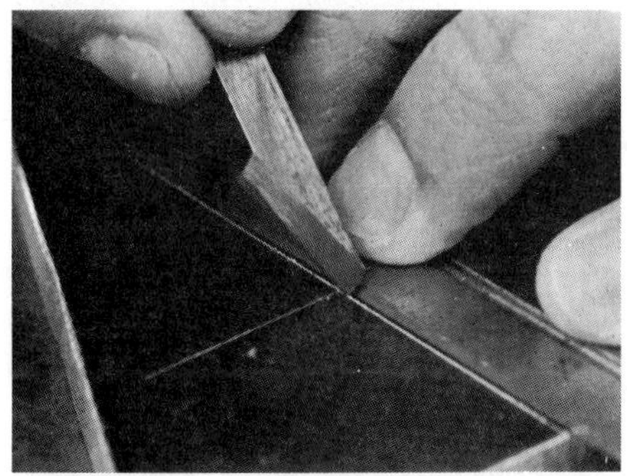

FIGURE E-7 Rule scribe made from a high-speed tool bit.

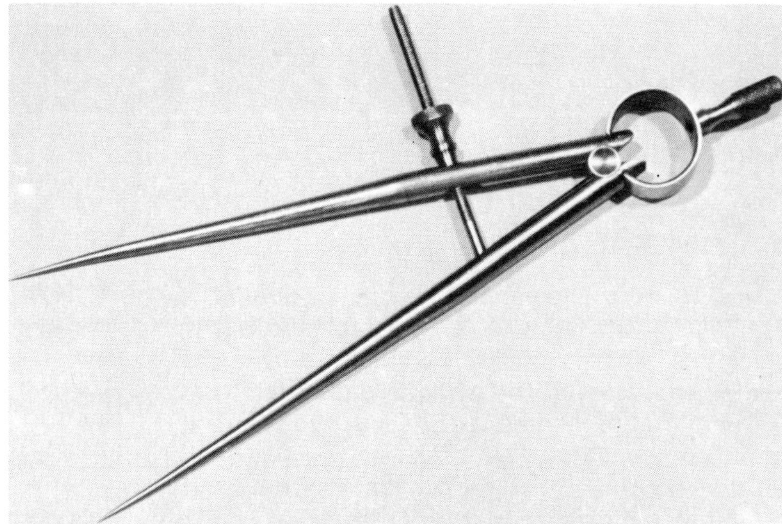

FIGURE E-8 Spring dividers.

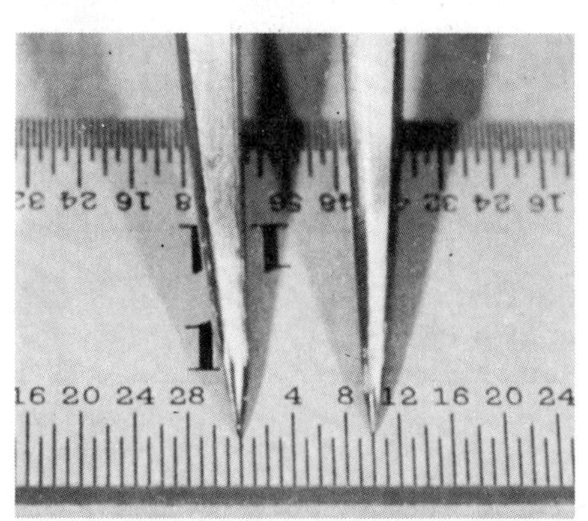

FIGURE E-9 Setting divider points to an engraved rule.

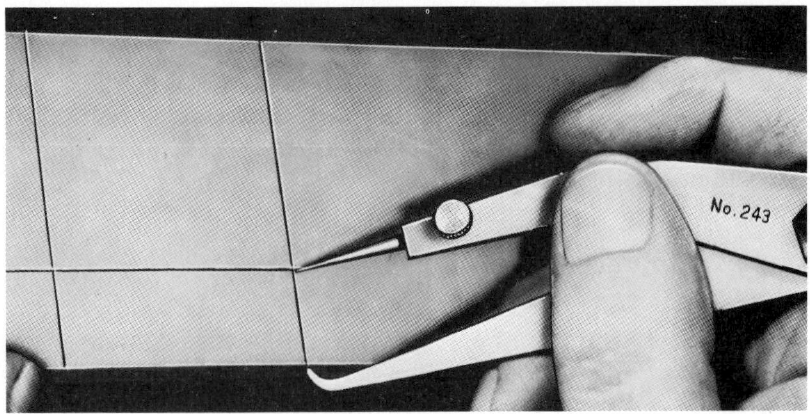

FIGURE E-10 Scribing a line parallel to an edge using a hermaphrodite caliper (L.S. Starrett Co.).

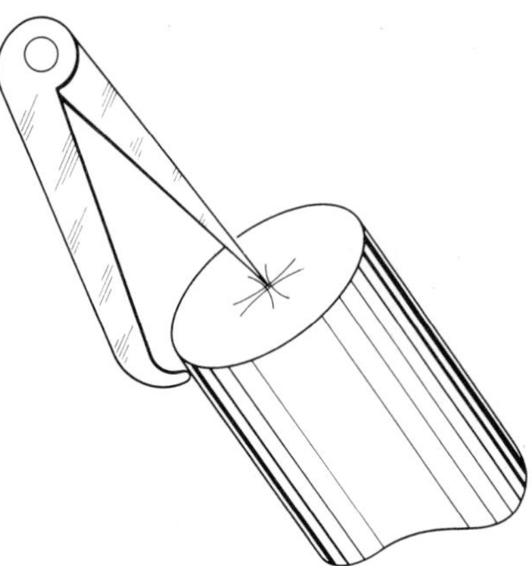

FIGURE E-11 Scribing the centerline of round stock with the hermaphrodite caliper.

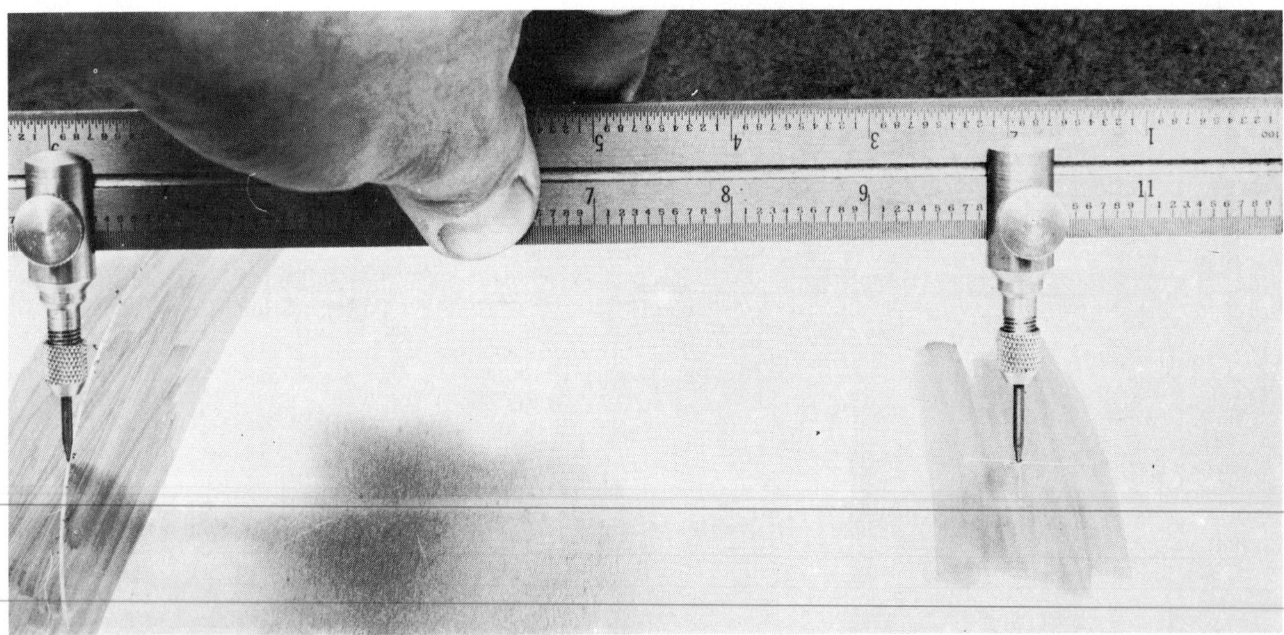

FIGURE E-12 Trammel point attached to a rule.

HERMAPHRODITE CALIPER

The hermaphrodite caliper has one leg similar to a regular divider. The tip is adjustable for length. The other leg has a hooked end that can be placed against the edge of the workpiece (Figure E-10). Hermaphrodite calipers can be used to scribe a line parallel to an edge.

The hermaphrodite caliper can also be used to lay out the center of round stock (Figure E-11). The hooked leg is placed against the round stock and an arc is marked on the end of the piece. By adjusting the leg spacing, tangent arcs can be laid out. By marking four arcs at 90 degrees, the center of the stock can be established.

TRAMMEL POINTS

Trammel points are used for scribing circles and arcs when the distance involved exceeds the capacity of the divider. Trammel points are either attached to a bar and set to circle dimensions or clamped directly to a rule where they can be set directly by rule graduations (Figure E-12).

LAYOUT HAMMERS AND PUNCHES

Layout hammers are usually lightweight (2 to 4 oz) machinist's ball peen hammers (Figure E-13). A heavy hammer should not be used in layout as it

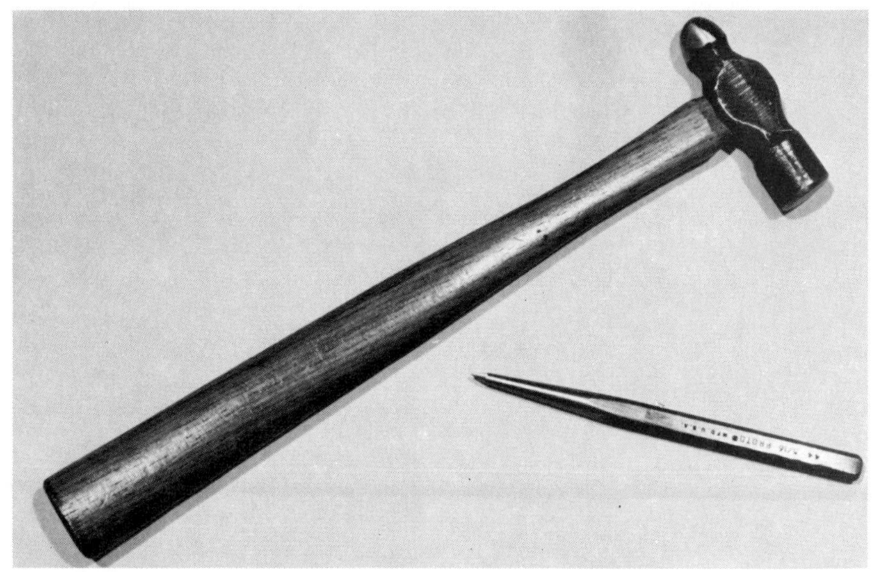

FIGURE E-13 Layout hammer and layout prick punch.

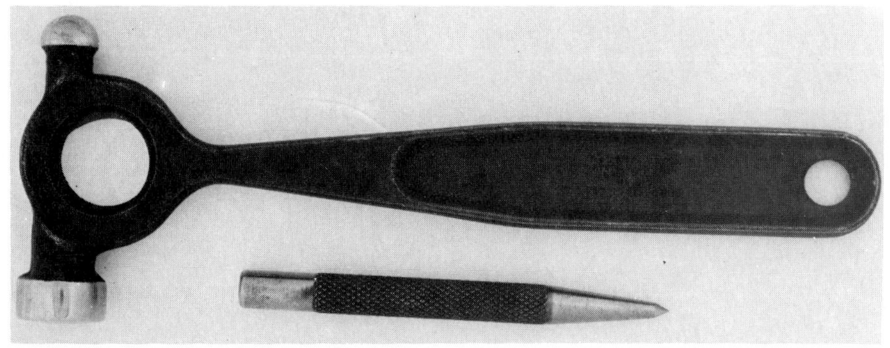

FIGURE E-14 Toolmaker's hammer.

tends to create unnecessarily large punch marks.

The toolmaker's hammer is also used (Figure E-14). This hammer is equipped with a magnifier that can be used to help locate a layout punch on a scribe mark (Figure E-15).

There is an important difference between a layout punch and a center punch. The **layout** or **prick punch** (Figure E-14) **has an included point angle of 30 degrees.** This is the only punch that should be used in layout. The slim point facilitates the locating of the punch on a scribe line. A prick punch mark is only used to preserve the location of a layout mark while doing minimum damage to the workpiece. On some workpieces, depending on the material used and the part application, layout punch marks are not acceptable as they create a defect in the material. A punch mark may affect surface finish or metallurgical properties. Before using a layout punch, you must make sure that it is acceptable. In all cases, layout punch marks should be of minimum depth.

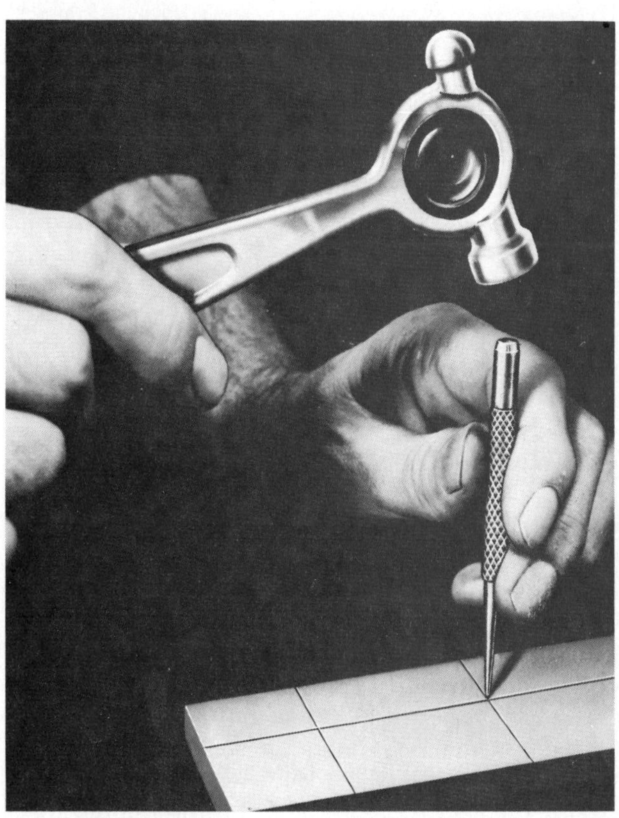

FIGURE E-15 Using a toolmaker's hammer and layout punch (L.S. Starrett Co.).

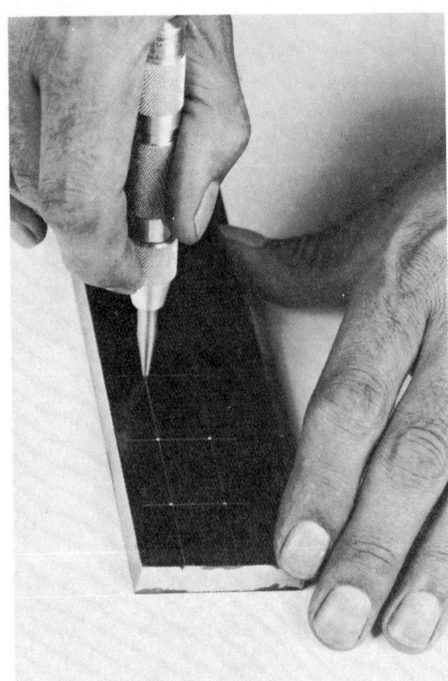

FIGURE E-16 Center punch.

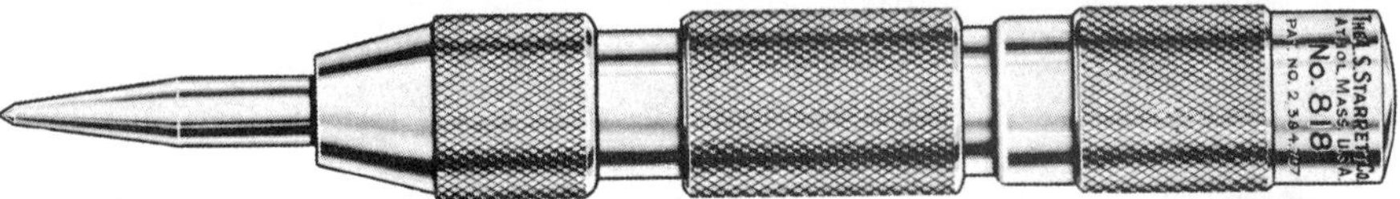

FIGURE E-18 Automatic center punch (L.S. Starrett Co.).

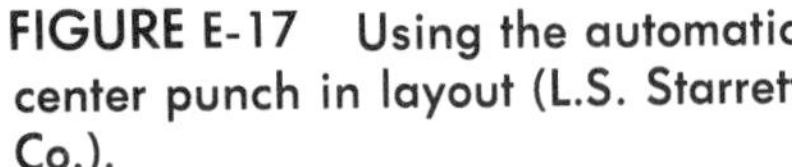

FIGURE E-17 Using the automatic center punch in layout (L.S. Starrett Co.).

FIGURE E-19 Optical center punch.

The **center punch** (Figure E-16) **has an included point angle of 90 degrees** and is used to mark the workpiece prior to such machining operations as drilling. A center punch should not be used in place of a layout punch. Likewise, a layout punch should not be used in place of a center punch. The center punch is used only to deepen the prick punch mark.

The automatic center punch (Figure E-17) requires no hammer. Although called a center punch, its tip is suitably shaped for layout applications (Figure E-18). Spring pressure behind the tip provides the required force. The automatic center punch may be adjusted for variable punching force by changing the spring tension. This is accomplished by an adjustment on the handle.

The optical center punch (Figure E-19) consists of a locator, optical alignment magnifier, and punch. This type of layout punch is extremely useful in locating punch marks precisely on a scribed line or line intersection. The locator is placed over the approximate location and the optical alignment magnifier is inserted (Figure E-20). The locator is magnetized so that it will remain in position when used on ferrous metals. The optical alignment magnifier has crossed lines etched on its lower end. By looking through the magnifier, you can move the locator about until the cross lines are matched to the scribe lines on the workpiece. The magnifier is then removed and the punch is inserted into the locator (Figure E-21). The punch is then tapped with a layout hammer (Figure E-22).

FIGURE E-20 Locating the punch holder with the optical alignment magnifier.

FIGURE E-21 Inserting the punch into the punch holder.

FIGURE E-22 Tapping the punch with a layout hammer.

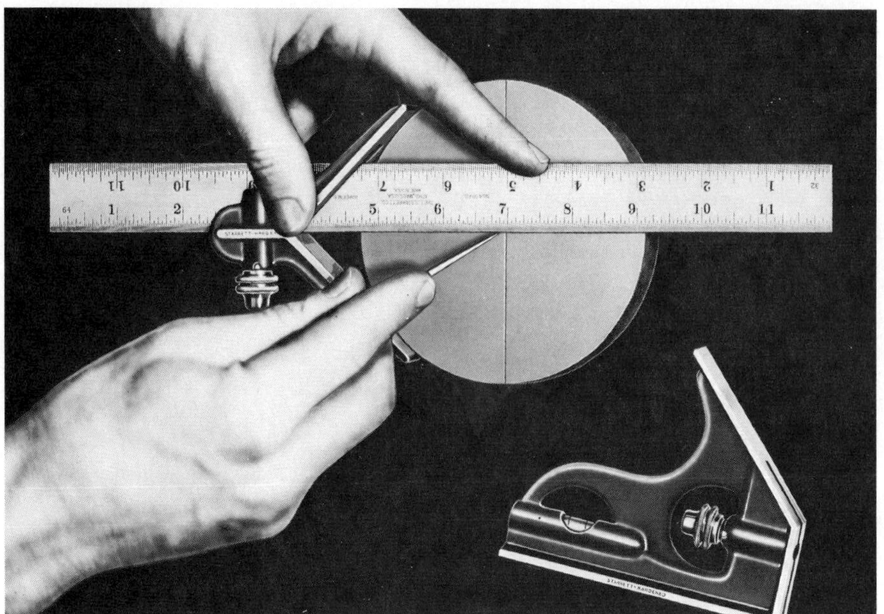

FIGURE E-23 Using the centerhead to lay out a centerline on round stock (L.S. Starrett Co.).

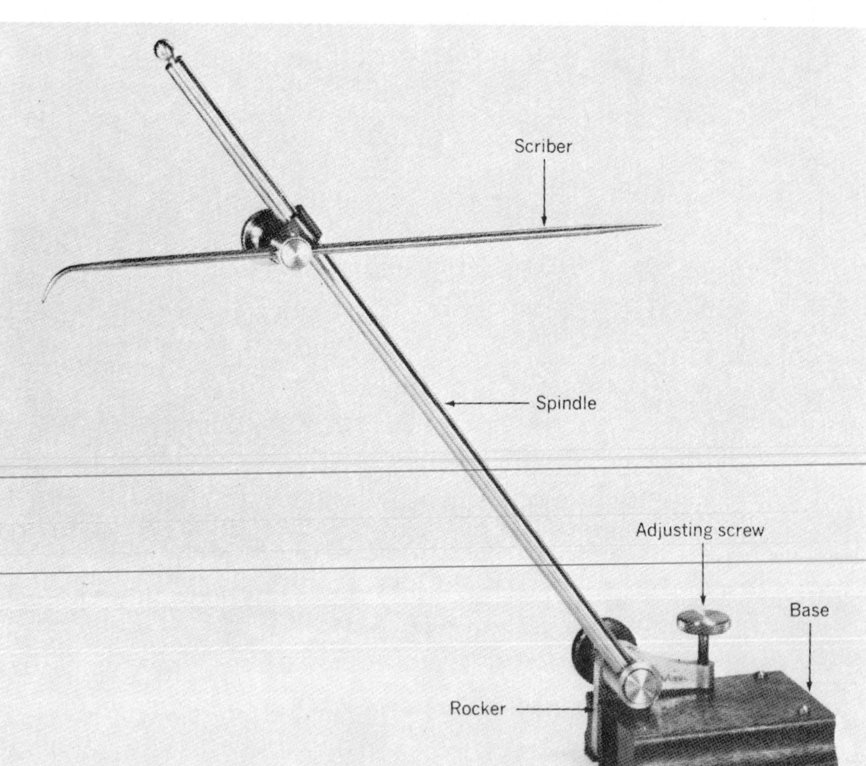

FIGURE E-24 Parts of the surface gage (Rank Scherr-Tumico, Inc.).

CENTERHEAD

The centerhead is part of the machinist's combination set. Centerheads are used to lay out centerlines on round workpieces (Figure E-23). When the centerhead is clamped to the combination set rule, the edge of the rule is in line with a circle center.

The other parts of the combination set are useful in layout. These include the **rule, square head,** and **bevel protractor**.

SURFACE GAGE

The surface gage consists of a base, rocker, spindle adjusting screw, and scriber (Figure E-24). The spindle of the surface gage pivots on the base and can be moved with the adjusting screw. The scriber can be moved along the spindle and locked at any desired position. The scriber can also swivel in its clamp. A surface gage may be used as a height transfer tool. The scribe is set to a rule dimension (Figure E-25) and then transferred to the workpiece.

FIGURE E-25 Setting a surface gage to a rule.

FIGURE E-26 Finding the centerline of the workpiece using the surface gage.

FIGURE E-27 Adjusting the position of the scribe line to center by inverting the workpiece and checking the existing differences in scribe marks.

FIGURE E-28 Mechanical dial height gage (Southwestern Industries, Inc.).

The hooked end of the surface gage scriber may be used to mark the centerline of a workpiece. The following procedure should be followed when doing this layout operation. The surface gage is first set as nearly as possible to a height equal to one half of the part height. The workpiece should be scribed for a short distance at this position. The part should be turned over and scribed again (Figure E-26). If a deviation exists, there will be two scribe lines on the workpiece. The surface gage scriber should then be adjusted so that it splits the difference between the two marks (Figure E-27). This ensures that the scribed line is in the center of the workpiece. This principle of inverting and thus splitting the errors can be applied to many other layout procedures.

HEIGHT GAGES

Height gages are some of the most important instruments for precision layout. The most common layout height gage is the vernier type. Use of this instrument will be discussed in the unit on precision layout. As a machinist you may use several other types of height gages for layout applications.

MECHANICAL DIAL AND ELECTRONIC DIGITAL HEIGHT GAGES Mechanical dial (Figure E-28) and electronic digital (Figures E-29

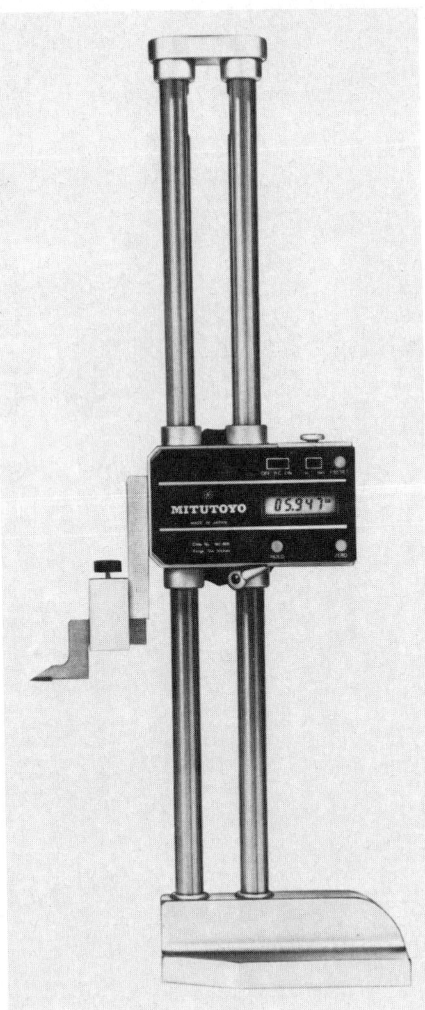

FIGURE E-29 Electronic digital height gage (MTI Corporation).

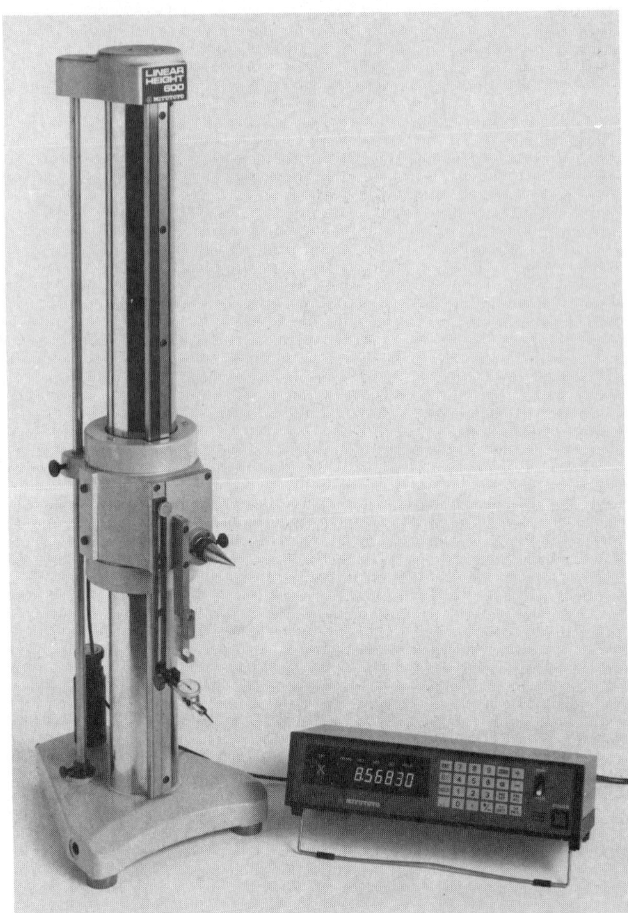

FIGURE E-30 The electronic height gage is useful for layouts and as a test indicator holder for inspection measurements (MTI Corporation).

and E-30) height gages eliminate the need to read a vernier scale. Often these height gages do not have beam graduations. Once set to zero on the reference surface, the total height reading is cumulative on the digital display. This makes beam graduations unnecessary. The electronic digital height gage will discriminate to .0001 in.

GAGE BLOCK HEIGHT GAGES Gage block height gages may be assembled from wrung stacks of gage blocks and accessories (Figure E-31). These height gages are extremely precise as they make use of the inherent accuracy of the gage blocks from which they are assembled.

THE PLANER GAGE AS A HEIGHT GAGE The planer gage may be equipped with a scriber and used as a height gage (Figure E-32). Dimensions are set by comparison to a precision height gage or height transfer micrometer. The planer gage can also be set with an outside micrometer.

LAYOUT MACHINES

The layout machine (Figure E-33) consists of a vertical column with a horizontal crossarm that can move up and down, in and out. The vertical column also moves horizontally across the layout table. From a single setup, the layout machine can accomplish layout on all sides, bottom, top, and inside of the workpiece. The instrument is equipped with an electronic digital display discriminating to .0001 in.

Layout Accessories

Layout accessories are tools that will aid you in accomplishing layout tasks. They are not specifically layout tools as they are used for many other purposes. The layout plate or surface plate used for layout is the most common accessory as it provides the reference surface from which to work. Other common accessories include vee blocks and angle plates that hold the workpiece during layout operations (Figure E-34).

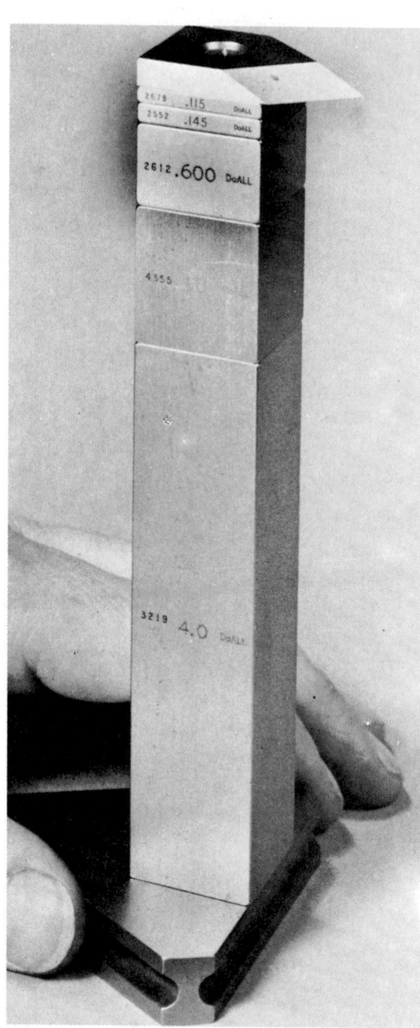

FIGURE E-31 Height gage assembled from gage blocks (DoALL Company).

FIGURE E-32 Using the planer gage as a height gage in layout.

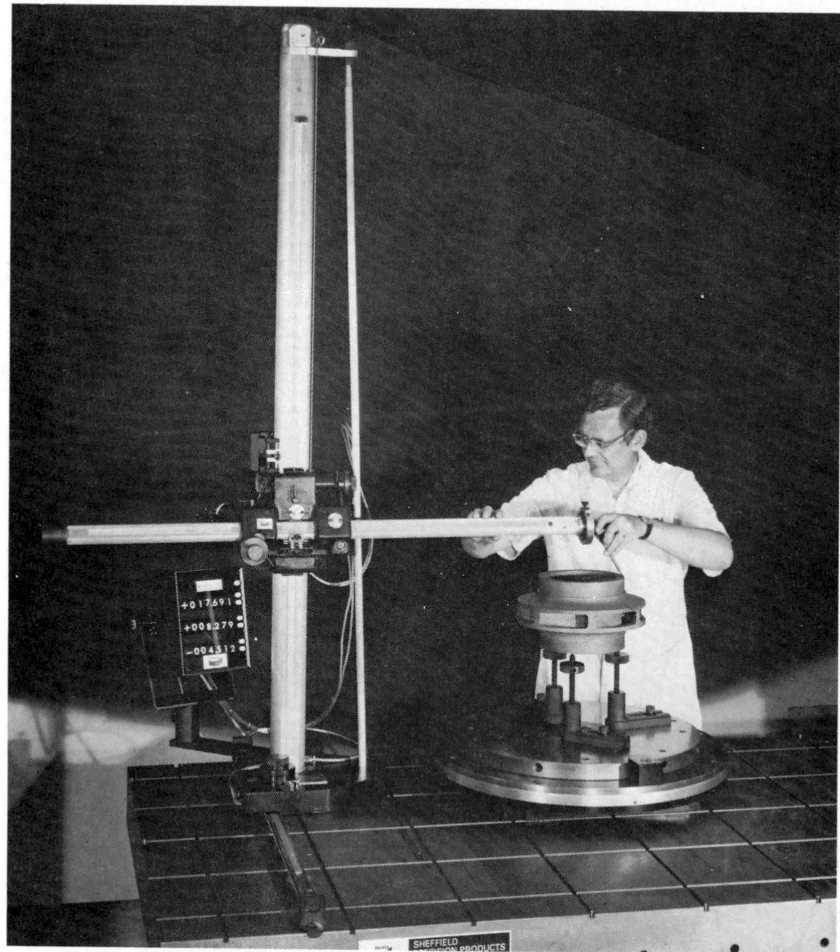

FIGURE E-33 Layout machine (Automation and Measurement Division-Bendix Corporation).

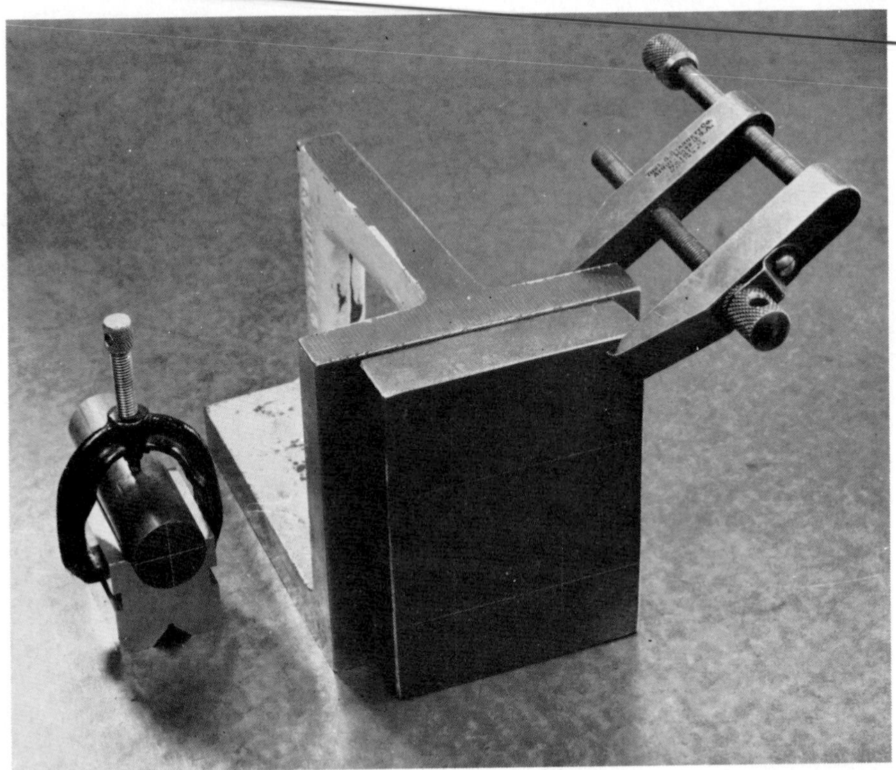

FIGURE E-34 Universal right angle plate and vee block used as layout accessories.

UNIT 1

Basic Semiprecision Layout Practice

Before you can cut material for a certain job, you must perform a layout operation. Layout for stock cutoff may involve a simple chalk, pencil, or scribe mark on the material. No matter how simple the layout job may be, you should strive to do it neatly and accurately. In any layout, semiprecision or precision, accuracy is the watchword. Up to this point, you have been introduced to a large number of measuring and layout tools. It is now up to you to put these tools to work in the most productive manner possible. In this unit, you proceed through a typical semiprecision layout task that will familiarize you with basic layout practice.

Objectives

After completing this unit, you should be able to:

1. Prepare the workpiece for layout.
2. Measure for and scribe layout lines on the workpiece outlining the various features.
3. Locate and establish hole centers using a layout prick punch and center punch.
4. Lay out a workpiece to a tolerance of $\pm\frac{1}{64}$ in.

Preparing the Workpiece for Layout

After the material has been cut, all sharp edges should be removed by grinding or filing before placing the stock on the layout table. Place a paper towel under the workpiece to prevent layout dye from spilling on the layout table (Figure E-35). Apply a thin even coat of layout dye to the workpiece. You will need a drawing of the part in order to do the required layout (Figure E-36). Avoid breathing the vapors.

Study the drawing and determine the best way to proceed. The order of steps depends on the layout task. Before some features can be laid out, certain reference lines may have to be established. Measurements for other layouts are made from these lines.

FIGURE E-35 Applying layout dye with workpiece on a paper towel (Lane Community College).

Layout of the Drill and Hole Gage

If possible, obtain a piece of material the same size as indicated on the drawing. Depending on the part to be made, you may be able to use material that is the same size as the finished job. However, certain parts may require that the edges be machined to finished dimensions. This may necessitate using material that is larger than the finished part in order to allow for machining of edges. Follow through each step as described in the text. Refer to the layout drawings to determine where layout is to be done. The pictures will help you in selecting and using the required tools.

The first operation is to establish the width of the gage. Measure a distance of $1\frac{1}{8}$ in. from one edge of the material. Use the combination square and rule. Set the square at the required dimension and scribe a mark at each end of the stock (Figures E-37 and E-38).

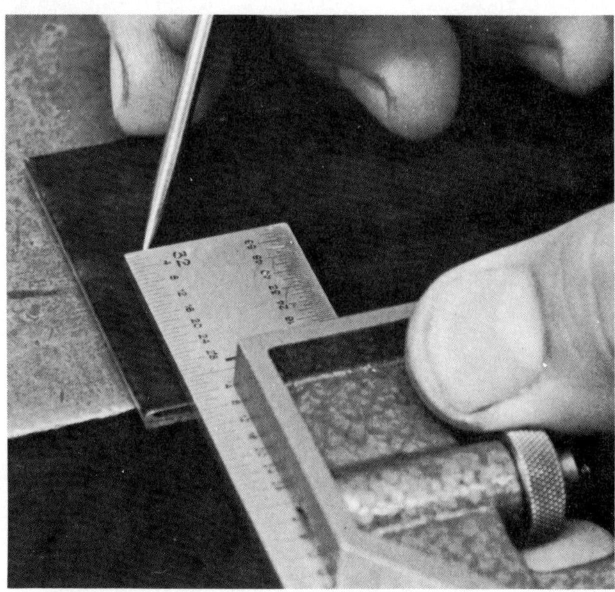

FIGURE E-37 Measuring and marking the width of the gage using the combination square and rule (Lane Community College).

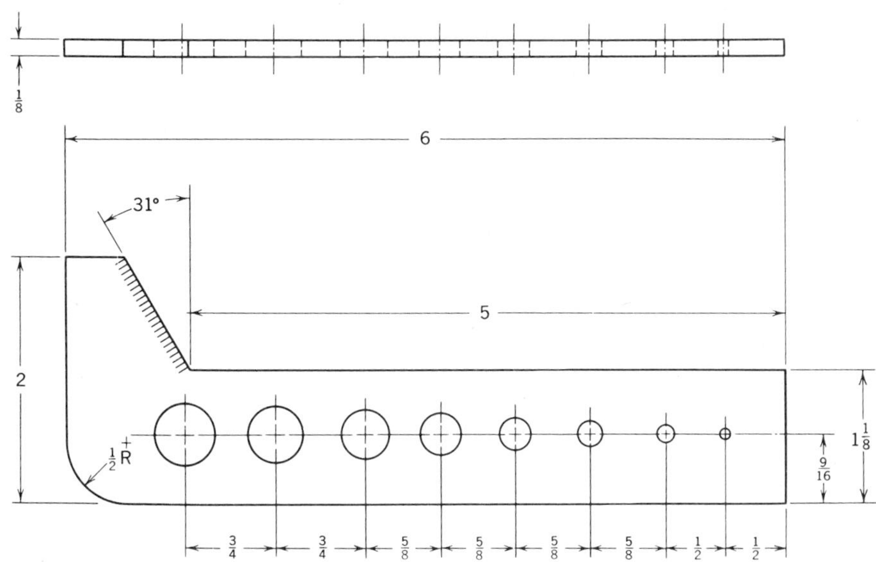

FIGURE E-36 Drill and hole gage (Lane Community College).

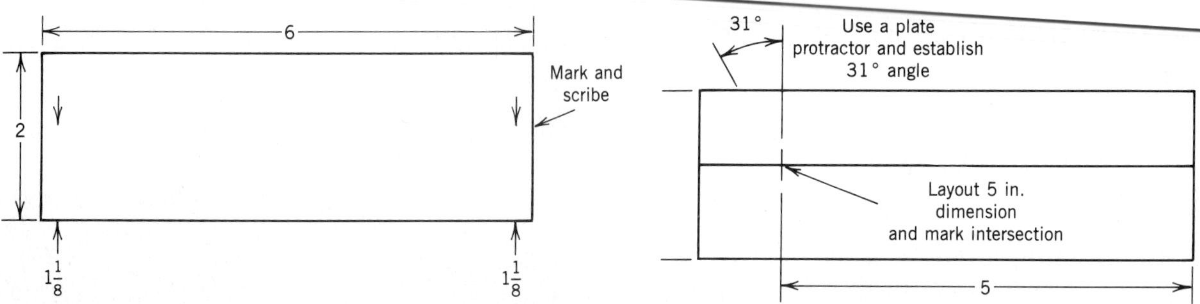

FIGURE E-38 Width line.

FIGURE E-41 Angle line.

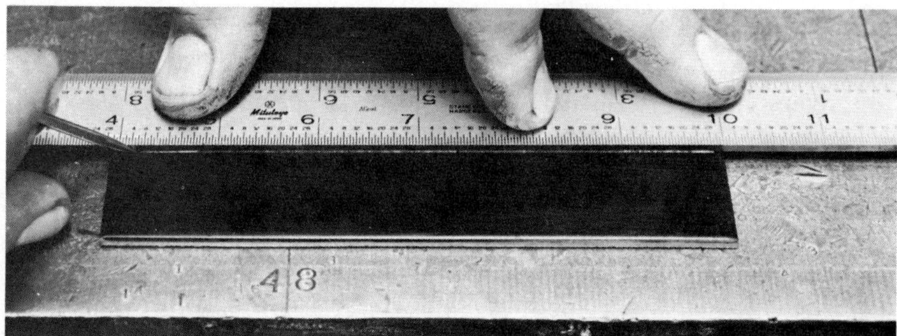

FIGURE E-39 Scribing the width line (Lane Community College).

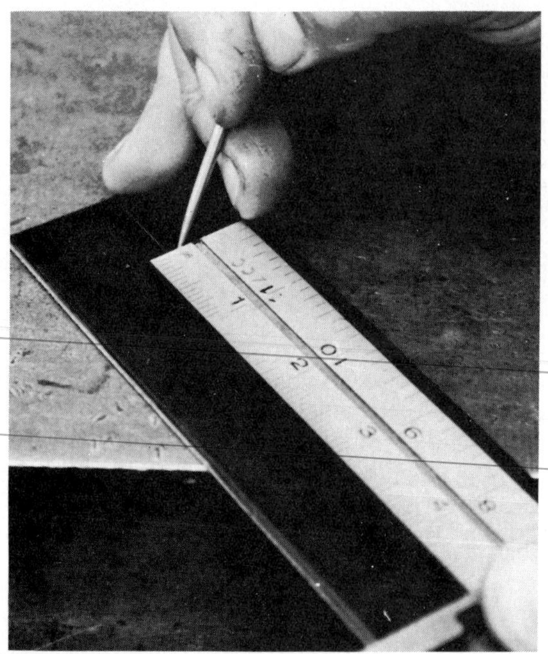

FIGURE E-40 Measuring the 5-in. dimension from the end to the angle vertex (Lane Community College).

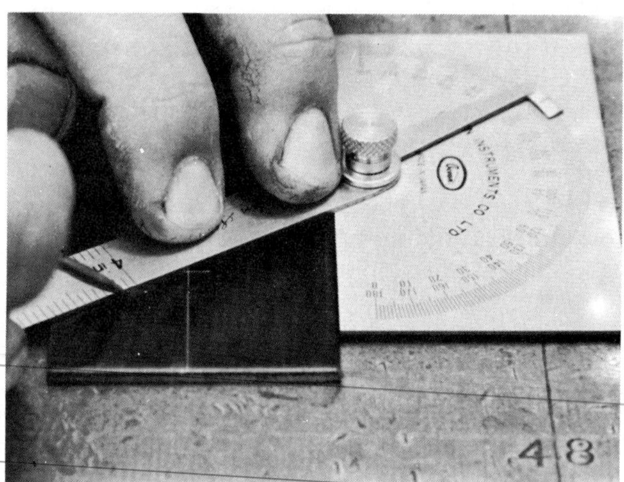

FIGURE E-42 Scribing the angle line (Lane Community College).

Remove the square and place the rule carefully on the scribe marks. Hold the rule firmly and scribe the line the full length of the material (Figures E-38 and E-39). Be sure to use a sharp scribe and hold it so that the tip is against the rule. If the scribe is dull, regrind or stone it to restore its point. Scribe a clean visible line. Lay out the 5-in. length from the end of the piece to the angle vertex. Use the combination square and rule (Figures E-40 and E-41).

Use a plate protractor to lay out the angle. The bevel protractor from the combination set is also a suitable tool for this application. Be sure that the protractor is set to the correct angle. The edge of the protractor blade must be set exactly at the 5-in. mark (Figures E-41 and E-42). The layout of the 31-degree angle establishes its complement of 59 degrees on the drill gage. The correct included angle for general-purpose drill points will be 118 degrees, or twice 59 degrees. The corner radius is

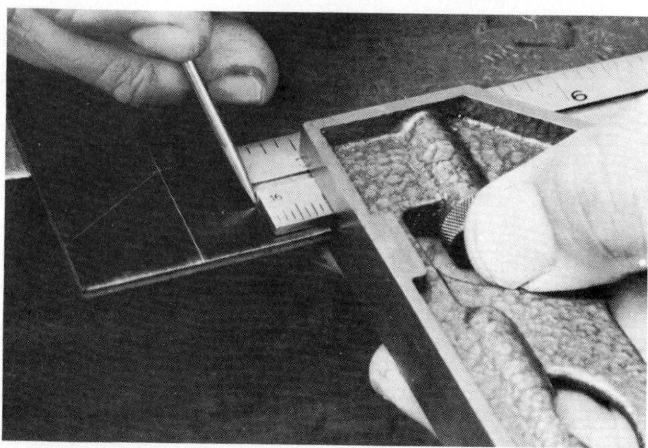

FIGURE E-43 Establishing the center point of the corner radius (Lane Community College).

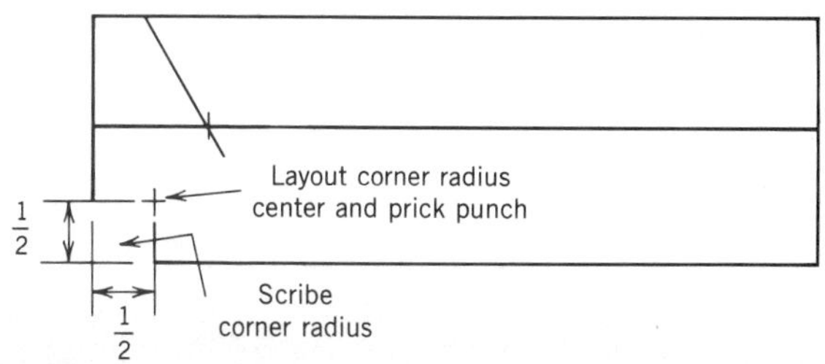

FIGURE E-44 Corner radius.

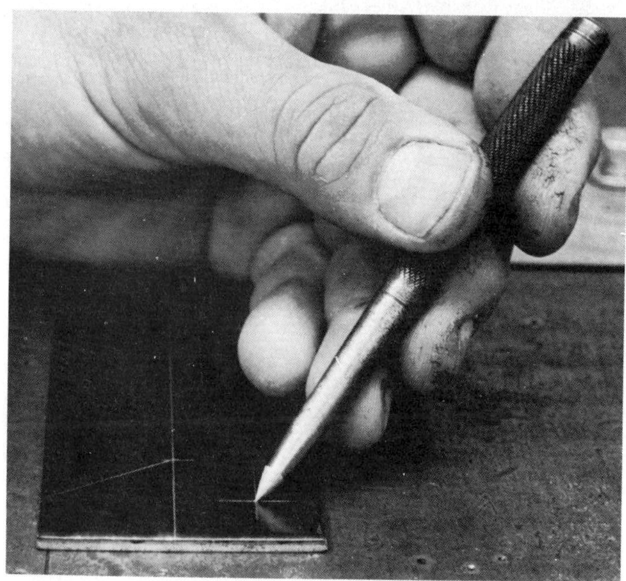

FIGURE E-45 Setting the layout punch on the center point of the corner radius (Lane Community College).

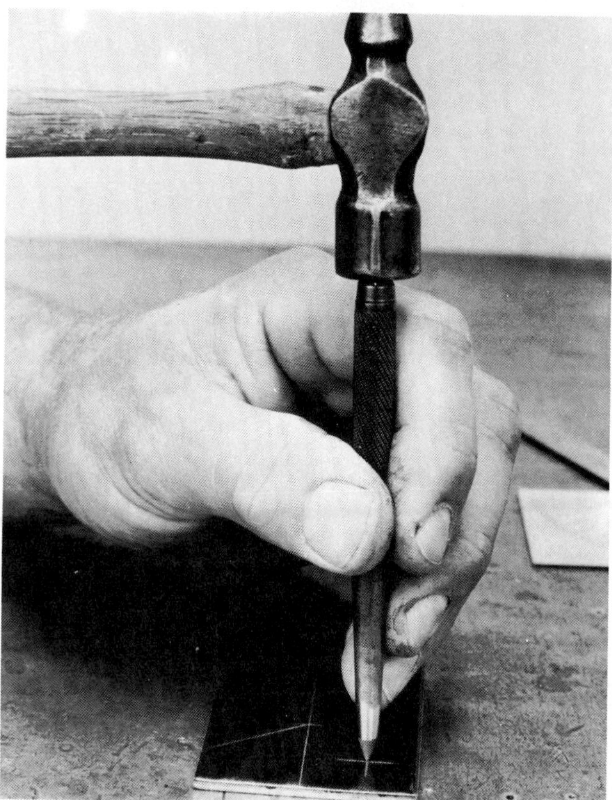

FIGURE E-46 Punching the center point of the corner radius (Lane Community College).

FIGURE E-47 Setting the dividers to the rule engravings (Lane Community College).

$\frac{1}{2}$ in. Establish this dimension using the square and rule. Two measurements will be required. Measure from the side and from the end to establish the center of the circle (Figures E-43 and E-44). Prick punch the intersection of the two lines with the 30-degree included point angle layout punch. Tilt the punch so that it can be positioned exactly on the scribe marks (Figure E-45). A magnifier will be useful here. Move the punch to its upright position and tap it lightly with the layout hammer (Figures E-44 and E-46).

Set the dividers to a dimension of $\frac{1}{2}$ in. using the rule. For maximum reliability, use the 1-in. graduation for a starting point. Adjust the divider spacing until you feel the tips drop into the rule engravings (Figure E-47). Place one divider tip into

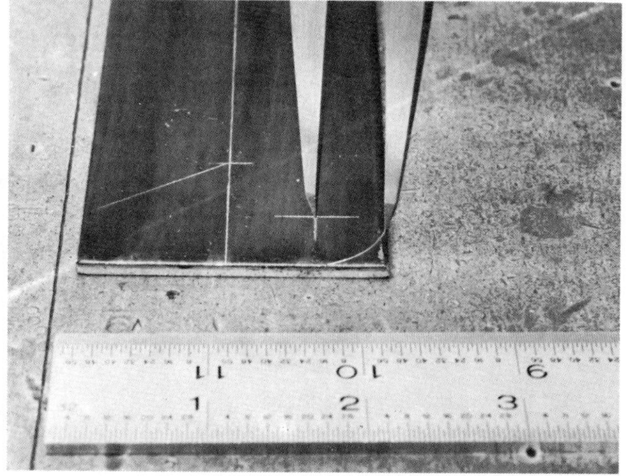
FIGURE E-48 Scribing the corner radius (Lane Community College).

FIGURE E-50 Center punching the hole centers (Lane Community College).

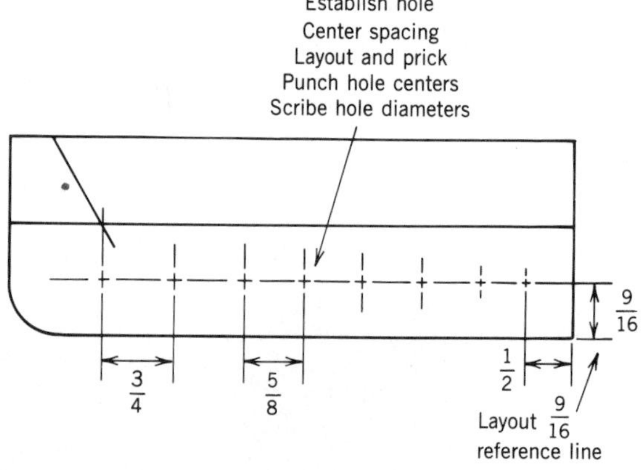

FIGURE E-49

FIGURE E-51 Completed layout for the drill and hole gage (Lane Community College).

the layout punch mark and scribe the corner radius (Figures E-44 and E-48).

The centerline of the holes is $\frac{9}{16}$ in. from the edge. Use the square and rule to measure this distance. Mark at each end and scribe the line full length (Figure E-49). Measure and lay out the center of each hole (Figure E-49). Use the layout punch and mark each hole center. After prick punching each hole center, set the dividers to each indicated radius and scribe all hole diameters (Figure E-49).

The last step is to center punch each hole center to deepen prick punch marks prior to drilling. Use a 90-degree included point angle center punch (Figure E-50). Layout of the drill and hole gage is now complete (Figure E-51).

SELF-TEST

1. How should the workpiece be prepared prior to layout?

2. What is the reason for placing the workpiece on a paper towel?

3. Describe the technique of using the layout punch.

4. Describe the use of the combination square and rule in layout.

5. Describe the technique of setting a divider to size using a rule.

UNIT 2

Basic Precision Layout Practice

Precision layout is generally more reliable and accurate than layout by semiprecision practice. On any job requiring maximum accuracy and reliability, precision layout practice should be used.

Objectives

After completing this unit, you should be able to:

1. Identify the major parts of the vernier height gage.
2. Describe applications of the vernier height gage in layout.
3. Read a vernier height gage in both metric and inch dimensions.
4. Accomplish layout using the vernier height gage.

The Vernier Height Gage

The fundamental precision layout tool is the height gage. The vernier height gage is the most common type found in the machine shop. This instrument will discriminate to .001 in. With this ability, a much higher degree of accuracy and reliability is added to a layout task. Whenever possible, you should apply the height gage in all precision layout requirements. Major parts of the height gage include the **base, beam, vernier slide,** and **scriber** (Figure E-52). The size of height gages is measured by the maximum height gaging ability of the instrument. Height gages range from 10 to 72 in.

Height gage scribers are made from tool steel or tungsten carbide. **Carbide scribers are subject to chipping and must be treated gently.** They do, however, retain their sharpness and scribe very clean narrow lines. Height gage scribers may be sharpened if they become dull. It is important that any **sharpening be done on the slanted surface so that the scriber dimensions will not be changed.**

The height gage scriber is attached to the vernier slide and can be moved up and down the beam. Scribers are either straight (Figure E-53) or offset (Figure E-54). The offset scriber permits direct readings with the height gage. The gage reads zero when the scriber rests down on the reference surface. With the straight scriber, the workpiece will have to be raised accordingly, if direct readings are to be obtained. This type of height gage scriber is less convenient.

Reading the Vernier Height Gage

On an inch height gage, the beam is graduated in inches with each inch divided into 10 parts. The tenth inch graduations are further divided into two or four parts depending on the divisions of the vernier. On the 25-division vernier, used on many older height gages, the $\frac{1}{10}$ in. divisions on the beam are graduated into four parts. The vernier permits discrimination to .001 in. Many newer height gages are making use of the 50-division vernier, which permits easier reading. On a height gage with a 50-division vernier, the $\frac{1}{10}$ in. graduations on the beam will be divided into two parts. Discrimination of this height gage is also .001 in. The metric vernier height gage has the beam graduated in millimeters. The vernier contains 50 divisions

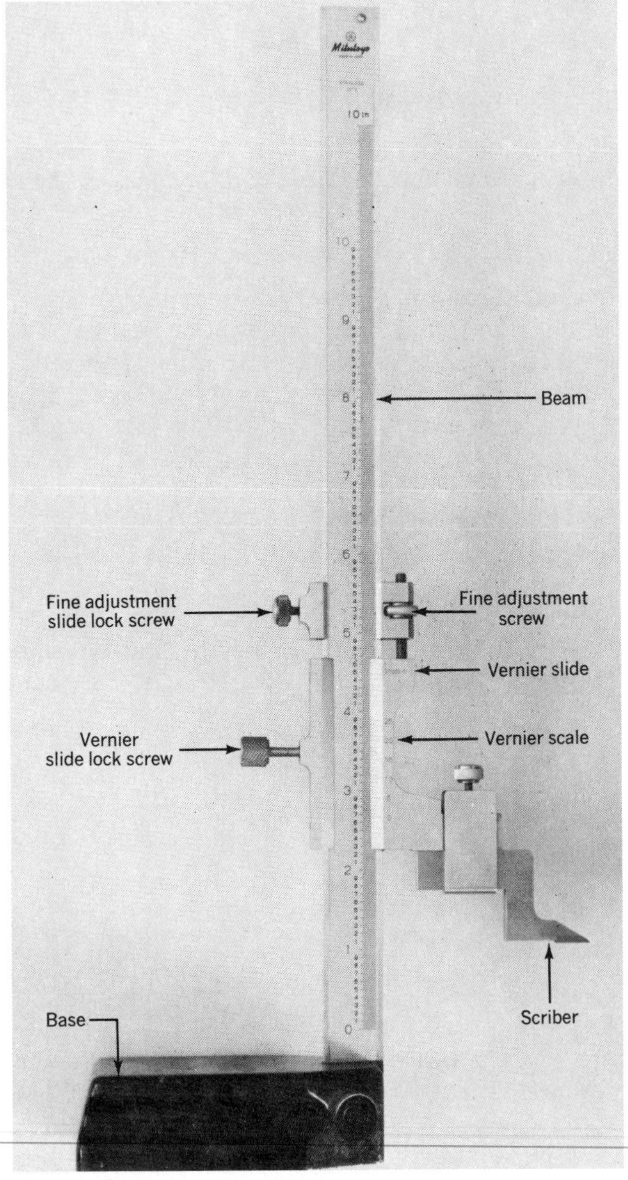

FIGURE E-52 Parts of the vernier height gage.

permitting the instrument to discriminate to $\frac{1}{50}$ mm.

The vernier height gage is read like any other instrument employing the principle of the vernier. The line on the vernier scale that is coincident with a beam scale graduation must be determined. This value is added to the beam scale reading to make up the total reading. The inch vernier height gage with a 50-division vernier is read as follows (Figure E-55, right-hand scale):

Beam reading	5.300
Vernier is coincident at 12 or .012 in.	.012
Total reading	5.312 in.

On the 50-division inch vernier height gage, the beam scale is graduated in $\frac{1}{10}$-in. graduations. Each $\frac{1}{10}$-in. increment is further divided into two parts. If the zero on the vernier is past the .050-in. mark on the beam, .050 in. must be added to the reading.

The metric vernier scale also has 50 divisions, each equal to $\frac{1}{50}$ or .02 mm (Figure E-55, left-hand scale):

Beam scale reading	134 mm
Vernier coincident at line 49 49 × .02 =	.98 mm
Total reading	134.98 mm

On the 25-division inch vernier height gage the $\frac{1}{10}$ in. beam graduations are divided into four parts, each equal to .025 in. Depending on the location of the vernier zero mark, .025, .050, or .075 in. may have to be added to the beam reading. The inch vernier height gage with a 25-division vernier is read as follows (Figure E-56):

Beam	5.0
Vernier coincident at 17 or .017	.017
Total reading	5.017 in.

FIGURE E-53 Straight vernier height gage scriber.

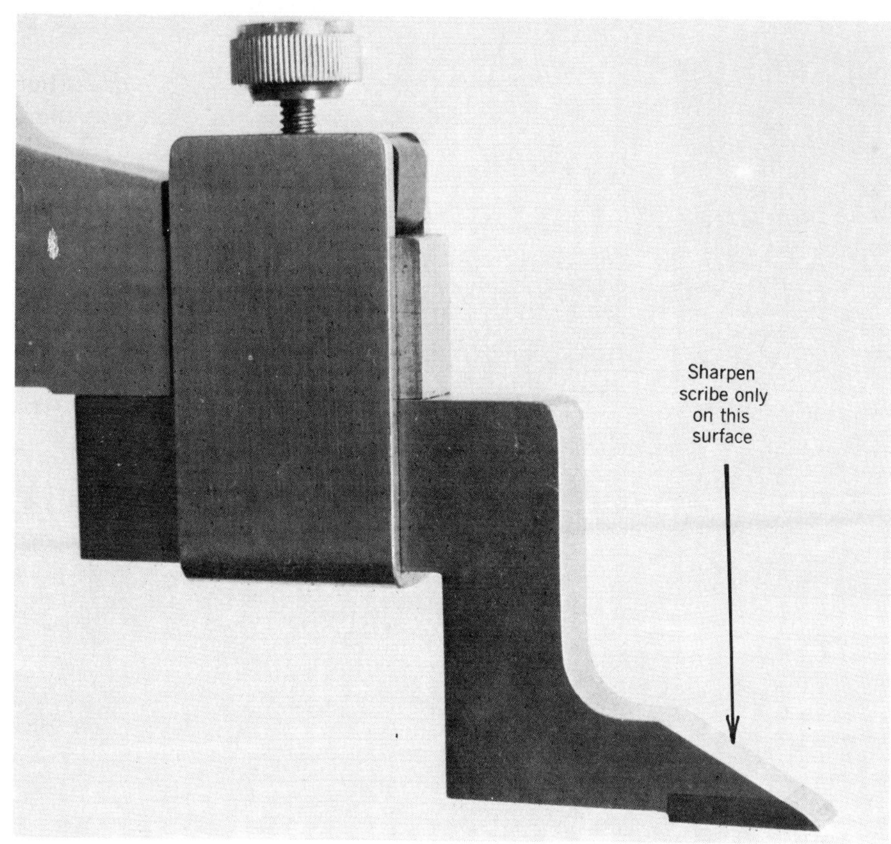

FIGURE E-54 Offset vernier height gage scriber.

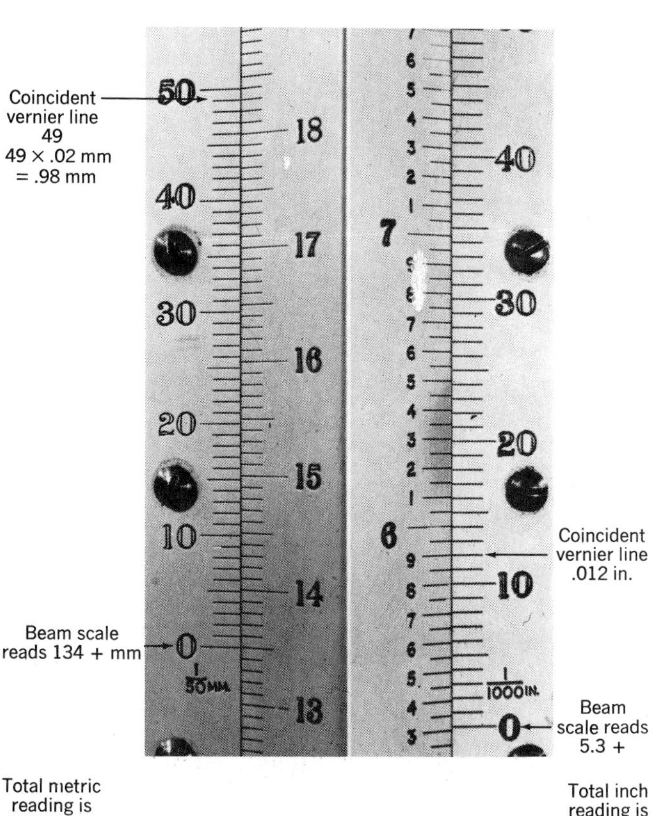

FIGURE E-55 Reading the 50-division inch/metric vernier height gage.

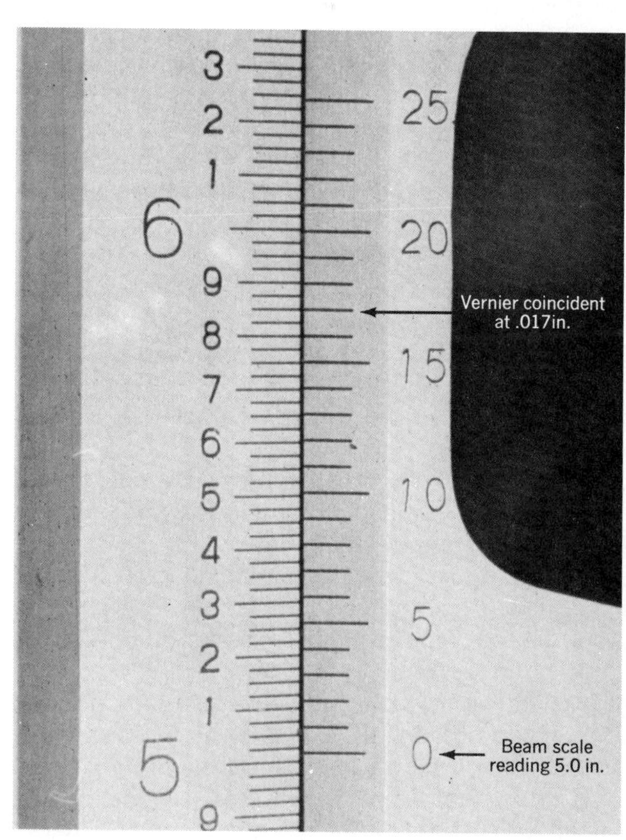

FIGURE E-56 Reading the 25-division inch vernier height gage.

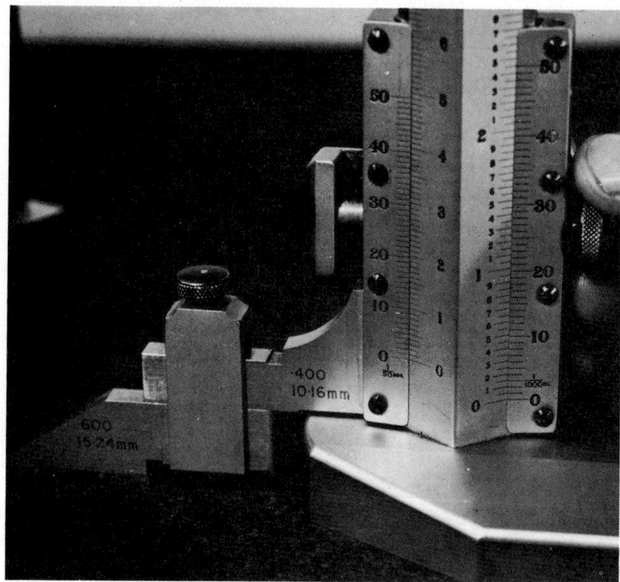

FIGURE E-57 Checking the zero reference.

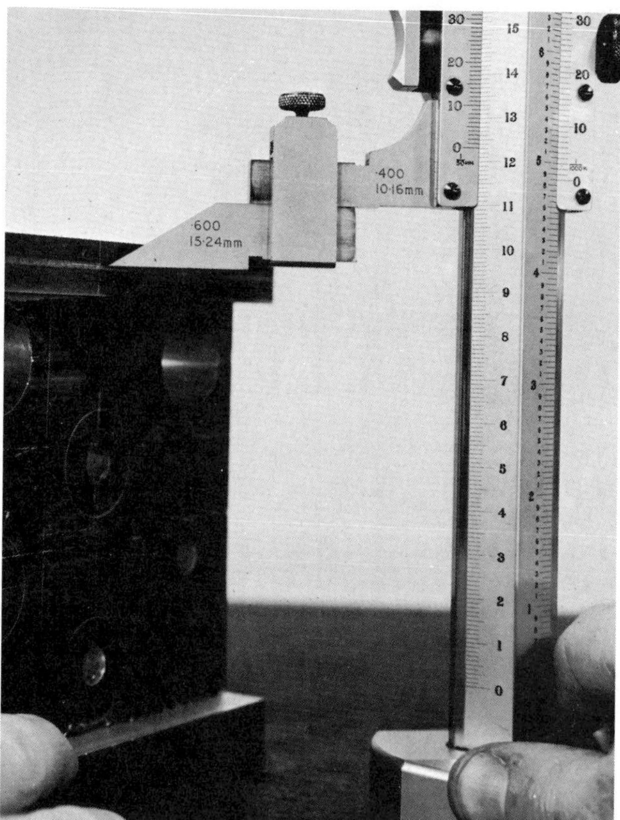

FIGURE E-58 Scribing height lines with the vernier height gage.

Checking the Zero Reference on the Vernier Height Gage

The height gage scriber must be checked against the reference surface before attempting to make any height measurements of layouts. Clean the surface of the layout table and the base of the gage. Slide the scriber down until it just rests on the reference surface. Check the alignment of the zero mark on the vernier scale with the zero mark on the beam scale. The two marks should coincide exactly (Figure E-57). Hold the height gage base firmly against the reference surface. Be sure that you do not tilt the base of the height gage by sliding the vernier slide past the zero point on the beam scale. If the zero marks on the vernier and beam do not coincide after the scriber has contacted the reference surface, an adjustment of the vernier scale is required.

Some height gages do not have an adjustable vernier scale. A misalignment in the vernier and the beam zero marks may indicate a loose vernier slide, an incorrect scriber dimension, or a beam that is out of perpendicular with the base. Loose vernier slides may be adjusted and scriber dimensions can be corrected. However, if the beam is out of perpendicular with the base, the instrument is unrealiable because of cosine error. A determination of such a condition can be made by an appropriate calibration process. All height gages, particularly those with nonadjustable verniers, must be treated with the same respect as any precision instrument that you will use.

Applications of the Vernier Height Gage in Layout

The primary function of the vernier height gage in layout is to measure and scribe lines of known height on the workpiece (Figure E-58). Perpendicular lines may be scribed on the workpiece by the following procedure. The work is first clamped to a right angle plate if necessary and the required lines are scribed in one direction. The height gage should be set at an angle to the work and the corner of the scriber pulled across while keeping the height gage base firmly on the reference surface (Figure E-59). Only enough pressure should be applied with the scriber to remove the layout dye and not actually remove material from the workpiece.

After scribing the required lines in one direction, turn the workpiece by 90 degrees. Setup is critical if the scribe marks are to be truly perpendicular. A square (Figure E-60) or a dial test indicator may be used (Figure E-61) to establish the work at right angles. In both cases the edges of the workpiece must be machined smooth and square. After the clamp has been tightened, the perpendicular lines may be scribed at the required height (Figure E-62).

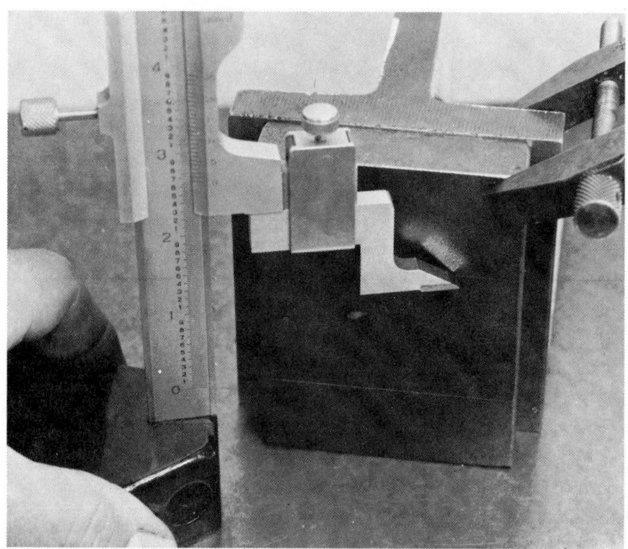

FIGURE E-59 Scribing layout lines with the workpiece clamped to a right angle plate.

FIGURE E-60 After turning the workpiece 90 degrees, it can be checked with a square.

FIGURE E-61 Checking the work using a dial test indicator.

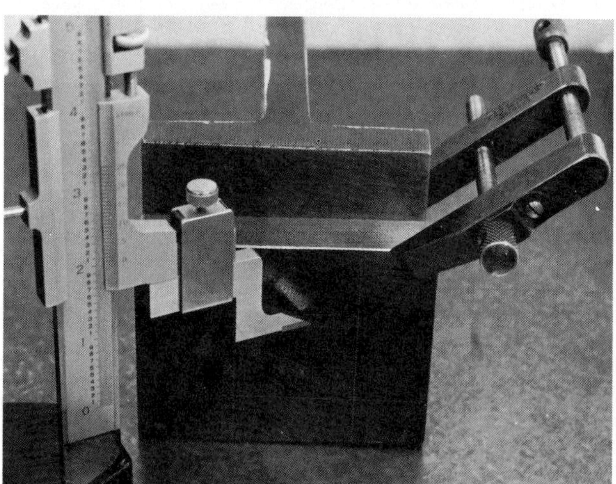

FIGURE E-62 Scribing perpendicular lines.

FIGURE E-63 Scribing centerlines on round stock clamped in a vee block.

The height gage may be used to lay out centerlines on round stock (Figure E-63). The stock is clamped in a vee block and the correct dimension to center is determined. This can be done with the dial test indicator attached to the height gage. However, it must not be done with the height gage scriber.

Parallel bars (Figure E-64) are a valuable and useful layout accessory. These bars are made from hardened steel or granite, and they have extremely accurate dimensional accuracy. Parallel bars are available in many sizes and lengths. In layout with the height gage they can be used to support the workpiece (Figure E-65). Angles may be laid out by placing the workpiece on the sine bar (Figure E-66).

Basic Precision Layout Practice

The workpiece should be prepared as in semiprecision layout. Sharp edges must be removed and a thin coat of layout dye applied. You will need a drawing of the part to be laid out (Figure E-67). The order of steps will depend on the layout task.

FIGURE E-64 Hardened-steel parallel bars.

FIGURE E-65 Using parallel bars in layout.

POSITION ONE LAYOUTS

In position one (Figure E-68) the clamp frame is on edge. In any position, all layouts can be defined as heights above the reference surface. Refer to the drawing on position one layouts and determine all of the layout that can be accomplished there.

Start by scribing the $\frac{3}{4}$-in. height that defines the width of the clamp frame. Set the height gage to .750 in. (Figure E-69). Attach the scriber (Figure E-70). Be sure that the scriber is sharp and properly installed for the height gage that you are using. Hold the workpiece and height gage firmly and pull the scriber across the work in a smooth motion (Figure E-71). The height of the clamp screw hole can be laid out at this time. Refer to the part drawing and determine the height of the hole. Set the height gage at 1.625 in. and scribe the line on the end of the workpiece (Figure E-72). The line may be projected around on the side of the part.

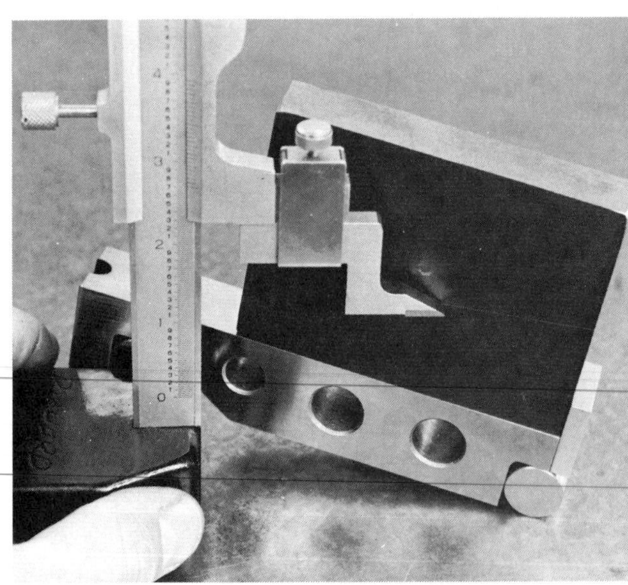

FIGURE E-66 Laying out angle lines using the sine bar.

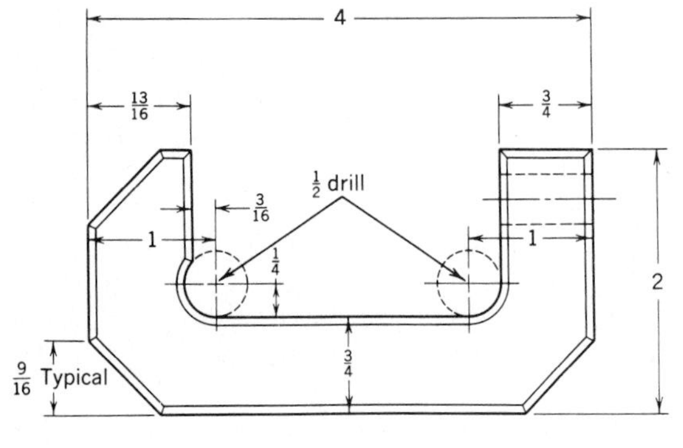

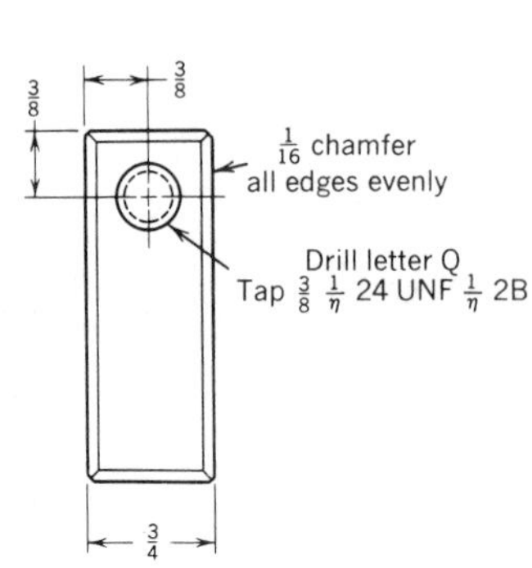

FIGURE E-67 Clamp frame (Lane Community College).

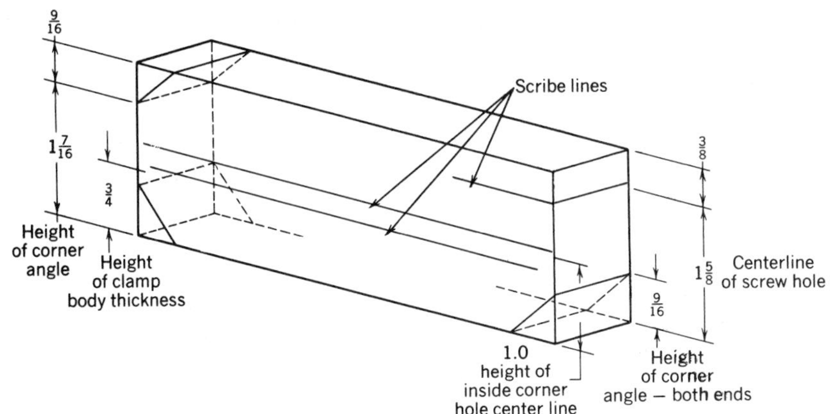

FIGURE E-68 Clamp frame—position one layouts.

This will facilitate setup in the drill press. Other layouts that can be accomplished at position one include the height equivalent of the inside corner hole centerlines.

The starting points of the corner angles on both ends may also be laid out. Refer to the drawing on position one layouts.

FIGURE E-70 Attaching the scriber (Lane Community College).

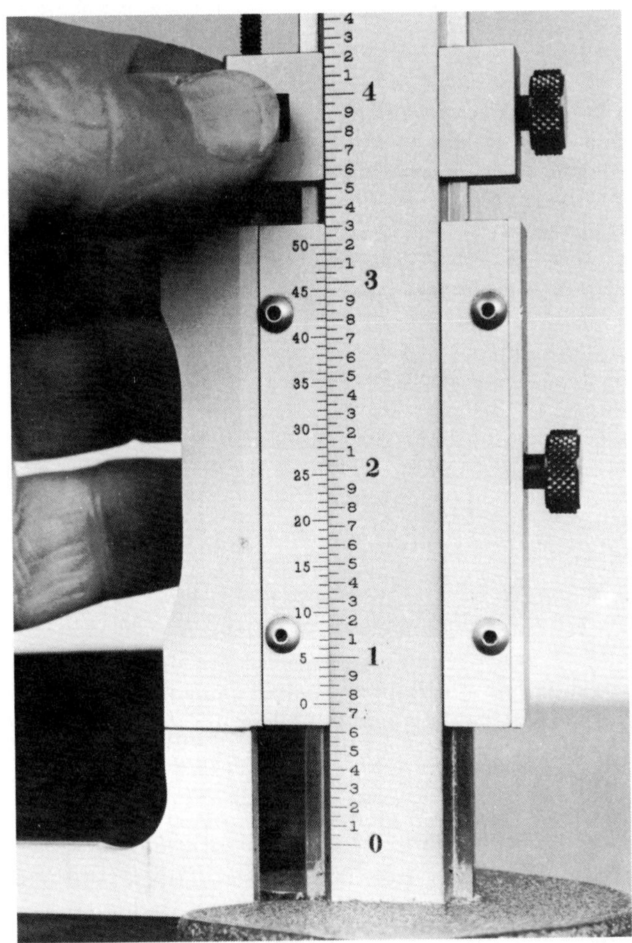

FIGURE E-69 Setting the height gage to a dimension of .750 in. (Lane Community College).

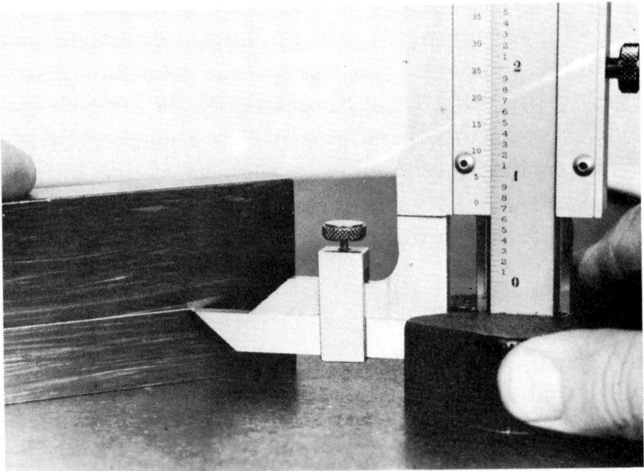

FIGURE E-71 Scribing the height equivalent of the frame thickness (Lane Community College).

FIGURE E-72 Scribing the height equivalent of the clamp screw hole (Lane Community College).

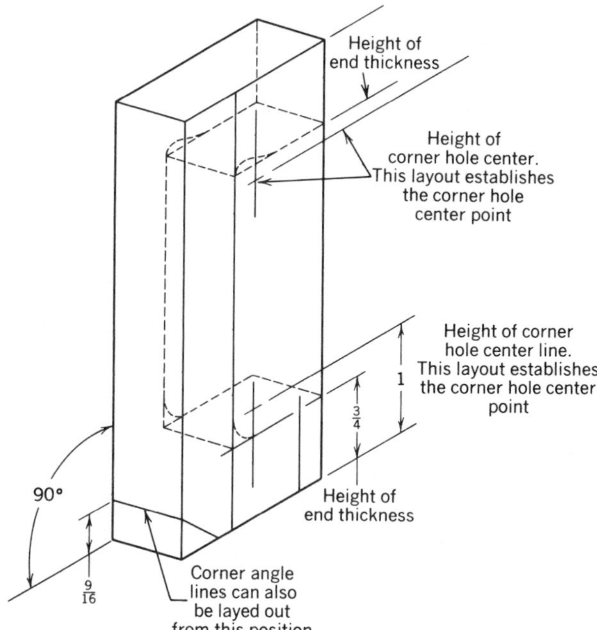

FIGURE E-74 Clamp frame—position three layouts.

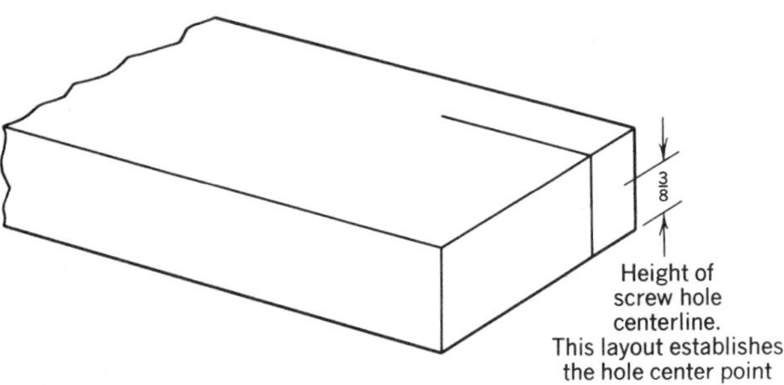

FIGURE E-73 Clamp frame—position two layouts.

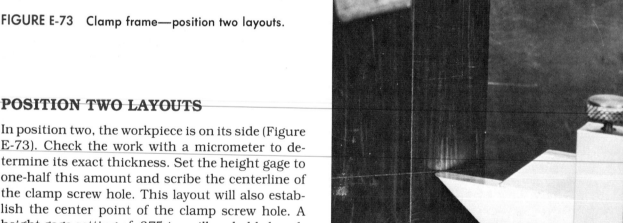

FIGURE E-75 Scribing the height equivalent of the end thickness (Lane Community College).

POSITION TWO LAYOUTS

In position two, the workpiece is on its side (Figure E-73). Check the work with a micrometer to determine its exact thickness. Set the height gage to one-half this amount and scribe the centerline of the clamp screw hole. This layout will also establish the center point of the clamp screw hole. A height gage setting of .375 in. will probably be adequate provided that the stock is .750 in. thick. However, if the thickness varies above or below .750 in. the height gage can be set to one-half of whatever the thickness is. This will insure that the hole is in the center of the workpiece.

POSITION THREE LAYOUTS

In position three, the workpiece is on end clamped to an angle plate (Figure E-74). The work must be established perpendicular using a square or dial test indicator. Set the height gage to .750 in. and scribe the height equivalent of the frame end thickness (Figure E-75). Other layouts that can be done at position three include the height equivalent of the inside corner hole centerlines. This layout will also locate the center points of the inside corner holes (Figure E-76). The height equivalent of the end thickness as well as the ending points of the corner angles can be scribed at position three.

FIGURE E-76 Completed layout of the clamp frame (Lane Community College).

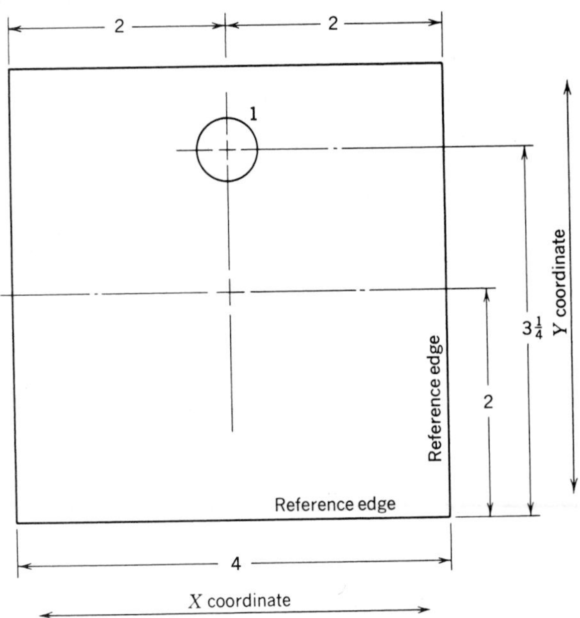

FIGURE E-78 Coordinate position of hole one.

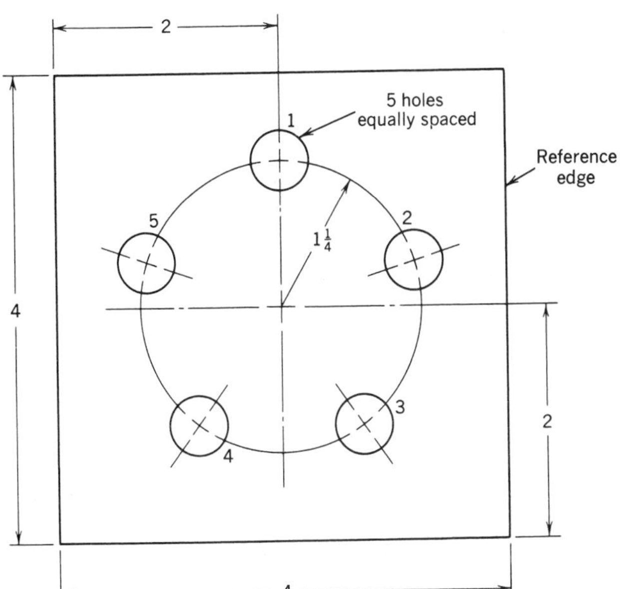

FIGURE E-77 Equally spaced five-hole circle.

Height Gage Layout by Coordinate Measure

Many layouts can be accomplished by calculating the coordinate position of the part features. Coordinate position simply means that each feature is located a certain distance from adjacent perpendicular reference lines. These are frequently known as the *X* and *Y* coordinates. You should begin to think of coordinates in terms of *X* and *Y*, as this terminology will be important, especially in the area of numerical control machining. The *X* coordinate on a two-dimensional drawing is horizontal. The *Y* coordinate is perpendicular to *X* and in the same plane. On a drawing, *Y* is the vertical coordinate. The *X* and *Y* coordinate lines can be and often are the edges of the workpiece, provided the edges have been machined true and square to each other.

Coordinate lengths can be calculated by the application of appropriate trigonometric formulas. They may also be determined from tables of coordinate measure. Such tables appear in most handbooks for machinists.

CALCULATING COORDINATE MEASUREMENTS

The drawing (Figure E-77) shows a five-hole equally spaced pattern centered on the workpiece. Since hole one is on the centerline, its coordinate position measured from the reference edges can be easily determined (Figure E-78). The *X* coordinate (horizontal) is 2 in. The *Y* coordinate (vertical is 2 in. plus the radius of the hole circle). This would be $3\frac{1}{4}$ in.

The coordinate position of hole two can be calculated by the following: Since there are five equally spaced holes, the central angle is $\frac{360}{5}$ or 72 degrees. Right triangle *ABC* (Figure E-79) is formed by constructing a perpendicular line from point B to point C. Angle A equals 18 degrees ($90 - 72 = 18$). To find the X coordinate, apply the following formula:

$$X_C = \text{circle radius} \times \cos 18^\circ$$
$$= 1.250 \times .951$$
$$= 1.188 \text{ in.}$$

The X-coordinate length from the reference edge is found by

$$2.0 - 1.188 = .812 \qquad \text{(Figure E-77)}$$

The Y coordinate is found by the following formula:

$$Y_C = \text{circle radius} \times \sin 18^\circ$$
$$= 1.250 \times .309$$
$$= .386$$

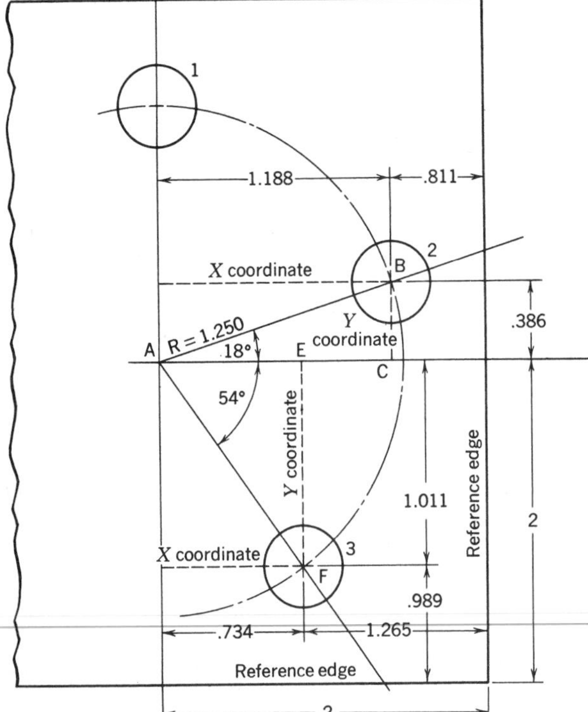

FIGURE E-79 Coordinate positions of holes two and three.

The Y-coordinate length from the reference edge is found by

$$2.0 + .386 = 2.386 \text{ in.}$$

The coordinate position of hole three is calculated in a similar manner. Right triangle AEF is formed by constructing a perpendicular line from point F to point E (Figure E-79). Angle FAE equals 54 degrees ($72 - 18 = 54$). To find the X coordinate, apply the following formula:

$$X_C = \text{circle radius} \times \cos 54^\circ$$
$$= 1.250 \times .587$$
$$= .734$$

The X-coordinate length from the reference edge is found by

$$2.0 - .734 = 1.265 \qquad \text{(Figure E-79)}$$

To find the Y coordinate, apply the following formula:

$$Y_C = \text{radius} \times \sin 54^\circ$$
$$= 1.250 \times .809$$
$$= 1.011 \text{ in.}$$

The Y-coordinate length from the reference edge is found by

$$2.0 - 1.011 = .989 \text{ in.} \qquad \text{(Figure E-79)}$$

The coordinate positions of holes four and five are the same distance from the centerlines as holes two and three. Their positions from the reference edges can be calculated easily.

Since this layout involves scribing perpendicular lines, the workpiece must be turned 90 degrees. If the edges of the work are used as reference, they must be machined square. Either coordinate may be laid out first. The workpiece is then turned 90 degrees to the adjacent reference edge. This permits the layout of the perpendicular lines (Figure E-80).

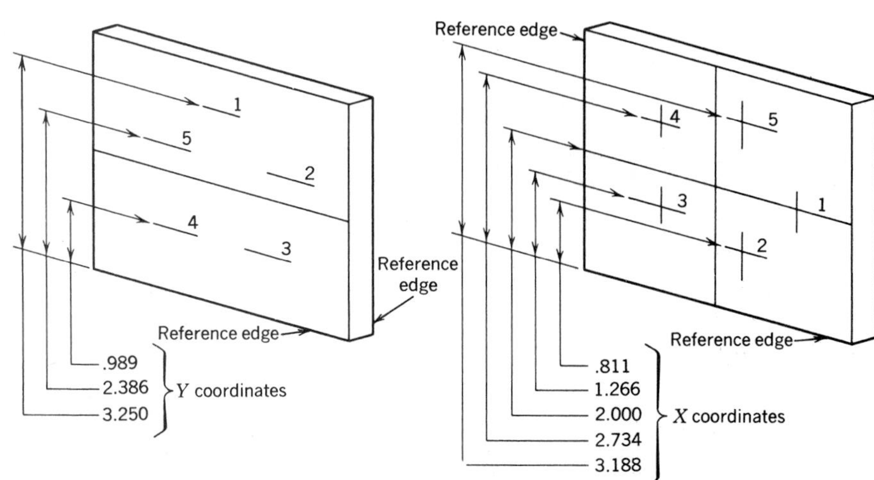

FIGURE E-80 Height equivalents of coordinate positions for all holes.

Laying Out Angles

Angles may be laid out using the height gage by calculating the appropriate dimensions using trigonometry. In the example (Figure E-81), the layout of height A will establish angle B at 36 degrees. Height A is calculated by the following formula:

$$\begin{aligned} \text{height } A &= 1.25 \times \tan B \\ &= 1.25 \times .726 \\ &= .908 \text{ in.} \end{aligned}$$

After scribing a height of .908 in., the workpiece is turned 90 degrees and the starting point of the angle established at point B. Scribing from point A to point B will establish the desired angle.

The sine bar can also be used in angular layout. In the example (Figure E-82), the sine bar is elevated for the 25 degree angle. Sine bar elevation is calculated by the formula

$$\text{bar elevation} = \text{bar length} \times \text{sine of required angle}$$

If we assume a 5-in. sine bar,

$$\begin{aligned} \text{elevation} &= 5 \times \sin 25^\circ \\ &= 5 \times .422 \\ &= 2.113 \text{ in.} \end{aligned}$$

A gage block stack is assembled and placed under the sine bar. Now that the bar has been elevated, the vertical distance CD from the corner to the scribe line AB must be determined. To find distance DC, a perpendicular line must be constructed from point C to point D. Angle A is 65 degrees (90 − 25 = 65). Length CD is found by the following formula:

$$\begin{aligned} CD &= .500 \times \sin B \\ &= .500 \times .906 \\ &= .453 \text{ in.} \end{aligned}$$

The height of the corner must be determined and the length of CD subtracted from this dimension. This will result in the corect height gage setting for scribing line AB. The corner height should be determined using the height gage and dial test indicator. The corner height must not be determined using the height gage scriber.

SELF-TEST

1. Read and record the 50-division inch/metric height gage readings in Figures E-83*a* to E-83*c*.

2. Read and record the 25-division inch height gage readings in Figures E-84*a* and E-84*b*.

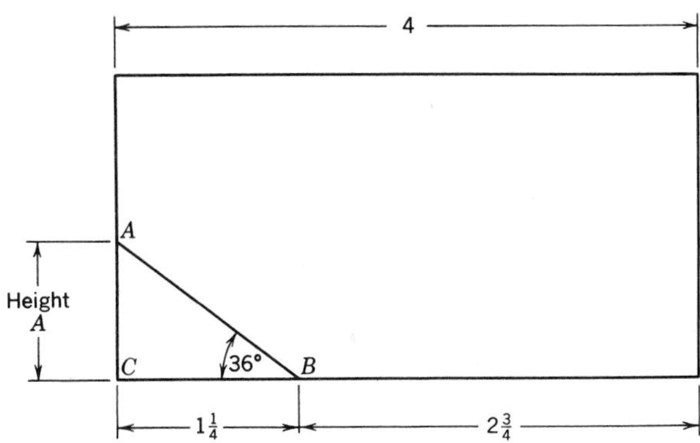

FIGURE E-81 Laying out a 36-degree angle.

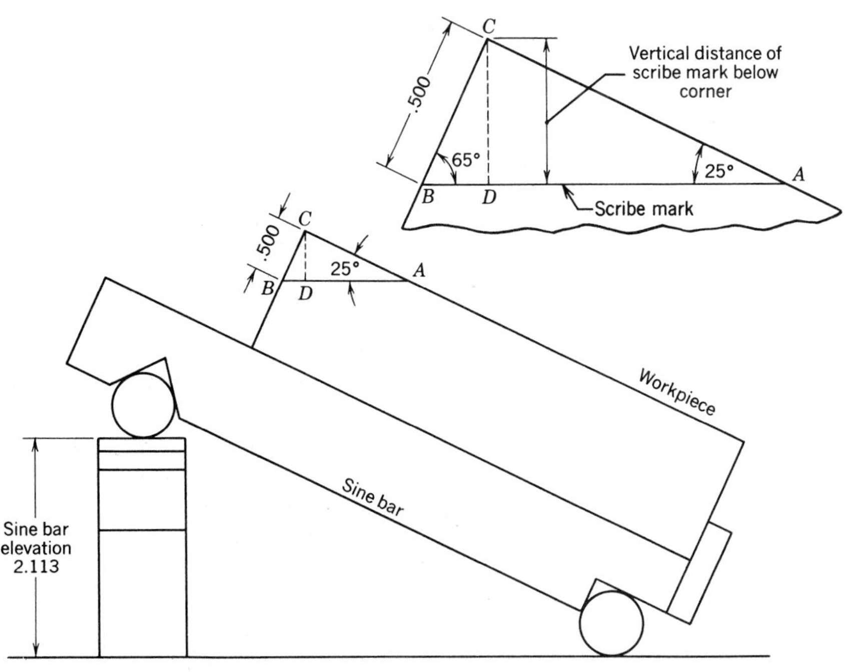

FIGURE E-82 Laying out an angle using the sine bar.

3. Describe the procedure for checking the zero reference.

4. How can the zero reference be adjusted?

5. How are perpendicular lines scribed with a height gage?

6. What is the measuring range of a typical height gage?

7. When laying out angles, what tool is used in conjunction with the height gage?

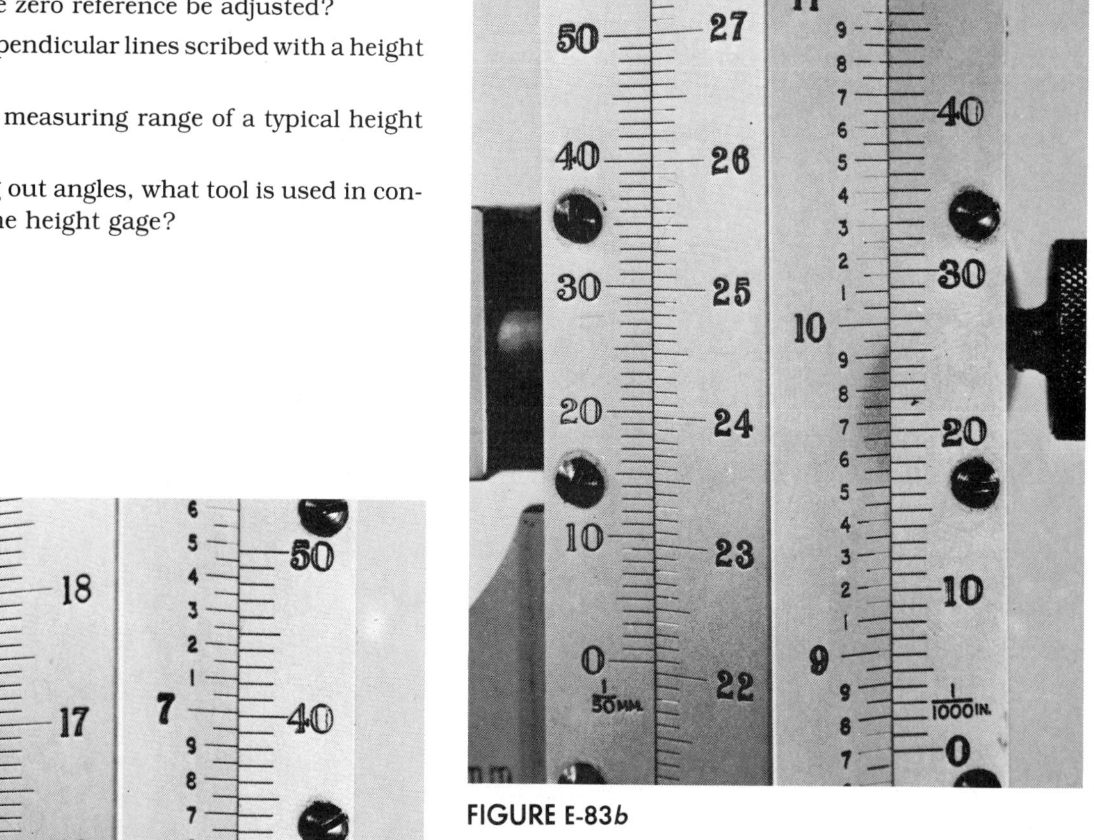

FIGURE E-83*b*

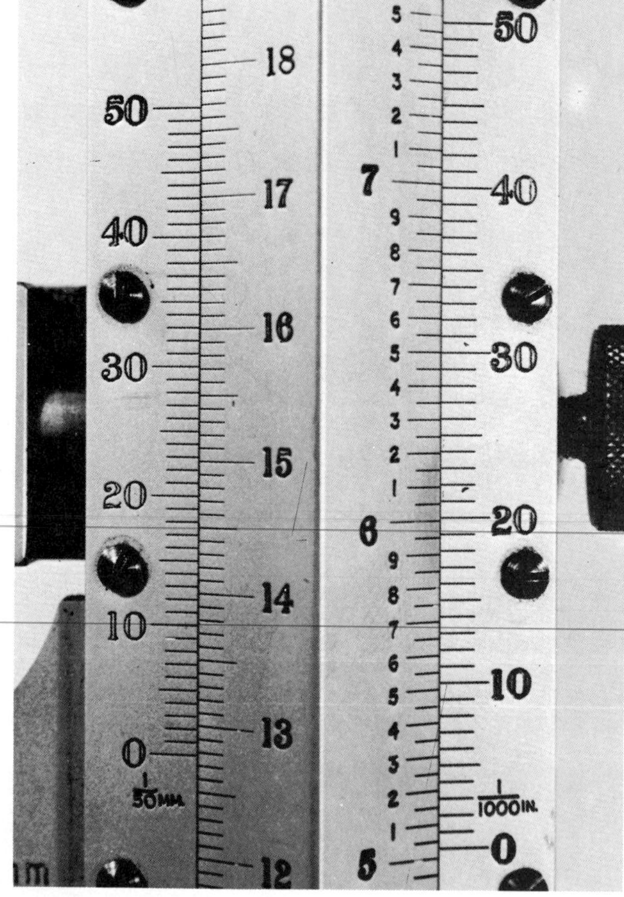

FIGURE E-83*a*

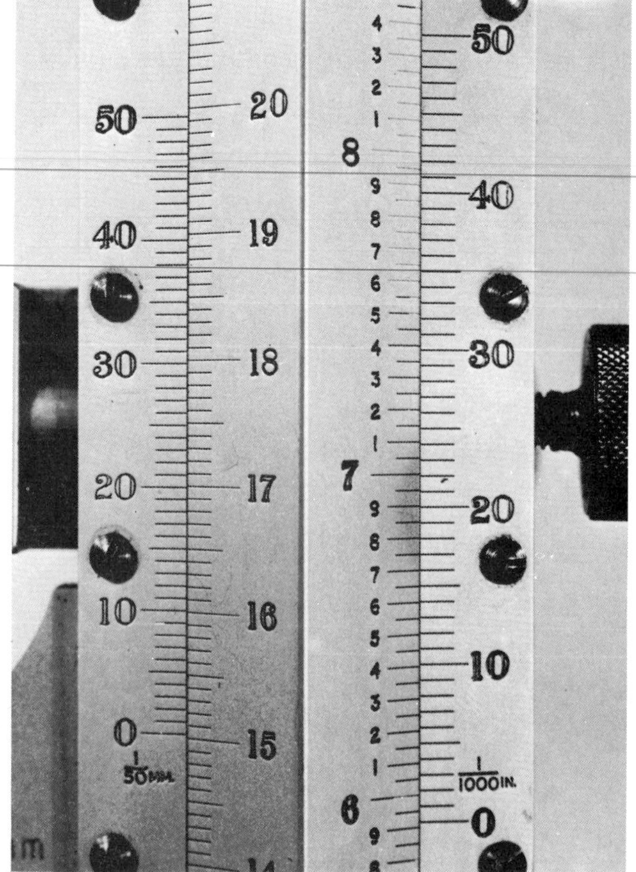

FIGURE E-83*c*

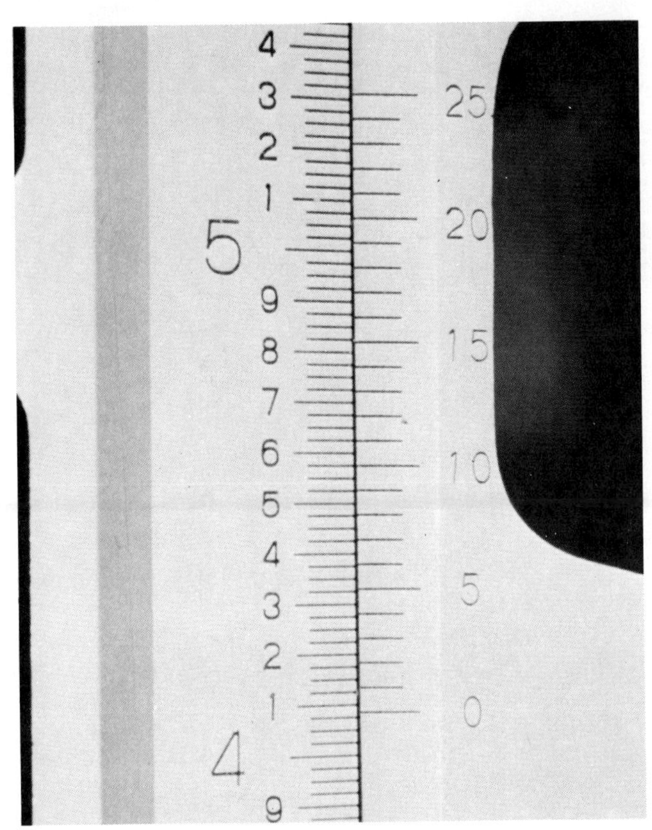

FIGURE E-84*a*

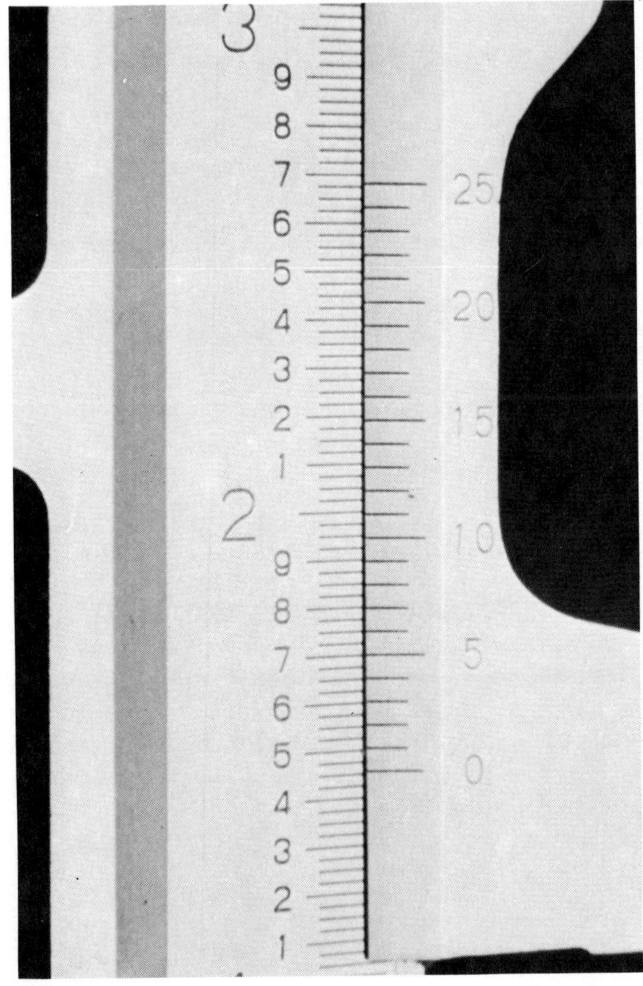

FIGURE E-84*b*

SECTION F

PREPARATION FOR MACHINING OPERATIONS

Cutting tool materials tend to get hot during the process of machining, that is, in cutting metals and producing chips. The heat rise on the cutting edge of a tool can be sufficient to burn the sharp edge and dull it (Figure F-1). The tool must then be resharpened or a new sharp edge must be indexed in place. A certain amount of wear and tool breakdown is to be expected, but lost machining time while replacing or sharpening the tool must be considered. Generally, a trade-off between maximum production and tool breakdown or loss is accepted in machining operations.

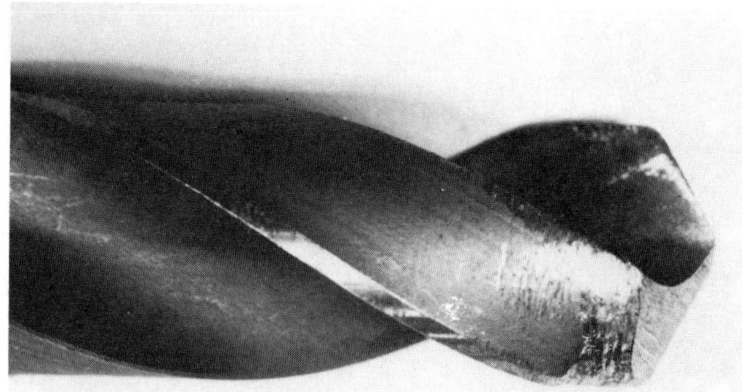

FIGURE F-1 The end of this twist drill has been burned due to excessive speed.

There are several factors that cause or reduce tool heating. Some tool materials can withstand higher temperatures than others. The hardness and toughness of workpiece materials are also factors in tool breakdown caused by heating. Tool breakage can be caused by heavy cuts (excessive feed) and interrupted cuts or by accidentally reversing the machine movement or rotation while a cut is being made.

The speed of the cutting tool point or edge moving along the workpiece or of the workpiece moving past the tool is a major factor in controlling tool breakdown and must always be considered when cutting metal with a machine tool. This is true of all machining operations—sawing, turning, drilling, milling, and grinding. In all of these, cutting speed is of utmost importance.

The feed on a machine is the movement that forces the tool into or along the workpiece. Increments of feed are usually measured in thousandths of an inch. Excessive feed usually results in broken tools (Figure F-2). Often this breakage ruins the tool, but in some cases the tool can be reground.

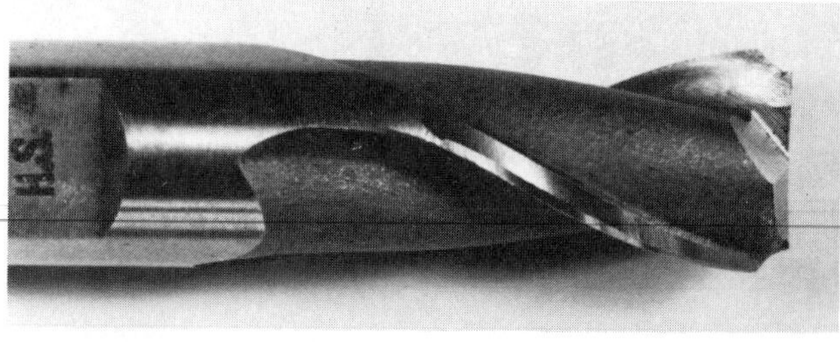

FIGURE F-2 The sharp corners of this end mill have been broken off due to excessive feed.

Cutting fluids are used in machining operations for two basic purposes: to cool the cutting tool and workpiece and to provide lubrication for easy chip flow across the toolface.

Some cutting fluids are basically coolants. These are usually water-based soluble oils or synthetics and their main function is to cool the work–tool interface. Cutting oils provide some cooling action but also act as lubricants and are mainly used where cutting forces are high as in die threading or tapping operations.

Tool materials range from hardened carbon steel to diamond, each type having its proper use, whether in traditional metal cutting in machine tools or for grinding operations. This section will enable you to use correctly all of these cutting tools, to avoid undue damage to tools and workpieces, and to use them to the greatest advantage.

Safety in Chip Handling

CAUTION **Certain metals, when divided finely as a powder or even as coarse as machining chips, can ignite with a spark or just by the heat of machining. Magnesium and zirconium are two such metals. Such fires are difficult to extinguish, and if water or a water or water-based fire extinguisher is used, the fire will only increase in intensity. The greatest danger of fire occurs when a machine operator fails to clean up zirconium or magnesium chips on a machine when the job is finished. The next operator may then cut alloy steel, which can produce high temperatures in the chip or even sparks that can ignite the magnesium chips. Such fires may destroy the entire machine if not the shop. Chloride-based powder fire extinguishers are commercially available. These are effective for such fires since they prevent water absorption and form an air-excluding crust over the burning metal. Sand is also used to smother fires in magnesium.**

Metal chips from machining operations are very sharp and are a serious hazard. They should not be handled with bare hands. Gloves may be worn only when the machine is not running.

UNIT 1

Machinability and Chip Formation

Machinability is the relative difficulty of a machining operation with regard to tool life, surface finish, and power consumption. In general, softer materials are easier to machine than harder ones. Chip removal and control are also major factors in the machining industry. These and other considerations concerning the use of machine tools are covered in this unit.

Objectives

After completing this unit, you should be able to:

1. Determine how metal cutting affects the surface structures of metals.
2. Analyze chip formation, structures, and chip breakers.
3. Explain machinability ratings and machining behavior of metals.

Principles of Metal Cutting

In machining operations, either the tool material rotates or moves in a linear motion or the workpiece rotates or moves (Table F-1). The moving or rotating tool must be made to move into the work material in order to cut a chip. This procedure is called *feed*. The amount of machine feed controls the thickness of the chip. The depth of cut is often called *infeed*. Besides the use of single point tools with one cutting edge, as they are called in lathe, shaper, and planer operations, there are multiple point tools such as milling machine cutters, drills, and reamers. In one sense, grinding wheels could be considered to be multiple-point cutters, since very small chips are removed by many tiny cutting points on grinding wheels.

TABLE F-1 Machining Principles and Operations

Operation	Diagram	Characteristics	Type of Machines
Turning	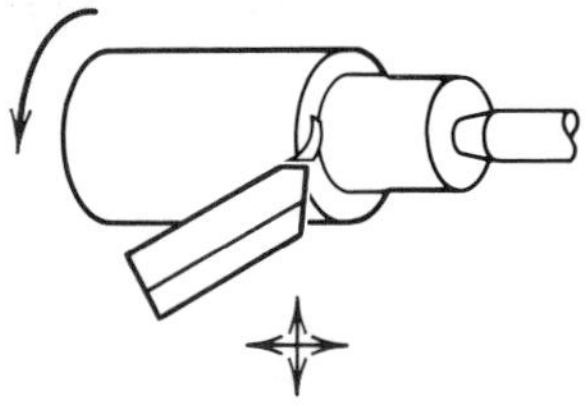	Work rotates, tool moves for feed	Lathe and vertical boring mill
Milling (horizontal)	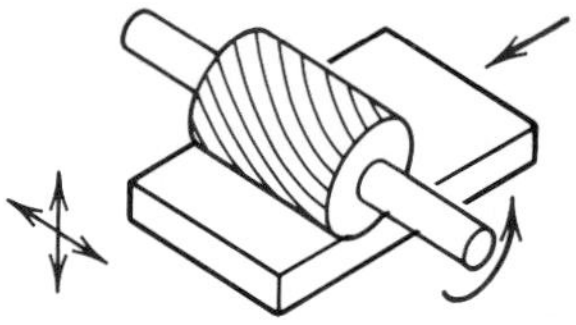	Cutter rotates and cuts on periphery; work feeds into cutter and can be moved in three axes	Horizontal milling machine
Face milling	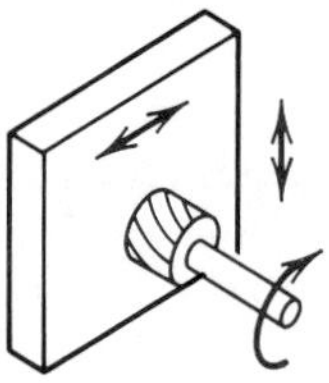	Cutter rotates to cut on its end and periphery of verical workpiece	Horizontal mill, profile mill, and machining center
Vertical (end) milling	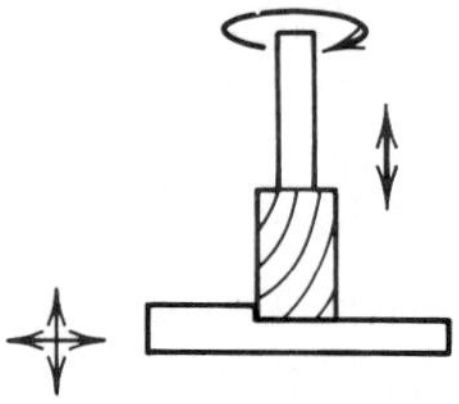	Cutter rotates to cut on its end and periphery, work moves on three axes for feed or position; spindle also moves up or down	Vertical milling machine, die sinker, machining center
Shaping	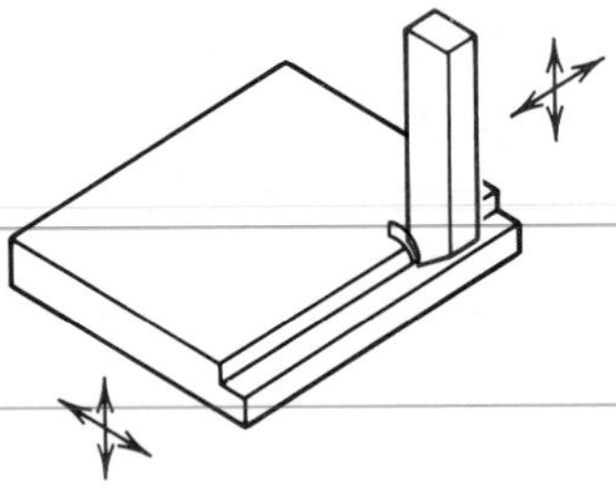	Work is held stationary and tool reciprocates; work can move in two axes; toolhead can be moved up or down	Horizontal and vertical shapers
Planing	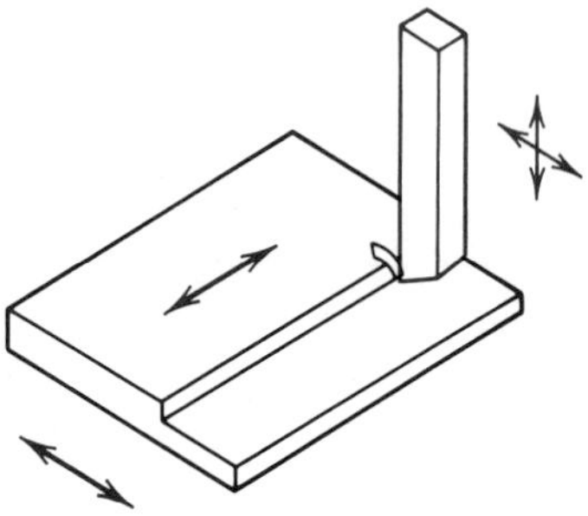	Work reciprocates while tool is stationary; tool can be moved up, down, or crosswise; worktable cannot be moved up or down	Planer
Horizontal sawing (cutoff)	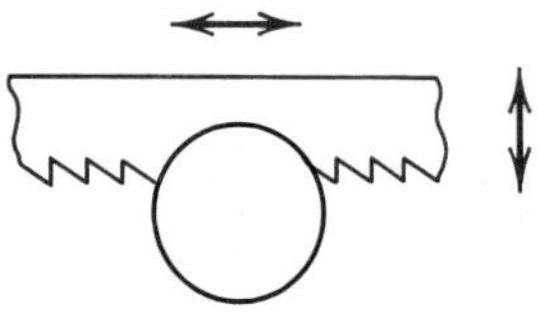	Work is held stationary while the saw cuts either in one direction as in band sawing or it reciprocates while being fed downward into the work	Horizontal band saw, reciprocating cutoff saw

(*continued*)

TABLE F-1 Machining Principles and Operations (*continued*)

Operation	Diagram	Characteristics	Type of Machines
Vertical band sawing (contour sawing)	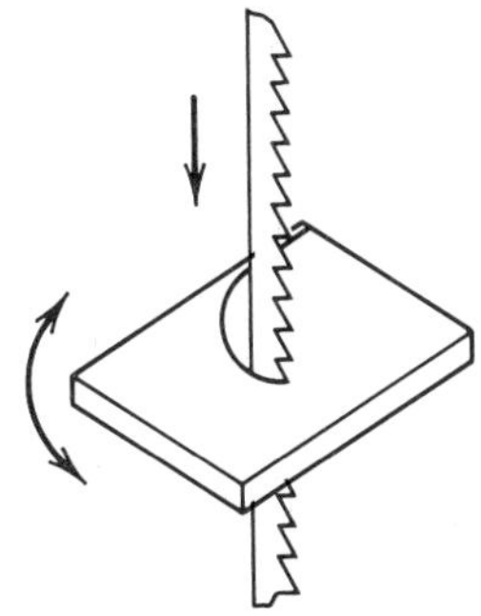	Endless band moves downward, cutting a kerf in the workpiece which can be fed into the saw on one plane at any direction	Vertical band saw
Broaching	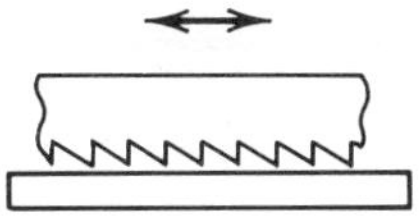	Workpiece is held stationary while a multitooth cutter is moved across the surface; each tooth in the cutter cuts progressively deeper	Vertical broaching machine, horizontal broaching machine
Horizontal spindle surface grinding	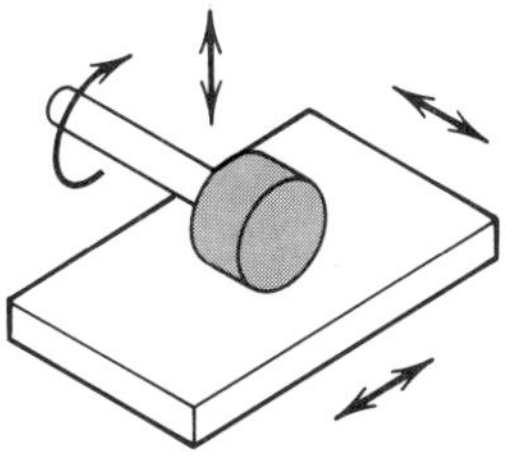	The rotating grinding wheel can be moved up or down to feed into the workpiece; the table, which is made to reciprocate, holds the work and can also be moved crosswise	Surface grinders, specialized industrial grinding machines
Vertical spindle surface grinding	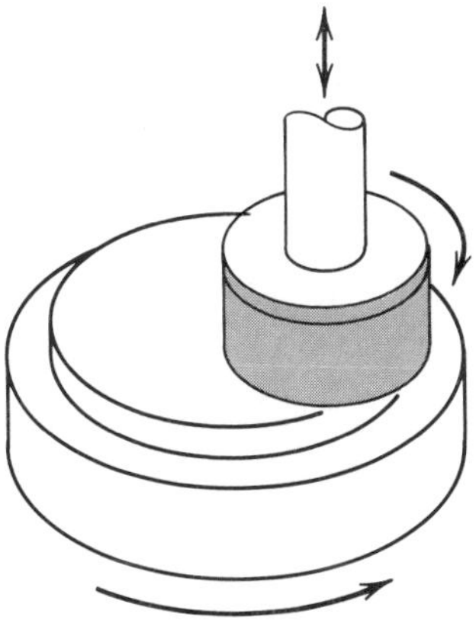	The rotating grinding wheel can be moved up or down to feed into the workpiece; the circular table rotates	Blanchard-type surface grinders
Cylindrical grinding	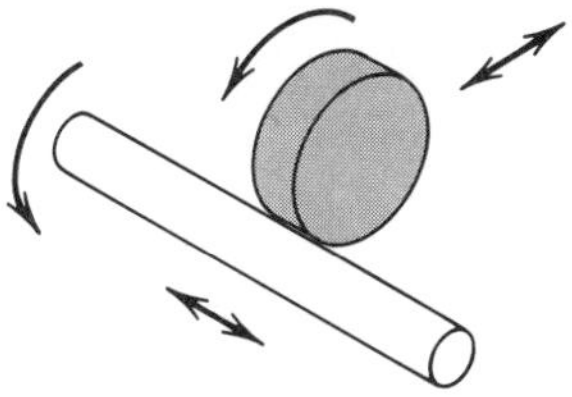	The rotating grinding wheel contacts a turning workpiece that can reciprocate from end to end; the wheelhead can be moved into the work or away from it	Cylindrical grinders, specialized industrial grinding machines
Centerless grinding	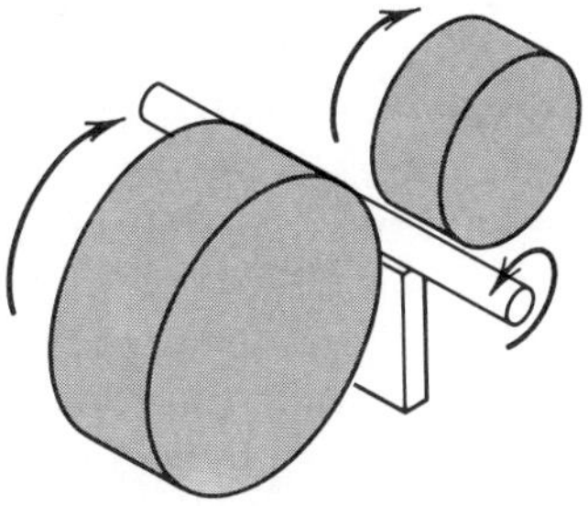	Work is supported by a workrest between a large grinding wheel and a smaller feed wheel	Centerless grinder

(*continued*)

TABLE F-1 Machining Principles and Operations (*continued*)

Operation	Diagram	Characteristics	Type of Machines
Drilling and reaming		Drill or reamer rotates while work is stationary	Drill presses, vertical milling machines
Drilling and reaming		Work turns while drill or reamer is stationary	Engine lathes, turret lathes, automatic screw machines
Boring		Work rotates, tool moves for feed on internal surfaces	Engine lathes, horizontal and vertical turret lathes, and vertical boring mills (on some horizontal and vertical boring machines the tool rotates and the work does not)

SOURCE: J. E. Neely and R. R. Kibbe, *Modern Materials and Manufacturing Processes*, John Wiley & Sons, Inc., New York, Copyright 1987.

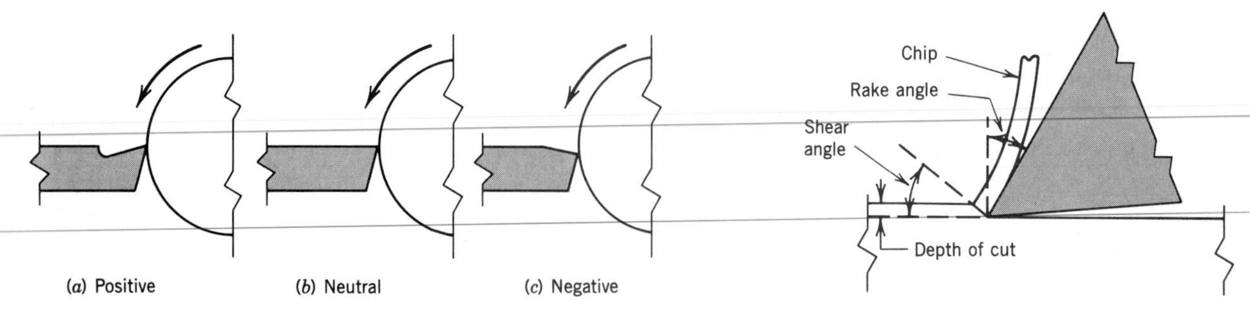

FIGURE F-3 Side view of rake angles.

FIGURE F-4 Metal cutting diagram.

One of the most important aspects of cutting tool geometry is rake (Figure F-3). Tool rake ranging from negative to positive has an effect on the formation of the chip and on the surface finish. Zero and negative rake tools are stronger and have a longer working life than positive rake tools. Negative rake tools produce poor finishes at low cutting speeds, but give a good finish at high speeds. Positive rake tools are freer cutting at low speeds and can produce good finishes when they are properly sharpened.

A common misconception is that the material splits ahead of the tool as wood does when it is being split with an axe. This is not true with metals; the metal is sheared off and does not split ahead of the chip (Figures F-4 to F-7). The metal is forced along in the direction of the cut and the grains are elongated and distorted ahead of the tool and forced along a shear plane, as can be seen in the micrographs. The surface is disrupted more with the tool having the negative rake than with the tool with the positive rake in this case, because it is moved at a slow speed. Negative rake tools require more power than positive rake tools.

Higher speeds give better surface finishes and produce less disturbance of the grain structure.

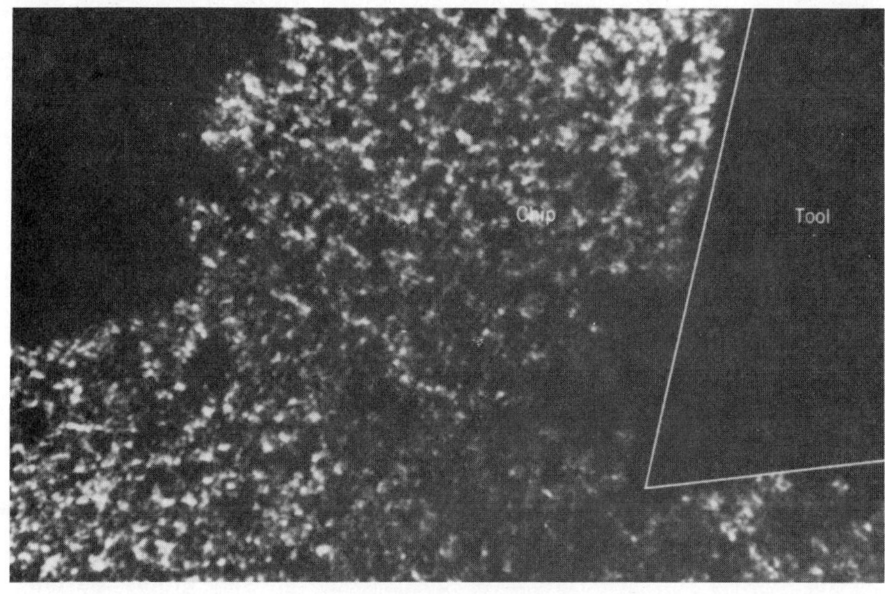

FIGURE F-5 A chip from a positive rake tool magnified 100 diameters at the point of the tool. The grain distortion is not as evident as in Figures 4 and 5. This is a continuous chip (Lane Community College).

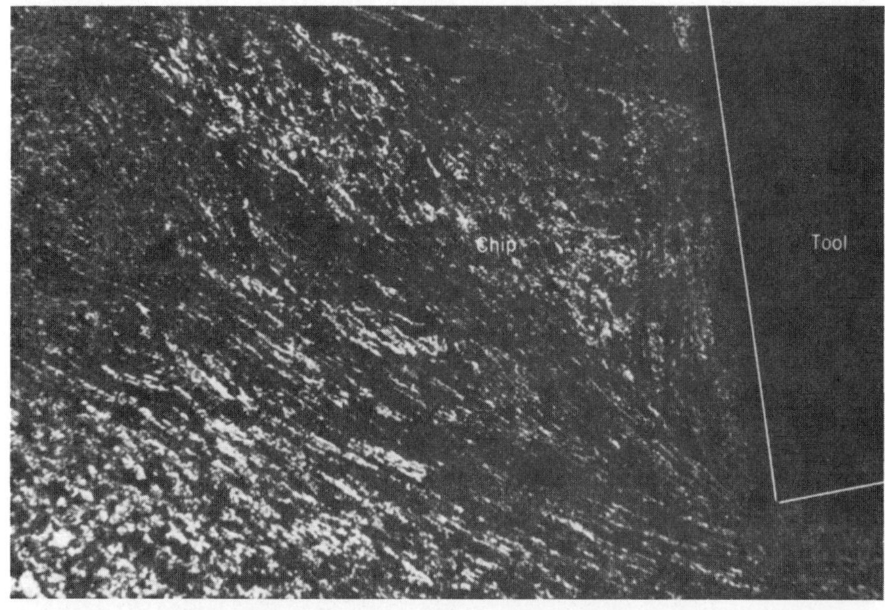

FIGURE F-6 Point of a negative rake tool magnified 100 diameters at the point of the tool (Lane Community College).

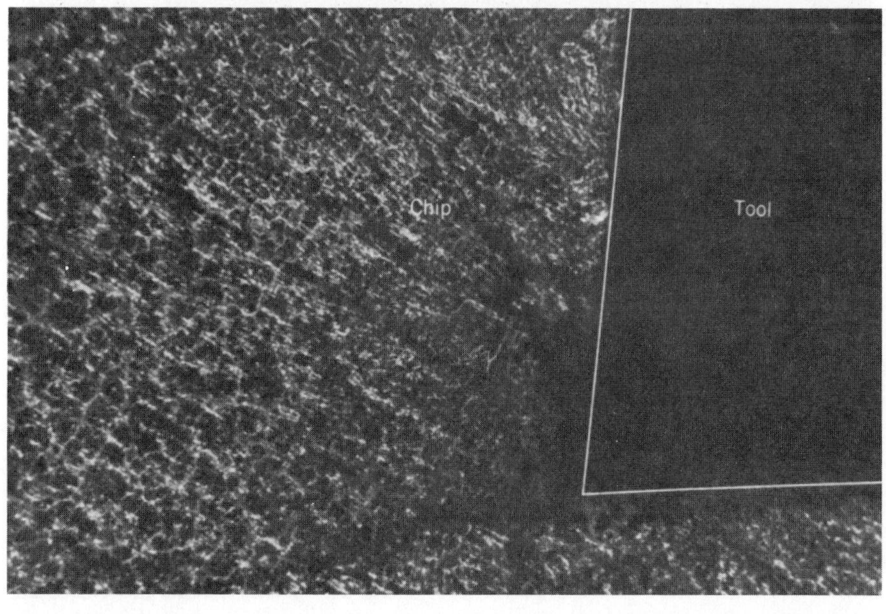

FIGURE F-7 A zero rake tool at 100 diameters shows similar grain flow and distortion to the negative rake tool (Lane Community College).

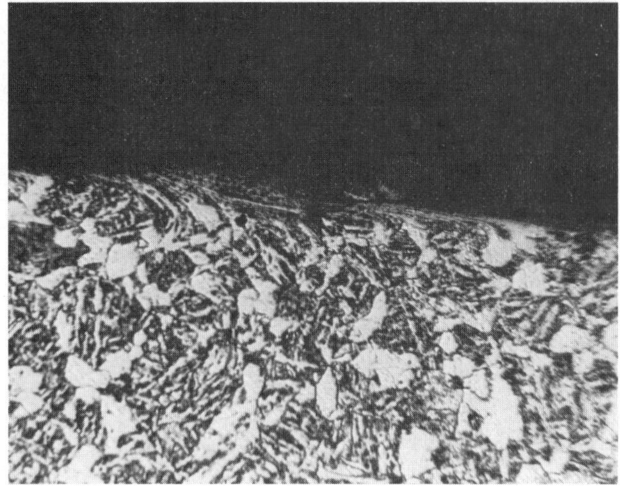

FIGURE F-8 This micrograph shows the surface of the specimen that was turned at 100 sfm. The surface is irregular and torn, and the grains are distorted to a depth of approximately .005 to .006 in. (250×) (Lane Community College).

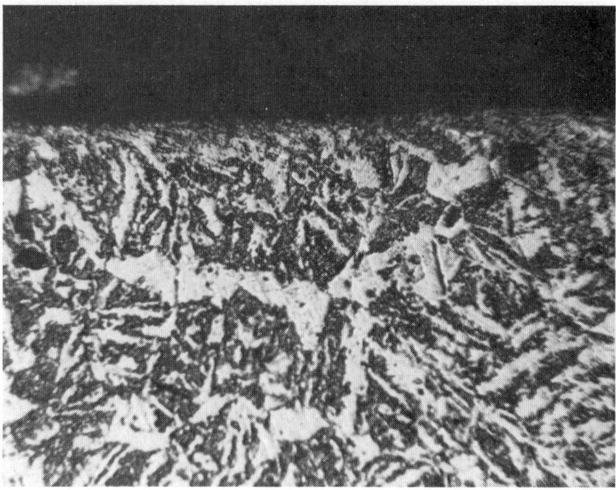

FIGURE F-9 At 400 sfm, this micrograph reveals that the surface is fairly smooth and the grains are only slightly distorted to a depth of approximately .001 in. (250×) (Lane Community College).

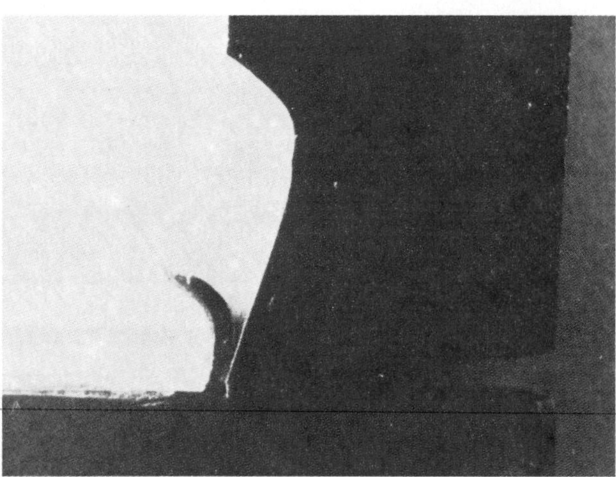

FIGURE F-10 A continuous form chip is beginning to curl away from this positive rake tool (Lane Community College).

FIGURE F-11 A thick discontinuous chatter chip being formed at slow speed with a negative rake tool (Lane Community College).

This can be seen in Figures F-8 and F-9. At a lower speed of 100 surface feet per minute (sfpm), the metal is disturbed to a depth of .005 to .006 in., and the grain flow is shown to be moving in the direction of the cut. The grains are distorted and in some places the surface is torn. This condition can later produce fatigue failures and a shorter working life of the part than would a better surface finish. Tool marks, rough surfaces, and an insufficient internal radius at shaft shoulders can also cause early fatigue failure where there are high stress and frequent reversals of the stress. At 400 sfpm, the surface is less disrupted and the grain structure is altered only to a depth about .001 in. When the cutting speeds are increased to 600 sfpm and above, little additional improvement is noted.

There is a great difference between the surface finish of metals cut with coolant or lubricant and metals that are cut dry. This is because of the cooling effect of the cutting fluid and the lubricating action that reduces friction between tool and chip. The chip tends to curl away from the tool more quickly and there is a more uniform chip when a cutting fluid is used. Also, the chip becomes thinner and pressure welding is reduced. When metals are cut dry, pressure welding is a definite problem, especially in the softer metals such as 1100 aluminum and low-carbon steels. Pressure welding

FIGURE F-12 A discontinuous, thick chip is being formed with a zero rake tool (Lane Community College).

FIGURE F-13 The crater on the cutting edge of this tool was caused by chip wear at high speeds. The crater often helps to cool the chip (Lane Community College).

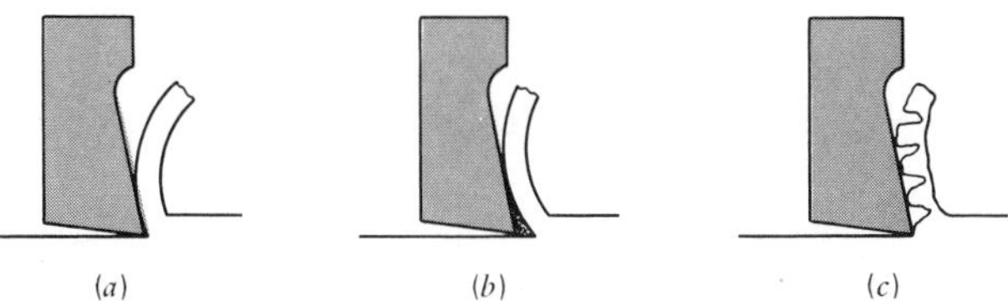

FIGURE F-14 The three types of chip formation: (*a*) continuous; (*b*) continuous with built-up edge; (*c*) discontinuous (segmented) (J. E. Neely, *Practical Metallurgy and Materials of Industry*, 2nd ed., John Wiley & Sons, Inc. Copyright © 1984).

produces a built-up edge (BUE) that causes a rough finish and a tearing of the surface of the workpiece. Built-up edge is also caused by speeds that are too slow; this often results in broken tools from excess pressure on the cutting edge. Figures F-10 to F-12 show chips formed with tools having positive, negative, and zero rakes. These chips were all formed at low surface speeds and consequently are thicker than they would have been at higher speeds. There is also more distortion at low speeds. The material was cut dry.

At high speeds, cratering begins to form on the top surface of the tool because of the wear of the chip against the tool; this causes the chip to begin to curl (Figure F-13). The crater makes an airspace between the chip and the tool, which is an ideal condition since it insulates the chip from the tool and allows the tool to remain cooler; the heat goes off with the chip. The crater also allows the cutting fluid to get under the chip.

Various metals are cut in different ways. Softer, more ductile metals produce a thicker chip and harder metals produce a thinner chip. A thin chip indicates a clean cutting action with a better finish.

A machine operator often notices that using certain tools having certain rake angles at greater speeds (several hundred surface feet per minute as compared with 50 to 100 sfpm) produces a better finish on the surface. The operator may also notice that when speeds are too low, the surface finish becomes rougher. Not only do speeds, feeds, tool shapes, and depth of cut have an effect on finishes, but the surface structure of the metal itself is disturbed and altered by these factors.

Tool materials are high-speed steel, cast alloy, carbide, ceramic, and diamond. Most manufacturing today is done with carbide tools. Greater amounts of materials may be removed and tool life extended considerably when carbides, rather than high-speed steels, are used. Much higher speeds can also be used with carbide tools than with high-speed tools.

Analysis of Chip Structures

Machining operations performed on various machine tools produce chips of three basic types: the continuous chip, the continuous chip with built-up edge on the tool, and the discontinuous chip. The formation of the three basic types of chips can be seen in Figure F-14. Various kinds of chip formations are shown in Figure F-15. High cutting speeds produce thin chips, and tools with a large positive rake angle favor the formation of the continuous chip. Any circumstances that lead to a reduction of friction between the chip–tool interface, such as the use of cutting fluid, tend to produce a continuous chip. The continuous chip usually produces the best surface finish and has the greatest efficiency in terms of power consumption. Continuous chips create lower temperatures at the

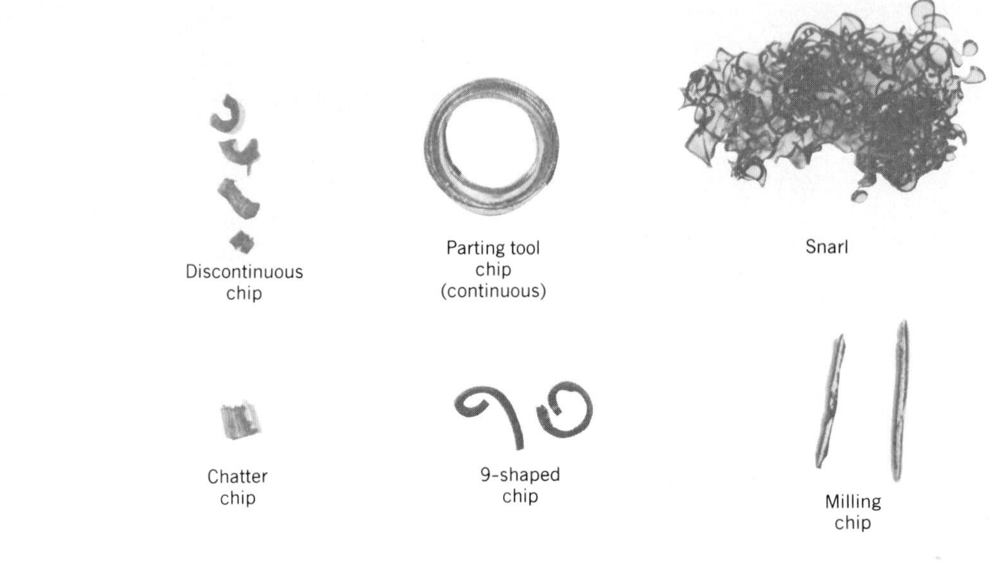

FIGURE F-15 **Some of the kinds of chips that are formed in machining operations (Lane Community College).**

cutting edge, but at very high speeds there are higher cutting forces and very high tool pressures. Since there is less strength at the point of positive rake angle tools than in the negative rake tools, tool failure is more likely with large positive rake angles at high cutting speeds or with intermittent cuts.

Negative rake tools are most likely to produce a built-up edge with a rough continuous chip and a rough finish on the work, especially at lower cutting speeds and with soft materials. Positive rake angles and the use of cutting fluid plus higher speeds decrease the tendency for a built-up edge on the tool. However, most carbide tools for turning machines have a negative rake. There are several reasons for this. Carbides are always used at high cutting speeds, lessening the tendency for a built-up edge. Better finishes are obtained at high speeds even with negative rake and the tool can withstand greater shock loads than positive rake tools. Another advantage of negative rake tools is that an indexable insert can have 90-degree angles between its top surface and flank, making more cutting edges on both sides. With a negative rake the flank of the tool has relief even though the tool has square edges. More horsepower is needed for negative rake tools than for positive rake tools when all other factors are the same.

The discontinuous or segmented chip is produced when a brittle metal, such as cast iron or hard bronze, is cut. Some ductile metals can form a discontinuous chip when the machine tool is old or loose and a chattering condition is present, or when the tool form is not correct. The discontinuous chip is formed as the cutting tool contacts the metal and compresses it to some extent; the chip then begins to flow along the tool and, when more stress is applied to the brittle metal, it tears loose and a rupture occurs. This causes the chip to separate from the work material. Then a new cycle of compression, the tearing away of the chip, and its breaking off begins.

Low cutting speeds and a zero or negative rake angle can produce discontinuous chips. The discontinuous chip is more easily handled on the machine since it falls into the chip pan.

The continuous chip sometimes produces snarls or long strings that are not only inconvenient but dangerous to handle. The optimum kind of chip for operator safety and producing a good surface finish is the 9-shaped chip that is usually produced with a chip breaker.

Chip breakers take many forms and most carbide tool holders either have an inserted chip breaker or the chip breaker is formed in the insert tool itself (Figure F-16). The action of a chip breaker is designed to curl the chip against the work and then to break it off to produce the proper type of chip.

Machine operators usually must form the tool

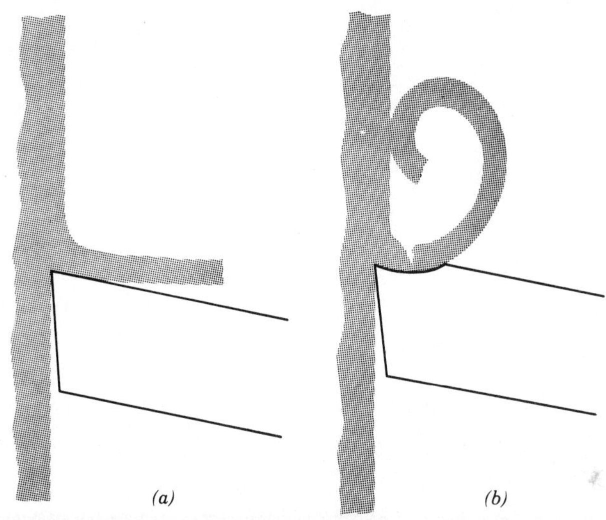

FIGURE F-16 The action of a chip breaker to curl a chip and cause it to break off: (*a*) plain tool; (*b*) tool with chip breaker.

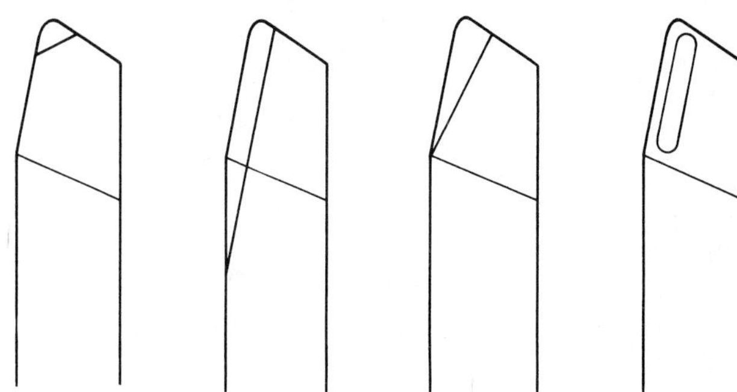

FIGURE F-17 Types of chip breakers in high-speed tools.

shape themselves on a grinder when using high-speed tools and may or may not grind a chip breaker on the tool. If they do not, a continuous chip is formed that usually makes a wiry tangle, but if they grind a chip breaker in the tool, depending on the feed and speed, they produce a more acceptable type of chip. A high rate of feed will produce a greater curl in the chip, and, often even without a chip breaker, a curl can be produced by adjusting the feed of the machine properly. The depth of cut also has an effect on chip curl and breaking. A larger chip is produced when the depth of cut is greater. This heavier chip is less springy than light chips and tends to break up into small chips more readily. Figure F-17 shows how chip breakers may be formed in a high speed tool. See Unit 4 for carbide chip breakers.

Machinability of Metals

Various ratings are given to metals of different types on a scale based on properly annealed carbon steel containing about .1 percent carbon (which is 100 on the scale). Much information has been published on the relative machinability of the various grades of carbon steel. They have been compared almost universally with B-1112, a free-machining steel, which was once the most popular screw machining stock. Machinability ratings by some manufacturers are now based on AISI 1212, which is rated at 100 percent and at 168 sfpm. Metals that are more easily machined have a number higher than 100 and materials that are more difficult to machine are numbered lower than 100. Table F-2 gives the machinability of various alloys of steel based on AISI B-1112 as 100 percent. Of course, the machinability is greatly affected by the cutting tool material, tool geometry, and the use of cutting fluids.

In general, the machinist must select the type of tool, speeds, feeds, and kind of cutting fluid for the material being cut. The most important material property, however, is hardness. A machinist often shop tests the hardness of material with a file to determine relative machinability since hardness is related to machinability. Hardness is also a factor in producing good finishes. Soft metals such as copper, 1100 series aluminum, and low-carbon hot-rolled steel tend to have poor finishes when cut with ordinary tooling. Cutting tools tend to tear the metal away rather than cut these soft

TABLE F-2 Machinability Ratings for Some Commonly Used Steels

AISI Number	Machinability Index (% Relative Speed Based on AISI B1112 as 100%)
B1112	100
C1120	81
C1140	72
C1008	66
C1020	72
C1030	70
C1040	64
C1060 (annealed)	51
C1090 (annealed)	42
3140 (annealed)	66
4140 (annealed)	66
5140 (annealed)	70
6120	57
8620	66
301 (stainless)	36
302 (stainless)	36
304 (stainless)	36
420 (stainless)	36
440A (stainless)	24
440B (stainless)	24
440C (stainless)	24

metals; however, very sharp tools and larger rake angles plus the use of cutting fluids can help to produce better finishes. Harder, tougher alloys almost always produce better finishes even with negative rake carbide tools. For example, an AISI 1020 hot-rolled steel bar and an AISI 4140 steel bar of the same-diameter cut at the same speed and feed with the same tooling would have markedly different finishes. The low-carbon HR steel would have a poor finish compared to the alloy steel.

The operator must also understand the effects of heat on normally machinable metals such as alloy steel, tool steel, or gray cast iron. Welding on any of these metals may harden the base metal near the weld and make it difficult to cut, even with carbide tools. To soften these hard areas, the entire workpiece must first be annealed.

Most alloy steel can be cut with high-speed tools but at relatively low cutting speeds, which produce a poor finish unless some back rake is used. These steels are probably best machined by using carbide tools at higher cutting speeds ranging from 200 to 350 sfpm, or even higher in some cases. Some nonferrous metals, such as aluminum and magnesium, can be machined with much higher speeds ranging from 500 to 8000 sfpm when the correct cutting fluid is used. Some alloy steels, however, are more difficult to machine because they tend to work harden. Examples of these are austenitic manganese steels (those used for wear resistance), Inconel, stainless steels (SAE 301 and others), and some tool steels.

SELF-TEST

1. In what way do tool rakes, positive and negative, affect surface finish?

2. Soft materials tend to pressure weld on the top of the cutting edge of the tool. What is this condition called and what is its result?

3. Which indicate a greater disruption of the surface material: thin uniform chips or thick segmented chips?

4. In metal cutting does the material split ahead of the tool? If not, what does it do?

5. Which tool form is stronger, negative or positive rake?

6. What effect does cutting speed have on surface finish? On surface disruption of grain structure?

7. How can surface irregularities caused by machining later affect the usefulness of the part?

8. Which property of metals is directly related to machinability? How do machinists usually determine this property?

9. Describe the type of chip produced on a machine tool that is the safest and easiest to handle.

10. Define *machinability*.

UNIT 2

Speeds and Feeds for Machine Tools

Modern machine tools are very powerful and with modern tooling they are designed for a high rate of production. However, if the operator uses the incorrect feeds and speeds for the machine operation, a much lower production rate will be realized resulting in an inferior product. Sometimes an operator will take a very light roughing cut when it is possible to take a greater depth of cut. Often the cutting speeds are too low to produce good surface finishes and for optimum metal removal. Because of the cost of labor, time is very important and removal of a larger amount of metal can shorten the time needed to produce a part and often improve its physical properties.

The importance of cutting speeds and machine feeds in machining operations cannot be overemphasized. The right speed can mean the difference between burning the end of a drill or other tool, causing lost time or having many hours of cutting time between sharpenings or replacement. The correct feed can also lengthen tool life. Excessive feeds often result in tool breakage. This unit will prepare you to think in terms of correct feeds and speeds.

Objectives

After completing this unit, you should be able to:

1. Calculate correct cutting speeds for various machine tools and grinding machines.

2. Determine correct feeds for various machining operations.

Cutting Speeds

Cutting speeds (CS) are normally given in tables for cutting tools and are based on surface feet per minute (fpm or sfpm). Surface feet per minute means that either the tool moves past the work or the work moves past the tool at a rate based on the number of feet that pass the tool in one minute. It can be determined from a flat surface or on the periphery of a cylindrical tool or workpiece. A small cylinder will have more revolutions per minute (rpm) than a large one with the same surface speed. For example, if a thin wire were pulled off a 1-in.-diameter spool at the rate of 20 fpm, the spool would rotate three times faster than one 3 in. in diameter with the wire being pulled off at the same rate of 20 fpm.

Since machine spindle speeds are given in revolutions per minute (rpm), they can be derived in the following manner:

$$\text{rpm} = \frac{\text{cutting speed (CS; in feet per minute)} \times 12}{\text{diameter of cutter } (D\text{; in inches}) \times \pi}$$

If you use 3 to approximate π (3.1416), the formula becomes

$$\frac{\text{CS} \times 12}{D \times 3} = \frac{\text{CS} \times 4}{D}$$

This simplified formula is certainly the most common one used in machine shop practice and it applies to the full range of machine tool operations, which include the lathe and the milling machine, as well as the drill press. The simplified formula

$$\text{rpm} = \frac{\text{CS} \times 4}{D}$$

is used throughout this book. The formula is used, for example, as follows, where D = the diameter of the drill or other rotating tool in inches and CS = an assigned cutting speed for a particular material. For a $\frac{1}{2}$-in. drill in low-carbon steel the speed would be

$$\frac{90 \times 4}{\frac{1}{2}} = 720 \text{ rpm}$$

Table F-3 gives cutting speeds and starting values for common materials. These may have to be varied up or down depending on the specific machining task. Always observe the cutting action carefully and make appropriate speed corrections as needed. Until you gain some experience in machining, use the lower values in the table when selecting cutting speeds. As you can see by the hardness values in the table, there is a general relationship between hardness and cutting speed.

TABLE F-3 Cutting Speeds and Starting Values for Some Commonly Used Materials

		Tool Material	
Work Material	**Harness (BHN)**	**High-Speed Steel**	**Cemented Carbide**
Aluminum	60–100	300–800	1000–2000
Brass	120–220	200–400	500–800
Bronze (hard drawn)	220	65–130	200–400
Gray cast iron (ASTM 20)	110	50–80	250–350
Low-carbon steel	220	60–100	300–600
Medium-carbon alloy steel (AISI 4140)	229[a]	50–80	225–400
High-carbon steel	240[a]	40–70	150–250
Carburizing-grade alloy steel (AISI 8620)	200–250[a]	40–70	150–350
Stainless steel (type 410)	120–200[a]	30–80	100–300

[a] Normalized.

This is not always true, however. Some stainless steel is relatively soft, but it must be cut at a low speed.

Cutting speeds/rpm tables for various materials are available in handbooks and on wall charts. The cutting speed for carbide tools is normally three to four times that of high-speed cutting tools.

You will learn more about cutting tool materials in later units in this section, but a good rule to remember concerning cutting speeds is: generally, cutting tools will wear out quickly at too-high cutting speeds. Carbide cutting tools will chip or break up quickly at too-low cutting speeds. When steel chips become blue colored, it is a sign of a higher temperature caused by high cutting speed and/or dull tools. Blue chips are acceptable and in fact desirable when using carbides, but not when using high-speed steel tools. The chips should not be discolored at all with high-speed tools, especially when using cutting fluids.

Cutting speed constants are influenced by the cutting tool material, workpiece material, rigidity of the machine setup, and the use of cutting fluids. As a rule, lower cutting speeds are used to machine hard or tough materials or where heavy cuts are taken and it is desirable to minimize tool wear and thus maximize tool life. Higher cutting speeds are used in machining softer materials in order to achieve better surface finishes.

Cutting speeds for drills are the same for drilling operations both where the drill rotates and where the work rotates and the drill remains stationary as in lathe work. For a step drill where there are two or more diameters, the largest diameter is used in the formula.

Where the workpiece rotates and the tool is stationary, as in lathes, the outer diameter (D) of the workpiece is used in the speed formula. As the workpiece is reduced to a smaller diameter, the speed should be increased. However, there is some latitude in speed adjustments and 5 to 10 sfpm one way or another is usually not significant. Therefore, if a machine cannot be set at the calculated rpm, use the next-lower speed. If there is vibration or chatter, a lower speed will often eliminate it.

For multiple point cutting, as in milling machines, D is the diameter of the milling cutter in inches. Cutting speed in milling is the rate at which a point on the cutter passes by a point on the workpiece in a given period of time.

Many grinding machines and portable grinders have a fixed rpm, so it is relatively easy to select a grinding wheel for that speed. However, tool and cutter grinders have variable speeds, so a range of wheels from tiny, mounted wheels to large, straight wheels is available. The rpm for these machines must be correct. Every grinding wheel, wherever used, has a safe maximum speed, and this should never be exceeded.

FIGURE F-18 The blotter on the wheel, besides serving as a buffer between the flange and the rough abrasive wheel, provides information as to the dimensions and the composition of the wheel, plus its safe speeds in rpm. This wheel can be run safely up to 3600 rpm (Bay State Abrasives Division, Dresser Industries, Inc.).

Any grinding machine in the shop, properly handled, is a safe machine. It has been designed that way and it should be maintained to keep it safe. The negative image of grinding wheels and breakage comes primarily from portable grinders, which often are not well maintained and sometimes are operated by unskilled and careless persons.

You must always stay within the safe speeds, which are shown on the blotter or label on every wheel of any size (Figure F-18). Vitrified wheels generally have a maximum safe speed of 6500 sfpm and organic wheels (resinoid, rubber, or shellac), 16,000 sfpm. These speeds are generally set by the grinding wheel manufacturer.

Because of the importance of wheel speed in grinding wheel safety, it is important to know how it is calculated. This speed is expressed in terms of surface feet per minute (sfpm), which simply expresses the distance that a given spot on a wheel periphery travels in a minute. It is calculated by multiplying the diameter (in inches) by 3.1416, dividing the result by 12 to convert it to feet, and multiplying that result by the number of revolutions per minute (rpm) of the wheel. Thus a 10-

in.-diameter wheel traveling at 2400 rpm would be rated at approximately 6283 sfpm, under the safe speed of most vitrified wheels of 6500 sfpm.

To find the safe speed in rpm of a 10-in.-diameter wheel, the formula becomes

$$\text{rpm} = \frac{\text{CS} \times 4}{D} = \frac{6500 \times 4}{10} = 2600$$

Most machine shop–type flat surface or cylindrical grinders are preset to operate at a safe speed for the largest grinding wheel that the machine is designed to hold. As long as the machine is not tampered with and no one tries to mount a larger wheel on the machine than it is designed for, there should be no problem.

It should be clear that with a given spindle speed (rpm) the speed in sfpm increases as the wheel diameter is increased, and it decreases with a decrease in wheel diameter. Maximum safe speed may be expressed either way, but on the wheel blotter it is usually expressed in rpm. However, if the cutting speed is sought and the rpm known, the following formula may be used:

$$\text{CS} = \frac{\text{rpm} \times D}{4}$$

Feeds

The machine movement that causes a tool to cut into or along the surface of a workpiece is called feed. The amount of feed is usually measured in thousandths of an inch in metal cutting. Feeds are expressed in slightly different ways on various types of machine tools.

Drills on drill presses rotate, causing the cutting edges on the end of the drill to cut into the workpiece. When metal is cut, considerable force is required to feed the cutting edges of the drill into the workpiece. Drilling machines that have power feeds are designed to advance the drill a given amount for each revolution of the spindle. Therefore, a .006-in. feed means that the drill advances .006 in. for every revolution of the machine spindle. Thus feeds for drill presses are expressed in inches per revolution (ipr). The amount of feed varies with the drill size and the kind of work material (Table F-4).

When drill presses have no power feed, the operator provides the feed with a handwheel or hand lever. These drilling machines are called sensitive drill presses because the operator can "sense" or feel the correct amount of feed. Also, the formation of the chip as a tightly rolled helix instead of a stringy chip is an indicator of the correct feed. Excessive feed on a drill can cause the machine to jam or stop, drill breakage, and a hazardous situation if the workpiece is not properly clamped.

TABLE F-4 Drilling Feed Table

Drill Size Diameter (in.)	Feeds per Revolution (in.)
Under $\frac{1}{8}$	.001 to .002
$\frac{1}{8}$ to $\frac{1}{4}$	.002 to .004
$\frac{1}{4}$ to $\frac{1}{2}$	.004 to .007
$\frac{1}{2}$ to 1	.007 to .015
Over 1	.015 to .025

Feeds on turning machines such as lathes are also expressed in inches per revolution (ipr) of the lathe spindle. The tool moves along the rotating workpiece to produce a chip. The depth of the cut is not related to feed. The quick-change gearbox on the lathe makes possible the selection of feeds in ipr in a range from approximately .001 in. to about .100 in., depending on size and machine manufacturer. Feed rates for small (10-in. swing) lathe operations can be as high as .015 ipr for roughing operations and usually .003 to .005 ipr for finishing. However, some massive turning machines may use .030 to .050 ipr feed with a .750-in. depth of cut. Coarser feeds, depending on the machine size, horsepower, and rigidity, necessary for rapid stock removal, are called roughing. As long as the roughing dimension is kept well over the finish size, feeds should be set to the maximum the machine will handle. In roughing operations, finish is not important and no attempt should be made to obtain a good finish; the only consideration is stock removal and, of course, safety. It is in the finishing cuts with fine feeds that dimensional accuracy and surface finish is of the greatest importance and stock removal is of little consideration.

Feeds on milling machines refer to the rate at which the workpiece material is advanced into the cutter by table movement or where the cutter is advanced into the workpiece in the manner of a drill cutting on its end. Since each tooth on a multitooth milling cutter makes a chip, the chip thickness depends on the amount of table feed. Feed rate in milling is measured in inches per minute (ipm) and is calculated by the formula

$$\text{ipm} = F \times N \times \text{rpm}$$

where ipm = feed rate, in inches per minute
F = feed per tooth
N = number of teeth in the cutter being used
rpm = revolutions per minute of the cutter

Feeds for end mills used in vertical milling machines range from .001 to .002 in. feed per tooth for very small diameter cutters in steel work material to .010 in. feed per tooth for large cutters in aluminum workpieces. Since the cutting speed for mild steel is 90, the rpm for a $\frac{3}{8}$-in. high-speed, two-flute end mill is

$$\text{rpm} = \frac{\text{CS} \times 4}{D} = \frac{90 \times 4}{3/8} = \frac{360}{.375} = 960 \text{ rpm}$$

To calculate the feed rate, we will select .002 in. feed per tooth. (See Section J, Table J-2, for a table for end mill feed rates.)

$$\text{ipm} = F \times N \times \text{rpm} = .002 \times 2 \times 960 = 3.84$$

The cutter should rotate at 960 rpm and the table feed adjusted to approximately 3.84 ipm.

Horizontal milling machines have a horizontal spindle instead of a vertical one, but the feed rate calculations are the same as for vertical mills. Horizontal milling cutters usually have many more teeth than end mills and are much larger. Cutting speeds and rpm are also calculated in the same way. (See Section K, Table K-2, for cutting speeds and feed rates for horizontal milling.)

Feeds for surface grinding are set for the crossfeed so that the workpiece moves at the end of each stroke of the table. The crossfeed dial is graduated in thousandths, but the feed is not usually calculated very precisely. As a rule, the crossfeed is set to move from one-quarter to one-third of the wheel width at each stroke. This coarse feed allows the wheel to break down (wear) more evenly than do fine feeds, which cause one edge of the wheel face to break down.

In plain cylindrical grinding where the rotating workpiece is traversed back and forth against a large grinding wheel, table feeds are used, as in surface grinding. The feed or traverse setting must be adjusted for roughing or finishing to get optimum results.

SELF-TEST

1. If the cutting speed (CS) for low-carbon steel is 90 and the formula for rpm is (CS × 4)/D, what should the rpm of the spindle of a drill press be for a $\frac{1}{8}$-in.-diameter high-speed twist drill? For a $\frac{3}{4}$-in.-diameter drill?

2. If the two drills in question 1 were used in a lathe instead of a drill press, what should the rpm of the lathe spindle be for both drills?

3. If, in question 2, the small drill could not be set at the calculated rpm because of machine limitations, what could you do?

4. An alloy steel 2-in.-diameter cylindrical bar having a cutting speed of 50 is to be turned on a lathe using a high-speed tool. At what rpm should the lathe be set?

5. A formula for setting the safe rpm for an 8-in. grinding wheel, when the cutting speed is known, is rpm = (CS × 4)/D. If the safe surface speed of a vitrified wheel is 6000 sfpm, what should the rpm be?

6. Are feeds on a drill press based on inches per minute or inches per revolution of the spindle?

7. Which roughing feed on a small lathe would work best, .100 or .010 in. per revolution?

8. What kind of machine tool bases feed on inches per minute instead of inches per revolution?

9. Approximately what should the feed rate be on a $\frac{3}{4}$-in.-wide grinding wheel on a surface grinder?

UNIT 3

Cutting Fluids

Since the beginning of the Industrial Revolution, when metals first began to be cut on machines, cutting fluids have been an important aspect of machining operations. Lubricants in the form of animal fats were first used to reduce friction and cool the workpiece. These straight fatty oils tended to become rancid, had a disagreeable odor, and often caused skin rashes on machine operators. Although lard oil alone is no longer used to any great extent, it is still used as an additive in cutting oils. Plain water was sometimes used to cool workpieces, but water alone is a corrosive liquid and tends to rust machine parts and workpieces. Water is now combined with oil in an emulsion that cools but does not corrode. Many new chemical and petroleum-based cutting fluids are in use today, making possible the high rate of production in machining and manufacturing of metal products we presently enjoy. The reasons for this are that cutting fluids reduce machining time by allowing higher cutting speeds and that they reduce tool breakdown and down time. This unit will prepare you to use cutting fluids properly when you begin to operate machine tools.

After completing this unit, you should be able to:

1. List the various types of cutting fluids.
2. Explain the correct uses and care in using several cutting fluids.
3. Describe several methods of cutting fluid application.

Effects of Cutting Fluids

Cutting fluids is a generic term that covers a number of different products used in cutting operations. Basically, there are two major effects derived from cutting and grinding fluids: cooling and lubrication. Although water-based fluids are most effective for cooling the tool and workpiece, and oil-based fluids are the better lubricants, there is a considerable overlap between the two.

Cutting force and temperature rise are not generated so much by the friction of the chip sliding over the surface of the tool as by the shear flow of the workpiece metal just ahead of the tool (Figure F-19). Some tool materials, such as carbide and ceramic, are able to withstand high temperatures. When these materials are used in high-speed cutting operations, the use of cutting fluids may actually be counterproductive. The increased temperatures tend to promote an easier shear flow and thus reduce cutting force. High-speed, high-temperature cutting also produces better finishes and disrupts the surface less than does lower-speed machining. For this reason, machining with extremely high cutting speeds using carbide tools is often done dry. The use of coolant may cause a carbide tool to crack and break up due to thermal shock, since the flooding coolant can never reach the hottest point at the tip of the tool and so will not maintain an even temperature on the tool.

It is at the lower cutting speeds in which high-speed tools and even carbides are used that cutting fluids are used to the greatest advantage. Also, cutting fluids are a necessity in most precision grinding operations. Most milling machine, drilling, and many lathe operations require the use of cutting fluids.

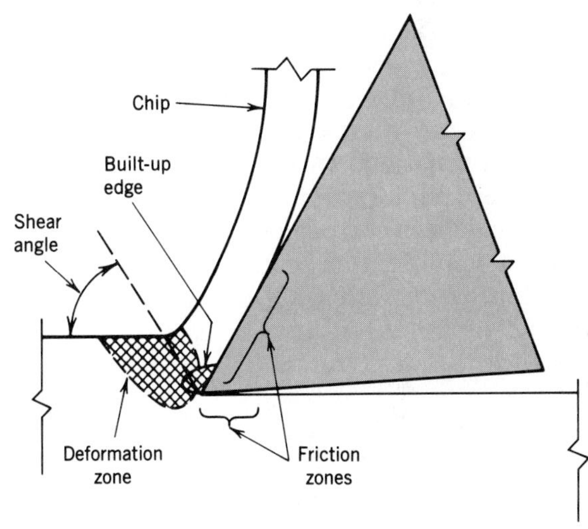

FIGURE F-19 Metal-cutting diagram showing the shear flow ahead of the tool.

TYPES OF CUTTING FLUIDS

Lubricants as well as coolants (which also provide some lubrication) are able to reduce cutting forces at lower cutting speeds (allowing somewhat higher speeds to be used). They also reduce or remove heat generation that tends to break down high-speed tools. Both lubricants and coolants extend tool life and improve workpiece finish. Another advantage in using cutting fluids is that the workpiece dimensions can be accurately measured on the cool workpiece. A hot workpiece must be cooled to room temperature to obtain a correct measurement.

Several types of cutting fluids are in use and their nomenclature is by no means universal, since a single fluid is often called by many different names. Next we describe the names used in this book.

SYNTHETIC FLUIDS (sometimes called chemical fluids). These can be *true solutions*, consisting of inorganic substances dissolved in water, which include nitrites, nitrates, borates, or phosphates. As their name implies, these chemical fluids form true solutions in water unlike the oils in emulsions. These chemical substances are water miscible and vary in color from milky to transparent when mixed with water. This type is superior to others for machining and grinding titanium and for grinding operations in general. The *surface-active* type is a water solution that contains additives that lower the surface tension of the water. Some types have good lubricity and corrosion inhibitors.

Some of the advantages of these chemical fluids are their resistance to becoming rancid (especially those that contain no fats), good detergent (cleansing) properties, rapid heat dissipation, and the fact that they are easy to mix, requiring little agitation. There are also some disadvantages. A lack of lubrication (oiliness) in most types may cause sticking in some machine parts that depend on the cutting fluid for lubrication. Also, the high detergency has a tendency to remove the normal skin oils and irritate workers' hands where there is a long continual exposure. Synthetic fluids generally provide less corrosion resistance than oilier types and some tend to foam to a certain extent. When improved lubricating qualities are needed for synthetic fluids, sulfur, chlorine, or phosphorus is added.

SEMISYNTHETIC FLUIDS (sometimes called semichemical fluids). These cutting fluids are essentially a combination of synthetic fluids and emulsions. They have a lower oil content than the straight emulsion fluids, and the oil droplets are smaller because of a higher content of emulsifying molecules, thus combining the best qualities of both types. Since a small amount of mineral oil is added to these semisynthetic fluids, they possess enhanced lubrication qualities. In addition, they may contain other additives such as chlorine, sulfur, or phosphorus.

EMULSIONS (also called water-miscible fluids or water-soluble oils). Ordinary soap is an emulsifying agent, causing oil to combine with water in a suspension of tiny droplets. Special soaps and other additives such as amine soaps, rosin soaps, petroleum sulfonates, and naphthenic acids are blended with a naphtha-based or paraffin mineral oil to emulsify it. Other additions, such as bactericides, help to reduce bacteria, fungi, and mold and to extend emulsion life. Without these additives, an emulsion tends to develop a strong, offensive odor because of bacterial action and must be replaced with a new mix.

These ordinary emulsified oils contain oil particles large enough to reflect light, and therefore they appear opaque or milky when combined with water. In contrast, many of the synthetic and semisynthetic emulsion particles are so small that the water remains clear or translucent when mixed with the chemical. The ratio of mixing oil and water varies with the job requirement and can range from 5 to 70 parts water to 1 part oil. However, for most machining and grinding operations, a mixture of 1 part oil to 20 parts water is generally used. A mixture that is too rich for the job can be needlessly expensive and a mixture that is too lean may cause rust to form on the workpiece and machine parts. When mixing, oil should be poured into the water.

Some emulsions are designed for greater lubricating value, with animal or vegetable fats or oils providing a "superfatted" condition. Sulfur, chlorine, and phosphorus provide even greater lubricating value for metal cutting operations where extreme pressures are encountered in chip forming. These fluids are mixed in somewhat rich ratios: 1 part oil with 5 to 15 parts water. All of these water-soluble cutting fluids are considered to be in the category of coolants even though they do provide some lubrication.

CUTTING OILS These are fluids that may be animal, vegetable, petroleum, or marine (fatty tissue of fish and other marine animals) oil, or a combination of two or more of these. The plain cutting oils (naphthenic and paraffinic) are considered to be lubricants and are useful for light-duty cutting on metals of high machinability, such as free-machining steels, brass, aluminum, and magnesium. Water-based cutting fluids should never be used for machining magnesium because of the fire hazard. The cutting fluid recommended for magnesium is an *anhydrous* (without water) oil that has a very low acid content. Fine magnesium chips can burn if ignited, and water tends to make it burn even more fiercely. A mineral–lard oil combination is often used in automatic screw machine practice.

Where high cutting forces are encountered and extreme pressure lubrication is needed, as in threading and tapping operations, certain oils, fats, waxes, and synthetic materials are added. The addition of animal, marine, or vegetable oil to petroleum oil improves the lubricating quality and the wetting action. Chlorine, sulfur, or phosphorus additives provide better lubrication at high pressures and high temperatures. These chlorinated or sulfurized oils are dark in color and are commonly used for thread-cutting operations.

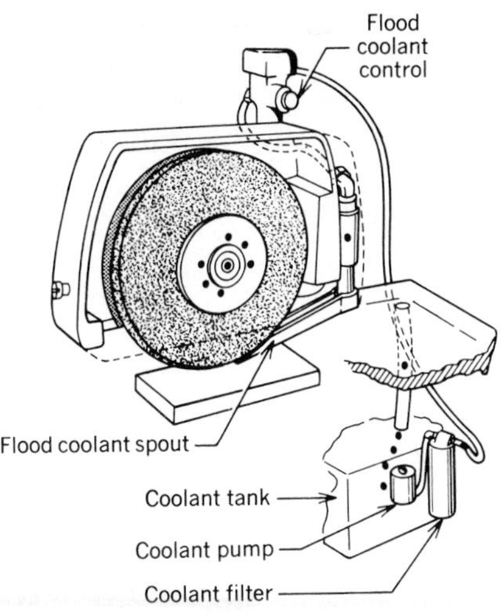

FIGURE F-20 Fluid recirculates through the tank, piping, nozzle, and drains in flood grinding system (DoALL Company).

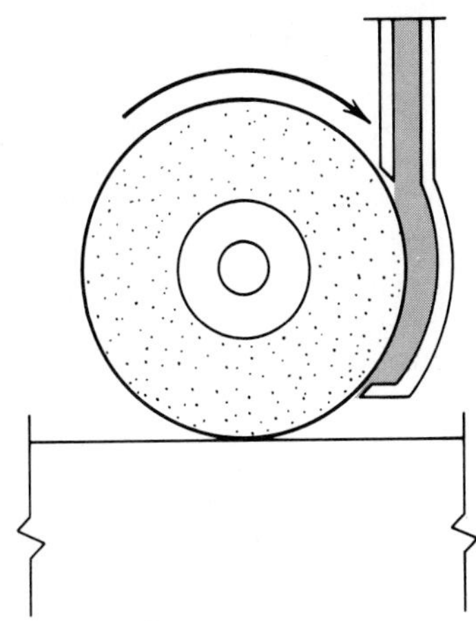

FIGURE F-21 Special wraparound nozzle used for surface grinding operations.

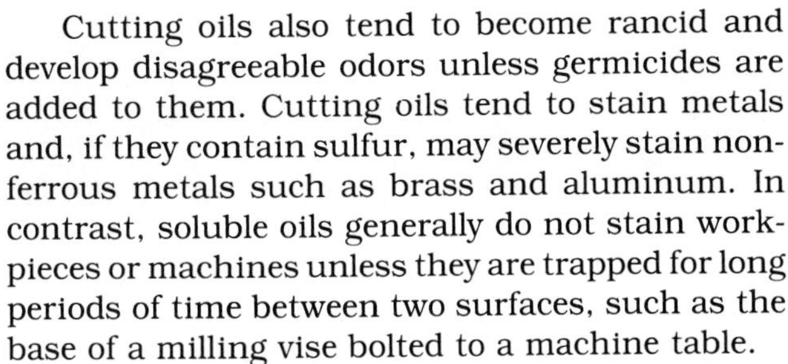

Cutting oils also tend to become rancid and develop disagreeable odors unless germicides are added to them. Cutting oils tend to stain metals and, if they contain sulfur, may severely stain nonferrous metals such as brass and aluminum. In contrast, soluble oils generally do not stain workpieces or machines unless they are trapped for long periods of time between two surfaces, such as the base of a milling vise bolted to a machine table.

Tanks containing soluble oil on lathes, milling, and grinding machines tend to collect tramp oil and dirt or grinding particles. For this reason, the fluid should be removed periodically, the tank cleaned, and a clean solution put in the tank. Water-based cutting fluid can be contaminated quickly when a machinist uses a pump oilcan to apply cutting oil to the workpiece instead of using the coolant pump and nozzle on the machine. The tramp oil goes into the tank and settles on the surface of the coolant, creating an oil seal where bacteria can grow. This causes an odorous scum to form that quickly contaminates the entire tank.

GASEOUS FLUIDS Fluids such as air are sometimes used to prevent contamination on some workpieces. For example, some reactive metals such as zirconium may be contaminated by water-based cutting fluids. The atmosphere is always, to some extent, a cutting fluid and is actually a coolant when cutting dry. Air can provide better cooling when a jet of compressed air is directed at the point of cut. Other gases such as argon, helium, and nitrogen are sometimes used to prevent oxidation of the chip and workpiece in special applications.

Methods of Cutting Fluid Application

The simplest method of applying a cutting fluid is manually with a pump oilcan. This method is sometimes used on small drill presses that have no coolant pumping system. Another simple method is with a small brush that is dipped into an open pan of cutting oil and applied to the workpiece, as in lathe threading operations.

Most milling and grinding machines, large drill presses, and lathes have a built-in tank or reservoir containing a cutting fluid and a pumping system to deliver it to the cutting area. The cutting fluid used in these machines is typically a water-soluble synthetic or soluble oil mix, with some exceptions, as in machining magnesium or in certain grinding applications where special cutting oils or fluids are used. The cutting fluid in the tank is picked up by a low-pressure (5 to 20 psi) pump and delivered through a tube or hose to a nozzle where the machining or grinding operation is taking place (Figure F-20).

When this pumping system is used, the most common method of application is by flooding. This is done by simply aiming the nozzle at the work-tool area and applying copious amounts of cutting fluid. This system usually works fairly well for most turning and milling operations, but not so well for grinding, since the rapid spinning of the grinding wheel tends to blow the fluid away from the work–tool interface. Special wraparound nozzles (Figure F-21) are often used on surface grinders to ensure complete flooding and cooling of the workpiece. On cylindrical grinders, a fan-shaped nozzle the width

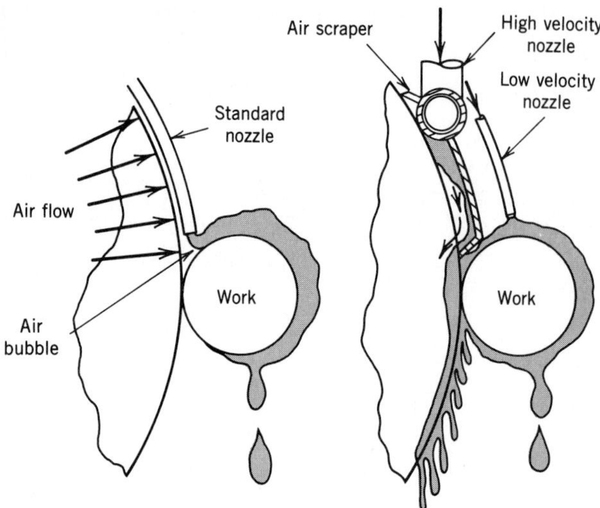

FIGURE F-22 This specially designed nozzle helps to keep the fan-like effect of the rapidly rotating wheel from blowing cutting fluid away from the wheel–work interface (Cincinnati Milacron).

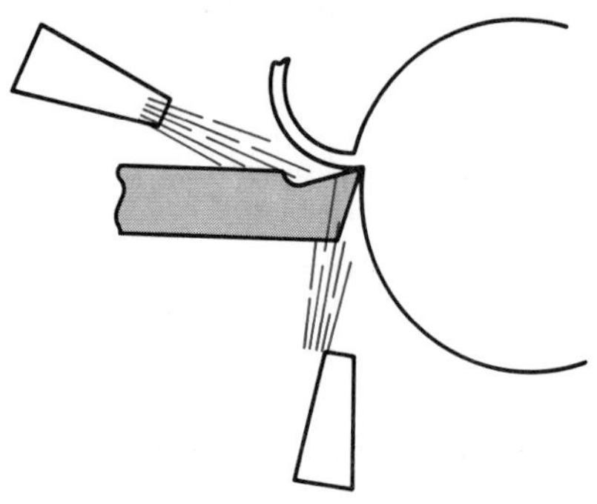

FIGURE F-23 Although not always possible, this is the ideal way of applying cutting fluid to a cutting tool.

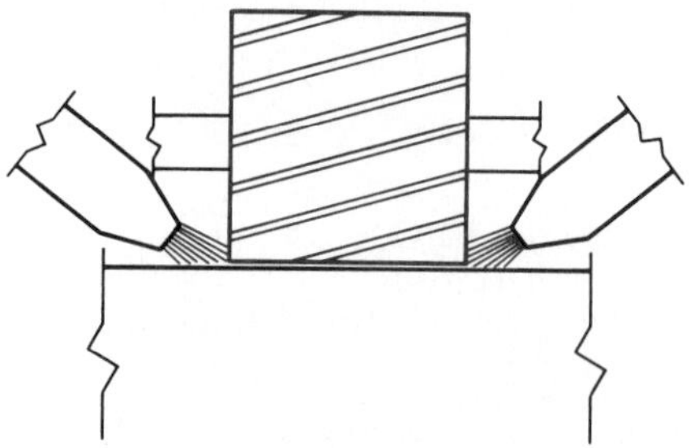

FIGURE F-24 The use of two nozzles ensures flooding of the cutting zone.

FIGURE F-25 Mist grinding fluid application on a surface grinder (DoALL Company).

of the wheel is normally used, but a specially designed nozzle (Figure F-22) is better.

In lathe operations, the nozzle is usually above the workpiece pointed downward at the tool. A better method of application would be above and below the tool as shown in Figure F-23. Internal lathe work, such as boring operations and close chuck work, cause the coolant flow to be thrown outward by the spinning chuck. In this case, a chip guard is needed to contain the spray. Although a single nozzle is normally used to direct the coolant flow over the cutter and on the workpiece on milling machines, a better method is to use two nozzles (Figure F-24) to make sure the cutting zone is completely flooded.

When a cutting fluid is properly applied, it can have a tremendous effect on finishes and surfaces. A finish may be rough when machined dry, but may be dramatically improved with lubrication. When carbide tools are used, whether on turning or milling work, and coolant is used, the work–tool area must be flooded to avoid intermittent cooling and heating cycles that create thermal shock and consequent cracking of the tool. The operator should not shut off the coolant while the cut is in progress to "see" the cutting operation.

There is no good way to cool and lubricate a twist drill, especially when drilling horizontally in a lathe. In deep holes, cutting forces and temperatures increase, often causing the drill to expand and bind. Even in vertical drilling on a drill press, the helical flutes, designed to lift out the chips, also pump out the cutting fluid. Some drills are made with oil holes running the length of the drill to the cutting lips that help to offset the problem. The fluid is pumped under relatively high pressure through a rotating gland. The cutting oil not only lubricates and cools the drill and workpiece, it also flushes out the chips. Gun drills designed to make very accurate and very deep holes have a similar arrangement, but the fluid pressures are as high as 1000 psi. Shields must be provided for operator safety when using these extremely high pressure coolant systems.

Air-carried mist systems (Figure F-25) are popular for some drilling and end milling operations. Usually, shop compressed air is used to make a spray of coolant drawn from a small tank on or near the machine. Since only a small amount of liquid ever reaches the cutting area, very little lubrication is provided. These systems are chosen for their ability to cool rather than to lubricate. This factor may reduce tool life. However, mist cooling has many advantages. No splash guards, chip pans, and return hoses are needed and only small amounts of liquid are used. The high-velocity air stream also cools the spray further by evaporation. There are some safety hazards in using mist spray equipment from the standpoint of the operator's health. Conventional coolants are not highly toxic, but when sprayed in a fine mist they can be inhaled; this could affect certain people over a period of time. However, many operators find that breathing any of the mist is uncomfortable and offensive. Good ventilation systems or ordinary fans can remove the mist from the operator's area.

SELF-TEST

1. What are the two basic functions of cutting fluids?

2. Name the four types of cutting fluids (*liquids*).

3. When soluble oil coolants become odorous or cutting oil becomes rancid in the reservoir, what can be done to correct the problem?

4. When a machine such as a lathe has a coolant pump and tank that contains a soluble oil–water mix, why should a machine operator not use a pump can containing cutting oil on a workpiece?

5. Some of the synthetic cutting fluids tend to irritate some workers' hands. What causes this?

6. What method of applying cutting fluid is often used on a small drill press?

7. What kinds of cutting fluid accessories are provided on most machine tools, lathes, milling machines, drill presses, and grinding machines?

8. Describe the most common method of cutting fluid application from a nozzle.

9. Why is a single, round nozzle rather inefficient when used on a grinding wheel?

10. Spray-mist cooling systems work well for cooling purposes, but do not lubricate the work area and tool very well. Why is this?

UNIT 4

Using Carbides and Other Tool Materials

The cutting tool materials such as carbon steels and high-speed steel that served the needs of machining in the past years are not suitable in many applications today. Tougher and harder tools are required to machine the tough, hard, space-age metals and new alloys. The constant demand for higher productivity led to the need for faster stock removal and quick-changing tooling. You, as a machinist, must learn to achieve maximum productivity at minimum cost. Your knowledge of cutting tools and ability to select them for specific machining tasks will affect your productivity directly.

Objectives

After completing this unit, you should be able to:

1. List six different cutting tool materials and compare some of their machining properties.
2. Select a carbide tool for a job by reference to operating conditions, carbide grades, nose radii, tool style, rake angles, shank size, and insert size, shape, and thickness.
3. Identify carbide inserts and toolholders by number systems developed by the American Standards Association.

The various tool materials used in today's machining operations are high-carbon steel, high-speed steel, cemented carbides, ceramics, and diamond.

High-Carbon Steels

High-carbon tool steels are used for hand tools such as files and chisels, and only to a limited extent for drilling and turning tools. They are oil or water hardening plain carbon steels with .9 to 1.4 percent carbon content. These tools maintain a keen edge and can be used for metals that produce low tool–chip interface temperatures; for example, aluminum, magnesium, copper, and brass. These tools, however, tend to soften at machining speeds above 50 feet per minute (fpm) in mild steels.

High-Speed Steels

High-speed steels (HSS) may be used at higher speeds (100 fpm in mild steels) without losing their hardness. The relationship of cutting speeds to the approximate temperature of tool–chip interface is as follows:

100 fpm	1000°F (538°C)
200 fpm	1200°F (649°C)
300 fpm	1300°F (704°C)
400 fpm	1400°F (760°C)

High-speed steel is sometimes used for lathe tools when special tool shapes are needed, especially for boring tools. However, high-speed steel is extensively used for milling cutters for both vertical and horizontal milling machines. Since milling cutters are usually used at cutting speeds of 100 sfpm in steels with cutting fluid, they never reach 1000°F (538°C) and so have a relatively long working life.

Cemented Carbides

A carbide, generally, is a chemical compound of carbon and a metal. The term *carbide* is commonly used to refer to cemented carbides, the cutting tools composed of tungsten carbide, titanium carbide, or tantalum carbide and cobalt in various combinations. A typical composition of cemented carbide is 85 to 95 percent carbides of tungsten and the remainder a cobalt binder for the tungsten carbide powder.

Cemented carbides are made by compressing various metal powders (Figure F-26) and sintering (heating to weld particles together without melting them) the briquettes. Cobalt powder is used as a binder for the carbide powder used, either tungsten, titanium, or tantalum carbide powder or a combination of these. Increasing the percentage of cobalt binder increases the toughness of the tool material and at the same time reduces its hardness or wear resistance. Carbides have greater hardness at both high and low temperatures than do high-speed steel or cast alloys. At temperatures of 1400°F (760°C) and higher, carbides maintain the hardness required for efficient machining. This makes possible machining speeds of approximately 400 fpm in steels. The addition of tantalum increases the red hardness of a tool material. Cemented carbides are extremely hard tool materials (above RA 90), have a high compressive strength and resist wear and rupture.

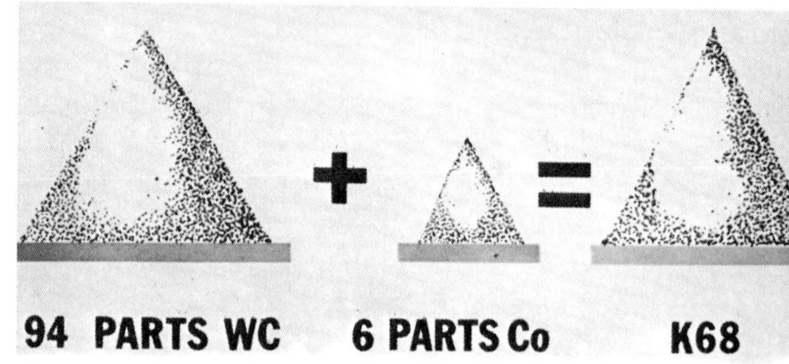

FIGURE F-26 The two basic materials needed to produce the straight grades of carbide tools are tungsten carbide (WC) and cobalt (CO) (Kennametal, Inc., Latrobe, Pa.).

Coated carbide inserts are often used to cut hard or difficult-to-machine workpieces. Titanium carbide (TiC) coating offers high wear resistance at moderate cutting speeds and temperatures. Aluminum oxide (Al_2O_3) coating resists chemical reactions and maintains its hardness at high temperatures. Titanium nitride (TiN) coating has high resistance to crater wear and reduces friction between the tool face and the chip, thereby reducing the tendency for a built-up edge.

Cemented carbides are the most widely used tool materials in the machining industry. They are particularly useful for cutting tough alloy steels which quickly break down high-speed tool steels. A very large percentage of machining is on alloy steels for automotive and other industrial machine parts. Various carbide grades and insert shapes are available and the correct selection should be made for machining a particular material.

Selecting Carbide Tools

The following steps may be used in selecting the correct carbide tool for a job.

1. Establish the operating conditions.
2. Select the cemented carbide grade.

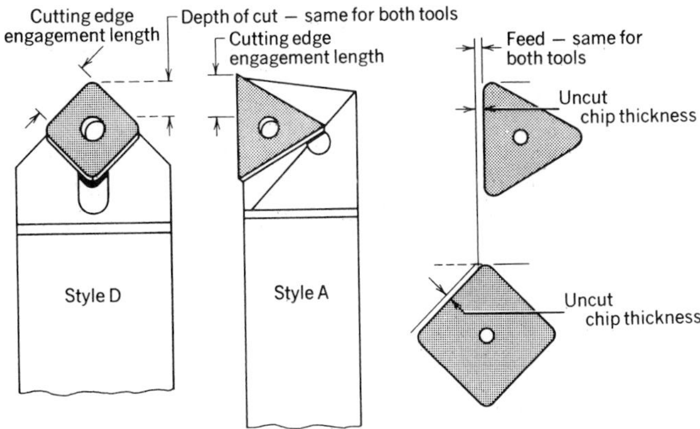

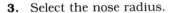

FIGURE F-27 The difference in Style A and Style D holders for depth of cut and cutting edge engagement length (Copyright © General Electric Company).

3. Select the nose radius.
4. Select the insert shape.
5. Select the insert size.
6. Select the insert thickness.
7. Select the tool style.
8. Select the rake angle.
9. Select the shank size.
10. Select the chip breaker.

STEP 1. ESTABLISHING THE OPERATING CONDITIONS

The tool engineer or machinist must use three variables to establish metal removal rate: speed, feed, and depth of cut. Cutting speed has the greatest effect on tool life. A 50 percent increase in cutting speed will decrease tool life by 80 percent. A 50 percent increase in feed will decrease tool life by 60 percent. The cutting edge engagement or depth of cut (Figure F-27) is limited by the size and thickness of the carbide insert and the hardness of the workpiece material. Hard workpiece materials require decreased feed, speed, and depth of cut.

Figure F-27 shows how the lead angle affects both cutting edge engagement length and chip thickness by comparing a style D square insert tool, using a 45-degree lead angle, to a style A triangular insert tool, using a 0-degree lead angle. The two tools are shown making an identical depth of cut at an identical feed rate. The feed rate is the same as the chip thickness with the style A tool or any tool with a 90-degree lead angle. The chip would be wider but thinner using the style D with the same feed rate. When large lead angles are used, the style D tool has a much greater strength than the style A tool since the cutting forces are directed into the solid part of the holder. Large lead angles can cause chatter to develop, however, if the setup is not rigid.

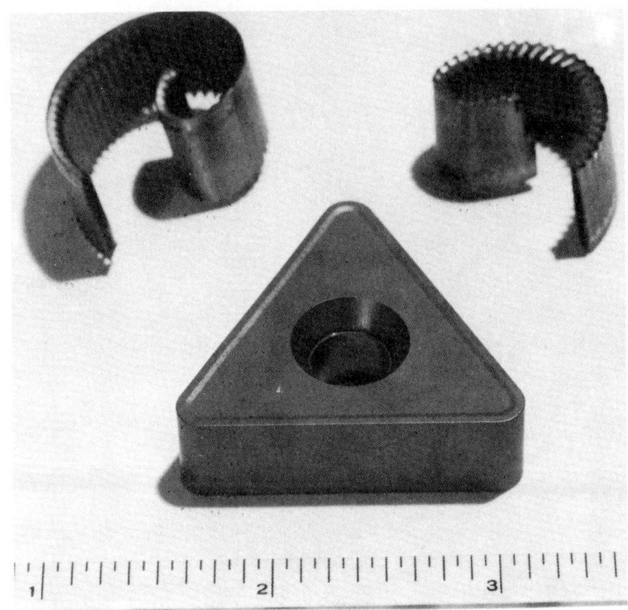

FIGURE F-28 Large, well-formed chips were produced by this tool with built-in chip breaker (Kennametal, Inc., Latrobe, Pa.).

FIGURE F-29 Normal edge wear.

The depth of cut is limited by the strength and thickness of the carbide insert, the rigidity of the machine and setup, the horsepower of the machine, and, of course, the amount of material to be removed. An example of a relatively large depth of cut with an insert that produced large "9"-shaped chips is shown in Figure F-28.

Edge wear and cratering are the most frequent tool breakdowns that occur. Edge wear (Figure F-29) is simply the breaking down of the tool relief

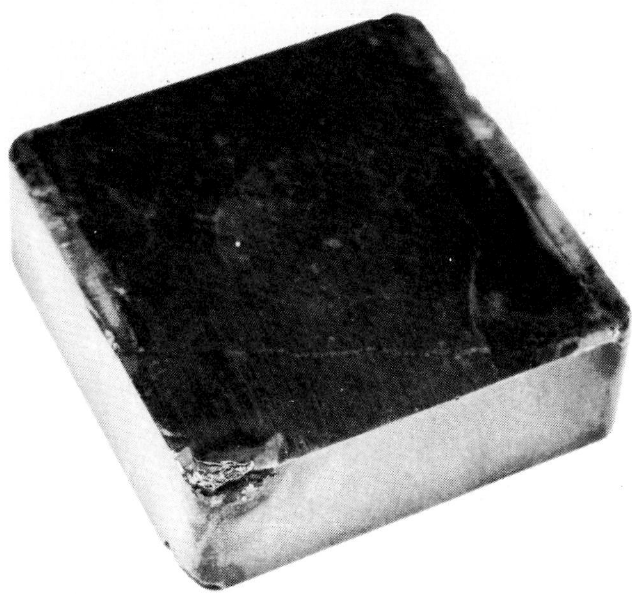

FIGURE F-30 Tool point breakdown caused by a built-up edge.

FIGURE F-31 Chipped or broken inserts. The triangular insert shows the typical breakage on straight tungsten carbide inserts. The square titanium-coated insert shows a stratified failure because of an edge impact.

surface caused by friction and abrasion and is considered normal wear. Edge breakdown is also caused by the tearing away of minute carbide particles by the built-up edge (Figure F-30). The cutting edge is usually chipped or broken in this case. Lack of rigidity, too much feed, or too slow a speed results in chipped or broken inserts (Figure F-31).

Tiny projections on the cutting edges of new inserts often break down and reduce tool life. Some machinists hone the edges of new inserts to increase tool life. However, if this practice is not carried out correctly, the tool can be damaged and actually have a lowered tool life. For this reason, some insert tool manufacturers provide prehoned carbide tools. Machine shop students should not hone the edges of their tools until they have gained more experience in the field.

Thermal shock, caused by sudden heating and cooling, is the cracking and checking of a tool that leads to breakage (Figure F-32). This condition is most likely to occur when an inadequate amount

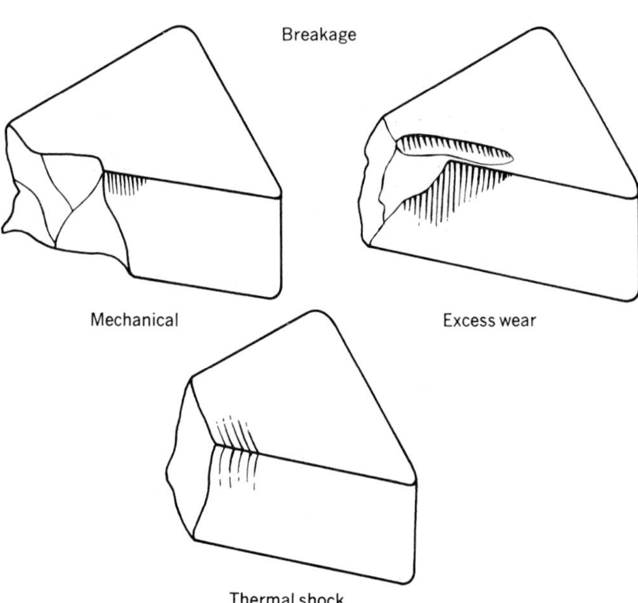

FIGURE F-32 Three causes of tool breakage (Kennametal, Inc., Latrobe, Pa.).

of coolant is used. It is better to machine dry if the work and tool cannot be kept flooded with coolant.

If edge wear occurs:

1. Decrease the machining speed.
2. Increase the feed.
3. Change to a harder, more wear-resistant carbide grade.

If the cutting edge is chipped or broken:

1. Increase the speed.
2. Decrease the feed and/or depth of cut.
3. Change to a tougher grade carbide insert.
4. Use a negative rake.
5. Hone the cutting edge before use.
6. Check the rigidity and tool overhang.

When there is a buildup on the cutting edge:

1. Increase the speed.
2. Change to a positive rake tool.
3. Change to a grade containing titanium.

For cutting edge notching (flank wear):

1. Increase the side cutting edge angle.
2. Decrease the feed.

Cratering (Figure F-33) is the result of high temperatures and pressures that cause the steel chip to weld itself to the tungsten carbide and tear

FIGURE F-33 Cratering on a carbide tool.

out small particles of the tool material. The addition of titanium carbide to the mixture of tungsten carbide and cobalt provides an antiweld quality, but there is some loss in abrasive wear and strength in these tools.

STEP 2. SELECT THE CEMENTED CARBIDE GRADES

There are two main groups of cemented carbides from which to select most grades: first, the straight carbide grades composed of tungsten carbide and cobalt binder, which are used for cast iron, nonferrous metals, and nonmetallics where resistance to edge wear is the primary factor; and second, grades composed of tungsten carbide, titanium carbide, and tantalum carbide plus cobalt binder, which are usually used for machining steels. Resistance to cratering and deformation is the major requirement for these steel grades.

Cemented carbides have been organized into grades. Properties that determine grade include hardness, toughness, and resistance to chip welding or cratering. The properties of carbide tools may be varied by the percentages of cobalt and titanium or tantalum carbides. Increasing the cobalt content increases toughness but decreases hardness. Properties may also be varied during the processing by the grain size of carbides, density, and other modifications. Some tungsten carbide inserts are given a titanium carbide coating (about .0003 in. thick) to resist cratering and edge breakdown. Tantalum carbide is added to sintered carbide principally to improve not hardness characteristics. This increases the composition's resistance to deformation at cutting temperatures.

The grades of carbides have been organized according to their suitable uses by the Cemented Carbide Producers Association (CCPA). It is recommended that carbides be selected by using such a table rather than by their composition. Cemented carbide grades with specific chip removal appliations are:

C-1 Roughing cuts (cast iron and nonferrous materials)
C-2 General purpose (cast iron and nonferrous materials)
C-3 Light finishing (cast iron and nonferrous materials)
C-4 Precision boring (cast iron and nonferrous materials)
C-5 Roughing cuts (steel)
C-6 General purpose (steel)
C-7 Finishing cuts (steel)
C-8 Precision boring (steel)

The hardest of the nonferrous/cast iron grades is C-4 and the hardest of the steel grades is C-8.

This system does not specify the particular materials or alloy, and the particular machining operations are not specified. For example, for turning a chromium–molybdenum steel, factors to be considered would include the difficulty of machining such an alloy because of its toughness. Given this example, a grade of C-5 or C-6 carbide would probably be best suited to this operation. The proof of the selection would come only with the actual machining. Cemented carbide tool manufacturers often supply catalogs designating uses and machining characteristics of their various grades. See Table F-5.

Grade classification–comparison tables that convert each manufacturer's carbide designations to CCPA "C" numbers are available. See Table F-6. There is, however, one major caution. The tables are intended to correlate grades on the basis of composition, not according to tested performance. Grades from different manufacturers having the same "C" number may vary in performance. Some general guidelines to grade selection are as follows:

1. Select the grade with the highest hardness with sufficient strength to prevent breakage.

2. Select straight grades of tungsten carbide for the highest resistance to abrasion.

STEP 3. SELECT NOSE RADIUS

Selecting the nose radius can be important because of tool strength, surface finish, or, perhaps, the forming of a fillet or radius on the work. To determine the nose radius according to strength

TABLE F-5 Grade and Machining Applications

Group	Subgroup	Grade	Hardness R_A	Typical Machining Applications
Combined Crater and Edge-Wear Resistance	Maximum Crater Resistance	CO6	Ceramic[a]	The hardest of this group. For finishing most ferrous and nonferrous alloys and nonmetals as in high-speed, light-chip-load precision maching, or for use at moderate speeds and chip loads where long tool life is desired.
		K165	93.5	Titanium carbide for finishing steels and cast irons at high to moderate speeds and light chip loads.
		K7H	93.5	For finishing steels at higher speeds and moderate chip loads.
		K5H	93.0	For finishing and light roughing steels at moderate speeds and chip loads through light interruptions.
		K45	92.5	The hardest of this group. General-purpose grade for light roughing to semifinishing of steels at moderate speeds and chip loads and for many low-speed, light-chip-load applications.
		K4H	92.0	For light roughing to semifinishing of steels at moderate speeds and chip loads and for form tools and tools that must dwell.
		K2S	91.5	For light to moderate roughing of steels at moderate speeds and feeds through medium interruptions.
	KC75			**For general-purpose use in machining of steels over a wide range of speeds in moderate roughing to semifinishing applications.**
		K21	91.0	For moderate to heavy roughing of steels at moderate speeds and heavy chip loads through medium interruptions where mechanical and thermal shock are encountered.
		K42	91.3	For heavy roughing of steels at low to moderate speeds and heavy chip loads through interruptions where mechanical and severe thermal shocks are encountered.
	Maximum Edge-Wear Resistance	K11	93.0	The hardest of this group. For precision finishing of cast irons, nonferrous alloys, nonmetals at high speeds and light chip loads, and for finishing many hard steels at low speeds and light chip loads.
		K68	92.6	General-purpose grade for light roughing to finishing of most high-temperature alloys, refractory metals, cast irons, nonferrous alloys, and nonmetals at moderate speeds and chip loads through light interruptions.
		K6	92.0	For moderate roughing of most high-temperature alloys, cast irons, nonferrous alloys, and nonmetals at moderate to low speeds and moderate to heavy chip loads through light interruptions.
		K1	90.0	The most shock resistant of this group. For heavy roughing of most high-temperature alloys, cast irons, and nonferrous alloys at low speeds and heavy chip loads through heavy interruptions.

[a] The hardness of CO6 is 91 Rockwell 45N (or about $94R_A$).

SOURCE: *Kentrol Inserts (Supplement 5 to Catalog 73*, "Kennametal Grade Systems and Machining Applications," 1975 (data courtesy of Kennametal, Inc., Latrobe, PA).

requirements, use the nomograph in Figure F-34. Consider that the feed rate, depth of cut, and workpiece condition determine strength requirements, since a larger nose radius makes a strong tool.

Large radii are strongest and can produce the best finishes, but they also can cause chatter between tool and workpiece. For example, the dashed line on the figure indicates that a $\frac{1}{8}$-in. radius would be required for turning with a feed rate of .015 in., and to obtain a 100-microinch finish, a $\frac{1}{4}$-in. radius would be required with a .020-in. feed rate.

STEP 4. SELECT INSERT SHAPES

Indexable inserts (Figure F-35), also called throwaway inserts, are clamped in toolholders of various design. These inserts provide a cutting tool with several cutting edges. After all edges have been used, the insert is discarded.

The round inserts have the greatest strength and, as with large radius inserts, make possible higher feed rates with equal finishes. Round inserts also have the greatest number of cutting edges possible, but are limited to workpiece con-

TABLE F-6 Carbide Grade Classification–Comparison Table with CCPA "C" Numbers and Manufacturer's Designations

Application			New-comer	Adamas	Atrax	Car-boloy	Carmet	Ex-cell-o	Firth Sterling	Green-leaf	Kenna-metal	Metal Carbides	Sand-vik	Valenite	V-R Wesson	Walmet	Wendt-Sonis
Cast irons	Roughing cuts	C-1	N10	B	FA5	44A	CA3	E8	H HB	G10	K1	C89	H20	VC-1	VR54 2A68	WA-1 WA-159	CQ12 CQ22
Nonferrous, nonmetallic, high-temperature alloys	General purpose	C-2	N20 N22	A AM	FA6 FA-62	883 860	CA4 CA443	E6 XL620	HA HTA	G20 G25	K6 K68	C91	H20	VC-2 VC-28	2A5 VR82	WA-69 WA-2	CQ2 CQ23
	Light finishing	C-3	N30	PWX	FA7	905	CA7	E5	HE HTA	G30	K8 K68	C93	R1P	VC-3	2A7 VR82	WA-35 WA3	CQ3 CQ23
200 and 300 series stainless	Precision boring	C-4	N40	AAA	FA8	999 895	CA8	E3	HF	G40	K11	C95	H1P HO5	VC-4	2A7	WA4	CQ4
Carbon steels	Roughing cuts	C-5	N50 N52	499 434	FT-3 FT-35	370 78B	CA721 CA740	10A 945	NTA TXH	G50 G55	K42 K21	S-880	S-6	VC-55 VC-125	VR77 WM	WA5 WA55	CY12 CY17
	General purpose	C-6	N60	6X T-60	FT-4 FT-6	78B	CA720	BA 606	T22 T25	G60	K25 K21	S-900 S-901	S-4	VC-6	26 VR75	WA6	CY5 CY16
Alloy steels	Finishing cuts	C-7	N70 N72	495 548	FT-6 FT-62	78 350	CA711	6A XL70 6AX	725 T31	G70 G74	K45 K5H	S-92 S-900	SM	VC7 VC-76	WH VR73	WA7 WA168	CY2 CY14
400 Series stainless	Precision boring	C-8	N80 N93	490 T-80	FT-7 FT-71	330 210	CA704	6AX XL88	T31	G80	K7H K165	S-94	FO2	VC-8 VC-83	VR71 VR65	WA8 WA800	CY31 Ti8
	High-velocity	C-80	N95			0–30					CO6				VR97		

NOTE: The grades listed are those usually recommended by the manufacturer for the categories shown.

SOURCE: Newcomer Products, Inc. *Reference card.*

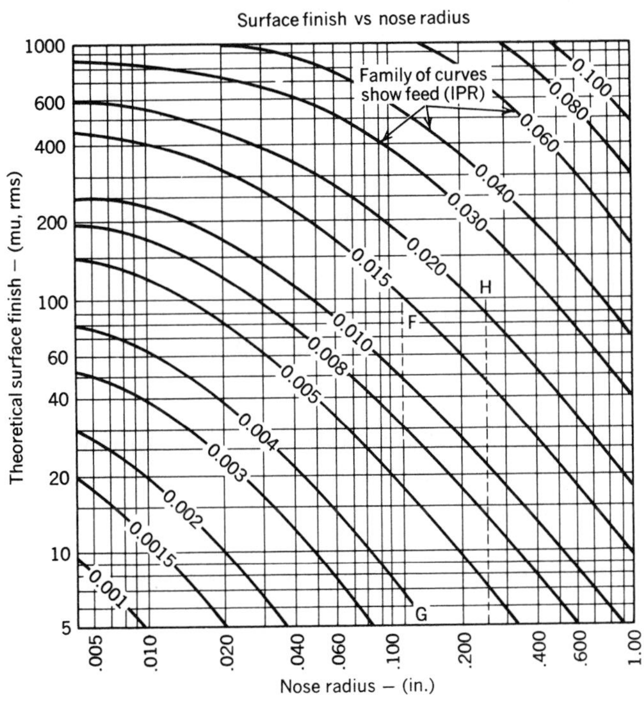

FIGURE F-34 Surface finish vs. nose radius (Copyright © General Electric Company).

figurations and operations that are not affected by a large radius. Round inserts would be ideally suited, for example, to straight turning operations.

Square inserts have lower strength and fewer possible cutting edges than round tools, but are much stronger than triangular inserts. The included angle between cutting edges (90 degrees) is greater than that for triangular inserts (60 degrees), and there are eight cutting edges possible, compared to six for the triangular inserts.

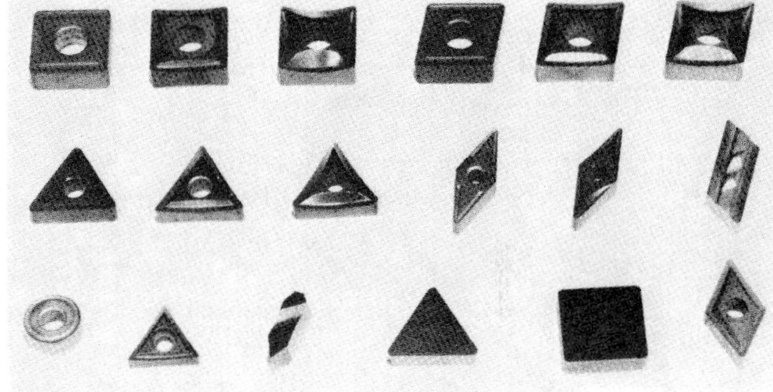

FIGURE F-35 Insert shapes for various applications (Kennametal, Inc., Latrobe, Pa.).

Triangular inserts have the greatest versatility. They can be used, for example, for combination turning and facing operations, while round or square inserts are often not adaptable to such combinations. Because the included angle between cutting edges is less than 90 degrees, the triangular inserts are also capable of tracing operations. The disadvantages include their reduced strength and fewer cutting edges per insert.

For tracing operations where triangular inserts cannot be applied, diamond-shaped inserts with smaller included angles between edges are available. The included angles on these diamond-shaped inserts range from 35 to 80 degrees. The smaller angle inserts in particular may be plunged

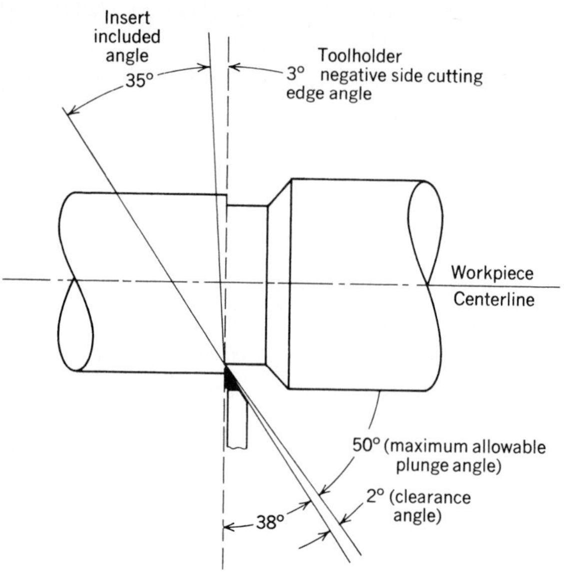

FIGURE F-36 A 38-degree triangular insert used for a tracing operation (Copyright General Electric Company).

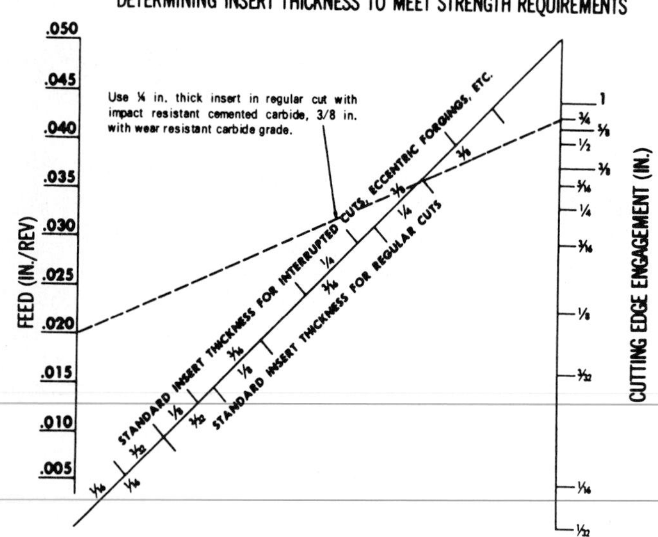

FIGURE F-37 Insert thickness as determined by length of cutting edge engagement and feed rate (Copyright General Electric Company).

into the workpiece as required for tracing. A typical setup for tracing is shown in Figure F-36. Note that small clearance angles are used for each cutting edge to permit plunging.

STEP 5. SELECT INSERT SIZE

The insert selected should be the smallest insert capable of sustaining the required depth of cut and feed rate. The depth of cut should always be as great as possible. A rule of thumb is to select an insert with cutting edges $1\frac{1}{2}$ times the length of

FIGURE F-38 Several of the many tool styles available (Kennametal, Inc., Latrobe, Pa.).

cutting edge engagement. The feed for roughing mild steel should be approximately $\frac{1}{10}$ the depth of cut.

STEP 6. SELECT INSERT THICKNESS

Insert thickness is also important to tool strength. The required depth of cut and feed rate are criteria that determine insert thickness. The nomograph in Figure F-37 simplifies the relationship of depth of cut and feed rate to insert thickness. Lead angle is also important in converting the depth of cut to a length of cutting edge engagement.

The dashed line on the nomograph in Figure F-37 represents an operation with a $\frac{3}{4}$-in. length of engagement (not depth of cut) and a feed rate of .020 ipr. Depending on the grade of carbide used (tough or hard), an insert of $\frac{1}{4}$- or $\frac{3}{8}$-in. thickness should be used.

STEP 7. SELECT TOOL STYLE

Tool style pertains to the configuration of toolholder for a carbide insert. To determine style, some familiarity with the particular machine tool and the operations to be performed is required. Figure F-38 shows some of the styles available for toolholders.

STEP 8. SELECT RAKE ANGLE

When selecting the rake angles, you need to consider the machining conditions. Negative rake should be used where there is maximum rigidity of the tool and work and where high machining speeds can be maintained. More horsepower is needed when using negative rake tools. Under

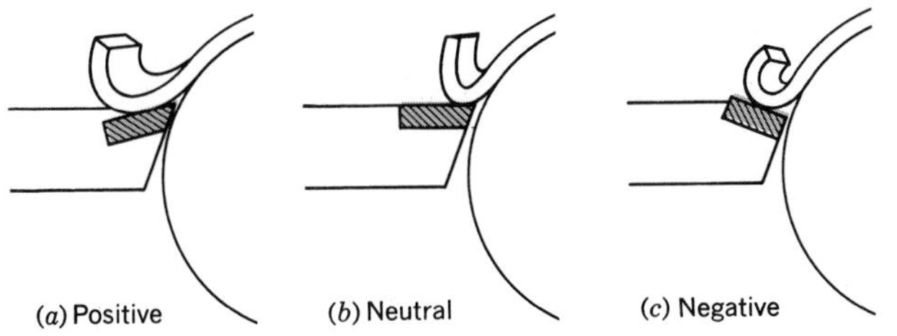

FIGURE F-39 Side view of back rake angles.

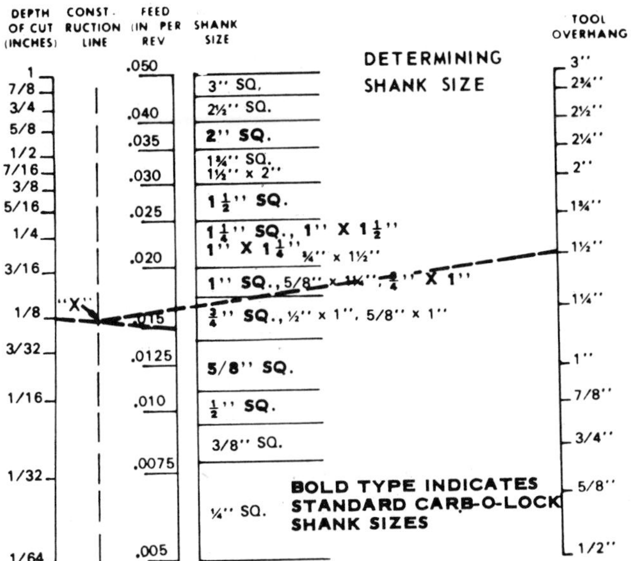

FIGURE F-40 Determining shank size according to depth of cut, feed rate, and tool overhang (Copyright © General Electric Company).

these conditions, negative rake tools are stronger and produce satisfactory results (Figure F-39).

Negative rake inserts may also be used on both sides, doubling the number of cutting edges per insert. This is possible because end and side relief are provided by the angle of the toolholder rather than by the shape of the insert.

Positive rake inserts should be used where rigidity of the tool and work is reduced and where high cutting speeds are not possible; for example, on a flexible shaft of smaller diameter. Positive rake tools cut with less force so deflection of the work and toolholder would be reduced. High cutting speeds (sfpm) are often not possible on small diameters because of limitations in spindle speeds.

STEP 9. SELECT SHANK SIZE

As with insert thickness and rake angles, the rate of feed and depth of cut are important in determining shank size. Overhang of the tool shank is extremely important for the same reason. Heavy

FIGURE F-41 A boring bar with various interchangeable adjustable heads (Kennametal, Inc., Latrobe, Pa.).

cuts at high feed rates create high downward forces on the tool. These downward forces acting on a tool with excessive overhang would cause tool deflection that would make it difficult or impossible to maintain accuracy or surface finish quality.

Having established the feed rate, depth of cut, and tool overhang, use of the nomograph in Figure F-40 to determine the shank size. Begin by drawing a line from the depth of cut scale to the feed rate scale. From the point where this line intersects the vertical construction line, draw a line to the correct point on the tool overhang scale. Determine the shank size from where this last line crosses through the shank size scale.

In the example shown on the graph, a line has been drawn from the $\frac{1}{8}$ in. depth of cut point to the .015 ipr feed rate point. Another line has been drawn from the point of intersection on the vertical construction line to the amount of tool overhang. The second line drawn passes through the shank size scale at the $\frac{3}{4}$ in.^{2} (cross-sectional area) point. Toolholders $\frac{1}{2} \times 1$ in. or $\frac{5}{8} \times 1$ in. would meet the requirements. Throwaway carbide inserts are also used for boring bars (Figure F-41) of various shapes and sizes. Brazed carbide tips are sometimes put on the end of a boring bar for special applications.

STEP 10. SELECT CHIP BREAKER

Chip control is a very important aspect of machining operations, especially on the lathe. Cutting tools on the lathe that have no chip breakers tend to produce long, stringy, or coiled chips which because they are very sharp on the edges and can cause severe cuts are a hazard to the operator. The ideal shape for chips is shown in Figure F-28, the compact "9"-shaped chip. Such chip formation is a necessity in manufacturing operations, where the chips are removed by a conveyor. Most cemented carbide inserts are provided with a chip

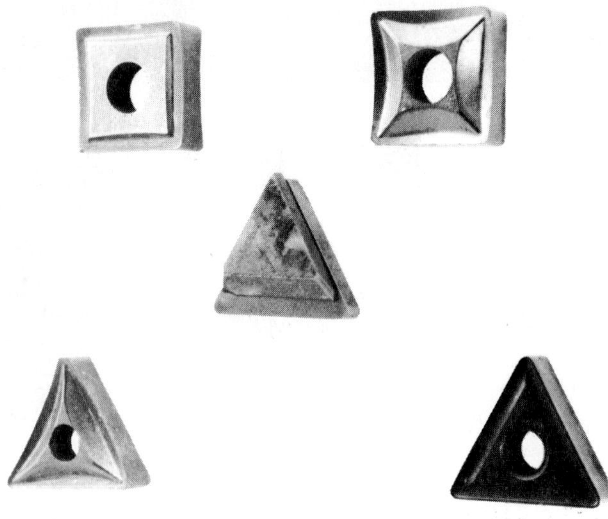

FIGURE F-42 Chip breakers used are the adjustable chip deflector (center) with a straight insert and the type with the built-in chip control groove.

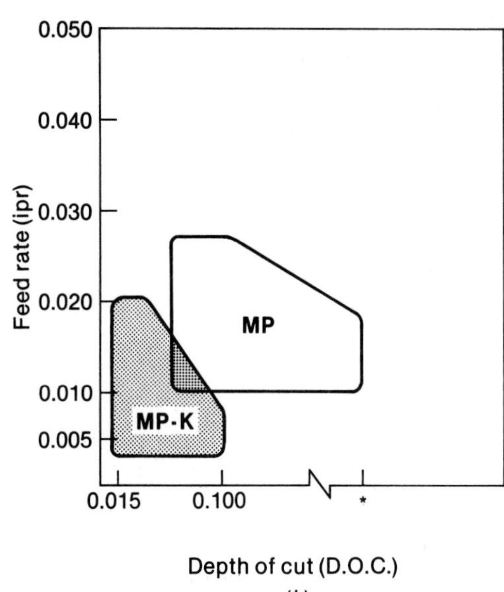

(b)

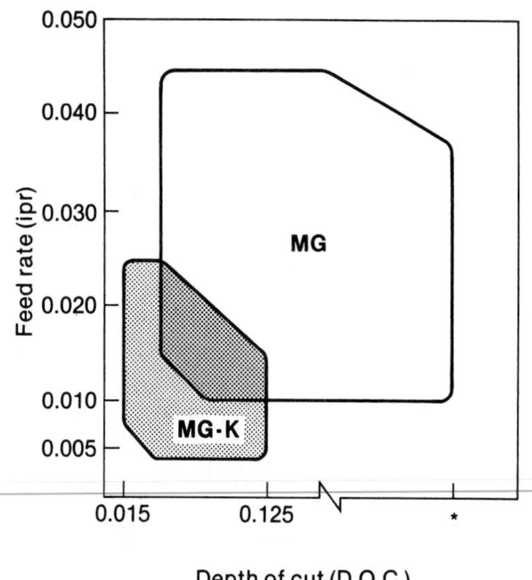

(a)

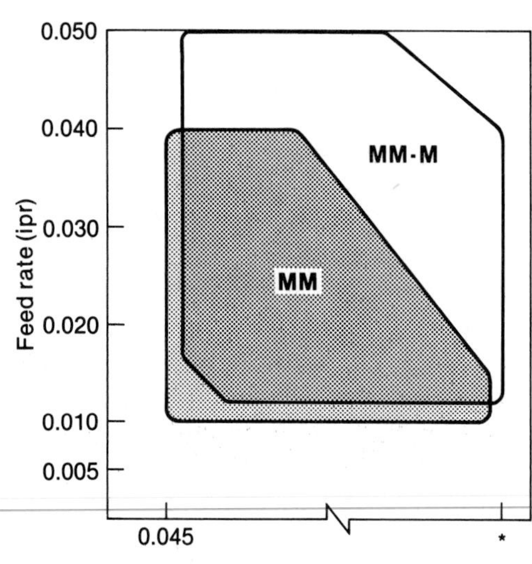

(c)

FIGURE F-43 (*a*) Negative rake two-sided Kenloc inserts; (*b*) negative/positive two-sided Kenloc inserts; (*c*) negative rake one-sided Kenloc inserts. *Maximum D.O.C. and feed rates (ipr) are limited by the insert thickness and cutting edge length. Application ranges are for AISI 1045 steel at 180 to 220 BHN (Kennametal, Inc., Latrobe, Pa.).

breaker, whose object is to form the desirable chip formation (Figure F-42). However, other factors, such as feed rate, depth of cut, and workpiece material, affect the chip curl and its final shape. A higher feed rate and deeper cut tend to curl the chip more, whereas a light cut and small feed rate will often produce a long stringy chip even if the tool has a chip breaker. A tough or hard workpiece will curl the chip more than a soft material. These factors are taken into account in Figure F-43 and in Table F-7. These ranges for chip breaker application are for Kennametal inserts. However, every tool manufacturer provides its own special shapes and specifications for their use with various materials and applications. Catalogs containing tool specifications are generally available from tool distributors.

Toolholder Identification

The carbide manufacturers and the American Standards Association (ASA) have adopted a system of identifying toolholders for inserted car-

TABLE F-7 Guide to Effective Application of Kenloc Chip-Control Inserts

Kenloc Series	Effective Rake	Number of Cutting Surfaces	Application
MP-K	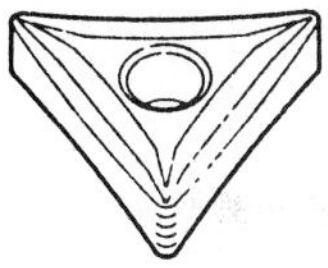 Positive +5°	Two-sided	For chip control when depths of cut are .100 or less and/or fine feed rates are required. Nose radius (in.) / Feed rate range[a] (ipn) .015 / .003–.010 .031 / .005–.015 .047 / .005–.015 .062 / .008–.020
MG-K	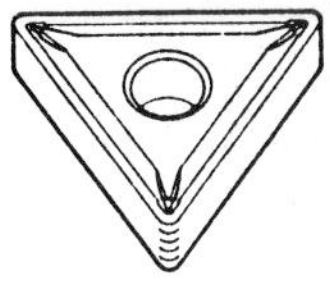 Negative −5°	Two-sided	For chip control in light to moderate cuts. The effective feed range extends from .005 to .025 ipr, depending on size of insert and nose radius. Excellent for work-hardening materials, using light feeds and light depths of cut.
MP	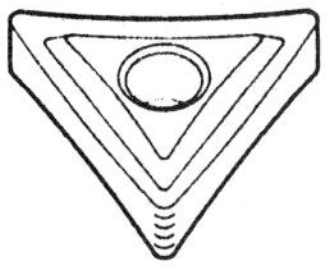 Positive +5°	Two-sided[b]	Used to minimize forces and/or workpiece deflection. Also, excellent for machining work-hardening materials. Minimum depth of cut for effective chip control is .060 in.[a]
MG	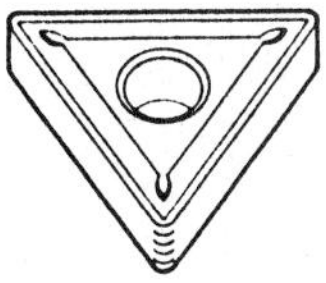 Negative −5°	Two-sided	General-purpose machining with wide range of effective chip control. The feed range extends from a low of .010–.015 to a high of .030–.050 ipr, depending on size of insert and nose radius.[a]
MM-M	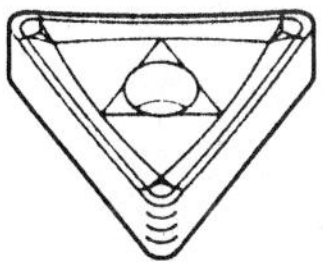 Negative −5°	One-sided	For maximum metal removal rates, and for additional strength in interrupted cutting. Effective positive cutting action reduces cutting forces and extends tool life.
MS	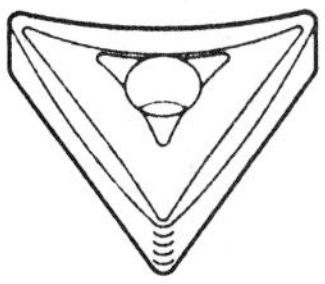 High positive +15°	One-sided	A supplemental geometry to minimize forces and workpiece deflection in light finishing cuts where horsepower is limited.
MM	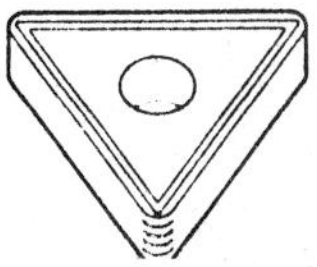 Negative −5°	One-Sided	Supplemental geometry for chip control in applications that demand heavy depth of cut with light to moderate feed rates.

[a] Parameters based on AISI 1045 steel 180–220 BHN; feed rates and depth of cut will vary for harder and higher-stength materials.

[b] Parameters based on AISI 1045 Steel 180–225 BHN, feed rates & depth of cuts will vary for harder, and higher strength materials.

SOURCE: Kennametal, Inc., Latrobe, Pa.

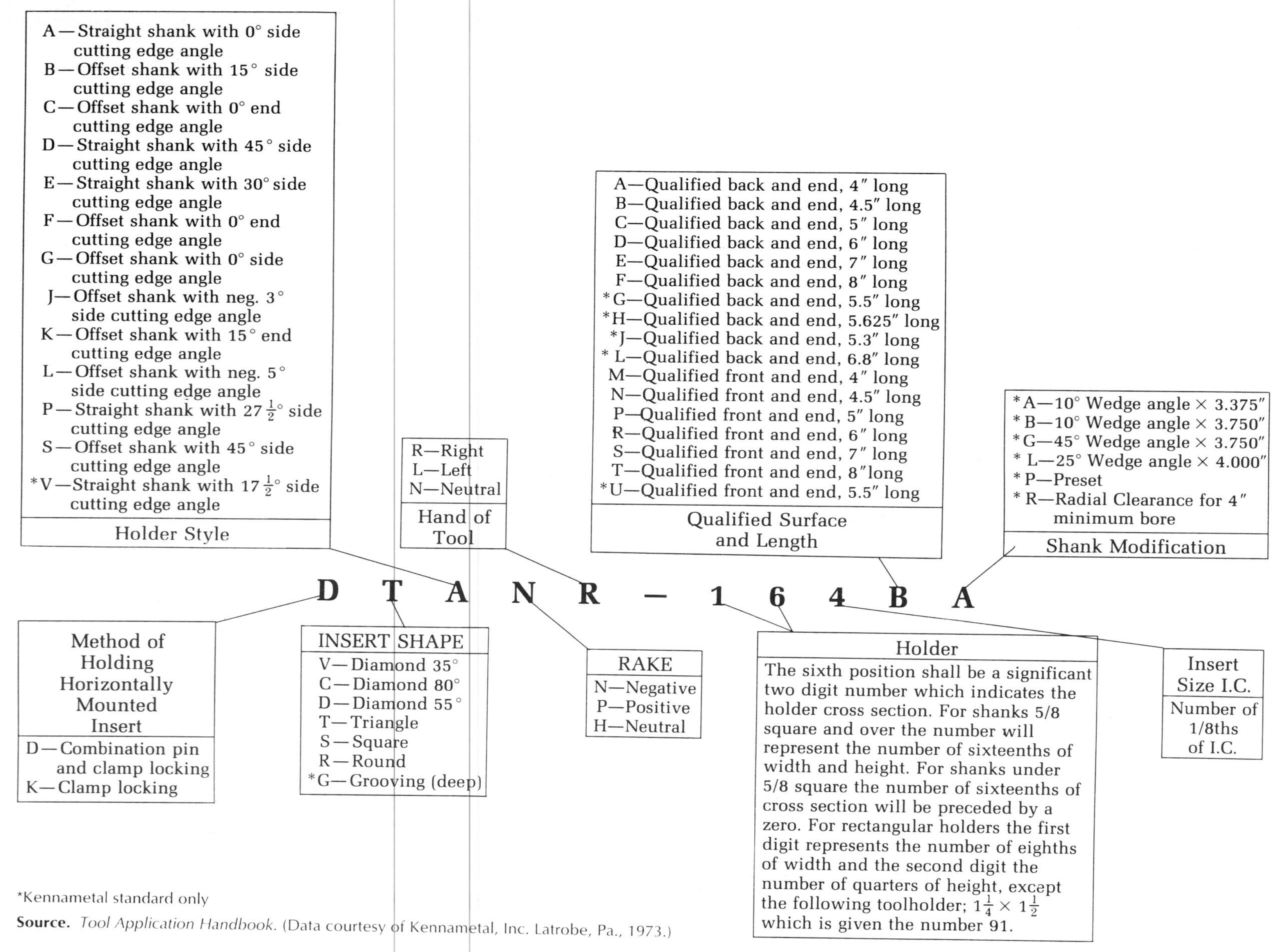

*Kennametal standard only

Source. *Tool Application Handbook.* (Data courtesy of Kennametal, Inc. Latrobe, Pa., 1973.)

FIGURE F-44 **ASA tool identification system (*Tool Application Handbook*; data courtesy of Kennametal, Inc., Latrobe, Pa., 1973).**

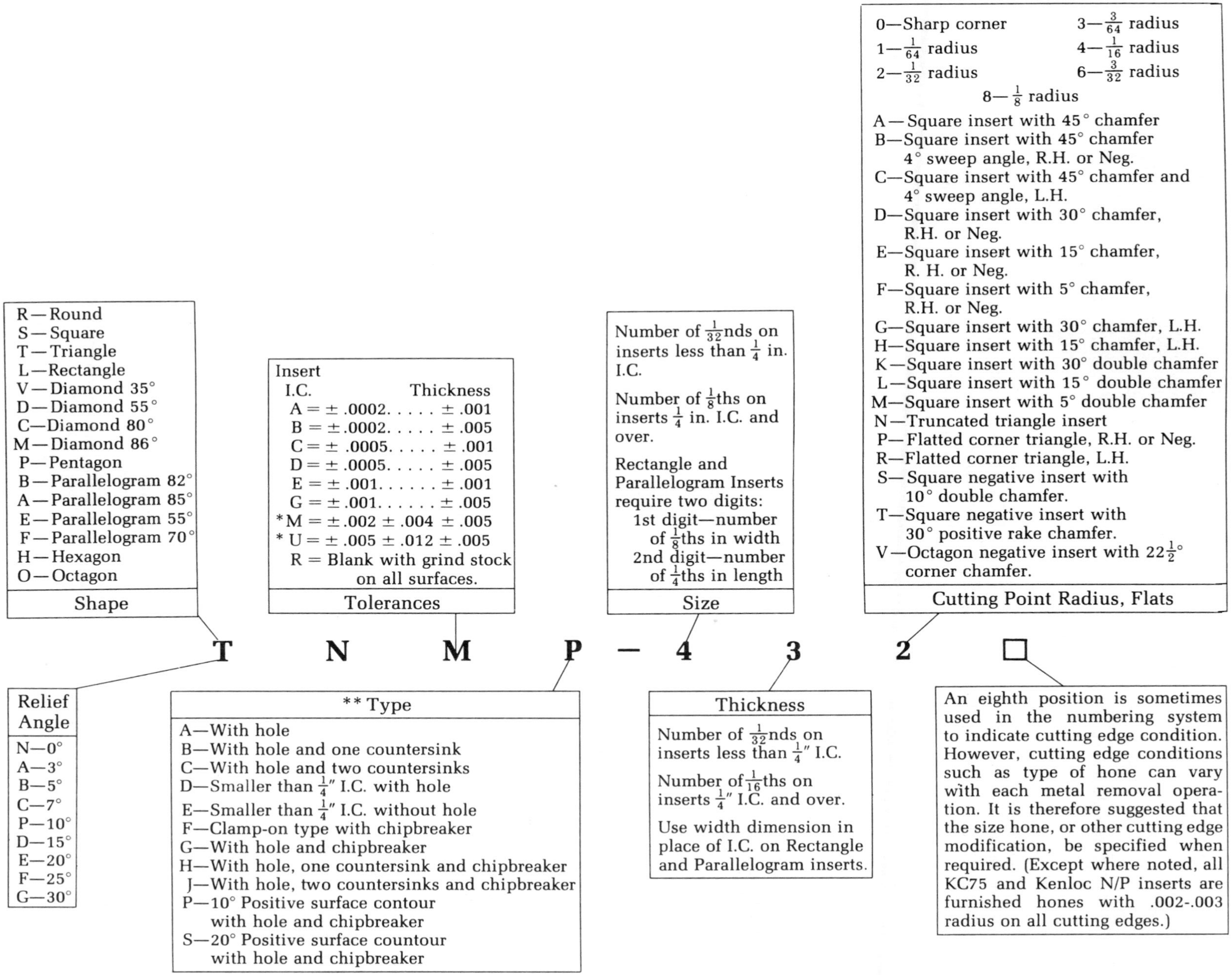

T N M P — 4 3 2 □

Shape

R—Round
S—Square
T—Triangle
L—Rectangle
V—Diamond 35°
D—Diamond 55°
C—Diamond 80°
M—Diamond 86°
P—Pentagon
B—Parallelogram 82°
A—Parallelogram 85°
E—Parallelogram 55°
F—Parallelogram 70°
H—Hexagon
O—Octagon

Relief Angle

N—0°
A—3°
B—5°
C—7°
P—10°
D—15°
E—20°
F—25°
G—30°

Tolerances

Insert I.C.	Thickness
A = ± .0002	± .001
B = ± .0002	± .005
C = ± .0005	± .001
D = ± .0005	± .005
E = ± .001	± .001
G = ± .001	± .005
*M = ± .002 ± .004	± .005
* U = ± .005 ± .012	± .005
R = Blank with grind stock on all surfaces.	

**** Type**

A—With hole
B—With hole and one countersink
C—With hole and two countersinks
D—Smaller than $\frac{1}{4}$" I.C. with hole
E—Smaller than $\frac{1}{4}$" I.C. without hole
F—Clamp-on type with chipbreaker
G—With hole and chipbreaker
H—With hole, one countersink and chipbreaker
J—With hole, two countersinks and chipbreaker
P—10° Positive surface contour with hole and chipbreaker
S—20° Positive surface countour with hole and chipbreaker

Size

Number of $\frac{1}{32}$nds on inserts less than $\frac{1}{4}$ in. I.C.

Number of $\frac{1}{8}$ths on inserts $\frac{1}{4}$ in. I.C. and over.

Rectangle and Parallelogram Inserts require two digits:
1st digit—number of $\frac{1}{8}$ths in width
2nd digit—number of $\frac{1}{4}$ths in length

Thickness

Number of $\frac{1}{32}$nds on inserts less than $\frac{1}{4}$" I.C.

Number of $\frac{1}{16}$ths on inserts $\frac{1}{4}$" I.C. and over.

Use width dimension in place of I.C. on Rectangle and Parallelogram inserts.

Cutting Point Radius, Flats

0—Sharp corner
1—$\frac{1}{64}$ radius
2—$\frac{1}{32}$ radius
3—$\frac{3}{64}$ radius
4—$\frac{1}{16}$ radius
6—$\frac{3}{32}$ radius
8—$\frac{1}{8}$ radius

A—Square insert with 45° chamfer
B—Square insert with 45° chamfer 4° sweep angle, R.H. or Neg.
C—Square insert with 45° chamfer and 4° sweep angle, L.H.
D—Square insert with 30° chamfer, R.H. or Neg.
E—Square insert with 15° chamfer, R. H. or Neg.
F—Square insert with 5° chamfer, R.H. or Neg.
G—Square insert with 30° chamfer, L.H.
H—Square insert with 15° chamfer, L.H.
K—Square insert with 30° double chamfer
L—Square insert with 15° double chamfer
M—Square insert with 5° double chamfer
N—Truncated triangle insert
P—Flatted corner triangle, R.H. or Neg.
R—Flatted corner triangle, L.H.
S—Square negative insert with 10° double chamfer.
T—Square negative insert with 30° positive rake chamfer.
V—Octagon negative insert with $22\frac{1}{2}$° corner chamfer.

An eighth position is sometimes used in the numbering system to indicate cutting edge condition. However, cutting edge conditions such as type of hone can vary with each metal removal operation. It is therefore suggested that the size hone, or other cutting edge modification, be specified when required. (Except where noted, all KC75 and Kenloc N/P inserts are furnished hones with .002-.003 radius on all cutting edges.)

*Exact tolerance is determined by size of insert **Shall be used only when required

Source. *Tool Application Handbook.* (Data courtesy of Kennametal, Inc., Latrobe, Pa., 1973.

FIGURE F-45 ASA carbide insert identification (*Tool Application Handbook*, data courtesy of Kennametal, Inc., Latrobe, Pa., 1973).

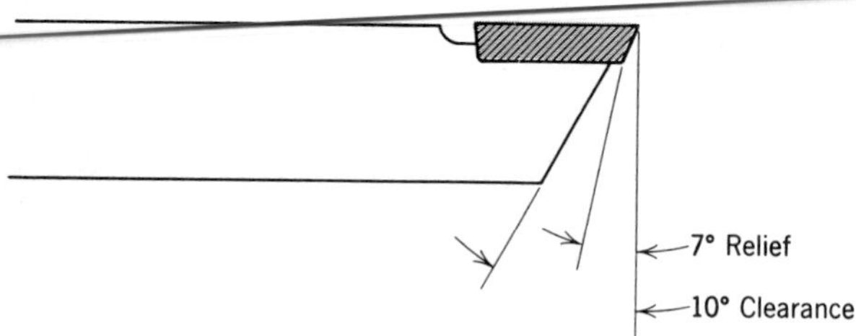

FIGURE F-46 Relief and clearance compared.

FIGURE F-47 This milling cutter uses carbide inserts that can be indexed. When one edge is dull, inserts are turned to a new sharp edge (Lane Community College).

bides. This system is used to call out the toolholder geometry and for ordering tools from manufacturers or distributors. The system is shown in Figure F-44.

As an example, use this system to determine the geometry of a toolholder called out as TANR-8:

- T Triangular insert shape
- A 0-degree side cutting edge angle
- N Negative rake
- R Right-hand turning (from left to right)
- 8 Square shank, $\frac{8}{16}$ in. per side

Carbide Insert Identification

As with toolholder identification, a system has been adopted by the carbide manufacturers and the American Standard Association for identifying inserts (see Figure F-45).

For example, use this system to determine the specifications for an insert called out as T N M G-323E:

- T Triangular shape
- N 0-degree relief (relief provided by holder)
- M Plus or minus .005 in. tolerance on thickness
- G With hole and chip breaker
- 3 $\frac{3}{8}$-in. inscribed circle (inside square and triangular insert)
- 2 $\frac{2}{16}$-in. thickness
- 3 $\frac{3}{64}$-in. radius
- E Unground honed

Brazed carbide tools have been used for many years and are still used on many machining jobs. These tools can be sharpened by grinding many times, while the insert tool is thrown away after its cutting edges are dull. Grinding the tool often causes thermal shock by the sudden heating of the carbide surface. This is evident when "crazing" or several tiny checks appear on the edge or when a large crack can be seen in the tool. To prevent crazing when grinding these tools:

1. Avoid the use of aluminum oxide wheels (except to rough grind the steel shank for clearance). Use only silicon carbide wheels with a soft bond.
2. Avoid excessive grinding pressure in a small area.
3. Avoid poor cutting action of a low concentration diamond wheel.
4. Avoid dry grinding.

Chip breakers may be ground on the edge of brazed carbide tools with diamond wheels on a surface grinder.

The side and end clearance angles are not to be confused with relief angles on carbide tools (Figure F-46). Clearance refers to the increased angle ground on the shank of a carbide tipped tool. Clearance provides for a narrow flank on the carbide and for regrinding of the carbide without contacting steel.

Since heavy forces and high speeds are involved in lathe work when using carbides, safety considerations are essential. Chip forms such as the ideal "9" or C-shape are convenient to handle, but shields, chip deflectors, or guards are required to direct chips away from workers in the shop.

Make sure that the setup is secure, that centers are large enough to support the work, and that work cannot slip in the chuck. If a long, slender work is machined at high speeds and there is a possibility of it being thrown out, guards should be provided.

Some milling cutters (Figure F-47) like lathe tools, are made to use indexable, throwaway inserts. Carbide cutters used on vertical and horizontal milling machines can increase stock removal and production for certain applications. The same considerations for use with certain workpiece materials is given to the selection of insert carbides as with lathe tools.

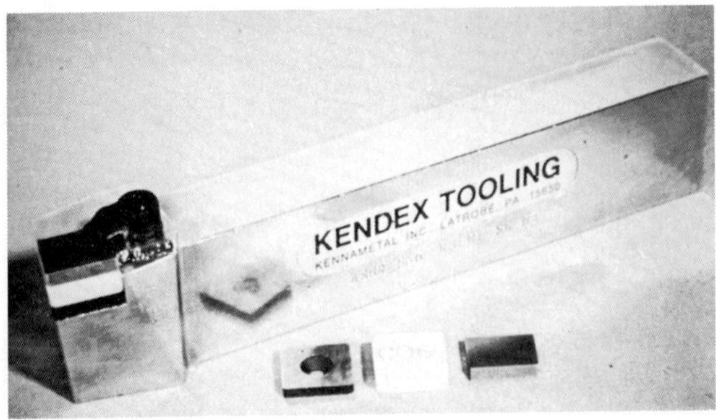

FIGURE F-48 Ceramic tool with carbide seat and chip deflector shown assembled in toolholder behind the parts (Kennametal, Inc., Latrobe, Pa.).

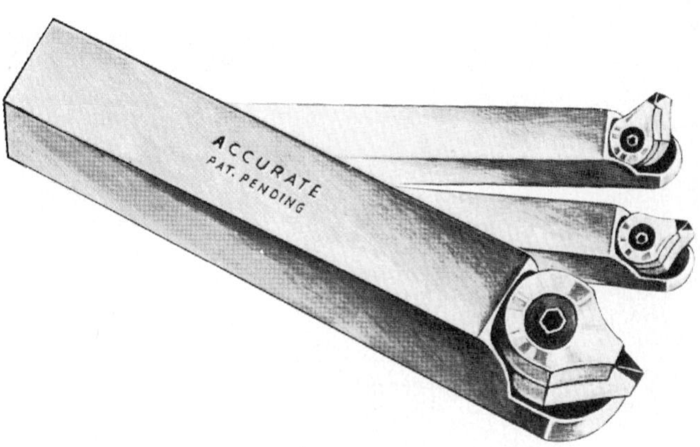

FIGURE F-49 Diamond tools are mounted in round or square shank holders as replaceable inserts (Accurate Diamond Tool Corp.). (The Universal Turning and Boring Tool is patented; the patent is owned by Accurate Diamond Tool Corp.).

Ceramic Tools

Ceramic or "cemented oxide" tools (Figure F-48) are made primarily from aluminum oxide. Some manufacturers add titanium, magnesium, or chromium oxides in quantities of 10 percent or less. The tool materials are molded at pressures over 4000 psi and sintered at temperatures of approximately 3000°F (1649°C). This process partly accounts for the high density of hardness of cemented oxide tools.

Cemented oxides are brittle and require that machines and setups are rigid and free of vibration. Some machines cannot obtain the spindle speeds required to use cemented oxides at their peak capacity in terms of sfpm. When practicable, however, cemented oxides can be used to machine relatively hard materials at high speeds.

Ceramic tools are either cold pressed or hot pressed, that is, formed to shape. Some hot-pressed, high-strength ceramics are termed *cermets* because they are a combination of ceramics and metals. They possess the high shear resistance of ceramics and the toughness and thermal shock resistance of metals. Cermets were originally designed for use at high temperatures such as those found in jet engines. These materials were subsequently adapated for cutting tools. The word *cermet* is a combination of the words "ceramic" and "metal." An insert tool composed of titanium carbide and titanium nitride is an example. In machining of steels, these tools have a low coefficient of friction and therefore less tendency to form a built-up edge. They are used at high cutting speeds. A common combination of materials in a cermet is aluminum oxide and titanium carbide. Another method of gaining the high-speed cutting characteristics of ceramics and the toughness of metal tools is to coat a carbide tool with a ceramic composite. Ceramic tools should be used as a replacement for carbide tools that are wearing rapidly, but not to replace carbide tools that are breaking.

Cubic Boron Nitride (CBN) Tools

CBN is next to diamond in hardness and therefore can be used to machine plain carbon steels, alloy steels, and gray cast irons with hardnesses of 45 Rc and above. Formerly, steels over 60 Rc would have to be abrasive machined, but with the use of CBN they can often be cut with single-point tools.

CBN inserts consist of a cemented carbide substrate with an outside layer of CBN formed as an integral part of the tool. Tool life, finishes, and resistance to cracking and abrasion make CBN a superior tool material to both carbides and ceramics.

Diamond Tools

Industrial diamonds are sometimes used to machine extremely hard workpieces. Only relatively small removal rates are possible with diamond tools (Figure F-49), but very high speeds are used and good finishes are obtained (Figure F-50). Nonferrous metals are turned at 2000 to 2500 fpm, for

FIGURE F-50 Turning at 725 sfm .010-in. stock is removed on each of two passes at a rate of $5\frac{1}{2}$ ipm (.0023 ipr) along the 29-in. length of the casting. A coolant is not required for this turning (Accurate Diamond Tool Company).

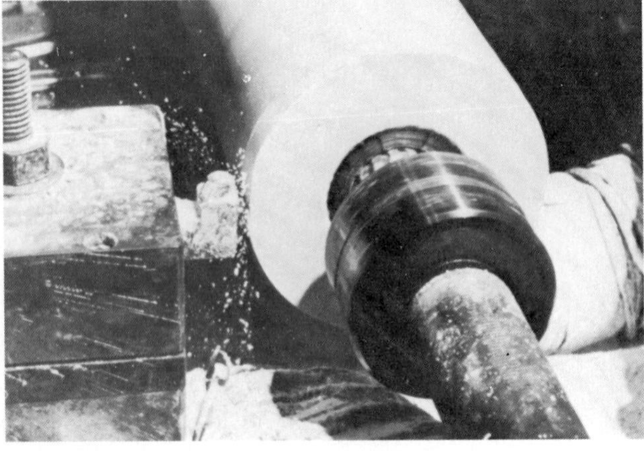

FIGURE F-52 To make abrasive fused silica tubes absolutely round where they fit into tundish valves, FloCon turns the tubes with a sintered diamond tool insert. After limited experience, the Megadiamond sintered diamond inserts appear to give about 200 times as much wear as carbides in the same application (Megadiamond Industries).

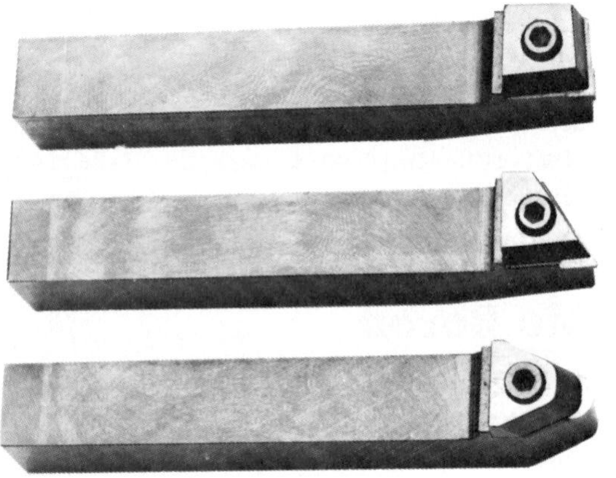

FIGURE F-51 Sintered diamond inserts are clamped in toolholders (Megadiamond Industries).

TABLE F-8 Hardness of Cutting Materials at High and Low Temperatures

Tool Material	Hardness at Room Temperature	Hardness at 1400°F (760°C)
High-speed steel	RA 85	RA 60
Carbide	RA 92	RA 82
Cemented oxide (ceramic)	RA 93–94	RA 84
Cubic boron nitride		Near diamond hardness
Diamond		Hardest known substance

example. Sintered polycrystalline diamond tools (Figure F-51), available in shapes similar to those of ceramic tools, are used for materials (Figure F-52) that are abrasive and difficult to machine. Polycrystalline tools consist of a layer of randomly oriented synthetic diamond crystals brazed to a tungsten carbide insert.

Diamond tools are particularly effective for cutting abrasive materials that quickly wear out other tool materials. Nonferrous metals, plastics, and some nonmetallic materials are often cut with diamond tools. Diamond is not particularly effective in carbon steels or superalloys that contain cobalt or nickel. Ferrous alloys chemically attack single- or polycrystalline diamonds, causing rapid tool wear. Because of the high cost of diamond tool material, it is usually restricted to those applications where other tool materials do not cut very well or break down quickly.

Diamond or ceramic tools should never be used for interrupted cuts such as on splines or keyseats because they could chip or break. They must never be used at low speeds or on machines that are not capable of attaining the higher speeds at which these tools should operate.

Each of the cutting tool materials varies in hardness. The differences in hardness tend to become more pronounced at high temperatures, as can be seen in Table F-8. Hardness is related to the wear resistance of a tool and its ability to machine materials that are softer than it is. Temperature change is important since the increase of temperature during machining results in the softening of the tool material.

SELF-TEST

1. List the major materials used in "straight" cemented carbides.

2. What effect does increasing the cobalt content have on cemented carbides?

3. How can you identify normal wear on a carbide tool?

4. Is chip thickness the same as feed on a style A tool?

5. Is the cutting edge engagement length the same as depth of cut on style B tools?

6. What effect does the addition of titanium carbides have on tool performance?

7. What effect does the addition of tantalum carbides have on tool performance?

8. What change in tool gemetry can make possible an increased rate of feed with equal surface finish quality in turning operations?

9. How does increasing the nose radius affect tool strength?

10. What is the hazard in using too large a nose radius?

11. In respect to carbide turning tools, how is clearance different from relief?

12. What are ceramic tools made of?

13. When should ceramic tools be used?

14. According to the CCPA chart, what designation of carbide would be used for finish-turning aluminum?

15. According to the CCPA chart, what designation of carbide is used for roughing cuts on steel?

16. According to the grade classification chart, what Carboloy designation of carbide would be used for rough cuts in cast iron?

17. According to the grade classification chart, what number on the CCPA table would a Kennametal K5H have?

18. Extremely hard or abrasive materials are machined with diamond tools. Is the material removal rate very high? What kind of finishes are produced?

19. Polycrystalline diamond tools are similar in some ways to ceramic inserts. What is their major advantage?

20. When brazed carbide tools are sharpened, should you use an aluminum oxide wheel or a silicon carbide wheel?

SECTION G

SAWING MACHINES

Sawing machines constitute some of the most important machine tools found in the machine shop. These machines can generally be divided into two classifications. The first class is **cutoff machines**. Common types of cutoff machines include reciprocating saws, horizontal endless band saws, universal tile frame band saws, abrasive saws, and cold saws. Of the cutoff machines, the **horizontal band saw** is the most important type. The second class is the **vertical band machine** that can be used as a band saw or with other band tools. The vertical band machine is most commonly used as a **band saw**.

The first machine tool that you will probably encounter is a cutoff machine. In the machine shop, cutoff machines are generally found near the stock supply area. The primary function of the cutoff machine is to reduce mill lengths of bar stock material into lengths suitable for holding in other machine tools. In a large production machine shop where stock is being supplied to many machine tools, the cutoff machine will be constantly busy cutting many materials. The cutoff machine that you will most likely see in the machine shop will be some type of saw.

Types of Cutoff Machines

RECIPROCATING SAWS

The **reciprocating saw** is often called a power hacksaw. Early saws were hand operated by a reciprocating or back-and-forth motion. It was logical that this principle be applied to power saws. The reciprocating saw is still used in the machine shop. However, they are giving way to the horizontal band machine. The reciprocating saw is built much like the metal cutting hand hacksaw. Basically, the machine consists of a frame that holds a blade. Reciprocating hacksaw blades are wider and thicker than those used in the hand hacksaw. The reciprocating motion is provided by hydraulics or a crankshaft mechanism.

Reciprocating saws are either the hinge type (Figure G-1) or the column type (Figure G-2). The

FIGURE G-1 Hinge-type reciprocating cutoff saw (Kasto-Racine Inc.).

saw frame on the hinge type pivots around a single point at the rear of the machine. On the column type, both ends of the frame rise vertically. Column-type reciprocating saws can accommodate larger sizes of material. The size of a reciprocating saw is determined by the largest piece of square material that can be cut. Sizes range from about 5 by 5 in. to 24 by 24 in. Large capacity reciprocating saws are often of the column design.

FIGURE G-2 Column-type reciprocating cutoff saw (Kasto-Racine Inc.).

HORIZONTAL BAND CUTOFF MACHINE

One disadvantage of the reciprocating saw is that it only cuts in one direction of the stroke. The band machine uses a steel band blade with the teeth on one edge. The band machine has a high cutting efficiency because the band is cutting at all times with no wasted motion. Band saws are the mainstay of production stock cutoff in the machine shop (Figure G-3).

A modern band saw may be equipped with a variable speed drive. This permits the most efficient cutting speed to be selected for the material being cut. The feed rate through the material may also be varied. The size of the horizontal band machine is determined by the largest piece of square material that the machine can cut. Large capacity horizontal band saws (Figure G-4) are designed to handle large dimension workpieces that can weigh

FIGURE G-3 Horizontal endless band cutoff saw (DoALL Company).

FIGURE G-4 Large capacity horizontal band saw (DoALL Company).

as much as 10 tons. With a wide variety of band types available, plus many special workholding devices, the band saw is an extremely valuable and versatile machine tool.

UNIVERSAL TILE FRAME CUTOFF

The universal tilt frame band saw is much like its horizontal counterpart. This machine has the band blade vertical, and the frame can be tilted from side to side (Figure G-5). The tile frame machine is particularly useful for making angle cuts on large structural shapes such as I-beams or pipe.

ABRASIVE CUTOFF MACHINE

The abrasive cutoff machine (Figure G-6) uses a thin circular abrasive wheel for cutting. Abrasive saws are very fast cutting. They can be used to cut a number of nonmetallic materials such as glass, brick, and stone. The major advantages of the abrasive cutoff machine are speed and the ability to cut nonmetals. Each particle of abrasive acts as a small tooth and actually cuts a small bit of materialo. Abrasive saws are operated at very high speeds. Blade speed can be as high as 10,000 to 15,000 surface feet per minute.

COLD SAW CUTOFF MACHINES

The cold saw (Figure G-7) uses a circular metal saw with teeth. These machine tools can produce extremely accurate cuts and are useful where length tolerance of the cut material must be held as close as possible. A cold saw blade that is .040 to .080 in. thick can saw material to a tolerance of plus or minus .002 in. Large cold saws are used to cut structural shapes such as angle and flat bar. These are also fast cutting machines.

FIGURE G-5 Tilt frame band saw (DoALL Company).

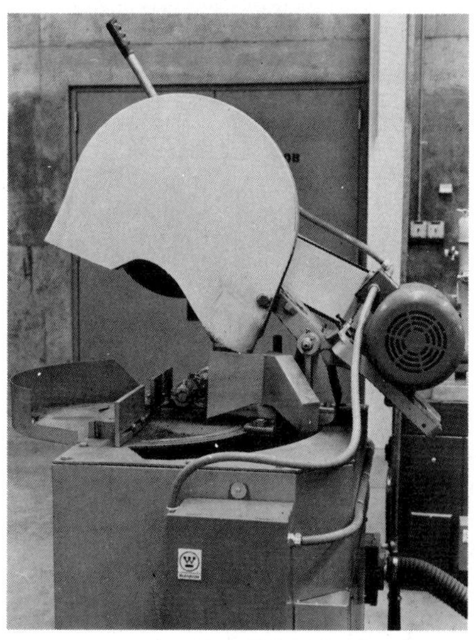

FIGURE G-6 Typical abrasive cutoff machine.

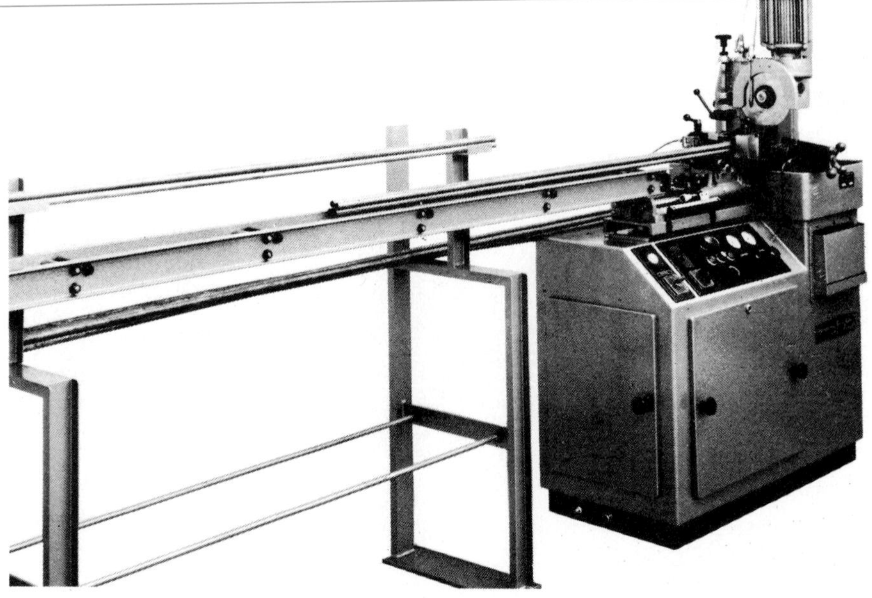

FIGURE G-7 Precision cold saw cutoff machine (Ameropean Industries, Inc.).

Cutoff Machine Safety

RECIPROCATING SAWS

Be sure that all guards around moving parts are in place before starting the machine. The saw blade must be properly installed with the teeth pointed in the right direction. Check for correct blade tension. Be sure that the width of the workpiece is less than the distance of the saw stroke. The frame will be broken if it hits the workpiece during the stroke. This will damage the machine. Be sure that the speed of the stroke and the rate of feed is correct for the material being cut.

When operating a saw with coolant, see that the coolant does not run on the floor during the cutting operation. This can cause an extremely dangerous slippery area around the machine tool.

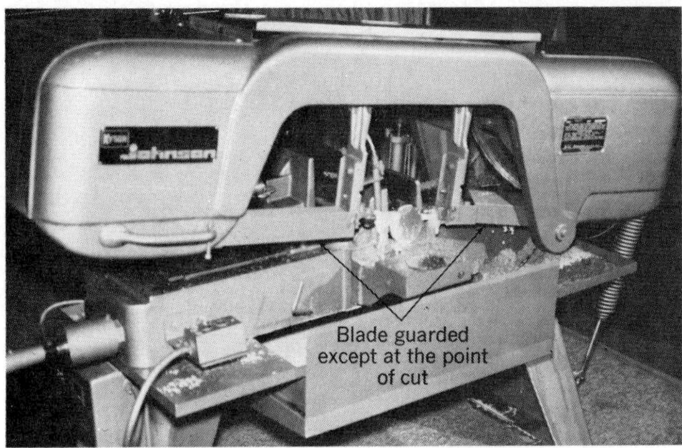

FIGURE G-8 The horizontal band blade is guarded except in the immediate area of the cut.

FIGURE G-9 Support both ends of the vise when cutting short material.

HORIZONTAL BAND SAWS

Recent regulations require that the blade of the horizontal band machine be fully guarded except at the point of cut (Figure G-8). Make sure that blade tensions are correct on reciprocating and band saws. Check band tensions, especially after installing a new band. New bands may stretch and loosen during their run-in period. Band teeth are sharp. When installing a new band, it should be handled with gloves. This is one of the few places that gloves may be worn around the machine shop. They must not be worn when operating any machine tool.

Endless band blades are often stored in double coils. Be careful when unwinding them, as they are under tension. The coils may spring apart and could cause an injury. Make sure that the band is tracking properly on the wheels and in the blade guides. If a band should break, it could be ejected from the machine and cause an injury.

Make sure that the material being cut is properly secured in the work-holding device. If this is a vise, be sure that it is tight. If you are cutting off short pieces of material, the vise jaw must be supported at both ends (Figure G-9). It is not good practice to attempt to cut pieces of material that are quite short. The stock cannot be secured properly and may be pulled from the vise by the pressure of the cut (Figure G-10). This can cause

FIGURE G-10 Result of cutting stock that is too short.

FIGURE G-11 The material is brought into the saw on the roll-case (opposite side) and, when pieces are cut off, they are supported by the stand (this side of the saw). The stand prevents the part from falling to the floor (DoALL Company).

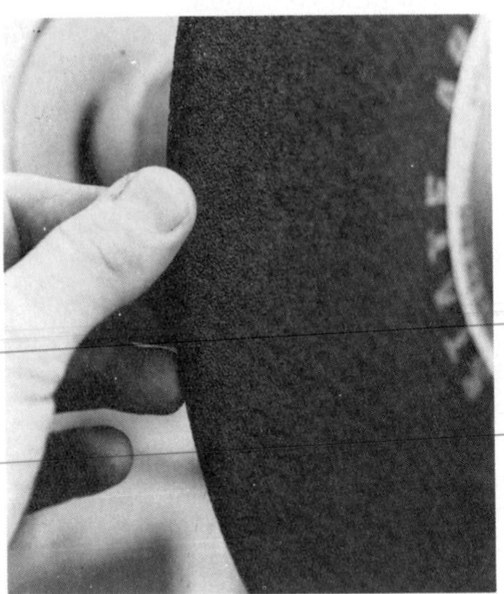

FIGURE G-12 Inspecting the abrasive wheel for chips and cracks.

damage to the machine as well as possible injury to the operator. Stock should extend at least halfway through the vise at all times.

Many cutoff machines have a rollcase that supports long bars of material while they are being cut. The stock should be brought to the saw on a rollcase (Figure G-11) or a simple rollstand. The pieces being cut off can sometimes be several feet long and should be similarly supported. Sharp burrs left from the cutting should be removed immediately with a file. You can acquire a nasty cut by sliding your hand over one of these burrs.

FIGURE G-13 Abrasive wheel must be operated at the correct rpm.

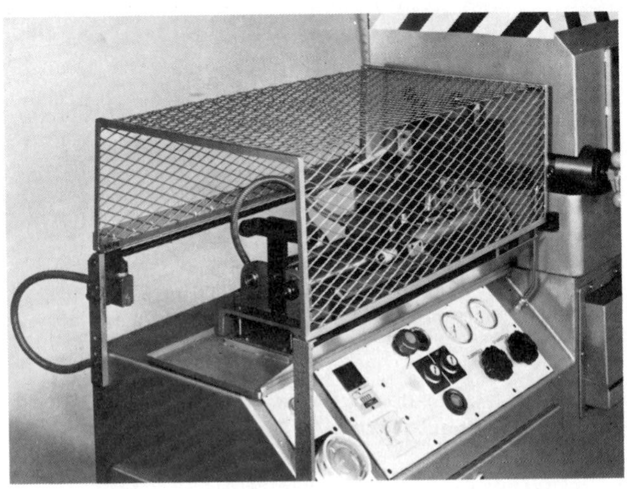

FIGURE G-14 Wire mesh guard on the cold saw material feeding mechanism (Ameropean Industries, Inc.).

Be careful around a rollcase, since bars of stock can roll, pinching fingers and hands. Also, be careful that heavy pieces of stock do not fall off the stock table or saw and injure feet or toes. Get help when lifting heavy bars of material. This will save your back and possibly your career.

ABRASIVE SAWS

On an abrasive saw, inspect the cutting edge of the blade for cracks and chips (Figure G-12). Replace the blade if it is damaged. Always operate an abrasive saw blade at the proper rpm (Figure G-13). Overspeeding the blade can cause it to fly apart. If an abrasive saw blade should fail at high speed, pieces of the blade can be thrown out of the machine at extreme velocities. A very serious injury indeed can result if you happen to be in the path of these bulletlike projectiles.

COLD SAWS

In terms of cold saws, safety is generally the same as with all cutoff machines. Guards must be in place around the saw and the feeding mechanism (Figure G-14). Before starting the saw, check to see that speeds and feeds are correct and that the workpiece is properly secured.

Safety extends to the machine as well as the operator. Never abuse any machine tool. They cost a great deal of money and in many cases are purchased with your tax dollars.

FIGURE G-15 Leighton A. Wilkie bandsaw of 1933 (DoALL Company).

Vertical Band Machines

The vertical band machine (Figure G-15) is often called the handiest machine tool in the machine shop. Perhaps the reason for this is the wide variety of work that can be accomplished on this versatile machine tool. The vertical band machine or vertical band saw is similar in general construction to its horizontal counterpart. Basically, it consists of an endless band blade or other band tool that runs on a driven and idler wheel. The band tool runs vertically at the point of the cut where it passes through a worktable on which the workpiece rests. The workpiece is pushed into the blade and the direction of the cut is guided by hand or mechanical means.

Advantages of Band Machines

Shaping of material with the use of a saw blade or other band, tool is often called **band machining**. The reason for this is that the band machine can perform other machining tasks aside from simple sawing. These include band friction sawing, band filing, and band polishing.

In any machining operation, a piece of stock material is cut by various processes to form the final shape and size of the part desired. In most machining operations, all of the unwanted material must be reduced to chips in order to uncover the final shape and size of the workpiece. With a band saw, only a small portion of the unwanted material must be reduced to chips in order to uncover the final workpiece shape and size (Figure G-16). A piece of stock material can often be shaped to final size by one or two saw cuts. A further advantage is gained in that the band saw cuts a very narrow kerf. A minimum amount of material is wasted.

A second important advantage in band sawing machines is **contouring ability.** Contour band

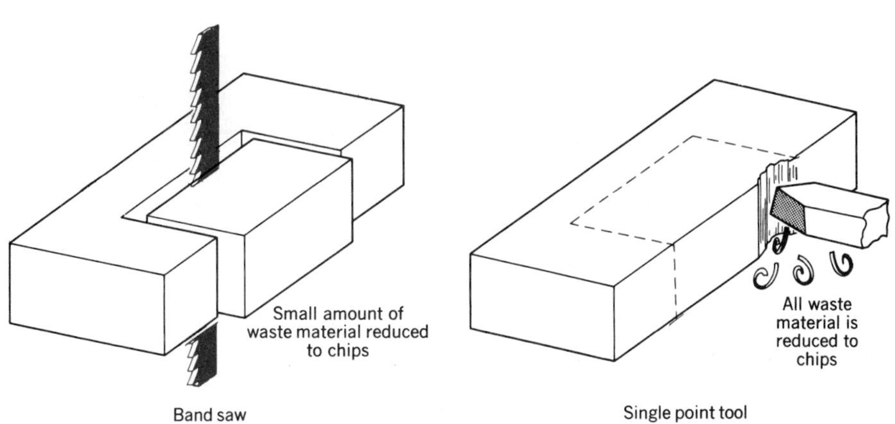

FIGURE G-16 Sawing can uncover the workpiece shape in a minimum number of cuts.

FIGURE G-17 Curved or contour band sawing can produce part shapes that would be difficult to machine by other methods (DoALL Company).

FIGURE G-18 Splitting a large diameter ring on the vertical band machine (DoALL Company).

FIGURE G-19 Workpieces larger than the machine tool can be cut (DoALL Company).

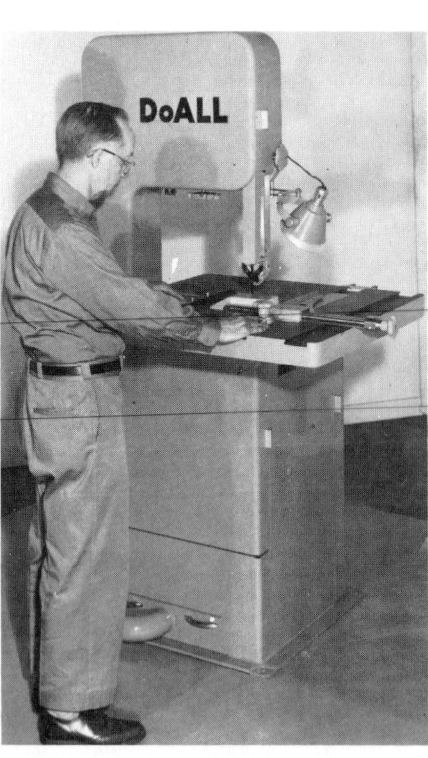

FIGURE G-20 General-purpose vertical band machine (DoALL Company).

sawing is the ability of the saw to cut intricate curved shapes that would be nearly impossible to machine by other methods (Figure G-17). The sawing of intricate shapes can be accomplished by a combination of hand and power feeds. On vertical band machines so equipped, the workpiece is steered by manual operation of the handwheel. The hydraulic table feed varies according to the

FIGURE G-21 Vertical band machine worktable can be tilted 10 degrees left.

FIGURE G-22 Vertical band machine worktable tilted 45 degrees right.

saw pressure on the workpiece. This greatly facilitates contour sawing operations.

Band sawing and band machining have several other advantages. There is no limit to the length, angle, or direction of the cut (Figure G-18). However, the throat capacity of the sawing machine will affect this depending on the dimension of the parts being sawed. Workpieces larger than the band machine can be cut (Figure G-19). Since the band tool is fed continuously past the work, cutting efficiency is high. A band tool, whether it be a saw blade band file, or grinding band, has a large number of cutting points passing the work. In most other machining operations, only one or a fairly low number of cutting points pass the work. With the band tool, wear is distributed over these many cutting points. Tool life is prolonged.

Types of Band Machines

GENERAL-PURPOSE BAND MACHINE WITH FIXED WORKTABLE

The general-purpose band machine is found in most machine shops (Figure G-20). This machine tool has a nonpower-fed worktable that can be tilted in order to make angle cuts. The table may be tilted 10 degrees left (Figure G-21). Tilt on this side is limited by the saw frame. The table may be tilted 45 degrees right (Figure G-22). On large machines, table tilt left may be limited to 5 degrees.

The workpiece may be pushed into the blade by hand. Mechanical (Figure G-23) or mechanical-hydraulic feeding mechanisms are also used. A band machine may be equipped with a hydraulic

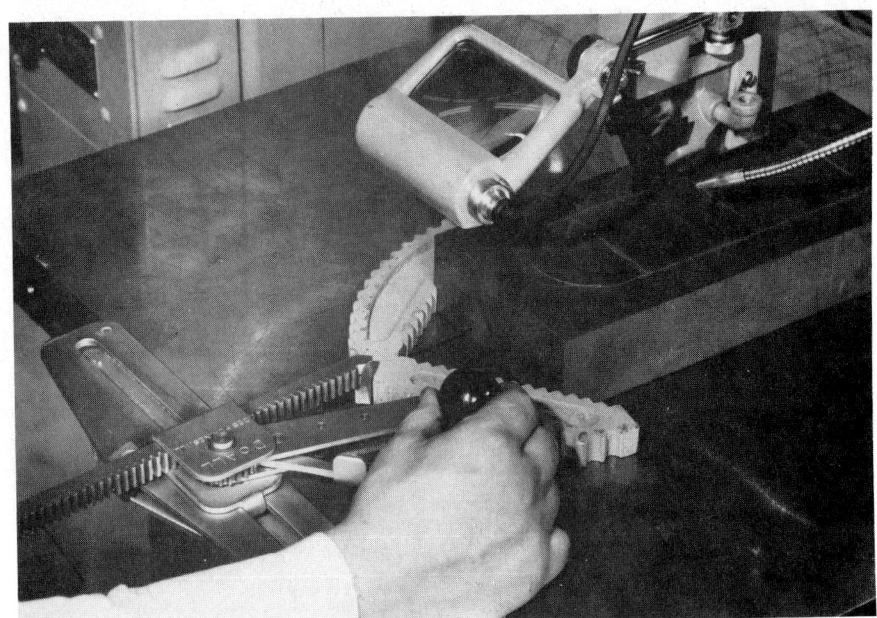

FIGURE G-23 Mechanical work feeding mechanism (DoALL Company).

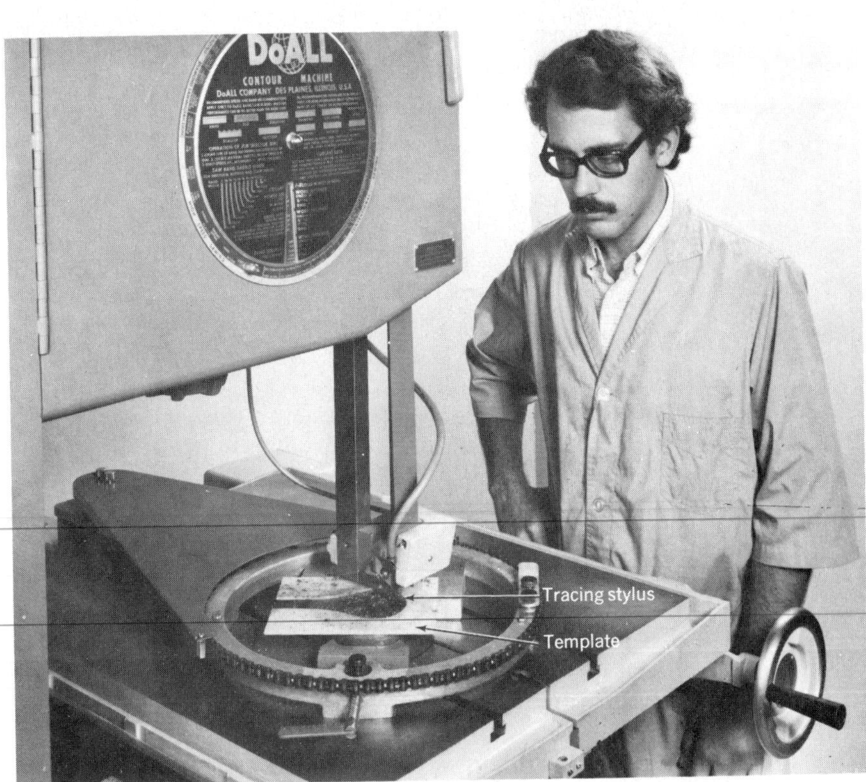

FIGURE G-24 Hydraulic tracing accessory (DoALL Company).

tracing attachment. This accessory uses a stylus contacting a template or pattern. The tracing accessory guides the workpiece during the cut (Figure G-24).

BAND MACHINES WITH POWER-FED WORKTABLES

Heavier construction is used on these machine tools. The worktable is moved hydraulically. The operator is relieved of the need to push the workpiece into the cutting blade. The direction of the cut can be guided by a steering mechanism (Figure G-25). A roller chain wraps around the workpiece and passes over a sprocket at the back of the worktable. The sprocket is connected to a steering wheel at the front of the worktable. The operator can then guide the workpiece and keep the saw cutting along the proper lines. The workpiece rests on roller bearing stands. These permit the workpiece to turn freely while it is being steered.

FIGURE G-25 Heavy-duty vertical band machine with power-fed worktable (DoALL Company).

FIGURE G-26 High-tool-velocity vertical band machine (DoALL Company).

FIGURE G-27 Trimming plastic laminates on the high-tool-velocity band machine (DoALL Company).

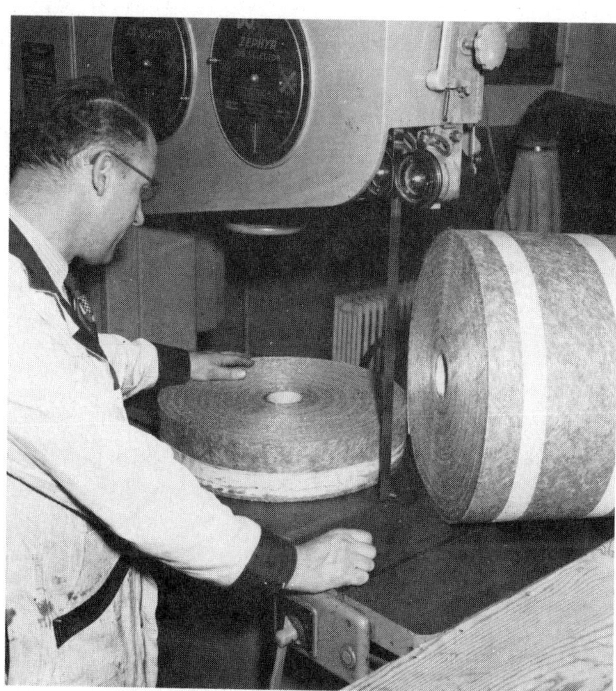

FIGURE G-28 Cutting fiber material on the high-tool-velocity band machine (DoALL Company).

HIGH TOOL VELOCITY BAND MACHINES

On the high-tool-velocity band machine (Figure G-26), band speeds can range as high as 10 to 15,000 feet per minute (fpm). These machine tools are used in many band machining applications. They are frequently found cutting nonmetal products. These include applications such as trimming plastic laminates (Figure G-27) and cutting fiber materials (Figure G-28).

LARGE-CAPACITY BAND MACHINES

This type of band machine is used on large workpieces. The entire saw is attached to a swinging column. The workpiece remains stationary and the saw is moved about to accomplish the desired cuts (Figure G-29).

Applications of the Vertical Band Machine

CONVENTIONAL AND CONTOUR SAWING

Vertical band machines are used in many conventional sawing applications. They are found in the

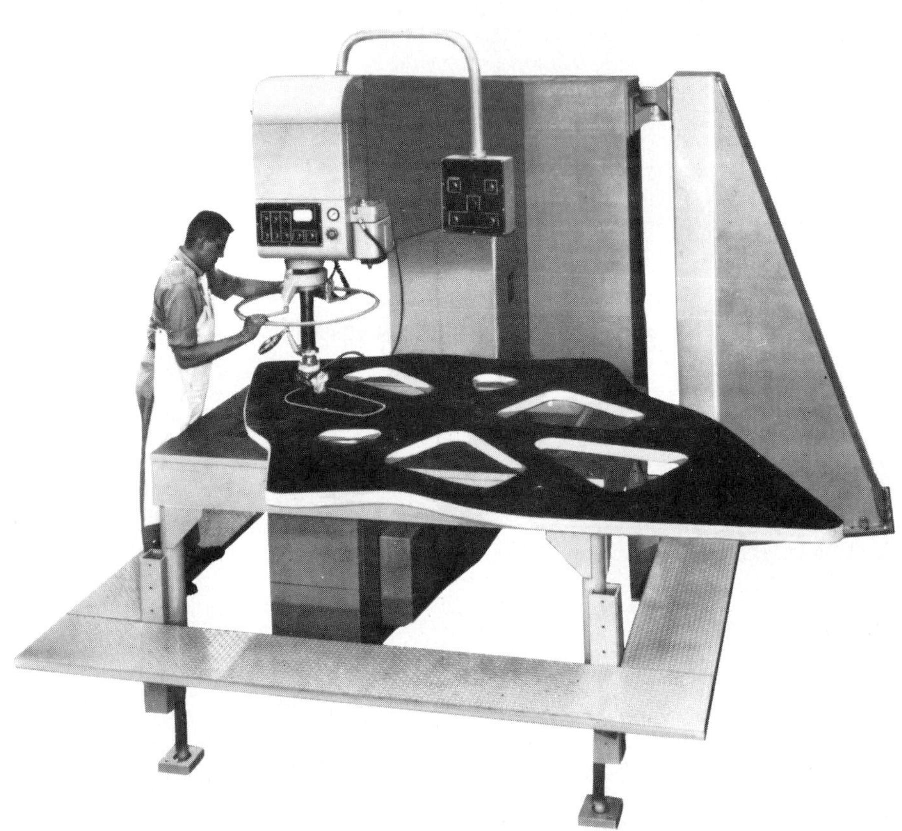

FIGURE G-29 Large-capacity vertical band machine (DoALL Company).

FIGURE G-30 Trimming casting sprues and risers on the vertical band machine (DoALL Company).

FIGURE G-31 Production trimming of castings on the vertical band machine (DoALL Company).

foundry trimming sprues and risers from castings. The band machine can accommodate a large casting and make widely spaced cuts (Figure G-30). Production trimming of castings is easily accomplished with the high-tool-velocity band machine (Figure G-31). Band saws are also useful in ripping operations (Figure G-32). In the machine shop, the vertical band machine is used in general purpose, straight line, and contour cutting mainly in sheet and plate stock.

FRICTION SAWING

Friction sawing can be used to cut materials that would be impossible or very difficult to cut by other means. In friction sawing, the workpiece is heated by friction created between it and the cutting blade. The blade melts its way through the work. Friction sawing can be used to cut hard materials such as files. Tough materials such as stainless steel wire brushes can be trimmed by friction sawing (Figure G-33). Friction sawing can only be done on machines with sufficiently high band speeds.

Vertical Band Machine Safety

CAUTION **The primary danger in operating the vertical band machine is accidental contact with the cutting blade. Workpieces are often hand guided. One advantage in sawing machines is that the pressure of the cut tends to hold the workpiece against the saw table. However, hands are often in close proximity to the blade. If you should contact the blade accidentally, an injury is almost sure to occur. You will not have time even to think about withdrawing your fingers before they are cut. Keep this in mind at all times when operating a band saw.**

Always use a pusher against the workpiece whenever possible. This will keep your fingers away from the blade. Be careful as you are about to complete a cut; as the blade clears through the work, the pressure that you are applying is suddenly released and your hand or finger could be carried into the blade. As you approach the end of the cut, reduce the feeding pressure as the blade cuts through.

The vertical band machine is generally not used to cut round stock. This can be extremely hazardous and should be done on the horizontal band machine, where round stock can be secured in a vice. Handheld round stock will turn if it is cut on the vertical band machine. This can cause an injury and may damage the blade as well. If round stock must be cut on the vertical band saw, it must be clamped securely in a vise, vee block, or other suitable workholding fixture.

Be sure to select the proper blade for the sawing requirements. Install it properly and apply the correct blade tension. Band tension should be rechecked after a few cuts. New blades will tend to stretch to some degree during their break-in period. Band tension may have to be readjusted.

The entire blade must be guarded except at the point of the cut. This is effectively accomplished by enclosing the wheels and blade behind guards that are easily opened for adjustments to the machine. Wheel and blade guard must be closed at all

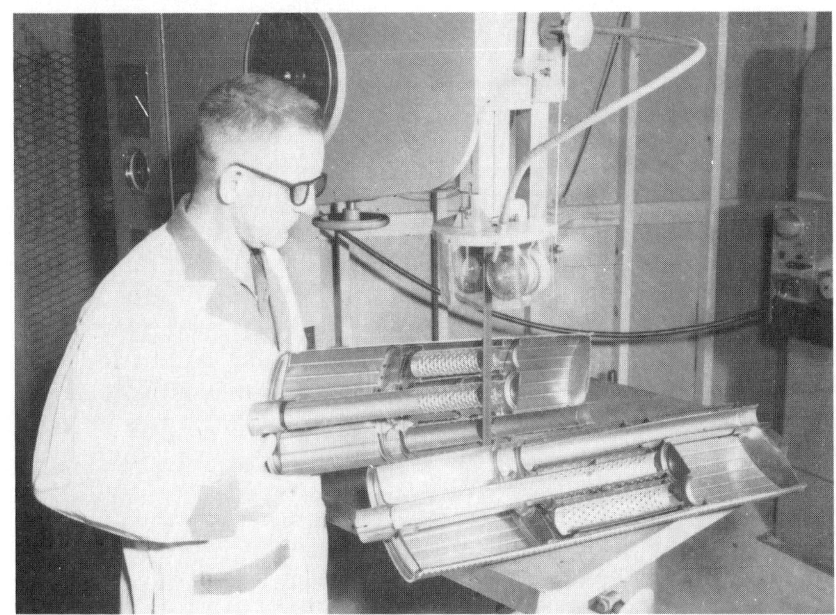

FIGURE G-32 Ripping on the vertical band machine (DoALL Company).

FIGURE G-33 Trimming stainless steel wire brushes by friction sawing (DoALL Company).

FIGURE G-34 Guidepost guard.

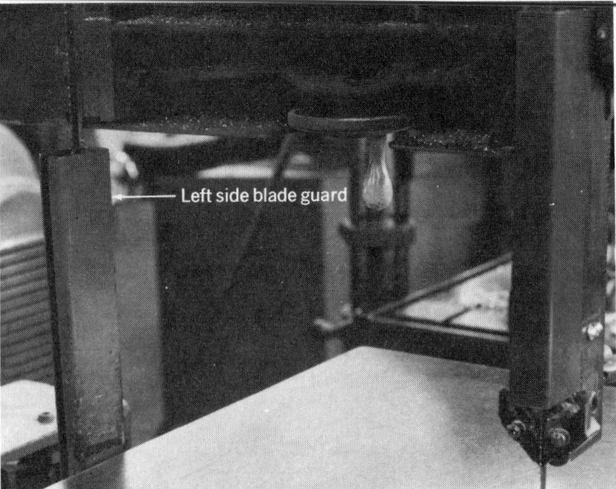

FIGURE G-35 Left-side blade guard when using a short blade over one idler wheel.

FIGURE G-36 Roller blade guide shield.

times during machine operation. The guidepost guard moves up and down with the guidepost (Figure G-34). The operator is protected from an exposed blade at this point. For maximum safety, set the guidepost $\frac{1}{8}$ to $\frac{1}{4}$ in. above the workpiece.

Band machines may have one or two idler wheels. On machines with two idler wheels, a short blade running over only one wheel may be used. Under this condition an additional blade guard at the left side of the wheel is required (Figure G-35). This guard is removed when operating over two idler wheels as the blade is then behind the wheel guard.

Roller blade guides are used in friction and high-speed sawing. A roller guide shield is used to provide protection for the operator (Figure G-36). Depending on the material being cut, the entire cutting area may be enclosed. This would apply to the cutting of hard, brittle materials such as granite and glass. Diamond blades are frequently used in cutting these materials. The clear shield protects the operator while permitting him or her to view the operation. Cutting fluids are also prevented from spilling on the floor. In any sawing operation making use of cutting fluids, see that they do not spill on the floor around the machine. This creates an extremely dangerous situation, not only for you but for others in the shop as well.

Gloves should not be worn around any machine tool. An exception to this is for friction or high-speed sawing or when handling band blades. Gloves will protect hands from the sharp saw teeth. If you wear gloves during friction or high-speed sawing, be extra careful that they do not become entangled in the blade or other moving parts. Be prepared for teeth flying off the band. If this happens, gloves will not necessarily protect your hands.

UNIT 1

Using Reciprocating and Horizontal Band Cutoff Machines

The reciprocating saw and the horizontal endless band saw are the most common cutoff machines you will encounter. Their primary function is to cut long lengths of material into lengths suitable for other machining operations on other machine tools. The cutoff machine is often the first step in machining a part to its final shape and size. In this unit, you are introduced to saw blades and the applications and operation of these important sawing machines.

Objectives

After completing this unit, you should be able to:

1. Use saw blade terminology.
2. Describe the conditions that define blade selection.
3. Identify the major parts of the reciprocating and horizontal band cutoff machine.
4. Properly install blades on reciprocating and horizontal band machines.
5. Properly use reciprocating and horizontal band machines in cutoff applications.

Cutting Speeds

An understanding of cutting speeds is one of the most important aspects of machining. Many years of machining experience have shown that certain tool materials are most effective if passed through workpiece materials at optimum speeds. If a tool material passes through the work too quickly, the heat generated by friction can rapidly dull the tool or cause it to fail completely. Too slow a passage of the tool through a material can result in premature dulling and low productivity.

A cutting speed refers to the amount of workpiece material that passes by a cutting tool in a given amount of time. Cutting speeds are measured in feet per minute (fpm). In some machining operations, the tool can pass the work. Sawing is an example. The work may pass the tool as in the lathe. In both cases, the fpm is the same. The shape of the workpiece does not affect the fpm. The circumference of a round part passing a cutting tool is still in fpm. In Section F, fpm was discussed in terms of revolutions per minute of a round workpiece.

In sawing, fpm is simply the speed of each saw tooth as it passes through a given length of material in one minute. If one tooth of a band saw passes through one foot of material in one minute, the cutting speed is one foot per minute. This is true of reciprocating saws as well. However, remember that this saw only cuts in one direction of the stroke. Cutting speeds are a critical factor in tool life. Productivity will be low if the sawing machine is stopped most of the time because a dull or damaged blade must be replaced frequently. The additional cost of replacement cutting tools must also be considered. In any machining operation, always keep cutting speeds in mind.

On a sawing machine, blade fpm is a function of rpm (revolutions per minute) of the saw drive.

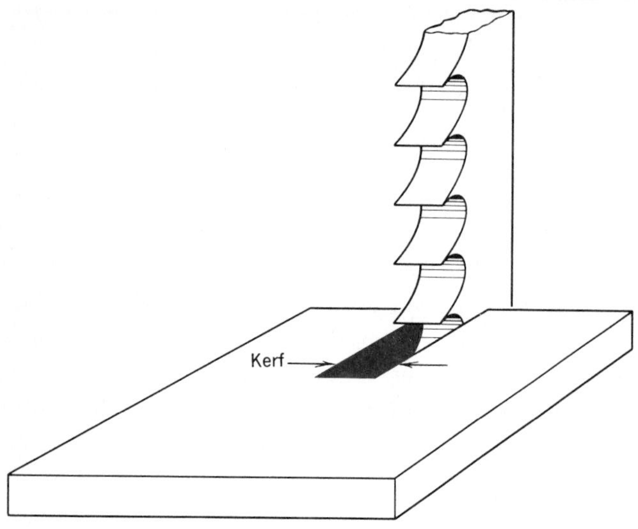

FIGURE G-37 Kerf.

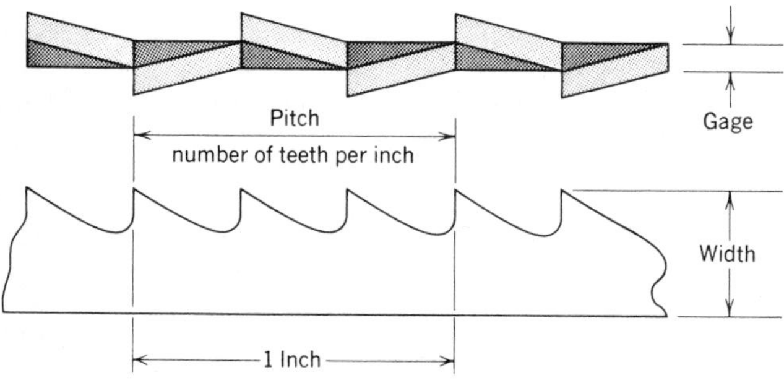

FIGURE G-38 Gage, pitch, and width.

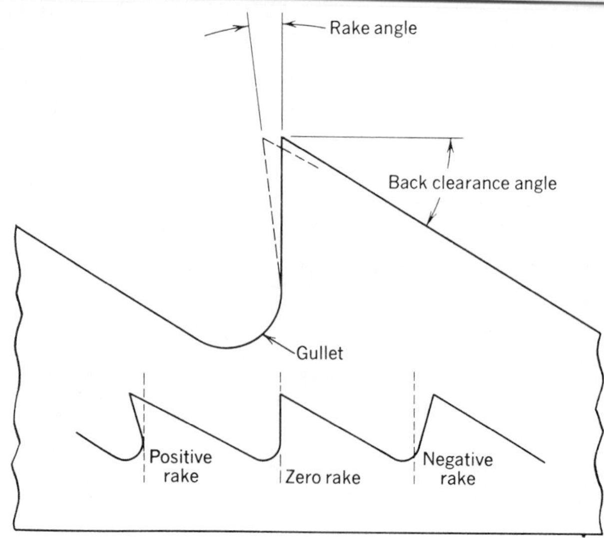

FIGURE G-39 Saw tooth terminology.

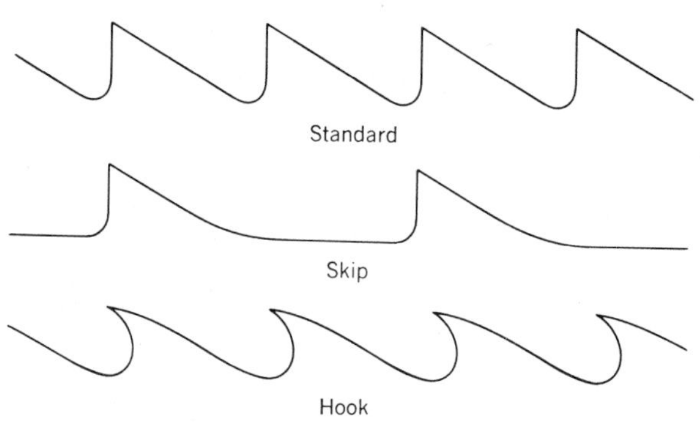

FIGURE G-40 Tooth forms.

That is, the setting of a specific rpm on the saw drive will produce a specific fpm of the blade. Feet per minute is also related to the material being cut. Generally, hard, tough materials have low cutting speeds. Soft material has higher cutting speeds. In sawing, cutting speeds are affected by the material, size, and cross section of the workpiece.

Saw Blades

The blade is the cutting tool of the sawing machine. In any sawing operation, at least three teeth on the saw blade must be in contact with the work at all times. This means that thin material requires a blade with more teeth per inch, while thick material can be cut with a blade having fewer teeth per inch. You should be familiar with the terminology of saw blades and saw cuts.

BLADE MATERIALS Saw blades for reciprocating and band saws are made from carbon steels and high-speed alloy steels. Blades may also have tungsten carbide tipped teeth. Some blades are bimetallic.

BLADE KERF The kerf of a saw cut is the width of the cut as produced by the blade (Figure G-37).

BLADE WIDTH The width of a saw blade is the distance from the tip of the tooth to the back of the blade (Figure G-38).

BLADE GAGE Blade gage is the thickness behind the set of the blade (Figure G-38). Reciprocating saw blades on large machines can be as thick as .250 in. Common band saw blades are .025 to .035 in. thick.

BLADE PITCH The pitch of a saw blade is the number of teeth per inch (Figure G-38). An eight-pitch blade has eight teeth per inch (a tooth spacing of $\frac{1}{8}$ in.). Blades of variable pitch are also used.

Saw Teeth

You should be familiar with saw tooth terminology (Figure G-39).

TOOTH FORMS Tooth form is the shape of the saw tooth. Sawtooth forms are either standard, skip, or hook (Figure G-40). Standard form gives accurate cuts with a smooth finish. Skip tooth gives additional chip clearance. Hook form pro-

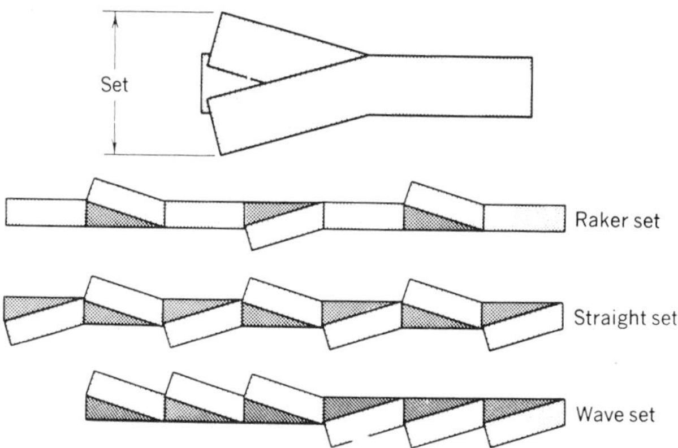

FIGURE G-41 Set and set patterns.

FIGURE G-42 Job selector on a horizontal band cutoff machine (DoALL Company).

vides faster cutting because of the positive rake angle, especially in soft materials.

SET The teeth of a saw blade must be offset on each side to provide clearance for the back of the blade. This offset is called set (Figure G-41). Set is equal on both sides of the blade. The set dimension is the total distance from the tip of a tooth on one side to the tip of a tooth on the other side.

SET PATTERNS Set forms include raker, straight, and wave (Figure G-41). Raker and wave are the most common. Raker set is used in general sawing. Wave set is useful where the cross-sectional shape of the workpiece varies.

Selecting a Blade for Reciprocating and Band Saws

Blade selection will depend upon the material, thickness, and cross-sectional shape of the workpiece. Some band cutoff machines have a job selector (Figure G-42). This will aid you greatly in

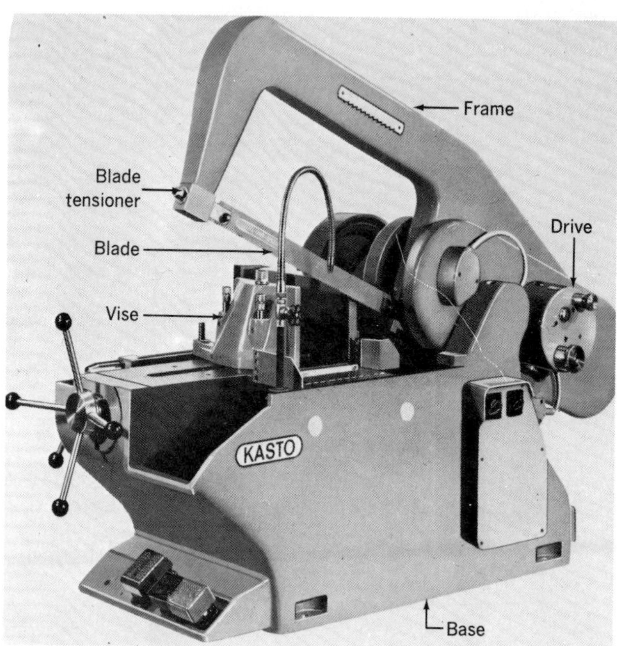

FIGURE G-43 Reciprocating cutoff saw (Kasto-Racine, Inc.).

selecting the proper blade for your sawing requirement. On a machine without a job selector, analyze the job and then select a suitable blade. For example, if you must cut thin tube, a fine pitch blade will be needed so that three teeth are in contact with the work. A particularly soft material may require a zero rake angle tooth form. Sawing through a workpiece with changing cross section may require a blade with wavy set to provide maximum accuracy.

Using Cutting Fluids

Cutting fluids are an extremely important aid to sawing. The heat produced by the cutting action can become so great that the metallurgical structure of the blade teeth can be affected. Cutting fluids will dissipate much of this heat and greatly prolong the life of the blade. Besides their function as a coolant, they also lubricate the blade. Sawing with cutting fluids will produce a smoother finish on the workpiece. One of the most important functions of a cutting fluid is to transport chips out of the cut. This allows the blade to work more efficiently. Common cutting fluids are oils, oils dissolved in water or soluble oils, and synthetic chemical cutting fluids.

Operating the Reciprocating Cutoff Machine

The reciprocating cutoff machine (Figure G-43) is often known as the power hacksaw. This machine is an outgrowth of the hand hacksaw. Basically,

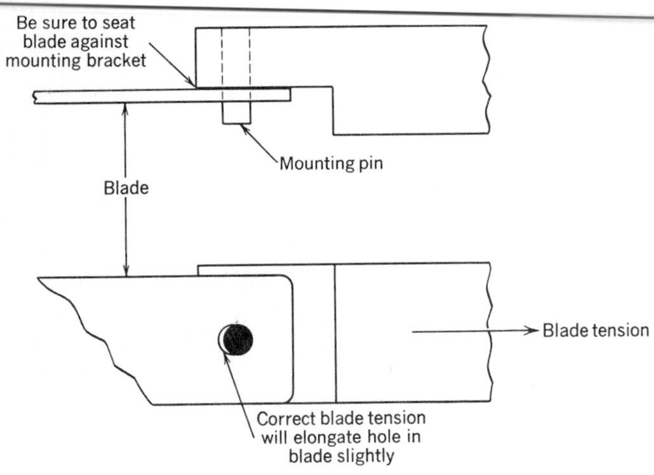

FIGURE G-44 Blade mounting on the reciprocating cutoff saw.

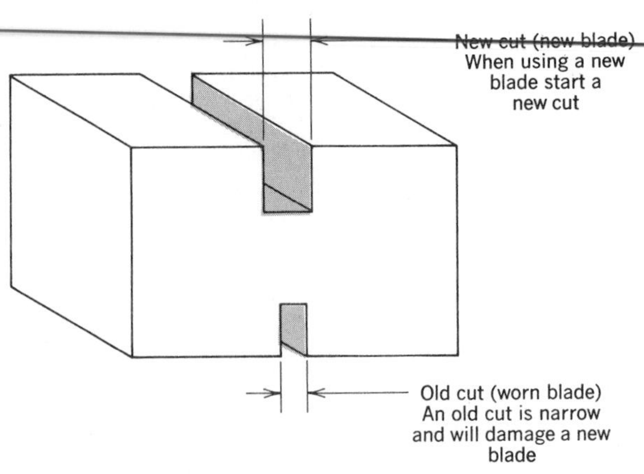

FIGURE G-46 If the blade is changed, begin a new cut on the other side of the workpiece.

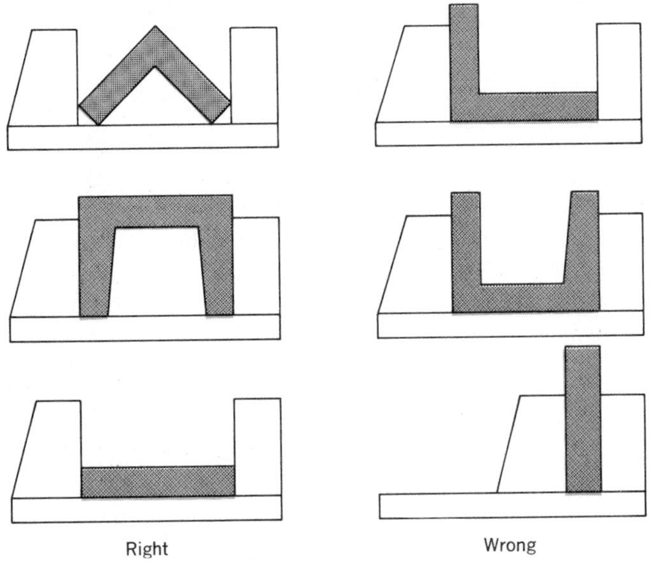

FIGURE G-45 Cutting workpieces with sharp corners.

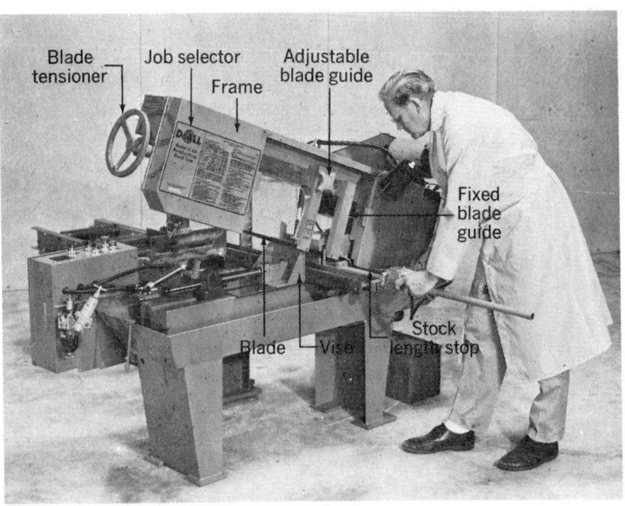

FIGURE G-47 Horizontal endless band cutoff machine (DoALL Company).

the machine consists of a frame-supported blade that is operated in a back-and-forth motion. Power hacksaws may be hydraulically driven or driven by a crankshaft mechanism.

INSTALLING THE BLADE ON THE POWER HACKSAW

Obtain a blade of the correct length and make sure that the teeth are pointed in the direction of the cut. This will be on the back stroke. Make sure that the blade is seated against the mounting plates (Figure G-44). Apply the correct tension. The blade may be tightened until a definite ring is heard when the blade is tapped. Do not overtighten the blade, as this may cause the pin holes to break out. If a new blade is installed, the tension should be rechecked after making a few cuts.

MAKING THE CUT

Select the appropriate strokes per minute speed rate for the material being cut. Be sure to secure the workpiece properly. If you are cutting material with a sharp corner, begin the cut on a flat side if possible. Note that angle material presents a sharp corner to the blade. Start the saw gently until a small flat is established (Figure G-45). Before making the cut, go over the safety checklist. Make sure that the length of the workpiece does not exceed the capacity of the stroke. This can break the frame if it should hit the workpiece. Bring the saw gently down until the blade has a chance to start cutting. Apply the proper feed. On reciprocating saws, feed is regulated with a sliding weight or feeding mechanism. If chips produced in the cut are blue, too much feed is being used. The blade will be damaged rapidly. Very fine powderlike chips indicate too little pressure. This will dull the blade. If the

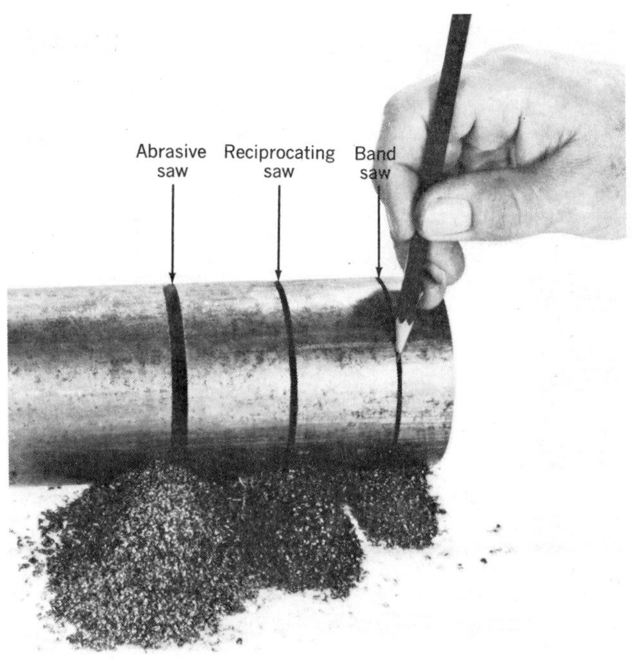

FIGURE G-48 Comparison of kerf widths from band, reciprocating, and abrasive cutoff machines (DoALL Company).

blade is replaced after starting a cut, turn the workpiece over and begin a new cut (Figure G-46). Do not attempt to saw through the old cut. This will damage the new blade. After a new blade has been used for a short time, recheck the tension and adjust if necessary.

Operating the Horizontal Band Cutoff Machine

The horizontal band cutoff machine (Figure G-47) is the most common stock cutoff machine found in the machine shop. This machine tool uses an endless steel band blade with teeth on one edge. Since the blade passes through the work continuously, there is no wasted motion. Cutting efficiency is greatly increased over the reciprocating saw.

The kerf from the band blade is quite narrow as compared to the reciprocating hacksaw or abrasive saw (Figure G-48). This is an added advantage in that minimum amounts of material are wasted in the sawing operation.

The size of the horizontal band saw is determined by the largest piece of square material that can be cut. Speeds on the horizontal band machine may be set by manual belt change (Figure G-49), or a variable-speed drive may be used. The variable-speed drive permits an infinite selection of band speeds within the capacity of the machine.

FIGURE G-49 Changing speeds by shifting belts on a horizontal band saw (DoALL Company).

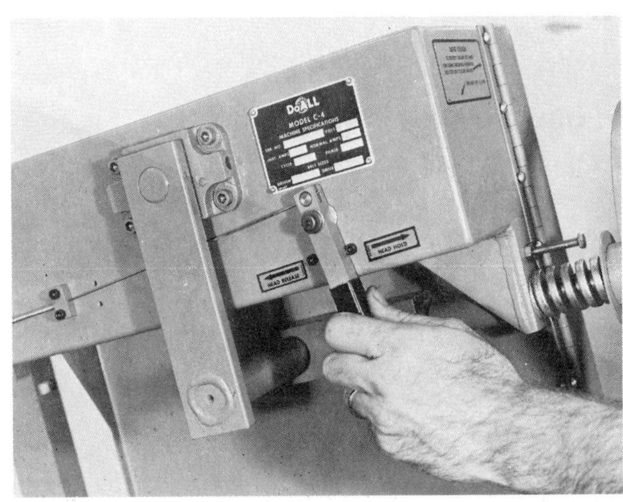

FIGURE G-50 Horizontal band saw head release lever (DoALL Company).

Cutting speeds can be set precisely. Many horizontal band machines are of the hinge design. The saw head, containing the drive and idler wheels, hinges around a point at the rear of the machine.

The saw head may be raised and locked in the up position while stock is being placed into or removed from the machine (Figure G-50). Feeds are accomplished by gravity of the saw head. The feed rate can be regulated by adjusting the spring ten-

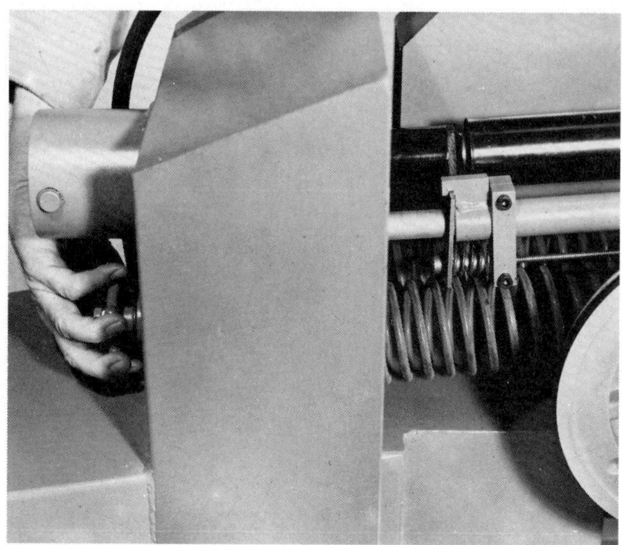

FIGURE G-51 Adjusting the head tension on the horizontal band saw (DoALL Company).

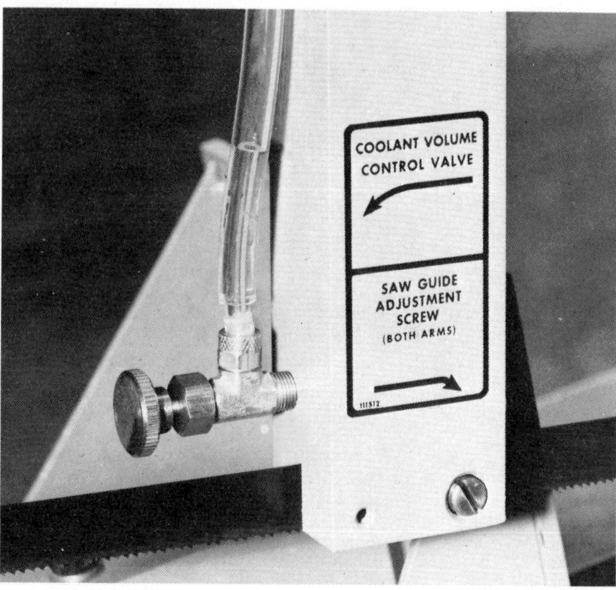

FIGURE G-53 Coolant volume control valve on the horizontal band saw (DoALL Company).

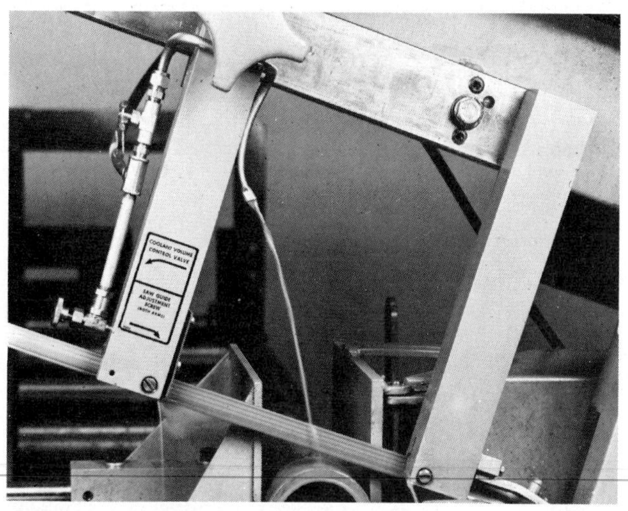

FIGURE G-52 Coolant system on the horizontal band saw (DoALL Company).

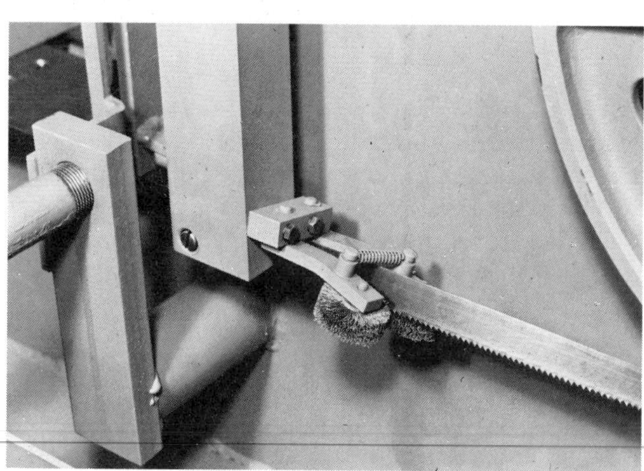

FIGURE G-54 Rotary chip brushes on the band saw blade (DoALL Company).

sion on the saw head (Figure G-51). Some sawing machines use a hydraulic cylinder to regulate the feed rate. The head is held in the up position by the cylinder. A control valve permits oil to flow into the reservoir as the saw head descends. This permits the feed rate to be regulated.

Cutting fluid is pumped from a reservoir and flows on the blade at the forward guide. Additional fluid is permitted to flow on the blade at the point of the cut (Figure G-52). The saw shown does not have the now-required full blade guards installed. Cutting fluid flow is controlled by a control valve (Figure G-53). Chips can be cleared from the blade by a rotary brush that operates as the blade runs (Figure G-54).

Common accessories used on many horizontal band saws include workpiece length measuring equipment (Figure G-55) and roller stock tables. The stock table shown has a hinged section that permits the operator to reach the rear of the machine (Figure G-56).

WORK-HOLDING ON THE HORIZONTAL BAND SAW

The vise is the most common work-holding fixture. Rapid adjusting vises are very popular (Figure G-57). These vises have large capacity and are quickly adjusted to the workpiece. After the vise jaws have contacted the workpiece, the vise is

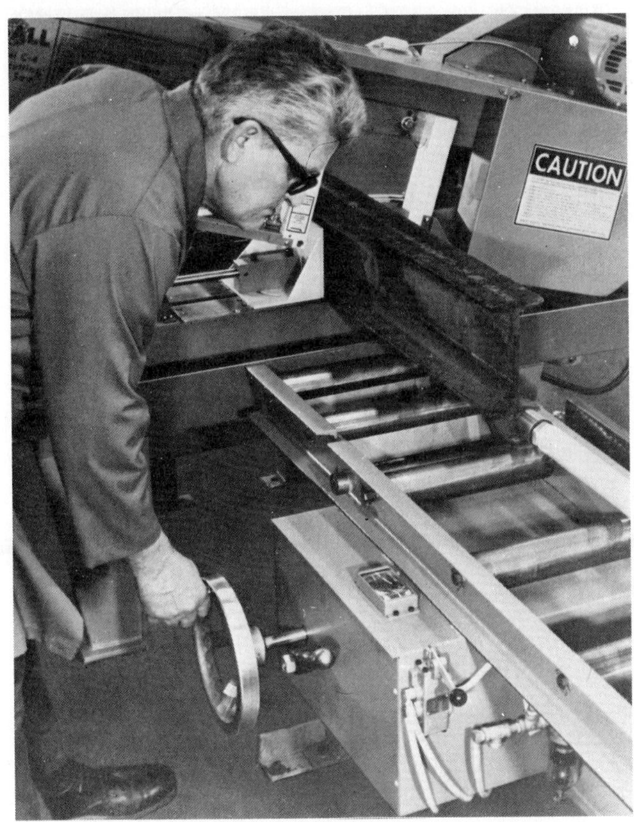

FIGURE G-55 Setting the stock length gage (DoALL Company).

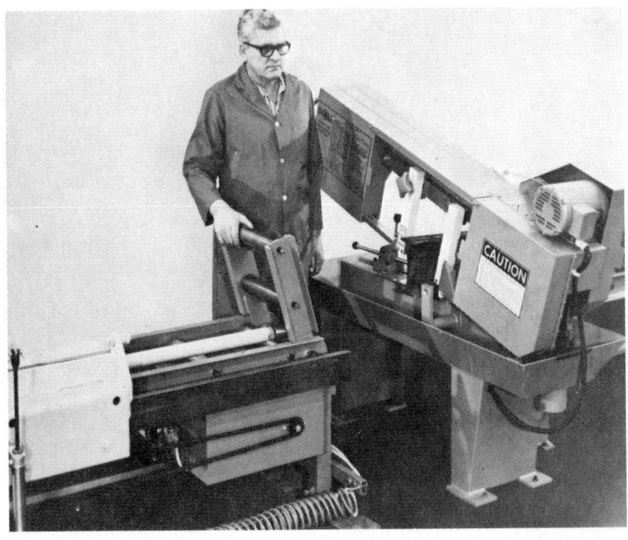

FIGURE G-56 Roller stock table with access gate (DoALL Company).

locked by operating the lock handle. The vise may be swiveled for miter or angle cuts (Figure G-58). On some horizontal band cutoff machines, the entire saw frame swivels for making angle cuts (Figure G-59).

FIGURE G-57 Quick-setting vise on the horizontal band saw (DoALL Company).

FIGURE G-58 Band saw vise swiveled for angle cutting (DoALL Company).

FIGURE G-59 Horizontal cutoff saw frame swiveled for angle cuts (DoALL Company).

FIGURE G-60 Nesting fixture for sawing multiple workpieces (DoALL Company).

FIGURE G-61 Make sure that the blade is tracking properly on the band wheel.

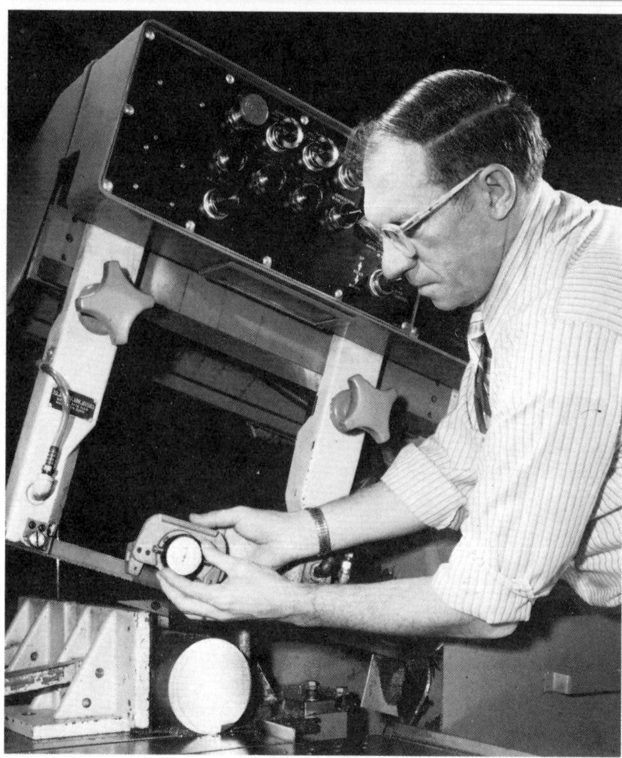

FIGURE G-62 Dial band tension indicator gage (DoALL Company).

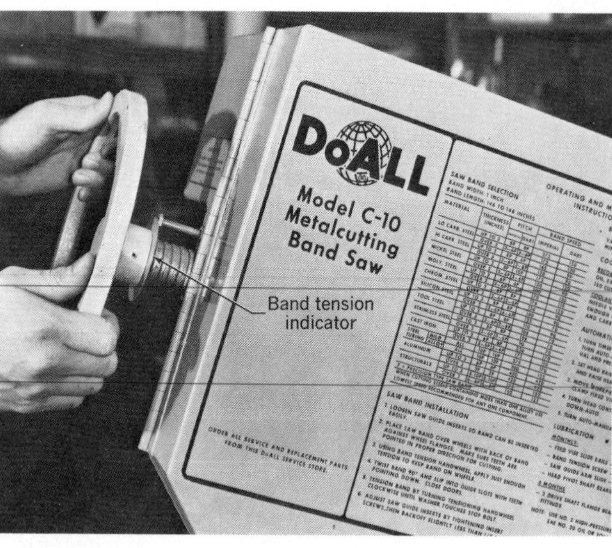

FIGURE G-63 Band tension indicator attached to the machine (DoALL Company).

The horizontal band saw is often used to cut several pieces of material at once. Stock may be held or nested in a special vise or nesting fixture (Figure G-60).

INSTALLING BLADES ON THE HORIZONTAL BAND MACHINE

Blades for the horizontal band saw may be ordered prewelded in the proper length for the machine. Band blade may also be obtained in rolls. The required length is then cut and welded at the sawing machine location.

To install the band, shut off power to the machine and open the wheel guards. Release the tension by turning the tension wheel. Place the blade around the drive and idler wheels. Be sure that the teeth are pointed in the direction of the cut. This will be toward the rear of the machine. See that the blade is tracking properly on the idler and drive wheels (Figure G-61). The blade will have to be twisted slightly to fit the guides. Guides should be adjusted so that they have .001 to .002 in. clearance with the blade. Adjust the blade tension using

FIGURE G-64 Blade guide should be set as close to the workpiece as possible.

FIGURE G-65 Stock length stop in position.

FIGURE G-66 Stock length stop must be removed before beginning the cut.

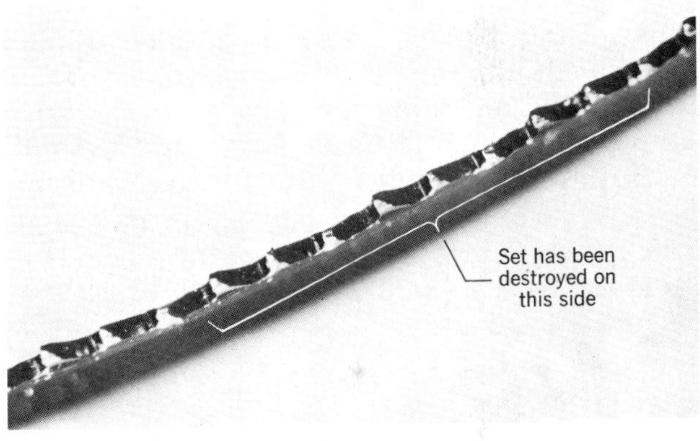

FIGURE G-67 Band saw blade with the set worn on one side.

the tension gage (Figure G-62) or the manual tension indicator built into the saw (Figure G-63). The blade is tightened until the flange on the tension wheel contacts the tension indicator stop.

MAKING THE CUT

Set the proper speed according to the blade type and material to be cut. If the workpiece has sharp corners, it should be positioned in the same manner as in the reciprocating saw (Figure G-45). The blade guides must be adjusted so that they are as close to the work as possible (Figure G-64). This will ensure maximum blade support and maximum accuracy of the cut. Sufficient feed should be used to produce a good chip. Excessive feed can cause blade failure. Too little feed can dull the blade prematurely. Go over the safety checklist for horizontal band saws. Release the head and lower it by hand until the blade starts to cut. Most saws are equipped with an automatic shutoff switch. When the cut is completed, the machine will shut off automaticallly.

Sawing Problems on the Horizontal Band Machine

You may use the stock stop to gage the length of duplicate pieces of material (Figure G-65). It is important to swing the stop clear of the work after the vise has been tightened and before the cut is begun (Figure G-66). A cutoff workpiece can bind between the stop and blade. This will destroy the blade set (Figure G-67). A blade with a tooth set worn on one side will drift in the direction of the side that has a set still remaining (Figure G-68). This is the principal cause of band breakage. As the saw progresses through the cut, the side draft of the blade will place the machine under great

FIGURE G-68 Using a band with the set worn on one side will cause the cut to drift toward the side of the blade with the set remaining.

stress. The cut may drift so far as to permit the blade to cut into the vise (Figure G-69).

Very thin and quite parallel workpieces can be cut by using a sharp blade in a rigid and accurate sawing machine (Figure G-70). Chip removal is important to accurate cutting. If chips are not cleared from the blade prior to it entering the guides, the blade will be scored. This will make it brittle and subject to breakage.

SELF-TEST

1. Name the most common saw blade set patterns.

2. Describe the conditions that define blade selection.

3. On a reciprocating saw, what is the direction of the cut?

4. What is set, and why is it necessary?

5. What are common tooth forms?

6. What can happen if the stock stop is left in place during the cut?

FIGURE G-69 The blade may drift far enough to damage the vise.

FIGURE G-70 Thin and parallel cuts may be made with a sharp blade and rigid sawing machine.

7. What type of cutoff saw will most likely be found in the machine shop?

8. Of what value are cutting fluids?

9. What can result if chips are not properly removed from the cut?

10. If a blade is replaced after a cut has been started, what must be done with the workpiece?

UNIT 2

Abrasive and Cold Saws

The abrasive saw is seldom used for stock cutoff in the machine shop. Small metal chips and abrasive wheel particles produced by the abrasive saw can damage other machine tools. However, the abrasive saw may be used in or around the machine shop or grinding room for the purpose of cutting hardened materials. Abrasive saws are very common in fabrication and welding shops. The abrasive saw has an advantage over other cutoff machines in that it can cut a number of nonmetallic materials such as slate, stone, brick, and glass.

The cold saw is seen in the machine shop. This type of cutoff machine uses a circular blade with teeth. Cold saws are useful in stock cutoff where length tolerances must be held as close as possible. You may see and possibly use a cold saw in stock cutoff applications.

Objectives

After completing this unit, you should be able to:

1. Identify abrasive and cold saws.
2. Identify abrasive wheel materials.
3. Identify abrasive wheel bonds.
4. Describe the operation of abrasive cold sawing machines.

Abrasive Sawing Machines

The abrasive saw (Figure G-71) consists of a high-speed motor-driven abrasive wheel mounted on a swing arm. Abrasive saws are very fast cutting. The reason for this is that each particle of the abrasive wheel acts to cut a small bit of material. Since the abrasive wheel has a large number of abrasive particles, many "teeth" are involved in the cutting action. Abrasive wheels may operate at cutting speeds of 15,000 to 20,000 feet per minute. A large number of abrasive particles delivered past the work at a high foot-per-minute rate makes for very fast cutting. One disadvantage of the abrasive saw is that common blades cut with a wide kerf. Considerable material may be wasted in the cut. However, some abrasive wheel materials permit a very thin wheel to be used. Abrasive sawing produces a large amount of heat in the workpiece. Cutting fluids act as coolants and are often used in abrasive sawing.

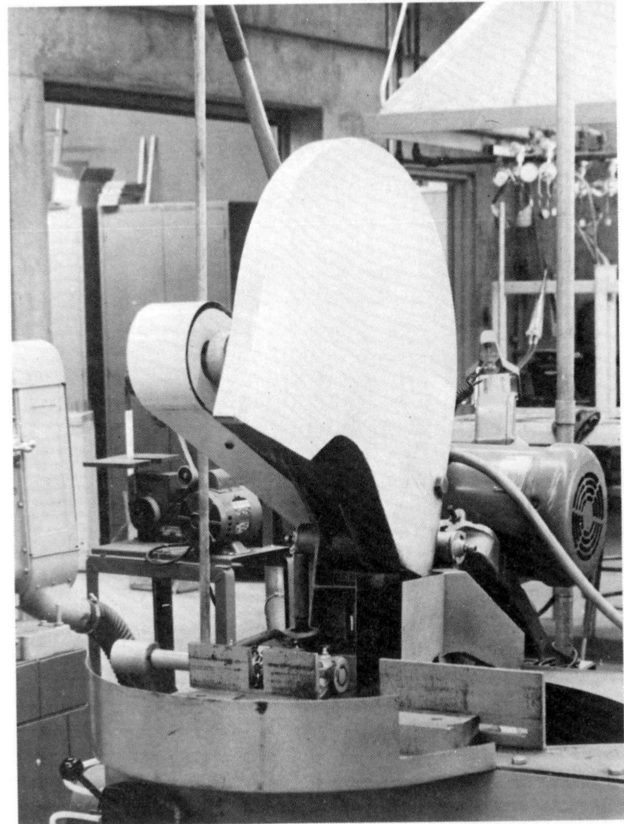

FIGURE G-71 Typical abrasive cutoff machine.

ABRASIVES

Abrasive saw wheels are made from aluminum oxide and silicon carbide. Aluminum oxide abrasives are used for most metals, including steel. Silicon carbide abrasive is used for nonmetallic materials such as stone. Diamond abrasives are used for extremely hard materials, such as glass.

ABRASIVE SAW WHEEL BONDS

The **bond** of an abrasive saw wheel is the material that holds the abrasive particles together. The bond must be strong enough to withstand the large force placed on the wheel at high revolutions per minute as well as the heat generated in the cut. However, the bond must also be able to break down as the wheel cuts. This is necessary to expose new abrasive particles to the cut.

Shellac bond abrasive wheels are suited for cutting hard steels. **Resinoid bone** wheels are suitable for cutting structural shapes and bar stock. **Rubber bond** wheels produce clean cuts when used with cutting fluids. Rubber bond wheels can be made as thin as .006 in.

ABRASIVE SAW WHEEL SPEEDS

Abrasive saw wheel speeds vary as to the diameter of the wheel. A 12-in.-diameter wheel can be operated at about 14,000 surface feet per minute (sfpm). Twenty-inch wheels should be run at about 12,000 sfpm. Surface feet is the speed of the wheel as measured at the circumference. Recommended wheel speeds are marked on each wheel. **Abrasive saw wheels must not be operated at speeds faster than recommended.**

SELECTING ABRASIVE SAWS

The following factors affect the selection of an abrasive saw.

1. Material to be cut
2. Size of material
3. Cut to be made dry or with cutting fluid
4. Degree of finish desired and acceptable burr (sharp edge left by the saw)

QUALITY OF ABRASIVE SAW CUTS

Excessive heat generated in the cut may discolor the material or possibly affect metalllurgical properties. This problem may be solved by selection of a different wheel. The use of a cutting fluid will produce a smooth cut with less of a burr. In some wet abrasive saw applications, the workpiece and part of the blade may be totally submerged in the cutting fluid. Fluids include a soda solution of plain water.

FIGURE G-72 Quick-setting chain vise.

ABRASIVE SAW FEED RATES

An extremely light feed will heat the wheel excessively. Heavy feeding may result in excessive wheel wear or the wheel may break causing an extreme hazard. Feed rates depend on many factors, including the mterial to be cut, wheel type, speeds, and the condition of the abrasive sawing machine.

Operating the Abrasive Sawing Machine

Inspect the abrasive wheel for chips and cracks. Be sure that it is rated at the proper rpm for the machine. Work holding on the abrasive saw may be a simple vise or a quick-setting production fixture such as an air-operated chain vise. Here the chain passes over the workpiece and a link hooks to the chain catch (Figure G-72). The chain is tightened by a winder operated by an air cylinder (Figure G-73).

Go over the safety checklist for abrasive saws. Start the machine and bring the abrasive wheel down until the cut begins. Use a feed rate that is suitable for the material being cut and the wheel being used (Figure G-74). The abrasive saw may be swiveled to the side for miter cuts (Figure G-75).

FIGURE G-73 Chain vise operating mechanism.

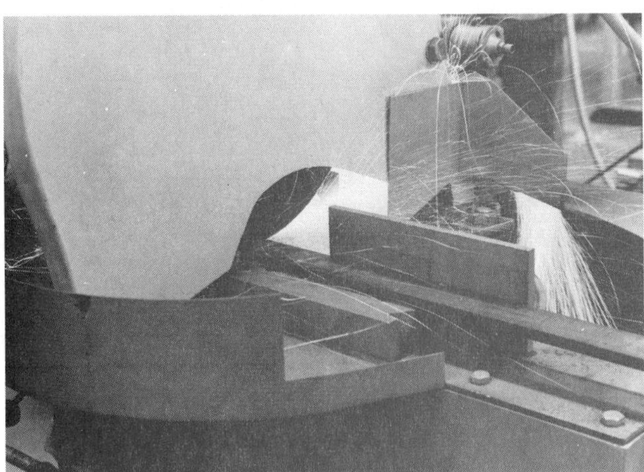

FIGURE G-74 Making a cut with the abrasive saw.

FIGURE G-75 The abrasive saw may be swiveled for angle cutting.

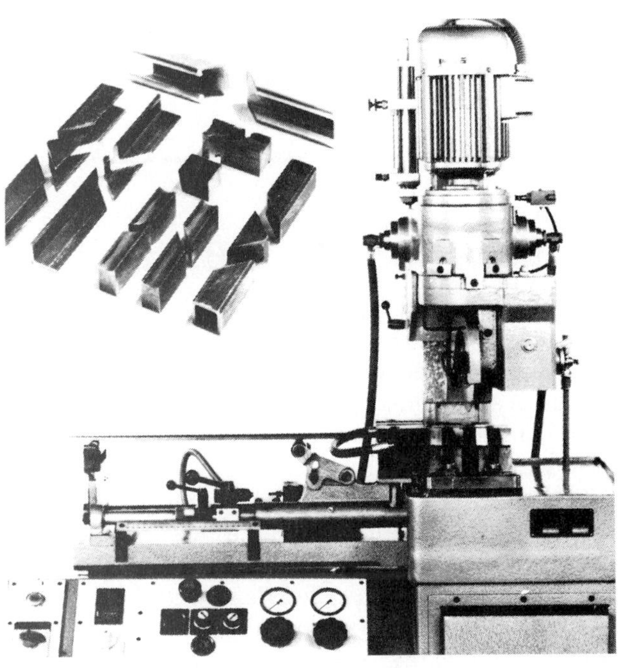

FIGURE G-76 The precision cold saw can make straight and angle cuts in materials with different cross sections (Ameropean Industries, Inc.).

Cold Saws

The **cold saw** (Figure G-76) uses a circular metal blade with teeth. Cold saws are very useful in precision cutoff applications where length tolerance must be held as close as possible. A cold saw blade on a precision cutoff machine may be 7 to 8 in. in diameter with a thickness of .040 to .080 in. The blade cuts a narrow kerf. A minimum amount of material is wasted in the cut. This is important when cutting expensive materials. Length tolerance and parallelism of the cut can be held to plus or minus .002 in. To obtain this degree of accuracy, hold the stock on both sides of the blade. A precision cold saw blade may be operated at speeds of 25 to 1200 rpm, depending on the size, shape, and material of the workpiece. Cold saws can be used to make straight and angle cuts in material with different cross sections (Figure G-76). Cold saws with blade diameters of about 24 in. are also used for stock cutoff in machine and fabrication shops.

SELF-TEST

1. Are you likely to find an abrasive saw in a machine shop?
2. What are the advantages of the abrasive saw?
3. What types of abrasives are used on what types of materials?
4. Name two types of abrasive wheel bonds.
5. What is one advantage of a precision cold saw?

UNIT 3

Preparing to Use the Vertical Band Machine

A machine tool can perform at maximum efficiency only if it has been properly maintained, adjusted, and set up. Before the vertical band machine can be used for a sawing or other band machining operation, several important preparations must be made. These include welding saw blades into bands and making several adjustments to the machine tool.

Objectives

After completing this unit, you should be able to:

1. Weld band saw blades.
2. Prepare the vertical band machine for operation.

Welding Band Saw Blades

Band saw blade stock is frequently supplied in rolls. The required length is measured and cut and the ends are welded together to form an endless band. Most band machines are equipped with a band welding attachment. These are frequently attached to the machine tool. They may also be separate pieces of equipment (Figure G-77).

The **band welder** is a resistance-type butt welder. They are often called **flash** welders because of the bright flash and shower of sparks created during the welding operation. The metal in the blade material has a certain resistance to the flow of an electric current. This resistance causes the blade metal to heat as the electric current flows during the welding operation. The blade metal is heated to a temperature that permits the ends to be forged together under pressure. When the forging temperature is reached, the ends of the blade are pushed together by mechanical pressure. They fuse, forming a resistance weld. The band weld is then annealed or softened and dressed to the correct thickness by grinding.

Welding band saw blades is a fairly simple operation and you should master it as soon as possible. Blade welding is frequently done in the

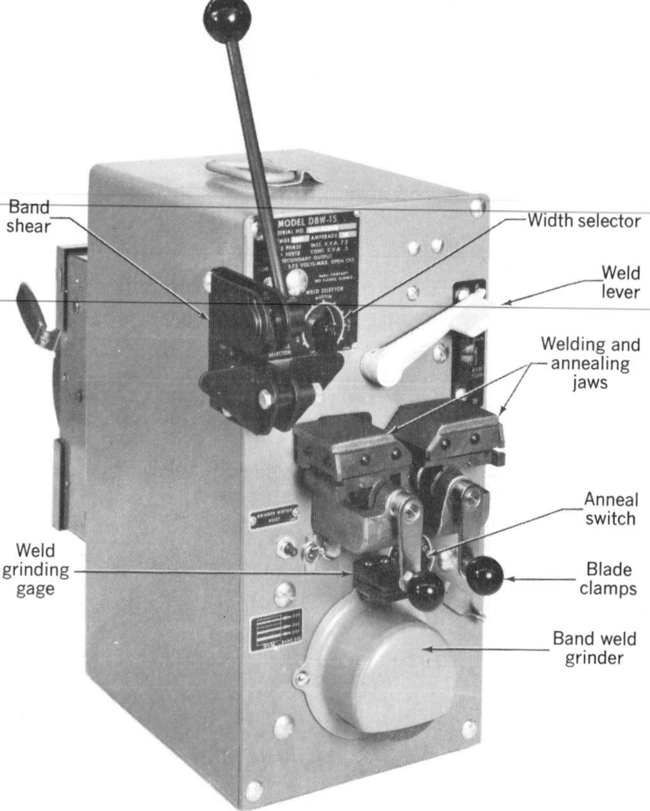

FIGURE G-77 Band blade welder (DoALL Company).

FIGURE G-78 Blade shear.

FIGURE G-79 Placing the blade ends together with the teeth opposed.

FIGURE G-80 End grinding the blade on the pedestal grinder.

machine shop. New blades are always being prepared. Sawing operations, where totally enclosed workpiece features must be cut, require that the blade be inserted through a starting hole in the workpiece and then welded into a band. After the enclosed cut is made, the blade is broken apart and removed.

Preparing the Blade for Welding

The first step is to cut the required length of blade stock for the band machine that you are using. Blade stock can be cut with snips or with the band shear (Figure G-78). Start the cut on the side of the band opposite the teeth. Many band machines have a blade shear near the welder. The required length of blade will usually be marked on the saw frame. Blade length, B_L, for two wheel sawing machines can be calculated by the formula

$$B_L = \pi D + 2L$$

where D is the diameter of the band wheel and L is the distance between band wheel centers. Set the tension adjustment on the idler wheel about midrange so that the blade will fit after welding. Most machine shops will have a permanent reference mark, probably on the floor, that can be used for gaging blade length.

After cutting the required length of stock, the ends of the blade must be ground so that they are square when positioned in the welder. Place the ends of the blade together so that the teeth are opposed (Figure G-79). Grind the blade ends in this position. The grinding wheel on the blade welder may be used for this operation. Blade ends may also be ground on the pedestal grinder (Figure G-80). Grinding the blade ends with the teeth opposed will ensure that the ends of the blade are square when the blade is positioned in the welder. Any small error in grinding will be canceled when the teeth are placed in their normal position.

Proper grinding of the blade ends permits correct tooth spacing to be maintained. After the blade has been welded, the tooth spacing across the weld should be the **same** as any other place on the band. Tooth set should be aligned as well. A certain amount of blade material is consumed in the welding process. Therefore, the blade must be ground correctly if tooth spacing is to be maintained. The amount consumed by the welding process may vary with different blade welders. You will have to determine this by experimentation. For example, if $\frac{1}{4}$ in. of blade length is consumed in weld-

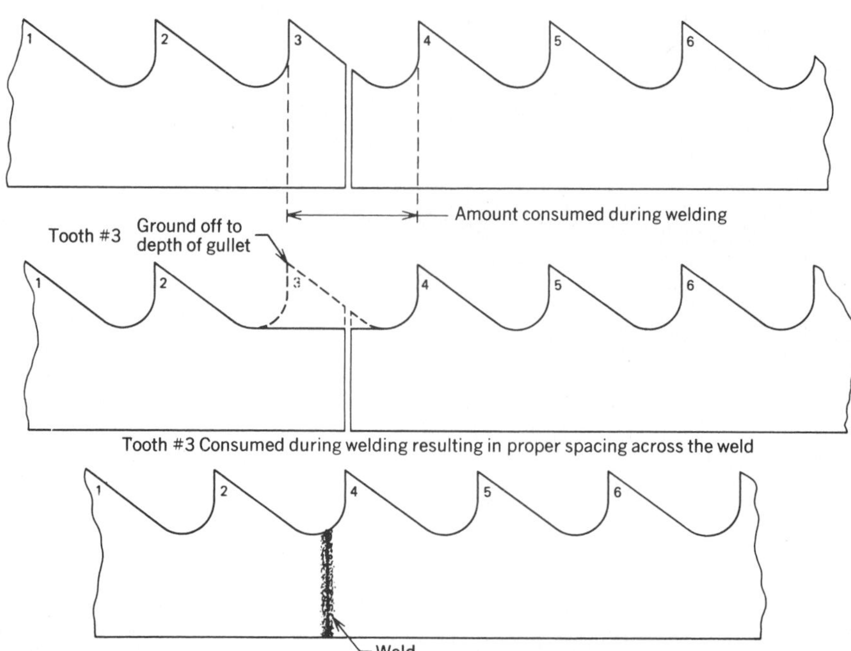

FIGURE G-81 The amount of blade lost in welding.

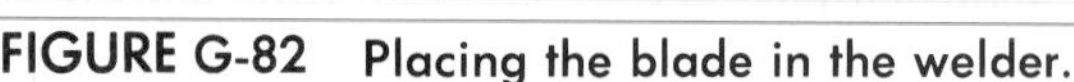

FIGURE G-82 Placing the blade in the welder.

FIGURE G-83 Welding the blade into a band (DoALL Company).

ing, this would amount to about one tooth on a four-pitch blade. Therefore, one tooth should be ground from the blade. This represents the amount lost in welding (Figure G-81). Be sure to grind only the tooth and not the end of the blade. The number of teeth to grind from a blade will vary according to the pitch and amount of material consumed by a specific welder. The weld should occur at the bottom of the tooth **gullet**. Exact tooth spacing can be somewhat difficult to obtain. You may have to practice end grinding and welding several pieces of scrap blades until you are familiar with the proper welding and tooth grinding procedure.

The jaws of the blade welder should be clean before attempting any welding. Position the blade ends in the welder jaws (Figure G-82). The saw teeth should point toward the back. This prevents scoring of the jaws when welding blades of different widths. A uniform amount of blade should extend from each jaw. The blade ends must contact squarely in the center of the gap between the welder jaws. Be sure that the blade ends are not offset or overlapped. Tighten the blade clamps.

Welding the Blade into an Endless Band

Adjust the welder for the proper width of blade to be welded. Wear eye protection and stand to one side of the welder during the welding operation. Depress the weld lever. A flash with a shower of sparks will occur (Figure G-83). In this brief operation, the mov-

FIGURE G-84 Weld flash should be evenly distributed after welding.

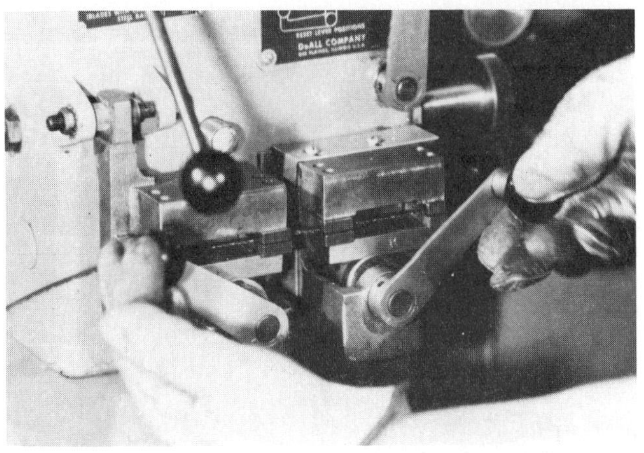

FIGURE G-85 Positioning the band for annealing.

able jaw of the welder moved toward the stationary jaw. The blade ends were heated to forging temperature by a flow of electric current, and the molten ends of the blades were pushed together forming a solid joint.

The blade clamps should be loosened before releasing the weld lever. This prevents scoring of the welder jaws by the now welded band. A correctly welded band will have the weld flash evenly distributed across the weld zone (Figure G-84). Tooth spacing across the weld should be the same as the rest of the band.

Annealing the Weld

The metal in the weld zone is hard and brittle immediately after welding. For the band to function, the weld must be **annealed** or **softened**. This improves strength qualities of the weld. Place the band in the annealing jaws with the teeth pointed out (Figure G-85). This will concentrate annealing heat away from the saw teeth. A small amount of compression should be placed on the movable welder jaw prior to clamping the band. This permits the jaw to move as the annealing heat expands the band.

It is most important not to overheat the weld during the annealing process. Overheating can destroy an otherwise good weld, causing it to become brittle. The correct annealing temperature is determined by the color of the weld zone during annealing. This should be a dull red color. Depress the anneal switch and watch the band heat. When the dull red color appears, release the anneal switch immediately and let the band begin to cool. As the weld cools, depress the anneal switch briefly several times to slow the cooling rate. Too rapid cooling can result in a band weld that is not properly annealed.

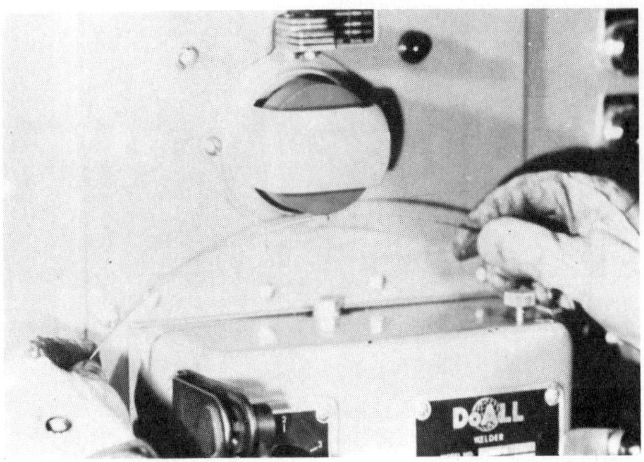

FIGURE G-86 Grinding the band weld.

Grinding the Weld

Some machinists prefer to grind the band weld prior to annealing. This permits the annealing color to be seen more easily. More often, the weld is ground after the annealing process. However, it is good practice to anneal the blade weld further after grinding. This will eliminate any hardness induced during the grinding operation. The grinding wheel on the band welder is designed for this operation. The top and bottom of the grinding wheel are exposed so that both sides of the weld can be ground (Figure G-86). **Be careful not to grind the teeth** when grinding a band weld. This will destroy the tooth set. Grind the band weld

FIGURE G-87 The saw teeth must not be ground while grinding the band weld.

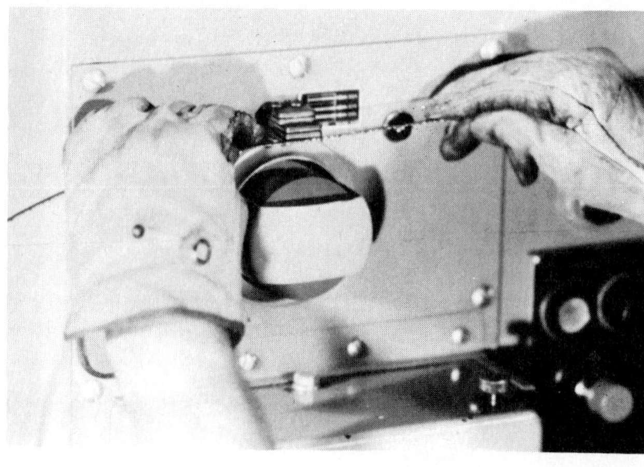

FIGURE G-89 Gaging the weld thickness in the grinding gage.

FIGURE G-88 Band weld thickness gage.

Too much welding heat

Too little welding heat

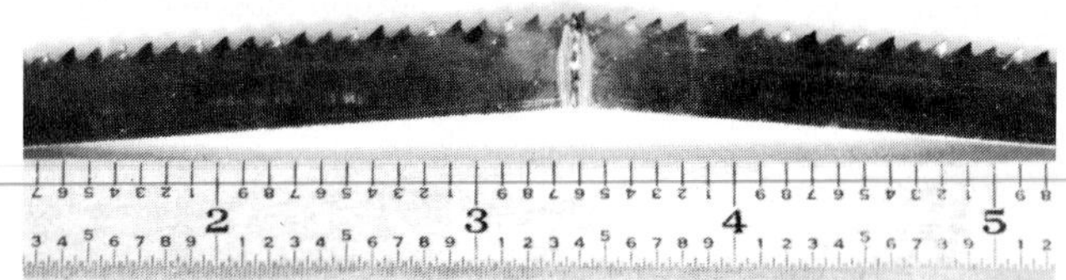

Blade misalignment

FIGURE G-90 Problems in band welding.

evenly on both sides (Figure G-87). The weld should be ground to the same thickness as the rest of the band. If the weld area is ground thinner, the band will be weakened at that point. As you grind, check the band thickness in the gage (Figure G-88) to determine proper thickness (Figure G-89).

Problems in Band Welding

Several problems may be encountered in band welding (Figure G-90). These include misaligned pitch, blade misalignment, insufficient welding heat, or too much welding heat. You should learn to recognize and avoid these problems. The best way to do this is to obtain some scrap blades and practice the welding and grinding operations.

Installing and Adjusting Band Guides on the Vertical Band Machine

Band guides must be properly installed if the band machine is to cut accurately and if damage to the band is to be prevented. Be sure to use the **correct**

FIGURE G-91 Band guides must fully support the band but must not extend over the saw teeth.

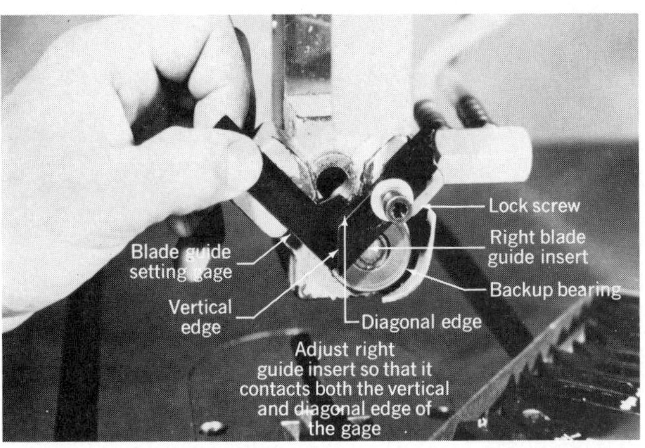
Lock screw
Right blade guide insert
Blade guide setting gage
Backup bearing
Vertical edge
Diagonal edge
Adjust right guide insert so that it contacts both the vertical and diagonal edge of the gage

FIGURE G-92 Using the saw guide setting gage.

width guides for the band (Figure G-91). The band must be fully supported except for the teeth. Using wide band guides with a narrow band will destroy tooth set as soon as the machine is started.

Band guides are set with a **guide setting gage**. Install the right-hand band guide and tighten the lock screw just enough to hold the guide insert in place. Place the setting gage in the left guide slot and adjust the position of the right guide insert so that it is in contact with both the vertical and diagonal edges of the gage (Figure G-92). Check the **backup bearing** at this time. Clear any chips that might prevent it from turning freely. If the backup bearing cannot turn freely, it will be scored by the band and damaged permanently.

Install the right-hand guide insert and make the adjustment for band thickness using the same setting gage (Figure G-93). The thickness of the band will be marked on the tool. Be sure that this is the same as the band that will be used. The lower band guide is adjusted in a like manner. Use the setting gage on the same side as it was used when adjusting the top guides (Figure G-94).

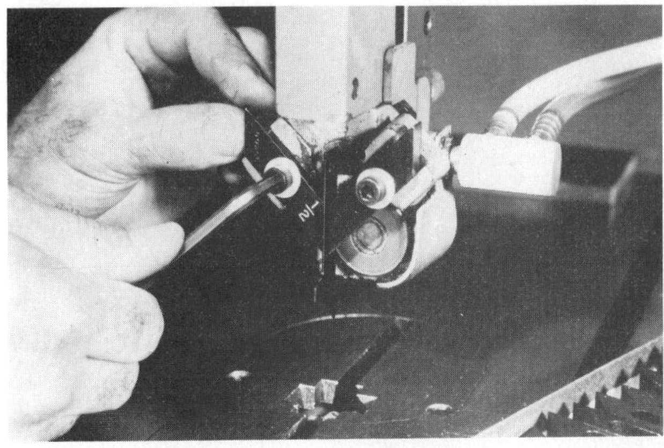

FIGURE G-93 Adjusting the band guides for band thickness.

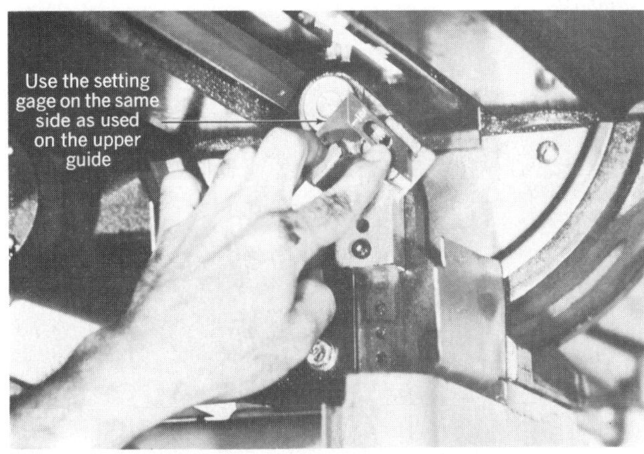
Use the setting gage on the same side as used on the upper guide

FIGURE G-94 When adjusting the lower guide, use the setting gage on the same side as the upper guide.

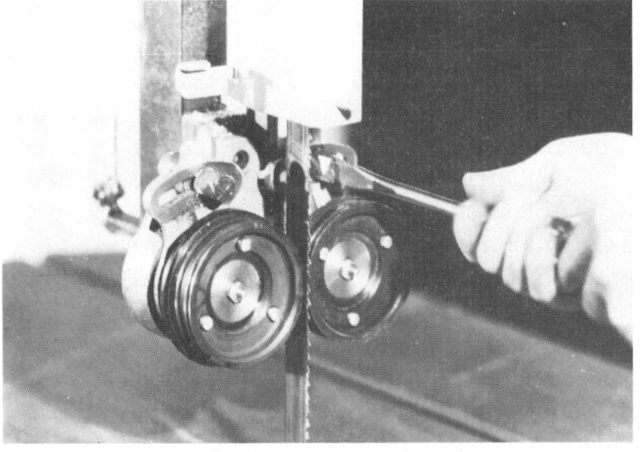

FIGURE G-95 Adjusting roller band guides.

Roller band guides are used in high speed sawing applications where band velocities exceed 2000 fpm. They are also used in friction sawing operations. The roller guide should be adjusted so that it has .001 to .002 in. clearance with the band (Figure G-95).

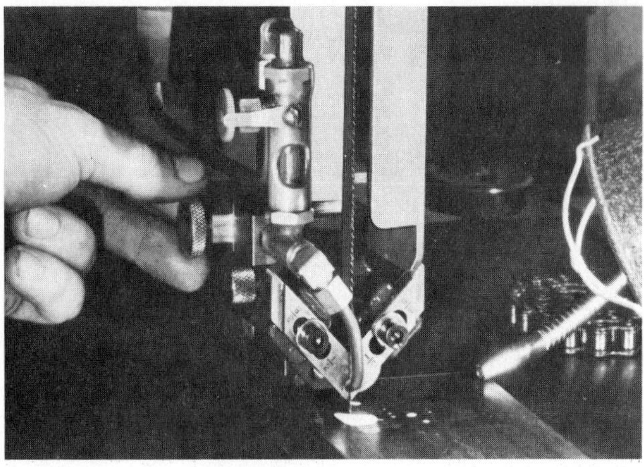

FIGURE G-96 Coolant may be introduced directly ahead of the band (DoALL Company).

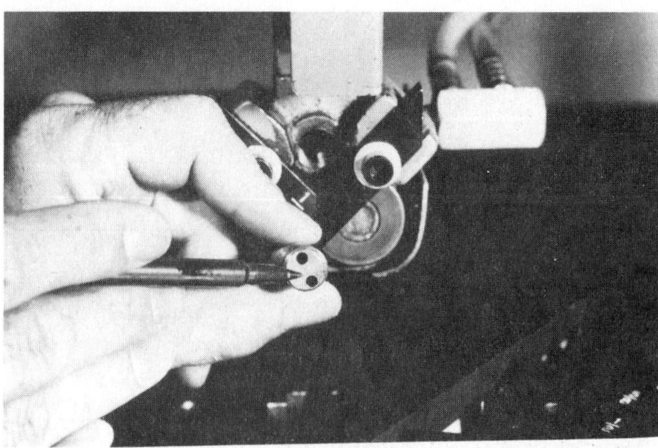

FIGURE G-98 Inlet side of the coolant nozzle.

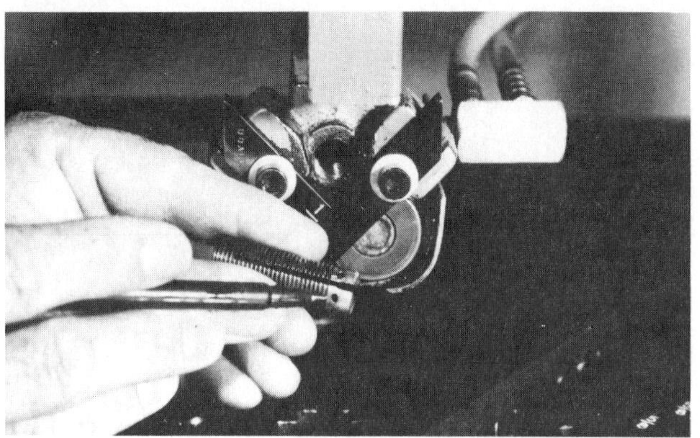

FIGURE G-97 Mist and flood coolant nozzles.

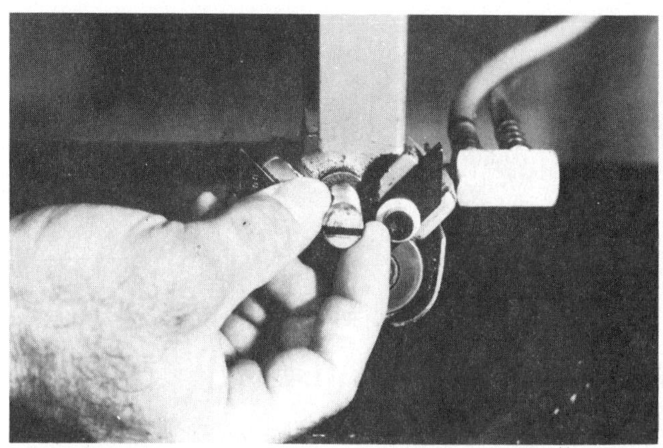

FIGURE G-99 Installing the coolant nozzle.

Adjusting the Coolant Nozzle

A band machine may be equipped with a flood or mist coolant. Mist coolant is liquid coolant mixed with air. Certain sawing operations may require only small amounts of coolant. With the mist system, liquid coolant is conserved and is less likely to spill on the floor. When cutting with floor coolant, be sure that the runoff returns to the reservoir and does not spill on the floor. Flood coolant may be introduced directly ahead of the band (Figure G-96).

Floor or mist coolant may be introduced through a nozzle in the upper guidepost assembly (Figure G-97). Air and liquid are supplied to the inlet side of the nozzle by two hoses (Figure G-98). The coolant nozzle must be installed (Figure G-99) and preset (Figure G-100) prior to installing the band. For mist coolant set the nozzle end $\frac{1}{2}$ in. from the face of the band guide. The setting for flood is $\frac{3}{8}$ in.

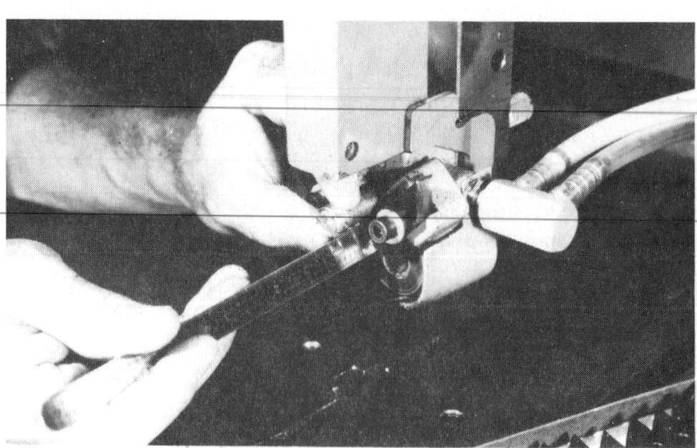

FIGURE G-100 Presetting the coolant nozzle position before installing the band guides.

Installing the Band on the Vertical Band Machine

Open the upper and lower wheel covers and remove the filler plate for the worktable. It is **safer to handle the band with gloves to protect your hands**

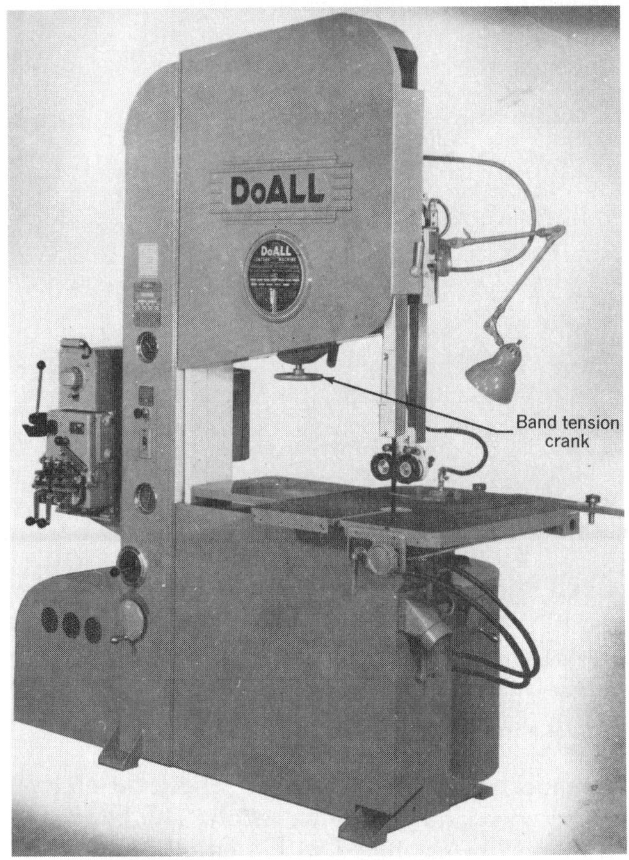

FIGURE G-101 Band tension crank (DoALL Company).

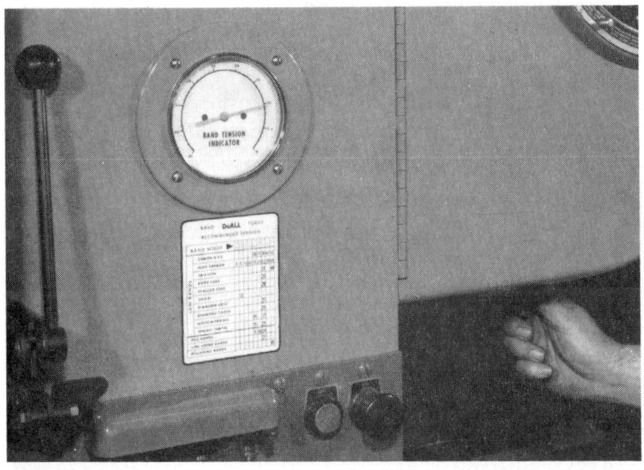

FIGURE G-102 Band tension dial (DoALL Company).

from the sharp saw teeth. The hand tension crank is attached to the upper idler wheel (Figure G-101). Turn the crank to lower the wheel to a point where the band can be placed around the drive and idler wheels. Be sure to install the band so the teeth point in the direction of the cut. This is always in a **down direction toward the worktable**. If the saw teeth seem to be pointed in the wrong direction, the band may have to be turned inside out. This can be done easily. Place the band around the drive and idler wheels and turn the tension crank so that tension is placed on the band. Be sure that the band slips into the upper and lower guides properly. Replace the filler plate in the worktable.

FIGURE G-103 Adjusting the band tracking position by tilting the idler wheel.

ADJUSTING BAND TENSION

Proper **band tension** is important to accurate cutting. A high tensile strength band should be used whenever possible. Tensile strength refers to the strength of the band to withstand stretch. The correct band tansion is indicated on the **band tension dial** (Figure G-102). Adjust the tension for the width of band that you are using. After a new band has been run for a short time, recheck the tension. New bands tend to stretch during their initial running period.

ADJUSTING BAND TRACKING

Band tracking refers to the position of the band as it runs on the idler wheel tires. On the vertical band machine, the idler wheel can be tilted to adjust tracking position. The band tracking position should be set so that the back of the band just touches the backup bearing in the guide assembly. Generally, you will not have to adjust band tracking very often. After you have installed a blade, check the tracking position. If it is incorrect, consult your instructor for help in adjusting the tracking position.

The tracking adjustment is made with the motor off and the speed range transmission in neutral. This permits the band to be rolled by hand. Two knobs are located on the idler wheel hub. The outer knob (Figure G-103) tilts the wheel.

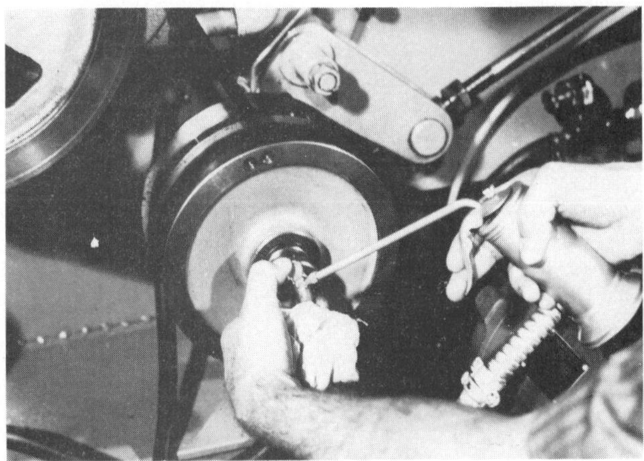

FIGURE G-104 Lubricating the variable-speed pulley hub.

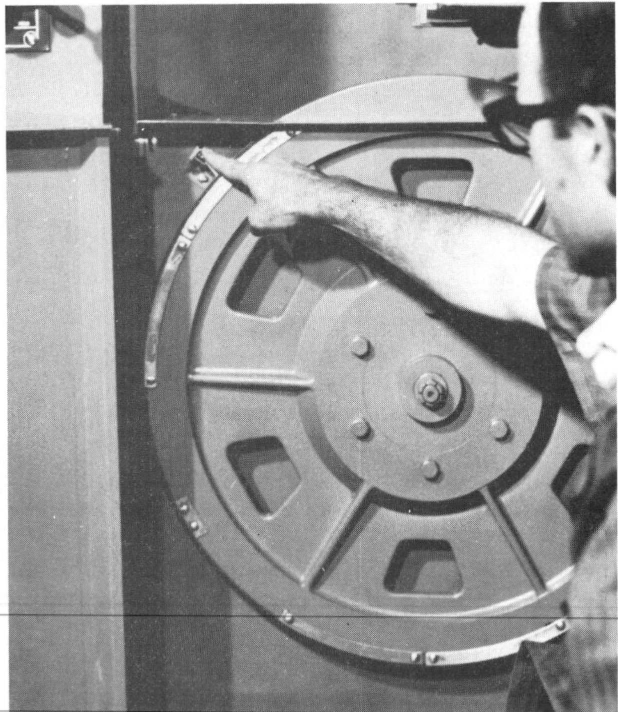

FIGURE G-105 Band wheel chip brush.

The inner knob is the tilt lock. Loosen the lock knob and adjust the tile of the idler wheel while rolling the band by hand. When the correct tracking position is reached, lock the inner knob. If the band machine has three idler wheels, adjust band tracking on the top wheel first. Then adjust tracking position on the back wheel.

Other Adjustments on the Vertical Band Machine

The hub of the variable speed pulley should be lubricated twice weekly (Figure G-104). While the drive mechanism guard is open, check the oil level in the speed range transmission. The band machine may be equipped with a chip brush on the band wheel (Figure G-105). This should be adjusted frequently. Chips that are transported through the band guides can score the band and make it brittle. The hydraulic oil level should be checked daily on band machines with hydraulic table feeds (Figure G-106).

FIGURE G-106 Checking the hydraulic oil level on the vertical band machine.

SELF-TEST

1. Describe the blade end grinding procedure.

2. Describe the band welding procedure.

3. Describe the weld grinding procedure.

4. What is the purpose of the band blade guide?

5. Why is it important to use a band guide of the correct width?

6. What tool can be used to adjust band guides?

7. What is the function of annealing the band weld?

8. Describe the annealing process.

9. What is band tracking?

10. How is band tracking adjusted?

UNIT 4

Using the Vertical Band Machine

After a machine tool has been properly adjusted and set up, it can be used to accomplish a machining task. In the preceding unit, you had an opportunity to prepare the vertical band machine for use. In this unit you will be able to operate this versatile machine tool.

Objectives

After completing this unit, you should be able to:

1. Use the vertical band machine job selector.
2. Operate the band machine controls.
3. Perform typical sawing operations on the vertical band machine.

Selecting a Blade for the Vertical Band Machine

Blade materials include standard carbon steel where the saw teeth are fully hardened but the back of the blade remains soft. The standard carbon steel blade is available in the greatest combination of width, set, pitch, and gage.

The carbon alloy steel blade has hardened teeth and a hardened back. The harder back permits sufficient flexibility of the blade, but, because of increased tensile strength, a higher band tension may be used. Because of this, cutting accuracy is greatly improved. The carbon alloy blade material is well suited to contour sawing.

High-speed steel and bimetallic high-speed steel blade materials are used in high-production and severe sawing applications where blades must have long wearing characteristics. The high-speed steel blade can withstand much more heat than the carbon or carbon alloy materials. On the bimetallic blade, the cutting edge is made from one type of high-speed steel, while the back is made from another type of high-speed steel that has been selected for high flexibility and high tensile strength. High-speed and bimetallic high-speed blades can cut longer, faster, and more accurately.

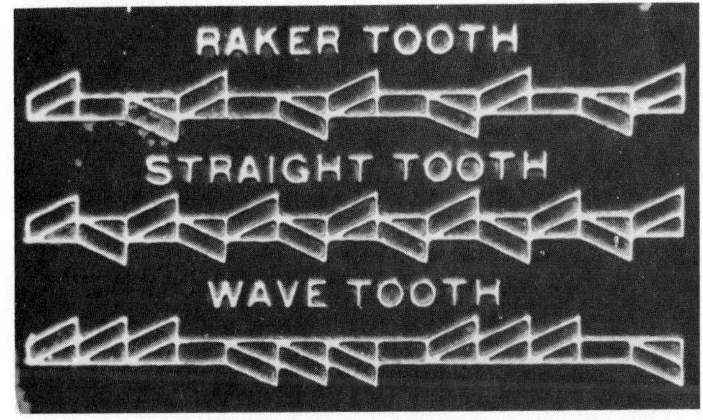

FIGURE G-107 Blade set patterns.

Band blade selection will depend on the sawing task. You should review saw blade terminology discussed in the cutoff machine unit. The first consideration is blade pitch. The pitch of the blade should be such that at least two teeth are in contact with the workpiece. This generally means that fine pitch blades with more teeth per inch will be used in thin materials. Thick material requires coarse pitch blades so that chips will be more effectively cleared from the kerf.

Remember that there are three tooth sets that can be used (Figure G-107). Raker and wave set are the most common in the metalworking indus-

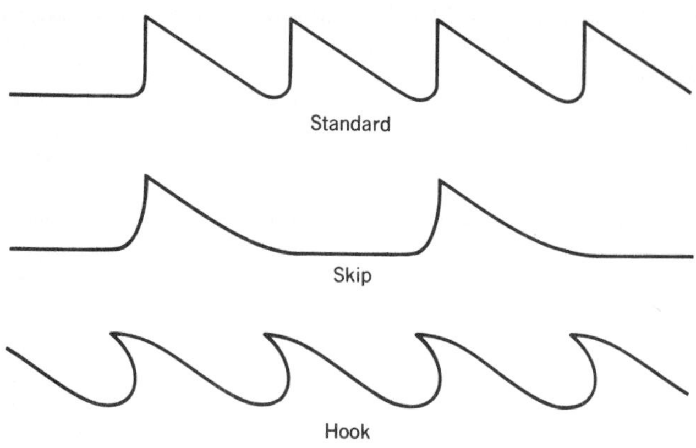

FIGURE G-108 Saw tooth forms.

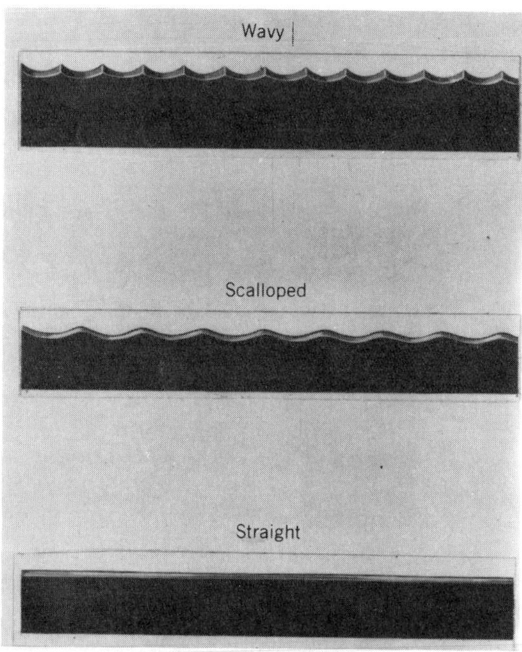

FIGURE G-109 Straight, scalloped, and wavy edge bands (DoALL Company).

FIGURE G-110 Continuous edge diamond band (DoALL Company).

FIGURE G-111 Segmented diamond edge band (DoALL Company).

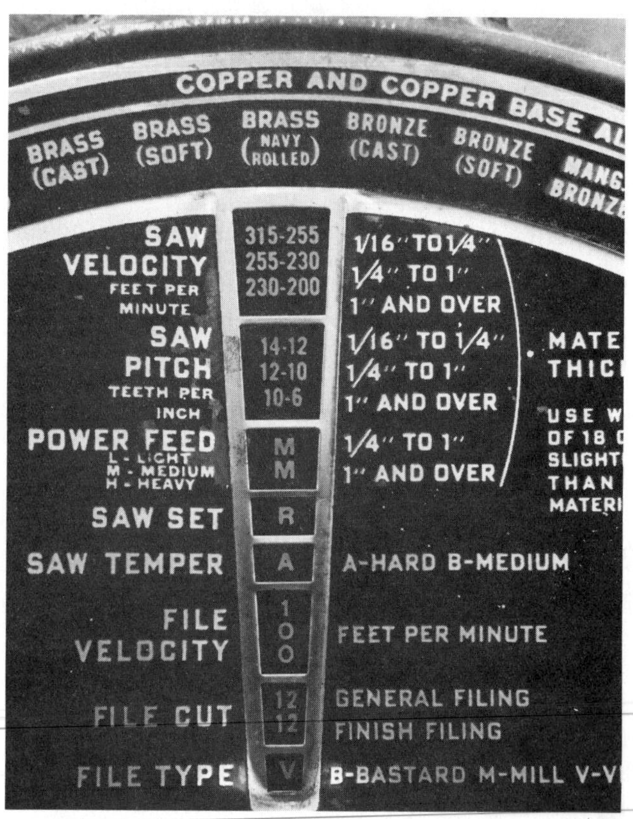

FIGURE G-112 Job selector on the vertical band machine.

tries. Straight set may be used for cutting thin materials. Wave set is best for accurate cuts through materials with variable cross sections. Raker set may be used for general-purpose sawing.

You also have a choice of **tooth forms** (Figure G-108). **Precision** or **regular** tooth form is best for accurate cuts where a good finish may be required. **Hook** form is fast cutting but leaves a rougher finish. **Skip** tooth is useful on deep cuts where additional chip clearance is required.

Several special bands are also used. **Straight, scalloped**, and **wavy edges** are used for cutting nonmetallic substances where saw teeth would tear the material (Figure G-109). **Continuous** (Figure G-110) and **segmented** (Figure G-111) **diamond edged band** are used for cutting very hard nonmetallic materials.

Using the Job Selector on the Vertical Band Machine

Most vertical band machines are equipped with a **job selector**. This device will be of great aid to you in accomplishing a sawing task. Job selectors are usually attached to the machine tool. They are frequently arranged by material. The material to be cut is located on the rim of the selector. The selector disk is then turned until the sawing data for the material can be read (Figure G-112).

The job selector yields much valuable information. Sawing velocity in feet per minute is the most important. The band must be operated at the

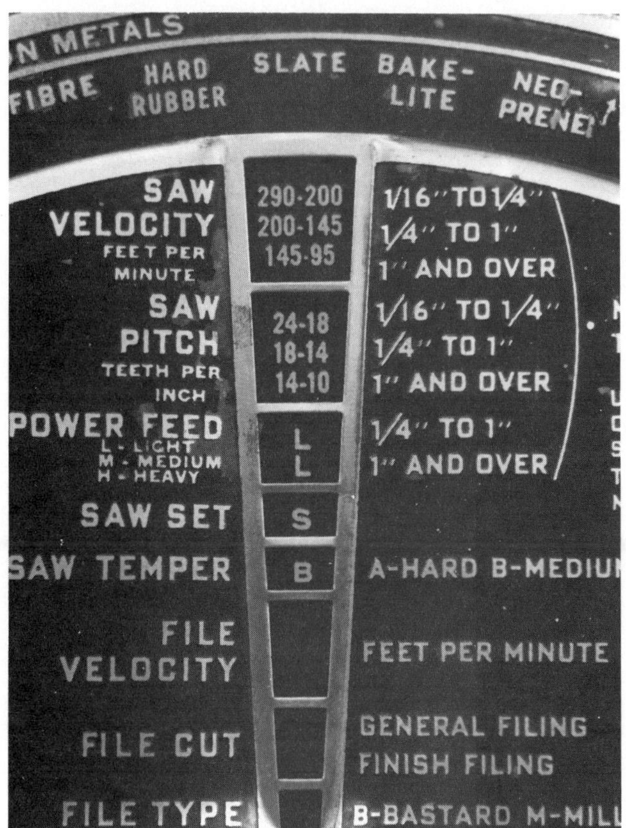

FIGURE G-113 The job selector set for a nonferrous material.

FIGURE G-114 Vertical band machine variable-speed drive (DoALL Company).

correct cutting speed for the material. If it is not, the band may be damaged or productivity will be low. Saw velocity is read at the top of the column and is dependent on material thickness. The job selector also indicates recommended pitch, set, feed, and temper. The job selector will provide information on sawing of nonmetallic materials (Figure G-113). Information on band filing can also be determined from the job selector.

FIGURE G-115 Band speed indicator.

Setting Saw Velocity on the Vertical Band Machine

Most vertical band machines are equipped with a variable-speed drive that permits a wide selection of band velocities. This is one of the factors that make the band machine such a versatile machine tool. Saw velocities can be selected that permit successful cutting of many materials.

The typical variable-speed drive uses a split-flange pulley to vary the speed of the drive wheel (Figure G-114). As the flanges of the pulley are spread apart by adjusting the speed control, the belt runs deeper in the pulley groove. This is the same as running the drive belt on a smaller-diameter pulley. Slower speeds are obtained. As the flanges of the pulley are adjusted for less spread, the belt runs toward the outside. This is equivalent to running the belt on a larger diameter pulley. Faster speeds are obtained.

SETTING BAND VELOCITY

Band velocity is indicated on the **band velocity indicator** (Figure G-115). Remember that band velocity is measured in **feet per minute**. The inner scale indicates band velocity in the low-speed range. The outer scale indicates velocity in the high-speed range. Band velocity is regulated by ad-

FIGURE G-116 Speed range and band speed controls.

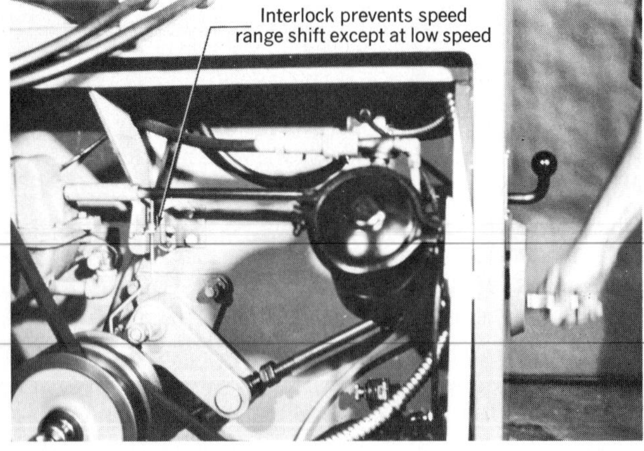

FIGURE G-117 An interlock prevents shifting speed ranges except at low speed.

justing the speed control (Figure G-116). Adjust this control only while the motor is running, as this adjustment moves the flanges of the variable speed pulley.

SETTING SPEED RANGES

Most band machines with a variable-speed drive have both a high and a low speed range, selected by operating the speed range shift lever (Figure G-116). This setting must be made while the band is stopped or is running at the lowest speed in the range. If the machine is set in high range and it is desired to go to low range, turn the band velocity control wheel until the band has slowed to the lowest speed possible. The speed range shift may now be changed to low speed. If the machine is in low speed and it is desired to shift to high range, slow the band to the lowest speed before shifting speed ranges. A speed range shift made while the band is running at a fast speed may damage the speed range transmission gears. Some band machines are equipped with an interlock to prevent speed range shifts except at low band velocity (Figure G-117).

FIGURE G-118 Adjusting the upper guide post (DoALL Company).

Straight Cutting on the Vertical Band Machine

Adjust the upper guidepost so that it is as close to the workpiece as possible (Figure G-118). This will maximize safety by properly supporting and guarding the band. Accuracy of the cut will also be aided. The guidepost is adjusted by loosening the clamping knob and moving the post up or down according to the workpiece thickness.

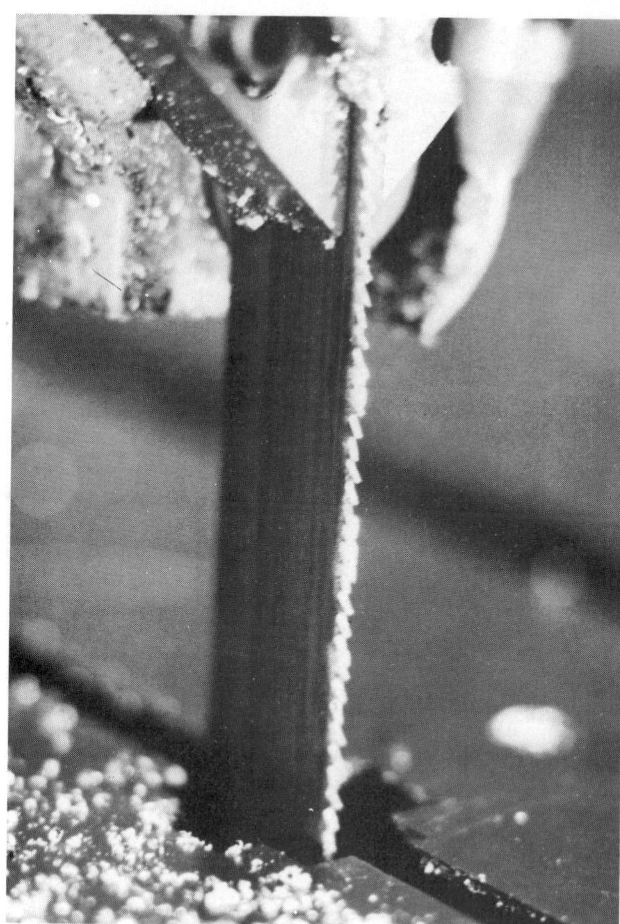

FIGURE G-119 Using too fine a pitch blade results in clogged teeth.

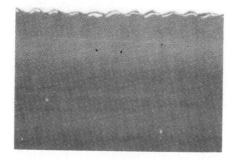

FIGURE G-120 Stripped and broken teeth resulting from overloading the saw (DoALL Company).

FIGURE G-121 Saw cut with a fine pitch blade in thick material.

FIGURE G-122 Saw cut with correct pitch band for thick material.

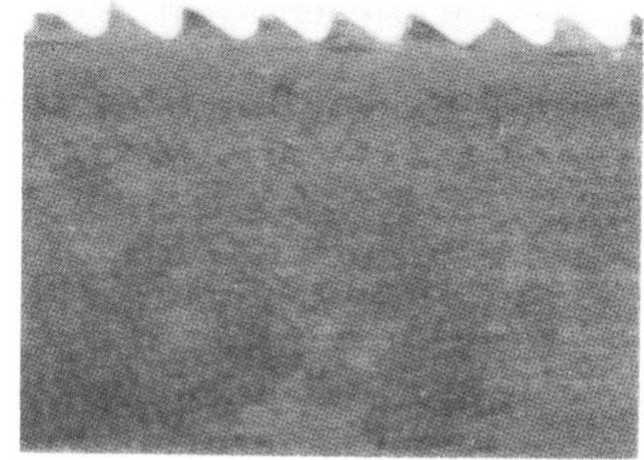

FIGURE G-123 Chipped and fractured teeth resulting from shock and vibration (DoALL Company).

Be sure to use a band of the **proper pitch** for the thickness of the material to be cut. If the band pitch is too fine, the teeth will clog (Figure G-119). This can result in stripping and breakage of the saw teeth due to overloading (Figure G-120). Cutting productivity will also be reduced. Slow cutting will result from using a fine pitch band on thick material (Figure G-121). The correct pitch for thick material (Figure G-122) results in much more efficient cutting in the same amount of time and at the same feeding pressure.

As you begin a cut, feed the workpiece gently into the band. A sudden shock will cause the saw teeth to chip or fracture (Figure G-123). This will quickly reduce band life. See that chips are cleared from the band guides. These can score the band (Figure G-124), making it brittle and subject to breakage.

FIGURE G-124 Scored bands can become brittle and lose flexibility (DoALL Company).

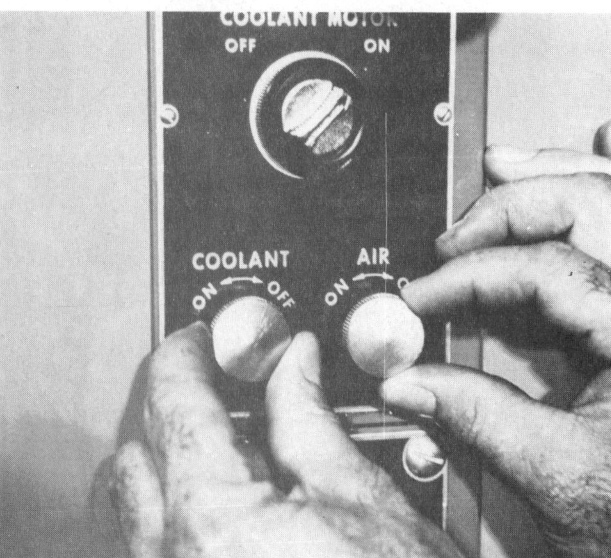

FIGURE G-125 Adjusting the coolant and air mix on the vertical band machine.

CUTTING FLUIDS

Cutting fluids are an important aid to sawing many materials. They **cool** and **lubricate** the band and **remove chips** from the kerf. Many band machines are equipped with a mist coolant system. Liquid cutting fluids are mixed with air to form a mist. With mist, the advantages of the coolant are realized without the need to collect and return large amounts of liquids to a reservoir. If your band machine uses mist coolant, set the liquid flow first and then add air to create a mist (Figure G-125). Do not use more coolant than is necessary. Overuse of air may cause a mist fog around the machine. This is both unpleasant and hazardous, as coolant mist should not be inhaled.

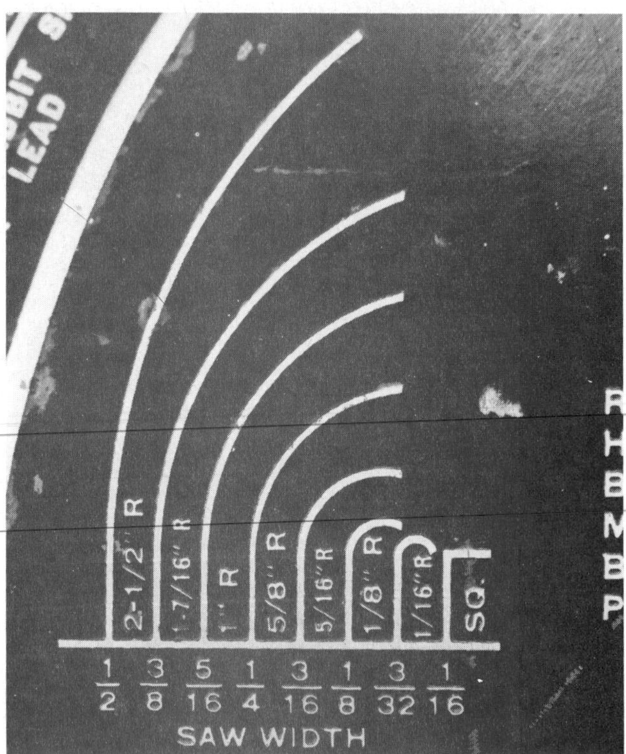

FIGURE G-126 Minimum radius per saw width chart on the job selector.

Contour Cutting on the Vertical Band Machine

Contour cutting is the ability of the band machine to cut around corners and produce intricate shapes. The ability of the saw to cut a specific radius depends on its **width**. The job selector will provide information on the minimum radius that can be cut with a blade of a given width (Figure G-126). As you can see, a narrow band can cut a smaller radius than a wide band.

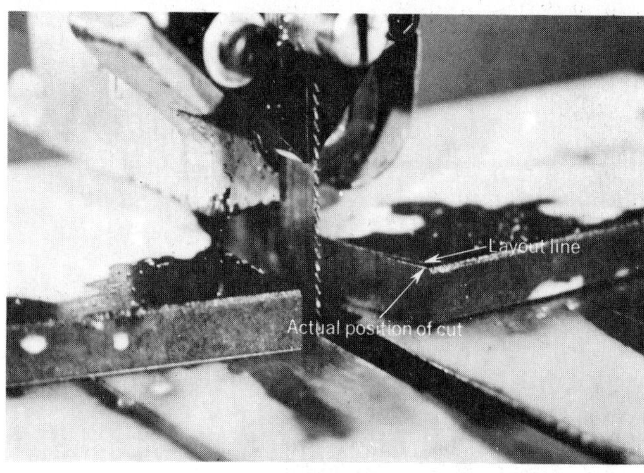

FIGURE G-127 **Band set must be adequate for the band width if the laid out radius is to be cut.**

Whenever possible, use the widest band available for straight cuts. A narrow band will tend to wander, making straight cuts more difficult. For small radius cutting, a narrow band will be required in order to follow the required radius. This band must have a full tooth set on each side for maximum efficiency in both cutting and following the desired radius curvature.

The set of the saw needs to be adequate for the corresponding band width. It is a good idea to make a test contour cut in a piece of scrap material. This will permit you to determine if the **saw set** is adequate to cut the desired radius. If the saw set is not adequate, you may not be able to keep the saw on the layout line as you complete the radius cut (Figure G-127).

If you are cutting a totally enclosed feature, be sure to insert the saw blade through the starting hole in the workpiece before welding it into a band. Also, be sure that the teeth are pointed in the right direction.

SELF-TEST

1. Name three saw blade sets and describe the applications of each.

2. When might scalloped or wavy edged bands be used?

3. What information is found on the job selector?

4. In what units of measure are band velocities measured?

5. Explain the operating principle of the variable speed pulley.

6. Explain the selection of band speed ranges.

7. What machine safety precaution must be observed when selecting speed ranges?

8. Describe the upper guidepost adjustment.

9. What does band pitch have to do with sawing efficiency?

10. What does band set have to do with contouring?

SECTION H

DRILLING MACHINES

A modern machinist must be able to use several types of complex and powerful drilling machines. Safety is the first thing you should learn about drilling machines. Chips are produced in great quantities and must be safely handled. The operator must also be protected from these chips as they fly from the machine as well as from the rotating parts of the machine. It is most important for you to become familiar with the operation and major parts of drilling machines as soon as possible. Because of the great power exerted by these machines, work-holding devices must be used to secure the work and to keep the operator safe.

There are three major types of drilling machines found in machine shops. The sensitive drill (Figure H-1) is used for light drilling on small parts. The upright drill press (Figure H-2) is used for heavy-duty drilling. The radial drill press (Figure H-3) is used for drilling large, heavy workpieces that are difficult to move.

There are a number of special-purpose drilling machines, ranging from the microscopic drilling

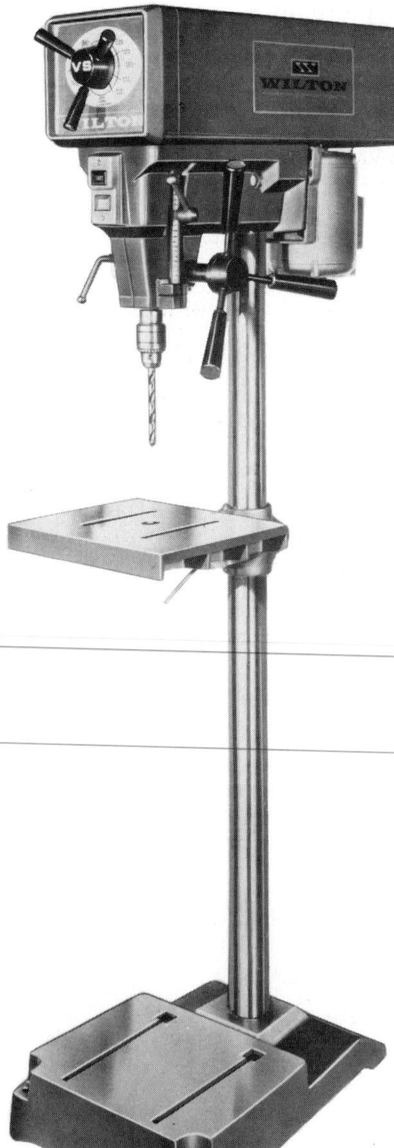

FIGURE H-1 The sensitive drill press (Wilton Corporation).

FIGURE H-2 Upright drill press (Giddings & Lewis, Inc.).

machine (Figure H-4), which can drill a hole smaller than a human hair, to deep-hole drilling machines and turret head drills, which are often CNC controlled for automatic operation.

The gang drilling machine (Figure H-5) is used when several successive operations must be performed on a workpiece. It has several drilling heads mounted over a single table, with each spindle tooled for drilling, reaming, or counterboring. Hand tooling, such as jigs and fixtures, is typically used on gang drilling machines. Tool guidance is provided by the jigs while the workpiece is clamped in the fixture. When one operation is completed, the workpiece is advanced to the next spindle.

FIGURE H-3 Radial drill press (LeBlond, Inc., Cincinnati, Ohio).

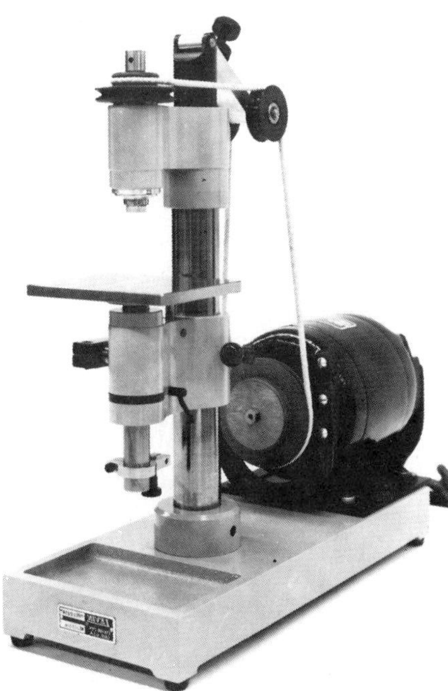

FIGURE H-4 Microdrill press for drilling very small holes in miniature and microminiature parts (Louis Levin & Sons, Inc.).

When a part requires the drilling of many holes, especially ones close together, a multispindle drilling machine (Figure H-6) is used. This machine has a number of spindles connected to the main spindle through universal joints.

FIGURE H-5 Gang drilling machine (Giddings & Lewis, Inc.).

FIGURE H-6 Multispindle drilling machine (Jarvis Products Corporation).

Turret drilling machines (Figure H-7) have several drill tools on a turret, which allows a needed tool to be rotated into position for operating on a workpiece. Many of these machines are CNC controlled so that the table position, spindle speed, and turret position are programmed to operate automatically. The operator simply clamps the workpiece in position and starts the machine.

Deep-hole drilling machines (Figure H-8) are used to drill precision holes many times longer than the bore diameter. These machines are usually horizontal with lathe-type ways and drives. The workpiece is held in clamps and is fed into a rotating drill through which coolant is pumped at high pressure. Some machines use gun drills. The deep drilling machine has the advantage of a very high metal removal rate.

FIGURE H-7 **Turret drilling machine with multispindle heads (Jarvis Products Corporation).**

In the following units, you will learn about tooling, how to use basic drilling machines, and how to select drills and reamers. Drills typically cut rough holes, and reamers are used to finish the holes. Reaming in the drill press is one way of producing a precision hole with a good finish. Countersinking and counterboring are also important tooling operations that will be introduced to you.

Since drills are used on a variety of materials and tend to take relatively heavy cuts, they often get dull or chipped. Sharpening can be done either by hand on a pedestal grinder or in a special machine. You will learn these sharpening methods in this section. Resharpening is kept to a minimum when the proper speeds, feeds, and coolants are used.

We have come a long way from the bow string and arrow drill. New ways of making holes in difficult materials are still being developed and used.

FIGURE H-8 **Deep-hole drilling machine (Giddings & Lewis, Inc.).**

Deep-hole drilling machine (Giddings & Lewis, Inc.).

Drilling Machine Safety

There is a tendency among students to dismiss the dangers involved in drilling since most drilling done in a school shop is performed in small sensitive drill presses with small diameter drills. This tendency, however, only increases the potential danger to the operator and has turned otherwise harmless situations into serious injuries. Clamps should be used to hold down workpieces, since hazards are always present even with small diameter drills.

One example of a safety hazard on even a small-diameter drill is the "grabbing" of the drill when it breaks through the hole. If you are holding the workpiece with your hands and the piece suddenly begins to spin, your hand will probably be injured, especially if the workpiece is thin.

Even if you were strong enough to hold the workpiece, the drill could break, continue turning, and with its jagged edge become an immediate hazard to your hand. The sharp chips that turned with the drill could also cut the hand that holds the workpiece.

CAUTION **Poor work habits produce many injuries. Chips flying into unprotected eyes, dropping heavy tooling, parts falling from the drill press onto toes, slipping on oily**

FIGURE H-9 Drilling operation on upright drill press with tools properly located on adjacent worktable (Lane Community College).

FIGURE H-10 Properly clamped drill press vise for holding work for drilling (Lane Community College).

floors, and getting hair or clothing caught in a rotating drill are all hazards that can be avoided by safe work habits.

SAFETY RULES FOR ALL TYPES OF DRILL PRESSES

1. Tools to be used while drilling should never be left lying on the drill press table, but should be placed on an adjacent worktable (Figure H-9).

2. Get help when lifting heavy vises or workpieces.

3. Workpieces should always be secured with bolts and strap clamps, C-clamps, or fixtures. A drill press vise should be used when drilling small parts (Figure H-10). If a clamp should come loose and a "merry-go-round" result, don't try to stop it from turning with your hands. Turn off the machine quickly; if the drill breaks or comes out, the workpiece may fly off the table.

4. Never clean the taper in the spindle when the drill is running, since this could result in broken fingers or worse injuries.

5. Always remove the chuck key immediately after using it. A key left in the chuck will be thrown out at high velocity when the machine is turned on. It is a good practice to never let the chuck key leave your hand when you are using it. It should not be left in the chuck even for a moment. Some keys are spring loaded so that they will automatically be ejected from the chuck when released. Unfortunately, very few of these keys are in use in the industry.

6. Never stop the drill press spindle with your hand after you have turned off the machine. Sharp chips often collect around the chuck or spindle. Do not reach around, near, or behind a revolving drill.

7. When removing taper shank drills with a drift, use a piece of wood under the drills so they will not drop on your toes. This will also protect the drill points from being damaged by striking the machine base.

8. Interrupt the feed occasionally when drilling to break up the chip so it will not be a hazard and will be easier to handle.

9. Use a brush instead of your hands to clean chips off the machine. Never use an air jet for removing chips as this will cause the chips to fly at a high velocity and cuts or eye injuries may result. Do not clean up chips or wipe up oil while the machine is running.

10. Keep the floor clean. Immediately wipe up any oil that spills, or the floor will be slippery and unsafe.

11. Remove burrs from a drilled workpiece as soon as possible, since any sharp edges or burrs can cause severe cuts.

12. When you are finished with a drill or other cutting tool, wipe it clean with a shop towel and store it properly.

13. Oily shop towels should be placed in a closed metal container to prevent a cluttered work area and to avoid a fire hazard.

14. When moving the head or table on sensitive drill presses, make sure a safety clamp is set just below the table or head on the column; this will prevent the table from suddenly dropping if the column clamp is prematurely released.

UNIT 1

The Drill Press

Before operating any machine, a machinist must know the names and the functions of all its parts. In this unit, therefore, you should familiarize yourself with the operating mechanisms of several types of drilling machines.

Objectives

After completing this unit, you should be able to:

1. Identify three basic drill press types and explain their differences and primary uses.
2. Identify the major parts of the sensitive drill press.
3. Identify the major parts of the radial arm drill press.

Drilling holes is one of the most basic of machining operations and one that is very frequently done by machinists. Metal cutting requires considerable pressure of feed on the cutting edge. A drill press provides the necessary feed pressure either by hand or power drive. The primary use of the drill press is to drill holes, but it can be used for other operations such as countersinking, counterboring, spot facing, reaming, and tapping, which are processes that modify the drilled hole.

There are three basic types of drill presses used for general drilling operations: the sensitive drill press, the upright drilling machine, and the radial arm drill press. The sensitive drill press (Figure H-11), as the name implies, allows you to "feel" the cutting action of the drill as you hand feed it into the work. These machines are either bench or floor mounted. Since these drill presses are used for light-duty applications only, they usually are used to drill holes up to $\frac{1}{2}$ in. in diameter. Machine capacity is measured by the diameter of work that can be drilled (Figure H-12).

The sensitive drill press has four major parts, not including the motor: the head, column, table, and base. Figure H-13 labels the parts of the drill press that you should remember. The spindle rotates within the quill, which does not rotate but carries the spindle up and down. The spindle shaft

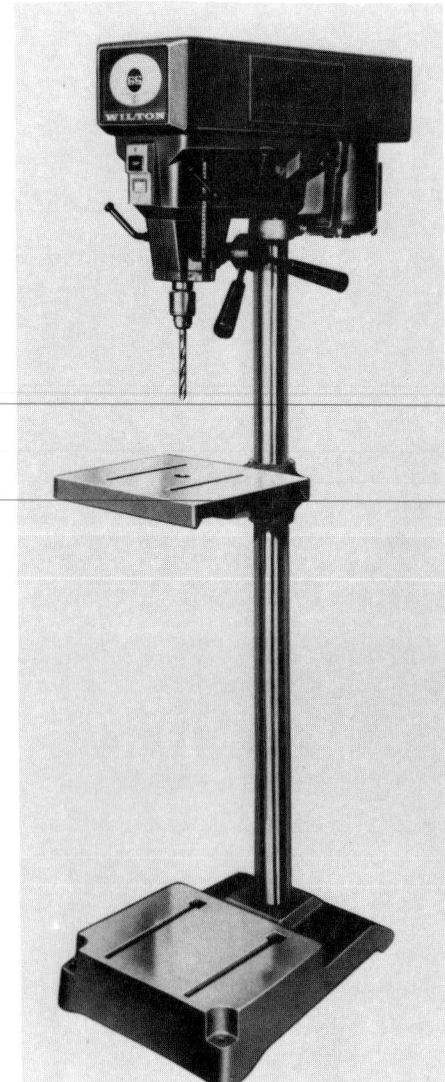

FIGURE H-11 A sensitive drill press. These machines are used for light-duty application (Wilton Corporation).

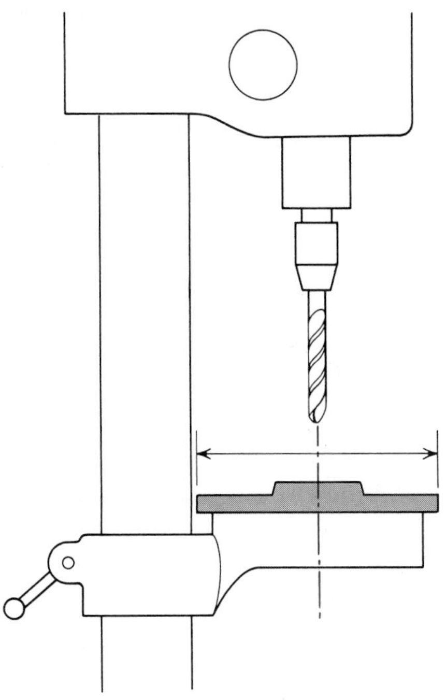

FIGURE H-12 Drill presses are measured by the largest diameter of a circular piece that can be drilled in the center.

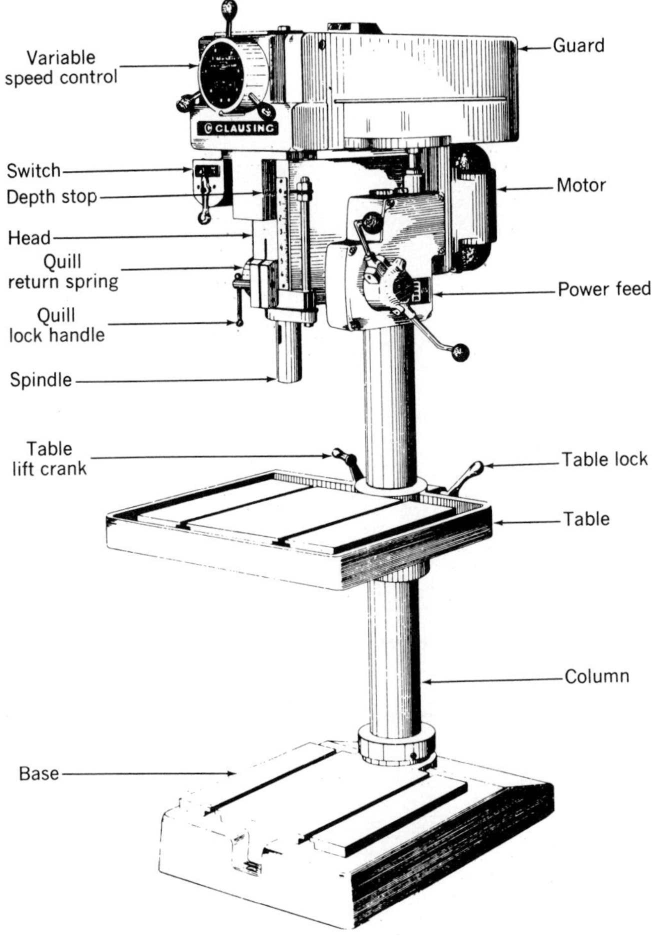

FIGURE H-13 Drill press showing the names of major parts (Clausing Corporation).

FIGURE H-14 View of a vee-belt drive. Spindle speeds are highest when the belt is in the top steps and lowest at the bottom steps (Clausing Corporation).

FIGURE H-15 View of a variable-speed drive. Variable speed selector should only be moved when the motor is running. The exact speed choice is possible for the drill size and material with this drive (Clausing Corporation).

is driven by a stepped-vee pulley and belt (Figure H-14) or by a variable-speed drive (Figure H-15). The motor must be running and the spindle turning when changing speeds with a variable speed drive.

The upright drill press is very similar to the sensitive drill press, but it is made for much heavier work (Figure H-16). The drive is more powerful and many types are gear driven, so they are capable of drilling holes to 2 in. or more in diameter. The motor must be stopped when changing speeds on a gear drive drill press. If it doesn't shift into the selected gear, turn the spindle by hand until it meshes. Since power feeds are needed to drill these large size holes, these machines are equipped with power feed mechanisms that can be adjusted by the operator. The operator may either feed manually with a lever or hand wheel or may engage the

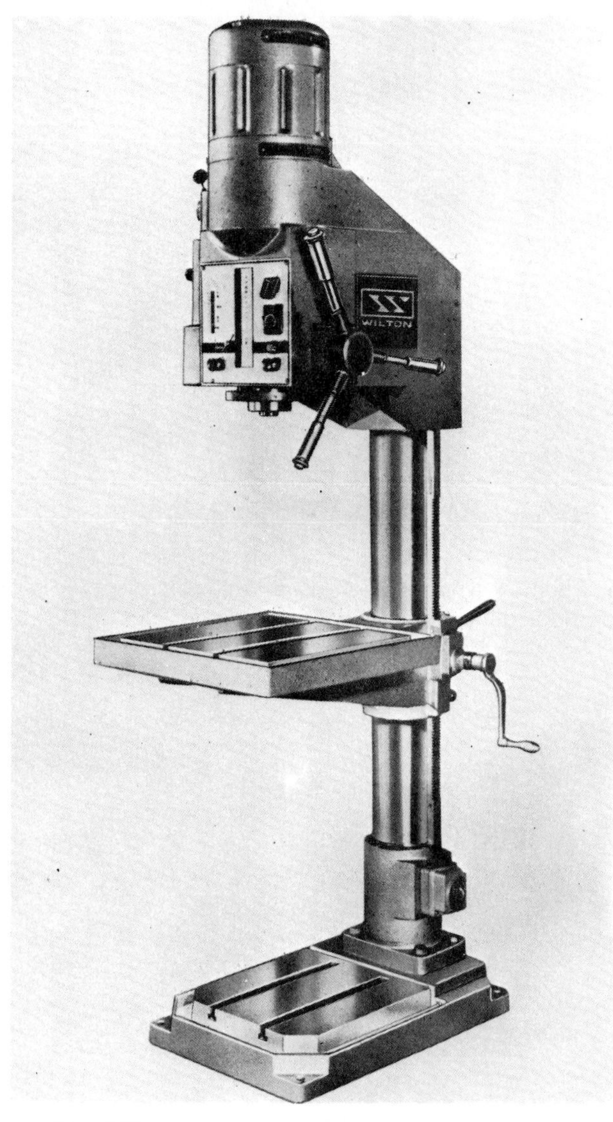

FIGURE H-16 Upright drill press (Wilton Corporation).

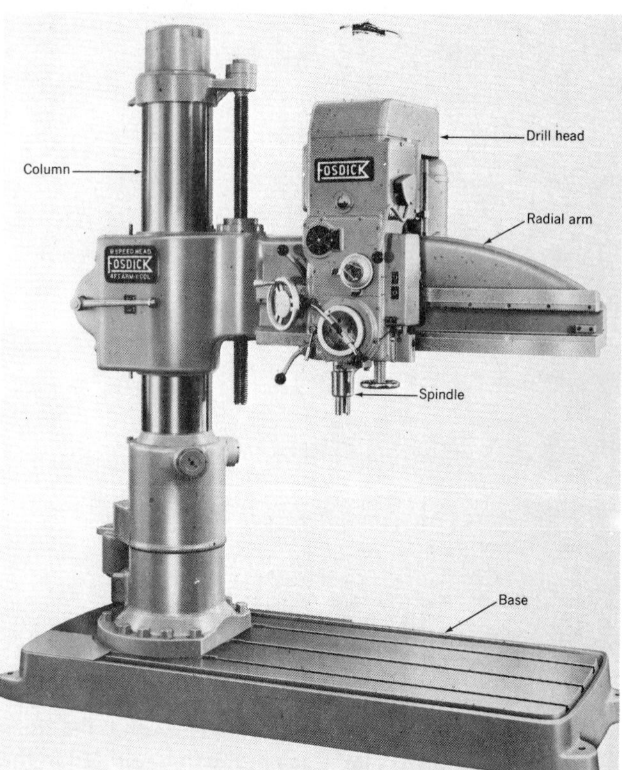

FIGURE H-17 Radial arm drill press with names of major parts (LeBlond Inc., Cincinnati, Ohio).

power feed. A mechanism is provided to raise and lower the table.

As Figures H-17 to H-19 show, the radial arm drill press is the most versatile drilling machine. Its size is determined by the diameter of the column and the length of the arm measured from the center of the spindle to the outer edge of the column. It is useful for operations on large castings that are too heavy to be repositioned by the operator for drilling each hole. The work is clamped to the table or base, and the drill can then be positioned where it is needed by swinging the arm and moving the head along the arm. The arm and head can be raised or lowered on the column and then locked in place. The radial arm drill press is used for drilling small to very large holes and for boring, reaming, counterboring, and countersinking. Like the upright machine, the radial arm drill press has a power feed mechanism and a hand feed lever.

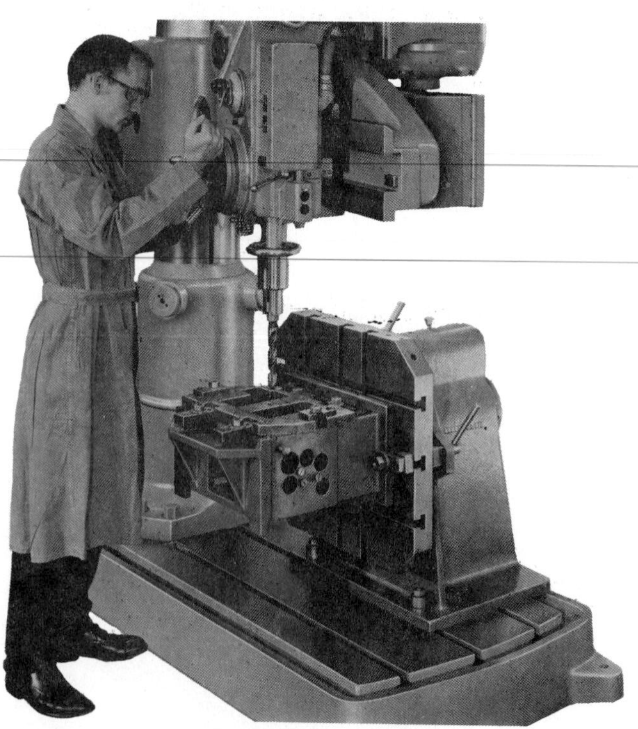

FIGURE H-18 Heavy workpiece is mounted on a trunnion-type worktable that can be rotated for positioning (LeBlond, Inc., Cincinnati, Ohio).

FIGURE H-19 Small holes are usually drilled by hand feeding on a sensitive radial drill. A workpiece is clamped on the tilting table so that a hole may be drilled at an angle (LeBlond, Inc., Cincinnati, Ohio).

SELF-TEST

1. List three basic types of drill presses and briefly explain their differences. Describe how the primary uses differ in each of these three drill press types.

2. *Sensitive drill press.* Match the correct letter with the name of that part shown on Figure H-20.

Spindle	Base
Quill lock handle	Power feed
Column	Motor
Switch	Variable speed control
Depth stop	Table lift crank
Head	Quill return spring
Table	Guard
Table lock	

3. *Radial drill press.* Match the correct letter with the name of that part shown on Figure H-21.

Column	Base
Radial arm	Drill head
Spindle	

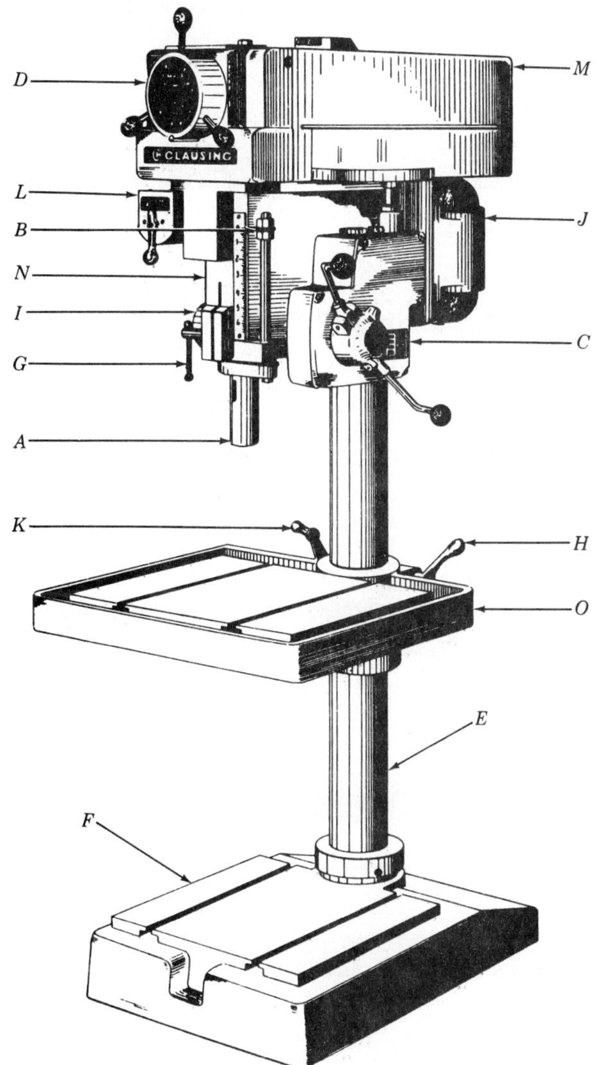

FIGURE H-20 (Clausing Corporation).

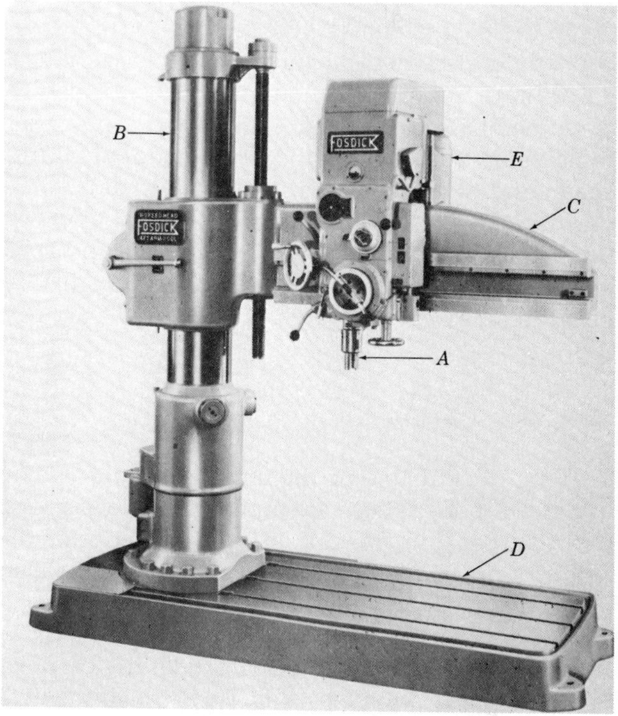

FIGURE H-21 (LeBlond, Inc., Cincinnati, Ohio).

UNIT 2

Drilling Tools

Before you learn to use drills and drilling machines, you will have to know of the great variety of drills and tooling available to the machinist. This unit will acquaint you with these interesting tools as well as show you how to select the one you should use for a given operation.

Objectives

After completing this unit, you should be able to:

1. Identify the various features of a twist drill.

2. Identify the series and size of 10 given decimal equivalent drill sizes.

The drill is an end-cutting rotary-type tool having one or more cutting lips and one or more flutes for the removal of chips and the passage of coolant. Drilling is the most efficient method of making a hole in metals softer than Rockwell 30. Harder metals can be successfully drilled, however, by using special drills and techniques.

In the past, all drills were made of carbon steel and would lose their hardness if they became too hot from drilling. Today, however, most drills are made of high-speed steel. High-speed steel drills can operate up to 1100°F (593°C) without breaking down, and, when cooled, will be as hard as before. Carbide-tipped drills are used for special applications such as drilling abrasive materials and very hard steels. Other special drills are made from cast heat-resistant alloys.

Twist Drills

The twist drill is by far the most common type of drill used today. These are made with two or more flutes and cutting lips and in many varieties of design. Figure H-22 illustrates several of the most

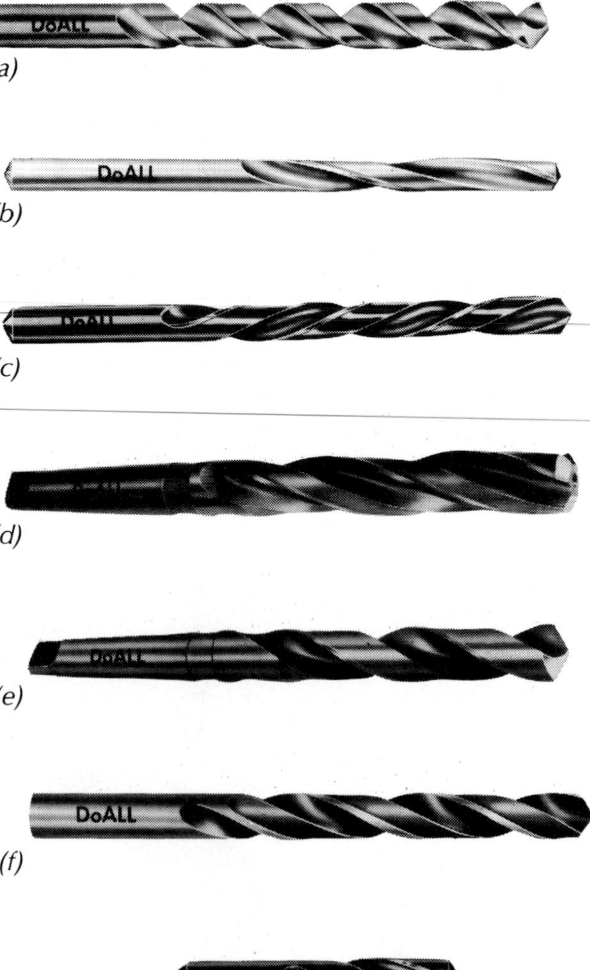

FIGURE H-22 Various types of twist drills used in drilling machines: (*a*) high-helix drill; (*b*) low-helix drill; (*c*) left-hand drill; (*d*) three-flute drill; (*e*) taper shank twist drill; (*f*) standard helix jobber drill; (*g*) center or spotting drill (DoALL Company).

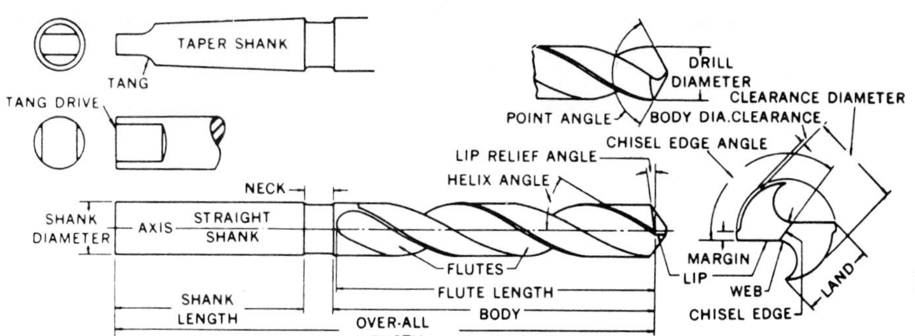

FIGURE H-23 Features of a twist drill (Bendix Industrial Tools Division).

FIGURE H-24 Drill chucks such as this one are used to hold straight shank drills (Lane Community College).

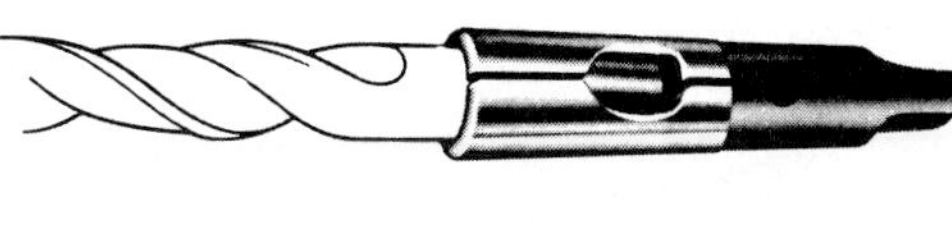

FIGURE H-25 Tang drive drill will fit into a Morse taper adapter shown here (Illinois/Eclipse, a Division of Illinois Tool Works, Inc., Chicago, Illinois 60639).

commonly used types of twist drills. The names of parts and features of a twist drill are shown in Figure H-23.

The twist drill has either a straight or tapered shank. The taper shank has a Morse taper, a standard taper of about $\frac{5}{8}$ in. per foot, which has more driving power and greater rigidity than the straight shank types. Ordinary straight shank drills are typically held in drill chucks (Figure H-24). This is a friction drive, and slipping of the drill shank is a common problem. Straight shank drills with tang drives have a positive drive and are less expensive than tapered shank drills. These are held in special drill chucks with a Morse taper (Figure H-25).

Jobbers drills have two flutes, a straight shank design, and a relatively short length-to-diameter ratio that helps to maintain rigidity. These drills are used for drilling in steel, cast iron, and nonferrous metals. Center drills and spotting drills are used for starting holes in workpieces. Oil hole drills are made so that coolant can be pumped through the drill to the cutting lips. This not only cools the cutting edges, but also forces out the chips along the flutes. Core drills have from three to six flutes, making heavy stock removal possible. They are generally used for roughing holes to a larger diameter or for drilling out cores in castings. Left-handed drills are mostly used on multispindle drilling machines where some spindles are rotated in reverse of normal drill press rotation. The step drill generally has a flat or an angular cutting edge, and can produce a hole with several diameters in one pass with either flat or countersunk shoulders.

HIGH- AND LOW-HELIX DRILLS

High-helix drills, sometimes called fast spiral drills, are designed to remove chips from deep

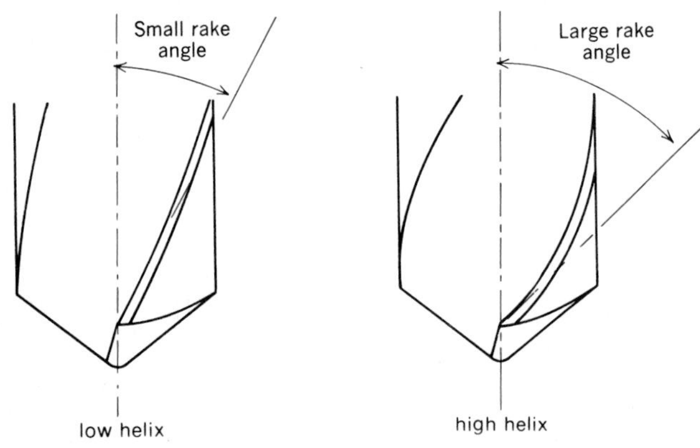

FIGURE H-26 Low and high rake angles on drills.

FIGURE H-27 Spotting drill (DoALL Company).

FIGURE H-28 Center drill.

FIGURE H-29 Step and subland drills (Mohawk Tools, Inc., Machine Tool Division).

holes. The large rake angle (Figure H-26) makes these drills suitable for soft metals such as aluminum and mild steel.

Straight fluted drills are used for drilling brass and other soft materials because the zero rake angle eliminates the tendency for the drill to "grab" on breakthrough. For the same reason they are used on thin materials.

Low-helix drills, sometimes called slow spiral drills, are more rigid than standard helix drills and can stand more torque. Like straight-fluted drills, they are less likely to grab when emerging from a hole because of the small rake angle. For this reason the low helix and straight flute drills are used primarily for drilling in brass, bronze, and some other nonferrous metals. Because of the low-helix angle, the flutes do not remove chips very well from deep holes, but the large chip space allows maximum drilling efficiency in shallow holes.

CENTER DRILLS

Spotting drills (Figure H-27) are used to position holes accurately for further drilling with regular drills. Spotting drills are short and have little or no dead center. These characteristics prevent the drills from wobbling. Lathe center drills (Figure H-28) are often used as spotting drills.

STEP DRILLS

Step and subland drills (Figure H-29) are used when two or more holes of different diameters are to be made in one drilling operation. Often, one diameter makes a tap drill hole and the larger diameter makes a countersink or counterbore for a bolt or screw head.

SPADE AND GUN DRILLS

Special drills such as spade and gun drills are used in many manufacturing processes. A spade drill is simply a flat blade with sharpened cutting lips. The spade bit, which is clamped in a holder (Figure H-30), is replaceable and can be sharpened many times. Some types provide for coolant flow to the cutting edge through a hole in the holder or shank for the purpose of deep drilling. These drills are made with very large diameters of 12 in. or more (Figure H-31) but can also be found as microdrills, smaller than a hair. Twist drills by comparison are rarely found with diameters over $3\frac{1}{2}$ in. Spade drills are usually ground with a flat top rake and with chip-breaker grooves on the end. A chisel edge and thinned web are ground in the dead center (Figure H-32).

Some spade drills are made of solid tungsten carbide, usually only in a small diameter. Twist drills with carbide inserts (Figure H-33) require a

FIGURE H-31 Large hole being drilled with spade drill. This 8-in.-diameter hole $18\frac{7}{8}$ in. deep was spade drilled in solid SAE No. 4145 steel rolling mill drive coupling housing with a Brinell hardness of 200 to 240. The machine that did the job is a 6 ft 19 in. Chip Master radial with a 25-hp motor (Giddings & Lewis, Inc.).

FIGURE H-30 Spade drill clamped in holder (DoALL Company).

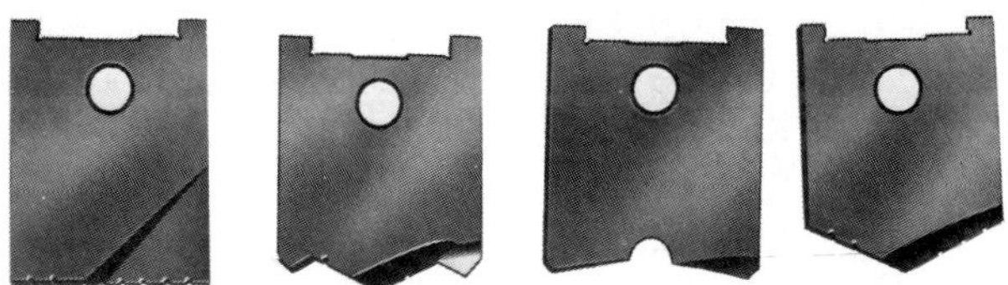

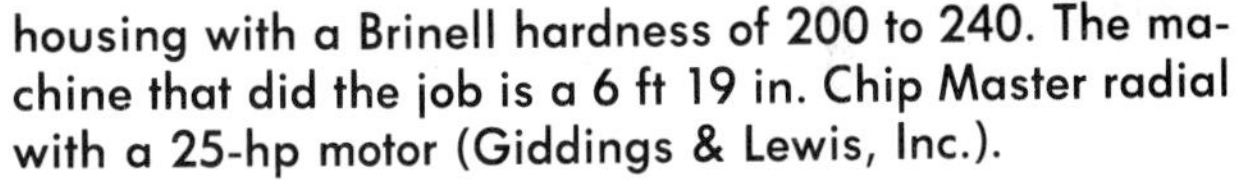

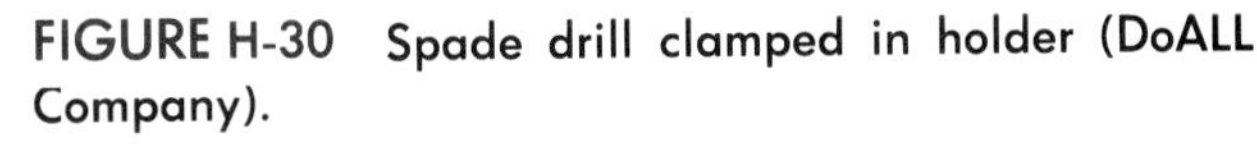

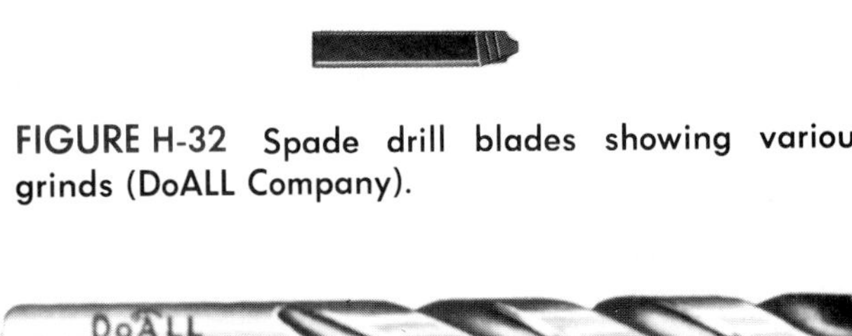

FIGURE H-32 Spade drill blades showing various grinds (DoALL Company).

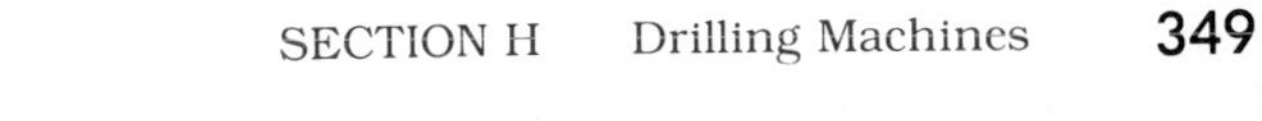

FIGURE H-33 Carbide-tipped twist drill (DoALL Company).

FIGURE H-34 Single-flute gun drill with insert of carbide cutting tip (DoALL Company).

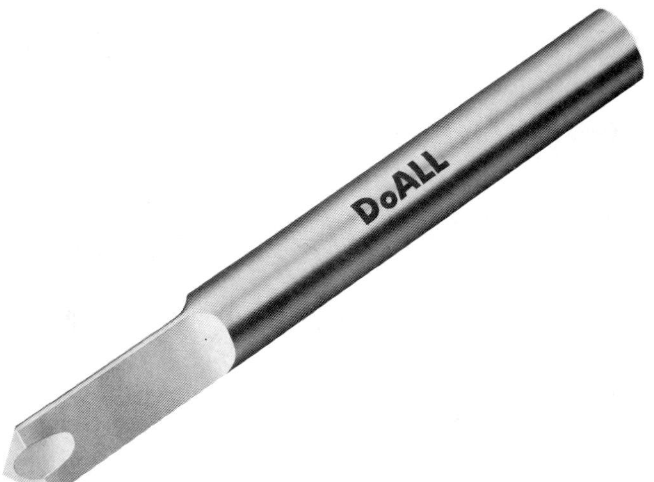

FIGURE H-35 Hard steel drill (DoALL Company).

rigid drilling setup. Gun drills (Figure H-34) are also carbide tipped and have a single V-shaped flute in a steel tube through which coolant is pumped under pressure. These drills are used in horizontal machines that feed the drill with a positive guide. Extremely deep precision holes are produced with gun drills.

Another special drill, used for drilling very hard steel, is the hard steel drill (Figure H-35). These drills are cast from a heat-resistant alloy, and the fluted end is ground to a triangular point. These drills work by heating the metal beneath the drill point by friction and then cutting out the softened metal as a chip. (Figures H-36 to H-39 show this drill in use.)

Drill Selection

The type of drill selected for a particular task depends upon several factors. The type of machine being used, rigidity of the workpiece, setup, and

FIGURE H-36 Hole started in file with a hard steel drill.

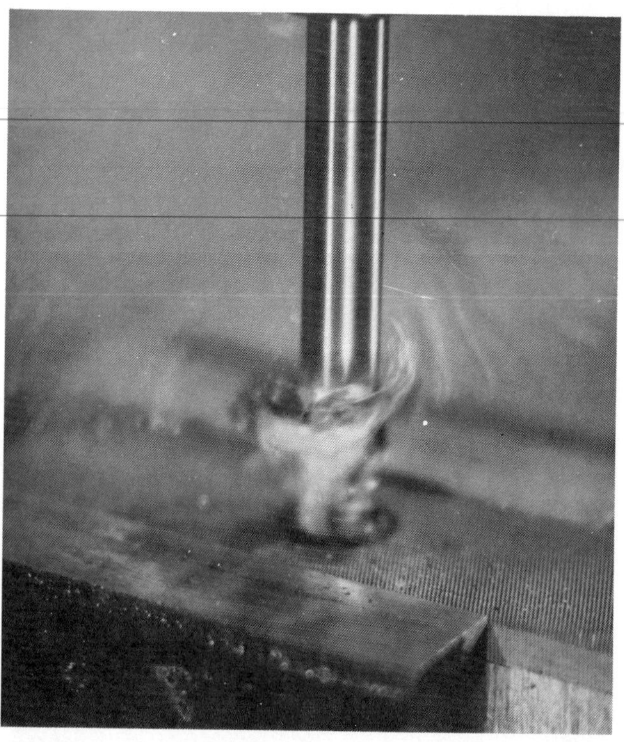

FIGURE H-37 Drilling the hole.

FIGURE H-38 Drill removed from file showing chip form.

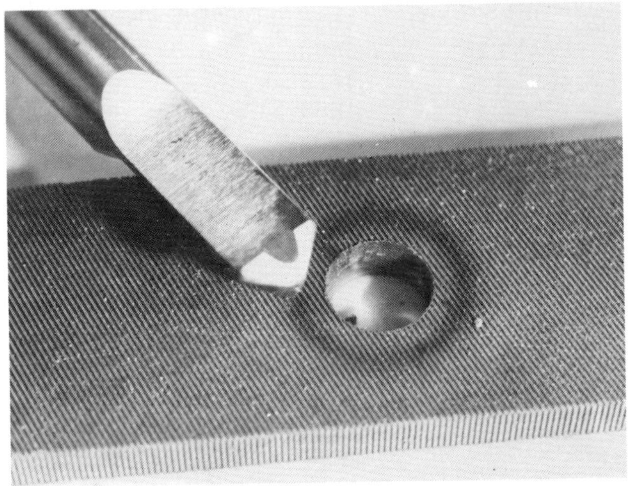

FIGURE H-39 Finished hole in file.

size of the hole to be drilled all are important. The composition and hardness of the workpiece are especially critical. The job may require a starting drill or one for secondary operations such as counterboring or spot facing, and it might need to be a drill for a deep hole or a shallow one. If the drilling operation is too large for the size or rigidity of the machine, there will be chatter and the work surface will be rough or distorted.

A machinist also must select the size of the drill, the most important dimension of which is the diameter. Twist drills are measured across the margins near the drill point (Figure H-40). Worn

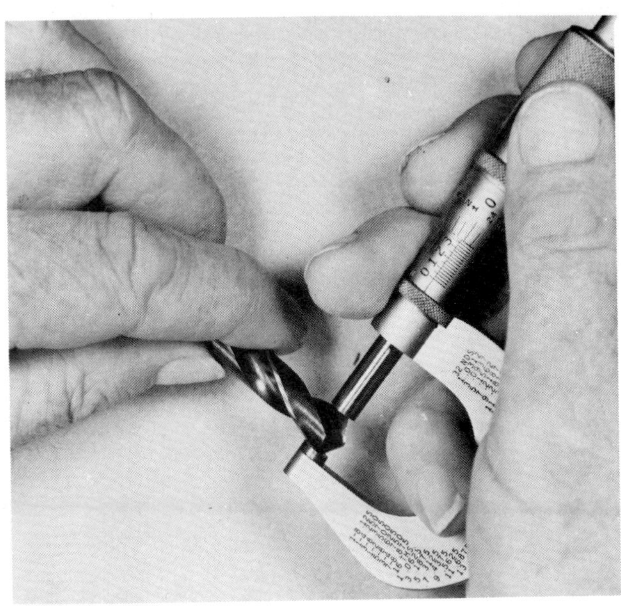

FIGURE H-40 Drill being measured across the margins.

FIGURE H-41 Morse taper drill sleeve (DoALL Company).

drills measure slightly smaller here. Drills are normally tapered back along the margin so that they will measure a few thousandths of an inch smaller at the shank. There are four drill size series: fractional, number, letter, and metric sizes. The fractional divisions are in $\frac{1}{64}$-in. increments, while the number, letter, and metric series have drill diameters that fall between the fractional inch measures. Together, the four series make up a series in decimal equivalents, as shown in Table H-1.

Identification of a small drill is simple enough as long as the number or letter remains on the shank. Most shops, however, have several series of drills, and individual drills often become hard to identify since the markings become worn off by the drill chuck. The machinist must then use a decimal equivalent table such as Table H-1. The drill in question is first measured by a micrometer, the decimal reading is then located in the table, and the equivalent fraction, number, letter, or metric size is found and noted.

Morse taper shanks on drills and Morse tapers in drill press spindles vary in size and are numbered from 1 to 6; for example, the smaller light-duty drill press has a No. 2 taper. Steel sleeves (Figure H-41) have a Morse taper inside and outside

TABLE H-1 Decimal Equivalents for Drills

Decimals of an Inch	Inch	Number/ Letter	Millimeter	Decimals of an Inch	Inch	Number	Millimeter	Decimals of an Inch	Inch	Number	Millimeter
.0135		80		.0630			1.6	.1299			3.3
.0145		79		.0635		52		.1339			3.4
.0156	$\frac{1}{64}$			.0650			1.65	.1360		29	
.0157			.4	.0669			1.7	.1378			3.5
.0160		78		.0670		51		.1405		28	
.0180		77		.0689			1.75	.1406			
.0197			.5	.0700		50		.1417			3.6
.0200		76		.0709			1.8	.1440		27	
.0210		75		.0728			1.85	.1457			3.7
.0217			.55	.0730		49		.1470		26	
.0225		74		.0748			1.9	.1476			3.75
.0236			.6	.0760		48		.1495		25	
.0240		73		.0768			1.95	.1496			3.8
.0250		72		.0781	$\frac{5}{64}$			.1520		24	
.0256			.65	.0785		47		.1535			3.9
.0260		71		.0787			2	.1540		23	
.0276			.7	.0807			2.05	.1563	$\frac{5}{32}$		
.0280		70		.0810		46		.1570		22	
.0293		69		.0820		45		.1575			4
.0295			.75	.0827			2.1	.1590		21	
.0310		68		.0846			2.15	.1610		20	
.0313	$\frac{1}{32}$			.0860		45		.1614			4.1
.0315			.8	.0866			2.2	.1654			4.2
.0320		67		.0886			2.25	.1660		19	
.0330		66		.0890		43		.1673			4.25
.0335			.85	.0906			2.3	.1693			4.3
.0350		65		.0925			2.35	.1695		18	
.0354			.9	.0935		42		.1719	$\frac{11}{64}$		
.0360		64		.0938	$\frac{3}{32}$			.1730		17	
.0370		63		.0945			2.4	.1732			4.4
.0374			.95	.0960		41		.1770		16	
.0380		62		.0966			2.45	.1772			4.5
.0390		61		.0980		40		.1800		15	
.0394			1	.0984			2.5	.1811			4.6
.0400		60		.0995		39		.1820		14	
.0410		59		.1015		38		.1850		13	
.0413			1.05	.1024			2.6	.1850			4.7
.0420		58		.1040		37		.1870			4.75
.0430		57		.1063			2.7	.1875	$\frac{3}{16}$		
.0433			1.1	.1065		36		.1890			4.8
.0453			1.15	.1083			2.75	.1890		12	
.0465		56		.1094	$\frac{7}{64}$			.1910		11	
.0469	$\frac{3}{64}$			.1100		35		.1929			4.9
.0472			1.2	.1102			2.8	.1935		10	
.0492			1.25	.1110		34		.1960		9	
.0512			1.3	.1130		33		.1969			5
.0520		55		.1142			2.9	.1990		8	
.0531			1.35	.1160		32		.2008			5.1
.0550		54		.1181			3	.2010		7	
.0551			1.4	.1200		31		.2031	$\frac{13}{64}$		
.0571			1.45	.1220			3.1	.2040		6	
.0591			1.5	.1250	$\frac{1}{8}$			.2047			5.2
.0595		53		.1260			3.2	.2055		5	
.0610			1.55	.1280			3.25	.2067			5.25
.0625	$\frac{1}{16}$			.1285		30		.2087			5.3

(*continued*)

TABLE H-1 Decimal Equivalents for Drills (*continued*)

Decimals of an Inch	Inch	Number/ Letter	Millimeter	Decimals of an Inch	Inch	Letter Size	Millimeter	Decimals of an Inch	Inch	Millimeter
.2090		4		.3150			8	.5156	33/64	
.2126			5.4	.3160		O		.5313	17/32	
.2130		3		.3189			8.1	.5315		13.5
.2165			5.5	.3228			8.2	.5469	35/64	
.2188	7/32			.3230		P		.5512		14
.2205			5.6	.3248			8.25	.5625	9/16	
.2210		2		.3268			8.3	.5709		14.5
.2244			5.7	.3281	21/64			.5781	37/64	
.2264			5.75	.3307			8.4	.5906		15
.2280		1		.3320		Q		.5938	19/32	
.2283			5.8	.3346			8.5	.6094	39/64	
.2323			5.9	.3386			8.6	.6102		15.5
.2340		A		.3390		R		.6250	5/8	
.2344	15/64			.3425			8.7	.6299		16
.2362			6	.3438	11/32			.6406	41/64	
.2380		B		.3445			8.75	.6496		16.5
.2402			6.1	.3465			8.8	.6563	21/32	
.2420		C		.3480		S		.6693		17
.2441			6.2	.3504			8.9	.6719	43/64	
.2460		D		.3543			9	.6875	11/16	
.2461			6.25	.3580		T		.6890		17.5
.2480			6.3	.3583			9.1	.7031	45/64	
.2500	1/4	E		.3594	23/64			.7087		18
.2520			6.4	.3622			9.2	.7188	23/32	
.2559			6.5	.3642			9.25	.7283		18.5
.2570		F		.3661			9.3	.7344	47/64	
.2598			6.6	.3680		U		.7480		19
.2610		G		.3701			9.4	.7500	3/4	
.2638			6.7	.3740			9.5	.7656	49/64	
.2656	17/64			.3750	3/8			.7677		19.5
.2657			6.75	.3770		V		.7812	25/32	
.2660		H		.3780			9.6	.7874		20
.2677			6.8	.3819			9.7	.7969	51/64	
.2717			6.9	.3839			9.75	.8071		20.5
.2720		I		.3858			9.8	.8125	13/16	
.2756			7	.3860		W		.8268		21
.2770		J		.3898			9.9	.8281	53/64	
.2795			7.1	.3906	15/64			.8438	27/32	
.2810		K		.3937			10	.8465		21.5
.2812	9/32			.3970		X		.8594	55/64	
.2835			7.2	.4040		Y		.8661		22
.2854			7.25	.4063	13/32			.8750	7/8	
.2874			7.3	.4130		Z		.8858		22.5
.2900		L		.4134			10.5	.8906	57/64	
.2913			7.4	.4219	27/64			.9055		23
.2950		M		.4331			11	.9063	29/32	
.2953			7.5	.4375	7/16			.9219	59/64	
.2969	19/64			.4528			11.53	.9252		23.5
.2992			7.6	.4531	29/64			.9375	15/16	
.3020		N		.4688	15/32			.9449		24
.3031			7.7	.4724			12	.9531	61/64	
.3051			7.75	.4844	31/64			.9646		24.5
.3071			7.8	.4921			12.5	.9688	31/32	
.3110			7.9	.5000	1/2			.9843		25
.3125	5/16			.5118			13	.9844	63/64	

Source: *Bendix Cutting Tool Handbook*, "Decimal Equivalents—Twist Drill Sizes," Bendix Corporation, Industrial Tools Division, 1972.

FIGURE H-42 Morse taper drill socket (DoALL Company).

FIGURE H-43 A drift is being used to remove a sleeve from a drill (Lane Community College).

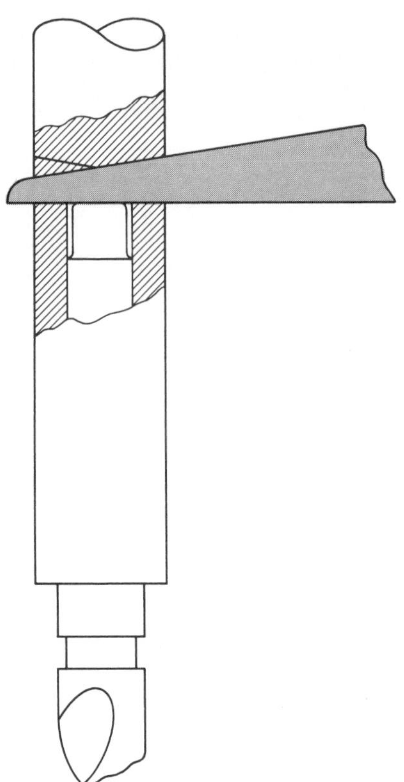

FIGURE H-44 Cutaway of a drill and sleeve showing a drift in place.

with a slot provided at the end of the inside taper to facilitate removal of the drill shank. A sleeve is used for enlarging the taper end on a drill to fit a larger spindle taper. Steel sockets (Figure H-42) function in the reverse manner of sleeves, as they adapt a smaller spindle taper to a larger drill. The tool used to remove a taper shank drill is called a drift (Figure H-43), made in several sizes and used to remove drills or sleeves. The drift is placed round side up, flat side against the drill (Figure H-44), and is struck a light blow with a hammer. A block of wood should be placed under the drill to keep it from being damaged and from being a safety hazard.

ACCURACY OF DRILLED HOLES

When twist drills have been sharpened by hand on a pedestal grinder, holes drilled by them are often rough and oversize. This is because it is not possible to make precision angles and dimensions on a drill point by offhand grinding. However, rough drilled holes are often sufficient for such purposes as bolt holes, preparation for reaming, and for tap drills. Drills almost never make a hole smaller than their measured diameter, but it is possible if the drill and workpiece become very hot during the drilling operation. When the work cools, the hole

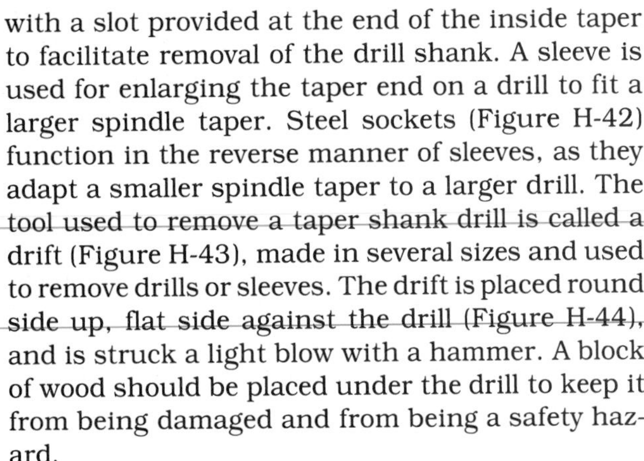

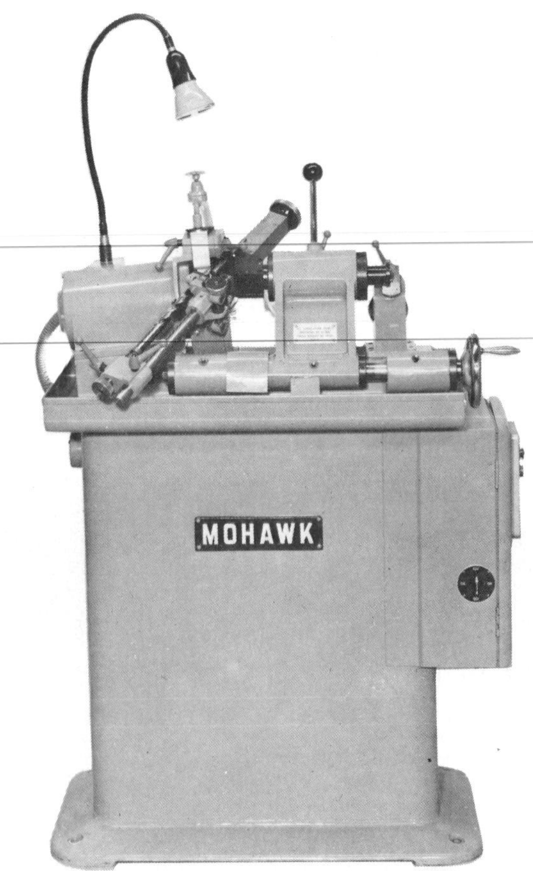

FIGURE H-45 Dupoint drill pointer (Mohawk Tools, Inc., Machine Tool Division).

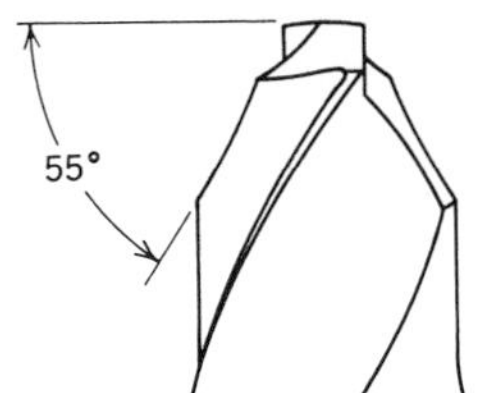

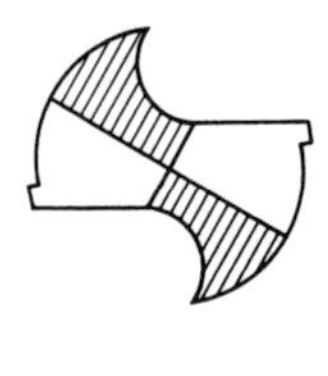

FIGURE H-46 Split-point design of a drill point (Bendix Industrial Tools Division).

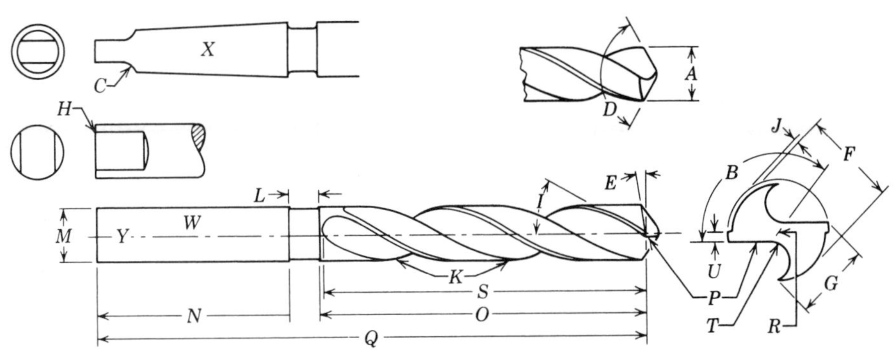

FIGURE H-47 (Bendix Industrial Tools Division).

becomes slightly smaller. Normally, the hole is 5 to 10 percent oversize if the drill has been hand sharpened.

DRILL POINTING MACHINES

A more precision method of sharpening drills is by machine (Figure H-45). Drill point grinding machines are available in several varieties and levels of complexity, ranging from those used for microdrilling to the automatic machines used by drill manufacturers. Some machines use a pivot so that the operator can shape the drill point to the proper geometry and others use cams to produce a generated drill point geometry. One kind uses the side of a grinding wheel, while another uses the circumference.

All these point grinding machines have one thing in common: precision. The precision drill point is the major advantage of the drill sharpening machine. Other advantages are that drill failures are reduced, drills stay sharp longer because both flutes always cut evenly, and the hole size is controllable within close limits.

Split-point design (Figure H-46) is often used for drilling crankshafts and tough alloy steels. The shape of the point is quite critical and too difficult to grind by hand; it should be done on a machine.

Large shops and factories that have toolrooms are more likely to use drill grinding machines than are small job shops and maintenance machine shops. Also drills used for NC and CNC machines should be machine sharpened. Step and subland drills used in production work should be sharpened only on a machine. Split-point or four-facet drill points used for deep-hole drilling as well as three- and four-flute core drills should be ground only on a machine, not by hand.

SELF-TEST

1. Match the correct letter from Figure H-47 to the list of drill parts.

Web	Body
Margin	Lip relief angle
Drill point angle	Land
Cutting lip	Chisel edge angle
Flute	Body clearance
Helix angle	Tang
Axis of drill	Taper shank
Shank length	Straight shank

2. Determine the letter, number, fractional, or metric equivalents of the 10 following decimal measurements of drills: .0781, .1495, .272, .159, .1969, .323, .3125, .4375, .201, and .1875.

	Decimal Diameter	Fractional Size	Number Size	Letter Size	Metric Size
a.	.0781				
b.	.1495				
c.	.272				
d.	.159				
e.	.1969				
f.	.323				
g.	.3125				
h.	.4375				
i.	.201				
j.	.1875				

UNIT 3

Hand Grinding of Drills on the Pedestal Grinder

Hand sharpening of twist drills has been until recent times the only method used for pointing a drill. Of course, various types of sharpening machines are now in use that can give a drill an accurate point. These precision machines are not found in every shop, however, so it is still necessary for a good machinist to learn the art of offhand drill grinding.

Objective

After completing this unit, you should be able to:

Properly hand sharpen a twist drill on a pedestal grinder so that it will drill a hole not more than .005 to .010 in. oversize.

One of the advantages of hand grinding drills on the pedestal grinder is that special alterations of the drill point such as web thinning and rake modification can be made quickly. The greatest disadvantage to this method of drill sharpening is the possibility of producing inaccurate, oversize holes (Figure H-48). If the drill has been sharpened with unequal angles, the lip with the large angle will do most of the cutting (Figure H-48*a*), and will force the opposite margin to cut into the wall of the hole. If the drill has been sharpened with unequal lip lengths, both will cut with equal force, but the drill will wobble and one margin will cut into the hole wall (Figure H-48*b*). When both conditions exist (Figure H-48*c*), holes drilled may be out of round and oversize. When drilling with inaccurate points, a great strain is placed on the drill and on the drill press spindle bearings. The frequent use of a drill point gage (Figure H-49) during the sharpening process will help to keep the point

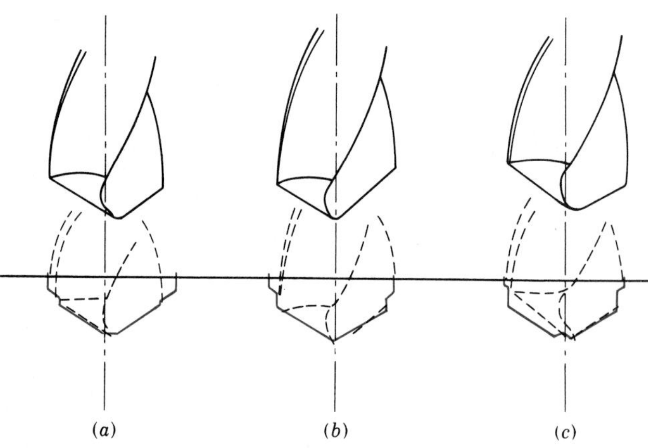

FIGURE H-48 Causes of oversize drilling: (*a*) drill lips ground to unequal lengths; (*b*) drill lips ground to unequal angles; (*c*) unequal angles and lengths.

FIGURE H-49 Using a drill point gage.

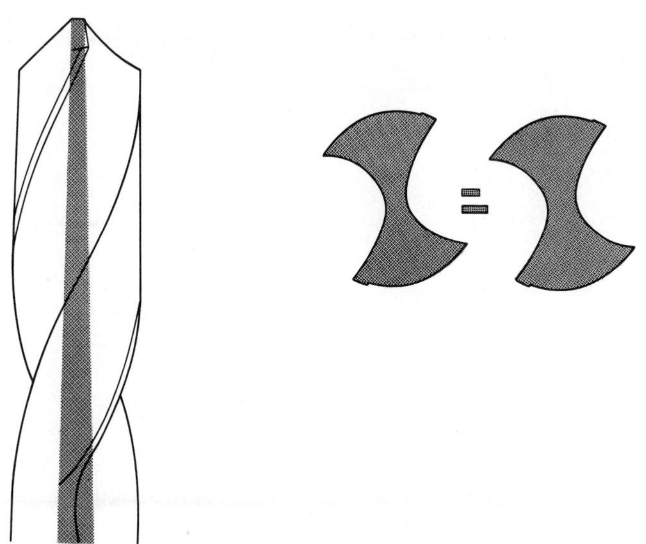

FIGURE H-50 Tapered web of twist drills (Bendix Industrial Tools Division).

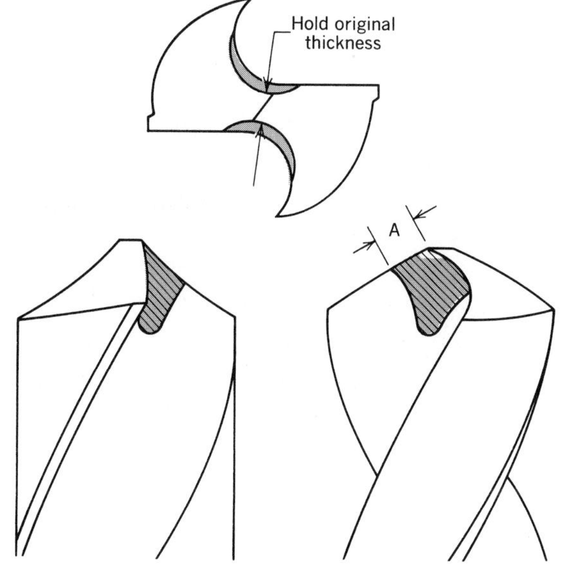

FIGURE H-51 The usual method of thinning the point on a drill. The web should not be made thinner than it was originally when the drill was new and full length (Bendix Industrial Tools Division).

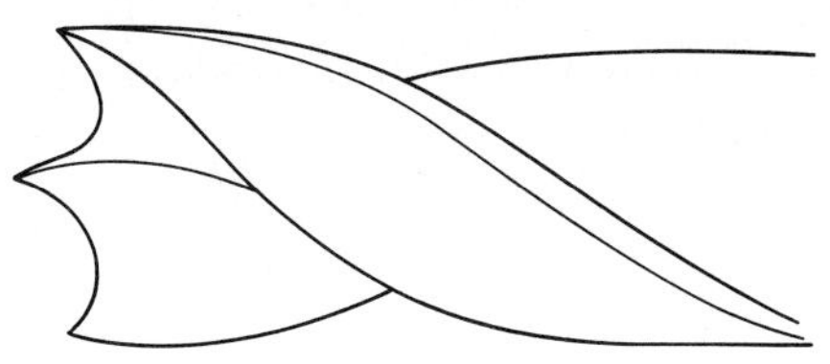

FIGURE H-52 Sheet metal drill point.

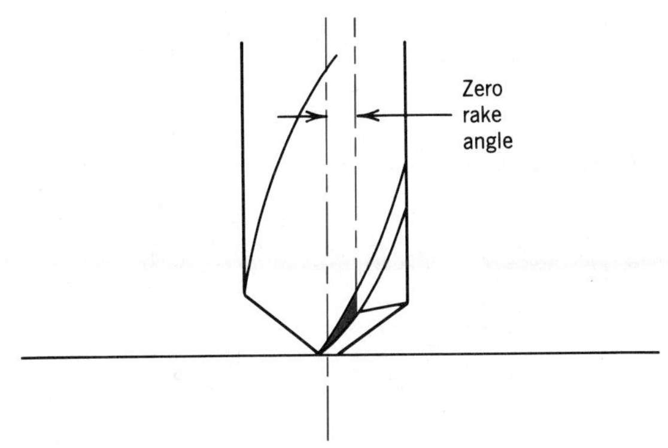

FIGURE H-53 Modification of the rake angle for drilling brass.

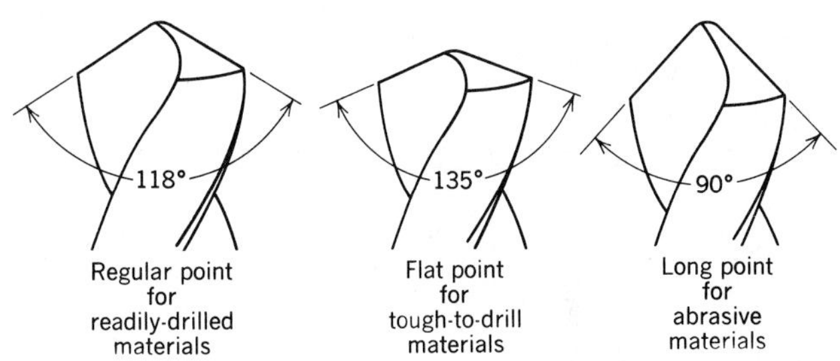

FIGURE H-54 Drill point angles (Bendix Industrial Tools Division).

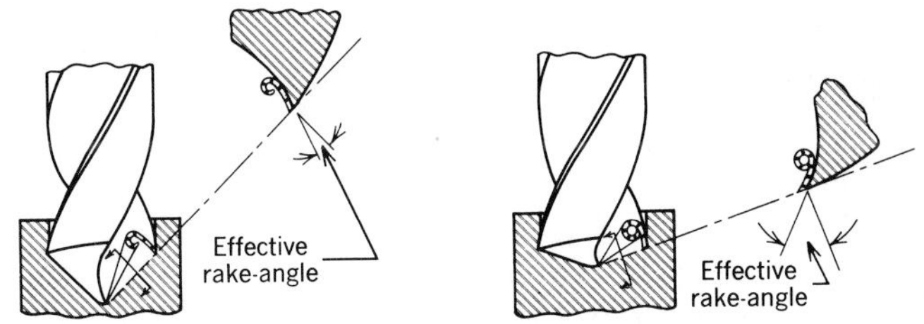

FIGURE H-55 Effective rake angles (Bendix Industrial Tools Division).

accurate and avoid such drilling problems.

The web of a twist drill (Figure H-50) is thicker near the shank. As the drill is ground shorter a thicker web results near the point. Also, the dead center or chisel point of the drill is wider and requires greater pressure to force it into the workpiece, thus generating heat. Web thinning (Figure H-51) is one method of narrowing the dead center in order to restore the drill to its original efficiency.

A sheet metal drill point (Figure H-52) may be ground by an experienced hand. The rake angle on a drill can be modified for drilling brass as shown in Figure H-53.

The standard drill point angle is an 118-degree included angle, while for drilling hard materials point angles should be from 135 to 150 degrees. A drill point angle from 60 to 90 degrees should be used when drilling soft materials, cast iron, abrasive materials, plastics, and some nonferrous metals (Figure H-54). Too great a decrease in the included point angle is not advisable, however, because it will result in an abnormal decrease in effective rake angle (Figure H-55). This will increase

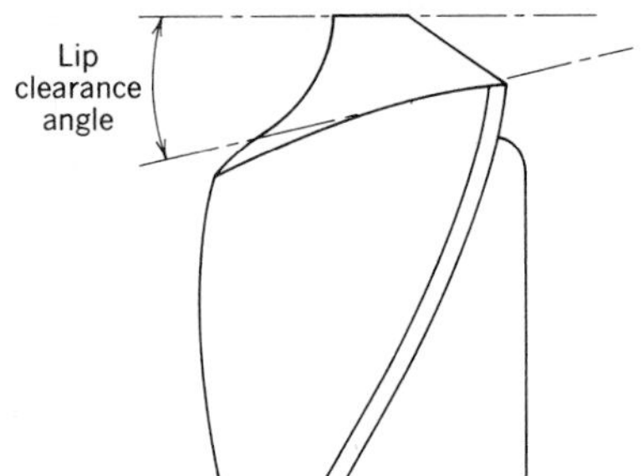

FIGURE H-56 Clearance angles on a drill point.

FIGURE H-57 Grinding wheel that is dressed and ready to use.

the required feed pressure and change the chip formation and chip flow in most steels. Clearance angles (Figure H-56) should be 8 to 12 degrees for most drilling.

FIGURE H-58 Trueing up a grinding wheel with a wheel dresser. The method shown here is often used in shops because it takes less time. The preferred method is to move the tool rest outward and to hook the lugs of the wheel dresser behind the tool rest, using it as a guide.

FIGURE H-59 Starting position showing 59-degree angle with the wheel and the cutting lip horizontal (Lane Community College).

Drill Grinding Procedure

Check to see if both the roughing and finishing wheels are true (Figure H-57). If not, move a wheel dresser across them (Figure H-58). If the end of the drill is badly damaged, use the coarse wheel to remove that part. If you overheat the drill, let it cool in air; do not cool high-speed steel drills in water.

The following method of grinding a drill is suggested:

STEP 1. Hold the drill shank with one hand and the drill near the point with the other hand. Rest your fingers that are near the point on the grinder tool rest. Hold the drill lightly at this point so you can manipulate it from the shank end with the other hand (Figure H-59).

STEP 2. Hold the drill approximately horizontal with the cutting lip (Figure H-60) that is being ground level. The axis of the drill should be at 59 degrees from the face of the wheel.

STEP 3. Using the tool rest and fingers as a pivot, slowly move the shank downward and slightly to the left (Figure H-61). The drill must be free to slip forward slightly to keep it against the wheel (Figure H-62). Rotate the drill very slightly. It is the most common mistake of the beginner to rotate the drill until the opposite cutting edge has been ground off. Do not rotate small drills at all—only larger ones. As you continue the downward movement of the shank, crowd the drill into the wheel so that it will grind lightly all the way from the lip to the heel (Figure H-63). This should all be one smooth movement. It is very important at this point to allow proper clearance (8 to 12 degrees) at the heel of the drill.

FIGURE H-60 Drill being held in the same starting position, approximately horizontal (Lane Community College).

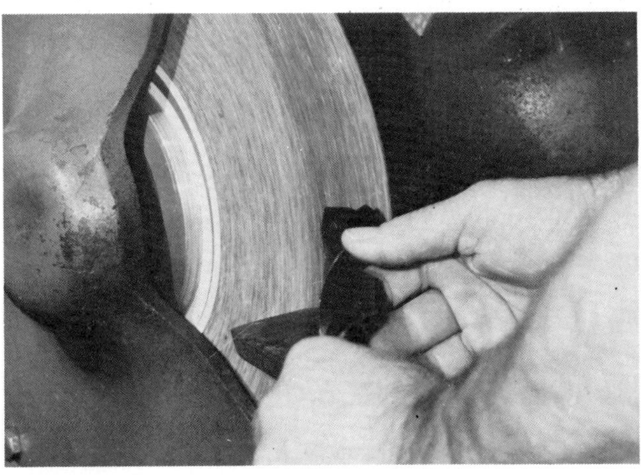

FIGURE H-61 Drill is now moved very slightly to the left with the shank being moved downward (Lane Community College).

STEP 4. Without changing your body position, pull the drill back slightly and rotate 180 degrees so that the opposite lip is now in a level position. Repeat step 3.

STEP 5. Check both cutting lips with the drill point gage (Figure H-64):

a. For correct angle.
b. For equal length.
c. Check lip clearances visually. These should be between 8 and 12 degrees.

If errors are found, adjust the regrind until they are correct.

STEP 6. When you are completely satisfied that the drill point angles and lip lengths are correct, drill a hole in a scrap metal that has been set aside for this purpose. Consult a drill speed

FIGURE H-62 Another view of the same position shown in Figure H-60 (Lane Community College).

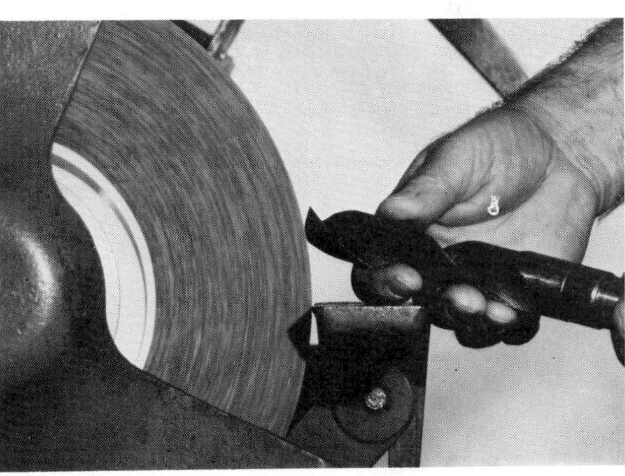

FIGURE H-63 Drill is now almost to final position of grinding. It has been rotated downward slightly from the starting position (Lane Community College).

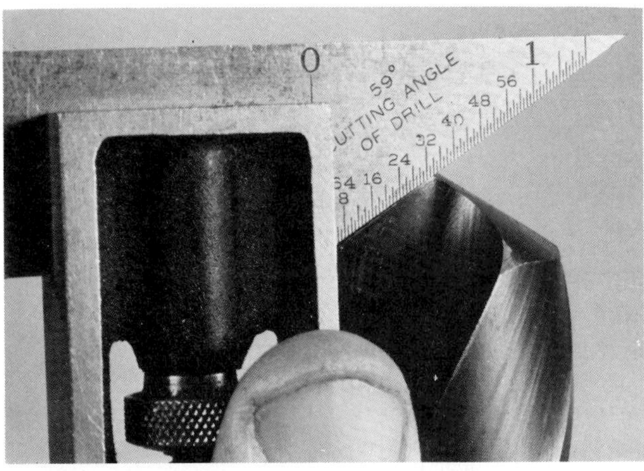

FIGURE H-64 After the sequence has been completed on both cutting lips, they are checked with a drill point gage for length and angle.

table so you will be able to select the correct rpm. Use cutting fluid.

STEP 7. Check the condition of the hole. Did the drill chatter and cause the start of the hole to be misshapen? This could be caused by too much lip clearance. Is the hole oversize more than .005 to .010 in.? Are the lips uneven or the lip angles off, or both? Running the drill too slowly for its size will cause a rough hole; too fast will overheat the drill. If the hole size is more than .01 in. × diameter over the drill size, resharpen and try again.

STEP 8. When you have a correctly sharpened drill, show the drill and the workpiece to your instructor for his evaluation.

SELF-TEST

Sharpen drills on the pedestal grinder and submit them to your instructor for approval.

UNIT 4

Operating Drilling Machines

You have already learned many things about drilling machines and tooling. You should now be ready to learn some very important facts about the use of these machines. How fast should the drill run? How much feed should be applied? Which kind of cutting fluid should be used? Workpieces of a great many sizes and shapes are drilled by machinists. In order to hold these parts safely and securely while they are drilled, several types of work-holding devices are used. In this unit you will learn how to correctly operate drilling machines.

Objectives

After completing this unit, you should be able to:

1. Determine the correct drilling speeds for five given drill diameters.

2. Determine the correct feed in steel by chip observation.

3. Set up the correct feed on a machine by using a feed table.

4. Identify and explain the correct uses for several work-holding and locating devices.

5. Set up and drill holes in two parts of a continuing project; align and start a tap using the drill press.

Cutting Speeds

Speeds (rpm) for drilling are calculated using the simplified formula

$$\text{rpm} = \frac{\text{CS} \times 4}{D}$$

where CS = an assigned surface speed for a given material, and D = the diameter of the drill. For a $\frac{1}{2}$-in. drill in low-carbon steel, the cutting speed would be 90. See Table H-2 for cutting speeds for some metals. Cutting speeds/rpm tables for various materials are available in handbooks and as wall charts. The rpm for a $\frac{1}{2}$-in. drill in low-carbon steel would be

$$\frac{90 \times 4}{\frac{1}{2}} = 720 \text{ rpm}$$

TABLE H-2 Drilling Speed Table

Material	Cutting Speed, CS
Low-carbon steel	90
Aluminum	300
Cast iron	70
Alloy steel	50
Brass and bronze	120

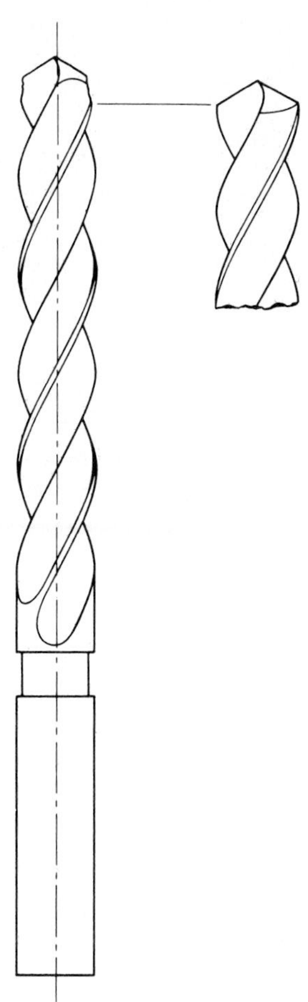

FIGURE H-65 Broken-down drill corrected by grinding back to full-diameter margins and regrinding cutting lips.

FIGURE H-66 Properly formed chip (Lane Community College).

TABLE H-3 Drilling Feed Table

Drill Size Diameter (in.)	Feeds per Revolution (in.)
Under $\frac{1}{8}$	.001 to .002
$\frac{1}{8}$ to $\frac{1}{4}$	.002 to .004
$\frac{1}{4}$ to $\frac{1}{2}$	.004 to .007
$\frac{1}{2}$ to 1	.007 to .015
Over 1	.015 and over

Excessive speeds can cause the outer corners and margins of the drill to bind in the hole, even if the speed is corrected and more cutting oil is applied. The only cure is to grind the drill back to its full diameter (Figure H-65) using methods discussed in Section B, Unit 9.

A blue chip from steel indicates the speed is too high. The tendency with very small drills, however, is to set the rpm of the spindle too slow. This gives the drill a very low cutting speed and very little chip is formed unless the operator forces it with an excessive feed. The result is often a broken drill.

Controlling Feeds

The feed may be controlled by the "feel" of the cutting action and by observing the chip. A long, stringy chip indicates too much feed. The proper chip in soft steel should be a tightly rolled helix in both flutes (Figure H-66). Some materials such as cast iron will produce a granular chip.

Drilling machines that have power feeds are arranged to advance the drill a given amount for each revolution of the spindle. Therefore, .006 in. feed means that the drill advances .006 in. every time the drill makes one full turn. The amount of feed varies according to the drill size and the work material. See Table H-3.

It is a better practice to start with smaller feeds than those given in tables. Materials and setups vary, so it is safer to start low and work up to an optimum feed. You should stop the feed occasionally to break the chip and allow coolant to flow to the cutting edge of the drill.

There is generally no breakthrough problem when using power feed, but when hand feeding, the drill may catch and grab while coming through the last $\frac{1}{8}$ in. or so of the hole. Therefore, the operator should let up on the feed handle near this point and ease the drill through the hole. This grabbing tendency is especially true of brass and some plastics, but it is also a problem in steels and other materials. Large upright drill presses and radial arm drills have power feed mechanisms with feed clutch handles (Figure H-67) that also can be used for hand feeding when the power feed is disengaged. Both feed and speed controls are set by

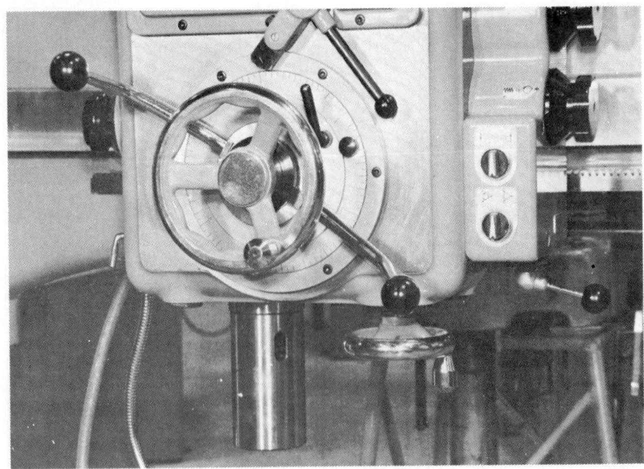

FIGURE H-67 Feed clutch handle. The power feed is engaged by pulling the handles outward. When the power feed is disengaged, the handles may be used to hand feed the drill (Lane Community College).

FIGURE H-69 Large speed and feed plates on the front of the head of the upright drill press can be read at a glance (Giddings & Lewis, Inc.).

FIGURE H-68 Speed and feed control dials (Lane Community College).

FIGURE H-70 Tapping attachment (Lane Community College).

levers or dials (Figure H-68). Speed and feed tables on plates are often found on large drilling machines (Figure H-69).

Tapping with small taps is often done on a sensitive drill press with a tapping attachment (Figure H-70) that has an adjustable friction clutch and reverse mechanism that screws the tap out when you raise the spindle. Large-size taps are power driven on upright or radial drill presses. These machines provide for spindle reversal (sometimes automatic) to screw the tap back out.

Cutting Fluids

A large variety of coolants and cutting oils are used for drilling operations on the drill press. Emulsifying or soluble oils (either mineral or synthetic) mixed in water are used for drilling holes where the main requirement is an inexpensive cooling medium. Operations that tend to create more friction and, hence, need more lubrication to prevent galling (abrasion due to friction), require a cutting oil. Animal or mineral oils with sulfur or chlorine added are often used. Reaming, counterboring, countersinking, and tapping all create friction and require the use of cutting oils, of which the sulfurized type is most frequently used. Cast iron and brass are usually drilled dry, but water-soluble oil can be used for both. Aluminum can be drilled with water-soluble oil or kerosene for a better finish. Both soluble and cutting oils are used for steel.

Drilling Procedures

Deep hole drilling requires sufficient drill length and quill stroke to complete the needed depth. A

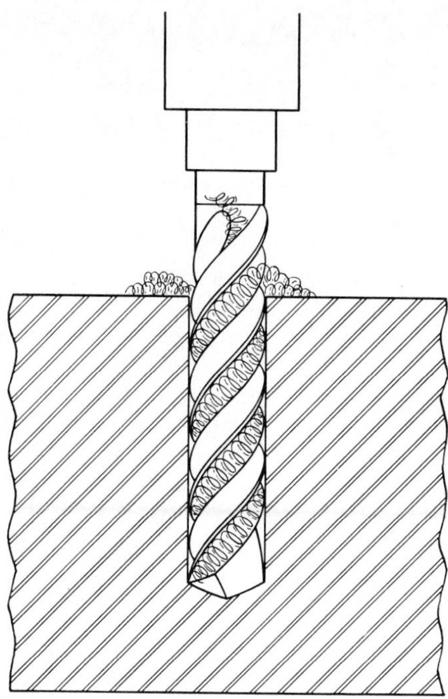

FIGURE H-71 Drill jammed in hole because of packed chips.

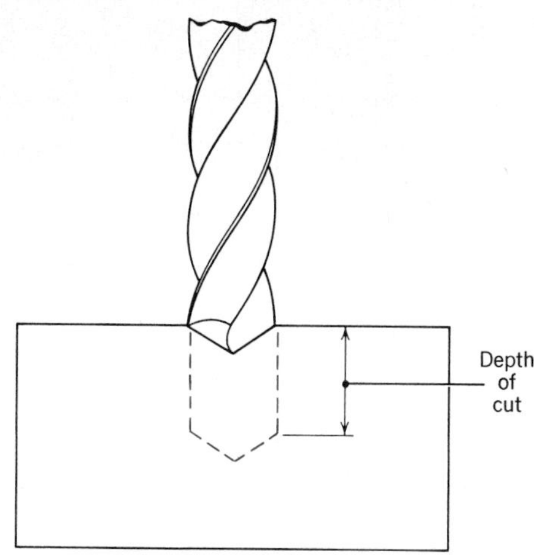

FIGURE H-73 Measuring the depth of a drilled hole.

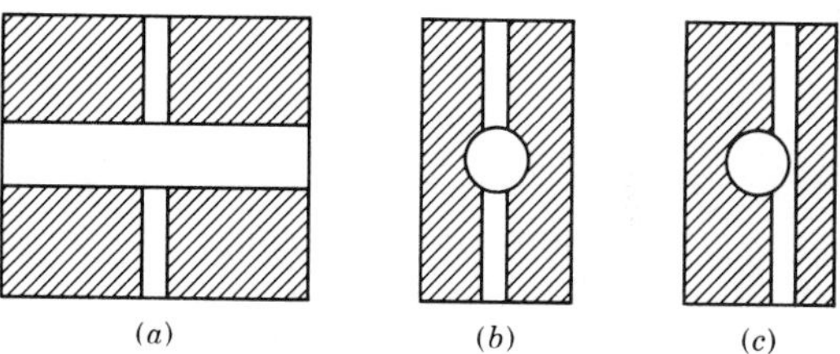

FIGURE H-74 Hole drilled at 90-degree angle into existing hole. Cross-drilling is done off center as well as on center: (*a*) side view; (*b*) end view drilled on center; (*c*) end view drilled off center.

FIGURE H-72 Using the depth stop (Lane Community College).

high helix drill helps to remove the chips, but sometimes the chips bind in the flutes of the drill and, if drilling continues, cause the drill to jam in the hole (Figure H-71). A method of avoiding this problem is called "pecking"; that is, when the hole is drilled a short distance, the drill is taken out from the hole, allowing the accumulated chips to fly off. The drill is again inserted into the hole, a similar amount is drilled, and the drill is again removed. Pecking is repeated until the required depth is reached.

A depth stop is provided on drilling machines to limit the travel of the quill so that the drill can be made to stop at a predetermined depth (Figure H-72). The use of a depth stop makes drilling several holes to the same depth quite easy. Spotfacing and counterboring should also be set up with the depth stop. Blind holes (holes that do not go through the piece) are measured from the edge of the drill margin to the required depth (Figure H-73). Once measured, the depth can be set with the stop and drilling can proceed. One of the most important uses of a depth stop, from a maintenance standpoint, is that of setting the depth so that the machine table or drill press vise will not be drilled full of holes.

Holes that must be drilled partly into or across existing holes (Figure H-74) may jam or bind a drill

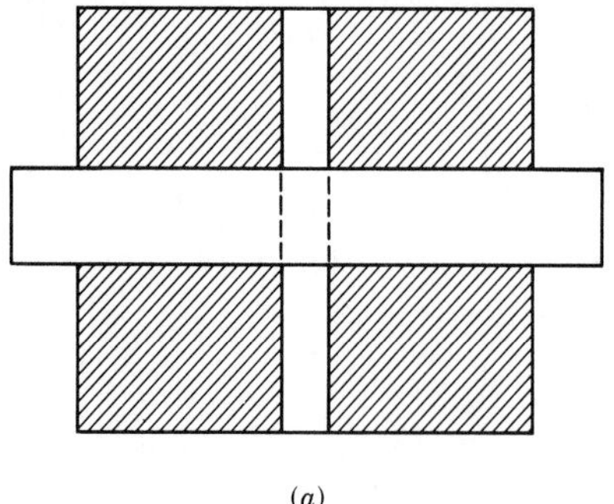

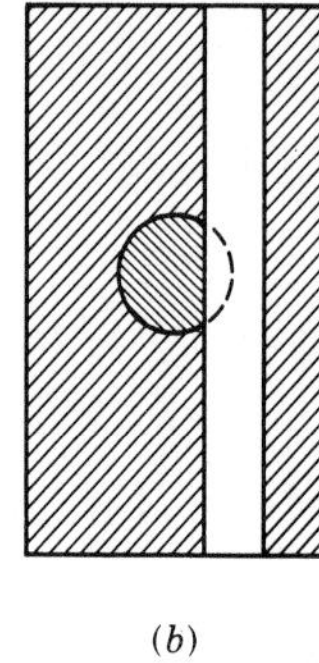

FIGURE H-75 Existing hole is plugged to make the hole more easily drilled: (*a*) hole is plugged with the same material that the workpiece consists of; (*b*) end view showing hole drilled through plug.

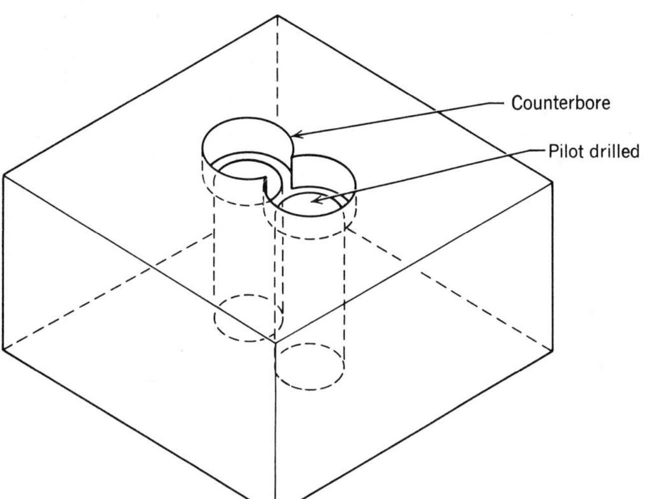

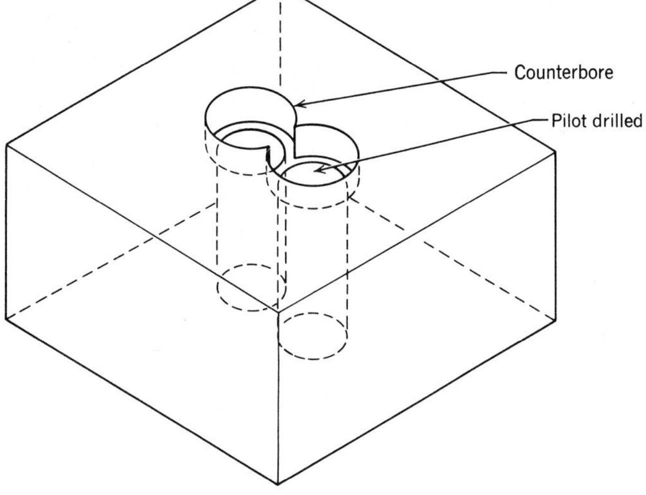

FIGURE H-76 Deep holes that overlap are difficult to drill. The holes are drilled alternately with a smaller pilot drill and a counterbore for final size.

unless special precautions are taken. A special drill with a double margin and high helix could be used directly. However, an ordinary jobber's drill will do a satisfactory job if a tight plug made of the same material as the work is first tapped into the cross hole (Figure H-75). The hole may then be drilled in a normal manner and the plug removed.

Holes that overlap may be made on the drill press if care is taken and a set of counterbores with interchangeable pilots is available. First, pilot drill the holes with a size drill that does not overlap. Then counterbore to the proper size with the appropriate pilot on the counterbore (Figure H-76). However, such an operation as this is best done with an end mill on a milling machine.

Heavy-duty drilling should be done on an upright or radial drill press (Figure H-77). The workpiece should be made very secure since high drilling forces are used with the larger drill sizes.

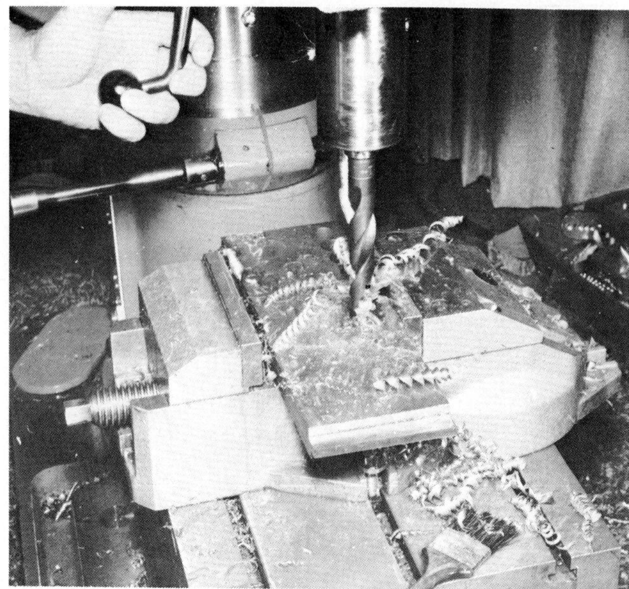

FIGURE H-77 Heavy-duty drilling on a radial drill press.

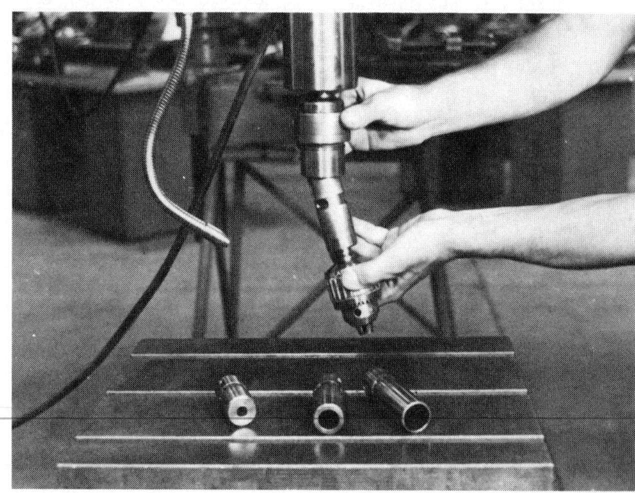

FIGURE H-78 Quick-change tooling for a drill press (Lane Community College).

The work should be well clamped or bolted to the work table. The head and column clamp should always be locked when drilling is being done on a radial drill press. Cutting fluid is necessary for all heavy-duty drilling.

QUICK-CHANGE TOOLING

Quick-change tooling (Figure H-78) is used on drill presses to facilitate speedy changes of tooling. Different size Morse taper sockets are used to hold drills, reamers, drill chucks, and other tools. These sockets can be quickly installed or removed on the spindle by sliding them into the adapter. They are automatically locked in place and can be removed by lifting up the collar.

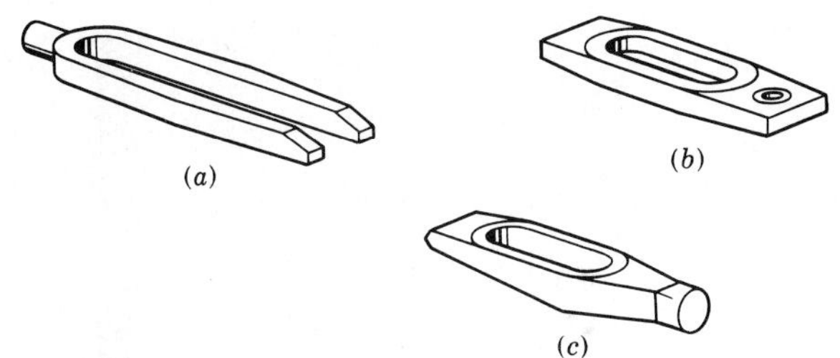

FIGURE H-79 Strap clamps: (*a*) U-clamp; (*b*) straight clamp; (*c*) finger clamp.

FIGURE H-80 T-bolts.

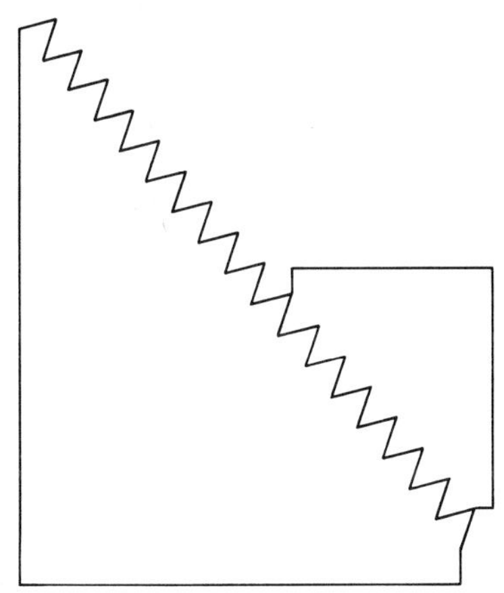

FIGURE H-81 Adjustable step blocks.

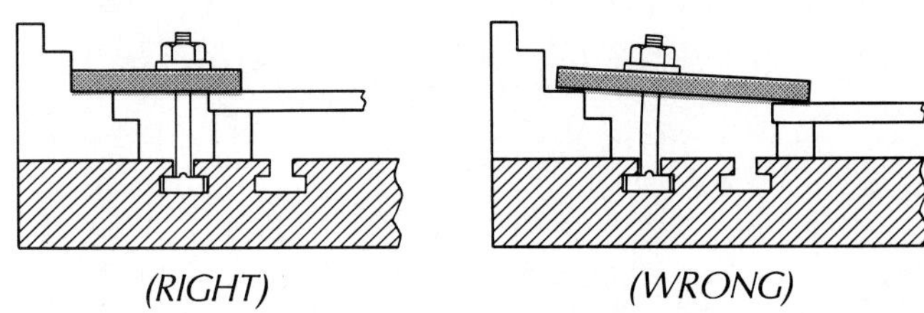

FIGURE H-82 Right and wrong setup for strap clamps. The clamp bolt should be as close to the workpiece as possible. The strap should also be level to avoid bending the bolt.

FIGURE H-83 Parallels of various sizes (Lane Community College).

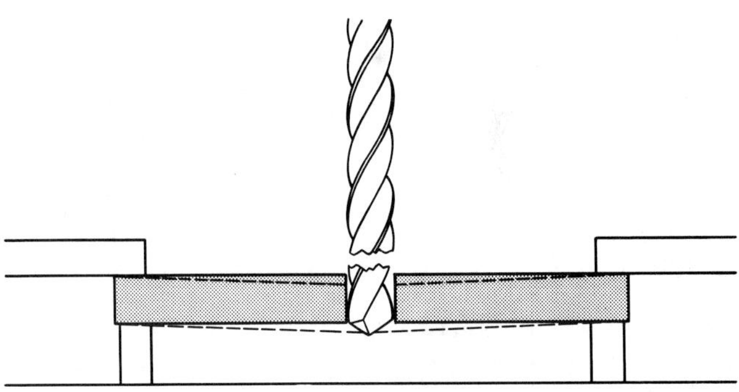

FIGURE H-84 Thin, springy material is supported too far from the drill. Drilling pressure forces the workpiece downward until the drill breaks through, relieving the pressure. The work then springs back and the remaining "fin" of material is more than the drill can cut in one revolution. The result is drill breakage.

Work Holding

Because of the great force applied by the machines in drilling, some means must be provided to keep the workpiece from turning with the drill or from climbing up the flutes after the drill breaks through. This is necessary not only for safety's sake but also for workpiece rigidity and good workmanship.

One method of work holding is to use strap clamps (Figure H-79) and T-bolts (Figure H-80). The clamp must be kept parallel to the table by the use of step blocks (Figure H-81) and the T-bolt should be kept as close to the workpiece as possible (Figure H-82). Parallels (Figure H-83) are placed under the work at the point where the clamp is holding. This provides a space for the drill to break through without making a hole in the table. A thin or narrow workpiece should not be supported too far from the drill, however, since it will spring down under the pressure of the drilling. This can cause the drill on breakthrough to suddenly grab more material than it can handle. The result is often a broken drill (Figure H-84). Thin workpieces

FIGURE H-85 C-clamp being used on an angle plate to hold work that would be difficult to support safely in other ways (Lane Community College).

FIGURE H-87 Part set up in vise with parallels under it (Lane Community College).

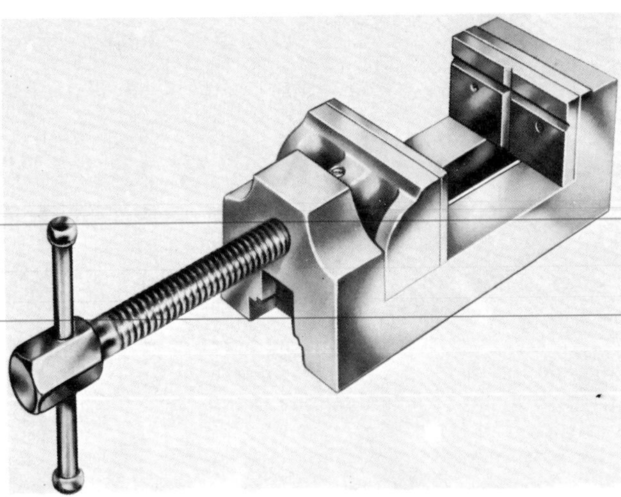

FIGURE H-86 Drill press vise. Small parts are held for drilling and other operations with the drill press vise (Wilton Corporation).

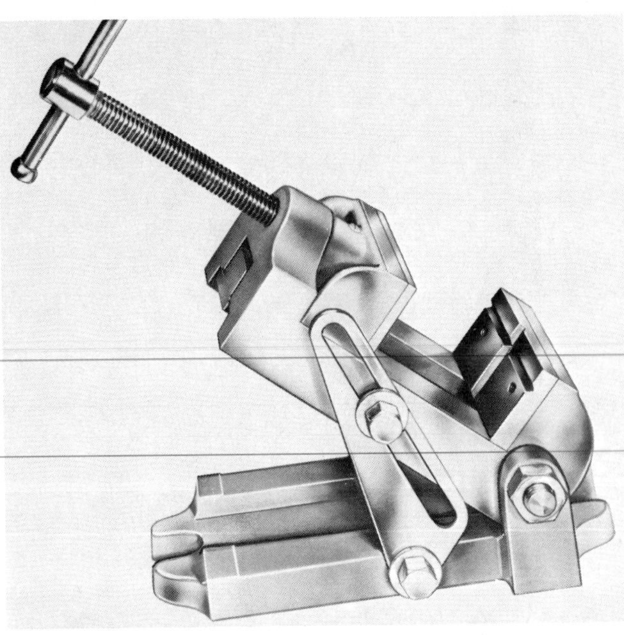

FIGURE H-88 Angle vise. Parts that must be held at an angle to the drill press table while being drilled are held with this vise (Wilton Corporation).

or sheet metal should be clamped over a wooden block to avoid this problem. C-clamps of various sizes are used to hold workpieces on drill press tables and on angle plates (Figure H-85).

Angle plates facilitate the holding of odd-shaped parts for drilling. The angle plate is either bolted or clamped to the table and the work is fastened to the angle plate. For example, a gear or wheel that requires a hole to be drilled into a projecting hub could be clamped to an angle plate.

Drill press vises (Figure H-86) are very frequently used for holding small workpieces of regular shape and size with parallel sides. Vises provide the quickest and most efficient setup method for parallel work, but should not be used if the work does not have parallel sides. The groove in the vise jaws is used to hold small cylindrical parts while larger cylindrical parts can rest on the

FIGURE H-89 View of an adjustable angle plate on a drill press table using a protractor to set up (Lane Community College).

FIGURE H-90 Checking the level of the table (Lane Community College).

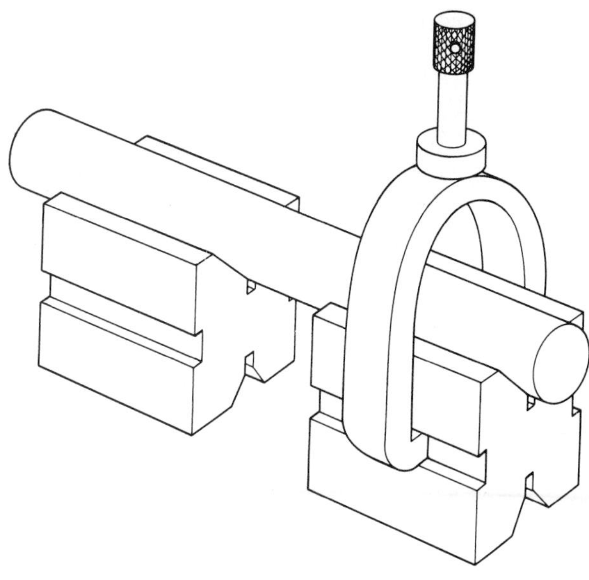

FIGURE H-91 Set of vee blocks with a vee block clamp.

FIGURE H-92 Set up of two vee blocks and round stock with strap clamp (Lane Community College).

bottom surface of the vise. The workpiece must be supported so the drill will not go into the vise. If precision parallels are used for support, they and the drill can be easily damaged since they are both hardened (Figure H-87). For rough drilling, however, cold-finished (CF) keystock would be sufficient for supporting the workpiece. Angular vises can pivot a workpiece to a given angle so that angular holes can be drilled (Figure H-88). Another method of drilling angular holes is by tilting the drill press table. If there is no angular scale on the vise or table, a protractor head with a level may be used to set up the correct angle for drilling. Of course, this method will not be accurate if the drill press is not level. A better way is to measure the angle between the side of the quill and the table. Angle plates are also sometimes used for drilling angular holes (Figure H-89). The drill press table must be level (Figure H-90).

Vee blocks come in sets of two, often with clamps for holding small-size rounds (Figure H-91). Larger-size round stock is set up with a strap clamp over the vee blocks (Figure H-92). The hole to be cross-drilled is first laid out and center punched. The workpiece is lightly clamped in the vee blocks and the punch mark is centered as

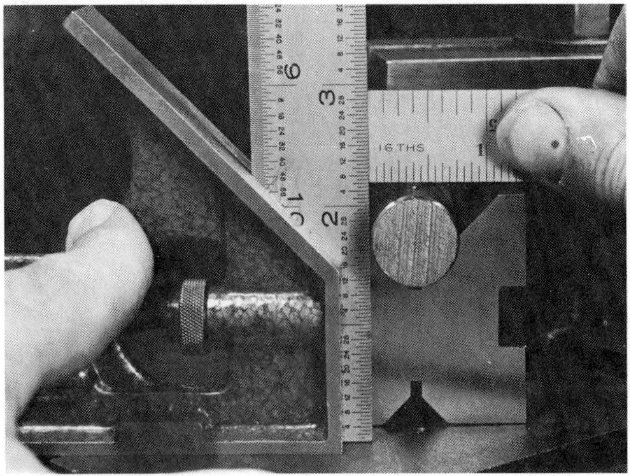

FIGURE H-93 Round stock in vee blocks; one method of centering layout line or punch mark using a combination square and rule (Lane Community College).

FIGURE H-94 Wiggler set in offset position (Lane Community College).

shown in Figure H-93. The clamps are tightened, and the drill is located precisely over the punch mark by means of a wiggler. A wiggler is a tool that can be put into a drill chuck to locate a punch mark to the exact center of the spindle.

The wiggler is clamped into a drill chuck and the machine is turned on (Figure H-94). Push on the knob near the end of the pointer with a 6-in. rule or other piece of metal until it runs with no wobble (Figure H-95). With the machine still running, bring the pointer down into the punch mark. If the pointer begins to wobble again, the mark is not centered under the spindle and the workpiece will have to be shifted. When the wiggler enters the punch mark without wobbling, the workpiece is centered.

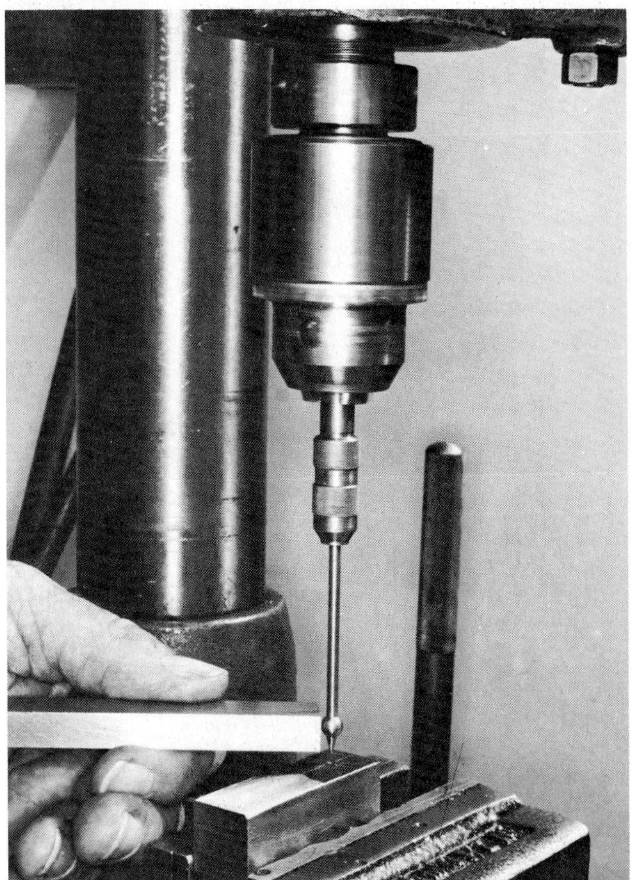

FIGURE H-95 Wiggler centered (Lane Community College).

The wiggler is not commonly used by general machinists, who often use a quicker but less accurate method of centering a drill. They simply set the dead center of the drill in the punch mark and slowly rotate the drill by hand with the workpiece not clamped, moving it into position. Once the drill is centered in this fashion, the workpiece is clamped and the hole drilled.

After the work is centered, use a spotting or center drill to start the hole. Then, for larger holes, use a pilot drill, which is always a little larger than the dead center of the next drill size used. Pilot drilled holes are made to reduce feeding pressure caused by the larger dead center. In some industrial applications, a split-point drill is used and no spotting or pilot drill is needed because split-point drills have no dead center. Use the correct cutting speed and set the rpm on the machine. Chamfer

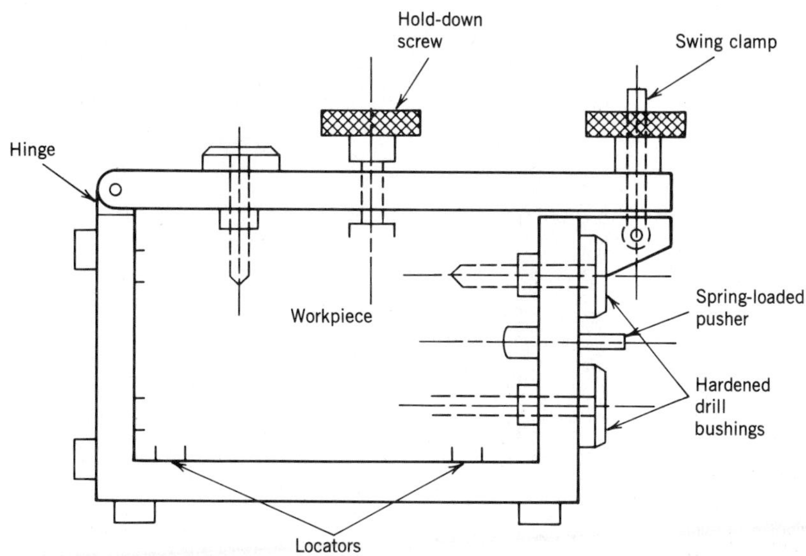

FIGURE H-96 Typical box jig. Hardened bushings guide the drill to precise locations on the workpiece.

both sides of the finished hole with a countersink or chamfering tool.

A tap may be started straight in the drill press by hand. After tap drilling the workpiece, and without removing any clamps, remove the tap drill from the chuck and replace it with a straight shank center. (An alternate method is to clamp the shank of the tap directly in the drill chuck and turn the spindle by hand two or three turns.) Insert a tap in the work and attach a tap handle. Then put the center into the tap, but do not turn on the machine. Apply sulfurized cutting oil and start the tap by turning the tap handle a few turns with one hand while feeding down with the other hand. Release the chuck while the tap is still in the work and finish the job of tapping the hole.

Jigs and fixtures are specially made tooling for production work. In general, fixtures reference a part to the cutting tool. Drill bushings guide the cutting tool (Figure H-96). The use of a jig assures exact positioning of the hole pattern in duplicate and eliminates layout work on every part.

Drilling Procedure C-Clamp Body

Given a combination square, wiggler, Q drill, countersink, $\frac{3}{8}$-24 tap, cutting oil, set of vee blocks with clamps, drill press vise, the C-clamp body as it is finished up to this point, and a piece of $\frac{1}{2}$-in.-diameter CF round stock $4\frac{1}{8}$ in. long:

STEP 1. Set up a workpiece square with the drill press table.

STEP 2. Locate the punch mark and center drill in the mark.

STEP 3. Pilot drill and then tap drill.

FIGURE H-97 Setup of C-clamp project in vise by squaring it with combination square (Lane Community College).

STEP 4. Hand start the tap in the drill press.

STEP 5. Set up round stock in vee blocks, locate the center, then clamp in place and drill a $\frac{3}{16}$-in. hole.

DRILLING THE CLAMP BODY

STEP 1. Clamp the C-clamp body in the vise as shown in Figure H-97 so that the back side extends from the vise jaws about $\frac{1}{16}$ in. Square it with the table by using the combination square. Tighten the vise.

FIGURE H-98 Using a center drill to start the hole (Lane Community College).

FIGURE H-99 Making the tap drill hole. Note the correct chip formation (Lane Community College).

STEP 2. Put a wiggler into the chuck and align the center as explained previously. Clamp the vise to the table, taking care not to move it.

STEP 3. Using the center drill or spotting drill, start the hole (Figure H-98). Change to a $\frac{1}{8}$- to $\frac{3}{16}$-in. pilot drill and make a hole clear through. Now change to the Q drill and enlarge the hole to size (Figure H-99). Chamfer the drilled hole with a countersink tool (Figure H-100). The chamfer should measure about $\frac{3}{8}$ in. across. Use cutting oil or coolant for drilling.

STEP 4. Place a straight shank center in the chuck and tighten it. Insert a $\frac{3}{8}$-24 tap and tap handle in the tap-drilled hole and support the other end on the center. Apply sulfurized cutting oil. **Do not turn on the machine.** Feed lightly downward with one hand while hand turning the tap handle with the other hand (Figure H-101). The tap will be started straight when partway into the work. Release the chuk and finish tapping with the tap handle.

CROSS DRILLING THE $\frac{1}{2}$-IN. ROUND

STEP 1. Take the $\frac{1}{2}$-in. CF round; lay out the hole location and punch. Place it in one or two vee blocks, depending on their size. Lightly clamp with about 1 in. extended from one end.

STEP 2. Set up the punch mark so that it is centered and on top by using the combination square and a rule. Refer to Figure H-91.

STEP 3. With a wiggler or using another method, locate the punch mark directly under the spindle.

STEP 4. Center drill; change to a $\frac{3}{16}$-in. drill and drill through. Chamfer both sides lightly. This part is now ready for lathe work.

SELF-TEST

1. Name three important things to keep in mind when using a drill press (in addition to operator safety and clamping work).

2. If rpm $= \dfrac{CS \times 4}{D}$ and the cutting speed for low-carbon steel is 90, what would the rpm be for the following drills: $\frac{1}{4}$, 2, $\frac{3}{4}$, $\frac{3}{8}$, $1\frac{1}{2}$ in. diameter?

3. What are some of the results of excessive drilling speed? What corrective measures can be taken?

4. Explain what can happen to small-diameter drills when the cuting speed is too slow.

5. How can an operator tell by observing the chip if the feed is about right?

6. In what way are power feeds designated?

7. Name two different cutting fluid types.

8. Such operations as counterboring, reaming, and tapping create friction that can cause heat. This can ruin a cutting edge. How can this situation be helped?

FIGURE H-100 Chamfering the drilled hole (Lane Community College).

FIGURE H-101 Hand tapping in the drill press to assure good alignment (Lane Community College).

9. How can jamming of a drill be avoided when drilling deep holes?

10. Name three uses for the depth stop on a drill press.

11. What is the main purpose for using work-holding devices on drilling machines?

12. List the names of all the work-holding devices that you can remember.

13. Explain the uses of parallels for drilling setups.

14. Why should the support on a narrow or thin workpiece be as close to the drill as possible?

15. Angle drilling can be accomplished in several ways. Describe two methods. How would this be done if no angular measuring devices were mounted on the equipment?

16. What shape of material is the vee block best suited to hold for drilling operations? What do you think its most frequent use would be?

17. What is the purpose of using a wiggler?

18. Why would you ever need an angle plate?

19. What is the purpose of starting a tap in the drill press?

20. Do you think jigs and fixtures are used to any great extent in small machines shops? Why?

UNIT 5

Counter-Sinking and Counterboring

In drill press work it is often necessary to make a recess that will leave a bolt head below the surface of the workpiece. These recesses are made with countersinks or counterbores. When holes are drilled into rough castings or angular surfaces, a flat surface square to these holes is needed, and spot facing is the operation used. This unit will familiarize you with these drill press operations.

Objectives

After completing this unit, you should be able to:

1. Identify tools for countersinking and counterboring.
2. Select speeds and feeds for countersinking and counterboring.

Countersinks

A countersink is a tool used to make a conical enlargement of the end of a hole. Figure H-102 illustrates a common countersink designed to produce smooth surfaces, free from chatter marks. A countersink is used as a chamfering or deburring tool to prepare a hole for reaming or tapping. Unless a hole needs to have a sharp edge, it should be chamfered to protect the end of the hole from nicks and burrs. A chamfer from $\frac{1}{32}$ to $\frac{1}{16}$ in. wide is sufficient for most holes.

A hole made to receive a flathead screw or rivet should be countersunk deep enough for the head to be flush with the surface or up to .015 in. below the surface. A flathead fastener should never project above the surface. The included angles on commonly available countersinks are 60, 82, 90, and 100 degrees. Most flathead fasteners used in metal working have an 82-degree head angle, except for in the aircraft industry, where the 100-degree angle is prevalent. The cutting speed used when countersinking should always be slow enough to avoid chattering.

A combination drill and countersink with a 60-degree angle (Figure H-103) is used to make center holes in workpieces for machining on lathes and grinders. The illustration shown is a bell-type center drill that provides an additional angle for a chamfer of the center, protecting it from damage. The combination drill and countersink, known as a center drill, is also used for spotting holes when using a drill press or milling machine, since it is extremely rigid and will not bend under pressure.

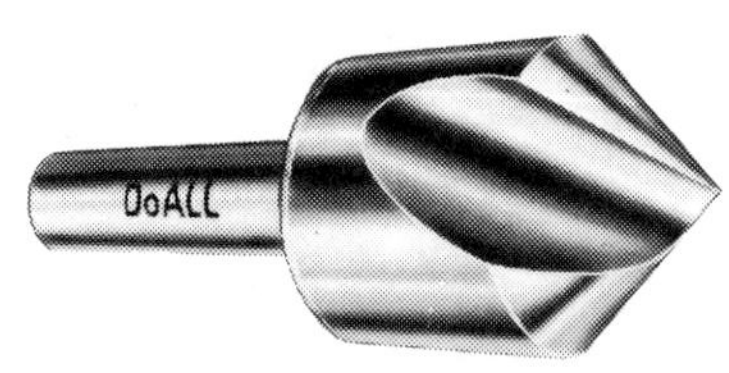

FIGURE H-102 Single-flute countersink (DoALL Company).

FIGURE H-103 Center drill or combination drill and countersink (DoALL Company).

Counterbores

Counterbores are tools designed to enlarge previously drilled holes, much like countersinks, and are guided into the hole by a pilot to assure the concentricity of the two holes. A multiflute counterbore is shown in Figure H-104, and a two-flute counterbore is shown in Figure H-105. The two-flute counterbore has more chip clearance and a larger rake angle than the counterbore in Figure H-104. Counterbored holes have flat bottoms, unlike the angled edges of countersunk holes, and are often used to recess a bolt head below the surface of a workpiece. Solid counterbores, such as those shown in Figures H-104 and H-105 are used to cut recesses for socket head cap screws or filister head

FIGURE H-104 Multiflute counterbore.

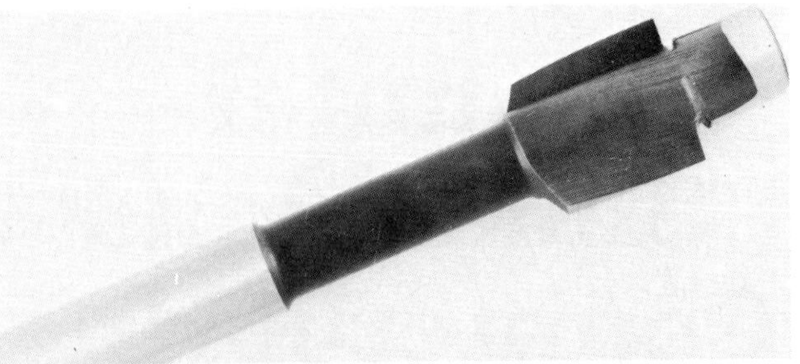

FIGURE H-105 Two-flute counterbore.

screws (Figure H-106). The diameter of the counterbore is usually $\frac{1}{32}$ in. larger than the head of the bolt.

When a variety of counterbore and pilot sizes is necessary, a set of interchangeable pilot counterbores is available. Figure H-107 shows a counterbore in which a number of standard or specially made pilots can be used. A pilot is illustrated in Figure H-108.

Counterbores are made with straight or tapered shanks to be used in drill presses and milling machines. When counterboring a recess for a hex head bolt, remember to measure the diameter of the socket wrench so the hole will be large enough to accommodate it. For most counterboring operations the pilot should have from .002 to .005 in. clearance in the hole. If the pilot is too tight in the hole, it may seize and break. If there is too much clearance between the pilot and the hole, the counterbore will be out of round and will have an unsatisfactory surface finish.

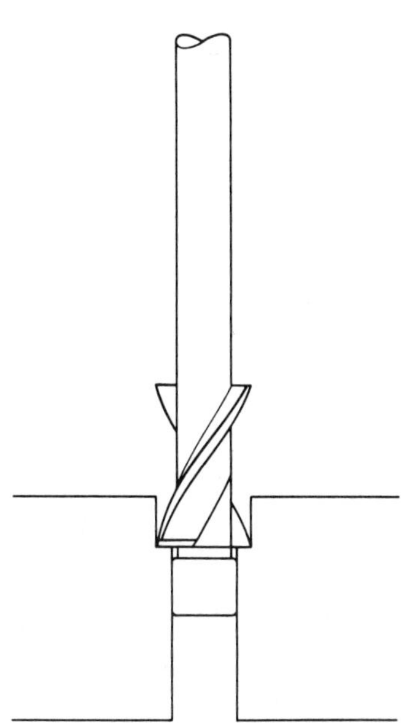

FIGURE H-106 The counterbore is an enlargement of a hole already drilled.

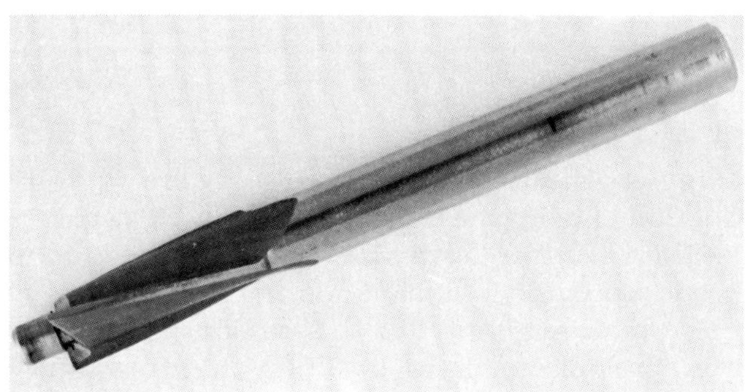

FIGURE H-107 Interchangeable pilot counterbore.

It is very important that the pilot be lubricated while counterboring. Usually, this lubrication is provided if a sulfurized cutting oil or soluble oil is used. When cutting dry, which is often the case with brass and cast iron, the hole and pilot should be lubricated with a few drops of lubricating oil.

Counterbores or spotfacers are often used to provide a flat bearing surface for nut or bolt heads on rough castings or a raised boss (Figure H-109). This operation is called spot facing. Because these rough surfaces may not be at right angles to the pilot hole, great strain is put on the pilot and counterbore and can cause breakage of either one. To avoid breaking the tool, be very careful when starting the cut, especially when hand feeding. Prevent hogging into the work by tightening the spindle clamps slightly to remove possible backlash. Back

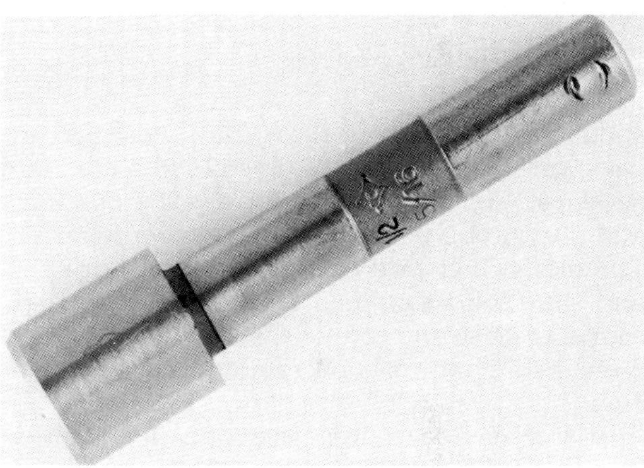

FIGURE H-108 Pilot for interchangeable pilot counterbore.

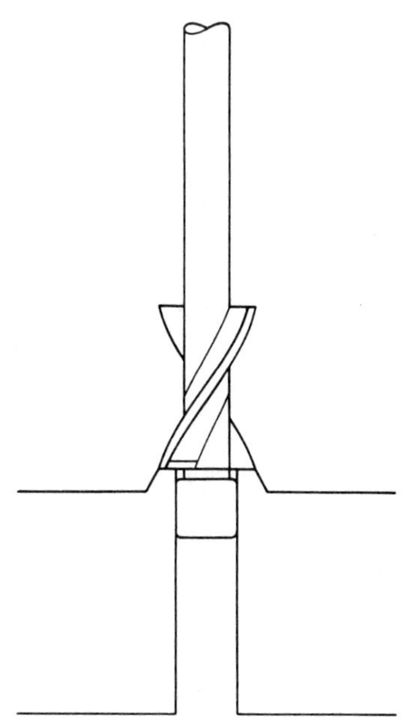

FIGURE H-109 Spotfacing on a raised boss.

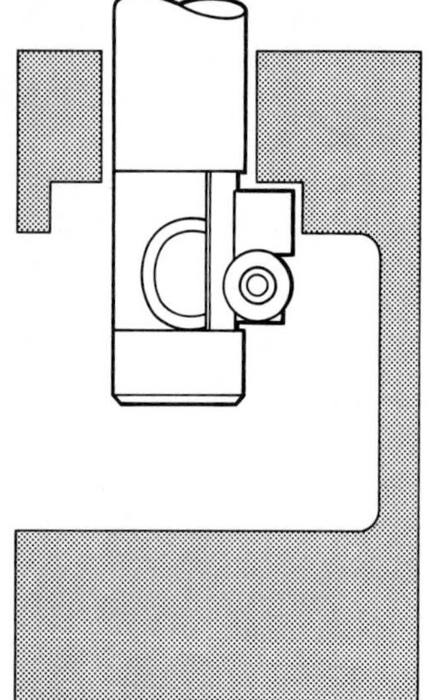

FIGURE H-110 Back spotfacing tool. (Illustration used by permission of Davis Tool, a Giddings & Lewis Company. Logo: Erix Tool AB, Sweden.)

spotfacing tool (Figure H-110) is used when the back side of a hole is inaccessible to a standard spotfacer. The tool is fastened in a drill press chuck and put through the hole with the tool retracted. The tool swings out when the spindle is rotated and the spotface is then cut. The spindle is reversed to close and retract the tool from the hole. Some deburring tools operate on the same principle.

Recommended power feed rates for counterboring are shown in Table H-4. The feed rate should be great enough to get under any surface scale quickly, thus preventing rapid dulling of the counterbore. The speeds used for counterboring are one third less than the speeds used for twist drills of corresponding diameters. The choice of speeds and feeds is very much affected by the condition of the equipment, the power available, and the material being counterbored.

TABLE H-4 Feeds for Counterboring

$\frac{3}{8}$ in. diameter up to .004 in. per revolution
$\frac{5}{8}$ in. diameter up to .005 in. per revolution
$\frac{7}{8}$ in. diameter up to .006 in. per revolution
$1\frac{1}{4}$ in. diameter up to .007 in. per revolution
$1\frac{1}{2}$ in. diameter up to .008 in. per revolution

Before counterboring a workpiece, it should be securely fastened to the machine table or tightly held in a vise because of the great cutting pressures encountered. Workpieces also should be supported on parallels to allow for the protrusion of the pilot. To obtain several equally deep countersunk or counterbored holes on the drill press or milling machine, the spindle depth stop can be set.

SELF-TEST

1. When is a countersink used?

2. Why are countersinks made with varying angles?

3. What is a center drill?

4. When is a counterbore used?

5. What relationship exists between pilot size and hole size?

6. Why is lubrication of the pilot important?

7. As a rule, how does cutting speed compare between an equal-size counterbore and twist drill?

8. What affects the selection of feed and speed when counterboring?

9. What is spotfacing?

10. What important points should be considered when a counterboring setup is made?

UNIT 6

Reaming in the Drill Press

Many engineering requirements involve the production of holes having smooth surfaces, accurate location, and uniform size. In many cases, holes produced by drilling alone do not entirely satisfy these requirements. For this reason, the reamer was developed for enlarging or finishing previously formed holes. This unit will help you properly identify, select, and use machine reamers.

Objectives

After completing this unit, you should be able to:

1. Identify commonly used machine reamers.
2. Select the correct feeds and speeds for commonly used materials.
3. Determine appropriate amounts of stock allowance.
4. Identify probable solutions to reaming problems.

Common Machine Reamers

Reamers are tools used mostly to precision finish holes, but they are also used in the heavy construction industry to enlarge or align existing holes. Machine reamers have straight or taper shanks; the taper usually is a standard Morse taper. The parts of a machine reamer are shown in Figure H-111; the cutting end of a machine reamer is shown in Figure H-112.

Chucking reamers (Figures H-113 to H-115) are efficient in machine reaming a wide range of materials and are commonly used in drill presses, turret lathes, and screw machines. Helical flute reamers have an extremely smooth cutting action that finishes holes accurately and precisely. Chucking reamers cut on the chamfer at the end of the flutes. This chamfer is usually at a 45-degree angle.

Jobber's reamers (Figure H-116) are used where a longer flute length than chucking reamers is needed. The additional flute length gives added

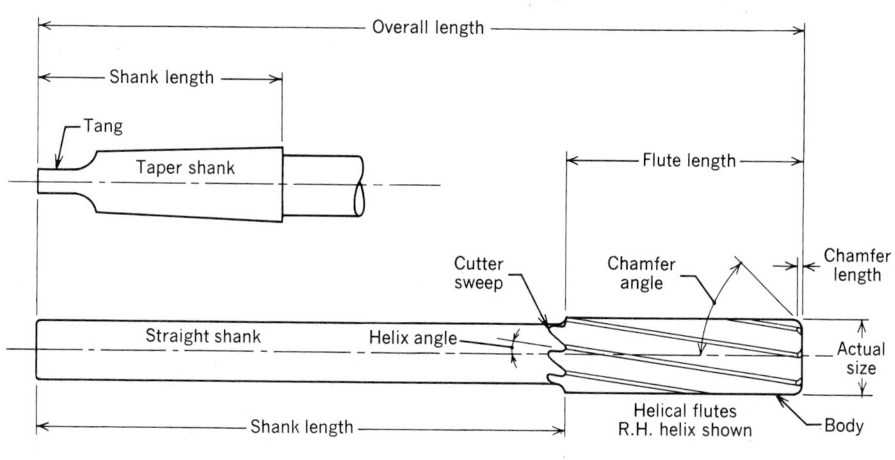

FIGURE H-111 The parts of a machine reamer (Bendix Industrial Tools Division).

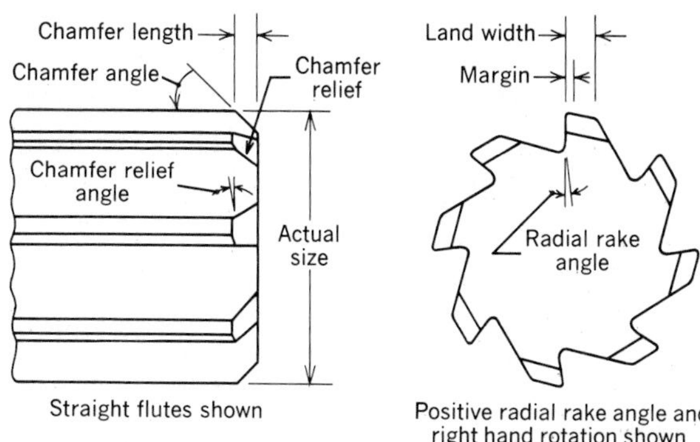

FIGURE H-112 Cutting end of a machine reamer (Bendix Industrial Tools Division).

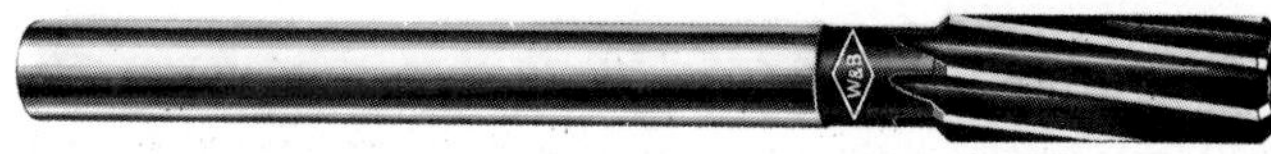

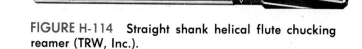

FIGURE H-114 Straight shank helical flute chucking reamer (TRW, Inc.).

FIGURE H-115 Taper shank helical flute chucking reamer (TRW, Inc.).

FIGURE H-113 Straight shank straight flute chucking reamer (TRW, Inc.).

FIGURE H-116 Taper shank straight flute jobber's reamer (TRW, Inc.).

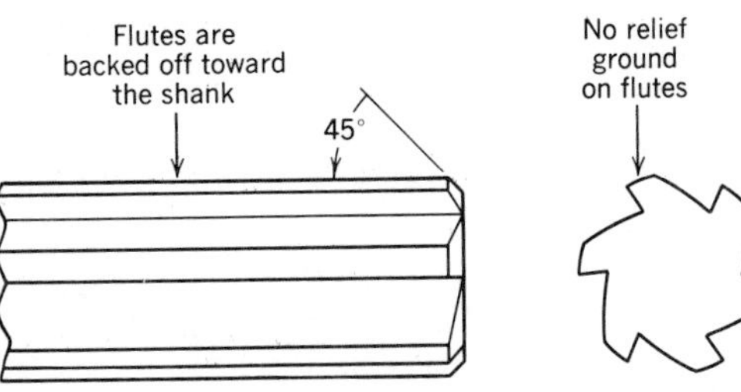

FIGURE H-117 Rose reamer.

FIGURE H-118 Shell reamer helical flute (TRW, Inc.).

FIGURE H-119 Taper shank shell reamer arbor (TRW, Inc.).

FIGURE H-120 Morse taper reamer (TRW, Inc.).

FIGURE H-121 Helical taper pin reamer (TRW, Inc.).

FIGURE H-122 Helical flute taper bridge reamer (TRW, Inc.).

FIGURE H-123 Carbide-tipped straight flute chucking reamer (TRW, Inc.).

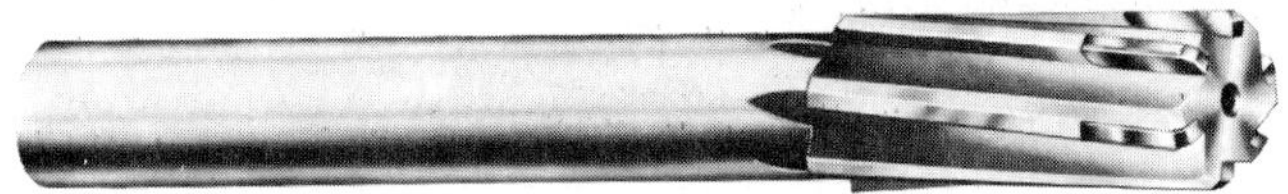

FIGURE H-124 Carbide-tipped helical flute chucking reamer, right-hand helix (TRW, Inc.).

guide to the reamer, especially when reaming deep holes.

The rose reamer (Figure H-117) is primarily a roughing reamer used to enlarge holes to within .003 to .005 in. of finish size. The rose reamer is typically followed by a fluted reamer to bring the hole to finish and size. The teeth are slightly backed off, which means that the reamer diameter is smaller toward shank end by approximately .001 in./in. of flute length. The lands on these reamers are ground cylindrically without radial relief, and all cutting is done on the end of the reamer. This reamer will remove a considerable amount of material in one cut.

Shell reamers (Figure H-118) are finishing reamers. They are more economically produced, especially in larger sizes, than solid reamers because a much smaller amount of tool material is used in making them. Two slots in the shank end of the reamer fit over matching driving lugs on the shell reamer or box (Figure H-119). The hole in the shell reamer has a slight taper ($\frac{1}{8}$ in./ft) in it to assure exact alignment with the shell reamer arbor. Shell reamers are made with straight or helical flutes and are commonly produced in sizes from $\frac{3}{4}$ to $2\frac{1}{2}$ in. in diameter. Shell reamer arbors come with matching straight or tapered shanks and are made in designated sizes from numbers 4 to 9.

Morse taper reamers (Figure H-120), with straight or helical flutes, are used to finish ream-tapered holes in drill sockets, sleeves, and machine tool spindles. Helical taper pin reamers (Figure H-121) are especially suitable for machine reaming of taper pin holes. There is no packing of chips in the flutes, as the chips are pushed forward through the hole during the reaming operation, which reduces the possibility of breakage. These reamers have a free-cutting action that produces a good finish at high cutting speeds. Taper pin reamers have a taper of $\frac{1}{4}$ in. per foot of length and are manufactured in 18 different sizes ranging from smallest number 8/0 (eight naught) to the largest at number 10.

Taper bridge reamers (Figure H-122) are used in structural iron or steel work, bridge work, and ship construction where extreme accuracy is not required. They have long tapered pilot points for easy entry in the out-of-line holes often encountered in structural work. Taper bridge reamers are made with straight and helical flutes to ream holes with diameters from $\frac{1}{4}$ to $1\frac{5}{16}$ in.

Carbide tipped chucking reamers (Figure H-123) are often used in production setups, particularly where abrasive materials or sand and scale (as in castings) are encountered. The right-hand helix chucking reamer (Figure H-124) is recom-

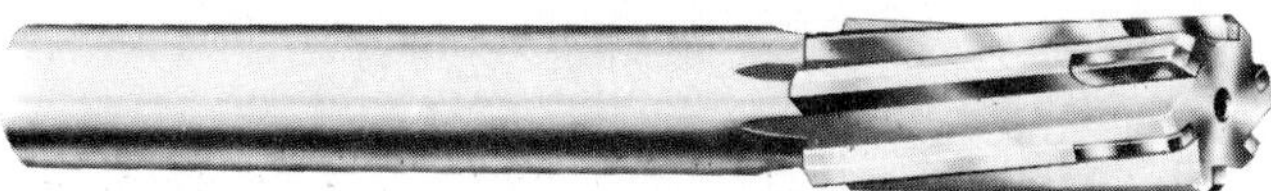

FIGURE H-125 Carbide-tipped helical flute chucking reamer, left-hand helix (TRW, Inc.).

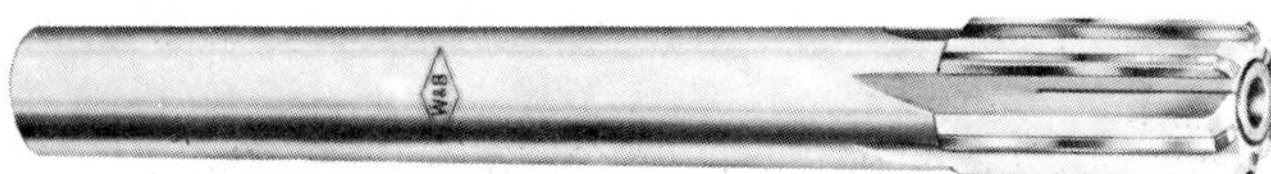

FIGURE H-126 Carbide-tipped expansion reamer (TRW, Inc.).

mended for ductile materials or highly abrasive materials or when machining blind holes. The carbide tipped left-hand helix chucking reamer (Figure H-125) will produce good finishes on heat-treated steels up to RC 40. They should be used only on through holes because the chips push out through the hole ahead of the reamer. All expansion reamers (Figure H-126) after becoming worn can be expanded and resized by grinding. This feature offsets normal wear from abrasive materials and provides for a long tool life. These tools should not be adjusted for reaming size by loosening or tightening the expansion plug but only by grinding.

Reaming is intended to produce accurate and straight holes of uniform diameter. The required accuracy depends on a high degree of surface finish, tolerance on diameter, roundness, straightness, and absence of bellmouth at the ends of holes. To make an accurate hole it is necessary to use reamers with adequate support for the cutting edges; an adjustable reamer may not be adequate. Machine reamers are often made of either high speed steel or cemented carbide. Reamer cutting action is controlled to a large extent by the cutting speed and feed used.

Speed

The most efficient cutting speed for machine reaming depends on the type of material being reamed, the amount of stock to be removed, the tool material being used, the finish required, and the rigidity of the setup. A good starting point, when machine reaming, is to use one-third to one-half of the cutting speed used for drilling the same materials. Table H-5 may be used as a guide.

Where conditions permit the use of carbide reamers, the speeds may often be increased over those recommended for HSS (high-speed steel) reamers. The limiting factor is usually an absence of rigidity in the setup. Any chatter, which is often caused by too high a speed, is likely to chip the cutting edges of a carbide reamer. Always select a speed that is slow enough to eliminate chatter. Close tolerances and fine finishes often require the use of considerably lower speeds than those recommended in Table H-5.

TABLE H-5 Reaming Speeds

Aluminum and its alloys	130–200[a]
Brass	130–200
Bronze, high tensile	50–70
Cast iron	
Soft	70–100
Hard	50–70
Steel	
Low carbon	50–70
Medium carbon	40–50
High carbon	35–40
Alloy	35–40
Stainless steel	
AISI 302	15–30
AISI 403	20–50
AISI 416	30–60
AISI 430	30–50
AISI 443	15–30

[a] Cutting speeds in surface feet per minute (fpm or sfm) for reaming with an HSS reamer.

Feeds

Feeds in reaming are usually two to three times greater than those used for drilling. The amount of feed may vary with different materials, but a good starting point would be between .0015 and .004 in. per revolution. Too low a feed may "glaze" the hole, which has the result of work hardening the material, causing occasional chatter and excessive wear on the reamer. Too high a feed tends to reduce the accuracy of the hole and the quality of the surface finish. Generally, it is best to use as high a feed as possible to produce the required finish and accuracy.

When a drill press that has only a hand feed is used to ream a hole, the feed rate should be estimated just as it would be for drilling. About twice the feed rate should be used for reaming as would be used for drilling in the same setup when hand feeding.

Stock Allowance

The stock removal allowance should be sufficient to assure a good cutting action of the reamer. Too small a stock allowance results in burnishing (a slipping or polishing action), or wedges the reamer in the hole causing excessive wear or breakage of the reamer. The condition of the hole before reaming also has an influence on the reaming allowance

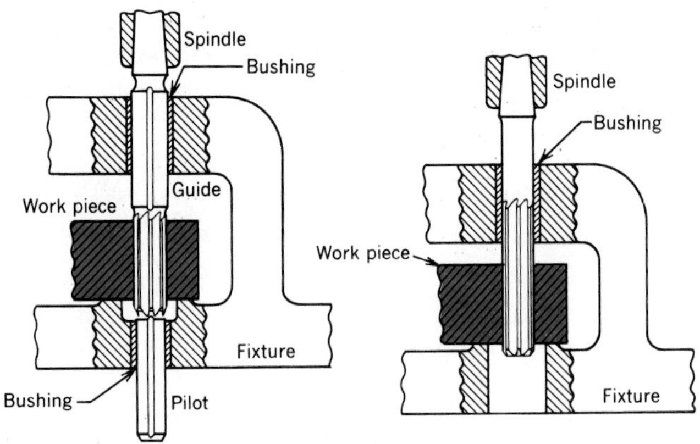

FIGURE H-127 Use of pilots and guided bushings on reamers. Pilots are provided so that the reamer can be held in alignment and can be supported as close as possible while allowing for chip clearance (Bendix Industrial Tools Division).

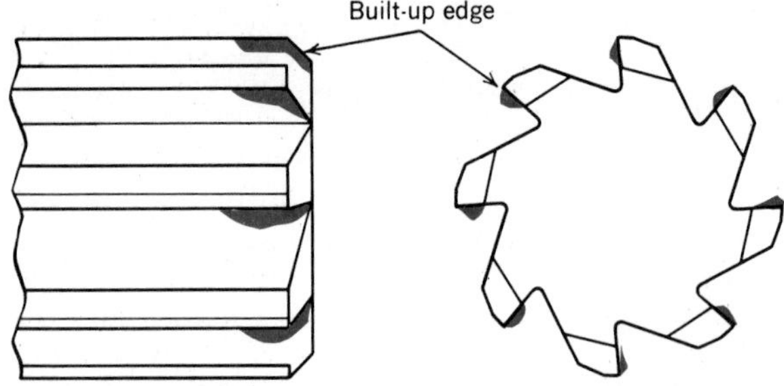

FIGURE H-128 Reamer teeth having built-up edges.

TABLE H-6 Stock Allowance for Machine Reaming

Reamer Size (in.)	Allowance (in.)
$\frac{1}{32}$ to $\frac{1}{8}$	.003 to .006
$\frac{1}{8}$ to $\frac{1}{4}$	.005 to .009
$\frac{1}{4}$ to $\frac{3}{8}$	.007 to .012
$\frac{3}{8}$ to $\frac{1}{2}$	.010 to .015
$\frac{1}{2}$ to $\frac{3}{4}$	$\frac{1}{64}$ to $\frac{1}{32}$
$\frac{3}{4}$ to 1	$\frac{1}{32}$

since a rough hole will need a greater amount of stock removed than an equal size hole with a fairly smooth finish. See Table H-6 for commonly used stock allowance for reaming. When materials that work harden readily are reamed, it is especially important to have adequate material for reaming.

Cutting Fluids

To ream a hole to a high degree of surface finish, a cutting fluid is needed. A good cutting fluid will cool the workpiece and tool and will also act as a lubricant between the chip and the tool to reduce friction and heat buildup. Cutting fluids should be applied in sufficient volume to flush the chips away. Table H-7 lists some cutting fluids used for reaming different materials.

Reaming Problems

Chatter is often caused by lack of rigidity in the machine, workpiece, or the reamer itself. Corrections may be made by reducing the speed, increasing the feed, putting a chamfer on the hole before reaming, using a reamer with a pilot (Figure H-127), or reducing the clearance angle on the cutting edge of the reamer. Carbide tipped reamers especially cannot tolerate even a momentary chatter at the start of a hole, as such a vibration is likely to chip the cutting edges.

Oversized holes can be caused by inadequate workpiece support, worn guide bushings, worn or loose spindle bearings, or bent reamer shanks. When reamers gradually start cutting larger holes, it may be because of the work material galling or forming a builtup edge on reamer cutting surfaces (Figure H-128). Mild steel and some aluminum alloys are particularly troublesome in this area. Changing to a different coolant may help. Reamers with highly polished flutes, margins, and relief angles or reamers that have special surface treatment may also improve the cutting action.

TABLE H-7 Cutting Fluids Used for Reaming

Material	Dry	Soluble Oil	Kerosene	Sulfurized Oil	Mineral Oil
Aluminum		×	×		
Brass	×	×			
Bronze	×	×			×
Cast iron	×				
Steels					
Low carbon		×		×	
Alloy		×		×	
Stainless		×		×	

Bellmouthed holes are caused by misalignment of the reamer with the hole. The use of accurate bushings or pilots may correct bellmouth, but in many cases the only solution is the use of floating holders. A floating holder will allow movement in some directions while restricting it in others. A poor finish can be improved by decreasing the feed, but this will also increase the wear and shorten the life of the reamer. A worn reamer will never leave a good surface finish as it will score or groove the finish and often produce a tapered hole.

Too fast a feed will cause a reamer to break. Too large a stock allowance for finish reaming will produce a large volume of chips with heat buildup, and will result in a poor hole finish. Too small a stock allowance will cause the reamer teeth to rub as they cut, and not cut freely, which will produce a poor finish and cause rapid reamer wear. Cutting fluid applied in insufficient quantity may also cause rough surface finishes when reaming.

SELF-TEST

1. How is a machine reamer identified?

2. What is the difference between a chucking and a rose reamer?

3. What is a jobber's reamer?

4. Why are shell reamers used?

5. How does the surface finish of a hole affect its accuracy?

6. How does the cutting speed compare between drilling and reaming for the same material?

7. How does the feed rate compare between drilling and reaming?

8. How much reaming allowance will you leave on a $\frac{1}{2}$-in. hole?

9. What is the purpose of using cutting fluid while reaming?

10. What can be done to overcome chatter?

11. What will cause oversize holes?

12. What causes a bellmouthed hole?

13. How can poor surface finish be overcome?

14. When are carbide-tipped reamers used?

15. Why is vibration harmful to carbide-tipped reamers?

SECTION I

TURNING MACHINES

The engine lathe is so named because it was originally powered by Watt's steam engine instead of by foot treadle or hand crank. With suitable attachments the engine lathe may be used for turning, threading, boring, drilling, reaming, facing, spinning, and grinding, although many of these operations are preferably done on specialized machinery. Sizes range from the smallest jeweler's or precision lathes (Figure I-1) to the massive lathes used for machining huge forgings (Figure I-2).

Engine lathes (Figures I-3 and I-4) are used by machinists to produce one-of-a-kind parts or a few

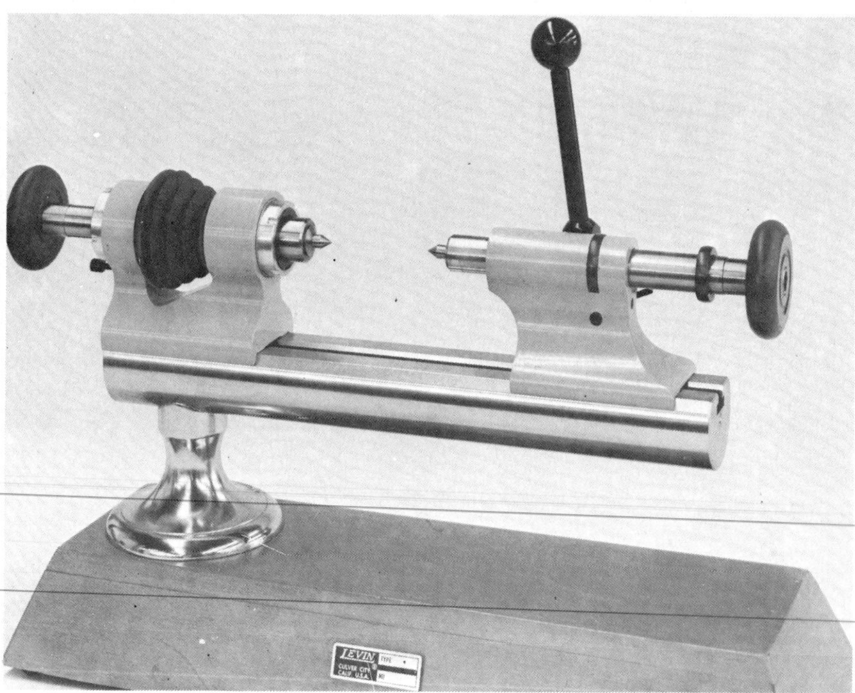

FIGURE I-1 Jeweler's or instrument lathe (Louis Levin & Son, Inc.).

FIGURE I-2 Massive forging being machined to exact specifications on a lathe (Bethlehem Steel Corporation, Bethlehem, Pa.).

FIGURE I-3 Heavy-duty engine lathe with $1\frac{1}{2}$-in.-diameter hole through the spindle. The carriage has rapid power traverse feature (Lodge & Shipley Company).

FIGURE I-4 Toolmaker's lathe (Monarch Machine Tool Company, Sidney, Ohio).

pieces for a short-run production. They are also used for toolmaking, machine repair, and maintenance.

Some lathes have a vertical spindle instead of a horizontal one, with a large rotating table on which the work is clamped. These huge machines, called vertical boring mills (Figure I-5), are the largest of our machine tools. A 25-ft-diameter table is not unusual. Huge turbines, weighing many tons, can be placed on the table and clamped in position to be machined. The machining of such castings would be impractical on a horizontal spindle lathe.

A more versatile and higher production version of the boring mill is the vertical turret (Figure I-6). It does similar work to the vertical boring mill, but on a smaller scale. It is arranged with toolholders and turret with multiple tools much like that on a turret lathe, which give it flexibility and relatively high production.

Turret lathes are strictly production machines. They are designed to provide short machining time and quick tool changes. Two types of semiautomatic turret lathes require an operator in constant attendance: the ram type (Figure I-7) and the saddle type (Figure I-8). Small, precision, hand-operated turret lathes (Figure I-9) are used to produce very small parts. Automatic bar chuckers require little operator action.

FIGURE I-5 Vertical boring mill (El-Jay Inc., Eugene, Ore.).

FIGURE I-6 Vertical turret lathe (Giddings & Lewis, Inc.)

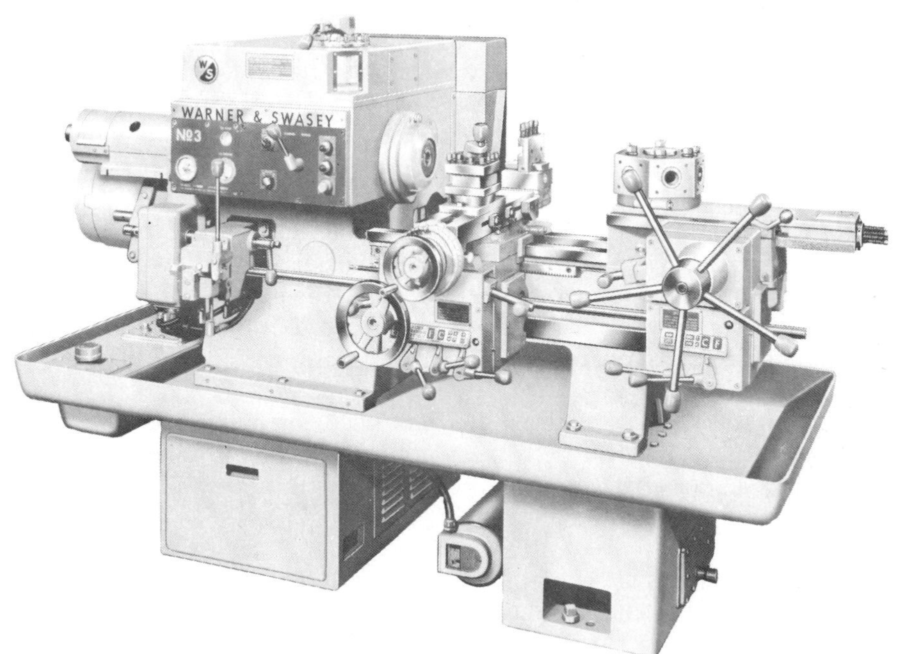

FIGURE I-7 Ram-type turret lathe (Warner & Swasey Company).

FIGURE I-8 Saddle-type turret lathe (Warner & Swasey Company).

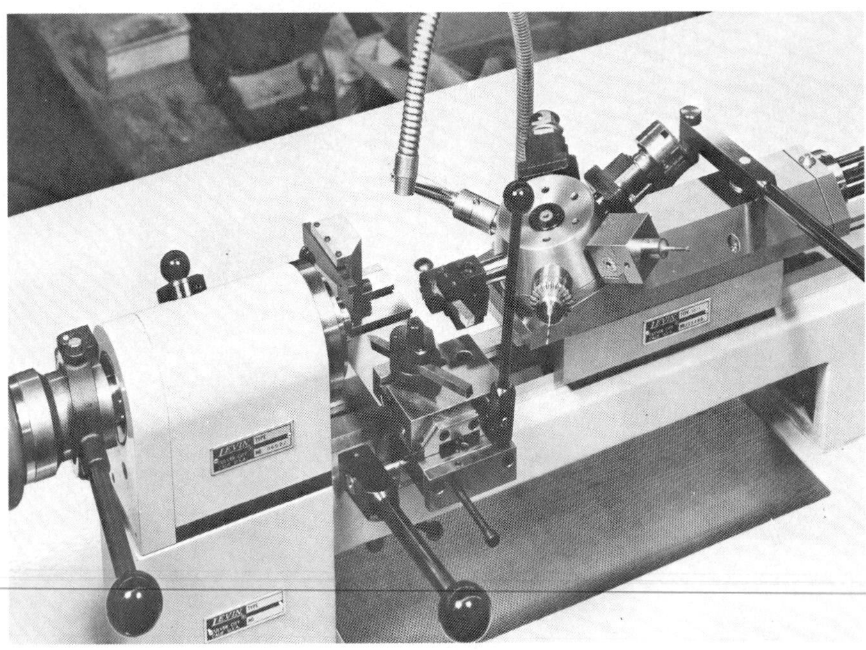

FIGURE I-9 Small precision manually operated turret lathe (Louis Levin & Sons, Inc.).

Fully automatic machines such as automatic turret lathes and automatic screw machines are programmed to do a sequence of machining operations to make a completed product. Automatic turret lathes usually are programmed by numerical control (NC) on punched tape. Automatic screw machines, used for high production of small parts, are typically programmed with cams, although some are numerically controlled. There are several types of automatic machines: the single spindle machine, the sliding head (or Swiss type), the multiple spindle, and the revolving head wire feed type. Multiple spindle bar and chucking machines are high-production turning machines that can do a variety of operations at different stations.

Tracer lathes follow a pattern or template to reproduce an exact shape on a workpiece. Tracing attachments (Figure I-10) are often used on engine or turret lathes. Tracer lathes have been largely replaced by NC or CNC lathes.

Lathes that are numerically controlled (NC) by programming and punching tape produce workpieces such as shafts with tapers and precision diameters. NC chuckers (Figure I-11) are high-production automatic lathes designed for chucking operations. Similar bar-feeding NC types take a full-length bar through the spindle and automatically feed it in as needed. Some automatic lathes operate as either chucking machines or bar-feed machines.

Lathes are also used for metal spinning. Reflectors, covers, and pans, for example, are made by this method out of aluminum, copper, and other metals.

Manufacturing methods for spinning heavy steel plate are entirely different. Hydraulically operated tools are used to form the steel (Figure I-12). The dimensions of the part shaped by this

FIGURE I-10 Tracer attachments on an engine lathe (Clausing Corporation).

FIGURE I-11 Numerically controlled chucking lathe with turret. Guards have been removed for the photograph. (Monarch Machine Tool Company, Sidney, Ohio).

FIGURE I-12 Floturn lathe. The photograph (right) shows the completion of the first two operations. Starting with the flat blank, the workpiece is shear formed (Floturn) to the shape shown on the mandrel. In the process the 1-in. thickness of the blank is reduced to .420 in. The second operation, although performed on a Floturn machine, is more properly spinning than shear forming. In this operation, the workpiece is brought to the finished shape as seen in the background (Floturn, Inc., Division of the Lodge-Shipley Co.).

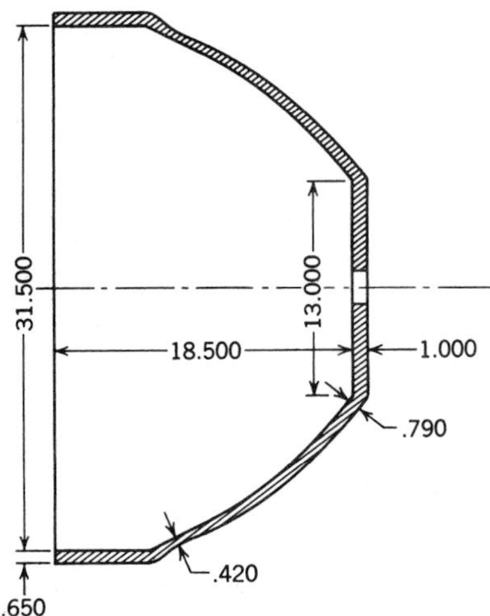

FIGURE I-13 The principle of the Floturn process (Floturn, Inc., Division of the Lodge & Shipley Co.).

Floturn process are shown in Figure I-13.

Digital readout systems for machine tools such as the engine lathe are becoming more common. This system features a completely self-guided rack and pinion that operates on the cross slide. The direct readout resolution is .001 in. on both the diameters and the cross slide movement. These systems can also be converted to metric measure.

In ordinary turning, metal is removed from a rotating workpiece with a single-point tool. The tool must be harder than the workpiece and held rigidly against it. Chips formed from the workpiece slide across the face of the tool. This essentially is the way chips are produced in all metal cutting operations. The pressures used in metal cutting can be as much as 20 tons per square inch. Tool geometry, therefore, is quite important to maintain the strength at the cutting edge of the tool bit. In this section you will learn how to use common lathes, how to grind high-speed tools, and how to select tools for lathe work.

■ ■ ■

Turning Machine Safety

CAUTION The lathe can be a safe machine only if the machinist is aware of the hazards involved in its operation. In the machine shop, as anywhere, you must always keep your mind on your work in order to avoid accidents. Develop safe work habits in the use of setups, chip breakers, guards, and other protective devices. Standards for safety have been established as guidelines to help you eliminate unsafe practices and procedures on lathes.

HAZARDS IN LATHE OPERATIONS

1. Pinch points due to lathe movement. A finger caught in gears or between the compound rest and a chuck jaw would be an example. The rule is to keep your hands away from such dangerous positions when the lathe is operating.

2. Hazards associated with broken or falling components. Heavy chucks or workpieces can be dangerous when dropped. Care must be used when handling them. If a threaded spindle is suddenly reversed, the chuck can come off and fly out of the lathe. A chuck wrench left in the chuck can become a missile when the machine is turned on. Always remove the chuck wrench immediately after using it (Figures I-14 and I-15).

FIGURE I-14 A chuck wrench left in the chuck is a danger to everyone in the shop (Lane Community College).

FIGURE I-15 A safety-conscious lathe operator will remove the chuck wrench when he or she finishes using it (Lane Community College).

3. Hazards resulting from contact with high-temperature components. Burns usually result from handling hot chips (up to 800°F or even more) or a hot workpiece. Gloves may be worn when handling hot

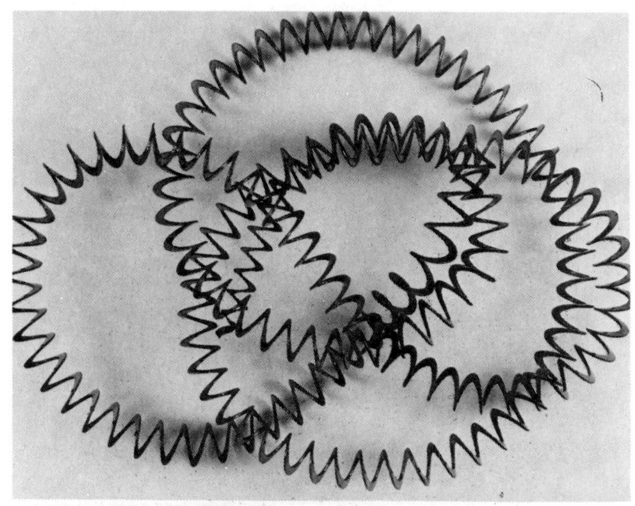

FIGURE I-16 **Unbroken lathe chips are sharp and hazardous to the operator (Lane Community College).**

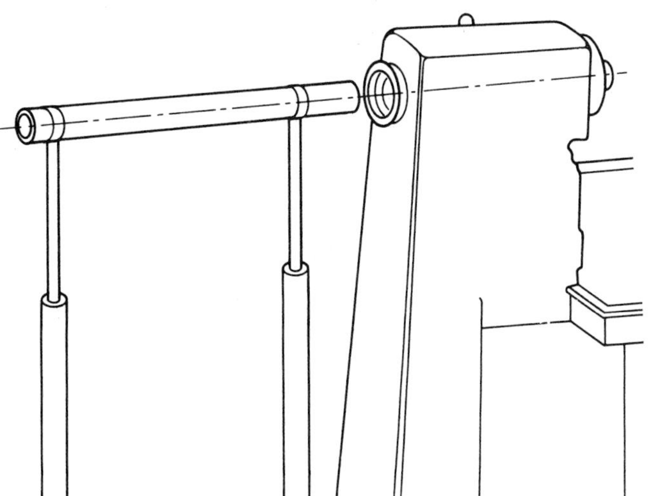

FIGURE I-17 **Stock tube is used to support long workpieces that extend out of the headstock of a lathe.**

chips or workpieces. Gloves should never be worn when you are operating the machine.

4. Hazards resulting from contact with sharp edges, corners, and projections. These are perhaps the most common cause of hand injuries in lathe work. Dangerous sharp edges may be found in many places: on a long stringy chip, on a tool bit, on a burred edge of a turned or threaded part. Shields should be used for protection from flying chips and coolant. These shields usually are made of clear plastic and are hinged over the chuck or clamped to the carriage of engine lathes. Even when shields are in place, safety glasses must be worn. Stringy chips must not be removed with bare hands; wear heavy gloves and use hook tools or pliers. Always turn off the machine before attempting to remove chips. Chips should be broken and "9"-shaped rather than in a stringy mass or a long wire (Figure I-16). Chip breakers on tools and correct feeds will help to produce safe, easily handled chips. Burred edges must be removed before the

FIGURE I-18 **Polishing in the lathe with abrasive cloth (Lane Community College).**

workpiece is removed from the lathe. Always remove the tool bit when setting up or removing workpieces from the lathe.

5. Hazards of workholding devices or driving devices. When workpieces are clamped, their components often extend beyond the outside diameter of the holding device. Guards, barriers, and warnings such as signs or verbal instructions are all used to make you aware of the hazards. On power chucking devices you should be aware of potential pinch points between workpiece and work-holding device. Make certain sufficient gripping force is exerted by the jaws to hold the work safely. Never run a geared scroll chuck without having something gripped in the jaws. Centrifugal force on the jaws can cause the scroll to unwind and the jaws to come out of the chuck. Keep tools, files, and micrometers off the machine. They may vibrate off into the revolving chuck or workpiece.

6. Spindle braking. The spindle or workpiece should never be slowed or stopped by hand gripping or by using a pry bar. Always use machine controls to stop or slow it.

7. Workpieces extending out of the lathe should be supported by a stock tube. If a slender workpiece is allowed to extend beyond the headstock spindle a foot or so without support, it can fly outward from centrifugal force. The piece will not only be bent, but it will present a very great danger to anyone standing near (Figure I-17).

OTHER SAFETY CONSIDERATIONS

Hold one end of abrasive cloth strips in each hand when polishing rotating work. Don't let either hand get closer than a few inches from the work (Figure I-18). Keep rags, brushes, and fingers away

FIGURE I-19 The skyhook in use bringing a large chuck into place for mounting (Syclone Products, Inc.).

FIGURE I-20 Left-hand filing in the lathe (Lane Community College).

from rotating work, especially when knurling. Roughing cuts tend to quickly drag in and wrap up rags, clothing, neckties, abrasive cloth, and hair. Move the carriage back out of the way and cover the tool with a cloth when checking boring work. When removing or installing chucks or heavy workpieces, use a board on the ways (a part of the lathe bed) so that the chuck can be slid into place. To lift a heavy chuck or workpiece (larger than an 8-in.-diameter chuck), get help or use a crane (Figure I-19). Remove the tool or turn it out of the way during this operation. Do not shift gears or try to take measurements while the machine is running and the workpiece is in motion. Never use a file without a handle as the file tang can quickly cut your hand or wrist if the file is struck by a spinning chuck jaw or lathe dog. Left-hand filing is considered safest in the lathe; that is, the left hand grips the handle while the right hand holds the tip end of the file (Figure I-20).

UNIT 1

The Engine Lathe

Modern lathes are highly accurate and complex machines capable of performing a great variety of operations. Before attempting to operate a lathe, you should familiarize yourself with its principal parts and their operation. Good maintenance is important to the life and accuracy of machine tools. A machinist depends on the lathe to make precision parts. A poorly maintained machine loses its usefulness to a machinist. This unit will show you how to make adjustments, lubricate, and properly maintain the machine that you use.

Objectives

After completing this unit, you should be able to:

1. Identify the most important parts of a lathe and their functions.
2. List all of the lubrication points for one lathe in your shop.
3. Determine the type of lubrication needed.
4. Adjust the cross slide, compound slide, and tailstock, and clamp the compound after rotating it.

One of the most important machine tools in the metal working industry is the lathe (Figure I-21). A lathe is a device in which the work is rotated against a cutting tool. As the cutting tool is moved lengthwise and crosswise to the axis of the workpiece, the shape of the workpiece is generated.

Figure I-22 shows a lathe and its most important parts. A lathe consists of the following major component groups: headstock, bed, carriage, tailstock, quick-change gearbox, and a base or pedestal. The headstock is fastened on the left side of the bed. It contains the spindle that drives the various work-holding devices. The spindle is supported by spindle bearings on each end. If they are sleeve-type bearings, a thrust bearing is also used to take up end play. Tapered roller spindle bearings are often used on modern lathes. Spindle speed changes are also made in the headstock, either with belts or with gears. Figure I-23 shows a geared-type headstock. Speed changes are made in

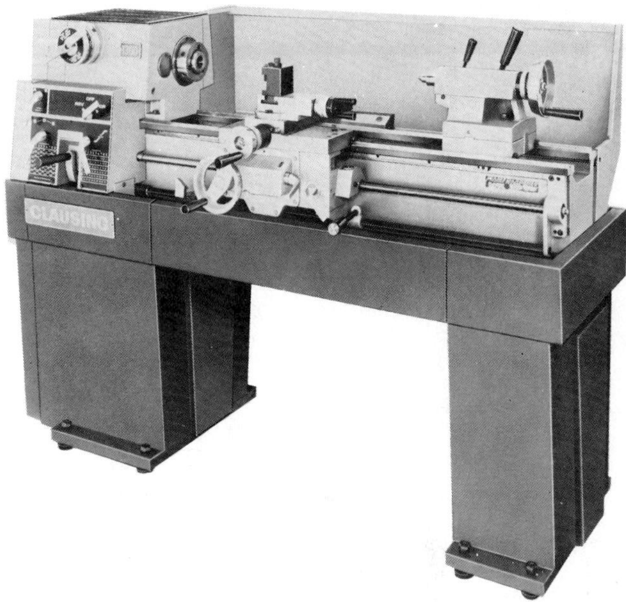

FIGURE I-21 The engine lathe (Clausing Corporation).

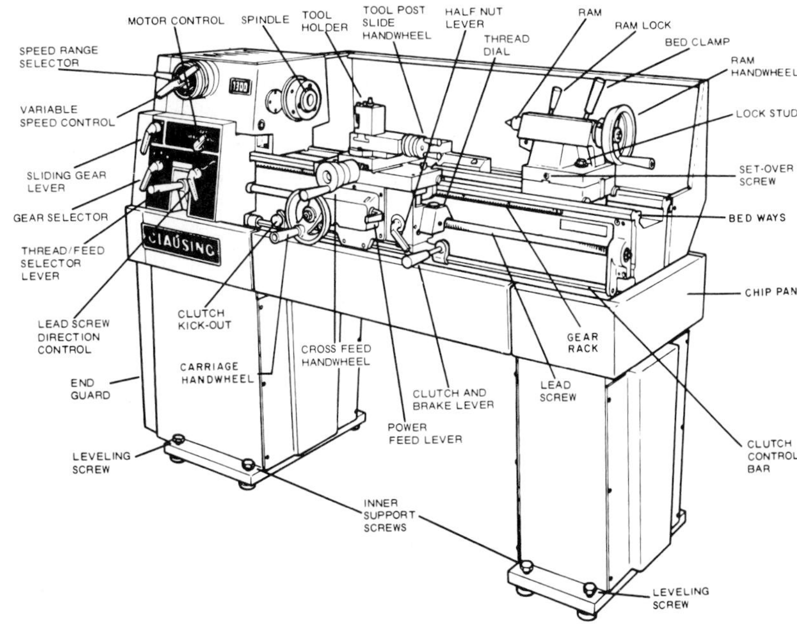

FIGURE I-22 Engine lathe with the parts identified (Clausing Corporation).

FIGURE I-23 Geared headstock for heavy-duty lathe (Lodge & Shipley Company).

FIGURE I-25 Speeds are changed on this lathe by moving the belt to various steps on the pulley (Lane Community College).

FIGURE I-24 Spindle drive showing gears and shifting mechanism located in the headstock (Lodge & Shipley Company).

FIGURE I-26 Long taper key drive spindle nose (Lane Community College).

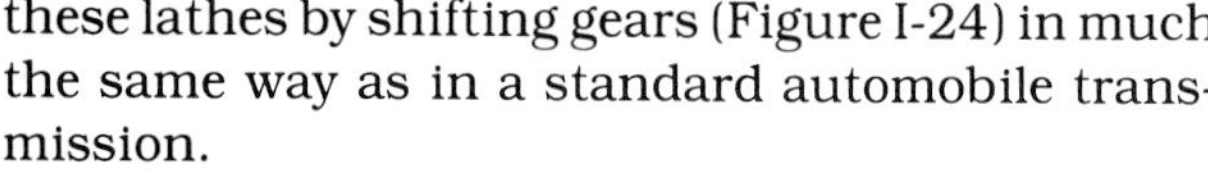

these lathes by shifting gears (Figure I-24) in much the same way as in a standard automobile transmission.

Most belt-driven lathes have a slow speed range when back gears are engaged. Figure I-25 shows a back-geared headstock. Usually, only older-type lathes have belt drives and back gears. See Unit 5 for operating details on these various drives.

A feed reverse lever, also called a leadscrew direction control, is located on the headstock. Its function is simply to control the direction of rotation of the leadscrew. This rotation determines the direction of feed and whether a thread cut on the lathe is left-hand or right-hand. The threading and feeding mechanisms of the lathe are also powered through the headstock.

The spindle is hollow to allow long slender workpieces to pass through. The spindle end facing the tailstock is called the spindle nose. Spindle noses usually are one of three designs: a long taper key drive (Figure I-26), a cam lock type (Figure I-27), or a threaded spindle nose (Figure I-28). Lathe chucks and other work-holding devices are fastened to and driven by the spindle nose. The hole in the spindle nose typically has a standard Morse taper. The size of this taper varies with the size of the lathe.

The bed (Figure I-29) is the foundation and backbone of a lathe. Its rigidity and alignment affect the accuracy of the parts machined on it. Therefore, lathe beds are constructed to withstand the stresses created by heavy machining cuts. On top of the bed are the ways, which usually consist of two inverted vees and two flat bearing surfaces. The ways of the lathes are very accurately machined by grinding or by milling and hand scraping. Wear or damage to the ways will affect the accuracy of workpieces machined on them. A gear

FIGURE I-27 Camlock spindle nose.

FIGURE I-28 Threaded spindle nose (Lane Community College).

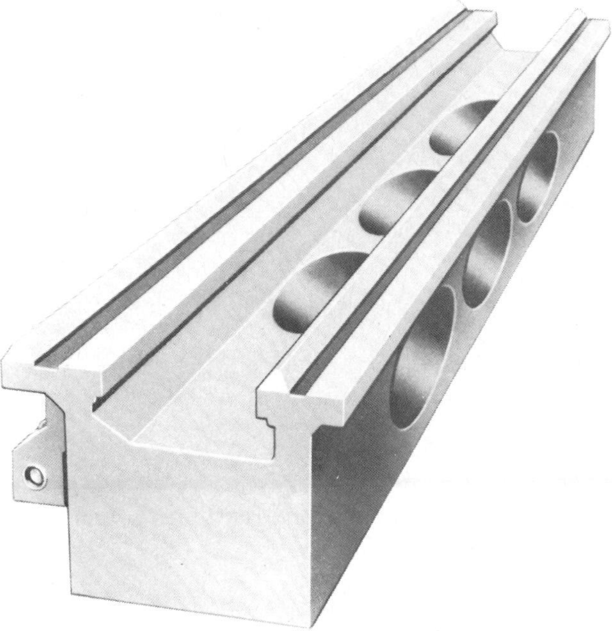

FIGURE I-29 Lathe bed (Clausing Corporation).

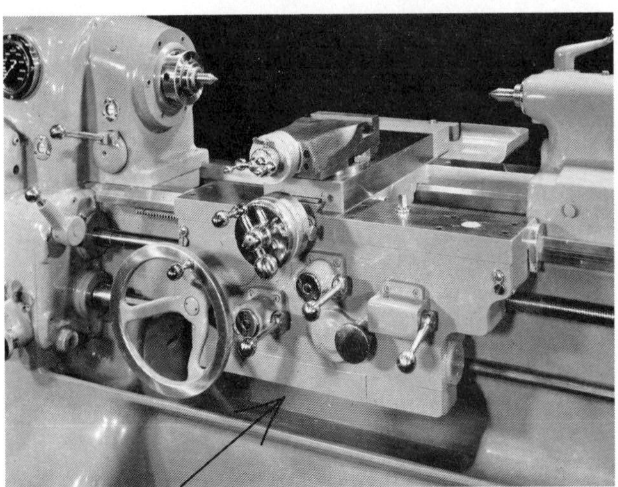

FIGURE I-30 Lathe carriage. The arrow is pointing to the apron (Monarch Machine Tool Company, Sidney, Ohio).

rack is fastened below the front way of the lathe. Gears that link the carriage handwheel to this rack make possible the lengthwise movement of the carriage by hand.

The carriage is made up of the saddle and apron (Figure I-30). The apron is the part of the carriage facing the operator; it contains the gears and feed clutches that transmit motion from the feed rod or leadscrew to the carriage and cross slide. The saddle slides on the ways and supports the cross slide and compound rest. The cross slide moves crosswise at 90 degrees to the axis of the lathe by manually turning the cross feed screw handle or by engaging the cross feed lever (also called power feed lever or on some lathes, the clutch knob), which is located on the apron for automatic feed. On some lathes a feed change lever (or plunger) on the apron is used to direct power from the feed mechanism to either the longitudinal (lengthwise) travel of the carriage or to the cross slide. On other lathes, two separate levers or knobs are used to transmit motion to the carriage and cross slide.

A thread dial is fastened to the apron (usually on the right side), which indicates the exact place to engage the half-nuts while cutting threads. The half-nut lever is used only for thread cutting and never for feeds for general turning. The entire carriage can be moved along the ways manually by turning the carriage handwheel or under power by engaging the power feed controls on the apron. Once in position, the carriage can be clamped to the bed by tightening the carriage lock screw.

The compound rest is mounted on the cross slide and can be swiveled to any angle horizontal

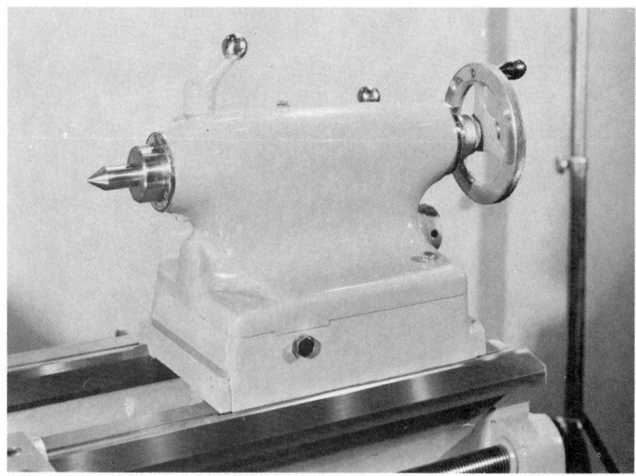

FIGURE I-31 Tailstock (Monarch Machine Tool Company, Sidney, Ohio).

FIGURE I-32 Quick-change gearbox showing index plate (Lane Community College).

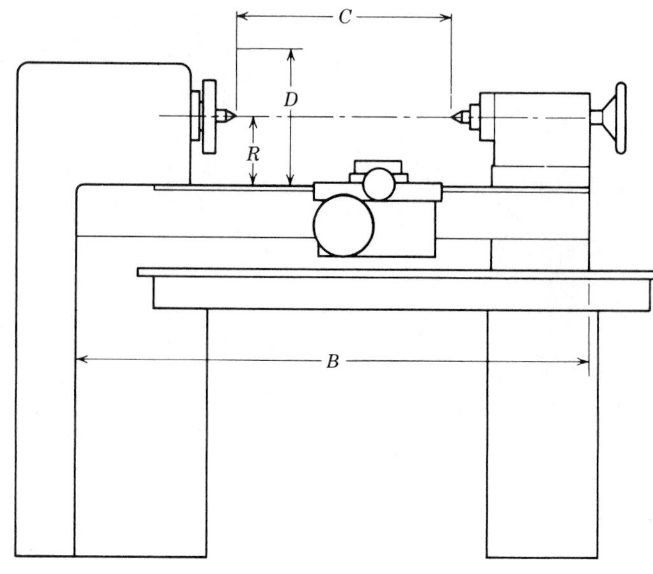

FIGURE I-33 Measuring a lathe for size. C = maximum distance between centers; D = maximum diameter of workpiece over ways (swing of lathe); R = radius, one-half swing; B = length of bed.

with the lathe axis in order to produce bevels and tapers. The compound rest can only be moved manually by turning the compound rest feed screw handle. Cutting tools are fastened on a tool post that is located on the compound rest.

The tailstock (Figure I-31) is used to support one end of a workpiece for machining or to hold various cutting tools such as drills, reamers, and taps. The tailstock slides on the ways and can be clamped in any position along the bed. The tailstock has a sliding spindle that is operated by a handwheel and locked in position with a spindle clamp lever. The spindle is bored to receive a standard Morse taper shank. The tailstock consists of an upper and lower unit and can be adjusted to make tapered workpieces by turning the adjusting screws in the base unit.

The quick-change gearbox (Figure I-32) is the link that transmits power between the spindle and the carriage. By using the gear shift levers on the quick-change gearbox, you can select different feeds. Power is transmitted to the carriage through a feed rod or, as on smaller lathes, through the leadscrew with a keyway in it. The index plate on the quick-change gearbox indicates the feed in thousandths of an inch or as threads per inch for the lever position.

The base of the machine is used to level the lathe and to secure it to the floor. The motor of the lathe is usually mounted in the base. Figure I-33 shows how the lathe is measured.

Engine Lathe Maintenance and Adjustments

The engine lathe, a precision machine tool, is perhaps the most abused of all shop equipment. With proper care the lathe will maintain its accuracy for many years, but its service life will be shortened severely if it is misused. Even small nicks or burrs on the ways can prevent the carriage or tailstock from seating properly. Fine chips, filings, or grindings combine with the oil to form an abrasive mixture that can score and wear sliding surfaces and bearings. Frequent cleaning of way surfaces is helpful, but do not use an air jet, since this will blow the abrasive sludge into the bearing surfaces. Use a brush to remove chips and wipe off with a cloth, then apply a thin film of oil.

Nicks or burrs are often caused by dropping chucks and workpieces on the ways and by laying tools such as files across them. This should never

FIGURE I-34 Wooden lathe board is used for handling chucks (Lane Community College).

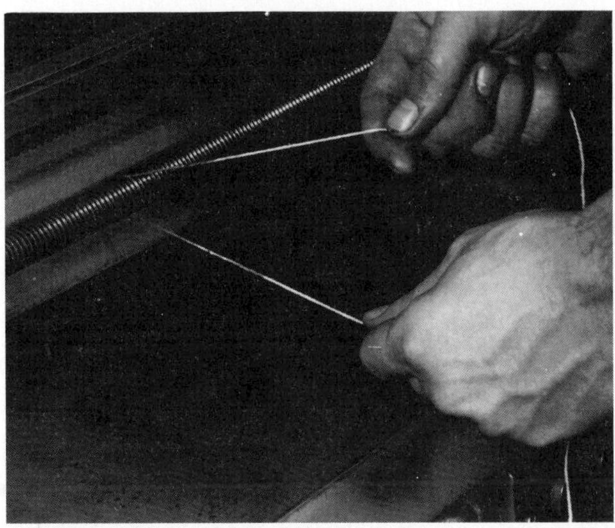

FIGURE I-36 Cleaning the leadscrew using a piece of string (Lane Community College).

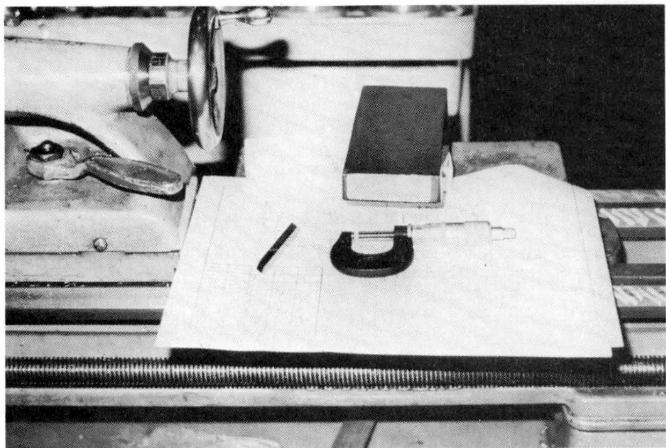

FIGURE I-35 Precision tools are protected from damage by keeping them in a safe place such as a tool tray while working on a lathe (Lane Community College).

FIGURE I-37 Gibs are adjusted by tightening screws while the compound is centered over its slide (Lane Community College).

be done. Wooden lathe boards are used for handling chucks and heavy work (Figure I-34). Larger boards are often used for tool trays that are placed on an unused portion of the lathe way (Figure I-35).

Occasionally, it is advisable to clean the leadscrew. To do this, loop a piece of string behind the leadscrew and hold each end of the string (Figure I-36). With the machine on and the leadscrew turning, draw the string along the threads. Never hold the string by wrapping it around your fingers; if the string grabs and begins winding on the leadscrew, let it go. Lathes should be completely lubricated daily or before using. Oil cups or holes should be given a few drops of oil (too much just runs out). Apron and headstock reservoirs should be checked for oil level. If they are low, use an oil recommended by the manufacturer or its equivalent.

Before operating any lathe, wipe the way surfaces, even if they look clean, as gritty dust can settle on them when the machine has been idle for a few hours. Wiped, the ways should then be given a thin film of way oil, which is specially compounded to remain on the way surfaces for a longer time than ordinary lubricating oil. After using the lathe, it should again be cleaned free of chips and grit, and the ways should be lightly oiled to prevent rusting. The chips and dirt on the surrounding floor space should be swept up and placed in a container.

Lathe maintenance and repair that requires extensive disassembly should only be done by qualified personnel. There are, however, many adjustments that a machinist should be able to perform. Adjustment of the gibs on the cross slide and compound slide may be needed if there is excess clearance between the gib and the dovetail slide. Adjust the gibs only when the slide is completely over its mating dovetail. Gibs adjusted without this backing may bend. Straight gibs are adjusted by tightening a number of screws (Figure I-37) until the

FIGURE I-38 Method of adjustment of tapered gibs. Thrust screw is being tightened. The lockscrew is on the other end of the gib (Lane Community College).

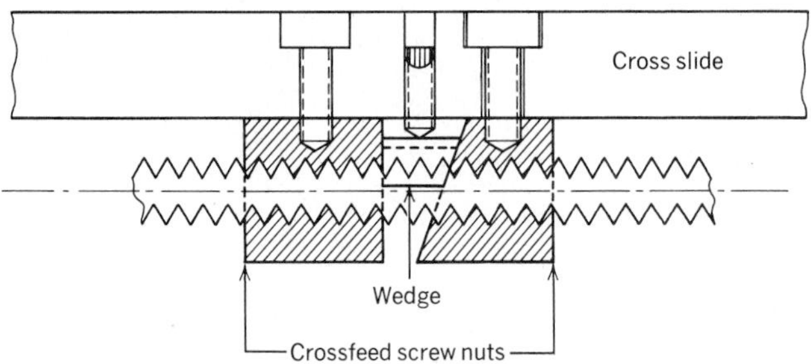

FIGURE I-39 Backlash compensating crossfeed nuts. This is one kind of backlash compensator in which a wedge is forced between two nuts on the same screw.

slide operates with just a slight drag. Too much tightening will cause binding. The gibs on the compound slide should be kept fairly tight when the compound is not being used. Locknuts should be tightened after making the adjustment. Tapered gibs are adjusted by first loosening the lockscrew and then adjusting the thrust screw. When the proper fit has been obtained, tighten the lockscrew (Figure I-38).

Lathes typically have some slack or backlash in the cross feed screw and compound screw. This slack must be removed before starting a cut by backing away from the work two turns and then bringing the tool to the work. Some lathes have backlash compensating crossfeed nuts (Figure I-39) that can be adjusted to remove most of the end play.

Sometimes a machinist must interchange gears in the gear train to a leadscrew on the headstock end of the lathe. When the proper gears have been set in place, the mounting or clamping bolts should be tightened lightly and a strip of paper placed between the gears. The gears should then be pushed together against the paper shim. This spaces the gears approximately .005 in. The clamping bolts should then be tightened (Figure I-40). If the gears are noisy, the adjustment must be made with more mating clearance between the gears.

FIGURE I-40 Adjustment of feed change gears. Gear backlash could be approximately .005 in. (Lane Community College).

FIGURE I-41 Tailstock is clamped to the ways by means of a bolt and crossbar (Lane Community College).

The method by which the tailstock is clamped to the ways is simple and self-adjusting in many lathes, especially older ones. A large bolt and nut draws a cross bar up into the bed when the nut is tightened (Figure I-41). A camlock type with a lever

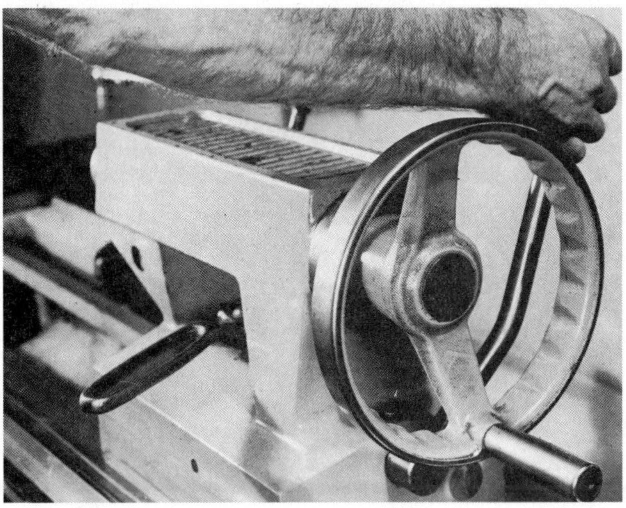

FIGURE I-42 Camlock-type tailstock with lever being tightened (Lane Community College).

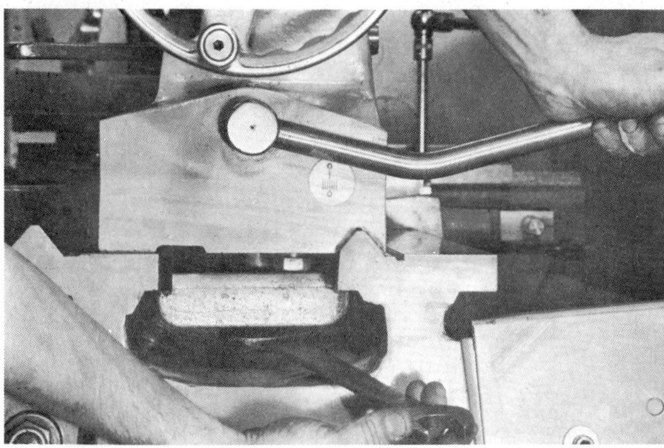

FIGURE I-43 Adjustment of camlock clamp. When the bolt on the crossbar is tightened, the cam will lock with the lever in a lower position (Lane Community College).

(Figure I-42) is sometimes used alone or on larger lathes in conjunction with a standard bolt and cross bar clamp. The clamping lever should never be forced. If it gets out of adjustment, it can be readjusted by means of a nut underneath the tailstock (Figure I-43).

The compound is made to rotate on all lathes so that an angle may be machined. Degrees marked on its base are read against an index mark on the carriage. Various methods are used to clamp the compound—usually, one or two bolts on each side (Figure I-44). Some types have setscrews.

SELF-TEST

I

Part A.

At a lathe in the shop, identify the following parts and describe their functions. **Do not** turn on the

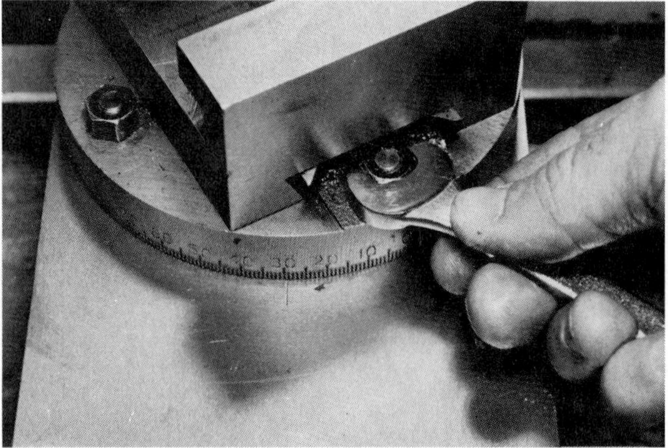

FIGURE I-44 Compound being tightened with wrench on clamping nut.

lathe until you get permission from your instructor.

The Headstock

1. Spindle
2. Spindle speed changing mechanism
3. Backgears
4. Bull gear lockpin
5. Spindle nose
6. What kind of spindle nose is on your lathe?
7. Feed reverse lever

The Bed

1. The ways
2. The gear rack

The Carriage

1. Cross slide
2. Compound rest
3. Saddle
4. Apron
5. Power feed lever
6. Feed change lever
7. Half-nut lever
8. Thread dial
9. Carriage handwheel
10. Carriage lock

The Tailstock

1. Spindle and spindle clamping lever
2. Tapered spindle hole and the size of its taper
3. Tailstock adjusting screws

The Quick-Change Gearbox

1. Leadscrew
2. Shift the levers to obtain feeds of .005 and .013 in. per revolution. Rotating the leadscrew with your fingers aids in shifting these levers.
3. Set the levers to obtain 4 threads/in. and then 12 threads/in.
4. Measure the lathe and record its size.

Part B.

1. Why should fine chips, filings, and grindings be cleaned from the ways and slides frequently?
2. How should cleaning of the ways and slides be done?
3. Nicks and scratches are very damaging to lathe ways. What means can be employed to avoid this?
4. How often should a lathe be lubricated?
5. Should you begin work immediately on a lathe that looks perfectly clean? Explain.
6. What should be done when you are finished using the lathe?
7. Name two types of gibs used on lathes.
8. How tight should the gibs be adjusted on the cross slide and on the compound?

UNIT 2

Toolholders and Tool Holding for the Lathe

For lathe work, cutting tools must be supported and fastened securely in the proper position to machine the workpiece. There are many different types of toolholders available to satisfy this need. Anyone working with a lathe should be able to select the best toolholding device for the operation performed.

Objectives

After completing this unit, you should be able to:

1. Identify standard, quick-change, and turret-type toolholders mounted on a lathe carriage.
2. Identify tool holding for the lathe tailstock.

A cutting tool is supported and held in a lathe by a toolholder that is secured in the tool post of the lathe with a clamp screw. A common tool post found on smaller or older lathes is shown in Figure I-45. Tool height adjustments are made by swiveling the rocker in the tool post ring. Making adjustments in this manner changes the effective back rake angle and also the front relief angle of the tool.

Many types of toolholders are used with the standard tool post. A straight shank turning toolholder (Figure I-46) is used with high-speed tool bits. The tool bit is held in the toolholder at a $16\frac{1}{2}$-degree angle, which provides a positive back rake angle for cutting. Straight shank toolholders are used for general machining on lathes. The type shown in Figure I-47 is used with carbide tools.

Offset toolholders (Figures I-48 and I-49) allow machining close to the chuck or tailstock of a lathe without tool post interference. The left-hand toolholder is intended for use with tools cutting from right to left or toward the headstock of the lathe.

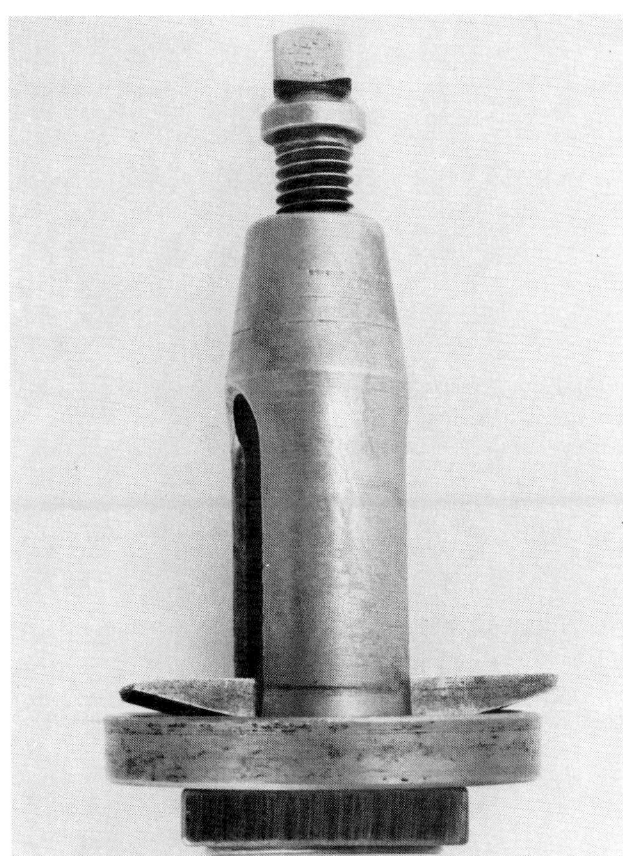

FIGURE I-45 Standard-type tool post with ring and rocker.

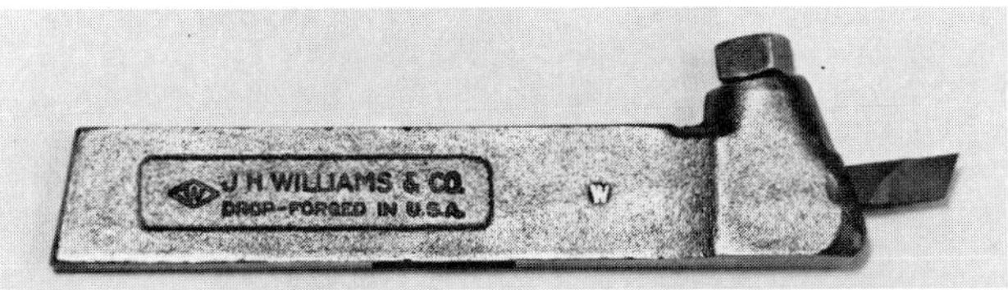

FIGURE I-46 Straight shank toolholder with built-in back rake holding a high-speed right-hand tool (Lane Community College).

FIGURE I-47 Right-hand toolholder for carbide tool bits without back rake.

FIGURE I-48 Left-hand toolholder with right-hand tool (Lane Community College).

FIGURE I-49 Right-hand toolholder with left-hand tool (Lane Community College).

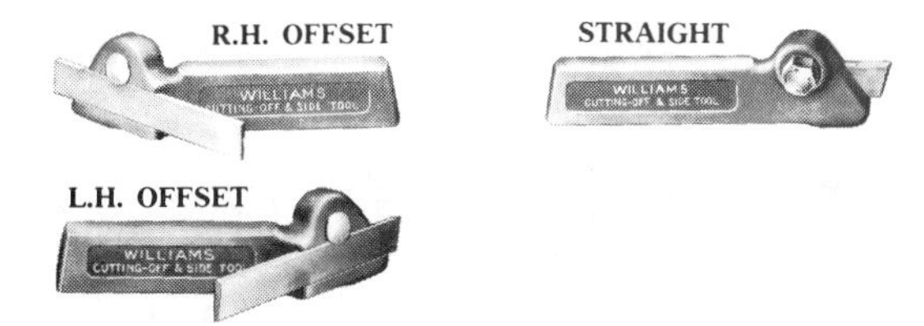

FIGURE I-50 Three kinds of cutoff toolholder with cutoff blades (TRW, Inc.).

FIGURE I-51 Knuckle-joint knurling tool (Lane Community College).

A toolholder should be selected according to the machining to be done. The setup should be rigid and the toolholder overhang should be kept to a minimum to prevent chattering. A variety of cutoff toolholders (Figure I-50) are used to cut off or make grooves in workpieces. Cutoff tools are available in a number of different thicknesses and heights. Knurling tools are made with one pair of rollers (Figure I-51) or with three pairs of rollers (Figure I-52) that make three different kinds of knurls. These toolholders as shown in Figures I-45 through I-52 and the so-called standard tool

FIGURE I-52 Triple-head knurling tool (Lane Community College).

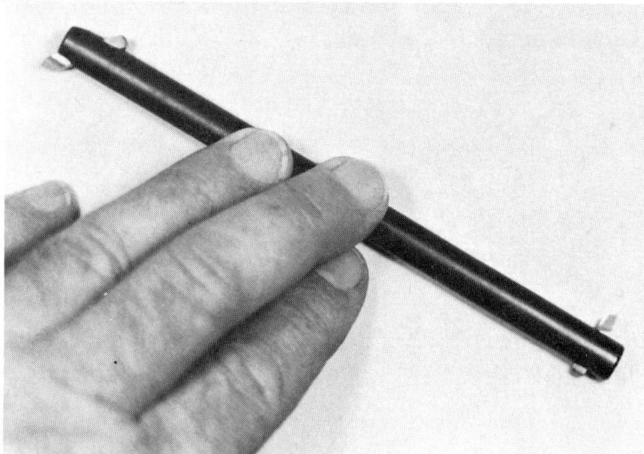

FIGURE I-53 Small inserted tool-type boring bar (Lane Community College).

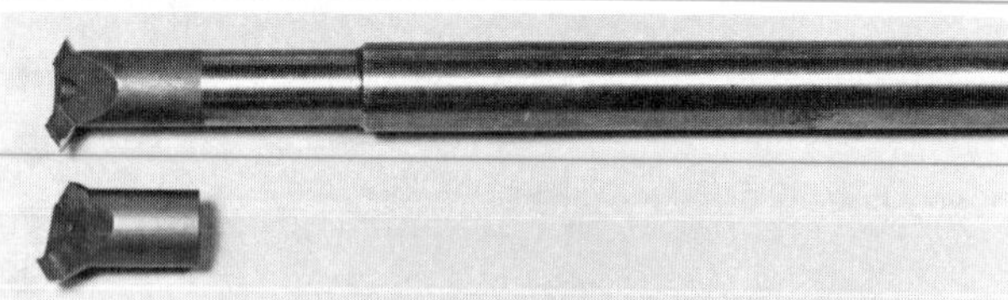

FIGURE I-54 Boring bar with form-relieved interchangeable cutter head. These are made in a range of sizes (Lane Community College).

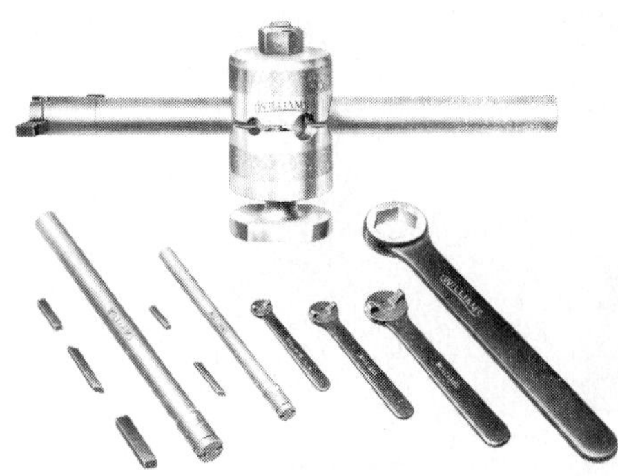

FIGURE I-55 Boring bar tool post and bars with special wrench (TRW, Inc.).

FIGURE I-56 Heavy-duty boring bar holder (TRW, Inc.).

FIGURE I-57 Quick-change tool post, dovetail type (Copyright © 1975, Aloris Tool Company, Inc.).

post are still used on some small lathes but are totally obsolete as far as manufacturing and industry are concerned.

Very small boring bars can be either the inserted tool type (Figure I-53) or the form-relieved interchangeable cutter head (Figure I-54). The boring bar tool post (Figure I-55) can be used with a number of different boring bar sizes. Another advantage of boring bars is the interchangeability of toolholding end caps. End caps hold the boring tool square to the axis of the boring bar or at a 45- or 60-degree angle to it. The heavy-duty boring bar holder in Figure I-56 is not as rigid as the holder in Figure I-54 because it is clamped in the tool post.

A quick-change tool post (Figure I-57), so called because of the speed with which tools can be interchanged, is more versatile than the standard post. The toolholders used on it are accurately held because of the dovetail construction of the post. This accuracy makes for more exact repetition of setups. Tool height adjustments are made with a micrometer adjustment collar, and the height alignment will remain constant through repeated

FIGURE I-58 Quick-change tool post (Lane Community College).

FIGURE I-59 Turning toolholder in use (Copyright © 1975, Aloris Tool Company, Inc.).

FIGURE I-60 Threading toolholder, using the top of the blade (Copyright © 1975, Aloris Tool Company, Inc.).

tool changes.

A three-sided quick-change tool post (Figure I-58) has the added ability to mount a tool on the tailstock side of the tool post. These tool posts are securely clamped to the compound rest. The tool post in Figure I-58 uses double vees to locate the toolholders, which are clamped and released from the post by turning the top lever.

FIGURE I-61 Threading is accomplished with the bottom edge of the blade with the lathe spindle in reverse. This assures cutting right-hand threads without hitting the shoulders (Copyright © 1975, Aloris Tool Company, Inc.).

FIGURE I-62 Drill toolholder in the tool post. Mounting the drill in the tool post makes drilling with power feed possible (Copyright © 1975, Aloris Tool Company, Inc.).

Toolholders for the quick-change tool posts include those for turning (Figure I-59), threading (Figures I-60 and I-61), and holding drills (Figure I-62). The drill holder makes it possible to use the carriage power feed when drilling holes instead of

FIGURE I-63 Boring toolholder. This setup provides good boring bar rigidity (Copyright © 1975, Aloris Tool Company, Inc.).

FIGURE I-64 Toolholders are made with wide or narrow slots to fit tools with various shank thicknesses (Lane Community College).

FIGURE I-65 Tailstock turret used in quick-change toolholder (Enco Manufacturing Company).

FIGURE I-66 Quick-change cutoff toolholder (Enco Manufacturing Company).

FIGURE I-67 Quick-change knurling and facing toolholder (Enco Manufacturing Company).

the tailstock hand feed. Figure I-63 shows a boring bar toolholder in use; the boring bar is very rigidly supported.

An advantage of the quick-change tool post toolholders is that cutting tools of various shank thicknesses can be mounted in the toolholders (Figure I-64). Shims are sometimes used when the shank is too small for the setscrews to reach. Another example of quick-change tool post versatility is shown in Figure I-65, where a tailstock turret is in use. Figure I-66 shows a cutoff tool mounted in a toolholder. Figure I-67 is a combination knurling tool and facing toolholder. A four-tool turret toolholder can be set up with several different tools such as turning tools, facing tools, or threading or boring tools. Often one tool can perform two or more operations, especially if the turret can be indexed in 30-degree intervals. A facing operation (Figure I-68), a turning operation (Figure I-69), and chamfering of a bored hole (Figure I-70) are all performed from this turret. Tool height adjustments are made by placing shims under the tool.

The toolholders studied so far are all intended for use on the carriage of a lathe. Toolholding is also done on the tailstock. Figure I-71 shows how the tailstock spindle is used to hold Morse taper shank tools. One of the most common toolholding devices used on a tailstock is the drill chuck (Figure I-72). A drill chuck is used for holding straight shank drilling tools. When a series of operations

FIGURE I-68 Facing cut with a four-tool turret-type toolholder (Enco Manufacturing Company).

FIGURE I-69 Turning cut with a turret-type toolholder (Enco Manufacturing Company).

FIGURE I-70 Chamfering cut with a turret-type toolholder (Enco Manufacturing Company).

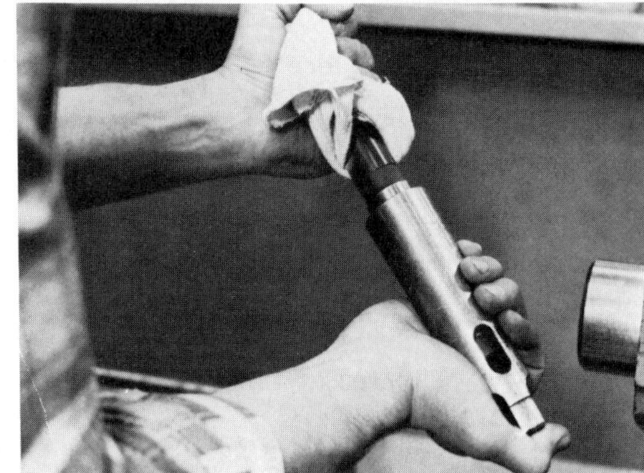
FIGURE I-71 Taper shank drill with sleeve ready to insert in tailstock spindle (Lane Community College).

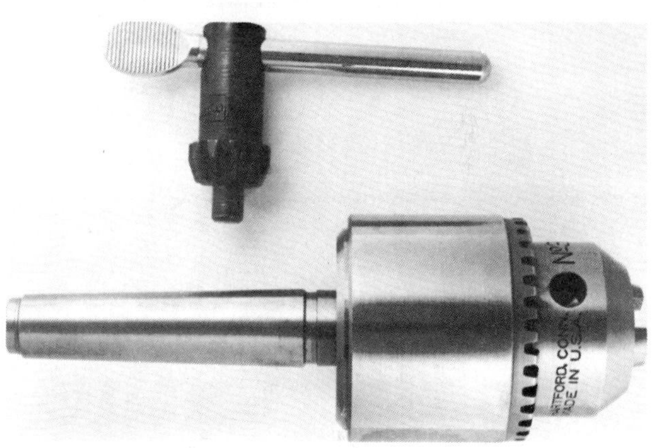
FIGURE I-72 Drill chuck with Morse taper shank.

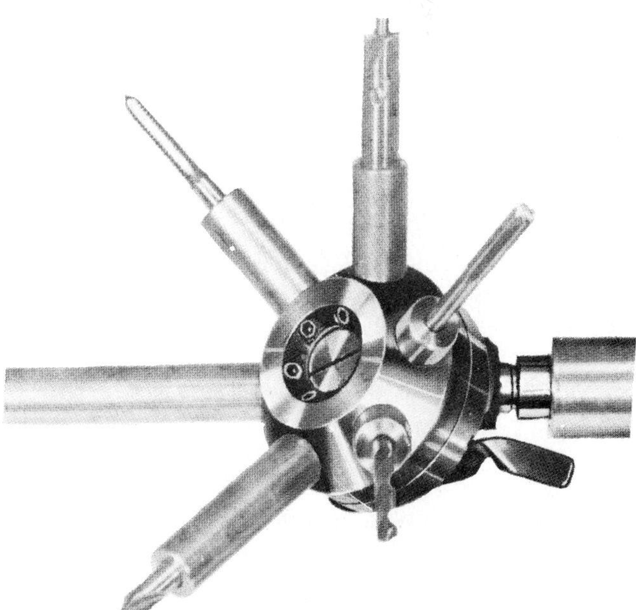
FIGURE I-73 Tailstock turret (Enco Manufacturing Company).

must be performed and repeated on several workpieces, a tailstock turret (Figure I-73) can be used. The illustrated tailstock turret has six tool positions, one of which is used as a workstop. The other positions are for center drilling, drilling, reaming, counterboring, and tapping. Tailstock tools are normally fed by turning the tailstock handwheel.

SELF-TEST

At a lathe in the shop, identify various toolholders and their functions.

1. What is the purpose of a toolholder?

2. How is a standard left-hand toolholder identified?

3. What is the difference between a standard-type toolholder for high-speed steel tools and for carbide tools?

4. Which standard toolholder would be best used for turning close to the chuck?

5. How are tool height adjustments made on a standard toolholder?

6. How are tool height adjustments made on a quick-change toolholder?

7. How are tool height adjustments made on a turret-type toolholder?

8. How does the toolholder overhang affect the turning operation?

9. What is the difference between a standard toolholder and a quick-change toolholder?

10. What kind of tools are used in the lathe tailstock?

11. How are tools fastened in the tailstock?

12. When is a tailstock turret used?

UNIT 3

Cutting Tools for the Lathe

A machinist must fully understand the purpose of cutting tool geometry, since it is the lathe tool that removes the metal from the workpiece. Whether or not this is done safely, economically, and with quality finishes depends to a large extent upon the shape of the point, the rake and relief angles, and the nose radius of the tool. In this unit, you will learn this tool geometry and also how to grind a lathe tool.

Objectives

After completing this unit, you should be able to:

1. Explain the purpose of rake and relief angles, chip breakers, and form tools.
2. Grind an acceptable right-hand roughing tool.

Cutting Tool Geometry

On a lathe, metal is removed from a workpiece by turning it against a single-point cutting tool. This tool must be very hard and it should not lose its hardness from the heat generated by machining. High-speed steel is used for many tools as it fulfills these requirements and is easily shaped by grinding. For this reason it is used in this unit to demonstrate tool geometry (see Table I-1). High speed tools have been largely replaced in production machining by carbide tools (Figure I-74) because of their higher metal removal rates. However, in general machining operations high-speed tools are still used for special tooling. High-speed steel tools are required for older lathes that are equipped with

TABLE I-1 Tool Geometry

Back rake	BR	12°
Side rake	SR	12°
End relief	ER	10°
Side relief	SRF	10°
End cutting edge angle	ECEA	30°
Side cutting edge angle	SCEA	15°
Nose radius	NR	$\frac{1}{32}$ in.

FIGURE I-74 Carbide insert tool and holder.

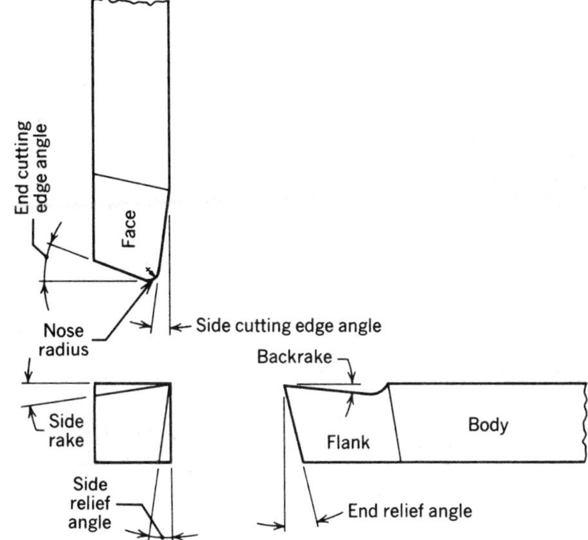

FIGURE I-75 The parts and angles of a tool.

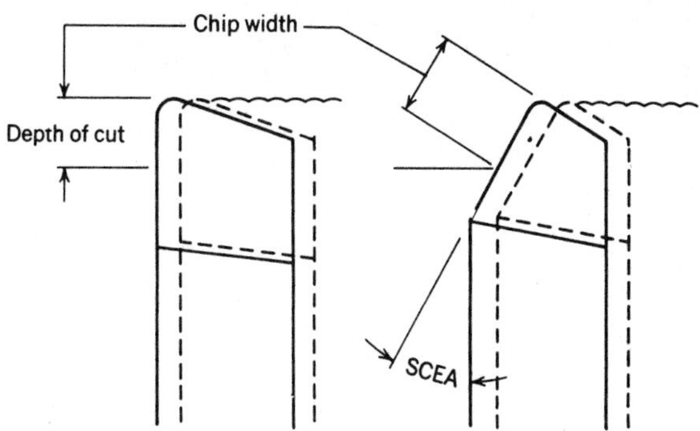

FIGURE I-76 The change in chip width with an increase of the side cutting edge angle. A large SCEA can sometimes cause chatter (vibration of work or tool).

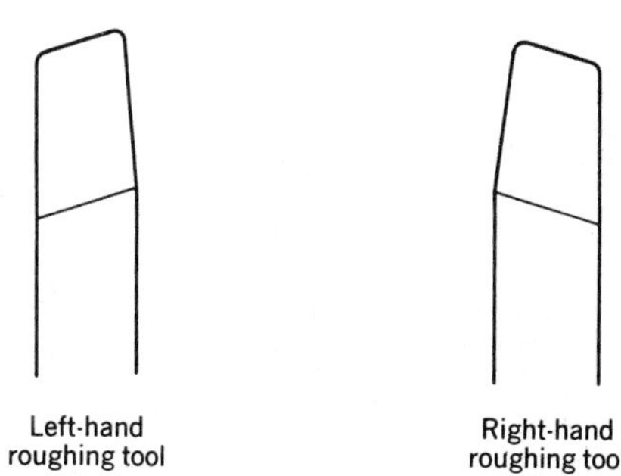

FIGURE I-77 Left-hand and right-hand roughing tools.

only low speed ranges. They are also useful for finishing operations, especially on soft metals.

The most important aspect of a lathe tool is its geometric form: the side and back rake, front and side clearance or relief angles, and chip breakers. Figure I-75 shows the parts and angles of the tool according to a commonly used industrial tool signature. The terms and definitions follow (the angles given are only examples and could vary according to the application).

1. The tool shank is that part held by the toolholder.

2. Back rake is very important to smooth chip flow, which is needed for a uniform chip and a good finish, especially in soft materials.

3. The side rake directs the chip flow away from the point of cut and provides for a keen cutting edge.

4. The end relief angle prevents the front edge of the tool from rubbing on the work.

5. The side relief angle provides for cutting action by allowing the tool to feed into the work material.

6. The side cutting edge angle (SCEA) may vary considerably. For roughing, it should be almost square to the work, usually about 5 degrees. Tools used for squaring shoulders or for other light machining could have angles from 5 to 32 degrees, depending on the application. This angle may be established by turning the toolholder or by grinding it on the toolbit or both. In finishing operations with a large nose radius and light cut, SCEA is not an important factor. The side cutting edge angle directs the cutting forces back into a stronger section of the tool point. It helps to direct the chip flow away from the workpiece. It also affects the thickness of the cut (Figure I-76).

7. The nose radius will vary according to the finish required. The smallest nose radius that will give the desired finish should be used.

Grinding a tool provides both a sharp cutting edge and the shape needed for the cutting operation. When the purpose for the rake and relief angles on a tool is clearly understood, then a tool suitable to the job may be ground. Left-hand tools are shaped just the opposite to right-hand tools (Figure I-77). The right-hand tool has the cutting edge on the left side and cuts to the left or toward the headstock. The hand of the lathe tool can be easily determined by looking at the end from the opposite side of the lathe; the cutting edge is to the right on a right-hand tool.

Tools are given a slight nose radius to strengthen the tip. A larger nose radius will give a

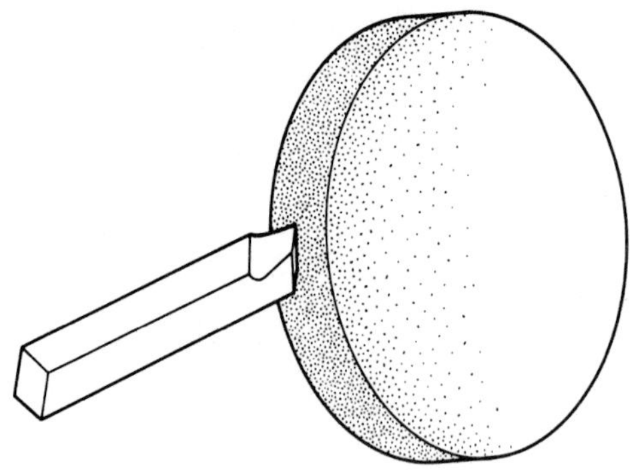

FIGURE I-78 One method of grinding the nose radius on the point of tool.

FIGURE I-81 Form tool being checked with radius gage (Lane Community College).

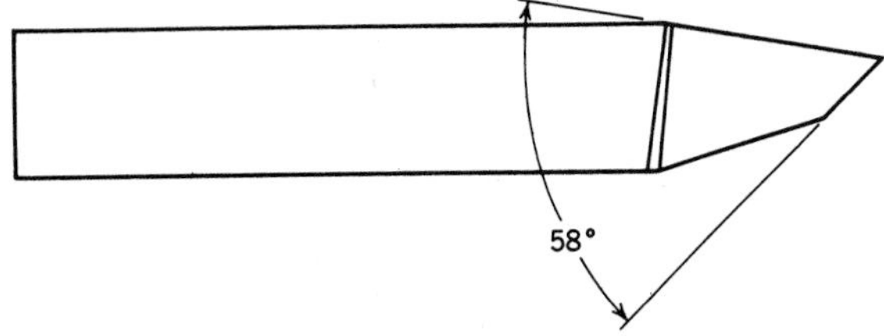

FIGURE I-79 Right-hand facing tool showing point angles. This tool is not suitable for roughing operations because of its acute point angle.

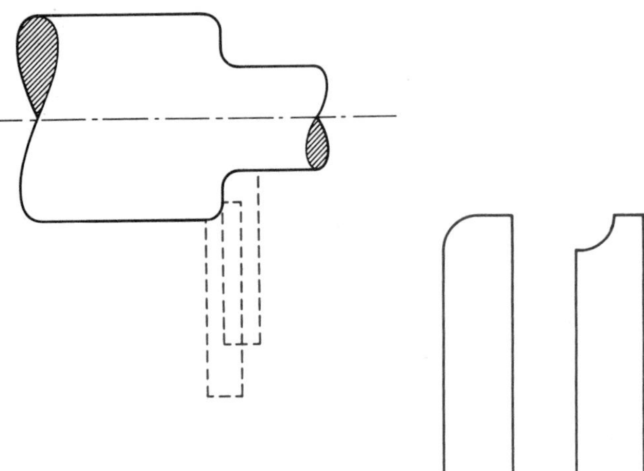

FIGURE I-82 Form tools are used to produce the desired shape in the workpiece. External radius tools, for example, are used to make outside corners round, while fillet radius tools are used on shafts to round the inside corners on shoulders.

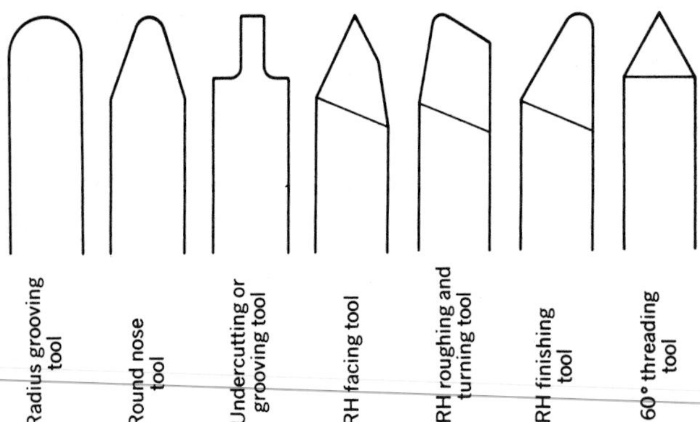

FIGURE I-80 Some useful tool shapes most often used. The first tool shapes needed are the three on the right, which are the roughing or general turning tool, finishing tool, and threading tool.

better finish (Figure I-78), but will also promote chattering (vibration) in a nonrigid setup. All lathe tools require some nose radius, however small. A sharp pointed tool is very weak at the point and will usually break off in use, causing a rough finish on the work. A facing tool (Figure I-79) for shaft ends and mandrel work has very little nose radius and an included angle of 58 degrees. This facing tool is not used for chucking work, however, as it is a relatively weak tool. A right-hand (RH) or left-hand (LH) roughing or finishing tool is often used for facing in chuck-mounted workpieces. Some useful tool shapes are shown in Figure I-80. These are used for general lathe work although they are not much used in industry.

FORM TOOLS

Tools that have specially shaped cutting edges are called form tools (Figure I-81). These tools are plunged directly into the cut in one operation. Form tools include those used for grooving, threading, cutting internal and external radii, and any other special shape needed for a particular operation. In contrast with single-point tools, these broad face tools have a tendency to chatter, so a lower cutting speed and cutting fluid is usually required when they are used.

The shape of any tool is dictated by the application. When cutting tool geometry is clearly understood, any specially shaped tool can be made for a specific purpose. For example, a tool with a nose radius can be used for making a fillet radius on a shaft (Figure I-82). Round corners can be formed with an external radius tool. A relief angle is ground on any kind of form tool to allow it to cut; side and back rakes are generally zero.

reverse jaws is used (Figure I-115) to hold pieces with larger diameters. The chuck and each of its jaws are stamped with identification numbers. Do not interchange any of these parts with another chuck or both will be inaccurate. Each jaw also is stamped 1, 2, or 3 to correspond to the same number stamped by the slot on the chuck. The jaws are removed from the chuck in the order 3, 2, 1 and should be returned in the reverse order 1, 2, 3.

A universal chuck with top jaws (Figure I-116) is reversed by removing the bolts in the top jaws and by reversing them. They must be carefully cleaned when this is done. Soft top jaws are frequently used when special gripping problems arise. Since the jaws are machined to fit the shape of the part (Figure I-117), they can grip it securely for heavy cuts (Figure I-118).

One disadvantage of most universal chucks is that they lose their accuracy when the scroll and jaws wear, and normally there is no compensation for wear other than regrinding the jaws. The three-jaw adjustable chuck in Figure I-119 has a compensating adjustment for wear or misalignment.

Combination universal and independent chucks also provide for quick opening and closing and have the added advantage of independent adjustment on each jaw. These chucks are like the universal type since three or four jaws move in or out equally, but each jaw can be adjusted independently as well.

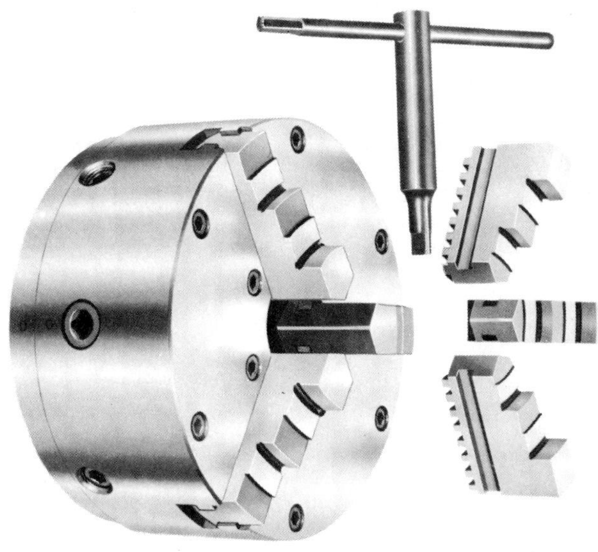

FIGURE I-115 Universal three-jaw chuck (Adjust-tru®) with a set of outside jaws (Buck Tool Company).

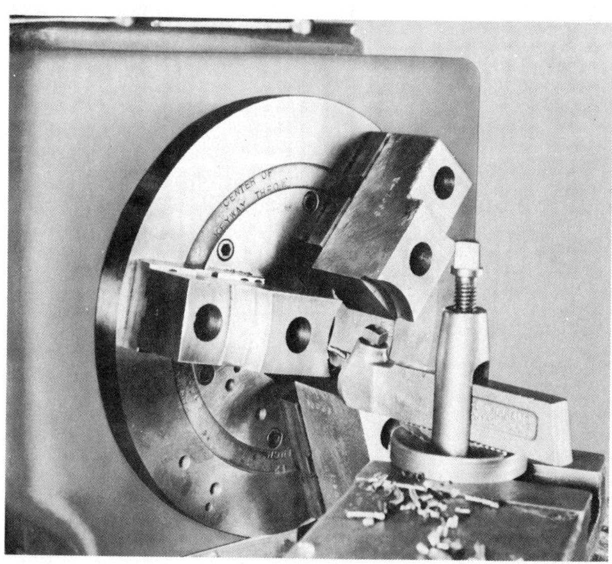

FIGURE I-117 Machining soft jaws to fit an odd-shaped workpiece on a jaw turning fixture (Warner & Swasey Company).

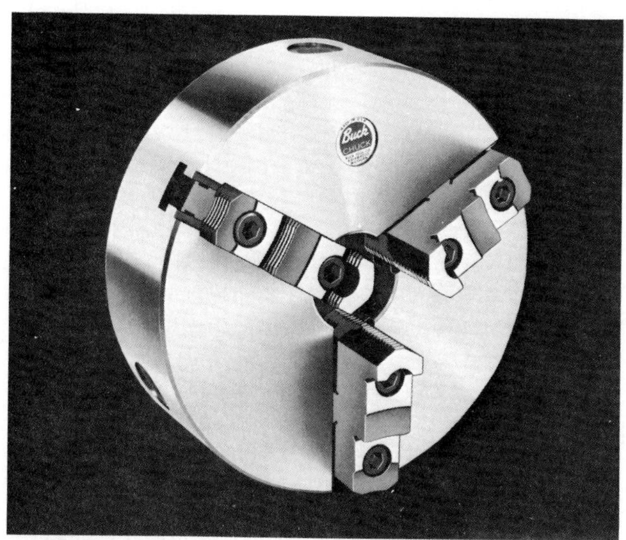

FIGURE I-116 Universal chuck with top jaws (Buck Tool Company).

FIGURE I-118 Soft jaws have been machined to fit the shape of this cast steel workpiece in order to hold it securely for heavy cuts (Warner & Swasey Company).

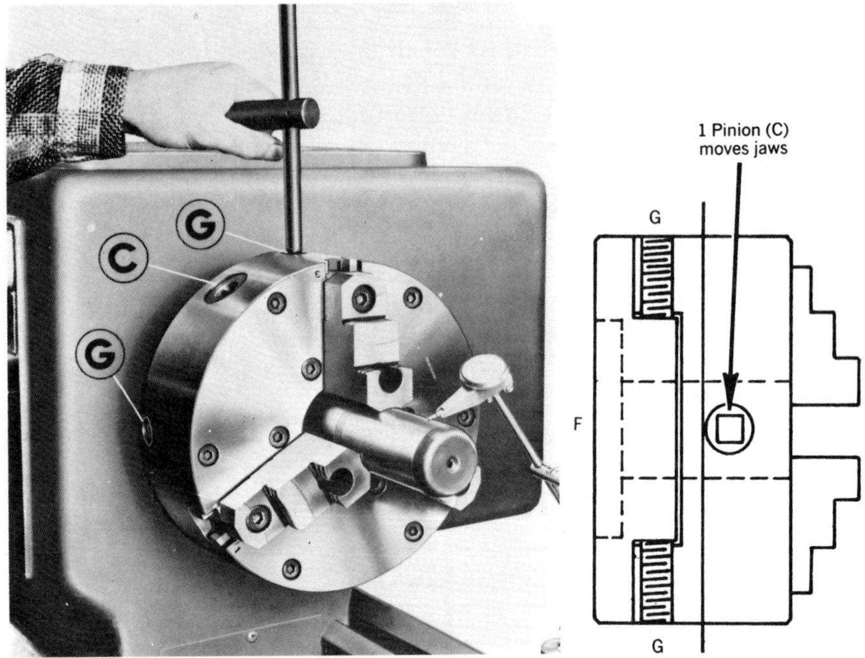

FIGURE I-119 Universal chuck (Adjust-tru®) with special adjustment feature (G) makes it possible to compensate for wear (Buck Tool Company).

FIGURE I-120 Magnetic chuck (Enco Manufacturing Company).

FIGURE I-121 Drive plate for turning between centers (Lane Community College).

Magnetic chucks (Figure I-120) are sometimes used for making light cuts on ferromagnetic material. They are useful for facing thin material that would be difficult to hold in conventional workholding devices. Magnetic chucks do not hold work very securely and so are not used much in lathe work.

All chucks need frequent cleaning of scrolls and jaws. These should be lightly oiled after cleaning, and chucks with grease fittings should be pressure lubricated. Chucks come in all diameters and are made for light-, medium-, and heavy-duty uses.

Drive plates are used together with lathe dogs to drive work mounted between centers (Figure I-121). The live center fits directly into the spindle taper and turns with the spindle. A sleeve is sometimes used if the spindle taper is too large in diameter to fit the center. The live center is usually made of soft steel so the point can be machined as needed to keep it running true. Live centers are removed by means of a knockout bar (Figure I-122).

Often when a machinist wants to machine the entire length of work mounted between centers without the interference of a lathe dog, special drive centers or face drivers (Figure I-123) can also be used to machine a part without interference. Quite heavy cuts are possible with these drivers, which are used especially for manufacturing purposes.

Face plates are used for mounting workpieces or fixtures. Unlike drive plates that have only slots, face plates have T slots and are more heavily built (Figure I-124). Face plates are made of cast iron and so must be operated at relatively slow speeds. If the speed is too high, the face plate could fly apart.

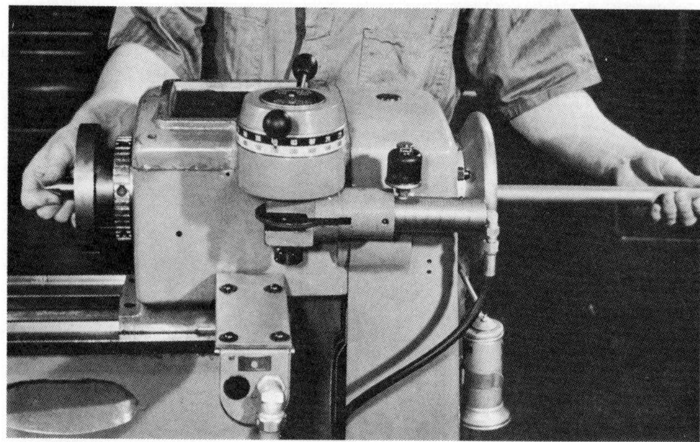

FIGURE I-122 Knockout bar is used to remove centers (California Community Colleges, IMC Project).

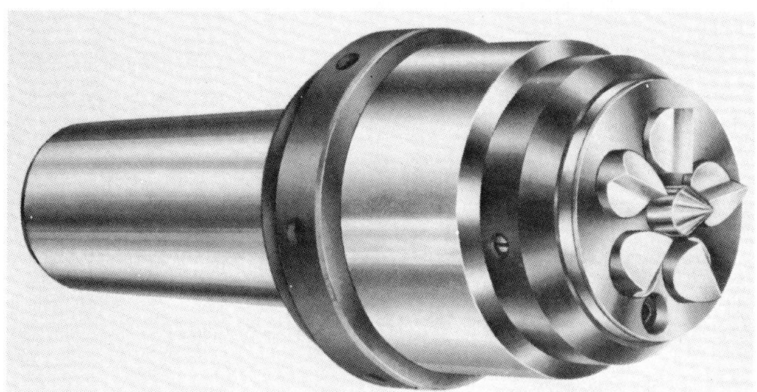

FIGURE I-123 Face driver is mounted in headstock spindle and work is driven by the drive pins that surround the deater (Madison-Kosta®, Madison Industries, a Division of Amtel, Inc.) (Madison-Kosta® is a registered trademark of Madison Industries, A Division of Amtel, Inc.).

Collet chucks (Figure I-125) are very accurate work-holding devices and are used in producing small high-precision parts. Steel spring collets are available for holding and turning hexagonal, square, and round workpieces. They are made in specific sizes (which are stamped on them) with a range of only a few thousandths of an inch. Workpieces to be gripped in a collet should not vary more than +.002 to −.003 in. from the collet size if the collets are to remain accurate. New collets can be expected to hold cylindrical work to .0005 in. eccentricity (indicator runout). Rough and inaccurate workpieces should not be held in the collet chuck since the gripping surfaces of the chuck would form an angle with the workpiece. The contact area would then be at one point on the jaws instead of along the entire length, and the piece would not be held firmly. If it is not held firmly, workpiece accuracy is impaired and the collet may be damaged. An adapter called a collet sleeve is fitted into the spindle taper and a draw bar is inserted into the spindle at the opposite end (Figure I-126). The collet is placed in the adapter, and the draw bar is rotated, which threads the collet into the taper and closes it. **Never tighten a collet without a workpiece in its jaws,** as this will damage it. Before collets and adapters are installed, they should be cleaned to ensure accuracy.

FIGURE I-124 T-slot face plate. Workpieces are clamped on the plate with T-bolts and strap clamps (Monarch Machine Tool Company, Sidney, Ohio).

FIGURE I-125 Side and end views of a spring collet for round work.

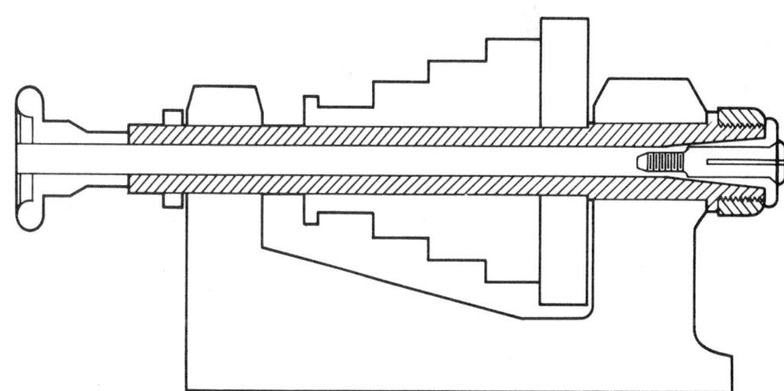

FIGURE I-126 Cross section of spindle showing construction of draw-in collet chuck attachment.

The rubber flex collet (Figure I-127) has a set of tapered steel bars mounted in rubber. It has a much wider range than the spring collet, each collet having a range of about $\frac{1}{8}$ in. A large handwheel is used to open and close the collets instead of a draw bar (Figure I-128).

The concentricity that you could expect from each type of work-holding device is shown in Table I-3.

FIGURE I-127 Rubber flex collet.

FIGURE I-128 Collet handwheel attachment for rubber flex collets (Monarch Machine Tool Company, Sidney, Ohio).

TABLE I-3 Accuracy of Holding Devices

Device	Centering Accuracy in Inches (Indicator Reading Difference)
Centers	Within .001
Four-jaw chuck	Within .001 (depending on the ability of machinist)
Collets	.0005 to .001
Three-jaw chuck	.001 to .003 (good condition) .005 or more (poor condition)

SELF-TEST

1. Briefly describe the lathe spindle. How does the spindle support chucks and collets?

2. Name the spindle nose types.

3. What is an independent chuck, and what is it used for?

4. What is a universal chuck, and what is it used for?

5. What chuck types make possible the frequent adjusting of chucks so that they will hold stock with minimum runout?

6. Workpieces mounted between centers are driven with lathe dogs. Which types of plate is used on the spindle nose to turn the lathe dog?

7. What is a live center made of? How does it fit in the spindle nose?

8. Describe a face driver.

9. On which type of plate are workpieces and fixtures mounted? How is it identified?

10. Name one advantage of using steel spring collets. Name one disadvantage.

UNIT 5

Operating the Machine Controls

Before using any machine, you must be able to use the controls properly, know what they are for, and understand how they work. You must also be aware of the potential hazards that exist for you and the machine if it is mishandled. This unit prepares you to operate lathes.

Objectives

After completing this unit, you should be able to:

1. Explain drives and shifting procedures for changing speeds on lathes.
2. Describe the use of various feed control levers.
3. Explain the relationship between longitudinal feeds and cross feeds.
4. State the differences in types of cross feed screw micrometer collars.

Most lathes have similar control mechanisms and operating handles for feeds and threading. Some machines, however, have entirely different driving mechanisms as well as different speed controls.

Drives

Spindle speed is controlled on some lathes by a belt on a step cone pulley in the headstock (Figures I-129 and I-130). The speed is changed by turning the belt tension lever to loosen the belt, moving the belt to the proper step for the desired speed, and then moving the lever to its former position. Several more lower speeds are made available by shifting to back gear. To do this, pull out or release the bull gear lockpin to disengage the spindle from the step cone pulley and engage the back gear lever as shown in Figure I-129. It may be necessary to rotate the spindle by hand slightly to bring the gear into mesh. The back gears must **never** be engaged when the spindle is turning with power. Do not forget to close the gear and pulley guards before starting the lathe.

FIGURE I-129 View of headstock showing back gear disengaged and lock pin engaged for direct belt drive (Lane Community College).

FIGURE I-130 View of headstock showing flat belt drive, the back gear engaged, and the lock pin disengaged (Lane Community College).

Another drive system uses a variable speed drive (Figure I-131) with a high and low range using a back gear. On this drive system, the motor must be running to change the speed on the varidrive, but the motor must be turned off to shift the back gear lever. Geared head lathes are shifted with levers on the outside of the headstock (Figure I-132). Several of these levers are used to set up the various speeds within the range of the machine. The gears will not mesh unless they are perfectly aligned, so that it is sometimes necessary to rotate the spindle by hand. **Never try to shift gears with the motor running and the clutch lever engaged.**

FIGURE I-131 Variable-speed control and speed selector (Clausing Corporation).

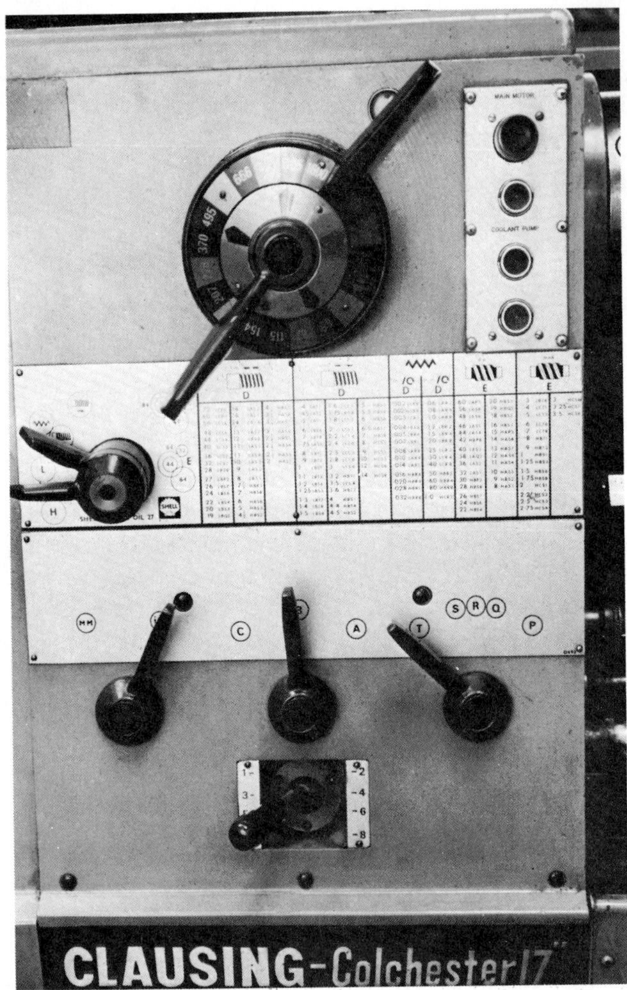

FIGURE I-132 Speed change levers and feed selection levers on a geared head lathe (Lane Community College).

FIGURE I-133 Quick-change gearbox with index plate (Lane Community College).

FIGURE I-134 Exposed quick-change gear mechanism for a large, heavy-duty lathe (Lodge & Shipley Company).

Feed Control Levers

The carriage is moved along the ways by means of the lead screw when threading, or by a separate feed rod when using feeds. On most small lathes, however, a leadscrew–feed rod combination is used. In order to make left-hand threads and reverse the feed, the feed reverse lever is used. This lever reverses the lead screw. It should never be moved when the machine is running.

The quick-change gearbox (Figures I-133 and I-134) typically has two or more sliding gear shifter levers. These are used to select feeds or threads per inch. On those lathes also equipped with metric selections, the threads are expressed in pitches (measured in millimeters). The carriage apron (Figure I-135) contains the handwheel for hand feeding and a power feed lever that engages a clutch to a gear drive train in the apron.

Hand feeding should not be used for long cuts as there would be lack of uniformity and a poor finish would result. When using power feed and approaching a shoulder or the chuck jaws, disengage the power feed and hand feed the carriage for the last $\frac{1}{8}$ in. or so. The handwheel is used to

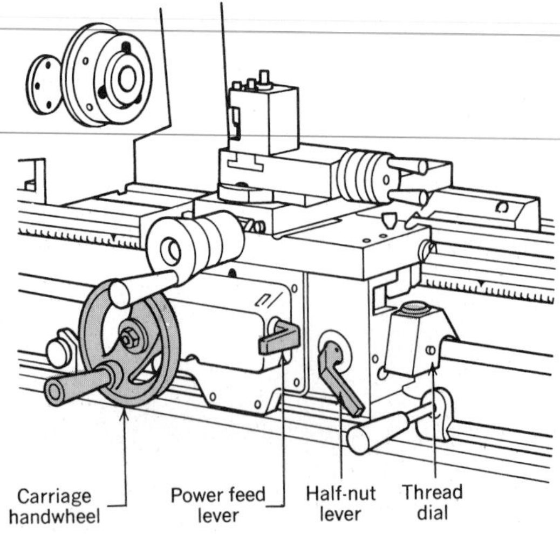

FIGURE I-135 View of carriage apron with names of parts (Clausing Corporation).

FIGURE I-136 Facing on a lathe. The guard is in place at left (Lane Community College).

FIGURE I-137 Micrometer collar on the crossfeed screw that is graduated in English units. Each division represents .001 in. (Lane Community College).

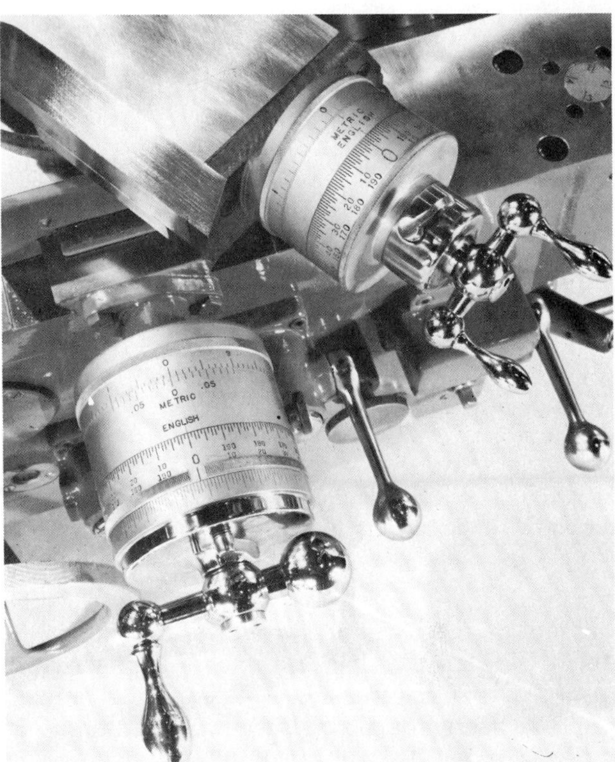

FIGURE I-138 Crossfeed and compound screw handles with metric–English conversion collars (the Monarch Machine Tool Company, Sidney, Ohio).

quickly bring the tool close to the work before engaging the feed and for rapidly returning to the start of a cut after disengaging the feed. A feed change lever diverts the feed either to the carriage for longitudinal movement or to the crossfeed screw to move the cross slide. There is generally some slack or backlash in the crossfeed and compound screws. As long as the tool is being fed to one direction against the work load, there is no problem, but if the screw is **slightly** backed off, the readings will be in error. Because of the backlash, no actual movement of the cross slide will take place even though the micrometer dial reading has been changed. When it is necessary to back the tool away from the cut a few thousandths of an inch, back off two full turns and then come back to the desired position on the micrometer dial.

Crossfeeds are geared differently than longitudinal feeds. On most lathes the crossfeed is approximately one-third to one-half that of the longitudinal feed, so a facing job (Figure I-136) with the quick-change gearbox set at about .012 in. feed would actually only be .004 in. for facing. The crossfeed ratio for each lathe is usually found on the quick-change gearbox index plate.

The half-nut or split-nut lever on the carriage engages the thread on the lead screw directly and is used **only** for threading. It cannot be engaged unless the feed change lever is in the neutral position.

Both the crossfeed screw handle and the compound rest feed screw handle are fitted with micrometer collars (Figure I-137). These collars traditionally have been graduated in English units, but metric conversion collars (Figure I-138) will help in transition to the metric system.

Some micrometer collars are graduated to read single depth; that is, the tool moves as much as the reading shows. When turning a cylindrical object such as a shaft, dials that read single depth

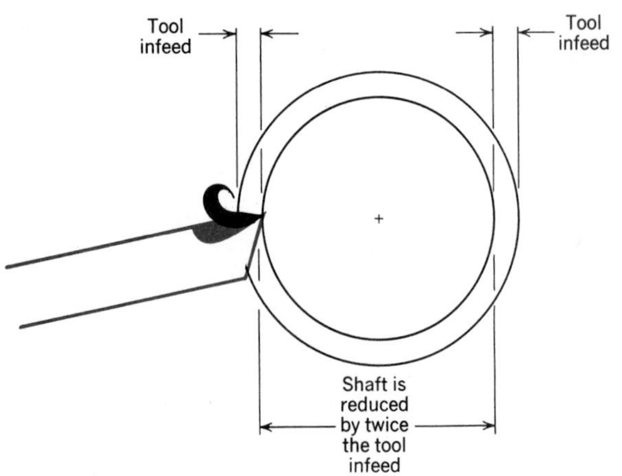

FIGURE I-139 The diameter of the workpiece is reduced twice the amount in which the tool is moved.

will remove twice as much from the diameter (Figure I-139). For example, if the crossfeed screw is turned in .020 in. and a cut is taken, the diameter will have been reduced by .040 in. Sometimes only the compound is calibrated in this way. Many lathes, however, are graduated on the micrometer dial to compensate for double depth on cylindrical turning. On this type of lathe, if the crossfeed screw is turned in until .020 in. shows on the dial and a cut is taken, the diameter will have been reduced .020 in. The tool would have actually moved into the work only .010 in. This is sometimes called radius and diameter reduction.

To determine which type of graduation you are using, set a fractional amount on the dial (such as .250 in. = $\frac{1}{4}$ in.) and measure on the cross slide with a rule. The actual slide movement you measure with the rule will be either the same as the amount set on the dial, for the single depth collar, or one half that amount, for the double depth collar.

Some lathes have a brake and clutch rod that is the same length as the leadscrew. A clutch lever connected to the carriage apron rides along the clutch rod (Figure I-140). The spindle can be started and stopped without turning off the motor by using the clutch lever. Some types also have a spindle brake that quickly stops the spindle when the clutch lever is moved to the stop position. An adjustable automatic clutch kickout is also a feature of the clutch rod.

When starting a lathe for the first time use the following checkout list:

STEP 1. Move carriage and tailstock to the right to clear work-holding device.

STEP 2. Locate feed clutches and half-nut lever and disengage before starting spindle.

FIGURE I-140 Clutch rod is actuated by moving the clutch lever. This disengages the motor from the spindle (Lane Community College).

STEP 3. Set up to operate at low speeds.

STEP 4. Read any machine information panels that may be located on the machine and observe precautions.

STEP 5. Note the feed direction; there are no built-in travel limits or warning devices to prevent feeding into the chuck or against the end of the slides.

STEP 6. When you are finished with a lathe, disengage all clutches, clean up chips, and remove any attachments or special setups.

SELF-TEST

1. How can the shift to the low speed range be made on the belt-drive lathes that have the step cone pulley?

2. Explain speed shifting procedure on the variable-speed drive.

3. In what way can speed changes be made on gear-head lathes?

4. What lever is shifted in order to reverse the leadscrew?

5. The sliding gear shifter levers on the quick-change gearbox are used for just two purposes. What are they?

6. When is the proper time to use the carriage handwheel?

7. Why will you not get the same surface finish (tool marks per inch) on the face of a workpiece as you would get on the outside diameter when on the same power feed?

8. The half-nut lever is not used to move the carriage for turning. Name its only use.

9. Micrometer collars are attached on the crossfeed handle and compound handle. In what ways are they graduated?

10. How can you know if the lathe you are using is calibrated for single or double depth?

UNIT 6

Facing and Center Drilling

Facing and center drilling the workpiece are often the first steps taken in a turning project to produce a stepped shaft or a sleeve from solid material. Much lathe work is done in a chuck requiring considerable facing and some center drilling. These important lathe practices will be covered in detail in this unit.

Objectives

After completing this unit, you should be able to:

1. Correctly set up a workpiece and face the ends.
2. Correctly center drill the ends of a workpiece.
3. Determine the proper feeds and speeds for a workpiece.
4. Explain how to set up to make facing cuts to a given depth and how to measure them.

Setting up for Facing

Facing is done to obtain a flat surface on the end of cylindrical workpieces or on the face of parts clamped in a chuck or face plate (Figures I-141 and I-142). The work most often is held in a three- or four-jaw chuck. If the chuck is to be removed from the lathe spindle, a lathe board must first be placed on the ways. Figure I-143 shows a camlock-mounted chuck being removed. The correct procedure for installing a chuck on a camlock spindle nose is shown in Figures I-144 to I-149. The cams should be tightly snugged (Figure I-149) for one or two revolutions around the spindle.

Setting up work in an independent chuck is simple, but mastering the setup procedures takes

FIGURE I-141 Facing a workpiece in a chuck (Lane Community College).

FIGURE I-142 Facing the end of a shaft (Lane Community College).

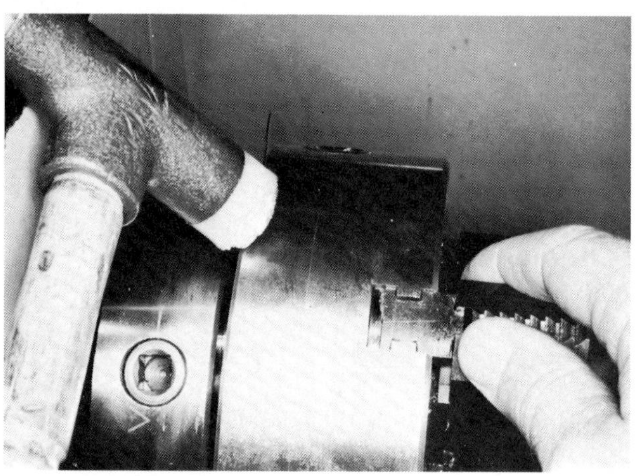

FIGURE I-143 Removing a camlock chuck that is mounted on a lathe spindle.

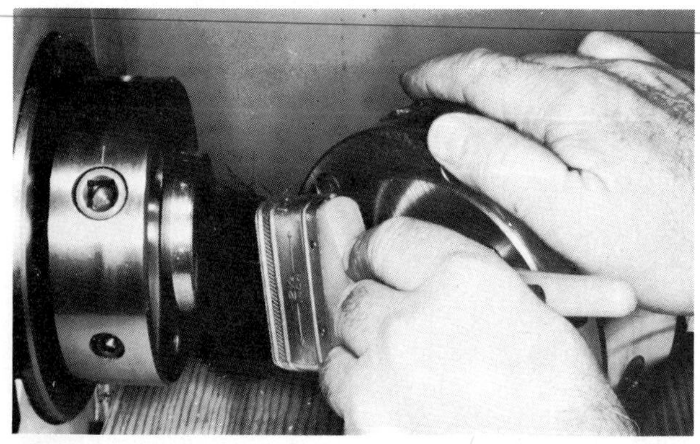

FIGURE I-144 Chips are cleaned from spindle nose with a brush.

some practice. Round stock can be set up by using a dial indicator (Figure I-150). Square or rectangular stock can be set up either with a dial indicator or by using a toolholder turned backward (Figures I-151 and I-152).

FIGURE I-145 Cleaning the chips from the chuck with a brush.

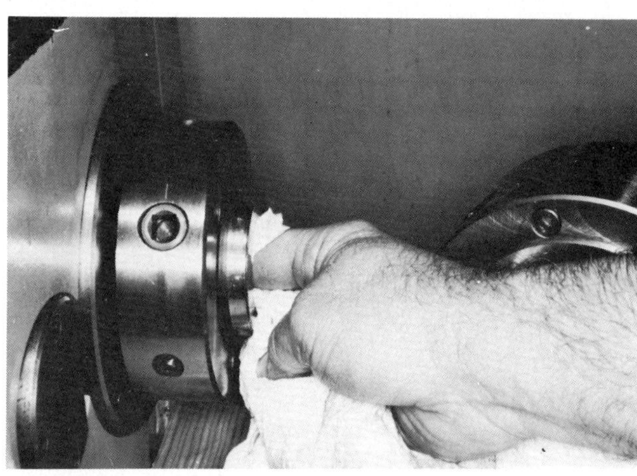

FIGURE I-146 Spindle nose is thoroughly cleaned with a soft cloth.

FIGURE I-147 Chuck is thoroughly cleaned with a soft cloth.

Begin the setup by aligning two opposite jaws with the same concentric ring marked in the face of the chuck while the jaws are near the workpiece. This will roughly center the work. Set up the other two jaws with a concentric ring also when they are

FIGURE I-148 Chuck is mounted on spindle nose.

FIGURE I-149 All cams are turned clockwise until locked securely.

FIGURE I-150 Setting up round stock in an independent chuck with a dial indicator (Lane Community College).

FIGURE I-151 Rectangular stock being set up by using a toolholder turned backward. The micrometer dial is used to center the workpiece (Lane Community College).

FIGURE I-152 Adjusting the rectangular stock at 90 degrees from Figure I-151 (Lane Community College).

near the work. Next, bring all of the jaws firmly against the work. When using the dial indicator, zero the bezel at the lowest reading. Now rotate the chuck to the opposite jaw with the high reading and tighten it half the amount of the runout. It may be necessary to loosen the jaw on the low side slightly. Always tighten the jaws at the position where the dial indicator contacts the work since any other location will give erroneous readings. When using the back of the toolholder, the micrometer dial on the cross slide will show the difference in runout. Chalk is sometimes used for setting up rough castings and other work too irregular to be measured with a dial indicator. Work-

FIGURE I-153 Normal chucking position (Lane Community College).

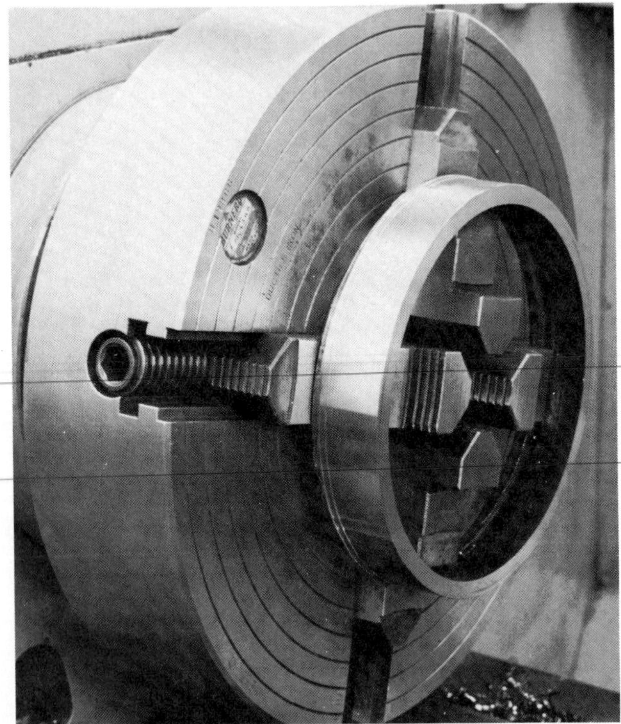

FIGURE I-154 Internal chucking position (Lane Community College).

pieces can be chucked either normally, internally, or externally (Figures I-153 to I-155).

FIGURE I-155 External chucking position (Lane Community College).

Facing

The material to be machined usually has been cut off in a power saw, so the piece is not square on the end or cut to the specified length. Facing from the center out (Figure I-156) produces a better finish, but it is difficult to cut on a solid face in the center. Facing from the outside (Figure I-157) is more convenient since heavier cuts may be taken and it is easier to work to the scribed lines on the circumference of the work. When facing from the center out, a right-hand turning tool in a left-hand toolholder is the best arrangement, but when facing from the outside to the center, a left-hand tool in a right-hand or straight toolholder can be used. Facing or other tool machining should not be done on workpieces extending more than five diameters from the chuck jaws.

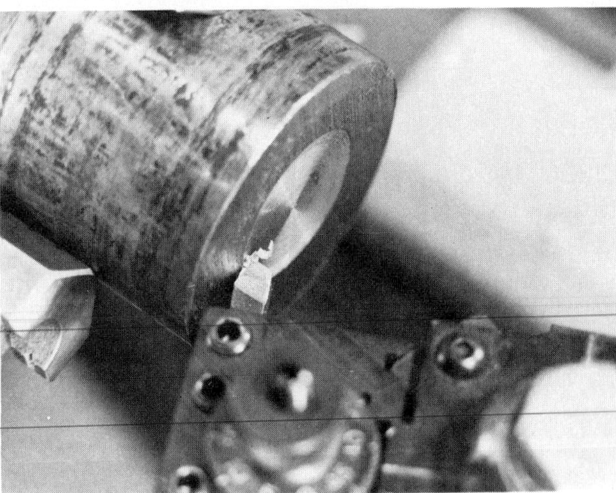

FIGURE I-156 Facing from the center to the outside of a workpiece (Lane Community College).

FIGURE I-157 Facing from the outside toward the center of the workpiece (Lane Community College).

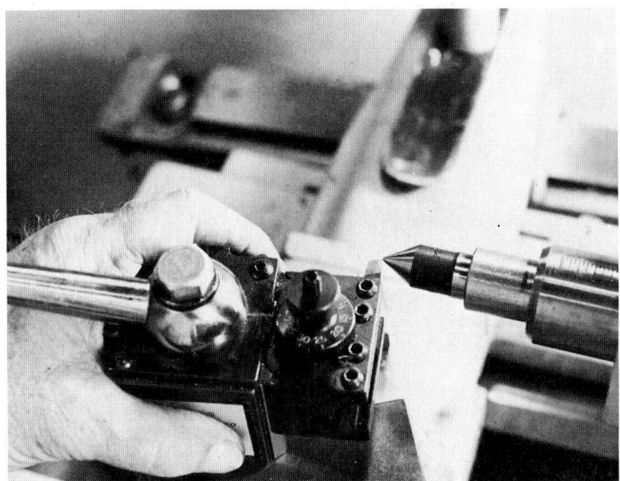

FIGURE I-158 Setting the tool to the center of the work using the tailstock center (Lane Community College).

The point of the tool should be set to the dead center (Figure I-158). This is done by setting the tool to the tailstock center point or by making a trial cut to the center of the work. If the tool is below center, a small uncut stub will be left. The tool can then be reset to the center of the stub.

The carriage must be locked (Figure I-159) when taking facing cuts as the cutting pressure can cause the tool and carriage to move away, which would make the faced surface curved rather than flat. Finer feeds should be used for finishing than for roughing. Remember, the cross feed is one half to one third that of the longitudinal feed. The ratio is usually listed on the index plate of the quick-change gearbox. A roughing feed could be from .005 to .015 in. and a finishing feed from .003 to .005 in. Use of cutting oils will help produce better finishes on finish facing cuts.

Facing to length may be accomplished by trying a cut and measuring with a hook rule (Figure I-160) or by facing to a previously made layout line. A more precise method is to use the graduations on the micrometer collar of the compound. The compound is set so its slide is parallel to the ways (Figures I-161 and I-162). The carriage is

FIGURE I-159 Carriage must be locked before taking a facing cut (Lane Community College).

FIGURE I-160 Facing to length using a hook rule for measuring (Lane Community College).

FIGURE I-161 The compound set at 90 degrees for facing operations (Lane Community College).

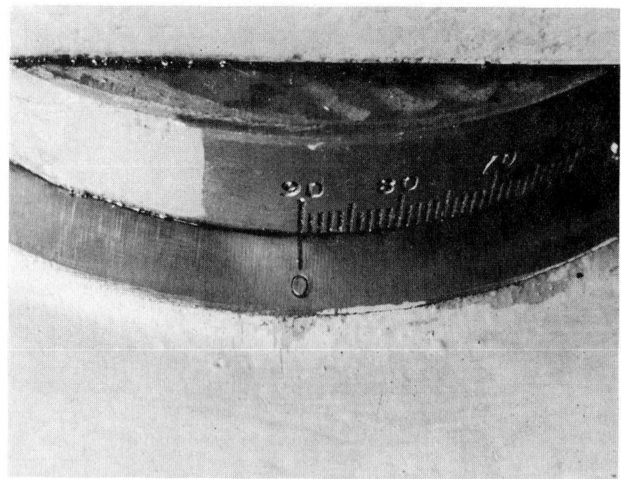

FIGURE I-162 Close-up of the compound set at 90 degrees (Lane Community College).

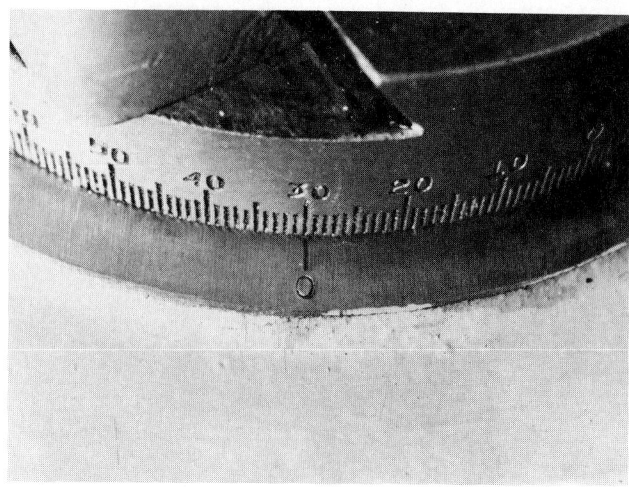

FIGURE I-164 Close-up of the compound set at 30 degrees (Lane Community College).

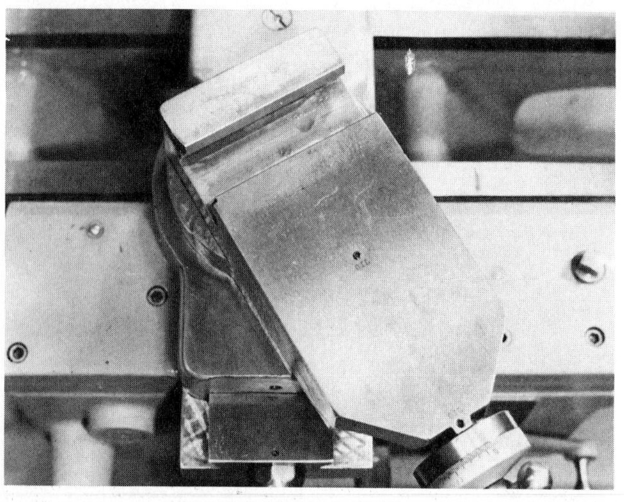

FIGURE I-163 The compound set at 30 degrees (Lane Community College).

FIGURE I-165 Turning and facing an offset with a tool that cuts on both the end and the side (Lane Community College).

locked in place and a trial cut is taken with the micrometer collar set on zero index. The workpiece is measured with a micrometer and the desired length is subtracted from the measurement; the remainder is the amount you should remove by facing. If more than .015 to .030 in. (depth left for finish cut) has to be removed, it should be taken off in two or more cuts by moving the compound micrometer dial the desired amount. A short trial cut (about $\frac{1}{8}$ in.) should again be taken on the finish cut and adjustment made if necessary. Roughing cuts should be approximately .060 in. in depth, but they can vary considerably, depending on machine size, horsepower, tooling, and setup.

Quite often the compound is kept at 30 degrees for threading purposes (Figures I-163 and I-164). It is convenient to know that at this angle the tool feeds into the face of the work .001 in. for every .002 in. that the slide is moved. For example, if you wanted to remove .015 in. from the workpiece, you would turn in .030 in. on the micrometer dial (assuming it reads single depth).

Turning to size and facing on a shoulder or offset requires a tool that can cut on both the end and the side (Figure I-165). Roughing should be done on both the diameter and face before finishing to size. The diameter is usually the critical dimension and should therefore be finished to size after the face is finished.

A specially ground tool is used to face the end of a workpiece that is mounted between centers. The right-hand facing tool is shaped to fit in the angle between the center and the face of the workpiece. Half centers (Figure I-166) are made to make

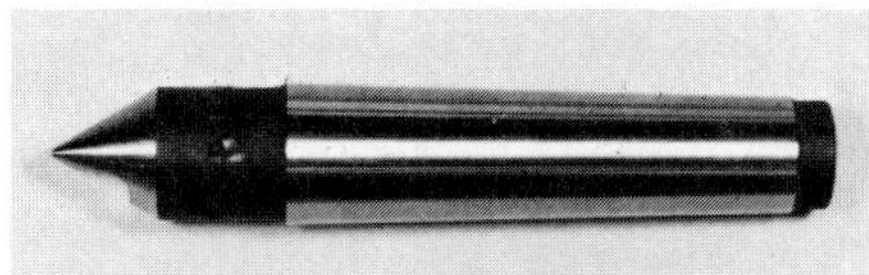

FIGURE I-166 Half centers make facing shaft ends easier (Lane Community College).

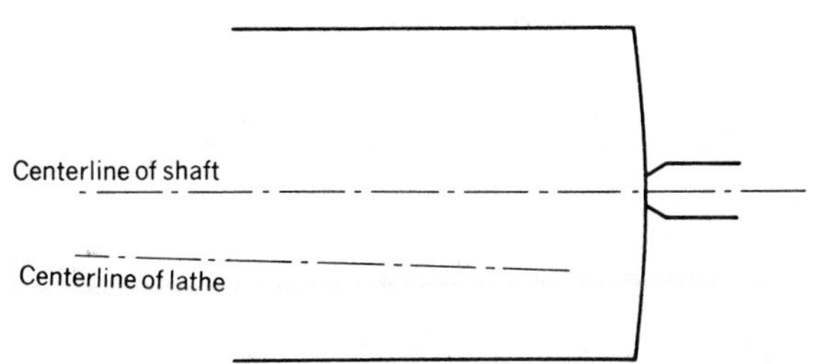

FIGURE I-167 Convex shaft ends caused by the tailstock being moved offcenter away from the operator.

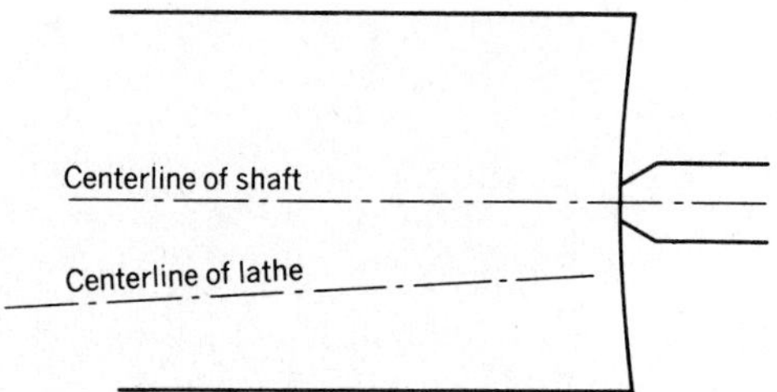

FIGURE I-168 Concave shaft ends caused by the tailstock being moved toward the operator.

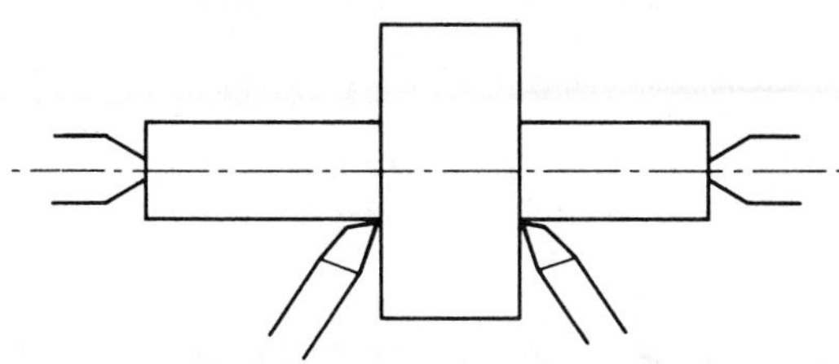

FIGURE I-169 Work that is held between centers on a mandrel can be faced on both sides with right-hand and left-hand facing tools.

the job easier, but they should be used only for facing, and not for general turning. If the tailstock is moved off center away from the operator, the shaft end will be convex; if it is moved toward the operator, it will be concave (Figures I-167 and I-168). Both right-hand and left-hand facing tools are used for facing work held on mandrels (Figure I-169). Care should be taken when machining pressure is toward the small end of a tapered mandrel (usually toward the tailstock). Excessive pressure may loosen the workpiece on the mandrel.

Speeds

Speeds (rpm) for lathe turning a workpiece are determined in essentially the same way as speeds for drilling tools. The only difference is that the diameter of the work is used instead of the diameter of the drill. In facing operations, the outside diameter of the workpiece has greater surface speed than its center. For this reason the rpm should vary as the tool is moved in or out. This is easily done with a variable speed control on the machine, but when belts or gears must be changed, the feed must be disengaged and the lathe stopped. Two or three speed changes on facing work may be required in order to get a uniform surface finish, depending on the size of the work. For facing work, the outside diameter is always used to determine maximum rpm. Thus

$$\text{rpm} = \frac{\text{CS} \times 4}{D}$$

where D = diameter of workpiece (where machining is done)

rpm = revolutions per minute

CS = cutting speed (surface feet per minute)

EXAMPLE 1

The cutting speed for low-carbon steel is 90 sfm (surface feet per minute) and the workpiece diameter to be faced is 6 in. Find the correct rpm.

$$\text{rpm} = \frac{90 \times 4}{6} = 60$$

EXAMPLE 2

A center drill has a $\frac{1}{8}$-in. drill point. Find the correct rpm to use on low-carbon steel (CS 90).

$$\text{rpm} = \frac{90 \times 4}{\frac{1}{8}} = \frac{360}{1} \times \frac{8}{1} = 2880$$

These are only approximate speeds and will vary according to the conditions. If chatter marks (vibration marks) appear on the workpiece, the rpm should be reduced. If this does not help, ask your instructor for assistance. For more information on speeds and feeds, see the next unit, "Turning Between Centers."

FIGURE I-170 Center drilling a workpiece held in a chuck (Lane Community College).

FIGURE I-171 Center drilling long material that is supported in a steady rest (Lane Community College).

Center Drills and Drilling

When work is held and turned between centers, a center hole is required on each end of the workpiece. The center hole must have a 60-degree angle to conform to the center and have a smaller drilled hole to clear the center's point. This center hole is made with a center drill, often referred to as a combination drill and countersink. These drills are available in a range of sizes from $\frac{1}{8}$ to $\frac{3}{4}$ in. body diameter and are classified by numbers from 00 to 8, which are normally stamped on the drill body. For example, a number 3 center drill has a $\frac{1}{4}$-in. body diameter and a $\frac{7}{64}$-in. drill diameter. Facing the workpiece is almost always necessary before center drilling because an uneven surface can push sideways on the fragile center drill point and break it.

Center drills are usually held in a drill chuck in the tailstock, while the workpieces are most often supported and turned in a lathe chuck for center drilling (Figure I-170). As a rule, center holes are drilled by rotating the work in a lathe chuck and feeding the center drill into the work by means of a tailstock spindle. Long workpieces, however, are generally faced by chucking one end and supporting the other in a steady rest (Figure I-171). Since the end of stock is not always sawed square, it should be center drilled only after spotting a small hole with the lathe tool. A slow feed is needed to protect the small, delicate drill end. Cutting fluid should be used and the drill should be backed out frequently to remove chips. The greater the work diameter and the heavier the cut, the larger the center hole should be. Table I-4 gives the center drill size that should be used.

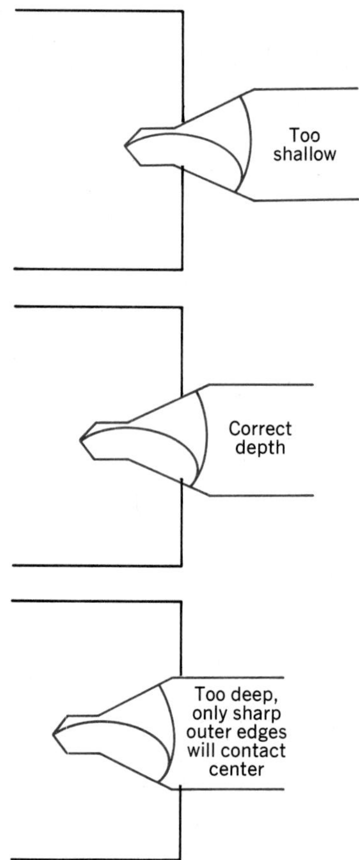

FIGURE I-172 Correct and incorrect depth for center drilling.

The size of the center hole can be selected by the center drill size and then regulated to some extent by the depth of drilling. You must be careful not to drill too deeply (Figure I-172), as this causes the center to contact only the sharp outer edge of the hole, which is a poor bearing surface. It soon becomes loose and out of round and causes such machining problems as chatter and roughness. Center drills often are broken from feeding the drill

FIGURE I-173 Center drill is brought up to work and lightly fed into material.

FIGURE I-174 Center drill is fed into work with a slow even feed.

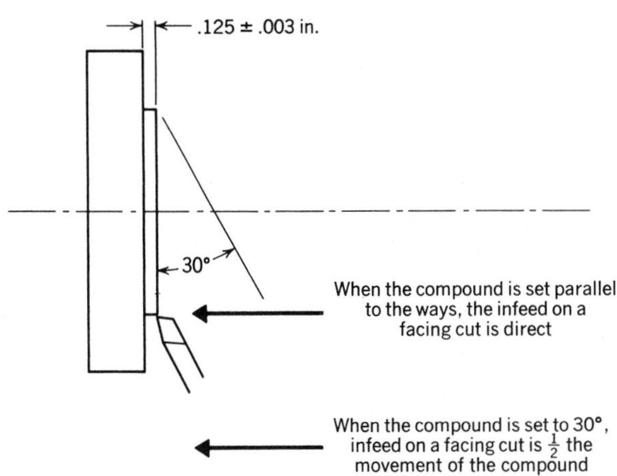
.125 ± .003 in.
30°
When the compound is set parallel to the ways, the infeed on a facing cut is direct
When the compound is set to 30°, infeed on a facing cut is $\frac{1}{2}$ the movement of the compound

FIGURE I-175

TABLE I-4 Center Drill Sizes

Center Drill Number	Drill Point Diameter (in.)	Body Diameter (in.)	Work Diameter (in.)
1	$\frac{3}{64}$	$\frac{1}{8}$	$\frac{3}{16}$ to $\frac{5}{16}$
2	$\frac{5}{64}$	$\frac{3}{16}$	$\frac{3}{8}$ to $\frac{1}{2}$
3	$\frac{7}{64}$	$\frac{1}{4}$	$\frac{5}{8}$ to $\frac{3}{4}$
4	$\frac{1}{8}$	$\frac{5}{16}$	1 to $1\frac{1}{2}$
5	$\frac{3}{16}$	$\frac{7}{16}$	2 to 3
6	$\frac{7}{32}$	$\frac{1}{2}$	3 to 4
7	$\frac{1}{4}$	$\frac{5}{8}$	4 to 5
8	$\frac{5}{16}$	$\frac{3}{4}$	6 and over

too fast with the lathe speed too slow or with the tailstock off-center.

Center drills are often used as starting or spotting drills when a drilling sequence is to be performed (Figures I-173 and I-174). This keeps the drill from "wandering" off the center and making the hole run eccentric. Spot drilling is done when work is chucked or is supported in a steady rest. Care must be taken that the workpiece is centered properly in the steady rest or the center drill will be broken.

SELF-TEST

1. You have a rectangular workpiece that needs a facing operation plus center drilling, and a universal chuck is mounted on the lathe spindle. What is your procedure to prepare for machining?

2. Should the point of the tool be set above, below, or at the center of the spindle axis when taking a facing cut?

3. If you set the quick-change gearbox to .012 in., would that be considered a roughing feed for facing?

4. An alignment step must be machined on a cover plate .125 in., plus or minus .003 in., in depth (Figure I-175). What procedure should be taken to face to this depth? How can you check your final finish cut?

5. What tool is used for facing shaft ends when they are mounted between centers? In what way is this tool different from a turning tool?

6. If the cutting speed of aluminum is 300 sfm and the workpiece diameter is 4 in., what is the rpm? The formula is

$$\text{rpm} = \frac{\text{CS} \times 4}{D}$$

7. Name two reasons for center drilling a workpiece in a lathe.

8. Name two causes of center drill breakage.

9. What happens when you drill too deeply with a center drill?

UNIT 7

Turning Between Centers

Since for a large percentage of lathe work the workpiece is held between centers or between a chuck and a center, turning between centers is a good way for you to learn the basic principles of lathe operation. The economics of machining time and quality will be detailed in this unit, as heavy roughing cuts are compared to light cuts, and speeds and feeds for turning operations are presented.

Objectives

After completing this unit, you should be able to:

1. Describe the correct setup procedure for turning between centers.
2. Select correct feeds and speeds for a turning operation.
3. Detail the steps necessary for turning to size predictably.
4. Turn a $1\frac{1}{4}$-in.-diameter shaft with six shoulders to a tolerance of plus .000 or minus .002 in.

Setup for Turning Between Centers

To turn a workpiece between centers, it is supported between the dead center (tailstock center) and the live center in the spindle nose. A lathe dog (Figure I-176) clamped to the workpiece is driven by a drive or dog plate (Figure I-177) mounted on the spindle nose. Machining with a single-point tool can be done anywhere on the workpiece except near or at the location of the lathe dog.

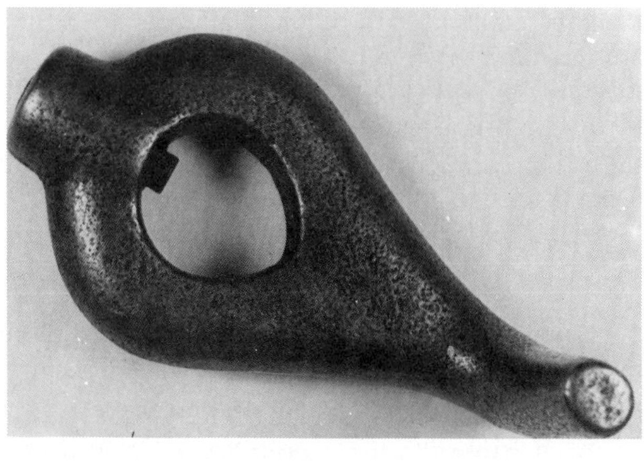

FIGURE I-176 Lathe dog.

Turning between centers has some disadvantages. A workpiece cannot be cut off with a parting tool while being supported between centers as this will bind and break the parting tool and ruin the workpiece. For drilling, boring, or machining the end of a long shaft, a steady rest is normally used to support the work. But these operations cannot very well be done when the shaft is supported only by centers.

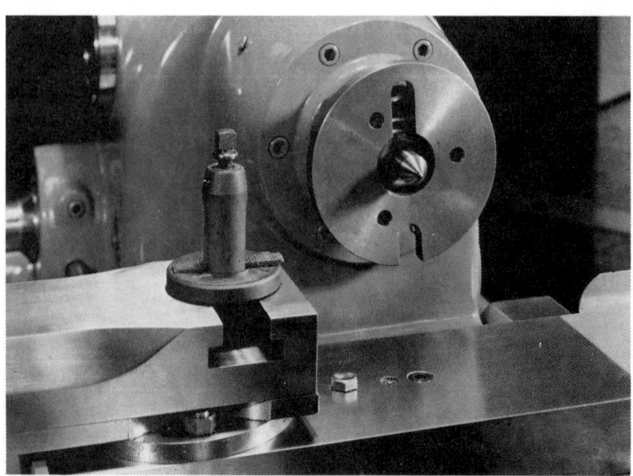

FIGURE I-177 Dog plate or drive plate on spindle nose of the lathe (Monarch Machine Tool Company, Sidney, Ohio).

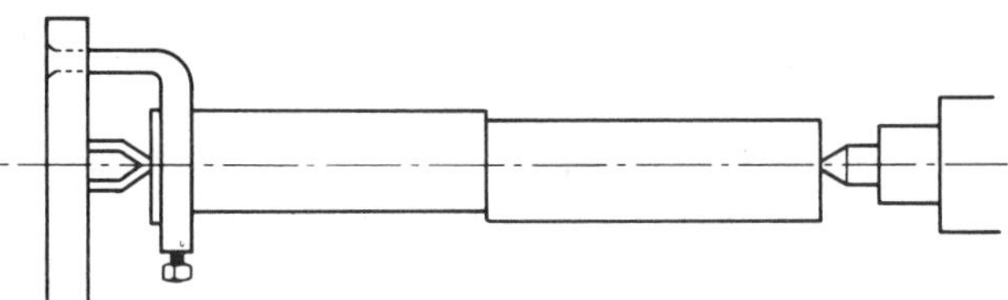

FIGURE I-178 Eccentricity in the center of the part because of the live center being off center.

FIGURE I-179 Make sure that the bushing is firmly seated in the taper (Lane Community College).

FIGURE I-180 Installing the center.

FIGURE I-181 Checking the live center for runout with a dial indicator.

The advantages of turning between centers are many. A shaft between centers can be turned end for end to continue machining without eccentricity if the live center runs true (Figure I-178). This is why shafts that are to be subsequently finish ground between centers must be machined between centers on a lathe. If a partially threaded part is removed from between centers for checking, and everything is left the same on the lathe, the part can be returned to the lathe and the threading resumed where it was left off.

LATHE CENTERS

The center for the headstock spindle has traditionally been called a live center in contrast to the dead center in the tailstock spindle. The live center rotates and the dead center does not. However, later innovations such as ball bearing tailstock centers that do rotate make this terminology confusing if not obsolete. Such ball bearing centers are commonly called live centers even though they are used in a tailstock spindle and not the headstock. To avoid difficulties in terminology in this text, the center in the headstock is referred to as a live center and a nonrotating tailstock center as a dead center. To avoid confusion, a ball bearing center will not be referred to as a live center even though that is common terminology in the shop.

The center for the headstock spindle is usually not hardened since its point frequently needs machining to keep it true. Thoroughly clean the inside of the spindle with a soft cloth and wipe off the live center. If the live center is too small for the lathe spindle taper, use a tapered bushing that fits the lathe. Seat the bushing firmly in the taper (Figure I-179) and install the center (Figure I-180). Set up a dial indicator on the end of the center (Figure I-181) to check for runout. If there is runout, remove the center by using a knockout bar through the spindle. Be sure to catch the center with one hand. Check the outside of the center for nicks or burrs. These can be removed with a file. Check the inside of the spindle taper with your finger for nicks or grit. If nicks are found, **do not** use a file

FIGURE I-182 Live center being machined in a four-jaw chuck. The lathe dog on the workpiece is driven by one of the chuck jaws (Lane Community College).

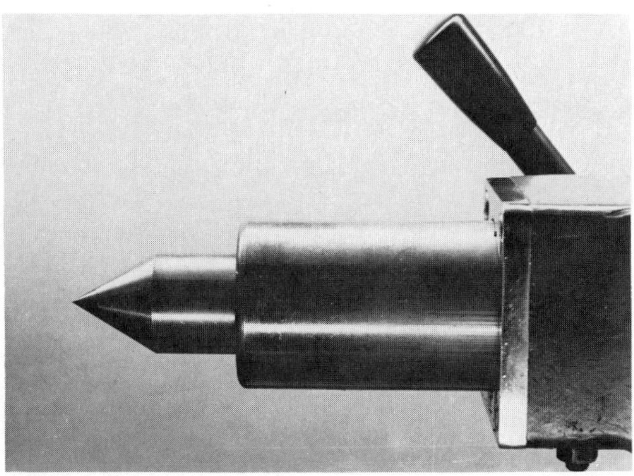

FIGURE I-183 The dead center is hardened to resist wear. It is made of high-speed steel with a carbide insert (California Community Colleges, IMC Project).

FIGURE I-184 Pipe center used for turning (Monarch Machine Tool Company, Sidney, Ohio).

FIGURE I-185 Antifriction ball bearing center (Monarch Machine Tool Company, Sidney, Ohio).

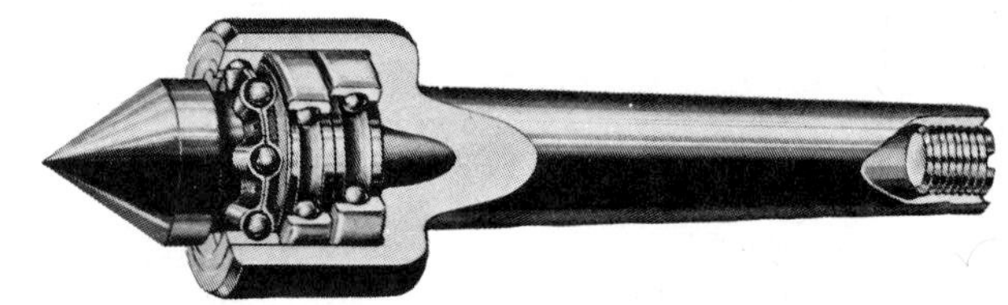

FIGURE I-186 Cutaway view of a ball bearing tailstock center (DoALL Company).

but check with your instructor. After removing nicks, if the center still runs out more than the acceptable tolerances (usually, .0001 to .0005 in.), a light cut by tool or grinding can be taken with the compound set at 30 degrees.

A live center is often machined from a short piece of soft steel mounted in a chuck (Figure I-182). It is then left in place and the workpiece is mounted between it and the tailstock center. A lathe dog with the bent tail against a chuck jaw is used to drive the workpiece. This procedure sometimes saves time on large lathes where changing from the chuck to a drive plate is cumbersome and the amount of work to be done between centers is small.

The tailstock (dead) center (Figure I-183) is hardened to withstand machining pressures and friction. Clean inside the taper and on the center before installing. Ball bearing antifriction centers are often used in the tailstock as they will withstand high-speed turning without the overheating problems of dead centers. Needle bearing and ball bearing centers are used extensively in machine shops and manufacturing operations. Dead centers are virtually obsolete but have the advantage of greater rigidity. Pipe centers are used for turning tubular material (Figures I-184 to I-186).

To set up a workpiece that has been previously center drilled, slip a lathe dog on one end with the bent tail toward the drive plate. Do not tighten the dog yet. Put antifriction compound into the center hole toward the tailstock and then place the workpiece between centers (Figure I-187). The tailstock spindle should not extend out too far as some rigidity in the machine would be lost and chatter or vibration might result. Set the dog in place and avoid any binding of the bent tail (Figure I-188). Tighten the dog and then adjust the tailstock so there is no end play, but so the bent tail of the dog moves freely in its slot. Tighten the tailstock bind-

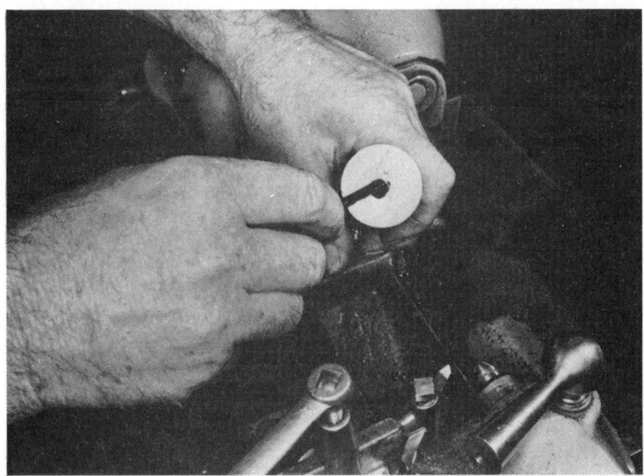

FIGURE I-187 Antifriction compound put in the center hole before setting the workpiece between centers. This step is not necessary when using an antifriction ball bearing center (Lane Community College).

FIGURE I-188 Lathe dog in position (Lane Community College).

FIGURE I-189 Incorrect position of a toolholder for roughing. If the toolholder should turn when using heavy feeds, the tool will gouge more deeply into the work (Lane Community College).

FIGURE I-190 Correct position of toolholder for roughing. The toolholder will swing away from the cut with excessive feeds (Lane Community College).

FIGURE I-191 Tools with excessive overhang. Both tool and toolholder extend too far from the tool post for roughing operations (Lane Community College).

ing lever. The heat of machining will expand the workpiece and cause the dead center to heat from friction. If overheated, the center may be ruined and may even be welded into your workpiece. Periodically, or at the end of each heavy cut, you should check the adjustment of the centers and reset if necessary.

When a tool post and toolholder are used, the toolholder must be positioned so it will not turn into the work when heavy cuts are taken (Figures I-189 and I-190). The tool and toolholder should not overhang too far (Figure I-191) for rough turn-

FIGURE I-192 Tool and toolholder in the correct position (Lane Community College).

FIGURE I-194 Work being turned between chuck and tailstock center. Note the chip formation (Lane Community College).

FIGURE I-193 Centering a tool by means of a steel rule (Lane Community College).

FIGURE I-195 Shoulder turned on a shaft to prevent it from sliding into the chuck (Lane Community College).

ing, but should be kept toward the tool post as far as practical (Figure I-192). Tools should be set on or slightly above the center of the workpiece for roughing and on center for finishing. The tool may be set to the dead center or to a steel rule on the workpiece (Figure I-193). This method should not be used with carbide tools because of their brittleness.

Turning Between Chuck and Center

A considerable amount of straight turning on shafts is done with the workpiece held between a chuck and tailstock center (Figure I-194). The advantages of this method are quick setup and a positive drive. One disadvantage is that eccentricities on the shaft can be caused by inaccuracies of the chuck jaws. Another problem is the tendency for the workpiece to slip endwise into the chuck jaws with heavy cuts, allowing the workpiece to come out of the tailstock center. This can present a very hazardous situation; the workpiece can fly out of the machine, or, if it is long and slender, it can act like a whip. Endwise movement of the shaft can be indicated by making a chalk mark next to the jaws. If the workpiece begins to loosen at the center, quickly shut off the machine and readjust the center. One way to prevent this movement is to first machine a shoulder on the shaft end that will be in the chuck (Figure I-195). Since the shoulder contacts the chuck jaws, the shaft cannot slide endwise into the chuck.

FIGURE I-196 Index chart on the quick-change gearbox (Lane Community College).

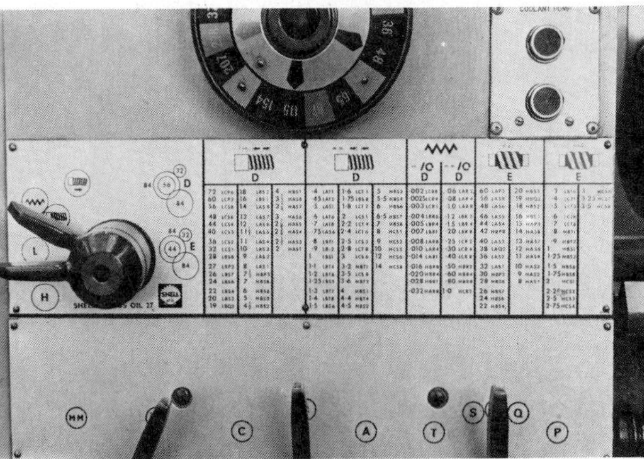

FIGURE I-197 Index chart for feed mechanism on a modern geared head lathe with both metric and inch thread and feed selections (Lane Community College).

Speeds and Feeds for Turning

Since machining time is an important factor in lathe operations, it is necessary for you to fully understand the principles of speeds and feeds in order to make the most economical use of your machine. Speeds are determined for turning between centers by using the same formula given for facing operations in the last unit:

$$\text{rpm} = \frac{\text{CS} \times 4}{D}$$

where rpm = revolutions per minute
D = diameter of workpiece
CS = cutting speed in surface feet per minute (sfm)

Cutting speeds for various materials are given in Table I-5.

EXAMPLE

If the cutting speed is 40 for a certain alloy steel and the workpiece is 2 in. in diameter, find the rpm.

$$\text{rpm} = \frac{40 \times 4}{2} = 80$$

After calculating the rpm, use the nearest or next-lower speed on the lathe and set the speed.

Feeds are expressed in inches per revolution (ipr) of the spindle. A .010-in. feed will move the carriage and tool .010 in. for one full turn of the headstock spindle. If the spindle speed is changed, the feed ratio still remains the same. Feeds are selected by means of an index chart (Figure I-196) found either on the quick-change gearbox or on the side of the headstock housing (Figure I-197). The sliding gear levers are shifted to different positions to obtain the feeds indicated on the index plate. The lower decimal numbers on the plate are feeds and the upper numbers are threads per inch.

Feeds and depth of cut should be as much as the tool, workpiece, or machine can stand without undue stress. The feed rate for roughing should be from one-fifth to one-tenth as much as the depth of cut. A small 10- or 12-in. swing lathe should hande a $\frac{1}{8}$-in. depth of cut in soft steel, but in some cases this may have to be reduced to $\frac{1}{16}$ in. If .100 in. were selected as a trial depth of cut, the feed could be anywhere from .010 to .020 in. If the machine seems to be overloaded, reduce the feed. Do not reduce the depth of cut. Finishing feeds can be from .003 to .005 in. for steel. Use a tool with a larger nose radius for finishing.

TABLE I-5 Cutting Speeds and Feeds for High-Speed Steel Tools

	Low-Carbon Steel	High-Carbon Steel Annealed	Alloy Steel Normalized	Aluminum Alloys	Cast Iron	Bronze
Speed (sfm)						
Roughing	90	50	45	200	70	100
Finishing	120	65	60	300	80	130
Feed (ipr)						
Roughing	.010–.020	.010–.020	.010–.020	.015–.030	.010–.020	.010–.020
Finishing	.003–.005	.003–.005	.003–.005	.005–.010	.003–.010	.003–.010

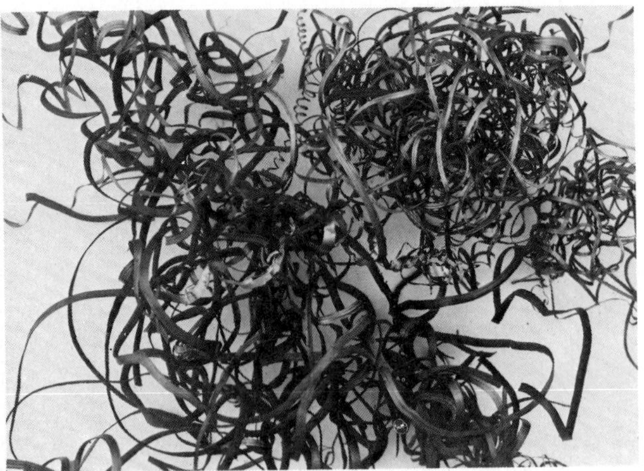

FIGURE I-198 A tangle of wiry chips. These chips can be hazardous to the operator (Lane Community College).

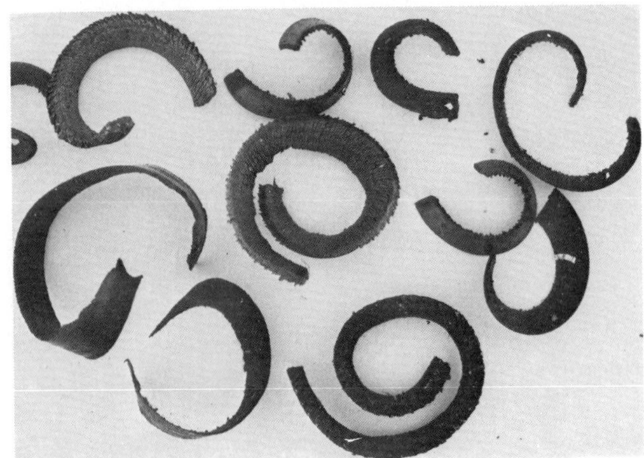

FIGURE I-199 A better formation of chips. This type of chip will fall into the chip pan and is more easily handled (Lane Community College).

Roughing and Finishing

Machining time is an exceedingly important aspect of lathe operations. The time required in finishing operations is generally governed by the surface finish requirements and dimensional tolerance limits. Shortcuts should never be made when finishing to size. Any time saved in machining a workpiece should be during the roughing operations. Coarser feeds, greater depth of cuts, higher rpm (when the tool material can withstand it), and cutting fluids all contribute to higher metal removal rates. The formula to determine time for turning, boring, and facing is

$$T = \frac{L}{fN}$$

where T = time, in minutes

L = length of cut, in inches

f = feed, in ipr

N = lathe spindle speed, in rpm

EXAMPLE 1

A 3-in.-diameter shaft is being turned at 130 rpm with a HS tool. The feed is .020 ipr and the depth of cut is .200 in. The length of cut is 4 in. The time required to make the cut is

$$\frac{4}{.02 \times 130} = 1.54 \text{ min}$$

EXAMPLE 2

The same 3-in.-diameter shaft on the same machine is being turned at 200 rpm with .005 ipr feed. The lathe horsepower limits the depth of cut to .050 in. because of the increased speed. The length of cut is 4 in. The time required to make the cut is

$$\frac{4}{.005 \times 200} = 4 \text{ min}$$

These examples show that light cuts and small feeds often waste time in roughing operations, even when higher cutting speeds are used. Example 1 not only takes less time than Example 2 for a 4-in.-long cut, but it has four times the depth of cut.

CHIP SAFETY

CAUTION **As in other lathe operations, chip formation and handling are important to safety. Coarser feeds, deeper cuts, and smaller rake angles all tend to increase chip curl, which breaks up the chip into small, safe pieces. Fine feeds and shallow cuts, on the other hand, produce a tangle of wiry, sharp hazardous chips (Figures I-198 and I-199) even with a chip breaker on the tool. Long strings may come off the tool, suddenly wrap in the work, and be drawn back rapidly to the machine. The edges are like saws and can cause very severe cuts.**

Turning to Size

The cut-and-try method of turning to size a workpiece, or making a cut and measuring how close you had come to the desired result, was used in the past when calipers and rule were used for measuring work diameters. A more modern method of turning to size predictably uses the compound and

FIGURE I-200 A trial cut is made to establish a setting of a micrometer dial in relation to the diameter of the workpiece (Lane Community College).

FIGURE I-201 Measuring the workpiece with a micrometer (Lane Community College).

cross feed micrometer collars and micrometer calipers for measurement. If the micrometer collar on the cross feed screw reads in single depth, it will remove twice the amount from the diameter of the work as the reading shows. A micrometer collar that reads directly or double depth will remove the same amount from the diameter that the reading shows, though the tool will actually move in only half that amount.

After taking one or several roughing cuts (depending on the diameter of the workpiece), .015 to .030 in. should be left for finishing. This can be taken in one cut if the tolerance is large, such as plus or minus .003 in. If the tolerance is small (plus or minus .0005 in.), two finish cuts should be taken. Between .005 and .010 in. should be left for the last finish cut. If insufficient material is left for machining—.001 in., for example—the tool will rub and will not cut.

The tool is advanced into the work and the first of the two finish cuts is made (Figure I-200). The tool is then returned to the start of the cut without moving the cross feed screw. The diameter of the workpiece is checked with a micrometer (Figure I-201) and the remaining amount to be cut is dialed on the cross feed micrometer dial. A short trial cut is taken (about $\frac{1}{8}$ in. long) and the lathe stopped. A final check with a micrometer is made to validate the tool setting, and then the cut is completed. If the lathe makes a slight taper, see the next unit on "Alignment of the Lathe Centers" to correct this problem.

Finishing of machined parts with a file and abrasive cloth should not be necessary if the tools are sharp and honed and if the feeds, speeds, and depth of cut are correct. A machine-finished part looks better than a part finished with a file and abrasive cloth. In the past, filing and polishing the precision surfaces of lathe workpieces were necessary because of lack of rigidity and repeatability of machines. In the same way, worn lathes are not dependable for close tolerances and so an extra allowance must be made for filing. The amount of surface material left for filing ranges from .0005 to .005 in., depending on the final finish and diameter. If more than the tips of the tool marks are removed with a file and abrasive cloth, a wavy surface will result. Also, stroking with the file causes a cylindrical part to become somewhat out of round. If short, rapid file strokes are used, the part may not only be out of round but it will have flat spots on the surface. The more material removed

FIGURE I-202 Filing in the lathe, left-handed (Lane Community College).

FIGURE I-204 Using a file for backing abrasive cloth for more uniform polishing (Lane Community College).

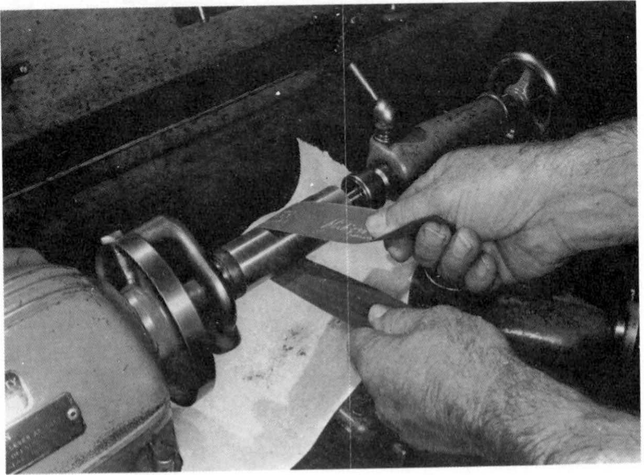

FIGURE I-203 Using abrasive cloth for polishing (Lane Community College).

FIGURE I-205 Measuring the workpiece length to a shoulder with a machinist's rule (Lane Community College).

by filing, the more likely it is that the finish size will not be cylindrical. For most purposes .002 in. is sufficient material to leave for file finishing.

When filing on a lathe, use a low speed and long strokes, and file left-handed (Figure I-202). For polishing with abrasive cloth, set the lathe for a high speed and move the cloth back and forth across the work. Hold an end of the cloth strip in each hand (Figures I-203 and I-204). Because of the damaging effect of abrasive grains on the sliding surfaces of machinery, some shops do not allow abrasive cloth to be used at all. When it is used for polishing in the lathe, the ways and saddle should always be covered with paper. Even if paper is used, abrasive cloth leaves grit on the ways of the lathe, so a thorough cleaning of the ways should be done after polishing.

TURNING TO LENGTH

Machining shoulders to specific lengths can be done in several ways. Using a machinist's rule (Figure I-205) to measure workpiece length to a shoulder is a very common but semiprecision method. Preset carriage micrometer stops (Figure I-206) can be used to limit carriage movement and establish a shoulder. This method can be very accurate if it is set up correctly. The compound rest can also be used to machine a given distance to establish a shoulder. The compound rest is set parallel to the ways and the micrometer collar indicates the distance moved. If the compound slide is not adjusted very accurately, a taper will be cut. Another very accurate means of machining shoulders is by using special dial indicators with long travel plunger rods. These indicators show the longitudinal position of the carriage. Whichever method is used, the power feed should be turned off $\frac{1}{8}$ in. short of the workpiece shoulder and the tool handfed to the desired length. If the tool should be accidentally fed into an existing shoul-

FIGURE I-206 Carriage micrometer stop set to limit tool travel in order to establish a shoulder (Lane Community College).

FIGURE I-207 Tapered mandrel or arbor.

FIGURE I-208 Tapered mandrel and workpiece set up between centers (Clausing Corporation).

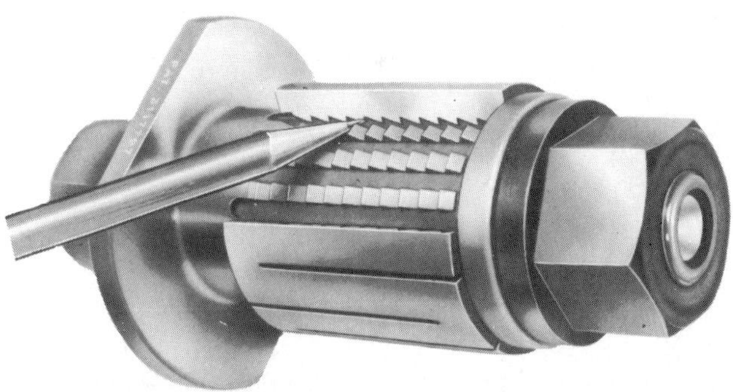

FIGURE I-209 Operating principle of a special type of an expanding mandrel. The "sabertooth" design provides a uniform gripping action in the bore (Buck Tool Company).

der, the feed mechanism may jam and be very difficult to release. A broken tool, toolholder, or lathe part may be the result.

Mandrels

Mandrels, sometimes called lathe arbors, are used to hold work that is turned between centers (Figures I-207 and I-208). Tapered mandrels are made in standard sizes and have a taper of only .006 in./ft. A flat is milled on one end of the mandrel for the lathe dog setscrew. High-pressure lubricant is applied to the bore of the workpiece and the mandrel is pressed into the workpiece with an arbor press. The assembly is mounted between centers and the workpiece is turned or faced on either side. Care must be taken when the feed is toward the small end of the taper. The workpiece may loosen on the mandrel.

Expanding mandrels (Figure I-209) have the advantage of providing a uniform gripping surface

FIGURE I-210 Expanding mandrel and workpiece set up between centers (Lane Community College).

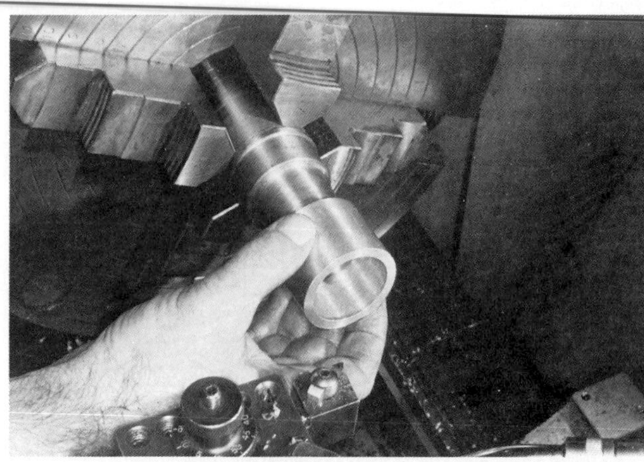

FIGURE I-213 Part to be machined being affixed to the mandrel. The mandrel is oiled so that the part can be easily removed (Lane Community College).

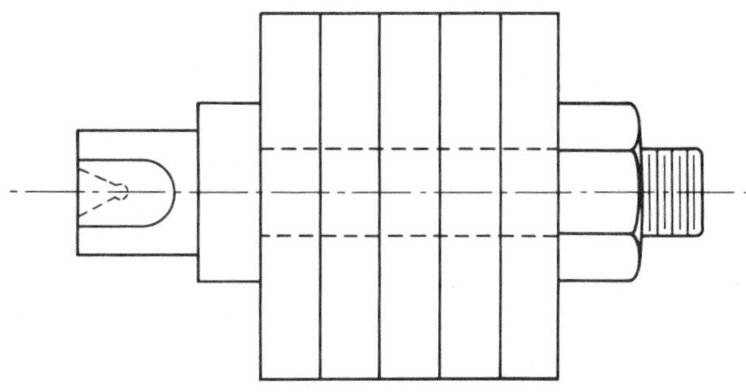

FIGURE I-211 Many similar parts are machined at the same time with a gang mandrel.

FIGURE I-214 Part being machined after assembly on mandrel (Lane Community College).

FIGURE I-212 Stub mandrel being machined to size with a slight taper, about .006 in./ft (Lane Community College).

FIGURE I-215 Special stub mandrel with Adjust-tru® feature (Buck Tool Company).

for the length of the bore (Figure I-210). A tapered mandrel grips tighter on one end than the other. Gang mandrels (Figure I-211) grip several pieces of similar size, such as disks, to turn their circumference. These are made with collars, thread, and nut for clamping.

Stub mandrels (Figures I-212 to I-214) are used in chucking operations. These are often

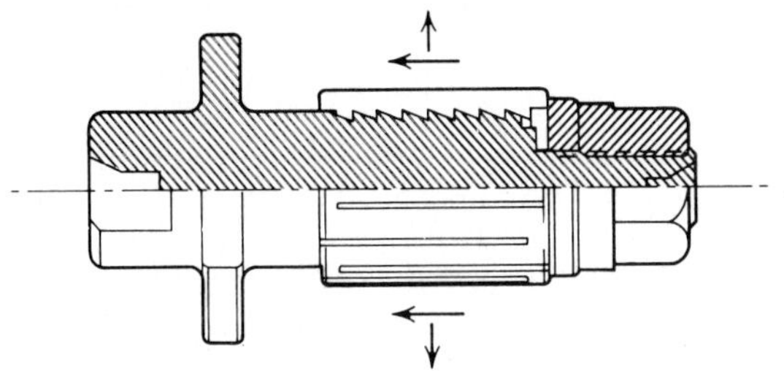

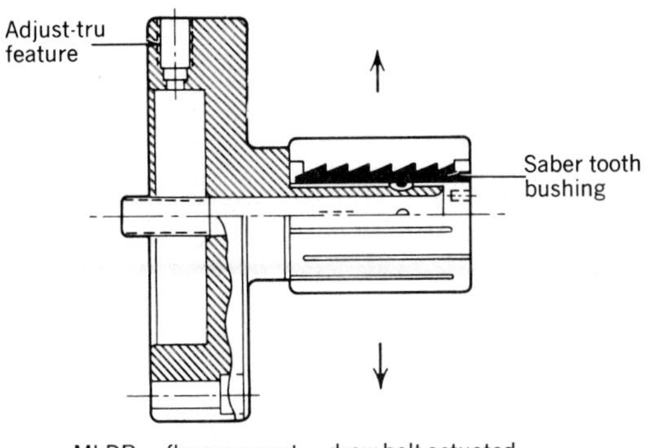

FIGURE I-216 Precision adjustment can be maintained on the expanding stub mandrels with the Adjust-tru® feature on the flange mount (Buck Tool Company).

quickly machined for a single job and then discarded. Expanding stub mandrels (Figures I-215 and I-216) are used when production of many similar parts is carried out. Threaded stub mandrels are used for machining the outside surfaces of parts that are threaded in the bore.

Coolants are used for heavy-duty and production turning. Oil–water emulsions and synthetic coolants are the most commonly used, while sulfurized oils usually are not used for turning operations except for threading. Most job work or single piecework is done dry. Many shop lathes do not have a coolant pump and tank, so, if any cutting fluid is used, it is applied with a pump oil can. Coolants and cutting oils for various materials are given in Table I-6.

SELF-TEST

1. Name two advantages and two disadvantages of turning between centers.

2. What other method besides turning between centers is extensively used for turning shafts and long workpieces that are supported by the tailstock center?

3. What factors tend to promote or increase chip curl so that safer chips are formed?

4. Name three kinds of centers used in the tailstock and explain their uses.

5. How is the dead center correctly adjusted?

6. Why should the dead center be frequently adjusted when turning between centers?

7. Why should you avoid excess overhang with the tool and toolholder when roughing?

8. Calculate the rpm for roughing a $1\frac{1}{2}$-in.-diameter shaft of machine steel.

9. What would the spacing or distance between tool marks on the workpiece be with a .010-in. feed?

10. How much should the feed rate be for roughing?

11. How much should be left for finishing?

12. Describe the procedure in turning to size predictably.

TABLE I-6 Coolants and Cutting Oils Used for Turning

Material	Dry	Water-Soluble Oil	Synthetic Coolants	Kerosene	Sulfurized Oil	Mineral Oil
Aluminum		×	×	×		
Brass	×	×	×			
Bronze	×	×	×			×
Cast iron	×					
Steel						
Low carbon		×	×		×	
Alloy		×	×		×	
Stainless		×	×		×	

UNIT 8

Alignment of the Lathe Centers

As a machinist, you must be able to check a workpiece for taper and properly set the tailstock of a lathe. Without these skills, you may lose much time in futile attempts to restore precision turning between centers when the workpiece has an unintentional taper. This unit will show you several ways to align the centers of a lathe.

Objectives

After completing this unit, you should be able to:

1. Check for taper with a test bar and restore alignment by adjusting the tailstock.
2. Check for taper by taking a cut with a tool and measuring the workpiece and restore alignment by adjusting the tailstock.

The tailstock will normally stay in good alignment on a lathe that is not badly worn. If a lathe has been used for taper turning with the tailstock offset, however, the tailstock may not have been realigned properly (Figure I-217). The tailstock also could be slightly out of alignment if an improper method of adjustment was used. It is therefore a good practice to occasionally check the center alignment of the lathe you usually use and to always check the alignment before using a different lathe.

It is often too late to save the workpiece by realigning centers if a taper is discovered while making a finish cut. A check for taper should be made on the workpiece while it is still in the roughing stage. You can do this by taking a light cut for some distance along the workpiece. Then check the diameter on each end with a micrometer; the difference between the two readings is the amount of taper in that distance.

Four methods are used for aligning centers on a lathe. In one method, the center points are brought together and visually checked for alignment (Figure I-218). This, of course, is not a precision method for checking alignment.

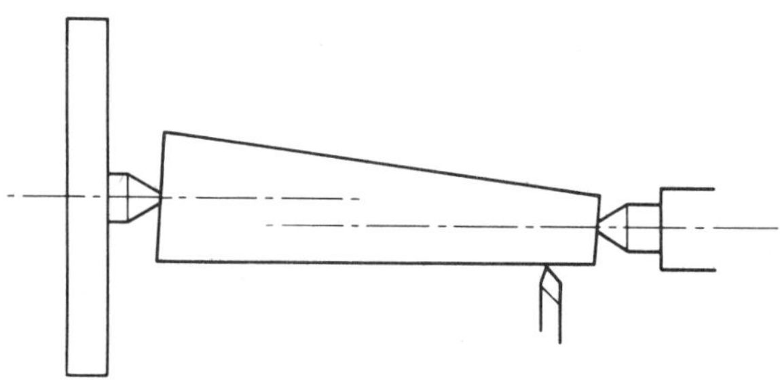

FIGURE I-217 Tailstock out of line causing a tapered workpiece.

FIGURE I-218 Checking alignment by matching center points.

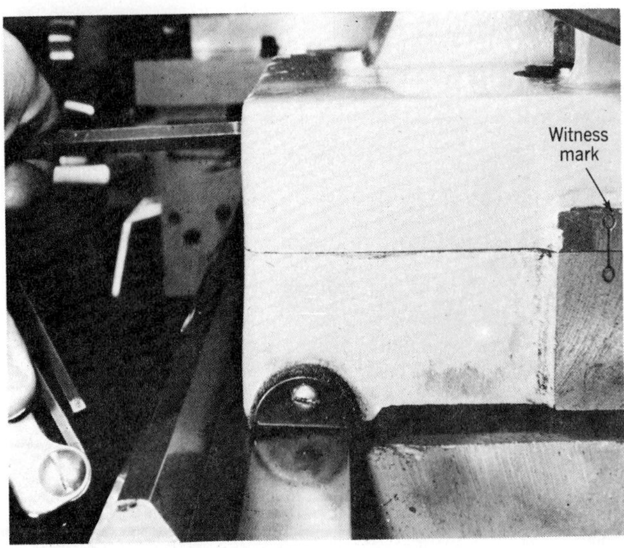

FIGURE I-219 Adjusting the tailstock to the witness marks for alignment.

FIGURE I-220 Hexagonal socket setscrew which, when turned, move the tailstock provided that the opposite one is loosened.

Another method of aligning centers is by using the tailstock witness marks. Adjusting the tailstock to the witness marks (Figure I-219), however, is only an approximate means of eliminating taper. The tailstock is moved by means of a screw or screws. A typical arrangement is shown in Figure I-220, where one set screw is released and the opposite one is tightened to move the tailstock on its slide (Figure I-221). The tailstock clamp bolt must be released before the tailstock is offset.

Two more accurate means of aligning centers are by using a test bar and by machining and measuring. A test bar is simply a shaft that has true centers (is not off center) and has no taper. Some test bars are made with two diameters for convenience. When checking alignment with a test bar, no dog is necessary as the bar is not rotated. A dial

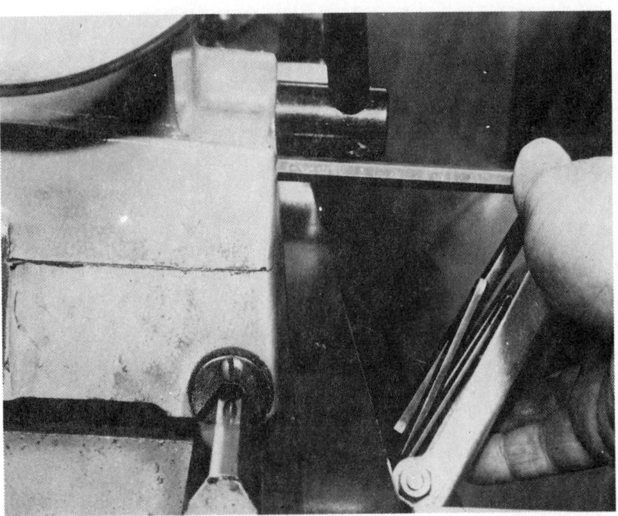

FIGURE I-221 The opposite setscrew being adjusted.

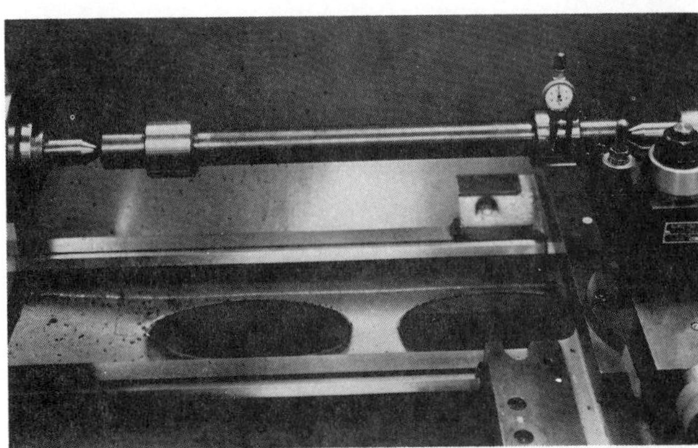

FIGURE I-222 Test bar setup between centers with a dial indicator mounted in the tool post.

FIGURE I-223 Indicator is moved to measuring surface at headstock end and the bezel is set on zero.

indicator is mounted, preferably in the tool post, so it will travel with the carriage (Figure I-222). Its contact point should be on the center of the test bar.

Begin with the indicator at the headstock end, and set the indicator bezel to zero (Figure I-223).

FIGURE I-224 The carriage with the dial indicator is moved to the measuring surface near the tailstock. In this case the dial indicator did not move, so the tailstock is on center.

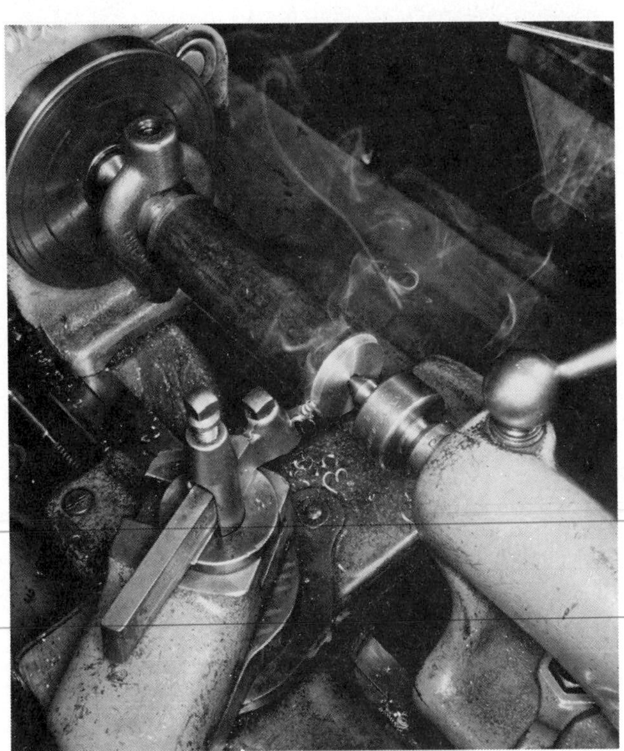

FIGURE I-225 Checking for taper by taking a cut on a workpiece. After the cut is made for the length of the workpiece, a micrometer reading is taken at each end to determine any difference in diameter (Lane Community College).

Now move the setup to the tailstock end of the test bar (Figure I-224) and check the dial indicator reading. If no movement of the needle has occurred, the centers are in line. If the needle has moved clockwise, the tailstock is misaligned toward the operator. This will cause the workpiece to taper with the smaller end near the tailstock. If the needle has moved counterclockwise, the tailstock is too far away from the operator and the workpiece will taper with the smaller end at the headstock. In either case, move the tailstock until both diameters have the same reading.

FIGURE I-226 Using a dial indicator to check the amount of movement of the tailstock when it is being realigned (Lane Community College).

Since usually only a minor adjustment is needed while a job is in progress, the most common method of aligning lathe centers is by cutting and measuring. It is also the most accurate. This method, unlike the bar test method, usually uses the workpiece while it is in the roughing stage (Figure I-225). A light cut is taken along the length of the test piece and both ends are measured with a micrometer. If the diameter at the tailstock end is smaller, the tailstock is toward the operator, and if the diameter at the headstock end is smaller, the tailstock is away from the operator. Set up a dial indicator (Figure I-226) and move the tailstock half the difference of the two micrometer readings. Make another light cut and check for taper.

SELF-TEST

1. What is the result on the workpiece when the centers are out of line?

2. What happens to the workpiece when the tailstock is offset toward the operator?

3. Name three methods of aligning the centers.

4. Which measuring instrument is used when using a test bar?

5. By what means is the measuring done when checking taper by taking a cut?

UNIT 9

Other Lathe Operations

Much of the versatility of the lathe as a machine tool is due to the variety of tools and work-holding devices used. This equipment makes possible the many special operations that you will begin to do in this unit.

Objectives

After completing this unit, you should be able to:

1. Explain the procedures for drilling, boring, reaming, knurling, recessing, parting, and tapping in the lathe.
2. Set up to drill, ream, bore, and tap on the lathe and complete each of these operations.
3. Set up for knurling, recessing, die threading, and parting on the lathe and complete each of these operations.

Drilling

The lathe operations of boring, tapping, and reaming usually begin with spotting and drilling a hole. The workpiece, often a solid material that requires a bore, is mounted in a chuck, collet, or faceplate, while the drill is typically mounted in the tailstock spindle that has a Morse taper. If there is a slot in the tailstock spindle, the drill tang must be aligned with it when inserting the drill.

Drill chucks with Morse taper shanks are used to hold straight shank drills and center drills (Figure I-227). Center drills are used for spotting or making a start for drilling (Figure I-228). When drilling with large size drills, a pilot drill should be used first. If a drill wobbles when started, place the

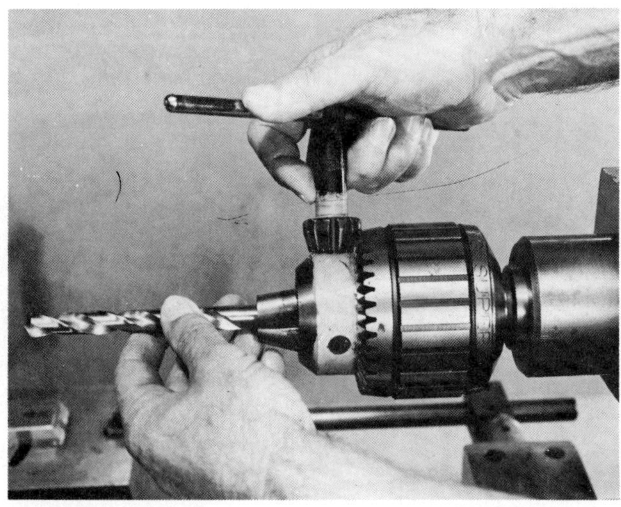

FIGURE I-227 Mounting a straight shank drill in a drill chuck in the tailstock spindle (Lane Community College).

FIGURE I-228 Center drilling is the first step prior to drilling, reaming, or boring.

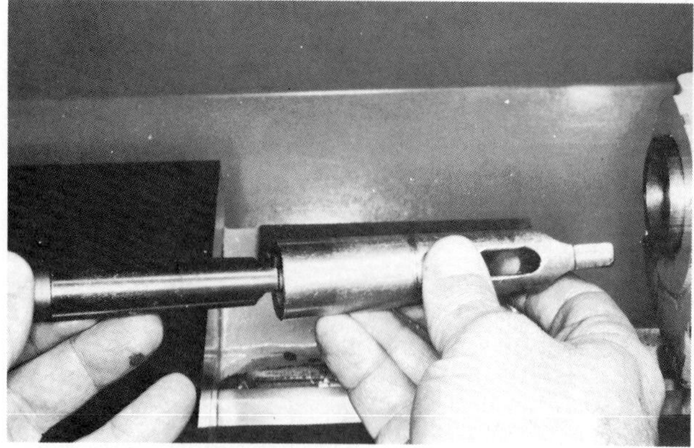

FIGURE I-229 A drill sleeve is placed on the drill so that it will fit the taper in the tailstock spindle.

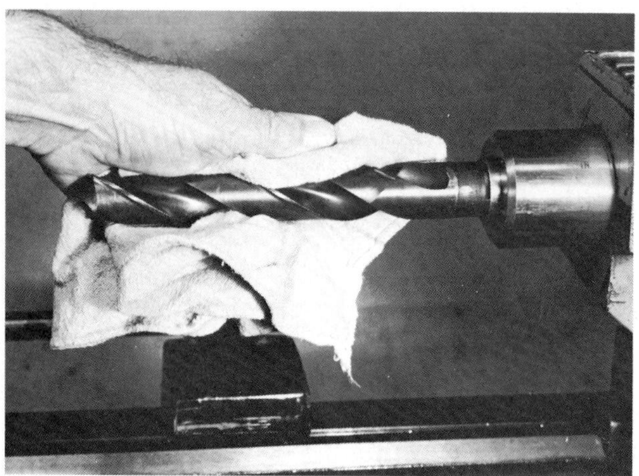

FIGURE I-230 The drill is then firmly seated in the tailstock spindle (Lane Community College).

heel of a toolholder against it near the point to steady the drill while it is starting in the hole. Taper shank drills (Figure I-229) are inserted directly into the tailstock spindle. The friction of the taper is usually all that is needed to keep the drill from turning while a hole is being drilled (Figures I-230 and I-231), but when using larger drills, the friction is not enough. A lathe dog is sometimes clamped to the drill just above the flutes (Figure I-232) with the bent tail resting on the compound. Hole depth can be measured with a rule or by means of the graduations on top of the tailstock spindle. The alignment of the tailstock with the lathe center line should be checked before drilling or reaming.

Drilled holes are not accurate enough for many applications, such as for gear or pulley bores, which should not be made over the nominal size more than .001 to .002 in. Drilling typically produces holes that are oversize and run eccentric to the center axis of the lathe (Figure I-233). This is not true in the case of some manufacturing types

FIGURE I-231 The feed pressure when drilling is usually sufficient to keep the drill seated in the tailstock spindle, thus keeping the drill from turning.

FIGURE I-232 A lathe dog is used when the drill has a tendency to turn in the tailstock spindle (Lane Community College).

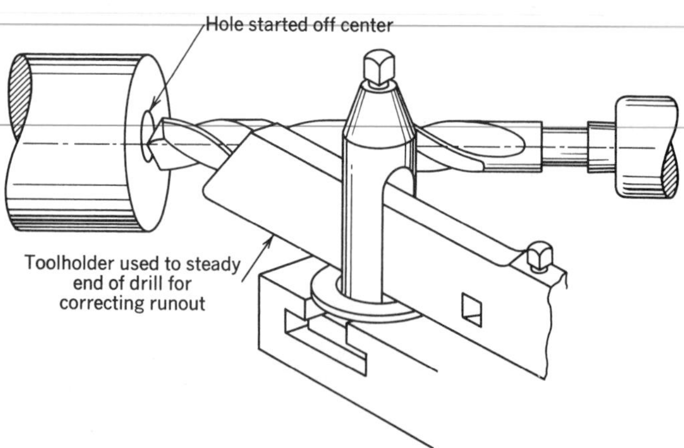

FIGURE I-233 Exaggerated view of the runout and eccentricity that is typical of drilled holes.

such as gun drilling. However, true axial alignment of holes is possible when the work is turned and the drill remains stationary, as in a lathe operation, in comparison to operations where the drill is turned and the work is stationary, as on a

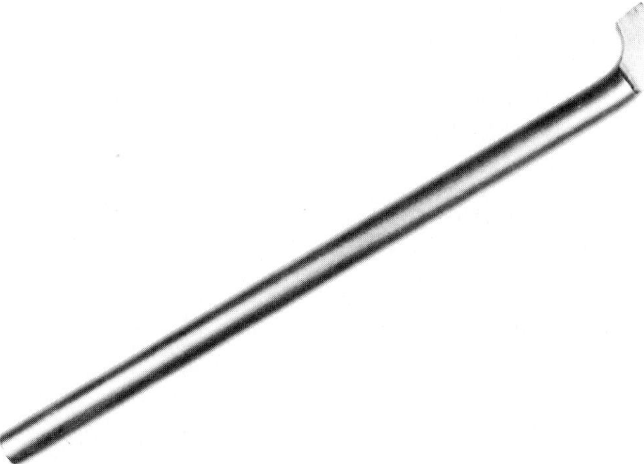

FIGURE I-234 Small forged boring bar made of high-speed steel.

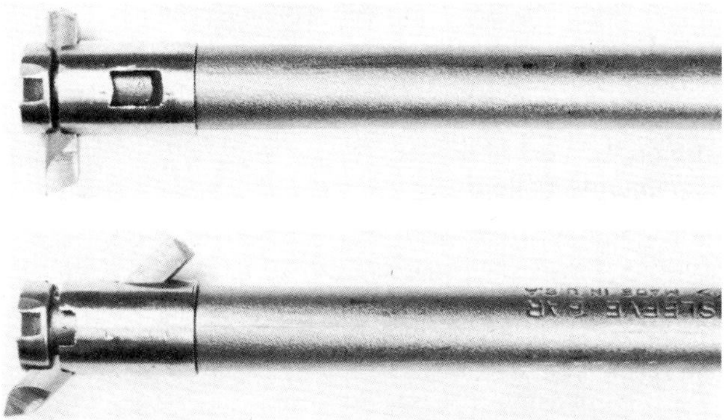

FIGURE I-235 Two boring bars with inserted tools set at different angles.

FIGURE I-236 Boring bar with carbide insert. When one edge is dull, a new one is selected (Kennametal Inc., Latrobe, PA).

FIGURE I-237 Boring bar setup with a large overhang for making a deep bore. It is difficult to avoid chatter with this arrangement.

drill press. Drilling also produces holes with rough finishes, which along with size errors can be corrected by boring or reaming. The hole must first be drilled slightly smaller than the finish diameter in order to leave material for finishing by either of these methods.

Boring

Boring is the process of enlarging and trueing an existing or drilled hole. A drilled hole for boring can be from $\frac{1}{32}$ to $\frac{1}{16}$ in. undersize, depending on the situation. Speeds and feeds for boring are determined in the same way as they are for external turning. Boring to size predictably is also done in the same way as in external turning except that the cross feed screw is turned counterclockwise to move the tool into the work.

An inside spring caliper and a rule are sometimes useful for rough measurement. Vernier calipers are also used by machinists for internal measuring, though the telescoping gage and outside micrometer are most commonly used for the precision measurement of small bores because they can take a more accurate measurement. Inside micrometers can be used for bores over $1\frac{1}{2}$ in. Precision bore gages are used where many bores are checked for similar size, such as for acceptable tolerance.

A boring bar is clamped in a holder mounted on the carriage compound. Several types of boring bars and holders are used. Boring bars designed for small holes ($\frac{1}{2}$ in. and smaller) are usually the forged type (Figure I-234). The forged end is sharpened by grinding. When the bar gets ground too far back, it must be reshaped or discarded. Boring bars for holes with diameters over $\frac{1}{2}$ in. (Figure I-235) use high-speed tool inserts, which are typically hand ground in the form of a left-hand turning tool. These tools can be removed from the bar for resharpening when needed. The cutting tool can be held at various angles to obtain different results, which makes the boring bar useful for many applications. Standard bars generally come with a tool angle of 30, 45, or 90 degrees. Some boring bars are made for carbide inserts (Figure I-236).

Chatter is the vibration between a workpiece and a tool because of the lack of rigid support for the tool. Chatter is a great problem in boring operations since the bar must extend away from the support of the compound (Figure I-237). For this reason boring bars should be kept back into their holders as far as practicable. Tuned boring bars

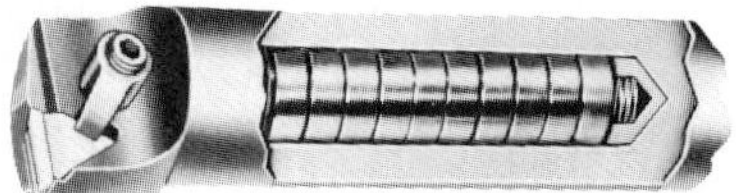

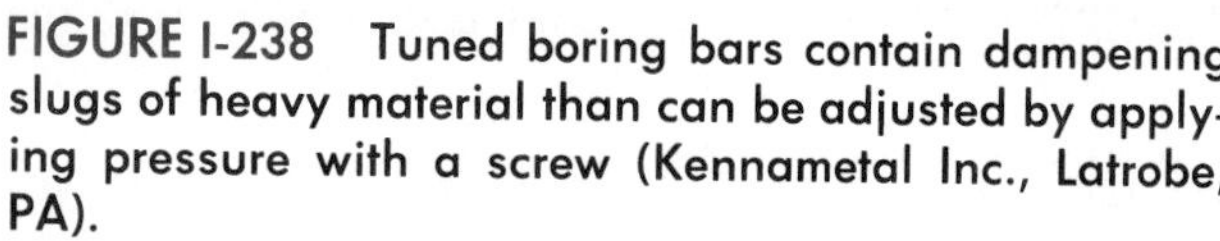

FIGURE I-238 Tuned boring bars contain dampening slugs of heavy material than can be adjusted by applying pressure with a screw (Kennametal Inc., Latrobe, PA).

FIGURE I-239 Boring tools must have sufficient side relief and side rake to be efficient cutting tools. Back rake is not normally used (Lane Community College).

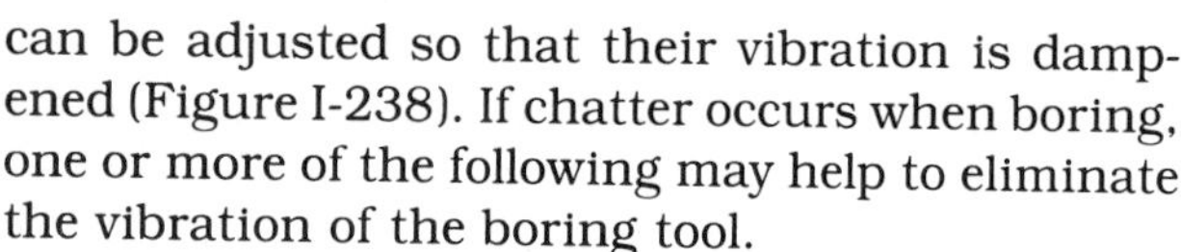

can be adjusted so that their vibration is dampened (Figure I-238). If chatter occurs when boring, one or more of the following may help to eliminate the vibration of the boring tool.

1. Shorten the boring bar overhang, if possible.
2. Increase feed.
3. Make sure that the tool is on center.
4. Reduce the spindle speed.
5. Use a boring bar as large in diameter as possible without it binding in the hole.
6. Reduce the nose radius on the tool.
7. Apply cutting oil to the bore.

Boring bars sometimes spring away from the cut and cause bellmouth, a slight taper at the front edge of a bore. One or two extra cuts (called free cuts) taken without increasing the infeed will usually eliminate the problem. A large variety of boring bar holders is used. Some types are designed for small, forged bars while others of more rigid construction are used for larger, heavier work.

Boring tools are made with side relief and end relief, but usually with zero back rake (Figure I-239). Insufficient end relief will allow the heel of the tool to rub on the workpiece. Figures I-240 to I-244 are views of bores that are looking outward from inside the chuck. The end relief should be between 10 and 20 degrees. The machinist must use judgment when grinding the end relief because the larger the bore, the less end relief is required. If the end of the tool is relieved too much, the cutting edge will be weak and break down.

FIGURE I-240 A tool with insufficient end relief will rub on the heel of the tool (arrow) and will not cut (Lane Community College).

FIGURE I-241 The point of the boring tool must be positioned on the centerline of the workpiece (Lane Community College).

The point of the cutting tool should be positioned on the centerline of the workpiece (Figures I-241 to I-243). There must be a space to allow the chips to pass between the bar and the surface being machined, or the chips will wedge and bind on the back side of the bar, forcing the cutting tool deeper into the work (Figure I-244).

FIGURE I-242 If the boring tool is too low, the heel of the tool will rub and the tool will not cut, even if the tool has the correct relief angle (Lane Community College).

FIGURE I-244 Allowance must be made so that the chips can clear the space between the bar and the surface being machined. This setup has insufficient chip clearance (where the arrow is pointing) (Lane Community College).

FIGURE I-243 If the tool is too high, the back rake becomes excessively negative and the tool point is likely to be broken off. A poor-quality finish is the result of this position (Lane Community College).

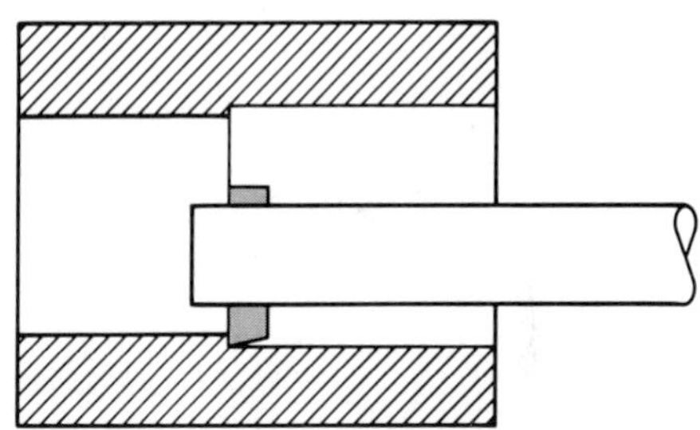

FIGURE I-245 Through boring showing bar and tool arrangement.

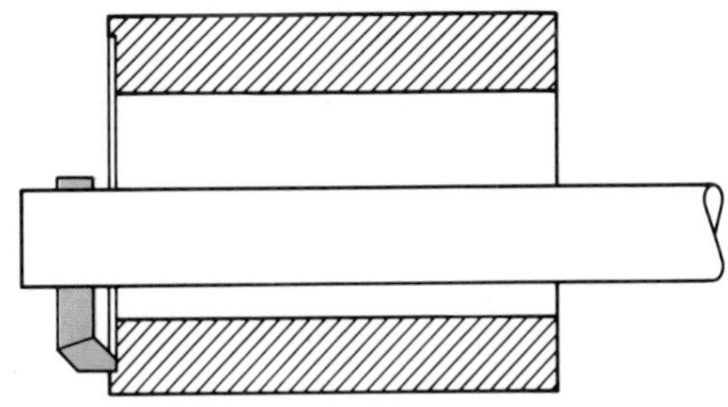

FIGURE I-246 Back facing using a boring bar.

Through boring is the boring of a workpiece from one end to the other or all the way through it. For through boring, the tool is held in a bar that is perpendicular to the axis of the workpiece. A slight side cutting edge angle is often used for through boring (Figure I-245). Back facing (Figure I-246) is sometimes done to true up a surface on the back side of a through bore. This is done with a specially ground right-hand tool also held in a bar perpendicular to the workpiece. The amount of facing that can be done in this way is limited to the movement of the bar in the bore.

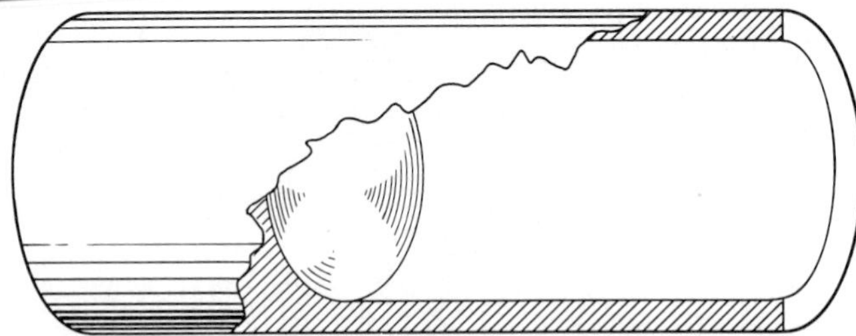

FIGURE I-247 A blind hole machined flat in the bottom.

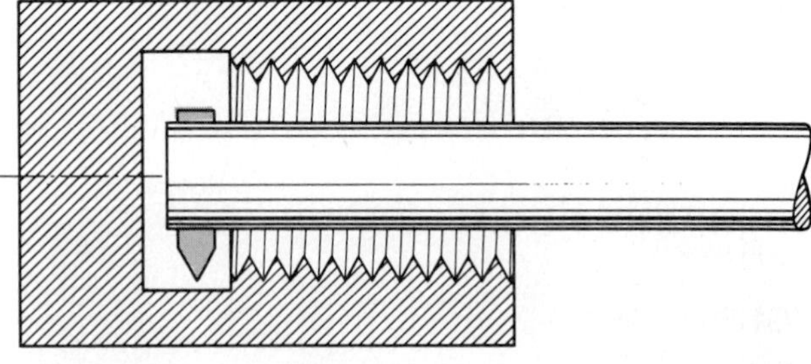

FIGURE I-250 Ample thread relief is necessary when making internal threads.

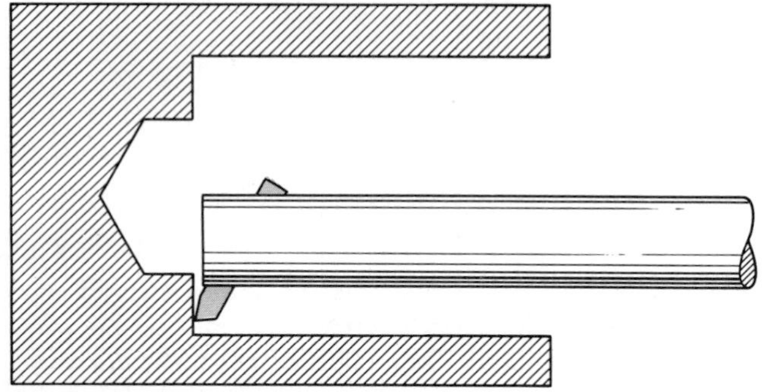

FIGURE I-248 A bar with an angled tool used to square the bottom of a hole with a drilled center.

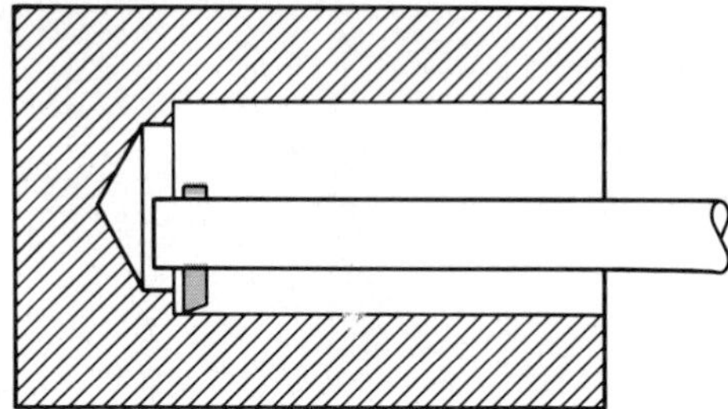

FIGURE I-251 The hole is drilled deeper than necessary to allow room for the boring bar.

FIGURE I-249 Workpiece clamped on face plate has been located, drilled, and bored (California Community Colleges IMC Project).

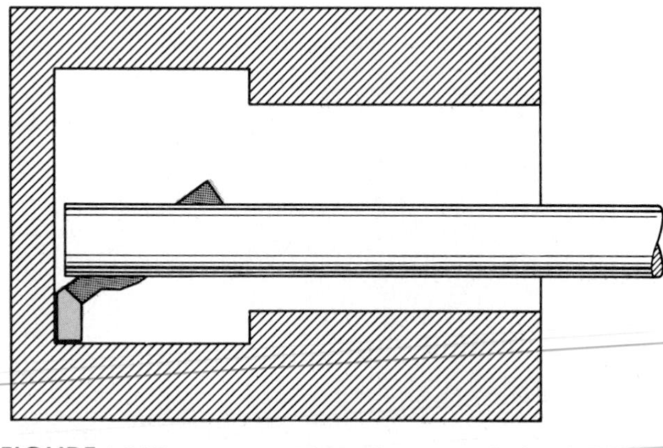

FIGURE I-252 A special tool is needed when the thread relief must be next to a flat bottom.

A blind hole is one that does not go all the way through the part to be machined (Figure I-247). Machining the bottom or end of a blind hole to a flat is easier when the drilled center does not need to be cleaned up. A bar with the tool set at an angle, usually 30 or 45 degrees, is used to square the bottom of a hole with a drilled center (Figure I-248).

Most boring is performed on workpieces mounted in a chuck. But it is also done in the end of workpieces supported by a steady rest. Boring and other operations are infrequently done on workpieces set up on a face plate (Figure I-249).

A thread relief is an enlargement of a bore at the bottom of a blind hole. The purpose of a thread relief is to allow a threading tool to disengage the work at the end of a pass (Figure I-250). When the work will allow it, a hole can be drilled deeper than necessary. This will give the end of the boring bar enough space so that the tool can reach into the area to be relieved and still be held at a 90-degree angle (Figure I-251). When the work will not allow for the deeper drilling, a special tool must be ground (Figure I-252).

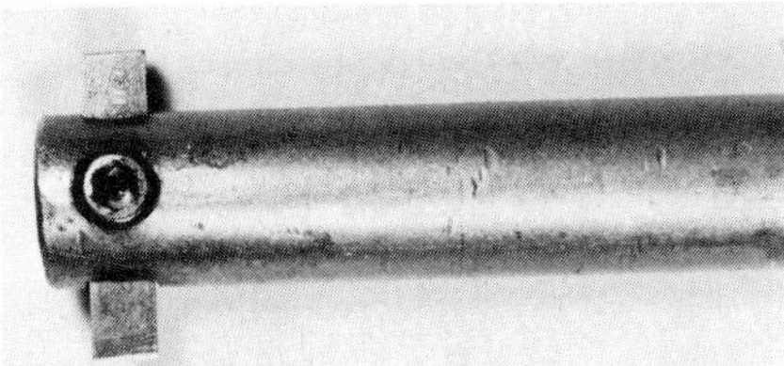

FIGURE I-253 A tool that is ground to the exact width of the desired groove can be moved directly into the work to the correct depth (Lane Community College).

FIGURE I-254 A square shoulder is made with a counterboring tool (Lane Community College).

Grooves in bores are made by feeding a form tool (Figure I-253) straight into the work. Snap ring, O-ring, and oil grooves are made in this way. Cutting oil should always be used in these operations.

Counterboring in a lathe is the process of enlarging a bore for definite length (Figure I-254). The shoulder that is produced in the end of the counterbore is usually made square (90 degrees) to the lathe axis. Boring and counterboring are also done on long workpieces that are supported in a steady rest. All boring work should have the edges and corners broken or chamfered.

Reaming

Reaming is done in the lathe to quickly and accurately finish drilled or bored holes to size. Machine reamers, like drills, are held in the tailstock spindle of the lathe. Floating reamer holders are sometimes used to assure alignment of the reamer, since the reamer follows the eccentricity of drilled holes. This helps eliminate bellmouth bores that result from reamer wobble, but does not eliminate the hole eccentricity. Only boring will remove the bore runout.

FIGURE I-255 Hand reaming in the lathe (Lane Community College).

Roughing reamers (rose reamers) are often used in drilled or cored holes, followed by machine or finish reamers. When drilled or cored holes have excessive eccentricity, they are bored .010 to .015 in. undersize and machine reamed. If a greater degree of accuracy or a better finish is required, the hole is bored to within .003 to .005 in. of finish size and hand reamed (Figure I-255). For hand reaming, the machine is shut off and the hand reamer is turned with a tap wrench. Precision dimensions are usually bored in diameters over $\frac{1}{2}$ in. An experienced machinist can make bores on a lathe that are within a tolerance of plus or minus .0005 in. Machine reamers may or may not produce an accurate hole depending on their sharpness and whether they have a built-up edge. However, small hand reamers can produce very accurate holes. Types of machine reamers are shown in Section H, Unit 6, "Reaming in the Drill Press." Hand reamers are shown in Section B, Unit 5, "Hand Reamers."

Cutting fluids used in reaming are similar to those used for drilling holes (see Table H-7). Cutting speeds are dependent on machine and workpiece material finish requirements. A rule of thumb for reaming speeds is to use one half the speed used for drilling (see Table H-5).

Feeds for reaming are about twice that used for drilling. The cutting edge should not rub without cutting as it causes glazing, work hardening, and dulling of the reamer.

A simple machine reaming sequence would be as follows:

STEP 1. Assuming that the hole has been drilled $\frac{1}{64}$ in. undersize, a taper shank machine reamer is seated in the taper by hand pressure (Figure I-256).

FIGURE I-256 The reamer must be seated in the taper.

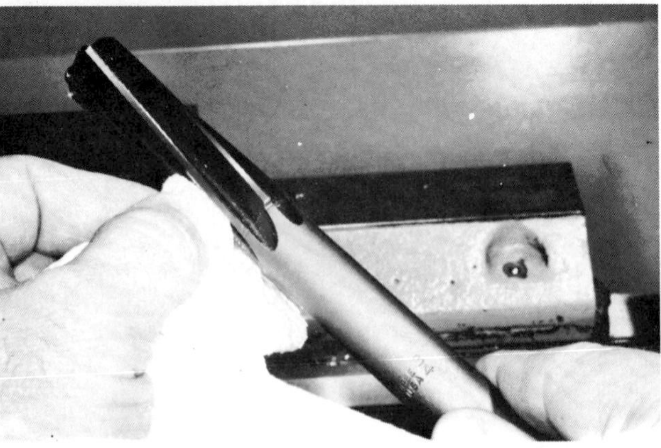

FIGURE I-258 The reamer should be cleaned and put away after using it.

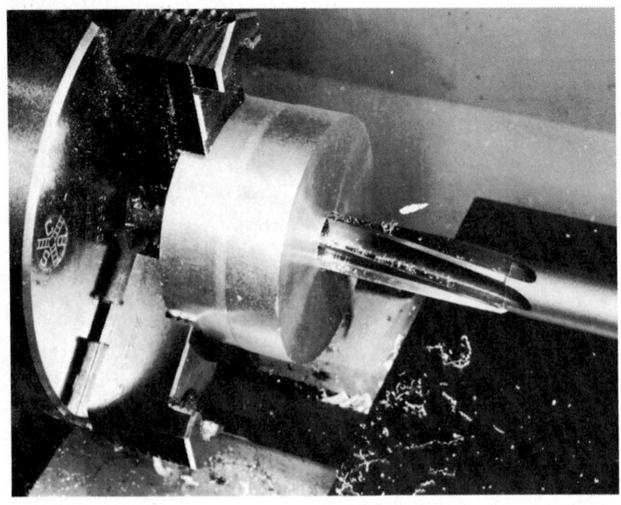

FIGURE I-257 Starting the reamer in the hole. Kerosene is being used as a cutting fluid.

FIGURE I-259 The chips are brushed into the chip pan.

STEP 2. Cutting fluid is applied to the hole and the reamer is started into the hole by turning the tailstock handwheel (Figure I-257). The hole is completed and the reamer is removed from the hole. Never reverse the machine when reaming.

STEP 3. The reamer is removed from the tailstock spindle and cleaned with a cloth (Figure I-258). The reamer is then returned to the storage rack.

STEP 4. The lathe is cleaned with a brush (Figure I-259).

Tapping

The tapping of work mounted in a chuck is a quick and accurate means of producing internal threads. Tapping in the lathe is similar to tapping in the drill press but it is generally reserved for small size holes, as tapping is the only way they can be internally threaded. Large internal threads are made in the lathe with a single-point threading tool held in a boring bar. A large tap requires considerable force or torque to turn, more than can be provided by hand turning. A tap that is aligned by the dead center will make a straight tapped hole that is in line with the lathe axis.

A plug tap or spiral point tap may be used for tapping through holes. When tapping blind holes, a plug tap could be followed by a bottoming tap, if threads are needed to the bottom of the hole. A good practice is to drill a blind hole deeper than the required depth of threads.

Two approaches may be taken for hand tapping. Power is not used in either case. One method is to turn the tap by means of a tap wrench or

FIGURE I-267 The finished groove (Lane Community College).

FIGURE I-269 Parting tool making a cut (Lane Community College).

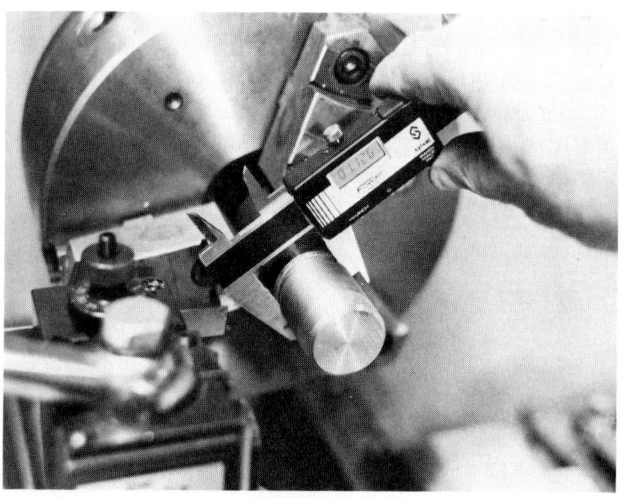

FIGURE I-268 Groove being measured with a sliding caliper with digital readout (Lane Community College).

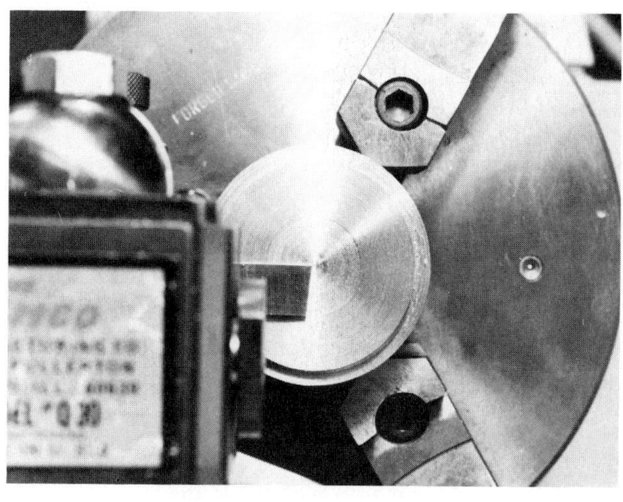

FIGURE I-270 Parting tools must be set to the center of the work (Lane Community College).

for thread relief, snap rings, and O-rings. Special tools are ground for both external and internal grooves and recesses. Parting tools are sometimes used for external grooving and thread relief. Grooves are sometimes measured with sliding-type calipers (Figure I-268).

Parting or cutoff tools (Figure I-269) are designed to withstand high cutting forces, but if chips are not sufficiencly cleared or cutting fluid is not used, these tools can quickly jam and break. Parting tools must be set on center and square with the work (Figure I-270). Lathe tools are often specially ground as parting tools for small or delicate parting jobs (Figure I-271). Diagonally ground parting tools leave no burr.

Parting alloy steels and other metals is some-

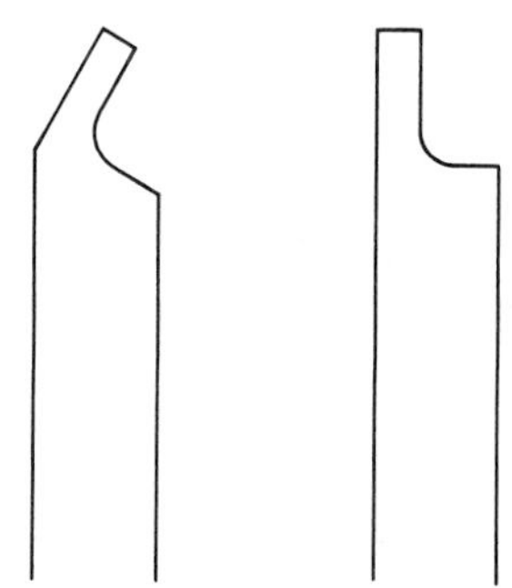

FIGURE I-271 Special parting tools that have been ground from lathe tools for small or delicate parting jobs. These tools should be ground slightly wider at the cutting edge to clear the tool in the cut.

FIGURE I-272 Step parting (Lane Community College).

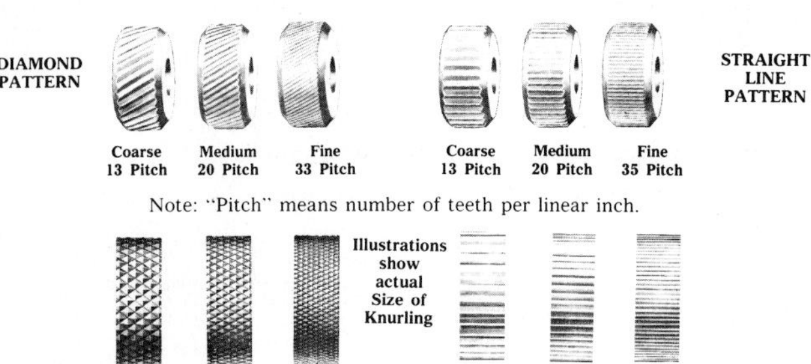

FIGURE I-273 Set of straight knurls and diagonal knurls (TRW, Inc.).

FIGURE I-274 Kuckle-joint knurling toolholder.

FIGURE I-275 Revolving head knurling toolholder.

times difficult, and step parting (Figure I-272) may help in these cases. When deep parting difficult material, extend the cutting tool from the holder a short distance and part to that depth. Then back off the cross feed and extend the tool a bit farther; part to that depth. Repeat the process until the center is reached. Sulfurized cutting oil works best for parting unless the lathe is equipped with a coolant pump and a steady flow of soluble oil is available. Parting tools are made in either straight or offset types. A right-hand offset cutoff tool is necessary when parting very near the chuck.

All parting and grooving tools have a tendency to chatter; therefore any setup must be as rigid as possible. A low speed should be used for parting; if the tool chatters, reduce the speed. A feed that is too light can cause chatter, but a feed that is too heavy can jam the tool. The tool should always be making a chip. Hand feeding the tool is best at first. Work should not extend very far from the chuck when parting or grooving, and no parting should be done in the middle of a workpiece or at the end near the dead center. This is because the tool will bind in the cut when the material is almost cut through. When the workpiece is cut off near the dead center, the same binding problem exists with the additional problem of the work climbing on the tool when it is cut off. This could break the tool and possibly damage the machine.

Knurling

A knurl is a raised impression on the surface of the workpiece produced by two hardened rolls, and is usually one of two patterns: diamond or straight (Figure I-273). The diamond pattern is formed by a right-hand and a left-hand helix mounted in a self-centering head. The straight pattern is formed by two straight rolls. These common knurl patterns can be either fine, medium, or coarse.

Diamond knurling is used to improve the appearance of a part and to provide a good gripping surface for levers and tool handles. Straight knurling is used to increase the size of a part for press fits in light-duty applications. A disadvantage to this use of knurls is that the fit has less contact area than a standard fit.

Three basic types of knurling toolholders are used: the knuckle-joint holder (Figure I-274), the revolving head holder (Figure I-275), and the straddle holder (Figure I-276). The straddle holder allows small diameters to be knurled with less distortion. This principle is used for knurling on production machines.

Knurling works best on workpieces mounted between centers. When held in a chuck and supported by a center, the workpiece tends to crawl back into the chuck and out of the supporting cen-

FIGURE I-276 Straddle knurling toolholder (Ralmike's Tool-A-Rama).

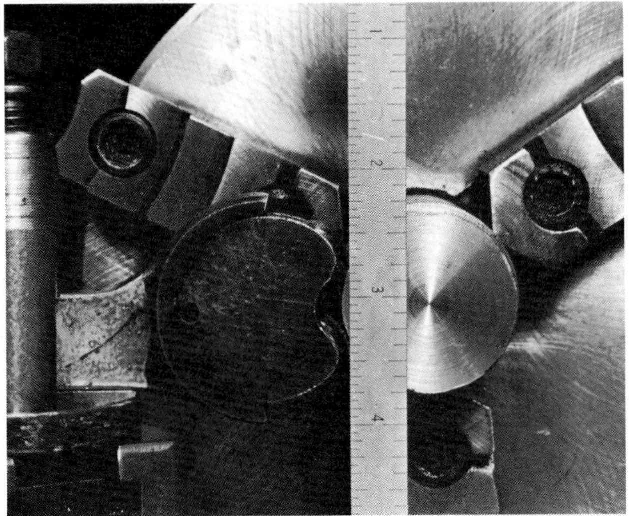

FIGURE I-277 Knurls are centered on the workpiece (Lane Community College).

ter with the high pressure of the knurl. This is especially true when the knurl is started at the tailstock end and the feed is toward the chuck. Long slender pieces push away from the knurl and will stay bent if the knurl is left in the work after the lathe is stopped.

Knurls do not cut, but displace the metal with high pressures. Lubrication is more important than cooling, so a cutting oil or lubricating oil is satisfactory. Low speeds (about the same as for threading) and a feed of about .010 to .015 in. are used for knurling.

The knurls should be centered on the workpiece vertically (Figure I-277) and the knurl toolholder should be square with the work, unless the knurl pattern is difficult to establish, as it often is in tough materials. In that case, the toolholder

FIGURE I-278 Angling the toolholder 5 degrees often helps establish the diamond pattern (Lane Community College).

FIGURE I-279 Double impression on the left is the result of the rolls not tracking evenly (Lane Community College).

should be angled about 5 degrees to the work so that the knurl can penetrate deeper (Figure I-278).

A knurl should be started in soft metal about half depth and the pattern checked. An even diamond pattern should develop. But, if one roll is dull or placed too high or too low, a double impression will develop (Figure I-279) because the rolls are not tracking evenly. If this happens, move the knurls

FIGURE I-280 More than one pass is usually required to bring the knurl to full depth in hard or springy material (Lane Community College).

FIGURE I-281 A knurling tool that cuts a knurl rather than forming it by pressure.

FIGURE I-282 A knurl being cut showing formation of chip (Lane Community College).

to a new position along the workpiece, readjust up or down, and try again. If possible, the knurl should be made in one pass. However, this is not always possible with ordinary knurling toolholders because of the extreme pressure bearing on one side of the workpiece. Several passes may be required on a slender workpiece to complete a knurl because the tool tends to push it away from the knurl. The knurls should be cleaned with a wire brush between passes.

Material that hardens as it is worked, such as high carbon or spring steel, should be knurled in one pass if at all possible, and in not more than two passes. Even in ordinary steel, the surface will harden after a diamond pattern has developed to points. It is best to stop knurling just before the points are sharp (Figure I-280). Metal flaking off the knurled surface is evidence that work hardening has occurred. Avoid knurling too deeply as it produces an inferior knurled finish.

Knurls are also produced with a type of cutting tool (Figure I-281) similar in appearance to a knurling tool. The serrated rolls form a chip on the edge (Figure I-282). Material difficult to knurl by pressure rolling, such as tubing and work hardening metals, can be knurled by this cutting tool. Sulfurized cutting oil should be used when knurling steel with this kind of knurling tool.

SELF-TEST

1. Why are drilled holes not used for bores in machine parts such as pulleys, gears, and bearing fits?

2. Describe the procedure used to produce drilled holes on workpieces in the lathe with minimum oversize and runout.

3. What is the chief advantage of boring over reaming in the lathe?

4. List five ways to eliminate chatter in a boring bar.

5. Explain the differences between through boring, counterboring, and boring blind holes.

6. By what means are grooves and thread relief made in a bore?

7. Reamers will follow an eccentric drilled hole, thus producing a bellmouth bore with runout. What device can be used to help eliminate bellmouth? Does it help remove the runout?

8. Machine reamers produce a better finish than is obtained by boring. How can you get an even better finish with a reamer?

9. Cutting speeds for reaming are (twice, half) that used for drilling; feeds used for reaming are (twice, half) that used for drilling.

10. Which would be best for making a 6-in.-diameter internal thread using an engine lathe: a tap or a single-point threading tool? Explain.

11. How can you avoid drilling oversize with a tap drill?

12. Standard plug or bottoming taps can be used when hand tapping in the lathe. If power is used, what kind of tap works best?

13. Why would threads cut with a hand die in a lathe not be acceptable for using on a feed screw with a micrometer collar?

14. By what means is thread relief on external grooves produced?

15. If cutting fluid is not used on parting tools or chips do not clear out of the groove because of a heavy feed, what is generally the immediate result?

16. How can you avoid chatter when cutting off stock with a parting tool?

17. State three reasons for knurling.

18. Ordinary knurls do not cut. In what way do they make the diamond or straight pattern on the workpiece?

19. If a knurl is producing a double impression, what can you do to make it develop a diamond pattern?

20. How can you avoid producing a knurled surface on which the metal is flaking off?

UNIT 10

Sixty-Degree Thread Information and Calculations

To cut threads, a good machinist must know more than how to set up the lathe. He or she must know the thread form, class of fit, and thread calculation. This unit prepares you for the actual cutting of threads, which you will do in the next unit.

Objectives

After completing this unit, you should be able to:

1. Describe the several 60-degree thread forms, noting their similarities and differences.

2. Calculate thread depth, infeeds, and minor diameters of threads.

The Sharp V Thread Form

Various screw thread forms are used for fastening and for moving or transmitting parts against loads.The most widely used of these forms are the 60-degree thread types. These are mostly used for fasteners. An early form of the 60-degree thread is the sharp V (Figure I-283). The sides of the thread form a 60-degree angle with each other. Theoretically, the sides and the base between two thread roots would form an equilateral triangle, but in practice this is not the case; it is necessary to make a slight flat on top of the thread in order to deburr it. Also, the tool will always round off and leave a slight flat at the thread root. The greatest drawback to this thread form is that it is so easily damaged while handling. The sharp V thread will fit closer and seal better than most threads, but is seldom used today. The depth (d) for the sharp V thread is calculated as follows:

$$d = \text{pitch} \times \cos 30 = .866\ \text{pitch}$$
$$= \frac{.866}{\text{number of threads per inch}}$$

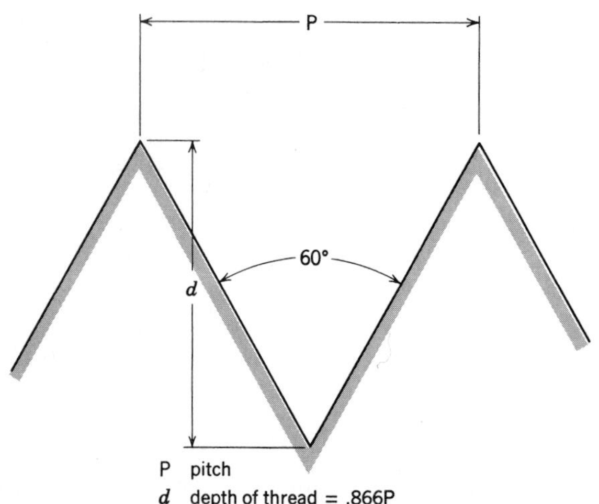

FIGURE I-283 The 60-degree sharp V thread.

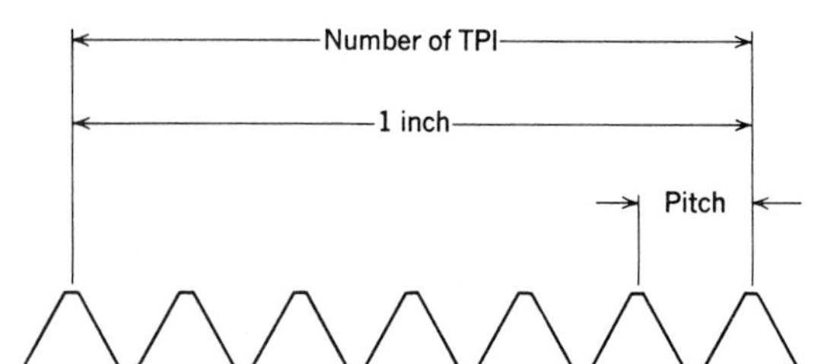

FIGURE I-284 Difference between threads per inch and pitch.

The relationship between pitch and threads per inch should be noted (Figure I-284). Pitch is the distance between a point on one screw thread and the corresponding point on the next thread, measured parallel to the thread axis. Threads per inch means the number of threads in 1 in. The pitch (P) may be derived by dividing the number of threads per inch (tpi) into one:

$$P = \frac{1}{\text{tpi}}$$

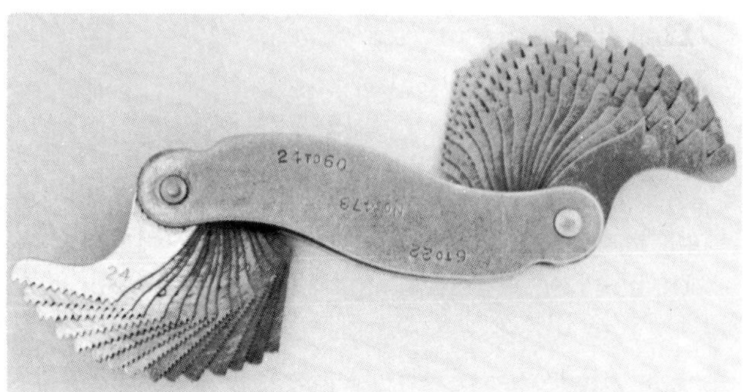

FIGURE I-285 Screw pitch gage for inch threads.

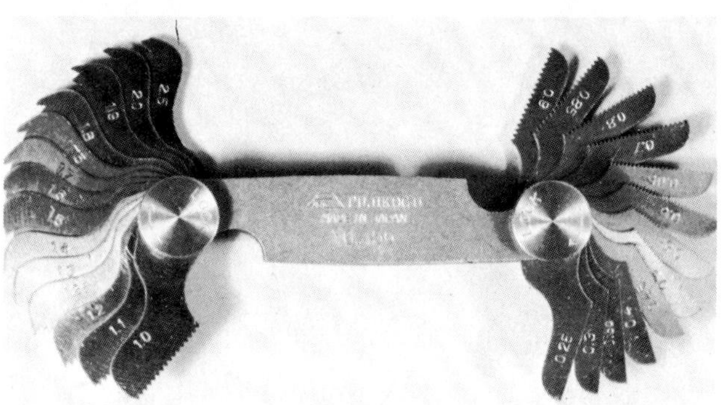

FIGURE I-286 Screw pitch gage for metric threads.

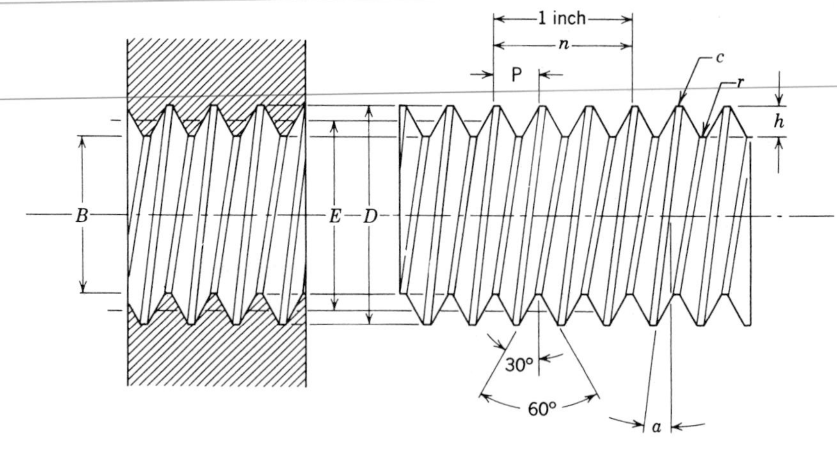

FIGURE I-287 General dimensions of screw threads.

EXAMPLE

Find the pitch of a ½-in.-diameter 20-tpi machine screw thread.

$$P = \frac{1}{20} = .050 \text{ in.}$$

Pitch is checked on a screw thread with a screw pitch gage (Figures I-285 and I-286). General dimensions and symbols for screw threads are shown in Figure I-287.

Unified and American National Forms

The American National form (Figure I-288), formerly United States Standard, was used for many years for screws, bolts, and other products. These National form threads are in either the national fine (NF) or the national coarse (NC) series. Other screw threads are listed in machinist's handbooks.

Taps and dies are marked with letter symbols to designate the series of the threads they form. For example, the symbol for American Standard Taper pipe thread is NPT; for Unified coarse thread it is UNC; and for Unified fine thread the symbol is UNF.

THREAD DEPTH

The American Standard for Unified threads (Figure I-289) is very similar to the American National Standard with certain modifications. The thread forms are practically the same and the basic 60-degree angle is the same. The depth of an external American National thread is .6495 × pitch, and the depth of the Unified thread is .6134 × pitch. The constant for American National thread depth may be rounded off from .6495 to .65, and the constant for Unified thread depth may be rounded to .613. The thread depth and the root truncation (tool flat) of the American National thread is fixed or definite, but these factors are variable within limits for Unified threads. A rounded root for Unified threads is desirable whether from tool wear or by design. A rounded crest is also desirable but not required. The constants .613 for thread depth and .125 for the flat on the end of the tool were selected for calculations on Unified threads in this unit.

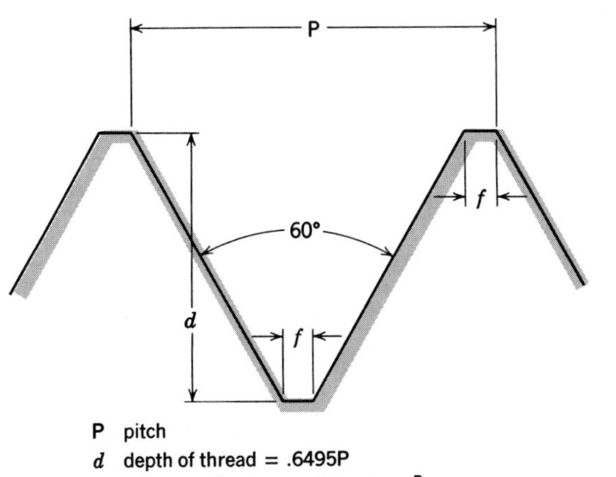

FIGURE I-288 The American National form of thread.

THREAD FIT CLASSES AND THREAD DESIGNATIONS

Unified and American National Standard form threads are interchangeable. An NC bolt will fit an UNC nut. The principal difference between the two systems is that of tolerances. The Unified system, a modified version of the old system, allows for more tolerances of fit. Thread fit classes 1, 2, 3, 4, and 5 were used with the American National Standard; 1 being a very loose fit, 2 a free fit, 3 a close fit, 4 a snug fit, and 5 a jam or interference fit. The Unified system expanded this number system to

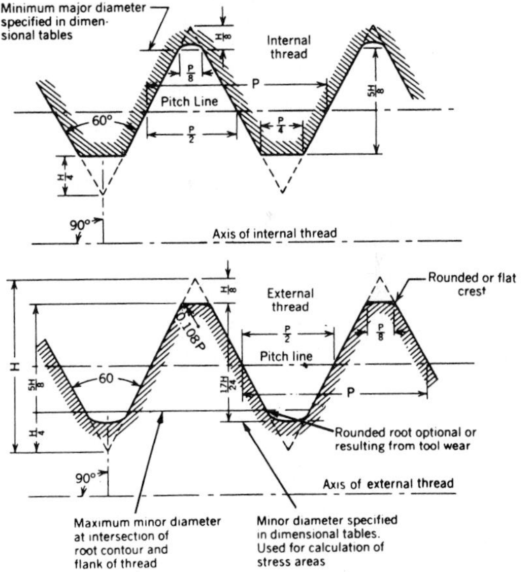

(H = height of sharp V thread = 0.86603 × pitch)

Unified Thread Form Data:

$$\text{Pitch} = \frac{1}{\text{Number of threads per inch}}$$

Depth, external thread = 0.613 × pitch
Depth, internal thread = 0.541 × pitch
Flat at crest, external thread = 0.125 × pitch
Flat at crest, internal thread = 0.25 × pitch
Flat at root, internal thread (tool flat) = 0.125 × pitch
Flat at root, external thread (permissible tool flat on new tool) = 0.125 × pitch

FIGURE I-289 Unified screw threads. (Reprinted from ASA B1.1—1960, Unified Screw Threads, with permission of the publisher, the American Society of Mechanical Engineers; data, John Neely).

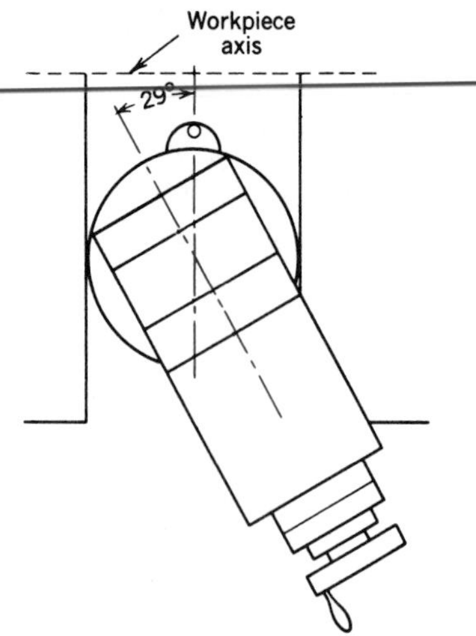

FIGURE I-290 Compound at 29 degrees for cutting 60-degree threads.

include a letter, so the threads could be identified as class 1A, 1B, 2A, 2B, and so on. "A" indicates an external thread and "B" an internal thread. Because of this expansion in the Unified system, tolerances are now possible on external threads and are 30 percent greater on internal threads. These changes make easier the manufacturer's job of controlling tolerances to insure the interchangeability of threaded parts. See *Machinery's Handbook* for tables of Unified thread limits. Limits are the maxium and minimum allowable dimensions of a part—in this case, internal and external threads.

Threads are designated by the nominal bolt size or major diameter, the threads per inch, the letter series, the thread tolerance, and the thread direction. Thus $1\frac{1}{4}$ in.—12 UNF-2BLH would indicate a $1\frac{1}{4}$-in. Unified nut with 12 threads per inch, a class 2 thread fit, and a left-hand helix.

Unified screw thread systems are the American Standard for fastening types of screw threads. Manufacturing processes where V threads are produced are based on the Unified system. Many job and maintenance machine shops, on the other hand, still use the American National thread system when chasing a thread with a single-point tool on an engine lathe.

TOOL FLATS AND INFEEDS FOR THREAD CUTTING

The flat on the crest of the thread on both the Unified and American National systems is P/8 or P × .125. The root flat (flat on the end of the external threading tool) is calculated P/8 or P × .125 for the American National system, but it varies in the Unified system. However, for purposes of convenience in cutting threads on a lathe, the same tool flat (P/8) may also be used for the Unified form thread.

To cut 60-degree form threads, the tool is fed into the work with the compound (Figure I-290). However, in practice, the compound is actually set to 29 degrees to provide a light finishing cut on the trailing side of the threading tool. The 1-degree difference will make no significant difference in calculations. The infeed depth along the flank of the thread at 30 degrees is greater than the depth at 90 degrees from the work axis. This depth may be calculated for American National threads by dividing the number of threads per inch (n) into .75:

$$\text{infeed} = \frac{.75}{n} \quad \text{or} \quad P \times .75$$

Thus for a thread with 10 threads per inch (.100-in. pitch),

$$\text{infeed} = \frac{.75}{10} = .075 \text{ in.}$$

or

$$\text{infeed} = .75 \times .100 = .075 \text{ in.}$$

For external Unified threads the infeed at 29 degrees may be calculated by the formula:

$$\text{infeed} = \frac{.708}{n} \quad \text{or} \quad .708P$$

Thus for a thread with 10 threads per inch (.100 in. pitch),

$$\text{infeed} = \frac{.708}{10} = .0708 \text{ in.}$$

or

$$\text{infeed} = .708 \times .100 = .0708 \text{ in.}$$

Pitch Diameter, Helix Angle, and Percent of Threads

The making of external and internal threads that are interchangeable depends upon the selection of thread fit classes. The clearances and tolerances for thread fits are derived from the pitch diameter. The pitch diameter on a straight thread is the diameter of an imaginary cylinder that passes through the thread profiles at a point where the width of the groove and thread are equal. The mating surfaces are the flanks of the thread.

The percent of thread has little do to with fit, but refers to the actual minor diameter of the internal thread. The typical nut for machine screws has 75 percent threads, which are easier to tap than 100 percent threads and retain sufficient strength for most thread applications.

The helix angle of a screw thread (Figure I-291) is larger for greater lead threads than for smaller

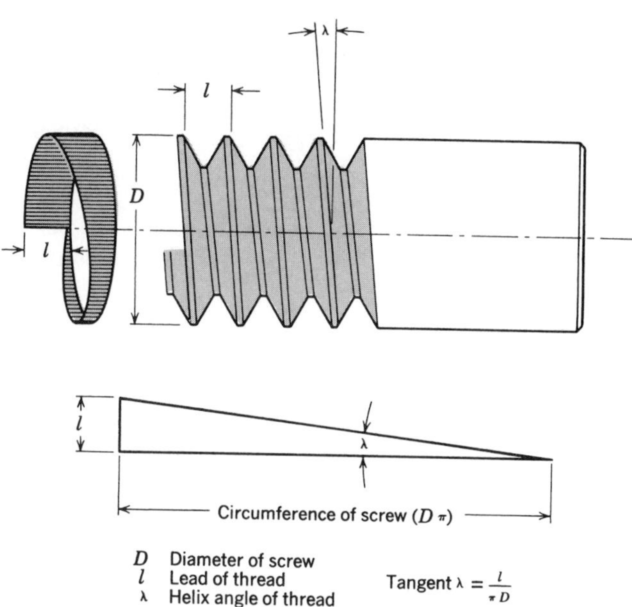

FIGURE I-291 Screw thread helix angle.

FIGURE I-292 Checking the relief and helix angle on the threading tool with a protractor (Lane Community College).

leads; and the larger the diameter of the workpiece, the smaller the helix angle for the same lead. Helix angles should be taken into account when grinding tools for threading. The relief and helix angles must be ground on the leading or cutting edge of the tool (Figure I-292). A protractor may be used to check this angle.

Helix angles may be determined by the following formula:

$$\text{tangent of helix angle} = \frac{\text{lead of thread}}{\text{circumference of screw}} = \frac{\text{lead of thread}}{\pi D}$$

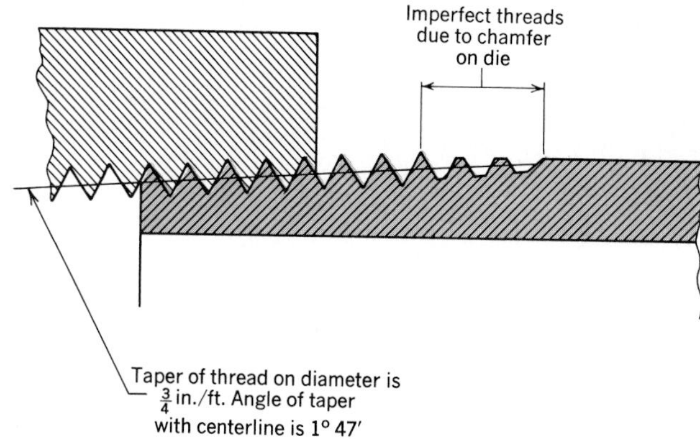

FIGURE I-293 American National standard taper pipe thread.

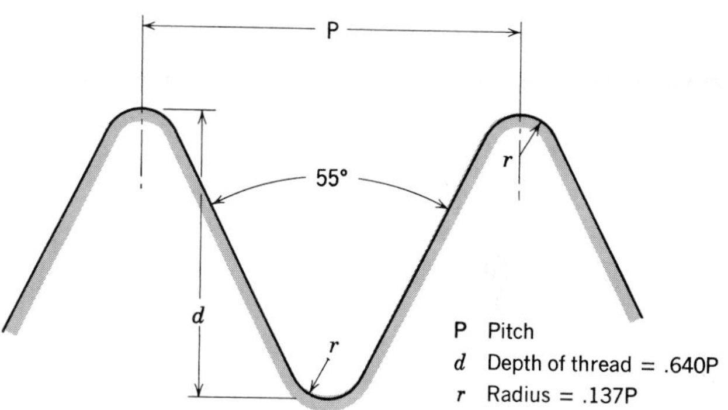

FIGURE I-294 Whitworth thread.

where $\pi = 3.1416$, D = the major diameter of the screw. (Also note that pitch and lead are the same for single lead screws.) Helix angles are given for Unified and other thread series in handbooks such as *Machinery's Handbook*.

A taper thread is made on the internal or external surface of a cone. An example of a 6-degree taper thread is the American National Standard pipe thread (Figure I-293). A line bisecting the 60-degree thread is perpendicular to the axis of the workpiece. On a taper thread the pitch diameter at a given position on the thread axis is the diameter of the pitch cone at that position.

The British Standard Whitworth thread (Figure I-294) has rounded crests and roots and has an included angle of 55 degrees. This thread form has been largely replaced by the Unified and metric thread forms.

Metric Thread Forms

Several metric thread systems such as the SAE standard spark plug threads and the British Standard for spark plugs are in use today. The Système

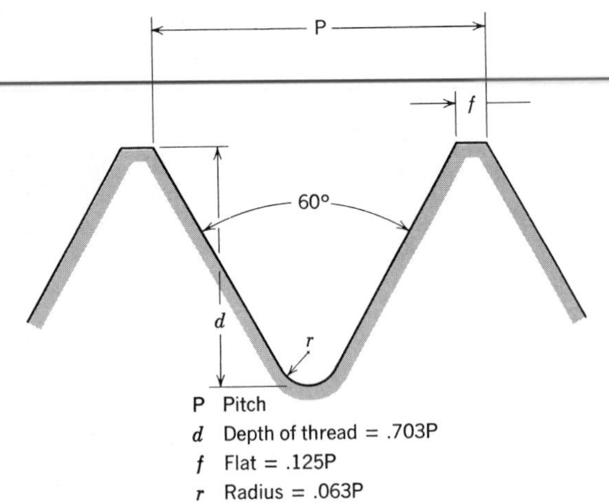

FIGURE I-295 SI metric thread form.

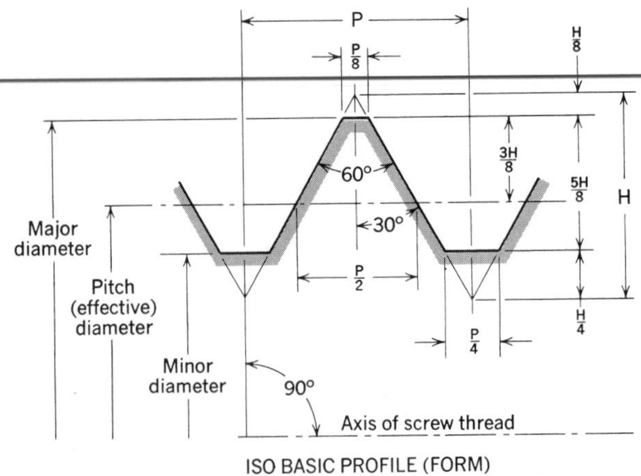

FIGURE I-296 ISO metric thread form (TRW, Inc.).

Internationale (SI) thread form (Figure I-295), adopted in 1898, is similar to the American National Standard. Metric bolt sizes differ slightly from one European country to the next. The British Standard for ISO (International Organization for Standardization) metric screw threads was set up to standardize metric thread forms. The basic form of the ISO metric thread (Figure I-296) is similar to the Unified thread form. These and other metric thread systems are listed in *Machinery's Handbook*. See Table 10 in Appendix 2 for ISO metric tap drill sizes.

SELF-TEST

1. Name one disadvantage of the sharp V thread.

2. Explain the difference between the threads per inch and the pitch of the thread.

3. Name two similarities and two differences between American National and Unified threads.

4. What is a major reason for thread allowances and classes of fits?

5. What does $\frac{1}{2}$–20 UNC–2A describe?

6. The root truncation for unified threads and for American National threads is found by .125*P*. What should the flat on the end of the threading tools be for both systems on a $\frac{1}{2}$–20 thread?

7. How far should the compound set at 30 degrees move to cut a $\frac{1}{2}$–20 Unified thread? (The formula is .708/*n*.) How far should the compound move to cut a $\frac{1}{2}$–20 American National thread? (The formula is .75/*n*.)

8. Explain the difference between the fit of threads and the percent of thread.

9. Name two metric thread standard systems.

UNIT 11

Cutting Unified External Threads

A machinist is frequently called upon to cut threads of various forms on the engine lathe. The threads most commonly made are the V form, American National, or Unified. This unit will show you how to make these threads on a lathe. You will need much practice to gain confidence in your ability to make external Unified threads on any workpiece.

Objectives

After completing this unit, you should be able to:

1. Detail the steps and procedures necessary to cut a Unified thread to the correct depth.
2. Set up a lathe for threading and cut several different thread pitches and diameters.
3. Identify tools and procedures for thread measurement.

How Threading is Done on a Lathe

Thread cutting on a lathe with a single-point tool is done by taking a series of cuts in the same helix of the thread. This is sometimes called chasing a thread. A direct ratio exists between the headstock spindle rotation, the leadscrew rotation, and the number of threads on the leadscrew. This ratio can be altered by the quick-change gearbox to make a variety of threads. When the half-nuts are clamped on the thread of the leadscrew, the carriage will move a given distance for each revolution of the spindle. This distance is the lead of the thread.

If the infeed of a thread is made with the cross slide (Figure I-297), equal-sized chips will be formed on both cutting edges of the tool. This causes higher tool pressures that can result in tool breakdown, and sometimes causes tearing of the threads because of insufficient chip clearance. A more accepted practice is to feed in with the compound, which is set at 29 degrees (Figure I-298) toward the right of the operator, for cutting right-hand threads. This assures a cleaner cutting action than with 30 degrees with most of the chip taken from the leading edge and a scraping cut from the following edge of the tool.

FIGURE I-297 An equal chip is formed on each side of the threading tool when the infeed is made with the cross slide (Lane Community College).

Setting up for Threading

Begin setup by obtaining or grinding a tool for cutting Unified threads of the required thread pitch. The only difference in tools for various pitches is the flat on the end of the tool. For Unified threads this is .125*P*, as discussed in Unit 10. If the toolholder you are using has no back rake, no grinding on the top of the tool is necessary. If the toolholder does have back rake, the tool must be ground to provide zero rake (Figure I-299).

FIGURE I-298 A chip is formed on the leading edge of the tool when the infeed is made with the compound set at 29 degrees to the right of the operator to make right-hand threads (Lane Community College).

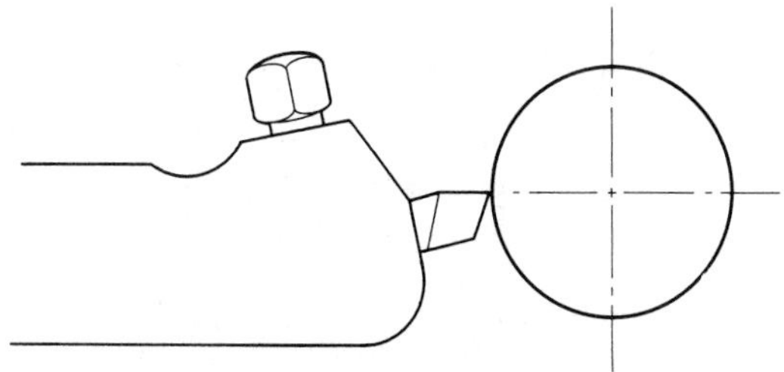

FIGURE I-299 The tool must have zero rake and be set on the center of the work in order to produce the correct form.

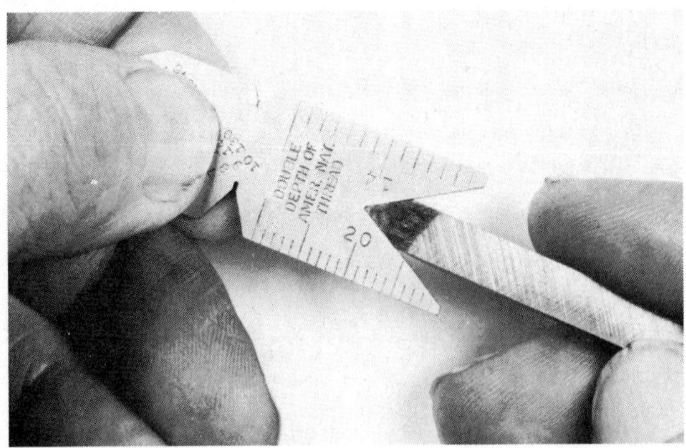

FIGURE I-300 Checking the tool angle with a center gage.

FIGURE I-301 The threading tool is placed in the holder and lightly clamped (Lane Community College).

A center gage (Figure I-300) or an optical comparator may be used to check the tool angle. An adequate allowance for the helix angle on the leading edge will assure sufficient side relief. (See Unit 10.)

The part to be threaded is set up between centers, in a chuck, or in a collet. An undercut of .005 in. less than the minor diameter should be made at the end of the thread. Its width should be sufficient to clear the tool.

The tool is clamped in the holder and set on the centerline of the workpiece (Figures I-301 and I-302). A center gage is used to align the tool to the workpiece (Figure I-303). The toolholder is clamped tightly after the tool is properly aligned.

SETTING DIALS ON THE COMPOUND AND CROSS FEED

The point of the tool is brought into contact with the work by moving the cross feed handle, and the micrometer collar is set on the zero mark (Figure I-304). The compound micrometer collar should also be set on zero (Figure I-305), but first be sure that all slack or backlash is removed by turning the compound feed handle clockwise.

FIGURE I-302 The tool is adjusted to the dead center for height. A tool that is set too high or too low will not produce a true 60-degree angle in the cut thread (Lane Community College).

FIGURE I-303 The tool is properly aligned by using a center gage. The toolholder is adjusted until the tool is aligned. The toolholder is then tightened (Lane Community College).

SETTING APRON CONTROLS

On some lathes a feed change lever, which selects either cross or longitudinal feeds, must be moved to a neutral position for threading. This action locks out the feed mechanism so that no mechanical interference is possible. All lathes have some interlock mechanism to prevent interference when the half-nut lever is used. The half-nut lever causes two halves of a nut to clamp over the leadscrew.

FIGURE I-304 After the tool is brought into contact with the work, the cross feed micrometer collar is set to the zero index (Lane Community College).

FIGURE I-306 The threads per inch selection is made on the quick-change gearbox (Lane Community College).

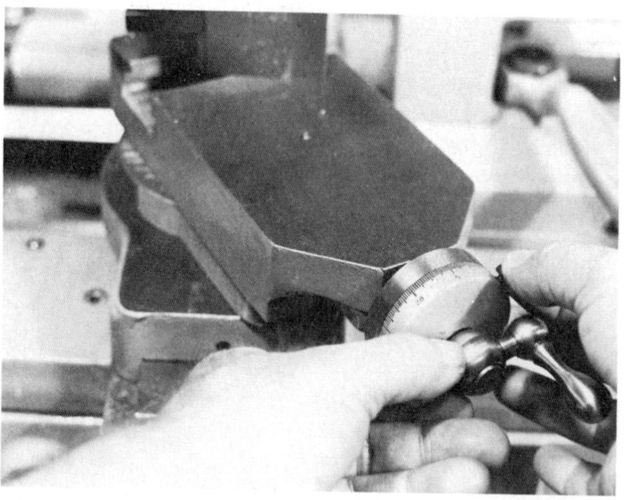

FIGURE I-305 The operator then sets the compound micrometer collar to the zero index (Lane Community College).

FIGURE I-307 Setting the correct speed for threading (Lane Community College).

The carriage will move the distance of the lead of the thread on the leadscrew for each revolution of the leadscrew.

Threading dials operate off the leadscrew and continue to turn when the leadscrew is rotating and the carriage is not moving. When the half-nut lever is engaged, the threading dial stops turning and the carriage moves. The marks on the dial indicate when it is safe to engage the half-nut lever. If the half-nuts are engaged at the wrong place, the threading tool will not track in the same groove as before but may cut into the center of the thread and ruin it. With any even number of threads such as 4, 6, 12, and 20, the half-nut may be engaged at any line. Odd numbered threads such as 5, 7, 13 may be engaged at any numbered line. With fractional threads it is safest to engage the half-nut at the same line every time.

THE QUICK-CHANGE GEARBOX

The settings for the gear shift levers on the quick-change gearbox are selected according to the threads per inch desired (Figure I-306). If the lathe has an interchangeable stud gear, be sure the correct one is in place.

SPINDLE SPEEDS

Spindle speeds for thread cutting are approximately one-fourth turning speeds. The speed should be slow enough so you will have complete control of the threadcutting operation (Figure I-307).

FIGURE I-308 The half-nut lever is engaged at the correct line or numbered line depending on whether the thread is an odd-, even-, or fractional-numbered thread (Lane Community College).

FIGURE I-310 The pitch of the thread is being checked with a screw pitch gage (Lane Community College).

FIGURE I-309 A light scratch cut is taken for the purpose of checking the pitch (Lane Community College).

FIGURE I-311 Cutting fluid is applied before taking the first cut (Lane Community College).

Cutting the Thread

The following is the procedure for cutting right-hand threads:

STEP 1. Move the tool off the work and turn the cross feed micrometer dial back to zero.

STEP 2. Feed it in .002 in. on the compound dial.

STEP 3. Turn on the lathe and engage the half-nut lever (Figure I-308).

STEP 4. Take a scratch cut without using cutting fluid (Figure I-309). Disengage the half-nut at the end of the cut. Stop the lathe and back out the tool using the cross feed. Return the carriage to the starting position.

STEP 5. Check the thread pitch with a screw pitch gage or a rule (Figure I-310). If the pitch is wrong, it can still be corrected.

STEP 6. Apply appropriate cutting fluid to the work (Figure I-311).

STEP 7. Feed the compound in .005 to .020 in. for the first pass, depending on the pitch of the thread. For a coarse thread, heavy cuts can be taken on the first few cuts. The depth of cut should be reduced for each pass until it is about .002 in. at the final passes. Bring the crossfeed dial to zero.

FIGURE I-312 The second cut is taken after feeding in the compound .010 in. (Lane Community College).

FIGURE I-313 The finish cut is taken with infeed of .001 to .002 in. (Lane Community College).

STEP 8. Make the second cut (Figure I-312).

STEP 9. Continue this process until the tool is within .010 in. of the finished depth (Figure I-313).

STEP 10. Brush the threads to remove the chips. Check the thread fit with a ring gage (Figure I-314), standard nut or mating part (Figure I-315), or screw thread micrometer (Figure I-316). The work may be removed from between centers and returned without dis-

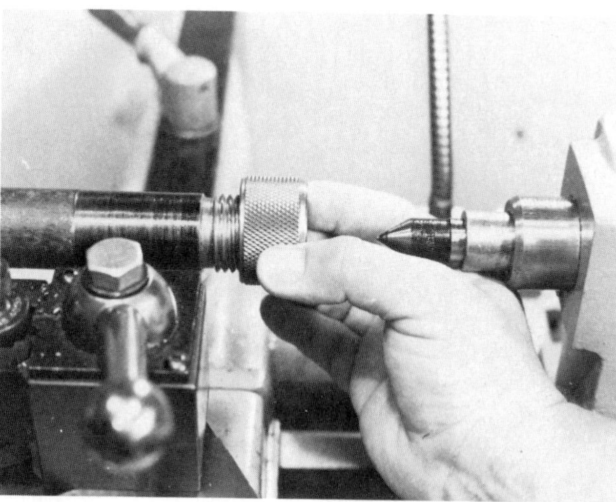

FIGURE I-314 The thread is checked with a ring gage (Lane Community College).

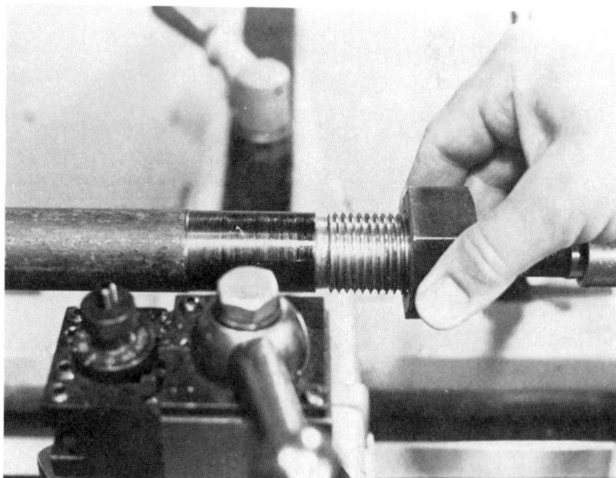

FIGURE I-315 A standard nut is often used to check a thread (Lane Community College).

FIGURE I-316 The screw thread micrometer may be used to check the pitch diameter of the thread (Lane Community College).

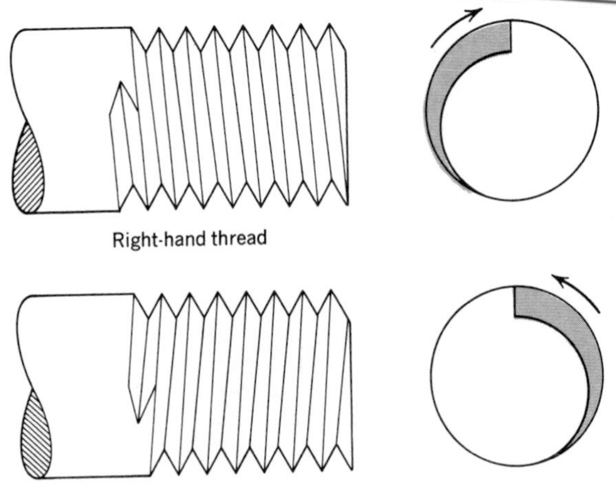

FIGURE I-317 The difference between right-hand and left-hand threads as seen from the side and end.

FIGURE I-318 Compound set for cutting a left-hand thread (Lane Community College).

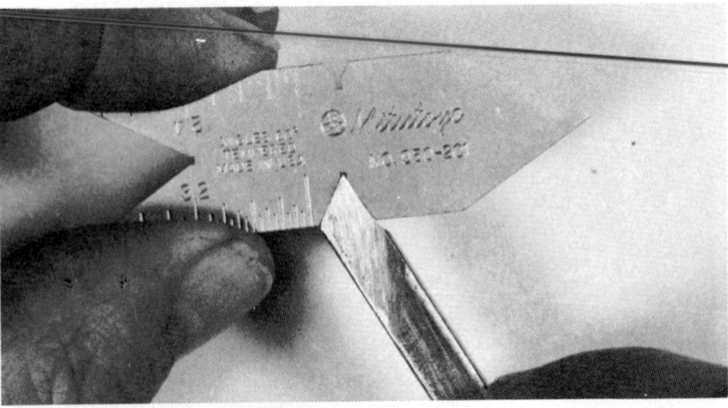
FIGURE I-319 A threading tool that is used for threading to a shoulder (Lane Community College).

FIGURE I-320 By placing the blade of the threading tool in the upper position, this tool can be made to thread on the bottom side. The lathe is reversed and the thread is cut from left to right making the job easier when threading next to a shoulder (Copyright © 1975, Aloris Tool Company, Inc.).

turbing the threading setup, provided that the tail of the dog is returned to the same slot.

STEP 11. Continue to take cuts of .001 or .002 in. (as shown in Figure I-313) and check the fit between each cut. Thread the nut with your fingers; it should go on easily but without end play. A class 2 fit is desirable for most purposes.

STEP 12. Chamfer the end of the thread to protect it from damage.

LEFT-HAND THREADS

The procedure for cutting left-hand threads (Figure I-317) is the same as that used for cutting right-hand threads with two exceptions. The compound is set at 29 degrees to the left of the operator (Figure I-318) and the leadscrew rotation is reversed so the cut is made from the left to the right. The feed reverse lever is moved to reverse the leadscrew. Sufficient undercut must be made for a starting place for the tool. Also, sufficient relief must be provided on the *right* side of the tool.

METHODS OF TERMINATING THREADS

Undercuts are often used for terminating threads. They should be made the single depth of the thread plus .005 in. The undercut should have a radius to lessen the possibility of fatigue failure resulting from stress concentration in the sharp corners.

Machinists sometimes simply remove the tool quickly at the end of the thread while disengaging the half-nuts. If a machinist misjudges and waits too long, the point of the threading tool will be broken off. A dial indicator is sometimes used to locate

FIGURE I-321 Carbide threading insert for external threading. Inserts are made for many different thread forms and for boring bars (Kennametal, Inc., Latrobe, Pa.).

FIGURE I-322 Threads being turned at high speed with carbide insert threading tool (Kennametal, Inc., Latrobe, Pa.).

FIGURE I-323 Uni-chaser—a single-tangent threading tool chaser that is a multiple-point tool and will produce a bottom and top radius on the thread (Geometric Tool Division of TRW, Inc.).

the exact position for removing the tool. When this tool withdrawal method is used, an undercut is not necessary.

Terminating threads close to a shoulder requires specially ground tools (Figure I-319). Sometimes, when cutting right-hand threads to a shoulder, it is convenient to turn the tool upside down and reverse the lathe spindle, which then moves the carriage from left to right. Some commercial threading tools are made for this purpose (Figure I-320).

OTHER TOOL TYPES

Since the crest of the Unified thread form should be rounded, some commercial threading tools provide this and other advantages (Figures I-321 and I-322). Another form of threading tool used in the lathe is shown in Figures I-323 and I-324. This is a multiple-point tool that will produce a full form

FIGURE I-324 A right-hand thread being cut with a uni-chaser (TRW, Inc.).

FIGURE I-325 Repositioning the tool.

FIGURE I-326 Thread ring gage (PMC Industries).

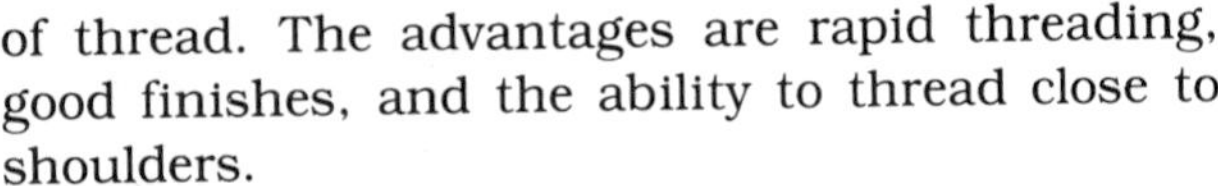

of thread. The advantages are rapid threading, good finishes, and the ability to thread close to shoulders.

PICKING UP A THREAD

It sometimes becomes necessary to reset the tool when its position against the work has been changed during a threading operation. This position change may be caused by removing the threading tool for grinding, by the work slipping in the chuck or lathe dog, or by the tool moving from the pressure of the cut.

To reposition the tool the following steps may be taken.

STEP 1. Check the tool position with reference to the work by using a center gage. If necessary, realign the tool.

STEP 2. With the tool backed away from the threads, engage the half-nuts with the machine running. Turn off the machine with the half-nuts still engaged and the tool located over the partially cut threads.

STEP 3. Position the tool in its original location in the threads by moving both the crossfeed and compound handles (Figure I-325).

STEP 4. Set the micrometer dial to zero on the crossfeed collar and set the dial on the compound to the last setting used.

STEP 5. Back off the crossfeed and disengage the half-nuts. Resume threading where you left off.

Basic External Thread Measurement

The simplest method for checking a thread is to try the mating part for fit. The fit is determined solely by feel with no measurement involved. While a loose, medium, or close fit can be determined by this method, the threads cannot be depended upon for interchangeability with others of the same size and pitch.

Thread ring gages are used to check the accuracy of external threads (Figure I-326). The outside of the ring gage is knurled and the no go gage can be easily identified by a groove on the knurled surface. When these gages are used, the go ring gage should enter the thread fully. The no go gage should not exceed more than $1\frac{1}{2}$ turns on the thread being checked.

Thread roll snap gages are used to check the

FIGURE I-327 Thread roll snap gage (PMC Industries).

FIGURE I-328 Thread comparator micrometer.

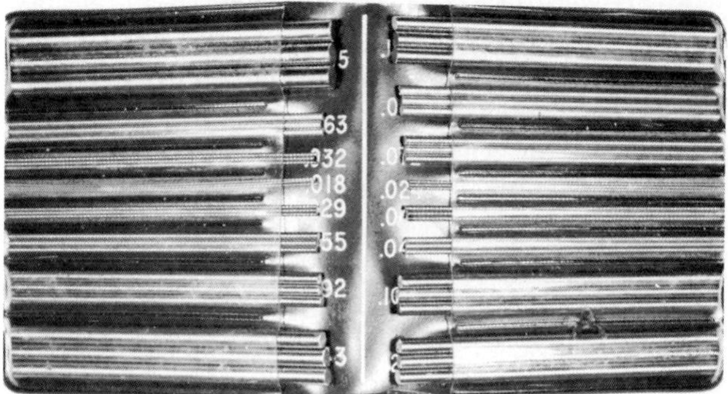
FIGURE I-329 Set of "best" three-wires for various thread sizes.

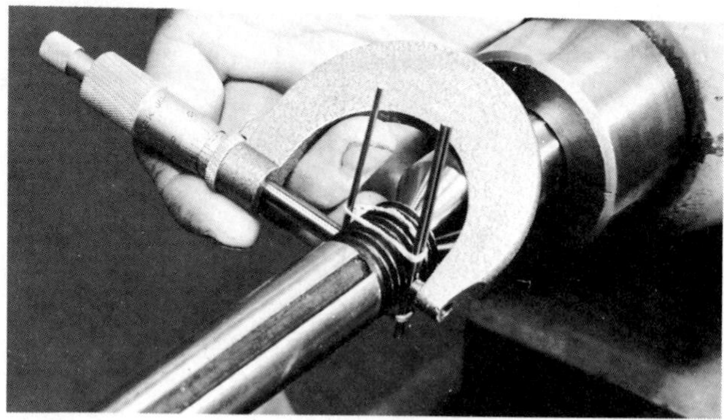
FIGURE I-330 Measuring threads with the three-wire and micrometer (Lane Community College).

accuracy of external screw threads (Figure I-327). These common measuring tools are easier and faster to use then thread micrometers or ring gages. The part size is compared to a preset dimension on the roll gage. The first set of rolls are the go and the second the no go rolls.

Thread roll snap gages, ring gages, and plug gages are used in production manufacturing where quick gaging methods are needed. These gaging methods depend on the operator's "feel," and the level of precision is only as good as the accuracy of the gage. The thread sizes are not measurable in any definite way.

The thread comparator micrometer (Figure I-328) has two conical points. This micrometer does not measure the pitch diameter of a thread, but is used only to make a comparison with a known standard. The micrometer is first set to the threaded part and then compared to the reading obtained from a thread plug gage.

Advanced Methods of Thread Measurement

The most accurate place to measure a screw thread is on the flank or angular surface of the thread at the pitch diameter. The outside diameter measured at the crest or the minor diameter measured at the root could vary considerably. Threads may be measured with standard micrometers and specially designated wires (Figure I-329) or with a screw thread micrometer. The pitch diameter is measured directly by these methods. The kit in Figure I-329 has wire sizes for most thread pitches. Most kits provide a number to be added to the nominal OD of the threads, which is the reading that should be seen on the micrometer if the pitch diameter of the thread is correct.

THE THREE-WIRE METHOD

The three-wire method of measuring threads is considered one of the best and most accurate. Figure I-330, shows three-wires placed in the threads with the micrometer measuring over them. Different sizes and pitches of threads require different-sized wires. For greatest accuracy a wire size that will contact the thread at the pitch diameter should be used. This is called the "best" wire size.

The pitch diameter of a thread can be calculated by subtracting the wire constant (which is the single depth of a sharp V thread, or $.866 \times P$) from the measurement over the three wires when the best wire size is used. The wires used for three-wire measurement of threads are hardened and lapped steel, and are available in sets that cover a large range of threaded pitches.

A formula by which the best size wire may be found is

$$\text{wire size} = \frac{.57735}{n}$$

where n = the number of threads per inch.

EXAMPLE

To find the best-size wire for measuring a $1\frac{1}{4}$ in.–12 UNC screw thread,

$$\text{wire size} = \frac{.57735}{12} = .048$$

The best wire size to use for measuring a 12-pitch thread would be .048 in.

If the best wire sizes are not available, smaller or larger wires may be used within limits. They should not be so small that they are below the major diameter of the thread, or so large that they do not contact the flank of the thread. Subtract the constant ($.866 \times P$) for best wire size from the pitch diameter and add three times the diameter of the available wire when the best wire size is not available.

EXAMPLE

The best wire size for $1\frac{1}{4}$–12 is .048 in., but only a $\frac{3}{64}$-in.-diameter drill rod is available, which has a diameter of .0469 in.

1.1959	Pitch diameter of $1\frac{1}{4}$–12
− .0722	Constant for best wire size
1.1237	
+ .1407	3 × .0469 available wire size
= 1.2644	Measurement over wires

The measurement over the wires will be slightly different than that of the best wire size because of the difference in wire size. After the best size wire is found, the wires are positioned in the threaded grooves as shown in Figure I-330. The anvil and spindle of a standard outside micrometer are then placed against the three wires and the measurement is taken.

To calculate what the reading of the micrometer should be if a thread is the correct finished size, use the following formula when measuring Unified coarse threads or American National threads:

$$M = D + 3W - \frac{1.5155}{n}$$

where M = micrometer measurement over wires
D = diameter of the thread
n = number of threads per inch
W = diameter of wire used

EXAMPLE

To find M for a $1\frac{1}{4}$–12 UNC thread proceed as follows. Where

$$W = .048$$

$$D = 1.250$$

$$n = 12$$

then

$$M = 1.250 + (3 \times .048) - \frac{1.5155}{12}$$

$$= 1.250 + 1.44 - .126$$

$$= 1.268 \text{ (micrometer measurement)}$$

When measuring a Unified fine thread, the same method and formula are used, except that the constant should be 1.732 instead of 1.5155.

The wire method of thread measurement is also used for other thread forms such as Acme and Buttress. Information and tables may be found in *Machinery's Handbook*. Another method of measuring threads with a standard micrometer is with precision ground triangular bars (Figure I-331). These also come in kits.

The screw thread micrometer (Figure I-332) may be used to measure sharp V, Unified, and American National threads. The spindle is pointed to a 60-degree included angle. The anvil, which swivels, has a double-V shape to contact the pitch diameter. The thread micrometer measures the pitch diameter directly from the screw thread. This reading may be compared with pitch diameters given in handbook tables. Thread micrometers have interchangeable anvils that will fit a wide range of thread pitches. Some are made in sets of four micrometers that have a capacity up to 1 in., and each covers a range of threads. The range of

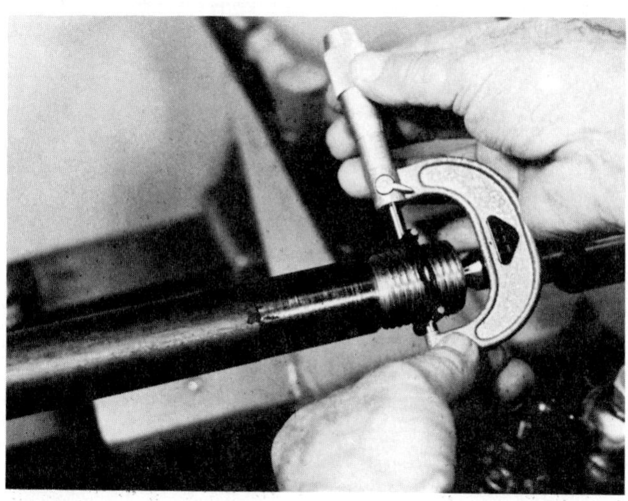

FIGURE I-331 Using precision ground triangular bars with a standard micrometer to measure threads (Lane Community College).

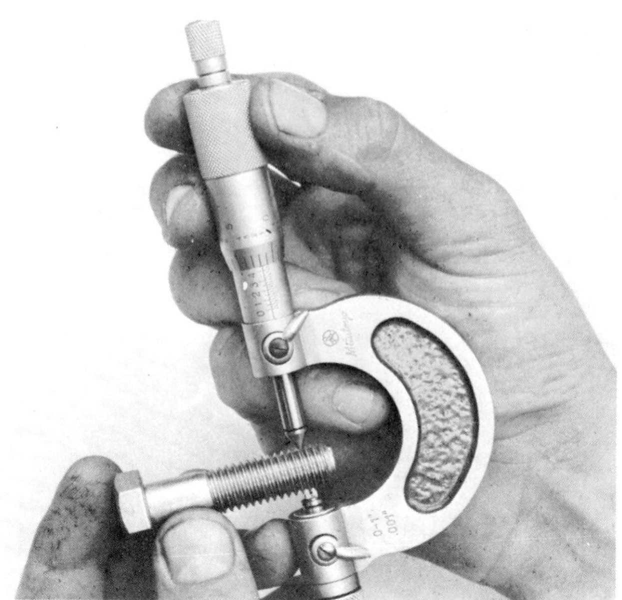

FIGURE I-332 Screw thread micrometer (Yuba College).

these micrometers depends on the manufacturer. Typical ranges are as follows:

No. 1	8 to 14 threads per inch
No. 2	14 to 20 threads per inch
No. 3	22 to 30 threads per inch
No. 4	32 to 40 threads per inch

The optical comparator is sometimes used to check thread form, helix angle, and depth of thread on external threads (Figure I-333). The part is mounted in a screw thread accessory that is adjusted to the helix angle of the thread so that the light beam will show a true profile of the thread.

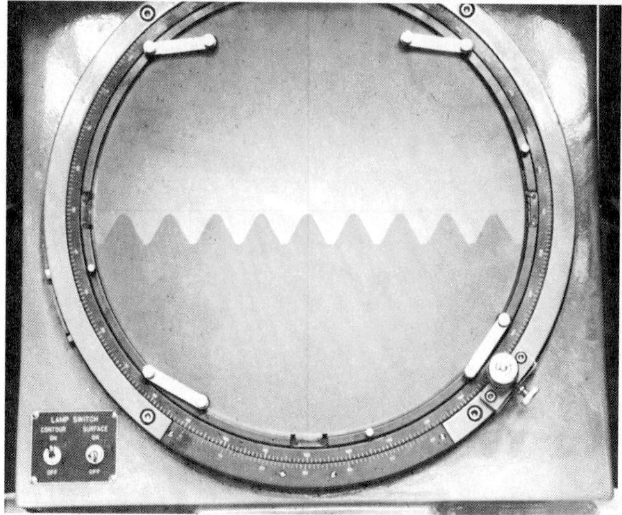

FIGURE I-333 Profile of thread as shown on the screen of the optical comparator (Lane Community College).

SELF-TEST

1. By what method are threads cut or chased with a single-point tool in a lathe? How can a given helix or lead be produced?

2. The better practice is to feed the tool in with the compound set at 29 degrees rather than with the cross slide when cutting threads. Why is this so?

3. By what means should a threading tool be checked for the 60-degree angle?

4. How can the number of threads per inch be checked?

5. How is the tool aligned with the work?

6. Is the carriage moved along the ways by means of gears when the half-nut lever is engaged? Explain.

7. Explain which positions on the threading dial are used for engaging the half-nuts for even-, odd-, and fractional-numbered threads.

8. How fast should the spindle be turning for threading?

9. What is the procedure for cutting left-hand threads?

10. If for some reason it becomes necessary for you to temporarily remove the tool or the entire threading setup before a thread is completed, what procedure is needed when you are ready to finish the thread?

UNIT 12

Cutting Unified Internal Threads

While small internal threads are tapped, larger sizes from 1 in. and up are often cut in a lathe. The problems and calculations involved with cutting internal threads differ in some ways from those of cutting external threads. This unit will help you understand these differences.

Objective

After completing this unit, you should be able to:

Calculate the dimensions of and cut an internal Unified thread.

Many of the same rules used for external threading apply to internal threading. The tool must be shaped to the exact form of the thread, and the tool must be set on the center of the workpiece. When cutting an internal thread with a single-point tool, the inside diameter (hole size) of the workpiece should be the minor diameter of the internal thread (Figure I-334). On the other hand, if the thread is made by tapping in the lathe, the hole size of the workpiece can be varied to obtain the desired percent of thread. See Section B, Unit 7 for determining percent of thread for tapping.

Full depth threads are very difficult to tap in soft metals and impossible in tough materials. Tests have proven that above 60 percent of thread very little additional strength is gained. Lower percentages, however, provide less flank surface for wear. Most commercial internal threads in steel are about 75 percent. Therefore, the purpose of making an internal thread with less than 100 percent thread depth is only for easier tapping. No such problem exists when making internal threads with a single-point tool on a lathe.

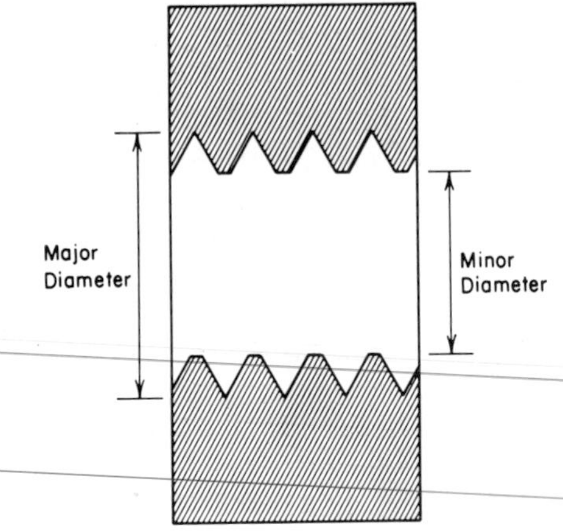

FIGURE I-334 View of internal threads showing major and minor diameters.

Single-Point Tool Threading

The advantages of making internal threads with a single-point tool are that large threads of various forms can be made and that the threads are concentric to the axis of the work. The threads may not be concentric when they are tapped. There are some difficulties encountered when making internal threads. The tool is often hidden from view and tool spring must be taken into account.

The hole to be threaded is first drilled to $\frac{1}{16}$ in. diameter less than the minor diameter. Then a boring bar is set up, and the hole is bored to the minor diameter of the thread. If the thread is to go completely through the work, no recess is nec-

FIGURE I-335 For right-hand internal threads, the compound is swiveled to the left (Lane Community College).

FIGURE I-336 The compound rest is swiveled to the right for left-hand internal threads (Lane Community College).

essary, but if threading is done in a blind hole, a recess must be made. The compound rest should be swiveled 29 degrees to the left of the operator for cutting right-hand threads (Figures I-335 and I-336). A threading tool is clamped in the bar and aligned by means of a center gage (Figure I-337).

The compound micrometer collar is moved to the zero index after the slack has been removed by turning the screw outwards or counterclockwise. The tool is brought to the work with the cross slide handle and its collar is set on zero. Threading may now proceed in the same manner as it is done with external threads. The compound is advanced outwards a few thousandths of an inch, a scratch cut is made, and the thread pitch is checked with a screw pitch gage.

The cross slide is backed out of the cut and reset to zero before the next pass. Cutting oil is

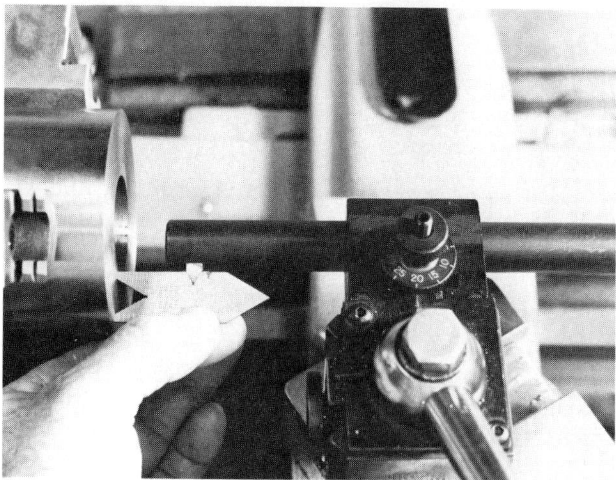

FIGURE I-337 Aligning the threading tool with a center gage (Lane Community College).

used. The compound is advanced a few thousandths (.001 to .010 in.).

The infeed on the compound is calculated in the same way as with external threads; for Unified internal threads, use the formula

$$\text{infeed} = P \times .625$$

for the depth of cut with the compound set at 29 degrees. Using the pitch from the previous example,

$$\text{infeed} = .1666 \times .625 = .104 \text{ in.}$$

The single depth of the American National thread equals $P \times .6495$ and the infeed on the compound set at 29 degrees is $P \times .75$. These figures may be substituted in the previous calculations to determine the single depth and infeed for a $1\frac{1}{2}$–6 NC internal thread.

Often it is necessary to realign an internal threading tool with the thread when the tool has been moved for sharpening or when the setup has moved during the cut. The tool is realigned in the same way as it is done for external threads: by engaging the half-nut and positioning the tool in a convenient place over the threads, then moving both the compound and cross slides to adjust the tool position.

The exact amount of infeed depends on how rigid the boring bar and holder are and how deep the cut has progressed. Too much infeed will cause the bar to spring away and produce a bellmouth internal thread.

If a slender boring bar is necessary or there is more than usual overhang, lighter cuts must be used to avoid chatter. The bar may spring away from the cut causing the major diameter to be less than the calculated amount, or that amount fed in on the compound. If several passes are taken

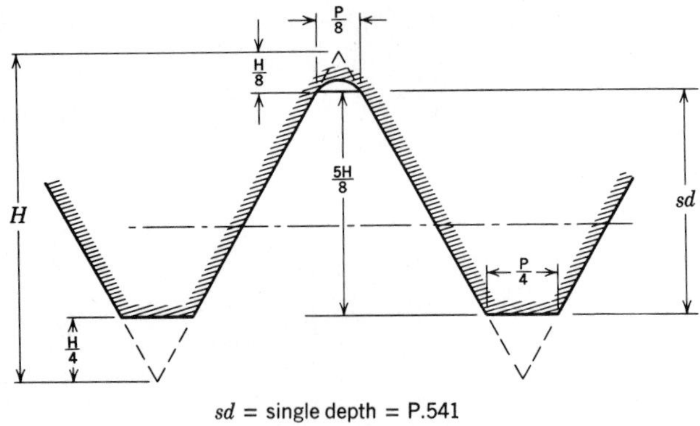

FIGURE I-338 Single depth of the Unified internal thread.

through the thread with the same setting on the compound, this problem can often be corrected.

The single depth of the Unified internal thread (Figure I-338) equals $P \times .541$. The minor diameter is found by subtracting the double depth of the thread from the major diameter. Thus if

$$D = \text{major diameter}$$

$$d = \text{minor diameter}$$

$$P = \text{thread pitch}$$

$$P \times .541 = \text{single depth}$$

the formula is

$$d = D - (P \times .541 \times 2)$$

EXAMPLE

A $1\frac{1}{2}$–6 UNC nut must be bored and threaded to fit a stud. Find the dimension of the bore.

$$P = \tfrac{1}{6} = .1666$$

$$d = 1.500 - (.1666 \times .541 \times 2)$$

$$d = 1.500 - .180$$

$$d = 1.320$$

Thus the bore should be made 1.320 in.

CUTTING THE INTERNAL THREAD

The following is the procedure for cutting right-hand internal threads.

STEP 1. After correctly setting up the boring bar and tool, touch the threading tool to the bore and set both crossfeed and compound micrometer dials to zero.

STEP 2. Feed in counterclockwise .002 in. on the compound dial.

STEP 3. Turn on the lathe and engage the half-nut lever.

STEP 4. Take a scratch cut. Disengage the half nut when the tool is through the workpiece.

STEP 5. Check the thread pitch with a screw pitch gage.

STEP 6. Apply cutting fluid to the work.

STEP 7. Feed the compound in an appropriate amount. (See Step 7 of "Cutting the Thread" in Unit 11 for an explanation of depth of cut when threading.) Slightly less depth for cutting internal threads than for similar external threads may be necessary because of the spring of the boring bar.

STEP 8. When nearing the calculated depth, test the thread with a plug gage between each pass. When the plug gage turns completely into the thread without being loose, the thread is finished.

STEP 9. Several free cuts (passes without infeed) should be taken when nearing the finish depth to compensate for boring bar spring. The test plug should be used between each free pass.

STEP 10. The inside and outside edges of the internal thread should be chamfered.

Basic Internal Thread Measurement

Since small internal threads are most often made by tapping, the pitch diameter and fit are determined by the tap used. However, internal threads cut with a single-point tool need to be checked. A precision thread plug gage (Figure I-339) is generally sufficient for most purposes. These gages are available in various sizes, which are stamped on the handle. An external screw thread is on each end of the handle. The longer threaded end is called the "go" gage, while the shorter end is called the "no go" gage. The no go end is made to a slightly larger dimension than the pitch diameter for the class of fit that the gage tests. To test an internal thread, both the go and no go gages should be tried in the hole. If the part is within the range or tolerance of the gage, the go end should turn in flush to the bottom of the internal thread, but the no go end should just start into the hole and become snug with no more than three turns. The gage should never be forced into the hole. If no gage is available, a shop gage can be made by cutting the required external thread to very precise dimen-

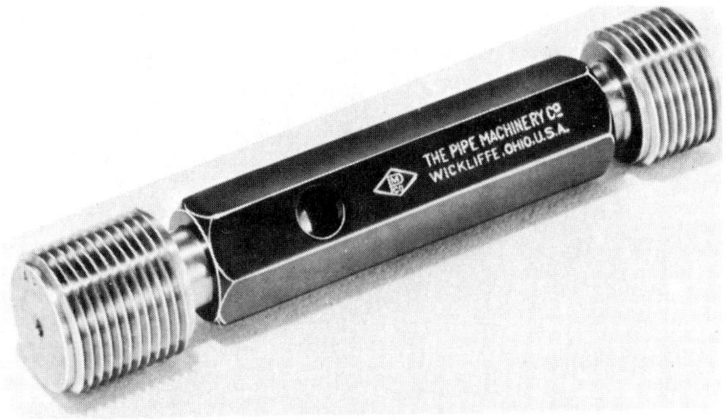

FIGURE I-339 Thread plug gage (PMC Industries).

sions. If only one threaded part of a kind is to be made and no interchangeability is required, the mating part may be used as a gage.

SELF-TEST

1. When internal threads are made with a tool, what should the bore size be?

2. In what way is percent of thread obtained? Why is this done?

3. What percent of thread are tap drill charts usually based on?

4. Drills often make an oversized hole that lowers the percent of thread. How can a more precise hole size be made for single-point tool thread cutting?

5. Name two advantages of making internal threads with a single-point tool on the lathe.

6. When making internal right-hand threads, which direction should the compound be swiveled?

7. After a scratch cut is made, what would be the most convenient method to measure the pitch of the internal thread?

8. What does deflection or spring of the boring bar cause when cutting internal threads?

9. Using $P \times .541$ as a constant for Unified single depth internal threads, what would the minor diameter be for a UNC 1–8 thread?

10. Name two methods of checking an internal thread for size.

UNIT 13

Cutting Tapers

Tapers are very useful machine elements that are used for many purposes. The machinist should be able to quickly calculate a specific taper and to set up a machine to produce it. The machinist should also be able to accurately measure tapers and determine proper fits. This unit will help you understand the various methods and principles involved in making a taper.

Objective

After completing this unit, you should be able to:

Describe different types of tapers and the methods used to produce and measure them.

Use of Tapers

Tapers are used on machines because of their capacity to align and hold machine parts and to realign when they are repeatedly assembled and disassembled. This repeatability assures that tools such as centers in lathes, taper shank drills in drill presses, and arbors in milling machines will run in perfect alignment when placed in the machine. When a taper is slight, such as a Morse taper that is about $\frac{5}{8}$ in. taper/ft, it is called a self-holding taper since it is held in and driven by friction (Fig-

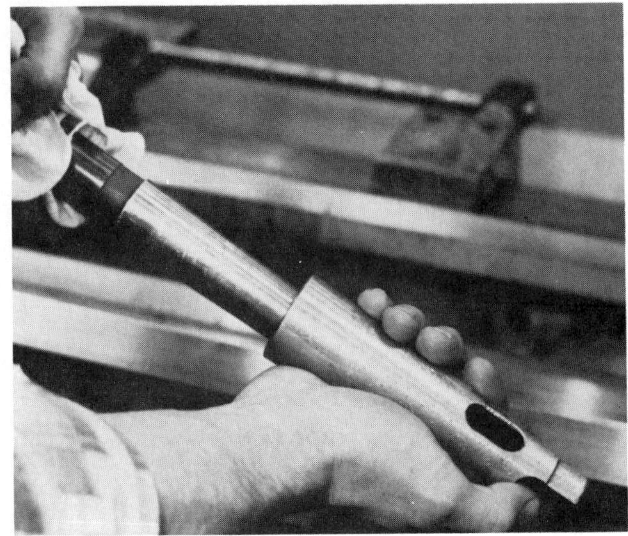

FIGURE I-340 The Morse taper shank on this drill keeps the drill from turning when the hole is being drilled (Lane Community College).

FIGURE I-341 The milling machine taper is driven by lugs and held in by a draw bolt (Lane Community College).

ure I-340). A steep taper, such as a quick-release taper of 3½ in./ft used on most milling machines, must be held in place with a draw bolt (Figure I-341).

A taper may be defined as a uniform increase in diameter on a workpiece for a given length measured parallel to the axis. Internal or external tapers are expressed in taper per foot (tpf), taper per inch (tpi), or in degrees. The tpf or tpi refers to the difference in diameters in the length of 1 ft or 1 in., respectively (Figure I-342). This difference is measured in inches. Angles of taper, on the other hand, may refer to the included angles or the angles with the centerline (Figure I-343).

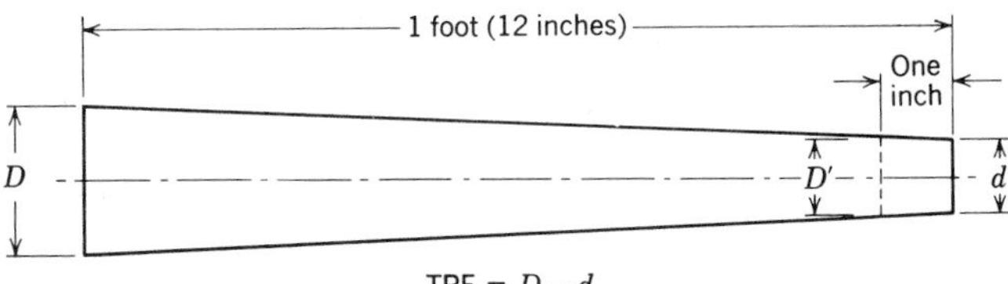

FIGURE I-342 The difference between taper per foot (tpf) and taper per inch (tpi).

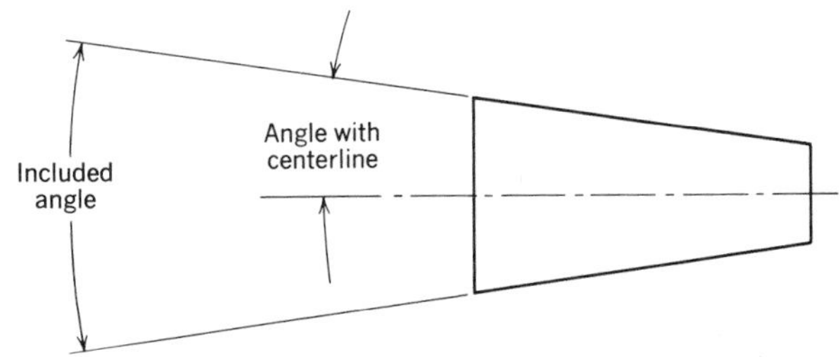

FIGURE I-343 Included angles and angles with centerline.

Some machine parts that are measured in taper per foot are mandrels (.006 in./ft), taper pins and reamers (¼ in./ft), the Jarno taper series (.600 in./ft), the Brown and Sharpe taper series (approximately ½ in./ft), and the Morse taper series (about ⅝ in./ft). Morse tapers include eight sizes that range from size 0 to size 7. Tapers and dimensions vary slightly from size to size in both the Brown and Sharpe and the Morse series. For instance, a No. 2 Morse taper has .5944 in./ft taper and a No. 4 has .6233 in./ft taper. See Table I-7 for more information on Morse tapers.

TABLE I-7 Morse Tapers Information

Number of Taper	Taper per Foot	Taper per Inch	P Standard Plug Depth	*D* Diameter of Plug at Small End	*A* Diameter at End of Socket	H Depth of Hole
0	.6246	.0520	2	.252	.356	$2\frac{1}{32}$
1	.5986	.0499	$2\frac{1}{8}$	.396	.475	$2\frac{3}{16}$
2	.5994	.0500	$2\frac{9}{16}$	.572	.700	$2\frac{5}{8}$
3	.6023	.0502	$3\frac{3}{16}$	.778	.938	$3\frac{1}{4}$
4	.6232	.0519	$4\frac{1}{16}$	1.020	1.231	$4\frac{1}{8}$
5	.6315	.0526	$5\frac{3}{16}$	1.475	1.748	$5\frac{1}{4}$
6	.6256	.0521	$7\frac{1}{4}$	2.116	2.494	$7\frac{3}{8}$
7	.6240	.0520	10	2.750	3.270	$10\frac{1}{8}$

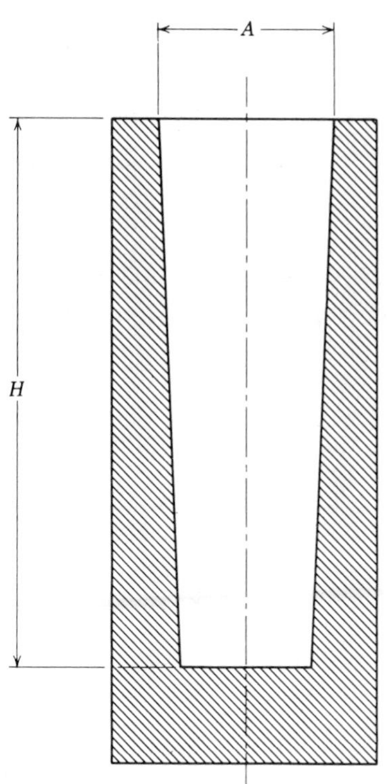

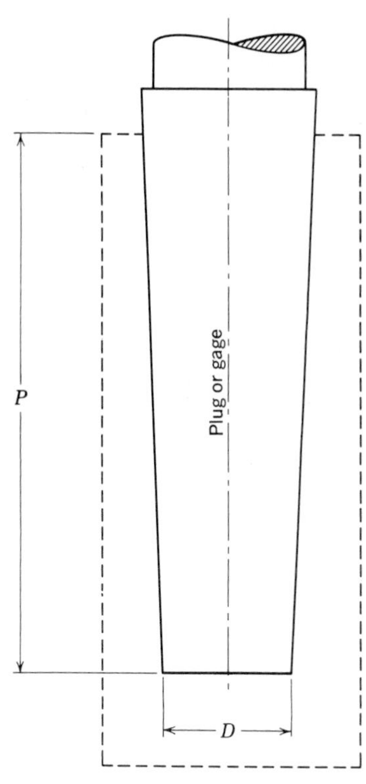

Methods of Making a Taper

There are several methods of turning a taper on a lathe. They are the compound slide method, the offset tailstock method, the taper attachment method, the use of a form tool, and the use of a tracer or CNC lathe. Each method has its advantages and disadvantages, so the kind of taper needed on a workpiece should be the deciding factor in the selection of the method that will be used.

THE COMPOUND SLIDE METHOD

Both internal and external short steep tapers can be turned on a lathe by hand feeding the compound slide (Figure I-344). The swivel base of the compound is divided in degrees. When the compound slide is in line with the ways of the lathe, the 0-degree line will align with the index line on the cross slide (Figure I-345). When the compound is swiveled off the index, which is parallel to the centerline of the lathe, a direct reading may be taken for the half angle or angle to centerline of the machined part (Figure I-346). When a taper is machined off the lathe centerline, its included angle will be twice the angle that is set on the compound. Not all lathes are indexed in this manner.

FIGURE I-344 Making a taper using the compound slide (Lane Community College).

FIGURE I-345 Alignment of the compound parallel with the ways (Lane Community College).

FIGURE I-346 An angle may be set off the axis of the lathe from this index (Lane Community College).

FIGURE I-347 The compound set 14½ degrees off the axis of the cross slide (Lane Community College).

When the compound slide is aligned with the axis of the cross slide and swiveled off the index in either direction, an angle is directly read off the cross slide centerline (Figure I-347). Since the lathe centerline is 90 degrees from the cross slide centerline, the reading on the lathe centerline index is the complementary angle. So if the compound is set off the axis of the cross slide 14½ degrees, the lathe centerline index reading is 90 − 14½ = 75½ degrees, as seen in Figure I-347.

Tapers of any angle may be cut by this method, but the length is limited to the stroke of the compound slide. Since tapers are often given in tpf, it is sometimes convenient to consult a tpf-to-angle conversion table, as in Table I-8. A more complete table may be found in *Machinery's Handbook*.

If a more precise conversion is desired, the following formula may be used to find the included angle: Divide the taper in inches per foot by 24; find the angle that corresponds to the quotient in a table of tangents; and double this angle. If the angle with centerline is desired, do not double the angle.

TABLE I-8 Tapers and Corresponding Angles

Taper per Foot	Including Angle		Angles with Centerline		Taper per Inch
	Degrees	Minutes	Degrees	Minutes	
1/8	0	36	0	18	.0104
3/16	0	54	0	27	.0156
1/4	1	12	0	36	.0208
5/16	1	30	0	45	.0260
3/8	1	47	0	53	.0313
7/16	2	5	1	2	.0365
1/2	2	23	1	11	.0417
9/16	2	42	1	21	.0469
5/8	3	00	1	30	.0521
11/16	3	18	1	39	.0573
3/4	3	35	1	48	.0625
13/16	3	52	1	56	.0677
7/8	4	12	2	6	.0729
15/16	4	28	2	14	.0781
1	4	45	2	23	.0833
1¼	5	58	2	59	.1042
1½	7	8	3	34	.1250
1¾	8	20	4	10	.1458
2	9	32	4	46	.1667
2½	11	54	5	57	.2083
3	14	16	7	8	.2500
3½	16	36	8	18	.2917
4	18	56	9	28	.3333
4½	21	14	10	37	.3750
5	23	32	11	46	.4167
6	28	4	14	2	.5000

EXAMPLE

What angle is equivalent to a taper of 3½ in./ft?

$$\frac{3.5}{24} = .14583$$

The angle of this tangent is 8 degrees 18 minutes, and the included angle is twice this, or 16 degrees 36 minutes.

When turning either an internal or external taper by any method on a lathe, the cutting tool must be on the exact centerline of the lathe. If the tool is either too high or too low, the taper will not correspond to any calculations and it will be in error.

The Offset Tailstock Method

Long, slight tapers may be produced on shafts and external parts between centers. Internal tapers

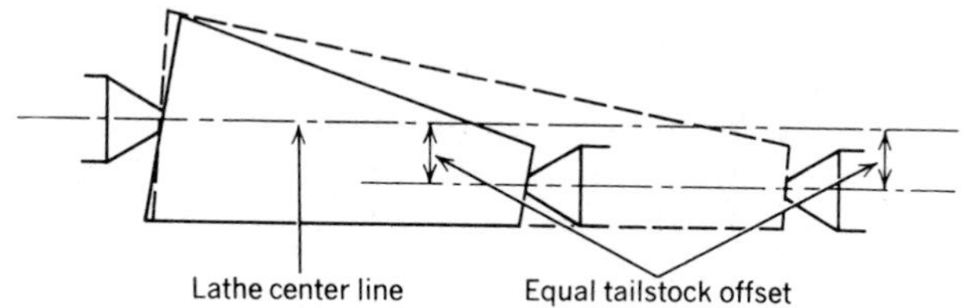

FIGURE I-348 When tapers are of different lengths, the tpf is not the same with the same offset.

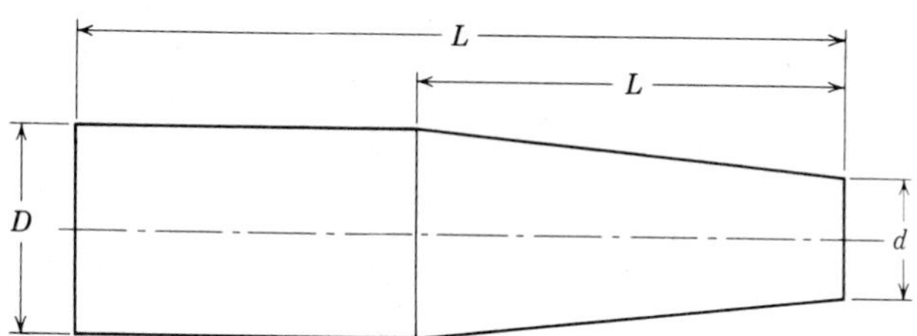

FIGURE I-349 Long workpiece with a short taper.

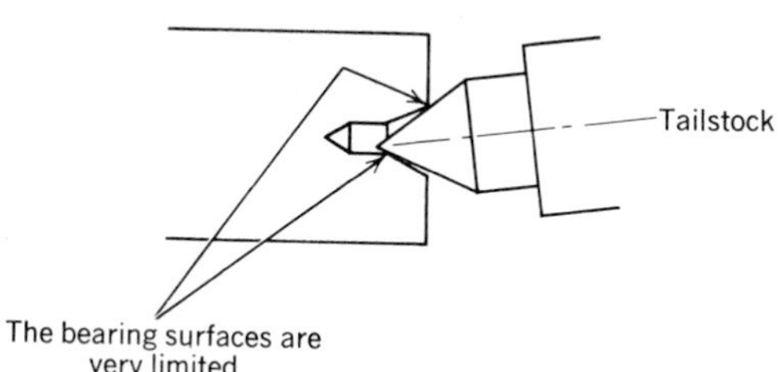

FIGURE I-350 The contact area between the center hole and the center is small.

cannot be made by this method. Power feed is used so good finishes are obtainable. Since most lathes are equipped with taper attachments, the offset tailstock method is not used much now. Its greatest advantage is that longer tapers can be made by this method than by any other. The taper per foot or taper per inch must be known so the amount of offset for the tailstock can be calculated. Since tapers are of different lengths, they would not be the same tpi or tpf for the same offset (Figure I-348). When the taper per inch is known, the offset calculation is as follows:

$$\text{offset} = \frac{\text{tpi} \times L}{2}$$

where tpi = taper per inch and L = length of workpiece. When the taper per foot is known, use the following formula:

$$\text{Offset} = \frac{\text{tpf} \times L}{24}$$

If the workpiece has a short taper in any part of its length (Figure I-349), and the tpi or tpf is not given, use the following formula:

$$\text{offset} = \frac{L \times (D - d)}{2 \times L_1}$$

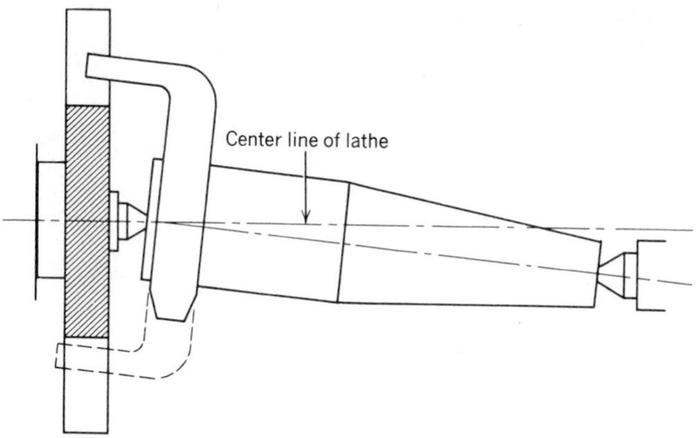

FIGURE I-351 The bent tail of the lathe dog should have adequate clearance.

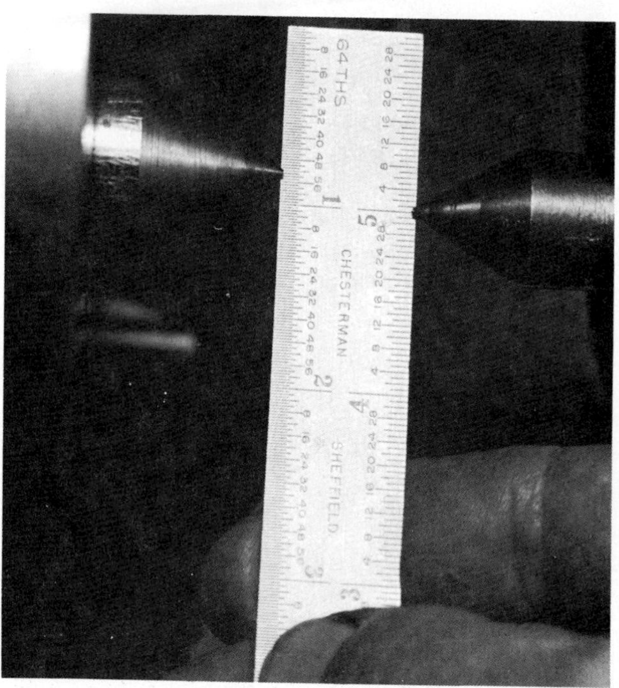

FIGURE I-352 Measuring the offset on the tailstock by use of the centers and the rule (Lane Community College).

where D = diameter at large end of taper
d = diameter at small end of taper
L = total length of workpiece
L_1 = length of taper

When you set up for turning a taper between centers, remember that the contact area between the center and the center hole is limited (Figure I-350). Frequent lubrication of the centers may be necessary.

You should also note the path of the lathe dog bent tail in the drive slot (Figure I-351). Check to see that there is adequate clearance.

To measure the offset on the tailstock, use either the centers and a rule (Figure I-352) or the

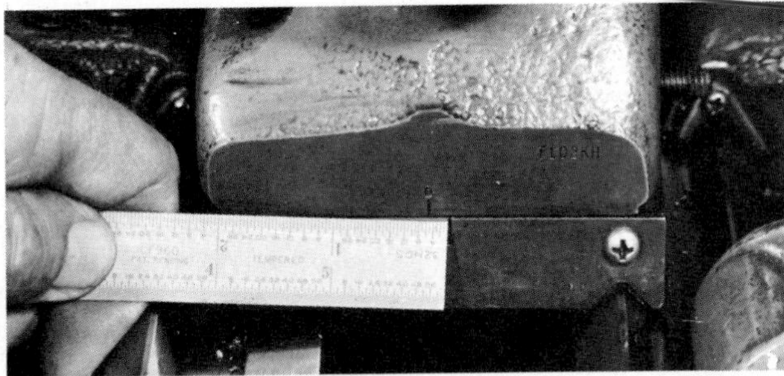

FIGURE I-353 Measuring the offset with the witness mark and a rule (Lane Community College).

FIGURE I-354 Using the dial indicator to measure the offset (Lane Community College).

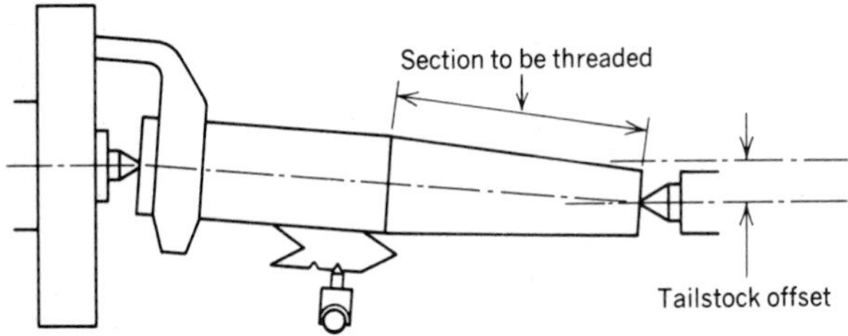

FIGURE I-355 Adjusting the threading tool for cutting tapered threads using the offset tailstock method. The tool is set square to the centerline of the work rather than the taper.

witness mark and a rule (Figure I-353); both methods are adequate for some purposes. A more precise measurement is possible with a dial indicator as shown in Figure I-354. The indicator is set on the tailstock spindle while the centers are still aligned. A slight loading of the indicator is advised since the first .010 or .020 in. of indicator movement may be inaccurate or the mechanism loose due to wear, causing fluctuating readings. The bezel is set at zero and the tailstock is moved toward the operator the calculated amount. Clamp the tailstock to the way. If the indicator reading changes, loosen the clamp and readjust.

When cutting tapered threads such as pipe threads, the tool should be square with the centerline of the workpiece, not the taper (Figure I-355). When you have finished making tapers by the offset tailstock method, realign the centers to .001 in. or less in 12 in. When more than one part must be turned by this method, all parts must have identical lengths and center hole depths if the tapers are to be the same.

FIGURE I-356 Plain taper attachment (Lane Community College).

THE TAPER ATTACHMENT METHOD

The taper attachment features a slide independent to the ways that can be angled and will move the cross slide according to the angle set. Slight to fairly steep tapers ($3\frac{1}{2}$ in./ft) may be made this way. Length of taper is limited to the length of the slide bar. Centers may remain in line without distortion of the center holes. Work may be held in a chuck and both external and internal tapers may be made, often with the same setting for mating parts. Power feed is used. Taper attachments are graduated in taper per foot (tpf) or in degrees.

There are two types of taper attachments: the plain taper attachment (Figure I-356) and the telescoping taper attachment (Figure I-357). The cross feed binding screw must be removed to free the nut when the plain type is set up. The depth of cut must then be made by using the compound feed screw handle. The crossfeed may be used for depth of cut when using the telescoping taper attachment since the crossfeed binding screw is not disengaged with this type.

When a workpiece is to be duplicated or an internal taper is to be made for an existing external taper, it is often convenient to set up the taper attachment by using a dial indicator (Figure I-358). The contact point of the dial indicator must

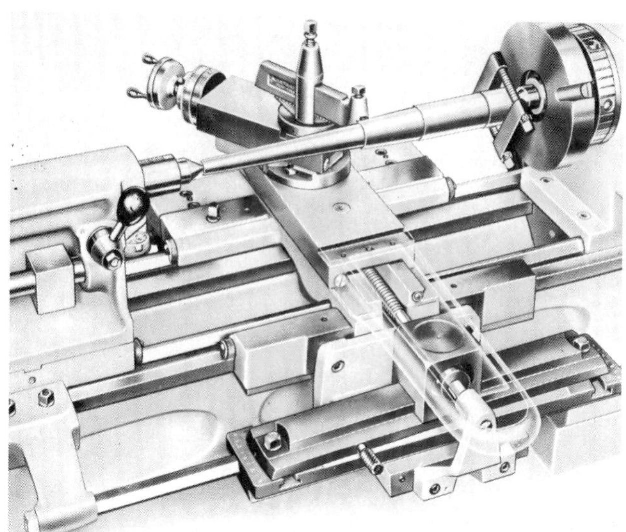
FIGURE I-357 Telescopic taper attachment (Clausing Corporation).

FIGURE I-358 Adjusting the taper attachment to a given taper with a dial indicator (Lane Community College).

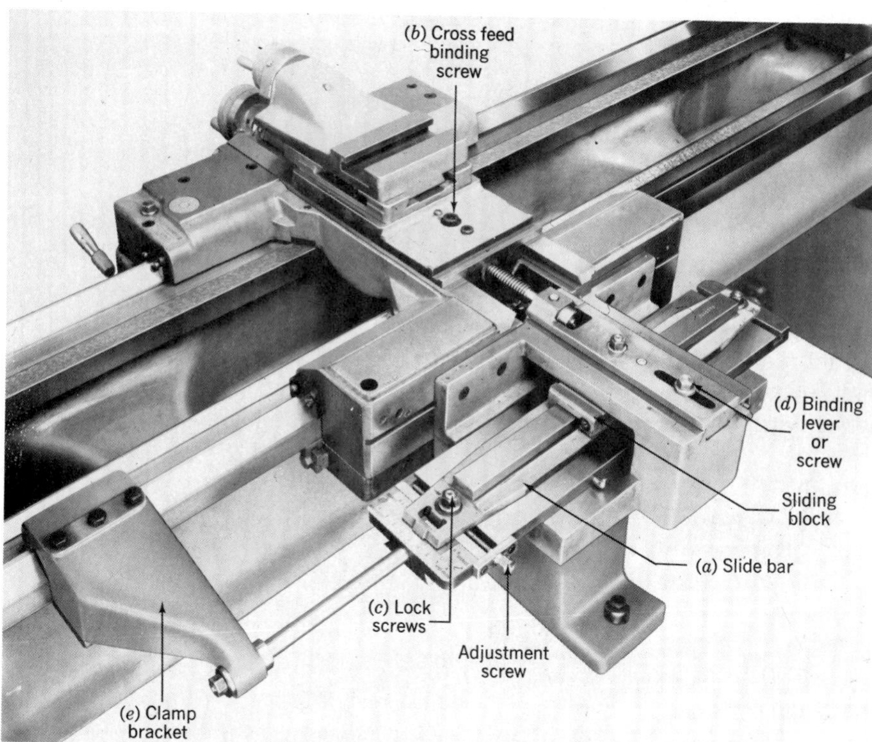

FIGURE I-359 Parts of the taper attachment (Lodge & Shipley Company).

be on the center of the workpiece. The workpiece is first set up in a chuck or between centers so there is no runout when it is rotated. With the lathe spindle stopped, the indicator is moved from one end of the taper to the other. The taper attachment is adjusted until the indicator does not change reading when moved.

The angle, the taper per foot, or the taper per inch must be known to set up the taper attachment to cut specific tapers. If none of these are known, proceed as follows: If the end diameters (D and d) and the length of taper (L) are given in inches, the following applies:

$$\text{taper per foot} = \frac{D - d}{L} \times 12$$

If the taper per foot is given, but you want to know the amount of taper in inches for a given length, use the following formula:

$$\text{amount of taper} = \frac{\text{tpf}}{12} \times \text{given length of tapered part}$$

When the tpf is known, to find tpi divide the tpf by 12. When the tpi is known, to find tpf multiply the tpi by 12.

TURNING A TAPER WITH A TAPER ATTACHMENT (Figure I-359)

STEP 1. Clean and oil the slide bar (a).

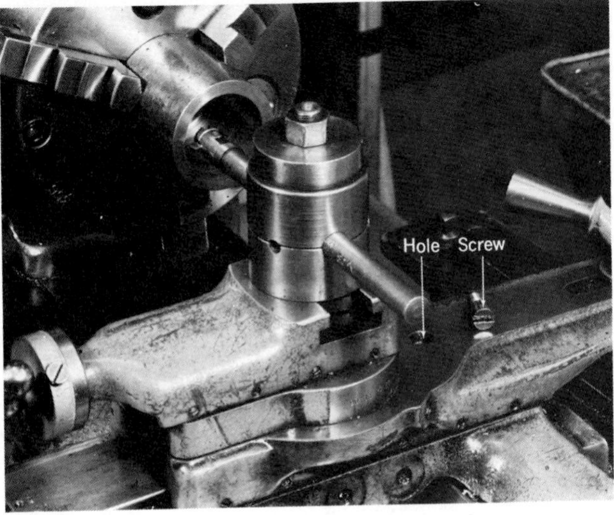

FIGURE I-360 Internal taper being made with a plain taper attachment. Note that the crossfeed nut locking screw has been removed and the hole has not been plugged. This hole *must* be plugged and the screw stored in a safe place (Lane Community College).

STEP 2. Set up the workpiece and the cutting tool on center. Bring the tool near the workpiece and to the center of the taper.

STEP 3. Remove the crossfeed binding screw (*b*) that binds the crossfeed screw nut to the cross slide. *Do not remove* this screw if you are using a telescoping taper attachment. The screw is removed *only* on the plain type. Put a temporary plug in the hole to keep chips out.

STEP 4. Loosen the lockscrews (*c*) on both ends of the slide bar and adjust to the required degree of taper.

STEP 5. Tighten the lockscrews.

STEP 6. Tighten the binding lever (*d*) on the slotted cross slide extension at the sliding block, *plain type only*.

STEP 7. Lock the clamp bracket (*e*) to the lathe bed.

STEP 8. Move the carriage to the right so that the tool is from $\frac{1}{2}$ to $\frac{3}{4}$ in. past the start position. This should be done on every pass to remove any backlash in the taper attachment.

STEP 9. Feed the tool in for the depth of the first cut with the cross slide unless you are using a plain-type attachment. Use the compound slide for the plain type.

STEP 10. Take a trial cut and check for diameter. Continue the roughing cut.

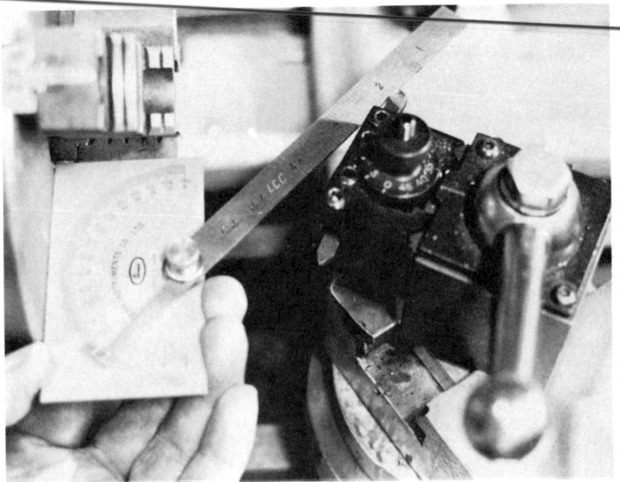

FIGURE I-361 Tool is set up with protractor to make an accurate chamfer or taper (Lane Community College).

FIGURE I-362 Making the chamfer with a tool (Lane Community College).

STEP 11. Check the taper for fit and readjust the taper attachment if necessary.

STEP 12. Take a light cut, about .010 in., and check the taper again. If it is correct, complete the roughing and final finish cuts.

Internal tapers (Figure I-360) are best made with the taper attachment. They are set up in the same manner as prescribed for external tapers.

OTHER METHODS OF MAKING TAPERS

A tool may be set with a protractor to a given angle (Figures I-361 and I-362) and a single plunge cut may be made to produce a taper. This method is

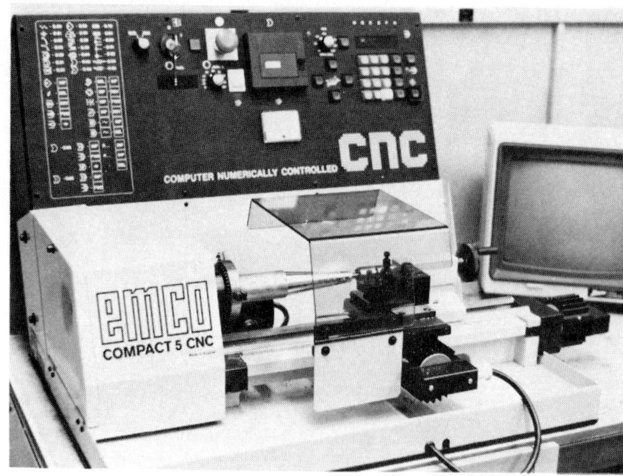

FIGURE I-363 CNC lathes such as this one can be easily programmed to produce an accurate taper (Lane Community College).

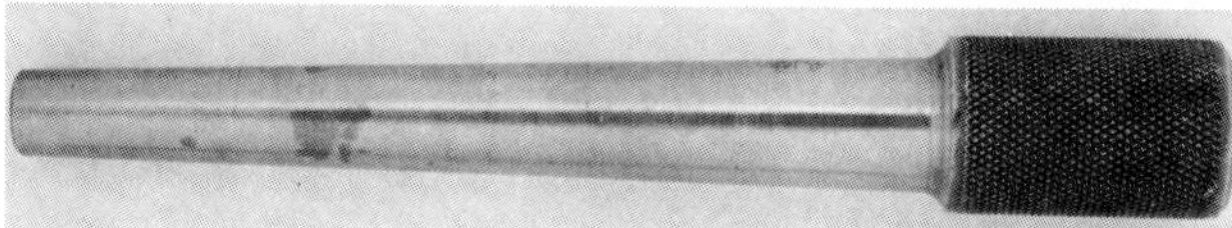

FIGURE I-364 Taper plug gage (Lane Community College).

FIGURE I-365 Taper ring gage (Lane Community College).

often used for chamfering a workpiece to an angle such as the chamfer used for hexagonal bolt heads and nuts. Tapered form tools sometimes are used to make V-shaped grooves. Only very short tapers can be made with form tools.

CNC machines (Figure I-363) can be programmed to make any taper, either external or internal. An advantage to this method is the precision repeatability when many tapers must be made with the same dimensions.

Tapered reamers are sometimes used to produce a specific taper such as a Morse taper. A roughing reamer is used first, followed by a finishing reamer. Finishing Morse taper reamers are often used to true up a badly nicked and scarred internal Morse taper.

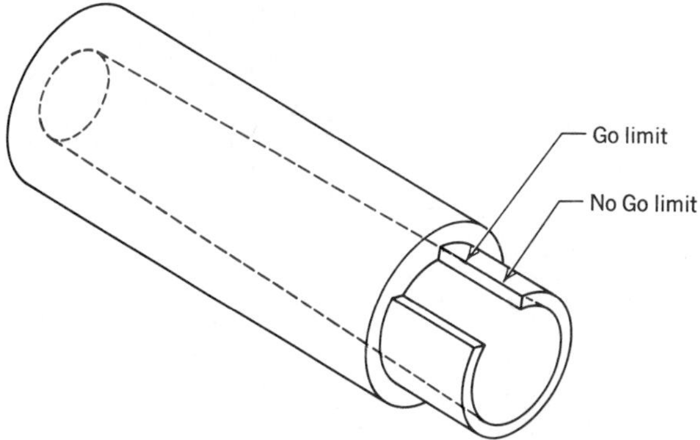

FIGURE I-366 Go/no go taper ring gage.

FIGURE I-367 Chalk mark is made along a taper plug gage prior to checking an internal taper (Lane Community College).

FIGURE I-368 The taper has been tested and the chalk mark has been rubbed off evenly, indicating a good fit (Lane Community College).

Methods of Measuring Tapers

The most convenient and simple way of checking tapers is to use the taper plug gage (Figure I-364) for internal tapers and the taper ring gage (Figure I-365) for external tapers. Some taper gages have go and no go limit marks on them (Figure I-366).

To check an internal taper, a chalk or Prussian blue mark is first made along the length of the taper plug gage (Figures I-367 and I-368). The gage

FIGURE I-369 The external taper is marked with chalk or Prussian blue before being checked with a taper ring gage (Lane Community College).

FIGURE I-371 The ring gage is removed and the chalk mark is rubbed off evenly for the entire length of the ring gage, which indicates a good fit (Lane Community College).

FIGURE I-370 The ring gage is placed on the taper snugly and is rotated slightly (Lane Community College).

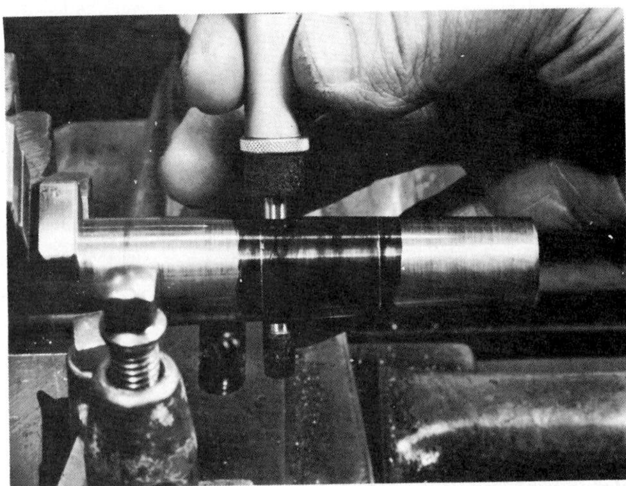

FIGURE I-372 Measuring the taper per inch (tpi) with a micrometer. The larger diameter is measured on the line with the edge of the spindle and the anvil of the micrometer contacting the line (Lane Community College).

is then inserted into the internal taper and turned slightly. When the gage is taken out, the chalk mark will be partly rubbed off where contact was made. Adjustment of the taper should be made until the chalk mark is rubbed off along its full length of contact, indicating a good fit. An external taper is marked with chalk to be checked in the same way with a taper ring gage (Figures I-369 to I-371).

The taper per inch may be checked with a micrometer by scribing two marks 1 in. apart on the taper and measuring the diameters (Figures I-372 and I-373) at these marks. The difference is the taper per inch. A more precise way of making this measurement is shown in Figures I-374 and I-375. A surface plate is used with precision parallels and drill rods. The tapered workpiece would have to be removed from the lathe if this method is used, however.

Perhaps an even more precise method of measuring a taper is with the sine bar and gage blocks on the surface plate (Figure I-376). When this is done, it is important to keep the centerline of the taper parallel to the sine bar and to read the indicator at the highest point.

Tapers may be measured with a taper micrometer. Refer to Section C, Unit 4 for a description of this instrument.

SELF-TEST

1. State the difference in use between steep tapers and slight tapers.

FIGURE I-373 The second measurement is taken on the smaller diameter at the edge of the line in the same manner (Lane Community College).

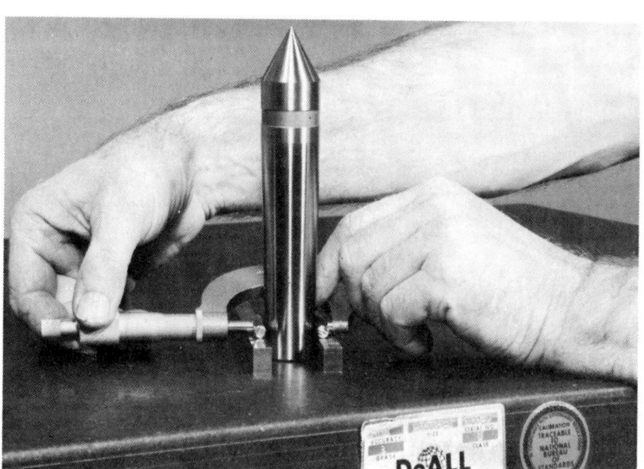

FIGURE I-374 Checking the taper on a surface plate with precision parallels, drill rod, and micrometer. The first set of parallels is used so that the point of measurement is accessible to the micrometer (Lane Community College).

2. In what three ways are tapers expressed (measured)?

3. Briefly describe the four methods of turning a taper in the lathe.

4. When a taper is produced by the compound slide method, is the reading in degrees on the compound swivel base the same as the angle of the finished workpiece? Explain.

5. If the swivel base is set to a 35-degree angle at the cross slide centerline index, what would the reading be at the lathe centerline index?

6. Calculate the offset for the taper shown in Figure I-377. The formula is

$$\text{offset} = \frac{L \times (D - d)}{2 \times L_1}$$

FIGURE I-375 When 1-in. wide parallels are in place, a second measurement is taken. The difference is the taper per inch (Lane Community College).

FIGURE I-376 Using a sine bar and gage blocks with a dial indicator to measure a taper (Lane Community College).

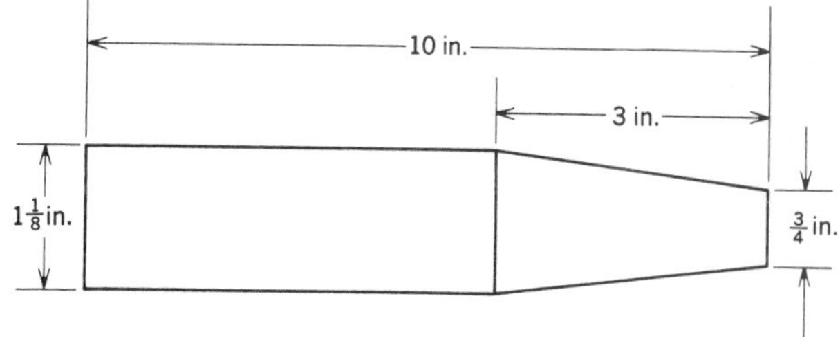

FIGURE I-377

7. Name four methods of measuring tapers.

8. What are the two types of taper attachments, and what are their advantages over other means of making a taper?

9. What is the most practical and convenient way to check internal and external tapers when they are in the lathe?

10. Describe the kinds of tapers that may be made by using a form tool or the side of a tool.

UNIT 14

Using Steady and Follower Rests

Many lathe operations would not be possible without the use of the steady and follower rests. These valuable attachments make internal and external machining operations on long workpieces possible on a lathe.

Objectives

After completing this unit, you should be able to:

1. Identify the parts and explain the uses of the steady rest.
2. Explain the correct uses of the follower rest.
3. Correctly set up a steady rest on a straight shaft.
4. Correctly set up a follower rest on a prepared shaft.

The Steady Rest

On a lathe, long shafts tend to vibrate when cuts are made, leaving chatter marks. Even light finish cuts will often produce chatter when the shaft is long and slender. To help eliminate these problems, use a steady rest to support workpieces that extend from a chuck more than four or five diameters of the workpiece for turning, facing, drilling, and boring operations.

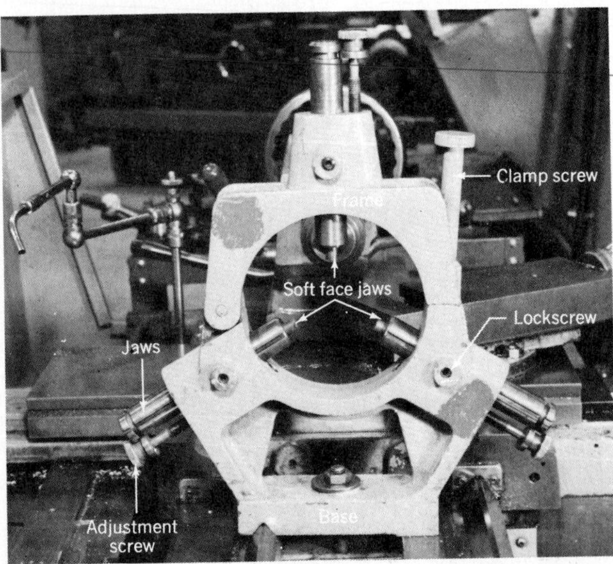

FIGURE I-378 The parts of the steady rest (Lane Community College).

The steady rest (Figure I-378) is made of a cast iron or steel frame that is hinged so that it will open to accommodate workpieces. It has three or more adjustable jaws that are tipped with bronze, plastic, or ball bearing rollers. The base of the frame is machined to fit the ways of the lathe and it is clamped to the bed by means of a bolt and crossbar.

A steady rest is also used to support long workpieces for various other machining operations such as threading, grooving, and knurling (Figure I-379). Heavy cuts can be made by using one or more steady rests along a shaft.

FIGURE I-379 A long, slender workpiece is supported by a steady rest near the center to limit vibration or chatter (Lane Community College).

FIGURE I-380 The jaws are adjusted to the shaft (Lane Community College).

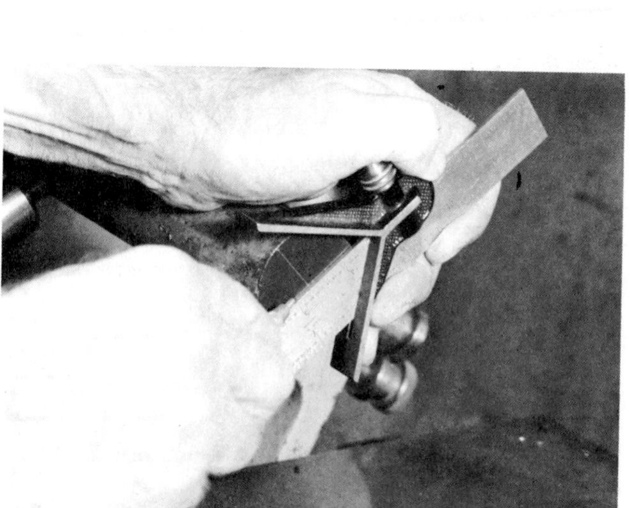

FIGURE I-381 Laying out the center of a shaft (Lane Community College).

ADJUSTING THE STEADY REST

Workpieces should be mounted and centered in a chuck whether a tailstock center is used or not. If the shaft has centers and finished surfaces that turn concentric (have no runout) with the lathe centerline, setup of the steady rest is simple. The steady rest is slid to a convenient location on the shaft, which is supported in the dead center and chuck, and the base is clamped to the bed. The two lower jaws are brought up to the shaft finger-tight only (Figure I-380). A good high-pressure lubricant is applied to the shaft and the top half of the steady rest is closed and clamped. The upper jaw is brought to the shaft finger-tight, and then all three lockscrews are tightened. Some clearance is necessary on the upper jaw to avoid scoring of

FIGURE I-382 Adjusting the steady rest jaws to a centered shaft (Lane Community College).

the shaft. As the shaft warms or heats up from friction during machining, readjustment of the upper jaw is necessary.

A finished workpiece can be scored if any hardness or grit is present on the jaws. To protect finishes, brass or copper strips or abrasive cloth is often placed between the jaws and the workpiece; with the abrasive cloth, the abrasive side is placed outward against the jaws. Steady or follower rest jaws should never support rough surfaces because they would soon be worn away, even if lubricant were used. Also, any machining done would be inaccurate because of the surface irregularities of the workpiece.

When there is no center in a finished shaft of the same diameter, setup procedure is as follows.

STEP 1. Position the steady rest near the end of the shaft with the other end lightly chucked in a three- or four-jaw chuck.

STEP 2. Scribe two cross centerlines with a center head on the end of the shaft and prick punch (Figure I-381).

STEP 3. Bring up the dead center near to the punch mark (Figure I-382).

STEP 4. Adjust the lower jaws of the steady rest to the shaft and lock.

STEP 5. Tighten the chuck. If it is a four-jaw chuck, check for runout with a dial indicator.

STEP 6. The steady rest may now be removed to any location along the shaft.

STEP 7. If needed, a center hole may now be drilled in the end of the shaft.

FIGURE I-383 Turning a concentric bearing surface on rough stock for the steady rest jaws (Lane Community College).

FIGURE I-385 Using a cat head for supporting a square piece. Drilling and boring in the end of a heavy square bar requires the use of an external cat head (Lane Community College).

FIGURE I-384 The rough shaft can now rotate in the steady rest jaws, which are being adjusted to the bearing surface in this view (Lane Community College).

FIGURE I-386 The cat head is adjusted with setscrews to center the shaft with the dead center (Lane Community College).

Stepped shafts may be set up by using a similar procedure, but the steady rest must remain on the diameter on which it is set up.

USING THE STEADY REST

A frequent misconception among students is that the steady rest may be set up properly by using a dial indicator near the steady rest on a rotating shaft. This procedure would never work since the indicator would show no offset or runout, no matter where the jaws were moved.

Steady rest jaws should never be used on rough surfaces. When a forging, casting, or hot-rolled bar must be placed in a steady rest, a concentric bearing with a good finish must be turned (Figures I-383 and I-384). Thick-walled tubing or other materials that tend to be out of round also should have bearing surfaces machined on them. The usual practice is to remove no more in diameter than necessary to clean up the bearing spot.

When the piece to be set up is very irregular, such as a square (Figure I-385), hexagonal part, HR bar, or rough casting, a cat head is used. The procedure to set up a cat head is not difficult. The piece is placed in the cat head and the cat head is mounted in the steady rest while the other end of the workpiece is centered in the chuck. The workpiece is made to run true near the steady rest by adjusting screws on the cat head (Figures I-386 to I-389). In most cases the workpiece is given a center to provide more support for turning operations. A centered cat head (Figures I-390 and I-391) is sometimes used when a permanent center is not required in the workpiece. Internal cat heads (Figure I-392) are used for truing to the inside diam-

FIGURE I-387 A spot is turned in the center of the shaft for center drilling (Lane Community College).

FIGURE I-388 The shaft is center drilled (Lane Community College).

FIGURE I-389 The center is put in place using center lube. the steady rest and cat head may now be removed, if necessary, for machining operations (Lane Community College).

FIGURE I-390 Using a centered cat head to provide a center when the end of the shaft or tube cannot be centered conveniently (Lane Community College).

FIGURE I-391 The cat head is adjusted over the irregular end of the shaft (Lane Community College).

FIGURE I-392 Tubing being set up with a cat head using a dial indicator to true the inside diameter (Lane Community College).

FIGURE I-393 A follower rest is used to turn this long shaft (Lane Community College).

FIGURE I-395 Long, slender Acme threaded screw being machined with the aid of a follower rest (Lane Community College).

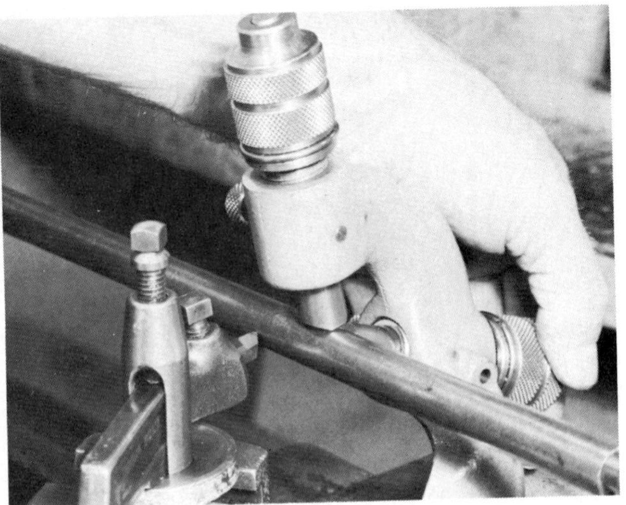

FIGURE I-394 Adjusting the follower rest (Lane Community College).

FIGURE I-396 Both steady and follower rests being used (Lane Community College).

eter of tubing that has an irregular wall thickness, so that a steady rest bearing spot can be machined on the outside diameter. These also have adjustment screws.

The Follower Rest

Long, slender shafts tend to spring away from the tool, vary in diameter, chatter, and often climb the tool. To avoid these problems when machining a slender shaft along its entire length, a follower rest (Figure I-393) is often used. Follower rests are bolted to the carriage and follow along with the tool. Most follower rests have two jaws placed to back up the work opposite to the tool thrust. Some types are made with different-size bushings to fit the work.

USING THE FOLLOWER REST

The workpiece should be 1 to 2 in. longer than the job requires to allow room for the follower rest jaws. The end is turned to smaller than the finish size. The tool is adjusted ahead of the jaws about $1\frac{1}{2}$ in. and a trial cut of 2 or 3 in. is made with the jaws backed off. Then the lower jaw is adjusted finger-tight (Figure I-394) followed by the upper jaw. Both locking screws are tightened. Oil should be used to lubricate the jaws.

The follower rest is often used when cutting threads on long, slender shafts, especially when cutting square or Acme threads (Figure I-395). Burrs should be removed between passes to prevent them from cutting into the jaws. Jaws with rolls are sometimes used for this purpose. On quite long shafts, sometimes both a steady rest and follower rest are used (Figure I-396).

SELF-TEST

1. When should a steady rest be used?
2. In what ways can a steady rest be useful?

3. How is the steady rest set up on a straight finished shaft when it has centers in the ends?

4. What precaution can be taken to prevent scoring of a finished shaft?

5. How can a steady rest be set up when there is no center hole in the shaft?

6. Is it possible to correctly set up a steady rest by using a dial indicator on the rotating shaft in order to watch for runout?

7. Should a steady rest be used on a rough surface? Explain.

8. How can a steady rest be used on an irregular surface such as square or hex stock?

9. When a long, slender shaft needs to be turned or threaded for its entire length, which lathe attachment could be used?

10. The jaws of the follower rest are usually 1 or 2 in. to the right of the tool on a setup. If the workpiece happens to be smaller than the dead center or tailstock spindle, how would it be possible to bring the tool to the end of the work to start a cut without interference by the follower rest jaws?

UNIT 15

Additional Thread Forms

Many thread forms other than the 60-degree types are to be found on machines. Each of the forms is unique and has a special use. To be able to recognize and measure these various thread forms will be very helpful to you as an apprentice machinist.

As a machinist, you may occasionally be called upon to make a multiple lead thread. When this happens, you should be prepared to select a method that is best for the job and proceed as efficiently as you would in cutting a single lead thread. This unit will acquaint you with the various methods and procedures for cutting multiple lead threads.

Objectives

After completing this unit, you should be able to:

1. Identify five different thread forms and explain their uses.
2. Calculate the dimensions needed to machine the five thread forms.
3. Describe the methods and procedures for machining multiple lead threads.

Transmitting (translating) screw threads are primarily used to transmit or impart power or motion to a mechanical part. Often these transmitting screws are of multiple lead to effect rapid motion. Bench vises and house jacks are familiar applications of single lead transmitting screws. The lead screw on a lathe and the table feed screws on milling machines are examples of these screw threads being used to impart power along the axis of the screw to move a part.

The earliest type of transmitting screws were of the square thread form (Figure I-397). Thrust

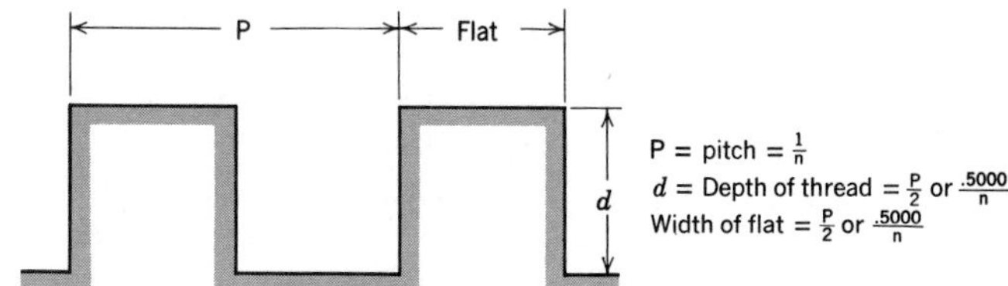

P
Flat
d
P = pitch = $\frac{1}{n}$
d = Depth of thread = $\frac{P}{2}$ or $\frac{.5000}{n}$
Width of flat = $\frac{P}{2}$ or $\frac{.5000}{n}$

FIGURE I-397 Square thread.

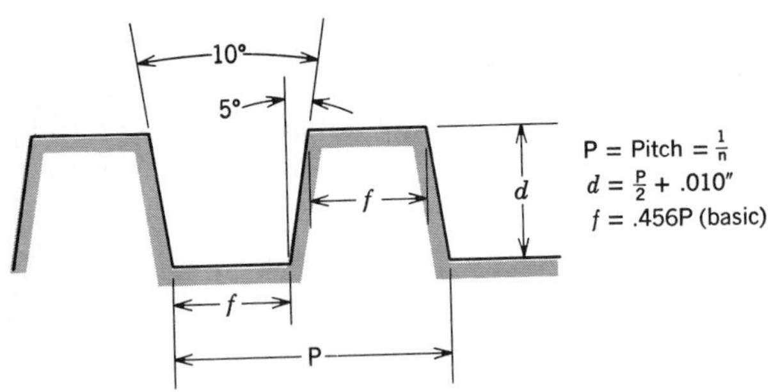

FIGURE I-398 Modified square thread. P = pitch = $\frac{1}{n}$; $d = \frac{P}{2} + \frac{.010}{\text{in.}}$; $f = .456P$ (basic).

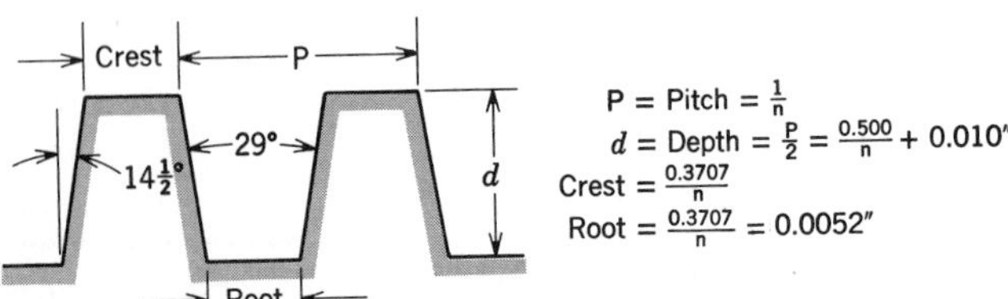

FIGURE I-399 Acme thread.

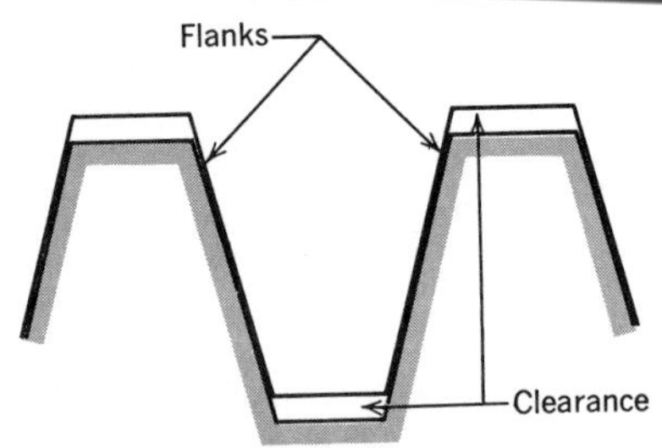

FIGURE I-400 General Purpose Acme threads bear on the flanks.

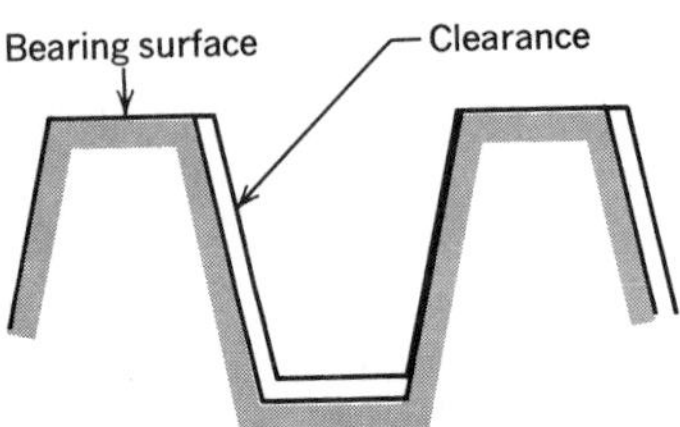

FIGURE I-401 Centralizing Acme threads bear at the major diameter

on the flanks of the thread is fully axial, thus reducing friction to a minimum. The square thread is more difficult to produce than other types and is not now widely used. The thread has a depth and thickness that is one-half the pitch. Clearance must be provided on the flanks and major diameter of the thread.

The modified square thread form (Figure I-398) was designed to replace the square thread. It is easier to produce than the square thread, yet it has all the advantages—and some of the drawbacks—of the square thread. It, like the square thread, is not widely used.

The Acme thread (Figure I-399) is generally accepted throughout the mechanical industries as an improved thread form over that of the square and modified square. The Acme thread is easier to machine and stronger than the square form thread since the Acme root cross section is thicker than its root clearance. Acme thread screws are used on milling machines and lathes. Like the square thread, the Acme has a basic depth equal to one-half the pitch; however, clearance is added both at the crest and the root of the thread for the general-purpose fit. The Acme general purpose threads bear on the flanks (Figure I-400). Centralizing fits bear at the major diameter (Figure I-401). For more detailed information on Acme thread fits, see *Machinery's Handbook*.

Three classes of General Purpose Acme threads 2G, 3G, and 4G are used. Class 2G is preferred for general purpose assemblies. If less backlash or end play is desired, classes 3G and 4G are used. Internal threads of any class may be combined with any external class to provide other degrees of fit. The included angle of General Purpose Acme threads is 29 degrees. Depth of thread is one-half the pitch plus .010 in. for 10 tpi and coarser. For finer pitches the depth of thread is one-half the pitch plus .005 in.

Stub Acme threads (Figures I-402 and I-403) are used where a coarse pitch thread with a shallow depth is required. The depth for the stub Acme is only $.3P$ and $.433P$ for the American Standard stub, as compared to $.5P$ for the standard Acme threads.

The Buttress thread (Figure I-404) is not usually used for translating motion. It is often used where great pressures are applied in one direction only, such as on vise screws and the breech of large guns. Acme threads are gradually replacing square and Buttress thread forms.

Acme threads may be measured by using the one-wire method (Figure I-405). If a wire with a diameter equal to $.48725 \times P$ is placed in the groove of an Acme thread, the wire will be flush with the top of the thread. For further information on one-wire and three-wire methods for checking Acme threads, see *Machinery's Handbook*.

Multiple Lead Threads

Multiple threads, though not often used, are usually found on industrial machines, valves, fire hydrants, and aircraft landing gear. They are also used on jars and other containers.

Most screws and bolts have single lead threads, which are formed by cutting one groove with a single-point tool. A double lead thread has two

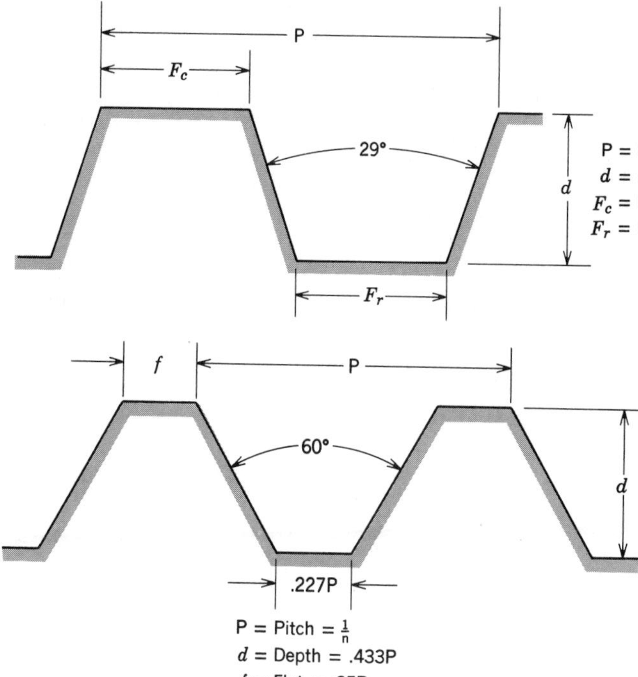

FIGURE I-402 American Standard Stub Acme thread. P = pitch = $1/n$; d = depth = $.3P$; Fc = basic flat at crest = $.4224P$; Fr = basic flat at root = $.4224P - .0052$.

FIGURE I-403 Stub Acme thread.

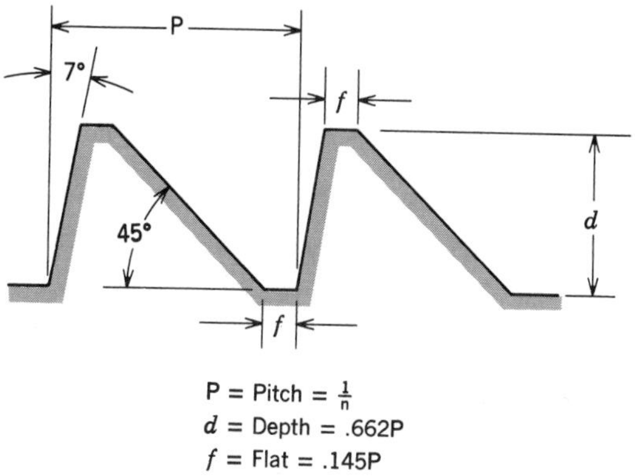

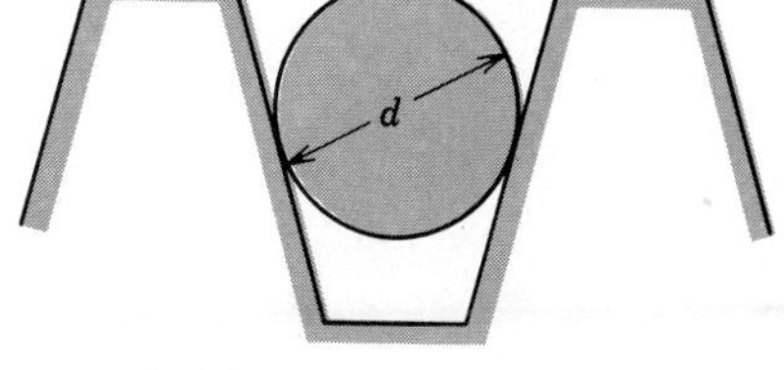

d = Diameter of wire = .48725P

FIGURE I-405 One-wire method for measuring Acme threads.

P
7°
f
45°
d
f

P = Pitch = $\frac{1}{n}$
d = Depth = .662P
f = Flat = .145P

FIGURE I-404 Buttress thread.

grooves, a triple lead thread has three grooves, and a quadruple lead thread has four grooves. Double, triple, and quadruple threads are also known as multiple lead threads. You may determine whether a bolt or screw is single or multiple threaded by looking at its end (Figure I-406) and counting the grooves that have been started. Multiple threads have less holding power, and less force is produced when these screws are tightened. A single lead thread should be used for fasteners where locking power is required.

Multiple lead threads offer several advantages:

1. They furnish more bearing surface area than single threads.

2. They have larger minor diameters and, therefore, a bolt is tronger than one with a single thread.

3. They provide rapid movement.

4. They are more efficient as they lose less power to friction than do single lead threads.

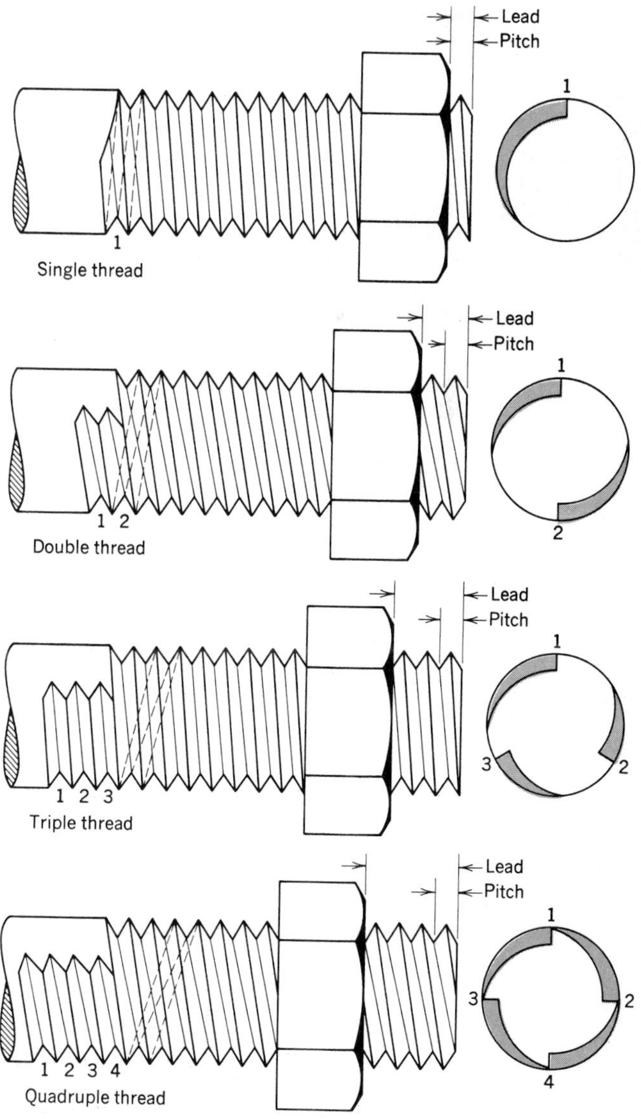

FIGURE I-406 Single, double, triple, and quadruple threads.

FIGURE I-407 Using the face plate and lathe dog method of indexing multiple lead threads (Lane Community College).

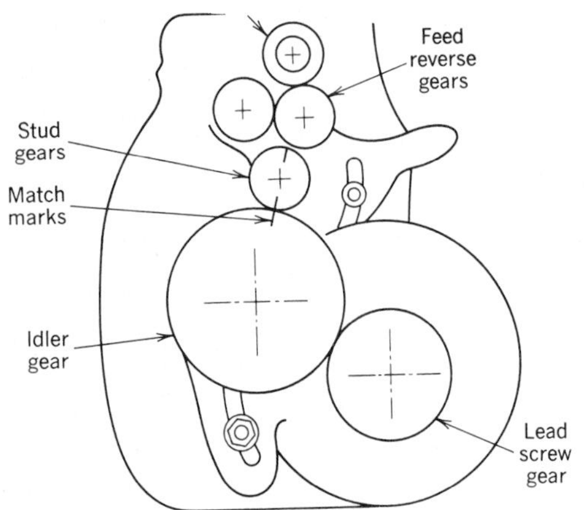

FIGURE I-408 Using the stud gear in the drive train for indexing.

The lead is the distance that a nut travels in one revolution. In one turn on a single lead screw, a nut moves forward the distance (pitch) of one thread; in one turn on a double lead thread it moves twice as far; on a triple lead thread it moves three times as far; and on a quadruple thread, it moves four times as far. On a single lead screw the lead and pitch are the same, but on a two lead screw, the lead is twice the pitch. Pitch is measured in the same way for single and multiple lead thread. It is the distance from a point on one thread to the corresponding point on the next thread.

The form and depth of thread for multiple lead threads can be based on any recognized thread form: Buttress, square, sharp V, Acme, American National, metric or Unified, both left- and right-hand. The thread depth is based on the pitch of the thread and not the lead.

FIGURE I-409 A thread chasing dial which may be used to cut double lead threads (Lane Community College).

Cutting Multiple Lead Threads

Several methods are used for indexing or dividing multiple lead threads. One method is to use an accurately slotted face plate (Figure I-407). The lathe dog is moved 180 degrees for two leads, 120 degrees for three leads, and 90 degrees for four leads. This method will only work on external threading. Another method is to mark the stud gear at 180 degrees for two leads, disengage the gear, rotate it 180 degrees, and reset it (Figure I-408). This procedure will work only if the spindle and stud gear have a 1:1 (1 to 1) ratio.

A thread chasing dial (Figure I-409) may be used to cut double threads if they are fractional or odd numbered. The first lead is cut on the numbered lines; the second lead is cut on the unnumbered lines. Some experimentation on various lathes will be useful to determine where the divisions will be found. If you cut the thread with the compound set at 29 degrees, be sure to back the compound out to its original position when you move to the next lead.

Many machinists prefer to index the thread with the compound set at 90 degrees, or parallel to the ways (Figure I-410). With this method, the tool must be fed straight into the work, so lighter cuts should be taken to keep from tearing the thread. Begin by taking up the slack in the compound feed screw and setting the tool to cut the first lead. Set the compound and cross feed micrometer dials to zero. Cut the first thread to the correct depth and move the compound forward one pitch of the thread using the micrometer dial or a dial indicator. Be sure that the slack is out of the

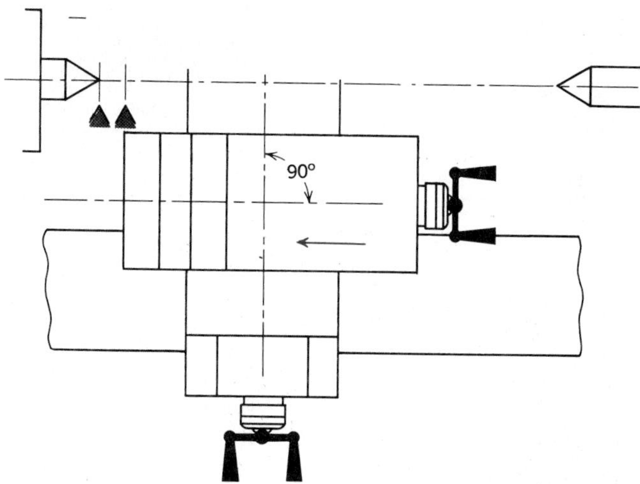

FIGURE I-410 Setting the compound parallel to the ways for cutting multiple lead threads. To index, move the compound a distance equal to the pitch, always in the same direction.

screw at all times. It is a good idea to tighten the compound gibs for this operation.

Make the cut for the second lead and check the fit. If it is a two-start thread, for example, use a gage or two-start nut to check it. If the thread is too tight, move the cross-slide in a few thousandths of an inch, take a cut, then advance the compound *forward* to the other position and take the same depth cut. This method may be used for a thread with any number of starts. On coarse threads (under 12 threads per inch), rough out both threads before taking the finishing cuts. Setups or tools may slip out of place slightly on heavy cuts, thus allowing for no adjustment if one thread is already finished.

An example of using this method to cut multiple lead threads is as follows: If a two-start, double lead unified screw thread with .100-in. pitch is required, the following steps can be used to make this thread.

CUTTING A DOUBLE LEAD THREAD

STEP 1. The lead is .200 in. or 5 threads per inch (tpi). The quick-change gearbox is set at 5 tpi.

STEP 2. The compound is set at 90 degrees, or parallel to the axis of the lathe, and the tool is set up.

STEP 3. The micrometer dials are zeroed and a scratch cut is taken. The compound is advanced .100 in. and the second scratch cut is taken. The pitch should now be .100 in. or 10 tpi when checked with a screw pitch gage.

STEP 4. The depth of the external unified thread is P × .613.

STEP 5. Feed in no more than .005 in. per pass but leave .010 in. for finishing. Now repeat this process in the second lead.

STEP 6. Both leads may now be cut to finish diameter.

If you use a dial indicator instead of a micrometer dial to measure the movement of the compound, the two lead threads can be within .001 in. of true position.

SELF TEST

1. For what two major purposes are translating-type screw threads used?

2. Name five thread forms used as translating screws.

3. What would the depth of thread be for a square thread that is 4 tpi?

4. What would the depth of thread be for a General Purpose Acme thread that is 4 tpi?

5. What is the main difference between General Purpose Acme threads and centralizing Acme threads?

6. What is the included angle of Acme threads?

7. Explain the general use of stub Acme threads.

8. Which thread form has a 10-degree included angle?

9. Of the translating thread forms, which type is most used and is easiest to machine?

10. What are Buttress threads mostly used for?

11. A $\frac{1}{8}$-in. pitch single lead thread will move .125 in. in one revolution of the nut. How far will a $\frac{1}{8}$-in. pitch, three-start thread move in one revolution?

12. Define "pitch."

13. Define "lead."

14. Which thread has more force or holding power for fasteners, a single or a multiple lead?

15. Name four methods of indexing the lathe for multiple threads.

16. Name three advantages offered by multiple lead threads.

17. What kinds of threads can be made multiple lead?

18. How can you determine the number of leads?

19. Why should both leads be roughed out before taking the finish cut on coarse threads?

20. Why are lighter cuts taken when using the compound slide method of indexing?

UNIT 16

Cutting Acme Threads on the Lathe

Machinists are sometimes required to cut internal and external Acme threads. These threads are, in most cases, larger and coarser than 60-degree form threads and require greater skill to produce them well. In this unit you will learn the procedure for cutting these threads.

Objective

After completing this unit, you should be able to:

Describe set up and procedure for making external and internal Acme threads.

Cutting Acme threads is similar to cutting 60-degree threads in many ways. The threads per inch and infeeds are calculated in the same way. Some calculations, tool form and relief angles, and finishes, however, involve different problems and procedures.

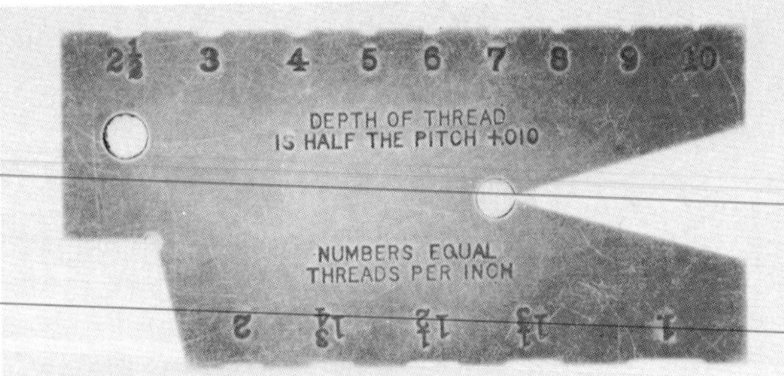

FIGURE I-411 Acme tool gage.

FIGURE I-412 Checking the tool angle with the Acme tool gage (Lane Community College).

Grinding the Tool for Acme Threads

The cutting tool form must be checked with the Acme tool gage (Figures I-411 and I-412) when the 29-degree included angle is ground. Side relief must be ground at the same time. The end of the tool is ground flat and perpendicular to the bisector of the angle. The flat is checked (Figure I-413)

FIGURE I-413 Checking the flat on the end of the tool with the Acme tool gage (Lane Community College).

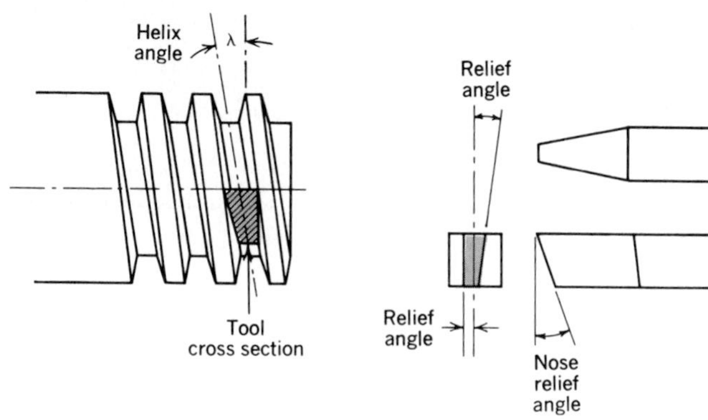

FIGURE I-414 Acme threads showing the importance of the relief angles on both sides of the tool.

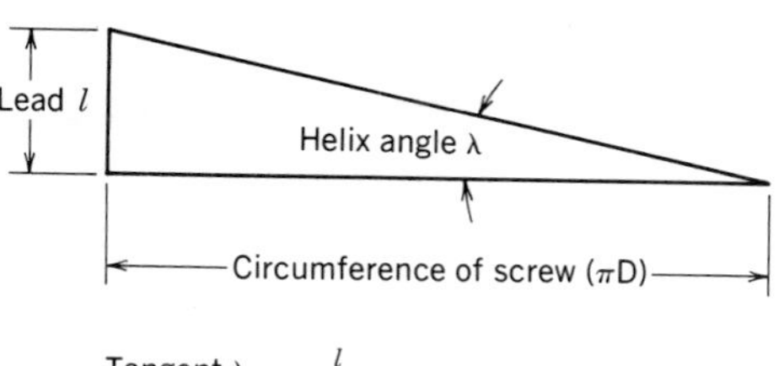

FIGURE I-415 The screw thread helix angle may be determined by dividing the lead by the circumference of the screw. The number thus obtained is the tangent of the helix angle, which can then be found in a table of tangents. D = diameter of screw, l = lead of thread, and λ = helix angle of thread.

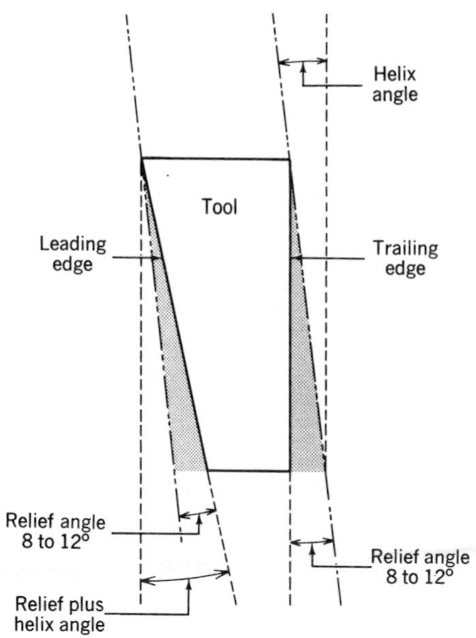

FIGURE I-416 The relationship of the helix angle of the thread to the relief angle of the tool.

FIGURE I-417 Aligning the tool with the work using the Acme gage (Lane Community College).

with the tool gage at the number corresponding to the threads per inch you will cut. It is very important to have the flat the exact width needed for the particular thread being cut.

The relief angles on the tool (Figures I-414 and I-415) are of greater importance when coarse threads are cut, since if the heel of the tool rubs on either side, a rough, inaccurate thread will result. When the helix angle has been determined, it should be added to the relief angle (8 to 12 degrees) of the tool on the leading edge and similar relief provided on the trailing edge of the tool (Figure I-416). As in other threading operatings, the tool must be ground for 0-degree rake and set on the center of the work to maintain the correct thread form.

Setting up to Cut External Acme Threads on the Lathe

The threads per inch are set up normally and the lead screw rotation is set for right- or left-hand threads. The gears, leadscrew, and carriage should be lubricated before cutting coarse threads. The compound is most often set at $14\frac{1}{2}$ degrees to the right for right-hand external threads. The workpiece must be set up and held very securely in the work-holding device; a four-jaw chuck and dead center would be most secure. The tool is aligned with the work by using the Acme gage (Figure I-417). With this setup, the tool is fed into the work by advancing the compound in small steps as with 60-degree threads. An undercut must be made at the end of the threads to clear the tool.

When Acme threads coarser than 5 tpi are cut, a square or round nose roughing tool that is smaller than the Acme tool should be used to remove up to 90 percent of the finished thread. The roughing tool does not cut to full depth or width and the Acme form tool is used to finish the thread. This procedure is also used when making threads in tough alloy materials.

FIGURE I-418 Taking the scratch cut (Lane Community College).

FIGURE I-419 Compound is set $14\frac{1}{2}$ degrees to the left of the operator for right-hand internal threads (Lane Community College).

Some machinists prefer to set the compound parallel to the ways so they can "shave" both flanks of the thread for a good finish. When this is done, the tool must be made a few thousandths of an inch narrower to allow for the "shaving" operation.

MAKING THE CUT

A scratch is taken (Figure I-418) and measured. The cross slide is moved out and the carriage returned. The cross slide is again set on zero and the compound is advanced .005 to .010 in., depending on what the lathe and setup will handle without chatter. Use sulfurized cutting oil. The total depth of the cut is $.5P + .010$ in. Feed in on the cross

FIGURE I-420 Aligning the Acme tool with a gage for cutting internal threads (Lane Community College).

slide for the last few thousandths of an inch so that the trailing flank will also receive a finish cut. For other Acme thread fits, see *Machinery's Handbook*.

Internal Threads

The bore size for making a General Purpose Acme internal thread is the major diameter of the screw minus the pitch. As with Unified threads, the actual minor diameter of an Acme thread is the inside diameter of the bore; thus the minor diameter of an Acme 1–5 thread would be $1 - .200 = .800$ in. The internal major diameter should be the major diameter of the screw plus .010 in. for 10 or more tpi and .020 in. or pitches less than 10 tpi.

The compound is set $14\frac{1}{2}$ degrees to the left for cutting right-hand internal threads (Figure I-419). An internal Acme threading tool is ground, checked (Figure I-420) and set up, and then fed into the work with the compound .002 to .005 in. for each pass. An Acme screw plug gage or the mating external thread should be used to check the fit as the internal thread nears completion (Figure I-421).

When internal Acme threads are too small in diameter to be cut with a boring bar and tool, an Acme tap (Figure I-422) is used to make the thread. Acme taps are made in sets of two or three; each tap cuts more of the thread, the last tap for the finishing cut. Two taps are sometimes made on the same shank, as in Figure I-422. The part is drilled or bored to the minor diameter of the thread and the Acme tap is turned in by hand. Use cutting oil when tapping steel, but cut threads in bronze dry.

A problem often encountered when threading coarse threads on small lathes is that of producing a thicker than normal last thread; the plug gage

FIGURE I-421 Completed internal Acme threads (Lane Community College).

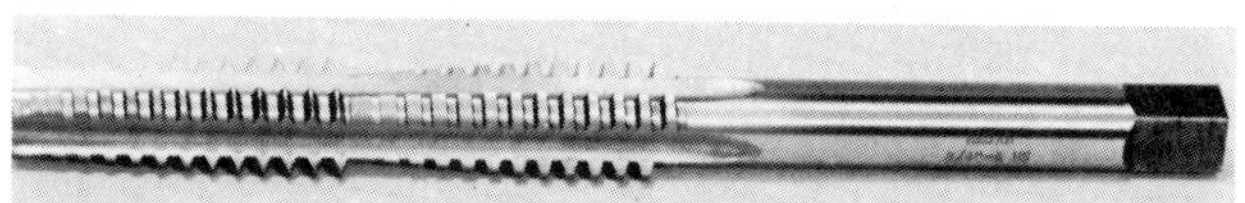

FIGURE I-422 Acme tap (Lane Community College).

will go in the nut all the way except for the final thread. This happens because the carriage is not heavy enough to provide sufficient drag to keep the tool cutting on its following side when the leading side of the tool is emerging from the cut. The result is a pitch error on the last thread. The slack in the half-nuts allows this to happen. Providing a slight drag on the carriage handwheel will eliminate this problem.

External Acme threads must often have a good finish. A final honing of the tool before the last few shaving passes will help. The setup must be very rigid and the gibs tight. Low speeds are essential. The grade of cutting oil is extremely important in this finishing operation.

The thread may be finished after it is cut by using a thin, safe edge file at low rpm and by using abrasive cloth at a higher rpm. A thin piece of wood is sometimes used to back up the abrasive cloth while each flank is being polished.

SELF-TEST

1. What is the major difference between V-form threads and Acme threads?

2. When grinding an Acme threading tool, what three important parts should be carefully measured?

3. Why should the gears, ways, leadscrew, and carriage be lubricated before cutting coarse threads?

4. Where is the compound most frequently set when cutting Acme threads? What setup is preferred by some machinists?

5. What is the depth of thread for a $\frac{3}{4}$-6 external General Purpose Acme thread?

6. How is the tool aligned with the workpiece?

7. Determine the bore size to make a $\frac{3}{4}$-6 general purpose internal Acme thread.

8. Which is the best way to make small internal Acme threads?

9. What can you use to check internal threads for fit?

10. Explain how a good tool finish may be obtained on an Acme thread.

SECTION J

VERTICAL MILLING MACHINES

The vertical milling machine is a relatively new development in comparison to the horizontal milling machine. The first vertical milling machines appeared in the 1860s. The vertical milling machine was a development more closely related to the drill press than to the horizontal spindle milling machine. The basic difference between drill presses and the earliest vertical milling machines was that the entire spindle assembly, pulleys and all, was moved vertically. This arrangement meant that the bearing that supported the cutting tool were always reasonably close to the tool, which permitted side thrust to be taken more readily.

The next significant step came in the mid-1880s with the adaptation of the "knee and column" from the horizontal milling machine. This allowed the milling table to be raised and lowered in relation to the spindle. Also, the spindle heads on some of these machines could be tilted to an angular relationship to the table.

Just after the turn of the twentieth century, vertical milling machines began to appear with power feeds on the spindle, housed in a heavy-duty quill. During that period, micrometers, and vernier scales had been applied to vertical milling machines to make them suitable for precise hole locating, known as jig boring.

Improvements in vertical milling machine design after 1910 related mostly to their drive and control mechanisms. Machines with automatic table cycles began to appear and by 1920 electrical servomechanisms were used on the vertical milling machine for operations like diesinking. By 1927 hydraulic tracing controls had been developed and applied to vertical milling machines (Figure J-1).

Control systems, not limited to vertical milling machines, have been developed that activate machine control movements from information stored on punched or magnetic tape (Figure J-2), called

FIGURE J-1 Hydraulic tracing controls on a vertical milling machine especially developed for die sinking (Cincinnati Milacron).

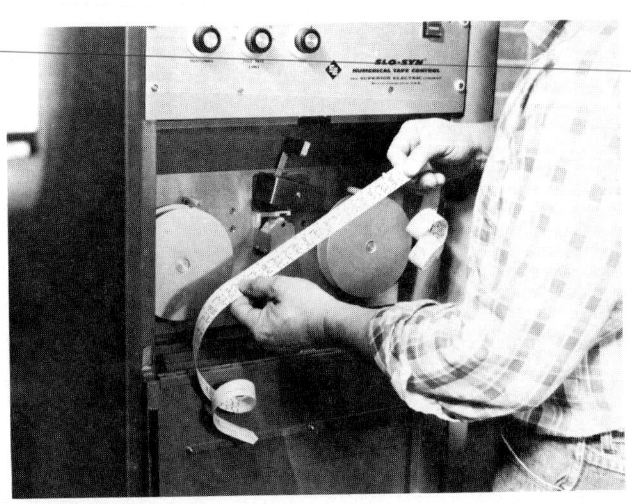

FIGURE J-2 Numerical control milling machine with information stored on punched tape (Lane Community College).

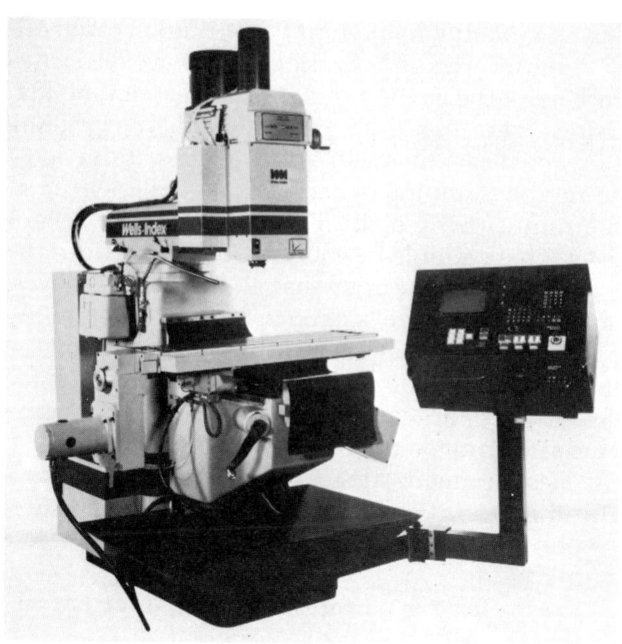

FIGURE J-3 Computer numerical control CNC vertical milling machine (Wells Manufacturing Corporation).

FIGURE J-4 Popular type of manually operated vertical milling machine with accessory slotting attachment. This is called a ram-type turret mill (Bridgeport Milling Machines, Division of Textron, Inc.).

numerical control (NC), or from computer numerical control (CNC) (Figure J-3).

The standard vertical milling machine (Figure J-4) is one of the most versatile machine tools found in the machine shop. In some respects, it is even more versatile than the lathe. In its various forms and with adaptations, it comes close to being a machine tool that can reproduce itself. This machine tool can accomplish a wide variety of machining tasks including milling, drilling, boring, and slotting.

■ ■ ■

Vertical Milling Machine Safety

Safe operation of the vertical milling machine requires that you learn about any hazards that exist so that you can protect yourself from them. It is very important that you are alert when you are working on a machine tool. Being sick, tired, or emotionally upset is dangerous when operating any equipment. Many prescription drugs should not be taken before driving a car because they affect your reflexes; the same is true of operating machinery. A safe machine operator thinks before doing anything. You need to know what is going to happen before you turn a lever or operate a control.

CAUTION **Proper dress is important for safe vertical milling machine operation. Short sleeves or tightly fitting sleeves are protection against being caught in a revolving spindle. Rings, bracelets, earrings, and even wristwatches can become dangerous if worn around machinery and they should be removed. Eye protection in the form of safety goggles and face shields should always be worn in a machine shop.**

Flying particles from the machine you are operating or even someone else's can blind you. Long hair should be safety covered under a cap or with a hair net. Heavy shoes should be worn to protect your feet from chips on the floor and from falling objects. Gloves should not be worn while operating machines because of the danger of being caught by the machine.

All machine guards should be in place prior to starting up a milling machine. Observe the other machines in operation around you to make sure that they are guarded properly. Report any unsafe or dangerous practices by those around you to the person in charge. A safe workplace depends on everyone.

Safety also involves keeping a clean machine and keeping the area surrounding it clean. Any oil or cutting fluid spills on the floor should be wiped up immediately to avoid slipping and falling. Chips should be swept up with a brush or broom and deposited in chip or trash containers. Do not handle chips with your bare hands. Cuts caused by contaminated chips can cause infections. Dirty and oily rags should be kept in closed containers and should never accumulate on the floor. Never use an airhose to clean a machine—flying chips are dangerous to you and others around you. If you need to lift a heavy workpiece or machine attachments, use a hoist or have someone help you.

Be careful when handling tools or sharp-edged workpieces to avoid getting cut; use a shop towel to protect your hands. Workpieces should be rigidly supported and tightly clamped to withstand the usually high cutting forces encountered in machining. If a workpiece comes loose while it is being machined, it is often ruined, and so is the cutter. The operator is also in danger from flying particles from a broken cutter on the workpiece.

The cutting tools need to be securely fastened in the machine spindle to prevent any movement during the cutting operation. Excessive feed rates can break the cutting tool. On vertical milling machines, care has to be exercised when swivelling the workhead to make angular cuts. After loosening the clamping bolts that hold the toolhead to the ram, retighten them slightly to create a light drag. There should be enough friction between the tool head and the ram that the toolhead swivels only when pressure is applied to it. If the clamping bolts are completely loosened, the weight of the heavy spindle motor will flip the toolhead upside down. This can cause serious injury to the operator and damage the machine table.

Measurements are frequently made during machining operations. Do not take any measurements until the spindle has come to a complete standstill.

Never leave a running machine unattended.

UNIT 1

The Vertical Spindle Milling Machine

The first step in efficient and safe operation of any machine tool is to know the names of the machine parts and its various controls. The next step is to know the function of each part and control so that you can operate the machine without damage to it, the workpiece, or possible injury to yourself. The purpose of this unit is to acquaint you with the nomenclature of the vertical milling machine and to identify the machine controls and their functions. Before starting any milling job, take a short time to operate all the machine controls and observe their functions.

Objectives

After completing this unit, you should be able to:

1. Identify the important components and controls on the vertical milling machine.
2. Describe the functions of machine parts and controls.
3. Perform routine maintenance on the machine.

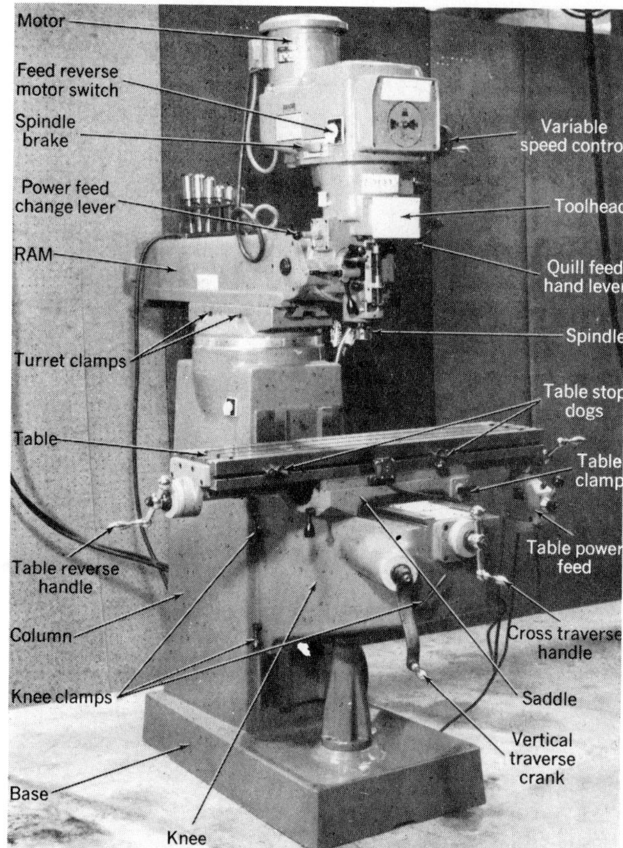

FIGURE J-5 The important parts of a vertical milling machine (Lane Community College).

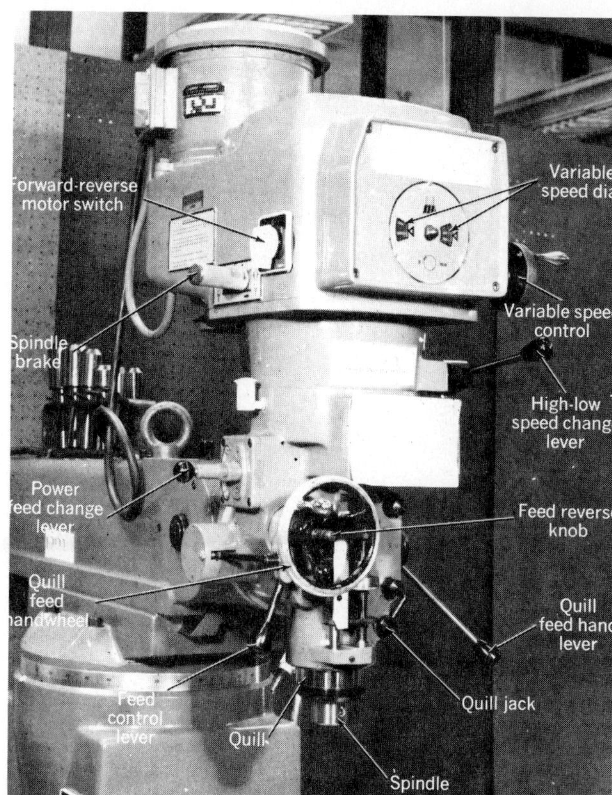

FIGURE J-6 The toolhead (Lane Community College).

Identifying Machine Parts, Controls, and their Functions

The major assemblies of the vertical milling machine are: base and column, knee and saddle, table, ram, and tool head (Figure J-5).

BASE AND COLUMN

The base and column are one piece and are the major structural component of this machine tool. A dovetail slide is machined on the face of the column to provide a vertical guide for the knee. A similar slide is machined on the top of the column to provide a guide for the ram. The top column slide and ram can be swiveled right and left of center to permit wide area positioning of the tool head over the table.

KNEE

The knee engages the slide on the face of the column and is moved up and down by turning the vertical traverse crank. The knee supports the saddle and table. Knee locks are provided that will securely lock the knee at a given position.

SADDLE

The saddle engages the slide on the top of the knee and can be moved in and out by turning the cross traverse handle. The saddle supports the table. Saddle locks are provided for the purpose of locking the saddle at a given position.

TABLE

The table engages the slide on the top of the saddle and is moved right and left by turning the table traverse handle. The workpiece or work-holding device is secured to the table. Table locks are provided so that the table may be locked at a given position. Most milling machines have a power feed mechanism on the table. This permits a variable table feed rate in either direction during a milling operation.

RAM

The ram engages the swiveling slide on the top of the column. The ram is moved in and out by turning the ram positioning pinion gear. Ram locks are provided to secure the ram position.

TOOLHEAD

The toolhead (Figure J-6) is attached to the end of the ram and contains the motor, which powers the

FIGURE J-7 Quill stop (Lane Community College).

FIGURE J-8 Clamping devices (Lane Community College).

spindle. The motor is turned on with a three-position switch (Forward/Off/Reverse). Be sure that the spindle is rotating in the proper direction when you turn on the machine. Speed changes are made with V-belts, gears, or variable-speed drives. When changing speeds into the high or low speed range, the spindle has to be stopped. The same is true for V-belt or gear-driven speed changes. On variable drives the spindle must be revolving while speed changes are being made. The quill is nonrotating and contains the rotating spindle. The quill can be extended and retracted into the tool head by a quill feed hand lever or handwheel. The quill feed hand lever is used to rapidly position the quill or to drill holes. The quill feed handwheel gives a controlled slow manual feed, as is needed when boring holes.

Power feed to the quill is obtained by engaging the feed control lever. Different quill feed rates, usually .0015, .003 and .006 in. per spindle revolution, are selected with the power feed change lever. The power feed is automatically disengaged when the quill dog contacts the adjustable micrometer depth stop (Figure J-7). When feeding upward, the power feed disengages when the quill reaches its upper limit. The micrometer dial allows depth stop adjustments in .001-in. increments. The quill clamp is used to lock the quill in a fixed position. To obtain maximum rigidity while milling, the quill clamp should be tightened.

The spindle lock or spindle brake is used to keep the spindle from rotating when installing or removing tools from it. Tools are usually held in collets, which are secured in the spindle with a drawbolt. The drawbolt threads into the upper end of the collet. When the drawbolt is tightened, the collet is drawn into the taper in the spindle. This aligns and also holds the tool. To release a tool from the spindle, use the following procedure:

STEP 1. Use the spindle brake to lock the spindle.

STEP 2. Raise the quill to its top position and lock.

STEP 3. Unscrew the drawbolt one turn.

STEP 4. With a lead hammer give the drawbolt a firm blow. This should release the collet from the spindle taper. At the same time, the tool is released from the collet. You should be holding the tool with a shop towel during this step to prevent the tool from falling on the workpiece or the machine table.

Locks, Adjustments, and Maintenance

The knee, saddle, table, and quill are all equipped with locks that will prevent movement of these parts (Figure J-8). During machining, all axes except the moving one should be locked. This will increase the rigidity of the setup. Do not use these clamping devices to compensate for wear on the machine slides. If the machine slides become

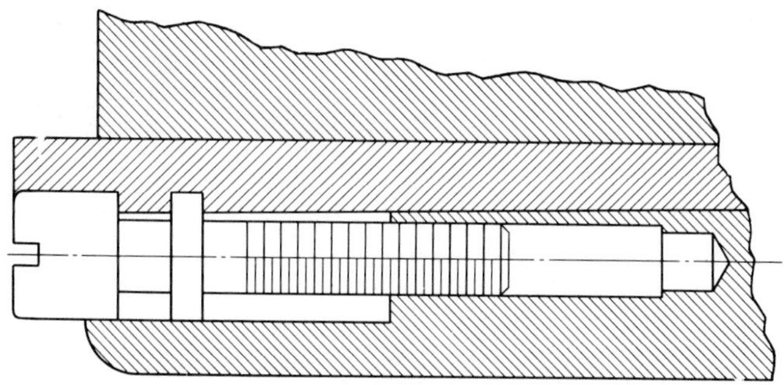

FIGURE J-9 Gib adjusting screw (Cincinnati Milacron).

loose, make adjustments with the gib adjustment screws (Figure J-9). Turning this screw in will tighten a tapered gib. Make a partial turn on the screw, then try moving the unit with the hand sheel. Repeat this operation until a free but not loose movement is obtained. Too tight an adjustment squeezes the lubricant from the slides, resulting in rapid wear.

All machine tools require periodic adjustment and lubrication. Many mills are equipped with "one shot" lubricators often located on the side of the knee. Oil from the lubricator is pressure fed to the knee, saddle, and table slides. Any other oil cups on the machine should be kept filled with light oil as specified by the manufacturer.

SELF-TEST

1. Name the six major components of a vertical milling machine.
2. Which parts are used to move the table longitudinally?
3. Which parts are used to move the saddle?
4. What moves the quill manually?
5. What is the purpose of the table clamp?
6. What is the purpose of the spindle brake?
7. What is important when changing the spindle speed range from high to low?
8. Why is the toolhead fastened to a ram?
9. How is a loose table movement adjusted?
10. What is the purpose of the quill clamp?

UNIT 2

Cutting Tools and Cutter Holders for the Vertical Milling Machine

The metal cutting versatility of the vertical milling machine can be fully realized and utilized by understanding, identifying, and selecting from the many types of milling cutters available for use on this machine tool. The purpose of this unit is to describe many of these common types of cutters and aid you in selecting the one that best fits your needs for a specific machining task.

Objectives

After completing this unit, you should be able to:

1. Identify common cutters for the vertical mill.
2. Select a proper cutter for a given machining task.

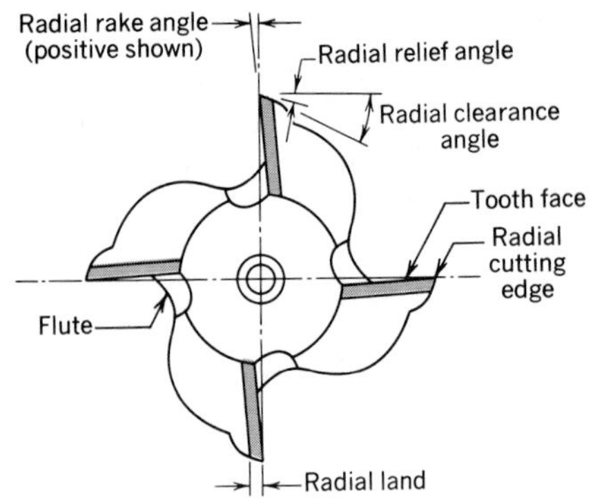

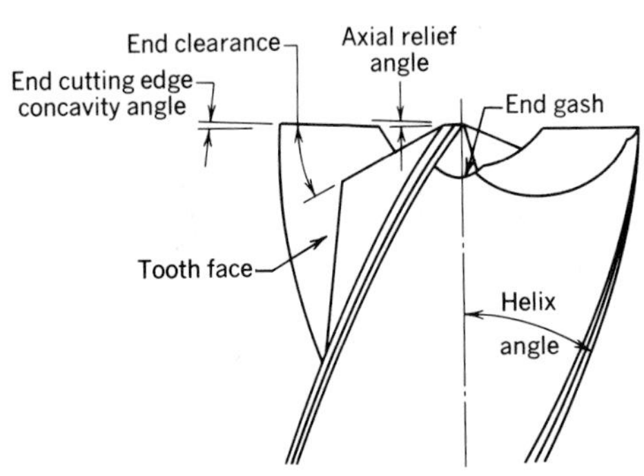

FIGURE J-10 End mill nomenclature. (Copyright © National Twist Drill & Tool Div., Lear Siegler, Inc.).

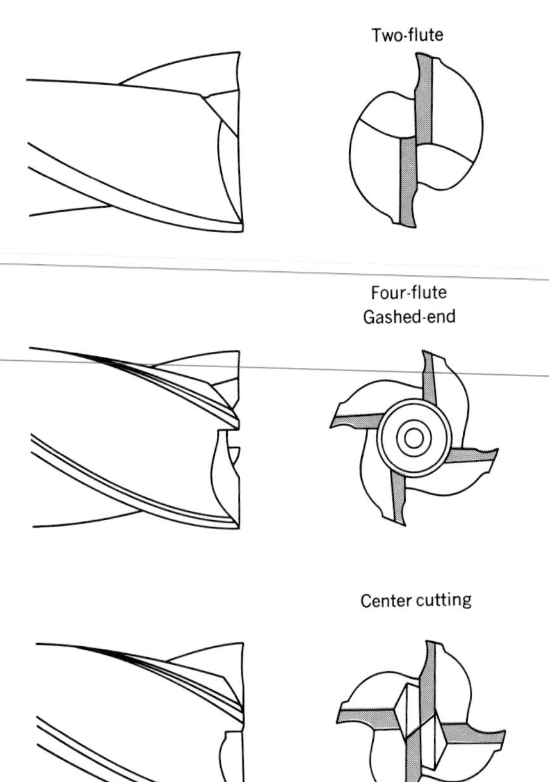

FIGURE J-11 Types of end teeth on end mills. (Copyright © National Twist Drill & Tool Div., Lear Siegler, Inc.).

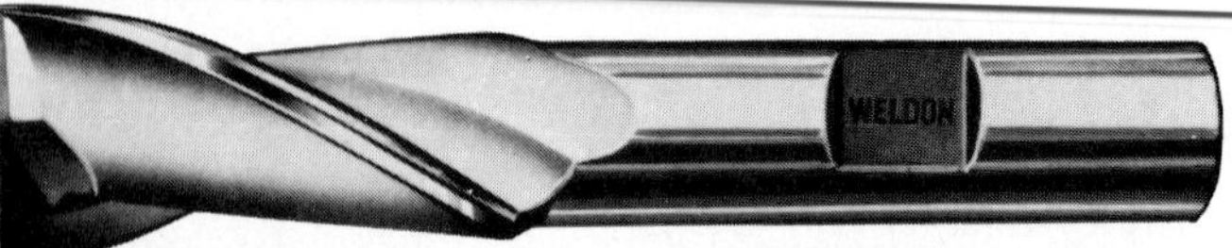

FIGURE J-12 Single-end helical teeth end mill (Weldon Tool Company, Cleveland, Ohio).

FIGURE J-13 Two-flute, double-end, helical teeth end mill (Weldon Tool Company, Cleveland, Ohio).

End Mills

The most frequently used cutting tool for the vertical milling machine is the end mill. End mills are so named because their primary cutting is done on their ends. End mills may have two, three, four, or more flutes and may be right-hand or left-hand cutting. To determine the cutting direction of an end mill, observe the cutter from its cutting end (Figure J-10). A right-handed cutter will cut while turning in a counterclockwise direction. A left-handed cutter will cut turning in a clockwise direction. The direction of flute twist or helix may also be right- or left-handed. For example, a right-handed helix twists to the right.

Two-flute end mills can be used for plunge cutting. These are called center cutting because they can make their own starting hole (Figure J-11). Four-flute end mills may also be center cutting. However, if these are center drilled or gashed on the end, they cannot start their own holes. This type of end mill will only cut on its periphery, but may be used in end milling provided the cut is begun off the workpiece or in a premachined hole or other cavity.

HIGH-SPEED STEEL HELICAL AND STRAIGHT-FLUTE END MILLS

The high-speed steel end mill is a very common cutter for the vertical mill. These cutters may be single ended (Figure J-12) or double ended (Figure J-13). They may also have straight flutes (Figure J-14). Slow, regular, and fast helix angles are also available. An example of slow helix is where the helix angle of the cutter is about 12 degrees. A regular helix angle may be 30 degrees and a fast helix 40 degrees or more. Selection of helix angle will

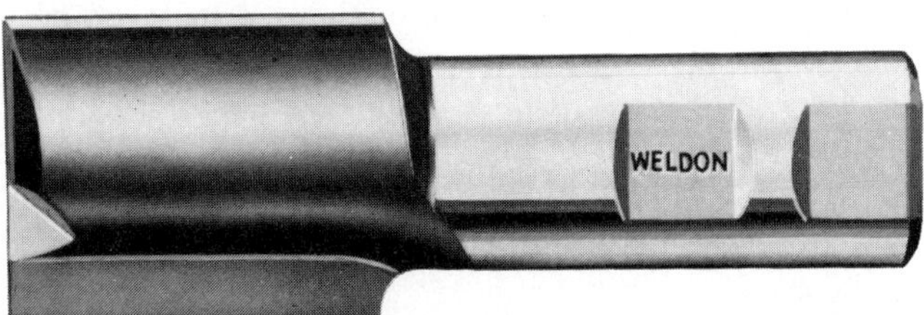

FIGURE J-14 Straight-tooth, single-end end mill (Weldon Tool Company, Cleveland, Ohio).

FIGURE J-15 Forty-five-degree helix angle aluminum cutting end mill (Weldon Tool Company, Cleveland, Ohio).

FIGURE J-16 Two-flute, carbide-tipped end mill (Brown & Sharpe Mfg. Co.).

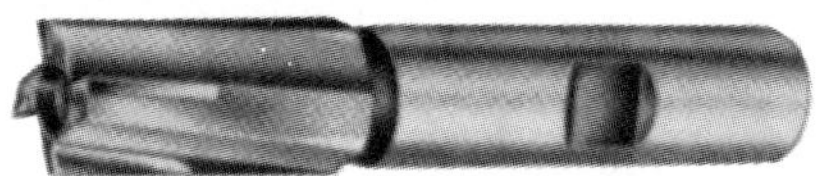

FIGURE J-17 Four-flute, carbide-tipped end mill (Brown & Sharpe Mfg. Co.).

FIGURE J-18 Disposable insert carbide end mill.

FIGURE J-19 Roughing mill. (Copyright © Illinois Tool Works, Inc., 1976).

FIGURE J-20 Three-flute, tapered end mill (Weldon Tool Company, Cleveland, Ohio).

depend on the machining task. For example, aluminum can be machined efficiently with a high helix angle cutter (45 degrees) and with a highly polished cutting face to minimize chip adherence (Figure J-15). Chips sticking to the cutting face can mar the surface finish of the part being machined. High-speed steel end mills are available at reasonable cost in a wide variety of styles, shapes, and sizes.

CARBIDE END MILLS

Carbide (tungsten carbide) end mills may be carbide tipped or made from solid carbide. These end mills are more expensive than high-speed types. However, they are extremely efficient in difficult-to-machine materials and for production machining applications. Two common carbide end mills are the two-flute cutter with a negative axial rake and slow helix (Figure J-16) designed to cut steel, and the four-flute cutter with a positive rake for nonferrous materials such as brass and aluminum (Figure J-17).

The disposable insert carbide end mill (Figure J-18) has carbide inserts providing the cutting edges. When the edges of the insert become dull, they are turned to expose a new cutting edge to the work. When all edges are dull the insert is discarded and a new one installed. No sharpening of the cutter is required. Various grades of carbide are also used depending on the material to be machined.

ROUGHING AND TAPERED END MILLS

The roughing end mill (Figure J-19) is used when large amounts of material must be quickly removed (roughed) from the workpiece. These end mills are also called hogging end mills and have a wavy tooth form cut on their periphery. These waved teeth form many individual cutting edges. The tip of each wave contacts the work and produces one short compact chip. Each succeeding wave tip is offset from the next one, which results in a relatively smooth surface finish. During the cutting operation, a number of teeth are in contact with the work. This reduces the possibility of vibration or chatter.

Tapered end mills (Figure J-20) are used in mold making, die work, and pattern making,

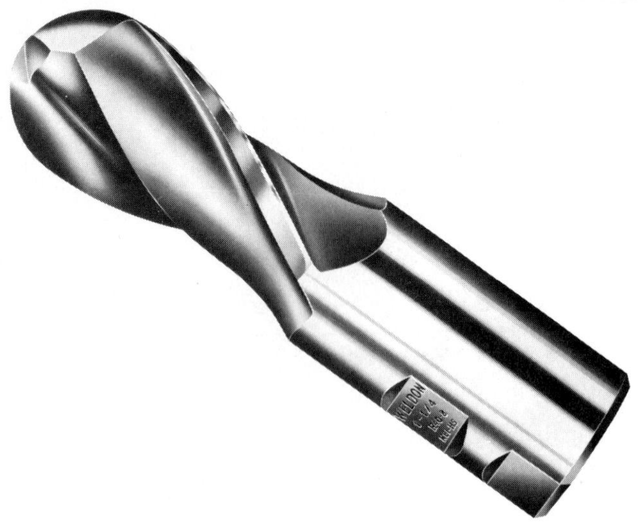

FIGURE J-21 Two-flute, single-end, ball-end mill. (Weldon Tool Company, Cleveland, Ohio).

FIGURE J-22 Corner rounding milling cutter. (Copyright © Illinois Tool Works, Inc., 1976.)

where precise tapered surfaces need to be made. Tapered end mills have included tapers ranging from 1 degree to over 10 degrees. Tapered end mills are also called diesinking mills.

GEOMETRY FORMING, DOVETAIL, T-SLOT, WOODRUFF KEY, AND SHELL END MILLS

Several types of end mills are used to form a particular geometry on the workpiece. Ball-end end mills (Figure J-21) have two or more flutes and form an inside radius or fillet between surfaces. Ball end end mills are used in tracer milling and in diesinking operations. Round-bottom grooves can also be machined with them. Precise convex radii can be machined on a milling machine with corner rounding end mills (Figure J-22). Dovetails are machined with single-angle milling cutters (Figure J-23). The two commonly available angles are 45 degrees and 60 degrees. T-slots in machine tables and work-holding devices are machined with T-slot cutters (Figure J-24). T-slot cutters are made in sizes to fit standard T-nuts.

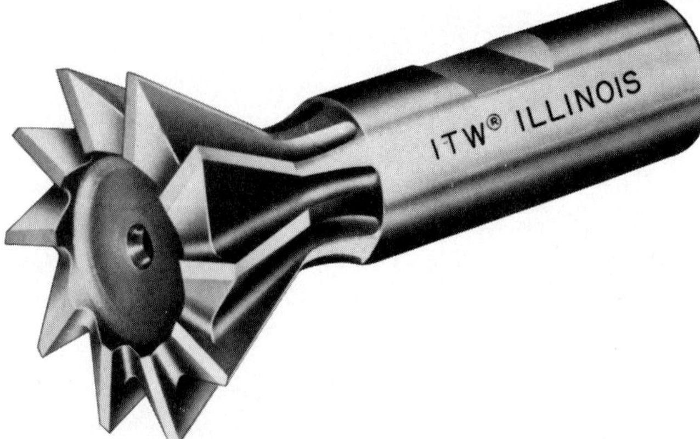

FIGURE J-23 Single-angle milling cutter. (Copyright © Illinois Tool Works, Inc., 1976.)

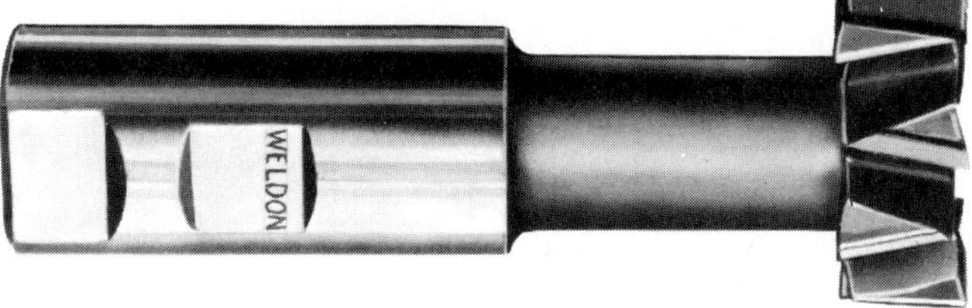

FIGURE J-24 T-slot milling cutter (Weldon Tool Company, Cleveland, Ohio).

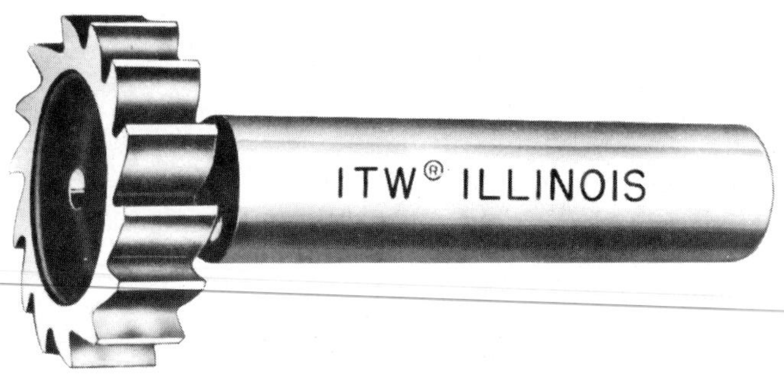

FIGURE J-25 Woodruff keyslot milling cutter. (Copyright © Illinois Tool Works, Inc., 1976.)

Woodruff keyseats are cut into shafts to retain a woodruff key as a driving and connecting member between shafts and pulleys or gears. Woodruff keyseat cutters (Figure J-25) come in many different standardized sizes. When larger flat surfaces need to be machined, a shell end mill (Figure J-26) can be used. Shell end mills are more economical to produce because less of the costly tool material is needed to make one than for a solid shank end mill of the same size. To obtain rapid metal re-

FIGURE J-26 Shell end mill (Copyright © Illinois Tool Works, Inc., 1976.)

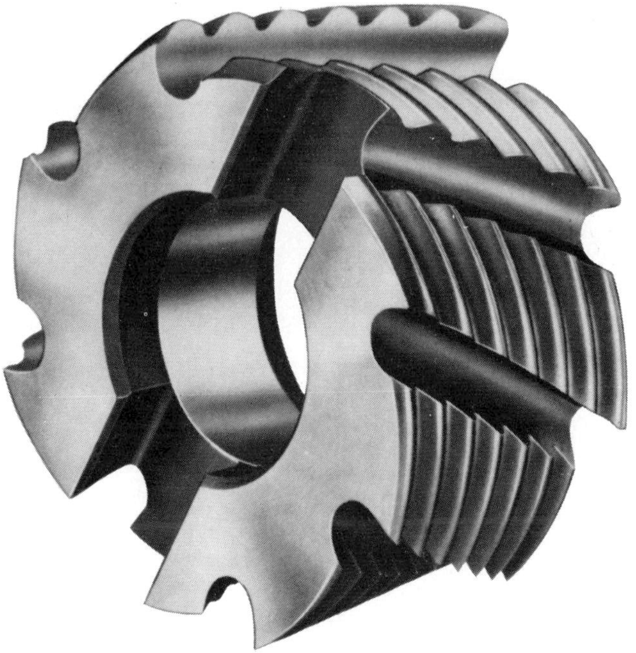

FIGURE J-27 Shell-type roughing mill. (Copyright © Illinois Tool Works, Inc., 1976.)

moval, shell end mills are made as a roughing type mill (Figure J-27) with a wavy thread forming many cutting edges. Shell mills are also made with carbide inserts (Figure J-28). The ease with which new sharp cutting edges can be installed makes this a very practical, efficient cutting tool. The great number of different carbide grades available

FIGURE J-28 Shell mill with carbide inserts (Lane Community College).

FIGURE J-29 Flycutter with an HSS tool installed (Lane Community College).

makes it possible to select a grade suitable for all work materials.

FLYCUTTERS

A flycutter is a single-point tool often consisting of a high-speed or carbide tool secured in an appropriate holder (Figure J-29). Although a flycutter is not truly an end mill, it is used for end milling applications. Flycutters are often used to take light face cuts from large surface areas. The tool bit in the flycutter must be properly ground to obtain the correct rake and clearance angles for the material being machined. Flycutters may also be used for boring operations. Care must be exercised when using a flycutter. When the tool is revolving the inserted cutting tool becomes almost invisible and could injure the operator.

Cutter Holding on the Vertical Mill

No matter which of the milling cutters discussed previously you might being using, they all must be securely mounted in the machine spindle before beginning a machining operation. Collet holders are widely used for this purpose.

The most rigid type of these is the solid collet (Figure J-30), sometimes called an end mill holder. The solid collet has a precision ground shank that fits the spindle on the milling machine. Most common vertical spindle milling machines have an R 8 spindle, meaning that they will accept all standard R 8 tooling. The solid collet has a hole that fits the shank of the end mill. The end mill is secured with setscrews that bear against a flat on the cutter shank. Solid collets will accommodate many different sizes of end mill shanks.

The split collet (Figure J-31) is widely used to hold cutters on the vertical mill. When the tapered part of the collet is pulled into the spindle taper by the drawbolt, the split in the collet permits it to squeeze tightly against the shank of the end mill. Although split collets are very effective cutter holding devices, it is possible for a cutter to be pulled from the collet because of heavy feed rates or the tool being dull. Helical flute end mills may tend to be pulled from the collet as well. In this respect the solid collet has an advantage over the split type since here the setscrews prevent slippage of the cutter.

QUICK-CHANGE SYSTEMS

To facilitate and speed tool changing, a quick-change tooling system may be used. Different tools can be mounted in their individual toolholders and preset to different lengths if desired. The toolholders are then mounted and removed from the master holder in the machine spindle by means of a clamping ring (Figure J-32).

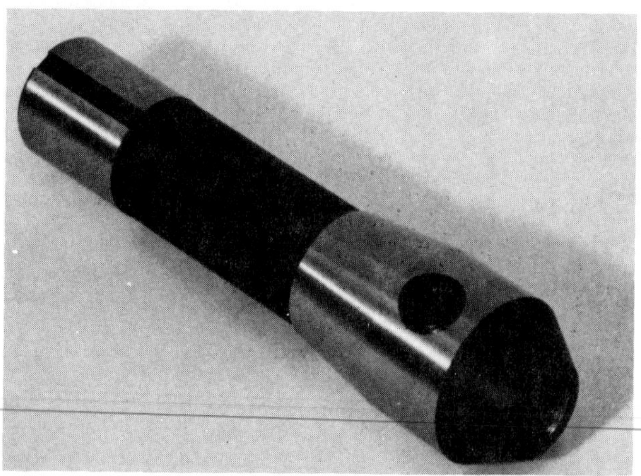

FIGURE J-30 Solid collet (Lane Community College).

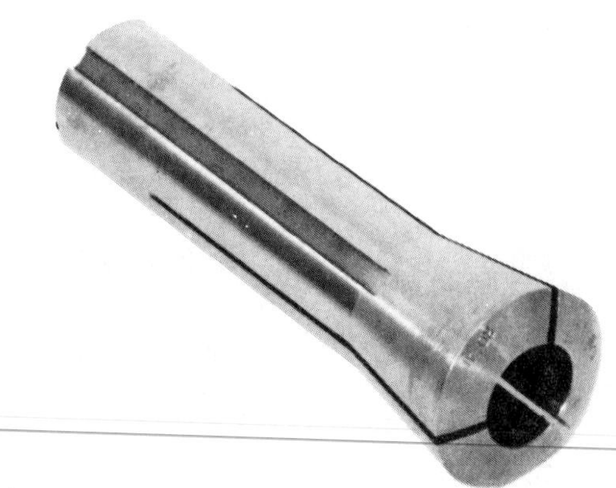

FIGURE J-31 Split collet (Lane Community College).

FIGURE J-32 Quick-change adapter and toolholders (Lane Community College).

FIGURE J-33 Shell mill arbor (Weldon Tool Company, Cleveland, Ohio).

ARBORS

Shell end mills or saws can be mounted on arbors where the cutter is secured by a nut. The shank of the arbor is the same as the shank of either the solid or split collet (Figure J-33).

SELF-TEST

1. How is a right-hand cut end mill identified?

2. What characteristic of end mills allows them to be used for plunge cutting?

3. What is the main difference between a general purpose end mill and one designed to cut aluminum?

4. When are carbide-tipped end mills chosen over high-speed steel end mills?

5. To remove a considerable amount of material, what kind of end mill is used?

6. Where are tapered end mills used?

7. Why are tools with carbide inserts used?

8. How are straight shank tools held in the machine spindle?

9. How are shell end mills driven?

10. Why are quick-change toolholders used?

UNIT 3

Setups on the Vertical Milling Machine

Before any machining can be done on the vertical mill, the toolhead must be squared to the table and saddle axes. Also, the workpiece or work-holding device must be secured to and aligned with the table and saddle axes. These common machine setups must be made or rechecked each time the machine is used. The purpose of this unit is to describe the procedure for squaring the toolhead and discuss common workpiece setups.

Objectives

After completing this unit, you should be able to:

1. Square the toolhead.

2. Set up and align a workpiece on the table.

3. Set up and align a mill vise.

4. Locate the edges of a workpiece relative to the spindle and position the spindle over a hole center.

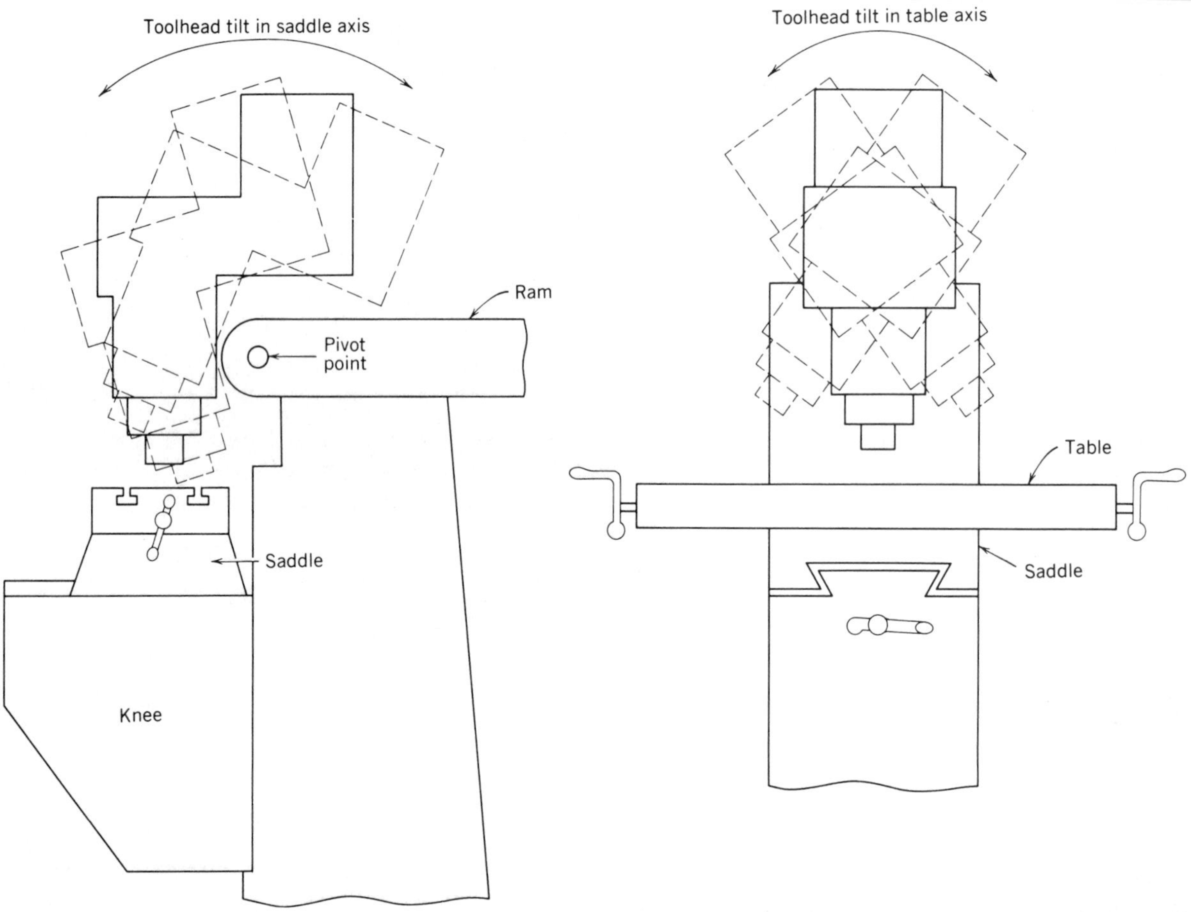

FIGURE J-34 **Vertical mill toolhead tilting capability.**

Squaring the Toolhead

On many vertical mills, although not all, the toolhead can be swiveled relative to the table and saddle axes (Figure J-34). This feature adds to the versatility of the machine tool since it permits drilling and milling on angled surfaces. However, by far the largest number of milling and drilling operations are done with the toolhead set square to the table and saddle axes. This alignment is quite critical as it is directly responsible for square mill cuts and straight drilled holes. The following procedure is used to square the toolhead.

FIGURE J-35 **Aligning the toolhead square to the table with a circular parallel and a dial indicator (Lane Community College).**

STEP 1. Fasten a dial indicator in the machine spindle (Figure J-35). The indicator should sweep a circle slightly smaller than the width of the table.

STEP 2. Lower the quill until the indicator contact point is depressed .015 to .020 in. Lock the quill in this position. The spindle indicator will be turned by hand and used to determine the position of the toolhead relative to the table and saddle. There is a tendency for the indicator tip to catch in the table T-slot as it is turned. To prevent this, an accurately machined ring can be used as an indicating surface. A large bearing race or the swivel base from a precision mill vise can be used for this purpose. If you use a ring for an indicating surface, be sure that the table is clean and that there are no burrs on the ring.

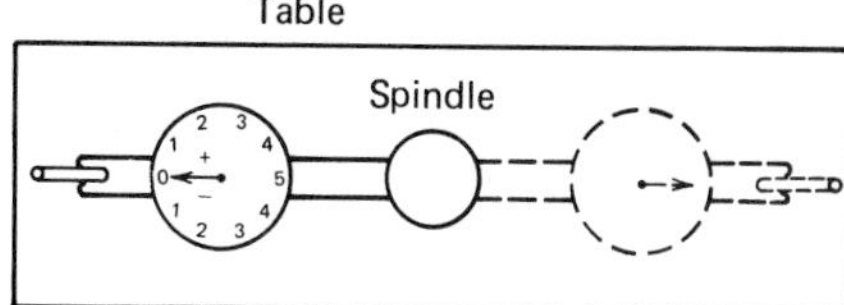

Set indicator to zero and rotate 180°

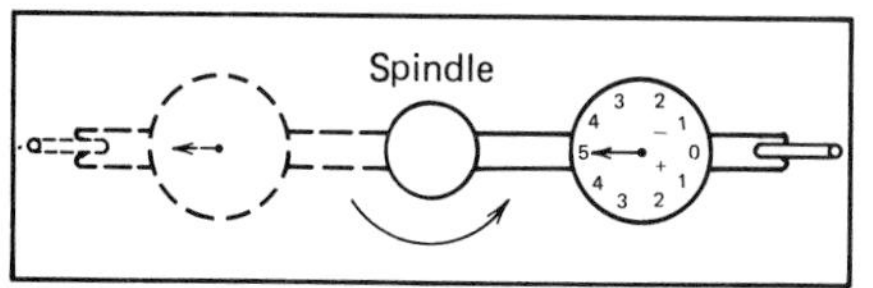

Observe reading (in this example +.005)

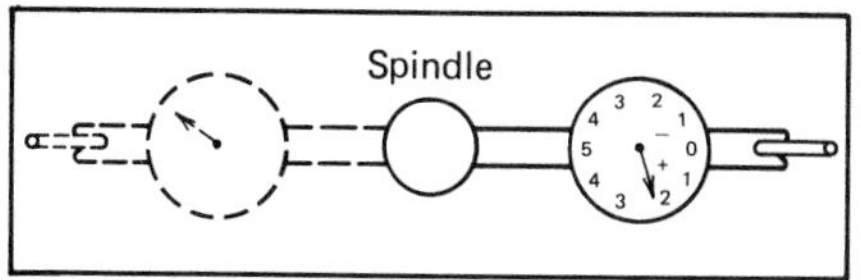

Adjust toolhead until indicator moves back toward zero ½ amount shown (.0025)

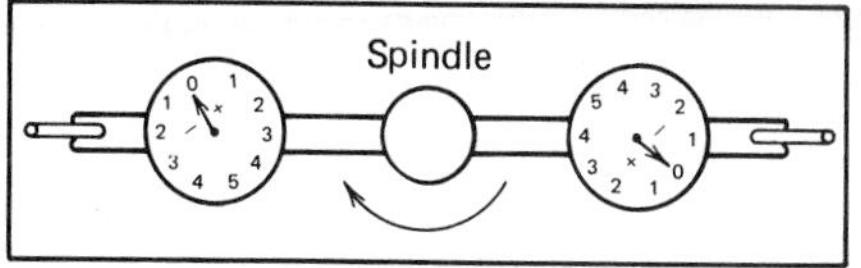

Reset indicator to zero and move back 180°. Readings should be the same on both sides

FIGURE J-36 Indicator readings while squaring the toolhead.

FIGURE J-37 Work aligned by locating against stops in T-slots (Lane Community College).

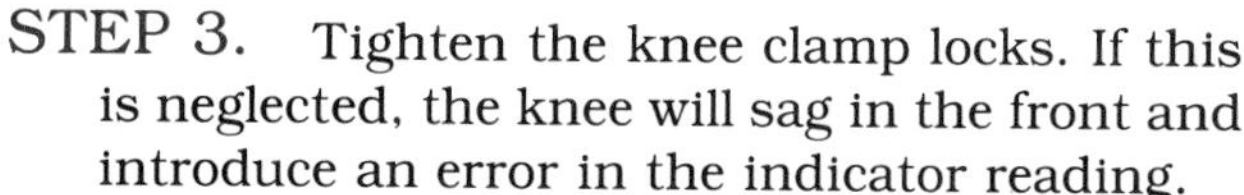

STEP 3. Tighten the knee clamp locks. If this is neglected, the knee will sag in the front and introduce an error in the indicator reading.

STEP 4. Loosen the toolhead clamping bolts one at a time and retighten them to provide a slight drag on the toolhead. Fine adjustments will be easier if the toolhead is just loose enough to be moved by slight pressure.

STEP 5. Rotate the spindle by hand until the indicator is to the left or right of the spindle and in line with the table axis.

STEP 6. Set the indicator bezel to zero. The example shown will aid you in the alignment procedure (Figure J-36).

STEP 7. Rotate the spindle 180 degrees so that the indicator is positioned on the opposite side and in line with the table axis. Note the reading at this position.

STEP 8. Tilt the toolhead using the tilt screw until the indicator moves back toward zero, one-half the amount showing.

STEP 9. Turn the indicator 180 degrees and see if the reading varies. If both readings are the same, tighten the toolhead clamping bolts and recheck the readings. Tightening the toolhead clamps will sometimes displace the head slightly, requiring a small additional adjustment.

STEP 10. Repeat this procedure for the toolhead alignment relative to the saddle axis.

STEP 11. After the toolhead has been squared in both axes, recheck its position and be sure that all clamping bolts are tight.

Work Holding on the Vertical Mill

The two most common work-holding methods on the vertical mill are securing parts directly to the table by means of clamps or by holding them in a mill vise.

MOUNTING DIRECTLY TO THE TABLE

Mounting a workpiece directly to the machine table is an excellent work-holding method. The same clamping techniques that you learned in drilling are applied in milling. Be careful not to distort the workpiece when applying clamping pressure.

In many cases, the workpiece must be aligned with the table or saddle axis to insure parallel or perpendicular cuts. A workpiece can be quite accurately aligned by placing it against stops that just fit the table T-slots (Figure J-37). Another

FIGURE J-38 Measuring the distance from the edge of the table to the workpiece (Lane Community College).

FIGURE J-39 Aligning a workpiece with the aid of a dial indicator (Lane Community College).

method is to measure from the edge of the table to the workpiece (Figure J-38).

Probably the most accurate method of workpiece alignment is to use a dial indicator fastened in the machine spindle or to the toolhead (Figure J-39). The workpiece is brought into contact with the indicator and the table or saddle is run back and forth while the workpiece position is adjusted so that the indicator reads zero.

USING A MILL VISE

A precision mill vise may also be used to hold the part being machined. The vise must also be aligned with the table or saddle axis. (See Section K, Unit 5). Once again, use a dial indicator for this purpose and always indicate a vise on its solid jaw.

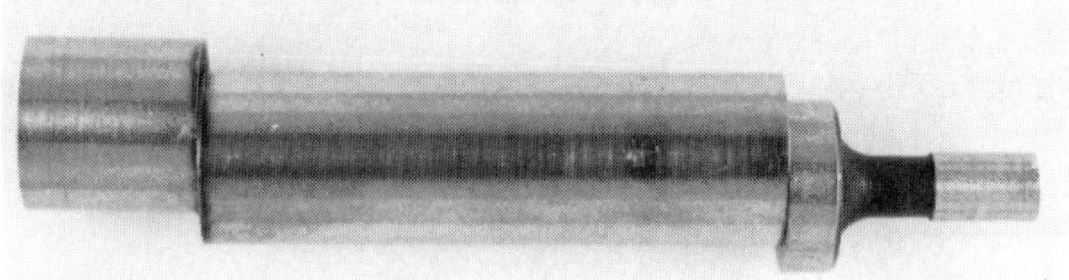

FIGURE J-40 Offset edgefinder (Lane Community College).

Mill vises are precision tools. Always treat them gently when setting them down on the machine table. Be sure that the table is clean and that there are no burrs on the bottom of the vise.

SOFT-VISE JAWS Sometimes a mill vise may be equipped with soft steel or aluminum jaws instead of the regular hardened steel types. After the vise has been bolted to the mill table and roughly aligned, a light cut can be taken on the soft jaws. The result of this procedure is a vise jaw that has been machined true to the axis of the saddle or table. Soft jaws are often used in production machining operations or where it might be desirable to shape the vise jaw in a certain way in order to hold a particular part to be machined.

Work Edge and Hole Centerline Locating

After a workpiece has been set up on the mill, it may be necessary to position the machine spindle relative to the edge of the part, or to center the spindle over an existing hole. Edge finding and centerline finding are very common operations that you will encounter almost daily in general milling machine operations.

USING AN EDGEFINDER

A very useful tool for edge finding is the offset edgefinder (Figure J-40). The edgefinder consists of a shank with a floating tip that is retained by an internal spring. The edgefinder tip is accurately machined to a known diameter, usually .200 or .500 in. Following this procedure for using the edgefinder.

STEP 1. Secure the edgefinder in a collet or chuck in the machine spindle.

STEP 2. Set the spindle speed to about 600 to 800 rpm, and slide the edgefinder tip over so that it is off-center.

STEP 3. Start the spindle and lower the quill or raise the knee so that the edgefinder tip can contact the edge of the part to be located.

FIGURE J-41 Work approaches the tip of the offset edgefinder (Lane Community College).

FIGURE J-42 Indicator used to locate the edge of a workpiece (Lane Community College).

STEP 4. Turn the table or saddle cranks and move the workpiece until it contacts the rotating edgefinder tip. Continue to slowly advance the workpiece against the edgefinder tip until the tip suddenly moves sideways. Stop movement at this moment (Figure J-41). The machine spindle is now positioned a distance equal to one-half of the edgefinder tip diameter from the edge of the workpiece. If you are using a .200-in.-diameter tip, the centerline of the spindle is .100 in. from the workpiece. Repeat the edgefinder-to-workpiece approach at least two times to confirm the position of the workpiece and the readings of the machine dials.

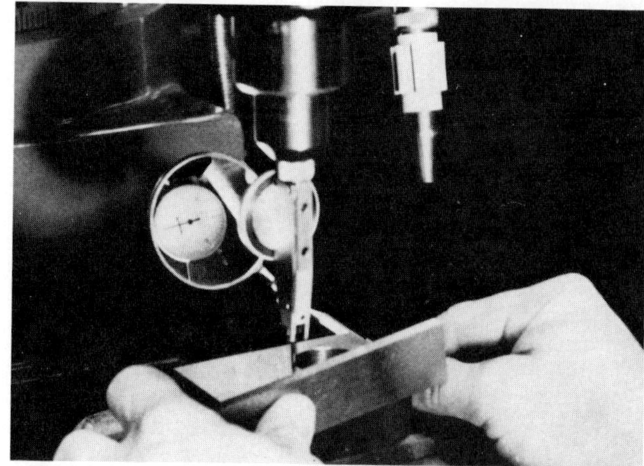

FIGURE J-43 Indicator against parallel to locate the edge of a workpiece (Lane Community College).

STEP 5. When you are sure that the positioning is correct, lower the workpiece or raise the spindle and set the table or saddle micrometer collars to zero. Then move the table or saddle the additional .100 in. in the same direction that it was moving as it approached the workpiece. This will prevent a backlash error from being introduced due to slack in the table or saddle nuts.

EDGE FINDING WITH A DIAL INDICATOR

If an edgefinder is not available, a workpiece edge can be located with the aid of a dial test indicator. The following procedure should be used.

STEP 1. Secure a dial test indicator in the machine spindle. Rotate the spindle by hand and set the indicator contact point as close to the spindle centerline as possible.

STEP 2. Lower the quill so that the indicator contact point touches the workpiece edge and registers a .010- to .015-in. deflection (Figure J-42). A slight rotating movement of the spindle forward and backward is used to locate the lowest reading on the dial test indicator. Now set the indicator bezel to register zero.

STEP 3. Raise the quill so that the indicator contact point is $\frac{1}{2}$ in. above the workpiece surface. Turn the spindle 180 degrees from the position it was in when the indicator was zeroed. Hold a precision parallel against the side of the workpiece so that it extends above the workpiece. Lower the quill until the indicator contact point is against the parallel (Figure J-43). Read the indicator value. Use a mirror to read the indicator when it faces away from you.

FIGURE J-44 Dial indicator used in locating the center of a hole (Lane Community College).

FIGURE J-45 Coaxial dial indicator used to locate a hole center (Lane Community College).

STEP 4. Move the table to where the indicator pointer is halfway between the reading against the parallel and the zero on the indicator dial.

STEP 5. Reset the bezel of the indicator to zero and recheck the position of the spindle as shown in Figure J-42.

STEP 6. Repeat this process until both readings on the indicator are the same.

STEP 7. This completes the locating of the spindle over the workpiece edge. The machine dial for this axis can now be set on zero.

HOLE CENTER LOCATING

Many vertical milling machine operations require that the machine spindle be positioned over the center of an existing hole in the workpiece. To locate a hole centerline, use the following procedure:

STEP 1. Mount a dial test indicator in the machine spindle.

STEP 2. Move the indicator contact point so it touches the side of the hole (Figure J-44). Set the indicator bezel to zero.

STEP 3. Rotate the spindle 180 degrees. Compare the two indicator readings and split the difference between them by moving the machine table. Locate the spindle center first in one axis, then in the second axis.

STEP 4. If the hole is not round or cylindrical, the spindle still has to be centered in both the table and saddle axis. In this case, the indicator readings would be identical 180 degrees apart, but would vary from axis to axis.

STEP 5. Always double-check the readings on the indicator and the machine dial settings before any additional machining is performed.

Another tool used to align a machine spindle with a hole or a round projection is the coaxial dial indicator (Figure J-45). With this indicator the spindle is turning at 500 to 600 rpm, but the face of the indicator is stationary. Since the dial indicator contact point is touching the surface of the hole being centered while the spindle is rotating, any table movement will either increase or decrease the indicator hand movement. The operator slowly moves the table, one axis at a time, until the indicator hand quits moving and stands still.

MACHINING HOLES IN VISE BODY (Figure J-46)

STEP 1. Align the workhead square to the machine table.

STEP 2. Align and fasten a machine vise on the table so that its jaw is parallel to the long axis of the table.

STEP 3. Mount the vise body in the machine vise with the bottom surface against the solid jaw of the machine vise.

STEP 4. Mount an edgefinder in a spindle collet and align the spindle axis with the base surface of the vise body.

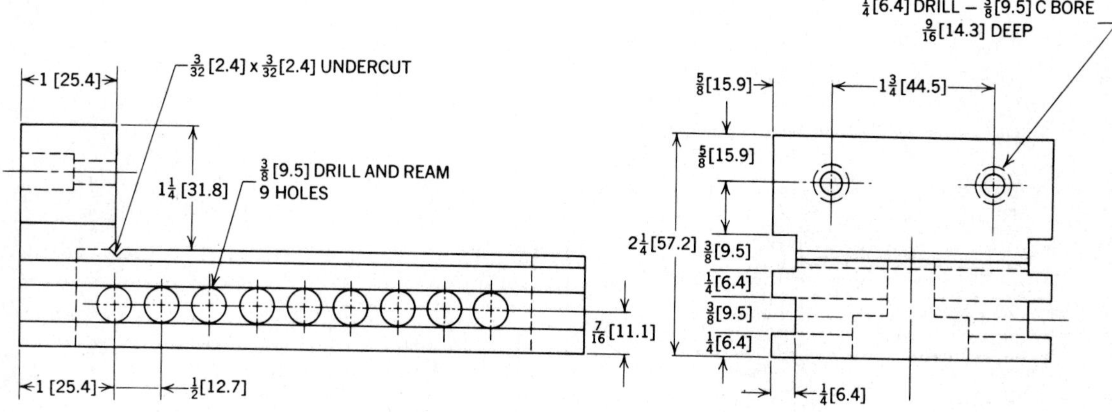

FIGURE J-46 Machining the holes in the vise body.

FIGURE J-47 Drilling $\frac{11}{32}$-in.-diameter holes in the vise body (Lane Community College).

STEP 5. Move the table the required .452-in. distance and lock the table cross slide.

STEP 6. Now pick up the outside of the solid jaw of the vise body.

STEP 7. Move to the first hole location 1.015 in. from the outside edge.

STEP 8. Center drill this hole.

STEP 9. Use a $\frac{1}{4}$-in.-diameter twist drill and drill this hole $1\frac{1}{2}$ in. deep.

STEP 10. Repeat steps 8 and 9 for the remaining eight holes. Accurate positioning is done with the micrometer dials.

STEP 11. Remove the workpiece from the machine vise. Turn it over so that the just-drilled holes are down and the bottom surface of the vise body is again against the solid jaw.

STEP 12. Use the edgefinder to pick up the two sides, as for the first drilling operation.

STEP 13. Position the spindle over the first hole location, again with the first hole on the solid jaw side.

STEP 14. Center drill this hole.

STEP 15. Drill this hole with a $\frac{1}{4}$-in.-diameter drill deep enough to meet the hole from below.

STEP 16. Switch to an $\frac{11}{32}$-in.-diameter drill and drill completely through the vise body (Figure J-47). The $\frac{1}{4}$-in. hole acts as a pilot hole to let the $\frac{11}{32}$-in. drill come out in the correct place on the bottom side.

STEP 17. Change from the $\frac{11}{32}$-in. drill to a $\frac{3}{8}$-in.-diameter machine reamer and ream this hole completely through also (Figure J-48).

STEP 18. Repeat steps 14 to 17 for the remaining eight holes.

STEP 19. Reposition the workpiece so that it is upright in the machine vise with the solid jaw of the vise body up.

STEP 20. With an edgefinder, pick up the edges of the workpiece.

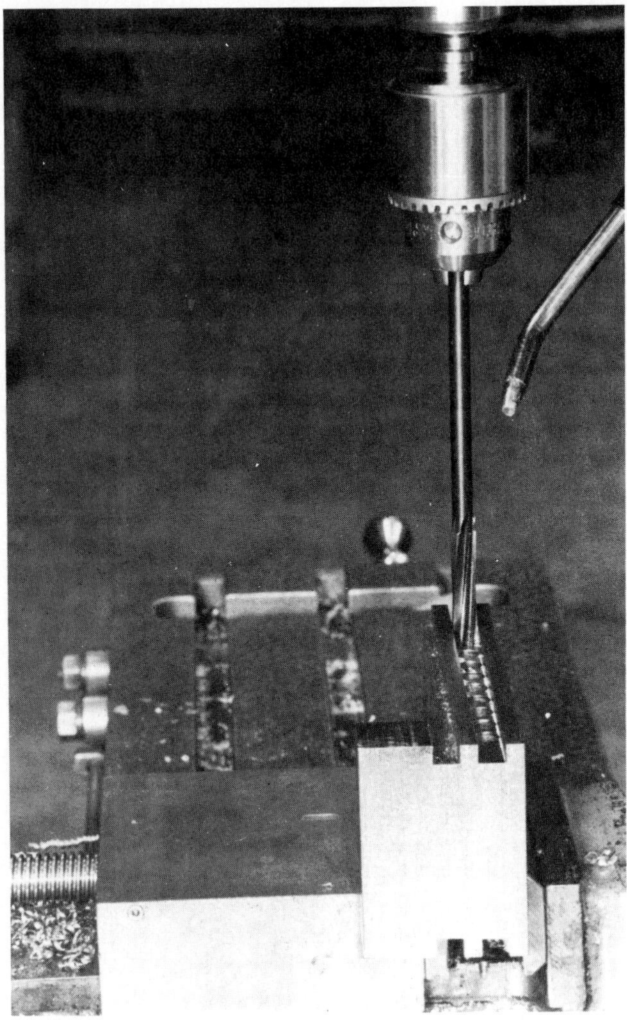

FIGURE J-48 Reaming $\frac{3}{8}$-in.-diameter holes in the vise body (Lane Community College).

FIGURE J-49 Drill and counterbore holes in the solid jaw of the vise body (Lane Community College).

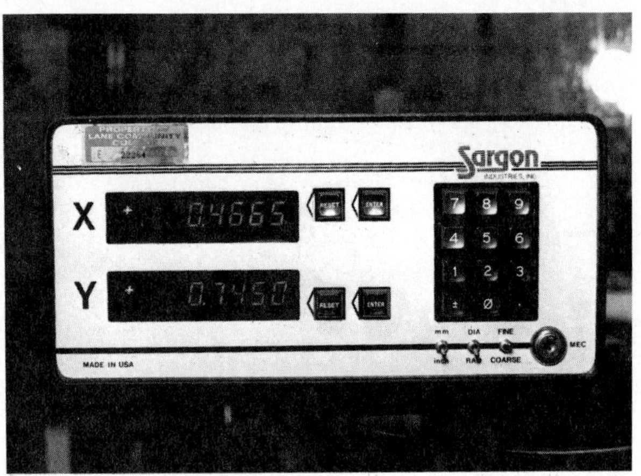

FIGURE J-50 Digital readout on a vertical milling machine (Lane Community College).

STEP 21. Position for the two hole locations and drill the $\frac{17}{64}$-in.-diameter holes with their $\frac{13}{32}$-in.-diameter counterbores (Figure J-49).

STEP 22. Remove all burrs.

One of the most desirable vertical milling machine attachments is a digital readout (DRO) (Figure J-50). A DRO increases the speed with which accurate dimensional movements can be made. Movements can be measured in either inches or millimeters in both positive or negative directions. Because actual machine movements are measured, backlash between feed screws and nuts can be ignored.

SELF-TEST

1. How can workpieces be aligned when they are clamped to the table?

2. How is a vise aligned on a machine table?

3. When is the toolhead alignment checked?

4. Why is it important that the knee clamping bolts are tight before aligning a toolhead?

5. Why does the toolhead alignment need to be checked again after all the clamping bolts are tightened?

6. How can the machine spindle be located exactly over the edge of a workpiece?

7. With a .200-in. edgefinder tip, when do you know that the spindle axis is .100 in. away from the edge of the workpiece?

8. What is the recommended rpm to use with an offset edgefinder?

9. When locating a number of positions on a workpiece, how can you eliminate the backlash in the machine screws?

10. How is the center of an existing hole located?

UNIT 4

Vertical Milling Machine Operations

The vertical milling machine is one of the most versatile machine tools found in the machine shop. The purpose of this unit is to explore some of this versatility and give you an idea of the wide scope of machining capability on this machine.

Objectives

After completing this unit, you should be able to:

1. Calculate cutting speeds and feeds for end milling operations.
2. Identify and select vertical milling machine setups and operations for a variety of machining tasks.

Climb and Conventional Milling

In milling, the direction that the workpiece is being fed can be either the same as the direction of cutter rotation or opposed to the direction of cutter rotation. When the direction of feed is opposed to the direction of rotation, this is said to be conventional or up milling (Figure J-51). When the direction of feed is the same as the direction of cutter rotation, this is said to be climb or down milling, because the cutter is attempting to climb onto the workpiece as it is fed into the cutter. If there is any large amount of backlash in the table or saddle nuts, the workpiece can be pulled into the cutter during climb milling. This can result in a broken cutter, damaged workpiece, and possible injury from flying metal. Climb milling should be avoided in most every case. However, in certain situations it may be desirable to climb mill. For example, if the milling machine has ball nuts and screws where backlash is virtually eliminated, climb milling is an acceptable technique. Even on conventional machines, climb milling with a very light cut can result in a better surface finish since chips are not swept back through the cut.

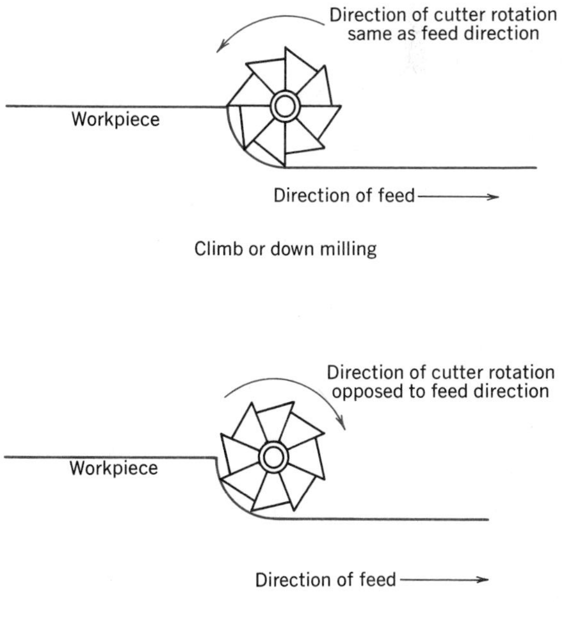

FIGURE J-51 Climb and conventional milling.

During any milling operation, all table movements should be locked except the one that is moving. This will insure the most rigid setup possible. Spiral fluted end mills may work their way out of a split collet when deep heavy cuts are made or when the end mill gets dull. As a precaution, to warn you that this is happening, you can make a mark with a felt-tip pen on the revolving end mill shank where it meets the collet face. Observing this mark during the cut will give you an early indication if the end mill is changing its position in the collet.

Cutting Performance Factors

Detailed information on machinability, cutting speeds, feeds, and cutting fluids is found in Section F.

CUTTING SPEEDS

Cutting speed (CS) in milling is the rate at which a point on the cutter passes by a point on the workpiece in a given period of time. Cutting speed is an extremely important factor in all machining operations. For rotating milling cutters, CS is expressed as a function of cutter rpm by the formula

$$\text{rpm} = \frac{\text{CS} \times 4}{D}$$

where rpm = revolutions per minute of the cutter

CS = cutting speed of the material being machined, in feet per minute

D = diameter of the cutter, in inches

Cutting speed constants (Table J-1) are influenced by the cutting tool material, workpiece material, rigidity of the machine setup, and the use of cutting fluids. As a rule, lower cutting speeds are used to machine hard or tough materials or where heavy cuts are taken and it is desirable to minimize tool wear and to extend tool life. Higher cutting speeds are used in machining softer materials in order to achieve better surface finishes. Higher speeds also apply when using small diameter cutters for light cuts on frail workpieces and in delicate setups. Table J-1 gives starting values for common materials. These values may have to be varied up or down depending on the specific machining task. Always observe the cutting action carefully and make appropriate speed corrections as needed. Until you gain some experience in milling, use the lower values in the table when selecting cutting speeds. Ball end mills, corner rounding mills, and angle milling cutters should be operated at one-half to two-thirds the speed of a comparable end mill.

TABLE J-1 Cutting Speeds for Some Commonly Used Materials

	Tool Material						
Work Material	**High-Speed Steel**	**Uncoated Carbide**	**Coated Carbide**	**Cermet**	**Ceramic**	**CBN**	**Diamond**
Aluminum							
Low silicon	300–800	700–1400					1000–5000
High silicon							500–2500
Bronze	65–130	500–700					1000–3000
Gray cast iron	50–80	250–450	350–500	400–1000	700–2000	700–1500	
Chilled cast iron					250–600	250–500	
Low-carbon steel	60–100	250–350	500–900	500–1300	1000–2500		
Alloy steel	40–70		350–600	300–100	500–1500	250–600	
Tool steel	40–70		250–500		500–1200	150–300	
Stainless steel							
200 and 300 series	30–80	100–250	400–650		300–1100		
400 and 500 series			250–350		400–1200		
Nonmetallics		400–600					400–2000
Superalloys		70–100	90–150		500–1000	300–800	

ADVANCED TOOL MATERIALS

The best machining results are achieved with very rigid machines and setups using a minimum of tool overhang and machine spindle extension. Vibration causing chatter will rapidly destroy tools. Many carbide grades are made for general-purpose machining within a range of different work materials. Others are designed for one specific application and they will fail if used for anything else. When using carbide, cermet, ceramic, CBN, or diamond inserts, it is absolutely essential to refer to the carbide tool manufacturer's catalog for specific application recommendations for a specific insert. Do not use any of the above-mentioned tools unless you know what the intended use and operating conditions are. General recommendations are:

Ceramics—used to machine all carbon, alloy, and stainless steels and superalloys.

CBN—used to machine hardened steel (RC 45 and above), chilled cast iron, and superalloys.

Diamonds—used to machine nonmetallic and nonferrous materials. Diamonds are not used to machine ferrous materials, because these alloys chemically attack the diamonds, which causes rapid tool wear. Diamonds are used most often in high-speed finishing and semifinishing operations.

FEED RATES

Another equally important factor in safe and efficient machining is the feed rate. Feed rate is the rate at which the material is advanced into the cutter or the cutter is advanced into the work material. Since each tooth of a multitooth milling cutter is cutting, a chip of a given thickness will be removed depending on the rate of feed. Chip thickness affects the life of the milling cutter. Excessive feed rates can cause a chipped cutting edge or a broken cutter. On the other hand, the highest practical feed rate per tooth will give the longest tool life between resharpenings. Feed rate in milling is measured in inches per minute, or ipm, and is calculated by the formula

$$\text{ipm} = F \times N \times \text{rpm}$$

where ipm = feed rate, in inches per minute

F = feed per tooth, in inches

N = number of teeth on the cutter being used

rpm = revolutions per minute of the cutter

Table J-2 gives starting values for chip loads in common materials.

CUTTING SPEED AND FEED RATE CALCULATIONS

The first step is to calculate the correct rpm for the cutter. Refer to Table J-1 for the cutting speed starting value.

EXAMPLE

Calculate the rpm for a $\frac{1}{2}$-in.-diameter HSS end mill machining aluminum.

$$\text{rpm} = \frac{\text{CS} \times 4}{D} = \frac{300 \times 4}{1/2} = \frac{1200}{.5} = 2400$$

The next step is to calculate the feed rate. Refer to Table J-2 for the starting values.

$$\text{ipm} = F \times N \times \text{rpm}$$

$$= .005 \text{ (feed per tooth, Table J-2)}$$

$$\times N \text{ (number of teeth)} \times \text{rpm}$$

$$= .005 \times 2 \times 2400 = 24 \text{ ipm}$$

Therefore, the cutter should revolve at 2400 rpm and the feed rate should be 24 ipm.

TABLE J-2 Feeds for High Speed Steel End Mills (Feed per Tooth in Inches)

Cutter Diameter	Aluminum	Brass	Bronze	Cast Iron	Low-Carbon Steel	High-Carbon Steel	Medium-Alloy Steel	Stainless Steel
$\frac{1}{8}$	.002	.001	.0005	.0005	.0005	.0005	.0005	.0005
$\frac{1}{4}$	.002	.002	.001	.001	.001	.001	.0005	.001
$\frac{3}{8}$	.003	.003	.002	.002	.002	.002	.001	.002
$\frac{1}{2}$	.005	.002	.003	.0025	.002	.002	.001	.002
$\frac{3}{4}$	.006	.004	.003	.003	.004	.003	.002	.002
1	.007	.005	.004	.0035	.005	.003	.002	.003
$1\frac{1}{2}$	.008	.005	.005	.004	.006	.003	.003	.004
2	.009	.006	.005	.005	.007	.004	.003	.004

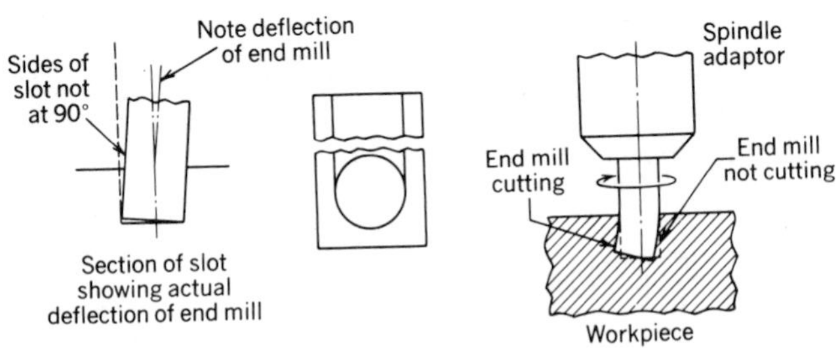

FIGURE J-52 The causes of a leaning slot in end milling.

FIGURE J-53 Using an end mill to mill steps (Lane Community College).

DEPTH OF CUT

The third factor to be considered in using end mills is the depth of cut. The depth of cut is limited by the amount of material that needs to be removed from the workpiece, by the power available at the machine spindle, and by the rigidity of the workpiece, tool, and setup. As a rule, the depth of cut for an end mill should not exceed one-half of the diameter of the tool in steel. In softer metals, depth of cut can be more. When using roughing end mills, the depth of cut can be as much as 1.5 times the cutter diameter and the width of cut one-half of the cutter diameter. But if deeper cuts need to be made, the feed rate needs to be reduced to prevent tool breakage. The end mill must be sharp and should run concentric in the end mill holder. The end mill should be mounted with no more tool overhang than necessary to do the job.

A problem that occasionally arises when using end mills to machine grooves or slots is a slot with nonperpendicular sides. Grooves with leaning sides are caused by worn spindles, excessive tool projection from the spindle, dull end mills, or excessive feed rates. The leaning slot is produced by an end mill that is deflected by high cutting forces (Figure J-52). To reduce the tendency of the tool to cut a leaning slot, reduce the feed rate, use end mills with only a short projection from the spindle, and use end mills with straight or low-helix angle flutes.

Another factor to consider is the horsepower rating of the machine. For horsepower calculations, see Section K, Unit 5.

CUTTING FLUIDS

Milling, as well as other machining operations, is often performed using cutting fluids. Cutting fluids serve to dissipate heat generated by the friction of the cutter against the workpiece. They also help to lubricate the interface between the cutting edge and work and to flush the chips away from the cutting area. Many machining operations are greatly improved through the use of cutting fluids. Cutting fluids increase productivity, extend tool life, and improve the surface finish of the workpiece. Cutting fluids are applied to the cut in a flood stream or by an air/cutting fluid mix mist. Cutting fluids generally reduce carbide tool life in milling cast iron and steel. The cutting fluid cannot get to the cutting edge while the insert is in the cut. After the extremely hot insert leaves the cut, the cutting fluid then has an extremely shocking cooling effect. These excessive temperature variations result in minute cracks along the cutting edge and premature tool failure. Cutting fluid recommendations of carbide tool manufacturers have to be followed to avoid tool failure caused by faulty coolant use.

Some materials, such as cast iron, brass, and plastics, are often machined dry. A stream of compressed air can be used to cool tools and to keep the cutting area clear of chips. A safety shield and protective clothing should always be used when compressed air is applied, for protection from flying hot chips.

FIGURE J-54 Using an end mill to square stock (Lane Community College).

FIGURE J-56 Plunge cutting with an end mill (Lane Community College).

FIGURE J-55 Roughing end mill used to remove a large volume of material (Lane Community College).

FIGURE J-57 Using an end mill to machine a pocket (Lane Community College).

Common Milling Operations

MACHINING STEPS AND SQUARING

Common milling operations on the vertical mill include machining steps (Figure J-53) and squaring or machining two surfaces perpendicular to each other (Figure J-54). The ends of the workpiece can be machined square and to a given length by using the peripheral teeth of an end mill.

If a large amount of material has to be removed, it is best to use a roughing end mill first (Figure J-55), then finish to size with a regular end mill.

On low-horsepower vertical mills plunge cutting is an efficient method of removing material quickly (Figure J-56). In this operation, the end mill is plunged a predetermined width and depth of cut, retracted, then advanced and plunged again repeatedly. In plunging, the maximum cutting force is in the direction in which the machine is the strongest—in the axial direction.

MILLING A CAVITY

Center-cutting end mills make their own starting hole when used to mill a pocket or cavity (Figure J-57). Prior to making any mill cuts, the outline of the cavity should be laid out on the workpiece. Only when finish cuts are made should these layout lines disappear. Good milling practice is to rough out the cavity to within .030 in. of finished size before making any finish cuts.

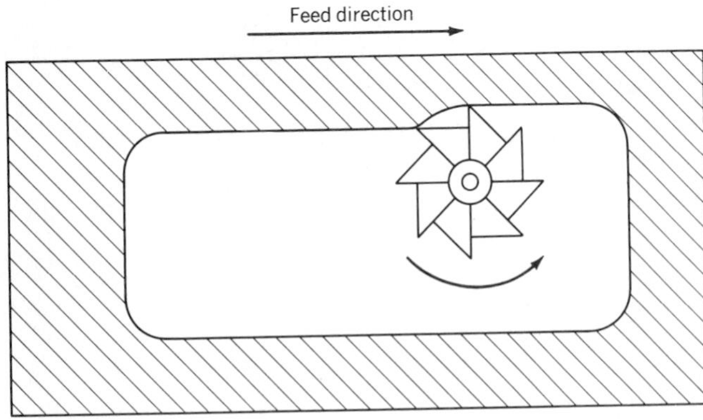

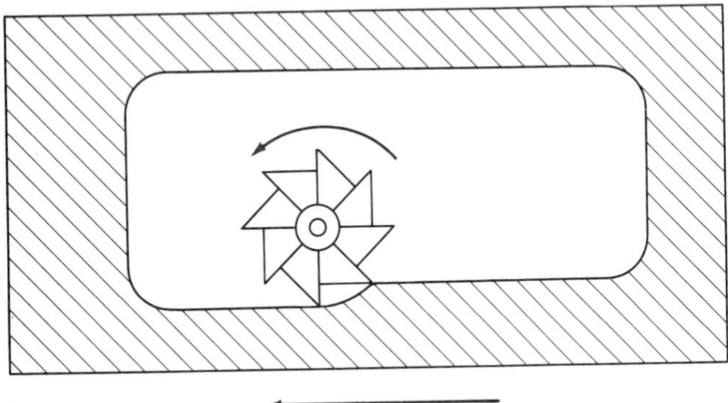

FIGURE J-58 Feed direction is against cutter rotation.

When you are milling a cavity, the direction of the feed should be against the rotation of the cutter (Figure J-58). This assures positive control over the distance the cutter travels and prevents the workpiece from being pulled into the cutter because of backlash. When you reverse the direction of table travel, you will have to compensate for the backlash in the table feed mechanism.

It is very important that the chips are removed from the cavity during the milling operation. Chips can be blown out of the cavity with compressed air or removed with a shop vacuum cleaner. If the chips are left to accumulate, the cutter will jam and very likely break. Safety guards around the machine are necessary when using compressed air.

END MILLING A SHAFT KEYSEAT

A common end milling operation on the vertical mill is keyseat milling. A shaft keyseat must be both centered on the shaft and cut to the correct depth. The following procedure may be used.

STEP 1. Secure the workpiece to the machine table or in an indicator aligned mill vise.

STEP 2. Select the correct size of cutter either in a two-flute or center cutting multiflute type and preferably one that has not been reduced in diameter by resharpenings. Install the cutter in the machine spindle.

FIGURE J-59 Setting an end mill to the side of a shaft with the aid of a paper feeler (Lane Community College).

STEP 3. Move the workpiece aside and lower the cutter beside the part. With the spindle motor off, insert a slip of paper between cutter and workpiece.

STEP 4. Use the saddle crank and move the workpiece toward the cutter until the paper feeler is pulled between cutter and work as you rotate the spindle by hand. At this point, the cutter is about .002 in. from the cutter (Figure J-59).

STEP 5. Set the saddle micrometer collar to zero, compensating for the .002 in. of paper thickness.

STEP 6. Add the diameter of the shaft and cutter and divide by two. This is the total distance to move the workpiece for centering.

STEP 7. Raise the cutter clear of the work and move the workpiece over the correct amount in the same direction that it was moving as it approached the cutter.

STEP 8. Raise the quill to its top position and lock the quill and the saddle in place.

STEP 9. Move the table crank to position the cutter at the point where the keyseat is to begin.

STEP 10. Start the spindle and raise the knee until the cutter makes a circular mark equal

FIGURE J-60 Cutter centered over the shaft and lowered to make a circular mark (Lane Community College).

FIGURE J-61 After centering and setting depth, keyseat is milled to required length (Lane Community College).

to the cutter diameter (Figure J-60). Set the knee micrometer collar to zero. Raise the knee a distance equal to one-half the cutter diameter plus .005 in. and lock the knee in this position. Using correct speeds and feeds mill the keyseat to the required length (Figure J-61).

MACHINING T-SLOTS, DOVETAILS, ANGLE MILLING, AND DRILLING

To machine a T-slot or a dovetail into a workpiece, two operations are performed. First, a slot is cut with a regular end mill, and then a T-slot cutter or a single angle milling cutter is used to finish the contour (Figures J-62 and J-63). Angular cuts on workpieces can be made by tilting the workpiece in a vise with the aid of a protractor (Figure J-64) and its built-in spirit level.

FIGURE J-62 Milling a slot and then the dovetail (Lane Community College).

FIGURE J-63 First a slot is milled and then the T-slot cutter makes the T-slot (Lane Community College).

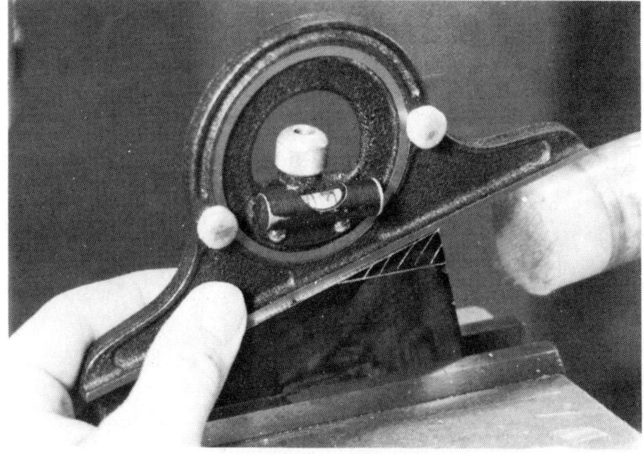

FIGURE J-64 Setting up a workpiece for an angular cut with a protractor (Lane Community College).

FIGURE J-65 Machining an angle with an end mill (Lane Community College).

FIGURE J-66 Using a shell mill to machine an angle (Lane Community College).

Machining the angle can be performed with an end mill (Figure J-65) or with a shell mill (Figure J-66). Another possibility for machining angles is the tilting of the workhead (Figures J-67 and J-68).

The head can be swiveled so that accurate angular holes can be drilled. These holes can be drilled by using the sensitive quill feed lever or the power feed mechanism (Figure J-69), or, in the case of vertical holes, the knee can be raised.

OTHER VERTICAL MILL OPERATIONS AND ACCESSORIES

Holes can be machine tapped by using the sensitive quill feed lever and the instant spindle reversal

FIGURE J-67 Cutting an angle by tilting the workhead and using the end teeth of an end mill (Lane Community College).

FIGURE J-68 Cutting an angle by tilting the workhead and using the peripheral teeth of an end mill (Lane Community College).

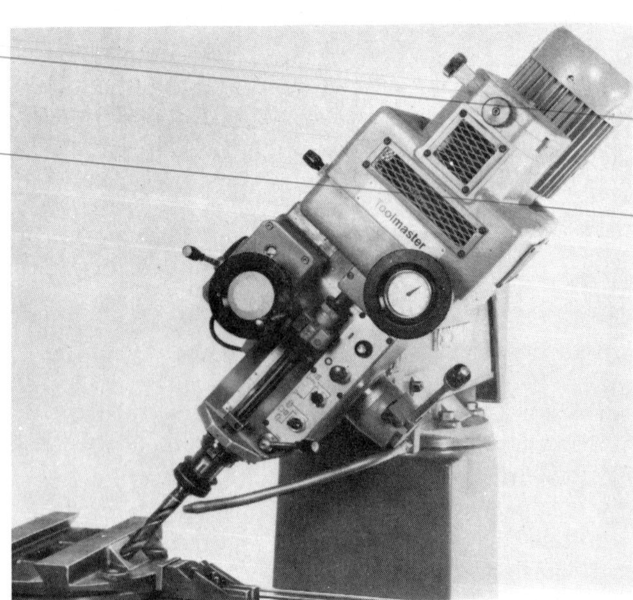

FIGURE J-69 Drilling of accurately located holes (Cincinnati Milacron).

FIGURE J-70 Tapping in a vertical milling machine (Cincinnati Milacron).

FIGURE J-71 Boring with an offset boring head (Cincinnati Milacron).

FIGURE J-72 Using a rotary table to mill a circular slot (Cincinnati Milacron).

FIGURE J-73 A dividing head in use (Cincinnati Milacron).

knob (Figure J-70). When an offset boring head is mounted in the spindle, precisely located and accurately dimensioned holes can be bored (Figure J-71). Circular slots can be milled when a rotary table is used (Figure J-72). Precise indexing can be performed when a dividing head is mounted on the milling machine. Figure J-73 shows the milling of a square on the end of a shaft. On many vertical milling machines a shaping attachment is mounted on the rear of the ram. This shaping attachment can be brought over the machine table by swiveling the ram 180 degrees. Shaping at-

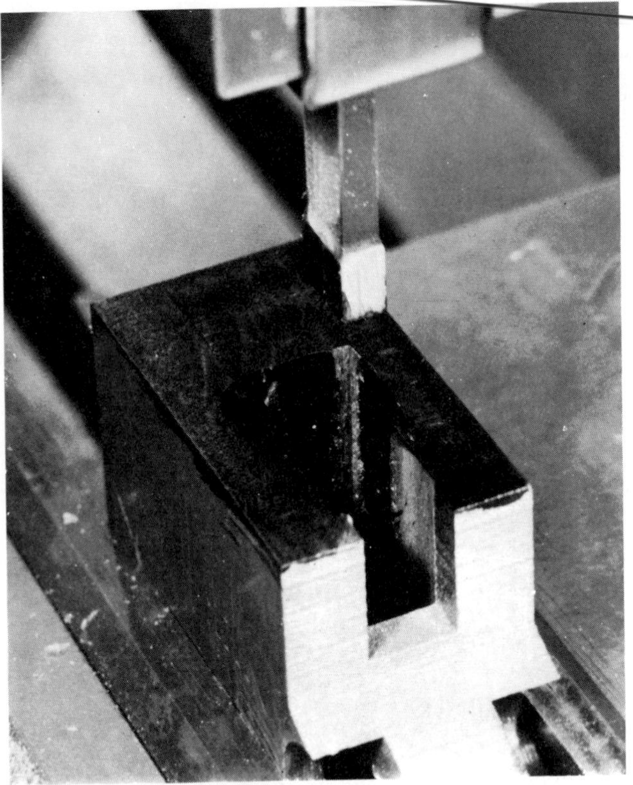

FIGURE J-74 A shaping head used to cut a square corner hole (Lane Community College).

FIGURE J-75 Using a right-angle milling attachment (Cincinnati Milacron).

tachments are used to machine irregular shapes on or in workpieces, such as the square corner hole shown in Figure J-74. When a right angle milling attachment (Figure J-75) is mounted on the spindle, it is possible to machine hard-to-get-at cavities at often difficult angles on workpieces.

SELF-TEST

1. When are the lower cutting speeds recommended?

2. When are the higher cutting speeds used?

3. Should you always use calculated rpm?

4. When are cutting fluids used?

5. When should machining be performed dry?

6. How is the tool life of an end mill affected by the chip thickness of a cut?

7. What is normally considered the maximum depth of cut for an end mill?

8. What are the limitations on the depth of cut?

9. Calculate the rpm for a $\frac{3}{4}$-in.-diameter HSS end mill to machine bronze.

10. Calculate the feed rate for a two-flute $\frac{1}{4}$-in.-diameter carbide end mill to machine low-carbon steel.

11. How is an end mill centered over a shaft prior to cutting a keyseat?

12. Why should the feed direction be against the cutter rotation when milling a cavity?

13. What can cause an end mill to work itself out of a collet while cutting?

14. Describe two methods of cutting angular surfaces in a vertical milling machine.

15. How are circular slots milled?

16. What milling machine attachment is used to mill a precise square or hexagon on a shaft?

17. How can a square hole in a workpiece be machined on the vertical milling machine?

18. When is a right-angle milling attachment used?

19. Why should workpieces be laid out before machining starts?

20. Why are two operations necessary to mill a T-slot?

UNIT 5

Using the Offset Boring Head

The offset boring head is used on the vertical mill to machine accurately located holes of precisely controlled diameters. Boring heads can also be used to machine external surfaces and to cut grooves and recesses. The purpose of this unit is to describe the setup and operation of this very useful accessory tool for the vertical mill.

Objective

After completing this unit, you should be able to:

Set up and use the offset boring head in common boring operations.

The Offset Boring Head

Most holes in workpieces are machined by drilling. When a better surface finish and better diameter accuracy is required, reaming may follow drilling. However, drilling and reaming are limited to standard sizes in which these tools are available. In addition, drilled holes may drift off position during a machining operation.

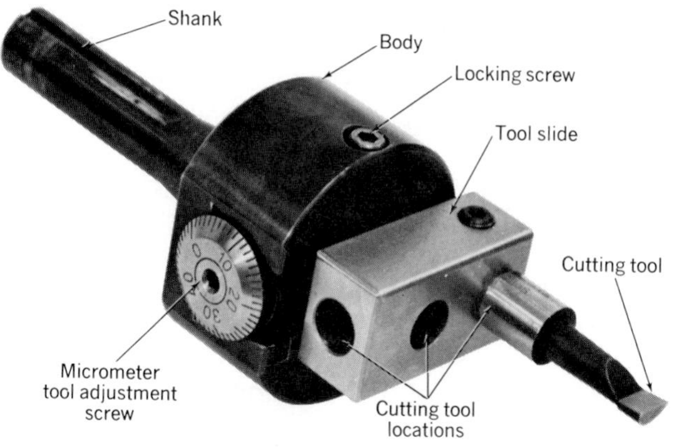

FIGURE J-76 Offset boring head (Lane Community College).

To machine holes of any size and at exact locations, the offset boring head may be used. A boring head can only be used to enlarge existing holes in the workpiece. The workpiece must be predrilled on the mill or drill press. If you are boring holes that are considerably larger than the largest-diameter drill available, time may be saved by using a high-speed hole saw to cut out most of the unwanted material. This will save many additional boring steps required to enlarge a hole to its finished size.

PARTS OF THE OFFSET BORING HEAD

The offset boring head consists of the body with shank and the tool slide (Figure J-76). The shank permits the body to be secured in the machine spindle. The tool slide contains several holes that will accommodate the several sizes of boring bars. Boring bar materials include HSS, brazed carbide, and disposable insert carbide. A typical boring bar set (Figure J-77) consists of several HSS or brazed carbide bars in assorted diameters and lengths. All bars have standard shanks that fit the holes in the tool slide.

FIGURE J-77 Set of boring tools for the offset boring head (Lane Community College).

FIGURE J-78 Workpiece supported on parallels. Note the clearance for the penetrating boring tool (Lane Community College).

Workpiece Preparation and Setup

The workpiece should be predrilled or otherwise machined to within about $\frac{1}{16}$ in. of finished size. If the hole in the part is rough, leave additional material so that cleanup machining will be ensured.

A workpiece to be bored may be clamped directly to the machine table by standard methods or held in a vise or other work-holding fixture. If the part is to be bored through, it must be supported on parallels so that the boring tool will not cut into the machine table (Figure J-78). The parallels must also be set far enough apart so that the tool will clear as it turns.

POSITIONING THE BORING HEAD

Before boring can begin, the boring head must be positioned over the hole to be bored. If the predrilled hole is accurately positioned, the boring head may be centered over the hole by using a dial indicator clamped to the boring head or held in the machine spindle. Use the procedure that you learned in Unit 3 to position the boring head over the hole.

If the predrilled hole is off location, the position of the final bored hole can still be correct if the machine spindle is first positioned in the saddle and table axes relative to the edge of the workpiece or other reference point or feature. Use the positioning techniques that you have already learned.

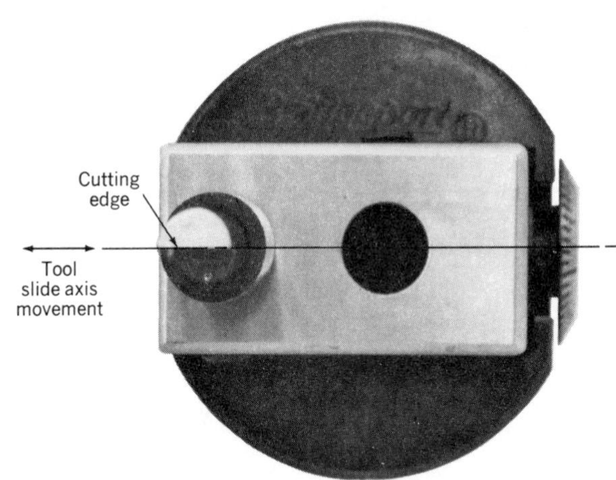

FIGURE J-79 Boring tool cutting edge is on the centerline of the boring head (Lane Community College).

Using the Offset Boring Head

BAR SETUP

When a boring tool is mounted in a boring head, it is very important that the cutting edge be on the centerline of the boring head and in line with the axis of the tool slide movement (Figure J-79). Only in this position are the rake angles and clearance angles correct as ground on the tool. This is also the only position in which the tool's cutting edge moves the same distance as the tool slide when adjustments are made. When selecting a boring bar, always pick the largest diameter bar with the shortest shank that will fit the hole to be bored. This ensures maximum rigidity of the setup.

When using any boring head, it is important to determine the amount of tool slide advance when the micrometer adjusting screw is rotated one graduation. On some boring heads the tool slide advances the same amount as on the micrometer dial. This will increase the diameter of the bore by twice the movement on the dial. Other boring heads are direct reading—the tool slide advance will increase the bore diameter by the same amount indicated on the dial. Movement of the tool slide is in thousandths of an inch and the ratio of tool slide advance to micrometer graduations will usually be indicated on the boring head.

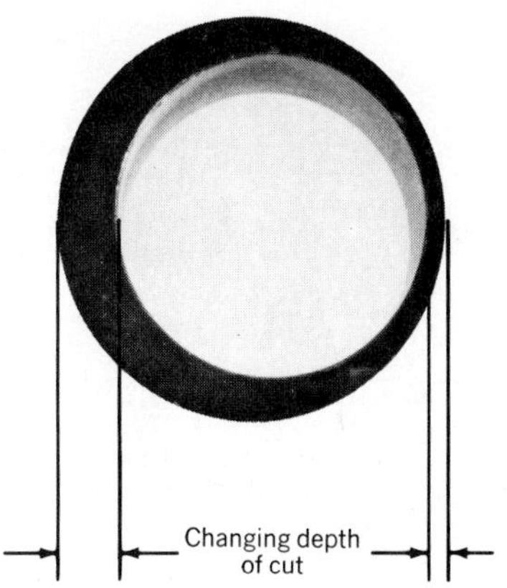

FIGURE J-80 When the hole is eccentric to the spindle centerline, it will cause a variable depth of cut for the boring tool (Lane Community College).

FEEDS AND SPEEDS IN BORING

The kind of tool material and the workpiece material determine the cutting speed that should be used. But the rigidity of the machine spindle and the setup often require a lower than calculated rpm because the imbalance of the offset boring head creates heavy machine vibrations.

The quill feed on many vertical milling machines is limited to .0015, .003, and .006 in. of feed per spindle revolution. Roughing cuts should be made at the higher figure and finishing cuts should be made with the two lower values. Roughing cuts are usually made with the tool feeding down into the hole. Finishing cuts are made with the tool feeding down and often the tool is fed back up through the hole by changing the feed direction at the bottom of the hole. Because of the tool deflection, a light cut will be made without resetting the tool on that second cut. When cuts are made with the tool only feeding down but not out, the spindle rotation is stopped before the tool is withdrawn from the hole. If the spindle rotates while the quill is raised, a helical groove will be cut into the wall of the just completed hole, possibly spoiling it.

CONTROLLING THE BORE DIAMETER

To obtain a predictable change in hole size for a given tool slide adjustment, certain conditions have to be met. The depth of cut of the boring tool needs to be the same around the circumference of the hole, not like the varying depth of cut illustrated in Figure J-80. Roughing cuts should be taken until the hole is round and concentric with the spindle centerline. The depth of cut needs to be equal for successive cuts. As an example, assume that a hole has been rough bored to be concentric with the spindle axis. The tool is now resharpened and fastened in the tool slide. The tool slide is advanced until the tool just touches the wall of the hole. After raising the tool above the work, the tool is moved 20 graduations, or a distance that should increase the hole diameter by .020 in. The spindle is turned on, and, with a feed of .003 in. per revolution, the cut is made through the hole. The spindle is stopped and the tool is withdrawn from the hole. Measuring the hole shows the diameter to have increased by only .015 in. What has happened is that the tool was deflected by the cutting pressure to produce a hole .005 in. smaller than expected. The tool is now advanced to again give an increase of .020 in. in the hole diameter. With the same feed as for the last cut, the hole is bored.

Measuring the hole again shows the hole to be .020 in. larger. Additional cuts made with the same depth of cut and the same feed will give additional .020-in.-diameter increases. If the depth of cut is increased, more tool deflection will take place, resulting in a smaller than expected diameter increase. If the depth of cut is decreased, the tool will be deflected less, resulting in a larger than expected diameter.

When the same depth of cut is maintained but the feed per revolution is increased, higher cutting pressures will produce more tool deflection and a smaller than expected hole diameter. With an equal depth of cut and a smaller feed per revolution, less cutting pressure will produce a larger than expected hole diameter. Another factor that affects the hole diameter with a given depth of cut is tool wear. As a tool cuts, it becomes dull. A dull tool will produce higher cutting pressures with a resultant larger tool deflection.

FIGURE J-81 A radius is machined on a workpiece with the offset boring head (Lane Community College).

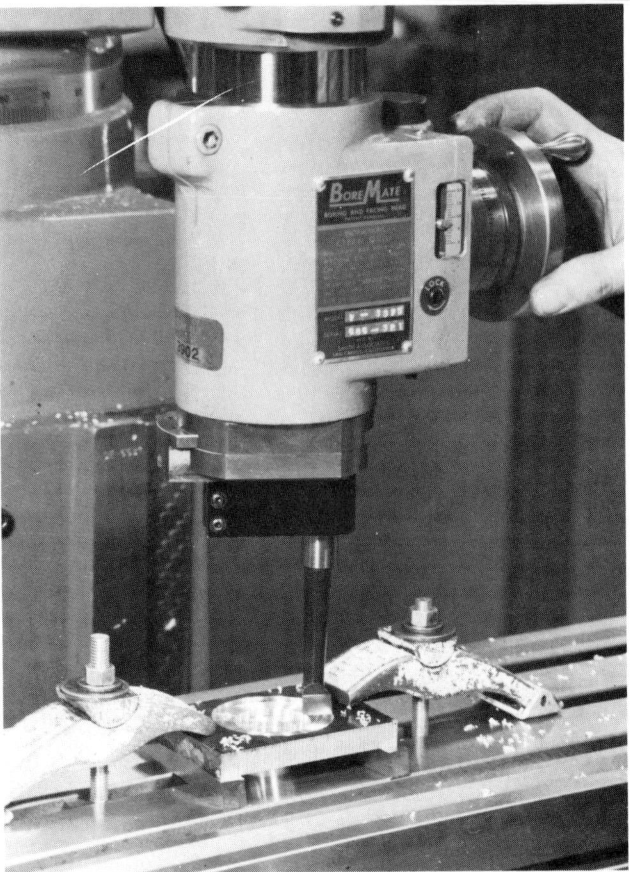

FIGURE J-82 A boring and facing head (Lane Community College).

MACHINING RADII AND FACING WITH THE BORING HEAD

An offset boring head can also be used to machine a precise radius on a workpiece (Figure J-81). The workpiece is positioned the specified distance from the spindle axis. A scrap piece of metal is clamped to the table opposite the workpiece. As cuts are being made with the offset boring head, the tool cuts on both the workpiece and the scrap piece. The diameter of the cuts is measured between the pieces being machined.

With a boring and facing head (Figure J-82), it is possible to machine flat surfaces with the same tool that was used to bore a hole size. The tool can be moved sideways while the spindle is rotating.

SELF-TEST

1. When is an offset boring head used?

2. Why is the workpiece normally placed on parallels?

3. Why is the locking screw tightened after tool slide adjustments have been made?

4. Why does the tool slide have a number of holes to hold boring tools?

5. Why is it important to determine the amount of tool movement for each graduation on the adjustment screw?

6. What would be the best boring tool to use on a job?

7. How important is the alignment of the tool's cutting edge with the axis of the tool slide?

8. What factors affect the size of the hole obtained for a given amount of tool adjustment?

9. Name three causes for changes in boring tool deflection.

10. What determines the cutting speed in boring?

SECTION **K**

HORIZONTAL MILLING MACHINES

In the preceding section you studied the vertical milling machine, where the machine spindle is primarily in the vertical axis. This section deals with the horizontal milling machine, where the spindle is primarily in the horizontal axis.

The horizontal milling machine, like its vertical counterpart, is an extremely versatile machine tool and capable of accomplishing a variety of machining tasks.

Types of Horizontal Milling Machines

BED TYPE

A bed-type milling machine is one in which the position of the milling spindle, but not the height of the table, can be changed. There are many configurations of bed-type milling machines. One type has the general appearance of a knee and column. One type of very large milling machine is known as a planer mill (Figure K-1) or adjustable rail milling machine. The example shown is equipped with two separate tables so that one table can be set up while machining is taking place on the other table. These machines often have 250 hp to their spindles. Other bed-type milling machines that employ two or more cutting heads are called duplex or triplex milling machines (Figure K-2). These are commonly used in high-production setups.

A particularly common form of bed milling machine is called the manufacturing milling machine (Figure K-3). On this machine the spindle assembly is positioned and secured at the correct height and parts are passed under the cutter. These machines are usually equipped with means for automatic cycling of the table, and they often have twin fixtures so that one part can be added while

FIGURE K-1 Large planer-type milling machine with two tables. A setup can be made on one table while machining is taking place on the other (Ingersoll Milling Machine Company).

FIGURE K-2 Triplex bed-type milling machine (Cincinnati Milacron).

FIGURE K-3 Plain manufacturing-type milling machine (Cincinnati Milacron).

FIGURE K-4 Bed-type horizontal milling machine with transverse table motion. This type of mill is also found in a vertical spindle design (Cincinnati Milacron).

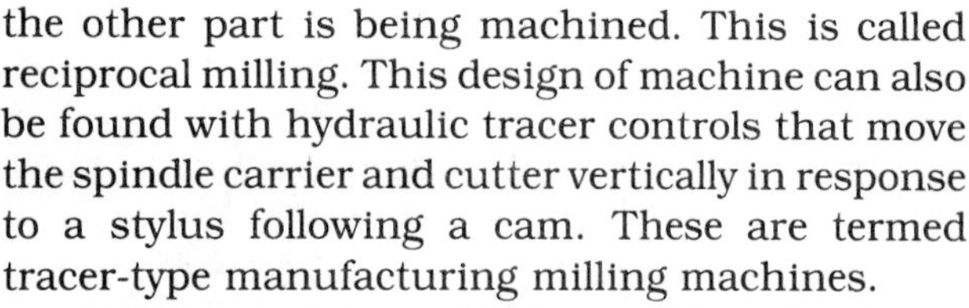

the other part is being machined. This is called reciprocal milling. This design of machine can also be found with hydraulic tracer controls that move the spindle carrier and cutter vertically in response to a stylus following a cam. These are termed tracer-type manufacturing milling machines.

Another bed-type milling machine also has a table traverse motion with a spindle assembly that can be moved vertically (Figure K-4). This type of machine is found with either horizontal or vertical head configuration and, by general appearance, is often mistaken for a knee and column type of milling machine.

Knee and Column Milling Machines

Knee and column milling machines are derived from the heritage of Joseph Brown's universal milling machines in the 1860s. Universal means that the machine table can swivel on its horizontal axis (Figure K-5) so that the work can be presented to the cutter at an angle in conjunction with a suitable indexing head permitting helical milling. The plain knee and column milling machine (Figure K-

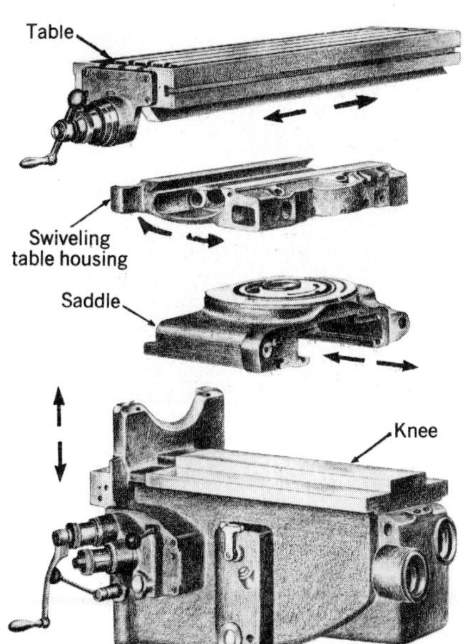

FIGURE K-5 The main features of the knee, saddle, and table assembly on the universal knee and column milling machine (Cincinnati Milacron).

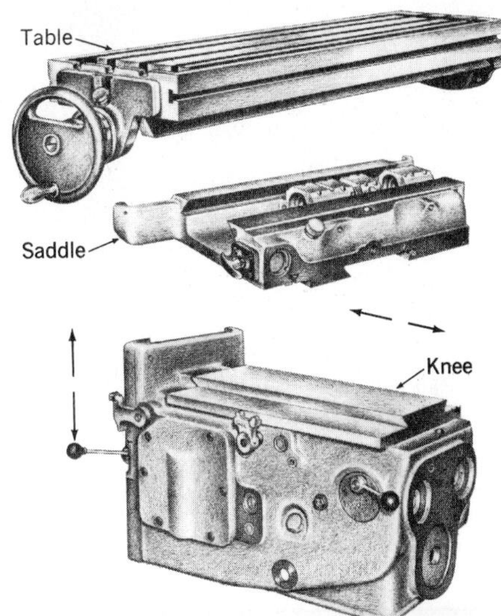

FIGURE K-6 The features of the knee, saddle, and column milling machine. The swiveling table housing is omitted in this design (Cincinnati Milacron).

FIGURE K-7 Small, plain, automatic knee and column milling machine (Cincinnati Milacron).

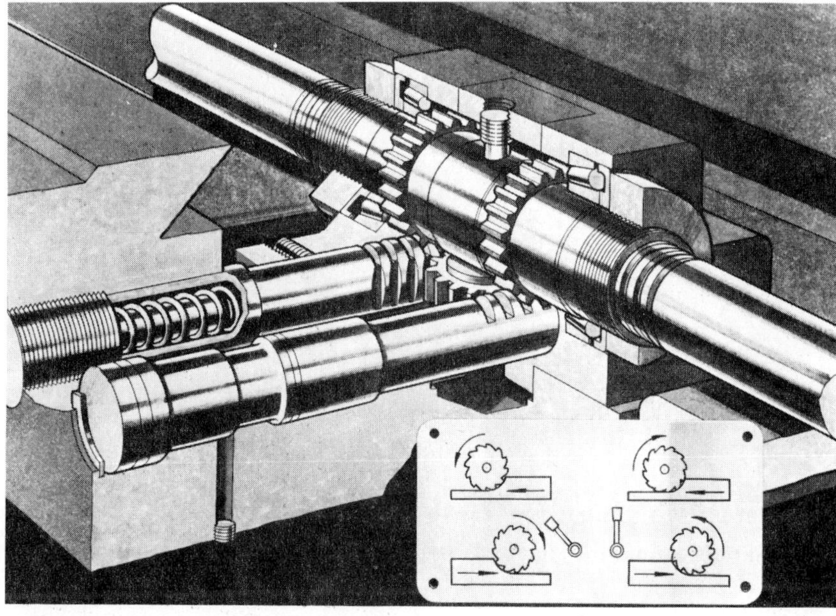

FIGURE K-8 Backlash eliminator to permit climb milling (Cincinnati Milacron).

6) omits the table swiveling feature in the interest of greater machine rigidity.

There is one type of knee and column milling machine used in manufacturing that is capable of vertical table positioning and longitudinal table travel, but it does not have transverse or cross feeding capability. On this type of machine (Figure K-7), the spindle bearings are carried in a quill so that the cutter can be positioned traversely over the part and locked into position. Eliminating the cross feeding saddle adds to the machine rigidity.

On the bed-type manufacturing milling machines, the table is moved by a hydraulic-mechanical means that includes backlash control. This control permits the cut to be made easily in either direction without the danger of having the milling cutter suddenly grab the work and take up the backlash, as can happen with most ordinary nut and screw table feeds. This sudden taking up of backlash results in many broken cutter teeth if it is not controlled by the method of milling or by special devices. One of these devices is called a backlash eliminator (Figure K-8), which automatically takes up the backlash by applying a preload between two nuts following the leadscrew. Another method, employed mainly with numerically con-

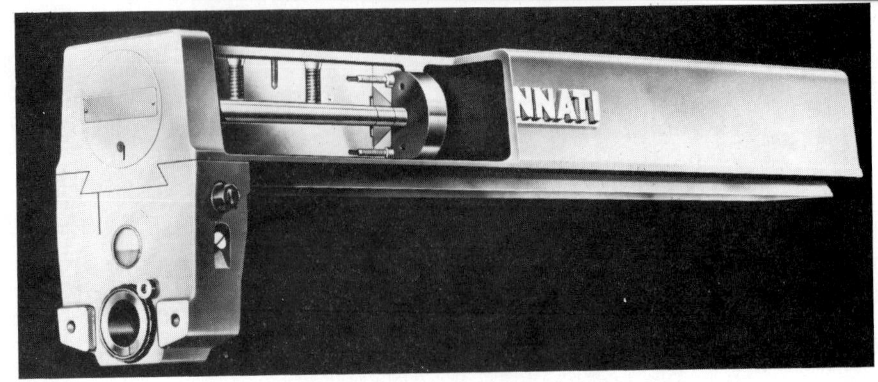

FIGURE K-9 The overarm can reduce cutting vibration (Cincinnati Milacron).

FIGURE K-10 Vertical milling attachment with quill feeding capability (Cincinnati Milacron).

trolled machine tools, is a ring ball nut, commonly called ball screw, which is essentially free of backlash.

Although much of the concern in the development of the milling machine and its cutters has been toward the highest possible machine rigidity by various means such as minimum number of moving components, overarm supports, and devices like backlash eliminators, another technique should be mentioned. Machine castings often vary in their ability to absorb vibration, even with the most careful design of internal webbing. Another means to attack the problem has been the tuning out of vibration by special vibration dampening devices. Figure K-9 shows the milling machine overarm equipped with a device to reduce the resonance of vibration passing through the casting. This capability permits increased cutting loads before chatter sets in.

Attachments and Accessories for the Horizontal Mill

A number of attachments are available to increase the capabilities of milling machines, particularly for toolroom applications where only a few parts are made or for limited production where the expense of a special machine would not be warranted. The vertical milling attachment (Figure K-10) is used on horizontal milling machines to ob-

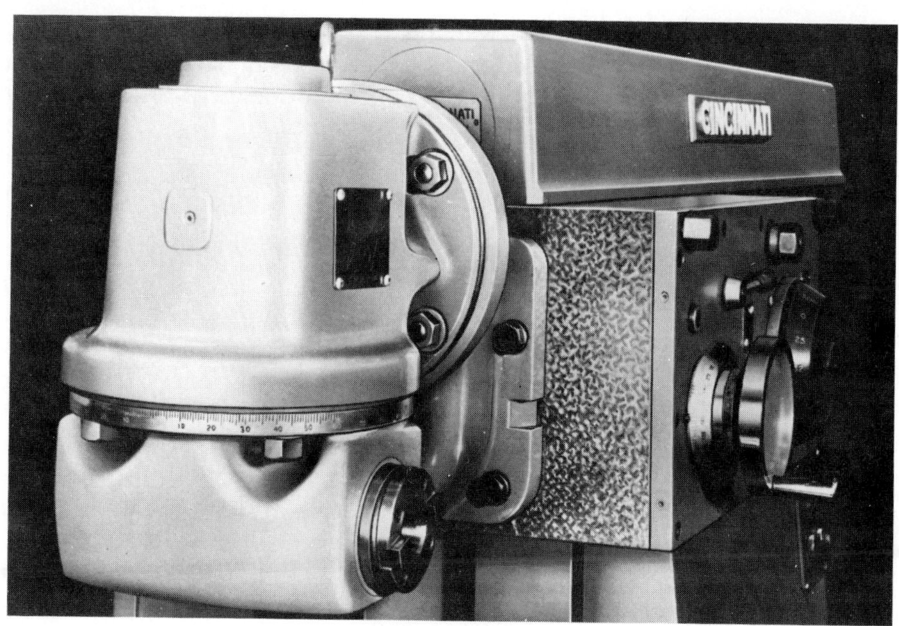

FIGURE K-11 Universal milling attachment (Cincinnati Milacron).

FIGURE K-12 Independent overhead spindle with angular head (Cincinnati Milacron).

tain the capability of both vertical and angular machining. The universal milling attachments (Figure K-11) permit an additional motion so that spiral milling may be done on a plain table milling machine in addition to vertical and angular cuts.

Another attachment is the independent overhead spindle with an angular swivel head (Figure K-12), which is powered separately from the horizontal machine spindle. This device replaces the standard overarm and can be used in conjunction with the horizontal spindle as needed to machine angular surfaces without moving the workpiece. When not needed, it can be swiveled out of the way, and the regular overarm brackets can be attached.

A slotting attachment (Figure K-13) is also available for horizontal milling machines to utilize a single-point tool for operations like internal keyseat cutting, where a vertical slotter is not available. It may be set at an angle as well as being set vertically. Devices for tool holding, such as arbors and adapters for horizontal milling, will be studied in this section. Table-mounted attachments such as rotary tables, indexing/dividing heads, and work-holding devices such as clamps and vises will also be studied.

FIGURE K-13 Slotting attachment for the horizontal milling machine (Cincinnati Milacron).

Milling machines can be equipped with accessory measuring equipment, called direct readouts (DROs), to reduce the chance of operator error when machining expensive complex parts (Figure K-14). These measuring systems can be switched to present information in either inch or metric form, which is a great timesaver and eliminates the possibility of making errors in conversion between the two systems.

It is important for you to learn to use vertical and horizontal milling machines competently because the cutting and locating principles apply to the most complex numerically controlled machine tool. Few companies are willing to risk damage to an expensive and complex numerically controlled machine by an operator without a background in

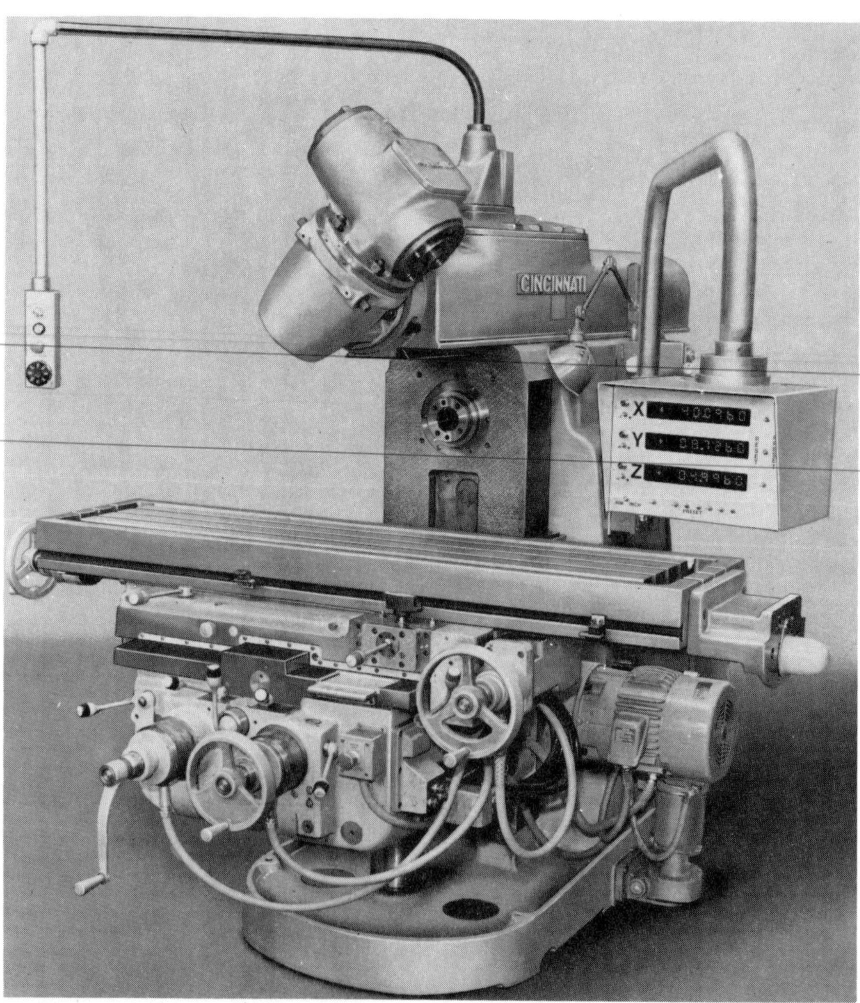

FIGURE K-14 Direct readout (DRO) fitted to a horizontal milling machine (Cincinnati Milacron).

conventional milling practice. An operator must be able to determine readily when there is something going wrong with the cutting operation and make appropriate corrections by replacing tools or manually overriding the machine control. Modern computer-controlled horizontal machining centers (Figure K-15) are high performance machines. As many as 120 preset tools are stored in tool-changing magazines, ready to be automatically inserted into the machine spindle. Many of these machines are equipped with multiple machine tables, so that one can be loaded while on the second one machining takes place.

In this section you will observe specific safety precautions that relate to horizontal milling, identify components and functions of horizontal mills, and perform routine maintenance. Various mounting systems used to drive milling cutters will be studied, and you will be able to match cutters to their respective applications. You will calculate rpm and feed rates for milling cutters and set the values into the machine controls. You will learn about a variety of work-holding methods and alignment procedures and how to mill a square workpiece. In addition, you will use side milling cutters in various combinations and you will use face milling cutters to machine flat surfaces.

Horizontal and vertical milling machines are as basic as the lathe, particularly where one-of-a-kind or small quantities of workpieces are involved. Both of these machine types will be in use for a long time to come. It is important for you to learn to set up and use these machines quickly, accurately, and safely.

HORIZONTAL MILLING MACHINE SAFETY

A safe worker is one who is properly dressed for the job. Loose-fitting clothing is dangerous—it can catch in rotating machinery. Sleeves should be rolled up. Rings, bracelets, and watches should be removed before operating machinery. Long or loose hair can be caught in a cutter or even be wrapped around a rotating smooth shaft. Persons with long hair should wear a cap or hairnet in the machine shop. A milling machine should not be operated while wearing gloves because of the danger of getting caught in the machine. When gloves are needed to handle sharp-edged materials, the machine has to be stopped. Eye protection should be worn at all times in the machine shop. Eye injuries can be caused by flying chips, tool breakage, or cutting fluid sprays. Keep your fingers away from the moving parts of the milling machine such as the cutter, gears, spindle, or arbor. Never reach over a rotating spindle or arbor. Safe operation of a machine tool requires that you think before you do something. Before starting up a machine, know the location and operation of its controls. Operate all controls on the machine yourself. Do not have another person start or stop the machine for you. Chances are good that he will turn a control at the wrong time. While operating a milling machine, observe the cutting action at all times so that you can stop the machine immediately when you see or hear something unfamiliar. Always stay within reach of the controls while the

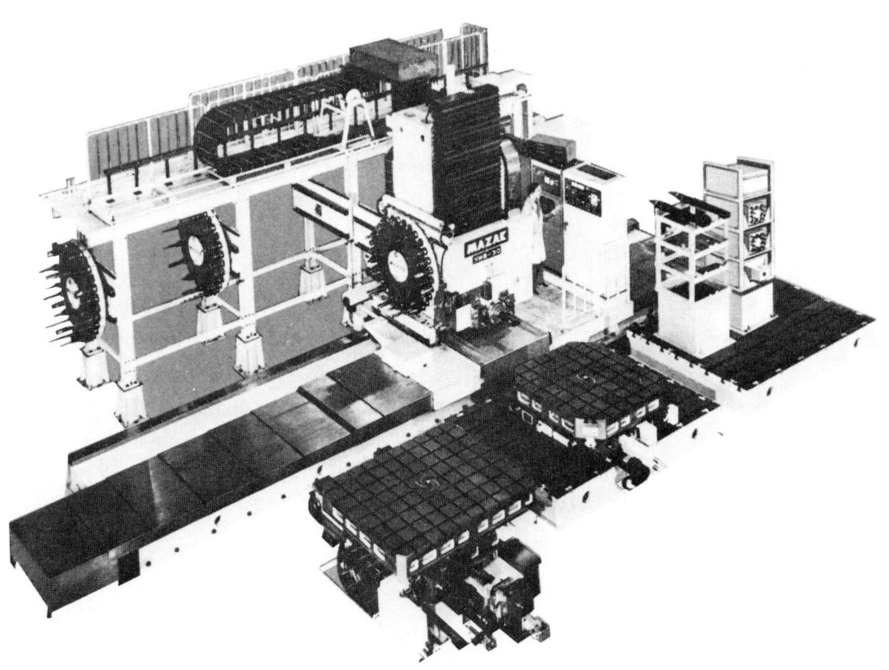

FIGURE K-15 Large CNC machining center (Mazak Corp.).

machine is running. An unexpected emergency may require quick action on your part. Never leave a running machine unattended.

Before operating the rapid traverse control on a milling machine, loosen the locking devices on the machine axis to be moved. Check that the handwheels or hand cranks are disengaged, or they will spin and injure anyone near them when the rapid traverse is engaged. The rapid traverse control will move any machine axis that has its feed lever engaged singularly or simultaneously. Do not try to position a workpiece too close to the cutter with this control, but approach the final 2 in. by using the handwheels or hand cranks.

Individuals concentrating on a machining operation should not be approached quietly from behind, since it may annoy and alarm them and they may ruin a workpiece or injure themselves. Do not lean on a running machine; moving parts can hurt you. Signs posted on a machine indicating a dangerous condition or a repair in progress should only be removed by the person making the repair or by a supervisor.

Measurements should only be taken on a milling machine after the cutter has stopped rotating and after the chips have been cleared away. Milling machine chips are dangerously sharp and often hot and contaminated with cutting fluids. They should not be handled with bare hands. Chips should be removed with a brush. Compressed air should not be used to clean off chips from a machine because it will make small missiles out of chips that can injure a person, even one who's quite a distance away. A blast of air will also force small chips into the ways and sliding surfaces of the milling machine where they will cause scoring and rapid premature wear. Cleaning chips and cutting fluids from the machine or workpiece should be done only after the cutter has stopped turning. Before and during the operation of a milling machine, keep the area around the machine clean of chips, oil spills, cutting fluids, and other obstructions to prevent the operator from slipping or stumbling.

Many milling machine attachments and workpieces are heavy; use a hoist to lift them on or off the table. Do not walk under a hoisted load. The hoist may release and drop the load on you. If a hoist is not available, ask for assistance.

Injuries can be caused by improper setups or the use of wrong tools. Use the correct-size wrench when loosening or tightening nuts or bolts, preferably a box wrench or a socket wrench. An oversized wrench will round off the corners on bolts and nuts and prevent sufficient tightening or loosening; a slipping wrench can cause smashed fingers or other injuries to the hands or arms. Milling machine cutters have very sharp cutting edges. Handling cutters carefully and with a cloth will prevent cuts on the hands.

All machine guards should be checked to see that they are in good condition and in place to increase milling machine safety. Workpieces should be centered in a vise with only enough extending out to permit machining. Clean the working area after a job is completed. A clean machine is safer than one buried under chips.

UNIT 1

Plain and Universal Horizontal Milling Machines

Before operating a horizontal mill, you must know the names of the machine parts, controls, and their functions. You must also know how to perform the routine maintenance on the machine required to preserve its accuracy and provide ease of operation. The purpose of this unit is to identify the parts, controls, and control functions of the horizontal mill and to describe routine maintenance procedures.

Objectives

After completing this unit, you should be able to:

1. Identify the important components and controls on the vertical milling machine.
2. Describe the functions of machine parts and controls.
3. Perform routine maintenance on the machine.

Plain Horizontal Milling Machine

DETERMINING THE SIZE OF THE MACHINE

The size of a horizontal milling machine is usually given as the range of movement possible and the power rating of the main drive motor of the machine. An example would be a milling machine with a 28-in. longitudinal travel, 10-in. cross travel, and 16-in. vertical travel with a 5-hp main drive motor. As the physical capacity of a machine increases, more power is also available at the spindle through a large motor.

IDENTIFYING MAJOR PARTS, CONTROLS, AND THEIR FUNCTIONS

The major assemblies of the horizontal mill are base and column, knee, saddle, table, spindle, and overarm (Figure K-16).

BASE AND COLUMN The base along with the column form the one-piece major structural component of the machine tool. A dovetail slide is machined on the vertical face of the column providing an accurate guide for the vertical travel of the knee. A dovetail slide is also machined on the top of the column providing a guide for the overarm. The column also contains the machine spindle, main

FIGURE K-16 Horizontal milling machine (Cincinnati Milacron).

drive motor, spindle speed selector mechanism, and spindle rotation direction selector.

KNEE The knee engages the slide on the face of the column and is moved vertically by turning the vertical hand feed crank. A slide on the top of the knee provides a guide for the saddle.

SADDLE The saddle engages the slide on the top of the knee and is moved horizontally toward or away from the face of the column by turning the cross feed handwheel. The saddle supports the table.

TABLE The table engages the slide on the top of the saddle and can be moved horizontally right and left by turning the table handwheel. The table is equipped with T-slots for direct mounting of the workpiece, vise, or other fixture.

SPINDLE The machine spindle is located in the upper part of the column and is used to hold, align, and drive the various cutters, chucks, and arbors. The front end or spindle nose has a tapered socket in a standard milling machine taper. This taper aligns the milling machine adapter or cutter arbor. Driving force is provided by two keys located on the spindle nose. These engage slots on the adapter or arbor.

Arbors and adapters are held in place by means of a drawbolt extending through the hollow center of the spindle to the rear of the machine. Like the vertical mill, the drawbolt is threaded on one end and is designed to screw into the thread in the end of the arbor or adapter shank. Tightening the drawbolt lock nut draws the taper shank of the arbor into the spindle taper.

OVERARM AND ARBOR SUPPORT The overarm engages the slide on the top of the column and may be moved in and out by loosening the overarm clamps and sliding this part to the desired position. The arbor support engages the dovetail on the overarm. The arbor support contains a bearing that is exactly in line with the spindle of the mill. The arbor support provides a rigid bearing support for the outer end of the mill arbor.

MACHINE CONTROLS

Most horizontal milling machines are equipped with power feeds for the table, saddle, and knee. This machine tool is also equipped with a rapid traverse feature that permits rapid positioning of a workpiece without the need to turn table, saddle, and knee cranks by hand.

CONTROLS FOR MANUAL MOVEMENTS Cranks and crank handwheels are provided to elevate the knee and move table and saddle. All of these controls are equipped with a micrometer collar graduated in .001-in. increments.

FIGURE K-17 Feed change crank (Cincinnati Milacron).

FEED RATE SELECTOR AND FEED ENGAGE CONTROLS The feed rate selector is located on the knee (Figure K-17) and is used to select the power feed rate for table, saddle, and knee in inches per minute (ipm). Power feeds are engaged by individual controls on the table, saddle, and knee. However, engaging all of these at the same time will cause all these components to move under power at the same time. On most mills, power feeds will not function unless the spindle is turning. Two safety stops at each axis travel limit prevent accidental damage to the feed mechanism by providing automatic kickout of the power feed. Adjustable power feed trip dogs are also provided so that you may preset the point at which the feed is to be disengaged.

USING THE RAPID TRAVERSE To expedite the positioning of the knee, saddle, and table in order to rapidly move the workpiece up to the cutter or clear of the overarm, a rapid traverse feature is provided. When the rapid traverse control is engaged (Figure K-18) it overrides the feed rate selector rate and rapidly moves the table, saddle, or knee depending on which feed control is engaged. The direction of rapid traverse is in the same direction as that of the feed. Furthermore, the rapid traverse will rapidly move the knee, saddle, and table all at the same time if all these parts should happen to have their feed controls engaged. Be careful when using the rapid traverse function not to run the work or table into the cutter or overarm. This will damage the work, cutter, and machine and could cause an injury.

SPINDLE CONTROLS Spindle controls include the main motor switch, clutch, spindle

FIGURE K-18 Rapid traverse lever (Cincinnati Milacron).

FIGURE K-19 Speed change levers (Cincinnati Milacron).

speed, and speed range controls. The motor switch will usually reverse the motor and spindle direction electrically. On some mills, spindle rotation is changed mechanically. A clutch is sometimes used to connect spindle and motor.

Spindle speeds are selected from the control on the side of the column (Figure K-19). Several spindle speeds are available in both a low and high speed range. Variable-speed controls are also used. The speed range selector is adjacent to the speed selector. This control has a neutral position between high- and low-speed settings. In the neutral

FIGURE K-20 Universal milling machine (Cincinnati Milacron).

position, the spindle may be turned by hand during machine setups. Spindle speed and direction must be selected and set while the motor and spindle are stopped. Shifting gears while the spindle is turning can damage the drive mechanism. However, on variable-speed drives, speeds must be set while the spindle is in motion.

LOCKS

Locks are provided on the table, saddle, and knee. These permit the components to be locked in place during a machining operation in order to increase rigidity of the setup. **All locks must be released before moving any part either by hand or under power.** During machining, locks should be set except the one on the moving axis. Locks should not be used to compensate for wear in the machine slides.

Universal Horizontal Milling Machine

The universal milling machine (Figure K-20) closely resembles a plain horizontal milling machine. The main difference between these machines is that the universal machine has an additional housing that swivels on the saddle and supports the table. This allows the table to be swiveled 45 degrees in either direction in a horizontal plane. The universal milling machine is specially designed to machine helical slots or grooves as in twist drills and milling cutters. Other than these special applications, a universal mill and a plain milling machine can perform the same operations.

Routine Maintenance on Mills

Before any machine tool is operated, it should be lubricated. A good starting point is to wipe clean all sliding surfaces and to apply a coat of a good way lubricant to them. Way lubricant is a specially formulated oil for sliding surfaces. Dirt, chips, and dust will act like an abrasive compound between sliding members and cause excessive machine wear. Most machine tools have a lubrication chart that outlines the correct lubricants and lubrication procedures. When no lubrication chart is available, check all oil sight gages for the correct oil level, and refill if necessary. Too much oil causes leakage. Lubrication should be performed progressively, starting at the top of the machine and working down. Machine points that are hand oiled should receive only a small amount of oil at any one time, but this should be repeated at regular intervals, at least daily. Motor or pulley bearings should not get too much grease, since this may destroy the seals.

SELF-TEST

1. On a plain or universal horizontal milling machine, locate the following parts:

Overarm
Column
Saddle clamping lever
Speed change lever
Powerfeed levers for longitudinal feed, crossfeed, and vertical feed
Rapid traverse lever
Switch for spindle on–off
Arbor support
Switch for coolant pump
Knee Table
Feed change lever
Spindle nose
Saddle
Knee clamping lever
Spindle forward-reverse switch
Trip dogs for all three axes

2. Lubricate a plain horizontal milling machine.

UNIT 2

Types of Spindles, Arbors, and Adapters

Several different devices are used to hold and drive cutters on the horizontal mill. As a machinist, part of your job is to know what these are and to select the one that best fits your needs. The purpose of this unit is to identify and describe types of spindles, arbors, and adapters used on the horizontal mill.

Objective

After completing this unit, you should be able to:

Identify machine spindles and set up different cutting tool mounting systems used to drive milling cutters.

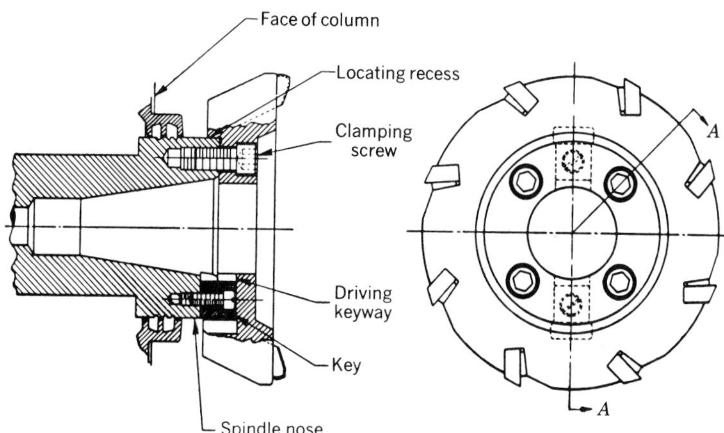

FIGURE K-21 Mounting a face mill on the spindle nose of a milling machine (Cincinnati Milacron).

Mill Spindle Tapers

The milling machine spindle provides the driving force for the milling cutter. A milling cutter may be attached directly to the spindle nose (Figure K-21) or held in a taper shank adapter or on a taper shank arbor. The arbor or adapter is then secured in the tapered socket located in the spindle nose.

Most modern milling machines have self-releasing tapers in their spindle sockets. These permit quick and easy installing and removal of tooling. The standard national milling machine taper is $3\frac{1}{2}$ in. per foot, which is an included angle of approximately $16\frac{1}{2}$ degrees. Since the standard milling machine taper is a locating or aligning taper only, tooling must be held in the spindle by a draw-in bolt or drawbolt extending through the center of the spindle. National milling machine tapers are available in four different sizes and are identified by the numbers 30, 40, 50, and 60.

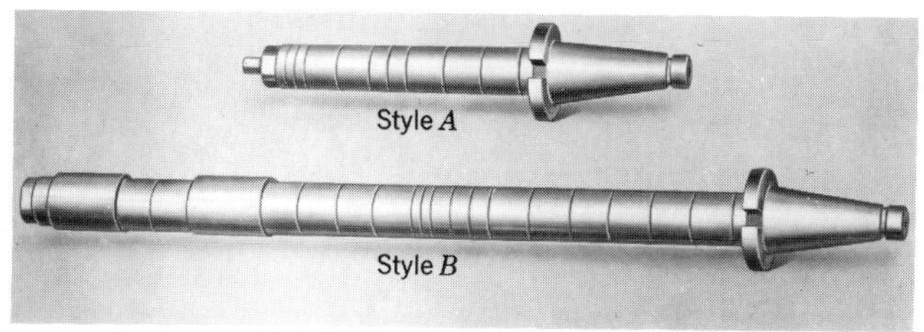

FIGURE K-22 Arbors, styles A and B (Cincinnati Milacron).

Number 50 is the most common. Positive drive of tapered shank tooling is provided through two keys on the spindle nose that engage keyseats in the flange on the arbor or adapter.

Types of Milling Machine Arbors

Two common arbor styles are shown in Figure K-22. A Style A arbor has a cylindrical pilot on the end opposite the shank. The pilot is used to support the free end of the arbor. Style A arbors are used mostly on small milling machines. But they are also used on larger machines when a style B arbor support cannot be used because of a small diameter cutter or interference between the arbor support and the workpiece.

Style B arbors are supported by one or more bearing collars and arbor supports. Style B arbors are used to obtain rigid setups in heavy-duty milling operations.

Style C arbors are also known as shell end mill arbors or as stub arbors (Figure K-23). Shell end milling cutters are face milling cutters up to 6 in. in diameter. Because of their relatively small diameter, these cutters cannot be counterbored so that they can be mounted directly on the spindle nose, as are face mills, but they are mounted on shell end mill arbors.

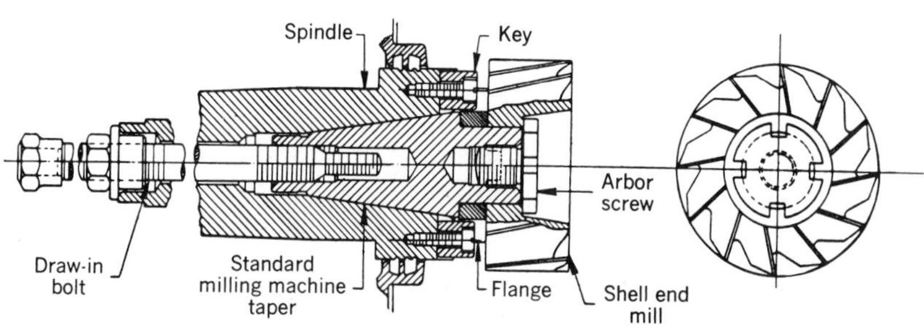

FIGURE K-23 Style C arbor; shell end mill arbor (Cincinnati Milacron).

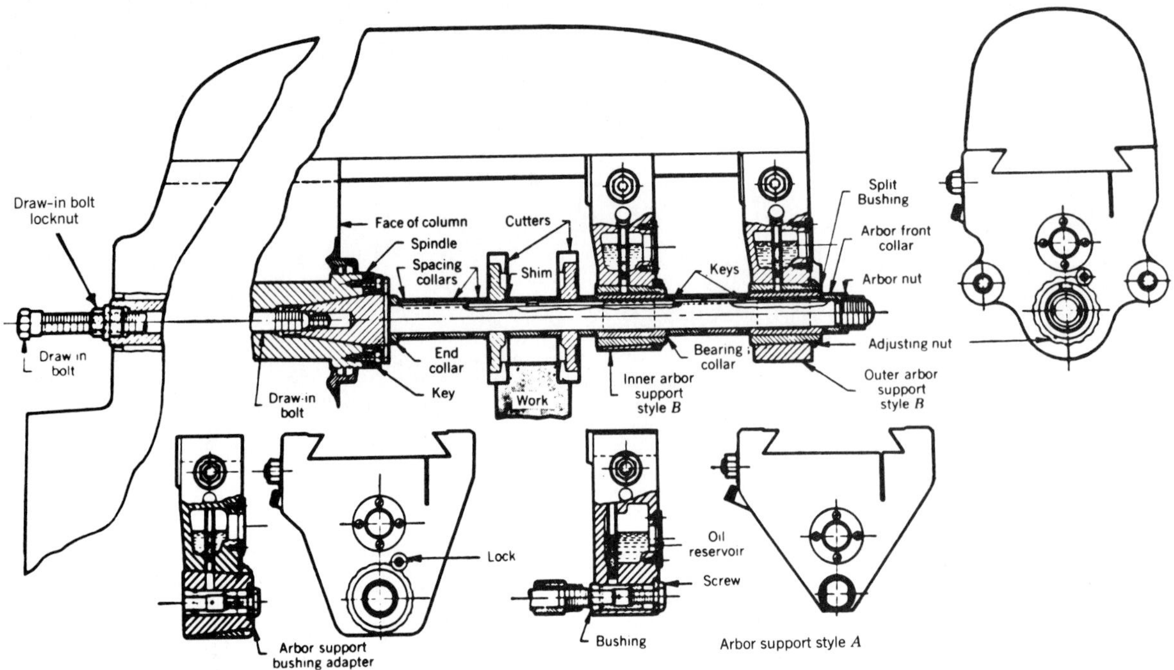

FIGURE K-24 Section through arbor showing location of arbor collars, keys, bearing collars, and various arbor supports (Cincinnati Milacron).

SPACING AND BEARING COLLARS

Precision spacing collars are used to take up the space between the cutter and the ends of the arbor. Shims may also be used to obtain exact cutter spacings in straddle milling operations. Bearing collars are larger in diameter than spacing collars. These collars ride in the arbor support bearing. On style A arbors, the end of the arbor rides in the arbor support bearing.

All collars are manufactured to very close tolerances with their ends or faces being parallel and also square to the hole. It is very important that the collars and other parts fitting on the arbor are handled carefully to avoid damaging the collar faces. **Any nicks, chips, or dirt between the collar faces wil misalign the cutter or deflect the arbor and cause cutter run-out.**

ARBOR SUPPORT BEARINGS

The arbor support bearing has a very important function in supporting the outer end of the arbor (Figure K-24). The arbor bearing collar or arbor pilot fits this bearing that is located in the arbor support. The arbor support bearing may be a sleeve bushing, or, on some mills, a sealed ball bearing may be used. On mills where a sleeve bushing is used for the arbor support bearing, the **fit of the arbor bearing collar or pilot can be quite critical.**

A provision for adjusting the fit of the arbor support bushing to the arbor collar or pilot is provided. Too loose a fit will cause inaccuracies in the mill cuts or permit chatter to occur. Too tight a fit will cause frictional heating and can damage the arbor collar, pilot, or the arbor support bushing. An arbor turning at high rpm will require more clearance than at slow rpm.

Arbor support bushings must be lubricated properly. Some mills have an oil reservoir in the arbor support for supplying oil to the arbor bushing. Check the oil level and method of lubrication before operating the machine, and consult with your instructor regarding the adjustment of the bushing to fit the arbor you are using.

Adapters, Collets, and Quick-Change Tooling Systems

Adapters are used on milling machines to mount cutters that cannot be mounted on arbors. Adapters can be used to hold and drive taper shank tools (Figure K-25). Collets used with these adapters increase the range of tools that can be used in a milling machine having a given-size spindle socket. The spring chuck adapter (Figure K-26), with different-size removable spring collets, is used with straight shank tools such as drills and end mills. With a quick-change adapter (Figure K-27) mounted on the spindle nose, a number of milling machine operations such as drilling, end milling, and boring can be performed without changing the setup of the part being machined. The different

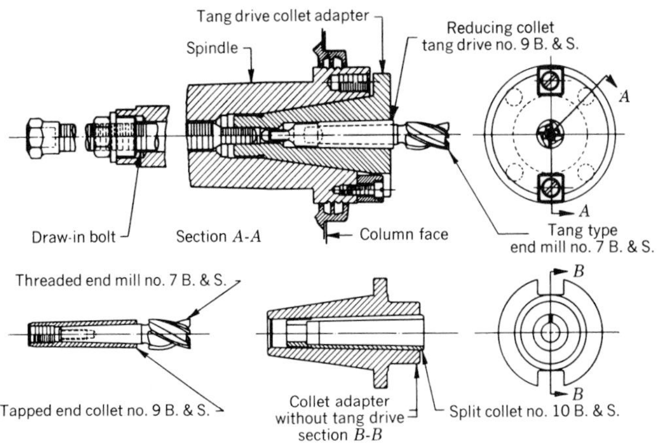

FIGURE K-25 Adapters and collets for self-releasing and self-holding tapers (Cincinnati Milacron).

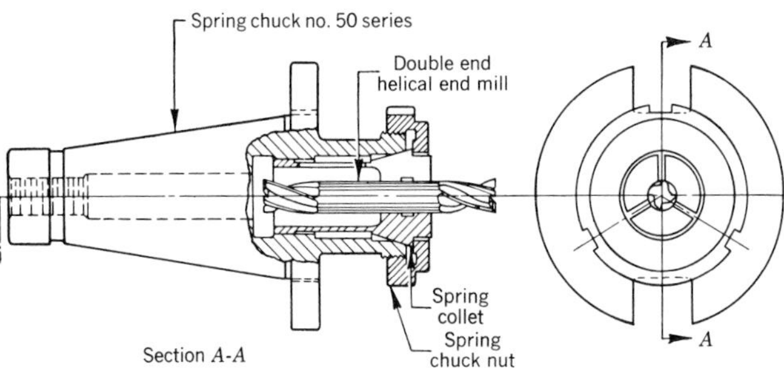

FIGURE K-26 Spring chuck adapter (Cincinnati Milacron).

FIGURE K-27 Quick-change adapter mounted on spindle nose (Cincinnati Milacron).

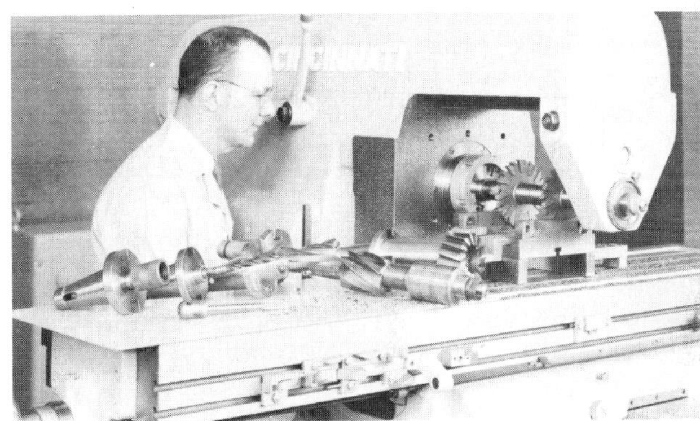

FIGURE K-28 A number of tools mounted on quick-change tool holders ready to use (Cincinnati Milacron).

tools are mounted on quick-change adapters that are ready for use (Figure K-28). The adapter is held in the spindle taper by a clamping ring (Figure K-29). Tool changing with this system is greatly facilitated.

SELF-TEST

1. What kind of cutters are mounted directly on the spindle nose?

2. Milling machine spindle sockets have two classes of taper. What are they?

3. What is the amount of taper on a national milling machine taper?

4. When is style A arbor used?

5. What is a style C arbor?

6. Why are milling machine adapters used?

7. Describe the function and care of spacing and bearing collars.

8. What is the result of dirty or nicked collars?

9. Describe clearances and lubrication requirements for the arbor support bearing and collar.

10. What is the advantage of quick-change tooling?

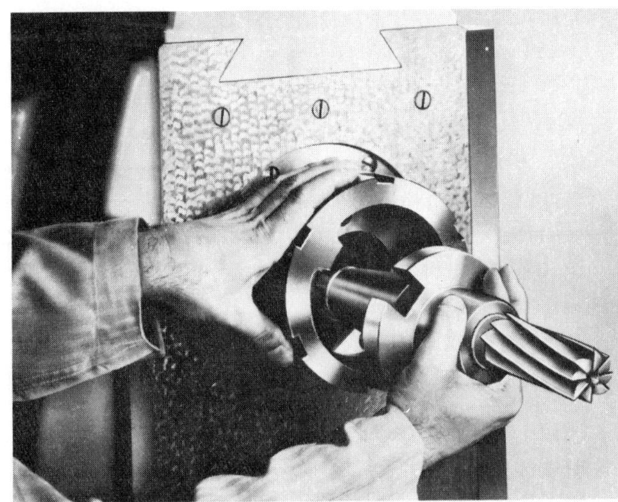

FIGURE K-29 Tools are locked into the spindle with a partial turn of the clamp ring (Cincinnati Milacron).

UNIT 3

Arbor-Driven Milling Cutters

The metal cutting versatility of the horizontal mill can be fully realized and utilized by understanding, identifying, and selecting from the many types of milling cutters available for use on this machine tool. The purpose of this unit is to describe many of these common milling cutters and aid you in selecting the one that best fits your needs for a specific machining task.

Objective

After completing this unit, you should be able to:

Identify common milling cutters, list their names, and select a suitable cutter for a given machining task.

Classifying Milling Cutters

Most milling cutters are designed to perform specific machining operations. You should learn to identify common types by sight and know their capabilities and limitations. Milling cutters can be generally classified as to:

1. Material of manufacture
2. Profile sharpened or form relieved
3. Arbor-driven or shank types
4. Rotation and helix hand

MATERIAL OF MANUFACTURE

Many milling cutters are made from high-speed steel (HSS). Some types have cemented carbide cutting edges. Large cutters have inserted blades or teeth.

PROFILE-SHARPENED OR FORM-RELIEVED

Profile-sharpened cutters are resharpened by grinding a narrow land (Figure K-30) back of the cutting edge. Form-relieved cutters are resharpened by grinding the face of the teeth parallel to the axis of the cutter.

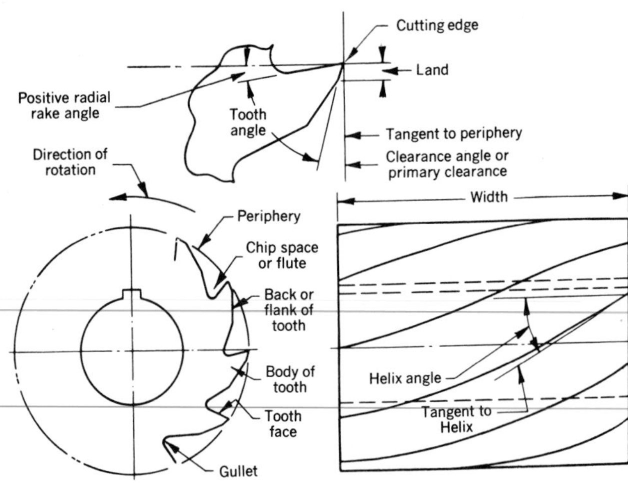

FIGURE K-30 Nomenclature of plain milling cutter (Cincinnati Milacron).

ARBOR-DRIVEN OR SHANK TYPES

Milling cutters for the horizontal mill are designed to be driven by the mill arbor, or they may have their own tapered shanks for direct mounting in the machine spindle socket. Although there are many instances where you will use a tapered shank mounted cutter in the horizontal mill, probably the most common cutters you will be using are arbor-mounted types.

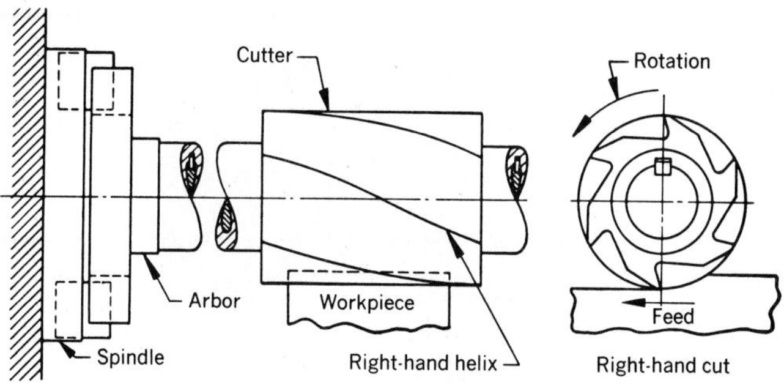

FIGURE K-31 Plain milling cutter with right-hand helix and right-hand cut (Cincinnati Milacron).

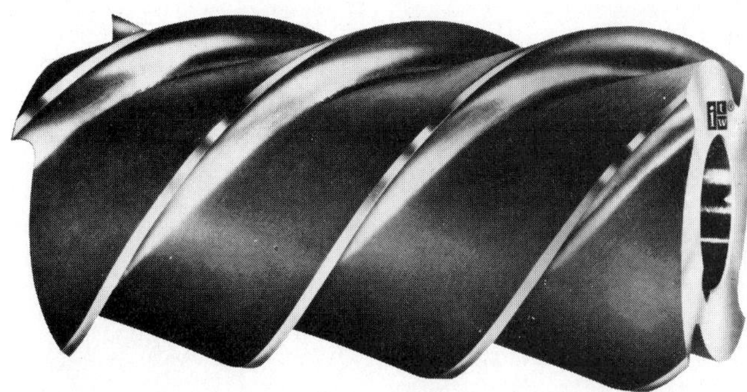

FIGURE K-33 Heavy-duty plain milling cutter (Illinois Tool Works Inc.).

FIGURE K-32 Light-duty plain milling cutter (Illinois Tool Works Inc.).

FIGURE K-34 Helical plain milling cutter (Illinois Tool Works Inc.).

ROTATION AND HELIX HAND

Milling cutters are either for right-hand rotation cutting or left-hand rotation cutting. The way in which a cutter is mounted on an arbor determines the hand or cut. Counterclockwise rotation determines right-hand cutting direction. Clockwise rotation determines a left-hand cutting direction (Figure K-31).

Helix hand is determined by looking at the end of the cutter and determining which direction the flutes twist. Flutes twisting to the right are right-hand helix and to the left are left-hand helix.

Plain Arbor-Driven Cutters

Plain milling cutters are designed for milling plain surfaces where the width of the work is narrower than the cutter (Figure K-32). Plain milling cutters less than $\frac{3}{4}$ in. wide have straight teeth. On straight tooth cutters, the cutting edge will cut along its entire length at the same time. Cutting pressure increases until the chip is completed. At this time the sudden change in tooth load causes a shock that is transmitted through the drive and often leaves chatter marks or an unsatisfactory surface finish. Light-duty milling cutters have a large number of teeth, which limits their use to light or finishing cuts because of insufficient chip space for heavy cutting. Heavy-duty plain mills (Figure K-33) have fewer coarse teeth, which makes for strong teeth with ample chip clearance. The helix angle of heavy-duty mills is about 45 degrees. The helical form enables each tooth to take a cut gradually, which reduces shock and lowers the tendency to chatter. Plain milling cutters are also called slab mills. Plain milling cutters with a helix angle over 45 degrees are known as helical mills (Figure K-34). These milling cutters produce a smooth finish when used for light cuts or on intermittent surfaces. Plain milling cutters do not have side cutting teeth and should not be used to mill shoulders or steps on workpieces.

Side Milling Cutters

Side milling cutters are used to machine steps or grooves. These cutters are made from $\frac{3}{16}$ to 1 in. in

FIGURE K-35 Side milling cutter (Lane Community College).

FIGURE K-36 Stagger tooth milling cutter (Illinois Tool Works Inc.).

width. Figure K-35 shows a straight tooth side milling cutter. To cut deep slots or grooves, a staggered tooth side milling cutter (Figure K-36) is preferred because the alternate right-hand and left-hand helical teeth reduce chatter and give more chip space or higher feeds than are possible with straight tooth side milling cutters. To cut slots over 1 in. wide, two or more side milling cutters may be mounted on the arbor simultaneously. Shims between the hubs of the side mills can be used to get any precise width cutter combination.

FIGURE K-37 Inserted tooth carbide cutter (Lane Community College).

FIGURE K-38 Half side milling cutter (Illinois Tool Works Inc.).

Inserted tooth carbide cutters (Figure K-37) are used to cut slots and grooves. Because of the inserted carbide teeth, these cutters can be set to cut variable-width slots. The common range of width adjustments on one cutter is .060 in. One big advantage of this type of cutter is that the inserts can be quickly changed when dull. The great number of carbide grades available allows the selection of the correct grade of carbide for the material being machined and the machining conditions being encountered.

Half side milling cutters are designed for heavy duty milling where only one side of the cutter is used (Figure K-38). For straddle milling, a right-hand and a left-hand cutter combination is used.

Plain metal slitting saws are designed for slotting and cutoff operations (Figure K-39). Their

FIGURE K-39 Plain metal slitting saw (Illinois Tool Works Inc.).

FIGURE K-41 Stagger tooth metal slitting saw (Illinois Tool Works Inc.).

FIGURE K-40 Side tooth metal slitting saw (Illinois Tool Works Inc.).

FIGURE K-42 Single-angle milling cutter (Illinois Tool Works Inc.).

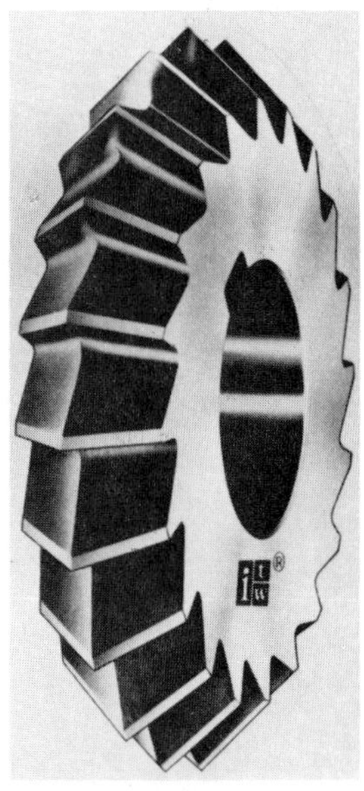

FIGURE K-43 Double-angle milling cutter (Illinois Tool Works Inc.).

sides are slightly relieved or dished to prevent binding in a slot. Their use is limited to a relatively shallow depth of cut. These saws are made in widths from $\frac{1}{32}$ to $\frac{5}{16}$ in.

To cut deep slots or when many teeth are in contact with the work, a side tooth metal slitting saw will perform better than a plain metal slitting saw (Figure K-40). These saws are made from $\frac{1}{16}$ to $\frac{3}{16}$ in. wide.

Extra deep cuts can be made with a staggered tooth metal slitting saw (Figure K-41). Staggered tooth saws have greater chip carrying capacity than other saw types. All metal slitting saws have a slight clearance ground on the sides toward the hole to prevent binding in the slot and scoring of the walls of the slot. Stagger tooth saws are made from $\frac{3}{16}$ to $\frac{5}{16}$ in. wide.

Angular milling cutters are used for angular milling such as cutting of dovetails, V-notches, and serrations. Single-angle cutters (Figure K-42) form an included angle of 45 to 60 degrees, with one side of the angle at 90 degrees to the axis of the cutter.

Double-angle milling cutters (Figure K-43) usually have an included angle of 45, 60, or 90 degrees. Angles other than those mentioned are special milling cutters.

FIGURE K-44 Convex milling cutter (Ilinois Tool Works Inc.).

FIGURE K-45 Concave milling cutter (Illinois Tool Works Inc.).

FIGURE K-46 Corner rounding milling cutter (Illinois Tool Works Inc.).

FIGURE K-47 Involute gear cutter (Lane Community College).

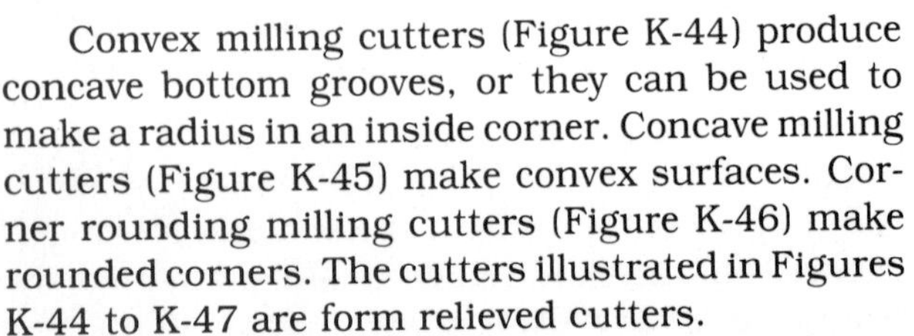

Convex milling cutters (Figure K-44) produce concave bottom grooves, or they can be used to make a radius in an inside corner. Concave milling cutters (Figure K-45) make convex surfaces. Corner rounding milling cutters (Figure K-46) make rounded corners. The cutters illustrated in Figures K-44 to K-47 are form relieved cutters.

Involute gear cutters (Figure K-47) are commonly available in a set of eight cutters for a given pitch, depending on the number of teeth for which the cutter is to be used. The ranges for the individual cutters are as shown in Table K-1.

These eight cutters are designed so that their forms are correct for the lowest number of teeth in each range. If an accurate tooth form near the upper end of a range is required, a special cutter is needed.

TABLE K-1 Involute Gear Cutters

Cutter Number	Range of Teeth
1	135 to rack
2	55 to 134
3	35 to 54
4	26 to 53
5	21 to 25
6	17 to 20
7	14 to 16
8	12 and 13

SELF-TEST

1. What are the two basic kinds of milling cutters with reference to their tooth shape?

2. What is the difference between a light-duty and a heavy-duty plain milling cutter?

3. Why are plain milling cutters not used to mill steps or grooves?

4. What kind of cutter is used to mill grooves?

5. How does the cutting action of a straight tooth side milling cutter differ from that of a stagger tooth side milling cutter?

6. Give an example of an application of half side milling cutters.

7. When are metal slitting saws used?

8. Give two examples of form-relieved milling cutters.

9. When are angular milling cutters used?

10. When facing the spindle, in which direction should the right-hand cutter be rotated in order to cut?

UNIT 4

Work-Holding Methods and Standard Setups

Before any machining can be done on the horizontal mill, the workpiece or work-holding fixture must be secured and aligned on the machine table. A specific setup will depend on the particular workpiece and the machining task to be done. Considerable ingenuity on your part is required to make a safe and secure setup that will not distort or damage the workpiece. Milling machine setups are almost infinite in number and it would not be possible to discuss them all. The purpose of this unit is to introduce you to the equipment and techniques for common setups on this machine.

Objectives

After completing this unit, you should be able to:

1. Select a work-holding method and device for common milling tasks.
2. Safely set up a workpiece on the machine using common standard techniques.

Direct Mounting to the Table

A large variety of milling clamps are available for direct table mounting of the workpiece. Always follow the rules of good clamping. Clamping studs or bolts should be fully screwed into their T-nuts. Clamp nuts should have full thread engagement on studs. Studs or clamp bolts should be located as close to the workpiece as possible and clamp support blocks arranged so that clamping pressure is applied to the workpiece (Figure K-48). Clamps should be located on both sides of the workpiece if possible. This will ensure maximum safety (Figure K-49). Clamp supports **must be the same height as the workpiece. Never use clamp supports that are lower than the workpiece.** Ad-

FIGURE K-48 Work clamped to the table with T-slot bolts and clamps (Cincinnati Milacron).

FIGURE K-49 Clamping bolt close to the work gives maximum clamping pressure; stop block prevents work slippage (Lane Community College).

FIGURE K-50 Highly finished surfaces should be protected from clamping damage (Lane Community College).

FIGURE K-51 Protect the machine table surface from rough workpieces (Lane Community College).

FIGURE K-52 Workpiece supported under the clamp.

FIGURE K-53 Examples of screw jacks (Cincinnati Milacron).

justable step blocks are extremely useful for this as the height of the clamp bar may be adjusted to ensure maximum clamping pressure.

To protect a soft or finished surface from damage by clamping pressure, place a shim between the clamp and work (Figure K-50). If you are machining a rough casting or weldment (Figure K-51), protect the machine table from damage by using parallels or a shim under the workpiece. Paper, plywood, and sheet metal are shim materials. Their selection depends on the accuracy of the machine cuts.

Workpieces can be easily distorted, broken, or otherwise damaged by excessive or improper clamping. One solution to this problem is to use a screw jack to support the workpiece (Figure K-52). The jack should be placed directly under the clamp. Screw jacks are available in a variety of sizes and styles (Figure K-53). In lieu of a screw jack, solid blocks of the correct size may be used as both workpiece supports and clamp supports (Figure K-54).

In machining operations where heavy cuts are involved, top clamps alone may not be sufficient to restrain the workpiece. A workpiece stop may be used to prevent slippage of the part being machined (Figure K-49).

FIGURE K-54 Work set up and clamped on table (Cincinnati Milacron).

FIGURE K-55 Quick-action jaws holding workpiece (Cincinnati Milacron).

Quick-action jaws (Figure K-55) may be used in direct table clamping. These can be positioned at any location on the table. Clamping action is independent of the clamp mounting, permitting the workpiece to be clamped or released without actually moving the clamp itself.

FIGURE K-56 Plain vise (Cincinnati Milacron).

FIGURE K-57 Swivel vise (Cincinnati Milacron).

Mill Vises

Probably the most common method of work holding on a milling machine is a vise. Vises are simple to operate and can quickly be adjusted to the size of the workpiece. A vise should be used to hold work with parallel sides if it is within the size limits of the vise, because it is the quickest and most economical work-holding method. The plain vise (Figure K-56) is bolted to the machine table. Alignment with the table is provided by two slots at right angles to each other on the underside of the vise. These slots are fitted with removable keys that align the vise with the table T-slots either lengthwise or crosswise. A plain vise can be converted to a swivel vise (Figure K-57) by mounting it on a swivel base. The swivel plate is graduated in degrees. This allows the upper section to be swiveled to any angle in the horizontal plane. When swivel bases are added to a plain vise, the versatility increases, but the rigidity is lessened.

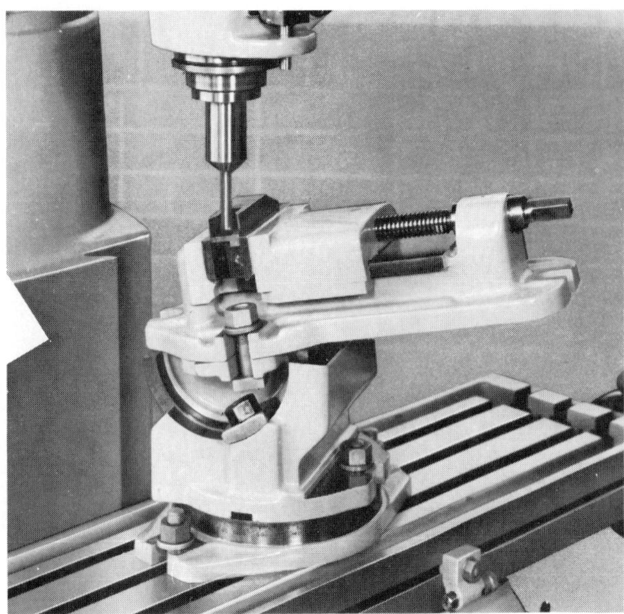

FIGURE K-58 Universal vise (Cincinnati Milacron).

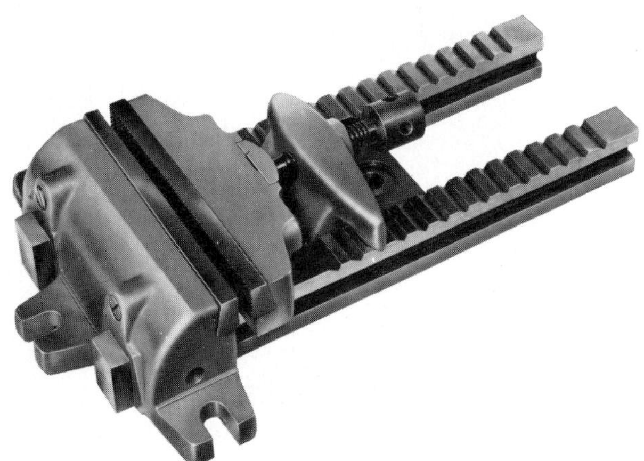

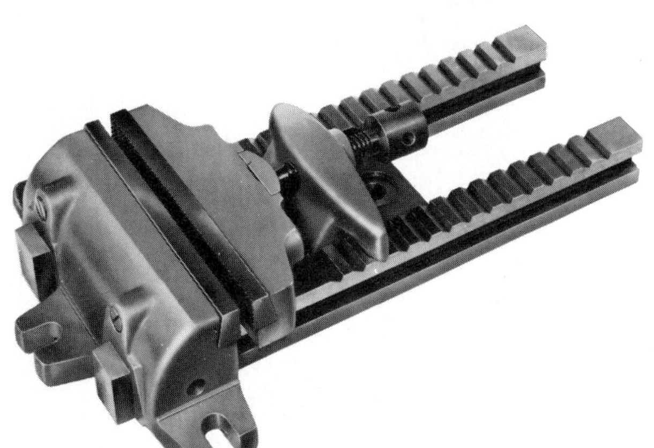

FIGURE K-59 All-steel vise (Cincinnati Milacron).

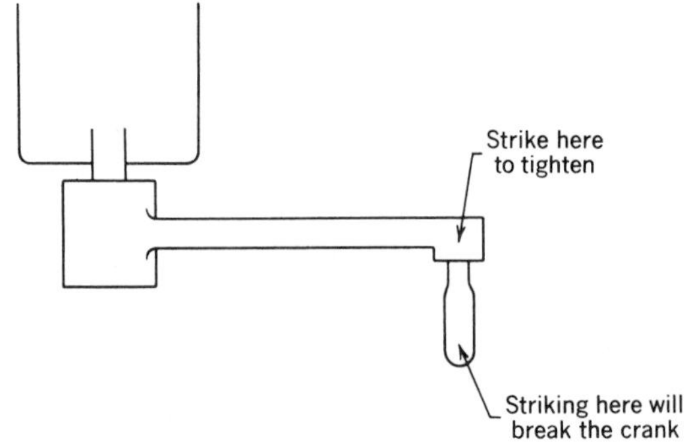

FIGURE K-60 Tightening a vise.

FIGURE K-61 Rotary table (Cincinnati Milacron).

For work involving compound angles, a universal vise (Figure K-58) is used. This vise can be swiveled 90 degrees in the vertical plane and 360 degrees in the horizontal plane.

The strongest setup is the one where the workpiece is clamped close to the table surface. Castings, forgings, or other rough workpieces can be securely fastened in an all-steel vise (Figure K-59). The movable jaw can be set in any notch on the two bars to accommodate different workpieces. The short clamping screw makes for a very strong and rigid setup. The hardened and serrated jaws grip the workpiece securely.

Air or hydraulically operated vises are often used in production work, but in general toolroom work, vises are opened and closed by cranks or levers. To hold workpieces securely without slipping under high cutting forces, a vise must be tightened by striking the crank with a lead hammer (Figure K-60) or other soft-face hammer.

Other Work-Holding Accessories for the Horizontal Mill

A rotary table or circular milling attachment (Figure K-61) is used to provide rotary movement to a workpiece. The rotary table can be used for angular indexing, milling circular grooves, or to cut radii. The rotary table is shown in Figure K-61 in a gear cutting operation.

The dividing head (Figure K-62) is used to divide the circumference of a workpiece into any number of equally spaced divisions. Work is held between centers, in collets, or in a chuck. The supporting member opposite the dividing head is the foot stock. The dividing head can be swiveled from below a horizontal line to beyond the vertical. The dividing head can also be used to drill equally

FIGURE K-62 Dividing head and foot stock (Cincinnati Milacron).

FIGURE K-63 Dividing head used to drill equally spaced holes (Cincinnati Milacron).

spaced holes in workpieces held in a chuck (Figure K-63). Round workpieces can be securely fastened in a set of vee blocks (Figure K-64). To prevent the shaft from bending under cutting pressure, a screw jack such as those shown in Figure K-53 can be used to support the shaft halfway between the vee blocks. If a number of identical workpieces are to be machined, a milling fixture (Figure K-65) may be the most efficient way of holding them. A fixture is used when the savings resulting from its use are greater than the cost of making the fixture.

FIGURE K-64 Round shaft being held in vee blocks.

FIGURE K-65 Milling fixture used for many identical parts (Cincinnati Milacron).

SELF-TEST

1. In relationship to the workpiece, where should the clamping bolt be located?

2. What precautions should be taken when clamping finished surfaces?

3. When are screw jacks used?

4. What is the reason for clamping a stop block to the table?

5. What are quick-action jaws?

6. What is the difference between a swivel vise and a universal vise?

7. When is an all-steel vise used?

8. When is a rotary table used?

9. When is a dividing head used?

10. When is a fixture used?

UNIT 5

Machine Setup and Plain Milling

Plain milling is the machine operation of milling a flat surface parallel to the cutter axis and machine table. Plain milling is accomplished most frequently on the plain horizontal milling machine and involves setting up the machine, selecting an appropriate arbor and cutter, and calculating correct speeds, feeds, and depth of cuts. Plain milling on the universal mill involves the additional operation of aligning the machine table. The purpose of this unit is to describe the procedure for preparing the machine tool and to acquaint you with the actual processes of plain milling.

Objectives

After completing this unit, you should be able to:

1. Select speeds and feeds for several different materials and milling cutters.
2. Set up the mill for plain milling.
3. Select and set up a work-holding system.
4. Select and set up an appropriate cutter and arbor.
5. Mill surfaces flat and square to each other.

Cutting Speeds

As in vertical milling, cutting speed (CS) is expressed as a function of cutter rpm by the general formula

$$\text{rpm} = \frac{\text{CS} \times 4}{D}$$

where rpm is the cutter speed in revolutions per minute, CS is the cutting speed of the material being machined expressed in surface feet per minute (sfpm), and D is the diameter of the milling cutter.

Cutting speeds (Table K-2) are influenced by the same factors as in vertical milling. Some additional factors relative to horizontal milling further influence their selection. Among these are cutting edges not remaining in the work continuously, a chip of variable thickness being formed, the type of milling being done (slab, face, or end), and the way in which heat is transferred from the cutting edge. Table values are approximate and may have to be varied up or down to fit specific machining tasks. Until you gain experience in horizontal milling, use the lower values when selecting cutting speeds. Computations are done in the same way as for vertical milling.

TABLE K-2 Cutting Speeds for Milling

	Cutting Speed (sfm)	
Material	High-Speed Steel Cutter	Carbide Cutter
Free-machining steel	100–150	400–600
Low-carbon steel	60–90	300–550
Medium-carbon steel	50–80	225–400
High-carbon steel	40–70	150–250
Medium-alloy steel	40–70	150–350
Stainless steel	30–80	100–300
Gray cast iron	50–80	250–350
Bronze	65–130	200–400
Aluminum	300–800	1000–2000

Feed Rates

Horizontal mill feed rates are expressed in inches per minute (ipm) and are calculated by the formula

$$\text{ipm} = F \times N \times \text{rpm}$$

where ipm is the inch/minute feed rate, F is the feed per tooth (Table K-3), N is the number of teeth on the milling cutter, and rpm is the revolutions per minute of the cutter.

To calculate a feed rate, first find rpm by the preceding formula, then refer to Table K-3 and calculate the feed rate by ipm $= F \times N \times$ rpm.

EXAMPLE

Find rpm and feed rate for a 3-in.-diameter high-speed helical mill cutting free-machining steel.

$$\text{rpm} = \frac{\text{CS} \times 4}{D}$$

$$= \frac{4 \times 100 \text{ fpm (Table K-2)}}{\text{3-in.-diameter cutter}}$$

$$= 134$$

$$\text{ipm} = F \times N \times \text{rpm}$$

$$= .002 \text{ in. feed/tooth (Table K-3)} \times 6 \text{ teeth} \times 134 \text{ rpm}$$

$$= 1.608 \text{ ipm feed}$$

To calculate the starting feed rate, use the low figure from the feed per tooth chart and, if conditions permit, increase the feed rate from there. The most economical cutting takes place when the most cubic inches of metal per minute are removed and a long tool life is obtained. The tool life is longest when a low speed and high feed rate are used. Try to avoid feed rates of less than .001 in. per tooth because this will cause rapid dulling of the cutter. Exceptions to this limit are small diameter end mills when used on harder materials. The depth and width of cut also affect the feed rate. Wide and deep cuts require a smaller feed rate than do shallow, narrow cuts. Roughing cuts are made to remove material rapidly. The depth of cut may be $\frac{1}{8}$ in. or more, depending on the rigidity of the machine, the setup, and the horsepower available. Finishing cuts are made to produce precise dimensions and acceptable surface finishes. The depth of cut on a finishing cut should be between .015 and .030 in. A depth of cut of .005 in. or less will cause the cutter to rub instead of cut and also results in excessive cutting edge wear.

HORSEPOWER CALCULATIONS

The rated horsepower (hp) of a milling machine is not the same as the hp available at the cutter. The hp at the cutter varies from 50 to 80 percent of the input hp. That indicates that a 5-hp milling machine has an available cutter hp from 2.5 to 4 hp. Another factor in hp calculation is the material removal constant expressed in in.^{3}/min/hp. In milling, the rule of thumb for hp requirement is

Steel	
150 BHN	1.0 in.^{3}/min/hp
350 BHN	.5 in.^{3}/min/hp
Alu	4.0 in.^{3}/min/hp

Hp at the cutter is calculated by the formula

$$\text{hp} = \frac{d \times w \times \text{ipm}}{C}$$

where d = depth of cut
w = width of cut
ipm = feed rate, in inches per minute
C = work material constant

TABLE K-3 Feed in Inches per Tooth (Instructional Setting)

	Aluminum		Bronze		Cast Iron		Free-Machining Steel		Alloy Steel	
Type of Cutter	HSS	Carbide	HSS	Carbide	HSS	Carbide	HSS	Carbide	HSS	Carbide
Face mills	.007 to .020	.007 to .020	.005 to .014	.004 to .012	.004 to .016	.006 to .020	.003 to .012	.004 to .016	.002 to .008	.003 to .014
Helical mills	.006 to .018	.006 to .016	.003 to .011	.003 to .010	.004 to .013	.004 to .016	.002 to .010	.003 to .013	.002 to .007	.003 to .010
Side cutting mills	.004 to .013	.004 to .012	.003 to .008	.003 to .007	.002 to .009	.003 to .012	.002 to .007	.003 to .009	.001 to .005	.002 to .008
End mills	.003 to .011	.003 to .010	.003 to .007	.002 to .006	.002 to .008	.003 to .010	.001 to .006	.002 to .008	.001 to .004	.002 to .007
Form-relieved cutters	.002 to .007	.002 to .006	.001 to .004	.001 to .004	.001 to .005	.002 to .006	.001 to .004	.002 to .005	.001 to .003	.001 to .004
Circular saws	.002 to .005	.002 to .005	.001 to .003	.001 to .003	.001 to .004	.002 to .006	.001 to .003	.001 to .004	.0005 to .002	.001 to .004

EXAMPLE

Calculate the hp requirement when using a 3-in.-diameter helical mill with 6 teeth, .005 in. feed per tooth, and a .200-in. depth of cut. The cutting speed is 90 sfm.

1. Calculate the rpm:

$$\text{rpm} = \frac{\text{CS} \times 4}{D} = \frac{90 \times 4}{3} = 120$$

2. Calculate the feed rate:

$$\text{ipm} = F \times N \times \text{rpm}$$

$$= .005 \times 6 \times 120 = 3.6$$

3. Calculate the hp:

$$\text{hp} = \frac{d \times w \times \text{ipm}}{C}$$

$$= \frac{.200 \times 4 \times 3.6}{1.0} = 2.88$$

From this example we see that we need a milling machine with a rating of at least 5 hp since it is the next larger size of electric motor from 3 hp. Because of friction losses in the power train, the machine would require much more than the calculated 2.88 hp.

Cutting Fluids

Cutting fluids should be used when machining most metals with high-speed steel cutters. A cutting fluid cools the tool and the workpiece. It lubricates, which reduces friction between the tool face and chip. Cutting fluids prevent rust and corrosion and, if applied in sufficient quantity, will flush away chips. Cutting fluids will, through these characteristics, increase production through higher speeds and produce better surface finishes. Most milling with carbide cutters is done dry unless a large constant flow of cutting fluid can be directed at the cutting edge. An interrupted coolant flow on a carbide tool causes thermal cracking and results in subsequent chipping of the tool.

Setting up the Mill

The machine table and all sliding surfaces should be cleaned prior to setup or operation. Any nicks or burrs found on the table or work-holding devices should be removed with a honing stone.

FIGURE K-66 Aligning the universal milling machine table (Lane Community College).

TABLE ALIGNMENT ON THE UNIVERSAL MILL

The table on the universal mill should be checked and adjusted if necessary for alignment prior to any machining. The following procedure may be used.

STEP 1. Clean the face of the column and the machine table.

STEP 2. Fasten a dial indicator to the table using a magnetic base or other mounting system (Figure K-66). **Never** try to align the table with the indicator mounted on the column since this would always show the table to be in alignment.

STEP 3. Preload the indicator about one-half revolution. If the table is set at an extreme angle, loosen the locking bolts and swing back to a position of approximate alignment.

STEP 4. Turn the handwheel and move the indicator until it is positioned near one edge of the column. Set the bezel to zero.

STEP 5. Move the table so that the indicator moves across the face of the column and observe the reading at the opposite edge.

STEP 6. Adjust the position of the table so that one-half of the total indicated runout is cancelled (back toward zero on the indicator).

STEP 7. Reset the bezel to zero and move the indicator back across the face of the column.

STEP 8. If an error is observed, cancel by one-half and recheck on the opposite edge of the column.

FIGURE K-67 Fixture alignment keys (Lane Community College).

FIGURE K-68 Aligning vise parallel to table (Lane Community College).

STEP 9. Tighten all locking bolts securely and **recheck the alignment.** A zero indicator reading should be observed at both edges of the column.

WORK HOLDING

Selection of a work-holding device and method will depend on the machining task to be done. Review Unit 4 and, if you are clamping directly to the mill table, be sure to follow the rules of good clamping. Use a work stop if necessary to prevent workpiece slippage from cutting pressure.

MILL VISES A mill vise is an accurate and dependable work-holding tool. When milling only on the top of the workpiece, it is not necessary to set up the vise square to the column or parallel to the table. However, if the workpiece has already machined outside surfaces, steps, or grooves, the vise must be precisely aligned to the machine table.

Many mill vises are equipped with keys that fit snugly into the table T-slots (Figure K-67). Once the keys are engaged, the vise is accurately aligned. Before mounting a vise or any other mill fixture, inspect the bottom for small chips, nicks, and burrs. Use a honing stone to remove these and make sure that the tooling and table are clean. Set the vise gently on the table and position it according to the job to be done. Install the hold-down bolts and tighten just enough to permit the vise to be moved by gentle tapping with a lead or other soft hammer. If the mill vise does not have alignment keys, the following procedures may be used to align it parallel or perpendicular to the machine table.

PARALLEL-TO-TABLE ALIGNMENT

STEP 1. Fasten a dial indicator on a magnetic base and attach to the arbor or overarm (Figure K-68). Be sure that overarm clamps are tight. Position the indicator to contact the solid jaw of the vise. Preload the indicator about one-half revolution.

STEP 2. Move the table by hand so that the indicator is positioned at one end of the solid jaw. Set the bezel to zero. Crank the table so that the vise jaw moves past the indicator tip and note the reading at the opposite end of the jaw.

STEP 3. Tap the vise gently with a soft hammer so that one-half of the total indicated runout is cancelled (back toward zero on the indicator). When tapping the vise, move it in such a direction that the solid jaw moves away from the indicator tip. Moving the jaw against the indicator tip can damage the delicate indicator because of the shock delivered to the indicator movement. Reset the bezel to zero.

STEP 4. Crank the table back and observe the indicator reading. If a zero reading is obtained, tighten the hold-down bolts securely and recheck the alignment.

PERPENDICULAR-TO-TABLE ALIGNMENT

A vise may be aligned at right angles to the table

FIGURE K-69 Aligning vise square to table travel (Lane Community College).

FIGURE K-71 Using a protractor to align a vise on the milling machine table (Lane Community College).

FIGURE K-70 Using a square to align a vise on the milling machine table (Lane Community College).

FIGURE K-72 Cutting pressure against solid jaw (Lane Community College).

(Figure K-69) by the technique discussed previously. Once again, always indicate on the solid jaw and move the saddle in order to carry the vise jaw past the indicator. Always recheck after tightening hold-down bolts.

SQUARING A VISE TO THE COLUMN A vise may be aligned by squaring the solid jaw to the column. Two paper strips may be used as feeler gages between the beam of the square and the vise jaw (Figure K-70). This technique will quite accurately align a vise, but it is not as reliable as the dial indicator methods previously discussed.

ANGULAR ALIGNMENT A protractor may be used to set a vise at an angle other than 90 degrees to the table (Figure K-71). The accuracy of the angle is dependent on the type of tool used and the technique. Errors can be introduced from the angle setting on the protractor and the relative alignment of the vise jaw along the protractor blade. Considerable care must be exercised in making a setup by this technique.

SECURING THE WORKPIECE The vise will effectively secure a workpiece in most cases. Whenever possible, set up the vise so that cutting pressure is applied to the solid jaw (Figure K-72). Avoid applying cutting pressure against the movable jaw. If the workpiece is sufficiently high, it may be seated on the bottom of the vise. If not, parallels may be used to elevate the work to a point where it can be machined. In many cases it will be necessary to apply the cutting pressure parallel to the vise jaws (Figure K-73). Remember that friction between the vise jaws and workpiece holds it in place.

FIGURE K-73 Workpiece held in vise (Lane Community College).

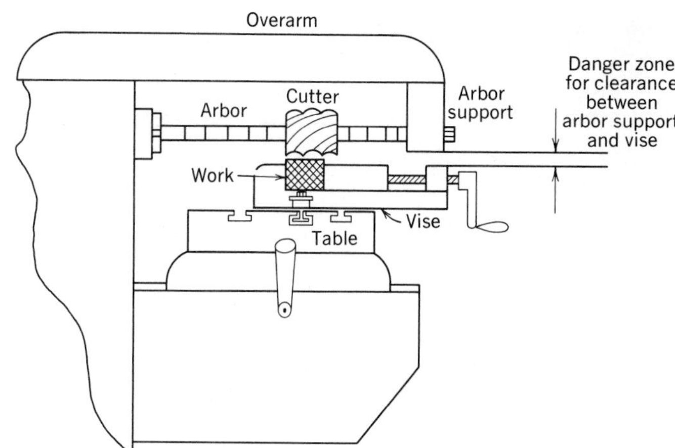

FIGURE K-75 Always check clearance between arbor support and vise when setting up the machine.

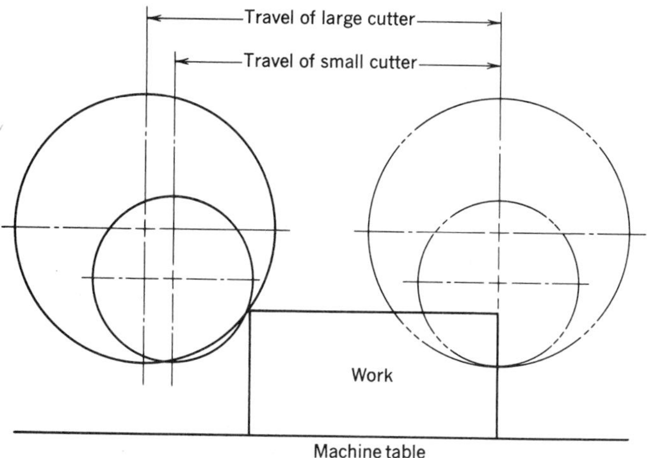

FIGURE K-74 Different travel distances between different diameters of cutters.

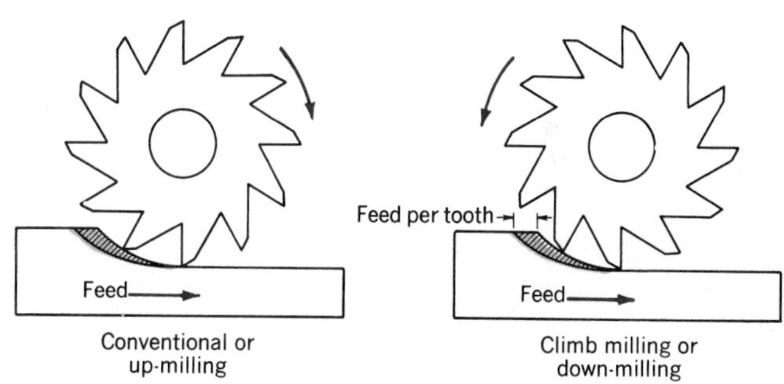

FIGURE K-76 Conventional and climb milling.

The more contact area there is, the better the holding power. When cutting pressure is applied parallel to the vise jaws, there is always a possibility that the part will be pushed from the vise. Therefore, if you must use parallels to elevate the workpiece, only raise it high enough to accomplish the required machining since this reduces contact area and the vise may not be able to hold the workpiece safely and securely.

SELECTING THE CUTTER

For a flat surface, use a plain milling cutter that is wider than the surface to be machined. The diameter of the milling cutter selected should be as small as practical. A large-diameter cutter must travel farther than one with a smaller diameter (Figure K-74). Therefore, a smaller diameter cutter is more efficient since time will be saved. Use a sharp cutter to minimize cutting pressure and to obtain a good surface finish.

Whatever diameter cutter is used, it is important to have sufficient clearance between the arbor supports and the vise or other work-holding fixture (Figure K-75). As material is machined from the workpiece, this clearance is reduced and it may become necessary to reset the workpiece if the vise and arbor supports will not clear each other before the final cut is taken.

CLIMB AND CONVENTIONAL MILLING

In milling, the direction that the workpiece is being fed either can be in the same direction as cutter rotation or can be opposed to the direction of cutter rotation (Figure K-76). When the direction of feed is opposed to the rotation direction of the cutter, this is said to be up or conventional milling. When the direction of feed is the same as the rotation direction, this is said to be down or climb milling because the cutter is attempting to climb onto the workpiece. If any appreciable amount of backlash exists in the table or saddle, the workpiece may be pulled into the cutter during climb milling. This can result in a bent arbor, broken cutter, damaged workpiece, and possible in-

FIGURE K-77 Clean the external and internal taper prior to installation (Lane Community College).

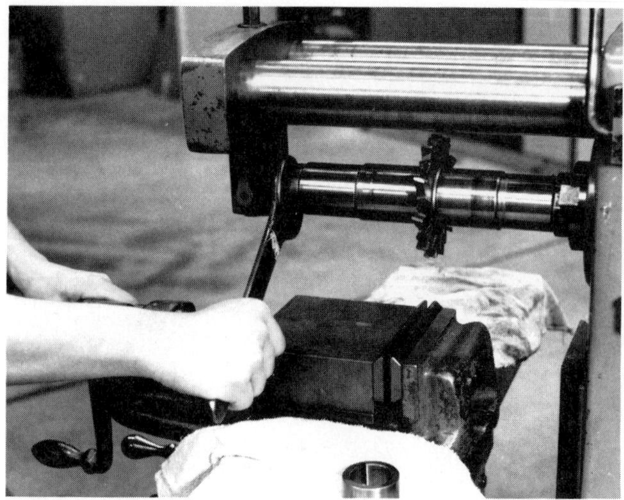

FIGURE K-79 Tighten the arbor nut (Lane Community College).

FIGURE K-78 Tighten the drawbolt locknut with a wrench (Lane Community College).

jury to the operator. Climb milling should be avoided unless the mill is equipped with adequate backlash control. Remember that any cutter can be operated in an up milling or down milling mode depending only on which side of the workpiece the cut is started.

SELECTING AND SETTING UP MILL ARBORS

When selecting a mill arbor, use one that has minimum overhang beyond the outer arbor support. Excessive overhang can cause vibration and chatter. After selecting the proper arbor, insert the tapered shank into the spindle socket. Be sure that the socket is clean and free from burrs or nicks (Figure K-77). Large mill arbors are heavy and you may need some help holding them in place until the drawbolt is engaged. Do not let the arbor fall out onto the machine table. Thread the drawbolt into the arbor shank all the way, then draw the arbor into the spindle taper by turning the drawbolt locknut. Tighten the locknut with a wrench (Figure K-78).

Remove the arbor nut and spacing collars. Place these on a clean surface so that their precision surfaces are not damaged. Position the cutter on the arbor as close to the spindle as the machining task will permit. Place a sufficient number of spacing collars on either side of the cutter so as to position it correctly. The cutter, spacing collars, and bearing collar should be a smooth sliding fit on the arbor.

A key is generally used to ensure a positive drive between cutter and arbor. However, a milling cutter can be driven without a key. Consult with your instructor and follow instructions regarding the use of keys.

Place the bearing collar on the arbor and locate as close to the cutter as the machining task will permit. Place the arbor support on the overarm and slide it in until the arbor bearing collar slips through the arbor support bearing. Tighten overarm and support clamps. Tighten the arbor nut only after the support is in place (Figure K-79). Tightening the arbor nut before the support is in place may bend the arbor. Do not overtighten the arbor nut and always use a wrench of the correct type and size.

FIGURE K-80 Loosening the locknut on the drawbolt (Lane Community College).

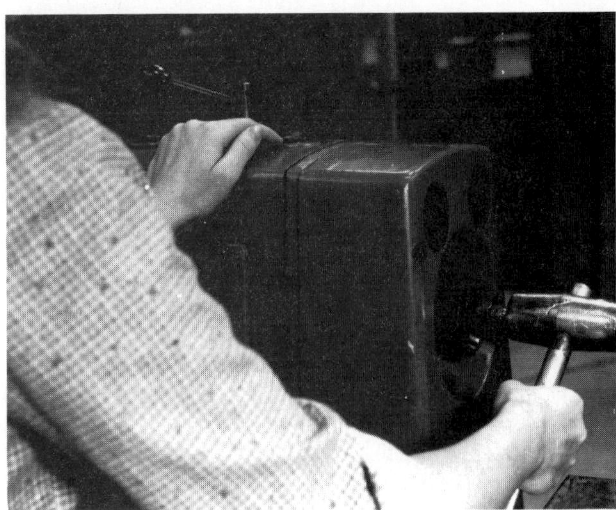

FIGURE K-81 Tapping the end of the drawbolt with a lead hammer (Lane Community College).

REMOVING AND STORING MILL ARBORS

Care should be exercised in removing the arbor from the spindle. Loosen the drawbolt locknut about one turn (Figure K-80). You may have to tap the drawbolt lightly to release the arbor taper shank from the spindle socket (Figure K-81). Hold the arbor in place or get help while you unscrew the drawbolt from the arbor shank (Figure K-82). Remove the arbor from the machine and store in an upright position. Long arbors stored in a flat position may bend.

FIGURE K-82 Have another person hold heavy arbors when you are loosening them (Lane Community College).

Plain Milling

Good milling practice is to take roughing cuts and then a finish cut. Better surface finish and higher dimensional accuracy are achieved when roughing and finishing cuts are made. The machining time of a workpiece is dependent to a large extent on how efficiently material is removed during the roughing operation. The depth of the roughing cut often is limited by the horsepower of the machine or the rigidity of the setup. A good starting depth for roughing is .100 to .200 in.

The finishing cut should be .015 to .030 in. deep. Depth of cut less than .015 in. should be avoided because a milling cutter, especially in conventional or up milling, has a strong rubbing action before the cutter actually starts cutting. This rubbing action causes a cutter to dull rapidly.

ROUGHING CUTS

Assuming that a roughing cut .100 in. deep is to be taken, the following procedure may be used.

STEP 1. Loosen the knee locking clamp and the cross slide lock.

STEP 2. Turn on the spindle and check its rotation.

STEP 3. Position the table so that the workpiece is under the cutter.

STEP 4. A paper strip may be used to determine when the cutter is about .002 in. away

FIGURE K-83 Using a paper strip to position the cutter on the top of the workpiece (Lane Community College).

FIGURE K-84 Coolant cools the cutter and flushes away chips (Lane Community College).

from the workpiece (Figure K-83). This should be done with the spindle off and turned by hand.

STEP 5. Set the micrometer dial on the knee crank on zero.

STEP 6. Lower the knee approximately one-half revolution of the hand feed crank. If the knee is not lowered, the cutter will leave tool marks on the workpiece in the following operation.

STEP 7. Move the table right or left until the cutter is clear of the workpiece. Start the cut on whichever side of the part results in up or conventional milling.

STEP 8. Raise the knee past the zero mark to the 100 mark on the micrometer dial.

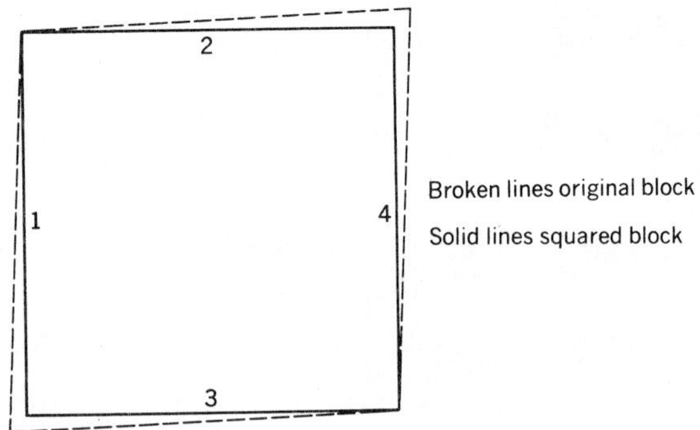

FIGURE K-85 Machining a square from a parallelogram.

STEP 9. Tighten the knee lock and the cross slide lock. Prior to starting a machining operation, **always** tighten all locking clamps except the one that would restrict table movement while cutting. This aids in making a rigid chatter-free setup.

STEP 10. The machine is now ready for the cut. Turn on the coolant (Figure K-84). Move the table slowly into the revolving cutter until the full depth of cut is obtained before engaging the power feed.

STEP 11. When the cut is completed, disengage the power feed, stop the spindle rotation, and turn off the coolant before returning the table to its starting position. If the revolving cutter is returned over the newly machined surface, it will leave cutter marks and mar the finish.

STEP 12. After brushing off the chips and wiping the workpiece clean, the part should be measured while it is still fastened in the machine. If the workpiece is parallel at this time, additional cuts can be made if more material needs to be removed.

SQUARING A WORKPIECE

Often a piece of stock shaped like a parallelogram (Figure K-85) must be machined square. This can be accomplished by machining all four sides in the correct sequence. The workpiece should be set up in a mill vise; it may be placed on parallels if necessary. This will facilitate measurements and will also raise a short part above the vise jaws. The following procedure may be used.

STEP 1. Set the workpiece in the vise and mill a reference surface on top.

FIGURE K-86 Setup to machine side 2 (Lane Community College).

FIGURE K-87 Setup to machine side 3 (Lane Community College).

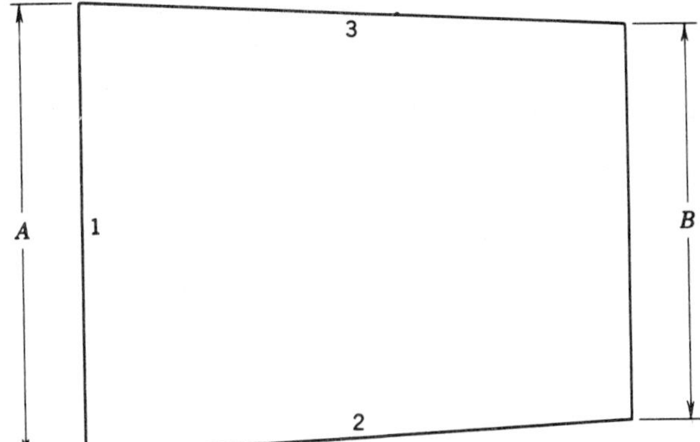

FIGURE K-88 Measuring for squareness.

FIGURE K-89 Location of shim to square up work (Lane Community College).

STEP 2. Deburr and place this surface (side 1, Figure K-86) against the solid jaw of the vise.

STEP 3. Place a short length of soft round rod between the work and the movable jaw of the vise. Tighten vise and seat workpiece down on the parallels with a lead hammer. The rod will ensure that the reference surface is pressed flat against the solid vise jaw.

STEP 4. Mill side 2 (Figure K-86).

STEP 5. Deburr and place side 2 down, keeping side 1 against the solid jaw. Be sure that side 2 is carefully seated against the parallels. Test by trying to move the parallels. If they can be moved, use the lead hammer and tap the workpiece down so that it will be firmly seated.

STEP 6. Mill side 3 (Figure K-87).

STEP 7. With a micrometer, measure dimensions *A* and *B* (Figure K-88). If dimension *A* is equal to dimension *B*, or within less than .004 in., the solid jaw is square to the base and you can skip step 8 and go on to step 9. If the difference is more than .004 in., go to step 8.

STEP 8. If dimension *A* is larger than *B*, use a thin shim and place it between the work and solid jaw (Figure K-89). With the shim in place and the workpiece securely fastened, take a cut .020 in. deep on side 2 and also on side 3. Measure dimensions *A* and *B* again to see if the shim corrected the previous difference. If necessary, repeat this operation with different size shims until the workpiece is parallel between sides 2 and 3.

STEP 9. Deburr and place side 1 down and sides 2 or 3 against the solid jaw of the vise. Be sure that side 1 is well seated on the parallels. Since sides 2 and 3 are parallel, the rod between the work and movable jaw will not be necessary.

FIGURE K-90 Setup to machine side 4 (Lane Community College).

FIGURE K-91 Setup of a workpiece to machine an end square (Lane Community College).

STEP 10. Mill side 4 (Figure K-90).

MILLING ENDS OF THE WORKPIECE

After the sides of the workpiece have been machined square, the end may also be finished by placing the work in a vise end up (Figure K-91). A square is used to position the workpiece. Tighten

FIGURE K-92 Work clamped off-center needs a spacer (Lane Community College).

the vise lightly and tap the work with a soft hammer to bring it into a square position. A dial indicator can also be used to position the workpiece in a perpendicular position. Attach the indicator to the overarm or arbor with the tip in contact with the workpiece. Run the knee up and down and adjust the workpiece until a zero indicator reading is obtained. If a workpiece needs to be fastened in a vise off-center (Figure K-92), it becomes very important that the opposite end of the vise jaw be supported with another piece of material that is the same thickness as the workpiece. Without this supporting spacer the vise cannot securely clamp the workpiece.

The end of a long workpiece may also be machined by clamping to an angle plate (Figure K-93) and adjusting the position of the workpiece with a square or dial indicator.

SELF TEST

1. What is *cutting speed?*

2. Why should cuts be started at the low end of the cutting speed range?

3. If the cutting speed is 100 fpm with an HSS tool, what would it be with a carbide tool for the same material?

FIGURE K-93 Use of angle plate to mill ends of workpieces (Lane Community College).

4. What is the effect of too low a cutting speed?

5. How is the feed rate expressed on a milling machine?

6. How is the feed rate calculated?

7. Why is a feed per tooth given instead of a feed per revolution?

8. What is the effect of too low a feed rate?

9. What rpm is used with a 3-in.-diameter HSS cutter on low carbon steel?

10. What is the feed rate for a 4-in.-diameter, five-tooth carbide face mill machining alloy steel?

11. Why is the solid jaw used to align a vise on a milling machine table?

12. Which is more accurate, using a precision square or a dial indicator to square a vise on a machine table?

13. What is the purpose of the keys used on the base of machine vises?

14. Should you mount the indicator on the column to align the table on a universal milling machine?

15. Should the solid vise jaw be in a specific position in relation to the direction of the cut?

16. Is a large- or small-diameter cutter more efficient?

17. What is the difference between conventional milling and climb milling?

18. How deep should a finish cut be?

19. Why are all table movements locked except the one being used during machining?

20. Why is the cutter rotation stopped while the table is returned over the newly cut surface to its starting position?

UNIT 6

Using Side Milling Cutters

Machining a flat surface is only one of the many machining operations that can be accomplished on the horizontal mill. Steps and grooves, straddle and gang milling all involve the use of side milling cutters. The purpose of this unit is to introduce you to these useful milling cutters and their applications.

Objectives

After completing this unit, you should be able to:

1. Set up side milling cutters and cut steps and grooves.
2. Use side milling cutters for straddle milling.
3. Use side milling cutters in gang milling.

FIGURE K-94 Full side milling cutter machining a groove (Lane Community College).

FIGURE K-95 Half side milling cutter machining a step (Lane Community College).

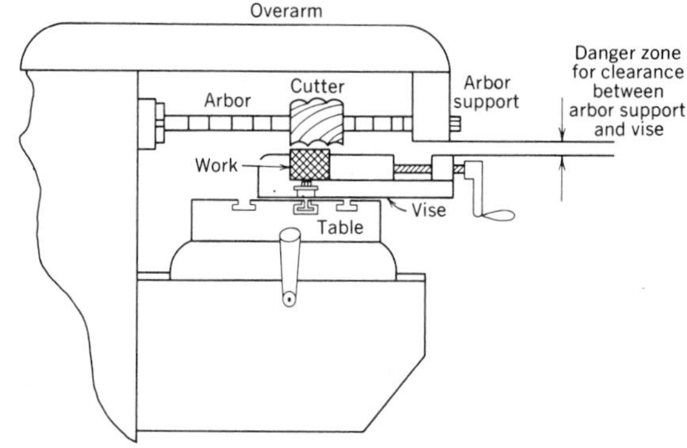

FIGURE K-96 Always check clearance between arbor support and vise when setting up the machine.

Side Milling Cutters and Side Milling

Milling cutters with side teeth are called side milling cutters. These are used to machine steps and grooves or, when only the sides of the workpiece are to be machined, in straddle setups. An example of this would be cutting hexes on bolt heads. Grooves are best machined in the workpiece with full side mills that have cutting teeth on both sides (Figure K-94). Steps may be cut with half side mills having cutting teeth only on one side (Figure K-95).

The size, type, and diameter of side milling cutter to use will depend on the machining task. As a rule, the smallest diameter cutter that will do the job should be used as long as sufficient clearance is maintained between the arbor support and work or vise (Figure K-96).

Frequently, side milling cutters will cut a slot or groove that is slightly wider than the nominal width of the cutter. Reasons for this include: cutter wobble due to chips or dirt between the cutter and arbor spacing collars, or making multiple cuts through the workpiece. Other factors contributing to wider than desired slots are the rate of feed and the type of material being machined. A slow feed will give the side mill more time to cut a slightly wider slot. A fast feed will crowd the cutter, resulting in a narrower slot. Wider slots may occur in softer materials more often than they do in harder materials.

If the specifications call for a slot width of .375 in., using a .375 ($\frac{3}{8}$) in. wide side mill may result in a slot width that ranges from .3755 to .376 in. In order to hold the required dimension, it may be necessary to use a narrower side mill, say .3125 ($\frac{5}{16}$) in., and make additional passes through the workpiece until the required width dimension is obtained.

PREPARING AND SETTING UP THE WORKPIECE

A good machinist will mark the workpiece with layout lines before securing it in the machine. The layout should be an exact outline of the final part

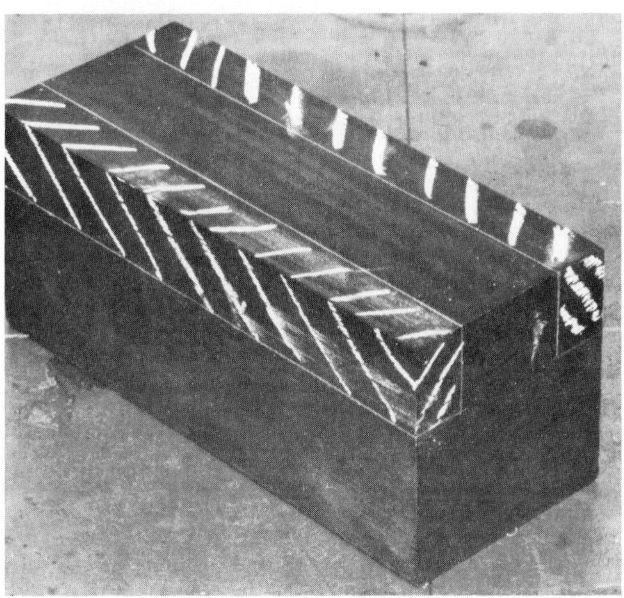

FIGURE K-97 Work laid out for milling (Lane Community College).

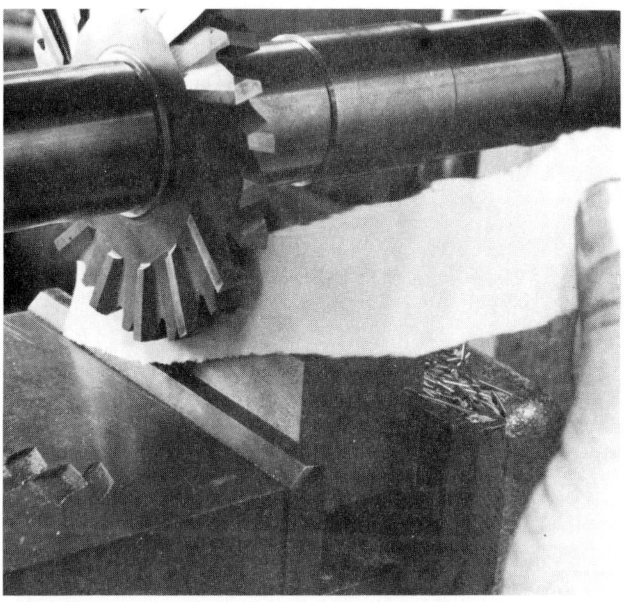

FIGURE K-98 Setting up a cutter by using a paper strip (Lane Community College).

shape and size. The reason for making the layout prior to beginning work is that reference surfaces are often removed during machining. After the layout has been made, diagonal lines should be chalked on the workpiece indicating the portions that are to be cut away. This helps to identify on which side of the layout lines the cut is to be made (Figure K-97).

The workpiece may be mounted by any of the traditional methods discussed previously. If you are using a vise, be sure it is aligned along with the table. If you are clamping directly to the table, follow the rules of good clamping.

FIGURE K-99 Positioning a cutter using a steel rule (Lane Community College).

POSITIONING A SIDE MILLING CUTTER

In order to machine a slot or groove at a particular location on the workpiece, the side mill must be positioned both horizontally (for location) and vertically (for depth of cut). To position the cutter for location, lower below the top surface of the workpiece. With the spindle off and free to turn by hand, insert a paper strip between cutter and work (Figure K-98) and move the workpiece toward the cutter until the paper is pulled between the work and cutter. At this point, the cutter is about .002 in. from the workpiece. Set the saddle micrometer collar to zero, compensating for the .002 in. of paper thickness.

Position for depth by lowering the knee and moving the workpiece under the cutter. Then, raise the knee and use the paper strip gage to determine when the cutter is about .002 in. above the workpiece. Set the knee micrometer collar to zero. Move the table until the cutter is clear of the workpiece and then raise the knee the amount required for the depth of the feature.

A less accurate, but quicker, method of cutter alignment is by direct measurement (Figure K-99). A rule may be used to measure the position of the cutter relative to the edge of the workpiece after which the saddle micrometer collar should be set to zero.

MAKING THE CUT After the cutter has been positioned for location, raise the knee the amount required for the depth of the cut. Machining to full depth may be accomplished in one pass depending

FIGURE K-100 Taking a trial cut (Lane Community College).

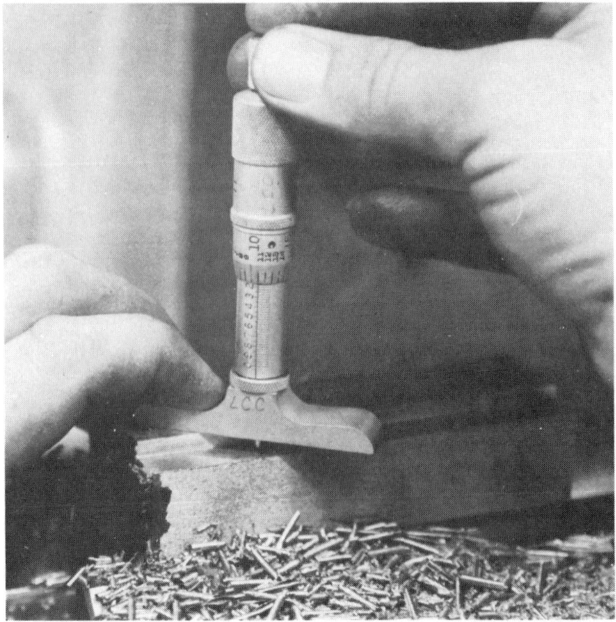

FIGURE K-102 Measuring depth of a step (Lane Community College).

FIGURE K-101 Measuring a slot with an adjustable parallel and a micrometer (Lane Community College).

FIGURE K-103 Setting the side mill alongside the shaft (Lane Community College).

on the width of the cut and the material being machined. A deep slot may have to be machined in more than one pass, each pass somewhat deeper than the one before. Until you gain some experience in milling, hold depth setting to about .100 in.

Set proper feeds and speeds and turn on the spindle. Approach the workpiece in an up milling mode. Hand feed the cutter into the work until a small nick is machined on the corner (Figure K-100, point X). Stop spindle, back away, and check the dimension relative to the edge of the workpiece. If you are machining a slot, check the width with an adjustable parallel and a micrometer (Figure K-101), or use a dial/vernier caliper. If the dimensions are correct, complete the required cuts. Precise depth can be measured with a depth micrometer (Figure K-102).

MILLING KEYSEATS

Side milling is an excellent way to machine a shaft keyseat, especially if the keyseat is quite long. The procedure for keyseat milling is much the same as in vertical milling. The cutter must be centered over the shaft and set for depth.

After selecting and mounting a cutter of the proper width, raise the workpiece beside the cutter (Figure K-103) and use the paper feeler technique to position the cutter alongside the shaft. Lower the knee and move the workpiece over a distance equal to one-half the total of cutter width and shaft diameter. Raise the knee until the cutter contacts the shaft and cuts a full width cut (Figure K-104). Set the knee micrometer collar to zero. Lower the

FIGURE K-104 After centering, cutter is brought into contact with shaft (Lane Community College).

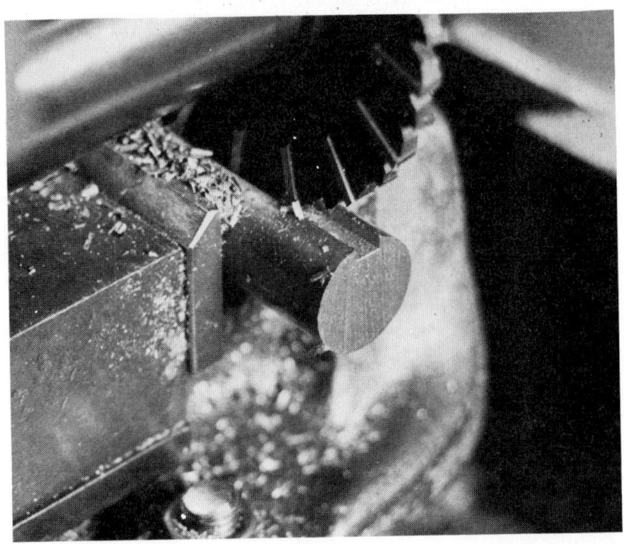

FIGURE K-105 Side milling a shaft keyseat (Lane Community College).

FIGURE K-106 Straddle milling (Lane Community College).

FIGURE K-107 Gang milling (Cincinnati Milacron).

knee and move the cutter clear of the workpiece in the table axis only. Raise the knee to obtain the correct keyseat depth. Using proper feeds and speeds, approach the workpiece in an up milling mode, and mill the keyseat to the proper length (Figure K-105).

STRADDLE AND GANG MILLING

Side milling cutters are combined to perform straddle milling. In straddle milling, two side milling cutters are mounted on an arbor and set at an exact spacing (Figure K-106). Two sides of the workpiece are machined simultaneously and final width dimensions are exactly controlled. Straddle milling has many useful applications in production machining. Parallel slots of equal depth can be milled by using straddle mills or equal diameters.

In gang milling, several cutters are mounted on the arbor and used to machine special shapes and contours on the workpiece (Figure K-107). The difference in cutter diameter determines the depths of steps and grooves. Cutter rpm is calculated for the largest diameter cutter in the gang.

INTERLOCKING TOOTH AND RIGHT- AND LEFT-HAND HELICAL SIDE MILLS

Wide slots and grooves can be machined with right- and left-hand helical side mills. The heavy

FIGURE K-108 Left-hand and right-hand helical flutes on wide cuts (Cincinnati Milacron).

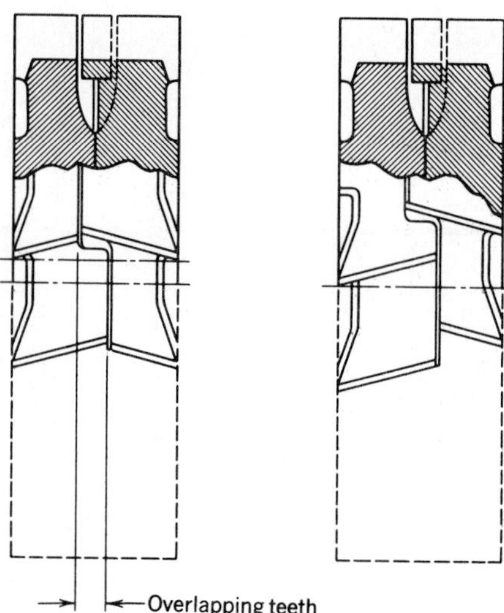

FIGURE K-109 Interlocking side milling cutters.

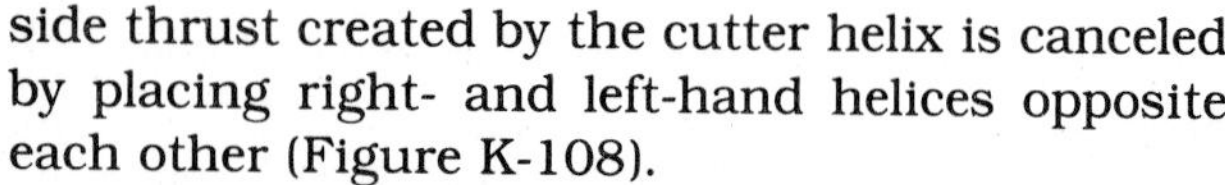

side thrust created by the cutter helix is canceled by placing right- and left-hand helices opposite each other (Figure K-108).

Interlocking side mills are used when grooves of a precise width are machined in one operation (Figure K-109). Shims inserted between individual cutters make precise adjustment possible. The overlapping teeth leave a smooth finish in the bottom of the groove. Cutter combinations that have become thinner from resharpenings can also be adjusted to their full width by adding shims.

SELF-TEST

1. When are full side milling cutters used?
2. When are half side milling cutters used?
3. What diameter side milling cutter is most efficient?
4. Is a groove the same width as the cutter that produces it?
5. Why should a layout be made on workpieces?
6. How can a side milling cutter be positioned for a cut without marring the workpiece surface?
7. Why should measurements be made before removing a workpiece from the work-holding device?
8. How is the width of a workpiece controlled in a straddle milling operation?
9. What determines the depth of the steps in gang milling?
10. When are interlocking side mills used?

UNIT 7

Using Face Milling Cutters on the Horizontal Milling Machine

Face milling cutters are used to machine flat surfaces at a right angle to the cutter axis. They are very efficient tools when large quantities of material are to be removed.

Objectives

After completing this unit, you should be able to:

1. Identify face milling cutters.
2. Use face milling cutters to machine flat surfaces.

Face milling cutters are the most commonly used milling cutters with inserted teeth. Face milling cutters up to 6 in. in diameter are called shell end mills. Face milling cutters are made with inserted teeth because it would be very difficult to make such large cutters in one piece, and because of the high cost of the cutting tool material. The cutter body, made from heat-treated alloy steel, can be used almost indefinitely and only the cutting inserts need to be changed. Face milling cutters are usually mounted directly on the spindle nose and held in position by four capscrews. The back of the cutter is counterbored to fit closely over the outer diameter of the spindle nose. The driving keys of the spindle nose fit the keyseats in the back of the face mill (Figure K-110). Another method of mounting face mills is the flat back mount (Figure K-111). It uses a centering plug located in the machine spindle taper to center the cutter. When a face mill is mounted on the machine spindle, it is very important that all mating surfaces between the cutter and the spindle are clean and free of any nicks or burrs. Even the smallest particle between the mating surfaces will make the required alignment of the cutter impossible.

After the cutter has been mounted on the machine, a check with a dial indicator should show the cutter to run true or to be out of alignment less than .001 in. Face mills are made as light-duty cutters with a large number of teeth for finishing cuts or as heavy-duty cutters with fewer teeth and heavier bodies for roughing operations.

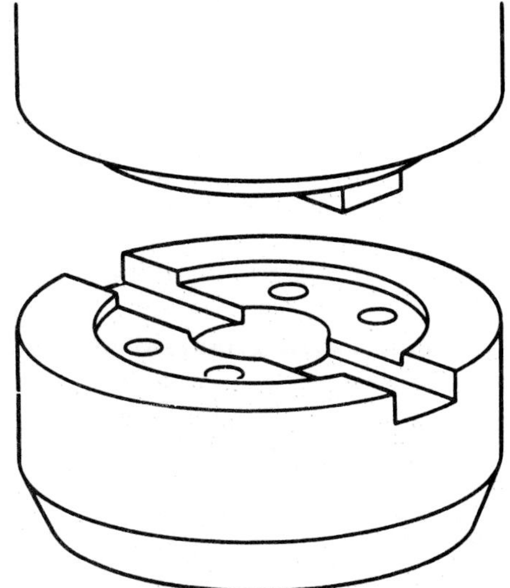

FIGURE K-110 National standard drive (Kennametal Inc., Latrobe, PA).

The cutting inserts are high speed steel in older face mills, but new face mills are made with carbide inserts. Major differences in face mills appear in the rake angles of the cutting surfaces. High speed milling cutters are normally used with positive rake angles (Figure K-112). Positive rake angles produce a good surface finish, increase cutter life, and use less power in cutting, both with HSS and carbide inserts. Positive rake angles are very effective in the machining of tough and work-hardening materials. Cutting pressures, which may deflect thin-walled workpieces, are smaller with positive rake cutters than with negative rake cutters under the same conditions. Zero or negative rake cutters are used only with carbide inserts These inserts are very strong and will give good

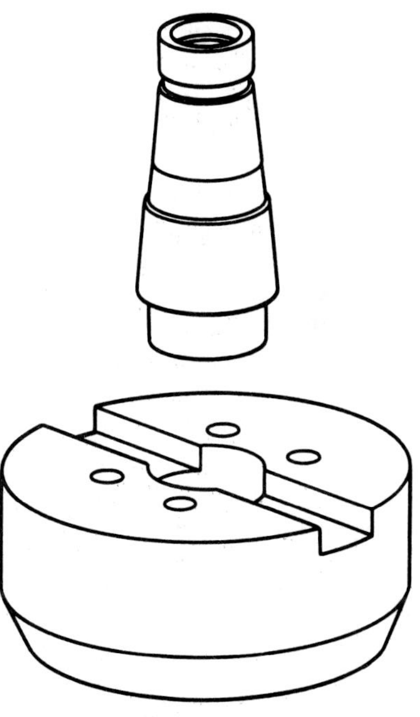

FIGURE K-111 Flat back mount (Kennametal Inc., Latrobe, PA).

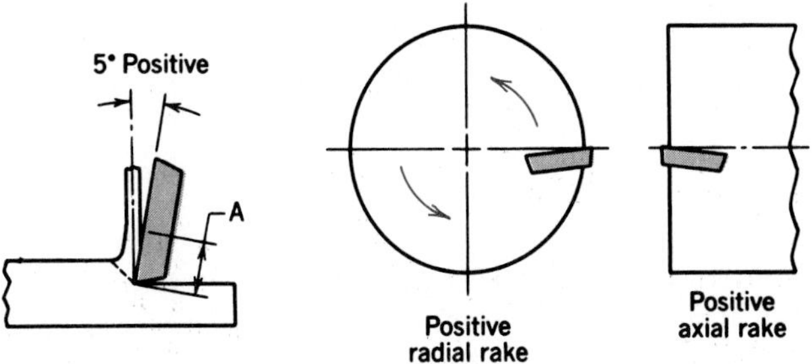

FIGURE K-112 Positive rake angles (Kennametal Inc., Latrobe, PA).

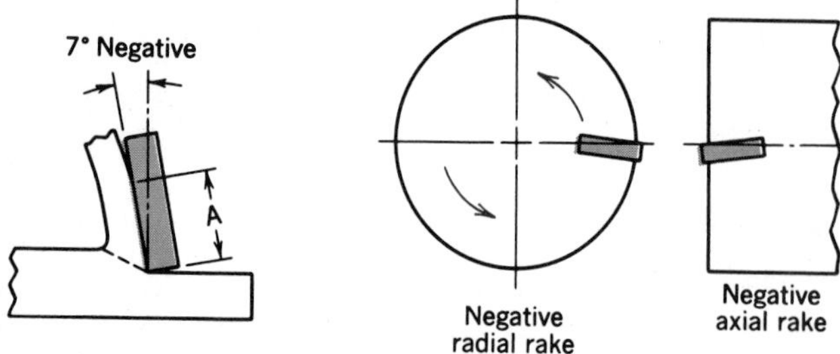

FIGURE K-113 Negative rake angles (Kennametal Inc., Latrobe, PA).

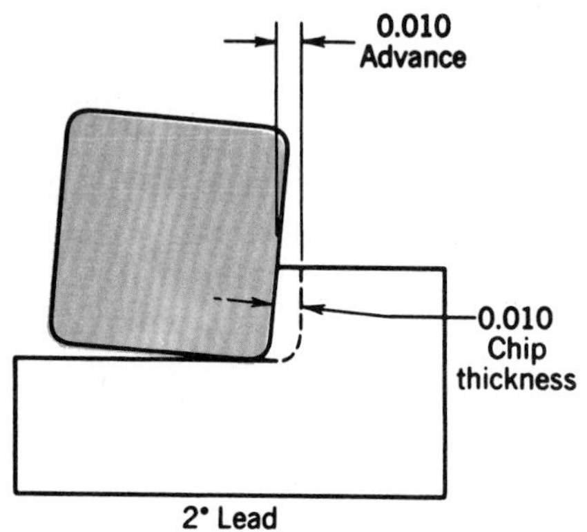

FIGURE K-114 Small lead angle (Kennametal Inc., Latrobe, PA).

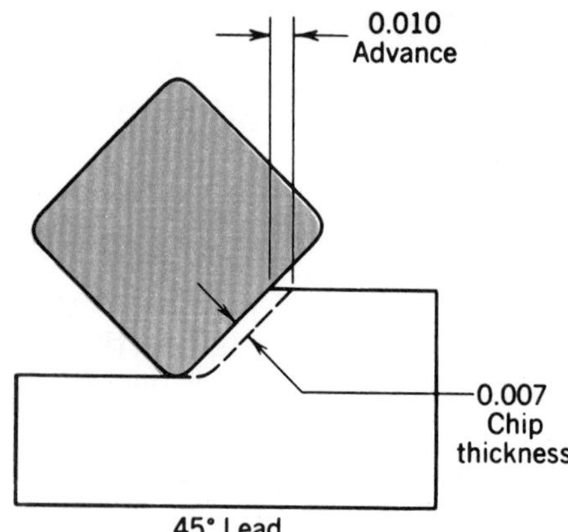

FIGURE K-115 Large lead angle (Kennametal Inc., Latrobe, PA).

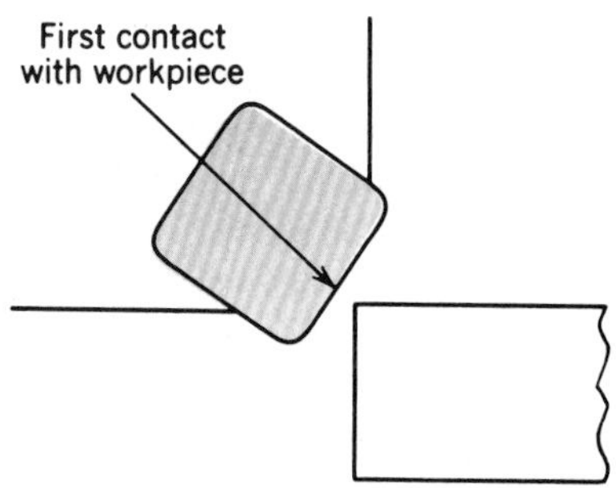

FIGURE K-116 Lead angle effect (Lane Community College).

service under heavy impact or interrupted cutting conditions (Figure K-113). Negative rake inserts create high cutting pressures, which tend to force the workpiece away from the cutter. Negative rake inserts should not be used on work hardening materials or ductile materials such as aluminum or copper.

The lead angle, which is the angle of the cutting edge measured from the periphery of the cutter, varies from 0 to 45 degrees, depending on the application. Small lead angles of 1 to 3 degrees can be used to machine close to square shoulders. A small lead angle cutter, when used with a square insert, will have sufficient clearance on the face of the cutter to prevent its rubbing. Figure K-114 shows the effect of a small lead angle on chip thickness. With a .010 in. feed per insert the chip is also .010 in. thick and as long as the depth of cut. Figure K-115 uses the same feed and depth of cut, but the chip is quite different. The chip now is longer than the depth of cut and its thickness is only .007 in. A thinner chip gives an increased cutting edge life, but it does limit the effective depth of cut. Feed rates of less than .004 in. per insert should be avoided. Figure K-116 shows another beneficial effect of a large lead angle; the cutter contacts the workpiece at a point away from the tip of the cutting edge. The cutter does not cut a full size chip on initial impact, but eases into the cut. The same thing happens on completion of the cut as the cutter eases out of the work.

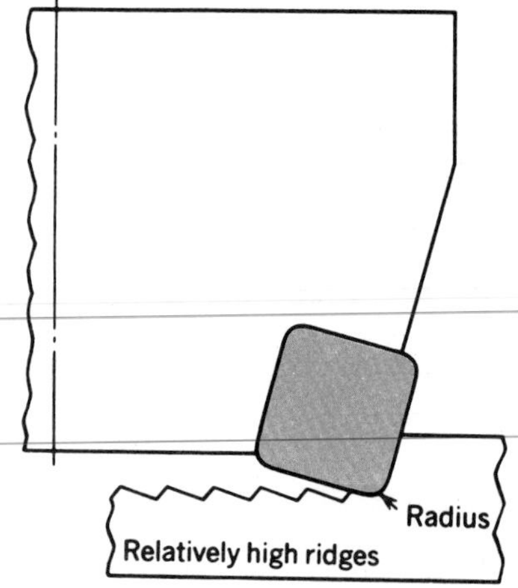

FIGURE K-117 Feed lines on surface (Kennametal Inc., Latrobe, PA).

The surface finish produced by a cutter depends largely on nose radius of the insert and the feed used. Figure K-117 is an exaggerated view of the ridges left by the cutter. The finish can be improved by the use of an insert with a wiper flat (Figure K-118).

For finishing operations use one wiper insert, the rest being regular inserts in the cutter body. The maximum feed rate should not exceed two-thirds of the width of the wiper flat. As an example

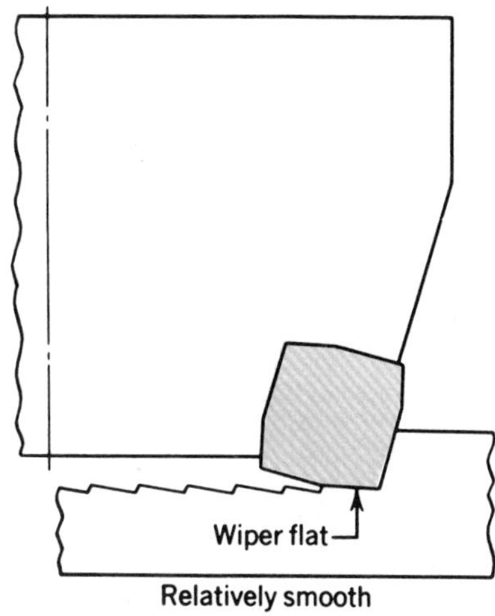

FIGURE K-118 Effect of wiper flat (Kennametal, Inc., Latrobe, PA).

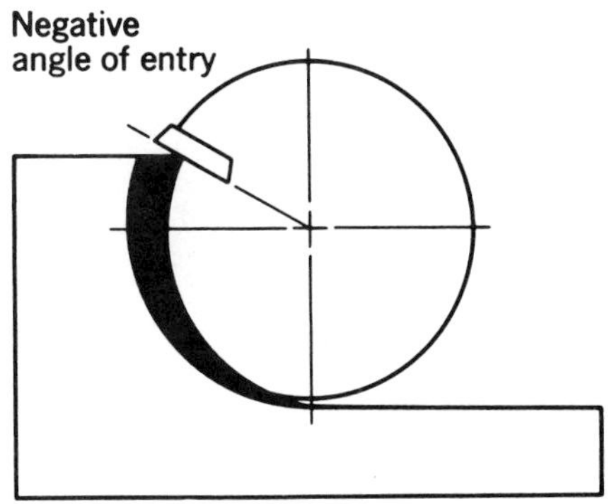

FIGURE K-119 Work contacts tool away from cutting tool tip (Kennametal, Inc., Latrobe, PA).

let's use a wiper insert with a .300-in.-wide flat and the cutter revolving at 200 rpm. Two-thirds of .300 in. equals .200 in. per spindle revolution. Multiply .200 in. times 200 rpm and we get 40 ipm. The wiper insert also extends axially .004 in. beyond the other inserts—this makes the wiper insert the finishing insert.

The cutting action of face mills is also affected by its diameter in relation to the width of the workpiece. When it is practical, a cutter should be chosen that is larger than the width of cut. A good ratio is obtained when the cut is two-thirds as wide as the cutter diameter. Cuts as wide as the cutter diameter should not be taken, because of the high friction and rubbing of the cutting edge before the material starts to shear. This rubbing action results in rapid cutting edge wear. Tool life can be extended when the cutting tool enters the workpiece at a negative angle (Figure K-119). The work makes its initial tool contact away from the cutting tip at a point where the insert is stronger. When possible, arrange the width of cut to obtain the cutting action in Figure K-119. Figure K-120 illustrates a workpiece where the cutter enters at a

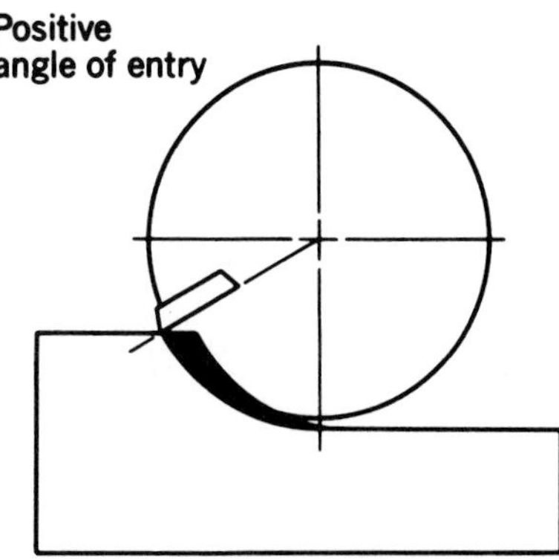

FIGURE K-120 Work contacts tool on the cutting tool tip (Kennametal, Inc., Latrobe, PA).

positive angle. Contact takes place at the weakest point of the insert, and the cutting edge may chip. Figures K-119 and K-120 show a climb milling operation. In climb milling, the cutter enters the work material and produces the thick part of the chip first, and then thins out toward the finished surface. Climb milling should be performed whenever possible because it reduces the tool wear caused in conventional milling by the rubbing action of the tool against the work before the actual cutting starts.

Cutting fluid is important when a face mill with high speed steel inserts is used. A good cutting fluid absorbs heat generated in cutting and provides a lubricant between the chips and the cutting tool. In face milling, cutting fluid should be applied where the cutting is being done, and it should flood the cutting area. Cooling a carbide face mill is very difficult because of the intermittent nature of the cutting process and the fanning action of the revolving cutter. If the cooling of the carbide insert is intermittent, thermal cracking, which causes rapid tool failure, takes place. Unless a sufficient coolant flow at the cutting edge is maintained, it is better to machine dry. A mist-type coolant application is sometimes used to provide lubrication between the chip and cutting tool. This reduces friction and generates less heat.

To machine a flat surface with a face mill, it is necessary that the face of the cutter and the travel of the table are exactly parallel to each other. If there is any looseness in the table movements or cutter run-out, it will appear either as back drag or as a concave surface. Back drag appears as light cuts taken behind the main cut by the trailing cutting edges of the cutter. A concave surface is generated when the cutter is tilted in relation to the table so that the trailing cutting edges are not in contact with the work surface. A slight tilt, where the trailing cutting edges are .001 to .002 in. higher than the leading cutting edges, gives satisfactory surfaces in most applications.

FIGURE K-121 **Handle tools with care—they are sharp (Cincinnati Milacron).**

FIGURE K-122 **Carbide face mill being used (Lane Community College).**

Effective face milling operations are the result of many often small factors. Rigidity of the setup is of main concern. The workpiece must be supported where the cutting takes place. Rigid stops prevent the workpiece from being moved by the cutting pressures. The table should be as close to the spindle as possible. The gibs on the machine slides need to be adjusted to prevent looseness. Locking devices should be used on machine slides. Coarse pitch cutters usually have an uneven spacing of the teeth, this helps to eliminate or reduce vibrations. Efficient machining requires sharp cutting edges. Sharp cutting edges are dangerous to the operator; handle cutters with a rag (Figure K-121). Dull cutting edges create excessive heat and leave a poor surface finish. Worn tools produce higher cutting pressures, which may deflect the workpiece. They also use additional power. When dull tools are used on work hardening materials, both the tool and workpiece may be ruined. The speed of the cutter affects the surface finish of the workpiece. Too low a speed leaves a poor surface because of the buildup on the cutting edge. Too high a speed results in excessive tool wear.

When using a carbide face mill, the color of the chips produced is an indicator of the machine conditions. When milling steel, chips should appear a light to medium blue on the outside chip surface. A straw to light brown color indicates that the surface speed is too low. A deep blue to black color indicates that the surface speed is too high. If the feed is too low, the rubbing action wears down the cutting edge. If the feed is too high, the cutting edges will chip or break. Milling cutters often have very high power requirements. It is good practice to calculate hp needs before starting a cutting operation. Unit 5 of this section shows how these calculations are done. Figure K-122 shows a carbide face mill in use. Metal removal is very efficient and rapid.

Figure K-123 is a shell end mill with inserted carbide teeth. This mill is held in the machine spindle with a standard milling machine shell mill arbor. When changing inserts it is very important that the seat and pocket for the insert be perfectly clean and not damaged. The most common reason for a bad surface finish is improper mounting of the inserts.

FIGURE K-123 **Inserted carbide tooth shell mill (Lane Community College).**

SELF-TEST

1. How are face mills mounted on a milling machine?

2. What is a shell end mill?

3. What is the difference between light-duty and heavy-duty face mills?

4. Give some reasons for using positive rake angles on face mills.

5. Give some reasons for using negative rake angles on face mills.

6. What is the effect of a large lead angle over a small one?

7. Why should the width of cut in face milling be narrower than the face mill diameter?

8. Should cutting fluid be used in face milling?

9. What causes a concave work surface in face milling?

10. Give four factors that help in efficient face milling.

SECTION L

ROTARY TABLES AND INDEXING DEVICES

Since the circle forms the basic geometric shape for so many mechanical devices, the ability to divide a circle circumference into almost any number of divisions is extremely useful. Furthermore, the ability to machine a circular feature on a workpiece has contributed greatly to the development of many useful devices.

The group of tools known as rotary tables or circular milling attachments divides the circle in angular terms—that is, by degrees and fractions of degrees. The group of tools known as indexing or dividing heads divides the circle in terms of the number of circumference divisions desired. Naturally, angles and numbers of circumference divisions can be equated to each other and the tools may often be applied interchangeably depending on the specific machining task to be done. However, a tool from one group or the other may be better suited to a specific application. Part of your job is to be able to select the one that best fits the need. The purpose of this section is to identify these useful accessories, describe their applications, aid you in selecting the one that best fits the job, and familiarize you with calculations relative to indexing requirements.

Rotary Tables

The rotary table is an extremely useful accessory for the milling machine. A rotary table consists of a precision worm and wheel unit that can be attached to the milling machine table (Figure L-1). One common application of the rotary table is machining a circular feature on the workpiece (Figure

FIGURE L-1 Circular milling table (Cincinnati Milacron).

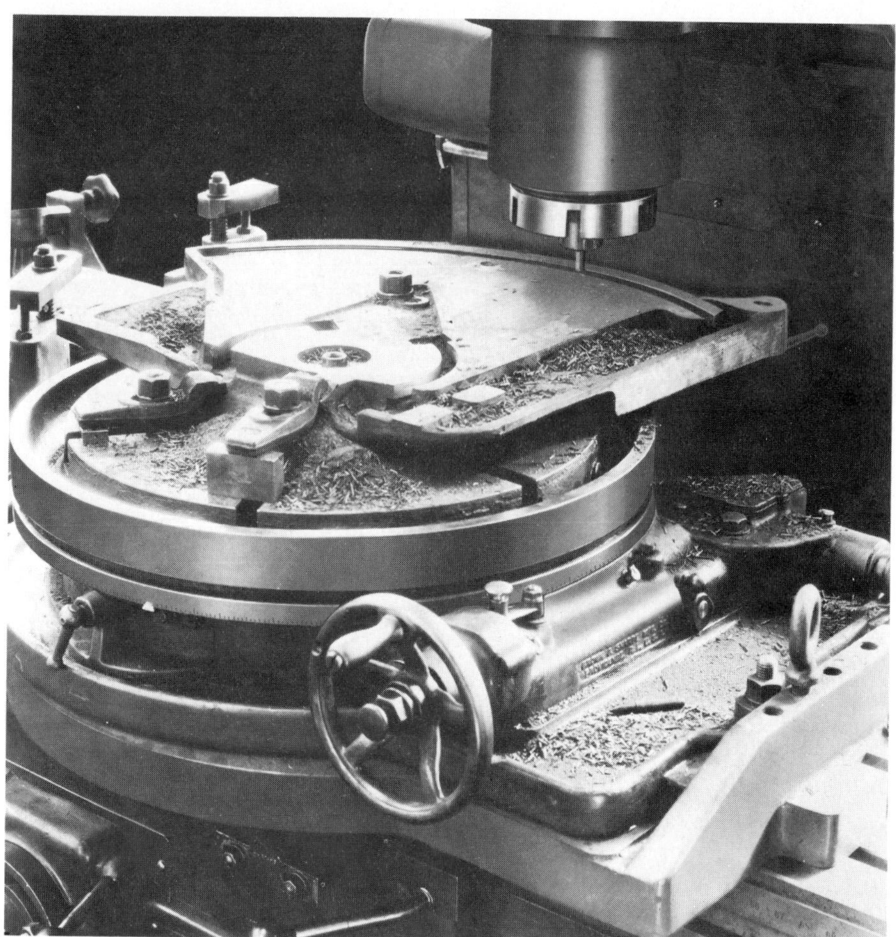

FIGURE L-2 Circular milling attachment being used to mill a circular T-slot (Cincinnati Milacron).

FIGURE L-3 A circular milling table with power feed is termed a circular milling attachment (Cincinnati Milacron).

FIGURE L-4 Circular milling attachment equipped with an indexing attachment (Cincinnati Milacron).

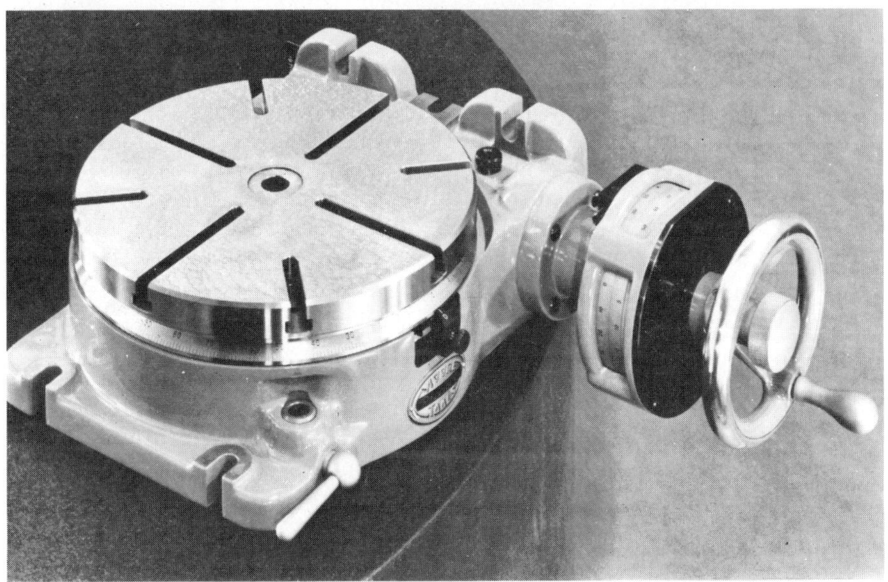

FIGURE L-5 Ultraprecise rotary table (Moore Special Tool Co., Inc., Bridgeport, Conn.).

L-2). Rotary tables may be power driven by mechanical connection to the mill table feed mechanism (Figure L-3). When using a rotary table for the more precise requirements of an indexing operation, as in cutting a gear, it should be equipped with an indexing attachment (Figure L-4). The typical rotary table is graduated in degrees and fractions. Discriminations range from 1 min of arc down to fractions of seconds (Figure L-5). An optical microscope may be used to further increase discrimination (Figure L-6). Rotary tables may be designed so they can be mounted on a machine tool in either a vertical or a horizontal position (Figure L-7). Inspection (Figure L-8) and calibration make use of high-resolution rotary tables discriminating to $\frac{1}{10}$ sec of arc (Figure L-9).

Indexing Devices

Indexing devices divide the circle in terms of the number of circumference spacings required. Indexing devices may also be called indexing heads or dividing heads. One of the simplest types is the

FIGURE L-6 Optical dividing table (American SIP Corporation).

FIGURE L-7 Using the rotary table on a jig boring machine (Moore Special Tool Co., Inc., Bridgeport, Conn.).

FIGURE L-8 Rotary table with tailstock and adjustable center being used for inspection (Moore Special Tool Co., Inc., Bridgeport, Conn.).

FIGURE L-9 Precision index—a master tool (Moore Special Tool Co.).

FIGURE L-11 Direct indexing head (Cincinnati Milacron).

FIGURE L-10 Collet index fixture (Hardinge Brothers, Inc.).

FIGURE L-12 Gear cutting attachment (Cincinnati Milacron).

collet index fixture (Figure L-10). This tool can be mounted horizontally or vertically and is suitable for indexing low numbers or divisions (2, 4, 6, 8, 10, 12, etc.). The workpiece is held in a split collet.

Many types of horizontal and universal indexing heads have been developed. The indexing capability is accomplished by making use of an index plate. An index plate consists of several circles of equally spaced holes into which the index crank pin can be engaged. Although the hole circles in the index plate are equally spaced, the number of holes varies in different circles, permitting many different numbers of circumference divisions to be obtained. Sets of index plates containing a variety of hole circles are also available, further increasing the range of circumference divisions possible.

The direct index head (Figure L-11) has the spindle connected directly to the index crank. The index plate is located on the back of the head. Indexing is accomplished by first setting the crank pin to the desired circle of holes, then moving the crank from hole to hole to achieve indexing of the workpiece. Direct index heads are limited to the maximum number of holes in the index plates, usually 50.

By adding a 40:1 worm and wheel to the index head, the ratio of crank turns to spindle turns can be greatly increased, thus permitting the circle circumference to be divided into many more parts. This type of plain index head is regularly used in spur gear cutting (Figure L-12).

The universal indexing head is much like the plain index head with the added feature that the spindle may be swiveled out of the horizontal. The workpiece may still be indexed, but milling can now be done on an angled surface (Figure L-13). Circular graduating on an angled surface is another operation that can be done with a universal index head (Figure L-14).

The wide range indexing head (Figure L-15) uses the standard 40:1 ratio plus an additional 100:1 ratio. The ratios are multiplied, resulting in a total of 4000:1. Divisions of 2 to 400,000 can be obtained with this tool.

Indexing heads may be applied in special inspection requirements (Figure L-16). For example, a cam lobe may be inspected by rotating the workpiece indexed according to the lobe positions.

INDEX HEADS AND HELICAL MILLING PROCESSES

Helical milling is a process where the workpiece is rotated at the same time that it is being fed into the cutter. The result is a helical cut such as the

FIGURE L-13 Milling a bevel gear with a universal index head (Cincinnati Milacron).

FIGURE L-15 Dividing head with wide range divider (Cincinnati Milacron).

FIGURE L-14 Dividing heads are also useful for making graduations on a conical surface (Cincinnati Milacron).

flute of a drill, helix of a milling cutter, or the helical teeth of a gear. By varying the rate of rotation relative to the distance the workpiece travels longitudinally, the amount of helical twist or lead can be changed depending on the machining requirements. For example, a drill would have a relatively long or slow helix whereas a high helix milling cutter would have a short or fast helix. You will probably not encounter a great deal of helical milling in routine machine shop work unless you happen to become involved with the more specialized manufacturing making use of this process. However, the process is quite interesting and the machine setup is probably one of the more involved processes done by the general machinist.

To accomplish helical milling, the spindle of the index head (Figure L-17) is coupled to the mill table screw. This will cause the index head spindle to rotate as the machine table is moved back and forth. The indexing capability of the index head permits the manufacture of a helical gear on the universal milling machine. However, it should be noted that milling of helical gears on a universal mill is not that common a process since the end result may not meet the exact specifications required on precision gears. Helical gear cutting for manufacturing purposes is done on sophisticated gear generators discussed in later sections.

Helical leads are obtained through the gear train connecting the table screw and index head spindle (Figure L-18). The short and long lead attachment (Figure L-19) permits the selection of over 13,000 leads ranging from .010 to 3000 in. Various accessories are combined to accomplish complex helical milling operations (Figure L-20).

Gear hobbing is another job that can be done by synchronizing an indexing device with the machine spindle (Figure L-21).

FIGURE L-16 Inspecting a cam using a dividing head (Cincinnati Milacron).

FIGURE L-17 Helical milling head (Cincinnati Milacron).

FIGURE L-18 Standard universal dividing head driving mechanism (Cincinnati Milacron).

FIGURE L-19 Short and long lead driving mechanism (Cincinnati Milacron).

FIGURE L-20 Combination of attachments for the milling of a leadscrew (Cincinnati Milacron).

FIGURE L-21 Special hobbing attachment for the horizontal milling machine (Cincinnati Milacron).

UNIT 1

Setup and Operation of Indexing Heads and Rotary Tables

The indexing head and rotary table are versatile and useful milling machine accessories. They serve to rotate the workpiece a full or partial turn for the purpose of indexing a specific number of divisions, an angle, or to machine a circular feature. The purpose of this unit is to discuss the operation and setup of common indexing heads and rotary tables.

Objectives

After completing this unit, you should be able to:

1. Identify the major parts of indexing heads and rotary tables.
2. Set up these tools on the milling machine.

Indexing Heads

Probably the most often used index head, also called dividing head, in the machine shop is either the plain or universal type. These tools consist of a precision worm and wheel unit housed in appropriate bearings (Figure L-22). The index head worm engages the spindle worm wheel and is turned by operating the index crank. The crank is geared to the worm usually in the ratio of 40:1. This means that 40 turns of the index crank are required to rotate the spindle one revolution.

On most indexing heads, the spindle can be disengaged from the crank worm, permitting the spindle to be turned by hand. This facilitates setup of the tool and also permits direct indexing. Direct indexing may be accomplished by engaging the direct index plunger and pin in holes on the spindle nose (Figure L-23). Smaller numbers of division can be indexed in this manner.

The index head spindle has a tapered bore that will accept tapered shank tooling. Some index head spindles are threaded to accept screw-on chucks. A lock is provided so that the spindle may be secured at a particular setting in order to increase rigidity of a setup.

The index crank contains a pin on a spring-loaded plunger that engages the holes in the index plate. Hole circles in the index plate provide the indexing function. The pattern on the plate consists of a number of equally spaced hole circles, and the crank pin is adjusted in and out so that the desired circle may be used. Some plates have additional circles of holes on the reverse side.

Two sector arms rotate about the crank hub and can be adjusted to indicate the correct number of holes for a partial turn of the crank. For example, if you are indexing one turn plus five holes for each division, the sector arms can be set so that you will not have to count the five extra holes each time around.

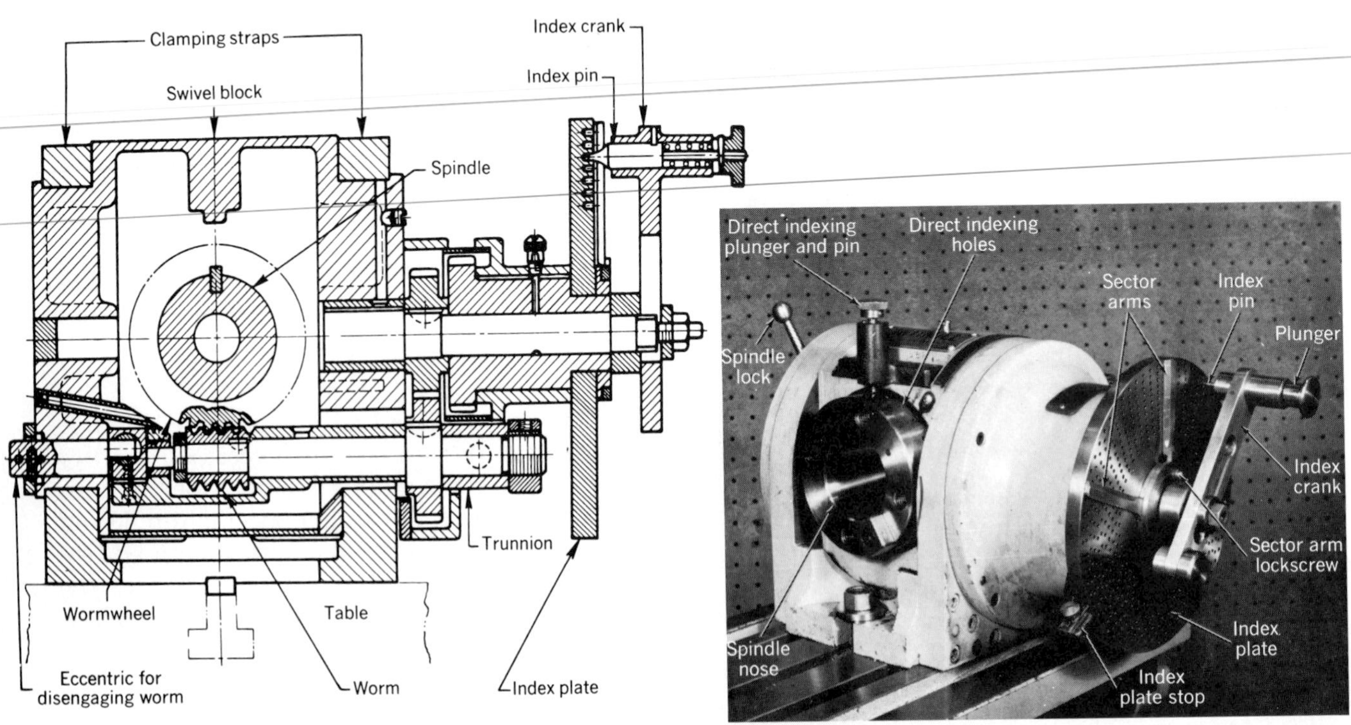

FIGURE L-22 Section through dividing head showing worm and worm shaft (Cincinnati Milacron).

FIGURE L-23 Indexing components of a dividing head (Lane Community College).

SETTING UP THE INDEX HEAD

The index head is secured to the mill table in the same manner as any other machine accessory. Many heads have alignment keys that fit the table T-slots. If the head has no keys, it must be aligned with the table using the dial indicator alignment techniques discussed in previous units. Furthermore, the universal indexing head must be aligned in the horizontal if it is used for parallel to table machining setups.

The ends of long workpieces or gear arbors are supported by a footstock (Figure L-24). Similar in function to the lathe tailstock, the footstock center can be adjusted in and out as well as up and down. Long or slender workpieces may be further supported by an adjustable center rest (Figure L-25).

Rotary Tables

The common rotary table is also a precision worm and wheel unit. Many rotary table setups use the tool in its horizontal position (Figure L-26). However, some rotary tables are designed to be used in both vertical and horizontal positions.

Rotary tables position the workpiece by degrees of arc, but may be used much like the index head with the addition of an indexing attachment. The rotary table is equipped with T-slots so that the workpiece may be secured.

Discriminations on rotary tables vary from 1 minute to fractions of seconds. Full degrees are graduated on the circumference of the table. Fractions of degrees are read on the worm crank scale, usually with the aid of a vernier. Ratios of common rotary tables are 40:1, 80:1, and 120:1. A lock is provided to increase the rigidity of the setup during machining.

SETTING UP AND USING ROTARY TABLES

If the rotary table is used to machine features equidistant from center, the tool must be accurately positioned under the machine spindle. This is done by the same techniques discussed in previous units. Use a dial test indicator and pick up center from the hole in the center of the rotary table. Once the rotary table has been centered under the spindle in both table and saddle axes, it may be moved off center in either axis an amount equal to the pitch circle radius of the features to be machined. It may also be necessary to center the workpiece on the rotary table so that concentric features may be obtained during a machining operation.

FIGURE L-25 Adjustable center rest (Lane Community College).

FIGURE L-24 Footstock (Lane Community College).

FIGURE L-26 Rotary table (Lane Community College).

SELF-TEST

1. When are indexing devices used?
2. What makes indexing devices so accurate?
3. When is the worm disengaged from the worm wheel?
4. When is the hole circle on the spindle nose used?
5. What is a commonly used index ratio on dividing heads?
6. Why does the index plate have a number of different hole circles?
7. What is the purpose of the sector arms?
8. What does the spindle lock do?
9. How can divisions be made that are not possible with a standard index plate?
10. Why should the index crank be rotated in one direction only while indexing?

UNIT 2

Direct and Simple Indexing

Most indexing for routine machining operations can be done by the direct and simple methods. Accuracy and the number of divisions required determine the method to use. Common machining operations that involve indexing include spline and gear cutting, keyseats, hexagons, octagons, squares, and other geometric shapes. The purpose of this unit is to familiarize you with the calculations required for indexing tasks.

Objective

After completing this unit, you should be able to:

Calculate full and fractional turns of the index head spindle to index the workpiece a required number of equal divisions.

Direct Indexing

Direct indexing is the simplest method of dividing a circle into a required number of equally spaced divisions. The number of divisions that can be obtained by the direct method is limited by the number of holes in the index plate or in the spindle nose. Direct index head plates contain 24-, 30-, and 36-hole circles. The following formula may be used to make the index calculation:

$$\text{number of holes per division} = \frac{\text{number of holes in the indexing circle}}{\text{required number of divisions}}$$

The result of this calculation must be a whole number.

EXAMPLES

The available hole circles for direct indexing may be evenly divided in the following manner.

24 divided by:	30 divided by:	36 divided by:
2 = 12 divisions	2 = 15 divisions	2 = 18 divisions
3 = 8	3 = 10	3 = 12
4 = 6	5 = 6	4 = 9
6 = 4		
8 = 3	6 = 5	6 = 6
12 = 2	10 = 3	9 = 4
24 = 1	15 = 2	12 = 3
	30 = 1	18 = 2
		36 = 1

Calculate direct indexing for 6 divisions:

$$\frac{24}{6} = 4 \text{ holes per division}$$

The 30- and 36-hole circles could also be used for 6 divisions. However, only the 30-hole circle could have been used for 5 divisions and only the 36-hole circle for 9 divisions.

When direct indexing, count the required number of new holes to the next division. Do not count the hole in which the index pin is engaged. After engaging the pin, always lock the spindle to increase rigidity of the setup. To avoid an indexing error, you may want to mark the holes that are to be used for a given indexing operation with a felt pen or chalk.

Direct indexing may be done on a plain index head by disengaging the worm from the spindle so that the spindles may be turned by hand. The hole circles in the spindle nose are used for the indexing circles.

Simple Indexing

Simple indexing, also known as plain indexing, makes use of the gear ratio between crank and spindle. On most common index heads this ratio is 40:1. Forty turns of the index crank are required to rotate the spindle one revolution.

Required numbers of circle divisions involving less than one full turn of the spindle now become fractions of 40 and the indexing calculation is made by applying the formula:

$$\text{index crank turns} = \frac{40}{N}$$

where N is the required number of divisions. Index plates with the following hole circles are available: 24, 25, 28, 30, 34, 37, 38, 39, 41, 42, 43, 46, 47, 49, 51, 53, 54, 57, 58, 59, 62, 66.

EXAMPLE

Calculate indexing for 20 divisions:

$$\frac{40}{N} = \frac{40}{20} = 2 \text{ full turns of the index crank}$$

In this case any index plate may be used since no partial turn is required. However, to calculate indexing for 52 divisions:

$$\frac{40}{N} = \frac{40}{52}$$

This means that less than one full turn is required for each division, 40/52 of a turn to be exact. A 52-hole circle is not available, so a proportional fraction must be generated. This can be done by the following procedure. Reduce the index fraction to its lowest terms:

$$\frac{40}{52} = \frac{10}{13}$$

Take the denominator of lowest terms, 13, and determine into which of the available hole circles it can be evenly divided. In this case, 13 may be divided into the available 39-hole circle exactly 3 times. Use this result 3 as a multiplier to generate the proportional fraction required.

$$\frac{10 \times 3}{13 \times 3} = \frac{30}{39}$$

Therefore, 30 holes on a 39-hole circle is the correct indexing for 52 divisions.

EXAMPLE

Calculate indexing for 27 divisions:

$$\frac{40}{N} = \frac{40}{27} = 1\,\frac{13}{27}$$

or 1 full turn plus $\frac{13}{27}$ fractional turn. If a 27-hole circle were available, then 13 holes of the 27 would give the fractional turn required. However, a 27-hole circle is not available so a proportional fraction must be generated containing a denominator corresponding to an available hole circle.

The fraction $\frac{13}{27}$ is already at lowest terms, so take the denominator and determine which of the available hole circles it will divide into evenly. When divided into the 54-hole circle, 27 goes exactly 2 times. This result becomes the multiplier used to generate the proportional fraction

$$\frac{13 \times 2}{27 \times 2} = \frac{26}{54}$$

Therefore, 26 holes on the 54-hole circle will give the correct partial turn fraction. Indexing will then be one full turn plus 26 of the 54 holes for each of the required divisions.

EXAMPLE

Calculate indexing for 35 divisions:

$$\frac{40}{N} = \frac{40}{35} = 1\,\frac{5}{35}$$

or 1 full turn plus $\frac{5}{35}$ fractional turn. Reducing $\frac{5}{35}$ to lowest terms equals $\frac{1}{7}$. The lowest-term denominator, 7, divides evenly into

28-hole circle (4 times)
42-hole circle (6 times) } multipliers
49-hole circle (7 times)

Therefore, any of these indexing circles may be used by generating the proportional fractions in the following manner:

$$\frac{1 \times 4}{7 \times 4} = \frac{4}{28} \quad \text{(4 holes on 28-hole circle)}$$

$$\frac{1 \times 6}{7 \times 6} = \frac{6}{42} \quad \text{(6 holes on 42-hole circle)}$$

$$\frac{1 \times 7}{7 \times 7} = \frac{7}{49} \quad \text{(7 holes on 49-hole circle)}$$

EXAMPLE

Calculate indexing for 51 divisions:

$$\frac{40}{N} = \frac{40}{51}$$

Since a 51-hole circle is available, 40 holes on a 51-hole circle will provide the correct indexing.

USING THE INDEX HEAD

It would be very time consuming to have to count numbers of holes for fractional turns each time around. To facilitate and to reduce possible errors in indexing, the sector arms are used. The sector arms rotate about the index crank hub and are locked in place by a lockscrew. The facing edges of the arms are beveled (Figure L-27) and the number of holes required for fractional turns is established between the beveled edges. Remember that the number of holes required for a fractional turn will be new holes. The hole where the crank pin is engaged is not counted.

To operate the index head, unlock the spindle and pull back the crank pin plunger. Rotate the crank the required number of turns plus fractions. Set the sector arms to keep track of additional holes beyond one full turn of the crank. Bring the crank pin around until it just drops into the required hole. If you overshoot the the position, back off well past the required hole and come up to it once again. This will eliminate any backlash in the worm and wheel. Always index only in one direction and always lock the spindle before beginning the machining operation.

After each indexing move, the sector arms must be rotated so that they will be properly positioned for the next move. For example, if you are indexing one full turn plus 11 holes, set the sector arms to include 12 holes between the beveled faces. Turn the index crank one full turn plus the 11 additional holes. Then, shift the sector arms around so that 11 new holes will be available for the next move.

FIGURE L-27 Dividing head sector arms set for indexing 11 spaces (Lane Community College).

Compound and Differential Indexing

To obtain indexing of divisions that cannot be done on standard index plates, high number plates or the wide range indexing head may be used. Nonstandard numbers of divisions can also be obtained by the processes of differential or compound indexing. This is accomplished by rotating the index plate backward or forward in order to increase the combinations of hole circles and obtain the required number of indexing divisions. Consult a machinist's handbook for more information regarding compound or differential indexing.

SELF-TEST

1. What is the difference between direct and simple indexing?
2. How can you avoid using a wrong hole in the index plate while direct indexing?
3. If the direct indexing hole circle has 24 holes, what are the different divisions you can make with it?
4. How are the sector arms used on a dividing head?
5. If both the 40- and the 60-hole circle can be used, which would be the better one to use?
6. Calculate how to index for 6 divisions. Use the hole circles given in the text.
7. Calculate how to index for 15 divisions.
8. Calculate how to index for 25 divisions.
9. Calculate how to index for 47 divisions.
10. Calculate how to index for 64 divisions.

UNIT 3

Angular Indexing

Circumference divisions on a circle can be defined in terms of specific numbers of equal spaces, or they may be defined in angular terms—that is, degrees or minutes of arc. Hole spacings in index plates can then become equivalent to angular measurement. The purpose of this unit is to define indexing plate hole circles in angular measure and familiarize you with the calculations for angular indexing.

Objective

After completing this unit you should be able to:

Perform calculations required for angular indexing.

Direct Angular Indexing

Since there are 360 degrees in a circle, one complete revolution of the dividing head spindle is 360 degrees. If the direct indexing plate has 24 holes, the angular spacing between each hole is 360/24 = 15 degrees. This allows us to make any division that requires 15 degree intervals or multiples of 15 degrees.

EXAMPLE

Calculate the indexing required to drill 2 holes at an angle of 75 degrees to each other:

$$\frac{75 \text{ deg}}{15 \text{ deg/hole}} = 5 \text{ holes in the index plate}$$

Simple Angular Indexing

On the simple index head making use of a 40:1 ratio, 40 turns of the index crank are required to turn the spindle 360 degrees or one full revolution. Therefore, one turn of the index crank is equal to

$$\frac{360 \text{ deg}}{40 \text{ turns}} = 9 \text{ deg per turn}$$

Any index plate circle that is divisible by nine can be used for angular indexing.

EXAMPLE

$$\frac{27\text{-hole circle}}{9} = 3 \text{ holes per deg}$$

$$(1 \text{ hole} = \tfrac{1}{3} \text{ deg or } 20 \text{ min})$$

$$\frac{36\text{-hole circle}}{9} = 4 \text{ holes per deg}$$

$$(1 \text{ hole} = \tfrac{1}{4} \text{ deg or } 15 \text{ min})$$

$$\frac{45\text{-hole circle}}{9} = 5 \text{ holes per deg}$$

$$(1 \text{ hole} = \tfrac{1}{5} \text{ deg or } 12 \text{ min})$$

$$\frac{54\text{-hole circle}}{9} = 6 \text{ holes per deg}$$

$$(1 \text{ hole} = \tfrac{1}{6} \text{ deg or } 10 \text{ min})$$

The index crank turns required for an angle can be calculated by the formula

$$\text{turns} = \frac{\text{degrees required}}{9 \text{ deg/turn}}$$

EXAMPLE

Calculate indexing for 37 degrees:

$$\frac{37}{9} = 4\frac{1}{9}$$

or 4 full turns plus $\frac{1}{9}$ additional turn. The $\frac{1}{9}$-partial turn is found by the method described in the preceding unit. Determine which of the available index circles the lowest term denominator 9 will divide into evenly.

$$\frac{\text{54-hole circle}}{9} = 6 \text{ holes}$$

Therefore, indexing is 4 full turns plus 6 holes on a 54-hole circle.

ANGULAR INDEXING IN MINUTES

Since one full turn of the index crank in a 40:1 ratio head is equal to 9 degrees, it is also equal to 9×60 or 540 minutes. Therefore, to index by minutes, apply the formula:

$$\text{crank turns} = \frac{\text{minutes required}}{\text{540 min/turn}}$$

EXAMPLES

Calculate indexing for 8 deg, 50 min. Convert angle to minutes:

$$\text{8 deg 50 min} = \text{530 min}$$

Apply the formula

$$\text{crank turns} = \frac{\text{minutes required}}{\text{540 min/turn}} = \frac{530}{540}$$

Reduce the fraction to lowest terms:

$$\frac{530}{540} = \frac{53}{54}$$

Take the lowest term denominator 54, and determine the available hole circle that it will divide into evenly. Since a 54-hole circle is available, 53 holes on a 54-hole circle will index 8 degrees and 53 minutes.

UNEVEN-MINUTE CALCULATIONS In some cases, it may not be possible to determine an exact partial turn on the index crank.

EXAMPLE

Calculate indexing for 1 deg and 35 min. Convert to minutes:

$$\text{1 deg 35 min} = \text{95 min}$$

Apply the formula:

$$\frac{\text{minutes required}}{\text{540 min/turn}} = \frac{95}{540}$$

Reduce fraction to lowest terms:

$$\frac{95}{540} = \frac{19}{108}$$

If a 108-hole circle were available, then 19 holes would index the required 95 minutes. However, 108 is not available so a proportional fraction must be generated containing a denominator equal to an available hole circle.

The method of creating a proportional fraction is not apparently obvious. The task is to determine which of the available hole circles will be closest to the requirements. A proportion may be established and evaluated for various hole circles:

$$\frac{19}{108} = \frac{X \text{ holes}}{\text{available circles}}$$

Evaluating this expression for a 51-hole circle, the result is

$$\frac{19}{108} = \frac{X}{51} \qquad X = \frac{19 \times 51}{108} = 8.972$$

This means that 8.972 holes on the available 51-hole circle will result in proper indexing. Obviously, it is not possible to have .972 hole, so the number may be rounded to the nearest full hole. In this case, 9 would be the closest. Since 9 holes on the 51-hole circle would be $\frac{9}{51}$ of 540 min, the actual indexed angle would be 95.294 min or 95′17″. The error is quite small. For more accuracy in angular indexing, it would be better to use the wide range indexing head.

SELF-TEST

1. If there are 24 holes in the direct indexing plate, how many degrees are between holes?

2. How many holes movement is necessary to index 45 degrees using the direct indexing method?

3. How many degrees are there in the movement produced by one complete turn of the index crank?

4. Which hole circles on the index plate can be used to index by whole degrees?

5. How many turns of the index crank are necessary to index 17 degrees?

6. What fraction of 1 degree is represented by one space on the 18-hole circle?

7. What fraction of 1 degree is represented by one space on the 36-hole circle?

8. What fraction of 1 degree is represented by one space on the 54-hole circle?

9. How many minutes movement is produced by one turn of the index crank?

10. How many turns of the index crank are necessary to index 54 deg 30 min using the 54-hole circle?

SECTION M

GEARS AND GEAR CUTTING

It would be difficult to envision a world without gears. Historically, the evolution of gears from pin cogs to the modern involute type is an interesting development in mechanical technology.

■ ■ ■

Purpose of Gears

Gears provide positive (nonslip) power transmission. They are also used to increase torque (turning effort) in many kinds of mechanical devices. Gears are used to increase and decrease speed in geared speeders and speed reducers. Mechanical timing requirements make extensive use of gears; they are widely used to establish set speed ratios between shafts.

Two or more gears running in mesh form a gear train. If the gears in a train are of different sizes, the smaller gear is called the pinion. Any two gears in mesh will always turn in opposite directions. To achieve same direction rotation, a third gear must be placed in mesh between two gears, or internal gears may be used.

Gears are often used to connect shafts at relatively short center distances. However, large marine reduction gear units may connect shafts that are several feet apart and are capable of transmitting many thousands of horsepower.

Gear Manufacturing Processes

Originally gears were milled with form cutters. This led to the involute geometry of modern gears and the method of gear cutting known as hobbing (Figures M-1 and M-2). Gear hobbers are a type of

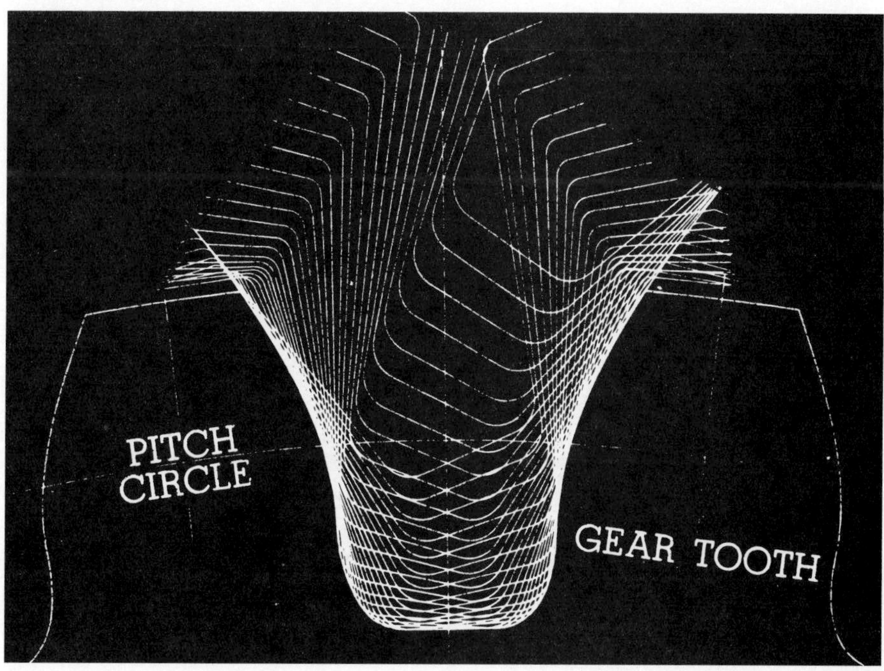

FIGURE M-1 Schematic view of the generating action of a hob (Barber-Colman Company).

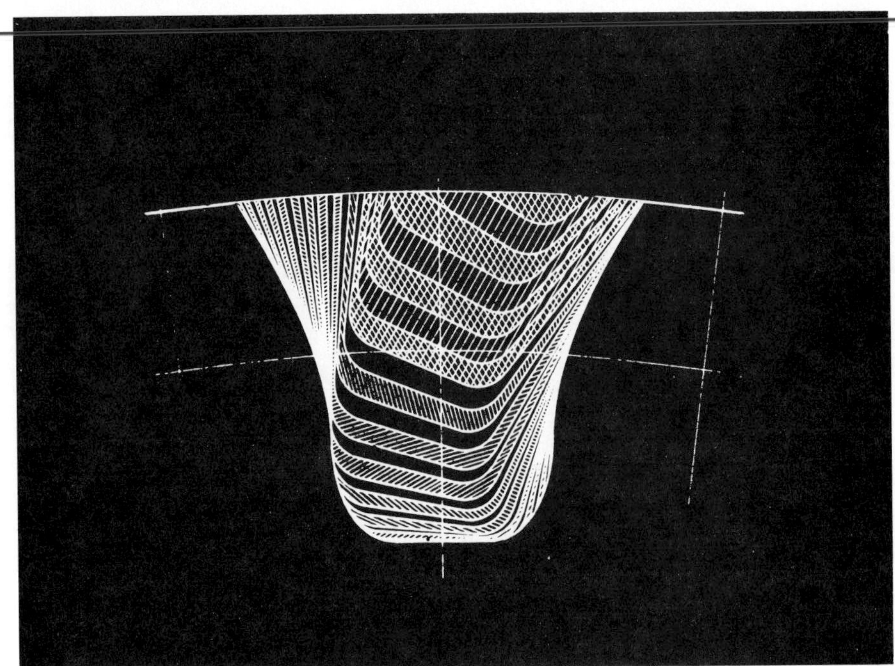

FIGURE M-2 Schematic view of the performance of an individual hob tooth (Barber-Colman Company).

FIGURE M-3 Small horizontal spindle hobbing machine.

milling machine designed with both horizontal (Figure M-3) and vertical spindles (Figure M-4).

Another important type of gear cutting machine is the gear shaper (Figure M-5). The shaper can make external, internal, spur, and helical gears. It can also form a gear next to a shoulder or next to another gear of the type found in many transmission gear clusters. Automotive differential gears that must be accurate, quiet, and long running are made on a hypoid gear generator (Figure M-6). The gear shaper will also form gears of very unusual geometry (Figure M-7).

As the need for high production of accurate gears increased, methods such as gear broaching were developed. Broaching is well suited to producing individual and nonclustered gears with straight teeth, or helical gears with low helix angles (Figure M-8).

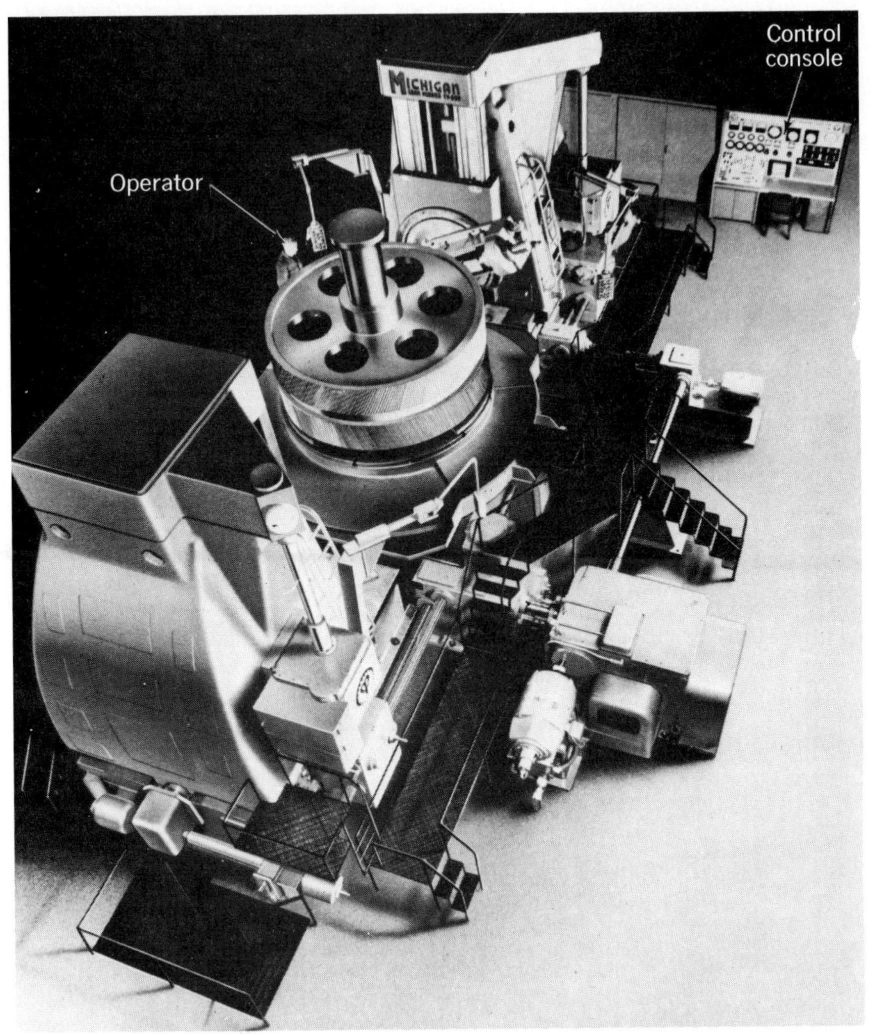

FIGURE M-4 Huge vertical spindle gear hobbing machine. Notice the operator with the pendant control at the upper left (Ex-Cell-O Corporation).

FIGURE M-5 Gear shaper. The cutter moves up and down with the same twist as the helix angle of the gear (Fellows Corporation).

FIGURE M-6 Hypoid gear generator (Gleason Works).

FIGURE M-7 The gear shaper can produce unusual gear shapes (Fellows Corporation).

FIGURE M-8 Pot broaching tool that produces a helical gear (National Broach and Machine Division, Lear Siegler, Inc.).

FIGURE M-9 Gear shaving (National Broach and Machine Division, Lear Siegler, Inc.).

Gear Finishing Processes

For commercial use, gear teeth are usually finished after they are cut and the gear has been heat-treated. The finishing process depends on the required accuracy of the gear, hardness, noise limitations, and where the gear will be used.

To finish gear teeth that are not harder than Rockwell C 30, the shaving process may be used. Shaving can result in surface finishes from 32 to 16 microinches. One shaving method uses a shaving cutter that is meshed with the gear to be finished. The shaving roll and gear are set at an angle (Figure M-9). This process is called diagonal shaving.

Roll finishing is used on unhardened gears (Figure M-10). This is a cold forming process that produces surface finishes to 10 microinches. Hardened gears may also be finished by form grinding (Figures M-11 and M-12). Finishes of 30 to 10 microinches can be obtained by this process.

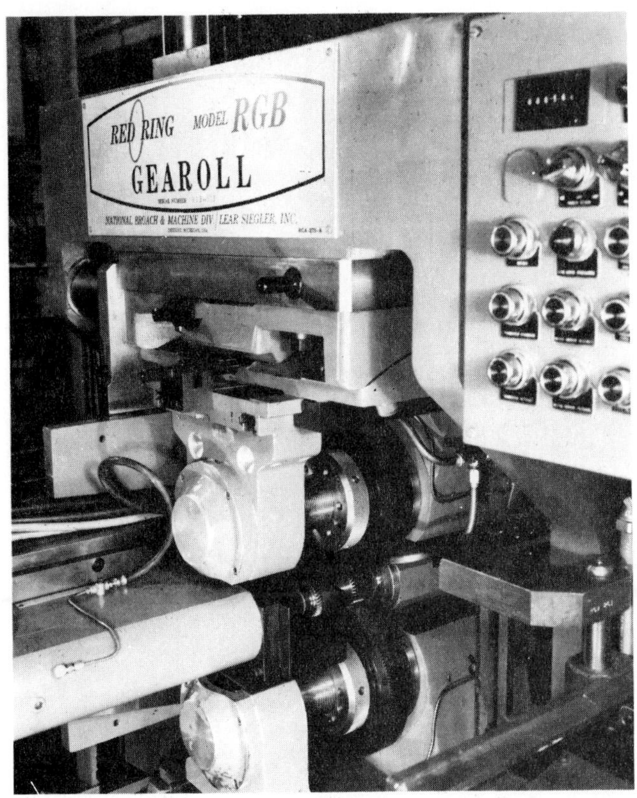

FIGURE M-10 Roll finishing of unhardened gears (Broach and Machine Division, Lear Siegler, Inc.).

FIGURE M-11 Finishing a gear by form grinding (National Broach and Machine Division, Lear Siegler, Inc.).

FIGURE M-12 Finishing a gear by generating the ground surface (Fellows Corporation).

FIGURE M-13 Finishing a hardened gear by honing. The honing tool has the helical teeth. (National Broach and Machine Division, Lear Siegler, Inc.)

FIGURE M-14 Inspecting gear profiles (Fellows Corporation).

Gear honing on hardened gears can obtain finishes of 10 to 5 microinches (Figure M-13). The gear hone is an abrasive impregnated mating gear that rolls in mesh with the gear to be finished. Gear honing is also applied to internal gears.

Gear Inspections

Due to the high degree of precision required in the manufacture and applications of gears, measurement and inspection is a highly refined technology (Figure M-14). Special testing and gear measuring tools are used to check the involute profile of a gear tooth, the tooth spacing, or the accuracy of the helix on a helical gear. Mesh is checked with a master gear engaged to the manufactured gear. In this section, you will have the opportunity to learn about common gear measuring techniques used in the machine shop.

Common Gear Cutting in the Machine Shop

Gear cutting in the job or school machine shop is done in order to make a part for a slow-speed drive or emergency situation, where an unhardened gear will meet needs. Although it would be possible to heat treat a gear properly, a required finishing process might not be available. In the school shop, gear cutting is often limited to spur types, although the milling of helical gears is sometimes done for instructional purposes.

Nonetheless, the processes of gear cutting make use of many of the setups and machining procedures that you have studied up to now. The work of gear cutting is very much in line with the overall work of the machine shop. Exposure to this technology will enhance both your experience and your knowledge in machining. In this section you will study types and applications of gears, machine setup and calculations for gear cutting, fundamental spur gear cutting techniques, and common methods for gear inspection and measurement.

UNIT 1

Introduction to Gears

Gears play an extremely important part in mechanical devices of all kinds. To meet a variety of gearing requirements, many types of gears have been developed. The purpose of this unit is to identify and familiarize you with several common types of gears and the materials from which they are made.

Objective

After completing this unit, you should be able to:

Identify common gear types and describe common gear materials.

Types of Gears

SPUR GEARS

Spur gears have straight teeth and are used to connect parallel shafts (Figure M-15). Spur gears are found in a wide variety of mechanisms ranging from watches and clocks to mechanical drives for large machinery.

INTERNAL GEARS

Internal gears may be spur or helical (Figure M-16) types. Two internal gears in mesh rotate in the same direction. Internal gears permit closer shaft center distances and these gears have stronger teeth than equivalent external gears.

HELICAL AND HERRINGBONE GEARS

When gear teeth are cut at an angle to the gear axis, a helical gear is formed (Figure M-17). Helical gears are smoother and quieter running than spur gears because several teeth are in mesh at any

FIGURE M-15 Spur gears in lathe headstock (B&K Mfg.).

FIGURE M-16 Helical internal gears (Lane Community College).

FIGURE M-17 Helical gears (Lane Community College).

FIGURE M-19 Crossed helical gear (Lane Community College).

FIGURE M-18 Double helical (herring bone) gears (Lane Community College).

FIGURE M-20 Bevel gears (Lane Community College).

given time. However, due to the helix of the teeth, an end thrust is created. Two helical gears in mesh will tend to push each other sideways. To overcome this tendency, herringbone or double helical gears are used. The right- and left-hand helixes cancel each other's end thrust (Figure M-18).

Crossed helical gears will connect shafts at a 90 degree angle, but not in the same plane (Figure M-19).

BEVEL GEARS

Bevel gears connect shafts at an angle (Figure M-20) in the same plane since the conical axis of the gears must intersect. Two meshed bevel gears of the same size connecting shafts at a 90-degree angle are called miter gears (Figure M-21). Shafts are connected at other angles by angular bevel gears.

FIGURE M-21 Miter gears (Lane Community College).

FIGURE M-23 Double enveloping worm and gear (Lane Community College).

FIGURE M-22 Worm and worm gear (Lane Community College).

WORM GEARS

Worm gears connect shafts at 90-degree angles, but not in the same plane. Worm gears are used extensively in speed reducers because of the large ratio differences that can be obtained using only two components, the worm and worm gear or wheel. In worm gear systems, the worm is the driving component while the gear or wheel is the driven component.

Two basic designs of worm gears are used. These are the single enveloping worm (Figure M-22) and the double enveloping worm (Figure M-23). In the double enveloping design, the worm is hourglass-shaped and envelops the worm wheel on both sides of center. Greater loads can be transmitted through this type of worm gear. Worms may be single or multiple leads depending on the speed ratio desired, and they will only mesh with corresponding wheels.

FIGURE M-24 Hypoid gears (Lane Community College).

HYPOID GEARS

A type of gear related to both the worm and crossed helical gear is the hypoid (Figure M-24). Hypoids connect shafts at right angles, but displaced in two planes. Hypoids are widely used in automotive differentials.

FIGURE M-25 Spur gear and gear rack (Lane Community College).

RACK AND PINION

Gear teeth cut side by side on a flat bar form a rack. A pinion gear meshes with the rack forming a rack and pinion (Figure M-25). The rack and pinion is used to convert rotary motion to linear motion or the other way around depending on which one is the driven component. A rack and pinion may be either spur or helical in tooth form; this device is found in numerous applications on machine tools and in many other mechanical devices.

Gear Materials

Gears are made from ferrous and nonferrous metals as well as nonmetallic materials.

FERROUS MATERIALS

Steel and cast iron are typical ferrous gear materials. High carbon steel gears can be hardened and tempered to exact specifications. The metallurgy of steels can be varied to suit gear specifications ranging from long wear to machinability.

Cast iron gears are easily machined, but this material is not as well suited for load carrying as steel. Cast iron gears have low impact strength and are poorly suited for shock loads. Ductile and malleable iron as well as sintered metals are also used as gear materials.

NONFERROUS MATERIALS

Gears for light applications such as clocks are often made from brass and aluminum alloys. Bronze is a superior gear material because of its tough and wear-resistant qualities.

Low-cost gears of nonferrous material may be made by die casting finished pieces. No machining is needed. Die cast gear materials include zinc, aluminum, magnesium, and copper-based alloys.

NONMETALLIC MATERIALS

A large number of gears for numerous applications are made from many types of plastic materials, including phenolic resins, nylon, and micarta. Nonmetallic gears have excellent wear resistance and often need no lubrication. Nonmetallic gears run very well meshed with steel or iron gears in many applications.

Gear Train Mechanics

In many instances, gear trains are made up with different gear materials. Many worm drives use a bronze worm gear with a hardened steel worm. Cast iron gears work well with steel gears. To equalize the wear in gear trains, the pinion is made harder than the gear. Even wear in gear trains can be obtained when a gear ratio is used that allows for a hunting tooth. For example, a gear train ratio of approximately 4:1 is needed. This is possible by using an 80-tooth gear in mesh with a 20-tooth gear. In this gear arrangement, the same tooth of the pinion will mesh with the same tooth of the gear in every revolution. If an 81-tooth gear were used, the teeth of the pinion would not equally divide into it, but each tooth of one gear would mesh with all of the mating teeth one after the other, distributing wear evenly over all teeth.

SELF-TEST

1. Name two types of gears used to connect parallel shafts.
2. What are some advantages and disadvantages of helical gears?
3. What is the direction of rotation of a pinion in relation to an internal gear when they are meshed together?
4. Two helical gears of the same hand and a 45 degree helix angle are in mesh. What relationship exists between the axis of the two shafts?
5. What is the gear reduction in a worm gear set when the worm has a double lead thread and the worm gear has 100 teeth?
6. Can a worm gear set ratio be changed by substituting a single start worm with a triple start worm?
7. What kind of gear material gives the greatest load carrying capacity for a given size?
8. What kind of material can be used for gears to run quietly at high speed?
9. How can the wear on the gear teeth be equalized when a large and a small gear are running together?
10. What kind of gear materials give corrosion resistance to gear sets?

UNIT 2

Spur Gear Terms and Calculations

Although most gear cutting is done on specialized machine tools, a spur gear may be cut on the milling machine by straightforward machining techniques. Gears and gear cutting in general involve a number of terms, numerous dimensions, and calculations. The purpose of this unit is to familiarize you with gear terminology and calculations involved in gear cutting.

Objective

After completing this unit, you should be able to:

Identify gear tooth parts and to calculate their dimensions.

Spur Gear Terminology

Spur gear terms are illustrated in Figure M-26. The definitions of these terms are as follows:

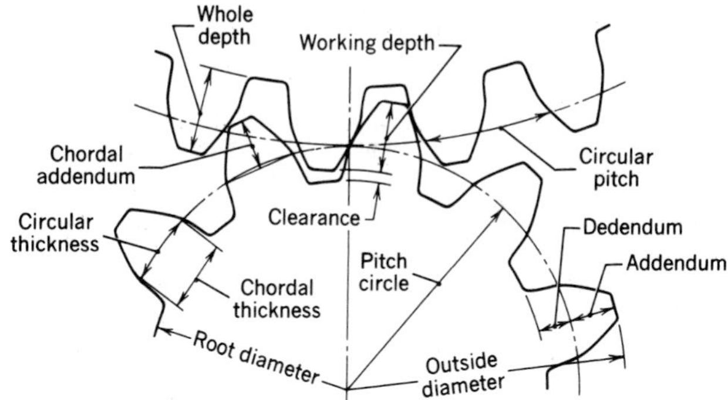

FIGURE M-26 Spur gear terms.

Addendum. The radial distance from the pitch circle to the outside diameter.

Dedendum. The radial distance between the pitch circle and the root diameter.

Circular thickness. The distance of the arc along the pitch circle from one side of a gear tooth to the other.

Circular pitch. The length of the arc of the pitch circle from one point on a tooth to the same point on the adjacent tooth.

Pitch diameter. The diameter of the pitch circle.

Outside diameter. The major diameter of the gear.

Root diameter. The diameter of the root circle measured from the bottom of the tooth spaces.

Chordal addendum. The distance from the top of the tooth to the chord connecting the circular thickness arc.

Chordal thickness. The thickness of a tooth on a straight line or chord on the pitch circle.

Whole depth. The total depth of a tooth space equal to the sum of the addendum and dedendum.

Working depth. The depth of engagement of two mating gears.

Clearance. The amount by which the tooth space is cut deeper than the working depth.

Backlash. The amount by which the width of a tooth space exceeds the thickness of the engaging tooth on the pitch circles.

Diametral pitch. The number of gear teeth to each inch of pitch diameter.

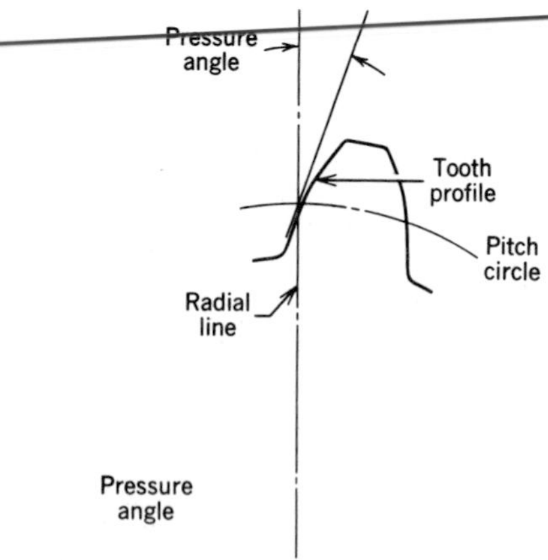

FIGURE M-27 Pressure angle.

Pressure angle. The angle between a tooth profile and a radial line at the pitch circle (Figure M-27).

Center distance. The distances between the centers of the pitch circles.

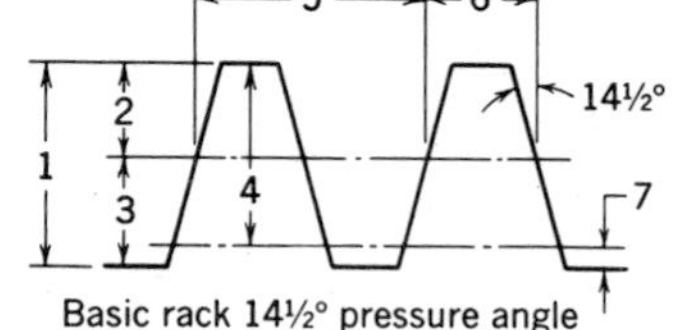

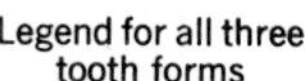

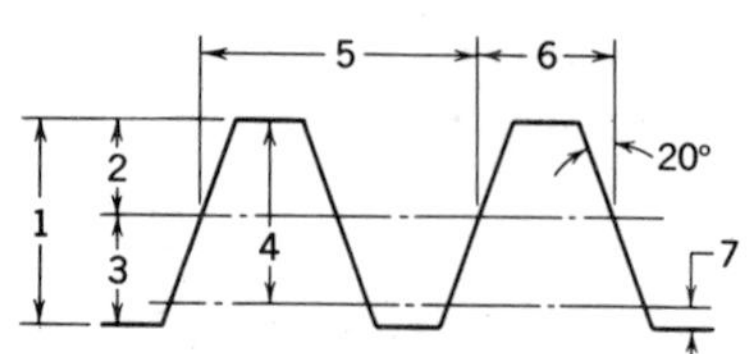

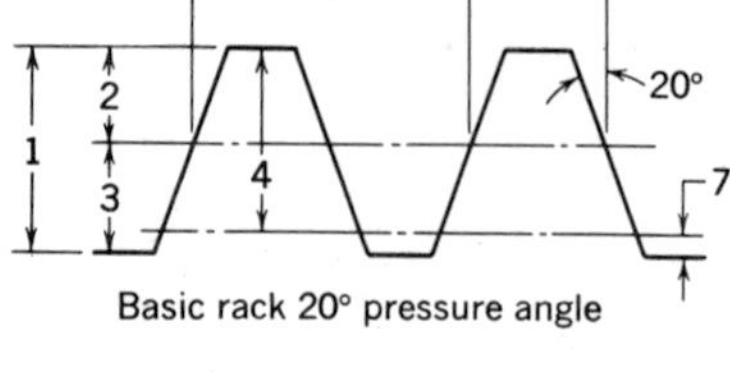

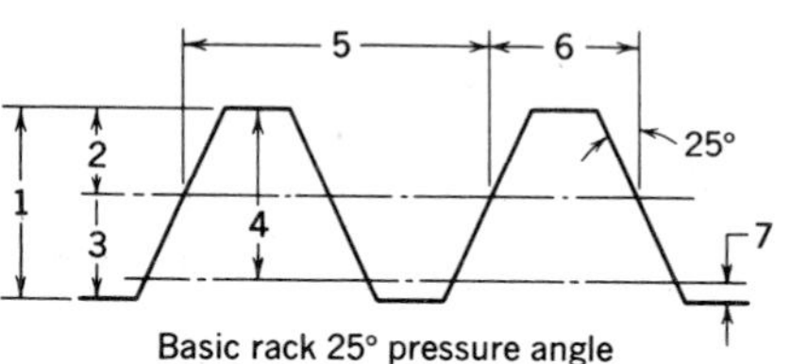

FIGURE M-28 Comparison of tooth shape on gear rack with different pressure angles.

Pressure Angles

Three spur gear tooth forms are generally used with pressure angles of 14½, 20, and 25 degrees. The 14½-degree tooth form is being replaced and made obsolete by the 20- and 25-degree forms. Figure M-28 illustrates these three pressure angles as applied to a gear rack with all teeth being the same depth. The larger pressure angle makes teeth with a much larger base, which also makes these teeth much stronger. The larger pressure angles also allow the production of gears with fewer teeth. Any two gears in mesh with each other must be of the same pressure angle and the same diametral pitch.

Information in this unit includes constants for 14½-degree pressure angle gear calculations because many existing gears and gear cutters are of this tooth form. When gear tooth measurements are to be made with gear tooth calipers, the chordal tooth thickness and the chordal addendum must be calculated. Most of the gear tooth dimensions that you will calculate in this unit can be found in tables in *Machinery's Handbook*.

Spur Gear Calculations

The following symbols are used to represent gear tooth terms in spur gear calculations.

P = diametral pitch
D = pitch diameter
D_o = outside diameter
N = number of teeth in gear
t = tooth thickness—circular
a = addendum
b = dedendum
c = clearance
C = center distance
h_k = working depth
h_t = total depth
t_c = tooth thickness—chordal
a_c = addendum—choral

Table M-1 gives the formulas to calculate the gear dimensions in the following examples.

EXAMPLE 1

Determine the dimensions for a 30-tooth gear, 14½-degree pressure angle, and a 2.500-in. pitch diameter.

1. Number of teeth $N = 30$.
2. Pitch diameter $D = 2.500$ in.
3. Diametral pitch $P = \frac{N}{D} = \frac{30}{2.500} = 12$
4. Addendum $a = \frac{1}{P} = \frac{1}{12} = .083$ in.
5. Dedendum $b = \frac{1.157}{P} = \frac{1.157}{12} = .096$ in.
6. Tooth thickness

$$t = \frac{1.5708}{P} = \frac{1.5708}{12} = .131 \text{ in.}$$

7. Clearance $c = \frac{.157}{P} = \frac{.157}{12} = .013$ in.

8. Total depth $h_t = \frac{2.157}{P} = \frac{2.157}{12} = .179$ in.

9. Working depth $h_k = \frac{2}{P} = \frac{2}{12} = .166$ in.

10. Chordal tooth thickness

$$t_c = D \sin\left(\frac{90 \text{ deg}}{N}\right)$$

$$= 2.500 \times \sin\left(\frac{90 \text{ deg}}{30}\right)$$

$$= 2.500 \times \sin 3 \text{ deg}$$

$$= 2.5 \times .0523 = .1307 \text{ in.}$$

11. Chordal addendum

$$a_c = a + \frac{t^2}{4D}$$

$$= .083 + \frac{.131^2}{4 \times 2.5}$$

$$= .083 + \frac{.017}{10}$$

$$= .083 + .0017 = .0847 \text{ in.}$$

12. Outside diameter

$$D_o = \frac{N + 2}{P}$$

$$= \frac{30 + 2}{12}$$

$$= \frac{32}{12} = 2.6666 \text{ in.}$$

TABLE M-1 Spur Gear Formulas

To Find:	Spur Gear Formulas – $14\frac{1}{2}$-Degree Pressure Angle	20- and 25-Degree Pressure Angles
Addendum, a	$a = \frac{1.0}{P}$	$a = \frac{1.0}{P}$
Dedendum, b	$b = \frac{1.157}{P}$	$b = \frac{1.250}{P}$
Pitch diameter, D	$D = \frac{N}{P}$	$D = \frac{N}{P}$
Outside diameter, D_O	$D_O = \frac{N + 2}{P}$	$D_O = \frac{N + 2}{P}$
Number of teeth, N	$N = D \times P$	$N = D \times P$
Tooth thickness, t	$t = \frac{1.5708}{P}$	$t = \frac{1.5708}{P}$
Whole depth, h_t	$h_t = \frac{2.157}{P}$	$h_t = \frac{2.250}{P}$
Clearance, c	$c = \frac{.157}{P}$	$c = \frac{.250K}{P}$
Center distance, C	$C = \frac{N_1 + N_2}{2 \times P}$	$C = \frac{N_1 \times N_2}{2 \times P}$
Working depth, h_k	$h_k = \frac{2}{P}$	$h_k = \frac{2}{P}$
Chordal tooth thickness, t_c	$t_c = D \sin \left(\frac{90 \text{ deg}}{N}\right)$	$t_c = D \sin \left(\frac{90 \text{ deg}}{N}\right)$
Chordal addendum, a_c	$a_c = a + \frac{t^2}{4D}$	$a_c = a + \frac{t^2}{4D}$
Diametral pitch, P	$P = \frac{N}{D}$	$P = \frac{N}{P}$
Center distance, C	$C = \frac{D_1 + D_2}{2}$	$C = \frac{D_1 + D_2}{2}$

EXAMPLE 2

Determine the gear dimensions for a 45-tooth gear, diametral pitch of 8, and a 20-degree pressure angle.

1. Number of teeth $N = 45$
2. Diametral pitch $P = 8$
3. Pitch diameter $D = \frac{N}{P} = \frac{45}{8} = 5.625$ in.
4. Addendum $a = \frac{1}{P} = \frac{1}{8} = .125$ in.
5. Dedendum $b = \frac{1.250}{P} = \frac{1.250}{8} = .1562$ in.
6. Tooth thickness

$$t = \frac{1.5708}{P} = \frac{1.5708}{8} = .1963 \text{ in.}$$

7. Clearance $c = \frac{.250}{P} = \frac{.250}{8} = .031$ in.
8. Total depth $h_t = \frac{2.250}{P} = \frac{2.250}{8} = .281$ in.
9. Working depth $h_k = \frac{2}{P} = \frac{2}{8} = .250$ in.
10. Outside diameter

$$D_o = \frac{N + 2}{P} = \frac{45 + 2}{8} = \frac{47}{8} = 5.875 \text{ in.}$$

11. Chordal tooth thickness

$$t_c = D \sin \frac{90 \text{ deg}}{N} = 5.625 \times \sin \frac{90 \text{ deg}}{45} = 5.625 \times \sin 2 \text{ deg.} = 5.625 \times .0349 = .1963 \text{ in.}$$

12. Chordal addendum

$$a_c = a + \frac{t^2}{4D} = .125 \text{ in.} + \frac{.1963^2}{4 \times 5.625} = .125 \text{ in.} + .0017 \text{ in.} = .1267 \text{ in.}$$

The center distance between gears can be calculated when the number of teeth in the gears and the diametral pitch is known. Two gears in mesh make contact at their pitch diameters.

EXAMPLE 3

Determine the center distance between gears with 25 and 40 teeth and a diametral pitch of 14.

$$\text{Center distance } C = \frac{N_1 + N_2}{2 \times P} = \frac{25 + 40}{2 \times 24} = \frac{65}{28} = 2.3214 \text{ in.}$$

Another method of finding the center distance if the pitch diameters are known is to add both pitch diameters and divide that sum by 2.

EXAMPLE 4

What is the center distance of two gears when their pitch diameters are 2.500 and 3.000 in., respectively?

$$\text{Center distance } C = \frac{D_1 + D_2}{2} = \frac{2.500 + 3.000}{2} = \frac{5.500}{2} = 2.750 \text{ in.}$$

SELF TEST

1. What are commonly found pressure angles for gear teeth?
2. What are larger pressure angles used on gear teeth?
3. What is the center distance between two gears with 20 and 30 teeth and a diametral pitch of 10?
4. What is the center distance between two gears with pitch, diameters of 3.500 and 2.500 in.?
5. What is the difference between the whole depth of a tooth and the working depth of a tooth?
6. What relationship do the addendum and the dedendum have with the pitch diameter on a tooth?
7. What is the outside diameter and the tooth thickness on a 50-tooth gear with a diametral pitch of 5?
8. What is the diametral pitch of a gear with 36 teeth and a pitch diameter of 3.000 in.?
9. What are the outside diameter, whole depth, pitch diameter, and dedendum for a 40-tooth, 8 diametral pitch, 20-degree pressure angle gear?
10. What are the outside diameter, clearance, whole depth, tooth thickness, and pitch diameter for a 48-tooth, 6 diametral pitch, $14\frac{1}{2}$-degree pressure angle gear?

UNIT 3

Cutting a Spur Gear

Spur gear cutting, like any machining job, includes setting up the machine and workpiece, selecting the correct cutter, and performing all calculations necessary to the task. The purpose of this unit is to describe and familiarize you with these procedures.

Objective

After completing this unit, you should be able to: Set up the mill and cut a spur gear.

Involute Gear Cutters

Spur gears are cut with involute gear cutters. The gear cutter set consists of eight cutters numbered 1 to 8, for typical pressure angles of $14\frac{1}{2}$ or 20 degrees. The eight gear cutters have eight different tooth forms for each diametral pitch, depending on the number of teeth on the gear to be cut. Each gear cutter can be used to cut a range of teeth (Table M-2). However, the cutter will only produce the exact tooth form for the lowest number of teeth in the range. For example, a No. 5 cutter will cut 21 to 25 teeth, but will produce the exact form only for 21 teeth. Other numbers of teeth will be less accurate in form.

Half-step cutters are also available. These serve to cut more accurate tooth forms within a cutter's range. An example would be a No. $5\frac{1}{2}$ for 19 and 20 teeth. Special cutters designed for specific numbers of teeth or for low numbers of teeth (6 to 11) are also used.

TABLE M-2

Number of Cutter	Number of Teeth Cut	Number of Cutter	Number of Teeth Cut
1	135 to rack	5	21 to 25
2	55 to 134	6	17 to 20
3	35 to 54	7	14 to 16
4	26 to 34	8	12 and 13

Gear cutters are identified and marked with the following information.

- Cutter number
- Diametral pitch
- Number of teeth
- Pressure angle
- Whole depth of tooth

Machining and Mounting the Gear Blank

The first step in gear cutting is to machine the gear blank. The outside diameter is the most critical dimension and should be held as close as possible to the calculated size. For example, a gear is to be cut to the following specifications:

Number of teeth—48

Diametral pitch—12

Pressure angle—$14\frac{1}{2}$ degrees

The gear blank is turned on the lathe to its correct outside diameter. This is calculated by the formula

$$\text{outside diameter } D_o = \frac{N + 2}{P}$$

where N = number of teeth and P = diametral pitch.

$$\frac{N + 2}{P} = \frac{48 + 2}{12} = 4.167 \text{ in.}$$

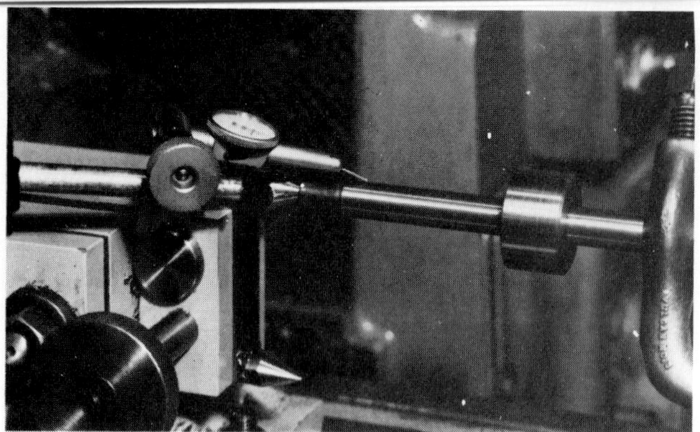

FIGURE M-29 Checking the axis of the mandrel to be parallel with the table surface (Lane Community College).

FIGURE M-30 Aligning the cutter centrally over the gear blank (Lane Community College).

If the gear has a hub, the outside diameter may be machined by chucking the gear on the hub. Better accuracy can be obtained by turning the outside diameter while the gear blank is mounted on the same mandrel that will be used while milling the teeth.

If more than one gear of the same size is to be made, a solid blank may be machined that is sufficiently long to accommodate the required number of gears. After teeth have been milled, individual gears may be cut off on the band saw or parted off in the lathe.

If you are using a mandrel to hold the gear blank during turning or milling, be sure to lubricate the bore of the blank before pressing it onto the mandrel.

Machine and Workpiece Setup

Mount the index head on the mill table. The index head spindle must be aligned with the table in both the horizontal and vertical axes. Secure the footstock to the mill table and use it to support the small end of the mandrel.

Attach a driving dog to the large end of the gear mandrel. Place the mandrel between the index head and footstock centers. Tighten the footstock center to hold the mandrel in place. Use a dial test indicator to check the height of the gear mandrel at both ends (Figure M-29). Allow for the taper of the gear mandrel and adjust the footstock center height if necessary. The gear blank must be parallel to the machine table.

From Table M-2, select the correct gear cutter for 48 teeth and a pressure angle of $14\frac{1}{2}$ degrees. This will be a No. 3 cutter (35 to 54 teeth). Keep the saddle as close to the column as possible and position the gear cutter on the mill arbor as close to the spindle as possible, allowing clear access to the gear blank. Install mill arbor spacing and bearing collars and the arbor support. Cutting pressure should always be applied toward the index head spindle holding the bit end of the mandrel.

CENTERING THE WORKPIECE

The gear blank must be exactly centered under the gear cutter. Several methods may be used for this. A direct measurement (Figure M-30) can be made by placing a square against the gear blank and measuring the distance to the cutter to determine the amount of saddle setover required. A more accurate method is to use the saddle micrometer collar to determine the setover dimension. Place a square against the gear blank and move the saddle until the cutter just touches the square (Figure M-31). Set the saddle micrometer to zero. Measure the width of the square with a micrometer and add this amount to the sum of half the cutter thickness plus half the diameter of the gear blank. Move the saddle over this amount to center the gear blank under the cutter.

CALCULATING INDEX

Calculate indexing for 48 teeth by the following formula:

$$\frac{40}{N} = \frac{40}{48} = \frac{5}{6}$$

Create the proportional fraction required by dividing the lowest term denominator, 6, into the available index plate hole circles to determine the multiplier. Any hole circle divisible by 6 can be used.

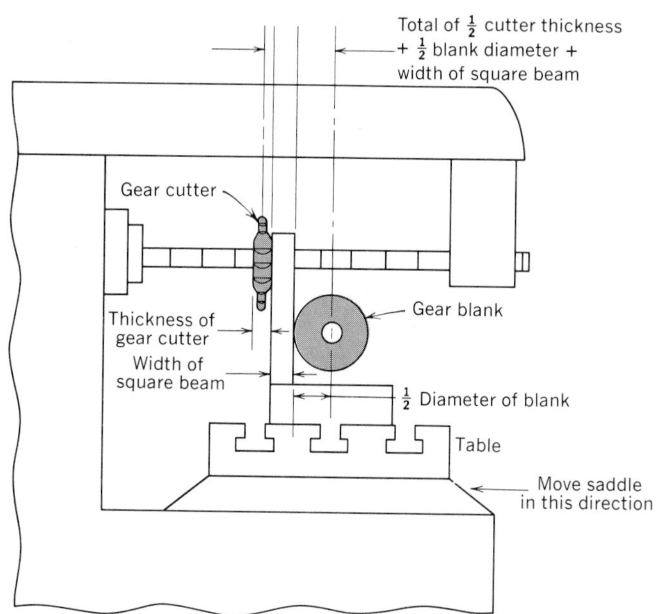

FIGURE M-31 **Determining the saddle setover dimension for centering the gear cutter.**

FIGURE M-32 **Marking all the tooth spaces around the circumference of the gear blank (Lane Community College).**

FIGURE M-33 **Finish cut is being taken on the gear teeth (Lane Community College).**

$$\frac{5 \times 5}{6 \times 5} = \frac{25 \text{ holes}}{30\text{-hole circle}}$$

$$\frac{5 \times 7}{6 \times 7} = \frac{35 \text{ holes}}{42\text{-hole circle}}$$

$$\frac{5 \times 9}{6 \times 9} = \frac{45 \text{ holes}}{54\text{-hole circle}}$$

Adjust the index crank pin to the correct hole circle and set the sector arms to include the correct number of new holes for the partial turn required. If you start the first cut with the index pin in the numbered hole of the index plate circle, you can double check the indexing by coming back to the starting location.

Cutting the Gear

Calculate the depth of cut by the formula

$$\text{whole depth} = \frac{2.157}{P} = \frac{2.157}{12} = .180 \text{ in.}$$

where P is the diametral pitch.

Set proper feeds and speeds on the machine. Start the mill spindle and raise the knee until the cutter just touches the gear blank. Set the knee micrometer collar to zero. A roughing cut of about .150 in. is recommended. This will leave .030 in. for finishing. Move the gear blank clear of the cutter and raise the knee .150 in. for roughing.

It is always a good idea to verify indexing by moving the gear blank up to the cutter until a small visible mark is made. Back away and index to the next tooth. Continue this process until all tooth locations have been marked and indexing has been verified (Figure M-32). The gear blank should return to the starting point exactly. If it does not, you may have made an indexing error. This can now be corrected before a gear blank is ruined. If indexing is correct, make roughing cuts. Be sure to lock the index head spindle each time. Table feed trip dogs may be set to stop feed at the completion of each cut. A center rest may be used under the gear blank to increase rigidity.

After completing all roughing cuts, raise the knee the remaining amount for the finish cuts (Figure M-33). Stay back from the full final depth until you have milled two adjacent teeth and made the tooth thickness measurement. After you have milled two adjacent teeth, deburr carefully and check tooth thickness with the gear tooth caliper. If you have stayed back from the full depth a few thousandths, the tooth thickness, should be a small amount over size. Raise the knee to the final full depth; make the cuts on two adjacent teeth. Recheck thickness measurement. If the tooth thickness is correct, mill the remaining teeth. Burrs may be removed after the completed gear has been removed from the mandrel.

SELF-TEST

1. What are two factors against cutting spur gears in the milling machine?
2. How many gear cutters are in a standard set?
3. Which cutter is used to cut a gear with 38 teeth?
4. Can a $14\frac{1}{2}$-degree pressure angle gear be in mesh with a 20-degree pressure angle gear?
5. A No. 6 gear cutter has the correct tooth shape for a gear with how many teeth?
6. What information is marked on the side of gear cutters?
7. Why should the work and cutter be mounted close to the column?
8. How many holes should be located within the sector arms?
9. Why is the gear blank marked with the cutter prior to cutting the teeth?
10. When is a center rest used?

UNIT 4

Gear Inspection and Measurement

Gears, like any machined parts, must be measured to determine if dimensions are within tolerance. Gears are somewhat special in that dimensions are numerous and vary from gear to gear. Many gear measurements require the use of highly specialized measuring tools and techniques. In common spur gear cutting, gear measurement and inspection can be accomplished with common tools designed for this purpose. The purpose of this unit is to describe the common tools and their application in spur gear measurements.

Objective

After completing this unit, you should be able to:

Measure a spur gear using common tools.

Methods of Gear Measurements

Common methods of gear measurement include the gear tooth vernier caliper, micrometer with two pins, and the optical comparator.

GEAR TOOTH VERNIER CALIPER

The gear tooth vernier caliper is a combination vernier depth gage and vernier caliper. This tool is designed to measure the thickness of a gear tooth along a chord drawn through the two points where the pitch circle intersects the side of the tooth (Figure M-34). In order to set the caliper jaws at this

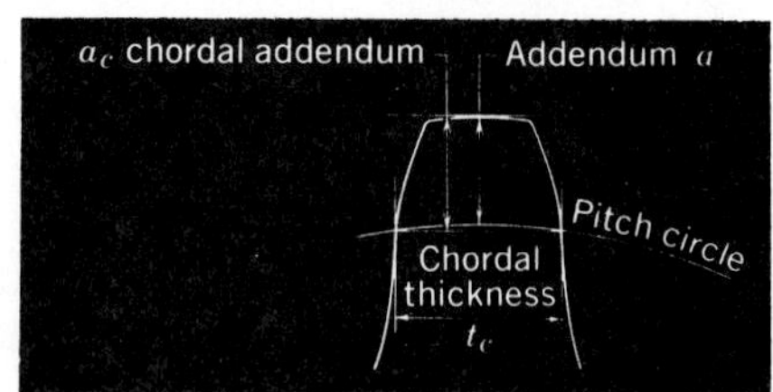

FIGURE M-34 Chordal addendum and chordal tooth thickness.

FIGURE M-35 Measuring a spur gear with a gear tooth vernier caliper (Lane Community College).

point, the depth gage on the gear tooth caliper must be set to the chordal or corrected addendum. This dimension is larger than the addendum. The required dimensions can be determined by the following calculations.

CALCULATING CHORDAL TOOTH THICKNESS

To determine the setting for the gear tooth caliper, it will be necessary to calculate several other dimensions first. Consider the following example. Calculate chordal tooth thickness for the following gear:

$$\text{Number of teeth } N = 12$$

$$\text{Diametral pitch } P = 6$$

$$\text{Outside diameter} = D_o$$

$$D_o = \frac{N+2}{P} = \frac{12+2}{6} = \frac{14}{6} = 2.3333$$

Addendum

$$a = \frac{1}{P} = \frac{1}{6} = .1666 \text{ in.}$$

Pitch diameter

$$D = \frac{N}{P} = \frac{12}{6} = 2.000 \text{ in.}$$

Circular tooth thickness

$$t = \frac{1.5708}{P} = \frac{1.5708}{6} = .2168 \text{ in.}$$

FIGURE M-36 Measuring a spur gear with two pins and a micrometer (Lane Community College).

Chordal addendum

$$a_c = a + \frac{t^2}{4D}$$

$$= .1666 + \frac{.2168^2}{4 \times 2}$$

$$= .1666 + \frac{.0685}{8} = .1751 \text{ in.}$$

Chordal tooth thickness

$$t_c = D \sin\left(\frac{90 \text{ deg}}{N}\right)$$

$$= 2 \sin\left(\frac{90}{12}\right)$$

$$= 2 \sin 7.5 = 2(.1305) = .261 \text{ in.}$$

MEASURING THE GEAR TOOTH

Set the gear tooth caliper depth gage to the chordal or corrected addendum of .175 in. Rest the blade of the depth gage on top of the gear tooth (Figure M-35). Use the vernier caliper to measure the chordal tooth thickness. This should be .261 in. If the outside diameter of the gear blank is undersized by .002 in., the depth gage setting should be reduced by half this amount of .001 in. This will assure that the tooth is being measured at the pitch diameter. If you are measuring a gear while still in the machine, be sure to remove all burrs on the teeth before making any measurements.

MICROMETER AND PINS

Another reliable method of gear measurement is by micrometer and pins (Figure M-36). The diameter of the pin to use is found by the formula

$$\text{pin diameter} = \frac{1.728}{D}$$

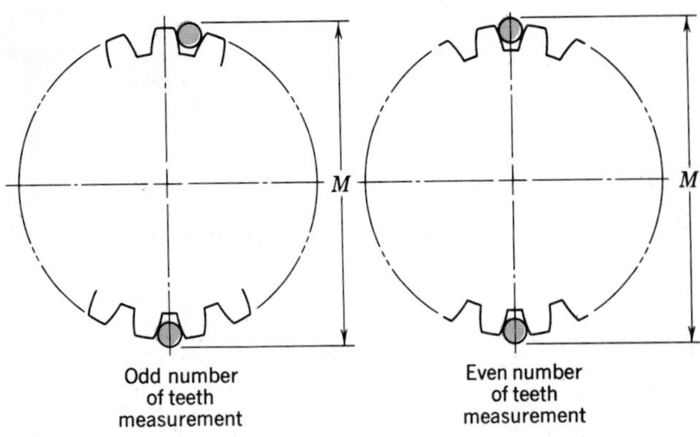

FIGURE M-37 Measuring gear dimensions over two pins.

where D is the diametral pitch. The constant 1.728 is most commonly used for external spur gears. The dimension M as measured across the pins (Figure M-37) is found in *Machinery's Handbook* depending on the pressure angle and number of teeth on the gear. Handbook tables for M dimension are given for a diametral pitch of 1. To apply the data to a specific gear, divide the table entry by the diametral pitch of the gear to be measured.

EXAMPLE

Number of teeth—45

Diametral pitch—10

Pressure angle—$14\frac{1}{2}$ degrees

$$\text{Pin diameter} = \frac{1.728}{10} = .1728 \text{ in.}$$

Measurement over pins M for a gear with a diametral pitch of 1 = 47.4437. Therefore,

$$M = \frac{47.4437}{10} = 4.7443 \text{ in.}$$

This calculated dimension assumes no backlash. However, in gear installations some backlash is required. If specifications call for .005 in. backlash, the M dimension must be recalculated. Handbook tables show that for a 45 tooth gear, each .001 in. of tooth thickness reduction at the pitch circle reduces M by .0031. Since .005 in. backlash is required,

$$.005 \times .0031 = .0155$$

The measurement over the pins must be reduced by this amount.

$$M = 4.74437 - .0155 = 4.72887$$

This result would be the required dimension over the pins to obtain the required backlash of .005 in.

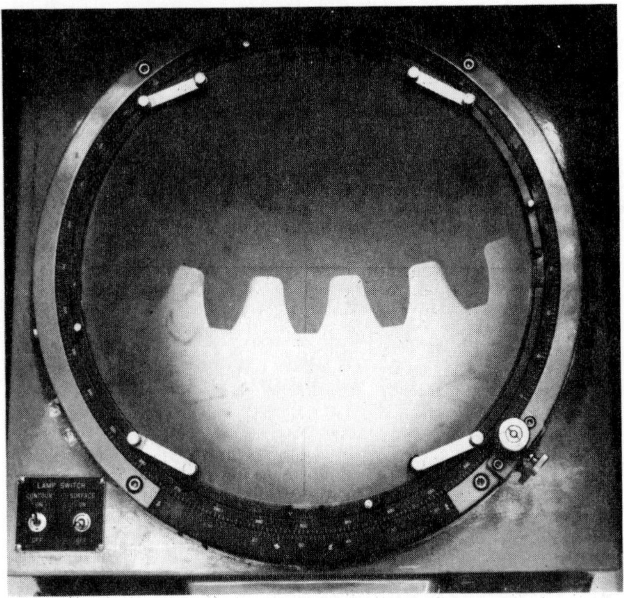

FIGURE M-38 Using an optical comparator to check gear tooth dimensions (Lane Community College).

OPTICAL COMPARATOR

The optical comparator can also be used for gear measurement. The comparator projects a greatly magnified shadow or reflected image of the gear onto a glass screen (Figure M-38). A transparent drawing of the gear to be measured is laid on the comparator screen and a direct comparison is made to the image of the gear being measured.

SELF-TEST

1. What are two common gear measurements performed by a machinist?
2. Which tooth thickness is measured by a gear tooth vernier caliper?
3. The vertical scale of a gear tooth vernier caliper is set to what dimension?
4. Which is larger, the chordal addendum or the standard addendum?
5. Which is larger, the chordal tooth thickness or the circular tooth thickness?
6. Name two ways of determining the chordal thickness of a gear tooth.
7. When a gear is measured with a micrometer and two pins, can any size pin be used?
8. How is the dimension determined when measuring over pins?
9. When a dimension is given for 1 diametral pitch but the gear measured is 12 diametral pitch, how is the measurement determined?
10. How is an optical comparator used in gear measurement?

SECTION N

GRINDING AND ABRASIVE MACHINING PROCESSES

The common machining processes of drilling, turning, and milling generally produce a fairly large chip in comparison to those produced by the processes discussed in this section. Abrasive materials are the cutting tools in the machining processes called grinding. Grinding machines and grinding processes make up one of the most important areas in all machining.

Grinding processes are chip-making metal cutting processes just like drilling, turning, and milling. However, grinding processes remove very small chips (called swarf) in very large numbers by the cutting action of many small individual abrasive grains. These abrasive grains are formed into a grinding wheel that is rotated against the workpiece at high speed. Each sharp corner of a grain cuts a small bit of material from the workpiece. When the corners become dull, heat and pressure increase, fracturing grains in the grinding wheel. This ability of the grains to microfracture and expose new sharp edges is termed *friability*. Some abrasive grains are produced to be tough (not break down readily). These are used in harsh grinding situations, such as the rough grinding (snagging) of castings in a foundry. Other grains are produced to break down readily and are used in grinding wheels where cool grinding is essential. An example of such an application would be the grinding of hardened steels. The characteristics of the bonding material are carefully matched to the abrasive grain by the producers of abrasive products. As the bonding material that holds the grains breaks down, new sharp grains are exposed, replacing the worn ones. This also provides a method of sharpening grinding wheels. Abrasive materials are also coated on sheets of cloth or paper in the form of sandpaper or sanding belts and disks. They also appear as solid blocks such as sharpening or honing stones or deburring media such as pellets.

In matchining, grinding processes are most often used as finish machining processes. The reason for this is that very small amounts (less than .001 in.) of material can be removed from the workpiece. This is extremely useful in finish machining a part to close dimensional accuracies. Furthermore, grinding processes result in very smooth surface finishes on the workpiece.

The abrasive materials that are the cutting tools in the grinding process are much harder than the equivalent materials used in common drills, lathe, and milling cutters. Therefore, these materials can be used to machine much harder materials than could ever be cut with high-speed-steel or even carbide. The grinding process can be applied to finish machine metals that have been hardened by heat treatment. For example, bearing races may be premachined to rough dimensions before heat treating. After hardening and tempering to exact specifications, they may be finish machined by grinding. In other instances, a hardened and tempered blank can be machined entirely with abrasives; this would be called abrasive machining. The term *abrasive machining* implies that the abrasive process is being used in a way that is competitive, on a cost basis, to some other metal removal process, such as turning or milling.

Grinding machines and grinding processes are essential to modern manufacturing industry. In this section, you will have the opportunity to study many of the types of grinding machines, other abrasive processes, and types of abrasives used. You will also set up and operate several common grinding machines found in machine shops.

■ ■ ■

Types of Grinding Machines

The versatility and wide application of the grinding processes have led to the development of many types of grinding machines. Your school shop will probably have two, three, or more of the common types. As a machinist, you should be familiar with the many types of these machine tools and the processes that they perform.

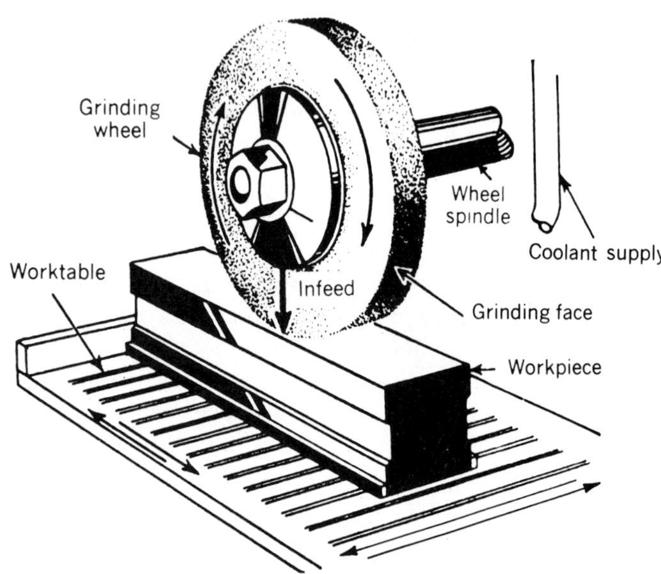

FIGURE N-1 Principle of the type I surface grinder (Bay State Abrasives, Dresser Industries, Inc.).

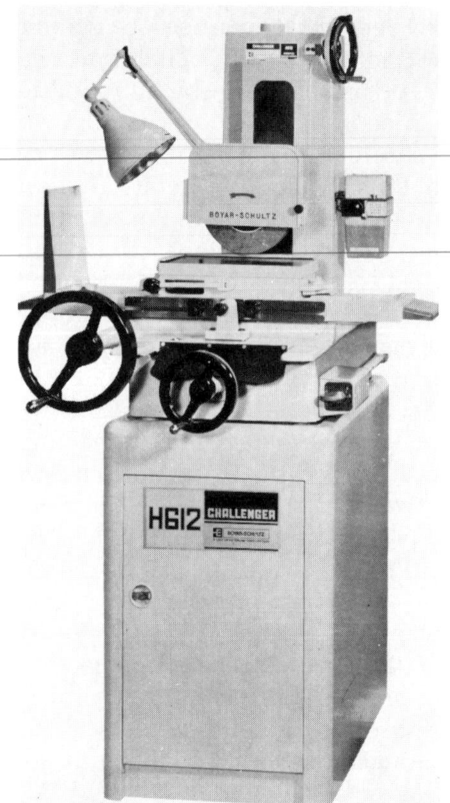

FIGURE N-2 Type I grinder (Boyar-Schultz Corp.).

Surface Grinders

One of the most common grinding machines is the surface grinder. Types of surface grinders include:

Type I—Horizontal spindle with reciprocating table

Type II—Horizontal spindle with rotary table

Type III—Vertical spindle with either reciprocating or rotary table

TYPE I

This type of surface grinder (Figure N-1) has the grinding wheel on a horizontal spindle. The face (periphery) of the wheel is presented to the workpiece. The workpiece, if of a ferrous material, is usually held on a magnetic chuck and is moved back and forth under the rapidly rotating grinding wheel. The chuck may also be moved in and out. Depth of cut is controlled by raising and lowering the wheel head. Sizes of surface grinders range from a small machine that will grind an area of 4 by 8 in. to large machines with the capacity to grind an area of 6 by 16 ft. The 6- by-18-inch-capacity machine is very common in machine shops (Figure N-2). The surface grinder in good condition can be expected to produce a flat surface on a workpiece within a tolerance of less than plus or minus .0001 in. By special shaping of the grinding wheel, contoured surfaces may be ground.

TYPE II

On the type II surface grinder, the face of the wheel is also brought into contact with the workpiece (Figure N-3). Parts to be ground are mounted on

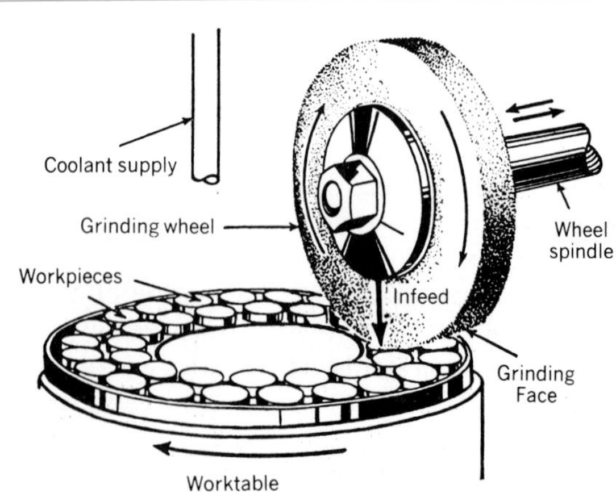

FIGURE N-3 Principle of the type II surface grinder. Sometimes circular parts are centered on this grinder; the resulting concentric scratch pattern is excellent for metal-to-metal seals of mating parts (Bay State Abrasives, Dresser Industries, Inc.).

a rotary table and run under the wheel. The table may also be tilted to provide special grinding geometry, such as the hollow grinding of a circular saw.

TYPE III

The type III surface grinder has a vertical spindle and the side (also called the grinding face) of the wheel is brought into contact with the workpiece. The worktable may be reciprocating (Figure N-4) or rotary (Figure N-5). Accessory spindles and special wheel shaping (dressing) add to the versatility of this grinding machine (Figure N-6). Vertical spindle grinding machines are often seen in large sizes (Figure N-7).

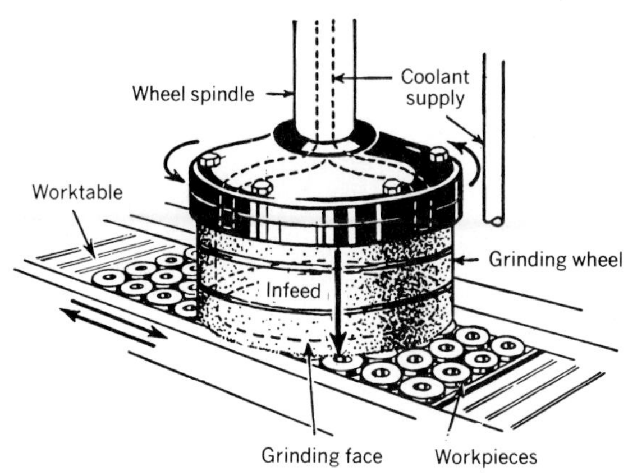

FIGURE N-4 Principle of the type III surface grinder, which has a vertical spindle and a reciprocating table (Bay State Abrasives, Dresser Industries, Inc.).

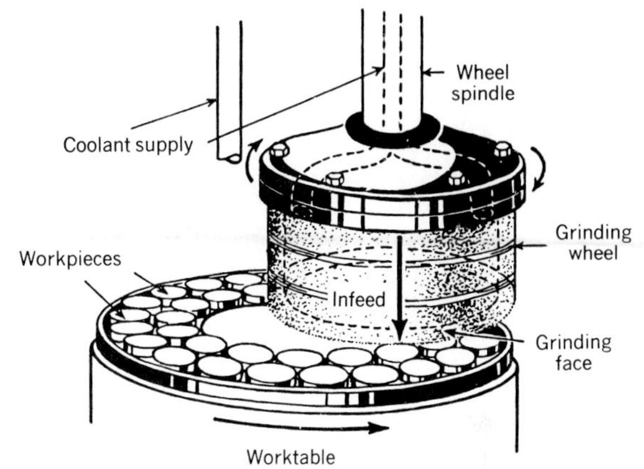

FIGURE N-5 Principle of the vertical spindle rotary grinder (Bay State Abrasives, Dresser Industries, Inc.).

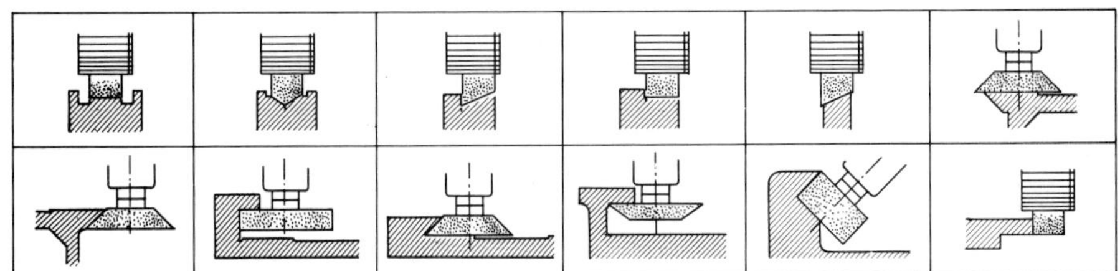

FIGURE N-6 A great variety of surfaces may be ground with an accessory spindle by dressing or tilting the wheel (Mattison Machine Works).

FIGURE N-7 Large vertical spindle rotary table grinding machine (Mattison Machine Works).

Cylindrical Grinders

Cylindrical grinders are used to grind the diameters of cylindrical workpieces. Types of cylindrical grinders include:

1. Center type
2. Roll type
3. Centerless
4. Internal cylindrical
5. Tool and cutter

CENTER-TYPE GRINDER

The center-type cylindrical grinder, sometimes called a center grinder, grinds the outside diameter of a cylindrical workpiece that is mounted between centers (Figure N-8). The workpiece is traversed past the grinding wheel while at the same time being rotated from 50 to 100 sfpm against the rotation of the grinding wheel (Figure N-9). By swiveling the table, tapers can be ground. This machine is also used for plunge grinding where the work and wheel are brought together without table traverse.

Steep tapers can be ground on the universal cylindrical grinder (Figure N-10). Long slender

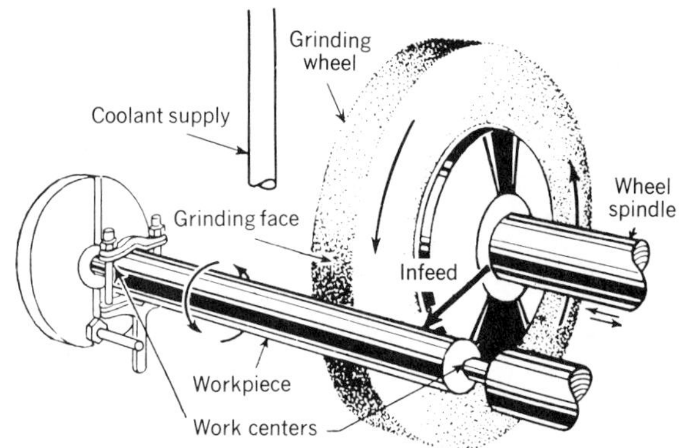

FIGURE N-8 Principle of the cylindrical grinding machine showing the workpiece and wheel motions (Bay State Abrasives, Dresser Industries, Inc.).

FIGURE N-9 Working area of a plain cylindrical grinder (Diamond Abrasives Corporation).

FIGURE N-10 Universal cylindrical grinder (Cincinnati Milacron).

workpieces can be ground using a steady rest (Figure N-11). The angular-type center grinder (Figure N-12) can grind into a shoulder (Figure N-13), and can form cylindrical geometry by using wheels (Figure N-14) that are accurately dressed to produce a mirror image of their form in the workpiece.

ROLL-TYPE GRINDER

Roll grinders are used to grind and to resurface large steel rolling mill rolls (Figure N-15). Since these rolls are usually very heavy, they are supported in bearings while in the grinding machine.

FIGURE N-11 The use of a steady rest for grinding slender workpieces (Cincinnati Milacron).

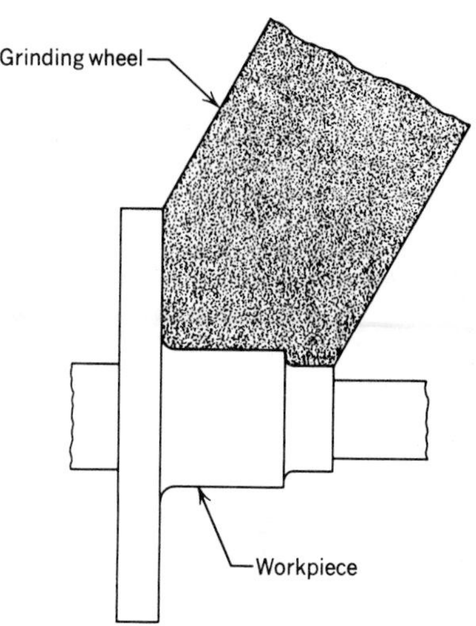

FIGURE N-13 Typical application of the angular center type grinding machine showing angled grinding wheel finishing both diameter and shoulder.

FIGURE N-12 Angular center-type grinder with automatic control (Cincinnati Milacron).

FIGURE N-14 Cylindrical from grinding (Bendix Automation and Measurement Division.

FIGURE N-15 **Roll grinding machine for grinding steel mill rolls (Landis Tool Div. Litton Industries).**

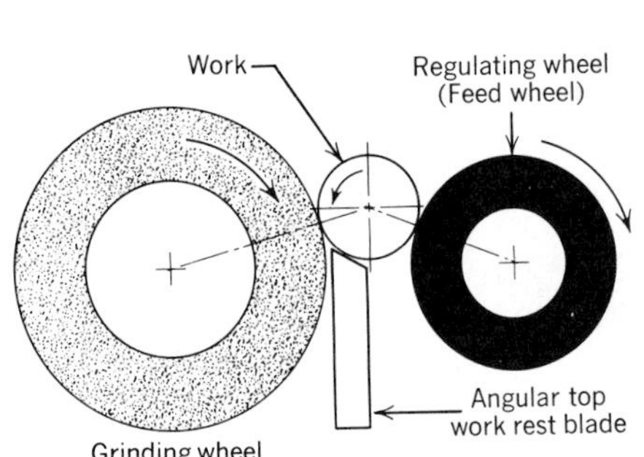

FIGURE N-16 **Principle of the centerless grinder. The grinding wheel travels at normal speeds and the regulating wheel travels at a slower speed to control the rate of spin of the workpiece (Bay State Abrasives, Dresser Industries, Inc.).**

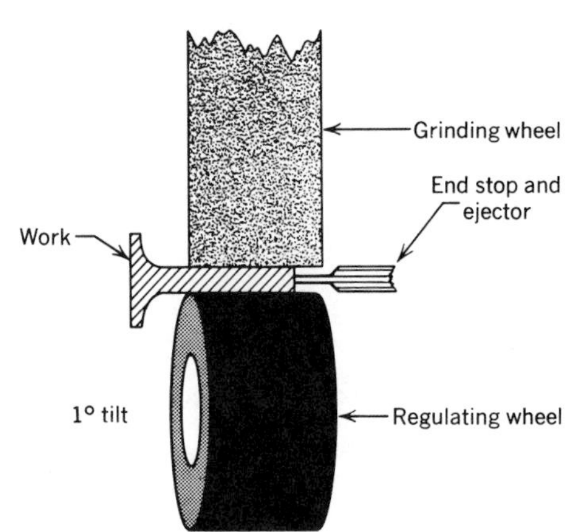

FIGURE N-17 **End feeding in the centerless grinder. The regulating wheel is set at a small tilt angle to keep the workpiece against the end stop (Bay State Abrasives, Dresser Industries, Inc.).**

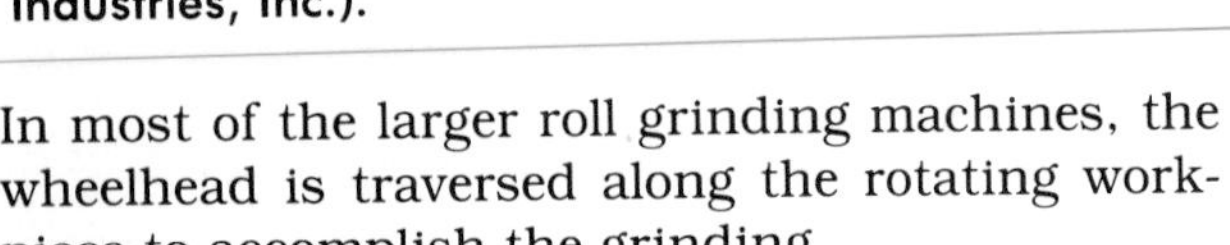

In most of the larger roll grinding machines, the wheelhead is traversed along the rotating workpiece to accomplish the grinding.

CENTERLESS GRINDER

The centerless grinder, as the name suggests, permits the grinding of cylindrical workpieces without the use of centers for support. It can be used for narrow parts, like the outside grinding of bearing races, or the grinding of long workpieces like lengths of drill rod. Machines may even be lined up in tandem, to permit roughing and finishing of parts in a continuous process, so that large numbers of parts can be quickly and accurately ground to close tolerances.

The workpiece is fed between the grinding wheel and regulating wheel (Figure N-16). The work rest located between the grinding and regulating wheel supports the workpiece. The regulating wheel is usually a rubber bonded wheel of the same grit size as the grinding wheel. It rotates in the same direction as the grinding wheel, but at a much slower speed, acting as a brake on the workpiece. The feeding action for moving the workpiece between the grinding wheel and the regulating wheel along the work rest blade is provided by a small tilt (typically 1 to 3 degrees) of the regulating wheel (Figure N-17). As shown in this figure, the same feeding action also may be used with a stop for the workpiece. A good example would be the grinding of the stems on automotive valves. Centerless grinding is also suited to the grinding of tapered parts (Figure N-18), and even parts with different diameters (Figure N-19). It is even possible to grind threads on headless set screws, em-

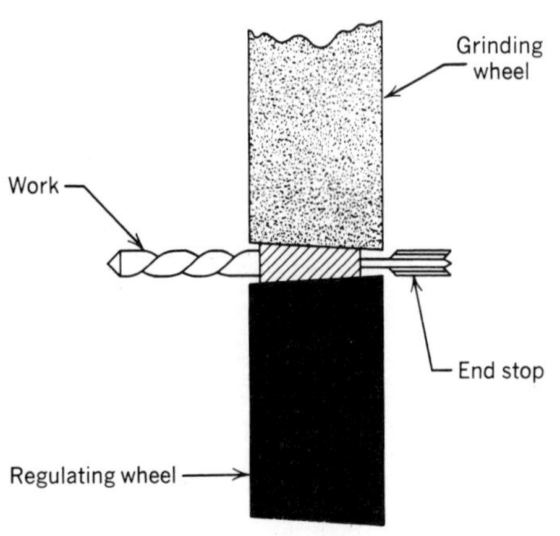

FIGURE N-18 Centerless form grinding (Bay State Abrasives, Dresser Industries, Inc.).

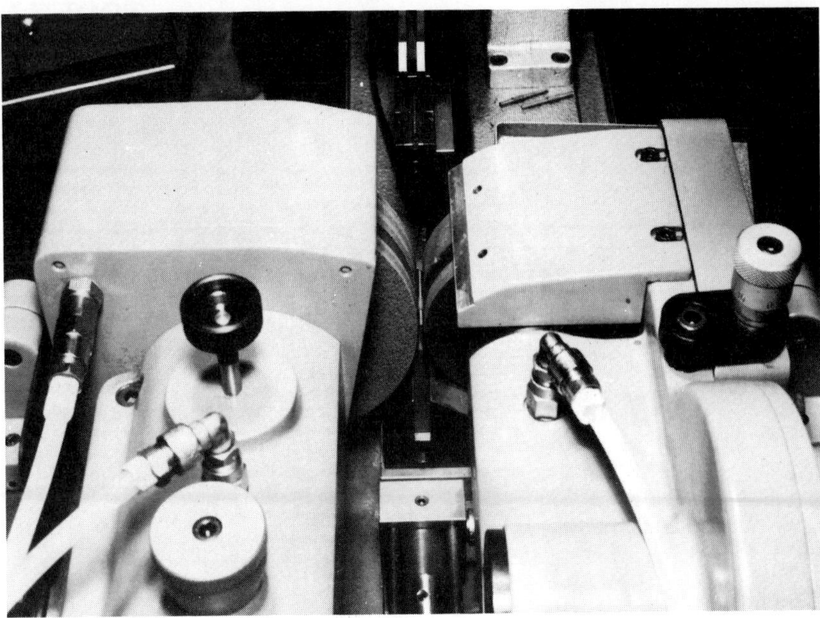

FIGURE N-19 Centerless grinding both ends of a part.

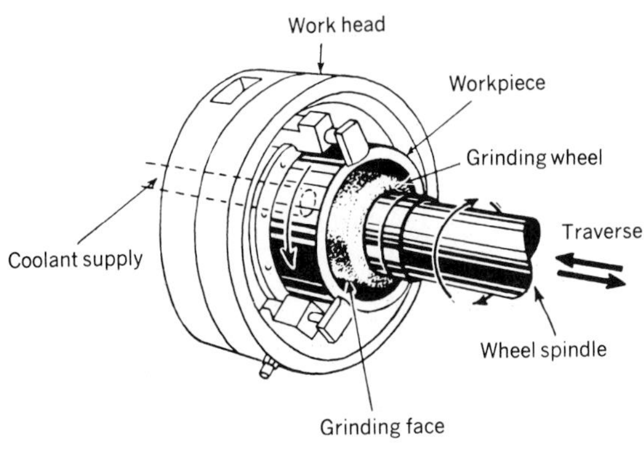

FIGURE N-20 Principles of internal cylindrical grinding (Bay State Abrasives, Dresser Industries, Inc.).

FIGURE N-21 Assortment of mounted wheels (Bay State Abrasives, Dresser Industries, Inc.).

ploying the through-feed method with the spiral thread form dressed into the grinding wheel. A very common through-feed centerless grinding operation is the grinding size of wrist pins for automotive pistons.

INTERNAL CYLINDRICAL GRINDER

Cylindrical grinding processes also apply to internal (Figure N-20) as well as external surfaces. Internal grinding uses mounted abrasive wheels (Figure N-21). These are grinding wheels cemented onto a shank for mounting into the machine's driving spindle (Figure N-22). Internal grinding can be done on concentric workpieces along the machine axis as shown, or for grinding of internal tapers. Internal grinders, when fitted with special tooling, can also be used for nonconcentric parts.

FIGURE N-22 Internal cylindrical grinding (Heald Machine Division/Cincinnati Milacron Company).

The centerless principle can also be applied to internal concentric parts, with the part held and driven between three rolls, while the inside is ground with a mounted wheel.

FIGURE N-23 Universal tool and cutter grinder (Cincinnati Milacron).

TOOL AND CUTTER GRINDER

The universal tool and cutter grinder (Figure N-23) is also a type of cylindrical grinder. However, this machine can also accomplish some limited surface grinding. On these machines the grinding head can be rotated around the vertical axis, and the workhead can be rotated and tilted in a variety of directions. On several designs the grinding spindle can also be tilted as shown in Figure N-23. A variety of grinding wheel shapes and a variety of abrasives can be used with tool and cutter grinders. The primary application of tool and cutter grinders is in the resharpening of milling cutters (Figures N-24 and N-25).

FIGURE N-24 Sharpening of a large brazed carbide-tipped face mill with a diamond cup wheel (Cincinnati Milacron).

FIGURE N-25 Sharpening a plain milling cutter held between centers on the cutter and tool grinder (Cincinnati Milacron).

FIGURE N-26 Form-type gear grinding machine (Ex-Cell-O Corporation).

Miscellaneous Grinding Machines

Gear grinders are another interesting specialized family of grinding machines. Basically these machines fall into two categories: form grinders (Figure N-26), where the grinding wheel is dressed to a precise "mirror image" of the shape of the tooth to be ground, and generating types (Figure N-27), where the form results from the combined action of the wheel and workpiece.

Thread grinders (Figure N-28) are very important in the production of precise lead screws and other precision screw forms.

Other Abrasive Processes

Some abrasive processes use abrasive material applied to the workpiece at low surface speeds. One type uses free abrasive grains flowing in a slurry on a hardened steel plate (Figure N-29) to generate a flat surface on the workpiece. This type is capable of comparatively large amounts of stock removal, but not of high finishes. This is because the abrasive grain tumbles along the hardened steel plate as it cuts the workpiece material. Another similar machine that works on a different principle is the

FIGURE N-27 Generating-type gear grinding machine (Reishauer AG, Wallisellen, Switzerland).

FIGURE N-28 Precision thread grinding machine (Reishauer AG, Wallisellen, Switzerland).

FIGURE N-29 Free abrasive grinding machine has a hard, water cooled plate on which the abrasive is fed in a slurry. The abrasive grains do not become embedded as they do in lapping. Hence the grains always roll around under the workpieces (Speedfam Corporation).

lapping machine (Figure N-30). Here abrasive grains are embedded into a relatively soft lapping plate. Since there is but a single layer of abrasive, very small amounts of material are removed in lapping, but extremely flat and very highly finished surfaces can be produced by this method. The surface of a gage block is an example of a finely lapped surface.

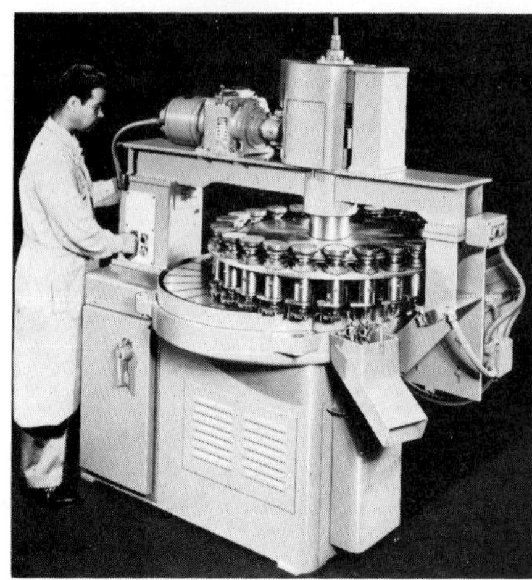

FIGURE N-30 Lapping machine. The pressure plate on top holds the parts to be lapped by the abrasive embedded in the plate beneath (Lapmaster Division of Crane Packing Company).

Parts such as engine cylinders may be finish machined by the honing process. The honing machine (Figure N-31) can be used for both external and internal parts.

Superfinishing is another slow speed abrasive process. A formed abrasive block (Figure N-32) is held against the workpiece as it turns. A small side motion is also applied. The result is a highly finished accurate surface.

Removal of sharp edges left from almost all machining processes can be accomplished by vibratory deburring (Figure N-33). The abrasive materials are in pellet form. Part and abrasive are vibrated together to remove burrs. This process can also be used to remove internal burrs that would be difficult or impossible to reach from the outside.

Grinding Machine Safety

WHEEL SPEED AND WHEEL GUARDS

The same general rules of safety apply to grinding machines as to other machine tools, but there are additional hazards because of the typically high speed of grinding

FIGURE N-31 Honing machine. Honing is a very popular method for finishing inside diameters of everything from bushings to cylinders of automobile engine blocks (Sunnen Products Company).

wheels, which can store a great deal of energy. If a grinding wheel becomes cracked, it can fly apart, ejecting chunks of wheel like missiles.

To reduce the safety hazards from wheel explosions, wheel guards (Figure N-34) are used on nearly every type of grinding machine.

All grinding wheels are rated at specific maximum rpm. Exceeding rated speeds can cause a wheel to fly apart. **Always** check rated speeds marked on the wheel blotter and **never** operate any grinding wheel beyond its stated maximum speed.

The American National Standards Institute (ANSI) has adopted a specific set of standards "Safety Requirements for the Construction, Care, and Use of Grinding Machines" (ANSI B11.9-1975). This document incorporates the earlier standards that specifically covered only grinding wheel safety.

FIGURE N-32 Superfinishing an automotive crank (Taft-Peirce Mfg. Co.).

FIGURE N-33 This high-production vibratory finisher shows the parts and abrasive coming from the vibrator to the front where the parts are unloaded; then the abrasive is returned to the vibrator by the conveyor (UltraMatic Equipment Co.).

RING TESTING A GRINDING WHEEL

A vitrified bond grinding wheel can be "ring tested" for possible cracks (Figure N-35). Hold the wheel on your finger or on a small pin. Tap the wheel lightly with a wooden mallet or screwdriver handle. A good wheel will give off a clear ringing sound, while a cracked wheel will sound dull. If you discover a cracked wheel, advise your instructor immediately. Large grinding wheels may be ring tested while resting on the floor, or while being supported by a sling (Figure N-36).

GRINDING WHEEL SAFETY RULES

1. Handle and store grinding wheels carefully (Figure N-37).

2. Wheels that have been dropped should never be used.

3. Inspect all wheels for cracks or chips before mounting. Ring test them.

4. Do not alter a wheel to fit the grinding machine and do not force it onto the machine spindle.

5. Operating speed should never exceed the maximum allowable operating speed of the wheel.

6. Mounting flanges (Figure N-38) must have equal and correct diameter. The bearing surfaces must be clean for using mounting flanges.

7. Mounting blotters should be used unless the wheel is designed for some other mounting method.

8. The mounting nut must not be tightened excessively.

9. If the machine is a pedestal or bench grinder, adjust the work rest properly, just clear, but not to exceed $\frac{1}{8}$ in. clearance.

10. Do not grind on the side of a straight wheel (Figure N-39). There are certain specific exceptions to this rule. It is recognized that in applications such as shoulder and form grinding, some amount of side grinding takes place.

11. Use a safety guard that covers at least half of the grinding wheel, and do not start the machine until the guard is in place.

12. Allow the grinding wheel to run at least one minute before using it to grind (with the guards in place, of course), and do not stand directly in line with the rotating grinding wheel.

FIGURE N-34 Safety guard on a surface grinder (DoALL Company).

FIGURE N-35 Making a ring test on a small wheel (Lane Community College).

FIGURE N-36 Large grinding wheels can be slung for ring testing (DoALL Company).

13. Always wear approved safety glasses or other eye protection.

14. Use the correct wheel for the material you are grinding. Check *Machining Data Handbook* for guidance.

15. Make sure there is no possibility of bump collision between the grinding wheel and the workpiece before starting the grinding machine. This is especially critical in cylindrical and surface grinding.

16. Turn off grinding fluid before stopping the wheel to avoid creating an out-of-balance condition. If the fluid is applied through the wheel, let the wheel run for at least a minute after the grinding fluid is stopped.

FIGURE N-37 Storing extra wheels at the machine on pegs is often convenient and practical. The main requirement is to keep the wheels separate or protected, and off the floor (DoALL Company).

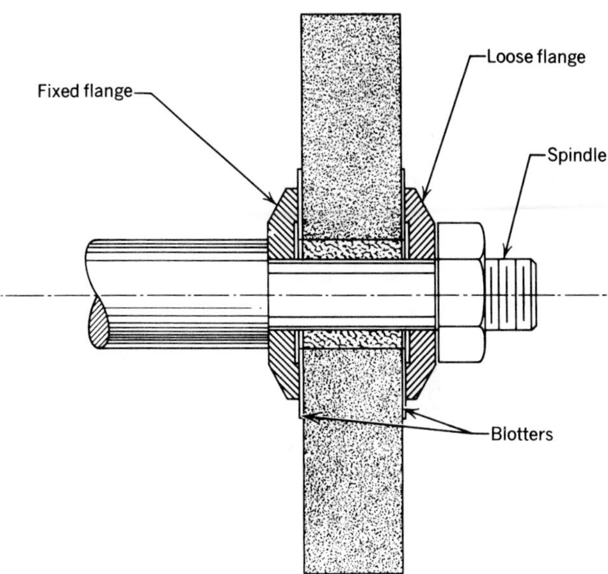

FIGURE N-38 Typical set of flanges with flat rims and hollow centers with blotters separating the wheel and the flanges. Tightening the nut too much could spring the flanges and perhaps even crack the wheel.

FIGURE N-39 Straight wheels are designed for grinding on the periphery. Never grind on the side. Cylinder wheels, cup wheels (both straight and flaring), and segments or segmental wheels (shown here without the holder) are all designed and safe for side grinding (Bay State Abrasives, Dresser Industries, Inc.).

UNIT 1

Selection and Identification of Grinding Wheels

Selecting a grinding wheel is much like selecting any other cutting tool. Many of the same factors apply, such as size and shape. However, in grinding wheels, additional factors influence the selection process. These include: types of abrasives, grit, grade, structure, and bond, as well as the particular grinding task to be done. The purpose of this unit is to familiarize you with the identification of grinding wheels and to assist you in selecting the one that best fits your needs.

Objectives

After completing this unit, you should be able to:

1. List five principal abrasives with their general areas of best use.
2. List four principal bonds with the types of applications where they are most used.
3. Identify by type number and name, from unmarked sketches, or from actual wheels, four commonly used shapes of grinding wheels.
4. Interpret wheel shape and size markings together with five basic symbols of a wheel specification into a description of the grinding wheel.
5. Given several standard, common grinding jobs, recommend the appropriate abrasive, approximate grit size and grade, and bond.

Types of Abrasives

ALUMINUM OXIDE

Aluminum oxide is the most widely used abrasive in manufacturing. The higher the level of purity, the more friable (less tough) the grit becomes. The purer grades are very often white in color, and are used mainly on hardened steels. The grinding of hard steel causes a fairly rapid breakdown of the individual grains to expose new sharp edges. A sharp wheel cuts easily, with less heat being produced. Less pure aluminum oxide usually is gray in appearance. It has a tougher grain structure that resists fracturing, and it is better for applications such as offhand grinding on a pedestal grinder. For extremely harsh applications, such as snagging in foundries, a crystalline combination of aluminum and zirconium oxide has been developed which results in a very tough abrasive.

While aluminum oxide works well on steels, it can work poorly on cast iron. This appears to be a matter of the aluminum oxide being somewhat soluble in cast iron at the high temperature of grinding. The wheel dulls quickly, apparently from the sharp edges being dissolved away into a solid solution with the iron.

SILICON CARBIDE

Silicon carbide (Figure N-40) is a somewhat harder abrasive than aluminum oxide, but with a sharper, more friable, and quite brittle crystalline structure. It works well on cast iron and nonferrous materials, such as aluminum and copper based alloys. It also works on carbide cutting tools, but not nearly so well as the cooler cutting diamond abrasives, which have largely replaced silicon carbide for this application. Silicon carbide is useful in the grinding of many of the titanium alloys, as well as austenic stainless steels.

CUBIC BORON NITRIDE

Cubic boron nitride (CBN) (Figure N-41) is a much harder abrasive than silicon carbide. This abrasive, called Borazon, was created by the General Electric Company in the late 1950s. It was intro-

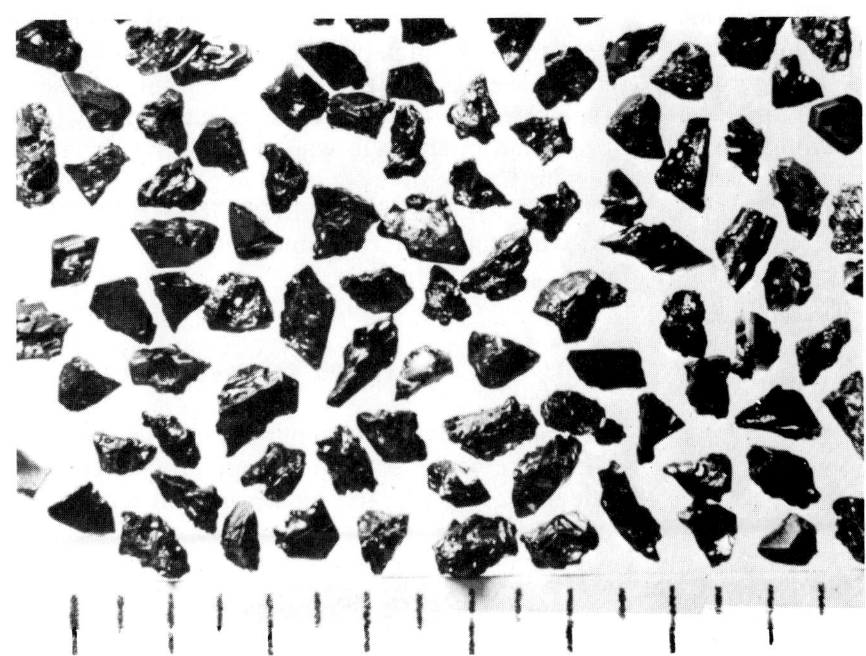

FIGURE N-40 Large grains of silicon carbide (Exolon Company).

FIGURE N-41 Cubic boron nitride (Borozon™ CBN) (Speciality Materials Dept., General Electric Company).

FIGURE N-42 Manufactured diamond abrasive (Speciality Materials Dept., General Electric Company).

duced as a commercial product in 1969. It works best on hardened ferrous alloys, especially the difficult-to-machine cobalt and nickel superalloys used widely in jet engine applications. Until 1981, this abrasive was available only as a cubic single crystal structure that would fracture only as large grains of the crystal (macrofracture). For this reason, most of the applications were for roughing, and only in materials harder than Rockwell C50. A microcrystalline version was introduced in 1981 which microfractures in micron-sized (.0000394 in.) particles to produce new sharp edges. This new form is also capable of operating at a much higher temperature than the original form. The chips produced by this superabrasive are well formed, like most milling chips. This sharp cutting action results in a much cooler operating temperature, which is important in the grinding of heat-sensitive materials. The abrasive is very expensive compared to either aluminum oxide or silicon carbide, but the time saved from faster cutting action, less dressing, consistent size control, and less frequent wheel changing makes cubic boron nitride the least expensive abrasive per part produced.

DIAMOND

Diamond (Figure N-42) is another superabrasive that can be obtained in either natural or manufactured form. It is the hardest substance known.

The single crystal diamond nib used for the trueing and dressing of grinding wheels is a common application. For several economic and strategic reasons, there was great interest in the development of synthetic diamond. This was accomplished by the General Electric Company in the early 1950s. The first commercial marketing was begun in 1957. Diamond abrasive is especially useful in the grinding of ceramics and tungsten carbide. It is not particularly effective in steels, or superalloys containing cobalt or nickel, probably because these materials pick up carbon from diamond very readily. Because of the cost of diamond, its use is frequently restricted to those applications where no other abrasive will work effectively. However, the same economic reasoning applied to diamond tooling as to cubic boron nitride. If it costs less per part with diamond, it should be the preferred tool for that operation.

Size and Shape of Wheels

The standard coding system uses numbers 1 to 28 to indicate wheel size and shape. You should be familiar with five of the most common types:

Type 1—Straight wheel (Figure N-43)

Type 2—Cylinder wheel (Figure N-44)

Type 6—Straight cup wheel (Figure N-45)

Type 11—Flaring cup wheel (Figure N-46) with grinding faces on both face and wall

Type 12—Shallow dish wheel (Figure N-47)

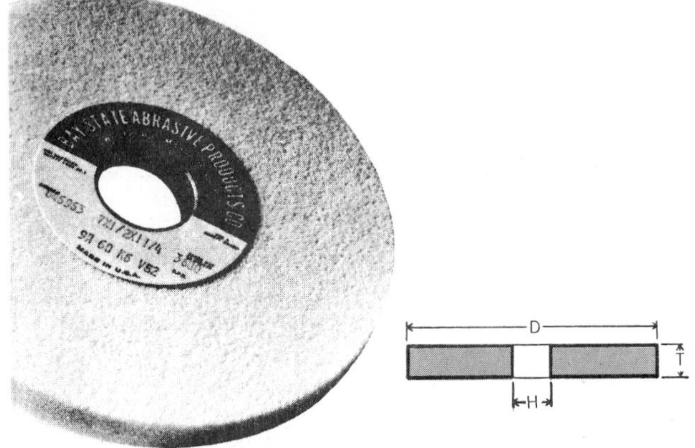

FIGURE N-43 Straight or type 1 wheel, whose grinding face is the periphery. Usually comes with the grinding face at right angles to the sides, in what is sometimes called an "A" face (Bay State Abrasives, Dresser Industries, Inc.).

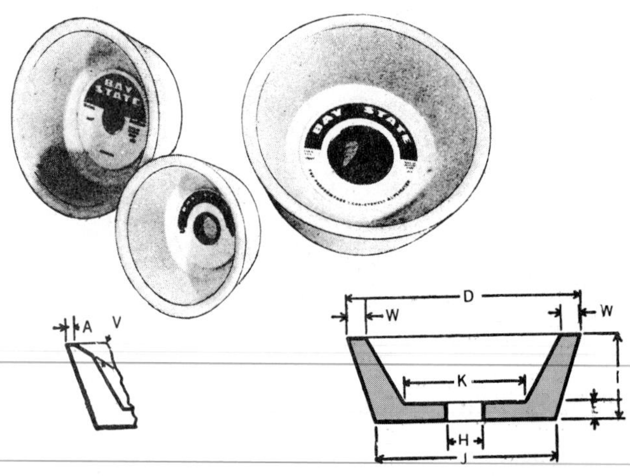

FIGURE N-46 Flaring cut or type 11 wheel, whose grinding face is also the flat rim or wall of the cup. Note that the wall of the cup is tapered (Bay State Abrasives, Dresser Industries, Inc.).

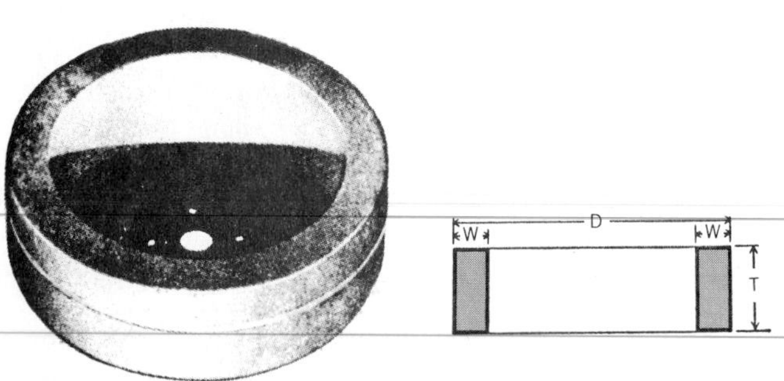

FIGURE N-44 Cylinder or type 2 wheel, whose grinding face is the rim or wall end of the wheel. Has three dimensions—diameter, thickness, and wall thickness (Bay State Abrasives, Dresser Industries, Inc.).

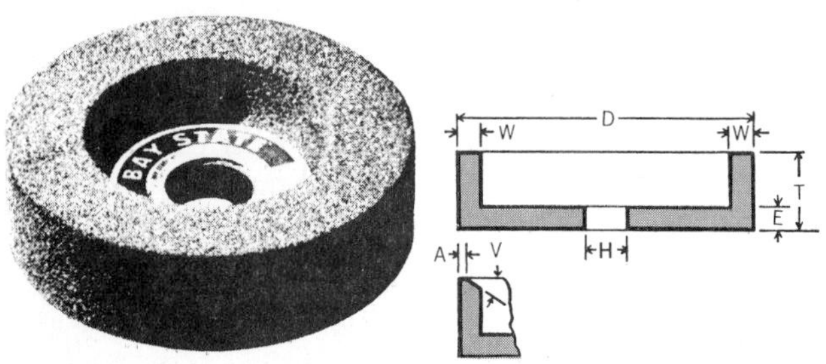

FIGURE N-45 Straight cut or type 6 wheel, whose grinding face is the flat rim or wall end of the cup (Bay State Abrasives, Dresser Industries, Inc.).

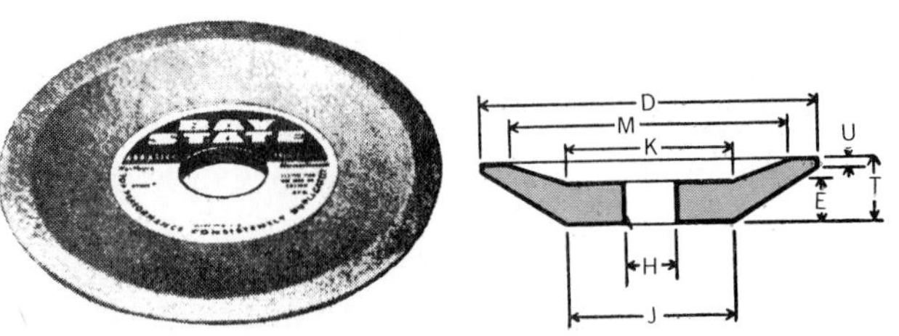

FIGURE N-47 Dish or type 12 wheel, similar to type 11, but with a narrow, straight peripheral grinding face in addition to the wall grinding face. Only wheel of those shown that is considered safe for both peripheral and wall or rim grinding (Bay State Abrasives, Dresser Industries, Inc.).

Standard Wheel Marking Systems

A standard wheel marking system (Figure N-48) is used for the purpose of identifying five major factors in grinding wheel selection:

1. Type of abrasive
2. Grit size
3. Grade or hardness
4. Structure
5. Bond

These factors are indicated on the grinding wheel blotter by a numeric and letter identification code. For example, a wheel marked

A 60 – J8V

would indicate the following:

FIRST SYMBOL—TYPE OF ABRASIVE (**A** 60 – J8V)

Five major abrasives are in common use:

1. A—Aluminum oxide
2. C—Silicon carbide
3. D—Natural diamond
4. MD or SD—Manufactured or synthetic diamond
5. B—Cubic boron nitride

SECOND SYMBOL—GRIT SIZE (A **60** – J8V)

Grit refers to the size of the abrasive grains. Grit size ranges from 4 to 8 (coarse) up to 500 or higher (fine). Grit numbering is derived from the screen openings used to sort the abrasive grains after manufacture. The following approximate scale can be used to determine grit:

4 36 46 100 120 240 500

←Coarse→ ←Medium→ ←Fine→

Selection of grit depends on the amount of stock to be removed and the surface finish requirements. Usually coarse grits are used for fast stock removal and on soft ductile materials. Fine grit is used for hard brittle materials. General usage calls for wheel grits ranging from 46 to 100.

THIRD SYMBOL—GRADE OF HARDNESS (A 60 – **J**8V)

Grade of hardness (Figure N-49) is a measure of the bond strength of the grinding wheel. The bond material holds the abrasive grains together in the wheel. The stronger the bond, the harder the

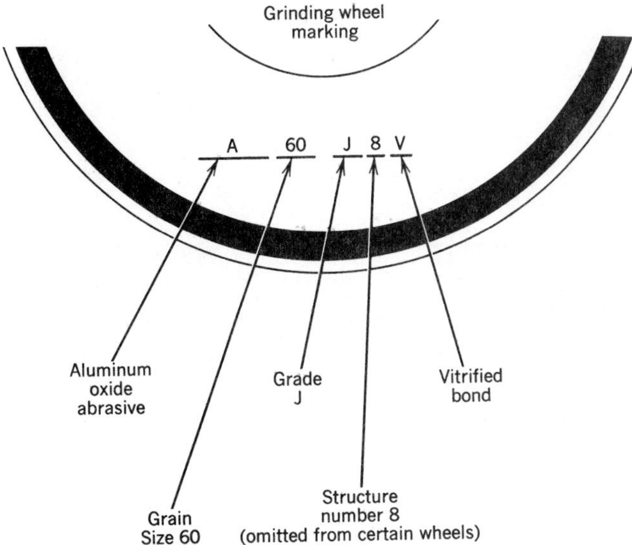

FIGURE N-48 Wheel specification.

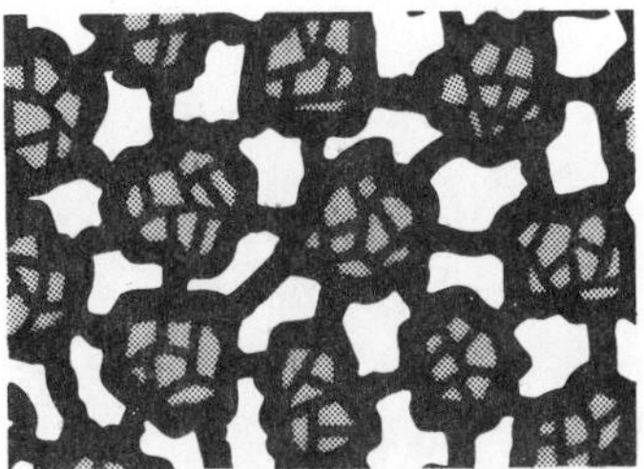

Weak holding power

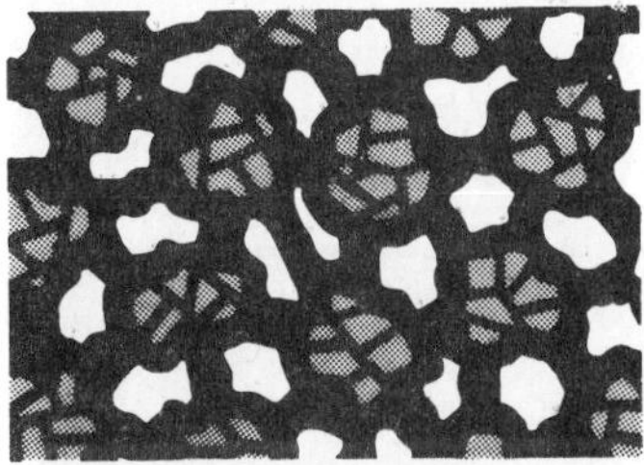

Medium holding power

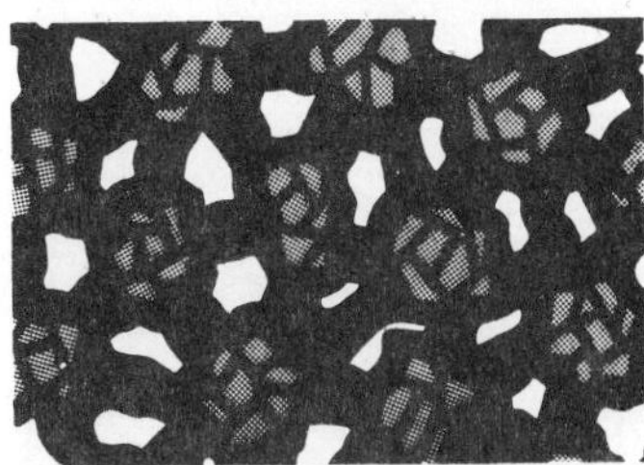

Strong holding power

FIGURE N-49 Three sketches illustrating (from top down) a soft, medium, and a hard wheel. This is the grade of the wheel. The white areas are voids with nothing but air; the black lines are the bond; and the others are the abrasive grain. The harder the wheel, the greater the proportion of bond and, usually, the smaller the voids (Bay State Abrasives, Dresser Industries, Inc.).

wheel. Precision grinding wheels tend to be softer grades because it is necessary to have dull abrasive grains pulled from the wheel as soon as possible. This will expose new sharp grains to the workpiece. If this does not happen, the wheel will become glazed with dull abrasive. Cutting efficiency and surface finish will be poor. Later alphabet letters indicate harder grades. For example, F to G are soft, where R to Z are very hard.

FOURTH SYMBOL— STRUCTURE (A 60 – J**8**V)

Structure, or the spacing of the abrasive grains in the wheel (Figure N-50), is indicated by the numbers 1 (dense) to 15 (open). Structure provides chip clearance so that chips may be thrown from the wheel by centrifugal force or washed out by the grinding coolant. If this does not happen, the wheel becomes loaded with workpiece particles (Figure N-51) and must be dressed.

Dense spacing

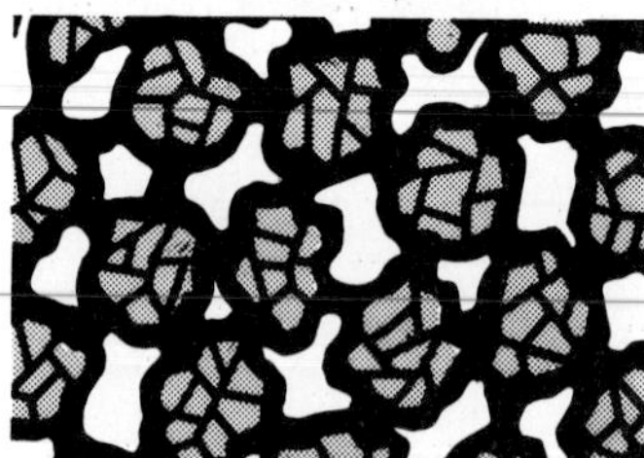

Medium spacing

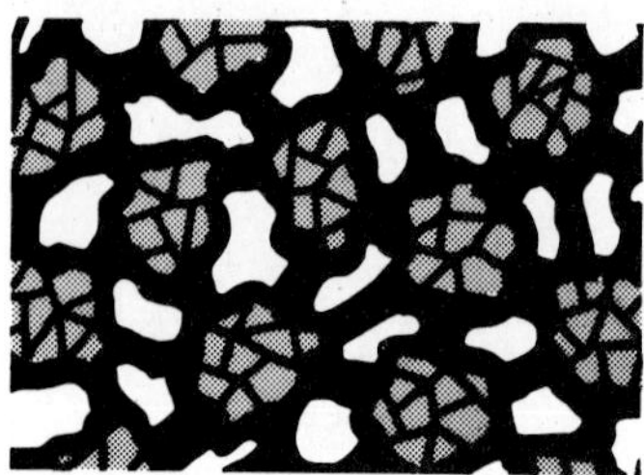

Open spacing

FIGURE N-50 Three similar sketches showing structure. From the top down, dense, medium, and open structure or grain spacing. The proportions of bond, grain, and voids in all three sketches are about the same (Bay State Abrasives, Dresser Industries, Inc.).

FIGURE N-51 The wheel at the left is called loaded, with small bits of metal embedded in its grinding face. It is probably too dense in structure or perhaps has too fine an abrasive grain. Same wheel at the right been dressed to remove all the loading (Desmond-Stephan Mfg. Co.).

FIFTH SYMBOL—BOND (A 60 – J8**V**)

Bond is identified by letter according to the following:

V Vitrified
B Resinoid
R Rubber
E Shellac
M Metal

In the machine shop, vitrified bonds are more common. Vitrified wheels are used for precision grinding. Resinoid wheels are used in rough grinding operations, such as "snagging" of castings in a foundry with high wheel speeds and heavy stock removal. Rubber is the usual bonding material for regulating wheels on centerless grinders, and shellac still finds some use in the finish grinding of surfaces like camshafts. Metal bonded wheels are necessary for such applications as electrochemical grinding, where the bonding material must be a good conductor of electricity. Bonds also affect wheel speeds. Vitrified wheels are rated up to 6500 sfpm. Resinoid wheels are rated to 16,000 sfpm or higher.

Other Factors in Wheel Selection

Grinding is probably the most challenging part of machine tool operation, because there are so many variables that must be balanced to get first-class results. Three of these factors relate to workpiece materials, which are constantly being changed in the shop environment. Five other factors in wheel selection are relatively fixed.

VARIABLE FACTORS

COMPOSITION OF THE WORKPIECE For most steels, and steel alloys, aluminum oxide grinding wheels are the usual choice. But with very high temperature alloys, particularly where the part will be operated under stress, cubic boron nitride may be far preferable. Now that microcrystalline cubic boron nitride is available, there are economical applications for this abrasive in softer steels, particularly for form grinding. For grinding cemented carbides, diamond is greatly preferred, as the use of silicon carbide often results in heat checking and premature tool failure. For most cast irons, silicon carbide is the wheel abrasive of choice, but for some roughing applications, aluminum oxide can be used. While ferritic and martensitic stainless steels are best with aluminum oxide, austenitic stainless steel is better ground with silicon carbide.

CUTTING FLUIDS For most toolroom grinding, a synthetic grinding fluid which avoids the use of soluble oil is generally preferred. But where heavy-duty grinding is done, soluble oils of differing characteristics are necessary, and in very high speed grinding, sulfo-chlorinated straight oils may be needed to avoid damage to the workpiece. In these latter cases, extensive exhaust ducting with oil precipitation equipment is needed.

MATERIAL HARDNESS This factor is of concern in the choice of both grit size and grade. Generally, for soft, ductile materials, the grit is more coarse, and the grade can be harder. For hard materials, finer grit and softer grades are the rule. In the machine shop a coarse abrasive would be considered to be in the range 36 to 60, and a fine abrasive in the range of 80 to 100. In this setting, grades F, G, and H would be considered soft, while J, K, and L would be considered relatively hard. Too coarse a grit tends to leave excessive scratching, which can be difficult to correct later. On very hard material, coarse grit may in fact remove less stock than a finer grit because of lack of workpiece penetration. Too soft a wheel may wear too fast to be economical, while an overly hard wheel may glaze quickly and cease to cut. In many cases, overly hard wheels are selected, when experimenting might show that a softer wheel would cut more freely, without significantly more wheel wear.

The structure of the grinding wheel also strongly relates to the material being ground. For very soft, free-cutting materials that produce relatively large chips, an open structure is essential to provide space for the chip to be carried through and be ejected. Otherwise, the wheel will become loaded.

WORKPIECE FINISH When production grinding is done on larger lot sizes, wheels can be very carefully selected for abrasive grit size and bonding material to perfectly match the finishing needs of the product. In the job shop setting, with typically small lot sizes (or even just one part), it is uneconomical to keep changing grinding wheels. In most such cases, a general purpose wheel is selected about midway between the usual roughing and finishing requirements. The skill of the operator in the dressing of the wheel then becomes the controlling factor in the finish obtained. The same wheel can be dressed either for aggressive rough grinding or for gentle finish grinding, to make a relatively coarse wheel behave as a finishing tool. This will be covered in the next unit.

FIXED FACTORS

HORSEPOWER OF THE MACHINE Machine horsepower tends to rise with each new generation of grinding machines. With increased horsepower goes the need for more machine rigidity, or the horsepower cannot be utilized. Higher power and increased rigidity permit the selection of harder wheels to go along with the capacity to do more work in a given time. As the superabrasives become more widely accepted, specific machine tools are being designed for their use.

SEVERITY OF THE GRINDING In a production setting, great care should be taken in matching the characteristics of the grinding wheel to the specific job to be done. This could involve trials with wheels that would come close to meeting the requirements, before selecting the one best wheel. *Machining Data Handbook* is an excellent source of starting recommendations on grit size and wheel grade. In the toolroom setting, where grinding pressures are typically low, a friable aluminum oxide wheel of relatively soft bond hardness is the typical choice.

AREA OF GRINDING CONTACT On any given design of grinding machine, this remains essentially constant. The rule is finer grit sizes and harder wheels for small areas of contact, and coarser grit sizes and softer wheels for larger areas of contact. In the machine shop, this means grits mostly in the 46 to 100 range, except for specialty work such as thread grinding that might use wheels as fine as 220 grit, or vertical spindle, rotary table grinding (often called Blanchard grinding), which often uses a grit size of 30, especially for soft steel applications. For horizontal spindle surface grinding the typical bond hardness selected lies in the I and J region, with the softer bond for the harder material. On cylindrical grinding, with line contact with the wheel, J and K are the most common hardnesses.

WHEEL SPEED This is usually fixed by the nature of the machine, but the operator has the responsibility to be sure that the wheel that is mounted is safe for the spindle speed of the machine. On vitrified wheels most have a maximum speed of 6500 sfpm. Organic bonded wheels (rubber, resinoid, or shellac) can be rated at higher speeds, up to 16,000 sfpm, and experimental work is being done with plated superabrasive media with speeds up in the 30,000 sfpm region.

USE OF GRINDING FLUID This is a relatively minor factor in grinding wheel selection, and typically results in only one step of grade change. With grinding fluids as opposed to dry grinding, one grade harder wheel is typically chosen.

In any shop, however, unless you are really just starting out, there should be information on what wheels have worked best on specific jobs. If there are changes to be made, these should be in small steps, rather than radical changes. Since most shops handle a variety of work, it usually doesn't pay to switch wheels all the time. A great deal can be done by the operator in his or her preparation procedures to adapt a general purpose wheel to specific workpiece grinding.

SELF-TEST

1. Select the preferred abrasive for each of the following materials: bronze, low carbon steel, carbide, and hard high temperature alloy steel.

2. What are the names of five common shapes for grinding wheels?

3. What shape of grinding wheel is considered safe for grinding both on the side and on the periphery of the wheel?

4. What result occurs from choosing a grinding wheel with a bond strength that is too hard?

5. What bond is essentially limited to 6500 SFPM?

6. What is meant by wheel grade?

7. You are to select a grinding wheel for horizontal surface finish grinding of 1020 steel. What would you choose for each of the following wheel specifications?

a. Abrasive.
b. Bonding material.
c. Abrasive grit size.
d. Bond hardness.

8. You are to select a grinding wheel for grinding a hardened and tempered lathe mandrel on a cylindrical grinder. What would you choose for each of the following wheel specifications?

a. Abrasive.
b. Bonding material.
c. Abrasive grit size.
d. Bond hardness.

9. You will be grinding the surface of tungsten carbide wear pads with a horizontal spindle surface grinding machine. What choices of abrasive would you consider for this job?

10. You have been given a straight grinding wheel to identify with the markings C 80 J8B. Describe the structure, bond hardness, grit size, and abrasive material and name the bonding material.

UNIT 2

Trueing, Dressing, and Balancing of Grinding Wheels

A grinding wheel must run true with every point on its cutting surface concentric with the machine spindle. As it becomes loaded with workpiece material, it must be dressed to restore sharpness. It must also run in balance because of its great speed. These procedures of trueing, dressing, and balancing are important parts of grinding operations and the purpose of this unit is to familiarize you with them.

Objectives

After completing this unit, you should be able to:

1. Describe trueing, dressing, and balancing of grinding wheels.
2. Distinguish the difference between the objectives of trueing and dressing a grinding wheel.
3. Correctly position a single-point diamond dresser in relation to the grinding wheel.

Trueing and Dressing

When a new wheel is installed on the grinder, it must be trued before use. The cutting surface of a new wheel will run out slightly due to the clearance between the wheel bore and machine spindle (Figure N-52). Trueing a wheel will bring every point on its cutting surface concentric with the machine spindle. This concentricity is very important for achieving smooth and accurate grinding conditions.

In most grinding operations, small chips of workpiece material become lodged in the cutting surface of the grinding wheel. In addition, if the wheel bonding hardness is excessive, dulled abrasive grains can remain in the grinding wheel. Both of these conditions will impair the cutting efficiency and they must be removed as needed to maintain proper cutting action. This process is

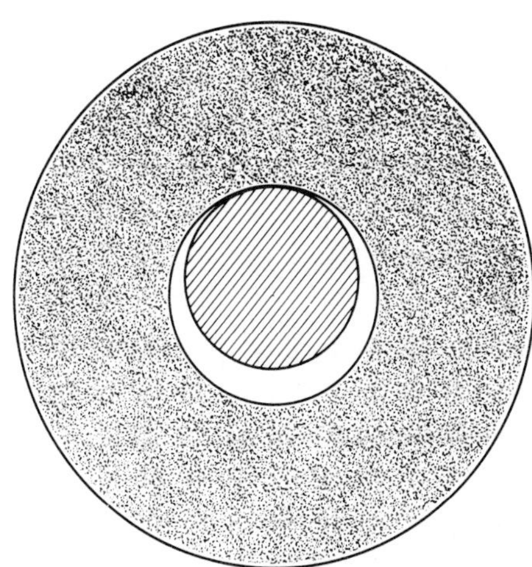

FIGURE N-52 Sketch, exaggerated for effect showing a grinding wheel on a spindle. Actually, the gap under the spindle may be only a few thousandths of an inch, but still enough to cause problems if the wheel is not trued to the center of the spindle.

FIGURE N-53 Single-point diamond dresser. The most important precaution in using such a dresser is to turn the diamond often to avoid grinding flats on it. This diamond is pointing to the right (Desmond-Stephen Manufacturing Co.).

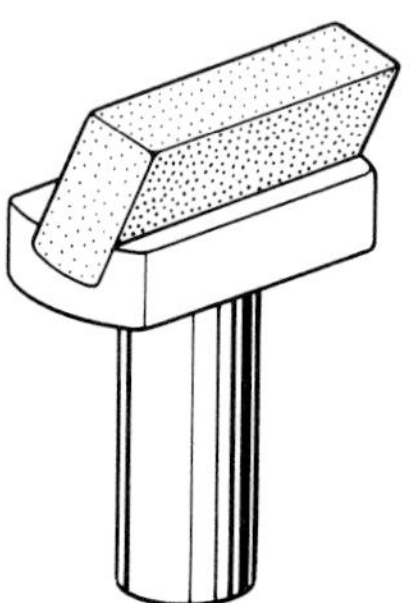

FIGURE N-54 Cluster-type dressers have come into use mostly because several smaller diamonds are cheaper than one large diamond (Desmond-Stephan Manufacturing Co.).

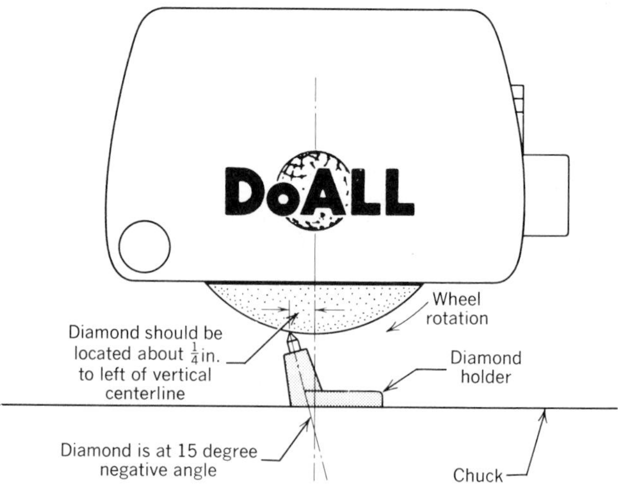

FIGURE N-55 This is one of several ways of mounting a dresser on a surface grinder. The dresser with its diamond is spotted on the clean magnetic chuck. Note that the diamond is slanted at a 15-degree angle and slightly past the vertical centerline of the wheel (DoALL Company).

called dressing and is highly important to good results in grinding. Dressing is the process of sharpening a grinding wheel.

Both trueing and dressing remove a certain amount of material from the grinding wheel. Wheels should be trued and dressed *only enough* to establish concentricity or to expose new sharp abrasive grains to the workpiece.

Trueing and dressing on precision grinders is done with single (Figure N-53) or multiple-point diamond dressers. The cluster dresser (Figure N-54) may be wide enough to reach across the entire cutting surface of the wheel.

Dressers are always solidly mounted. The diamond and its holder are mounted on the grinder chuck so that they can be traversed across the cutting surface of the wheel. The dresser must be positioned offcenter on the wheel on the outgoing rotation side (Figure N-55). This is to prevent the dresser from getting caught and being pulled under the wheel. Dressers are sometimes marked with arrows that are to be pointed in the same direction as wheel rotation.

When trueing, lower the wheel head while traversing the diamond with the table travel, until you find the highest point on the wheel. Continue to traverse the dresser across the wheel and feed down .001 in. after each pass. Each time, a bit more abrasive will be removed from the wheel. When the dresser is cutting all around, the wheel has been fully trued. Do not remove any more abrasive than is necessary to achieve concentric running of the cutting surface. Trueing may be done dry, if the diamond is given a few seconds to cool after several passes have been made. If the trueing is done with grinding fluid it is very important to flood the diamond continuously; otherwise, the diamond could be fractured by becoming hot and then having fluid splash against it. After trueing and dressing it is desirable to break the sharp corners of the wheel with a dressing stick, leaving a small radius. This will prevent the sharp corners from leaving undesirable feed line scratches on the workpiece.

After the wheel has been in use, the diamond will be used to remove embedded workpiece particles and dulled grains to expose new sharp abrasive grains. This is the process of dressing and is done in the same way as trueing. The speed of traverse can influence the surface finish obtainable on the workpiece. A slow dressing traverse will result in the diamond machining the abrasive grit smoother, which in turn will result in a smoother workpiece finish, but less efficient grinding. A rapid dressing traverse will leave a sharper wheel, but the surface finish will be rougher. In many cases the same wheel can be used for both roughing and finishing, a coarse dress for roughing a number of parts, followed by a fine dress for finishing the parts. It is important not to try to rough grind the next batch of parts without redressing the wheel. The effect of different dressing procedures can be substantial; for example, a slowly dressed 60-grit wheel can be made to behave like a rapidly dressed 120-grit wheel for finishing, but

FIGURE N-56 Built-in wheel dresser. Lever traverses dresser across wheel (DoALL Company).

FIGURE N-57 Since the hole in the core of a diamond or CBN wheel is machined, it can be fitted to much closer tolerances (K & M Industrial Tool, Inc.).

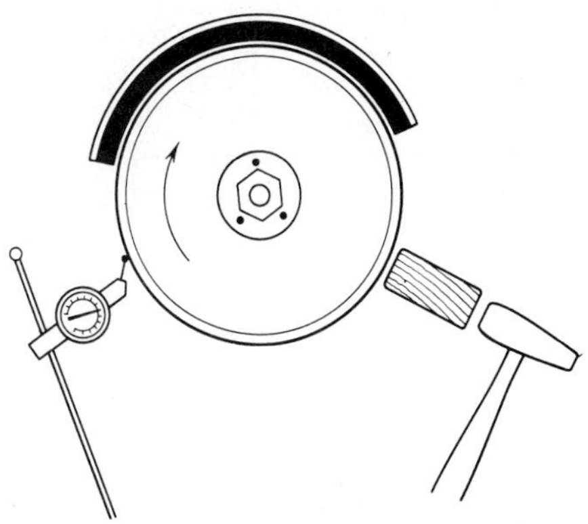

FIGURE N-58 Runout on a diamond wheel is checked with a dial indicator and must be within .0005 in. for resinoid wheels or .00025 in. (half as much) for metal bonded wheels. Tapping a wooden block held against the wheel to shift the wheel on the spindle is often enough to bring it within limits. Otherwise, it will have to be trued (Precision Diamond Tool Co.).

FIGURE N-59 Sometimes it is practical to use one grinding wheel to dress another. The abrasive wheel in the dresser is being used with a metal bonded diamond wheel on the grinding machine (Norton Company).

the reverse is never true. Grinding machines may have a built-in dresser with micrometer feed (Figure N-56).

Because boron nitride and diamond abrasives are expensive, it is not desirable to true or dress these wheels excessively. The bore of a diamond wheel is machined so that it will closely fit the grinder spindle (Figure N-57). When a diamond or CBN wheel is mounted, it may be adjusted to run true by using a dial indicator. The wheel is tapped lightly with a block of wood (Figure N-58). If the diamond wheel cannot be adjusted within tolerance, it may be trued or dressed. The single point diamond or cluster type dresser must **not** be used for this purpose. A better alternative is to use a brake type trueing device, with a 46- to 60-grit silicon carbide grinding wheel (Figure N-59). Trueing glazes superabrasive grinding wheels, so they must be conditioned afterwards in order to work. On wheels less than 12 in. in diameter, this can

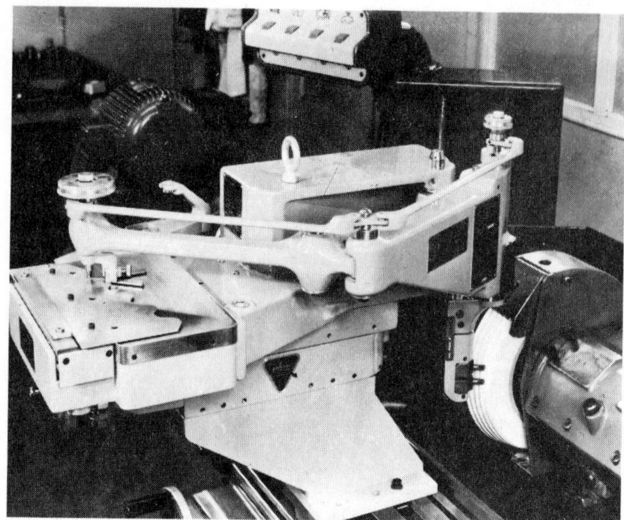
FIGURE N-60 A dressing unit such as this can be set to dress practically any desired shape in a wheel. It is very versatile (Engis Co., Diamond Tool Division).

FIGURE N-61 This is a crush roll, which literally crushes out the form in the wheel, instead of cutting it, as with diamond rolls or blocks (DoALL Company).

be done by hand, using a fine grit aluminum oxide dressing stick. This is done using a small amount of grinding fluid. The dressing stick is pushed into the wheel by hand, and gradually the wheel begins to consume the stick more rapidly. When the rate increases sharply the dressing is done. This proc-

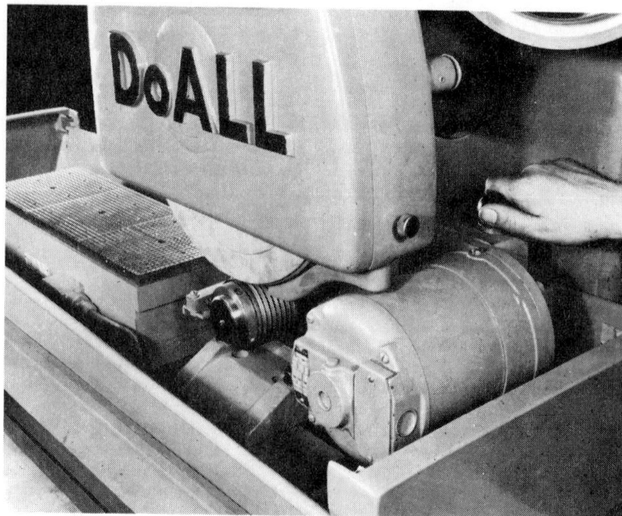

FIGURE N-62 The crush roll must be mounted to provide support against considerable force, either as here on a center type cylindrical grinder, or on a surface grinder (Bendix Automation and Measurement Division).

ess may use up 2 or 3 cubic inches of a dressing stick in the process. Electroplated superabrasive wheels should **not** be either trued or dressed, as the abrasive is only a single layer deep.

RADIUS AND FORM DRESSING

Contours, radii, and other special shapes can be ground by forming the reverse geometry on the grinding wheel. Various methods are used for this. A radius dresser that swings the single point diamond in a preset arc is one method. The wheel dressing pantograph (Figure N-60) can form dress a grinding wheel to almost any shape.

In crush roll dressing (Figure N-61) the form is literally crushed into the wheel by a carbide or diamond roll. Crush dressing is used on both surface and cylindrical grinders. Crush roll dressing results in a very sharp, free-cutting grinding wheel that leaves a less burnished finish than a diamond dressing. For this reason the wheel must be an accurate match for the grinding to be done. The roll must be solidly mounted to withstand the pressures involved (Figure N-62). The diamond dressing block is another method used to form dress a wheel (Figure N-63).

Balancing

Balancing may be required on large wheels (over 14 in. diameter) but is usually not required for smaller wheels. An out-of-balance wheel can cause chatter marks in the workpiece finish.

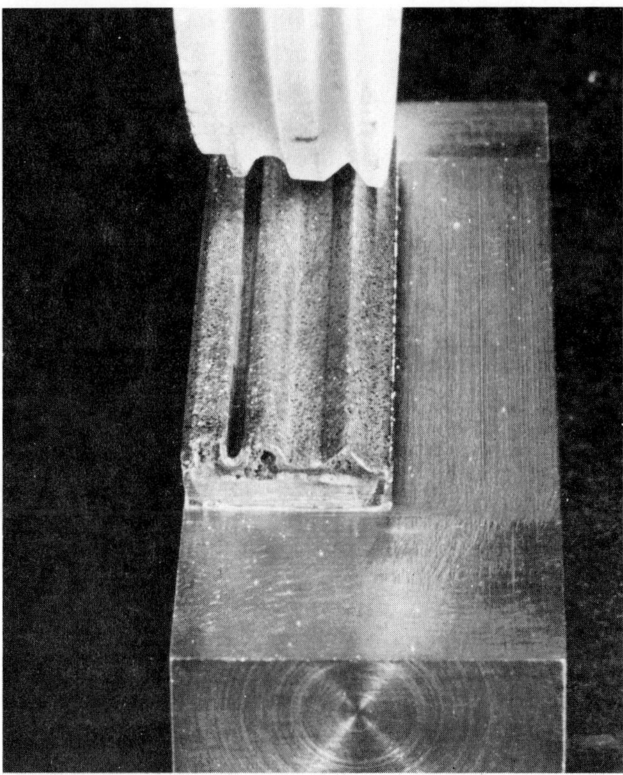

FIGURE N-63 Diamond-plated dressing block, intended for use on a surface grinder. The block is held flat on the magnetic chuck, and the wheel is traversed back and forth along it. The block is formed to the shape of the finished workpiece (Engis Co., Diamond Tool Division).

FIGURE N-64 This type of balancing device with overlapping wheels or disks is quite common. It has an advantage in that it need not be precisely leveled (Bay State Abrasives, Dresser Industries, Inc.).

FIGURE N-65 Balancing a wheel on two knife edges, as on this unit, is very accurate, because there is minimum friction. Of course, the unit must be perfectly level and true. Otherwise, the wheel may roll from causes other than out-of-balance (Bay State Abrasives, Dresser Industries, Inc.).

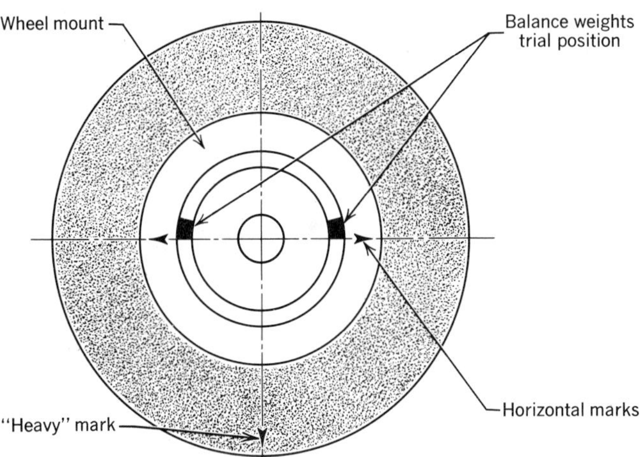

FIGURE N-66 With the weight at some point between the vertical and the horizontal centerlines, the wheel should be in proper balance, stationary in any position. If not, one or two other balance weights should be used (Bay State Abrasives, Dresser Industries, Inc.).

Wheels are balanced on the overlapping disk balancing ways (Figure N-64) or parallel ways (Figure N-65). These tools must be set to a precise level. The grinding wheel is mounted on a balancing arbor and placed on the ways. The heavy point will rotate to the lowest position. By adjusting weights in the flanges (Figure N-66), balance can be achieved. Balance should be carefully done, to the point where the weight of a postage stamp applied to the grinding surface in a horizontal position will cause the wheel to move. Some production grinders are equipped with devices used to balance a wheel while it is in motion. This is termed *dynamic balancing*.

SELF-TEST

1. What is meant by *trueing* a grinding wheel, and why is it important?

2. Explain the difference between dressing and trueing a wheel. How often should a wheel be dressed?

3. Define *form dressing*, mentioning at least two methods of form dressing.

4. Explain the placement of a single-point dresser.

5. List the procedure in trueing an aluminum oxide or silicon carbide wheel after mounting.

6. Explain the essential differences between dressing a grinding wheel for roughing as against finishing.

7. Under what circumstances might it be necessary to true a superabrasive wheel and how is it done?

8. Should electroplated superabrasive wheels be trued or dressed?

9. Generally, what determines whether a wheel needs to be balanced? Describe the procedure.

10. What advantages is there to using a balancing device with overlapping disks as compared with a knife edge balancing stand?

UNIT 3

Grinding Fluids

In grinding, a great deal of heat is generated from the friction between abrasive grains and the workpiece. Temperatures to 2000°F (1093°C) are not uncommon at the point where the grain is actually cutting. Even though much grinding is done with grinding fluids, the high speed of the grinding wheel creates enough of a fan effect to blow the grinding fluid away from the contact area. The actual cutting process may then be occurring in a dry and hot environment. Nonetheless, grinding fluids are almost essential in grinding. The purpose of this unit is to describe the purpose, types, application methods, and cleaning of grinding fluids.

Objectives

After completing this unit, you should be able to:

1. List reasons for using grinding fluids.

2. List three types of grinding fluids.

3. Describe methods and grinding fluid application.

4. Describe methods of cleaning grinding fluids.

Purpose of Grinding Fluids

The correct selection of grinding fluid can have a significant effect on the grinding process. But the correct grinding fluid, poorly applied, can in some cases be worse than none at all. All of these fluids serve a cooling function, but some emphasize lubrication, even at the expense of cooling capacity.

Grinding fluids are used to:

1. Reduce temperature in the workpiece, thus reducing warping in thin parts and aiding in size control of all shapes of parts.

2. Lubricate the contact area between wheel and workpiece, thus helping to prevent chips from sticking to the wheel, which aids surface finish.

3. Flush chips and abrasive (swarf) from the workpiece.

4. Help control grinding dust, which can be hazardous to the health of the operator.

Types of Grinding Fluids

WATER-SOLUBLE CHEMICAL TYPES

These types of fluids, called synthetic fluids, are oil free. They are typically transparent, which helps operator visibility. They are compounded to provide good rust control. They are not prone to bacterial growth. They provide excellent cooling capacity, but have less lubricating ability than the soluble or straight oils. A great deal of development is taking place in synthetic grinding fluids.

FIGURE N-67 This is a very common method of flood coolant application. For the photograph, the volume of coolant has been reduced (Lane Community College).

WATER-SOLUBLE OIL TYPES

Like the chemical fluids, this type uses water as a base with a water soluble oil mixed in. The fluid is the common milky substance often seen in the machine shop. For grinding, this fluid is good for medium stock removal operations. To avoid bacterial growth when the fluid is idle, adding bacterial growth inhibitors is necessary. There are various duty levels of this type fluid, some better adapted than others to grinding with superabrasives.

STRAIGHT OILS

Straight oils are excellent for lubricating characteristics, but their heat transfer properties are not as good as those of water based grinding fluids. Straight oils are used for heavy form grinding and thread grinding. When grinding superalloys, using superabrasives, sulfochlorinated mineral oils are preferred. Good positive ventilation of the work area is essential when straight or enhanced oils are being used.

Methods of Application

Most grinding fluids are applied in a high-volume, low-velocity flood stream (Figure N-67) by the machine coolant pump. The grinding fluid is constantly recirculated and cleaned of swarf (Figure N-68). The supply must be replenished from time to time with water and additives. It is important to apply grinding fluid at the right place (Figure N-69). This is usually at the immediate vicinity of the contact area. One of the greatest difficulties in

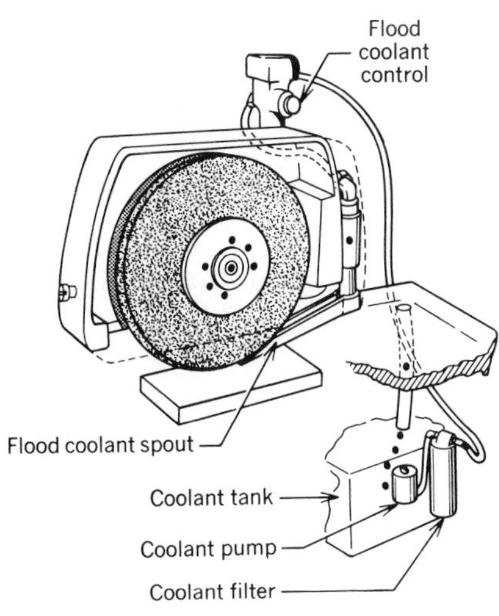

FIGURE N-68 Fluid recirculates through the tank, piping, nozzle, and drains in flood grinding system (DoALL Company).

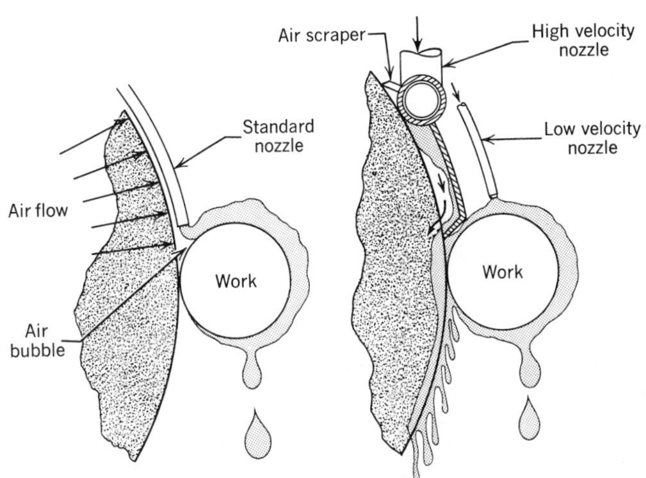

FIGURE N-69 A specially designed nozzle like this helps to keep the fanlike effect of the rapidly rotating wheel from blowing coolant away from the wheel–work interface (Cincinnati Milacron, Products Division).

both surface and cylindrical grinding is "grinding dry with fluid." The explanation for this is that on some heat sensitive steels inadequate grinding fluid application can result in grinding cracks developing from having the high temperature line contact with the grinding wheel quenched an instant later with the grinding fluid. The hot metal at the contact point tries to flow, but the abrupt quench arrests the flow, putting the metal into such stress that cracking often results. This type of grinding crack is parallel to the grinding wheel axis. Unless full flowing contact with the grinding fluid is assured, it is sometimes better to shut off the fluid to avoid damaging a sensitive workpiece. One good way to assure better fluid flow is to use a dummy block of the same height ahead of the workpiece, to keep the fluid flowing over the workpiece.

There is also another good way to avoid this problem. Since vitrified wheels are porous, grinding fluid can be applied through the wheel (Figure N-70). Grinding fluid is fed to a circular channel, then through holes into the wheel, and ultimately out through the porous material. Centrifugal force keeps the grinding fluid flowing outward. This method is very effective in getting the fluid into the contact area. However, the fluid requires filtration down to the 3- to 5-micron range (120 to 200 millionths of an inch) to keep the pores in the wheel open. Also, the wheel guards must be kept very clean to avoid having swarf washed onto the work.

Air/mist coolant is another application method (Figure N-71) in which air under pressure is used to spray a small volume of grinding fluid between the wheel and the workpiece. Mist systems use much less liquid coolant volume and need no coolant return system. The rapidly evaporating mist provides good cooling in the contact area and the view of the workpiece is not obstructed as in flood application. Good ventilation of the mist is essential when using mist coolant.

FIGURE N-70 Nozzle and flange design for through-the-wheel application. This allows coolant to filter through the wheel to the cutting area. It also sprays coolant over the inside of the wheel guard (DoALL Company).

FIGURE N-71 Mist grinding fluid application is sometimes used on a normally dry grinder. The cooling effect is excellent, lubrication practically nil, but no recirculating system is required (DoALL Company).

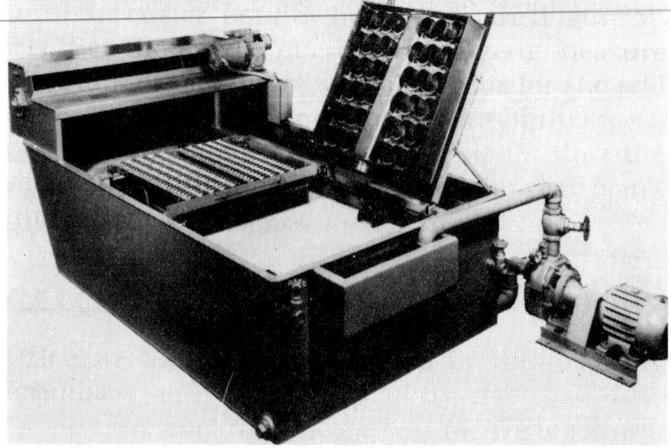

FIGURE N-72 Settling tank. The two requirements for cleaning coolant by settling are that the tank be big enough to allow the coolant to remain still for enough time to allow the swarf to settle out, and that there always be enough coolant in the system. This also means that the tank must be kept clean; otherwise, dirty coolant is recirculated (Carborundum Company).

Cleaning the Grinding Fluid

Grinding fluid must be cleaned of swarf before returning it to the workpiece. If this is not done, swarf will be circulated back through the contact area and result in surface finish problems (called fishtails) on the workpiece. Many shops use centrally supplied grinding fluid to all grinding machines. However, many grinding machines have self-contained grinding fluid systems. A number of methods are used for grinding fluid cleaning. Cleaning of the fluid should be carefully done and monitored.

A fundamental cleaning technique is settling. Grinding fluid is pumped to the settling tank where swarf settles out by gravity (Figure N-72). Since settling is slow, filtering may be used (Figure N-73). The filtering material is changed periodically. The type of filter used is limited as to the size of particles that can be trapped.

Other grinding fluid cleaning methods include the centrifugal separator (Figure N-74) where grinding fluid is directed into a rapidly rotating centrifuge. Swarf is moved to the outside and clean coolant remains in the center.

The cyclonic separator (Figure N-75) also uses centrifugal force. Clean grinding fluid exits out the top while swarf goes to the bottom. The magnetic separator (Figure N-76) will remove iron and steel grindings. A rotary magnetic drum rotates in the grinding fluid and attracts the grindings, which are later scraped off and removed.

FIGURE N-74 The centrifugal unit on top spins at high speed to remove swarf from the fluid. Dirty fluid, already partly cleaned by settling, is pumped into the unit by the large hose from the tank and recirculated through the smaller hose and piping (Barrett Centrifugals).

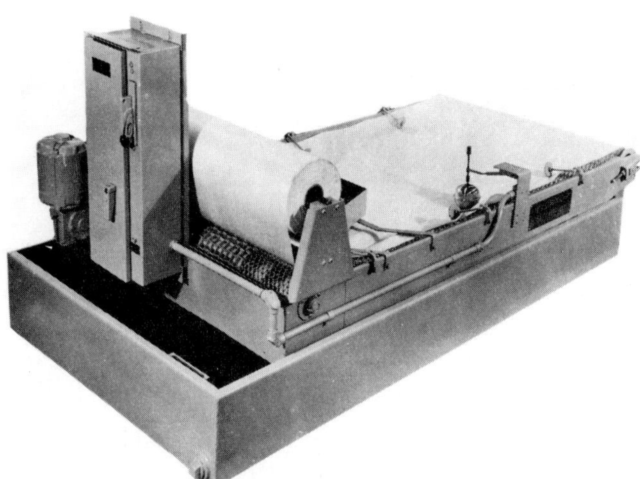

FIGURE N-73 A widely used cleaning method is to let the fluid fall through filter fabric directly into a tank. Then, when the load on the fabric becomes heavy enough, it actuates a switch that pulls the loaded fabric into a waste container and also pulls a fresh section of fabric into place (Carborundum Company).

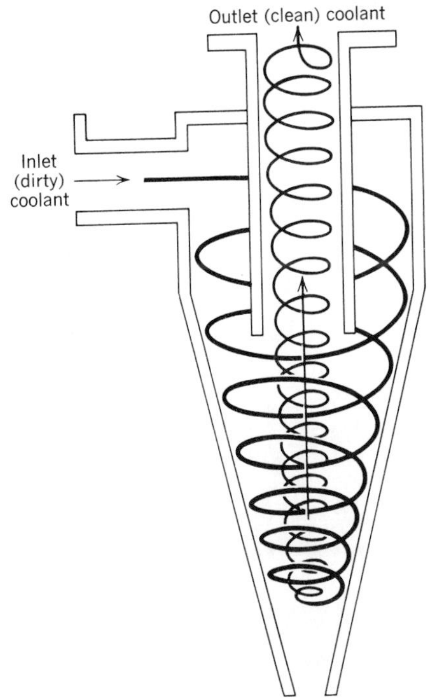

FIGURE N-75 Cyclonic filter. The dirty coolant is fed in from the side, swirls around in the cone as the clean coolant goes out through the top, and the swarf through the bottom of the cone (Barnes Drill Co., Rockford, Ill.).

Using Grinding Fluids

When using grinding fluids, be sure to observe the following.

1. Maintain proper coolant level.
2. Maintain proper balance of water and additives.
3. Check for contamination by lubricating oils; skim off oil droplets.
4. Make sure that coolant is supplied to the contact area in sufficient volume and at the right place.

FIGURE N-76 This magnetic separator removes iron and steel chips from water-soluble oils and cutting oils. The magnets are in the roll visible just past the electrical control box. Chips picked up by the roll are scraped off onto the slide and go down into the sludge box (Barnes Drill Co., Rockford, Ill.).

SELF-TEST

1. List at least four major functions of grinding fluids.
2. Where should the grinding fluid be applied?
3. What is mist coolant and how is it applied?
4. What are the major types of grinding fluids?
5. Describe at least three methods of cleaning grinding fluids.
6. What is a "dummy block," and how does its use help cool the workpiece?
7. What is the primary disadvantage in the use of a water-soluble oil grinding fluid?
8. What type of grinding fluid is used in thread grinding, and what precautions are needed to protect the operator?
9. What is one of the advantages of using water-soluble chemical-type grinding fluids?
10. What is meant by the term *grinding dry with fluid*?

UNIT 4

Horizontal Spindle Reciprocating Table Surface Grinders

The horizontal spindle reciprocating table surface grinder is probably the most common precision grinder found in the machine shop. The primary application of this machine is surface grinding. With accessories it can accomplish several other grinding tasks. The purpose of this unit is to familiarize you with the major parts of this machine, its controls, and its accessory capabilities.

Objectives

After completing this unit, you should be able to:

1. Name the components of the horizontal spindle surface grinder.
2. Define the functions of the various component parts of this grinder.
3. Name and describe the functions of at least two accessory devices used to increase the versatility of the surface grinder.

Identifying Machine Parts and Their Functions

The horizontal spingle reciprocating table surface grinder (Figure N-77), often simply called a surface grinder, presents the face of the grinding wheel to the workpiece. The workpiece is mounted on a chuck, which in turn is mounted on a table that reciprocates back and forth under the wheel. Major parts of this grinding machine include the following.

WHEELHEAD

The wheelhead contains the spindle, bearings, and drive motor. This assembly is mounted on the downfeed slide.

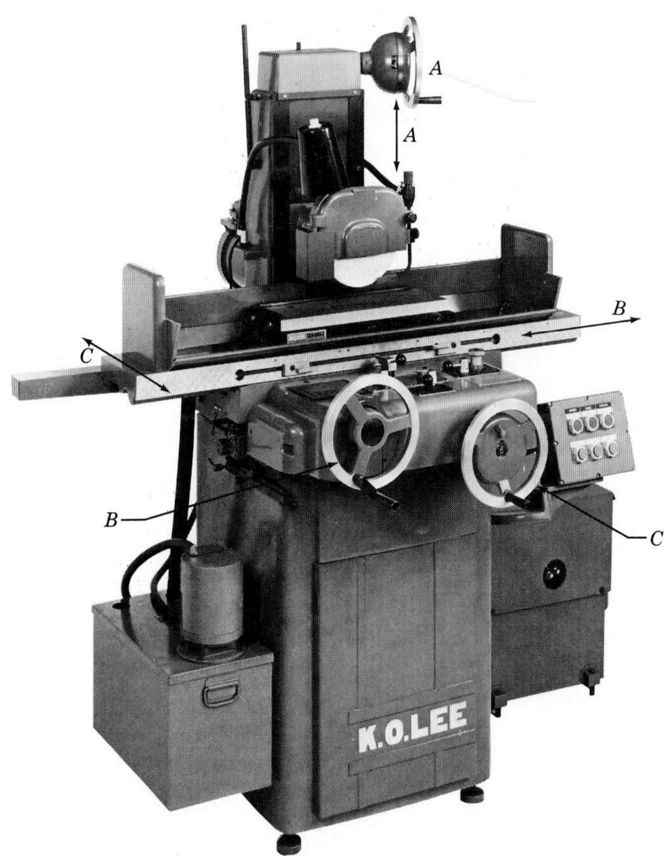

FIGURE N-77 Surface grinder with direction and control of movements indicated by arrows. Wheel *A* controls downfeed *A*. Large wheel *B* controls table traverse *B*. Wheel *C* controls crossfeed *C* (K. O. Lee Company).

TABLE WITH CHUCK

The table supports the chuck, which is the primary work-holding device on the grinder. The table reciprocates right and left to carry the workpiece under the grinding wheel. Table reciprocation may be done by hand or it may be a powered function. On powered tables, the length of stroke may be preset by positioning the table reverse trip dogs. The table is then reversed at the end of each stroke.

SADDLE

The table is supported by the saddle, which moves in and out in order to set the workpiece over after each reciprocation. This action is termed crossfeed. This is also an automatic powered function on many grinders.

CONTROLS

The downfeed handwheel (Figure N-78) located on the wheelhead raises and lowers the wheel with

FIGURE N-78 Close-up of downfeed handwheel. Moving from one mark to the next lowers or raises the grinding wheel .0001 in. (DoALL Company).

FIGURE N-79 Same control wheel with slip ring set to zero. Now it is simpler for operator to downfeed the grinding wheel as he or she grinds (DoALL Company).

FIGURE N-80 Centerless grinding attachment mounted on surface grinder with an assortment of parts that can be ground on it (Unison Corporation).

reference to the workpiece. Depth of cut is controlled by the downfeed. The downfeed handwheel is graduated to .0001-in. increments and usually has an arrow indicating the direction of wheel feed. The downfeed handwheel is equipped with a micrometer collar (Figure N-79) that may be adjusted to zero at any point where the wheel just begins to cut the workpiece.

The table feed handwheel reciprocates the table back and forth. The crossfeed handwheel moves the saddle in and out. The crossfeed handwheel usually has a dial graduated in .001-in. increments, permitting setover to be regulated so that the correct amount of overlap is obtained on each grinding pass.

Skilled diemakers sometimes prefer to use completely hand operated surface grinders, especially for the sharpening of dies. Others prefer to use hydraulic table traverse combined with manual crossfeeding action to obtain a feel for the process.

In general, large crossfeeding movements combined with small downfeeding movements are to be preferred. This combination keeps more of the wheel working, and tends to keep the wheel surface flat longer between dressings. Narrow crossfeeding movements tend to round the edges of the wheel excessively. As mentioned previously, a small amount of deliberate radiusing of the edge of the wheel with a dressing stick helps to prevent feed lines that mark the workpiece showing the amount of crossfeed for each pass. It is also desirable to allow enough longitudinal movement beyond the part to have time to complete the crossfeeding movement before the work is again brought under the grinding wheel, or unattractive angular lines on the workpiece can be the result.

Accessories

The versatility of the horizontal surface grinder can be extended by the use of accessory equipment. Several of these pertain to work holding and will be discussed in the next unit. Accessories that increase the grinding capability of the machine include the following.

CENTERLESS ATTACHMENT

Centerless grinding can be accomplished using the centerless attachment (Figure N-80). Parts are held between rollers and the grinding wheel. Diameters up to 5 in. can be centerless ground using this attachment.

CENTER-TYPE CYLINDRICAL ATTACHMENT

Cylindrical grinding can be accomplished with this accessory (Figure N-81). It can also be used to grind a flat on the workpiece.

FIGURE N-81 Center-type cylindrical attachment mounted on surface grinder. Attachment can be tilted for grinding a taper, as shown here, or set level for grinding a straight cylinder (Harig Mfg. Corp., Chicago, Ill.).

SELF-TEST

1. What is the function of the table?
2. What is the function of the saddle?
3. Explain the purpose of the downfeed handwheel, table, and saddle handwheel.
4. What is the usual discrimination of the downfeed handwheel graduations?
5. What is the typical crossfeed measuring increment on most surface grinders?
6. Why are small downfeed movements in combination with large crossfeeding movement preferable in surface grinding?
7. Why is it preferable to make a small radius on each edge of a surface grinding wheel?
8. Why is it desirable to set the longitudinal table to travel some distance beyond the workpiece on each end?
9. How can the surface grinder be used to grind small-diameter cylindrical parts (like dowel pins)?
10. How can the surface grinder be used to grind cylindrical parts mounted on centers?

UNIT 5

Work Holding on the Surface Grinder

A fundamental consideration in any machining operation is work holding. Grinding machines are no exception. In most common grinding, the amount of material that is removed from the workpiece on each pass of the wheel is quite small compared to other machining operations. The workpiece must be securely held, while permitting access to the grinding wheel. Large clamps or heavy vises are usually not required. Special chucks designed for surface grinder work holding are in common use. These will usually meet a majority of common work-holding requirements. The purpose of this unit is to describe the function and applications of these common grinder work-holding devices.

Objectives

After completing this unit, you should be able to:

1. Describe the basic operating principle of common grinder chucks.
2. Explain care of grinder chucks.
3. Describe methods of holding odd-shaped, nonmagnetic, and thin workpieces.

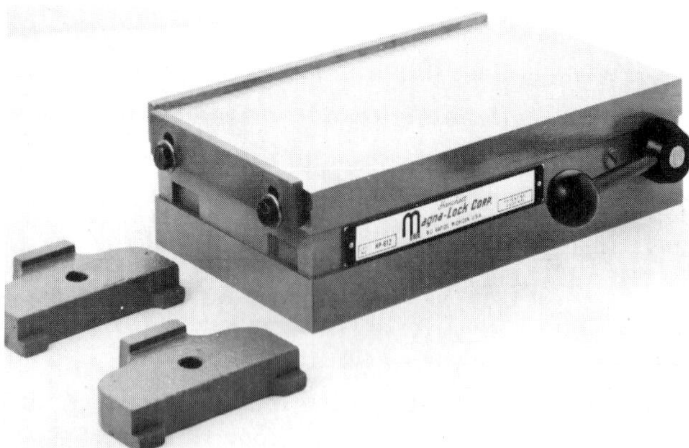

FIGURE N-82 The permanent-magnetic chuck looks about the same as any other chuck of the same shape and often is used where wiring would be inconvenient (Hitachi Magna-Lock Corp.).

FIGURE N-84 Magnetic sine chuck needed for grinding nonparallel surfaces (Hitachi Magna-Lock Corp.).

FIGURE N-83 Common type of magnetic chuck for reciprocating surface grinder. The guards at the back and left side are usually adjustable and help keep work from sliding off the chuck (DoALL Company).

Types of Grinder Chucks

MAGNETIC CHUCKS

The most common work-holding device for the surface grinder is the magnetic chuck. The three types are the permanent magnet (Figure N-82), the electromagnetic (Figure N-83), and the electro-permanent magnetic chuck. Permanent magnet chucks, which are the most widely used chucks in the job shop setting, are composed of a series of alternating plates consisting of powerful magnets, with pole pieces inside that can be moved to increase magnetic flux for firm work holding, or to diminish magnetic flux to help release the part.

The electromagnet chuck is magnetized when electrically energized. The chuck remains magnetized as long as power is applied. When power is removed, the chuck is demagnetized so that the workpiece can be removed. However, some residual magnetism often remains in the workpiece and can make removal from the chuck difficult. An electrical demagnetization process (degaussing) is used to remove any residual magnetism from the part. This is accomplished by a series of magnetic reversals activated by a special switch and takes only a few seconds.

The third type of magnetic chuck, the electro-permanent chuck, has some distinct advantages in two important areas. The first is safety. If the power fails during grinding with an electromagnetic chuck, parts can be violently ejected from the chuck, which can endanger the operator, can cause damage to the grinding wheel, and even cause spindle damage to the machine (if the part does not eject smoothly). The electro-permanent magnetic chuck retains holding ability with or without electrical power. Another advantage, especially for precision grinding, is that once the magnetism is established (which takes only a few seconds), it is not necessary to apply additional energy to the chuck—which in a purely electromagnetic chuck could cause a temperature rise of 30 to 40°F. If for some reason grinding fluid as coolant is not being applied, this can result in size variation for which it is difficult to compensate. Alnico is used in these electro-permanent magnet chucks, because it is easier to magnetize and demagnetize than the ceramic (ferrite) that is commonly used in permanent magnet chucks.

Magnetic chucks for surface grinders are usually rectangular. Those for vertical spindle rotary grinders are circular. Workpieces can be held on magnetic sine chucks for grinding angles or compound angles (Figure N-84).

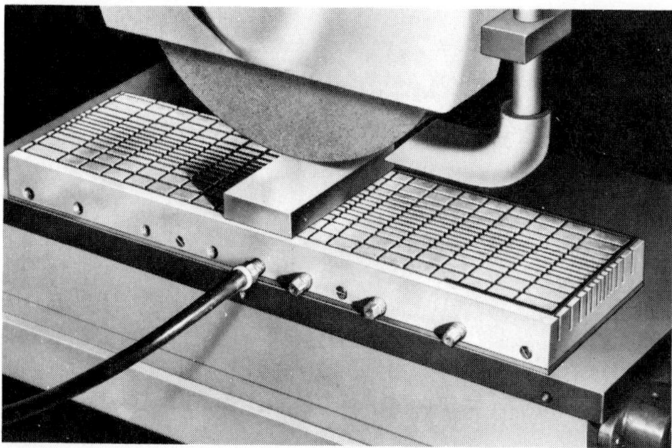

FIGURE N-85 Vacuum chucks such as this one hold practically anything and are considered good for thin work (Thompson Vacuum Co., Inc.).

VACUUM CHUCKS

The vacuum chuck (Figure N-85) is another popular tool for grinder work holding. By evacuating the air under the workpiece, the force of atmospheric pressure can be applied to hold the part on the chuck. This can be quite significant. On a chuck 6 by 12 in., an atmospheric pressure of over 1000 pounds is applied to the workpiece if a good vacuum is achieved. Vacuum chucks are also useful for holding thin workpieces.

Care of Grinder Chucks

Like much of the equipment found in the machine shop, grinder chucks are precision tools and deserve the same care and handling you would give your precision measuring instruments. With reasonable care they will maintain their accuracy over many years. Observe the following precautions when using grinder chucks.

STEP 1. Clean the chucking surface before mounting a workpiece.

STEP 2. Be sure that the workpiece is free from burrs.

STEP 3. On a magnetic chuck, be sure that small parts span as many magnetic poles as possible to insure maximum holding power.

STEP 4. Deburr the chucking surface from time to time (Figure N-86).

STEP 5. Check the chucking surface with a dial indicator if the chuck has been removed and reinstalled on the machine or a new chuck is being used for the first time.

FIGURE N-86 Periodic deburring of the chuck with a granite deburring stone like this, or with a fine grit oilstone, is a good practice (DoALL Company).

FIGURE N-87 Chuck setup for workpiece with projection on chucking side. Work is supported on laminated magnetic parallels (DoALL Company).

Grinder Chucking Setups

ODD-SHAPED WORKPIECES

A magnetic or vacuum chuck will exert the most reliable holding power on an odd-shaped workpiece if the area of contact is large enough. Whenever possible, place the workpiece in such a manner that the largest surface area possible is on the chuck. Sometimes it is necessary to block a small or thin part with additional material to prevent the part from moving on the chuck during the grinding process. The blocking, of course, needs to be thinner than the workpiece if it is not to be ground also. Sometimes it is necessary to chuck on a smaller than desirable area (Figure N-87). In

FIGURE N-88 A set of magnetic parallels and vee blocks can be very useful (Hitachi Magna-Lock Corp.).

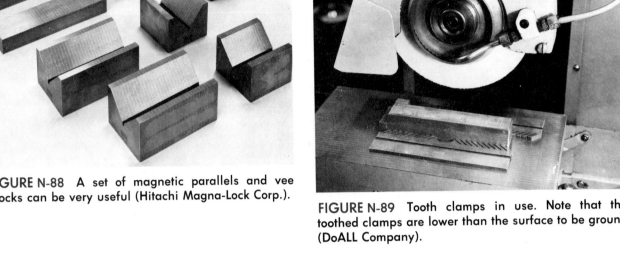

FIGURE N-89 Tooth clamps in use. Note that the toothed clamps are lower than the surface to be ground (DoALL Company).

these cases, the workpiece must be supported with laminated parallels or vee blocks (Figure N-88). These are specially designed for grinder work. Laminated accessories are made with nonmagnetic and soft steel inserts so that the lines of magnetic force will be conducted through to the workpiece. Laminated parallels must be treated with great care because they are soft for best magnetic permeability. They should be checked carefully for burrs each time they are used. Odd-shaped workpieces may also be held by traditional machine tool work-holding methods such as vises or clamps.

NONMAGNETIC WORK

Nonmagnetic materials may be held on a magnetic chuck by blocking the parts with steel (magnetic) tooth clamps (Figure N-89). The comblike teeth will keep a nonmagnetic workpiece from sliding off the chuck. As the magnetism is applied the comblike teeth of the clamps are pulled down against the nonmagnetic workpiece, to keep the part from sliding off the chuck. If you are using tooth clamps, be sure that they are lower than the workpiece so that they will not come in contact with the grinding wheel.

THIN WORKPIECES

Thin material already warped or twisted will probably be pulled flat on either a magnetic or vacuum chuck. However, these workpieces may assume their original distorted shape after grinding and removal from the chuck.

Thin material can be successfully held and ground by using a minimum amount of holding force. One method of accomplishing this is by using double-backed tape (with adhesive on both sides). One side of the tape is applied to the chuck while the other side holds the workpiece. This may be sufficient for light cuts. Avoid the use of grinding fluids when double-backed tape is used, as it will promote part slippage.

It is also possible to shim under a warped thin workpiece until one side has been ground true. The part may then be turned over and the true side placed on the chucking surface. The grinding of thin cold rolled steel stock that has not been annealed or normalized can be very frustrating. The outside of the material is in much greater compression than the core. As the highly compressed layer is ground away on one side, the compression of the layer on the opposite side can actually result in enough force to unseat the part from the chuck. It is important to grind uniform amounts from both sides of such materials and to be prepared to grind opposite sides more than a single time to get a flat, dimensionally stable part.

SELF-TEST

1. What is the most widely used work-holding device on the surface grinder?

2. How is residual magnetism removed from a workpiece?

3. What two advantages are there in using an electromagnetic chuck over the permanent-magnet chuck?

4. How does the electro-permanent chuck offer two important advantages over either the permanent-magnet chuck or the electromagnetic chuck?

5. Why must care be taken to span as many magnetic poles as possible when placing work on a magnetic chuck?

6. If a chuck has been removed and reinstalled on the surface grinder, what should be done to check that it is mounted accurately?

7. If the chuck has been found to have surface burrs, what steps should be taken?

8. What are laminated accessories, and how can they be used in grinder work holding?

9. Name at least two methods of work holding on the surface grinder that can be used to hold nonmagnetic workpieces?

10. Describe at least two methods of holding thin workpieces.

UNIT 6

Using the Surface Grinder

Using the surface grinder involves selecting and mounting the proper grinding wheel, surface grinding the chuck (grinding in) if necessary, and using the proper combination of table speed and crossfeeds. The surface grinder is a versatile and precision machine tool. Used correctly, it will finish your workpiece to a high degree of dimensional accuracy. The purpose of this unit is to describe the surface grinder and to familiarize you with its preparation and with the grinding process required to finish a common grinding project.

Objectives

After completing this unit, you should be able to:

1. Prepare the surface grinder for a typical grinding job.
2. Check and grind in the chuck if necessary.
3. Finish grind a set of vee-blocks to required specifications.

Horizontal spindle surface grinding is becoming ever more important as equipment and machined parts are miniaturized, and dimensional tolerances get ever smaller. A few years ago, plus or minus .0001 in. was considered a very close tolerance. Now half that amount (.000050 in.) is seen more frequently on drawings, especially in the advanced technology areas. These pressures for increasingly closer work are reflected in newer, more rigid machine tools and greater concern with surface texture specification and measurement.

Avoiding damage such as grinding cracks is critically important in ground parts that are subjected to high and variable loadings, such as those found in bearings, missiles, and aircraft applications.

These requirements for improved machine tools are also reflected in the development of improved superabrasives. These permit sharpness to be better controlled, resulting in less metallurgical damage and easier size control. A grinding wheel made with superabrasives will last much longer, while removing a given amount of work material, than one made with aluminum oxide or silicon carbide. With superabrasives, the volume of abrasive loss to workpiece removal volume (grinding ratio) is very low, and the ratio is considerably greater than 1000:1, in contrast to the 40:1 considered acceptable in using conventional abrasives. As there is movement toward automation of grinding processes, more and more research is being put into making the grinding processes more predictable.

All of this relates to the task of setting up and using the horizontal spindle surface grinder. It is at this point that you start putting together a large number of variables that through your skill and observation can result in an accurate workpiece. As the operator, you will have more control of the outcome in surface grinding than in nearly any other portion of the machinist's trade. This is also where the wheel selection and preparation process are put to the test.

In surface grinding, it is necessary to reverse one of the more usual principles of machining used in most of the other chip-making processes. In both turning and milling for metal cutting volume, the machinist tries to keep the depth of cut to a maximum, adjusting all the other variables such as speeds and feeds to match that concept. In surface grinding, the important thing is to match the actions of workpiece speed, crossfeed, and downfeed to keep the grinding action as consistent as possible, while at the same time avoiding damage to the workpiece. The example used in this unit will be in the grinding of a vee block, but before beginning with that, some other generalities about horizontal spindle surface grinding should be considered.

Starting with a broad surface of a soft steel like AISI 1018, a suitable selection of wheel and grinding variables would be as follows: An aluminum oxide abrasive, of 46 grit size in a J bond hardness. The bond itself is vitrified. The wheel speed for vitrified wheels is usually from 5500 to 6500 sfpm; the table speed would be 50 to 100 sfpm. The crossfeed rate would be .050 to .500 in. per pass, with a maximum being one-fourth the wheel width. For roughing, the downfeed would be .003 in., and for finishing, .001 in. maximum.

If the same AISI 1018 material were pack hardened (carburized) to about 55 Rockwell C, how would this change the initial recommendations? The wheel speed and the table speed would remain the same. The abrasive chosen and the grit size would also remain the same, but the bond selected would be one grade softer, or I bond. The roughing downfeed would change to .002 in., and the finishing to .0005 in. The greatest change in recommendation other than the grade of the wheel would be in the way that the crossfeeding is done. Here the recommendation would be .025 to .250 in. with one-tenth of the wheel width taken as a maximum. All of these recommendations are based on a sharp wheel prepared with a dresser that is in good condition. If the dressing is done incorrectly so that the wheel is dulled, it could make the recommendations useless.

A general observation in horizontal surface grinding is to keep the crossfeed travel as large as possible, and the downfeed adjustments matched so that you cannot hear a significant change in grinding wheel speed during the work's traverse under the wheel. This procedure helps to keep the surface of the wheel parallel to the spindle and avoids excessive dressing, providing the grade and abrasive friability are correctly matched to the workpiece needs.

At the surface grinder you will be using, familiarize yourself with the controls and tooling. If someone is using the machine, observe the operation for a time. Check the supply of grinding wheels and determine their possible uses by observing their color and by reading the printed symbol on the blotter.

Grinding in a Magnetic Chuck

The first step in surface grinding is to check, and if necessary grind in or true, the chuck. Trueing the chuck is usually not necessary unless it has been removed and reinstalled or a new chuck is to be used for the first time. Any grinder chuck should be checked occasionally and if necessary trued by grinding a minimum amount from the chucking surface. To true a grinder chuck, the following procedure may be used.

STEP 1. Mount the chuck and align it in both table and saddle axes using a dial indicator with .0001 in. resolution.

STEP 2. Use the indicator to check the chucking surface to determine if trueing will be necessary.

STEP 3. If so, select and mount a proper wheel for this grinding task. A good choice would be a friable grade of aluminum oxide, 46 grit, J grade, 8 structure with a vitrified bond.

STEP 4. Mark the entire surface of the chuck with pencil lines. As you are grinding, the removal of the last pencil lines will show you when the task is finished.

STEP 5. True and dress the wheel. Make the final dressing pass with a fast cross movement of the diamond, for a free-cutting grinding action.

STEP 6. Magnetize the chuck.

STEP 7. Bring the wheel down toward the chuck, using a feeler gage to set the wheel a few thousandths above the chucking surface.

STEP 8. Turn on the wheel and let it run for several minutes to stabilize the spindle and its bearings. Then start the grinding fluid. Set the table in motion with the table stop dogs set to overtravel about an inch at each end. Lower the wheel slowly the additional amount while you are moving the table saddle back and forth crosswise, until you just barely contact the highest point on the chuck's surface.

STEP 9. Set the downfeed handwheel collar to zero.

STEP 10. Using a downfeed of no more than .0002 in., a fairly rapid table speed (50 to 100 sfpm), and a crossfeed of about one-fourth the wheel width on each pass, grind to a cleanup condition on the chucking surface. Remove only enough material to clean the surface. (The total amount is usually no more than .001 in.)

STEP 11. Remove any remaining burrs with a fine grit stone.

Grinding Vee-Blocks

The vee-block (Figure N-90) is a common and useful machine shop tool. Precision types are finish machined by grinding after they have been rough machined oversize (about .015 in.) and heat treated. Finish grinding of vee-blocks will provide you with a well-rounded surface grinding experience.

FIGURE N-90 Finished, hardened, and ground precision vee-block.

Refer to a working drawing (Figure N-91) to determine dimensions and to plan the sequence of grinding operations.

SELECTING THE WHEEL

For these vee blocks, the best abrasive will be a friable grade of aluminum oxide. A vitrified wheel of 46 grit, with an I bond hardness would be a reasonable selection.

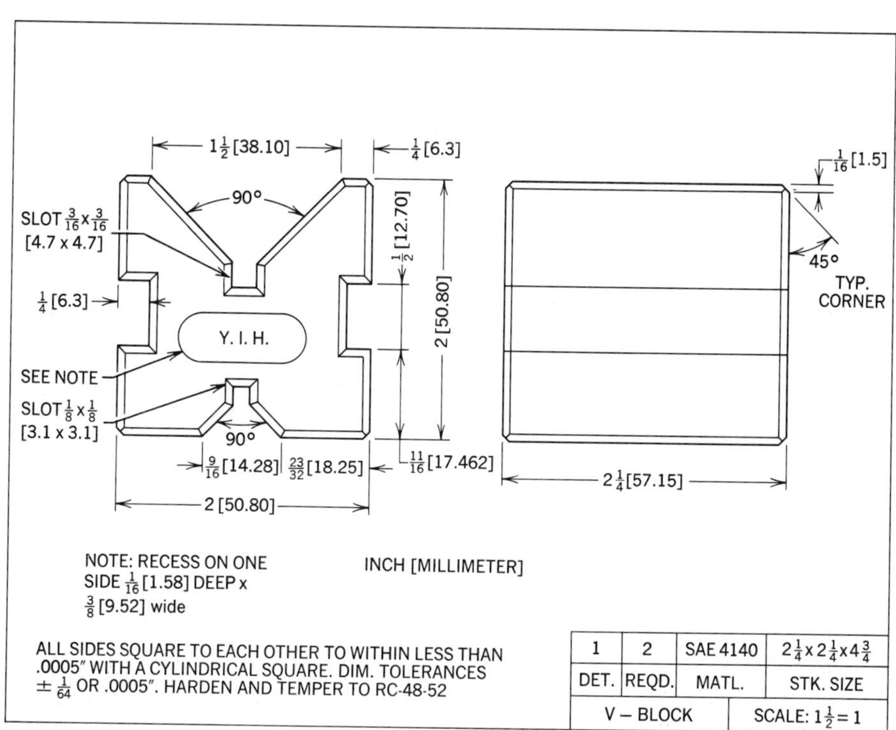

FIGURE N-91 Dimensions and information for grinding the vee block.

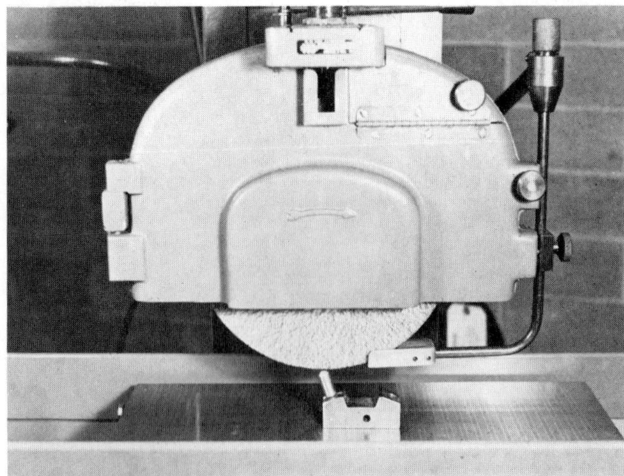
FIGURE N-92 Diamond dresser in the correct position (Lane Community College).

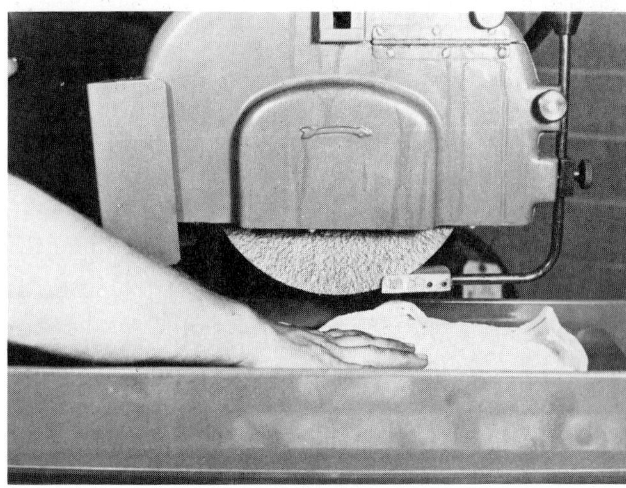
FIGURE N-94 Cleaning the magnetic chuck with a cloth. The wheel must be completely stopped before this is done (Lane Community College).

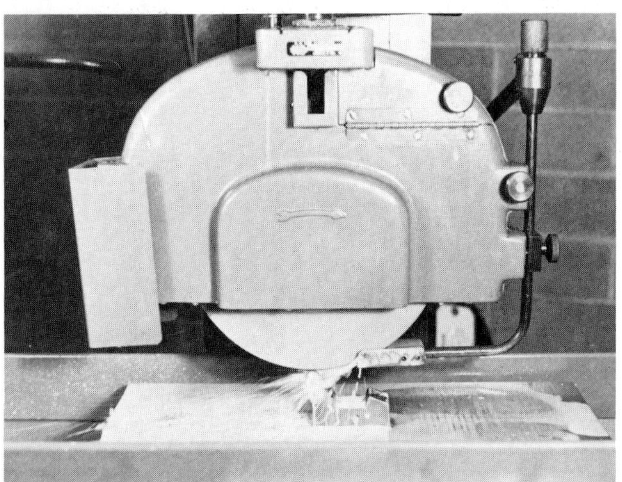
FIGURE N-93 Dressing the wheel using grinding fluid (Lane Community College).

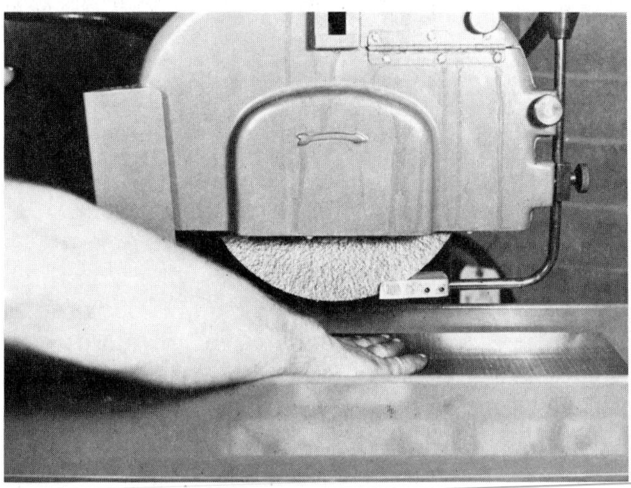
FIGURE N-95 Checking for burrs on the magnetic chuck (Lane Community College).

GRINDING MACHINE SETUP

The following procedure may be used in setting up the grinding machine for vee-block grinding.

STEP 1. Select a suitable wheel.

STEP 2. Clean the spindle. Use a cloth to remove any grit or dirt from the spindle. If you lay tools and wheels on the chuck, cover it with a cloth for protection.

STEP 3. Ring test the grinding wheel.

STEP 4. Mount the wheel. The wheel should fit snugly and the flanges should be the same size. Blotters will probably be attached to the wheel. If not, place a blotter between each of the flanges and the wheel.

STEP 5. Install the spindle nut and tighten firmly. Do not overtighten.

STEP 6. Replace or close the wheel guard.

STEP 7. Place the diamond dresser on the magnetic chuck in the proper position (Figure N-92). (Or use the built-in dresser if you have one.)

STEP 8. Dress the wheel, using fluid and a rapid crossfeed (Figure N-93). (Use fluid only if you can be sure of flooding the diamond.)

STEP 9. Remove the dresser and clean the chucking surface (Figure N-94).

STEP 10. Check chucking surface for nicks and burrs (Figure N-95). Use a deburring stone if necessary.

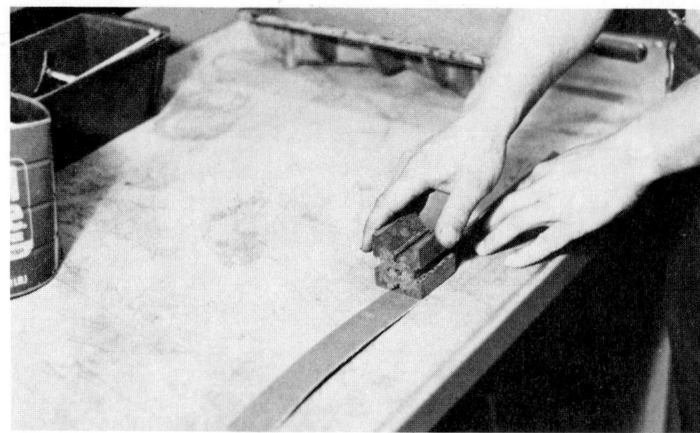

FIGURE N-96 Preparing the part to put on the magnetic chuck (Lane Community College).

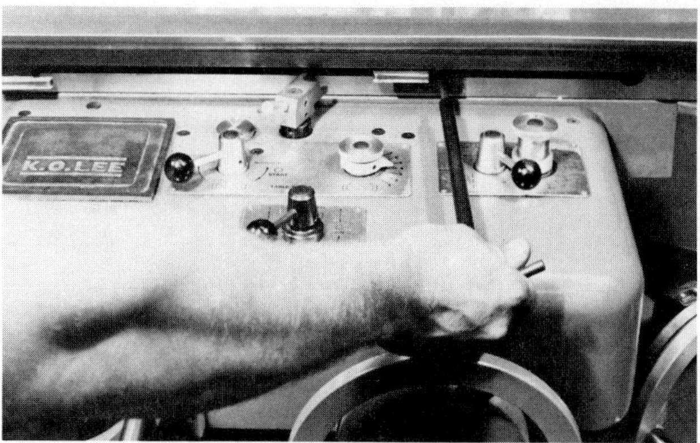

FIGURE N-98 Setting the table trip dogs (stop dogs) (Lane Community College).

FIGURE N-97 The rough blocks in place with the large vee side up ready to be ground (Lane Community College).

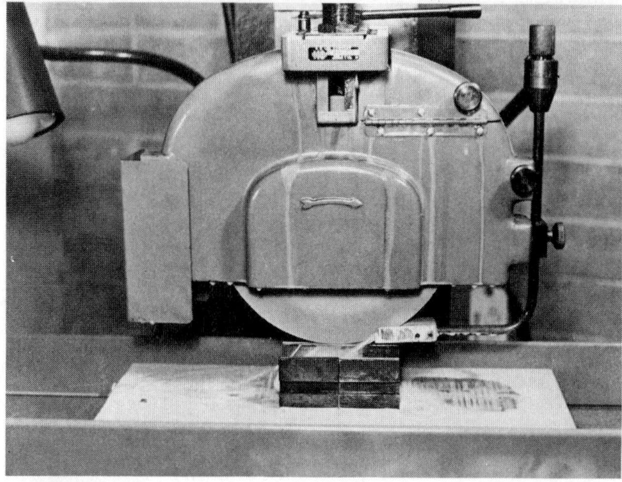

FIGURE N-99 Grinding the first surface until the surface is "cleaned up" (Lane Community College).

To avoid the necessity of having to protect the chuck surface with paper, prepare the first part surfaces by removing any heat treatment scale with the use of abrasive cloth on a flat surface (Figure N-96).

SIDE AND END GRINDING PROCEDURE

The vee-blocks should be match ground; that is, ground together in pairs so that they are exactly the same dimensions when completed. The following procedural steps and illustrations will describe the process for side and end grinding.

STEP 1. Place the blocks with the large vee side up on the grinder chuck (Figure N-97).

STEP 2. Magnetize the chuck.

STEP 3. Lower the wheelhead until it is about an inch above the workpiece. Adjust table and saddle position so that blocks are centered.

STEP 4. Adjust table feed reverse trips so that the workpiece has about 1 in. of overtravel at each end of the table stroke (Figure N-98).

STEP 5. Use a feeler gage or piece of paper and lower the wheelhead until it is a few thousandths of an inch above the surface to be ground.

STEP 6. Start the wheel and grinding fluid. Start table crossfeed and table travel. Carefully lower the wheel until it just begins to contact the high point of the workpiece. Set the downfeed micrometer collar to zero at this point.

STEP 7. The amount to grind from the workpiece will depend on the amount of extra material left from the original machining. Downfeed about .001 in. per pass and grind to a cleanup condition (Figure N-99). About .003 to .005 in. should be left for finish grinding on each surface.

FIGURE N-100 The blocks are turned over and the opposite sides (small vee) are ground (Lane Community College).

FIGURE N-101 A ground side of the vee block is clamped to a precision angle plate, and the part is made parallel, in preparation for grinding the end square (Lane Community College).

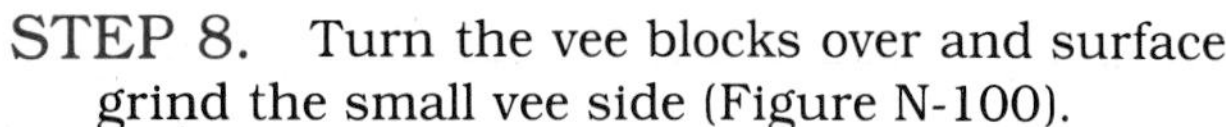

STEP 8. Turn the vee blocks over and surface grind the small vee side (Figure N-100).

STEP 9. The end of the block may be ground by clamping it to a precision angle plate (Figure N-101). The block is adjusted into alignment using a test indicator. The end of the block should extend slightly beyond the angle plate.

STEP 10. Set the angle plate on the chuck and end grind the vee block (Figure N-102).

STEP 11. For grinding the remaining side, clamp the ground end to the angle plate, leaving the side surface projecting above the angle plate (Figure N-103). Grind one remaining side square to the end and square to the first surface ground in step 7.

STEP 12. End grind opposite ends (Figure N-104). The vee blocks may be set up on the magnetic chuck without further support.

FIGURE N-102 The precision angle plate and vee-block setup is turned with the vee block end up on the magnetic chuck. The end of the vee block is ground square to a ground side (Lane Community College).

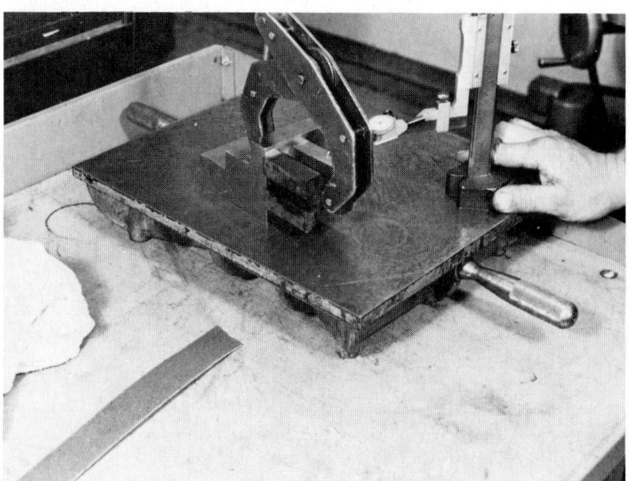

FIGURE N-103 Setting up to grind the third square surface (Lane Community College).

STEP 13. Grind remaining sides leaving .003 to .005 in. for finishing (Figure N-105).

STEP 14. Redress the wheel with a slow pass (or passes) of the diamond dresser for finish grinding. Use a light cut of about .0002 in.

STEP 15. Check all sides and ends for squareness (Figure N-106). Use a precision cylindrical square and dial test indicator. Small errors can be corrected by further grinding if necessary. Tissue-paper shims (.001 in.) may be used to achieve squareness. Take as little material as possible if corrections are needed.

STEP 16. Check all dimensions with a vernier micrometer (.0001 in. discrimination) or test indicator (Figure N-107).

FIGURE N-104 Grinding the opposite ends of both blocks in one setup to make them parallel (Lane Community College).

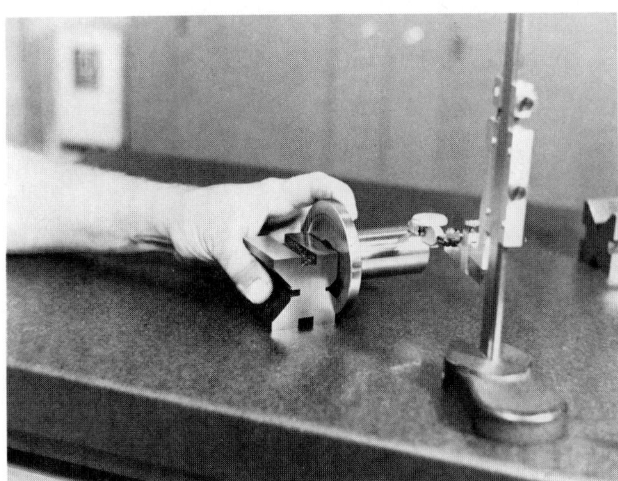

FIGURE N-106 A vee block now being checked for square on all sides using a precision cylindrical square, a dial indicator, and a height gage on a surface plate (Lane Community College).

FIGURE N-105 The sides that have not been ground are also being ground in one setup to make them parallel to the other sides (Lane Community College).

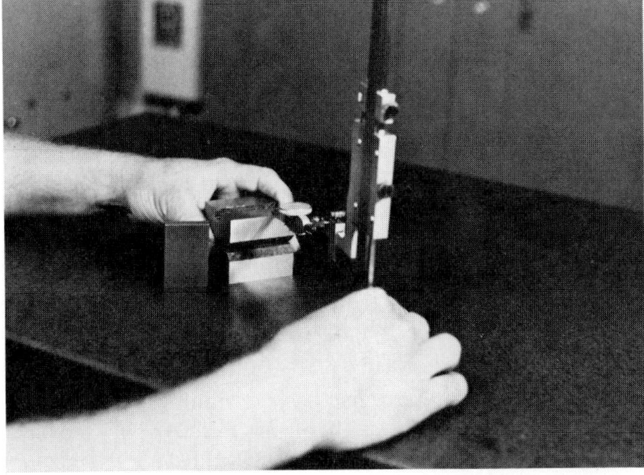

FIGURE N-107 Dimensions being checked using a .0001-in. reading dial indicator on a height gage. Precision gage blocks are used for comparison measurement for this operation (Lane Community College).

STEP 17. Finish grind all sides and ends to final dimensions. Both blocks should be match ground.

VEE-GRINDING

STEP 1. Grind one side of the large vee (Figure N-108). Set the blocks in a magnetic vee block and carefully align with table travel. **Do not make contact with the side of the wheel.**

STEP 2. After rough grinding one side of the vee, note the number on the downfeed micrometer collar or set it to zero index. Raise the wheel about $\frac{1}{2}$ in. Reverse the blocks in the magnetic vee block and grind the other side. Bring the wheel down to contact the work and make grinding passes until the micrometer collar is at the same position as it was when the first

FIGURE N-108 Setting up the magnetic vee block to grind the angular surfaces on the large vees (Lane Community College).

FIGURE N-109 Setting up the parts again in the magnetic vee block to grind the external angular surfaces (Lane Community College).

vee side was roughed. Now dress the wheel for finishing and repeat the procedure for both vees, removing only enough material necessary to obtain a finish. This procedure will ensure that the vee is accurately centered on the blocks.

STEP 3. Repeat the procedure for the remaining vee grooves.

STEP 4. Set up the parts again in the magnetic vee blocks to grind the external angular surfaces (chamfers) (Figure N-109). This operation must be done gently, as work holding from the accessory magnetic vee blocks is much less firm than from the surface of the magnetic chuck itself.

SELF-TEST

1. What is meant by the *grinding ratio*?

2. Why is capable precision surface grinding critically important in advanced technology applications?

3. Skillful surface grinding involves combining a large number of variables into a desirable result. What general observation applies in the practice of using the surface grinder?

4. If a workpiece is hardened, what variable in grinding wheel selection would typically be changed?

5. For grinding harder workpieces of the same material, what variation in crossfeeding often is applied?

6. If a "grinding in" or trueing of the grinding chuck becomes necessary, how do you determine when the surface has been trued?

7. When grinding scaly heat-treated parts, what preparation can you do to avoid having to protect the chuck surface with paper?

8. In dressing the grinding wheel prior to precision surface grinding, what caution should be observed relative to using grinding fluid during the dressing?

9. Why should just a minimum of stock be removed in grinding the first surface of a precision workpiece?

10. After the initial "rough grinding" of the part is accomplished, how is preparation made for finish grinding?

UNIT 7

Problems and Solutions in Surface Grinding

In any machining operation, there are many factors that can influence the final result. A good machinist knows what to look for; from experience, the machinist can identify a problem and implement a solution. The purpose of this unit is to help you identify common problems in surface grinding and recommended solutions.

Objective

After completing this unit, you should be able to:

Recognize the effects of common surface grinding problems and recommend solutions.

General Causes of Grinding Problems

Surface grinding problems appear in a variety of forms. In many cases they are related to acceptable surface finish, or even just the avoidance of the appearance of burning on the workpiece. But in some cases, particularly where safety is involved, grinding defects can be a very serious matter. Surface grinding problems can usualy be attributed to the general factors of machine condition and machine operation.

MACHINE CONDITION

Factors in machine condition include spindle bearings, way lubrication, the overall weight and quality of the machine, and machine location.

All mechanical functions of the grinder must function properly for best grinding results. Spindle bearings must be properly adjusted, and in most cases surface grinding machines produce the best results if the spindles have been warmed up by running for several minutes before grinding begins. Sliding components must be properly lubricated so they will move smoothly. More massive grinding machines are generally more rigid and will produce better surface finishes than will lightweight machines.

External sources of vibration can affect a grinding machine. A grinder located close to a railroad track or punch press may pick up external vibrations that can show up as surface finish defects on the workpiece.

OPERATIONAL CONDITIONS

Conditions that are variable because of the way in which the machine is operated can also be the cause of grinding problems. These include wheel selection, dressing technique, swarf in the coolant, loaded or glazed wheel, and the technique of the operator in regard to table and cross feed. Except for the irregular surface finish problems that come from poor maintenance of cutting fluid, cleanliness of area, and machine tool defects, the operator in selecting and preparing the grinding wheel has the most to do with the results in surface grinding. Most of these problems relate to the matter of controlling wheel sharpness. Many times, rather than change grinding wheels, the operator will try to obtain too fine a finish with too coarse a wheel. An overly fine dressing can lead to grinding a burnished surface, in which dull grains, rather than cutting, cause extensive cold working of the surface. In hardened materials, this additional working action can result in failure of the part for its design purpose.

The maintenance of wheel sharpness is greatly dependent on how the operator handles the table travel and cross feed actions. If this combination is poorly matched, as in too slow a table travel and too narrow a cross feed, the wheel will tend to not be self-dressing, and dulling (glazing) can again be the result.

Specific Problems and Solutions in Surface Grinding (Table N-1)

Effect—Chatter or vibration marks (Figure N-110)
Cause—Interrupted cutting due to wheel balance, wheel not concentric to spindle, external vibrations, or a skipping wheel that is loaded or glazed
Solution—Rebalance wheel; true wheel; redress wheel; relocate grinding machine and isolate from external vibration

Effect—Irregular scratches or fishtails (Figure N-111)
Cause—Swarf in the grinding fluid; swarf dripping from wheel guard; low grinding fluid; sliding workpiece off a dirty chuck

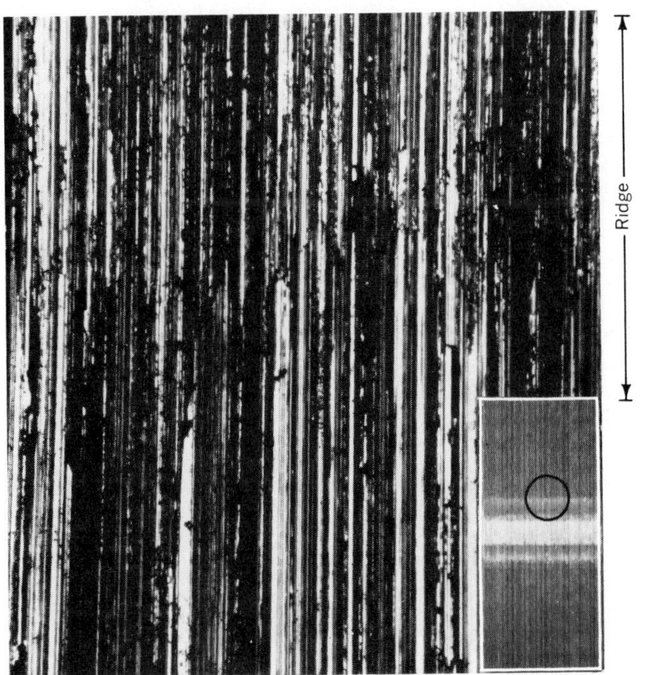

FIGURE N-110 Chatter marks enlarged × 80. Marked inset (× 1.25) shows area that is enlarged (Courtesy Surface Finishes, Inc.).

TABLE N-1 Summary of Surface Grinding Defects and Possible Causes (DoALL Company)

	Problems								
Causes	**Burning or Checking**	**Burnishing of Work**	**Chatter Marks**	**Scatches on Work**	**Wheel Glazing**	**Wheel Loading**	**Work Not Flat**	**Work Out of Parallel**	**Work Sliding on Chuck**
Machine operation									
Dirty coolant				×		×			
Insufficient coolant	×						×	×	
Wrong coolant					×	×			
Dirty or burred chuck				×			×	×	
Inadequate blocking									×
Poor chuck loading							×	×	×
Sliding work off chuck				×					
Dull diamond					×				
Too fine dress	×				×	×	×		
Too long a grinding stroke								×	
Loose dirt under guard				×					
Grinding wheel									
Too fine grain size	×				×	×			
Too dense structure					×	×			
Too hard grade	×	×	×		×	×	×		
Too soft grade			×	×					
Machine adjustment									
Chuck out of line								×	
Loose or cracked diamond				×			×	×	
No magnetism									×
Vibration			×						
Condition of work									
Heat-treat stresses							×		
Thin							×		

Solution—Clean inside of wheel guard; clear grinding fluid; clean chuck before removing workpiece and tilt work up on edge rather than sliding it off the chuck; maintain grinding fluid level

Effect—Discoloration or burning of workpiece (Figure N-112)
Cause—Insufficient grinding fluid; wheel too hard or too fine; concentration of heat because of too heavy a cut in a small area
Solution—Speed up table travel; lighten depth of cut; increase grinding fluid flow and volume for cooling

Effect—Work not parallel
Cause—Chuck out of line; chuck surface not parallel to table travel (not ground in); chuck dirty or burred
Solution—Align chuck with dial indicator; follow grind in procedure; deburr and clean

Effect—Workpiece not flat
Cause—Local overheating; internal stress relief; bent or twisted workpiece

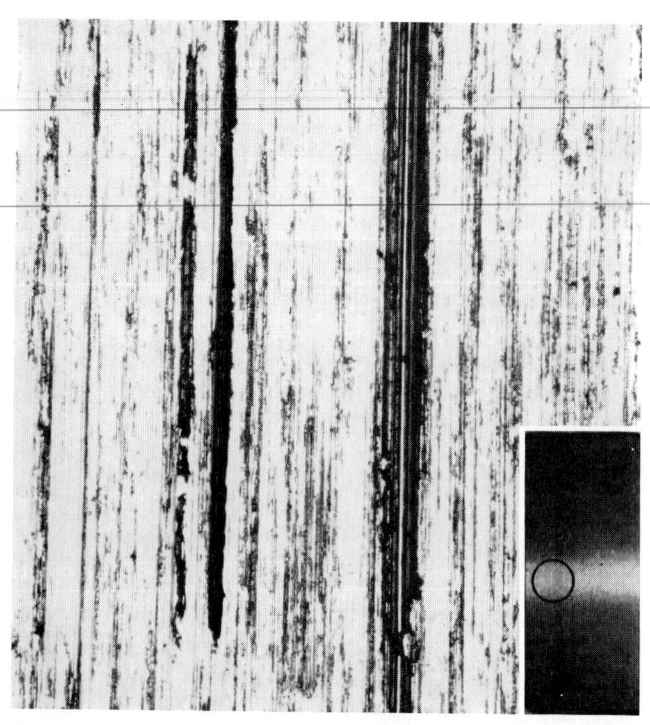

FIGURE N-111 Grinding marks or fishtails also enlarged ×80. Inset (enlarged ×1.25) shows the damaged area (Surface Finishes, Inc.).

FIGURE N-112 Discoloration or burning also enlarged × 80, with damaged area marked on inset (Surface Finishes, Inc.).

Solution—Use minimum amount of chuck-holding power; place workpiece on chuck bowed side up; turn workpiece over and shim with paper; take additional light cuts; reverse workpiece; remove and repeat procedure until flat

SELF-TEST

1. What grinding machine conditions can the operator readily affect?

2. If chatter or vibration marks are observed on the workpiece, what steps can the operator take to correct the problem?

3. If a "fishtail" appearance is seen on the ground surface, what steps can be taken by the operator?

4. If the workpiece shows burn marks, what can be done by the operator?

5. What if the work is not parallel?

6. List three causes of a workpiece not being ground flat.

7. If a thin cold-rolled steel workpiece warps after removal from the chuck, what is the probable cause, and what correction can be applied?

8. What sort of environmental circumstance can lead to grinding defects?

9. How can overly fine wheel dressing lead to part failure?

10. Most grinding problems relate to a matter of controlling wheel sharpness. How does the operator take part in this controlling process?

UNIT 8

Center-Type Cylindrical Grinders

The center-type cylindrical grinder, as its name implies, is used to grind cylinders or, more generally, the diameter of a round workpiece. The cylindrical grinder is a versatile machine tool capable of finish machining a round part to a high degree of dimensional accuracy. Various setups on the basic machine permit many different grinding tasks to be accomplished. The purpose of this unit is to familiarize you with the major parts of this machine and its general capabilities.

Objectives

After completing this unit, you should be able to:

1. Identify the major parts of the cylindrical grinder.
2. Describe the various movements of the major parts.
3. Describe the general capabilities of this machine.

Cylindrical Grinding

The cylindrical grinder grinds the outside or inside diameter of a cylindrical (round) part. The rotating abrasive wheel, moving from 4000 to 6500 sfpm, is brought into contact with a rotating workpiece moving 50 to 200 sfpm. The workpiece is traversed lengthwise against the wheel to reduce the outside diameter as in the case of external grinding. In the case of internal grinding, the workpiece diameter would be increasing. In cylindrical plunge grinding, the wheel is brought into contact with the work but without traverse. In all cylindrical grinding, the workpiece is rotated opposite to the rotation of the abrasive wheel.

Identifying Machine Parts and Their Functions

On the center-type cylindrical grinder (Figure N-113), the workpiece is mounted between centers much as it would be in the lathe. The plain cylindrical grinder (Figure N-114) has a fixed wheelhead that cannot be swiveled, only traversed. On the universal cylindrical grinder (Figure N-115), the wheelhead and table may be swiveled for taper grinding. All possible motions are illustrated in Figure N-116.

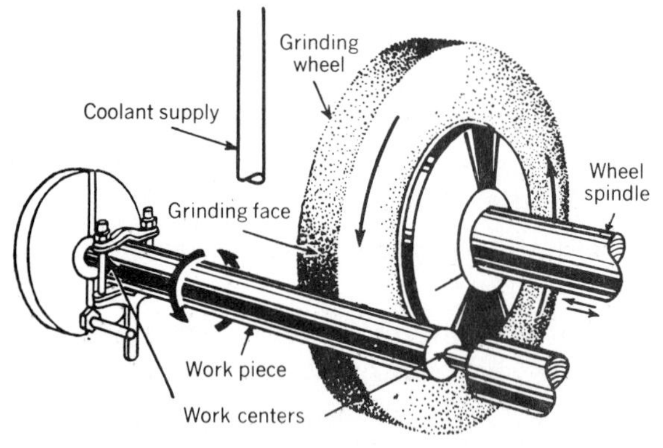

FIGURE N-113 Sketch of center-type cylindrical grinder set up for traverse grinding. Note particularly the direction of travel of the grinding wheel and the workpiece, and the method of rotating the workpiece (Bay State Abrasives, Dresser Industries, Inc.).

FIGURE N-115 Modern universal cylindrical grinder (Landis Tool Co., Division of Litton Industries).

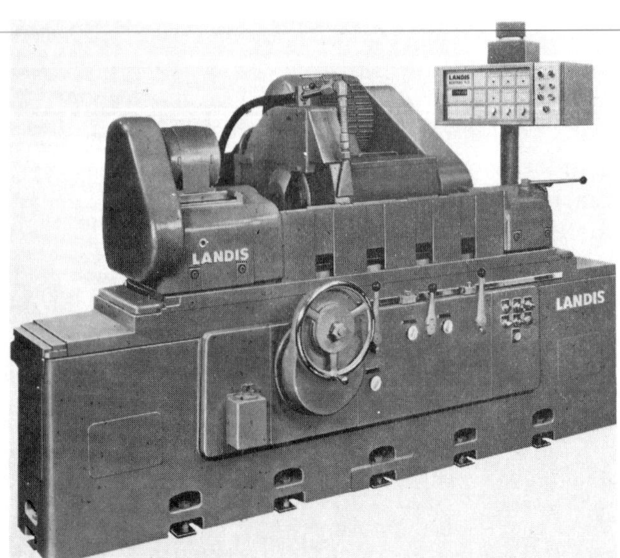

FIGURE N-114 Plain cylindrical grinder (Landis Tool Co., Division of Litton Industries).

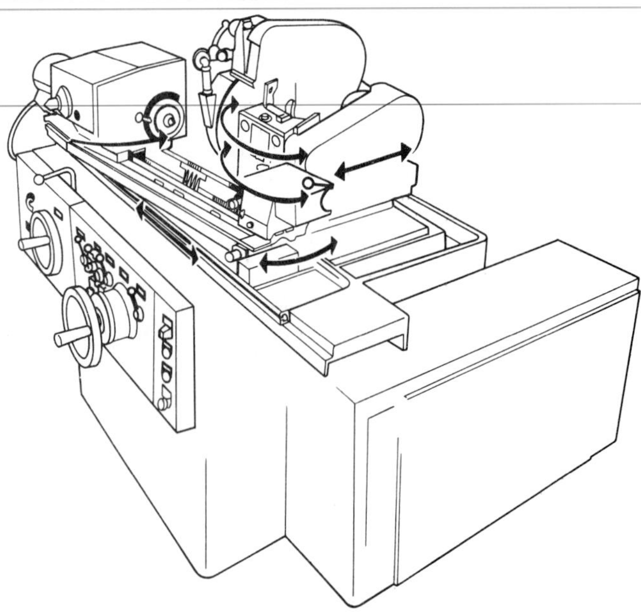

FIGURE N-116 View of universal cylindrical grinder with arrows indicating the swiveling capabilities of the various major components (Cincinnati Milacron).

MAJOR PARTS OF THE UNIVERSAL CENTER TYPE CYLINDRICAL GRINDER

Major parts of the machine include the bed, slide, swivel table, headstock, footstock, and wheelhead.

BED The bed is the main structural component and is responsible for the rigidity of the machine tool. The bed supports the slide, which in turn supports the swivel table.

SLIDE AND SWIVEL TABLE The slide carries the swivel table and provides the traverse motion to carry the workpiece past the wheel. The swivel table is mounted on the slide and supports the head and foot stocks. Graduated scales are located on the swivel table for establishing taper angles.

HEADSTOCK The headstock (Figure N-117) mounts on the swivel table and is used to support one end of the workpiece. The headstock also provides the rotating motion for the workpiece. The headstock spindle may be designed to accept a chuck or face plate. The headstock center is installed when workpieces are mounted between centers. Variable spindle speed selection is also available. For the most precise cylindrical grinding, the headstock center is held stationary, while the driving plate that rotates the part rotates concentric to the dead center. This procedure eliminates the possibility of duplicating headstock bearing eccentricity into the workpiece. For parts that can tolerate minor eccentricity, it is preferable to have the center turn with the driving plate.

FOOTSTOCK The footstock (Figure N-118) is also mounted on the swivel table and supports the opposite end of a workpiece mounted between centers. The footstock center does not rotate. It is spring loaded and is retracted by lever. This permits easy installation and removal of the workpiece. Compression loading on the spring is adjustable, and the tailstock spindle typically can be locked after it is adjusted. The footstock assem-

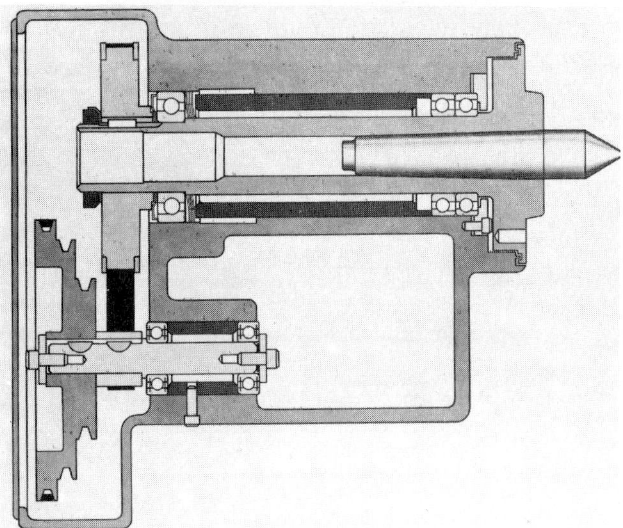

FIGURE N-117 Typical headstock of a center-type cylindrical grinder with cutaway sketch (Landis Tool Co., Division of Litton Industries).

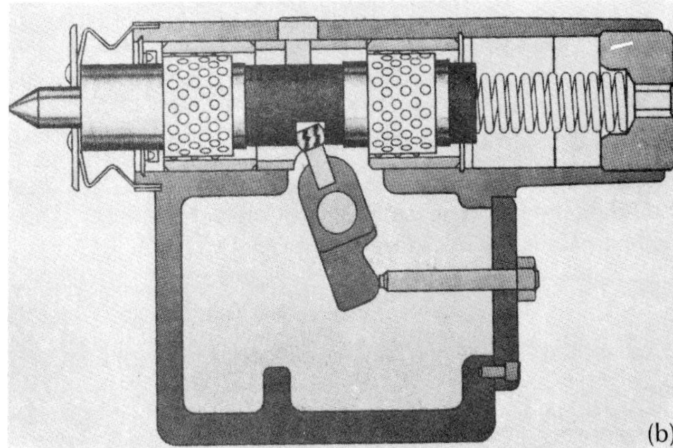

FIGURE N-118 (*a*) Typical footstock. The lever on top of the footstock retracts the work center so that the workpiece can be mounted on the grinder. (*b*) The spring (right end of sketch) provides tension to hold the workpiece in place (Landis Tool Co., Division of Litton Industries).

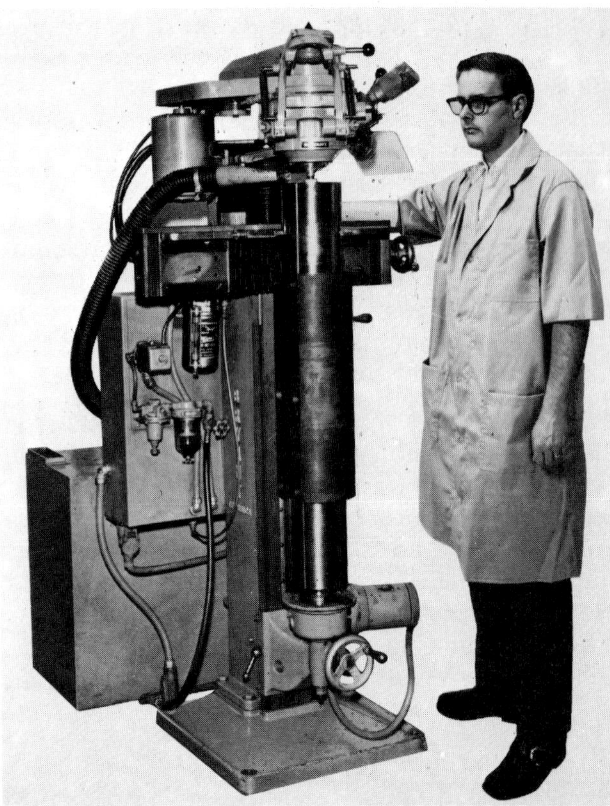

FIGURE N-119 Center-hole grinding machine (Bryant Grinder Corp.).

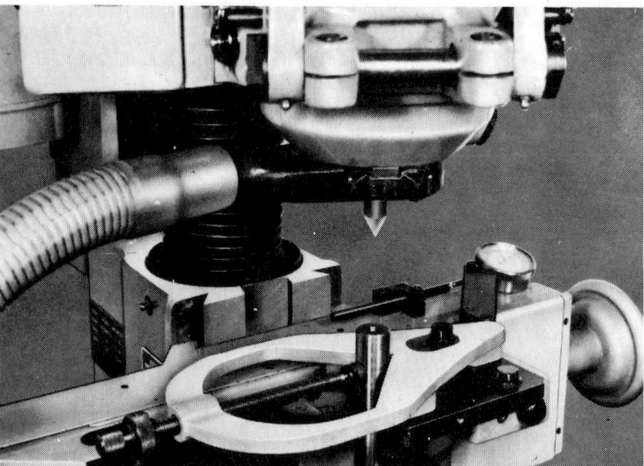

FIGURE N-120 Close-up of center-hole locating setup. Exact location of the center is a most critical step in the operation (Bryant Grinder Corp.).

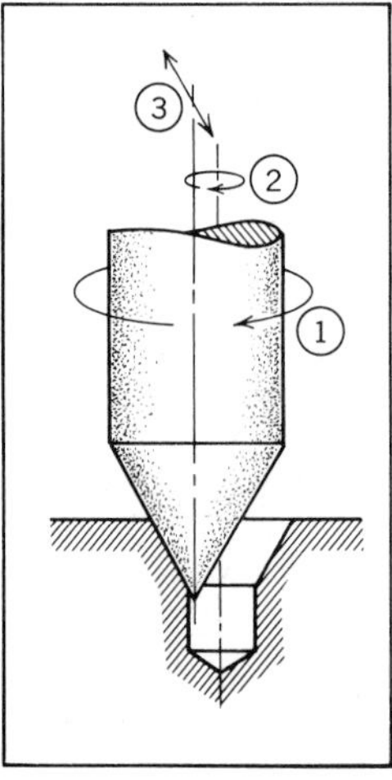

FIGURE N-121 Sketch showing the motions of the grinding wheel in center-hole grinding (Bryant Grinder Corp.).

bly is positioned on the swivel table at whatever points will accommodate the length of the workpiece.

WHEELHEAD The wheelhead, located at the back of the machine, contains the spindle, bearings, drive, and main motor.

Work Holding and Center-Hold Grinding

The workpiece in center cylindrical grinding is most frequently mounted between the head and footstock centers. This essentially provides a single point mounting on each end of the workpiece, permitting maximum accuracy to be achieved in the grinding operation.

The angle of center holes in the workpiece is extremely important. The final result of a cylindrical grinding job depends on this factor. Center holes in the workpiece are often precision ground to the correct geometry by a center-hold grinder (Figure N-119). Location of the center holes (Figure N-120) and producing the correct angle (Figure N-121) are critical operations in center hole grinding. Since a large proportion of cylindrical grinding is done on heat treated parts, the center hole grinder is especially useful for removing heat treatment scale and leaving a precise locating surface.

Capabilities of the Center-Type Cylindrical Grinder

Most common is simple traverse grinding (Figure N-122). This may be done on interrupted surfaces (Figure N-123) where the wheel face is wide enough to span two or more surfaces.

Multiple-diameter-form grinding (Figure N-124) is a type of plunge grinding where the wheel may be form dressed to a desired shape. In straight plunge grinding, the wheel is brought into contact with the workpiece, but the workpiece is not traversed (Figure N-125).

O.D. taper grinding (Figures N-126 and N-127) is done by swiveling the table. Steeper tapers may

FIGURE N-122 Traverse grinding, which is probably the most common type of cylindrical grinding (Cincinnati Milacron).

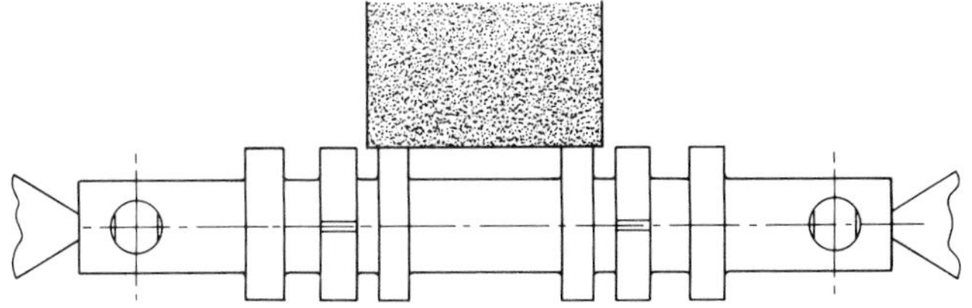

FIGURE N-123 Sketch of traverse grinding with interrupted surfaces. Wheel should always be thick enough to span two surfaces or more at once (Cincinnati Milacron).

FIGURE N-124 Multiple-diameter or form grinding. This is usually a plunge grinding operation with the form dressed in the wheel face (Cincinnati Milacron).

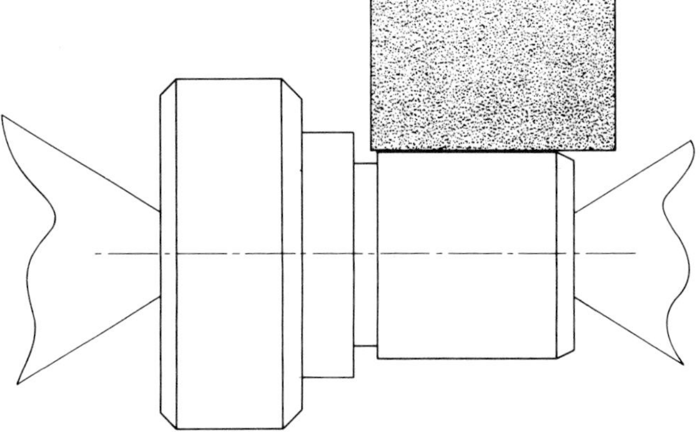

FIGURE N-125 Straight plunge grinding, where the wheel is usually thicker than the length of the workpiece, except where the intent is to take a series of overlapping plunge cuts across a longer piece and finish with several traverses along the entire length (Cincinnati Milacron).

FIGURE N-126 O.D. taper grinding, which may be done to produce a tapered finished workpiece or to straighten up a rough piece that was previously tapered (Cincinnati Milacron).

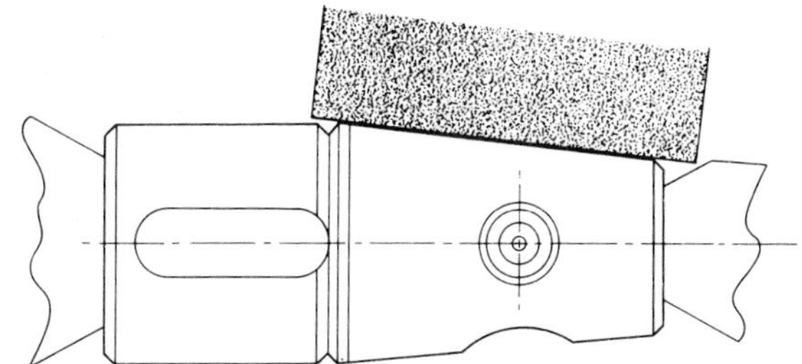

FIGURE N-127 Taper grinding with the workpiece swiveled to the desired angle. For a steeper angle, the wheel might also have to be dressed at an angle less than 90 degrees (Cincinnati Milacron).

require that both table and wheelhead be swiveled in combination (Figure N-128).

Angular shoulder grinding (Figure N-129) and angular plunge grinding may require that the wheel be dressed at an appropriate angle (Figure N-130).

Internal cylindrical grinding can be straight (Figure N-131) or tapered (Figure N-132). The workpiece is held in a chuck or fixture so that the inside diameter can be accessed by the grinding wheel.

FIGURE N-128 Steep taper grinding. Here the wheelhead has been swiveled to grind the workpiece taper (Cincinnati Milacron).

FIGURE N-129 Angular-shoulder grinding. This is very often a production-type operation, but it is shown here on a universal grinder (Cincinnati Milacron).

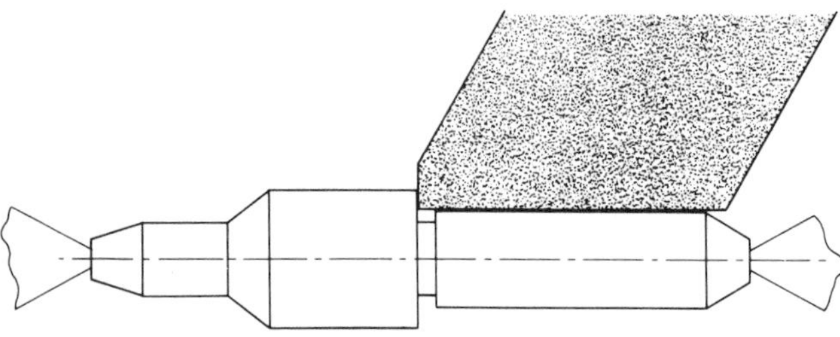

FIGURE N-130 Angular plunge grinding with shoulder grinding. Note the dressing of the grinding wheel (Cincinnati Milacron).

Accessories for Center-Type Grinders

Accessories for center-type cylindrical grinders include the high speed internal attachment, headstock chucks, dressers, and the back or steady rest for grinding slender workpieces (Figure N-133).

SELF-TEST

1. Explain the differences in construction between plain and universal cylindrical grinders.

2. Why is the most critical cylindrical grinding done on nonrotating centers?

3. Why should the footstock adjustment be locked after the workpiece has been positioned?

4. Can the plain cylindrical grinder produce a tapered (conical) surface?

FIGURE N-131 Internal grinding, which requires a special high speed attachment that is mounted on the wheelhead so that it can be swung up out of the way when not in use. It also requires either a chuck or a face plate on the headstock (Cincinati Milacron).

5. Since many cylindrically ground workpieces are hardened, what can be done for center-hole preparation to prevent heat treatment scale from causing inaccurate cylindrical grinding?

6. What is *traverse grinding*?

7. What is *plunge grinding*?

8. Describe the basic method by which a workpiece is held on a center-type cylindrical grinder.

9. If you are traverse grinding a workpiece with interrupted surfaces, what is the major limitation that must be observed in the workpiece?

10. What two additional components are required to do internal grinding on a universal cylindrical grinder?

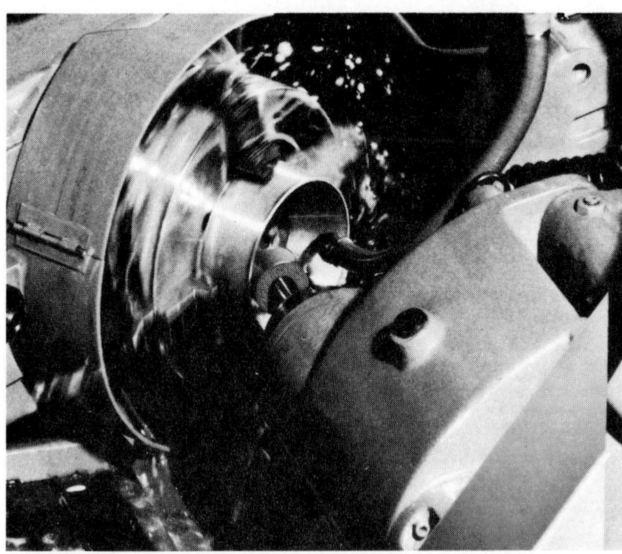

FIGURE N-132 I.D. taper grinding. This is the same sort of operation as shown in Figure 19, but with the workpiece swiveled at the required angle (Cincinnati Milacron).

FIGURE N-133 Backrest or steady rest used to support a thin piece for grinding. If only one is used, as here, it is in the center of the workpiece. If more are used, always an odd number, and they are equally spaced along the workpiece (Cincinnati Milacron).

UNIT 9

Using the Cylindrical Grinder

Using the cylindrical grinder involves many of the same steps that you learned in surface grinding. The cylindrical grinder is a versatile and precision machine tool. Used correctly, it will finish machine your workpiece to a high degree of dimensional accuracy. The purpose of this unit is to familiarize you with the preparation of the machine and the grinding process required to finish machine a typical cylindrical grinding project.

Objectives

After completing this unit, you should be able to:

1. Prepare the cylindrical grinder for a typical grinding job.
2. Finish grind a lathe mandrel to required specifications

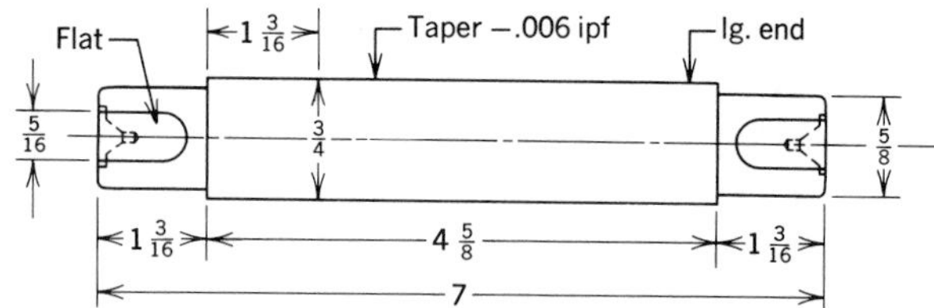

FIGURE N-134 Sketch of tapered lathe mandrel.

FIGURE N-135 Setting up the diamond for wheel dressing (Lane Community College).

FIGURE N-136 Dressing the wheel with the coolant on (Lane Community College).

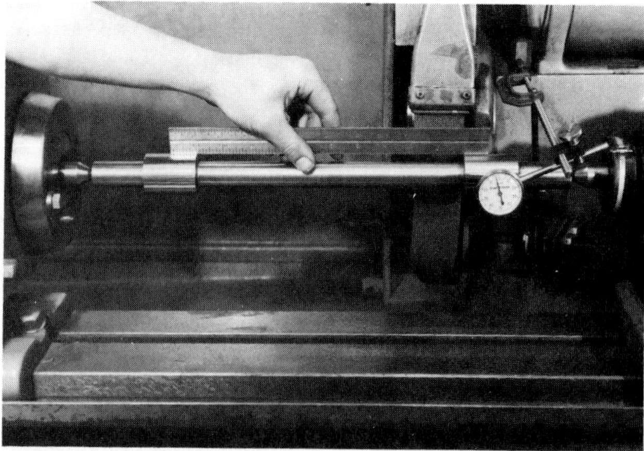

FIGURE N-137 Setting up the parallel test bar (Lane Community College).

FIGURE N-138 Second reading of the dial indicator (Lane Community College).

Cylindrical Grinding a Lathe Mandrel

A lathe mandrel is a common and useful machine shop tool. Precision mandrels are finished by cylindrical grinding after they have been machined and heat treated. Finish grinding the mandrel will provide you with a well-rounded experience in cylindrical grinding. Refer to a working drawing (Figure N-134) to determine the required dimensions.

SELECTING THE WHEEL

For a hardened steel mandrel, the best abrasive will be aluminum oxide of 60 grit and a J bond hardness with a somewhat dense structure.

GRINDING MACHINE SETUP

STEP 1. Set up the diamond dresser (Figure N-135). Diamond dressers have a height to match the wheel centerline. On many machines the dresser is built into the footstock assembly.

STEP 2. Dress the wheel using cutting fluid. If you cannot obtain continuous fluid coverage, then dress dry. Use a rapid traverse of the diamond, with about .001 infeed per pass, until the wheel is true and sharp (Figure N-136).

FIGURE N-139 Adjusting the swivel table (Lane Community College).

FIGURE N-141 Checking the work dog clearance and driver pin contact (Lane Community College).

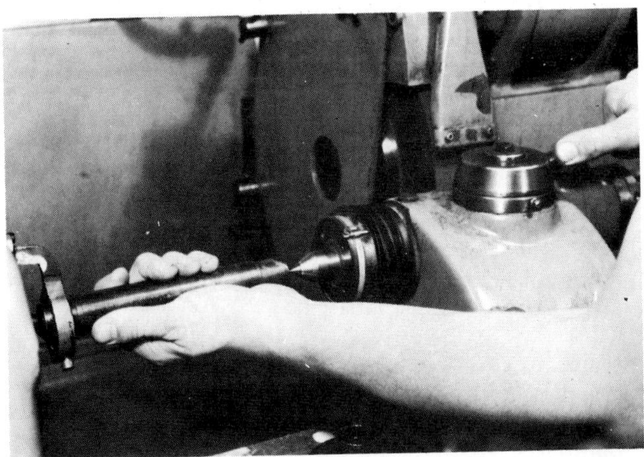

FIGURE N-140 Clamping the footstock (Lane Community College).

FIGURE N-142 Adjusting the wheel overrun (Lane Community College).

STEP 3. Place a parallel test bar between centers (Figure N-137). Mount a dial indicator on the wheelhead, and set to zero at one end of the test bar.

STEP 4. Move the table 12 in. and read the indicator at the other end of the bar. The indicator should read .003 or a total of .006 in. per foot of taper (Figure N-138).

STEP 5. Adjust swivel table to obtain correct amount of taper (Figure N-139).

STEP 6. Lubricate the center hole in each end of the mandrel, using a high pressure lubricant specially prepared for use with centers.

STEP 7. Place the dog on the mandrel and insert it between the headstock and footstock centers. Move the footstock up so that some tension is on the footstock center (Figure N-140).

STEP 8. The work must have no axial movement, but be free to rotate on the centers. The dog must be clamped firmly to the work and contact the drive pin (Figure N-141). If your machine has a slotted driver, be sure that the dog does not touch the bottom of the slot, or it could force the part off the center.

STEP 9. Set the table stops (Figure N-142) so that the wheel can be traversed some distance beyond the ground surface, but at least $\frac{1}{4}$ in. away from contacting the driving dog. If there is adequate clearance on the footstock, set the stop to permit at least one-third of the wheel width to go beyond the surface that is being ground. This overtravel will insure that the ends of the mandrel are ground.

FIGURE N-143 Turning on the wheel (Lane Community College).

FIGURE N-145 Measuring taper 4 in. down the part (Lane Community College).

FIGURE N-144 Confirming taper (Lane Community College).

FIGURE N-146 Grinding the mandrel to size (Lane Community College).

MANDREL GRINDING PROCEDURE

STEP 1. Move the wheel close to the small end of the workpiece. You may use a feeler gage to position the wheel very close to the mandrel.

STEP 2. Turn on the wheel and headstock spindle. Adjust the workspeed to 70 to 100 sfpm.

STEP 3. Start the table traverse. The table traverse rate should be about one-quarter of the wheel width for each revolution of the workpiece. On unhardened work a more rapid traverse would be suitable, up to one-half the wheel width. If your machine is equipped with a Tarry control, set the adjustment for a slight dwell at the end of the traverse, so that the table does not "bounce" on the table stops.

STEP 4. Infeed the wheel until it contacts the rotating mandrel. Sparks will begin to show (Figure N-143). Now start the grinding fluid, and infeed about .001 in. per traverse until the surface of the part is cleaned up. At this point, set the infeed dial to read zero.

STEP 5. Stop the machine completely, but do not disturb the last infeed setting. Measure the mandrel in two places, 4 in. apart (Figures N-144 and N-145), to ensure that your taper setting was correctly done. There should be a .002 in. difference in diameter between the larger and smaller diameters. If adjustment of the swivel table is necessary, be sure that the grinding head is retracted from the work while you make the adjustment.

STEP 6. Rough and finish grind the mandrel to final size. The grinder may have an auto-

FIGURE N-147 The finished mandrel (Lane Community College).

matic infeed stop that will stop the infeed when final diameter is reached. On a manual machine, infeed the wheelhead about .001 in. off the diameter per pass. The finishing passes should be about .0002 in. off the diameter.

STEP 7. Check taper and diameter periodically while making the finishing passes. The table traverse rate should be reduced to about one-eighth the width of the wheel per workpiece revolution. When the final size is reached, allow the work to spark out, by traversing several times without additional infeed (Figure N-146).

Only after careful measurement should the workpiece be removed from between the centers (Figure N-147), as returning the work to the centers to remove very small amounts of material is usually not successful, because of minor eccentricities that occur. The smallest amount of grit can cause large differences.

SELF-TEST

1. Why is the wheel used on a cylindrical grinder typically of a more dense structure than a surface grinding wheel for the same workpiece material condition?

2. What advantage is there in having the diamond wheel dresser integral with the footstock?

3. What is the purpose of a parallel test bar on a cylindrical grinder?

4. What amount of end play should there be when mounting workpieces on centers?

5. What special care must be taken with the driver dog in cylindrical grinding?

6. When setting the traversing stops, how much of the wheel width should be permitted to overtravel the ground part length?

7. What purpose is served by the tarry control?

8. For rough cylindrical grinding, about how much of the wheel width should be traversed for each rotation of the workpiece?

9. For finish grinding, about how much of the wheel width should be traversed for each rotation of the workpiece?

10. Finishing passes on the cylindrical grinder should be about ______ in. off of the diameter.

UNIT 10

Universal Tool and Cutter Grinder

By now, you are well aware of the variety of cutting tools used in the machine shop. Like any cutting tools, milling cutters become dull or occasionally broken during normal use. No production machine shop can function at peak efficiency unless it can keep all of its cutting tools sharp. The tool and cutter grinder is an essential tool for sharpening milling cutters and reamers. In a large machine shop several tool and cutter grinders will be kept continuously busy sharpening and reconditioning a variety of cutting tools. Having at least one tool and cutter grinder, or immediate access to a competent sharpening service, is essential to any machine shop that expects its machinists to turn out quality work.

The operation of the tool and cutter grinder in a large machine shop is often delegated to specialists in tool grinding. These individuals may be assigned to the shop's toolmaking and tool grinding department. Even though tool grinding is a somewhat specialized area of machining, any well-rounded machinist should be familiar with the machines and processes used.

Objectives

After completing this unit, you should be able to:

1. Identify a tool and cutter grinder and its parts.
2. Briefly describe the function of this machine tool.
3. Under guidance from your instructor, sharpen common cutting tools

The tool and cutter grinder is constructed much like a center-type cylindrical grinder (Figure N-148). It has the additional ability to swivel and tilt the workhead in two axes. Some designs, as shown here, have the ability to tilt the wheelhead. Some of these machines can be fitted with a powered table traverse (Figure N-149) and a power-driven workhead, making cylindrical grinding possible.

Cutter Sharpening

The major application of the tool and cutter grinder is in sharpening milling cutters of all types. Cutter sharpening requires an understanding of milling cutters and the grinding machine setups required to corectly produce cutting edge geometry.

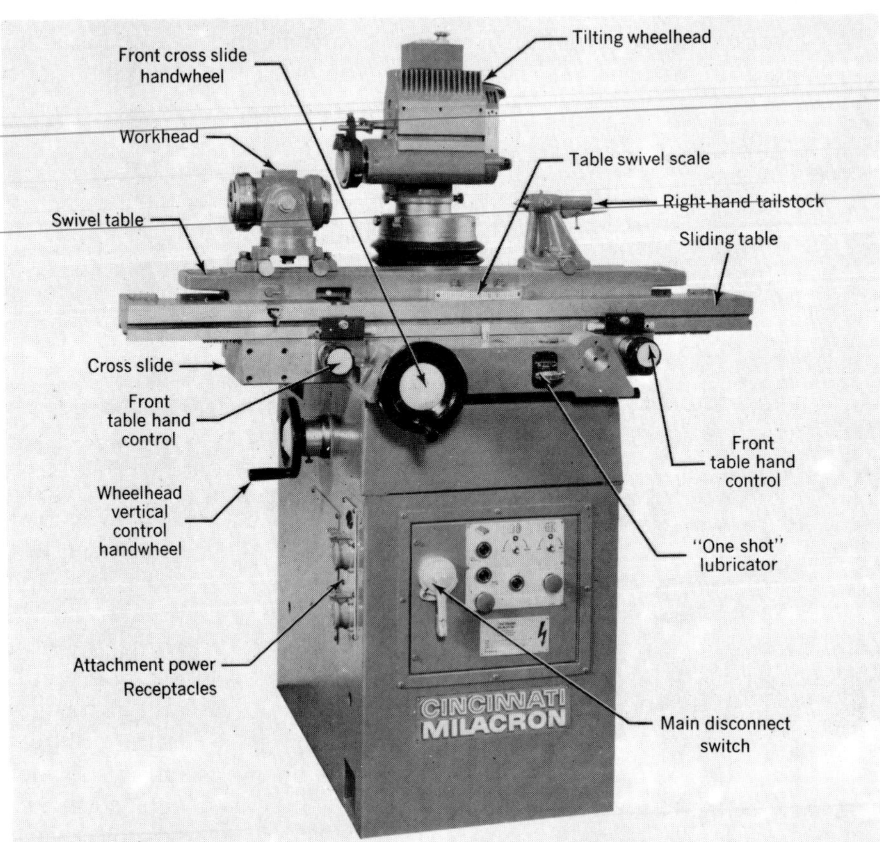

FIGURE N-148 Components of the cutter and tool grinder (Cincinnati Milacron).

PRIMARY AND SECONDARY CLEARANCE ANGLES

In order to cut, a milling cutter must have clearance behind its cutting edge. The required clearance actually consists of two angles forming the primary and secondary clearances.

The surface immediately behind the cutting edge is called the land. The angle formed by the land and a line tangent to the cutter at the tooth tip creates the primary clearance (Figure N-150). The angle between the back of the land and the heel of the tooth forms the secondary clearance.

Primary clearance is extremely important to the performance and life of a milling cutter. If this clearance is excessive, the cutting edge will be insufficiently supported and will chip from the pressure of the cut (Figure N-151).

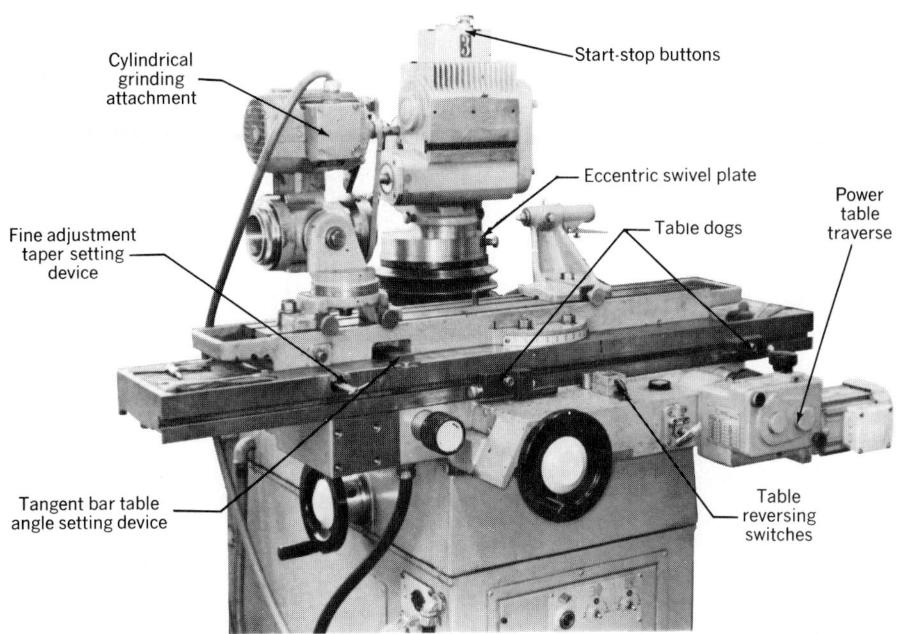

FIGURE N-149 Tool and cutter grinder equipped with power table traverse and with a cylindrical grinding attachment mounted on the workhead (Industrial Plastic Products, Inc.).

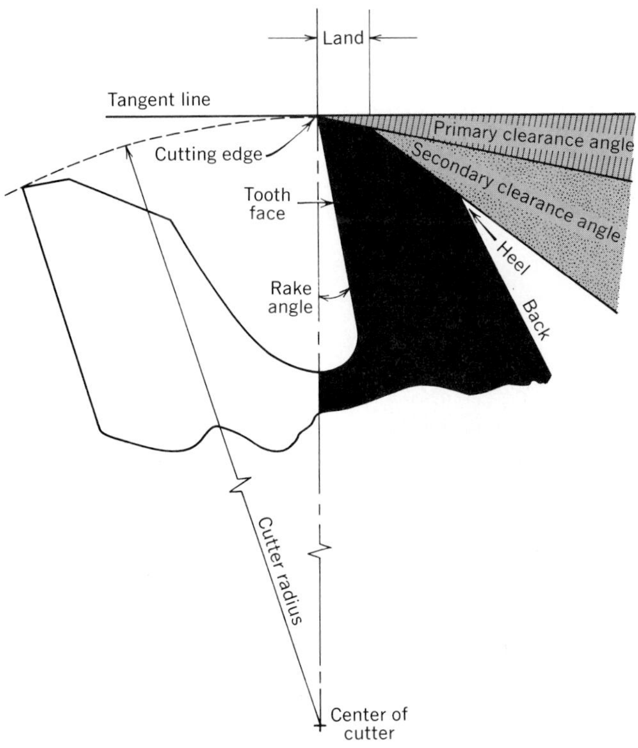

FIGURE N-150 Primary and secondary clearance angles.

FIGURE N-151 This cutter was weakened by incorrect excessive clearance.

TABLE N-2 Primary and Secondary Clearance Angles Suggested for High-Speed Steel Cutters

Material	Primary Clearance (deg)	Secondary Clearance (deg)
Carbon steels	3–5	8–10
Gray cast iron	4–7	9–12
Bronze	4–7	7–12
Brasses and other copper alloys	5–8	10–13
Stainless steels	5–7	11–15
Titanium	8–12	14–18
Aluminum and magnesium alloys	10–12	15–17

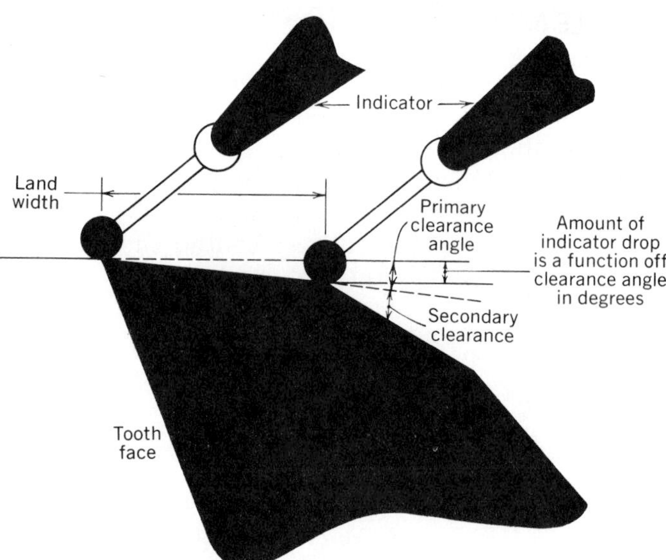

FIGURE N-153 Measuring primary clearance by indicator drop.

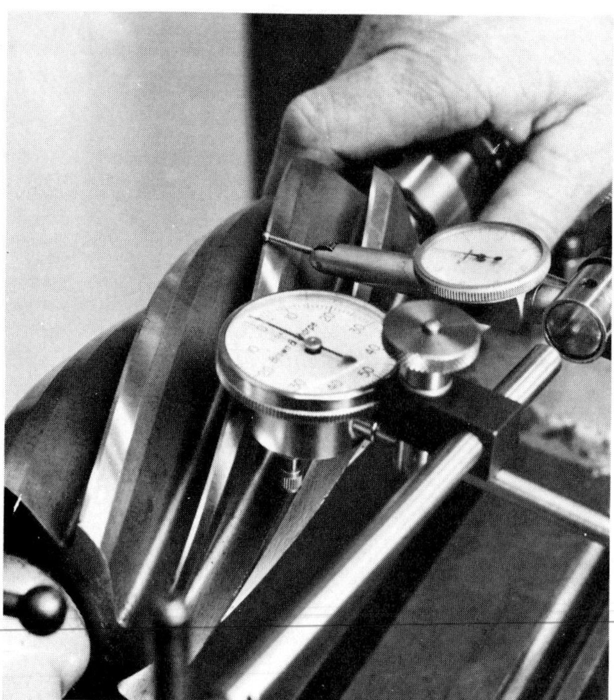

FIGURE N-152 Setting up indicator for the indicator drop method of checking clearances (Lane Community College).

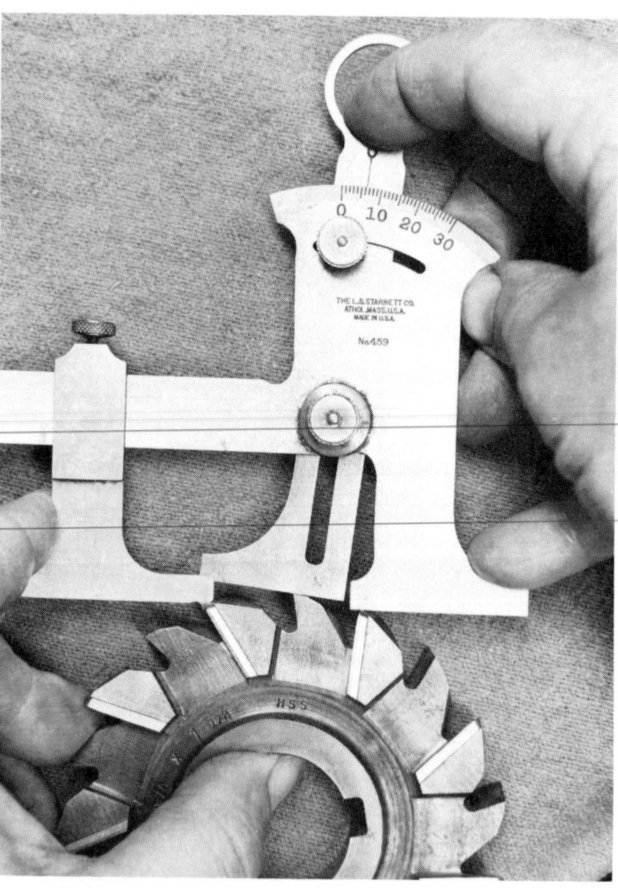

FIGURE N-154 Checking the primary relief of a stagger tooth milling cutter with a Starrett Cutter clearance gage (K & M Tool, Inc.).

Primary and secondary clearance angles for larger diameter (arbor mounted) cutters machining various materials are detailed in Table N-2. When a milling cutter is sharpened, the primary clearance is ground first. The secondary clearance is ground later so that the land may be brought to the recommended width. Land width will vary depending on the cutter diameter.

MEASURING CLEARANCE ANGLES

Two common methods are used to determine the clearance angle of a milling cutter. The indicator drop method (Figure N-152) uses a dial test indicator to measure the difference in height across the width of the land (Figure N-153). This amount is expressed in terms of the angle formed by primary clearance (Table N-3).

TABLE N-3 Indicator Drop Method of Determining Primary Clearance Angles

Diameter of Cutter (in.)	Average Range of Primary Clearance (deg)	Indicator Drop for Range of Primary Clearance Shown Minimum	Maximum	Radial Movement for Checking
$\frac{1}{16}$	20–25	.0018	.0027	.010
$\frac{1}{8}$	15–19	.0021	.0032	.015
$\frac{3}{16}$	12–16	.0020	.0034	.020
$\frac{1}{4}$	10–14	.0019	.0033	.020
$\frac{5}{16}$	10–13	.0020	.0033	.020
$\frac{7}{16}$	9–12	.0025	.0038	.025
$\frac{1}{2}$	9–12	.0027	.0040	.025
$\frac{5}{8}$	8–11	.0028	.0045	$\frac{1}{32}$
$\frac{7}{8}$	8–11	.0033	.0049	$\frac{1}{32}$
1	7–10	.0028	.0045	$\frac{1}{32}$
$1\frac{1}{4}$	6–9	.0025	.0042	$\frac{1}{32}$
$1\frac{1}{2}$	6–9	.0026	.0043	$\frac{1}{32}$
$1\frac{3}{4}$	6–9	.0027	.0044	$\frac{1}{32}$
2	6–9	.0028	.0045	$\frac{1}{32}$
$2\frac{1}{2}$	5–8	.0024	.0040	$\frac{1}{32}$
3	5–8	.0024	.0041	$\frac{1}{32}$
4	5–8	.0025	.0042	$\frac{1}{32}$
5	4–7	.0020	.0037	$\frac{1}{32}$
6	4–7	.0021	.0037	$\frac{1}{32}$
8	4–7	.0021	.0037	$\frac{1}{32}$

The cutter clearance gage (Figure N-154) is somewhat more convenient as clearance angles can be measured directly in degrees.

WHEEL STYLES FOR CUTTER SHARPENING

Milling cutters may be sharpened with a flaring cup wheel or straight wheel. The flaring cup wheel will produce a flat land (Figure N-155). A straight wheel will produce a slight hollow grind behind the cutting edge with curvature dependent on the diameter of the grinding wheel used. Form-relieved cutters are sharpened with a shallow dish wheel.

WORKHOLDING

Arbor driven milling cutters are held on a grinding arbor for sharpening (Figure N-156). An expanding bushing arbor may also be used (Figure N-157). Shank cutters like end mills are held in an accessory spindle or the universal workhead.

TOOTHRESTS

Since the cutter is free to turn on the grinding arbor, a provision must be made so that each tooth to be sharpened can be solidly positioned. The toothrest is used for this purpose. Types of tooth-

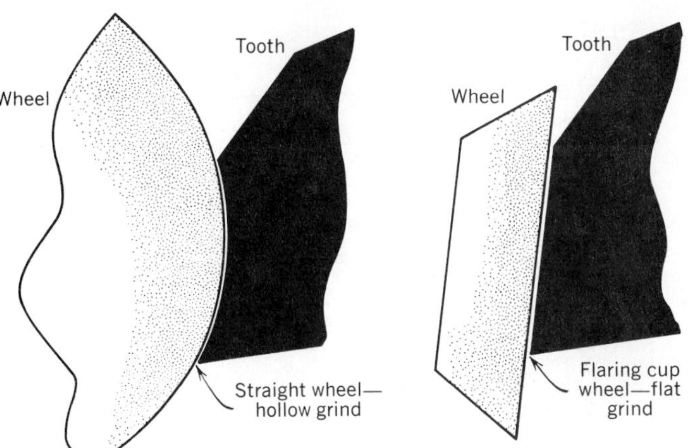

FIGURE N-155 Wheel type will produce flat or hollow form.

FIGURE N-156 Slitting saw being mounted on grinding arbor (Lane Community College).

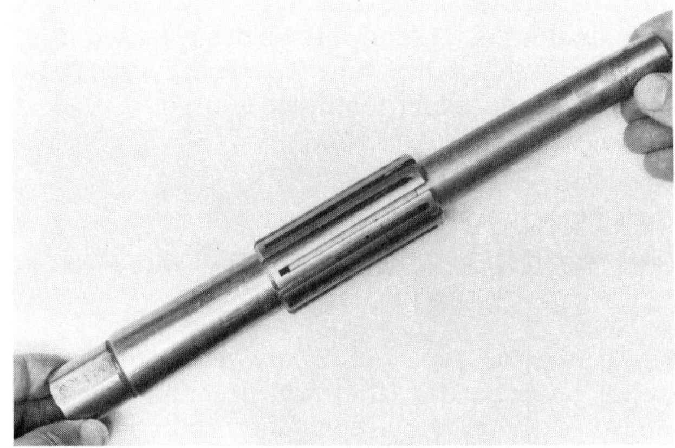

FIGURE N-157 Adjustable grinding mandrel. The slotted bushing is moved along the mandrel to adjust for the I.D. of the tool to be mounted (Lane Community College).

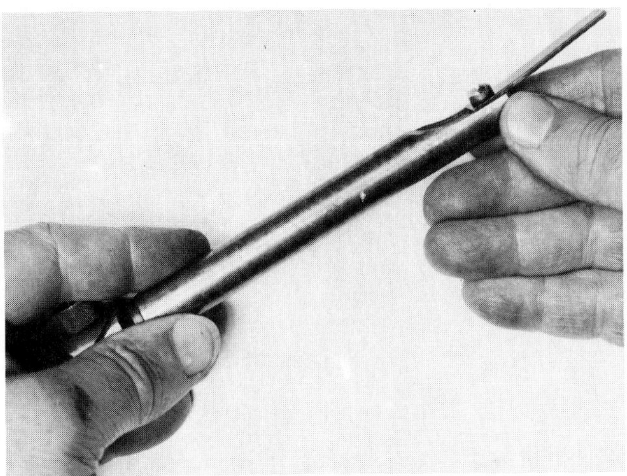

FIGURE N-158 Plain-type toothrest support (Lane Community College).

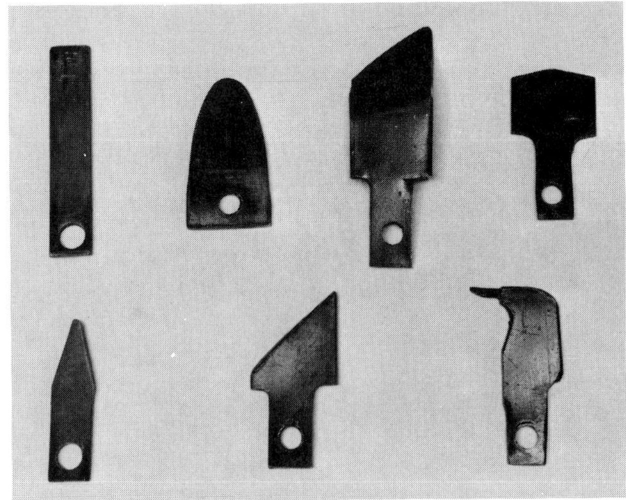

FIGURE N-160 Various designs of toothrest blades (Lane Community College).

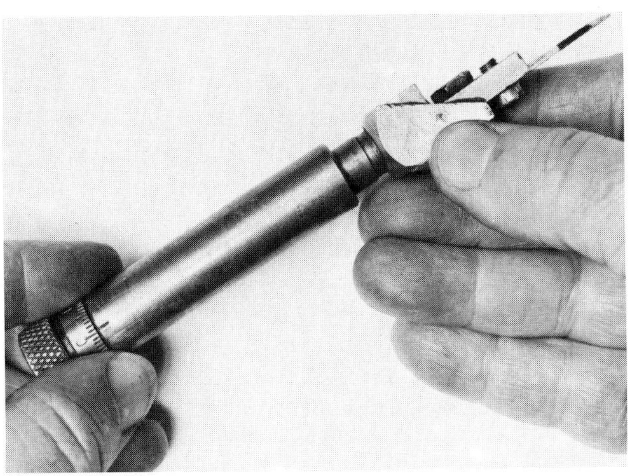

FIGURE N-159 Micrometer-type toothrest support. This example also has provision for spring loading of the finger to permit ratcheting of the cutter tooth, called a flicker finger.

FIGURE N-161 Offset toothrest blade for use in grinding helical milling cutters.

rests include the plain design (Figure N-158) and the micrometer adjustable type (Figure N-159). Various designs of toothrest blades are used (Figure N-160) depending on the type of cutter to be sharpened. An offset toothrest is used for helical milling cutters (Figure N-161).

SETTING UP THE GRINDING MACHINE FOR CUTTER SHARPENING

The centers (Figure N-162) are mounted on the swivel table so that they will accommodate the length of the cutter arbor. One center is spring loaded to facilitate installation and removal of the cutter arbor.

The arbor must be checked for runout and the table must be aligned parallel with the wheelhead.

FIGURE N-162 Tailstock centers are basic to a large portion of cutter grinding (Industrial Plastic Products, Inc.).

FIGURE N-163 After checking the grinding arbor for runout of the centers at each end, the bezel should be zeroed and table alignment checked (Lane Community College).

Attach a dial indicator to the wheelhead (Figure N-163) and use a parallel grinding arbor or test bar held between centers to determine table alignment. Make the necessary adjustments and lock the table in place (Figure N-164).

ESTABLISHING CLEARANCE ANGLES

The type of cutter to be sharpened will determine where the toothrest is to be mounted. For a straight tooth cutter such as a slitting saw, the toothrest wil be mounted on the grinder table. When sharpening this type of cutter it is only necessary to rest each tooth solidly against the toothrest blade.

A center height gage that is adjusted to the arbor centerline is used to set the toothrest blade and the grinder spindle to the arbor centerline height.

Since the toothrest is mounted to the table, clearance angles are established by raising the

FIGURE N-164 Adjusting and locking the swivel table in alignment (Lane Community College).

wheelhead above the center of the cutter. This will cause the grinding to occur back of the cutting edge, thus establishing the correct clearance angle. The amount to raise the wheelhead (W_r) is calculated by the following formula:

$$W_r = \text{sin of clearance angle} \times \text{radius of grinding wheel}$$

Helical cutters, because of their helix angle, must be rotated during grinding so that each point on the tooth is presented to the grinding wheel. It would not be possible to rotate the cutter if the toothrest were mounted on the table. However, by mounting the toothrest on the wheelhead, the cutter tooth may be constantly kept in contact with the toothrest blade as the cutter rotates through its helix.

The clearance angle is established by lowering the wheelhead. This causes the toothrest to lower also and causes the cutter tooth to be rotated below cutter centerline. The grinding then occurs back of the cutting edge with the correct clearance angle. The amount to lower the wheelhead (W_l) is calculated by the following formula:

$$W_l = \sin \text{(clearance angle)} \times \text{radius of cutter}$$

FLAT GRINDING A PLAIN HELICAL MILLING CUTTER

The following procedure may be used to sharpen a plain helical milling cutter.

STEP 1. Align the table.

STEP 2. Attach toothrest to wheelhead.

STEP 3. Set wheelhead at 89 degrees or so that the face of the flaring cup wheel will be set at a 1-degree angle to the cutter. Dress the wheel

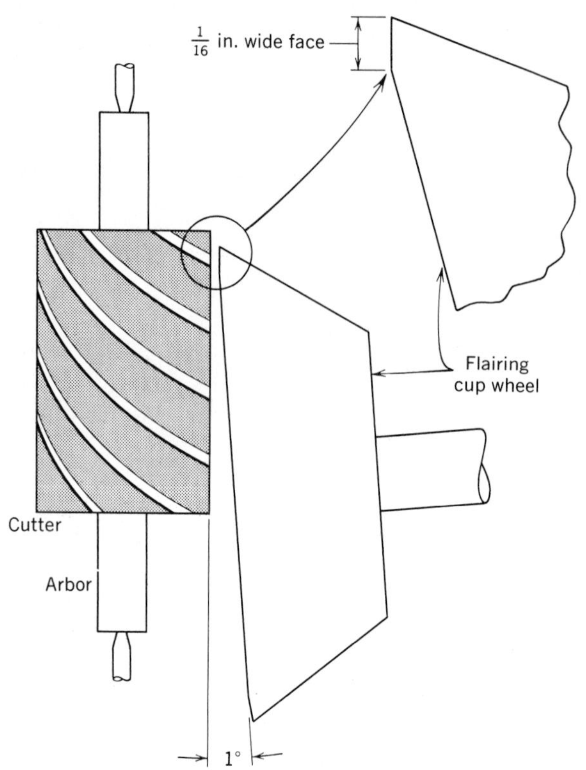

FIGURE N-165 Wheel set at 1 degree to cutter and dressed to narrow land.

FIGURE N-166 Use the center gage to set wheel and toothrest blade height (Lane Community College).

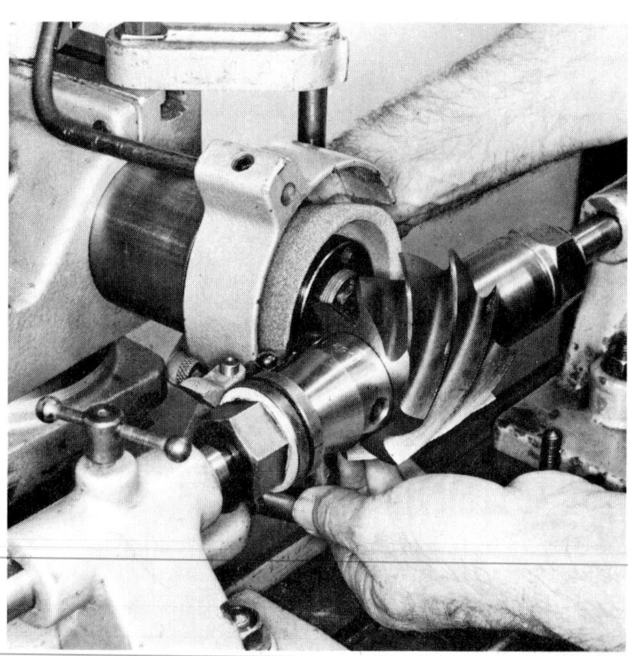

FIGURE N-167 Recheck tooth level and adjust toothrest blade to level if necessary (Lane Community College).

to a $\frac{1}{16}$-in.-wide face. This will ensure that only the edge of the wheel will contact the cutter (Figure N-165).

STEP 4. Lubricate the arbor centers and install arbor between centers.

STEP 5. Use the center gage and position grinder centers, wheel, and toothrest to the same height (Figure N-166).

STEP 6. Rotate cutter firmly against the toothrest and recheck center height. Adjust if necessary (Figure N-167).

STEP 7. Calculate required wheelhead drop to determine primary clearance.

STEP 8. Lower wheelhead by this amount.

STEP 9. Approach the cutter carefully and grind a narrow land. Hold the cutter firmly against the toothrest as the table moves (Figure N-168). Note the cross slide reading.

STEP 10. Move wheel back and rotate the cutter 180 degrees.

STEP 11. Approach cutter and grind according to the same cross slide setting noted in step 9.

STEP 12. Measure cutter for taper with a micrometer. Correct table alignment and arbor runout if necessary. Taper should not exceed .001 in.

STEP 13. Grind all remaining teeth to a spark-out condition (Figure N-169). Spark-out is when no more sparks are visible from the wheel after grinding at a given setting.

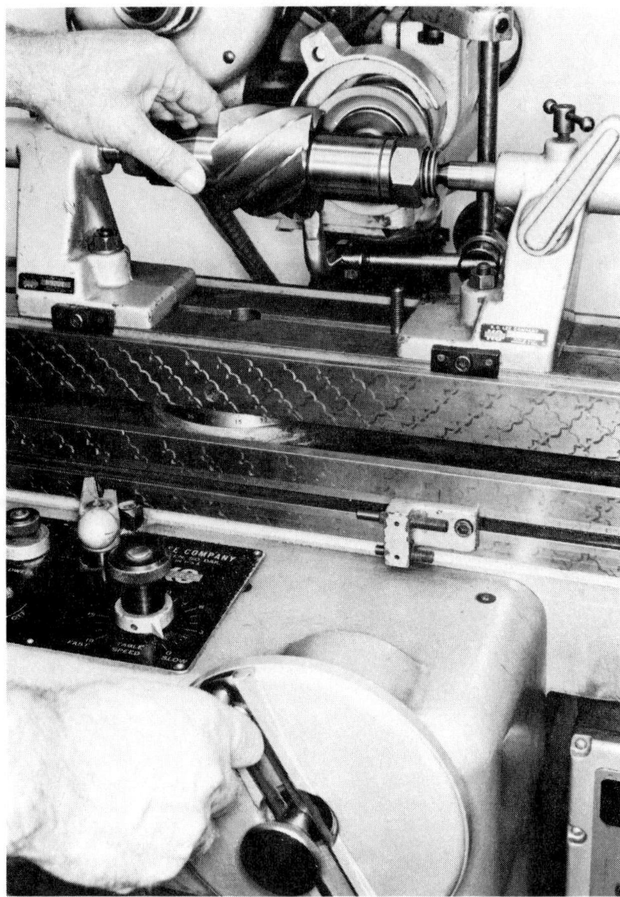

FIGURE N-168 Make the grinding passes (Lane Community College).

FIGURE N-169 The completed grind should have this appearance (K & M Tool, Inc.).

STEP 14. Reset for secondary clearance.

STEP 15. Grind the secondary clearance. When this is complete the cutter should have the appearance shown in Figure N-169.

STEP 16. Check the clearances by indicator drop or clearance gage.

FIGURE N-170 Traversing the table while maintaining cutter contact with the support finger (K & M Tool, Inc.).

HOLLOW GRINDING A PLAIN HELICAL MILL

A straight wheel will produce the hollow grind form. The larger the wheel diameter, the less hollow will be ground. The same procedure will be used for setup as described in flat grinding. The wheel will be positioned perpendicular to the cutter (Figure N-170). Since the toothrest is attached to the wheelhead, it will be necessary to **lower** the wheel to obtain the primary and secondary clearance.

SHARPENING A SLITTING SAW

For sharpening slitting saws, the toothrest is mounted on the swivel table since it is only necessary to hold the cutter in firm contact. Rotation of the cutter during the grinding process is not necessary as it is for helical cutters. A flicker-type toothrest is very useful in this operation. After one tooth is ground, the table with the cutter is moved aside to clear the wheel, and the cutter rotated up

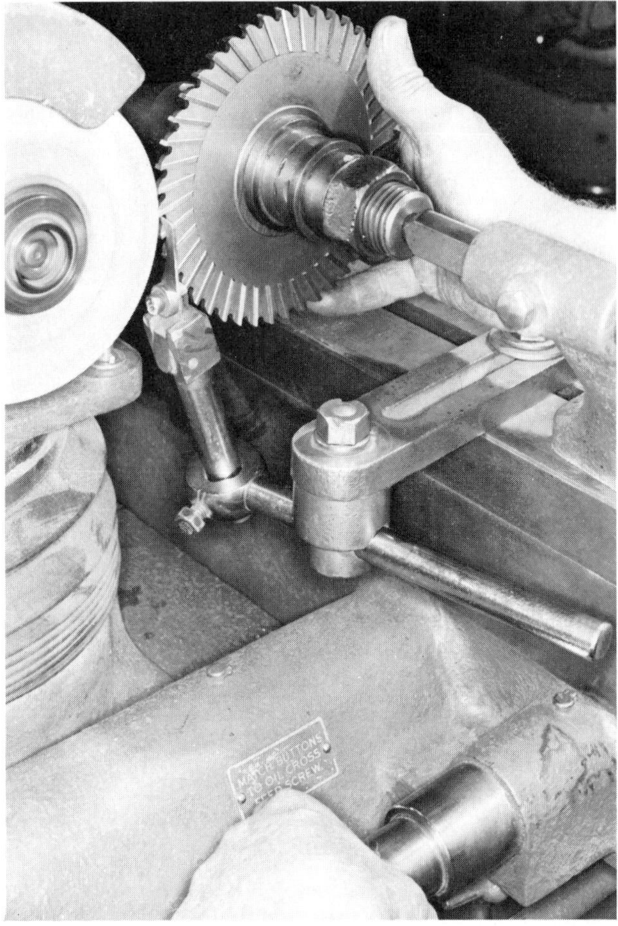

FIGURE N-171 Grinding the primary relief (K & M Tool, Inc.).

FIGURE N-172 Grinding two successive teeth to check for centering of the toothrest (K & M Tool, Inc.).

one tooth. The spring-loaded toothrest snaps aside to index for the next tooth (Figure N-171).

STAGGER TOOTH CUTTERS

The stagger tooth cutter is a helical cutter with the helix angle alternating direction on each tooth. The grinding wheel must be dressed to a shallow V form, and the toothrest must be mounted on the wheelhead and carefully centered to the apex of the V form dressed into the wheel (Figure N-172). This is necessary so that the primary clearance on each opposing helix will be a uniform distance from the cutter axis (Figure N-173). If this is not done correctly, when a slot is made with the stagger tooth cutter, a step is left in the work.

An additional procedure is required to grind the side teeth of a stagger tooth cutter. Primary and secondary clearances are obtained by tilting the workhead (Figure N-174). On some machines the wheelhead can be tilted. Note that the side cutting tooth must be in a level position to grind.

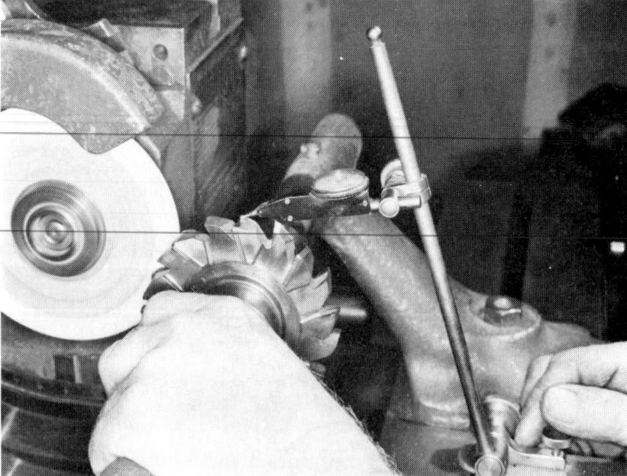

FIGURE N-173 Checking for concentricity on the two teeth of opposite helix (K & M Tool, Inc.).

FORM-RELIEVED CUTTERS

Form-relieved cutters are designed to machine a specific workpiece. Included among these are involute gear cutters and convex and concave milling cutters. Form cutters are sharpened only on the

FIGURE N-174 The workhead is given additional tilt to provide for secondary clearance (K & M Tool, Inc.).

FIGURE N-175 The wheel is positioned relative to the face of the tooth (K & M Tool, Inc.).

face of the tooth (Figure N-175) in order that the particular geometry be maintained. It is important to have the face that is being ground in line with the cutter axis, unless otherwise marked on the side of the cutter. The shallow dish-shaped grinding wheel must be properly dressed and positioned relative to the face of the tooth.

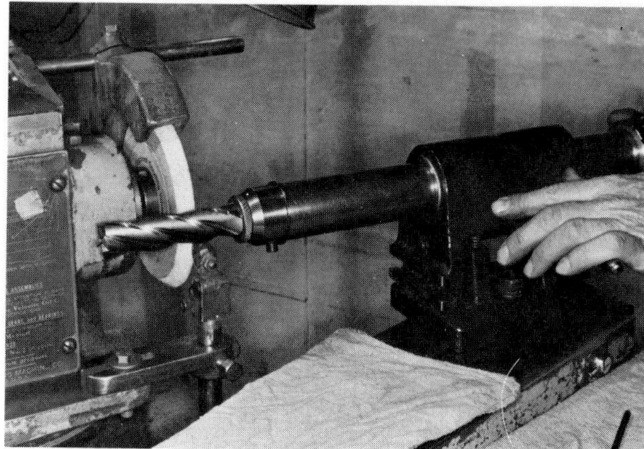

FIGURE N-176 The fixture is rocked away from the wheel and the cutter is moved forward (K & M Tool, Inc.).

TABLE N-4 Primary Clearance Angles for High-Speed Steel End Mills

Workpiece Material	End Mill Diameter (in.)							
	$\frac{1}{8}$	$\frac{1}{4}$	$\frac{3}{8}$	$\frac{1}{2}$	$\frac{3}{4}$	1	$1\frac{1}{2}$	2
Carbon steels[a]	16°	12°	11°	10°	9°	8°	7°	6°
Nonferrous metals	19°	15°	13°	13°	12°	10°	8°	7°

[a] For tool steels, decrease indicated values about 20 percent. For secondary clearance angle, increase value by about $\frac{1}{3}$.

END MILLS

Although end mills can be sharpened on both the side and end, often only end sharpening is required since most wear and damage occurs there. When side wear requires peripheral sharpening, the tool loses its nominal size and cannot be used to cut accurate slots such as keyseats in one pass. However, the reground end mill can have new cutter performance if the regrinding is done expertly. The machinist compensates for the reduced diameter to generate specific slot widths. Where earlier numerically controlled milling machines required end mills that were accurately to size, most currently manufactured machine controllers can accommodate nonstandard diameter cutters. When the side of an end mill requires sharpening, it may be held in a free-turning accessory spindle. The spindle may be moved lengthwise in an ordinary sleeve bearing or in a type that floats in an air bearing (preferred) providing a high degree of concentricity and friction-free movement. The accessory spindle can also be tilted away from the grinding wheel to facilitate the grinding procedure (Figure N-176).

Grinding of the primary and secondary clearances on the side of an end mill is much the same procedure as any plain helical cutter. Primary and secondary angles are specified in Table N-4. The

FIGURE N-177 Passes are made with additional infeed until the primary relief is complete (K & M Tool, Inc.).

FIGURE N-178 Correct appearance of the completed side grinding of the end mill (K & M Tool, Inc.).

FIGURE N-179 Cutting off the damaged end (K & M Tool, Inc.).

best practice is to sharpen from the shank end toward the tip (Figure N-177). The toothrest will be mounted on the wheelhead and the wheelhead will be lowered to provide the proper clearances (Figure N-178). The formula for the wheelhead drop is

$$W_1 = \sin \text{(primary relief angle)} \times \text{radius of the cutter}$$

END SHARPENING WITH THE UNIVERSAL WORKHEAD

The end of an end mill may be sharpened by using the universal workhead. If the end of the cutter has been damaged, it must first be cut off with an abrasive cut-off wheel (Figure N-179). The workhead spindle is tilted (Figure N-180) to obtain the proper primary and secondary clearances (Figure N-181).

CENTER GASHING

Center gashing is necessary to permit a center cutting end mill to make its own starting hole. A straight wheel is dressed to a sharp beveled edge. The cutter is then gashed (Figure N-182) to center. After grinding each tooth, burrs may be removed with a suitable stone (Figure N-183).

It should be noted that the gashing techniques shown present a significant hazard. The cutter is

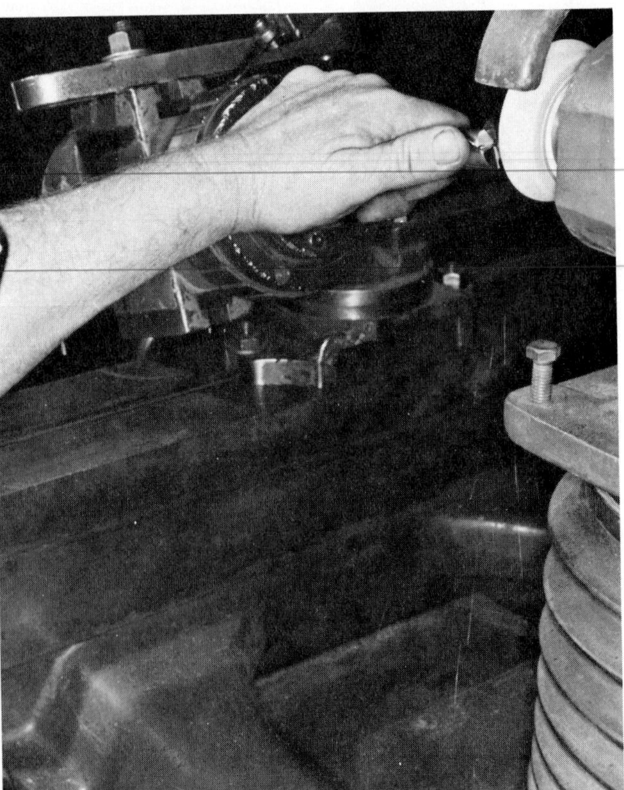

FIGURE N-180 Grinding the secondary clearance (K & M Tool, Inc.).

hand-held with fingers close to the high-speed wheel. Wrapping a cutter in a shop cloth also presents a significant hazard near the revolving equipment. **You must be extremely careful if you are gashing end mills by this technique.** Until you become experienced, ask your instructor for help in performing this cutter grinding operation.

Additional Capabilities of the Tool and Cutter Grinder

Additional capabilities of this machine include sharpening of reamers, taps, drills, and single-point cutting tools. The power-driven workhead permits cylindrical grinding (Figure N-184). The workpiece may be held in a three-jaw chuck or magnetic grinder chuck (Figures N-185 and N-

FIGURE N-183 Remove the burrs. Note the completed appearance (K & M Tool, Inc.).

FIGURE N-181 Appearance of the completed end grinding of the end mill.

FIGURE N-184 Resizing the pilot portion of the counterbore by cylindrical grinding (Industrial Plastics Products, Inc.).

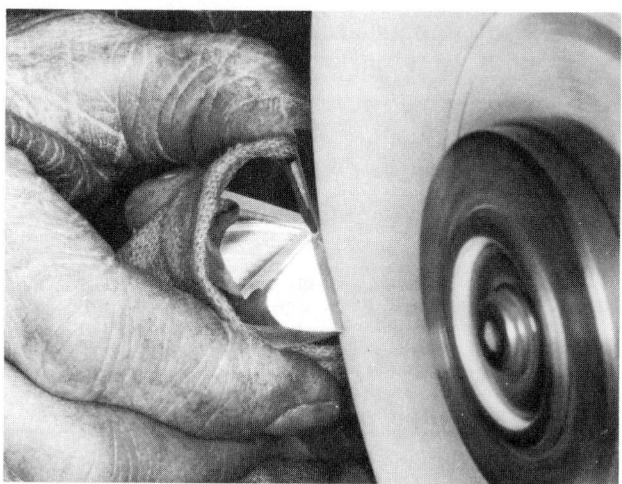

FIGURE N-182 Grinding across the face of the tooth to center (K & M Tool, Inc.).

FIGURE N-185 Cylindrical grinding attachment being used with adjustable scroll chuck to set up part for cylindrical grinding (Industrial Plastics Products, Inc.).

FIGURE N-186 A permanent magnet chuck is also very useful for cylindrical grinding (Industrial Plastics Products, Inc.).

FIGURE N-187 Internal grinding attachment set up for use with a mounted grinding wheel. Only the forward portion of the wheel is used; the rest is dressed away for clearance (Industrial Plastics Products, Inc.).

186). The internal grinding attachment permits internal cylindrical grinding (Figure N-187). The swiveling capability of the workhead permits regrinding of lathe centers (Figure N-188).

FIGURE N-188 Another use for the cylindrical grinding attachment is the reconditioning of centers (Industrial Plastics Products, Inc.).

SELF-TEST

1. What machine capabilities differentiate the tool and cutter grinder from other designs of universal cylindrical grinders?

2. If the cutter clearances are excessive, what is the result in the use of the cutter?

3. Why is it desirable to sharpen different sets of milling cutters for aluminum alloys as compared to steels?

4. Why is the primary clearance ground first on a milling cutter?

5. Why is the toothrest attached to the grinding head when sharpening to helical cutters is done?

6. In grinding cutters, why should the second trial cut be made 180 degrees from the first grind?

7. In grinding stagger tooth milling cutters, why must the toothrest be carefully centered to the grinding wheel?

8. On what surface are form-relieved cutters sharpened?

9. What grinding accessory is essential for the sharpening of the sides of end mills?

10. What tool and cutter grinder accessory is necessary for grinding the clearances on both the ends of end milling cutters and the side teeth of stagger tooth cutters?

SECTION **O**

ADVANCED MACHINING PROCESSES

As the hardware of an advanced technology becomes more complex, new and visionary approaches to the processing of materials into useful products come into common use. This has been the trend in machining processes in recent years. Advanced methods of machine control as well as completely different methods of shaping materials have permitted the mechanical designer to proceed in directions that would have been totally impossible only a few years ago.

Parallel development in other technologies such as electronics and computers have made available to the machine tool designer methods and processes that can permit a machine tool to far exceed the capabilities of the most experienced machinist. Since these advanced technologies are constantly changing and expanding, the purpose of this section is to provide only an overview of two major areas:

1. Numerical control of machine tools
2. Electrical machining processes

Electronic Control of Machine Tools

The purpose of this section is to give you some information on the impact numerical control (NC) has in the field of machining. It is only an overview and cannot take the place of a programming manual for a specific machine tool. Because of the tremendous growth in numbers and capability of computers, changes in machine controls are rapidly and constantly taking place. The exciting part of this evolution in machine controls is that programming becomes easier with each new advance in this technology. Many of the NC machines with older style controllers which use punched tape would benefit from a conversion to a new computer numerical control (CNC) unit. Most CNC machines have the ability to accept input from punched tape, but one double density floppy disk can hold about as much machine control information as 3000 ft of punched tape and the transmission of information is many times faster. On many CNC machines, while one part is being machined, a new part can be programmed.

Advantages of Numerical Control

A manually operated machine tool may have the same physical characteristics as a CNC machine, such as size and horsepower. The principles of metal removal are the same. The big gain comes from the computer controlling the machine axes' movements. The cutting tools that remove materials are standard tools such as milling cutters, drills, boring tools, or lathe tools depending on the type of machine used. Cutting speeds and feeds need to be correct as in any other machining operation. The greatest advantage in CNC machining comes from the unerring and rapid positioning movements possible. A CNC machine does not stop at the end of a cut to plan its next move; it does not get fatigued; it is capable of uninterrupted machining, error free, hour after hour. A machine tool is productive only while it is making chips.

Complex contoured shapes were extremely difficult to produce prior to CNC machining. CNC has made the machining of these shapes economically feasible. Design changes on a part are relatively easy to make by changing the program that directs the machine tool.

A CNC machine produces parts with high dimensional accuracy and close tolerances without taking extra time or special precautions. CNC machines generally need less complex work-holding fixtures, which saves time by getting the parts machined sooner. Once a program is ready and producing parts, each part will take exactly the same amount of time as the previous one. This repeatability allows for a very precise control of production costs. Another advantage of CNC machining is the elimination of large inventories; parts can be machined as needed. In conventional production often a great number of parts must be made at the same time to be cost effective. With CNC even one piece can be machined economically. In many instances, a CNC machine can perform in one setup the same operations that would require several conventional machines.

When the term *NC* is used, it usually applies to both NC and CNC machining. CNC is rapidly expanding and the programming of CNC machine tools is becoming easier all the time. A commonly used programming aid is the use of a CAM system. Part drawing information is entered into a computer as lines, arcs, or points. As this information is entered, it is graphically displayed on the computer screen. Errors show up immediately as each input is displayed and can be corrected with a few keyboard strokes. When the part geometry input is complete, information on the cutting tool, feed, and speed is entered. These inputs are manipulated in the computer and converted in the postprocessor into machine-ready code. With modern CNC machine tools a trained machinist can program and produce even a single part economically. CNC machine tools are used in small and large machining facilities and range in size from tabletop models to huge machining centers. In a facility with many CNC tools, programming is usually done by CNC programmers away from the CNC tools. The machine control unit (MCU) on the machine is then used mostly for small program changes or corrections. Manufacturing with CNC tools usually requires three categories of persons. The first is the programmer, who is responsible for developing machine-ready code. The next person involved is the setup person, who loads the program into the MCU, checks that the correct tools are loaded, and makes the first part. The third person is the machine operator, who loads the raw stock into the machine and unloads the finished parts.

CNC controls are generally divided into two basic categories. One uses a word address format with coded inputs such as G and M codes. The other uses a conversational input; conversational input is also called user-friendly or prompted input. Later in this section examples of each of these programming formats in machining applications will be described.

Computer-aided design (CAD) is widely used, and in many instances this information stored in a computer can be processed to directly control machine tools. This activity is called computer-aided manufacturing (CAM). Under this general heading, robots can be linked to machine tools to load and unload machines or to be used in the assembly of products. When designing, machining, and assembly operations are linked to the management functions of manufacturing, the process is termed computer-integrated manufacturing (CIM).

CNC Machine Safety

All safety rules that apply to conventional machine tools also apply to CNC machines. Always wear appropriate eye protection. Rings, bracelets, watches, necklaces, and loose clothing can be caught by moving or revolving machine or work parts. Items like these should be removed before any machine is operated. When working with large machines or heavy workpieces, wear safety shoes. Long hair should be adequately covered.

The work area around the machine needs to be kept clean and clear of obstructions to prevent slipping or tripping. Machine surfaces should not be used as worktables. Use proper lifting methods to handle heavy workpieces, fixtures, or heavy cutting tools. Make measurements only when the spindle has come to a complete standstill. Chips should never be handled with bare hands.

Before starting the machine make sure that the work-holding device and the workpiece are securely fastened. When changing cutting tools, protect the workpiece being machined from damage, and protect your hands from sharp cutting edges. Use only sharp cutting tools. Check that cutting tools are installed correctly and securely.

Do not operate any machine controls unless you understand their function and what they will do. All CNC machines have electrical compartments; keep the doors to these compartments closed to keep dirt and chips out. Only authorized technicians should perform maintenance or repairs in these enclosures.

To avoid damage to the machine or workpiece, check the tool path by making a dry run through the program, that is, letting the machine make all the positioning and cutting moves without a cutting tool installed, one move at a time. If the controller is equipped with one of the new CRT video displays, the path of each tool can be graphically shown and then compared to the expected tool path.

UNIT 1

CNC Machine Tools

Before any CNC machine tool can be used productively, its capabilities must be understood. This unit shows you what some CNC machine tools can do, how machine tool axes are designated, and the importance of efficient tooling.

Objectives

After completing this unit, you should be able to:

1. Identify a variety of machining operations performed on CNC machine tools.
2. Explain the benefits of adaptive control.
3. Identify the *X*, *Y*, *Z*, and rotational axes on a machine tool.
4. Explain the reason for precision tool presetting on CNC machine tools.

Machine Control by Numerical Instructions

Numerical control (NC) is a system of controlling a machine or process by instructions in the form of numbers and letters. NC is a term often used to describe a machine control system that uses punched tape to feed information into the controller memory. Computer numerical control (CNC) stores and processes machining information directly in a machine's built-in computer. Initially, DNC stood for *direct numerical control.* Early DNC used a mainframe computer to feed individual NC controllers with one program block or a buffer load at a time. Because of the high transmission speed of these computers, this computer could keep as many as 50 NC tools in one facility supplied with machine code on a continuous basis. The problem was that a power interruption shut down all of the machines relying on that computer. Today DNC stands for *distributed numerical control.* The mainframe computer still stores all of a facility's CNC programs, but these programs are downloaded to an MCU in their entirety rather than a block at a time. The MCU on each CNC machine tool then supplies control information as needed during the machining cycle.

Vertical Spindle NC and CNC Machining Centers

Vertical spindle CNC machining centers patterned after the vertical milling machine are very popular. Many vertical milling machines have been converted with retrofit machine controls to CNC operation. Control units such as the Centurion IV (Figure O-1) can take a conventional machine tool and make a high output CNC tool out of it. Smaller vertical CNC machines usually do not have auto-

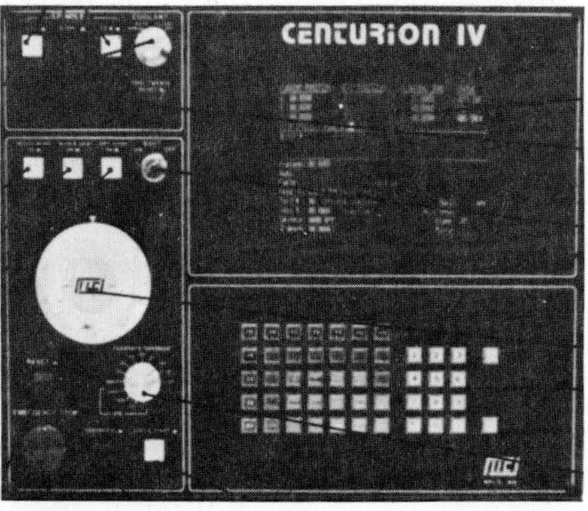

FIGURE O-1 CNC adaptable to conventional machines (Industrial Information Controls, Inc.).

matic tool changers but require the operator to exchange cutting tools in the spindle. Larger machines are equipped with tool magazines holding a number of tools. A carousel-type toolholder is shown in Figure O-2. Tool changing is directed from the machine control unit. The vertical spindle design also includes large capacity multispindle milling machines such as the gantry-type profiler (Figure O-3). Several workpieces can be machined at the same time on these machine tools.

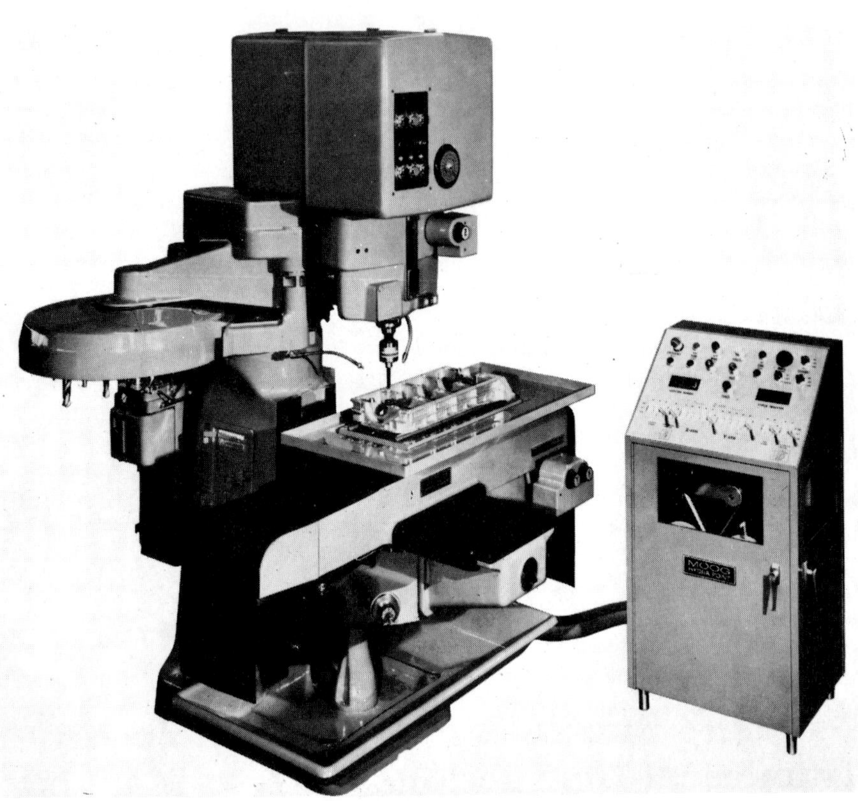

FIGURE O-2 Vertical CNC milling machine with side-mounted tool drum (Hydra-Point Division, Moog Inc.).

FIGURE O-3 Gantry-type multiple-spindle NC profiler (Cincinnati Milacron).

Horizontal Spindle NC and CNC Machining Centers

Horizontal spindle CNC machining centers are patterned after the horizontal boring machine. With a four-axis horizontal spindle machining center (Figure O-4), workpieces that previously could be produced only with a profiling or tracing machine can now be programmed and machined.

AUTOMATIC TOOL CHANGING AND WORK LOCATING

CNC toolholders can be side- or top-mounted tool carousels (Figure O-5). Tool carousels, also called tool drums, move the correct tool to the position where the tool changing arm can make the tool exchange. A workpiece can be mounted on a rotary

FIGURE O-4 Four-axis horizontal spindle machining center (WCI Machine Tools & Systems Company).

FIGURE O-5 Horizontal spindle, CNC machining center with side-mounted tool drum (Cincinnati Milacron).

FIGURE O-6 Rotary table on a horizontal spindle machining center (Cincinnati Milacron).

FIGURE O-7 CNC horizontal spindle machining center (Kearney & Trecker Corporation).

FIGURE O-8 Pivoting presenter arm on a side-mounted tool changer (Heald Machine Division/Cincinnati Milacron).

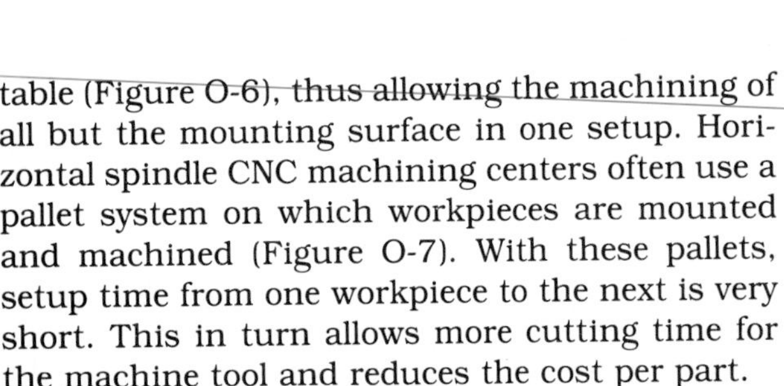

table (Figure O-6), thus allowing the machining of all but the mounting surface in one setup. Horizontal spindle CNC machining centers often use a pallet system on which workpieces are mounted and machined (Figure O-7). With these pallets, setup time from one workpiece to the next is very short. This in turn allows more cutting time for the machine tool and reduces the cost per part.

Both vertical and horizontal machining centers can accomplish a wide variety of machining tasks. With the tool drum mounted on the side of the machining center (Figure O-8), tool changing can be a two-setup operation. After the tool drum rotates the required tool into position it is pivoted down and grasped by the changing arm. The tool is then placed into the spindle (Figure O-9). At the same time, the tool already in the spindle is returned to the pivot arm and replaced in the tool drum. The tools are held in a self-releasing taper shank toolholder. The toolholder is fastened in the machine by a locking mechanism contained within the spindle (Figure O-10). Tool changing methods vary on different machines; machine tool manufacturers use the methods best suited to the machines they make. The greater the number of preloaded and preset tools available to the programmer, the greater the variety of machining operations that can be performed. Whereas a tool changer capacity of 30 tools was considered to be large a few years ago, some machining centers now store as many as 120 tools (Figure O-11).

FIGURE O-9 Tool changing arm (Heald Machine Division/Cincinnati Milacron).

FIGURE O-10 Spindle tool-holding mechanism (Hydra-Point Division, Moog Inc.).

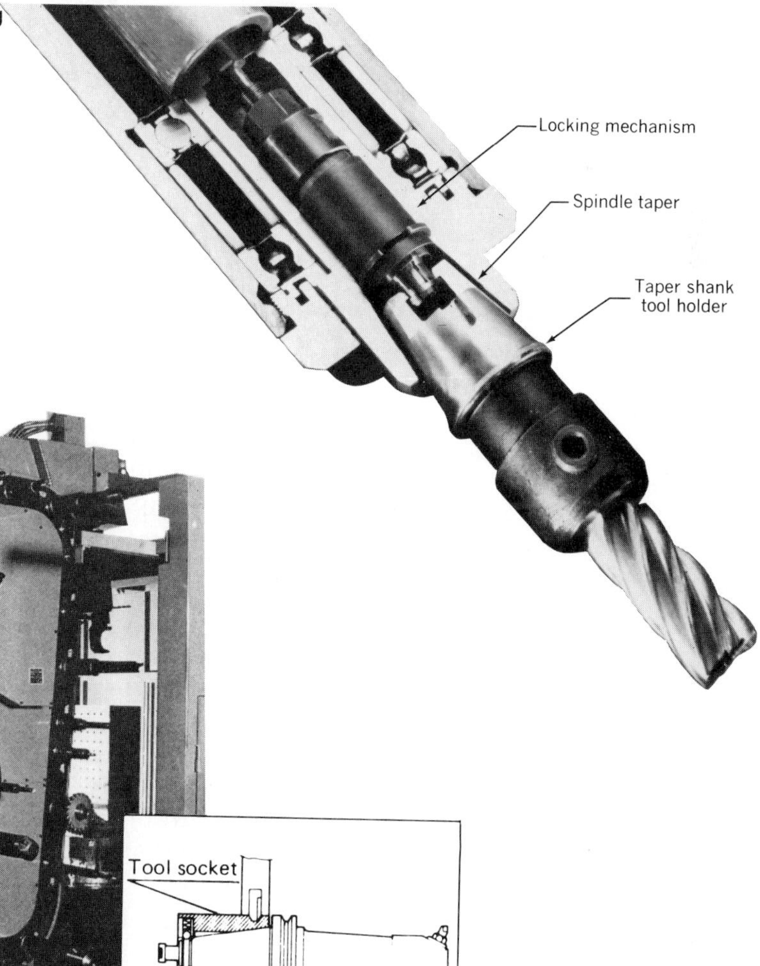

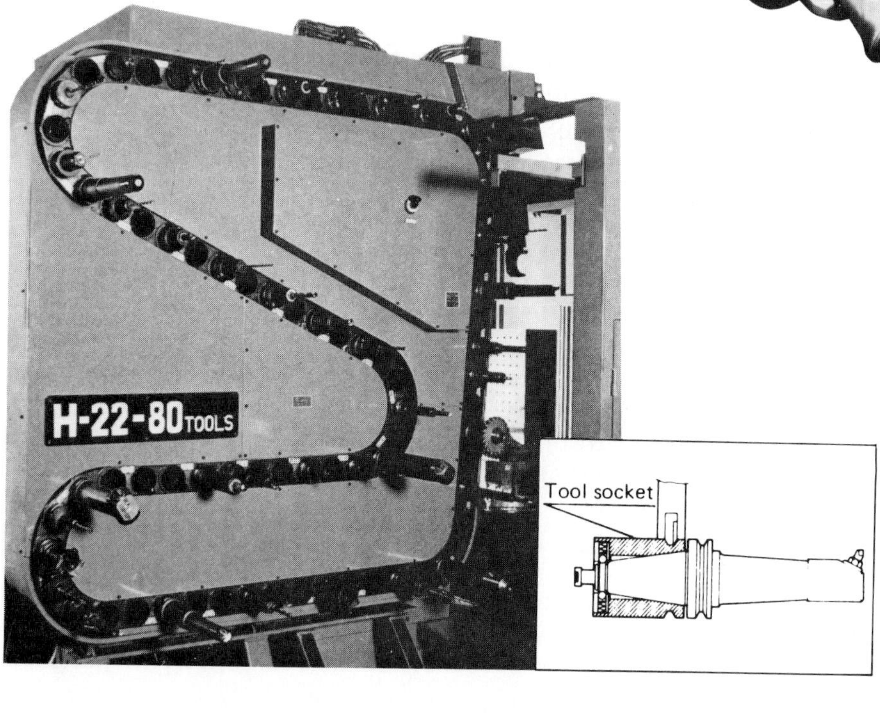

FIGURE O-11 Large-capacity tool magazine (Mazak Corporation).

FIGURE O-12 CNC drilling (Hydra-Point Division, Moog Inc.).

FIGURE O-14 CNC boring (Hydra-Point Division, Moog Inc.).

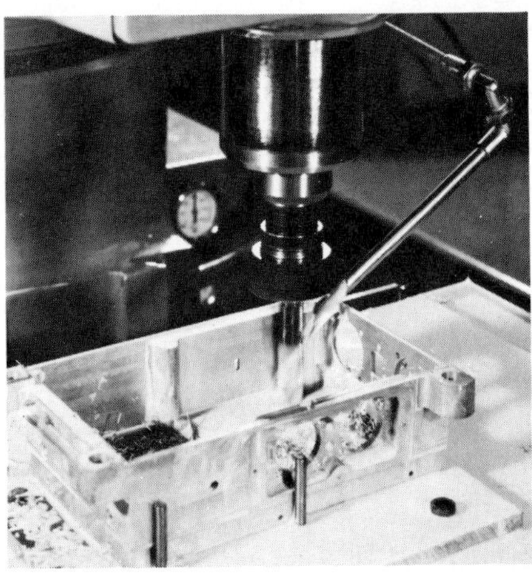

FIGURE O-13 CNC pocket milling (Hydra-Point Division, Moog Inc.).

FIGURE O-15 CNC leadscrew tapping (Hydra-Point Division, Moog Inc.).

CNC DRILLING

Drilling (Figure O-12) is a common CNC machining operation. Spindle speed, feed rate, and hold depth are controlled by CNC instructions. A peck drilling cycle is used for deep hole or small hole drilling. In a pecking cycle, the drill feeds a short distance into the workpiece and is then withdrawn rapidly to clear chips from the hole. The drill then returns rapidly to the point from which it was withdrawn. Then the drill advances another short distance at the preset feed rate before the next cycle begins. Pecking cycles can be programmed as a given number of pecks per hole or as a distance drilled between cycles.

CNC MILLING

Milling a cavity or pocket milling (Figure O-13) is a frequently performed machining operation. Spindle speeds, milling feed rates, and direction and distance of the cuts are controlled from the machine control unit. Many CNC controllers have a canned cycle function for pocket milling.

CNC BORING

Close tolerance boring (Figure O-14) is often done on CNC machines. The boring tool can be withdrawn rapidly from the hole during rough boring operations where tool marks are not critical. For close tolerance bores, withdrawing the tool at the same feed rate that was used for boring into the workpiece leaves an excellent surface finish. Accurate bores require the use of boring tools preset for roughing, semifinishing, and finishing cuts.

CNC TAPPING

Tapping operations are well suited for CNC machines. One such method is leadscrew tapping (Figure O-15). The leadscrew causes the machine spindle to feed the same distance per revolution as the lead of the tap. The leadscrews are changed to correspond to the different tap leads. Tapping heads can also be used; in these the lead of the tap provides the feed.

On CNC controls the lead of the tap is calculated and then programmed as a feed per revolution of the machine spindle. A control unit with a tapping canned cycle will reverse the spindle rotation when the correct tapping depth is reached and will unscrew the tap from the hole.

CNC CONTOURING

Probably the greatest machining capability of the CNC machine tool is that of contouring or continuous path machining. This includes circular, angular, and irregular shape cuts in two or three dimensions. Figure O-16 is an example of a four-axis machine milling a cam; the fourth axis is the rotary table. The CNC machine tool can produce shapes that would be impossible with conventional machines.

FIGURE O-16 Four-axis vertical spindle machining center machining a cam (Mazak Corporation).

AUTOMATIC CENTERING

High-precision machining requires accurate workpiece centering and positioning. The setup procedure to establish the *X–Y* coordinate axes requires the use of edgefinders and dial indicators by a highly skilled machine operator. The workpiece has to be aligned on the machine table with its axes parallel with the machine axes, often a very time-consuming, difficult process. When a touch sensor, often called a probe, is used, this process is greatly simplified. The workpiece is secured on the machine table without the need of exact alignment of the workpiece and machine axes. The probe is used to record the location of specific reference points on the workpiece relative to the machine datum. The MCU will then rearrange its tool path database to accurately machine the workpiece where it is located. When a probe is used to locate and plot the periphery of a casting, the CNC control can use that information as the starting dimension for its programmed rough machining cycle. A large stock allowance will result in additional roughing cuts; a small stock allowance will result in fewer passes. When a touch sensor (Figure O-17) makes contact with a surface, it registers this coordinate point in the controller's memory. Probes can be mounted in a lathe turret or fixed in an automatic tool changer. A probe can be used for in-cycle gaging and measuring. If the probe senses that a machined surface is dimensionally different from the programmed size, it can alert the MCU to make tool offset changes to correct the discrepancy.

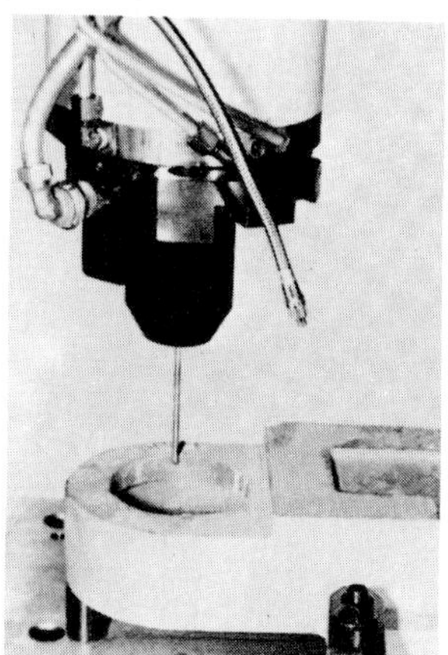

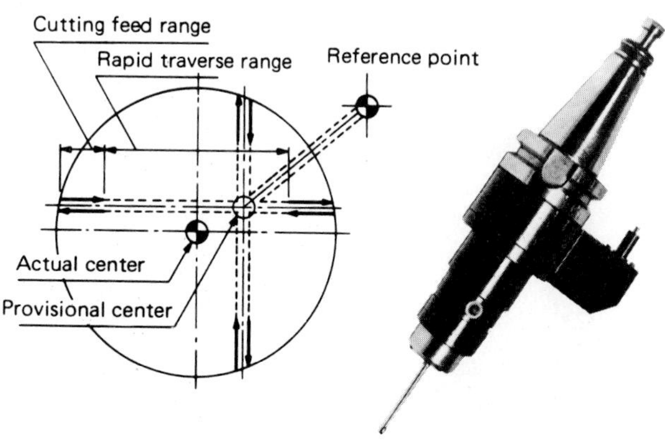

FIGURE O-17 Touch sensor locating a bore center (Mazak Corporation).

ADAPTIVE CONTROL

AC (adaptive control) is continual and automatic process monitoring and constant compensation adjustment without operator intervention. The benefit of an AC system lies in its ability to sense and adjust the feed rate depending on workpiece hardness and cutting tool conditions, varying depth of cut, deep holes with chips clogging the flutes of the drill, and so on. As an example, handbook data on drilling feed rates take into consideration that a drill wears while drilling. Consequently, the handbook feed rate recommendations are only 50 percent of what they could be under ideal conditions. The value of an AC system lies in its ability to sense cutting variations and to make adjustments, rather than using a constant machining rate that is "safe" for all machining conditions. Adaptive controls in metal-cutting operations differ from each other depending on what

FIGURE O-18 CNC turning center (Hardinge Brothers, Inc., Elmira, N.Y.).

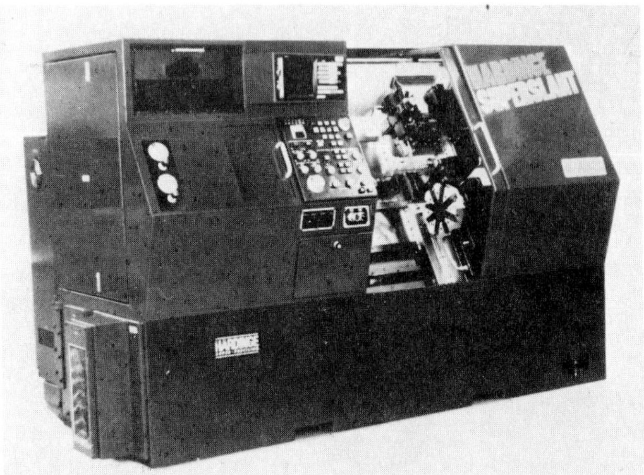

FIGURE O-20 CNC four-axis chucker and bar machine (Hardinge Brothers, Inc., Elmira, N.Y.).

FIGURE O-19 Inside- and outside-diameter NC turning center turrets (Cincinnati Milacron).

they monitor and what adjustments they make. Most often power consumption is monitored. In a rough machining operation, the desired feed rate, spindle speed, and target horsepower are programmed. The horsepower target is based on the expected load at the tool. Since horsepower is directly related to metal removal in cubic inches per minute, the control will adjust the feed rate as required to maintain the target horsepower. This maximizes the metal removal rate and produces a more uniform chip load. Another type of AC measures changes in tool or spindle deflection forces. This control continuously monitors the torque at the cutting tool and the power consumed at the motor. Whenever the torque at the tool exceeds a preset limit, the feed rate is lowered. This assures that at all times the cutting torque and power are limited to levels that protect the tool, motor, and setup. One type of AC measures vibrations at specific points on the machine. When this type of control detects a tool failure on a lathe, it is capable of stopping the feed movement in a time period shorter than it takes to make one spindle revolution. This fast reaction time prevents a broken tool from damaging the workpiece.

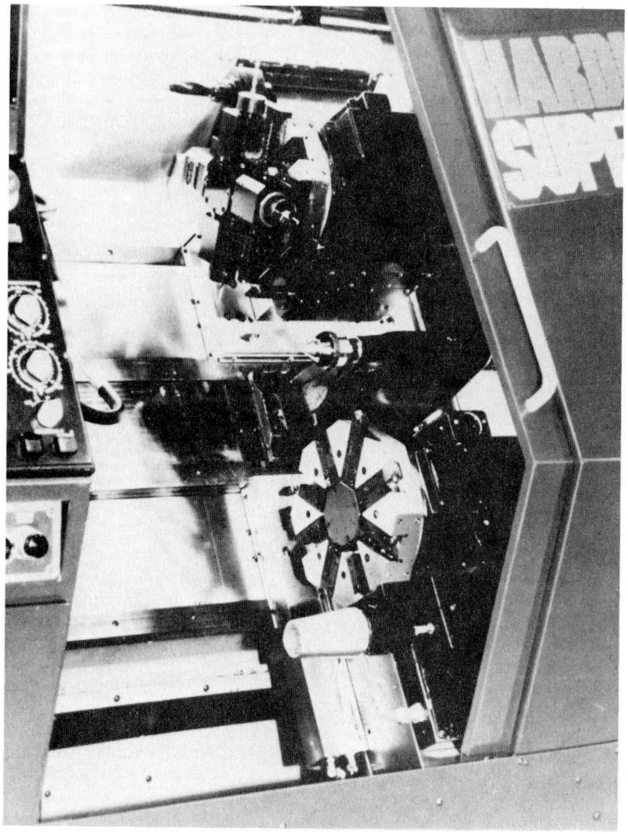

FIGURE O-21 Twin turrets; top turret with live tooling (Hardinge Brothers, Inc., Elmira, N.Y.).

NC and CNC Turning Centers

A sophisticated CNC lathe or turning center (Figure O-18) uses one or more numerically controlled turret toolholders. With the workpiece held in a chuck, a variety of cutting tools can be used for inside and outside turning (Figure O-19). The machine is able to drill, bore, tap, thread, and machine internal and external contours and shapes in one setup. A four-axis turning machine (Figure O-20) can make parts that used to require multiple machine setups. One of the two turrets shown in Figure O-21 has live tooling, that is, rotating tools. With live tooling, milling a hexagon on a part previously turned is simple (Figure O-22). Cross drilling a hole (Figure O-23), tapping, or other milling machine functions can be performed on such a machine. When an electronic sensing probe is part

FIGURE O-22 Milling a hexagon with an indexable endmill (Hardinge Brothers, Inc., Elmira, N.Y.).

FIGURE O-23 Cross-drilling a hole (Hardinge Brothers, Inc., Elmira, N.Y.).

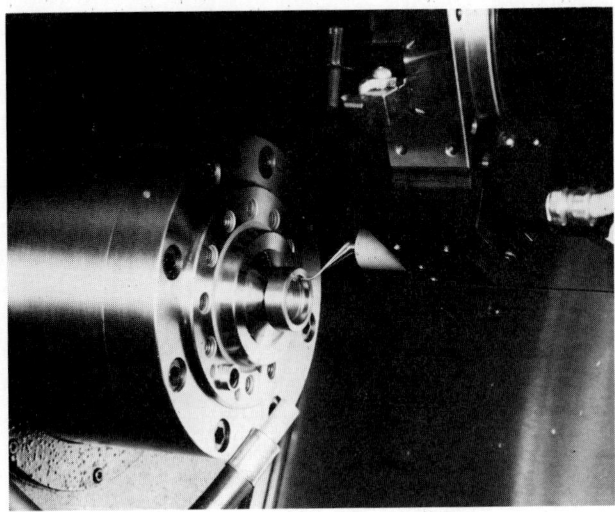

FIGURE O-24 Electronic sensing probe checking workpiece size (Hardinge Brothers, Inc., Elmira, N.Y.).

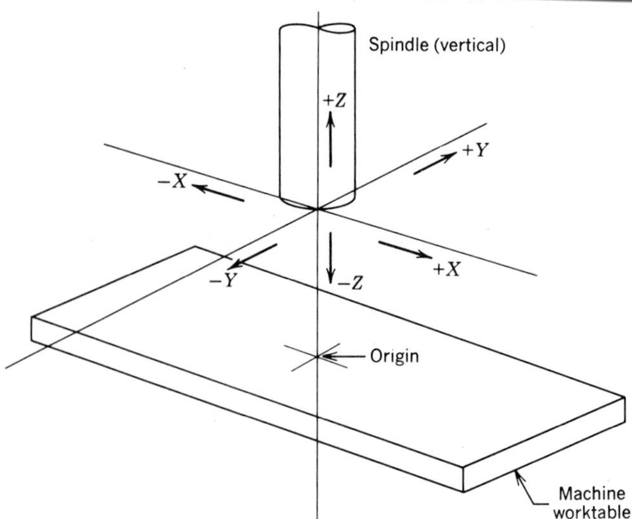

FIGURE O-25 Directions of vertical spindle movement.

of the available machine options, workpiece dimensions can be measured while the part is still in the machine (Figure O-24). This guarantees precision and extreme accuracy of the machined parts.

CNC turning machines have a feature that allows the programming of a constant surface speed. This feature gives an excellent surface finish in facing cuts or in the machining of tapers. A constant surface speed also helps in obtaining maximum cutting tool life.

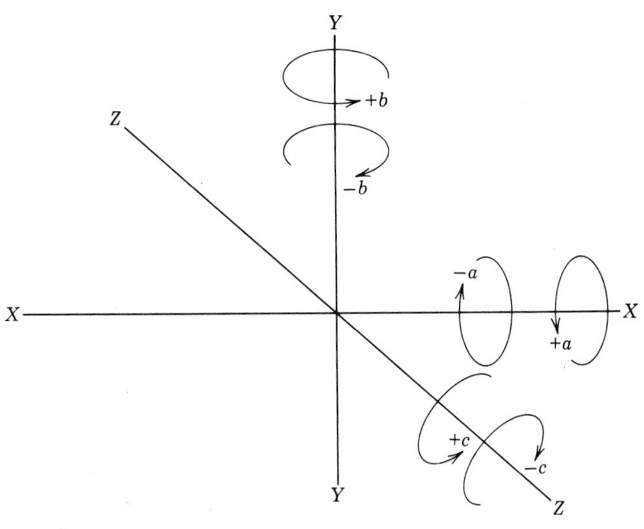

FIGURE O-26 Rotational axes.

Directions of Machine Tool Spindle Travel

The MCU tells the machine spindle or worktable to move in a certain direction or to a specified location. This is done by instructions that indicate positive and negative movement directions. CNC programming is done as if the tool were actually moving even though the worktable may be the moving component. Figure O-25 shows the directions of a three-axis machine tool. The direction of spindle movement is expressed by noting its direction of travel along a specified axis. Spindle movement in the *Z* axis is also defined in terms of positive and negative directions. A longitudinal movement of the machine table is an *X* axis move. When you are facing the machine, a positive command in *X* moves the tool to the right. The saddle on this machine moves in the *Y* axis. A positive *Y* command moves the tool away from you. A movement in the *Z* axis directs the spindle to raise or lower. A positive command in *Z* raises the spindle. So remember, a negative *Z* axis command will bring the tool closer to the machine table. The directions for positive and negative commands for the rotational axes of *a*, *b* and *c* are illustrated in Figure O-26.

MACHINE TOOL AXES

BASIC AXES The rectangular coordinate system consists of the two perpendicular axes of *X* and *Y*. The *X* and *Y* axes lie in the same plane and are known as coordinate axes. With the addition of a third axis, *Z*, that is perpendicular to the *X-Y* plane, a three-dimensional volume of space can be described and identified. The point at which the axes intersect is called the origin and has a numeric value of zero.

These notations are applied to CNC machine tools in order to identify the basic machine axes.

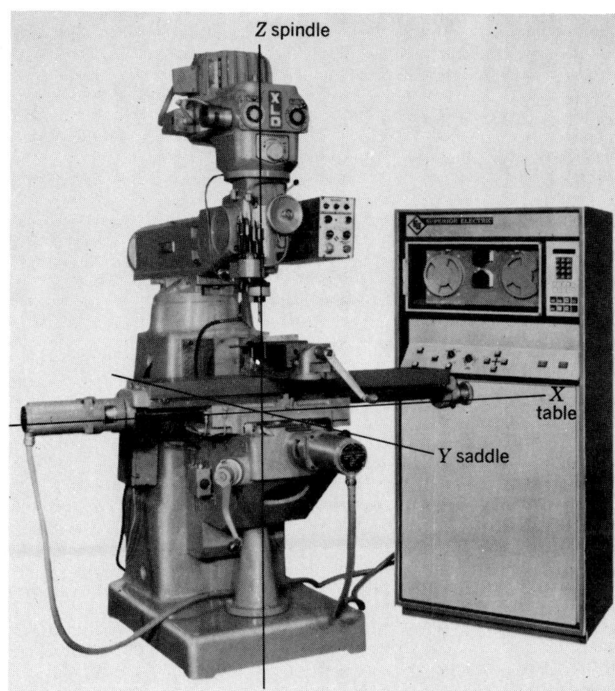

FIGURE O-27 Basic axes of a vertical spindle CNC machine tool (Superior Electric Company).

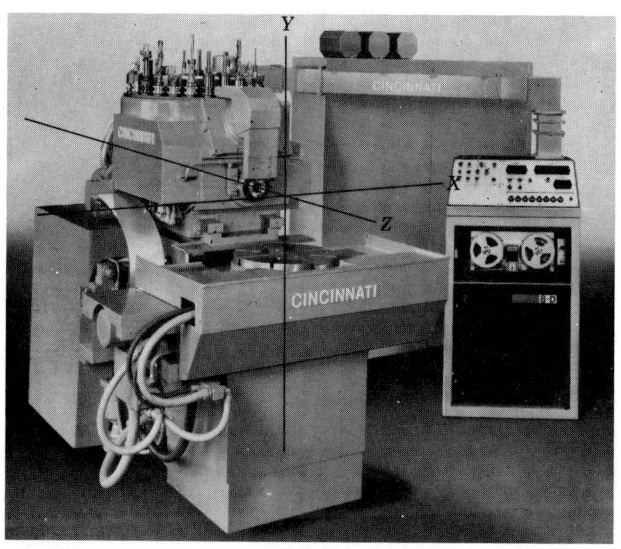

FIGURE O-28 Basic axes of a horizontal spindle CNC machine tool (Cincinnati Milacron).

The Z axis is always the spindle axis, even though the machine spindle may be horizontal or vertical. On a typical vertical spindle machine tool, such as a vertical mill, the spindle axis is Z. The X axis is the table and the Y axis is the saddle (Figure O-27). The knee is in the Z axis.

On a horizontal spindle machine tool, Z remains the spindle axis while Y becomes vertical and X remains horizontal (Figure O-28). The CNC lathe is also a horizontal spindle machine tool. The spindle axis is Z and the cross slide is X, or the horizontal axis perpendicular to Z (Figure O-29).

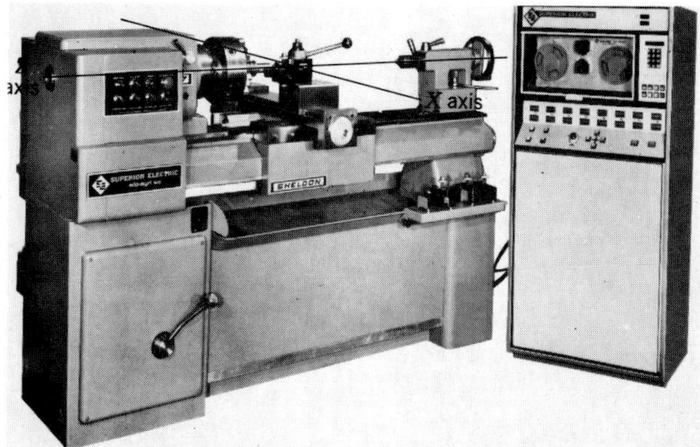

FIGURE O-29 Basic axes of a CNC lathe (Superior Electric Company).

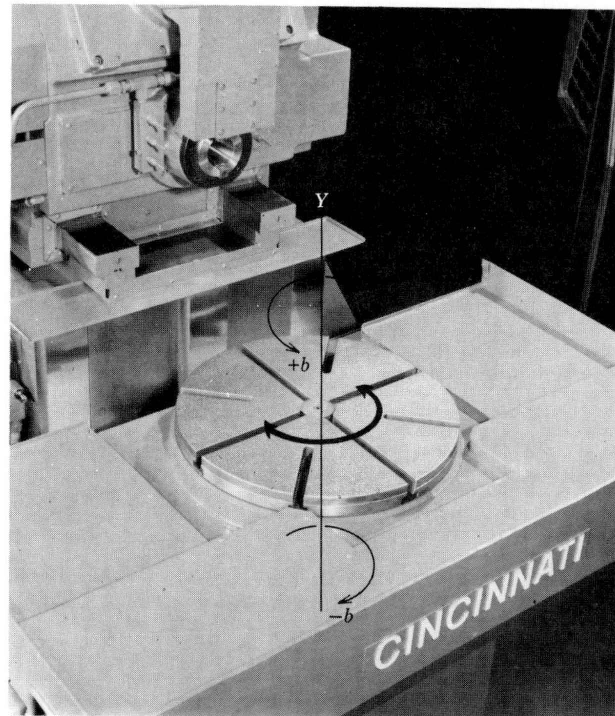

FIGURE O-30 Rotational axis b defining rotary table motion (Cincinnati Milacron).

Since the lathe toolholder is not moved vertically, the Y axis is not used.

ROTATIONAL AXES Rotational axes define numerically controlled motion around the X, Y, and Z basic axes. A CNC machine tool may have a rotary table or part indexer (Figure O-30).

These rotating auxiliary axes are also computer controlled. The direction of rotation, as well as the basic axes around which rotation occurs, must be identified. Rotational axes are identified as a, b, and c (Figure O-26). Discussions about four and five axis CNC machine tools refer to the basic axes of X, Y, and Z and the rotational axes of a, b, and c.

FIGURE O-31 Micrometer tool length gage (Hydra-Point Division, Moog Inc.).

TOOLING PARAMETERS

Tool length is a very critical component of programming and it varies generally from one tool to the next. Presetting tools in their respective toolholders gives the programmer information on the exact tool length for each tool used in machining. All commonly used tools are listed for the programmer's use by their type, size, and length. From this chart the programmer sees that a tool identified as T17 is a $\frac{1}{4}$-in. drill with a preset length of 3 in. Tool T35 is a 4-in.-diameter carbide tipped shell mill with six teeth and a preset length of 4.500 in. From this information the programmer knows then how to program the depth of the drilled hole without ever having seen the actual tool. These tool standard charts are usually made up by companies to meet their individual requirements.

The newer CNC machines have built-in automatic tool length measurement and tool length compensation features. Most CNC controllers have a cutter diameter compensation feature. This capability allows the use of an over- or undersized cutter, different from the one originally intended when the part was programmed. The CNC operator only has to call up the tool number to be adjusted and enter the new tool diameter to replace the original size in the controller's memory, and the machine will compensate for the cutter diameter change. Because the cutter diameter compensation affects only a specific tool, the other tools and their programmed cutting path need not be changed.

CNC machines usually are equipped with automatic tool change (ATC) systems. ATC makes tool changes very rapidly, usually in less than 10 seconds. Depending on the ATC system used, tools are selected either sequentially or randomly. In sequential tool selection, the tools need to be loaded into the tool magazine in the exact order in which they will be used during the machining cycle. Random selection of tools is more widely used, because of the versatility it provides. In one system of random selection, the machine control has been programmed to know which tool is in a specific tool pocket in the tool magazine. As the program calls for a specific tool by tool number, the tool magazine moves the correct tool to the place where the tool change arm grips it and makes the exchange. Another system selects the right tool by actually reading tool code identification markings on the toolholder. With this system tools can be loaded into or interchanged in the tool magazine at will. Since the controller reads a number of program blocks ahead, the ATC moves the next tool to be used into the tool change position while the tool in the spindle is still cutting, thereby reducing tool change time.

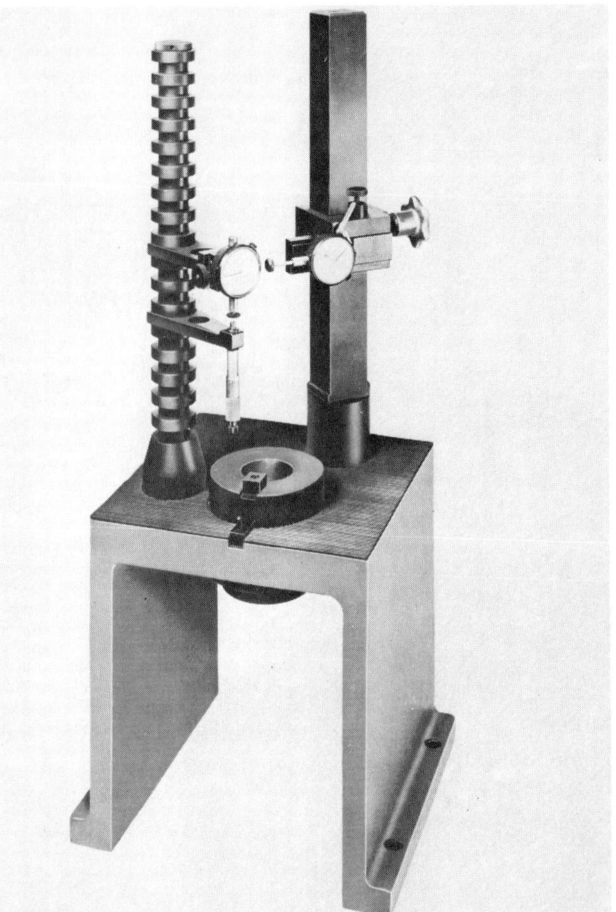

FIGURE O-32 Dial indicator gage for adjusting length and diameter (Cincinnati Milacron).

TOOL PRESETTING

A CNC machining task usually requires a number of different tools. A drilling operation may consist of center drilling, drilling, and reaming. Each of the three tools needed is of a different length. The programmer has to know the diameter and length

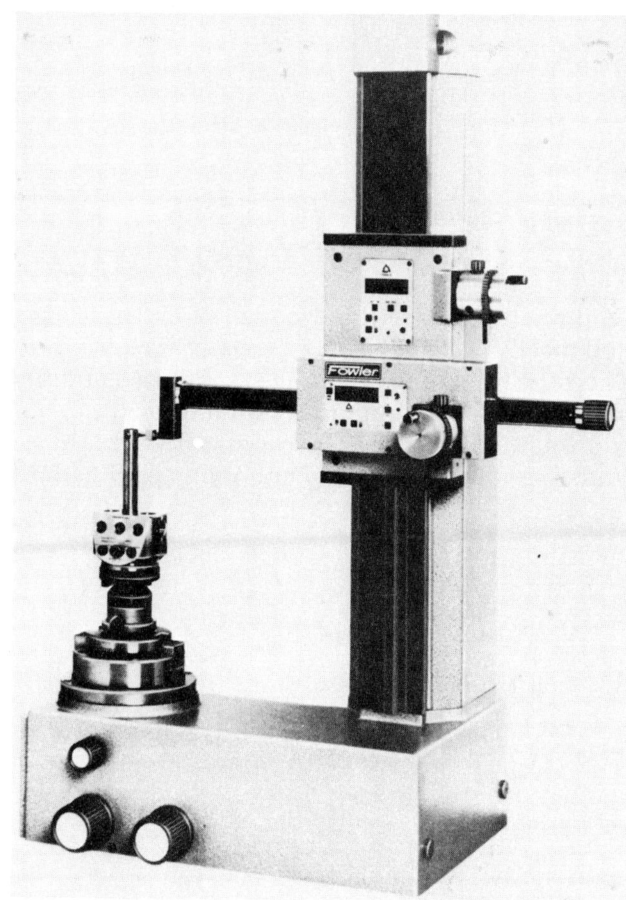

FIGURE O-33 Electronic tool length gage (Fred V. Fowler Co., Inc.).

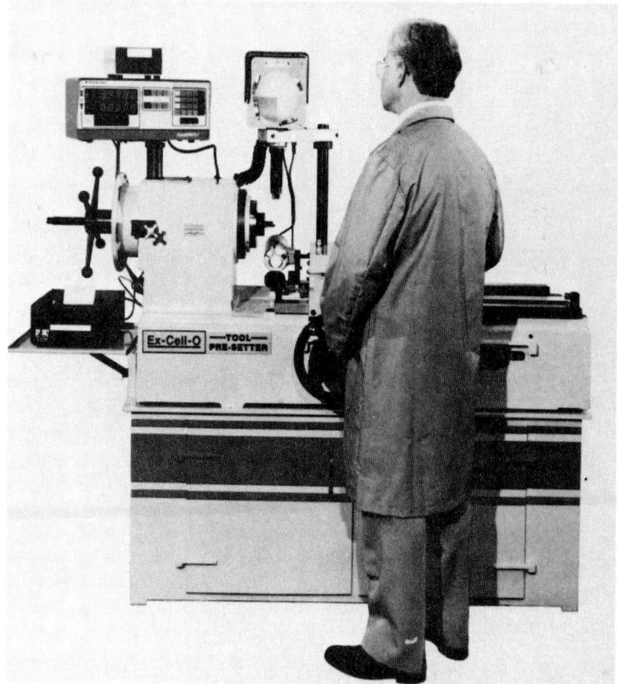

FIGURE O-34 Electronic-optical tool presetter (Excello Corporation).

of each tool to be able to program a correct tool path. Most CNC machine tools have probes that measure tool diameter and length of each tool prior to this tool being used for the first time in a machining operation. These dimensions, for each tool, are stored in the computer memory. The programmer defines the part profile and specifies which tool to use. The controller then uses these measured tool values to create a tool path that produces the programmed part profile. Without these built-in probes, tools need to be preset. A tool length gage is used to measure the projection of cutting tools from their toolholders. Some tool length gages consist of a series of accurately spaced rings mounted on a column (Figure O-31). The ring spacing is usually 1 in. and a micrometer head with a 1-in. travel spans the distance between the rings. The micrometer head can be placed at any desired height within the range of the gage. The cutting tool that is to be preset is placed in its holder and its projection adjusted until its measured length is equal to the required tool length.

Some tool setting gages use dial indicators to measure tool projection (Figure O-32). Both length and diameters can be set with this tool. High-discrimination electronic tool length gages are equipped with digital readouts (Figure O-33). One advantage of an electronic gage is its ability to measure in either inch or metric dimensions. A more sophisticated tool presetting device uses a combination of electronic and optical features (Figure O-34). Modern electronic measuring devices such as this usually include a printed record of the measured tool dimensions.

SELF-TEST

1. What is the difference between NC, CNC, and DNC?

2. Describe the movement of the axes on a five-axis vertical spindle machining center.

3. Why are pallets used on machining centers?

4. Why is a rotary table used on a machining center?

5. Why are automatic tool changers used?

6. What is a pecking cycle?

7. Give two factors that help in the machining of a precision bore.

8. What is adaptive control?

9. Describe live tooling on a CNC chucker.

10. What machine movement takes place when a negative *Z*-axis move is programmed on a vertical spindle machine?

11. What are the designations of the two axes on a lathe?

12. Why is tool presetting necessary?

UNIT 2

CNC Positioning and Input

To direct the efficient movement of machine axes and machining operations the programmer has to know the kind of positioning system used and the format of information input into the machine control unit. This unit will provide you with a basic understanding of these programming parameters.

Objectives

After completing this unit, you should be able to:

1. Explain the difference between Cartesian and polar coordinates.
2. Explain the difference between incremental and absolute positioning.
3. List tape code systems.
4. Describe the words in a word address format.

Numerical Control Systems

CLOSED-LOOP SYSTEMS

Closed-loop systems use servomotors which can be dc, ac, or hydraulic. Servomotors can be turned on and off, speeded up or slowed down, and are instantly reversible. To control precise positioning, a servomotor is coupled with a position-sensing resolver. Two types of resolvers are commonly used. The first, a rotary resolver, measures angular or rotary motion. It can be located on an axis motor or leadscrew, where a signal is generated that tells the MCU the actual axis position. The second type of resolver is the linear resolver. It measures actual linear movements of the machine axes. Both types compare the desired and actual distance moved and make automatic corrections of any discrepancy between the two.

OPEN-LOOP SYSTEMS

In the less expensive open-loop system, no feedback signal is used. The open-loop system may use an electric stepping motor to control the movement of machine components. The stepping motor is used on many numerical controls that are added to existing machine tools.

A stepping motor operates on a pulse of electric current supplied by the MCU. Each current pulse causes the motor rotor to turn or step a fraction of a revolution. When the motor is coupled to a mill table, lathe cross slide, or lathe leadscrew, it can act to move the screw specific amounts according to the number of pulses received from the MCU. Stepping motors are often designed to move machine tool tables a distance of .001 in. per pulse. They are reversible and can move a component in either direction.

CARTESIAN COORDINATE SYSTEMS

When using rectangular coordinates, any specific point can be described in mathematical terms. Figure O-35 illustrates the Cartesian coordinate system. This system uses the *X* axis as its horizontal line, and the *Y* axis is represented by a vertical line. The point formed at the intersection of the *X* and *Y* axes is called the origin or zero point. The perpendicular coordinate axes *X* and *Y* form four quadrants. Quadrants are numbered in a counterclockwise direction beginning at the upper right. All points to the right of zero along the *X* axis have positive values. All points to the left of zero have negative values. All points on the *Y* axis

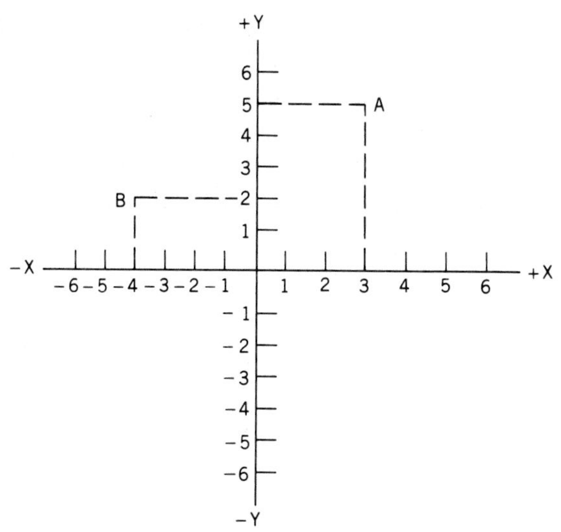

FIGURE O-35 Cartesian coordinate points.

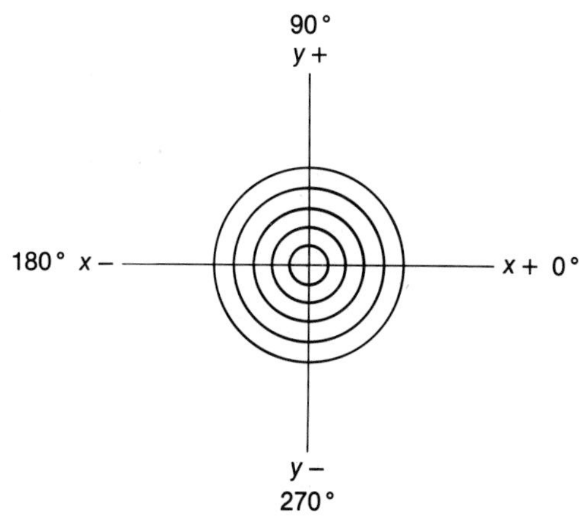

FIGURE O-36 Polar coordinate points.

above the *X* axis have positive values and points below that axis are negative. Any positive number can be identified without a plus sign, but negative numbers must have a minus sign preceding them. Figure O-35 shows where point *A* (*A* 3, *Y* 5) is located. Another point identified is point *B* (*X* −4, *Y* 2).

POLAR COORDINATE SYSTEMS

In the polar coordinate system the position of a point is defined by its direction and distance from a point that has been defined as a zero point. Polar coordinates can be envisioned by looking at a world map with a view of one of the polar regions. Polar coordinates use concentric circles as their reference lines, with the zero point at the center (Figure O-36). The other necessary factor in identifying a point on a plane is the angular distance from the positive *X* axis. The angular reference of zero degrees is at the 3 o'clock position. Any point on a

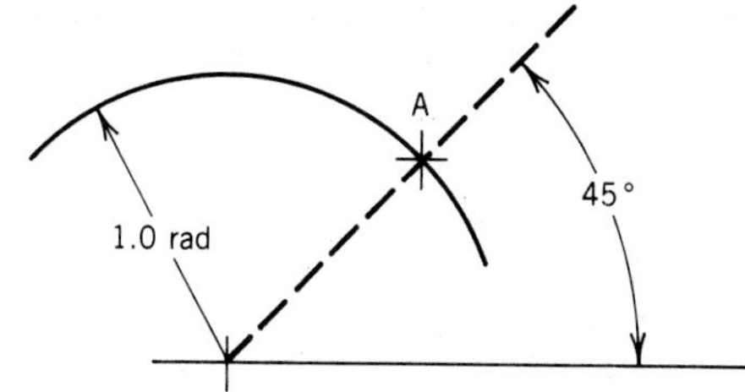

FIGURE O-37 Point *A* plotted with polar coordinates.

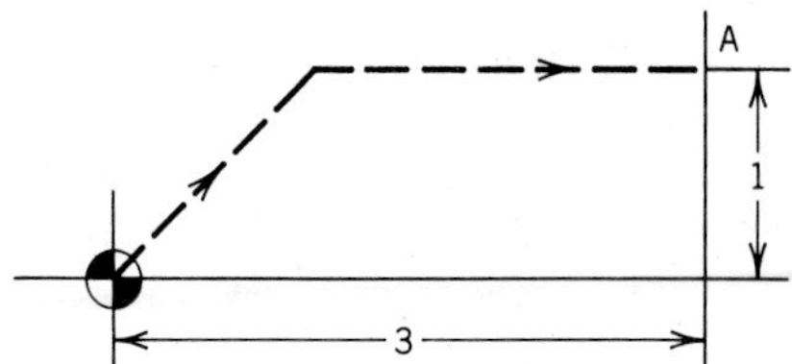

FIGURE O-38 Tool path on a point-to-point machine.

plane can be identified when the radius and angle are known. Figure O-37 is an example, where point *A* is identified as a radius of 1 in. and an angle of 45 degrees. A positive angular value is in a counterclockwise direction from the angular reference line. A negative angular value describes a clockwise movement from the angular reference line.

Positioning and Continuous Path Systems

CNC systems fall into two basic categories, positioning and continuous path systems. Positioning or point-to-point systems are usually used to perform operations such as drilling, reaming, tapping, and boring. These processes require that the machine table movements be completed before the machine spindle lowers and does its task. Many point-to-point systems have the ability to make straight line milling cuts. Forty-five degree angles can be machined if two axes can be moved simultaneously and at precisely the same speed. Figure O-38 illustrates the path that the tool takes. The distance that the tool travels is *X* 3 and *Y* 1, but it is not a straight line move. The reason for this is that both axes are moving at the same speed, so that the axis with the shortest distance to move arrives first while the second axis is still moving to its destination.

In a continuous path system any machine movement is under constant controller direction. This system allows the machining of exact angles and curves. One important characteristic of continuous path machining is the ability to move each machine axis simultaneously at different speeds.

The continuous path system can perform any of the operations of a point-to-point system. The example shown in Figure O-39 is the tool path taken under continuous path control.

INCREMENTAL POSITIONING SYSTEMS

Many CNC machines use the incremental positioning system to move the tool from one machining point to another. In incremental positioning the spindle measures the distance to its next location from the position at which it was last located. Incremental positioning requires positive and negative travel directions. An example of incremental positioning is given in Figure O-40. The workpiece shown has to have two holes drilled at locations 1 and 2. The workpiece is set up so that the spindle start point or zero point is over the left-hand corner and the edges of the part are parallel with the coordinate axes. The CNC program instructs the machine spindle to move in a $+X$ direction a distance of 1 in. and in a $+Y$ direction a distance of 4 in. This will position the spindle over drilling location 1. To reach location 2, the spindle moves in the $+X$ direction an additional distance of 2 in. However, it must move in a $-Y$ direction a distance of 1 in. To return the spindle from location 2 to the start point a $-X$ move of 3 in. and a $-Y$ move of 3 in. is required. Once again, location 2 became a new origin from which the distance back to the starting point was measured.

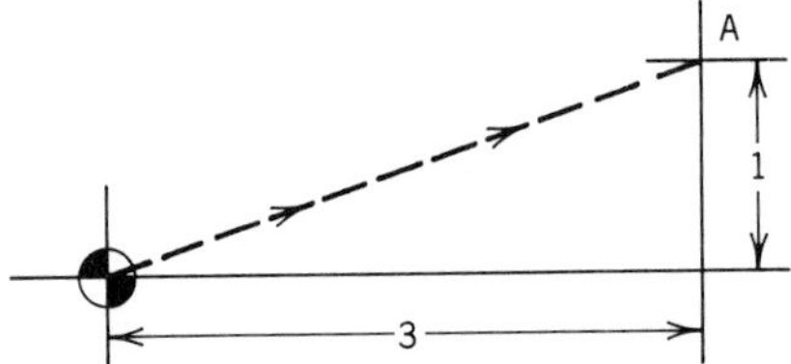

FIGURE O-39 Tool path on a continuous path machine.

ABSOLUTE POSITIONING SYSTEMS

In absolute positioning, all machine movements are measured from one fixed zero point. Two basic categories of absolute positioning systems are in use. One uses a fixed zero position, the other a floating zero position. With fixed zero machines the zero location is permanently established and cannot be moved. This absolute zero point is usually at the point where the spindle has reached the end of travel in the $-X$ and $-Y$ axes. With a fixed zero, the spindle is operating in quadrant 1 and all points have positive values. Figure O-41 is an example of absolute positioning. This workpiece is located with its left corner at the fixed zero position of the machine table. Each drilling location is expressed as a dimension from the absolute zero point. Drilling location 1 is at point (X 1, Y 3) from zero. Location 2 is at point (X 3, Y 3) from zero. The machine goes directly from location 1 to location 2 without first returning to zero. The return of the machine spindle to the start point is accomplished by directing the spindle to position at (X 0, Y 0).

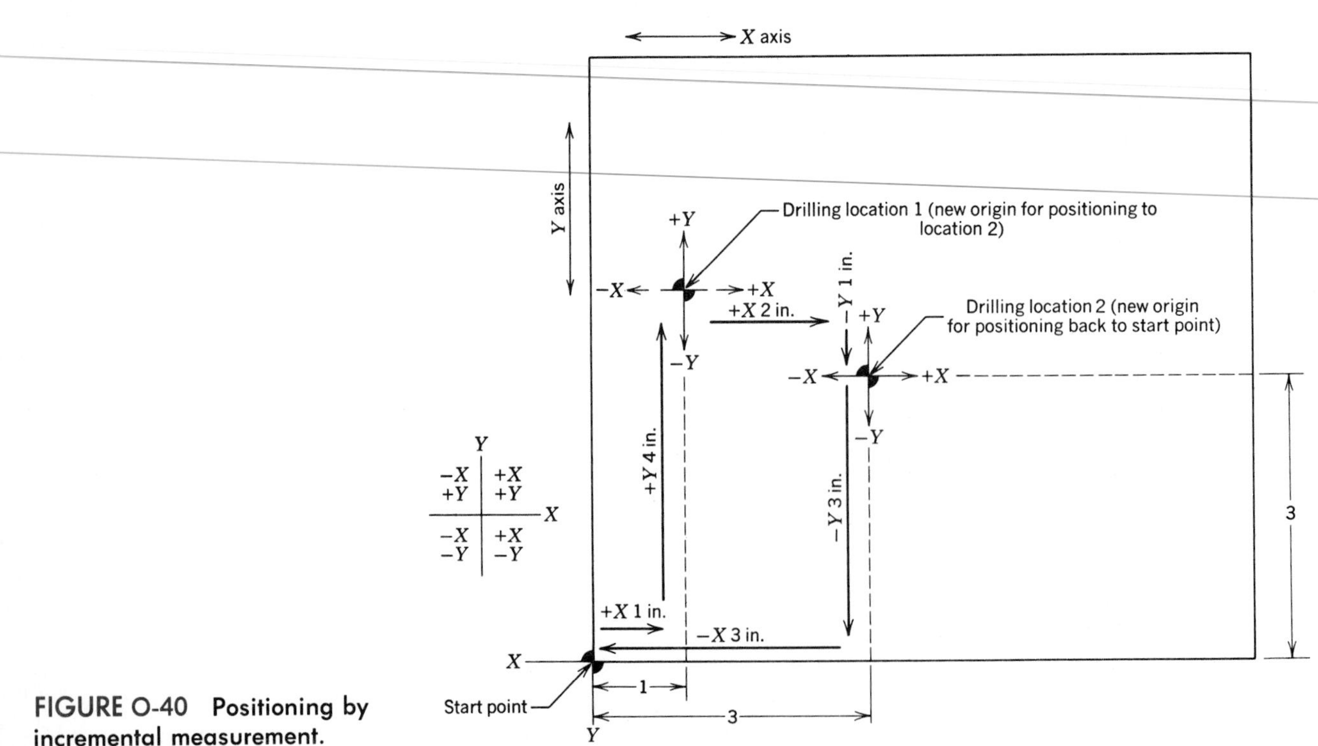

FIGURE O-40 Positioning by incremental measurement.

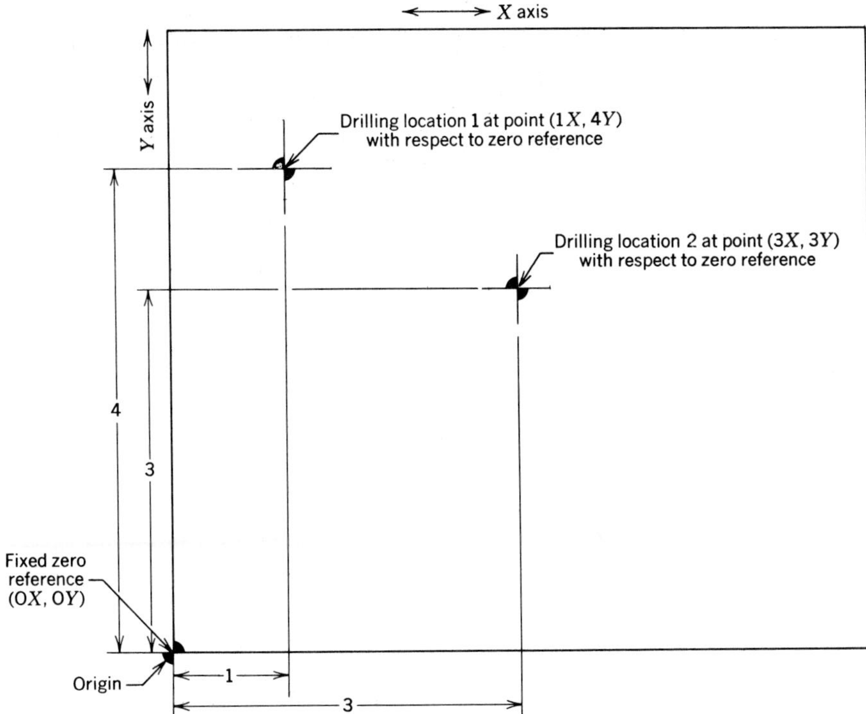

FIGURE O-41 Positioning by fixed zero absolute for coordinate measurement.

When using a floating zero system, the workpiece can be positioned at any convenient location on the machine table. The machine operator then moves the machine spindle to the position on the workpiece that is designated as the zero position, also called the setup point. At this location the machine zero point is established and the machine can start to perform its programmed operations. Figure O-42 shows a workpiece with the zero position at the intersection of its centerlines. To move from the setup point to the hole 1 location the command would be (*X* 1.25, *Y* 0). From hole 1 to hole 2 it is (*X* 0, *Y* 1.25). To go to hole 3 it is (*X* −1.25, *Y* 0). From hole 3 to hole 4 it is (*X* 0, *Y* −1.25). From hole 4 back to he setup point it is (*X* 0, *Y* 0).

One advantage of absolute positioning systems is that an error made in positioning is limited to one location only, because the next location is again a reference to the zero position. If a positioning error is made on an incremental system, all subsequent positions are affected and all the following locations are incorrect.

Most modern CNC machines allow the programmer to mix incremental and absolute programming. This mixing capability coupled with a floating zero point makes programming very efficient. Since a floating zero point can, and probably will, be relocated during the programming process, it is critical that the programmer be aware of where the presently active zero location is.

Circular Interpolation

With circular interpolation the tool moves in a circular path. The length of the arc made can be anything from a part of a degree to a full 360-degree circle. The benefits of circular interpolation can be seen by comparing the program input to generate a circular cut. The circle cut with linear interpolation requires 1000 or more blocks of data, each blocking defining a small span. With circular interpolation a single block of data will make the

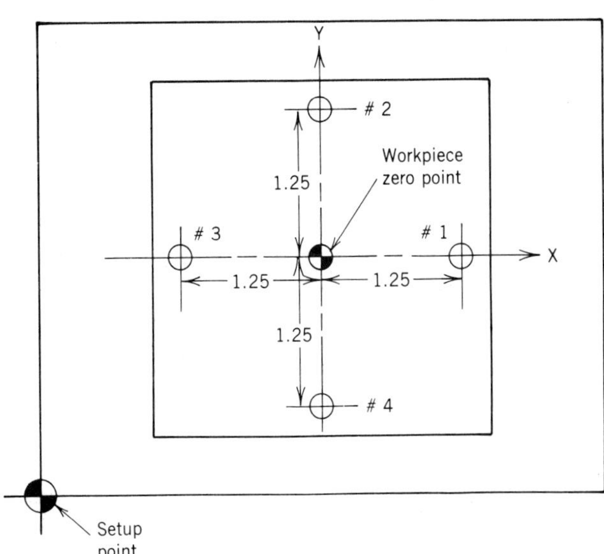

FIGURE O-42 Positioning with a floating zero system.

same cut. Circular interpolation is limited to a two-axis plane.

Other types of interpolation, such as helical, parabolic, and cubic, are available on some MCUs. Helical interpolation combines a two-axis circular movement with a linear third-axis move. Helical interpolation can be used for thread milling. Parabolic interpolation uses three points on a curved line (two end points and one midpoint) to define what can be a complete or a partial parabola. Because of increased computer power at lower cost, this form of interpolation has been overtaken by cubic interpolation. These advanced interpolation forms are usually used to create freeform, three-

8	7	6	5	4		3	2	1	
		•	•		·				0
•		•	•		·			•	1
•		•	•		·		•		2
		•	•		·		•	•	3
•		•	•		·	•			4
		•	•		·	•		•	5
		•	•		·	•	•		6
•		•	•		·	•	•	•	7
•		•	•	•	·				8
		•	•	•	·		•		9
	•				·			•	A
	•				·		•		B
•	•				·		•	•	C
	•				·	•			D
•	•				·	•		•	E
•	•				·	•	•		F
	•				·	•	•	•	G
	•			•	·				H
•	•			•	·			•	I
•	•			•	·		•		J
	•			•	·		•	•	K
•	•			•	·	•			L
	•			•	·	•		•	M
	•			•	·	•	•		N
•	•			•	·	•	•	•	O
	•		•		·				P
•	•		•		·			•	Q
•	•		•		·		•		R
	•		•		·	•	•		S
•	•		•		·	•			T
	•		•		·	•		•	U
	•		•		·	•	•		V
•	•		•		·	•	•	•	W
•	•		•	•	·				X
	•		•	•	·			•	Y
	•		•	•	·		•		Z
•	•	•	•	•	·	•	•	•	Delete
•				•	·				Back Space
				•	·		•		Horiz. Tab
				•	·	•			Line Feed
•				•	·	•		•	Carr. Ret. (EOB)
•		•			·				Space
•		•			·	•		•	%
		•		•	·				((Open Paren.)
•		•		•	·			•	) (Close Paren.)
		•		•	·		•	•	+
		•		•	·	•		•	−

(a)

8	7	6	5	4		3	2	1	
		•			·				0
					·			•	1
					·		•		2
			•		·		•	•	3
					·	•			4
			•	•	·			•	5
			•	•	·		•		6
					·	•	•	•	7
				•	·				8
			•	•	·			•	9
	•	•			·			•	a
	•	•			·		•		b
	•	•	•		·		•	•	c
	•	•			·	•			d
	•	•	•		·	•		•	e
	•	•	•		·	•	•		f
	•	•			·	•	•	•	g
	•	•		•	·				h
	•	•	•	•	·			•	i
	•		•		·			•	j
	•		•		·		•		k
	•				·		•	•	l
	•		•		·	•			m
	•				·	•		•	n
	•				·	•	•		o
	•		•		·	•	•	•	p
	•		•	•	·				q
	•			•	·			•	r
		•	•		·		•		s
		•			·		•	•	t
		•	•		·	•			u
		•			·	•		•	v
		•			·	•	•		w
		•	•		·	•	•	•	x
		•	•	•	·				y
		•		•	·			•	z
	•	•		•	·		•	•	. (Period)
		•	•	•	·		•	•	, (Comma)
		•	•		·			•	/
	•	•	•		·				+
	•				·				−
			•		·				Space
	•	•	•	•	·	•	•	•	Delete
•									Carr. Ret. (EOB)
		•		•	·		•		Back Space

(b)

FIGURE O-43 Tape formats: (*a*) ASCII RS-358; (*b*) EIA RS-244.

dimensional shapes such as those used in forming dies for automotive parts. Using cubic interpolation, the sophisticated cutter paths needed to machine such dies can be programmed with relatively few data point inputs.

Linear Interpolation

When programmed points are connected by straight-line movements it is defined as linear interpolation. Linear interpolation takes place when two machine axes move simultaneously, but at different speeds, creating a straight-line movement. A G 01 preparatory function code used on a milling machine will make straight line or angular cuts; used on a turning machine it will make cylindrical or tapered cuts. If linear interpolation is used to make a circular contouring cut, it consists of a number of straight line moves. The shorter these straight line moves are, and the more there are of them in a given circle, the closer the resulting tool path will be to a true circular move.

Numerical Control and the Computer

The computer is an important and valuable component of the modern numerical control system. The computer can do mathematics with great speed and accuracy. This has made it an important tool in the CNC programming of complex machining operations. Mathematical computation necessary for continuous path machining is difficult and time consuming. The computer can accommodate this kind of calculation easily.

A computer can interpret direct descriptions of workpiece geometry, machine control functions, and machining operations. This permits the CNC programmer to program much in the same way that one would verbally describe the machining task to be done. The computer can respond to direct statements such as GO/TO, MILL, DRILL, or BORE. These direct descriptions are translated by the computer into appropriate instructions for a specific machine tool.

The information from the programming manuscript is put into the computer memory on a keyboard. This information is examined either on a video screen or on a printout by the programmer. If there are errors, these are corrected by editing the information in the memory of the computer. The programming information in the computer memory can be stored on a disk or tape, or it can be transmitted through an electrical cable to a tape punch to make a punched tape that will operate a machine tool.

MCU INPUT

When workpiece geometry and machining parameters are converted into machine-ready code, this information must be provided to the machine tool. Input to the MCU can be made in different ways. One input method uses punched tape. On this tape, machining information is translated into specific patterns of holes. The information contained in these hole patterns is then scanned and interpreted by a tape reader and converted into signals to affect machine axis movements and other functions. An example of the two most common standardized tape formats are shown in Figure O-43. The EIA (Electronic Industries Association) code is known as RS-244. The ASCII (American Standard Code for Information Interchange) code, RS-358, is also called the ISO (International Standards Organization) code. The EIA code provides that each character be made up of an odd number of holes in the tape. The ASCII code uses an even number of holes for every character represented. This characteristic is called parity. MCUs sense which tape format is being received and initiate an internal parity check. A tape is either odd-parity (EIA) or even-parity (ASCII). If the MCU detects even parity on an EIA transmission, it will give an error signal and stop machine operation until a correction is made. This machining code can be punched on tape with a special typewriter tape punch machine Figure O-44. The tape punch typewriter has the same letters and numbers as those used on a standard typewriter plus several extra symbols and control keys for the tape punch. The tape punch is activated as each typewriter key is depressed. As the tape is punched, a printed record is typed on paper in the typewriter carriage. Modern tape preparation is performed by using the

FIGURE O-44 Typewriter tape punch (California State University at Fresno).

high-speed data transmission of a computer coupled with a high-speed tape punch and printer. Corrections and changes are made in computer memory, with the program being displayed on the CRT. These tape preparation systems have tape reading capability, which allows an existing tape to be read into memory, edited, and a corrected tape made on the tape punch. The printer will make a written copy of the information on the punched tape, called a "hard copy," used to verify what is on the tape. Punched tapes are generally read by using a photoelectric reader (Figure O-45). Photoelectric readers operate by translating the beams of light that pass through the holes in the tape striking a photocell into electrical impulses. These readers have a reading speed from 300 to 500 characters per second. This high speed is necessary when making contouring cuts to produce smooth and continuous machine movements. Such contours consist of a great number of minute machine movements. Because of these small movements, the machine may actually have to wait for the reader to catch up and to provide more machining information. To prevent this from happening, many controllers are built with a buffer storage. This buffer allows the MCU to accept and store information from a tape while at the same time previously received directions are being acted on by the machine. One drawback of photoelectric readers is the chance of errors due to dirty or oily tapes and imperfections in the punched holes.

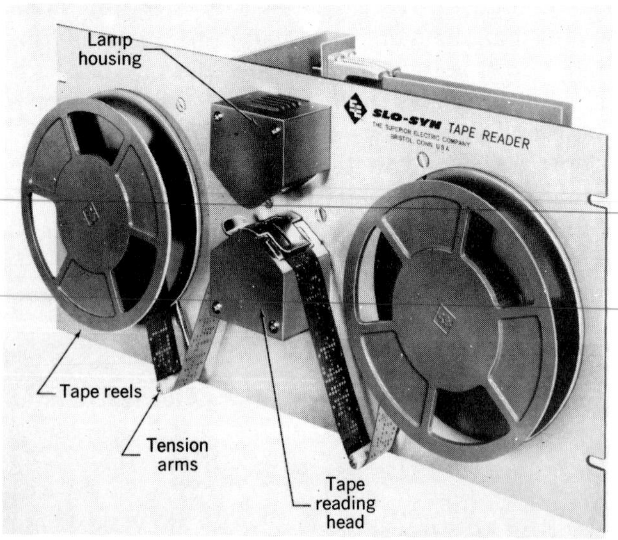

FIGURE O-45 Photoelectric MCU tape reader (Superior Electric Company).

A CNC control unit can also receive machining information from a punched tape. In this application the complete tape is read and stored into the MCU memory. The machining operation is now controlled and directed from this built-in computer. To shorten the time required to load a machining program into the computer memory, once a tape has been read into memory, it is saved on a floppy or fixed disk for future use. Programs saved on disks use much less space than does storing punched tapes. A program can also be loaded into the MCU through the controller's keyboard. This is fine as long as the CNC machine is able to keep machining while a new program is being loaded. If an expensive machine tool stands idle while a program is being loaded, one command at a time, machining costs become excessive. A very efficient method of program input to the MCU is to download from another computer. The coded program is in either EIA or ASCII format and is transmitted through a communications link between the two computers. Uploading, transferring a program from the MCU to another computer for editing or storage, is also possible with this communications link. DNC (distributed numerical control) uses a mainframe computer to store the CNC program files for all the CNC machine tools in a shop. The MCU on each CNC machine can access the large computer to download a program file to produce machined parts.

PROGRAM FORMAT

A CNC program is made up of a series of commands arranged so that the controller can understand and execute them. The program format is the arrangement of machining data in the correct order for a specific machine tool. This means that a program prepared for a CNC machining center cannot be used to operate a CNC turning center. Each of these different types of machine tools recognizes and acts only on the words specifically assigned to them. Programs for CNC machine tools often use the word address format. An example of a word address format program sheet is shown in Figure O-46. The word address format uses a letter to

Sequence number O/N	Prep function G	X (U)	Z (W)	I	K	R	F/E	S	T	Misc function M	EOB
O....	G..	X ±	Z ±	I ± ...	K ± ...	R...	F...	S....	T....	M...	CR

FIGURE O-46 Word address format.

identify and separate words. The identifying letter of a word is followed by a numeric entry. A program line, consisting of a series of words ending with an EOB (End Of Block) symbol, is called a block. The word address format is usually called a variable block format, because the program blocks can vary in length. The block length varies because many entries in one block stay effective in the following blocks unless changed or canceled. Words within a block need not be in a specific sequence, since the MCU reads each identifying letter and then directs its numeric value to the corresponding machine register. But common programming practice is to give an *X* value before a *Y* or a *Z* value. Following is a description of words in a program sheet and their purpose.

SEQUENCE NUMBER

The sequence number is usually identified by the letter N and a number such as N001. The primary purpose of the sequence number is the identification of each block of information. Knowing the sequence number of specific blocks helps the programmer or machine operator to locate problem areas. The MCU can be directed to find and go to a specific block in a program. The first entry in the sequence number column is normally used to identify a program by a number. The letter O (not number zero) is a prefix to a program number, such as O 113.

PREPARATORY FUNCTIONS

The preparatory function is identified by the letter G, followed by a one- or two-digit number. It is commonly referred to as a G code. Many G codes are standardized, where one G code will perform the same function on any machine control unit. However, there is also a great variation in nonstandard G codes that are used by machine tool manufacturers to perform a certain function only in their brand of MCU. To write a program for a CNC machine, the programmer has to consult the programming manual for a specific machine and machine controller. MCUs are being upgraded and improved constantly. An MCU upgrade will generally increase the MCU functions and make programming easier because of the improved built-in computer capability. The following descriptions are an example of some G codes. Complete G code listings and explanations are found in CNC programming manuals and CNC textbooks. Most G codes are modal, which means that this code stays in effect in the following program blocks until changed or canceled. A nonmodal code is effective only in the block where it is programmed.

G 00 Axis movements are made at rapid traverse speed; only positioning, noncutting moves
G 01 Linear (straight line) movements of the machine axes are made at programmed feed rates; cutting moves
G 02 Circular interpolation; clockwise arc moves at programmed feed rates
G 03 Circular interpolation; counterclockwise
G 04 Dwell; a programmed time delay
G 33 Thread cutting; coordinates the revolving spindle with the movement of the carriage so that successive cuts are made in the same groove
G 70 Programming in inch units
G 71 Programming in metric units
G 90 Absolute positioning
G 91 Incremental positioning
G 92 Preloading the machine registers with a dimension that specifies the physical location of the cutting tool
G 94 Feed rate in inches (meters if G 71) per minute
G 95 Feed rate in inches (millimeters if G 71) per spindle revolution
G 96 CSS (constant surface speed); the spindle rpm will increase or decrease as the workpiece diameter decreases or increases in turning, to maintain the programmed cutting speed; when a G 96 is programmed, it is also necessary to program the maximum spindle rpm that should not be exceeded
G 97 Spindle speed in rpm

AXES INPUT

Machine movements and their direction are preceded by an X, Y, or Z word. Rotational axes are identified by A, B, or C. The movement direction is either positive (+) or negative (−). Positive entries do not need a + (plus) sign to identify these; it is usually omitted. A negative entry always has a − (minus) sign preceding a numeric entry. Some controllers allow decimal-point entries; others assume that the decimal point would be between the fourth and fifth digits from the right. A 1-in. negative axis move would be written as X-1. if a decimal point is allowed. This same input without the decimal point would appear as X-10000. On most controllers, the smallest axis entry is .0001 in. or .0001 mm. Some controllers identify an axis entry as either absolute positioning or incremental positioning by the letter prefix of the word, where

X . . . (absolute input) or U . . . (incremental *X*-axis input)
Y . . . (absolute input) or V . . . (incremental *Y*-axis input)
Z . . . (absolute input) or W . . . (incremental *Z*-axis input)

ARC CENTER LOCATION

When programmed with an arc movement (circular interpolation) commanded by a G 02 or G 03, many MCUs require an entry defining the center of such an arc. An arc command on a turning center, for example, requires entry of the endpoint of the arc as an X and Z or an U and W location, with the center of that arc defined by an I and a K word entry. I defines the arc center distance in the *X* axis, while K is the arc center distance in the *Z* axis. I and K are incremental values, even if the arc endpoint is programmed in absolute dimensions. On a three-axis machine, the arc center location is given as I in the *X* axis, as J in the *Y* axis, and as K in the *Z* axis. If I, J, or K is zero, most controllers do not require that entry. I and K values are not used if the R word specifies the size of the radius of the arc. When an arc command is programmed, the cutting tool is already at the start point of the arc; then only the endpoint and the arc radius need to be entered. This information is sufficient to let the controller calculate the center point and perform the arc movement.

FEED RATES

Feed rates are measured in inches per minute (ipm) or as inches per revolution (ipr) of the spindle. Either ipm or ipr is selected with a G code. Most controllers will accept F . . . commands with two digits to the left and four digits to the right of the decimal point (Fxx.xxxx). To increase the accuracy of the thread lead during threading operations, an E address word can be used. Entries for E.. usually have a limit of one digit to the left and six digits to the right of the decimal point (Ex.xxxxxx).

SPINDLE SPEED

Spindle speeds are programmed as rpm if a G 97 is active; here an S3210 would give a 3210-rpm spindle speed. If a G 96 is programmed, S550 would give a constant surface speed of 550 surface feet per minute at the contact point of the cutting tool and workpiece. This means that the spindle rpm will change as the workpiece diameter changes.

TOOL DESIGNATION

The tool code number identifies the tool that is used for a machining operation. Most tool codes use four digits (Txxxx) to identify tools, but two- or six-digit codes are also used. A T0204 tool designation on a turning machine shows that this is the tool in turret position number 2. The last two digits 04 are the tool offset number. The programmer selects the particular number from a tool offset chart. The information on a tool offset chart consists of the *X* and *Z* dimensions of a specific turning tool, the tool nose radius of this tool, and the direction of the cutting tip of this tool. During the initial setup to machine a part, the machine operator has to load this tool offset information into the MCU.

MISCELLANEOUS FUNCTIONS

Miscellaneous function (M) codes activate auxiliary functions such as turning on the spindle or the start and stop of coolant flow. Here is a description of some M codes:

M 00	Program stop; machining is halted until restarted; often this code is used to allow measuring of a workpiece dimension or to check the surface finish
M 03	Spindle start command, forward
M 04	Spindle start in reverse
M 05	Spindle stop
M 06	Tool change command; manual or automatic, depending on the machine tool
M 07	Coolant on, flood
M 08	Coolant on, mist
M 09	Coolant off
M 02 or M 30	End of program; reset program to its starting point.

CANNED CYCLES

Some G codes activate canned cycles. A canned cycle is a combination of machine moves that results in a machining function such as drilling, tapping, boring, or threading. As an example, let's look at a drilling cycle. The G 81 drill cycle will:

1. Position the spindle at a rapid feed rate to the programmed *X* and *Y* location.

2. The spindle, the *Z* axis, will advance at a rapid feed rate to a programmed point. This point is usually .100 in. above the workpiece and is called the gage height.

3. The spindle will advance the drilling tool to the programmed depth at the programmed feed rate.

4. The spindle will retract the drill from the hole to the gage height at a rapid feed rate.

Use of this canned cycle compresses into one program block what would otherwise have been programmed in four blocks. A canned threading cycle for a CNC turning center combines in a single program block a great number of different machine moves. For a threading cycle, the block of information usually contains:

1. The major diameter of the thread
2. The depth of the thread
3. The beginning of the thread
4. The ending of the thread
5. The lead of the thread
6. The depth of the first pass
7. The depth of the final pass
8. The spindle speed

From this information, the controller directs all machine moves to cut a complete thread. Many different canned cycles are available on machine controllers. Canned cycles require programmer input, which differs from one controller to another. It is absolutely essential to use the programming manual for a specific CNC machine in order to write a correct program. Programming manuals are generally very descriptive of all programming functions. These manuals contain program examples with illustrations and explanations of each program entry.

SELF-TEST

1. What is the difference between an open-loop and a closed-loop system?
2. How is a point identified in a Cartesian coordinate system?
3. How is a point identified in a polar coordinate system?
4. Where is a point-to-point positioning system usually used?
5. What is a continuous path positioning system?
6. What is incremental positioning?
7. What is absolute positioning?
8. What is a floating zero system?
9. What are linear and circular interpolation?
10. Name two standard tape code systems.

UNIT 3

CNC Applications

Programming for different CNC machine tools requires that the programmer understand the capabilities of each machine tool. The intent of this unit is to familiarize you with several programming formats for different machine tools.

Objectives

After completing this unit, you should be able to:

1. Explain the difference between prompted and word address programs.
2. Explain the purpose of a canned cycle.
3. Explain the difference between programs that use Cartesian and polar coordinates.

Operating the CNC Machine

Most CNC machining operations during setup require the machine operator to manually (not under tape control) move the machine elements. Manual data input (MDI) means telling the control system what to do by moving dials and switches or by pushing buttons. Because the CNC machine is under the complete control of the operator, he or she must have a complete understanding of the function of the MDI controls. MDI is used to interrupt a program in order to make corrections or

FIGURE O-47 Dyna Myte 2,400 CNC milling machine (Lane Community College).

FIGURE O-48 Machining the 11-hole pattern (Lane Community College).

to add an additional operation to an existing program. MDI is commonly used during setup operations to position the workpiece or the machine spindle at the correct location relative to the program reference point.

CNC Programming

The CNC programmer needs to be able to visualize the actual cutting motions that are taking place on the machine. The programmer needs to understand the information on a blueprint, the sequence of machining operations, and how a workpiece is setup and secured during the machining process. The programmer is responsible for the selection of the cutting tools, feeds, and speeds. Another important aspect of programming is to be familiar with the operating characteristics and requirements of the particular machine being programmed. The programmer, at any time, must know where the tool is, where it is going next, and what path it is taking.

CNC Programming for a Vertical Milling Machine

The programming format for this machine is conversational or prompted. Information is entered into the MCU from the keyboard on the machine (Figure O-47). On this machine the MCU is portable, so that this small controller can be taken to a desk for part programming. Information input can be absolute or incremental or a combination of both. The control will accept Cartesian or polar dimensions. The floating zero point can be moved

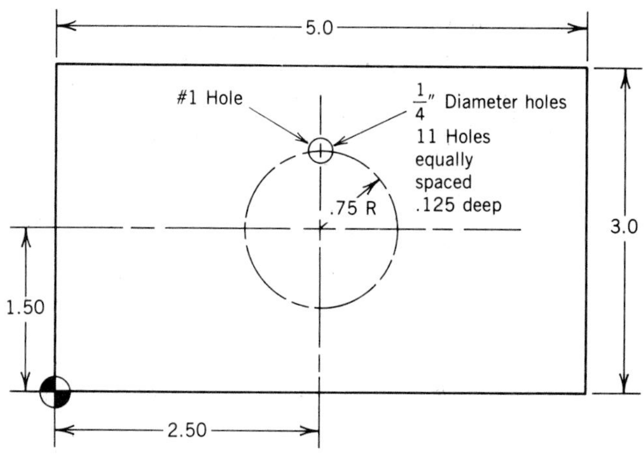

FIGURE O-49 Drawing of the plate with the 11 holes.

to any point within the working parameters of the machine. Canned cycles such as rectangle pocket, circle pocket, bolt circle, etc., are available to the programmer. To show the versatility of CNC machining, three different methods of making the part shown in Figure O-48 will be demonstrated. A drawing of this part is given in Figure O-49.

These three examples will show how much of the work of the programmer can now be performed by computer. The time required to chart the coordinate points for the first program is nonproductive. The time to enter these programs into the machine control also varies as can be seen in the different program lengths.

EXAMPLE ONE

The first example uses Cartesian coordinates to establish the *X* and *Y* positions for each hole location. Point-to-point positioning is used. To make a chart for the 11 hole locations consult a

```
700 START INS 03
701  TO=  0·2500
702 FR XY  =20·0
703 FR   Z =02·0
704 SETUP →zcxyu
705 SPINDLE   ON
706 GR X  2·5000
707     Y  2·2500
708 GOcZ- 0·1250
709 GR X- 0·4056
710     Y- 0·1198
711 GOcZ- 0·1250
712 GR X- 0·2758
713     Y- 0·3195
714 GOcZ- 0·1250
715 GR X- 0·0601
716     Y- 0·4183
717 GOcZ- 0·1250
718 GR X  0·1755
719     Y- 0·3843
720 GOcZ- 0·1250
721 GR X  0·3555
722     Y- 0·2286
723 GOcZ- 0·1250
724 GR X  0·4226
725     Y  0·0000
726 GOcZ- 0·1250
727 GR X  0·3555
728     Y  0·2286
729 GOcZ- 0·1250
730 GR X  0·1756
731     Y  0·3843
732 GOcZ- 0·1250
733 GR X- 0·0601
734     Y  0·4183
735 GOcZ- 0·1250
736 GR X- 0·2768
737     Y  0·3185
738 GOcZ- 0·1250
739 GR X- 2·9056
740     Y  2·1302
741 SPINDLE  OFF
742 END NEWPART
```

FIGURE O-50 Program for the 11 holes with Cartesian coordinates.

```
600 START INS 02
601  TO=  0·2500
602 FR XY  =20·0
603 FR   Z =02·0
604 SETUP →zcxyu
605 SPINDLE   ON
606 GO X  2·5000
607     Y  1·5000
608 ZERO     XY
609 GO r  0·7500
610     a  90·000
611 REPEAT    11
612 GOcZ- 0·1250
613 GR a  32·727
614 REPEAT END
615 SPINDLE  OFF
616 END NEWPART
```

FIGURE O-51 Program for the 11 holes with polar coordinates.

table of hole coordinate dimension factors in *Machinery's Handbook*.

#	1	2	3	4	5	6
X	0.0	−.4056	−.2768	−.0601	.1756	.3555
Y	0.0	−.1198	−.3185	−.4183	−.3843	−.2286

#	7	8	9	10	11
X	.4226	.3555	.1756	−.0601	−.2768
Y	0.0	.2286	.3843	.4183	.3185

With this information and the setup point located on the corner of the workpiece programming can begin. A printout of this program is shown in Figure O-50. This program is 42 lines long. A program can be placed in any empty space within the 900 lines of available memory. Programs can be saved on cassette or erased if no longer needed. The start of this program is on line 700. It is in inches and the program number is 03. Line 701 is the diameter of the tool used. This entry is critical if cutter diameter compensation is required. The feed rate given in line 702 for the *X* and *Y* axes is for positioning moves only. Line 703 is the feed rate of 2 ipm for the *Z* axis. The setup command in line 704 requires the operator to physically position each axis on the zero reference point for this program and lock it into the controller memory. If this reference point is not established, the controller uses the machine's home position as the program zero point. Machine home is the position where the table is in its maximum negative *X* and *Y* location and the spindle is at maximum positive *Z*. The steps to this point are the start of any program and the information programmed stays in effect until revoked by different commands.

Line 705 turns on the spindle. Line 706 gives the first machining directions. The first two letters determine the machining mode. GR means "go relative," a move relative from where you are to where you are going. This is incremental positioning. The *X* movement in line 706 is tied to the *Y* movement because no GR appears in line 707. With *X* and *Y* axes moving at the same time, an angular move is made. The prefix in line 708 is GO, which means "go in an absolute mode." This is a move in the *Z* axis. The *Z* zero point was established at the workpiece surface. This makes the $Z\ -0.1250$ dimension a .125-in.-deep cut into the workpiece. The letter "c" is a "come back" command that will raise the spindle after the depth of cut is reached. The spindle or *Z* axis will raise to the height above the work surface that was established in setup as the clearance height. The following steps to line 740 are positioning and drilling moves to complete the 11-hole pattern.

In line 741 the spindle is turned off. Line 742 is the program end. At this point the *X* and *Y* axes are at their zero reference point and the *Z* axis goes to maximum positive *Z*. This extra clearance under the tool makes workpiece changeover easier. The control also goes back to line 705, ready to machine a new part as soon as the start button is activated.

EXAMPLE TWO

The second example uses polar coordinates to drill the 11 hole pattern. When programming with polar coordinates, coordinate points are specified as distances from the existing reference or zero point and identified by the letter "r" (radius). The second value needed to establish a coordinate point is an angle identified by the letter "a." Straight lines, full circles, arcs, and chords can be machined by using the polar coordinate prefixes of r and a. Figure O-51 is the printout of the program.

Line 600 is the start of this program. Again the dimensions are in inches, and the program number is 02. The program start through the setup instructions and the spindle on command are the same as for the first example. In line 606 the GO specifies that this move is an absolute dimension, measured from the reference zero point. The *X* command is tied to the *Y* command in line 607 because no prefix precedes the *Y* axis value. This machine, with its linear interpolation capability, describes an angular path if both the *X* and *Y* axis move at the same time. This first move has positioned the spindle over the hole circle center.

The command in line 608 directs the controller to establish a new zero point, called a local zero, at the coordinate location of the *X* and *Y* axes. This new zero point is now at the hole circle center. Lines 609 and 610 are an absolute command in polar coordinates. The r 0.7500 directs the spindle to move .750 in. and the a 90.000 gives the angle of the move. In absolute positioning, zero degrees are on the *X* axis or at a 3 o'clock position. This would put a positive 90 degree angle at a 12 o'clock location. This move, then, has located the spindle over the hole number 1 position.

Line 611 gives instructions to the controller to repeat 11 times the following directions, ending with the repeat end statement in line 614. These directions are, in line 612, to make a negative 0.1250 in. move in the *Z* axis (lower spindle) combined with a c (come back, raise spindle) command; this drills the hole. In line 613, the GR a 32.727 is a "go relative angle" move. This means that the angle move is incremental from its present position and, being a positive 32.727 degrees, it is a counterclockwise movement. This positions the spindle over the next hole location. Lines 612 and 613 are repeated 11 times to drill the 11 holes. In line 615 the spindle is turned off, and the end of the program is line 616. The END NEWPART moves are the same as those in the first example.

```
500 START INS 01
501  TO=  0·2500
502 FR XY  =20·0
503 FR   Z =02·0
504 SETUP →zcxyu
505 SPINDLE   ON
506 BOLT PECK=00
507  ZH=  0·0000
508  Zd=  0·1250
509  XC=  2·5000
510  YC=  1·5000
511  a1=  90·000
512   N=      11
513   r=  0·7500
514 SPINDLE  OFF
515 END NEWPART
```

FIGURE O-52 Program for the 11 holes using a canned cycle.

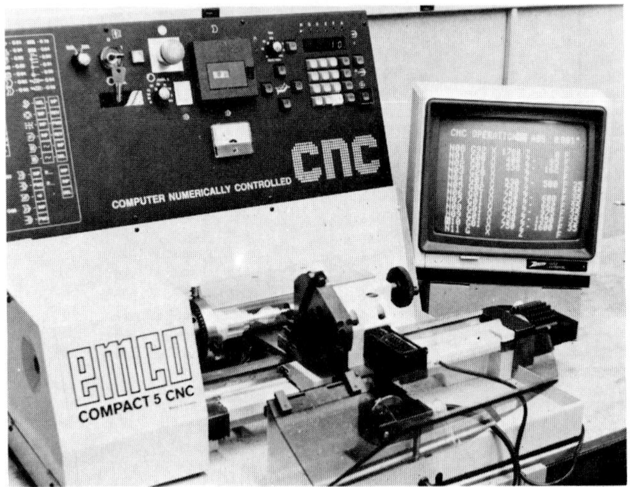

FIGURE O-53 **EMCO CNC Lathe (Lane Community College).**

EXAMPLE THREE

The third example uses a canned cycle to produce the 11 hole pattern. In a canned cycle, the programmer is prompted to give answers to questions asked by the machine control. Figure O-52 is the printout of the program that machines the 11 hole pattern.

The start of the program through line 505 is identical for all three examples. The BOLT CIRCLE canned cycle is started by touching a key on the controller keyboard. The response in line 506 is BOLT (for bolt circle) and PECK = ?. In prompted programming a question mark requires a response. Here the question asks for the number of pecks wanted to drill each hole. The holes to be drilled are only .125 in. deep, so pecking is not required. The programmer's entry then is PECK 00. Line 507 asks for the *Z*-axis reference surface location, which in this case is zero (0.0). In line 508 the prompt Zd asks for the hole depth. Lines 509 and 510 ask for the center location of the bolt circle, relative to the reference zero point; here it is *X* 2.500 and *Y* 1.500. Line 511 asks for the angular direction of hole number 1; here it is 90 degrees. The *X* axis, *Y* axis, and angle positions are absolute coordinate locations. Line 512 asks for the number of holes in the bolt circle. In line 513 the question as to the radius of the bolt circle is answered by an r = 0.7500 entry. This is all the information needed to machine this part. The last two lines make up the end of this program and are the same as the endings of the earlier two programs.

CNC Programming for a Lathe

Programming for this lathe is done using the word address format with G and M codes. The keyboard on the MCU is used to put information into the machine memory (Figure O-53). The shape of the part to be machined is shown on the drawing (Figure O-54). The machining consists of cylindrical surfaces, tapers, radii, and threads. The program sheet is shown in Figure O-55. To keep this ex-

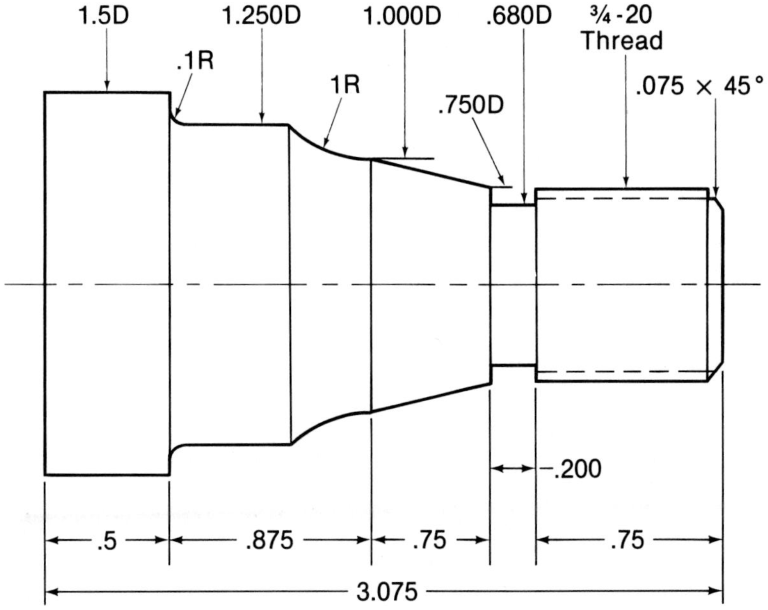

FIGURE O-54 Part drawing for CNC lathe programming example.

PROGRAM SHEET EMCO COMPACT 5 CNC

N	G (M)	X (I)	Z (K)	F (L)(K)(T)	H	Remarks
00	90					Tool I-DNMG-2211-30L
01	92	1700	50			S-2500 RPM
02	M 03					
03	04	200				
04	00	600	50			
05	01	600	0	30		
06	01	750	-75	30		
07	01	750	-950	30		
08	01	1000	-1700	30		
09	02	1250	-2184	20		
10	M 99	I1000	K 0			
11	01	1250	-2475	30		
12	02	1450	-2575	20		
13	01	1480	-2575	30		
14	03	1500	-2585	30		.01" R on Corner
15	00	1550	-2585			
16	00	1700	50			
17	M 05					
18	M 06	109	-64	T02		.075"w Groove tool
19	M 00					S-1000 RPM
20	M 03					
21	00	770	-750			
22	86	680	-950	20	75	
23	00	1700	50			
24	M 05					
25	M 06	126	98	T04		Threading tool
26	M 00					S-320 RPM
27	M 03					
28	00	790	100			
29	78	734	-850	K 50		
30	78	720	-850	K 50		
31	78	708	-850	K 50		
32	78	698	-850	K 50		
33	78	690	-850	K 50		
34	78	686	-850	K 50		
35	78	684	-850	K 50		
36	00	1700	50			
37	M 05					Return turret to start
38	M 06	0	0	T00		
39	M 30					

FIGURE O-55 Program for part in Figure O-54.

ample simple, only the finish pass, grooving, and threading are shown. The dimensions in the program ignore the nose radius of the turning tool, assuming a sharp point. In this example the programmer selected the threaded end of the workpiece as the Z 0 location. Since this controller has a floating zero capability, the programmer could also have picked the chuck face as the absolute zero point. In line 00, the G 90 code specifies absolute programming mode. The G 92 in line 01 establishes in the MCU the location of the tool presently in the cutting position. The X 1700, Z 50 dimensions are achieved during the initial setup.

This control unit does not use a decimal point; the smallest increment that can be programmed is .001 in. The X 1700 gives the tool point location as being at a diameter of 1.700 in. The Z 50 puts the tool point at .050 in. to the right of Z 0, which is the end of the workpiece. X dimensions in absolute programming are diameter values. With Z 0 being the end of the workpiece, any cutting requires a Z negative entry. Line 02 is an M 03 spindle on, clockwise command. Line 03 contains a G 04 code, this is a dwell command 2 seconds in length. The reason for this dwell is a time delay to allow the spindle to reach its full rpm before the tool starts to cut.

The G 00 in line 04 rapidly positions the tool near the beginning of the cut. Line 05 is the beginning of a number of machine movements at a programmed feed rate of 3.0 ipm. G 01 axes moves are metal-cutting movements. The coordinate points described by the X and Z values in each line define the endpoints of each move. The endpoint of one move becomes the start point for the following move. The G 02 in line 09 is a counterclockwise circular interpolation move. The X and Z values are the coordinates of the endpoint of the arc. The drawing in Figure O-54 does not give the endpoint of the arc in the *Z* axis. Figure O-56 illustrates how this dimension is calculated using trigonometry.

The M 99 in line 10 informs the controller that additional information is needed to complete this machine movement. The I and K values describe the centerpoint of the arc in line 09. This MCU requires an I and K entry if an arc of less than 90 degrees is to be made. Line 12 is a full 90-degree arc with a radius of .100 in. Line 14 has the cutting tool break a sharp corner by machining a .010-in. radius. Line 15 pulls the tool away from the work. Line 16 returns the tool to the starting point at a rapid feed rate. This point also is used as the tool change point. Automatic tool changes have to be made at a location where there is no tool–workpiece interference.

In line 17 the M 05 turns off the spindle. Line 18 has the first tool change activated by the M 06. The X and Z values are tool offsets. This means that T 02 (tool 2) was different in length and width by these amounts from tool 1. As part of this tool change, as soon as the new tool is in position, both *X* and *Z* axes will move by the tool offset amounts. Tool offsets make it possible that a 1-in. diameter programmed for tool 1 is still to be programmed as 1 in. with tool 2, even if these tools are physically different from each other. The M 00 in line 19 is a programmed Halt. The M 00 can be used to inspect the surface finish or make a measurement. In line 21 the grooving tool (tool 2) is positioned in line with the edge of the groove and .010 in. away from the .750-in. diameter of the workpiece. In line 22 we see a G 86 code. G 86 on this controller is a grooving cycle.

The inputs for this canned cycle are the bottom diameter of the groove, the endpoint in width of the groove, the feed rate, and the grooving tool width. The controller will take this information and make as many plunge cuts side by side as needed until the width of the groove is achieved. Since the controller knows the width of the grooving tool, successive passes are made at 90 percent of the tool width or less. At the end of a canned cycle, the controller will reposition the cutting tool at the place where the cycle started. Line 25 has another M 06 tool change, this time to bring a threading tool into cutting position. A new set of tool offsets will reposition this new tool so that an existing 1-in. diameter is still programmed as 1 in.

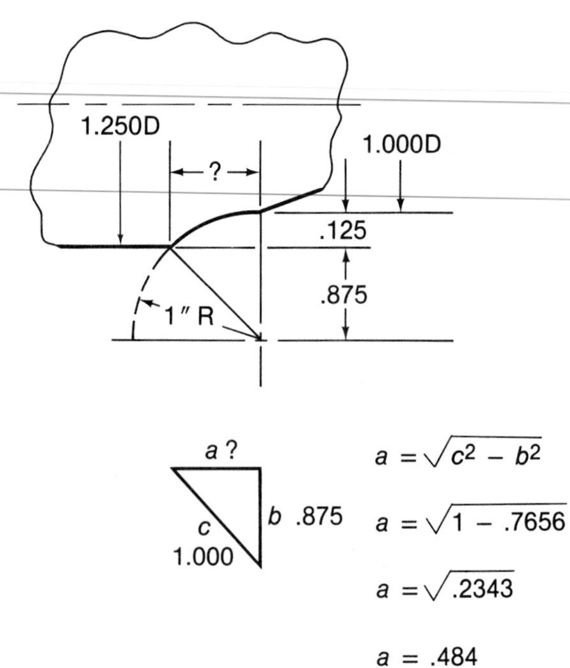

FIGURE O-56 Calculations to find *Z*-axis endpoint for the 1-in. radius in Figure O-54.

Most controllers have spindle rpm limitations while threading. This controller specifies a maximum of 300 rpm when cutting a thread with 20 threads per inch. Threading also requires that the threading cut start a certain distance before the part and end past the threaded piece. This gives the carriage an opportunity to accelerate from a stop at the beginning and decelerate to a stop at the end without any change in the spindle speed. Program manuals give formulas that are used to calculate these distances.

Line 28 is the coordinate point to which the threading tool moves to from the tool change point at rapid traverse speed. This is also the starting and ending point for the next seven canned cycle threading movements. Figure O-57 illustrates the four machine movements in this G78 code. Moves 1, 3, and 4 are made at rapid traverse speed. Move 2 is the threading cut; in this example the K 50 entry in the F column is the thread lead of .050 in. per spindle revolution. The cut in line 29 is .008 in. deep. The following cuts are .007, .006, .005, .004, .002 in., with a final pass at .001 in. deep. The depth of cut for each threading pass and the number of threading passes for a given thread are listed in carbide tool manufacturers' catalogs or in a machining data handbook. The final line with the M 30 stops the machining operation and resets the control to the starting point, ready to repeat this machining sequence.

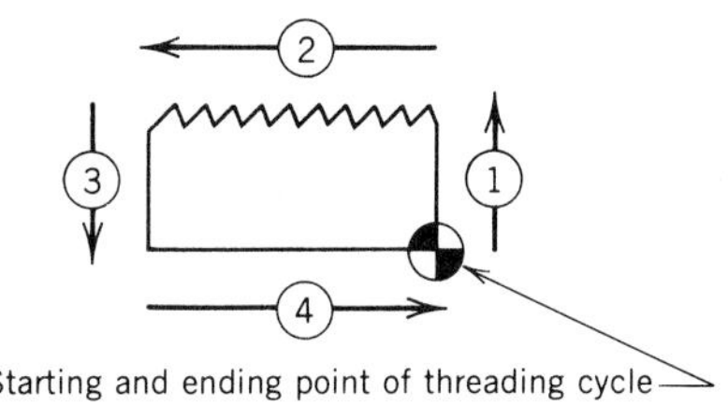

FIGURE O-57 Movements in a G78 threading cycle.

PROMPTED PROGRAMMING

A good example of prompted programming can be illustrated by the MAZATROL M-2 control from the Mazak Corporation. The computer in this control has been programmed with basic machining information such as work materials, cutting tool shapes, and cutting tool applications. The major steps required to program the machining of the part shown in Figure O-58 are as follows:

STEP 1. From the material selection menu (Figure O-59), select the appropriate workpiece material. The computer is already programmed with the cutting speed and feed rate for each material.

STEP 2. From the three basic machining patterns displayed (point machining, line machining, and face machining), select the face

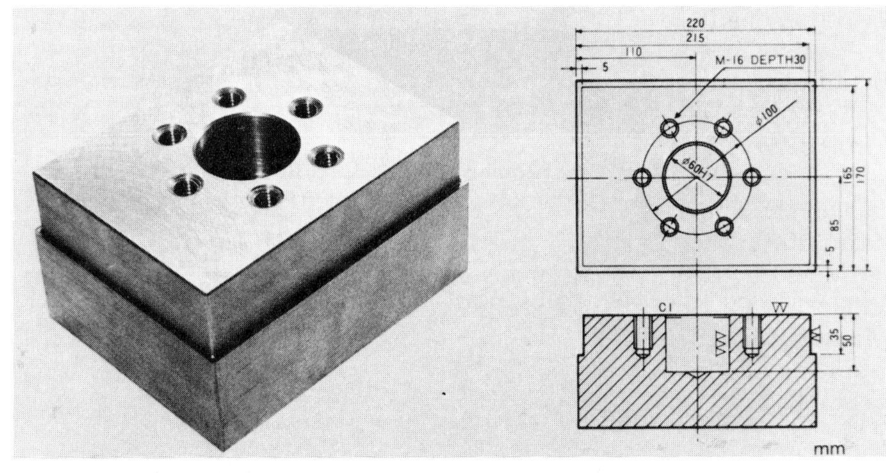

FIGURE O-58 Workpiece and part drawing (Mazak Corporation).

Material Menu	Cast Iron	Ductile Cast Iron	Carbon Steel	Alloy Steel	Stainless Steel	Aluminum	Copper Alloy		Other
Surface Roughness Menu	▽ 1	▽ 2	▽▽ 3	▽▽ 4	▽▽▽ 5	▽▽▽ 6	▽▽▽ 7	▽▽▽▽ 8	▽▽▽▽ 9

FIGURE O-59 Material selection menu (Mazak Corporation).

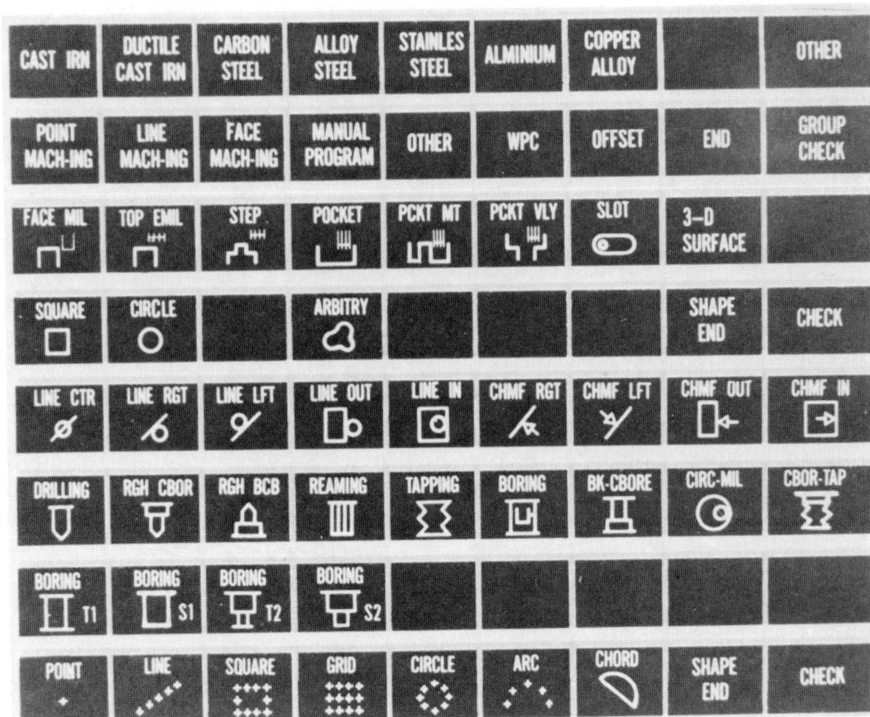

FIGURE O-60 Video display of material selections and machining functions (Mazak Corporation).

milling mode by pressing the menu key (Figure O-60). From the face milling mode choose the square symbol, which represents the shape of the workpiece (Figure O-61). For this machining operation the control selects the face milling cutter and displays a cutter picture on the CRT screen. After the cutter diameter is entered, by using the automatic selection of cutting conditions function, the depth of cut, cutting width, and feed rate are automatically calculated and entered. After questions displayed on the CRT about the finished workpiece contour are answered by entering coordinate values for the *X* and *Y* axis, the computer will calculate start and end points for the cutter and automatically determine the required number of tool passes. Figure O-62 is the surface roughness selection chart. When selecting the number 4 from the surface roughness chart, the maximum acceptable roughness would be a RMS of 125. After this selection is made, the computer, with the dimension, shape, and corner radius of the cutter in memory, calculates the correct feed rate and speed to achieve this surface finish.

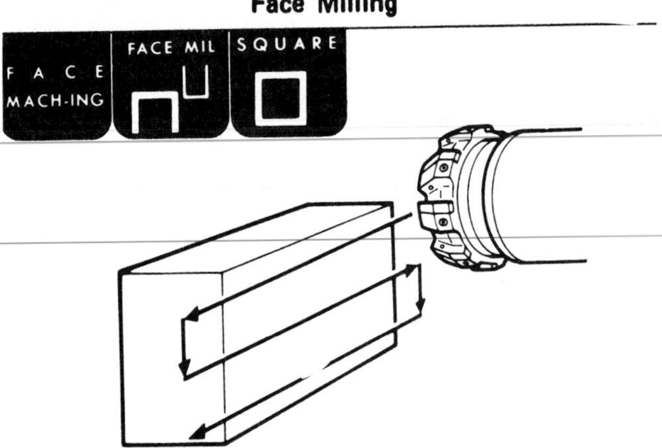

FIGURE O-61 Face milling mode (Mazak Corporation).

Surface Roughness Menu	▽ 1	▽ 2	▽▽ 3	▽▽ 4	▽▽▽ 5	▽▽▽ 6	▽▽▽ 7	▽▽▽▽ 8	▽▽▽▽ 9
RMS m-in	1000	500	250	125	63	32	16	8	4

FIGURE O-62 Surface roughness selection chart.

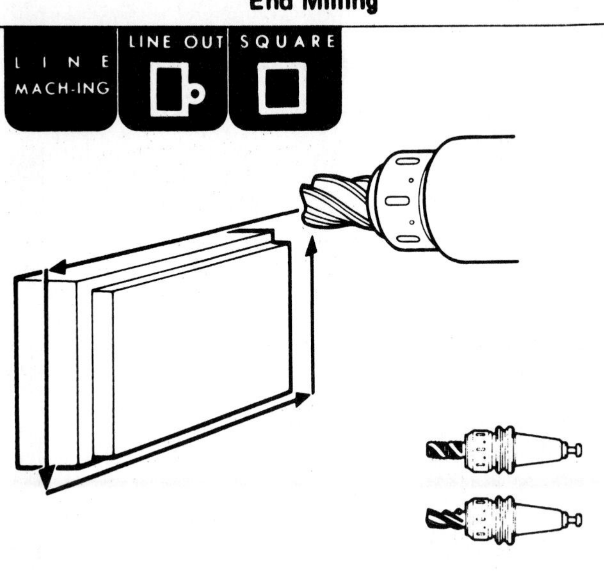

FIGURE O-63 Line machining mode (Mazak Corporation).

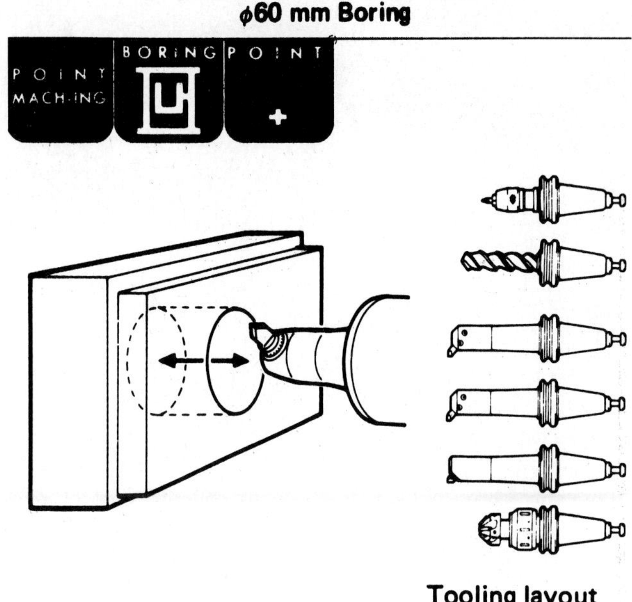

FIGURE O-64 Point machining mode (Mazak Corporation).

STEP 3. The next step is the end milling of the periphery. From the three machining patterns select the line machining mode (Figure O-63). Select the line outside and square shape symbols. Enter the cutter diameter, cutting contour, and the workpiece dimension as the CRT displays the appropriate questions. With this information entered, the computer again will calculate all necessary feeds, speeds and positioning moves.

STEP 4. At this time the hole needs to be machined. From the three patterns the point machining mode is needed (Figure O-64). From the menu the boring and the one point only selections are made. Questions from the computer need to be answered with the hole depth, hole diameter, surface finish symbol, inside chamfer, and coordinate values for the *X* and *Y* axis. All the required tools for this machining operation are determined again by the automatic tooling layout function and displayed on the CRT. The tools are displayed in the proper machining order; center drill, drill, rough boring bar, semi-finish boring bar, finish boring bar, and chamfering tool. Optimum speed and feed rate are calculated automatically for each tool.

STEP 5. The last operation is the drilling and tapping of the six holes (Figure O-65). From the machining menu, select point machining, tapping, and the hole circle pattern. The computer asks for the tap size, depth of threaded holes, number of tapped holes, and the location of the center for the hole circle. Again, all the required tools are automatically selected: center drill, tap drill, and tap. The computer calculates the lead of the tap and then sets the feed rate per revolution of the spindle to exactly the same amount. With the spindle speed and feed rate exactly synchronized and the depth of the tapped hole determining the point of spindle reversal, exact threads are simple to produce.

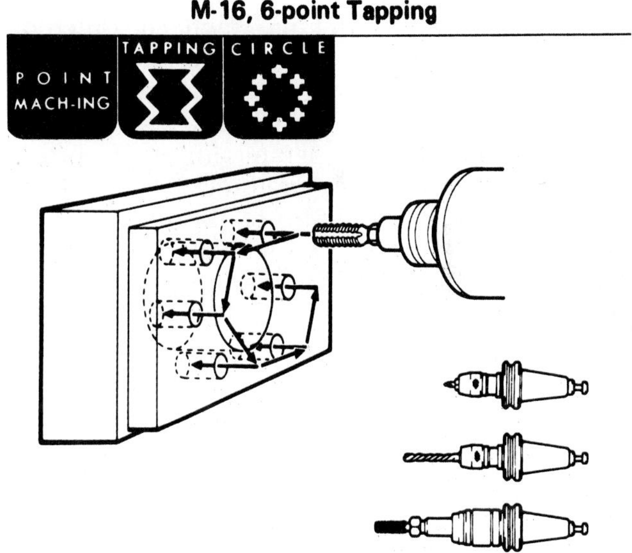

FIGURE O-65 Point machining mode—tapping (Mazak Corporation).

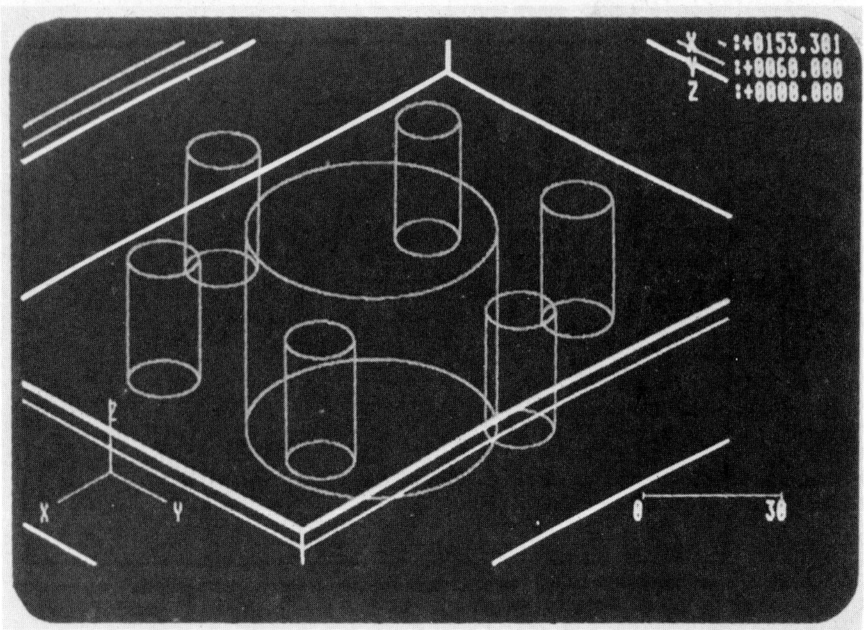

FIGURE O-66 **Programming check display (Mazak Corporation).**

At any time during the programming a programming check display (Figure O-66) can be requested. This graphic display will show the path of each tool that is being programmed. Different tools can be shown with lines of different colors. To see small details more clearly, any part of the screen display can be magnified to full screen size. If the display shows a programming error, corrections can be made easily and quickly.

After the programming of the part is completed, the CRT will show the total time required to machine the part. This example used here illustrates how relatively simple programming a sophisticated CNC machining center can be.

SELF-TEST

1. Give two examples where MDI is used.
2. What knowledge is needed by a good CNC programmer?
3. What is the difference between prompted input and word address input?
4. Why are canned cycles used?

UNIT 4

Nontraditional Machining Processes

Most of this book deals with traditional machining processes such as turning, milling, drilling, and grinding. In these processes, the cutting tool directly contacts the workpiece and removes chips of the workpiece material. Although these processes are widely used and relatively efficient they do have some limitations. For example, machining certain materials by the traditional processes may be extremely difficult or even impossible. Furthermore, the traditional processes do not always lend themselves to machining parts for certain designs that are necessary to the manufacturing of exotic products in aerospace and other high technology fields. To solve these and other problems in machining manufacturing, the engineer and industrial technologist have developed other types of machining processes that are quite different from the classical chip-producing methods. These processes are sometimes known as nontraditional in that they remove workpiece material by applications of electrical energy, laser light energy, intensely hot plasmas, electrochemical processes, high-pressure water/abrasive jets, and ultrasound.

Included in nontraditional machining processes are:

1. Electrodischarge machining (EDM)
2. Electron beam machining (EBM)
3. Electrolytic grinding (ELG)
4. Electrochemical machining and deburring (ECM, ECDB)
5. Ultrasonic machining
6. Hydrojet machining
7. Plasma beam machining
8. Laser machining

Objective

After completing this unit, you should be able to:

Identify common nontraditional machining processes and generally describe how these processes work and where they might be applied.

Electrodischarge Machining

Electrodischarge machining, commonly know as **EDM,** removes workpiece material by an electrical spark erosion process. The process is accomplished by establishing a large potential (voltage) difference between the workpiece to be machined and an electrode. A large burst (spark) of electrons travels from the electrode to the workpiece. When the electrons impinge on the workpiece, some of the workpiece material is eroded away. Thus the machining is accomplished by electrical spark erosion.

The EDM process (Figure O-67) takes place in a dielectric (nonconducting) oil bath. The dielectric bath concentrates the spark and also flushes away the spark-eroded workpiece material. A typical EDM system consists of a power supply, dielectric reservoir, electrode, and workpiece. The EDM machine tool has many of the same features as its conventional counterpart. These include worktable positioning mechanisms and measurement devices. Many EDM machine tools are equipped with computer numerical control systems. Thus, the versatility of CNC for workpiece positioning and tool control functions can be effectively used in the process.

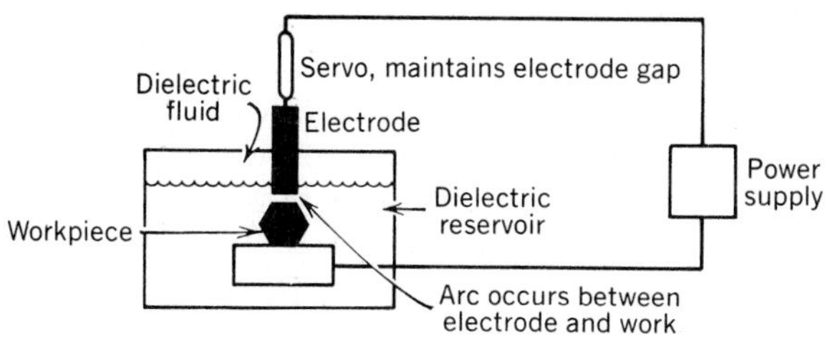

FIGURE O-67 EDM system.

EDM ELECTRODES

In the EDM process, the shape of the electrode controls the shape of the machined feature on the workpiece. EDM electrodes may be made from metal or carbon (graphite), and shaped by molding or machining to the desired geometry. Erosion of the electrode takes place as well during the EDM process. In time it will become unusable. To circumvent this problem, a roughing electrode may be used to generally shape the workpiece and then a finish electrode may be applied to complete the process and establish final dimensions and geometry.

WIRE-CUT EDM

Wire-cut EDM uses a slender wire as an electrode (Figure O-68). It is extremely useful for narrow slots and detailed internal features in the workpiece (Figures O-69 and O-70). Its capability for detailed tool and die making is very useful. Wire-cut EDM is similiar in some ways to band sawing in conventional machining. Whereas the saw teeth do the cutting in band sawing, in wire-cut EDM the spark is occurring between the wire electrode and the workpiece. Wire feed is continuous during the process and the wire may not be reused.

Different metals are used for wire electrodes, brass being one popular material. EDM wire can

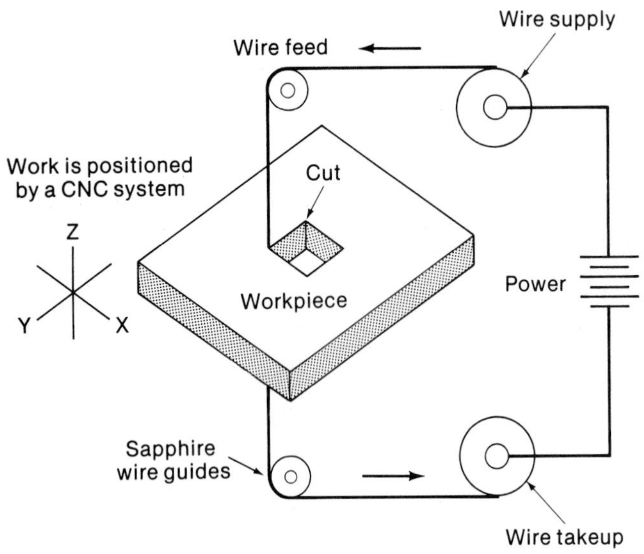

FIGURE O-68 Wire-cut EDM system.

FIGURE O-69 Blanking dies for watch parts (Japax Inc.).

be of very small diameter on the order of .005 in. Sapphire guides maintain the alignment of the wire as it passes through the work. Wire tension is maintained by weighted pulleys or other mechanical tension schemes.

Wire-cut EDM, because it uses such small diameter wires, permits extremely narrow slots to be machined in the workpiece. The kerf is only slightly wider than the wire diameter. This process is highly suited to intricate slot features in tools and dies (Figures O-71 and O-72).

FIGURE O-70 Machining a turbine disk by wire-cut EDM (Japax Inc.).

FIGURE O-71 Plastic mold for squirrel cage fan (Japax Inc.).

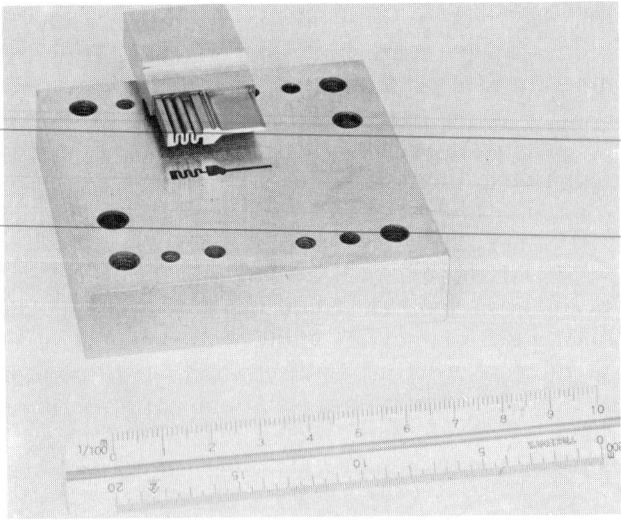

FIGURE O-72 Blanking punch and die set (Japax Inc.).

ADVANTAGES AND APPLICATIONS OF EDM

EDM processes can accomplish machining that would be impossible by the traditional methods. Any odd electrode shape will be reproduced in the workpiece. Thus, fine detail is possible, making the process very useful in tool, die and mold work. When EDM is coupled to numerical control, an excellent machining system for the tool and die maker is available.

Since machining is accomplished by spark erosion, the electrode does not actually touch the workpiece and is therefore not dulled by hard workpiece materials as would be the case with conventional cutters. Thus, the EDM process can spark erode very hard metals and has found wide application in removal of broken taps without destroying an expensive workpiece in the process.

DISADVANTAGES OF EDM

The EDM process is quite slow from a metal removal standpoint as compared to conventional machining processes involving direct cutter contact. There is also a possibility of overcutting and local area heat treating of the parts being machined.

However, EDM can accomplish many machining tasks that could never be done by conventional machining. Thus, the EDM process has become a well-established, very versatile manufacturing process in modern industrial applications.

Electron Beam Machining

Electron beam machining (EBM) is in some ways related to EDM and to electron beam welding. In EDM, however, a burst of electrons (spark) impinges on the workpiece, whereas electrons in a continuous beam are used in the EBM process. The workpiece material is heated and vaporized by an intense electron beam. The process, like that of electron beam welding, must be carried on in a vacuum chamber and appropriate shielding must be employed to protect personnel from X-ray radiation.

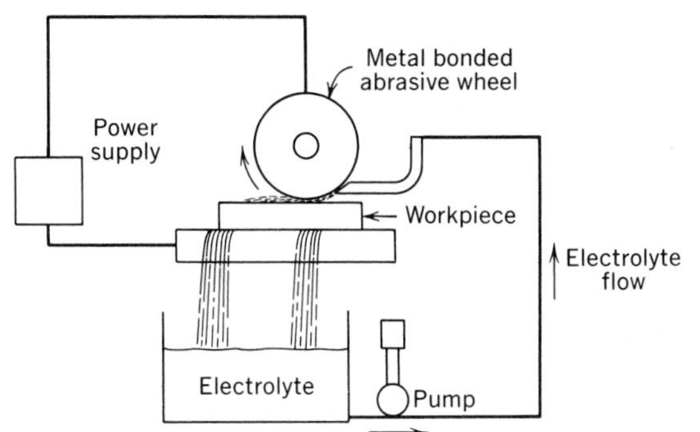

FIGURE O-73 **ELG system.**

Electrolytic Grinding

In the process of electrolytic grinding (ELG), an abrasive wheel much like a standard grinding wheel is used. The abrasive wheel bond is metal, thus making it a conducting medium. The abrasive grains in the grinding wheel are nonconducting and aid in removing oxides from the workpiece while at the same time helping to maintain the gap between wheel and work. ELG, like ECM, is a deplating process and workpiece material is carried away by the circulating electrolyte.

THE ELG SYSTEM

The basic ELG system (Figure O-73) consists of the appropriate power supply, the electrode (metal bonded grinding wheel), work-holding equipment, and the electrolyte supply and filtration system. Workpiece material is deplated and goes into the electrolyte solution.

ADVANTAGES AND APPLICATIONS OF ELG

Since this process is primarily electrochemical and not mechanical as is conventional grinding, the abrasive wheel in ELG wears little in the process (Figure O-74). ELG is burr-free and will not distort or overheat the workpiece. The process is therefore very useful for small precision parts and thin or fragile workpieces.

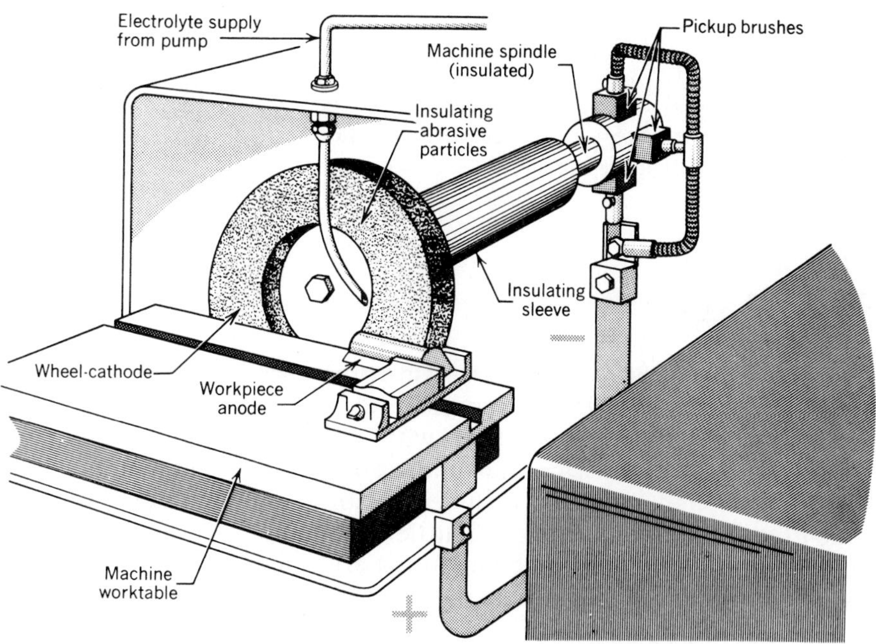

FIGURE O-74 **Electrolytic surface grinder (Machinery Builders Inc.).**

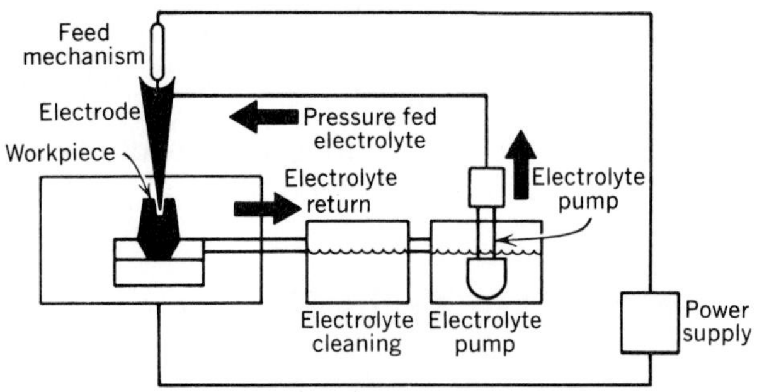

FIGURE O-75 **ECM system.**

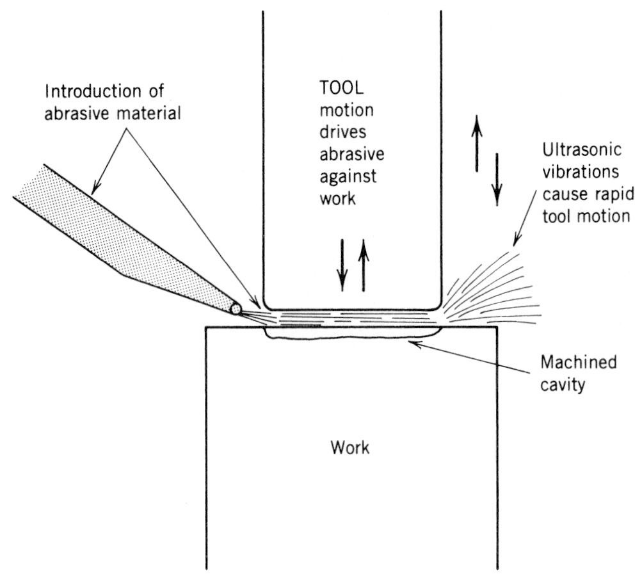

FIGURE O-76 **Ultrasonic machining.**

Electrochemical Machining and Electrical Chemical Deburring

Electrochemical machining (ECM) is essentially a reverse metal plating process (Figure O-75). The process takes place in a conducting fluid or electrolyte that is pumped under pressure between electrode and workpiece. As workpiece material is deplated, it is flushed away by the flow of electrolyte. Workpiece material is removed from the electrolyte by a filtration system.

ADVANTAGES AND APPLICATIONS OF ECM

As in EDM, ECM can accomplish machining of intricate shapes in hard-to-machine material. The process is also burr-free and does not subject the workpiece to distortion and stress as do conventional machining processes. This makes it useful for work on thin or fragile workpieces. ECM is also used for part deburring in the process called electrochemical deburring (ECDB). This process is very useful for deburring internal workpiece features that are inaccessible to traditional mechanical deburring processing.

Ultrasonic Machining

Ultrasonic machining is akin to abrasive processes like sand blasting. High-frequency sound (Figure O-76) is used as the motive force with which to propel abrasive particles against the workpiece. Advantages of this process include the ability to machine very hard material with little distortion. Good surface finishes may be obtained and part features of many different shapes can be machined.

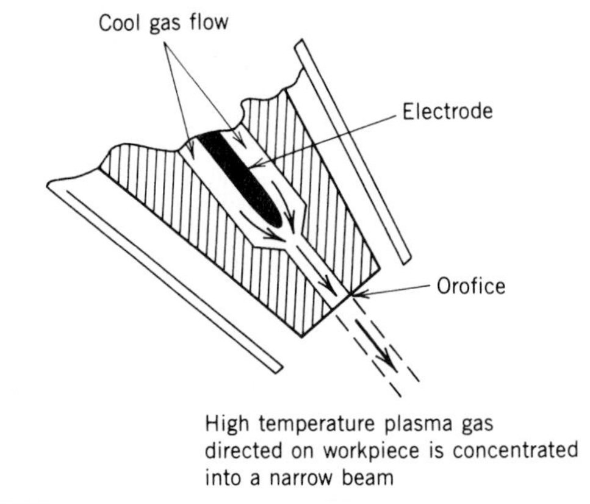

FIGURE O-77 **Plasma torch.**

Hydrojet Machining

The cutting tool in this process is an extremely high pressure water jet. Pressures of several thousand pounds per square inch are used to pump water through a small nozzle, thus directing a narrow stream at the workpiece. The efficiency of the process is enhanced by adding an abrasive material to the water. Thus pressure and mechanical abrasion join forces to accomplish cutting and machining tasks. The process may be used for hard metallic materials. However, there are limitations as to the thickness of the material that can be machined. In the fabric industry, the process has efficiently been applied to fabric cutting applications.

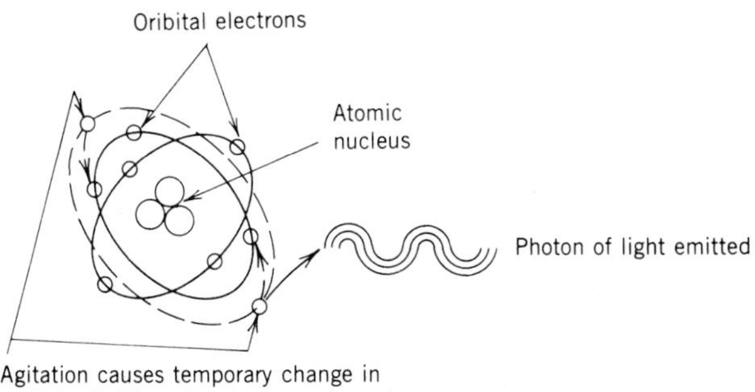

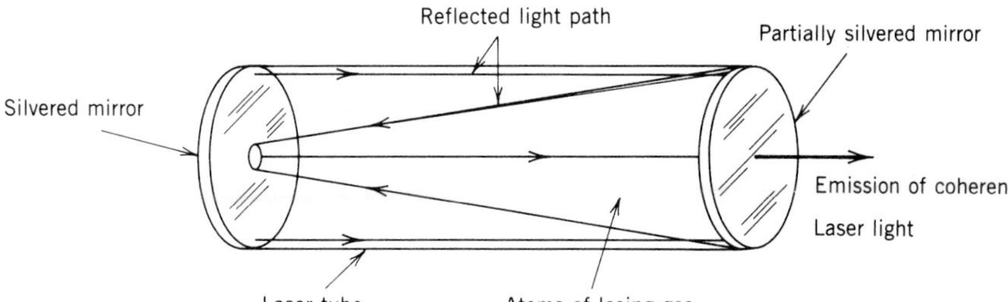

FIGURE O-78 **Gas laser generator.**

Plasma Beam Machining

A great deal of interest and development of plasma beam has occurred in recent years. Plasma is created by passing a gas through an electric arc. The gas is ionized by the arc and an extremely high temperature results. Temperatures in plasma arcs can exceed 40,000°F (22,204°C), many times hotter than any arc or flame temperature produced by other processes (Figure O-77).

Plasma arcs of such high temperatures have a multitude of uses not only in machining, but also for other applications such as ore smelting and waste metal recovery operations. Ongoing research is studying the applications of high temperature plasmas for incinerating industrial wastes.

The process is useful for cutting, welding, and machining nonferrous metals and stainless steel. Plasma torch machining is extremely fast and will produce smooth surface finishes to close dimensions.

Lasers and Laser Machining

LASERS

The applications of lasers in manufacturing are widespread. Laser energy finds many applications other than the use of its energy for a cutting tool. LASER is an acronym for *light amplification by stimulated emission of radiation*. By electrically stimulating the atoms of certain materials such as various crystals and certain gases, the electrons of these atoms can be temporarily displaced to higher electron shell energy level positions within the atomic structure. When the electrons fall back to their originally stable levels, photons of light energy are released (Figure O-78). This light energy can be enhanced and focused into a coherent beam and then used in many manufacturing, medical, measurement, and other useful applications.

LASER MACHINING

In laser machining, the coherent laser light becomes the cutting tool. When the laser is integrated with CNC machine tool positioning, an extremely versatile system with many capabilities is available.

Applications of laser machining include cutting plate for shapes such as saw blades, slotting stainless tubing, and in marking systems.

SELF-TEST

1. Describe in general terms how the following processes work: EDM, ECM, ELG, laser, ultrasonic, hydrojet, plasma beam, and electron beam machining.

2. Which processes might be used for deburring applications?

3. Which process would be used for detail die slotting?

4. What temperatures can be reached in a plasma system?

5. Which processes are essentially deplating?

6. What is the primary function of the grinding wheel in ELG?

7. What provides the energy in ultrasonic machining and what does the cutting?

8. Which process is accomplished in a dielectric fluid?

9. Which process uses a conducting fluid?

10. Describe in general terms how a laser functions.

APPENDIX 1

ANSWERS TO SELF-TESTS

Section A / Unit 1 / Shop Safety

SELF-TEST ANSWERS

1. Eye protection equipment.

2. Wear safety goggles or a full face shield. Prescription glasses may be made as safety glasses.

3. Shoes, short sleeves, short or properly secured hair, no rings and wristwatches, shop apron or shop coat with short sleeves.

4. Use of cutting fluids and vacuum dust collectors.

5. They may cause skin rashes or infections.

6. Bend knees, squat, and lift with your legs, keeping your back straight.

7. Compressed air can propel chips through the air, implant dirt into skin, and possibly injure eardrums.

8. Good housekeeping includes cleaning up oil spills, keeping material off the floor, and keeping aisles clear of obstructions.

9. In the vertical position or with a person on each end.

10. Do I know how to operate this machine? What are the potential hazards involved?
Are all guards in place?
Are my procedures safe?
Am I doing something I probably should not do?
Have I made all proper adjustments and tightened all locking bolts and clamps?
Is the workpiece secured properly?
Do I have proper safety equipment?
Do I know where the stop switch is?
Do I think about safety in everything I do?

Section A / Unit 2 / Mechanical Hardware

SELF-TEST ANSWERS

1. A bolt goes through parts being assembled and is tightened with a nut. A screw is used where a part is internally threaded and no nut is needed.

2. The minimum recommended thread engagement for a screw in an assembly is as much as the screw diameter; a better assembly will result when $1\frac{1}{2}$ times the screw diameter is used.

3. Class 2 threads are found on most screws, nuts, and bolts used in the manufacturing industry. Car and machine tools would be good examples.

4. Machine bolts are not machined to the precise dimensions of capscrews. Machine bolts have coarse threads, whereas capscrews may have coarse or fine threads. Machine bolts have many uses in the construction industry, and capscrews are usually used in precision assemblies.

5. The formula is D = number of the machine screw times .013 in. plus .060 in. $D = 8 \times .013$ in. plus $.060 = .164$.

6. Setscrews are used to secure gears or pulleys to shafts.

7. Stud bolts can be used instead of long bolts. Stud bolts are used to aid in the assembly of heavy parts by acting as guide pins.

8. Thread-forming screws form threads by displacing material. Thread-cutting screws produce threads by actually cutting grooves and making chips.

9. Castle nuts can be secured on a bolt with a cotter pin to prevent their accidental loosening.

10. Cap nuts are used because of their neat appearance. They also protect projecting threads from damage.

11. Flat washers protect the surface of parts from being marred by the tightening of screws or nuts. Flat washers also provide a larger contact area than nuts and screw heads to distribute the clamping pressure over a larger area.

12. A helical spring lock washer prevents the unplanned loosening of nut and bolt or screw assemblies. Spring lock washers will also provide for a limited amount of takeup when expansion or contraction takes place.

13. Internal–external tooth lock washers are used on oversized holes or to provide a large bearing surface.

14. Dowel pins are used to achieve accurate alignment between two or more parts.

15. Taper pins give accurate alignment to parts that have to be disassembled frequently.

16. Roll pins are used to align parts. Holes to receive roll pins do not have to be reamed, which is necessary for dowel pins and taper pins.

17. Retaining rings are used to hold bearings or seals in bearing housings or on shafts. Retaining rings have a spring action and are usually seated in grooves.

18. Keys transmit the driving force between a shaft and pulley.

19. Woodruff keys are used where only light loads are transmitted.

20. Gib head keys are used to transmit heavy loads. These keys are installed and removed from the same side of a hub and shaft assembly.

Section A / Unit 3 / Reading Drawings

SELF-TEST ANSWERS

1. This is how Figure A-61 (page 33) should appear as a three-view orthographic drawing.

For answers to Questions 2–10, refer to Figure A-62 (page 34).

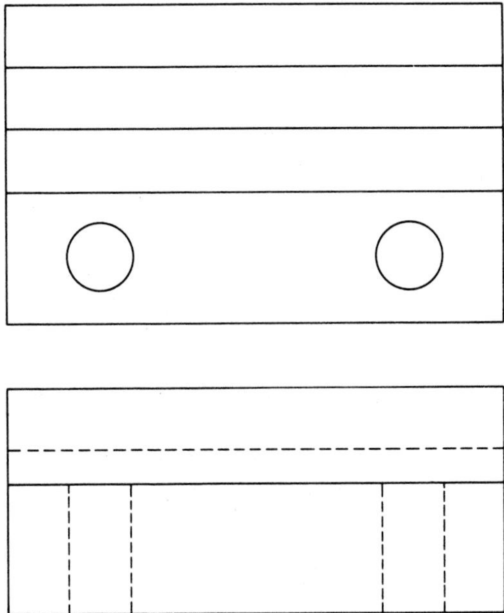

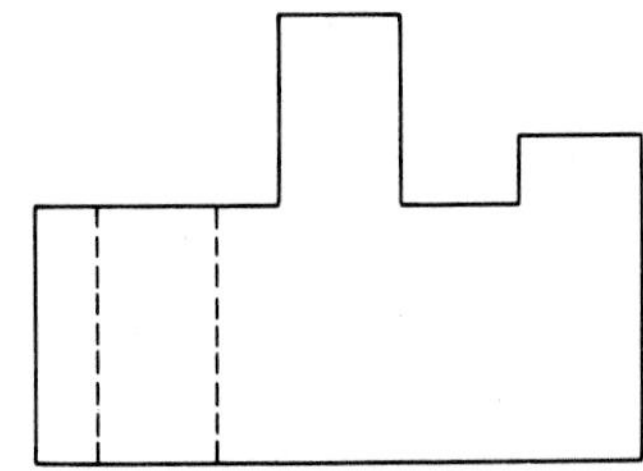

2. The tolerance of the hole is specified as ± .005. Therefore, the minimum size of the hole would be .750 − .005 or .745.

3. $2\frac{1}{4}$ in.

4. Since the slot conforms to standard tolerances, width is ± $\frac{1}{64}$.

5. $\frac{1}{8}$ in.

6. 6 in.

7. 1 in.

8. Drilling and reaming.

9. $\frac{1}{8}$ in. by 45 degrees.

10. Note 1 indicates that all sharp corners are to be broken, which means that all burrs and sharp edges left from machining are to be smoothed off.

Section B / Unit 1 / Arbor and Shop Presses

SELF-TEST ANSWERS

1. To use the arbor press without instruction is unsafe for the operator; also, very expensive equipment and materials may be damaged or ruined.

2. Arbor presses are hand powered and can be mechanical or hydraulic. Large power-driven presses do not provide the "feel" needed when pressing delicate parts.

3. The arbor press is used for installing and removing mandrels, bushings, and ball bearings. The hydraulic shop press is also used for straightening and bending.

4. The shaft has seized or welded in the bore because it was not lubricated with pressure lubricant.

5. A loose and rounded ram could cause a bushing to tilt or twist side-ways while pressing and thus be ruined. In any case, the operator should always check to see if a bushing is going in straight. A pressing plug with a pilot would be helpful here.

6. Just enough pressure should be applied to press in the bushing, and, when it stops moving, more pressure will be sensed. At that point it is time to stop.

7. The bearing should be supported on the inner race.

8. Ordinary shafts with press fits are not tapered but have the same dimension along the pressing length. Mandrels taper .060 in./ft, which causes them to tighten in the bore somewhere along their length.

9. The two most important steps, assuming that the dimensions are all correct, are to:

Make sure that the bore has a good chamfer; the bushing should also have a chamfer or "start."

Apply high-pressure lubricant to the bore and the bushing.

10. Five ways to avoid tool breakage and other problems when broaching keyseats in the arbor press:

Make sure that the press ram is not loose and check to see that the proper hole in the press plate is under the work so that the broach has clearance to go through the work.

Clean and lubricate the broach, especially the back edge between each cut.

Do not use a broach on hard materials (over RC 35).

Use the right-size bushing for the bore and broach.

Make sure at least two teeth are continuously engaged in the work.

Section B / Unit 2 / Work-Holding and Hand Tools

SELF-TEST ANSWERS

1. The vise should be positioned so that a long piece can be held vertically in the jaws without interference from the workbench.

2. The solid base and the swivel base types.

3. By the width of the jaws.

4. Insert jaws are hardened and have diamond pattern or criss-cross serrations.

5. Copper, soft metal, or wood may be used to protect finishes from insert jaw serrations.

6. "Cheater" bars should never be used on the handle. The movable jaw slide bar should never be hammered upon, and excessive heat should never be applied to the jaws.

7. The vise should be taken apart, cleaned, and the screw and nut cleaned in solvent. A heavy grease should be packed on the screw and thrust collars before reassembly.

8. True.

9. The purpose of soft jaws is to protect finished work from being damaged by the hard serrated jaws. Also, when gripping soft metals, soft jaws of leather or a softer metal are used.

10. False. C-clamps are used for clamping work. Some heavy-duty types can hold many hundreds of pounds.

11. False. This practice can quickly ruin good machinery so that the proper tool can never be used.

12. The principal advantage of the lever-jawed wrench is its great holding power. Most types have hard serrated jaws and so should not be used on nuts and bolt heads.

13. No. Hammers used for layout work range from 4 to 10 oz in weight. A smaller one should be used with a prick punch and for delicate work.

14. Soft hammers and mallets are made for this purpose. When setting down work in a drill press or milling machine vise, for instance, a lead hammer is best because it has no rebound.

15. The box type, either socket or end wrench, would be best as it provides contact on the six points of the capscrew, thus avoiding the damage and premature wear that would be caused by using an adjustable wrench.

16. The hard serrated jaws will damage machine parts. Pipe wrenches should be used on pipe and pipe fittings only.

17. Standard screwdrivers should have the right-width blade to fit the screw head. They should be shaped correctly and, if worn, reground and shaped properly.

Section B / Unit 3 / Hacksaws

SELF-TEST ANSWERS

1. The kerf is the groove produced in the work by a saw blade.

2. The set on a saw blade is the width of the teeth that are bent out from the blade back.

3. The pitch of a hacksaw blade refers to the number of teeth per inch on a saw blade.

4. The first consideration in the selection of a saw blade is the kind of material being cut. For soft materials use a coarse tooth blade and for harder materials use a fine pitch blade. The second point to watch is that at least three teeth should be cutting at the same time.

5. The two basic kinds of saw blades are the all-hard blade and the flexible blade.

6. Generally, a speed between 40 and 60 strokes per minute is suggested. It is best to use long and slow strokes utilizing the full length of the blade.

7. Excessive dulling of saw blades is caused by pressure on the saw blade on the return stroke, sawing too fast, letting the saw slide over the workpiece without any cutting pressure, or applying too much pressure.

8. Saw blades break if too much pressure is used or if the blade is not sufficiently tightened in the saw frame.

9. When the saw blade is used, the set wears and makes the kerf cut narrower with a used blade than the kerf cut with a new blade. If a cut started with a used blade can't be finished with that blade, but has to be completed with a new blade, the workpiece should be turned over and a new cut started from the opposite end of the original cut. A new blade used in a kerf started with a used blade would lose its set immediately and start binding in the groove.

10. When a blade breaks, it shatters, causing blade particles to fly quite a distance with the possibility of injuring someone. Should the blade break while sawing, it may catch the operator off balance and cause him to push his hand into the workpiece while following through with his sawing stroke. Serious cuts or abrasions can be the result.

Section B / Unit 4 / Files

SELF-TEST ANSWERS

1. By its length, shape, cut, and coarseness.

2. Single cut, double cut, curved cut, and rasp.

3. Four out of these: rough, coarse, bastard, second cut, smooth, and dead smooth.

4. The double cut file.

5. To make it possible to file a flat surface by offsetting the tendency to rock a file. To compensate for the slight downward deflection when pressure is applied while filing. To concentrate pressure on fewer teeth for deeper penetration.

6. A blunt file has the same cross-sectional area from heel to point, whereas a tapered file is larger at the heel than at the point.

7. A mill file is thinner than a comparable-sized flat file.

8. A warding file is rectangular in shape and it tapers to a small point, making filing possible in narrow slots and grooves.

9. Swiss pattern files are more precise in construction; they are more slender and have teeth to the extreme edges.

10. Coarseness is identified by numbers from 00 (fine) to 6 (coarse).

11. Where files touch each other, teeth break off or become dull.

12. Too much pressure will break teeth off a file. It also will cause pinning and scratching of the work surface.

13. Files in contact with each other, files rubbing over the work without any pressure being applied, filing too fast, or filing on hardened materials cause files to dull.

14. As a safety precaution; an unprotected tang can cause serious injury.

15. Measuring the workpiece for flatness and size assures the worker that the filing is done in the right place and that the filing is stopped before a piece is undersize. Measuring often is not a waste of time.

16. Touching a workpiece is just like lubricating the workpiece or the file so that it slips over the work without cutting. This also causes a file to dull quickly.

17. A soft workpiece requires a file with coarser teeth because there is less resistance to tooth penetration. A fine-toothed file would clog up on soft materials. For harder materials use a fine-toothed file in order to have more teeth making smaller chips.

18. The drawfiling stroke should be short enough that the file never slips over the end of the workpiece. Care should be taken that no hollow surface is created through too short a stroke.

19. Pressure is only applied on the forward stroke, which is the cutting stroke.

20. Rotating a round file clockwise while filing makes the file cut better and improves the surface finish.

Section B / Unit 5 / Hand Reamers

SELF-TEST ANSWERS

1. Hand reamers have a square on the shank and a long starting taper on the fluted end.

2. A reamer does its cutting on the tapered portion. A long taper will help in keeping a reamer aligned with the hole.

3. Spiral-fluted reamers cut with a shearing action. They will also bridge over keyseats and grooves without chattering.

4. The shank diameter is usually a few thousandths of an inch smaller than the nominal size of the reamer. This allows the reamer to pass through the hole without marring it.

5. Expansion reamers are useful to increase hole sizes by a very small amount.

6. Expansion reamers can only be adjusted a small amount by moving a tapered internal plug. Adjustable reamers have a larger range of adjustments, from $\frac{1}{32}$ in. on small diameters to $\frac{5}{16}$ in. on large reamers. Adjustable reamers have removable blades. Size changes are made by moving these blades with nuts in external tapered slots.

7. Cutting fluids are used to dissipate the heat generated by the reaming process, but in reaming cutting fluids are more important in obtaining a high quality surface finish of the hole.

8. Reamers dull rapidly if they are rotated backward.

9. The hand reaming allowance is rather small, only between .001 and .005 in.

Section B / Unit 6 / Identification and Uses of Taps

Self-Test Answers

1. A set of taps consists of three taps with equal pitch diameters and major diameters with the difference being the number of chamfered threads on the cutting end. Serial taps have different pitch and major diameters within a nominal size designation. The smallest tap in the series is marked with a single ring on the shank near the square. The next larger tap has two rings on the shank and the tap that cuts the thread to its full size is marked with three rings.

2. Spiral-pointed taps are used on through holes or blind holes with sufficient chip space at the bottom of a hole.

3. Fluteless spiral-pointed taps are especially useful to tap holes in sheet metal or on soft, stringy materials where the thickness is no greater than one tap diameter.

4. Spiral-fluted taps draw the chips out of the hole and are useful when tapping a hole that has a keyseat in it bridged by the helical flutes.

5. Thread-forming or fluteless taps do not produce chips because they don't cut threads. Their action can be compared to thread rolling in that material is being displaced in grooves to form ridges shaped in the precise form of a thread.

6. Taper pipe taps are identified by the taper of the body of the tap, which is $\frac{3}{4}$ in. per foot of length; also by the size marked on the shank.

7. When an Acme thread is cut, the tap is required to cut too much material in one pass. To obtain a quality thread, a roughing pass and then a finishing pass are needed.

8. Rake angles vary on tools depending on the kind of work machined. In general, softer, more ductile materials require larger rake angles than do harder, less ductile materials.

9. Friction is reduced by back tapering, eccentric or coneccentric relief on the pitch diameter, concave groove land relief, making the tap in interrupted design, and by various surface treatments such as oxides and flash chrome plating.

Section B / Unit 7 / Tapping Procedures

SELF-TEST ANSWERS

1. Taps are driven with tap wrenches or T-handle tap wrenches.

2. A hand tapper is a fixture used to hold a tap in precise alignment while hand tapping holes.

3. A tapping attachment is used when tapping holes in a machine or for production tapping.

4. The strength of a tapped hole is determined by the kind of material being tapped, the percentage of thread used, and the length or depth of thread engagement.

5. Holes should be tapped deep enough to provide 1 to $1\frac{1}{2}$ times the tap diameter of usable thread.

6. Tap drilled holes should be reamed when close control over the percentage of thread produced is necessary and when fine pitches of thread are produced, because a small change in tap drilled hole size would mean a large change in the percentage of thread cut.

7. Taps break because holes are drilled too shallow, chips are packed tight in the flutes, hard materials or hard spots are encountered, inadequate or the wrong kind of lubricant is used, or the cutting speed used is too great.

8. Tapped holes that are rough and torn are often caused by dull taps, chips clogging and flutes, insufficient lubrication, wrong kind of lubrication, or already rough holes being tapped.

9. Oversize tapped holes can be caused by a loose machine spindle or a worn tap holder, misaligned spindle, oversized tap, a dull tap, chips packed in flutes, and buildup on cutting edges.

10. Broken taps can be removed by drilling them out after annealing the tap, or by using an electrical discharge machine to erode the tap.

Section B / Unit 8 / Thread-Cutting Dies and Their Uses

SELF-TEST ANSWERS

1. A die is used to cut external threads.

2. A diestock is used to hold the die when hand threading. A special die holder is used when machine threading.

3. The size of thread cut can only be changed a very small amount on round adjustable dies. Too much expansion or contraction may break the die.

4. The purpose of the guide is to align the die square to the workpiece to be threaded.

5. When assembling a two-piece die collet, be sure both die halves are marked with the same serial number and that the starting chamfers on the dies are toward the guide.

6. Hexagon rethreading dies are used to clean and recut slightly damaged or rusty threads. Only in emergencies should they be used to cut new threads.

7. The chamfer on the cutting end of a die distributes the cutting force over a number of threads and aids in starting the thread-cutting operation.

8. Cutting fluids are very important in threading to achieve threads with a good surface finish and close tolerance, and to give long tool life.

9. Before a rod is threaded, it should be measured to assure its size is no longer than its nominal size. Preferably, it is .002 to .005 in. undersized.

10. The chamfer on a rod before threading makes it easy to start a die. It also protects the starting thread on a finished bolt.

Section B / Unit 9 / Off-Hand Grinding

SELF-TEST ANSWERS

1. The sharpening of tool bits and drills.

2. The wheels tend to become misshapen and out-of-round.

3. A Desmond dresser.

4. Overheating, which ruins the tool.

5. Safety factors are: wear eye protection; do not grind nonferrous metals on an aluminum oxide wheel; keep workrest close to wheel; keep wheel guards in place; let newly installed wheel run idle for 1 minute; keep wheel dressed.

Section C / Unit 1 / Systems of Measurement

SELF-TEST ANSWERS

1. To find in. knowing mm, multiply mm by .03937: 35 × .03937 = 1.377 in.

2. To find mm knowing in., multiply in. by 25.4: .125 × 25.4 = 3.17 mm.

3. To find cm knowing in., multiply in. by 2.54: 6.273 × 2.54 = 15.933 cm.

4. To find mm knowing in., multiply in. by 25.4: .050 × 25.4 = 1.27 mm.

5. 10 mm = 1 cm; therefore, to find cm knowing mm, divide mm by 10.

6. To find in. knowing mm, multiply mm by .03937: .02 × .03937 = .0008 in. The tolerance would be ± .0008 in.

7. SI refers to the International System of Units.

8. Conversions between metric and inch systems can be accomplished by mathematical procedures, conversion charts, and direct-converting calculators.

9. The yard is presently defined in terms of the meter: 1 yard = 3600/3937 meter.

10. Yes, by the use of appropriate conversion dials or electronic digital readouts.

Section C / Unit 2 / Using Steel Rules

SELF-TEST ANSWERS

Fractional Inch Rules (Figure C-74)

$A = 1\frac{1}{4}$ in.
$B = 2\frac{1}{8}$ in.
$C = \frac{15}{16}$ in.
$D = 2\frac{5}{16}$ in.
$E = \frac{15}{32}$ in.
$F = 2\frac{25}{32}$ in.
$G = \frac{63}{64}$ in.
$H = 1\frac{59}{64}$ in.

Decimal Inch Rules (Figure C-75)

A = .300 in.
B = .510 in.
C = 1.020 in.
D = 1.200 in.
E = 1.260 in.

Metric Rules (Figure C-76)

A = 11 mm or 1.1 cm
B = 27 mm or 2.7 cm
C = 52 mm or 5.2 cm
D = 7.5 mm or .75 cm
E = 20.5 mm or 2.05 cm
F = 45.5 mm or 4.55 cm

Section C / Unit 3 / Using Vernier, Dial, and Digital Instruments for Direct Measurements

SELF-TEST ANSWERS

Reading Inch Vernier Calipers

Figure C-88*a*: 1.304 in.
Figure C-88*b*: .724 in.

Reading Metric Vernier Calipers

Figure C-89*a*: 20.26 mm
Figure C-89*b*: 35.62 mm

Reading Inch Vernier Depth Gages

Figure C-90*a*: .943 in.
Figure C-90*b*: 1.326 in.

Section C / Unit 4 / Using Micrometer Instruments

SELF-TEST ANSWERS

1. Anyone taking pride in his or her tools usually takes pride in his or her workmanship. The quality of a product produced depends to a large extent on the accuracy of the measuring tools used. Skilled workers protect their tools because they guarantee their products.

2. Moisture between the contact faces can cause corrosion.

3. Even small dust particles will change a dimension. Oil or grease attracts small chips and dirt. All of these can cause incorrect readings.

4. A measuring tool is no more discriminatory than the smallest division marked on it. This means that a standard micrometer can discriminate to the nearest thousandth. A vernier scale on a micrometer will make it possible to discriminate a reading to one ten-thousandth of an inch under controlled conditions.

5. The reliability of a micrometer depends on the inherent qualities built into it by its maker. Reliability also depends on the skill of the user and the care the tool receives.

6. The sleeve is stationary in relation to the frame and is engraved with the main scale, which is divided into 40 equal spaces, each equal to .025 in. The thimble is attached to the spindle and rotates with it. The thimble circumference is graduated with 25 equal divisions, each representing a value of .001 in.

7. There is less chance of accidentally moving the thimble when reading a micrometer while it is still in contact with the workpiece.

8. Measurements should be made at least twice. On critical measurements, checking the dimensions additional times will assure that the size measurement is correct.

9. As the temperature of a part is increased, the size of the part will increase. When a part is heated by the machining process, it should be permitted to cool down to room temperature before being measured. Holding a micrometer by the frame for an extended period of time will transfer body heat through the hand and affect the accuracy of the measurement taken.

10. The purpose of the ratchet stop or friction thimble is to enable equal pressure to be repeatedly applied between the measuring faces and the object being measured. Use of the ratchet stop or friction thimble will minimize individual differences in measuring pressure applied by different persons using the same micrometer.

Outside Micrometer Readings		**Depth Micrometer Readings**	
Figure C-134*a*:	.669 in.	Figure C-147*a*:	.535 in.
Figure C-134*b*:	.787 in.	Figure C-147*b*:	.815 in.
Figure C-134*c*:	.237 in.	Figure C-147*c*:	.732 in.
Figure C-134*d*:	.994 in.	Figure C-147*d*:	.535 in.
Figure C-134*e*:	.072 in.	Figure C-147*e*:	.647 in.

Vernier Micrometer Readings		**Inside Micrometer Readings**	
Figure C-154*a*:	.3749 in.	Figure C-141*a*:	1.617
Figure C-154*b*:	.5377 in.	Figure C-141*b*:	2.000
Figure C-154*c*:	.3123 in.	Figure C-141*c*:	2.254
Figure C-154*d*:	.1255 in.	Figure C-141*d*:	2.562
Figure C-154*e*:	.2498 in.	Figure C-141*e*:	2.784

Metric Micrometer Readings	
Figure C-152*a*:	21.21 mm
Figure C-152*b*:	13.27 mm
Figure C-152*c*:	9.94 mm
Figure C-152*d*:	5.59 mm
Figure C-152*e*:	4.08 mm

Section C / Unit 5 / Using Comparison Measuring Instruments

SELF-TEST ANSWERS

1. Comparison measurement is measurement where an unknown dimension is compared to a know dimension. This often involves a transfer device that represents the unknown and is then transferred to the known, where the reading can be determined.

2. Most comparison instruments do not have the capability to show measurement directly.

3. Cosine error is error incurred when misalignment exists between the axis of measurement and the axis of the measuring instrument.

4. Cosine error can be reduced by making sure that the axis of the measuring instrument is exactly in line with the axis of measurement.

5. Adjustable parallel: (c).

6. Dial test indicator in conjunction with a height transfer micrometer. Height transfer measurements can also be accomplished with the test indicator and planer gage: (f) and (n).

7. Opitcal comparator: (p).

8. A combination square can be checked against a precision square or, if the actual amount of deviation is required, the cylindrical square or micrometer square can be used: (i), (j), and (l).

9. Telescoping gage and outside micrometer: (b).

10. Thickness gage or, in the case of a thick shim or chock, the adjustable parallel: (c) and (e).

Section C / Unit 6 / Using Gage Blocks

SELF-TEST ANSWERS

1. The wringing interval in the space or interface between wrung gage blocks.

2. Wear blocks are made from very hard material such as tungsten carbide. Wear blocks are used in applications where direct contact with gage blocks might damage them.

3. If gage blocks should become heated or cooled above or below room temperature, normalizing is the process of returning them to room temperature.

4. Grade 1: ± .000002 in.
Grade 2: + .000004 in.
− .000002 in.
Grade 3: + .000008 in.
− .000002 in.

5. The conditioning stone is a highly finished piece of granite or ceramic material and is used to remove burrs from the wringing surface of a gage block.

6. A microinch is 1 millionth of an inch. On surface finish it refers to the deviation of a surface from a uniform plane.

7. Gage block accuracy depends on the following factors:

Extreme cleanliness.
No burrs.
Minimum use of the conditioning stone.
Leaving stacks assembled only for minimum amounts of time.
Cleaning before storage.
Application of gage block preservative.

8. 3.0213
.1003
2.9210
.121
2.800
.800
2.000
2.000
0.0000

9. 1.9643
.100 (wear blocks 2 × .050)
1.8643
.1003
1.7640
.114
1.6500
.650
1.0000
1.000
0.000

10. Gage blocks can be used to check other measuring instruments, to set sine bars for angles, as precision height gages for layout, in direct gaging applications, and for setting machine and cutting tool positions.

Section C / Unit 7 / Using Angular Measuring Instruments

SELF-TEST ANSWERS

1. Plate protractor and machinist's combination set bevel protractor.

2. Five minutes of arc.

3. The sine bar becomes the hypotenuse of a right triangle. Angles are measured or established by elevating the bar a specified amount or calculating the amount of bar elevation, knowing the angle.

4. Figure C-252*a*: 50°
Figure C-252*b*: 96° 15′
Figure C-252*c*: 34° 30′
Figure C-252*d*: 61° 45′
Figure C-252*e*: 56° 25′

5. Bar elevation = bar length × sine of angle desired

= 5 in. × sin 37°
= 5 × .6018
= 3.0090 in.

6. Sine of the angle desired = elevation/bar length

= 2.750/5
Sine of angle = .550
Angle = 33° 22′

Section C / Unit 8 / Tolerances and Fits

SELF-TEST ANSWERS

1. Tolerances are important because they control the size and therefore the ability of parts to fit together in complex assemblies.

2.

Fractional dimensions	$\pm \frac{1}{64}$ in.
Two-place decimals	± .010 in.
Three-place decimals	± .005 in.
Four-place decimals	± .0005 in.
Angles	$\pm \frac{1}{2}$ degree

3. Straightness, roundness, flatness, perpendicularity, parallelism, concentricity, runout.

4. Typical press fit allowance is calculated by the formula .0015 in. × the diameter of the part.

5. Shrink and expansion fits are accomplished by cooling (shrinking) or heating (expanding) the parts to be fitted together. After fitting in the shrunk or expanded state, the parts will securely hold together upon cooling or warming to ambient temperature.

Section D / Unit 1 / Selection and Identification of Steels

SELF-TEST ANSWERS

1. Carbon and alloy steels are designated by the numerical SAE or AISI system.

2. The three basic types of stainless steels are: martensitic (hardenable) and ferritic (nonhardenable), both magnetic and of the 400 series, and austenitic (nonmagnetic and nonhardenable, except by work hardening) of the 300 series.

3. The identification for each piece would be as follows:

a. AISI C1020 CF is a soft, low-carbon steel with a dull metallic luster surface finish. Use the observation test, spark test, and file test for hardness.

b. AISI B1140 (G and P) is a medium-carbon, resulfurized, free-machining steel with a shiny finish. Use the observation test, spark test, and machinability test.

c. AISI C4140 (G and P) is a chromium–molybdenum alloy of medium carbon content with a polished, shiny finish. Since an alloy steel would be harder than a similar carbon- or low-carbon-content steel, a hardness test should be used such as the file or scratch test to compare with known samples. The machinability test would be useful as a comparison test.

d. AISI 8620 HR is a tough low-carbon steel used for carburizing purposes. A hardness test and a machinability test will immediately show the difference from low-carbon hot-rolled steel.

e. AISI B1140 (ebony) is the same as the resulfurized steel in part b, but the finish is different. The test would be the same as for part b.

f. AISI D1040 is a medium-carbon steel. The spark test would be useful here as well as the hardness and machinability tests.

4. A magnetic test can quickly determine whether it is a ferrous metal or perhaps nickel. If the metal is white in color, a spark test will be needed to determine whether it is a nickel casting or one of white cast iron, since they are similar in appearance. If a small piece can be broken off, the fracture will show whether it is white or gray cast iron. Gray cast iron will leave a black smudge on the finger. If it is cast steel, it will be more ductile than cast iron and a spark test should reveal a smaller carbon content.

5. O1 refers to an alloy-type oil-hardening (oil-quench) tool steel. W1 refers to a water-hardening (water-quench) tool steel.

6. The 40-in.-long, $2\frac{7}{16}$ in.-diameter shaft weighs 1.322 lb/in. The cost is $.30/lb; $1.322 \times 40 \times .30 = \$15.86$ cost of the shaft.

7. a. No.
b. Hardened tool steel or case-hardened steel.

8. Austenitic (having a face-centered cubic unit cell in its lattice structure). Examples are chromium, nickel, stainless steel, and high-manganese alloy steel.

9. Nickel is a nonferrous metal that has magnetic properties. Some alloy combinations of nonferrous metals make strong permanent magnets: for example, the well-known Alnico magnet, an alloy of aluminum, nickel, and cobalt.

10. Some properties of steel to be kept in mind when ordering or planning for a job would be:

Strength.
Machinability.
Hardenability.
Weldability (if welding is involved).
Fatigue resistance.
Corrosion resistance (especially if the piece is to be exposed to a corrosive atmosphere).

Section D / Unit 2 / Selection and Identification of Nonferrous Metals

SELF-TEST ANSWERS

1. Since aluminum is about one-third lighter than steel, it is used extensively in aircraft. It also forms an oxide on the surface that resists further corrosion. The initial cost is much greater. Some higher-strength aluminum alloys cannot be welded.

2. The letter "H" following the four-digit number always designates strain or work hardening. The letter "T" refers to heat treatment.

3. Magnesium weighs approximately one-third less than aluminum and is approximately one-quarter the weight of steel. Magnesium will burn in air when finely divided.

4. Copper is most extensively used in the electrical industries because of its low resistance to the passage of current when it is unalloyed with other metals. Copper can be strain hardened or work hardened and certain alloys may be hardened by a solution heat-treat and aging process.

5. Bronze is basically copper and tin. Brass is basically copper and zinc.

6. Nickel is used to electroplate surfaces of metals for corrosion resistance, and as an alloying element with steels and nonferrous metals.

7. All three resist deterioration from corrosion.

8. Alloy.

9. Tin, lead, and cadmium.

10. Die-cast metals, sometimes called "pot metal."

Section D / Unit 3 / Hardening, Case Hardening, and Tempering

SELF-TEST ANSWERS

1. No hardening would result as 1200°F (649°C) is less than the lower critical point and no dissolving of carbon has taken place.

2. There would be almost no change. For all practical purposes in the shop these low-carbon steels are not considered hardenable.

3. They are shallow hardening, and liable to distortion and quench cracking because of the severity of the water quench.

4. Air and oil hardening steels are not so subject to distortion and cracking as W1 steels, and they are deep hardening.

5. 1450°F (788°C). 50°F (10°C) above the upper critical limit.

6. Tempering is done to remove the internal stresses in martensite, which is very brittle. The temperature used gives the best compromise between hardness and toughness or ductility.

7. Tempering temperature should be specified according to the hardness, strength, and ductility desired. Mechanical properties charts give these data.

8. 525°F (274°C). Purple.

9. 600°F (315°C). It would be too soft for any cutting tool.

10. Immediately. If you let it set for any length of time, it may crack from internal stresses.

11. The low-carbon steel core does not harden when quenched from 1650° (899°C), so it remains soft and tough, but the case becomes very hard. No tempering is therefore required as the piece is not brittle all the way through as a fully hardened carbon steel would be.

12. A deep case can be made by pack carburizing or by a liquid bath carburizing. A relatively deep case is often applied by nitriding or by similar procedures.

13. No. The base material must contain sufficient carbon to harden by itself without adding more for surface hardening.

14. Three methods of introducing carbon into heated steel are roll, pack, and liquid carburizing.

15. Nitriding.

16. Electric-, gas-, and oil-fired furnaces, and pot furnaces.

17. The surface decarburizes or loses surface carbon to the atmosphere as it combines with oxygen to form carbon dioxide.

18. Dispersion of carbon atoms in the solid solution of austenite may be incomplete and little or no hardening in the quench takes place as a result. Also, the center of a thick section takes more time to come to the austenitizing temperature.

19. Circulation or agitation breaks down the vapor barrier. This action allows the quench to proceed at a more rapid rate.

20. By furnace.

21. They run from the surface toward the center of the piece. The fractured surfaces usually appear blackened. The surfaces have a fine crystalline structure.

22. Overheating.
Wrong quench.
Wrong selection of steel.
Poor design.
Time delays between quench and tempering.
Wrong angle into the quench.
Not enough material to grind off decarburization.

23. Controlled atmosphere furnace.
Wrapping the piece in stainless steel foil.
Covering with cast iron chips.

24. Changes in hardness of the surface area and the development of high internal stresses during grinding.

25. An air-hardening tool steel should be used when distortion must be kept to a minimum.

Section D / Unit 4 / Annealing, Normalizing, and Stress Relieving

SELF-TEST ANSWERS

1. Medium-carbon steels that are not uniform and have hardened areas from welding or prior heat treating need to be normalized so they can be machined. Forgings, castings, and tool steel in the as-rolled condition are normalized before any further heat treatments or machining is done.

2. 1550°F (843°C). 50°F (10°C) above the upper critical limit.

3. The spheroidization temperature is quite close to the lower critical temperature line, about 1300°F (704°C).

4. The full anneal brings carbon steel to its softest condition as all the grains are reformed (recrystallized), and any hard carbide structures become soft pearlite as it slowly cools. Stress relieving will only recrystallize distorted ferrite grains and not the hard carbide structures or pearlite grains.

5. Stress relieving should be used on severely cold-worked steels or for weldments.

6. High-carbon steels (.8 to 1.7 percent C).

7. Process annealing is used by the sheet and wire industry and is essentially the same as stress relieving.

8. In still air.

9. Very slowly. Packed in insulating material or cooled in a furnace.

10. Low-carbon steels tend to become gummy when spheroidized, so the machinability is worse than in the as-rolled condition. Spheroidization sometimes is desirable when stress relieving weldments on low-carbon steels.

Section D / Unit 5 / Rockwell and Brinell Hardness Testers

SELF-TEST ANSWERS

1. Resistance to penetration is the one category that is utilized by the Rockwell and Brinell testers. The depth of penetration is measured when the major load is removed on the Rockwell tester and the diameter of the impression is measured to determine a Brinell hardness number.

2. As the hardness of a metal increases, the strength increases.

3. The A scale and a Brale marked "A" with a major load of 60 kgf should be used to test a tungsten carbide block.

4. It would become deformed or flattened and give an incorrect reading.

5. No. The Brale used with the Rockwell superficial tester is always marked or prefixed with the letter "N."

6. False. The ball penetrator is the same for all the scales using the same diameter ball.

7. The diamond spot anvil is used for superficial testing on the Rockwell tester. When used, it does not become indented, as is the case when using the spot anvil.

8. Roughness will give less accurate results than would a smooth surface.

9. The surface "skin" would be softer than the interior of the decarburized part.

10. A curved surface will give inaccurate readings.

11. A 3000-kg weight would be used to test specimens on the Brinell tester.

12. A 10-mm steel ball is usually used on the Brinell tester.

Section E / Unit 1 / Basic Semiprecision Layout Practice

SELF-TEST ANSWERS

1. The workpiece should have all sharp edges removed by grinding or filing. A thin, even coat of layout dye should be applied.

2. The towel will prevent spilling layout dye on the layout plate or layout table.

3. The punch should be tilted so that it is easier to see when the point is located on the scribe mark. It should then be moved to the upright position before it is tapped with the layout hammer.

4. The combination square can be positioned on the rule for measurements. The square head acts as a positive reference point for measurments. The rule may be removed and used as a straight edge for scribing.

5. The divider should be adjusted until you feel the tip drop into rule engraving.

Section E / Unit 2 / Basic Precision Layout Practice

SELF-TEST ANSWERS

1. Figure E-83a: Inch—5.030 in.; Metric—127.76mm
Figure E-83*b*: Inch—8.694 in.; Metric—220.82 mm
Figure E-83*c*: Inch—5.917 in.; Metric—150.30 mm

2. Figure E-84*a*: Inch—4.086 in.
Figure E-84*b*: Inch—1.456 in.

3. Zero reference is checked by bringing the scriber to rest on the reference surface and then checking the alignment of the beam and vernier zero lines.

4. The position of the vernier scale may be adjusted on height gages with this feature.

5. By turning the workpiece 90 degrees.

6. 10 to 72 in.

7. The sine bar.

Section F / Unit 1 / Machinability and Chip Formation

SELF-TEST ANSWERS

1. At lower speeds, negative rake tools usually produce a poorer surface finish than do positive rakes.

2. It is called a built-up edge and causes a rough, ragged cutting action that produces a poor surface finish.

3. Thin uniform chips indicate the least surface disruption.

4. No. It slides ahead of the tool on a shear plane, elongating and altering the grain structure of the metal.

5. Negative rake.

6. Higher cutting speeds produce better surface finishes. There also is less disturbance and disruption of the grain structure at higher speeds.

7. Surface irregularities such as scratches, tool marks, microcracks, and poor radii can shorten the working life of the part considerably by causing stress concentration that can develop into a metal fatigue failure.

8. The property of hardness is related to machinability. Machinists sometimes use a file to determine hardness.

9. The "9"-shaped chip.

10. Machinability is the relative difficulty of a machining operation with regard to tool life, surface finish, and power consumption.

Section F / Unit 2 / Speeds and Feeds for Machine Tools

SELF-TEST ANSWERS

1. The rpm for the $\frac{1}{8}$ in. twist drill should be 2880, and for the $\frac{3}{4}$-in. drill, 480.

2. They would be the same.

3. Use the next-lower speed setting or the highest setting on the machine if the calculated speed is higher than the top speed of the machine.

4. 100 rpm.

5. The rpm should be 3000.

6. Inches per revolution (ipr).

7. .010 in. per revolution.

8. Milling machines, both vertical and horizontal.

9. About $\frac{1}{4}$ in. per table stroke.

Section F / Unit 3 / Cutting Fluids

SELF-TEST ANSWERS

1. Cooling and lubrication.

2. Synthetic and semisynthetic fluids, emulsions, and cutting oils.

3. Remove the cutting fluid, clean the tank, and replace with new cutting fluid.

4. Because the tramp oil will contaminate the soluble oil-water mix.

5. They tend to remove skin oils. The "oilier" types do not cause this problem.

6. Using a pump oiler can containing the appropriate cutting fluid.

7. A coolant tank, pump, hose, and nozzle.

8. Flooding the workpiece–tool area with cutting fluid.

9. Because the rapidly spinning grinding wheel tends to blow it away.

10. Because only a very small amount of liquid which contains the fatty particles ever reaches the cutting area.

Section F / Unit 4 / Using Carbides and Other Tool Materials

SELF-TEST ANSWERS

1. Tungsten carbide and cobalt.

2. Decreased hardness and increased toughness.

3. Edge or flank wear is considered normal wear.

4. Yes, if it is set at 90 degrees to the axis of the work.

5. No. The cutting edge engagement length is greater than the depth of cut.

6. Antiweld, anticratering for machining steels.

7. Higher red hardness.

8. Increasing the nose radius will give good finishes even with an increased feed rate.

9. Tools are stronger.

10. Chatter may develop between tool and work.

11. Relief is ground just below the cutting edges of the carbide and clearance is ground primarily on the shank of the tool.

12. Aluminum oxide.

13. When considerable wear is evident on carbide tools. Equipment and setups must be very rigid for ceramics.

14. C-2.

15. C-5.

16. 44-A.

17. C-7.

18. The stock removal rate is low but very high finishes are obtained with these tools.

19. Very high tool life on some abrasive, difficult-to-machine materials.

20. Only silicon carbide (or diamond) wheels can be used for grinding carbide tools. The shank is made of steel, however, and clearance may be ground first on an aluminum oxide wheel.

Section G / Unit 1 / Using Reciprocating and Horizontal Band Cutoff Machines

SELF-TEST ANSWERS

1. Raker, wave, and straight.
2. Workpiece material, cross-section shape, and thickness.
3. On the back stroke.
4. The tooth offset on either side of the blade. Set provides clearance for the back of the blade.
5. Standard skip and hook.
6. Cutoff material can bind the blade and destroy the set.
7. The horizontal band saw.
8. Cooling, lubrication, and chip removal.
9. Scoring and possible blade breakage.
10. The workpiece must be turned over and a new cut started.

Section G / Unit 2 / Abrasive and Cold Saws

SELF-TEST ANSWERS

1. Probably not.
2. Fast cutting and they can be used to cut nonmetallic materials.
3. Aluminum oxide, silicon carbide, and diamond.
4. Shellac, resinoid, and rubber.
5. Length tolerance of cutoff stock can be held very close.

Section G / Unit 3 / Preparing to Use the Vertical Band Machine

SELF-TEST ANSWERS

1. The ends of the blade should be ground with the teeth opposed. This will ensure that the ends of the blade are square.
2. The blade ends are placed in the welder with the teeth pointed in. The ends must contact squarely in the gap between the jaws. The welder must be adjusted for the band width to be welded. You should wear eye protection and stand to one side during the welding operation. The weld will occur when the weld lever is depressed.
3. The weld is ground on the grinding wheel attached to the welder. Grind the weld on both sides of the band until the band fits the thickness gage. Be careful not to grind the saw teeth.
4. The guides support the band. This is essential to straight cutting.
5. Band guides must fully support the band except for the teeth. A wide guide used on a narrow band will destroy the saw set as soon as the machine is started.
6. The guide setting gage is used to adjust the band guides.
7. Annealing is the process of softening the band weld in order to improve strength qualities.
8. The band should be clamped in the annealing jaws with the teeth pointed out. A small amount of compression should be placed on the movable welder jaw prior to clamping the band. The correct annealing color is dull red. As soon as this color is reached, the anneal switch should be released and then operated briefly several times to slow the cooling rate of the weld.
9. Band tracking is the position of the band as it runs on the idler wheels.
10. Band tracking is adjusted by tilting the idler wheels until the band just touches the backup bearing.

Section G / Unit 4 / Using the Vertical Band Machine

SELF-TEST ANSWERS

1. The three sets are straight, wave, and raker. Straight set may be used for thin material, wave for material with a variable cross section, and raker for general-purpose sawing.
2. Scalloped and wavy edged bands might be used on nonmetallic material where blade teeth would tear the material being cut.
3. Band velocity is measured in feet per minute.
4. The variable speed pulley is designed so that the pulley flanges may be moved toward and away from each other. This permits the belt position to be varied, resulting in speed changes.
5. The job selector provides information about recommended saw velocity, saw pitch, power feed, saw set, and temper. Band filing information is also indicated.
6. Speed range is selected by shifting the transmission.
7. Speed range shift must be done with the band speed set at the lowest setting.

8. The upper guidepost must be adjusted so that it is as close to the workpiece as possible.

9. Band pitch must be correct for the thickness of material to be cut. Generally, a fine pitch will be used on thin material. Cutting a thick workpiece with a fine pitch band will clog saw teeth and reduce cutting efficiency.

10. Band set must be adequate for the thickness of the blade used in a contour cut. If set is insufficient, the blade may not be able to cut the desired radius.

Section H / Unit 1 / The Drill Press

SELF-TEST ANSWERS

1. The sensitive, upright, and radial-arm drill presses are the three basic types. The sensitive drill press is made for light-duty work and it provides the operator with a sense of "feel" of the feed on the drill. The upright is a similar, but heavy-duty, drill press equipped with power feed. The radial-arm drill press allows the operator to position the drill over the work where it is needed, rather than to position the work under the drill as with other drill presses.

The sensitive, upright, and radial-arm drill presses all perform much the same functions of drilling, reaming, counterboring, countersinking, spot facing, and tapping, but the upright and radial machines do heavier and larger jobs. The radial-arm drill can support large, heavy castings and work can be done on them without the workpiece being moved.

2. Sensitive drill press:

Spindle, *A*	Base, *F*
Quill lock handle, *G*	Power feed, *C*
Column, *E*	Motor, *J*
Switch, *L*	Variable-speed control, *D*
Depth stop, *B*	Table lift crank, *K*
Head, *N*	Quill return spring, *I*
Table, *O*	Guard, *M*
Table lock, *H*	

3. Radial drill press:

Column, *B*	Base, *D*
Radial arm, *C*	Drill head, *E*
Spindle, *A*	

Section H / Unit 2 / Drilling Tools

SELF-TEST ANSWERS

1.

Web, *T*	Chisel edge angle, *B*
Margin, *U*	Body clearance, *J*
Drill point angle, *D*	Helix angle, *I*
Cutting lip, *P*	Axis of drill, *Y*
Flute, *K*	Shank length, *N*
Body, *O*	Tang, *C*
Lip relief angle, *E*	Taper shank, *X*
Land, *G*	Straight shank, *W*

2.

	Decimal Diameter	Fractional Size	Number Size	Letter Size	Metric Size
a	.0781	$\frac{5}{64}$			
b	.1495		25		
c	.272			I	
d	.159		21		
e	.1969				5
f	.323			P	
g	.3125	$\frac{5}{16}$			
h	.4375	$\frac{7}{16}$			
i	.201		7		
j	.1875	$\frac{3}{16}$			

Section H / Unit 4 / Operating Drilling Machines

SELF-TEST ANSWERS

1. The three considerations would be speeds, feeds, and coolants.

2. The rpm rates of the drills would be:
$\frac{1}{4}$ in. diameter: 1440 rpm
2 in. diameter: 180 rpm
$\frac{3}{4}$ in. diameter: 480 rpm
$\frac{3}{8}$ in. diameter: 960 rpm
$1\frac{1}{2}$ in. diameter: 240 rpm

3. Worn margins and outer corners broken down. The drill can be ground back to its full size and resharpened.

4. The operator will increase the feed in order to produce a chip. This increased feed is often greater than the drill can stand without breaking.

5. The feed is about right when the chip rolls into a close helix. Long, stringy chips can indicate too much feed.

6. Feeds are designated by a small measured advance movement of the drill for each revolution. A .001-in. feed for a $\frac{1}{8}$-in.-diameter drill, for example, would move the drill .001 in. into the work for every turn of the drill.

7. The water-soluble oil types and the cutting oils, both animal and mineral.

8. Besides having the correct cutting speed, a sulfurized oil-based cutting fluid helps to reduce friction and cool the cutting edge.

9. Drill jamming can be avoided by a "pecking" procedure. The operator drills a small amount and pulls out the drill to remove the chips. This is repeated until the hole is finished.

10. The depth stop is used to limit the travel of the drill so that it will not go into the table or vise. The depth of blind holes is preset and drilled. Countersink and counterbore depths are set so that several can be easily made the same.

11. The purpose of using work-holding devices is to keep the workpiece rigid, to prevent it from turning with the drill, and for operator safety.

12. Included in a list of work-holding devices would be strap clamps, T-bolts, and step blocks. Also used are C-clamps, vee blocks, vises, jigs, and fixtures, and angle plates.

13. Parallels are mostly used to raise workpieces off the drill press table or to lift a workpiece higher in a vise, thus providing a space for the drill breakthrough. They are made of hardened steel so care should be exercised in their use.

14. Thin, limber materials tend to be sprung downward from the force of drilling until drill breakthrough begins. The drill then "grabs" as the material springs upward and a broken drill is often the result. This can be avoided by placing the support or parallels as near the drill as possible.

15. Angle drilling is done by tilting a drill press table (not all types tilt), or by using an angular vise. If no means of setting the exact angle is provided, a protractor head with level may be used.

16. Vee blocks are suited to hold round stock for drilling. The most frequent use of vee blocks is for cross-drilling holes in shafts, although many other setups are used.

17. The wiggler is used for locating a center punch mark under the center axis of a drill spindle.

18. Some odd-shaped workpieces, such as gears with extending hubs that need holes drilled for setscrews, might be difficult to set up without an angle plate.

19. One of the difficulties with hand tapping is the tendency for taps to start crooked or misaligned with the tap-drilled hole. Starting a tap by hand in a drill press with the same setup as used for the tap drilling assures a perfect alignment.

20. Since jigs and fixtures are mostly used for production manufacturing, small machine shops rarely have a use for them.

Section H / Unit 5 / Counter sinking and Counterboring

SELF-TEST ANSWERS

1. Countersinks are used to chamfer holes and to provide tapered holes for flat head fasteners such as screws and rivets.

2. Countersink angles vary to match the angles of different flat head fasteners or different taper hole requirements.

3. A center drill is used to make a 60-degree countersunk hole in workpieces for lathes and grinders.

4. A counterbore makes a cylindrical recess concentric with a smaller hole so that a hex head bolt or socket head capscrew can be flush mounted with the surface of a workpiece.

5. The pilot diameter should always be a few thousandths of an inch smaller than the hole, but not more than .005 in.

6. Lubrication of the pilot prevents metal-to-metal contact between it and the hole. It will also prevent the scoring of the hole surface.

7. A general rule is to use approximately one-third of the cutting speed when counterboring as when using a twist drill with the same diameter.

8. Feeds and speeds when counterboring are controlled to a large extent by the condition of the equipment, the available power, and the material being counterbored.

9. Spotfacing is performed with a counterbore. It makes a flat bearing surface, square with a hole to set a nut, washer, or bolt head.

10. Counterboring requires a rigid setup with the workpiece securely fastened and provisions made to allow the pilot to protrude below the bottom surface of the workpiece.

Section H / Unit 6 / Reaming in the Drill Press

SELF-TEST ANSWERS

1. Machine reamers are identified by the design of the shank—either a straight or tapered shank and usually a 45-degree chamfer on the cutting end.

2. A chucking reamer is a finishing reamer, the fluted part is cylindrical, and the lands are relieved. A rose reamer is a roughing reamer. It can remove a considerable amount of material. The body has a slight back taper and no relief on the lands. All cutting takes place on the chamfered end.

3. A jobber's reamer is a finishing reamer like a chucking reamer but it has a longer fluted body.

4. Shell reamers are more economical to produce than solid reamers, especially in larger sizes.

5. An accurate hole size cannot be obtained without a high-quality surface finish.

6. As a general rule the cutting speed used to ream a hole is about one-third to one-half of the speed used to drill a hole of the same size in the same material.

7. The feed rate, when reaming as compared to drilling the same material, is approximately two to three times as great. As an example, for a 1-in. drill the feed rate is about .010 to .015 in. per revolution. A 1-in. reamer would have a feed rate of between .020 and .030 in.

8. The reaming allowance for a $\frac{1}{2}$-in.-diameter hole would be $\frac{1}{64}$ in.

9. Cutting fluids cool the tool and workpiece and act as lubricants.

10. Chatter may be eliminated by reducing the speed, increasing the feed, or using a piloted reamer.

11. Oversized holes may be caused by a bent reamer shank or buildup on the cutting edges. Check also to see if there is a sufficient amount and if the correct kind of coolant is being used.

12. Bellmouthed holes are usually caused by a misaligned reamer and workpiece set up. Piloted reamers, bushings, or a floating holder may correct this problem.

13. Surface finish can be improved by decreasing the feed and checking the reaming allowance. Too much or not enough material will cause poor finish. Use a large volume of coolant.

14. Carbide tipped reamers are recommended for long production runs where highly abrasive materials are reamed.

15. Cemented carbides are very hard, but also very brittle. The slightest amount of chatter or vibration may chip the cutting edges.

Section I / Unit 1 / The Engine Lathe

SELF-TEST ANSWERS—PART B

1. Fine chips, filings, and grindings form an abrasive sludge that wears and scores the sliding surfaces. Frequent cleaning will help to prevent damage to the machine.

2. Heavier chips should be removed with a brush; never use an air jet. The ways should then be wiped clean with a cloth and lightly oiled.

3. Since most nicks come from dropping chucks and heavy workpieces on the lathe; a lathe board used every time a chuck is changed or heavy work is installed will prevent much of this. A tool board will help keep tools such as files from being laid across the ways.

4. Once daily.

5. No. The oil on the ways may have collected dirt or grit from the air to form an abrasive mixture. The ways should first be cleaned and oiled.

6. The chips should be cleaned from the lathe and swept up on the surrounding floor area. The lathe ways and slides should be wiped and oiled.

7. Straight gibs and tapered gibs.

8. The gib on the cross slide should be adjusted so it will have a slight drag, but the compound should be set up fairly tight when it is not being used.

Section I / Unit 2 / Toolholders and Tool Holding for the Lathe

SELF-TEST ANSWERS

1. A toolholder is needed to rigidly support and hold a cutting tool during the actual cutting operation. The cutting tool is often only a small piece of high speed steel or other cutting material that has to be clamped in a much larger toolholder in order to be usable.

2. On a left-hand toolholder, when viewed from above, the cutting tool end is bent to the right.

3. For high-speed tools the square tool bit hole is angled upward in relation to the base of the toolholder, where it is parallel to the base for carbide tools.

4. Turning close to the chuck is usually best accomplished with a left-handed toolholder.

5. Tool height adjustments on a standard-type toolholder are made by swiveling the rocker in the tool post ring.

6. Quick-change toolholders are adjusted for height with a micrometer collar.

7. Tool height on turret toolholders is adjusted by placing shims under the tool.

8. Toolholder overhang affects the rigidity of a setup; too much overhang may cause chatter.

9. The difference between a standard-type toolholder and a quick-change toolholder is in the speed with which tools can be interchanged. The tools are usually fastened more securely in a quick-change toolholder and height adjustments on a quick-change toolholder do not change the effective back rake angle of the cutting tool.

10. Drilling machine tools are used in a lathe tailstock.

11. The lathe tailstock is bored with a Morse taper hole to hold Morse taper shank tools.

12. When a series of repeat tailstock operations is to be performed on several different workpieces, a tailstock turret should be used.

Section I / Unit 3 / Cutting Tools for the Lathe

SELF-TEST ANSWERS

1. High-speed steel is easily shaped into the desired shape of cutting tool. It produces better finishes on low-speed machines and on soft metals.

2. Its geometrical form: the side and back rake, front and side relief angles, and chip breakers.

3. Unlike single-point tools, form tools produce their shape by plunging directly into the work.

4. When a "chip trap" is formed by improper grinding on a tool, the chip is not able to clear the tool; this prevents a smooth flow across the face of the tool. The result is tearing of the surface on the workpiece and possibly a broken tool.

5. Some toolholders provide a built-in back rake of about 16 degrees; to this is added any back rake on the tool to make a total back rake that is excessive.

6. A zero rake should be used for threading tools. A zero to slightly negative back rake should be used for plastics and brass, since they tend to "dig in."

7. The side relief allows the tool to feed into the work material. The end relief angle keeps the tool end from rubbing on the work.

8. The side rake directs the chip flow away from the cut and it also provides for a keen cutting edge. The back rake promotes smooth chip flow and good finishes.

9. The angles can be checked with a tool grinding gage, a protractor, or an optical comparator.

10. Long, stringy chips or those that become snarled on workpieces, tool post, chuck, or lathe dog are hazardous to the operator. Chip breakers and correct feeds can produce an ideal chip that does not fly off but will simply drop to the chip pan and is easily handled.

11. Chips can be broken up by using coarse feeds and maximum depth of cuts for roughing cuts and by using tools with chip breakers on them.

12. Overheating a tool causes small cracks to form on the edge. When a stress is applied, as in a roughing cut, the tool end may break off.

Section I / Unit 4 / Lathe Spindle Tooling

SELF-TEST ANSWERS

1. The lathe spindle is a hollow shaft that can have one of three mounting devices machined on the spindle nose. It has an internal Morse taper that will accommodate centers or collets.

2. The spindle nose types are the threaded, long taper key drive, and the camlock.

3. The independent chuck is a four-jaw chuck in which each jaw can move separately and independently of the others. It is used to hold odd-shaped workpieces.

4. The universal chuck is most often a three-jaw chuck, although these chucks are made with more or fewer jaws. Each jaw moves in or out by the same amount when the chuck wrench is turned. They are used to hold and quickly center round stock.

5. Combination chucks and Adjust-Tru three-jaw chucks.

6. A drive plate.

7. The live center is made of soft steel so it can be turned to true it up if necessary. It is made with a Morse taper to fit the spindle taper or special sleeve if needed.

8. Face drivers use a number of driving pins that dig into the work and are hydraulically compensated for irregularities.

9. Workpieces and fixtures are mounted on faceplates. These are identified by their heavy construction and the T-slots. Drive plates have only slots.

10. Collet chucks are very accurate work-holding devices. Spring collets are limited to smaller material and to specific sizes.

Section I / Unit 5 / Operating the Machine Controls

SELF-TEST ANSWERS

1. Very low speeds are made possible by disengaging the spindle by pulling out the lockpin and engaging the back gear.

2. The varidrive is changed while the motor is running, but the back gear lever is only shifted with the motor off.

3. Levers that are located on the headstock can be shifted in various arrangements to select speeds.

4. The feed reverse lever.

5. These levers are used for selecting feeds or threads per inch for threading.

6. The quick approach and return of the tool; this is used for delicate work and when approaching a shoulder or chuck jaw.

7. Since the crossfeed is geared differently (about one-third of the longitudinal feed), the outside diameter would have a coarser finish than the face.

8. The half-nut lever is used only for threading.

9. They are graduated in English units. Some metric conversion collars are being made and used that read in both English and metric units at the same time.

10. You can test with a rule and a given slide movement such as .125 or .250 in. If the slide moves one-half that distance, the lathe is calibrated for double depth and reads the same amount as that taken off the diameter.

Section I / Unit 6 / Facing and Center Drilling

SELF-TEST ANSWERS

1. A lathe board is placed on the ways under the chuck and the chuck is removed, since it is the wrong chuck to hold rectangular work. The mating parts of an independent chuck and the lathe spindle are cleaned and the chuck is mounted. The part is roughly centered in the jaws and adjusted to center by using the back of a tool holder or a dial indicator.

2. The tool should be on the center of the lathe axis.

3. No. The resultant facing feed would be approximately .003 to .005 in., which would be a finishing feed.

4. The compound must be swung to either 30 or 90 degrees so that the tool can be fed into the face of the work by a measured amount. A depth micrometer or a micrometer caliper can be used to check the trial finish cut.

5. A right-hand facing tool is used for shaft ends. It is different from a turning tool in that its point is only a 58-degree included angle to fit in the narrow space between the shaft face and the center.

6. $\text{rpm} = \dfrac{300 \times 4}{4} = 300.$

7. Center drilling is done to prepare work for turning between centers and for spotting workpieces for drilling in the lathe.

8. Center drills are broken as a result of feeding the drill too fast and having the lathe speed too slow. Breakage also can result from having the tailstock off-center, the work off-center in a steady rest, or a lack of cutting oil.

9. The sharp edge provides a poor bearing surface and soon wears out of round, causing machining problems such as chatter.

Section I / Unit 7 / Turning Between Centers

SELF-TEST ANSWERS

1. A shaft between centers can be turned end for end without loss of concentricity and it can be removed from the lathe and returned without loss of synchronization between thread and tool. Cutting off between centers is not done as it would break the parting tool and ruin the work. Steady rest work is not done with work mounted in a center in the headstock spindle.

2. The other method is one whereby the workpiece is held in a chuck on one end and in the tailstock center on the other end.

3. Coarser feeds, deeper cuts, and smaller rake angles all tend to increase chip curl. Chip breakers also make the chip curl.

4. Dead centers are hardened 60-degree centers that do not rotate with the work but require high-pressure lubricant. Ball bearing centers turn with work and do not require lubricant. Pipe centers turn with the work and are used to support tubular material.

5. With no end play in the workpiece and the bent tail of the lathe dog free to move in the slot.

6. Because of expansion of the workpiece from the heat of machining, it tightens on the center, thus causing more friction and more heat. This could ruin the center.

7. Excess overhang promotes lack of rigidity. This causes chatter and tool breakage.

8. $\text{Rpm} = \dfrac{90 \times 4}{1\frac{1}{2}} = 360 \div \dfrac{3}{2} = 360 \times \dfrac{2}{3} =$ 240 or $\dfrac{360}{1.5} = 240.$

9. The spacing would be .010 in. as the tool moves that amount for each revolution of the spindle.

10. The feed rate for roughing should be one-fifth to one-tenth as much as the depth of cut. This should be limited to what the tool, workpiece, or machine can stand without undue stress.

11. For most purposes where liberal tolerances are allowed, .015 to .030 in. can be left for finishing. When closer tolerances are required, two finish cuts are taken with .005 to .010 in. left for the last finish cut.

12. After roughing is completed, .015 to .030 in. is left for finishing. The diameter of the workpiece is checked with a micrometer and the remaining amount is dialed on the cross feed micrometer collar. A short trial cut is taken and the lathe is stopped. This diameter is again checked. If the diameter is within tolerance, the finish cut is taken.

Section I / Unit 8 / Alignment of the Lathe Centers

SELF-TEST ANSWERS

1. The workpiece becomes tapered.
2. The workpiece is tapered with the small end at the tailstock.
3. By the witness mark on the tailstock, by using a test bar, and by taking a light cut on a workpiece and measuring.
4. The dial indicator.
5. With a micrometer. The tailstock is set over with a dial indicator.

Section I / Unit 9 / Other Lathe Operations

SELF-TEST ANSWERS

1. Drilled holes are not sufficiently accurate for bores in machine parts as they would be loose on the shaft and would not run true.
2. The workpiece is center spotted with a center drill at the correct rpm and, if the hole is to be more than $\frac{3}{8}$ in. in diameter, a pilot drill is put through. Cutting fluid is used. The final size drill is put through at a slower speed.
3. The chief advantage of boring in the lathe is that the bore runs true with the centerline of the lathe and the outside of the workpiece, if the workpiece has been set up to run true (with no runout). This is not always possible when reaming bores that have been drilled, since the reamer follows the eccentricity or runout of the bore.
4. Ways to eliminate chatter in a boring bar are:
   Shorten the bar overhang, if possible.
   Reduce the spindle speed.
   Make sure the tool is on center.
   Use as large a diameter bar as possible without binding in the bore.
   Reduce the nose radius on the tool.
   Apply cutting fluid to the bore.
   Use tuned or solid carbide boring bars.
5. Through boring is making a bore the same diameter all the way through the part. Counterboring is making two or more diameters in the same bore, usually with 90-degree or square internal shoulders. Blind holes are bores that do not go all the way through.
6. Grooves and thread relief are made in bores by means of specially shaped or ground tools in a boring bar.
7. A floating reamer holder will help to eliminate the bell mouth, but it does not remove the runout.
8. Hand reamers produce a better finish than machine reaming.
9. Cutting speeds for reaming are one-half that used for drilling; feeds used for reaming are twice that used for drilling.
10. Large internal threads are produced with a boring tool. Heavy forces are needed to turn large hand taps, so it is not advisable to use large taps in a lathe.
11. A tap drill can be used as a reamer by first drilling with a drill that is $\frac{1}{32}$ to $\frac{1}{16}$ in. undersized. This procedure assures a more accurate hole size by drilling.
12. A spiral-point tap works best for power tapping.
13. The variations in pitch of the hand cut threads would cause the micrometer collar to give erroneous readings. A screw used for this purpose and for most machine parts must be threaded with a tool guided by the leadscrew on the lathe.
14. Thread relief and external grooves are produced by specially ground tools that are similar to internal grooving tools except that they have less end relief. Parting tools are often used for making external grooves.
15. Parting tools tend to seize in the work, especially with deep cuts or heavy feeds. Without cutting fluids seizing is almost sure to follow with the possibility of a broken parting tool and misaligned or damaged work.
16. You can avoid chatter when cutting off with a parting tool by maintaining a rigid setup and keeping enough feed to produce a continuous chip, if possible.
17. Knurling is used to improve the appearance of a part, to provide a good gripping surface, and to increase the diameter of a part for press fits.
18. Ordinary knurls make a straight or diamond pattern impression by displacing the metal with high pressures.
19. When knurls produce a double impression, they can be readjusted up or down and moved to a new position. Angling the toolholder 5 degrees may help.
20. You can avoid producing a flaking knurled surface by stopping the knurling process when the diamond points are almost sharp. Also, use a lubricant while knurling.

Section I / Unit 10 / Sixty-Degree Thread Information and Calculations

SELF-TEST ANSWERS

1. The sharp V thread can be easily damaged during handling if it is dropped or allowed to strike against a hard surface.

2. The pitch is the distance between a point on a screw thread to a corresponding point on the next thread measured parallel to the axis. "Threads per inch" is the number of threads in one inch.

3. American National Standard and Unified Standard threads both have the 60-degree included angle and are both based on the inch measure with similar pitch series. The depth of the thread and the classes of thread fits are different in the two systems.

4. To allow for tolerancing of external and internal threads to promote standardization and interchangeability of parts.

5. This describes a diameter of $\frac{1}{2}$ in., 20 threads per inch, and Unified coarse series external thread with a class 2 thread tolerance.

6. The flat on the end of the tool for 20 threads per inch should be P = .050 × .125 = .006 in. for American National threads and for Unified threads.

7. The compound at 30 degrees will move in .708/20 = .0354 in. for Unified threads.
The compound will move in .75/20 = .0375 in. for American National threads.

8. The fit of the thread refers to classes of fits and tolerances, while percent of thread refers to the actual minor diameter of an internal thread, a 100 percent thread being full depth internal threads.

9. The Système International (SI) thread and the British Standard ISO Metric Screw Threads are two metric thread systems in use.

Section I / Unit 11 / Cutting Unified External Threads

SELF-TEST ANSWERS

1. A series of cuts is made in the same groove with a single-point tool by keeping the same ratio and relative position of the tool on each pass. The quick-change gearbox allows choices of various pitches or leads.

2. The chips are less likely to bind and tear off when feeding in with the compound set at 29 degrees, and the tool is less likely to break.

3. The 60-degree angle on the tool is checked with a center gage or optical comparator.

4. The number of threads per inch can be checked with a screw pitch gage or by using a rule and counting the threads in one inch.

5. A center gage is used to align the tool to the work.

6. No. The carriage is moved by the thread on the leadscrew when the half-nuts are engaging it.

7. Even-numbered threads may be engaged on the half-nuts at any line and odd-numbered threads at any numbered line. It would be best to use the same line every time for fractional-numbered threads.

8. The spindle should be turning slowly enough for the operator to maintain control of the threading operation, usually about one-fourth turning speeds.

9. The leadscrew rotation is reversed, which causes the cut to be made from the left to the right. The compound is set at 29 degrees to the left. The threading tool and lathe settings are set up in the same way as for cutting right-hand threads.

10. Picking up the thread or resetting the tool is a procedure that is used to position a tool to existing threads.

Section I / Unit 12 / Cutting Unified Internal Threads

SELF-TEST ANSWERS

1. The minor diameter of the thread.

2. By varying the bore size, usually larger than the minor diameter. This is done to make tapping easier.

3. 75 percent.

4. By using a $\frac{1}{16}$-in. undersized drill and boring to the minor diameter.

5. Large internal threads of various forms can be made, and the threads are concentric to the axis of the work.

6. To the left of the operator.

7. A screw pitch gage should be used.

8. Boring bar and tool deflection cause the threads to be undersize from the calculations and settings on the micrometer collars.

9. The minor diameter equals D − (P × .541 × 2).
P = $\frac{1}{8}$ in. = .125 in.
D = 1 in. − (.125 × .541 × 2) = .8648 or .865 in.

10. A thread plug gage, a shop-made plug gage, or the mating part.

Section I / Unit 13 / Cutting Tapers

SELF-TEST ANSWERS

1. Steep tapers are quick-release tapers and slight tapers are self-holding tapers.

2. Tapers are expressed in taper per foot, taper per inch, and by angles.

3. Tapers are turned by hand feeding the compound slide, by offsetting the tailstock and turning between centers, or by using a taper attachment. A fourth method is to use a tool that is set to the desired angle and form cut the taper.

4. No. The angle on the workpiece would be the included angle, which is twice that on the compound setting. The angle on the compound swivel base is the angle with the work centerline.

5. The reading at the lathe centerline index would be 55 degrees, which is the complementary angle.

6. Offset $= \dfrac{10 \times (1.125 - .75)}{2 \times 3} = \dfrac{3.75}{6} =$ .625 in.

7. Four methods of measuring tapers are using the plug and ring gages; using a micrometer on layout lines; using a micrometer with precision parallels and drill rod on a surface plate; and using a sine bar, gage block, and a dial indicator.

8. The two types of taper attachments are the plain and the telescopic. Internal and external and slight to fairly steep tapers can be made. Centers remain in line, and power feed is used for good finishes.

9. The taper plug gage and the taper gage are the simplest and most practical means to check a taper.

10. Chamfers, V-grooves, and very short tapers may be made by the form tool method.

Section I / Unit 14 / Using Steady and Follower Rests

SELF-TEST ANSWERS

1. When workpieces extend from the chuck more than four or five workpiece diameters and are unsupported by a dead center; when workpieces are long and slender.

2. Since they are useful for supporting long workpieces, heavier cuts can be taken or operations such as turning, threading, and grooving may be performed without chattering. Internal operations such as boring may be done on long workpieces.

3. The steady rest is placed near the tailstock end of the shaft, which is supported in a dead center. The steady rest is clamped to the lathe bed and the lower jaws are adjusted to the shaft finger tight. The upper half of the frame is closed and the top jaw is adjusted with some clearance. The jaws are locked and lubricant is applied.

4. The jaws should be readjusted when the shaft heats up from friction in order to avoid scoring. Also, soft materials are sometimes used on the jaws to protect finishes.

5. A center punch mark is placed in the center of the end of the shaft. The lower two jaws on the steady rest are adjusted until the center punch mark aligns with the point of the dead center.

6. No.

7. No. When the surface is rough, a bearing spot must be turned for the steady rest jaws.

8. By using a cat head.

9. A follower rest.

10. The shaft is purposely made 1 or 2 in. longer and an undercut is machined on the end to clear the follower rest jaws.

Section I / Unit 15 / Additional Thread Forms

SELF-TEST ANSWERS

1. Translating type screws are mostly used for imparting motion or power and to position mechanical parts.

2. Square, modified square, Acme, stub Acme, and Buttress are five basic translating thread forms.

3. Since the pitch for 4 TPI would be .250 in., the depth of thread would be $P/2 = .250/2 = .125$ in.

4. Since the pitch for 4 TPI would be .250 in., the depth of the Acme thread would be $.5P + .010$ in. or $.5 \times .250$ in. $+ .010$ in. $= .135$ in.

5. General Purpose Acme threads bear on the flanks and centralizing Acme threads bear at the major diameter.

6. 29 degrees.

7. The general use for stub Acme threads is where a coarse pitch thread with a shallow depth is required.

8. The modified square thread.

9. The Acme thread form.

10. Buttress threads are used where great forces or pressures are exerted in one direction.

11. .375 in.

12. The distance from a point on one thread to a corresponding point on the next.

13. The distance the nut travels in one revolution.

14. Single lead.

15. Accurately slotted face plate.
Indexing a gear on the drive train.
Using the thread chasing dial.
The compound rest method.

16. They provide rapid traverse, are more efficient, have a larger minor diameter, are stronger, and furnish more bearing surface area than a single thread.

17. Most kinds. Sharp-V, American National, Unified, metric, Buttress, square, and Acme can be made multiple lead, either in right- or left-hand threads.

18. You can determine the number of leads by counting the number of starting grooves at the end of a bolt or screw.

19. Roughing of coarse threads may move the tool slight. If one thread has already been finished, there is no more allowance for adjustment. Finishing both threads consecutively, however, gives a much greater assurance that the setup will not move.

20. Lighter cuts should be taken to keep from tearing the threads.

Section I / Unit 16 / Cutting Acme Threads on the Lathe

SELF-TEST ANSWERS

1. The thread angle.

2. The included angle, the relief angle, and the flat on the end of the tool.

3. Coarse threading on a lathe imposes heavy loads on these parts. Lubrication prior to threading helps to reduce excess wear.

4. The compound is usually set at $14\frac{1}{2}$ degrees. Some machinists prefer to set the compound at 90 degrees.

5. $P = \frac{1}{6} = .1666$ in.; depth $- .5P + .010$ in. $=$ .93 in.

6. The tool is aligned by using the Acme tool gage.

7. $P = .1666$ in.; minor diameter $= .750$ in. $-$ .1666 in. $= .583$ in.

8. With an Acme tap set.

9. An Acme thread plug gage.

10. Light finishing cuts with a honed tool and a good grade of sulfurized cutting oil will help make good thread finishes. A rigid setup and low speeds will also help.

Section J / Unit 1 / The Vertical Spindle Milling Machine

SELF-TEST ANSWERS

1. The column, knee, saddle, table, ram, and toolhead.

2. The table traverse handwheel and the table power feed.

3. The cross traverse handwheel.

4. The quill feed hand lever and handwheel.

5. The table clamp locks the table rigidly and keeps it from moving while other table axes are in movement.

6. The spindle brake locks the spindle while tool changes are being made.

7. The spindle has to stop before speed changes from high to low are made.

8. The ram movement increases the working capacity of the toolhead.

9. Loose machine movements are adjusted with the slide gibs.

10. The quill clamp is tightened to lock the quill rigidly while milling.

Section J / Unit 2 / Cutting Tools and Cutter Holders for The Vertical Milling Machine

SELF-TEST ANSWERS

1. When viewed from the cutting end, a right-hand cut end mill will rotate counterclockwise.

2. An end mill has to have center cutting teeth to be used for plunge cutting.

3. End mills for aluminum usually have a fast helix angle and highly polished flutes and cutting edges.

4. Carbide end mills are very effective when milling abrasive or hard materials.

5. Roughing mills are used to remove large amounts of material.

6. Tapered end mills are mostly used in mold or die making to obtain precisely tapered side on workpieces.

7. Carbide insert tools are used because new cutting edges are easily exposed. They are available in grades to cut most materials and are very efficient cutting tools.

8. Straight shank mills are held in collets or adapters.

9. Shell end mills are mounted on shell mill arbors.

10. Quick-change toolholders make presetting of a number of tools possible and tools can be changed with a minimum loss of time.

Section J / Unit 3 / Setups on the Vertical Milling Machine

SELF-TEST ANSWERS

1. Workpieces can be aligned on a machine table by measuring their distance from the edge of the table, by locating against stops in the T-slots, or by indicating the workpiece side.

2. To align a vise on a machine table, the solid vise jaw needs to be indicated.

3. Toolhead alignment is checked when it is important that machining take place square to the machine table.

4. When the knee clamping bolts are loose, the weight of the knee makes it sag. But when the knee clamps are tightened, the knee is pulled into its normal position in relation to the column.

5. When the toolhead clamping bolts are tightened, it usually produces a small change in the toolhead position.

6. A machine spindle can be located over the edge of a workpiece with an edge finder or with the aid of dial indicator.

7. The spindle axis is one-half of the tip diameter away from the workpiece edge when the tip suddenly moves sideways.

8. An offset edge finder works best at 600 to 800 rpm.

9. To eliminate the effect of backlash, always position from the same direction.

10. The center of a hole is located with a dial indicator mounted in the machine spindle.

Section J / Unit 4 / Vertical Milling Machine Operations

SELF-TEST ANSWERS

1. Lower cutting speeds are used to machine hard materials, tough materials, and abrasive material; on heavy cuts; and to get maximum tool life.

2. Higher cutting speeds are used to machine softer materials, to obtain good surface finishes, with small diameter cutters, for light cuts, on frail workpieces, and on frail setups.

3. The calculated rpm is a starting point and may change depending on conditions illustrated in the answers to Problems 1 and 2.

4. Cutting fluids are used with HSS cutters except on materials such as cast iron, brass, and many nonmetallic materials.

5. Cutting with carbide cutters is performed without a cutting fluid, unless a steady stream of fluid can be maintained at the cutting edge of the tool.

6. The thickness of the chips affects the tool life of the cutter. Very thin chips dull a cutting edge quickly. Too thick chips cause tool breakage or the chipping of the cutting edge.

7. The depth of cut for an end mill should not exceed one-half of the diameter of the cutter in mild steel.

8. Limitations on the depth of cut for an end mill are the amount of material to be removed, the power available, and the rigidity of the tool and setup.

9. The rpm for a $\frac{3}{4}$-in.-diameter end mill to cut bronze is

$$\text{rpm} = \frac{CS \times 4}{D} = \frac{100 \times 4}{\frac{3}{4}} = 533 \text{ rpm}$$

10. The feed rate for a two-flute, $\frac{1}{4}$-in.-diameter carbide end mill in low carbon steel is

$$\begin{aligned} \text{feed rate} &= f \times \text{rpm} \times n \\ f &= .0005 \\ \text{rpm} &= \frac{CS \times 4}{D} = \frac{300 \times 4}{\frac{1}{4}} = 4{,}800 \\ n &= 2 \\ \text{feed rate} &= .0005 \times 4{,}800 \times 2 = 4.8 \text{ ipm} \end{aligned}$$

11. Accurate centering of a cutter over a shaft is done with the machine dials.

12. The feed direction against the cutter rotation assures positive dimensional movement. It also prevents the workpiece from being pulled into the cutter because of any backlash in the machine.

13. End mills can work themselves out of a split collet if the cut is too heavy or when the cutter gets dull.

14. Angular cuts can be made by tilting the workpiece or by tilting the workhead.

15. Circular slots can be milled by using a rotary table or an index head.

16. Squares, hexagons, or other shapes that require surfaces at precise angles to each other are made by using a dividing head.

17. Square holes or other internal hole shapes can be made by using a vertical shaping attachment.

18. With a right angle milling attachment, milling cuts can be made in very inaccessible places.

19. Layout lines are used as guides to indicate where machining should take place. Layouts should be made prior to machining any reference surfaces away.

20. A T-slot cutter only enlarges the bottom part of a groove. The groove has to be made before a T-slot cutter can be used.

Section J / Unit 5 / Using the Offset Boring Head

SELF-TEST ANSWERS

1. An offset boring head is used to produce standard and nonstandard size holes at precisely controllable hole locations.

2. Parallels raise the workpiece off the table or other work-holding device to allow through holes to be bored.

3. Unless the locking screw is tightened after toolslide adjustments are made, the toolslide will move during the cutting operation, resulting in a tapered or odd-sized hole.

4. The toolslide has a number of holes so that the boring tool can be held in different positions in relation to the spindle axis for different size bores.

5. It is important that you know if one graduation is one-thousandth of an inch or two-thousandths of an inch in hole size change.

6. The best boring tool to use is the one with the largest diameter that can be used and with the shortest shank.

7. It is very important that the cutting edge of the boring tool be on the centerline of the axis of the toolslide. Only in this position are the rake and clearance angles correct as ground on the tool.

8. The hole size obtained for given amount of depth of cut can change depending on the sharpness of the tool, the amount of tool overhang (boring bar length), and the amount of feed per revolution.

9. Boring tool deflection changes when the tool gets dull, the depth of cut increases or decreases, or the feed is changed.

10. The cutting speed is determined by the kind of tool material and the kind of work material, but boring vibrations set up through an unbalanced cutting tool or a very long boring bar may require a smaller than calculated rpm.

Section K / Unit 2 / Types of Spindles, Arbors, and Adaptors

SELF-TEST ANSWERS

1. Face mills over 6 in. in diameter.

2. The two classes of taper are self-holding, with a small included angle, and self-releasing, with a steep taper.

3. $3\frac{1}{2}$ ipf.

4. Where small diameter cutters are used, on light cuts, and where little clearance is available.

5. A style C arbor is a shell end mill arbor.

6. To increase the range of cutters that can be used on a milling machine with a given size spindle socket.

7. Spacing collars are used to take up the space between the cutter and the end of the arbor. They are also used to space straddle milling cutters. Bearing collars ride in the arbor support bearing; they provide support for the outer end of the arbor. Spacing and bearing collars are precision accessories and should be protected against nicks and burrs.

8. Dirt or burrs between collars can cause cutter runout and inaccurate machining.

9. Your instructor will show you how to adjust bearing clearance. Check to see that the oil reservoir is full or that sufficient oil is applied to keep the arbor bearing well lubricated during milling.

10. Quick-change tooling systems save time as many different tools can be set up and quickly inserted in the machine.

Section K / Unit 3 / Arbor-Driven Milling Cutters

SELF-TEST ANSWERS

1. Profile-sharpened cutters and form-relieved cutters.

2. Light-duty plain milling cutters have many teeth. They are used for finishing operations. Heavy-duty plain mills have few but coarse teeth, designed for heavy cuts.

3. Plain milling cutters do not have side cutting teeth. This would cause extreme rubbing if used to mill steps or grooves. Plain milling cutters should be wider than the flat surface they are machining.

4. Side milling cutters, having side cutting teeth, are used when grooves are machined.

5. Straight tooth side mills are used only to mill shallow grooves because of the limited chip space between the teeth and their tendency to chatter. Stagger tooth mills have a smoother cutting action because of the alternate helical teeth; more chip clearance allows deeper cuts.

6. Half side milling cutters are efficiently used when straddle milling.

7. Metal slitting saws are used in slotting or cutoff operations.

8. Gear tooth cutters and corner rounding cutters.

9. To mill V-notches, dovetails, or chamfers.

10. A right-hand cutter rotates counterclockwise when viewed from the outside end.

Section K / Unit 4 / Work-Holding Methods and Standard Setups

SELF-TEST ANSWERS

1. A clamping bolt should be close to the workpiece and a greater distance away from the support block.

2. Finished surfaces should be protected with shims from being marked by clamps or rough vise jaws.

3. Screwjacks are used to support workpieces or to support the end of a clamp.

4. A stop block prevents a workpiece from being moved by cutting pressure.

5. Quick-action jaws are two independent jaws mounted anywhere on a machine table to form a custom vise.

6. A swivel vise has one movement in a horizontal plane, whereas a universal vise swivels both horizontally and vertically.

7. All-steel vises are used to hold rough workpieces such as castings or forgings.

8. A rotary table is used to mill gears or circular grooves, or for angular indexing.

9. A dividing head is used to divide accurately the circumference of a workpiece into any number of equal divisions.

10. A fixture is used when a great number of pieces have to be machined in exactly the same way, and when the cost of making the fixture can be justified in savings resulting from its use.

Section K / Unit 5 / Machine Setup and Plain Milling

SELF-TEST ANSWERS

1. Cutting speed is the distance a cutting edge of a tool travels in one minute. It is expressed in feet per minute (fpm).

2. Starting a cut at the low end of the speed range will save the cutter from overheating.

3. Carbide tools are operated at two to six times the speed of HSS tools. If an HSS tool is 100 fpm, the carbide tool would be 200 to 600 fpm.

4. Too low a cutting speed is inefficient because the cutter could do more work in a given time period.

5. The feed rate on a milling machine is given in inches per minute (ipm).

6. The feed rate is the product of rpm times the number of teeth of the cutter times the feed per tooth.

7. Feed per revolution does not consider the number of teeth on different cutters.

8. Too low a feed rate causes the cutter to rub and scrape the surface of the work instead of cut. Because of the high friction, the tool will dull quickly.

9. $\text{Rpm} = \frac{CS \times 4}{D} = \frac{60 \times 4}{3} = 80$ rpm.

10. $\text{Rpm} = \frac{CS \times 4}{D} = \frac{150 \times 4}{4} = 150$ rpm.
Feed per tooth = .003 in.
Number of teeth = 5.
Feed rate = rpm × feed per tooth × number of teeth = $150 \times .003 \times 5 = 2.25 = 2\frac{1}{4}$ ipm.

11. Because the movable jaw is not solidly held. It can move and swivel slightly to align itself with the work to some extent.

12. The dial indicator is more accurate. It will show you the amount of misalignment when you make adjustments; you can see when alignment is achieved.

13. Keys are designed to align a vise or other attachments with the T-slots on a machine table.

14. No, indicating the table from the column only measures the table sliding in its ways; it does not show if it travels parallel to the column.

15. Yes, if possible the cutting tool pressure should be against the solid jaw. This makes the most rigid setup.

16. The smallest diameter cutter that will do the job will be the most efficient, because it requires a shorter movement to have the workpiece move clear of the cutter.

17. In conventional milling, the cutting pressure is against the feed direction and also up, whereas in climb milling the cutting pressure is down and the cutter tends to pull the workpiece under itself.

18. A good depth for a finish cut is .015 to .030 in. Less than .010 in. makes the cutter rub, which causes rapid wear.

19. Cutting vibrations and cutting pressures may make the table move when it should be rigidly clamped.

20. When a revolving cutter is moved over a just machined surface, it will leave tool marks.

Section K / Unit 6 / Using Side Milling Cutters

SELF-TEST ANSWERS

1. Full side milling cutters are used to cut slots and grooves and where contact on both sides of the cutter is made.

2. Half side milling cutters make contact on one side only, as in straddle milling where a left-hand and a right-hand cutter are combined to cut a workpiece to length.

3. The best cutter is the one with the smallest diameter that will work, considering the clearance needed under the arbor support.

4. Usually a groove is wider than the cutter.

5. A layout shows the machinist where the machining is to take place. It helps in preventing errors.

6. Accurate positioning is done with the help of a paper feeler strip. Adequate accuracy often is achieved by using a steel rule or by aligning by sight with the layout lines.

7. If a workpiece is measured while it is clamped in the machine, additional cuts can be made without additional setups being made.

8. Shims and spacers control workpiece width in straddle milling operations.

9. The diameters of the individual cutters determine the relationship of the depth of the steps in gang milling.

10. Interlocking side mills are used to cut slots over 1 in. wide and also when precise slot width is to be produced.

Section K / Unit 7 / Using Face Milling Cutters on the Horizontal Milling Machine

SELF-TEST ANSWERS

1. Face mills are mounted on the spindle nose, driven by two keys, and held with four capscrews.

2. A shell end mill is a face mill 6 in. in diameter and smaller.

3. Light-duty face mills have a great number of teeth and are used in finishing operations. Heavy-duty face mills have fewer, stronger teeth and are used in roughing operations.

4. Positive rake angles on cutters use less power, have less cutting pressures, generate less heat, and work well on soft and ductile materials.

5. Negative rake angles make cutters very strong. They will withstand heavy interrupted cuts.

6. A large lead angle on a cutter produces a thinner but wider chip under equal depth of cut and feed conditions as a small lead angle would. A large lead angle helps in reducing initial cutting pressure on the tool because it gradually eases the cutter into the work.

7. Cuts as wide as the diameter of the face mill result in excessive rubbing and friction of the cutting edge, which causes rapid tool wear.

8. Cutting fluids are used with high speed steel cutters. With carbide tools, it is better not to use coolant at all than to apply it intermittently to the cutting edge.

9. A concave surface results when the cutter is tilted so that the trailing edge is not in contact with the workpiece surface.

10. Effective face milling requires:
A rigid setup.
A sharp cutting tool.
The right cutting speed.
The correct feed.

Section L / Unit 1 / Setup and Operation of Indexing Heads and Rotary Tables

SELF-TEST ANSWERS

1. When accurate spacings are made, as with gears, splines, keyseats, or precise angular spacings.

2. The use of a worm and worm wheel unit.

3. To make the spindle freewheeling and when direct indexing.

4. For direct indexing.

5. The most common ratio is 40:1.

6. Different hole circles are used to make precise partial revolutions with the index crank.

7. The sector arms are set to the number of holes that the index crank is to move. They eliminate the counting of these holes for each spacing.

8. The spindle lock is tightened after each indexing operation and before a cut is made to prevent any rotary movement of the spindle.

9. With the use of high number index plates or with a wide range divider.

10. When the rotation of the index crank is reversed, the backlash between the worm and worm wheel affects the accuracy of the spacing.

Section L / Unit 2 / Direct and Simple Indexing

SELF-TEST ANSWERS

1. Direct indexing is performed from index holes on the spindle nose. Simple indexing uses the worm and worm wheel drive and the side index plate.

2. Use a marking pen or layout dye to mark the holes to be used in direct indexing.

3. Twenty-four index holes let you make equal divisions of 2, 3, 4, 6, 8, 12, and 24 parts.

4. The sector arms can be adjusted so that the number of holes to be indexed plus one are between the beveled edges.

5. For the highest degree of accuracy use the largest possible hole circle.

6. $\frac{40}{6} = \frac{20}{3} = 6\frac{2}{3}$ turns. Use the 57-hole circle for the $\frac{2}{3}$ turn. $\frac{2}{3} \times \frac{19}{19} = \frac{38}{57}$. The fraction is 38 holes in the 57-hole circle.

7. $\frac{40}{15} = \frac{8}{3} = 2\frac{2}{3}$ turns. Use the same hole circle as for problem 6. Two turns and $\frac{38}{57}$ turn.

8. $\frac{40}{25} = \frac{8}{5} = 1\frac{3}{5}$ turns. $\frac{3}{5} \times \frac{6}{6} = \frac{18}{30}$. One turn and 18 holes in the 30-hole circle.

9. $\frac{40}{47}$. There is a 47 hole circle, so use 40 holes in the 47-hole circle.

10. $\frac{40}{64} = \frac{5}{8}$ turns. The only hole circle divisible by 8 is 24. So $\frac{5}{8} \times \frac{3}{3} = \frac{15}{24}$ = 15 holes in the 24-hole circle.

Section L / Unit 3 / Angular Indexing

SELF-TEST ANSWERS

1. 15 degrees.

2. Three holes.

3. 9 degrees.

4. All those hole circles are divisible by 9.

5. $\frac{17}{9} = 1\frac{8}{9}$ turn of the index crank.

6. 30 min or $\frac{1}{2}$ degrees.

7. 15 min or $\frac{1}{4}$ degrees.

8. 10 min or $\frac{1}{6}$ degrees.

9. 540 min.

10. Converting 54°30′ into minutes = 3270 min

$$\frac{\text{Required minutes}}{540} = \frac{3270}{540} = 6\,\frac{30}{540} = 6\,\frac{3}{54}$$

= 6 turns and 3 holes in the 54-hole circle.

Section M / Unit 1 / Introduction to Gears

SELF-TEST ANSWERS

1. Spur gears and helical gears.

2. Helical gears run more smoothly than spur gears because more than one tooth is in mesh at all times. Helical gears, however, generate axial thrust that has to be offset with thrust bearings.

3. The pinion in mesh with an internal gear rotates in the same direction as the gear.

4. The shafts will be at 90 degrees to each other.

5. 50:1.

6. No, a single start worm can only be replaced with a single start worm. To change the gear ratio, the number of teeth in the gear must be changed.

7. Hardened steel.

8. Nonmetallic materials.

9. By making the pinion harder than the gear.

10. Nonferrous materials.

Section M / Unit 2 / Spur Gear Terms and Calculations

SELF-TEST ANSWERS

1. Pressure angles on gears vary from $14\frac{1}{2}$ to 20 to 25 degrees.

2. Larger pressure angles make stronger teeth. They also allow gears to be made with fewer teeth.

3. $C = \dfrac{N_1 + N_2}{2P} = \dfrac{20 \times 30}{2 \times 10} = \dfrac{50}{20} = 2.500$ in.

4. $C = \dfrac{D_1 + D_2}{2} = \dfrac{3.500 + 2.500}{2} = \dfrac{6.000}{2} =$ 3.000 in.

5. The whole depth of a tooth is how deep a tooth is cut. The working depth gives the distance the teeth from one gear enter the opposing gear in meshing.

6. The addendum is above the pitch diameter and the dedendum is below the pitch diameter of a gear.

7. $D_o = \dfrac{N + 2}{P} = \dfrac{50 + 2}{5} + \dfrac{52}{5} = 10.400$ in.

$t = \dfrac{1.5708}{P} = \dfrac{1.5708}{5} = .314$ in.

8. $P = \dfrac{N}{D} = \dfrac{36}{3} = 12.$

9. $D_o = \dfrac{N + 2}{P} = \dfrac{40 + 2}{8} = \dfrac{42}{8} = 5.250$ in.

$h_t = \dfrac{2.250}{P} = \dfrac{2.250}{8} = .2812$ in.

$D = \dfrac{N}{P} = \dfrac{40}{8} = 5.000$ in.

$b = \dfrac{1.250}{P} = \dfrac{1.250}{8} = .1562$ in.

10. $D_o = \dfrac{N + 2}{P} = \dfrac{48 + 2}{6} = \dfrac{50}{6} = 8.3333$ in.

$c = \dfrac{.157}{P} = \dfrac{.157}{6} = .0261$ in.

$h_t = \dfrac{2.157}{P} = \dfrac{2.157}{6} = .3595$ in.

$t = \dfrac{1.5708}{P} = \dfrac{1.5708}{8} = .2618$ in.

$D = \dfrac{N}{P} = \dfrac{48}{6} = 8.000$ in.

Section M / Unit 3 / Cutting a Spur Gear

SELF-TEST ANSWERS

1. Gears cut on a milling machine lack a high degree of accuracy and are expensive to make.

2. Eight.

3. No. 3.

4. No, the tooth profile differs with different pressure angles.

5. For a 17-tooth gear.

6. The number of the cutter, the diametral pitch, the number of teeth the cutter can cut, the pressure angle, and the whole depth of tooth.

7. To get a setup as rigid as possible.

8. The number of holes to be indexed plus one.

9. To check the correctness of the number of spaces required.

10. A center rest is used under gear blanks, mandrels, or shafts to help prevent chatter and deflection.

Section M / Unit 4 / Gear Inspection and Measurement

SELF-TEST ANSWERS

1. Gear measurements with a gear tooth vernier caliper and with a micrometer over two wires or pins.

2. The chordal tooth thickness.

3. The chordal addendum.

4. The chordal addendum.

5. The circular thickness.

6. By calculation or from *Machinery's Handbook* tables.

7. No, the pin sizes are specifically calculated for each differing diametral pitch.

8. By consulting a *Machinery's Handbook* table.

9. The dimension given is divided by the diametral pitch used; in this question, the divisor is 12.

10. The optical comparator magnifies the gear tooth profile and projects it on a screen. On the screen there is also a transparent drawing with the tooth profile; by aligning the shadow with the drawn profile, any variation between the two can be seen and measured.

Section N / Unit 1 / Selection and Identification of Grinding Wheels

SELF-TEST ANSWERS

1. Bronze—silicon carbide; low-carbon steel—aluminum oxide; carbide—diamond or silicon carbide; high-temperature alloy—cubic boron nitride.

2. Straight, cylinder, straight cup, flaring cup, shallow dish.

3. Shallow dish, and flaring cup (with tool and cutter grinding).

4. Vitrified bonds.

6. The proportion of bonding material to abrasive. The more bonding strength, the "harder" the wheel.

7. a. Aluminum oxide.
 b. Vitrified bond.
 c. 46–120.
 d. J bond hardness.

8. a. Aluminum oxide.
 b. Vitrified bond.
 c. About 60 grit.
 d. I bond hardness.

9. Either silicon carbide or, preferably, diamond.

10. Structure is 8 (open); bond, J; grit size, 80; abrasive, silicon carbide; and bonding material, resinoid.

Section N / Unit 2 / Trueing, Dressing, and Balancing of Grinding Wheels

SELF-TEST ANSWERS

1. Trueing makes a grinding wheel's working surface concentric to the machine spindle. It is important in obtaining accurate grinding results.

2. Dressing is sharpening, and trueing is making the tool concentric. The wheel should only be dressed when necessary to maintain free cutting action.

3. Form dressing is used to shape the wheel to the reverse of the geometry required in the part. It can be done by generating the form needed or by using a diamond dressing block.

4. The single-point diamond dresser should be about $\frac{1}{4}$ in. after the wheel center, and tipped about 15 degrees away from the wheel rotation.

5. Lower the revolving grinding wheel head while traversing the diamond until contact is made. Traverse the dresser across the grinding surface. Add about .001 in. depth per pass. Continue until the wheel is completely concentric.

6. A wheel dressing for roughing utilizes rapid traverse of the diamond to provide a sharp open surface. Slow traverses are used to obtain a smoother surface for finish grinding.

7. If the superabrasive wheel cannot be indicated in and adjusted to run true, then it must be trued. This is best done with a brake-type trueing device.

8. Electroplated superabrasive wheels must not be trued or dressed as they have only one layer depth of abrasive.

9. An unbalanced wheel leaves an irregular surface on the workpiece. Wheels need to be placed on an arbor and balanced on overlapping disk balancing ways, or on a carefully leveled pair of knife edges. Balance should be achieved to less than the weight of a postage stamp.

10. The overlapping disk type of balancer can work well without precision leveling.

Section N / Unit 3 / Grinding Fluids

SELF-TEST ANSWERS

1. They cool a workpiece for size control, lubricate to prevent chip adherence to the wheel, flush swarf, and control grinding dust.

2. As close as possible to the grinding contact area.

3. Mist coolant is sprayed between the wheel and workpiece with air pressure.

4. Water-soluble chemical types, water-soluble oil types, straight oils, and sulfochlorinated oils.

5. Settling, filtering, or using a centrifugal separator, cyclonic separator, or magnetic separator.

6. A block of nearly the same height as the workpiece, to help keep the grinding fluid flowing over the actual workpiece.

7. Bacterial growth occurs, which must be combated with bacterial growth inhibitors.

8. Straight oils are used. The work area must be carefully and thoroughly ventilated to protect the operator.

9. Synthetics are not prone to bacterial growth and they are transparent, which gives the operator better workpiece visibility.

10. If fluid is not applied accurately, it can be less harmful to the workpiece to grind without fluid being applied.

Section N / Unit 4 / Horizontal Spindle Reciprocating Table Surface Grinders

SELF-TEST ANSWERS

1. The table supports the work chuck and carries both back and forth under the grinding wheel.

2. The saddle supports the table and moves it in and out so that the workpiece may be set over for the next grinding pass.

3. The downfeed handwheel raises and lowers the wheel so that the depth of cut may be set. The table handwheel reciprocates the table when in manual traverse. The saddle handwheel moves the saddle in and out.

4. The downfeed handwheel graduations typically discriminate to .0001 in.

5. The typical crossfeed measuring increment is .001 in.

6. It tends to keep the grinding surface flat, which extends the time interval between needed wheel dressings.

7. It helps to prevent the appearance of "feed lines" in the surface of the work.

8. It gives the operator or cross feed mechanism time to complete the crossfeed, to avoid unattractive angular feed lines on the workpiece.

9. This can be done with a centerless grinding attachment.

10. This can be done by using a center-type cylindrical grinding attachment.

Section N / Unit 5 / Workholding on the Grinder

SELF-TEST ANSWERS

1. The permanent magnet chuck.

2. By reversing the magnetic field electrically in a process called degaussing.

3. The intensity of the part holding can be varied to reduce part distortion, and the part can be fully demagnetized by degaussing.

4. If power fails, the electro-permanent chuck retains its holding ability. Heat transfer to the part, which is a problem with electro-magnetic chucks, is minimal.

5. The work-holding ability of a magnetic chuck is enhanced by spanning as many poles as possible.

6. The parallelism should be checked with a dial indicator.

7. Burrs should be removed with a granite conditioning stone or a very fine grit oilstone.

8. Laminated accessories include parallels and vee blocks, made with alternating layers of steel and nonmagnetic material. They are used as accessory work-holding devices on magnetic chucks.

9. The vacuum chuck.
Magnetic tooth clamps with magnetic chucks.
Double-backed tape can also be used.

10. Thin workpieces may be shimmed to compensate for initial warpage; they may also be held with double-backed tape to avoid magnetic distortion. Grinding fluid should not be used on a double-backed tape setup.

Section N / Unit 6 / Using the Surface Grinder

SELF-TEST ANSWERS

1. This is the volume of abrasive used in proportion to the volume of workpiece material removed. For example, 1 unit of abrasive for 40 parts of workpiece material is common in conventional grinding.

2. Tolerances can be very small, with plus or minus 50 millionths of an inch (.000050 in.) often appearing on drawings. Highly stressed critical parts must be ground without metallurgical damage.

3. Keep the crossfeeding movements as large as possible by varying the downfeed so that little change in grinding wheel spindle speed occurs. This helps to avoid too frequent wheel dressings.

4. One-grade softer bond would be preferred, as this change helps keep the wheel self-sharpening.

5. The feeding rate is often reduced from $\frac{1}{4}$ of wheel width per pass, down to about $\frac{1}{10}$ of the wheel width.

6. Marking the surface with pencil lines and grinding until they disappear. Usually less than .001 in. is needed.

7. Use an abrasive cloth on a flat base surface (not on your shop surface plate!) to remove the heat-treating scale to leave a smooth initial surface for contact with the chuck.

8. Dress the wheel with fluid applied *only if* full flooding can be assured, or the diamond may crack from thermal stress.

9. It leaves you, the operator, the most latitude in bringing the part to size in the subsequent grinding operations.

10. .003 to .005 in. of stock is left for finishing. The wheel is redressed for finish grinding with a slow pass (or passes) of the diamond.

Section N / Unit 7 / Problems and Solutions in Surface Grinding

SELF-TEST ANSWERS

1. The operator can warm up the spindle before

grinding and be sure that the machine is correctly lubricated.

2. The wheel can be trued. In some cases where a wheel has been unbalanced by residual dried grinding fluid, it may be necessary to replace the grinding wheel.

3. First, examine the inside of the wheelguard for cleanliness. If the problem persists, check the grinding fluid for cleanliness.

4. Sharpen the wheel by dressing. Reduce the intensity of the grinding by reducing the downfeed, and possibly, increasing the table speed. If wheel is still not self-dressing, change to a softer wheel bond.

5. First, check for burrs on the chuck surface and dress those discovered with a deburring stone. Check the chuck for parallel with an indicator. Grind-in the chuck if necessary.

6. Local overheating; internal stress relieved by the grinding; bent or twisted workpiece.

7. The grinding has reduced the surface compression of the material. Turn the part over on the chuck and take an equal amount from the opposite side, if the part design permits. Otherwise, anneal the part before grinding to remove the stresses of the cold rolling, or use hot-rolled material if you have the choice of material.

8. If the grinder is located near sources of vibration such as punch presses, passing trains, and/or heavy truck traffic, surface finish defects can result.

9. It can lead to surface burnishing, which imparts extensive cold-working stresses in the workpiece. These stresses can lead to part failure.

10. To avoid frequent wheel changes, the operator will usually attempt to dress the wheel to match workpiece requirements. The way in which the downfeed and crossfeed are related also has a large effect on whether or not the grinding wheel tends to be self-dressing.

Section N / Unit 8 / Center-Type Cylindrical Grinders

SELF-TEST ANSWERS

1. On the plain cylindrical grinder the wheelhead does not swivel. The wheelhead may travel parallel to the rotating workpiece, in one design, or the workpiece can be traversed past the grinding wheel in lighter-duty plain cylindrical grinders. On the universal cylindrical grinder, all the major components are able to swivel, and the grinding head can be equipped with a chuck for internal grinding. The plain cylindrical grinder is usually a very rigidly constructed production machine, while the construction of the universal grinder is much less rigid because of all the motions that can be used. The universal cylindrical grinder is used mainly in toolroom applications where versatility is critical.

2. This prevents the duplication of headstock bearing eccentricity into the workpiece.

3. This prevents the force of the grinding toward the footstock end from deflecting the part away from the grinding wheel.

4. Yes, the table that carries the centers is usually capable of being swiveled for this purpose.

5. A center-hole grinding machine can be used to prepare workpieces for cylindrical grinding.

6. This is grinding in which the rotating workpiece is moved across the face of the grinding wheel, which generates the surface.

7. This is cylindrical grinding in which the rotating workpiece is moved directly into the grinding wheel. This action imparts a "mirror image" of the form of the face of the grinding wheel into the part.

8. The workpiece is supported on two conical work centers that project into matching conical holes in the ends of the workpiece.

9. The interruption should be narrower than the width of the grinding wheel face.

10. A chuck or other workholding device, like a faceplate, to hold and rotate the part relative to the grinding spindle. A "high-speed" attachment which mounts to the wheelhead assembly and carries a mounted internal grinding wheel.

Section N / Unit 9 / Using the Cylindrical Grinder

SELF-TEST ANSWERS

1. The combined curvature of the wheel and the workpiece create a very narrow line of contact from which the grinding swarf can easily escape.

2. It does not have to be mounted for each use, hence it is very convenient for the operator.

3. It is useful for aligning the swivel table to parallel, or for offsetting the table for accurate conical (tapered) surfaces.

4. None. The part should be adjusted with nil end play but be able to rotate freely on centers lubricated with a high-pressure center lubricant.

5. The "tail" of the driving dog must not touch the bottom of the driving slot, or the workpiece could be forced off its conical seat. This would result in poor accuracy.

6. At least $\frac{1}{3}$ of the wheel width, where possible. Half the wheel width would be optimal to permit full grinding to size.

7. The Tarry Control permits a dwell at the end of the traverse to stop "table bounce." (Without a Tarry Control, it is difficult to obtain the highest possible dimensional accuracy.)

8. About $\frac{1}{4}$ of the wheel width.

9. About $\frac{1}{8}$ of the wheel width.

10. .0002 in. (or less).

Section N / Unit 10 / Universal Tool and Cutter Grinder

SELF-TEST ANSWERS

1. Vertical height control of the wheelhead, and in some designs, the ability to tilt the wheelhead also.

2. The cutter edges break down quickly for lack of proper support.

3. The larger clearances that are optimal for aluminum would lead to early failure in steels.

4. The amount necessary to take off the cutter to restore the cutting edge can easily be seen, so less of the cutter material is lost in regrinding.

5. Helical cutters must rotate while being sharpened to follow the tooth form. Mounting the stop on the table works only for straight tooth cutters.

6. It permits the checking of the setup for parallel, and correcting if necessary. The measured difference should not exceed .001 in.

7. The distance from the cutting edge to the axis of the cutter must be uniform or the slot made by the cutter will be stepped.

8. Form relieved cutters are sharpened only on the face. To preserve the form of the cutter the grind should be radial to the cutter axis.

9. A free-turning accessory spindle. (Air bearing types are preferable because of their uniformly low friction.)

10. The workhead. It is capable of being both rotated and tilted for these clearances. A tilting spindle design of tool and cutter grinder is also desirable.

Section O / Unit 1 / CNC Machine Tools

SELF-TEST ANSWERS

1. CNC, often using punched tape to transmit information, is an early form of machine control using numbers and letters. CNC uses a computer built into the MCU for machine control. DNC uses a computer away from the machine for machine control.

2. The X axis is the long table movement. The Y axis is the table moving toward and away from the column. The Z axis is the spindle up and down movement. The a axis is a rotary table on the machine table. The b axis is the ability of the spindle to rotate.

3. Pallets are used to hold workpieces while they are being machined. They cut down setup time because they are loaded and unloaded during the machining of another workpiece.

4. A rotary table is used in contouring cuts and to machine cams.

5. Automatic tool changers cut short the time required to change cutting tools during the machining operation.

6. When drilling holes, a pecking cycle withdraws the drill from the hole in planned intervals to clear the chips from the flutes.

7. To use a number of boring tools from roughing to finishing tools and to make the finishing cut by boring out of the hole at the same feed rate as boring in.

8. With adaptive control, the MCU makes changes in the feed rate if increased cutting pressures signal changes in the cutting conditions.

9. With live tooling, the tools in a turret can be revolving tools such as drills or milling cutters.

10. A negative Z axis move on a vertical spindle machine is a tool movement down or toward the table.

11. The X axis is the cross-slide and the Z axis is the longitudinal axis.

12. Tool presetting makes it possible for the programmer to select tools, knowing their exact diameters and relative lengths, without seeing these tools.

Section O / Unit 2 / CNC Positioning and Input

SELF-TEST ANSWERS

1. In an open-loop system the MCU sends a signal to the axis drive unit, but no feedback signal informs the MCU of the actual distance moved. In a closed-loop system a feedback signal reports to the MCU that the directed distance was actually moved.

2. With an X and a Y value designation.

3. With a distance given as a radius and a direction given as an angle.

4. Mostly in drilling operations.

5. It allows the machining of contours.

6. In incremental positioning, the location of the tool becomes the reference for the next movement.

7. In absolute positioning, every movement is measured as a distance from the established reference zero point.

8. A floating zero system allows the programming of a new zero point away from the original reference zero point.

9. Linear interpolation is used in the machining of straight lines. Circular interpolation is used to machine circles and arcs.

10. ASCII and EIA.

Section O / Unit 3 / CNC Applications

SELF-TEST ANSWERS

1. MDI is used during setup procedures and when a program is interrupted to correct or add to a machining operation.

2. A programmer needs to be able to read and interpret blueprints, to plan the sequence of machining operations, and to understand work holding, tool selection, feeds and speeds, and the operating characteristics of the particular machine being programmed.

3. In prompted input, the controller usually directs the programmer by asking questions and giving choices from which to select. A word address input uses G and M words to direct machine functions.

4. A canned cycle is a program in the MCU memory of frequently performed machining operations. By programming a canned cycle, one command will direct a series of machine operations that would otherwise be a number of separate programmed commands.

Section O / Unit 4 / Nontraditional Machining Processes

SELF-TEST ANSWERS

1. EDM is electrodischarge machining, where workpiece material is removed by spark erosion.

ECM is electrochemical machining and ELG is electrolytic grinding. The processes take place in a conducting medium and workpiece material is deplated by electrochemical action.

Laser machining uses the engery from an intense coherent beam of light energy as a cutting tool.

Ultrasonic machining uses ultrasound to induce sympathetic vibrations in an anvil. The anvil's vibrations drive an abrasive material against the workpiece, providing the cutting action.

Hydrojet machining uses a high-pressure stream of water containing an abrasive material as a cutting tool.

Plasma beam machining uses a very high temperature gas or plasma as a cutting tool.

Electron beam machining is accomplished in a vacuum by an intense beam of electrons impinging on the workpiece.

2. ECM and ECDB.

3. Wire-cut EDM.

4. 40,000°F (22,204°C).

5. ECM and ELG.

6. To maintain the gap between work and wheel.

7. Ultrasound vibrations provide the energy. Abrasive materials do the cutting.

8. EDM.

9. ECM, ELG, and ECDB.

10. By electrically stimulating the atoms of a gas, photons of light energy are produced and are concentrated in a coherent (extremely narrow) beam as they exit the laser tube.

APPENDIX 2

GENERAL TABLES

Basic Designations

ISO Metric Threads are designated by the letter "M" followed by the *nominal size* in millimeters, and the *pitch* in millimeters, separated by the sign "X."

EXAMPLE

M16 × 1.5

Those numbers in the table marked with an asterisk are the commercially available sizes in the United States.

Tolerance Symbols

3 4 5 [6]

7 8 9

Numbers are used to define the amount of product tolerance permitted on either internal or external threads. Smaller grade numbers carry smaller tolerances, that is, grade 4 tolerances are smaller than grade 6 tolerances, and grade 8 tolerances are larger than grade 6 tolerances.

e [H] G [g]

Letters are used to designate the "position" of the product thread tolerances relative to basic diameters. Lower case letters are used for external threads, and capital letters for internal threads.

In some cases the "position" of the tolerance establishes an allowance (a definite clearance) between external and internal threads.

By combining the tolerance amount number and the tolerance position letter, the *tolerance symbol* is established that identifies the actual maximum and minimum product limits for external or internal threads. Generally the first number and letter refer to the pitch diameter symbol. The second number and letter refer to the crest diameter symbol (minor diameter of internal threads or major diameter of external threads).

EXAMPLE

5g 6g

Pitch Diameter Tolerance Symbol — 5g

Crest Diameter Tolerance Symbol — 6g

Where the pitch diameter and crest diameter tolerance symbols are the same, the symbol need only be given once.

EXAMPLE

6g

Pitch Diameter and Crest Diameter Tolerance Symbol

It is recommended that the *coarse series* be selected whenever possible, and that *general purpose* grade 6 be used for both internal and external threads.

Tolerance positions "g" for external threads and "H" for internal threads are preferred.

Other product information may also be conveyed by the ISO metric thread designations. Complete specifications and product limits may be found in the ISO Recommendations or in the B1 report "ISO Metric Screw Threads."

Some examples of ISO Metric Thread designations are as follows:

M10

M18 × 1.5

M6—6H

M4—6g

M12 × 1.25—6H

M20 × 2—6H/6g

M6 × 0.75—7g 6g

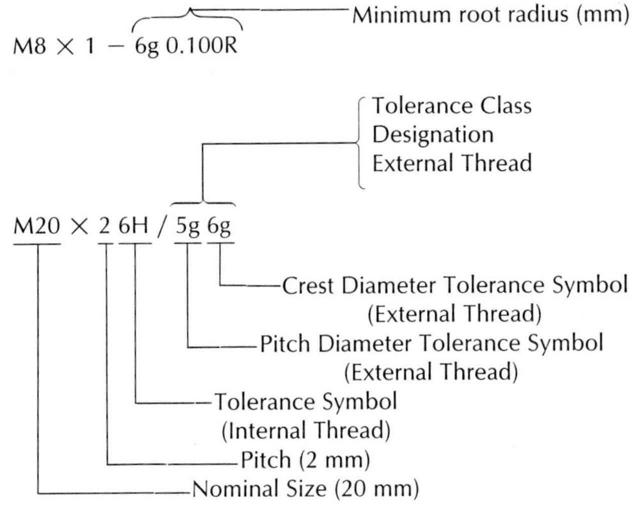

(SOURCE: Material courtesy of TRW Inc., *New Greenfield Geometric Screw ISO Metric Thread Manual*, 1973.)

TABLE 1 Decimal Equivalents of Fractional Inches

FRACTION INCH	DECIMAL INCH	DECIMAL MILLI-METERS	FRACTION INCH	DECIMAL INCH	DECIMAL MILLI-METERS
1/64	.015625	0.39688	33/64	.515625	13.09690
1/32	.03125	0.79375	17/32	.53125	13.49378
3/64	.046875	1.19063	35/64	.546875	13.89065
1/16	.0625	1.58750	9/16	.5625	14.28753
5/64	.078125	1.98438	37/64	.578125	14.68440
3/32	.09375	2.38125	19/32	.59375	15.08128
7/64	.109375	2.77813	39/64	.609375	15.47816
1/8	.1250	3.17501	5/8	.6250	15.87503
9/64	.140625	3.57188	41/64	.640625	16.27191
5/32	.15625	3.96876	21/32	.65625	16.66878
11/64	.171875	4.36563	43/64	.671875	17.06566
3/16	.1875	4.76251	11/16	.6875	17.46253
13/64	.203125	5.15939	45/64	.703125	17.85941
7/32	.21875	5.55626	23/32	.71875	18.25629
15/64	.234375	5.95314	47/64	.734375	18.65316
1/4	.2500	6.35001	3/4	.7500	19.05004
17/64	.265625	6.74689	49/64	.765625	19.44691
9/32	.28125	7.14376	25/32	.78125	19.84379
19/64	.296875	7.54064	51/64	.796875	20.24067
5/16	.3125	7.93752	13/16	.8125	20.63754
21/64	.328125	8.33439	53/64	.828125	21.03442
11/32	.34375	8.73127	27/32	.84375	21.43129
23/64	.359375	9.12814	55/64	.859375	21.82817
3/8	.3750	9.52502	7/8	.8750	22.22504
25/64	.390625	9.92189	57/64	.890625	22.62192
13/32	.40625	10.31877	29/32	.90625	23.01880
27/64	.421875	10.71565	59/64	.921875	23.41567
7/16	.4375	11.11252	15/16	.9375	23.81255
29/64	.453125	11.50940	61/64	.953125	24.20942
15/32	.46875	11.90627	31/32	.96875	24.60630
31/64	.484375	12.30315	63/64	.984375	25.00318
1/2	.5000	12.70003	1	1.0000	25.40005

TABLE 2 Inch/Metric Conversion Table

Drill No. or Letter		Inch	mm
		.001	0,0254
		.002	0,0508
		.003	0,0762
		.004	0,1016
		.005	0,1270
		.006	0,1524
		.007	0,1778
		.008	0,2032
		.009	0,2286
		.010	0,2540
		.011	0,2794
		.012	0,3048
80	.0135	.013	0,3302
79	.0145	.014	0,3556
		.015	0,3810
	1/64	.0156	0,3969
78		.016	0,4064
		.017	0,4318
77		.018	0,4572
		.019	0,4826
76		.020	0,5080
75		.021	0,5334
74	.0225	.022	0,5588
		.023	0,5842
73		.024	0,6096
72		.025	0,6350
71		.026	0,6604
		.027	0,6858
70		.028	0,7112
69	.0292	.029	0,7366
		.030	0,7620
68		.031	0,7874
	1/32	.0312	0,7937
67		.032	0,8128
66		.033	0,8382
		.034	0,8636
65		.035	0,8890
64		.036	0,9144
63		.037	0,9398
62		.038	0,9652
61		.039	0,9906
		.0394	1,0000
60		.040	1,0160
59		.041	1,0414
58		.042	1,0668
57		.043	1,0922
		.044	1,1176
		.045	1,1430
56	.0465	.046	1,1684
	3/64	.0469	1,1906
		.047	1,1938
		.048	1,2192
		.049	1,2446
		.050	1,2700
		.051	1,2954
55		.052	1,3208
		.053	1,3462
		.054	1,3716
54		.055	1,3970
		.056	1,4224
		.057	1,4478
		.058	1,4732
53	.0595	.059	1,4986
		.060	1,5240
		.061	1,5494
		.062	1,5748
	1/16	.0625	1,5875
		.063	1,6002
52	.0635	.064	1,6256
		.065	1,6510
		.066	1,6764
51		.067	1,7018
		.068	1,7272
		.069	1,7526
50		.070	1,7780
		.071	1,8034
		.072	1,8288
49		.073	1,8542
		.074	1,8796
		.075	1,9050
48		.076	1,9304
		.077	1,9558
47	.0785	.078	1,9812
	5/64	.0781	1,9844
		.0787	2,0000
		.079	2,0066
		.080	2,0320
46		.081	2,0574
45		.082	2,0828
		.083	2,1082
		.084	2,1336
		.085	2,1590
44		.086	2,1844
		.087	2,2098
		.088	2,2352
43		.089	2,2606
		.090	2,2860
		.091	2,3114
		.092	2,3368
42	.0935	.093	2,3622
	3/32	.0937	2,3812
		.094	2,3876
		.095	2,4130
41		.096	2,4384
		.097	2,4638
40		.098	2,4892
		.099	2.5146
39	.0995	.100	2.5400

Drill No. or Letter		Inch	mm
38	.1015	.101	2,5654
		.102	2,5908
		.103	2,6162
37		.104	2,6416
		.105	2,6670
36	.1065	.106	2,6924
		.107	2,7178
		.108	2,7432
		.109	2,7686
	7/64	.1094	2,7781
35		.110	2,7940
34		.111	2,8194
		.112	2,8448
33		.113	2,8702
		.114	2,8956
		.115	2,9210
32		.116	2,9464
		.117	2,9718
		.118	2,9972
		.1181	3,0000
		.119	3,0226
31		.120	3,0480
		.121	3,0734
		.122	3,0988
		.123	3,1242
		.124	3,1496
	1/8	.125	3,1750
		.126	3,2004
		.127	3,2258
		.128	3,2512
30	.1285	.129	3,2766
		.130	3,3020
		.131	3,3274
		.132	3,3528
		.133	3,3782
		.134	3,4036
		.135	3,4290
29		.136	3,4544
		.137	3,4798
		.138	3,5052
		.139	3,5306
28	.1405	.140	3,5560
	9/64	.1406	3,5719
		.141	3,5814
		.142	3,6068
		.143	3,6322
27		.144	3,6576
		.145	3,6830
		.146	3,7084
26		.147	3,7338
		.148	3,7592
25	1495	.149	3,7846
		.150	3,8100
		.151	3,8354
24		.152	3,8608
		.153	3,8862
23		.154	3,9116
		.155	3,9370
		.156	3,9624
	5/32	.1562	3,9687
22		.157	3,9878
		.1575	4,0000
		.158	4,0132
21		.159	4,0386
		.160	4,0640
20		.161	4,0894
		.162	4,1148
		.163	4,1402
		.164	4,1656
		.165	4,1910
19		.166	4,2164
		.167	4,2418
		.168	4,2672
		.169	4,2926
18	.1635	.170	4,3180
		.171	4,3434
	11/64	.1719	4,3656
		.172	4,3688
17		.173	4,3942
		.174	4,4196
		.175	4,4450
		.176	4,4704
16		.177	4,4958
		.178	4,5212
		.179	4,5466
15		.180	4,5720
		.181	4,5974
14		.182	4,6228
		.183	4,6482
		.184	4,6736
13		.185	4,6990
		.186	4,7244
		.187	4,7498
	3/16	.1875	4,7625
		.188	4,7752
12		.189	4,8006
		.190	4,8260
11		.191	4,8514
		.192	4,8768
10	.1935	.193	4,9022
		.194	4,9276
		.195	4,9530
9		.196	4,9784
		.1969	5,0000
		.197	5,0038
		.198	5,0292
		.199	5,0546
8		.200	5,0800

Drill No. or Letter		Inch	mm
7		.201	5,1054
		.202	5,1308
		.203	5,1562
	13/64	.2031	5,1594
6		.204	5,1816
5	.2055	.205	5,2070
		.206	5,2324
		.207	5,2578
		.208	5,2832
4		.209	5,3086
		.210	5,3340
		.211	5,3594
		.212	5,3848
3		.213	5,4102
		.214	5,4356
		.215	5,4610
		.216	5,4864
		.217	5,5118
		.218	5,5372
	7/32	.2187	5,5562
		.219	5,5626
		.220	5,5880
2		.221	5,6134
		.222	5,6388
		.223	5,6642
		.224	5,6896
		.225	5,7150
		.226	5,7404
		.227	5,7658
1		.228	5,7912
		.229	5,8166
		.230	5,8420
		.231	5,8674
		.232	5,8928
		.233	5,9182
A		.234	5,9436
	15/64	.2344	5,9531
		.235	5,9690
		.236	5,9944
		.2362	6,0000
		.237	6,0198
B		.238	6 0452
		.239	6,0706
		.240	6,0960
		.241	6,1214
C		.242	6,1468
		.243	6,1722
		.244	6,1976
		.245	6,2230
D		.246	6,2484
		.247	6,2738
		.248	6,2992
		.249	6,3246
E	1/4	.250	6,3500
		.251	6,3754
		.252	6,4008
		.253	6,4262
		.254	6,4516
		.255	6,4770
		.256	6,5024
F		.257	6,5278
		.258	6,5532
		.259	6,5786
		.260	6,6040
G		.261	6,6294
		.262	6,6548
		.263	6,6802
		.264	6,7056
		.265	6,7310
	17/64	.2656	6,7469
H		.266	6,7564
		.267	6,7818
		.268	6,8072
		.269	6,8326
		.270	6,8580
		.271	6,8834
I		.272	6,9088
		.273	6,9342
		.274	6,9596
		.275	6,9850
		.2756	7,0000
		.276	7,0104
J		.277	7,0358
		.278	7,0612
		.279	7,0866
		.280	7,1120
K		.281	7,1374
	9/32	.2812	7,1437
		.282	7,1628
		.283	7,1882
		.284	7,2136
		.285	7,2390
		.286	7,2644
		.287	7,2898
		.288	7,3152
		.289	7,3406
L		.290	7,3660
		.291	7,3914
		.292	7,4168
		.293	7,4422
		.294	7,4676
M		.295	7,4930
		.296	7,5184
	19/64	.2969	7,5406
		.297	7,5438
		.298	7,5692
		.299	7,5946
		.300	7,6200

Drill No. or Letter		Inch	mm
		.301	7,6454
N		.302	7,6708
		.303	7,6962
		.304	7,7216
		.305	7,7470
		.306	7,7724
		.307	7,7978
		.308	7,8232
		.309	7,8486
		.310	7,8740
		.311	7,8994
		.312	7,9248
	5/16	.3125	7,9375
		.313	7,9502
		.314	7,9756
		.3150	8,0000
		.315	8,0010
O		.316	8,0264
		.317	8,0518
		.318	8,0772
		.319	8,1026
		.320	8,1280
		.321	8,1534
		.322	8,1788
P		.323	8,2042
		.324	8,2296
		.325	8,2550
		.326	8,2804
		.327	8,3058
		.328	8,3312
	21/64	.3281	8,3344
		.329	8,3566
		.330	8,3820
		.331	8,4074
Q		.332	8,4328
		.333	8,4582
		.334	8,4836
		.335	8,5090
		.336	8,5344
		.337	8,5598
		.338	8,5852
R		.339	8,6106
		.340	8,6360
		.341	8,6614
		.342	8,6868
		.343	8,7122
	11/32	.3437	8,7312
		.344	8,7376
		.345	8,7630
		.346	8,7884
		.347	8,8138
S		.348	8,8392
		.349	8,8646
		.350	8,8900
		.351	8,9154
		.352	8,9408
		.353	8,9662
		.354	8,9916
		.3543	9,0000
		.355	9,0170
		.356	9,0424
		.357	9,0678
T		.358	9,0932
		.359	9,1185
	23/64	.3594	9,1281
		.360	9,1440
		.361	9,1694
		.362	9,1948
		.363	9,2202
		.364	9,2456
		.365	9,2710
		.366	9,2964
		.367	9,3218
U		.368	9,3472
		.369	9,3726
		.370	9,3980
		.371	9,4234
		.372	9,4488
		.373	9,4742
		.374	9,4996
	3/8	.375	9,5250
		.376	9,5504
V		.377	9,5758
		.378	9,6012
		.379	9,6266
		.380	9,6520
		.381	9,6774
		.382	9,7028
		.383	9,7282
		.384	9,7536
		.385	9,7790
W		.386	9,8044
		.387	9,8298
		.388	9,8552
		.389	9,8806
		.390	9,9060
	25/64	.3906	9,9219
		.391	9,9314
		.392	9,9568
		.393	9,9822
		.3937	10,0000
		.394	10,0076
		.395	10,0330
		.396	10,0584
X		.397	10,0838
		.398	10,1092
		.399	10,1346
		.400	10,1600

Drill No. or Letter		Inch	mm
		.401	10,1854
		.402	10,2108
		.403	10,2362
Y		.404	10,2616
		.405	10,2870
		.406	10,3124
	13/32	.4062	10,3187
		.407	10,3378
		.408	10,3632
		.409	10,3886
		.410	10,4140
		.411	10,4394
		.412	10,4648
Z		.413	10,4902
		.414	10,5156
		.415	10,5410
		.416	10,5664
		.417	10,5918
		.418	10,6172
		.419	10,6426
		.420	10,6680
		.421	10,6934
	27/64	.4219	10,7156
		.422	10,7188
		.423	10,7442
		.424	10,7696
		.425	10,7950
		.426	10,8204
		.427	10,8458
		.428	10,8712
		.429	10,8966
		.430	10,9220
		.431	10,9474
		.432	10,9728
		.433	10,9982
		.4331	11,0000
		.434	11,0236
		.435	11,0490
		.436	11,0744
		.437	11,0998
	7/16	.4375	11,1125
		.438	11,1252
		.439	11,1506
		.440	11,1760
		.441	11,2014
		.442	11,2268
		.443	11,2522
		.444	11,2776
		.445	11,3030
		.446	11,3284
		.447	11,3538
		.448	11,3792
		.449	11,4046
		.450	11,4300
		.451	11,4554
		.452	11,4808
		.453	11,5062
	29/64	.4531	11,5094
		.454	11,5316
		.455	11,5570
		.456	11,5824
		.457	11,6078
		.458	11,6332
		.459	11,6586
		.460	11,6840
		.461	11,7094
		.462	11,7348
		.463	11,7602
		.464	11,7856
		.465	11,8110
		.466	11,8364
		.467	11,8618
		.468	11,8872
	15/32	.4687	11,9062
		.469	11,9126
		.470	11,9380
		.471	11,9634
		.472	11,9888
		.4724	12,0000
		.473	12,0142
		.474	12,0396
		.475	12,0650
		.476	12,0904
		.477	12,1158
		.478	12,1412
		.479	12,1666
		.480	12,1920
		.481	12,2174
		.482	12,2428
		.483	12,2682
		.484	12,2936
	31/64	.4844	12,3031
		.485	12,3190
		.486	12,3444
		.487	12,3698
		.488	12,3952
		.489	12,4206
		.490	12,4460
		.491	12,4714
		.492	12,4968
		.493	12,5222
		.494	12,5476
		.495	12,5730
		.496	12,5984
		.497	12,6238
		.498	12,6492
		.499	12,6746
	1/2	.500	12,7000

TABLE 2 Continued

Inch	mm	Inch	mm	Inch	mm	Inch	mm	Inch	mm
.501	12,7254	.600	15,2400	.701	17,8054	.800	20,3200	.901	22,8854
.502	12,7508	.601	15,2654	.702	17,8308	.801	20,3454	.902	22,9108
.503	12,7762	.602	15,2908	.703	17,8562	.802	20,3708	.903	22,9362
.504	12,8016	.603	15,3162	45/64 .7031	17,8594	.803	20,3962	.904	22,9616
.505	12,8270	.604	15,3416	.704	17,8816	.804	20,4216	.905	22,9870
.506	12,8524	.605	15,3670	.705	17,9070	.805	20,4470	.9055	23,0000
.507	12,8778	.606	15,3924	.706	17,9324	.806	20,4724	.906	23,0124
.508	12,9032	.607	15,4178	.707	17,9578	.807	20,4978	29/32 .9062	23,0187
.509	12,9286	.608	15,4432	.708	17,9832	.808	20,5232	.907	23,0378
.510	12,9540	.609	15,4686	.7087	18,0000	.809	20,5486	.908	23,0632
.511	12,9794	39/64 .6094	15,4781	.709	18,0086	.810	20,5740	.909	23,0886
.5118	13,0000	.610	15,4940	.710	18,0340	.811	20,5994	.910	23,1140
.512	13,0048	.611	15,5194	.711	18,0594	.812	20,6248	.911	23,1394
.513	13,0302	.612	15,5448	.712	18,0848	13/16 .8125	20,6375	.912	23,1648
.514	13,0556	.613	15,5702	.713	18,1102	.813	20,6502	.913	23,1902
.515	13,0810	.614	15,5956	.714	18,1356	.814	20,6756	.914	23,2156
33/64 .5156	13,0968	.615	15,6210	.715	18,1610	.815	20,7010	.915	23,2410
.516	13,1064	.616	15,6464	.716	18,1864	.816	20,7264	.916	23,2664
.517	13,1318	.617	15,6718	.717	18,2118	.817	20,7518	.917	23,2918
.518	13,1572	.618	15,6972	.718	18,2372	.818	20,7772	.918	23,3172
.519	13,1826	.619	15,7226	23/32 .7187	18,2562	.819	20,8026	.919	23,3426
.520	13,2080	.620	15,7480	.719	18,2626	.820	20,8280	.920	23,3680
.521	13,2334	.621	15,7734	.720	18,2880	.821	20,8534	.921	23,3934
.522	13,2588	.622	15,7988	.721	18,3134	.822	20,8788	59/64 .9219	23,4156
.523	13,2842	.623	15,8242	.722	18,3388	.823	20,9042	.922	23,4188
.524	13,3096	.624	15,8496	.723	18,3642	.824	20,9296	.923	23,4442
.525	13,3350	5/8 .625	15,8750	.724	18,3896	.825	20,9550	.924	23,4696
.526	13,3604	.626	15,9004	.725	18,4150	.826	20,9804	.925	23,4950
.527	13,3858	.627	15,9258	.726	18,4404	.8268	21,0000	.926	23,5204
.528	13,4112	.628	15,9512	.727	18,4658	.827	21,0058	.927	23,5458
.529	13,4366	.629	15,9766	.728	18,4912	.828	21,0312	.928	23,5712
.530	13,4620	.6299	16,0000	.729	18,5166	53/64 .8281	21,0344	.929	23,5966
.531	13,4874	.630	16,0020	.730	18,5420	.829	21,0566	.930	23,6220
17/32 .5312	13,4937	.631	16,0274	.731	18,5674	.830	21,0820	.931	23,6474
.532	13,5128	.632	16,0528	.732	18,5928	.831	21,1074	.932	23,6728
.533	13,5382	.633	16,0782	.733	18,6182	.832	21,1328	.933	23,6982
.534	13,5636	.634	16,1036	.734	18,6436	.833	21,1582	.934	23,7236
.535	13,5890	.635	16,1290	47/64 .7344	18,6532	.834	21,1836	.935	23,7490
.536	13,6144	.636	16,1544	.735	18,6690	.835	21,2090	.936	23,7744
.537	13,6398	.637	16,1798	.736	18,6944	.836	21,2344	.937	23,7998
.538	13,6652	.638	16,2052	.737	18,7198	.837	21,2598	15/16 .9375	23,8125
.539	13,6906	.639	16,2306	.738	18,7452	.838	21,2852	.938	23,8252
.540	13,7160	640	16,2560	.739	18,7706	.839	21,3106	.939	23,8506
.541	13,7414	41/64 .6406	16,2719	.740	18,7960	.840	21,3360	.940	23,8760
.542	13,7668	.641	16,2814	.741	18,8214	.841	21,3614	.941	23,9014
.543	13,7922	.642	16,3068	.742	18,8468	.842	21,3868	.942	23,9268
.544	13,8176	.643	16,3322	.743	18,8722	.843	21,4122	.943	23,9522
.545	13,8430	.644	16,3576	.744	18,8976	27/32 .8437	21,4312	.944	23,9776
.546	13,8684	.645	16,3830	.745	18,9230	.844	21,4376	.9449	24,0000
35/64 .5469	13,8906	.646	16,4084	.746	18,9484	.845	21,4630	.945	24,0030
.547	13,8938	.647	16,4338	.747	18,9738	.846	21,4884	.946	24,0284
.548	13,9192	.648	16,4592	.748	18,9992	.847	21,5138	.947	24,0538
.549	13,9446	.649	16,4846	.7480	19,0000	.848	21,5392	.948	24,0792
.550	13,9700	.650	16,5100	.749	19,0246	.849	21,5646	.949	24,1046
.551	13,9954	.651	16,5354	3/4 .750	19,0500	.850	21,5900	.950	24,1300
.5512	14,0000	.652	16,5608	.751	19,0754	.851	21,6154	.951	24,1554
.552	14,0208	.653	16,5862	.752	19,1008	.852	21,6408	.952	24,1808
.553	14,0462	.654	16,6116	.753	19,1262	.853	21,6662	.953	24,2062
.554	14,0716	.655	16,6370	.754	19,1516	.854	21,6916	61/64 .9531	24,2094
.555	14,0970	.656	16,6624	.755	19,1770	.855	21,7170	.954	24,2316
.556	14,1224	21/32 .6562	16,6687	.756	19,2024	.856	21,7424	.955	24,2570
.557	14,1478	.657	16,6878	.757	19,2278	.857	21,7678	.956	24,2824
.558	14,1732	.658	16,7132	.758	19,2532	.858	21,7932	.957	24,3078
.559	14,1986	.659	16,7386	.759	19,2786	.859	21,8186	.958	24,3332
.560	14,2240	.660	16,7640	.760	19,3040	55/64 .8594	21,8281	.959	24,3586
.561	14,2494	.661	16,7894	.761	19,3294	.860	21,8440	.960	24,3840
.562	14,2748	.662	16,8148	.762	19,3548	.861	21,8694	.961	24,4094
9/16 .5625	14,2875	.663	16,8402	.763	19,3802	.862	21,8948	.962	24,4348
.563	14,3002	.664	16,8656	.764	19,4056	.863	21,9202	.963	24,4602
.564	14,3256	.665	16,8910	.765	19,4310	.864	21,9456	.964	24,4856
.565	14,3510	.666	16,9164	49/64 .7656	19,4469	.865	21,9710	.965	24,5110
.566	14,3764	.667	16,9418	.766	19,4564	.866	21,9964	.966	24,5364
.567	14,4018	.668	16,9672	.767	19,4818	.8661	22,0000	.967	24,5618
.568	14,4272	.669	16,9926	.768	19,5072	.867	22,0218	.968	24,5872
.569	14,4526	.6693	17,0000	.769	19,5326	.868	22,0472	31/32 .9687	24,6062
.570	14,4780	.670	17,0180	.770	19,5580	.869	22,0726	.969	24,6126
.571	14,5034	.671	17,0434	.771	19,5834	.870	22,0980	.970	24,6380
.572	14,5288	43/64 .6719	17,0656	.772	19,6088	.871	22,1234	.971	24,6634
.573	14,5542	.672	17,0688	.773	19,6342	.872	22,1488	.972	24,6888
.574	14,5796	.673	17,0942	.774	19,6596	.873	22,1742	.973	24,7142
.575	14,6050	.674	17,1196	.775	19,6850	.874	22,1996	.974	24,7396
.576	14,6304	.675	17,1450	.776	19,7104	7/8 .875	22,2250	.975	24,7650
.577	14,6558	.676	17,1704	.777	19,7358	.876	22,2504	.976	24,7904
.578	14,6812	.677	17,1958	.778	19,7612	.877	22,2758	.977	24,8158
37/64 .5781	14,6844	.678	17,2212	.779	19,7866	.878	22,3012	.978	24,8412
.579	14,7066	.679	17,2466	.780	19,8120	.879	22,3266	.979	24,8666
.580	14,7320	.680	17,2720	.781	19,8374	.880	22,3520	.980	24,8920
.581	14,7574	.681	17,2974	25/32 .7812	19,8433	.881	22,3774	.981	24,9174
.582	14,7828	.682	17,3228	.782	19,8628	.882	22,4028	.982	24,9428
.583	14,8082	.683	17,3482	.783	19,8882	.883	22,4282	.983	24,9682
.584	14,8336	.684	17,3736	.784	19,9136	.884	22,4536	.984	24,9936
.585	14,8590	.685	17,3990	.785	19,9390	.885	22,4790	.9843	25,0000
.586	14,8844	.686	17,4244	.786	19,9644	.886	22,5044	63/64 .9844	25,0031
.587	14,9098	.687	17,4498	.787	19,9898	.887	22,5298	.985	25,0190
.588	14,9352	11/16 .6875	17,4625	.7874	20,0000	.888	22,5552	.986	25,0444
.589	14,9606	.688	17,4752	.788	20,0152	.889	22,5806	.987	25,0698
.590	14,9860	.689	17,5006	.789	20,0406	.890	22,6060	.988	25,0952
.5906	15,0000	.690	17,5260	.790	20,0660	57/64 .8906	22,6219	.989	25,1206
.591	15,0114	.691	17,5514	.791	20,0914	.891	22,6314	.990	25,1460
.592	15,0368	.692	17,5768	.792	20,1168	.892	22,6568	.991	25,1714
.593	15,0622	.693	17,6022	.793	20,1422	.893	22,6822	.992	25,1968
19/32 .5937	15,0812	.694	17,6276	.794	20,1676	.894	22,7076	.993	25,2222
.594	15,0876	.695	17,6530	.795	20,1930	.895	22,7330	.994	25,2476
.595	15,1130	.696	17,6784	.796	20,2184	.896	22,7584	.995	25,2730
.596	15,1384	.697	17,7038	51/64 .7969	20,2402	.897	22,7838	.996	25,2984
.597	15,1638	.698	17,7292	.797	20,2438	.898	22,8092	.997	25,3238
.598	15,1892	.699	17,7546	.798	20,2692	.899	22,8346	.998	25,3492
.599	15,2146	.700	17,7800	.799	20,2946	.900	22,8600	.999	25,3746
								1.000	25,4000

TABLE 3 Tap Drill Sizes

Tap	Tap Drill	Decimal Equivalent of Tap Drill
0-80	56	.0465
	3/64	.0469
1-64	54	.0550
	53	.0595
1-72	53	.0595
	1/16	.0625
2-56	51	.0670
	50	.0700
	49	.0730
2-64	50	.0700
	49	.0730
3-48	48	.0760
	5/64	.0781
	47	.0785
	46	.0810
	45	.0820
3-56	46	.0810
	45	.0820
	44	.0860
4-40	44	.0860
	43	.0890
	42	.0935
	3/32	.0938
4-48	42	.0935
	3/32	.0938
	41	.0960
5-40	40	.0980
	39	.0995
	38	.1015
	37	.1040
5-44	38	.1015
	37	.1040
	36	.1065
6-32	37	.1040
	36	.1065
	7/64	.1094
	35	.1100
	34	.1110
	33	.1130
6-40	34	.1110
	33	.1130
	32	.1160
8-32	29	.1360
	28	.1405
8-36	29	.1360

Tap	Tap Drill	Decimal Equivalent of Tap Drill
	28	.1405
	9/64	.1406
10-24	27	.1440
	26	.1470
	25	.1495
	24	.1520
	23	.1540
	5/32	.1563
	22	.1570
10-32	5/32	.1563
	22	.1570
	21	.1590
	20	.1610
	19	.1660
12-24	11/64	.1719
	17	.1730
	16	.1770
	15	.1800
	14	.1820
12-28	16	.1770
	15	.1800
	14	.1820
	13	.1850
	3/16	.1875
1/4-20	9	.1960
	8	.1990
	7	.2010
	13/64	2031
	6	.2040
	5	.2055
	4	.2090
1/4-28	3	.2130
	7/32	.2188
	2	.2210
5/16-18	F	.2570
	G	.2610
	17/64	.2656
	H	.2660
5/16-24	H	.2660
	I	.2720
	J	.2770
3/8-16	5/16	.3125
	O	.3160
	P	.3230
3/8-24	21/64	.3281

Tap	Tap Drill	Decimal Equivalent of Tap Drill
	Q	.3320
	R	.3390
7/16-14	T	.3580
	23/64	.3594
	U	.3680
	3/8	.3750
	V	.3770
7/16-20	W	.3860
	25/64	.3906
	X	.3970
1/2-13	27/64	.4219
	7/16	.4375
1/2-20	29/64	.4531
9/16-12	15/32	.4688
	31/64	.4844
9/16-18	1/2	.500
	33/64	.5156
5/8-11	17/32	.5313
	35/64	.5469
5/8-18	9/16	.5625
	37/64	.5781
3/4-10	41/64	.6406
	21/32	.6563
3/4-16	11/16	.6875
7/8-9	49/64	.7656
	25/32	.7812
7/8-14	51/64	.7969
	13/16	.8125
1″-8	55/64	.8594
	7/8	.875
	57/64	.8906
	29/32	.9063
1-12	29/32	.9063
	59/64	.9219
	15/16	.9375
1-14	59/64	.9219
	15/16	.9375
1 1/8-7	31/32	.9688
	63/64	.9844
	1″	1.0000

TABLE 3 Continued

Tap	Tap Drill	Decimal Equivalent of Tap Drill
	$1\frac{1}{64}$	1.0156
$1\frac{1}{8}$-12	$1\frac{1}{32}$	1.0313
	$1\frac{3}{64}$	1.0469
$1\frac{1}{4}$-7	$1\frac{3}{32}$	1.0938
	$1\frac{7}{64}$	1.1094
	$1\frac{1}{8}$	1.1250
$1\frac{1}{4}$-12	$1\frac{5}{32}$	1.1563
	$1\frac{11}{64}$	1.1719
$1\frac{3}{8}$-6	$1\frac{3}{16}$	1.1875
	$1\frac{13}{64}$	1.2031
	$1\frac{7}{32}$	1.2188
	$1\frac{15}{64}$	1.2344
$1\frac{3}{8}$-12	$1\frac{9}{32}$	1.2813
	$1\frac{19}{64}$	1.2969
$1\frac{1}{2}$-6	$1\frac{5}{16}$	1.3125
	$1\frac{21}{64}$	1.3281
	$1\frac{11}{32}$	1.3438
	$1\frac{23}{64}$	1.3594
$1\frac{1}{2}$-12	$1\frac{13}{32}$	1.4063
	$1\frac{27}{64}$	1.4219

Pipe (American Standard) For use with Taper Pipe Taps

Nominal Pipe Size Inches	Threads per Inch	Tap Drill Size
$\frac{1}{16}$	..	R
$\frac{1}{8}$	27	$\frac{11}{32}$
$\frac{1}{4}$	18	$\frac{7}{16}$
$\frac{3}{8}$	18	$\frac{37}{64}$
$\frac{1}{2}$	14	$\frac{23}{32}$
$\frac{3}{4}$	14	$\frac{59}{64}$
1	$11\frac{1}{2}$	$1\frac{5}{32}$
$1\frac{1}{4}$	$11\frac{1}{2}$	$1\frac{1}{2}$
$1\frac{1}{2}$	$11\frac{1}{2}$	$1\frac{47}{64}$
2	$11\frac{1}{2}$	$2\frac{7}{32}$
$2\frac{1}{2}$	8	$2\frac{5}{8}$
3	8	$3\frac{1}{4}$
$3\frac{1}{2}$	8	$3\frac{3}{4}$
4	8	$4\frac{1}{4}$

TABLE 4 Wire Gages and Metric Equivalents

Gage No.	American or Brown & Sharpe's (in.)	(mm)	Gage No.	American or Brown & Sharpe's (in.)	(mm)
000000	.5800	14.732	21	.02846	.723
00000	.5165	13.119			
0000	.4600	11.684	22	.02535	.644
000	.4096	10.404	23	.02257	.573
00	.3648	9.266	24	.02010	.511
0	.3249	8.252	25	.01790	.455
			26	.01594	.405
1	.2893	7.348	27	.01420	.361
2	.2576	6.543	28	.01264	.321
3	.2294	5.827			
4	.2043	5.189	29	.01126	.286
5	.1819	4.620	30	.01003	.255
6	.1620	4.115	31	.008928	.227
7	.1443	3.665	32	.007950	.202
			33	.007080	.180
8	.1285	3.264	34	.006305	.160
9	.1144	2.906	35	.005615	.143
10	.1019	2.588			
11	.09074	2.305	36	.005000	.127
12	.08081	2.053	37	.004453	.113
13	.07196	1.828	38	.003965	.101
14	.06408	1.628	39	.003531	.090
			40	.003145	.080
15	.05707	1.450	41	.002800	.071
16	.05082	1.291	42	.002494	.063
17	.04526	1.150			
18	.04030	1.024	43	.002221	.056
19	.03589	.912	44	.001978	.050
20	.03196	.812			

TABLE 5A Tapers

Useful Information on Tapers

Taper Per Foot	Amount of Taper Length Tapered Portion (in.)													
	1/32	1/16	1/8	3/16	1/4	5/16	3/8	7/16	1/2	9/16	5/8	11/16	3/4	13/16
1/16	.0002	.0003	.0007	.0010	.0013	.0016	.0020	.0023	.0026	.0029	.0033	.0036	.0039	.0042
3/32	.0002	.0005	.0010	.0015	.0020	.0024	.0029	.0034	.0039	.0044	.0049	.0054	.0059	.0063
1/8	.0003	.0007	.0013	.0020	.0026	.0033	.0039	.0046	.0052	.0059	.0065	.0072	.0078	.0085
1/4	.0007	.0013	.0026	.0039	.0052	.0065	.0078	.0091	.0104	.0117	.0130	.0143	.0156	.0169
3/8	.0010	.0020	.0039	.0059	.0078	.0098	.0117	.0137	.0156	.0176	.0195	.0215	.0234	.0254
1/2	.0013	.0026	.0052	.0078	.0104	.0130	.0156	.0182	.0208	.0234	.0260	.0286	.0312	.0339
5/8	.0016	.0033	.0065	.0098	.0130	.0163	.0195	.0228	.0260	.0293	.0326	.0358	.0391	.0423
3/4	.0020	.0039	.0078	.0117	.0156	.0195	.0234	.0273	.0312	.0352	.0391	.0430	.0469	.0508
1	.0026	.0052	.0104	.0156	.0208	.0260	.0312	.0365	.0417	.0469	.0521	.0573	.0625	.0677
1 1/4	.0063	.0065	.0130	.0195	.0260	.0326	.0391	.0456	.0521	.0586	.0651	.0716	.0781	.0846

Taper Per Foot	Amount of Taper Length Tapered Portion (in.)													
	7/8	15/16	1	2	3	4	5	6	7	8	9	10	11	12
1/16	.0046	.0049	.0052	.0104	.0156	.0208	.0260	.0312	.0365	.0417	.0469	.0521	.0573	.0625
3/32	.0068	.0073	.0078	.0156	.0234	.0312	.0391	.0469	.0547	.0625	.0703	.0781	.0859	.0937
1/8	.0091	.0098	.0104	.0208	.0312	.0417	.0521	.0625	.0729	.0833	.0937	.1042	.1146	.1250
1/4	.0182	.0195	.0208	.0417	.0625	.0833	.1042	.1250	.1458	.1667	.1875	.2083	.2292	.2500
3/8	.0273	.0293	.0312	.0625	.0937	.1250	.1562	.1875	.2187	.2500	.2812	.3125	.3437	.3750
1/2	.0365	.0391	.0417	.0833	.1250	.1667	.2083	.2500	.2917	.3333	.3750	.4167	.4583	.5000
5/8	.0456	.0488	.0521	.1042	.1562	.2083	.2604	.3125	.3646	.4167	.4687	.5208	.5729	.6250
3/4	.0547	.0586	.0625	.1250	.1875	.2500	.3125	.3750	.4375	.5000	.5625	.6250	.6875	.7500
1	.0729	.0781	.0833	.1667	.2500	.3333	.4167	.5000	.5833	.6667	.7500	.8333	.9167	1.0000
1 1/4	.0911	.0977	.1042	.2083	.3125	.4167	.5208	.6250	.7292	.8333	.9375	1.0417	1.1458	1.2500

Taper Per Foot	Amount of Taper Length Tapered Portion (in.)													
	13	14	15	16	17	18	19	20	21	22	23	24	...	...
1/16	.0677	.0729	.0781	.0833	.0885	.0937	.0990	.1042	.1094	.1146	.1198	.1250	...	...
3/32	.1016	.1094	.1172	.1250	.1328	.1406	.1484	.1562	.1641	.1719	.1797	.1875	...	...
1/8	.1354	.1458	.1562	.1667	.1771	.1875	.1979	.2083	.2187	.2292	.2396	.2500	...	...
1/4	.2708	.2917	.3125	.3333	.3542	.3750	.3958	.4167	.4375	.4583	.4792	.5000	...	...
3/8	.4062	.4375	.4687	.5000	.5312	.5625	.5937	.6250	.6562	.6875	.7187	.7500	...	...
1/2	.5417	.5833	.6250	.6667	.7083	.7500	.7917	.8333	.8750	.9167	.9583	1.0000	...	...
5/8	.6771	.7292	.7812	.8333	.8854	.9375	.9896	1.0417	1.0937	1.1458	1.1979	1.2500	...	...
3/4	.8125	.8750	.9375	1.0000	1.0625	1.1250	1.1875	1.2500	1.3125	1.3750	1.4375	1.5000	...	...
1	1.0833	1.1667	1.2500	1.3333	1.4167	1.5000	1.5833	1.6667	1.7500	1.8333	1.9167	2.0000	...	...
1 1/4	1.3542	1.4583	1.5625	1.6667	1.7708	1.8750	1.9792	2.0833	2.1875	2.2917	2.3958	2.5000	...	...

TABLE 5B Tapers and Angles

Taper per Foot	Included Angle Deg.	Min.	Sec.	Angle with Center Line Deg.	Min.	Sec.	Taper per Inch	Taper per Inch from Center Line
$\frac{1}{8}$	0	35	47	0	17	54	.010416	.005208
$\frac{3}{16}$	0	53	44	0	26	52	.015625	.007812
$\frac{1}{4}$	1	11	38	0	35	49	.020833	.010416
$\frac{5}{16}$	1	29	31	0	44	46	.026042	.013021
$\frac{3}{8}$	1	47	25	0	53	42	.031250	.015625
$\frac{7}{16}$	2	5	18	1	2	39	.036458	.018229
$\frac{1}{2}$	2	23	12	1	11	36	.041667	.020833
$\frac{9}{16}$	2	41	7	1	20	34	.046875	.023438
$\frac{5}{8}$	2	59	3	1	29	31	.052084	.026042
$\frac{11}{16}$	3	16	56	1	38	28	.057292	028646
$\frac{3}{4}$	3	34	48	1	47	24	.062500	.031250
$\frac{13}{16}$	3	52	42	1	56	21	.067708	.033854
$\frac{7}{8}$	4	10	32	2	5	16	.072917	.036456
$\frac{15}{16}$	4	28	26	2	14	13	.078125	.039063
1	4	46	19	2	23	10	.083330	.041667
$1\frac{1}{4}$	5	57	45	2	58	53	.104166	.052084
$1\frac{1}{2}$	7	9	10	3	34	35	.125000	.062500
$1\frac{3}{4}$	8	20	28	4	10	14	.145833	.072917
2	9	31	37	4	45	49	.166666	.083332
$2\frac{1}{2}$	11	53	38	5	56	49	.208333	.104166
3	14	2	0	7	1	0	.250000	.125000
$3\frac{1}{2}$	16	35	39	8	17	49	.291666	.145833
4	18	55	31	9	27	44	.333333	.166666
$4\frac{1}{2}$	21	14	20	10	37	10	.375000	.187500
5	23	32	12	11	46	6	.416666	.208333
6	28	4	20	14	2	10	.500000	.250000

TABLE 6 General Measurements

Measurement Rules

Length

Side of square of equal periphery as circle = diameter × 0.7854.

Diameter of circle of equal periphery as square = side × 1.2732.

Length of arc = number of degrees × diameter × 0.008727

Area

Triangle = base × half perpendicular height.

Parallelogram = base × perpendicular.

Trapezoid = half the sum of the parallel sides × perpendicular height.

Trapezium, divide two triangles and find area of the triangles.

Parabola = base × $\frac{2}{3}$ height.

Ellipse = long diameter × short diameter × 0.7854.

Regular polygon = sum of sides × half perpendicular distance from center to sides.

Surface of cylinder = circumference × length + area of two ends.

Surface of pyramid or cone = circumference of base × $\frac{1}{2}$ of the slant height + area of the base.

Surface of a frustrum of a regular right pyramid or cone = sum of peripheries or circumferences of the two ends × half slant height + area of both ends.

Area of rectangle = length × breadth.

TABLE 6 Continued

General Information

To find the circumference of a circle, multiply diameter by 3.1416.

To find diameter of a circle, multiply circumference by .31831.

To find area of a circle, multiply square of radius by 3.1416.

Area of rectangle: Length multiplied by breadth. Doubling the diameter of a circle increases its area four times.

To find area of a triangle, multiply base by $\frac{1}{2}$ perpendicular height.

To find side of square inscribed in a circle, multiply diameter by 0.7071, or multiply circumference by 0.2251, or divide circumference by 4.4428.

To find diameter of circle circumscribing a square, multiply one side by 1.4142.

A side multiplied by 4.4428 equals circumference of its circumscribing circle.

A side multiplied by 1.128 equals diameter of a circle of equal area.

A side multiplied by 3.547 equals circumference of a circle of equal area.

Equivalent Measures

Measures of Length

1 Meter =

39.37	inches
3.28083	feet
1.09361	yards
1000.	millimeters
100.	centimeters
10.	decimeters
0.001	kilometers

1 Centimeter =

0.3937	inch
0.0328083	foot
10.	millimeters
0.01	meters

1 Millimeter =

39.370	mils
0.03937	inch (or $\frac{1}{25}$ inch nearly)
0.001	meter

Kilometer =

3280.83	feet
1093.61	yards
0.62137	mile
1000.	meters

Mil =

0.001	inch
0.02540	millimeter
0.00254	centimeter

1 Inch =

1000.	mils
0.0833	foot
0.02777	yard
25.40	millimeters
2.540	centimeters

1 Foot =

12.	inches
1.33333	yard
0.0001893	miles
0.30480	meter
30.480	centimeters

1 Yard =

36.	inches
3.	feet
0.0005681	mile
0.914402	meter

1 Mile =

63360.	inches
5280.	feet
1760.	yards
320.	rods
8.	furlongs
1609.35	meters
1.60935	kilometers

Measures of Volume and Capacity

1 Cubic Meter =

61023.4	cubic in.
35.3145	cubic feet
1.30794	cubic yd
1000.	liters
264.170	gallons U.S. liquid = 231 cubic in.

1 Cubic decimeter =

61.0234	cubic in.
0.0353145	cubic foot
0.26417	U.S. liquid gallon
1000.	cubic centimeters
0.001	cubic meter

1 Cubic Centimeter =

0.0000353	cubic foot
0.0610234	cubic inch
1000.0	cubic millimeters
0.001	liter

1 Cubic millimeter =

0.000061023	cubic inch
0.0000000353	cubic foot
0.001	cubic centimeter

1 Liter =

1.	cubic decimeter
61.0234	cubic inches
0.353145	cubic foot
1000.	cubic centimeters or centiliters
0.001	cubic meter
0.26417	U.S. gallon liquid
1.0567	U.S. quart
2.202	lbs. of water at 62 degrees Fahrenheit

1 Cubic Yard =

46656.	cubic inches
27.	cubic feet
0.76456	cubic meter

1 Cubic foot =

1728.	cubic inches
0.03703703	cubic yard
28.317	cubic decimeters or liters
0.028317	cubic meter
7.4805	gallons

1 Cubic inch =
16.3872 cubic centimeters

1 Gallon (British) =
4.54374 liters

1 Gallon (U.S.) =
3.78543 liters

Measures of Weight

1 Gram =

15.432	grains
0.0022046	lb (avoir.)
0.03527	oz (avoir.)

1 Kilogram =

1000.	grams
2.20462	lb (avoir.)
35.2739	oz (avoir.)

1 Metric ton =

2204.62	pounds
0.984206	ton of 2240 pounds
22.0462	cwt
1.10231	ton of 2000 pounds =
1000	kilograms

1 Grain =
0.064799 grams

1 Ounce =

437.5	grains
0.0625	pounds
28.3496	grams

1 Pound =

7000.	grains
16.	ounces
453.593	grams
0.453593	kilograms

1 Ton (2240 pounds) =

1.01605	metric tons
1016.05	kilograms

TABLE 7 Density and Melting Points for Metals and Other Materials

Density or Specific Gravity of Metals and Alloys

Material	Specific Gravity	Weight in Lb. Cu. Ft.	Weight in Lb. Cu. In.	Cu. In. in One Lb.
Aluminum Cast	2.569	160.	.093	10.80
Aluminum Wrought	2.681	167.	.097	10.35
Aluminum Bronze	7.787	485.	.281	3.56
Antimony	6.712	418.	.242	4.13
Arsenic	5.748	358.	.207	4.83
Bismuth	9.827	612.	.354	2.82
Benedict nickel	8.691	542.6	.3140	3.19
Gold (pure)	19.316	1203.	.696	1.44
Standard 22 carat fine	17.502	1090.	.631	1.59
Iron – cast	6.904	430.	.249	4.02
	7.386	499.	.266	3.76
	7.209	464.	.260	3.85
Iron – wrought	7.547	470.	.272	3.56
	7.803	486.	.281	3.68
	7.707	480.	.278	3.60
Lead – cast	11.368	708.	.410	2.44
Lead sheet	11.432	712.	.412	2.43
Manganese	8.012	499.	.289	2.46
Nickel – cast	8.285	516.	.299	3.35
Nickel rolled	8.687	541.	.313	3.19
Platinum	21.516	1340.	.775	1.29
Silver	10.517	655.	.379	2.64
Steel	7.820	487.	.282	3.55
	7.916	493.	.285	3.51
	7.868	490.	.284	3.53
Tin	7.418	462.	.267	3.74
White metal (Babbitt's)	7.322	456.	.264	3.79
Zinc – cast	6.872	428.	.248	4.05
Zinc sheet	7.209	449.	.260	3.84

Approximate Melting Points of Metals and Various Substances

Solid	Degrees Centigrade	Degrees Fahrenheit
Alloy – 3 Lead, 2 Tin, 5 Bismuth	100	212
Alloy – 1 Lead 1½ Tin	200	392
Alloy – 1 Lead 1 Tin	215	419
Aluminum	657.3	1215
Antimony	430 to 630	806 to 1166
Bismuth	269.2	517
Brass	1030	1886
Bronze	920	1688
Cadmium	320	608
Chromium	1487 to 1515	2709 to 2749
Cobalt	1463 to 1500	2665 to 2732
Copper	1054 to 1084	1929 to 1893
Gold	1045 to 1064	1913 to 1947
Iridium	1950 to 2500	3542 to 4532
Iron – Cast Gray	1220 to 1530	2228 to 2786
Iron – Cast White	1050 to 1135	1922 to 2075
Iron – Wrought	1500 to 1600	2732 to 1912
Lead	327	620
Magnesium	750	1382
Manganese	1207 to 1245	2205 to 2273
Mercury	−39.7	−39.5
Nickel	1435	2615
Osmium	2500	4532
Palladium	1546 to 1900	2815 to 3452
Platinum	1753 to 1780	3187 to 3276
Potassium	62	144
Rhodium	2000	3632
Ruthenium	2000+	3632
Silver	960	1760
Sodium	79 to 95	174.2 to 203
Steel	1300 to 1378	2372 to 2532
Steel – Hard	1410	2570
Steel – Mild	1475	2687
Tin	232	449
Titanium	1700	3092
Tungsten	3000	5432
Vanadium	1775	3227
Zinc	419	786
Phosphorous	44.4	112
Calcium	760	1400

TABLE 8 Right Triangle Solution Formulas

∠A	∠B	Side a	Side b	Side c
$\sin A = \frac{a}{c}$		$a = c \times \sin A$		$c = \frac{a}{\sin A}$
$\cos A = \frac{b}{c}$			$b = c \times \cos A$	$c = \frac{b}{\cos A}$
$\tan A = \frac{a}{b}$		$a = b \times \tan A$	$b = \frac{a}{\tan A}$	
$\cot A = \frac{b}{a}$		$a = \frac{b}{\cot A}$	$b = a \times \cot A$	
$\sec A = \frac{c}{b}$			$b = \frac{c}{\sec A}$	$c = b \times \sec A$
$\csc A = \frac{c}{a}$		$a = \frac{c}{\csc A}$		$c = a \times \csc A$
	$\sin B = \frac{b}{c}$		$b = c \times \sin B$	$c = \frac{b}{\sin B}$
	$\cos B = \frac{a}{c}$	$a = c \times \cos B$		$c = \frac{a}{\cos B}$
	$\tan B = \frac{b}{a}$	$a = \frac{b}{\tan B}$	$b = a \times \tan B$	
	$\cot B = \frac{a}{b}$		$b = \frac{a}{\cot B}$	
	$\sec B = \frac{c}{a}$	$a = \frac{c}{\sec B}$		$c = a \times \sec B$
	$\csc B = \frac{c}{b}$		$b = \frac{c}{\csc B}$	$c = b \csc B$

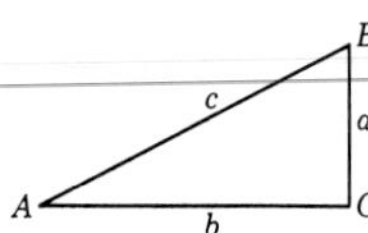

TABLE 9 Natural Trigonometric Functions

	0°						1°						2°						3°						
′	sin	cos	tan	cot	sec	cosec	sin	cos	tan	cot	sec	cosec	sin	cos	tan	cot	sec	cosec	sin	cos	tan	cot	sec	cosec	′
0	.00000	1.0000	.00000	Infinite	1.0000	Infinite	.01745	.99985	.01745	57.290	1.0001	57.299	.03490	.99939	.03492	28.636	1.0006	28.654	.05234	.99863	.05241	19.081	1.0014	19.107	60
1	.00029	.0000	.00029	3437.7	.0000	3437.7	.01774	.99984	.01775	56.350	.0001	56.359	.03519	.99938	.03521	28.399	.0006	28.417	.05263	.99861	.05270	18.975	.0014	19.002	59
2	.00058	.0000	.00058	1718.9	.0000	1718.9	.01803	.99984	.01804	55.441	.0001	55.450	.03548	.99937	.03550	28.166	.0006	28.184	.05292	.99860	.05299	18.871	.0014	18.897	58
3	.00087	.0000	.00087	1145.9	.0000	1145.9	.01832	.99983	.01833	54.561	.0002	54.570	.03577	.99936	.03579	27.937	.0006	27.955	.05321	.99858	.05328	18.768	.0014	18.794	57
4	.00116	.0000	.00116	859.44	.0000	859.44	.01861	.99983	.01862	53.708	.0002	53.718	.03606	.99935	.03608	27.712	.0006	27.730	.05350	.99857	.05357	18.665	.0014	18.692	56
5	.00145	1.0000	.00145	687.55	.0000	687.55	.01891	.99982	.01891	52.882	.0002	52.891	.03635	.99934	.03638	27.490	.0007	27.508	.05379	.99855	.05387	18.564	.0014	18.591	55
6	.00174	.0000	.00174	572.96	.0000	572.96	.01920	.99981	.01920	52.081	.0002	52.090	.03664	.99933	.03667	27.271	.0007	27.290	.05408	.99854	.05416	18.464	.0015	18.491	54
7	.00204	.0000	.00204	491.11	.0000	491.11	.01949	.99981	.01949	51.303	.0002	51.313	.03693	.99932	.03696	27.056	.0007	27.075	.05437	.99852	.05445	18.365	.0015	18.393	53
8	.00233	.0000	.00233	429.72	.0000	429.72	.01978	.99980	.01978	50.548	.0002	50.558	.03722	.99931	.03725	26.845	.0007	26.864	.05466	.99850	.05474	18.268	.0015	18.295	52
9	.00262	.0000	.00262	381.97	.0000	381.97	.02007	.99980	.02007	49.816	.0002	49.826	.03751	.99930	.03754	26.637	.0007	26.655	.05495	.99849	.05503	18.171	.0015	18.198	51
10	.00291	0.99999	.00291	343.77	.0000	343.77	.02036	.99979	.02036	49.104	.0002	49.114	.03781	.99928	.03783	26.432	.0007	26.450	.05524	.99847	.05532	18.075	.0015	18.103	50
11	.00320	.99999	.00320	312.52	.0000	312.52	.02065	.99979	.02066	48.412	.0002	48.422	.03810	.99927	.03812	26.230	.0007	26.249	.05553	.99846	.05562	17.980	.0015	18.008	49
12	.00349	.99999	.00349	286.48	.0000	286.48	.02094	.99978	.02095	47.739	.0002	47.750	.03839	.99926	.03842	26.031	.0007	26.050	.05582	.99844	.05591	17.886	.0016	17.914	48
13	.00378	.99999	.00378	264.44	.0000	264.44	.02123	.99977	.02124	47.085	.0002	47.096	.03868	.99925	.03871	25.835	.0007	25.854	.05611	.99842	.05620	17.793	.0016	17.821	47
14	.00407	.99999	.00407	245.55	.0000	245.55	.02152	.99977	.02153	46.449	.0002	46.460	.03897	.99924	.03900	25.642	.0008	25.661	.05640	.99841	.05649	17.701	.0016	17.730	46
15	.00436	.99999	.00436	229.18	.0000	229.18	.02181	.99976	.02182	45.829	.0002	45.840	.03926	.99923	.03929	25.452	.0008	25.471	.05669	.99839	.05678	17.610	.0016	17.639	45
16	.00465	.99999	.00465	214.86	.0000	214.86	.02210	.99975	.02211	45.226	.0002	45.237	.03955	.99922	.03958	25.264	.0008	25.284	.05698	.99837	.05707	17.520	.0016	17.549	44
17	.00494	.99999	.00494	202.22	.0000	202.22	.02240	.99975	.02240	44.638	.0002	44.650	.03984	.99921	.03987	25.080	.0008	25.100	.05727	.99836	.05737	17.431	.0016	17.460	43
18	.00524	.99999	.00524	190.98	.0000	190.99	.02269	.99974	.02269	44.066	.0002	44.077	.04013	.99919	.04016	24.898	.0008	24.918	.05756	.99834	.05766	17.343	.0017	17.372	42
19	.00553	.99998	.00553	180.93	.0000	180.93	.02298	.99974	.02298	43.508	.0003	43.520	.04042	.99918	.04045	24.718	.0008	24.739	.05785	.99832	.05795	17.256	.0017	17.285	41
20	.00582	.99998	.00582	171.88	.0000	171.89	.02326	.99973	.02327	42.964	.0003	42.976	.04071	.99917	.04075	24.542	.0008	24.562	.05814	.99831	.05824	17.169	.0017	17.198	40
21	.00611	.99998	.00611	163.70	.0000	163.70	.02356	.99972	.02357	42.433	.0003	42.445	.04100	.99916	.04104	24.367	.0008	24.388	.05843	.99829	.05853	17.084	.0017	17.113	39
22	.00640	.99998	.00640	156.26	.0000	156.26	.02385	.99971	.02386	41.916	.0003	41.928	.04129	.99915	.04133	24.196	.0008	24.216	.05872	.99827	.05883	16.999	.0017	17.028	38
23	.00669	.99998	.00669	149.46	.0000	149.47	.02414	.99971	.02415	41.410	.0003	41.423	.04158	.99913	.04162	24.026	.0009	24.047	.05902	.99826	.05912	16.915	.0017	16.944	37
24	.00698	.99997	.00698	143.24	.0000	143.24	.02443	.99970	.02444	40.917	.0003	40.930	.04187	.99912	.04191	23.859	.0009	23.880	.05931	.99824	.05941	16.832	.0018	16.861	36
25	.00727	.99997	.00727	137.51	.0000	137.51	.02472	.99969	.02473	40.436	.0003	40.448	.04217	.99911	.04220	23.694	.0009	23.716	.05960	.99822	.05970	16.750	.0018	16.779	35
26	.00756	.99997	.00756	132.22	.0000	132.22	.02501	.99969	.02502	39.965	.0003	39.978	.04246	.99910	.04249	23.532	.0009	23.553	.05989	.99820	.05999	16.668	.0018	16.698	34
27	.00785	.99997	.00785	127.32	.0000	127.32	.02530	.99968	.02531	39.506	.0003	39.518	.04275	.99908	.04279	23.372	.0009	23.393	.06018	.99819	.06029	16.587	.0018	16.617	33
28	.00814	.99997	.00814	122.77	.0000	122.78	.02559	.99967	.02560	39.057	.0003	39.069	.04304	.99907	.04308	23.214	.0009	23.235	.06047	.99817	.06058	16.507	.0018	16.538	32
29	.00843	.99996	.00844	118.54	.0000	118.54	.02589	.99966	.02589	38.618	.0003	38.631	.04333	.99906	.04337	23.058	.0009	23.079	.06076	.99815	.06087	16.428	.0018	16.459	31
30	.00873	.99996	.00873	114.59	.0000	114.59	.02618	.99966	.02618	38.188	.0003	38.201	.04362	.99905	.04366	22.904	.0009	22.925	.06105	.99813	.06116	16.350	.0019	16.380	30
31	.00902	.99996	.00902	110.89	.0000	110.90	.02647	.99965	.02648	37.769	.0003	37.782	.04391	.99903	.04395	22.752	.0010	22.774	.06134	.99812	.06145	16.272	.0019	16.303	29
32	.00931	.99996	.00931	107.43	.0000	107.43	.02676	.99964	.02677	37.358	.0003	37.371	.04420	.99902	.04424	22.602	.0010	22.624	.06163	.99810	.06175	16.195	.0019	16.226	28
33	.00960	.99995	.00960	104.17	.0000	104.17	.02705	.99963	.02706	36.956	.0004	36.969	.04449	.99901	.04453	22.454	.0010	22.476	.06192	.99808	.06204	16.119	.0019	16.150	27
34	.00989	.99995	.00989	101.11	.0000	101.11	.02734	.99963	.02735	36.563	.0004	36.576	.04478	.99900	.04483	22.308	.0010	22.330	.06221	.99806	.06233	16.043	.0019	16.075	26
35	.01018	.99995	.01018	98.218	.0000	98.223	.02763	.99962	.02764	36.177	.0004	36.191	.04507	.99898	.04512	22.164	.0010	22.186	.06250	.99804	.06262	15.969	.0019	16.000	25
36	.01047	.99994	.01047	95.489	.0000	95.495	.02792	.99961	.02793	35.800	.0004	35.814	.04536	.99897	.04541	22.022	.0010	22.044	.06279	.99803	.06291	15.894	.0020	15.926	24
37	.01076	.99994	.01076	92.908	.0000	92.914	.02821	.99960	.02822	35.431	.0004	35.445	.04565	.99896	.04570	21.881	.0010	21.904	.06308	.99801	.06321	15.821	.0020	15.853	23
38	.01105	.99994	.01105	90.463	.0001	90.469	.02850	.99959	.02851	35.069	.0004	35.084	.04594	.99894	.04599	21.742	.0010	21.765	.06337	.99799	.06350	15.748	.0020	15.780	22
39	.01134	.99993	.01134	88.143	.0001	88.149	.02879	.99958	.02880	34.715	.0004	34.729	.04623	.99893	.04628	21.606	.0011	21.629	.06366	.99797	.06379	15.676	.0020	15.708	21
40	.01163	.99993	.01164	85.940	.0001	85.946	.02908	.99958	.02910	34.368	.0004	34.382	.04652	.99892	.04657	21.470	.0011	21.494	.06395	.99795	.06408	15.605	.0020	15.637	20
41	.01193	.99993	.01193	83.843	.0001	83.849	.02937	.99957	.02939	34.027	.0004	34.042	.04681	.99890	.04687	21.337	.0011	21.360	.06424	.99793	.06437	15.534	.0021	15.566	19
42	.01222	.99992	.01222	81.847	.0001	81.853	.02967	.99956	.02968	33.693	.0004	33.708	.04711	.99889	.04716	21.205	.0011	21.228	.06453	.99791	.06467	15.464	.0021	15.496	18
43	.01251	.99992	.01251	79.943	.0001	79.950	.02996	.99955	.02997	33.366	.0004	33.381	.04740	.99888	.04745	21.075	.0011	21.098	.06482	.99790	.06496	15.394	.0021	15.427	17
44	.01280	.99992	.01280	78.126	.0001	78.133	.03025	.99954	.03026	33.045	.0004	33.060	.04769	.99886	.94774	20.946	.0011	20.970	.06511	.99788	.06525	15.325	.0021	15.358	16
45	.01309	.99991	.01309	76.390	.0001	76.396	.03054	.99953	.03055	32.730	.0005	32.745	.04798	.99885	.04803	20.819	.0011	20.843	.06540	.99786	.06554	15.257	.0021	15.290	15
46	.01338	.99991	.01338	74.729	.0001	74.736	.03083	.99952	.03084	32.421	.0005	32.437	.04827	.99883	.04832	20.693	.0012	20.717	.06569	.99784	.06583	15.189	.0022	15.222	14
47	.01367	.99991	.01367	73.139	.0001	73.146	.03112	.99951	.03113	32.118	.0005	32.134	.04856	.99882	.04862	20.569	.0012	20.593	.06598	.99782	.06613	15.122	.0022	15.155	13
48	.01396	.99990	.01396	71.615	.0001	71.622	.03141	.99951	.03143	31.820	.0005	31.836	.04885	.99881	.04891	20.446	.0012	20.471	.06627	.99780	.06642	15.056	.0022	15.089	12
49	.01425	.99990	.01425	70.153	.0001	70.160	.03170	.99950	.03172	31.528	.0005	31.544	.04914	.99879	.04920	20.325	.0012	20.350	.06656	.99778	.06671	14.990	.0022	15.023	11
50	.01454	.99989	.01454	68.750	.0001	68.757	.03199	.99949	.03201	31.241	.0005	31.257	.04943	.99878	.04949	20.205	.0012	20.230	.06685	.99776	.06700	14.924	.0022	14.958	10
51	.01483	.99989	.01484	67.402	.0001	67.409	.03228	.99948	.03230	30.960	.0005	30.976	.04972	.99876	.04978	20.087	.0012	20.112	.06714	.99774	.06730	14.860	.0023	14.893	9
52	.01512	.99988	.01513	66.105	.0001	66.113	.03257	.99947	.03259	30.683	.0005	30.699	.05001	.99875	.05007	19.970	.0012	19.995	.06743	.99772	.06759	14.795	.0023	14.829	8
53	.01542	.99988	.01542	64.858	.0001	64.866	.03286	.99946	.03288	30.411	.0005	30.428	.05030	.99873	.05037	19.854	.0013	19.880	.06772	.99770	.06788	14.732	.0023	14.765	7
54	.01571	.99988	.01571	63.657	.0001	63.664	.03315	.99945	.03317	30.145	.0005	30.161	.05059	.99872	.05066	19.740	.0013	19.766	.06801	.99768	.06817	14.668	.0023	14.702	6
55	.01600	.99987	.01600	62.499	.0001	62.507	.03344	.99944	.03346	29.882	.0005	29.899	.05088	.99870	.05095	19.627	.0013	19.653	.06830	.99766	.06846	14.606	.0023	14.640	5
56	.01629	.99987	.01629	61.383	.0001	61.391	.03374	.99943	.03375	29.624	.0006	29.641	.05117	.99869	.05124	19.515	.0013	19.541	.06859	.99764	.06876	14.544	.0024	14.578	4
57	.01658	.99987	.01658	60.306	.0001	60.314	.03403	.99942	.03405	29.371	.0006	29.388	.05146	.99867	.05153	19.405	.0013	19.431	.06888	.99762	.06905	14.482	.0024	14.517	3
58	.01687	.99986	.01687	59.266	.0001	59.274	.03432	.99941	.03434	29.122	.0006	29.139	.05175	.99866	.05182	19.296	.0013	19.322	.06918	.99760	.06934	14.421	.0024	14.456	2
59	.01716	.99985	.01716	58.261	.0001	58.270	.03461	.99940	.03463	28.877	.0006	28.894	.05204	.99864	.05212	19.188	.0013	19.214	.06947	.99758	.06963	14.361	.0024	14.395	1
60	.01745	.99985	.01745	57.290	.0001	57.299	.03490	.99939	.03492	28.636	.0006	28.654	.05234	.99863	.05241	19.081	.0014	19.107	.06976	.99756	06993	14.301	.0024	14.335	0
′	cos	sin	cot	tan	cosec	sec	cos	sin	cot	tan	cosec	sec	cos	sin	cot	tan	cosec	sec	cos	sin	cot	tan	cosec	sec	′
	89°						88°						87°						86°						

SOURCE: Courtesy of Bethlehem Steel.

TABLE 9 Continued

	4°						5°						6°						7°						
′	sin	cos	tan	cot	sec	cosec	sin	cos	tan	cot	sec	cosec	sin	cos	tan	cot	sec	cosec	sin	cos	tan	cot	sec	cosec	′
0	.06976	.99756	.06993	14.301	1.0024	14.335	.08715	.99619	.08749	11.430	1.0038	11.474	.10453	.99452	.10510	9.5144	1.0055	9.5668	.12187	.99255	.12278	8.1443	1.0075	8.2055	60
1	.07005	.99754	.07022	14.241	.0025	14.276	.08744	.99617	.08778	11.392	.0038	11.436	.10482	.99449	.10540	.4878	.0055	.5404	.12216	.99251	.12308	.1248	.0075	.1861	59
2	.07034	.99752	.07051	14.182	.0025	14.217	.08773	.99614	.08807	11.354	.0039	11.398	.10511	.99446	.10569	.4614	.0056	.5141	.12245	.99247	.12337	.1053	.0076	.1668	58
3	.07063	.99750	.07080	14.123	.0025	14.159	.08802	.99612	.08837	11.316	.0039	11.360	.10540	.99443	.10599	.4351	.0056	.4880	.12273	.99244	.12367	.0860	.0076	.1476	57
4	.07092	.99748	.07110	14.065	.0025	14.101	.08831	.99609	.08866	11.279	.0039	11.323	.10568	.99440	.10628	.4090	.0056	.4620	.12302	.99240	.12396	.0667	.0076	.1285	56
5	.07121	.99746	.07139	14.008	.0025	14.043	.08860	.99607	.08895	11.242	.0039	11.286	.10597	.99437	.10657	9.3831	.0057	9.4362	.12331	.99237	.12426	8.0476	.0077	8.1094	55
6	.07150	.99744	.07168	13.951	.0026	13.986	.08889	.99604	.08925	11.205	.0040	11.249	.10626	.99434	.10687	.3572	.0057	.4105	.12360	.99233	.12456	.0285	.0077	.0905	54
7	.07179	.99742	.07197	13.894	.0026	13.930	.08918	.99601	.08954	11.168	.0040	11.213	.10655	.99431	.10716	.3315	.0057	.3850	.12389	.99229	.12485	.0095	.0078	.0717	53
8	.07208	.99740	.07226	13.838	.0026	13.874	.08947	.99599	.08983	11.132	.0040	11.176	.10684	.99428	.10746	.3060	.0057	.3596	.12418	.99226	.12515	7.9906	.0078	.0529	52
9	.07237	.99738	.07256	13.782	.0026	13.818	.08976	.99596	.09013	11.095	.0040	11.140	.10713	.99424	.10775	.2806	.0058	.3343	.12447	.99222	.12544	.9717	.0078	.0342	51
10	.07266	.99736	.07285	13.727	.0026	13.763	.09005	.99594	.09042	11.059	.0041	11.104	.10742	.99421	.10805	9.2553	.0058	9.3092	.12476	.99219	.12574	7.9530	.0079	8.0156	50
11	.07295	.99733	.07314	13.672	.0027	13.708	.09034	.99591	.09071	11.024	.0041	11.069	.10771	.99418	.10834	.2302	.0058	.2842	.12504	.99215	.12603	.9344	.0079	7.9971	49
12	.07324	.99731	.07343	13.617	.0027	13.654	.09063	.99588	.09101	10.988	.0041	11.033	.10800	.99415	.10863	.2051	.0059	.2593	.12533	.99211	.12633	.9158	.0079	.9787	48
13	.07353	.99729	.07373	13.563	.0027	13.600	.09092	.99586	.09130	10.953	.0041	10.998	.10829	.99412	.10893	.1803	.0059	.2346	.12562	.99208	.12662	.8973	.0080	.9604	47
14	.07382	.99727	.07402	13.510	.0027	13.547	.09121	.99583	.09159	10.918	.0042	10.963	.10858	.99409	.10922	.1555	.0059	.2100	.12591	.99204	.12692	.8789	.0080	.9421	46
15	.07411	.99725	.07431	13.457	.0027	13.494	.09150	.99580	.09189	10.883	.0042	10.929	.10887	.99406	.10952	9.1309	.0060	9.1855	.12620	.99200	.12722	7.8606	.0080	7.9240	45
16	.07440	.99723	.07460	13.404	.0028	13.441	.09179	.99578	.09218	10.848	.0042	10.894	.10916	.99402	.10981	.1064	.0060	.1612	.12649	.99197	.12751	.8424	.0081	.9059	44
17	.07469	.99721	.07490	13.351	.0028	13.389	.09208	.99575	.09247	10.814	.0043	10.860	.10944	.99399	.11011	.0821	.0060	.1370	.12678	.99193	.12781	.8243	.0081	.8879	43
18	.07498	.99718	.07519	13.299	.0028	13.337	.09237	.99572	.09277	10.780	.0043	10.826	.10973	.99396	.11040	.0579	.0061	.1129	.12706	.99189	.12810	.8062	.0082	.8700	42
19	.07527	.99716	.07548	13.248	.0028	13.286	.09266	.99570	.09306	10.746	.0043	10.792	.11002	.99393	.11069	.0338	.0061	.0890	.12735	.99186	.12840	.7882	.0082	.8522	41
20	.07556	.99714	.07577	13.917	.0029	13.235	.09295	.99567	.09335	10.712	.0043	10.758	.11031	.99390	.11099	9.0098	.0061	9.0651	.12764	.99182	.12869	7.7703	.0082	7.8344	40
21	.07585	.99712	.07607	13.146	.0029	13.184	.09324	.99564	.09365	10.678	.0044	10.725	.11060	.99386	.11128	8.9860	.0062	.0414	.12793	.99178	.12899	.7525	.0083	.8168	39
22	.07614	.99710	.07636	13.096	.0029	13.134	.09353	.99562	.09394	10.645	.0044	10.692	.11089	.99383	.11158	.9623	.0062	.0179	.12822	.99174	.12928	.7348	.0083	.7992	38
23	.07643	.99707	.07665	13.046	.0029	13.084	.09382	.99559	.09423	10.612	.0044	10.659	.11118	.99380	.11187	.9387	.0062	8.9944	.12851	.99171	.12958	.7171	.0084	.7817	37
24	.07672	.99705	.07694	12.996	.0029	13.034	.09411	.99556	.09453	10.579	.0044	10.626	.11147	.99377	.11217	.9152	.0063	.9711	.12879	.99167	.12988	.6996	.0084	.7642	36
25	.07701	.99703	.07724	12.947	.0030	12.985	.09440	.99553	.09482	10.546	.0045	10.593	.11176	.99373	.11246	8.8918	.0063	8.9479	.12908	.99163	.13017	7.6821	.0084	7.7469	35
26	.07730	.99701	.07753	12.898	.0030	12.937	.09469	.99551	.09511	10.514	.0045	10.561	.11205	.99370	.11276	.8686	.0063	.9248	.12937	.99160	.13047	.6646	.0085	.7296	34
27	.07759	.99698	.07782	12.849	.0030	12.888	.09498	.99548	.09541	10.481	.0045	10.529	.11234	.99367	.11305	.8455	.0064	.9018	.12966	.99156	.13076	.6473	.0085	.7124	33
28	.07788	.99696	.07812	12.801	.0030	12.840	.09527	.99545	.09570	10.449	.0046	10.497	.11262	.99364	.11335	.8225	.0064	.8790	.12995	.99152	.13106	.6300	.0085	.6953	32
29	.07817	.99694	.07841	12.754	.0031	12.793	.09556	.99542	.09599	10.417	.0046	10.465	.11291	.99360	.11364	.7996	.0064	.8563	.13024	.99148	.13136	.6129	.0086	.6783	31
30	.07846	.99692	.07870	12.706	.0031	12.745	.09584	.99540	.09629	10.385	.0046	10.433	.11320	.99357	.11393	8.7769	.0065	8.8337	.13053	.99144	.13165	7.5957	.0086	7.6613	30
31	.07875	.99689	.07899	12.659	.0031	12.698	.09613	.99537	.09658	10.354	.0046	10.402	.11349	.99354	.11423	.7542	.0065	.8112	.13081	.99141	.13195	.5787	.0087	.6444	29
32	.07904	.99687	.07929	12.612	.0031	12.652	.09642	.99534	.09688	10.322	.0047	10.371	.11378	.99350	.11452	.7317	.0065	.7888	.13110	.99137	.13224	.5617	.0087	.6276	28
33	.07933	.99685	.07958	12.566	.0032	12.606	.09671	.99531	.09717	10.291	.0047	10.340	.11407	.99347	.11482	.7093	.0066	.7665	.13139	.99133	.13254	.5449	.0087	.6108	27
34	.07962	.99682	.07987	12.520	.0032	12.560	.09700	.99528	.09746	10.260	.0047	10.309	.11436	.99344	.11511	.6870	.0066	.7444	.13168	.99129	.13284	.5280	.0088	.5942	26
35	.07991	.99680	.08016	12.474	.0032	12.514	.09729	.99525	.09776	10.229	.0048	10.278	.11465	.99341	.11541	8.6648	.0066	8.7223	.13197	.99125	.13313	7.5113	.0088	7.5776	25
36	.08020	.99678	.08046	12.429	.0032	12.469	.09758	.99523	.09805	10.199	.0048	10.248	.11494	.99337	.11570	.6427	.0067	.7004	.13226	.99121	.13343	.4946	.0089	.5611	24
37	.08049	.99675	.08075	12.384	.0032	12.424	.09787	.99520	.09834	10.168	.0048	10.217	.11523	.99334	.11600	.6208	.0067	.6786	.13254	.99118	.13372	.4780	.0089	.5446	23
38	.08078	.99673	.08104	12.339	.0033	12.379	.09816	.99517	.09864	10.138	.0048	10.187	.11551	.99330	.11629	.5989	.0067	.6569	.13283	.99114	.13402	.4615	.0089	.5282	22
39	.08107	.99671	.08134	12.295	.0033	12.335	.09845	.99514	.09893	10.108	.0049	10.157	.11580	.99327	.11659	.5772	.0068	.6353	.13312	.99110	.13432	.4451	.0090	.5119	21
40	.08136	.99668	.08163	12.250	.0033	12.291	.09874	.99511	.09922	10.078	.0049	10.127	.11609	.99324	.11688	8.5555	.0068	8.6138	.13341	.99106	.13461	7.4287	.0090	7.4957	20
41	.08165	.99666	.08192	12.207	.0033	12.248	.09903	.99508	.09952	10.048	.0049	10.098	.11638	.99320	.11718	.5340	.0068	.5924	.13370	.99102	.13491	.4124	.0090	.4795	19
42	.08194	.99664	.08221	12.163	.0034	12.204	.09932	.99505	.09981	10.019	.0050	10.068	.11667	.99317	.11747	.5126	.0069	.5711	.13399	.99098	.13520	.3961	.0091	.4634	18
43	.08223	.99661	.08251	12.120	.0034	12.161	.09961	.99503	.10011	9.9893	.0050	10.039	.11696	.99314	.11777	.4913	.0069	.5499	.13427	.99094	.13550	.3800	.0091	.4474	17
44	.08252	.99659	.08280	12.077	.0034	12.118	.09990	.99500	.10040	9.9601	.0050	10.010	.11725	.99310	.11806	.4701	.0069	.5289	.13456	.99090	.13580	.3639	.0092	.4315	16
45	.08281	.99656	.08309	12.035	.0034	12.076	.10019	.99497	.10069	9.9310	.0050	9.9812	.11754	.99307	.11836	8.4489	.0070	8.5079	.13485	.99086	.13609	7.3479	.0092	7.4156	15
46	.08310	.99654	.08339	11.992	.0035	12.034	.10048	.99494	.10099	9.9021	.0051	9.9525	.11783	.99303	.11865	.4279	.0070	.4871	.13514	.99083	.13639	.3319	.0092	.3998	14
47	.08339	.99652	.08368	11.950	.0035	11.992	.10077	.99491	.10128	9.8734	.0051	9.9239	.11811	.99300	.11895	.4070	.0070	.4663	.13543	.99079	.13669	.3160	.0093	.3840	13
48	.08368	.99649	.08397	11.909	.0035	11.950	.10106	.99488	.10158	9.8448	.0051	9.8955	.11840	.99296	.11924	.3862	.0071	.4457	.13571	.99075	.13698	.3002	.0093	.3683	12
49	.08397	.99647	.08426	11.867	.0035	11.909	.10134	.99485	.10187	9.8164	.0052	9.8672	.11869	.99293	.11954	.3655	.0071	.4251	.13600	.99071	.13728	.2844	.0094	.3527	11
50	.08426	.99644	.08456	11.826	.0036	11.868	.10163	.99482	.10216	9.7882	.0052	9.8391	.11898	.99290	.11983	8.3449	.0071	8.4046	.13629	.99067	.13757	7.2687	.0094	7.3372	10
51	.08455	.99642	.08485	11.785	.0036	11.828	.10192	.99479	.10246	9.7601	.0052	9.8112	.11927	.99286	.12013	.3244	.0072	.3843	.13658	.99063	.13787	.2531	.0094	.3217	9
52	.08484	.99639	.08514	11.745	.0036	11.787	.10221	.99476	.10275	9.7322	.0053	9.7834	.11956	.99283	.12042	.3040	.0072	.3640	.13687	.99059	.13817	.2375	.0095	.3063	8
53	.08513	.99637	.08544	11.704	.0036	11.747	.10250	.99473	.10305	9.7044	.0053	9.7558	.11985	.99279	.12072	.2837	.0073	.3439	.13716	.99055	.13846	.2220	.0095	.2909	7
54	.08542	.99634	.08573	11.664	.0037	11.707	.10279	.99470	.10334	9.6768	.0053	9.7283	.12014	.99276	.12101	.2635	.0073	.3238	.13744	.99051	.13876	.2066	.0096	.2757	6
55	.08571	.99632	.08602	11.625	.0037	11.668	.10308	.99467	.10363	9.6493	.0053	9.7010	.12042	.99272	.12131	8.2434	.0073	8.3039	.13773	.99047	.13906	7.1912	.0096	7.2604	5
56	.08600	.99629	.08632	11.585	.0037	11.628	.10337	.99464	.10393	9.6220	.0054	9.6739	.12071	.99269	.12160	.2234	.0074	.2840	.13802	.99043	.13935	.1759	.0097	.2453	4
57	.08629	.99627	.08661	11.546	.0037	11.589	.10366	.99461	.10422	9.5949	.0054	9.6469	.12100	.99265	.12190	.2035	.0074	.2642	.13831	.99039	.13965	.1607	.0097	.2302	3
58	.08658	.99624	.08690	11.507	.0038	11.550	.10395	.99458	.10452	9.5679	.0054	9.6200	.12129	.99262	.12219	.1837	.0074	.2446	.13860	.99035	.13995	.1455	.0097	.2152	2
59	.08687	.99622	.08719	11.468	.0038	11.512	.10424	.99455	.10481	9.5411	.0055	9.5933	.12158	.99258	.12249	.1640	.0075	.2250	.13888	.99031	.14024	.1304	.0098	.2002	1
60	.08715	.99619	.08749	11.430	.0038	11.474	.10453	.99452	.10510	9.5144	.0055	9.5668	.12187	.99255	.12278	8.1443	.0075	8.2055	.13917	.99027	.14054	7.1154	.0098	7.1853	0
′	cos	sin	cot	tan	cosec	sec	cos	sin	cot	tan	cosec	sec	cos	sin	cot	tan	cosec	sec	cos	sin	cot	tan	cosec	sec	′
	85°						84°						83°						82°						

TABLE 9 Continued

	8°						9°						10°						11°						
′	sin	cos	tan	cot	sec	cosec	sin	cos	tan	cot	sec	cosec	sin	cos	tan	cot	sec	cosec	sin	cos	tan	cot	sec	cosec	′
0	.13917	.99027	.14054	7.1154	1.0098	7.1853	.15643	.98769	.15838	6.3137	1.0215	6.3924	.17365	.98481	.17633	5.6713	1.0154	5.7588	.19081	.98163	.19438	5.1445	1.0187	5.2408	60
1	.13946	.99023	.14084	.1004	.0099	.1704	.15672	.98764	.15868	.3019	.0125	.3807	.17393	.98476	.17663	.6616	.0155	.7493	.19109	.98157	.19468	.1366	.0188	.2330	59
2	.13975	.99019	.14113	.0854	.0099	.1557	.15701	.98760	.15898	.2901	.0125	.3690	.17422	.98471	.17693	.6520	.0155	.7398	.19138	.98152	.19498	.1286	.0188	.2252	58
3	.14004	.99015	.14143	.0706	.0099	.1409	.15730	.98755	.15928	.2783	.0126	.3574	.17451	.98465	.17723	.6425	.0156	.7304	.19166	.98146	.19529	.1207	.0189	.2174	57
4	.14032	.99010	.14173	.0558	.0100	.1263	.15758	.98750	.15958	.2665	.0126	.3458	.17479	.98460	.17753	.6329	.0156	.7210	.19195	.98140	.19559	.1128	.0189	.2097	56
5	.14061	.99006	.14202	7.0410	.0100	7.1117	.15787	.98746	.15987	6.2548	.0127	6.3343	.17508	.98455	.17783	5.6234	.0157	5.7117	.19224	.98135	.19589	5.1049	.0190	5.2019	55
6	.14090	.99002	.14232	.0264	.0101	.0972	.15816	.98741	.16017	.2432	.0217	.3228	.17537	.98450	.17813	.6140	.0157	.7023	.19252	.98129	.19619	.0970	.0191	.1942	54
7	.14119	.98998	.14262	.0117	.0101	.0827	.15844	.98737	.16047	.2316	.0128	.3113	.17565	.98445	.17843	.6045	.0158	.6930	.19281	.98124	.19649	.0892	.0191	.1865	53
8	.14148	.98994	.14291	6.9972	.0102	.0683	.15873	.98732	.16077	.2200	.0128	.2999	.17594	.98440	.17873	.5951	.0158	.6838	.19309	.98118	.19680	.0814	.0192	.1788	52
9	.14176	.98990	.14321	.9827	.0102	.0539	.15902	.98727	.16107	.2085	.0129	.2885	.17622	.98435	.17903	.5857	.0159	.6745	.19338	.98112	.19710	.0736	.0192	.1712	51
10	.14205	.98986	.14351	6.9682	.0102	7.0396	.15931	.98723	.16137	6.1970	.0129	6.2772	.17651	.98430	.17933	5.5764	.0159	5.6653	.19366	.98107	.19740	5.0658	.0193	5.1636	50
11	.14234	.98982	.14380	.9538	.0103	.0254	.15959	.98718	.16167	.1856	.0130	.2659	.17680	.98425	.17963	.5670	.0160	.6561	.19395	.98101	.19770	.0581	.0193	.1560	49
12	.14263	.98978	.14410	.9395	.0103	.0112	.15988	.98714	.16196	.1742	.0130	.2546	.17708	.98419	.17993	.5578	.0160	.6470	.19423	.98095	.19800	.0504	.0194	.1484	48
13	.14292	.98973	.14440	.9252	.0104	6.9971	.16017	.98709	.16226	.1628	.0131	.2434	.17737	.98414	.18023	.5485	.0161	.6379	.19452	.98090	.19831	.0427	.0195	.1409	47
14	.14320	.98969	.14470	.9110	.0104	.9830	.16045	.98704	.16256	.1515	.0131	.2322	.17766	.98409	.18053	.5393	.0162	.6288	.19480	.98084	.19861	.0350	.0195	.1333	46
15	.14349	.98965	.14499	6.8969	.0104	6.9690	.16074	.98700	.16286	6.1402	.0132	6.2211	.17794	.98404	.18083	5.5301	.0162	5.6197	.19509	.98078	.19891	5.0273	.0196	5.1258	45
16	.14378	.98961	.14529	.8828	.0105	.9550	.16103	.98695	.16316	.1290	.0132	.2100	.17823	.98399	.18113	.5209	.0163	.6107	.19537	.98073	.19921	.0197	.0196	.1183	44
17	.14407	.98957	.14559	.8687	.0105	.9411	.16132	.98690	.16346	.1178	.0133	.1990	.17852	.98394	.18143	.5117	.0163	.6017	.19566	.98067	.19952	.0121	.0197	.1109	43
18	.14436	.98952	.14588	.8547	.0106	.9273	.16160	.98685	.16376	.1066	.0133	.1880	.17880	.98388	.18173	.5026	.0164	.5928	.19595	.98061	.19982	.0045	.0198	.1034	42
19	.14464	.98948	.14618	.8408	.0106	.9135	.16189	.98681	.16405	.0955	.0134	.1770	.17909	.98383	.18203	.4936	.0164	.5838	.19623	.98056	.20012	4.9969	.0198	.0960	41
20	.14493	.98944	.14648	6.8269	.0107	6.8998	.16218	.98676	.16435	6.0844	.0134	6.1661	.17937	.98378	.18233	5.4845	.0165	5.5749	.19652	.98050	.20042	4.9894	.0199	5.0886	40
21	.14522	.98940	.14677	.8131	.0107	.8861	.16246	.98671	.16465	.0734	.0135	.1552	.17966	.98373	.18263	.4755	.0165	.5660	.19680	.98044	.20073	.9819	.0199	.0812	39
22	.14551	.98936	.14707	.7993	.0107	.8725	.16275	.98667	.16495	.0624	.0135	.1443	.17995	.98368	.18293	.4665	.0166	.5572	.19709	.98039	.20103	.9744	.0200	.0739	38
23	.14579	.98931	.14737	.7856	.0108	.8589	.16304	.98662	.16525	.0514	.0136	.1335	.18023	.98362	.18323	.4575	.0166	.5484	.19737	.98033	.20133	.9669	.0201	.0666	37
24	.14608	.98927	.14767	.7720	.0108	.8454	.16333	.98657	.16555	.0405	.0136	.1227	.18052	.98357	.18353	.4486	.0167	.5396	.19766	.98027	.20163	.9594	.0201	.0593	36
25	.14637	.98923	.14796	6.7584	.0109	6.8320	.16361	.98652	.16585	6.0296	.0136	6.1120	.18080	.98352	.18383	5.4396	.0167	5.5308	.19794	.98021	.20194	4.9520	.0202	5.0520	35
26	.14666	.98919	.14826	.7448	.0109	.8185	.16390	.98648	.16615	.0188	.0137	.1013	.18109	.98347	.18413	.4308	.0168	.5221	.19823	.98016	.20224	.9446	.0202	.0447	34
27	.14695	.98914	.14856	.7313	.0110	.8052	.16419	.98643	.16644	.0080	.0137	.0906	.18138	.98341	.18444	.4219	.0169	.5134	.19851	.98010	.20254	.9372	.0203	.0375	33
28	.14723	.98910	.14886	.7179	.0110	.7919	.16447	.98638	.16674	5.9972	.0138	.0800	.18166	.98336	.18474	.4131	.0169	.5047	.19880	.98004	.20285	.9298	.0204	.0302	32
29	.14752	.98906	.14915	.7045	.0111	.7787	.16476	.98633	.16704	.9865	.0138	.0694	.18195	.98331	.18504	.4043	.0170	.4960	.19908	.97998	.20315	.9225	.0204	.0230	31
30	.14781	.98901	.14945	6.6911	.0111	6.7655	.16505	.98628	.16734	5.9758	.0139	6.0588	.18223	.98325	.18534	5.3955	.0170	5.4874	.19937	.97992	.20345	4.9151	.0205	5.0158	30
31	.14810	.98897	.14975	.6779	.0111	.7523	.16533	.98624	.16764	.9651	.9139	.0483	.18252	.98320	.18564	.3868	.0171	.4788	.19965	.97987	.20375	.9078	.0205	.0087	29
32	.14838	.98893	.15004	.6646	.0112	.7392	.16562	.98619	.16794	.9545	.9140	.0379	.18281	.98315	.18594	.3780	.0171	.4702	.19994	.97981	.20406	.9006	.0206	.0015	28
33	.14867	.98889	.15034	.6514	.0112	.7262	.16591	.98614	.16824	.9439	.0140	.0274	.18309	.98309	.18624	.3694	.0172	.4617	.20022	.97975	.20436	.8933	.0207	4.9944	27
34	.14896	.98884	.15064	.6383	.0113	.7132	.16619	.98609	.16854	.9333	.0141	.0170	.18338	.98304	.18654	.3607	.0172	.4532	.20051	.97969	.20466	.8860	.0207	.9873	26
35	.14925	.98880	.15094	6.6252	.0113	6.7003	.16648	.98604	.16884	5.9228	.0141	6.0066	.18366	.98299	.18684	5.3521	.0173	5.4447	.20079	.97963	.20497	4.8788	0208	4.9802	25
36	.14953	.98876	.15123	.6122	.0114	.6874	.16677	.98600	.16914	.9123	.0142	5.9963	.18395	.98293	.18714	.3434	.0174	.4362	.20108	.97957	.20527	.8716	.0208	.9732	24
37	.14982	.98871	.15153	.5992	.0114	.6745	.16705	.98595	.16944	.9019	.0142	.9860	.18424	.98288	.18745	.3349	.0174	.4278	.20136	.97952	.20557	.8644	.0209	.9661	23
38	.15011	.98867	.15183	.5863	.0115	.6617	.16734	.98590	.16973	.8915	.0143	.9758	.18452	.98283	.18775	.3263	.0175	.4194	.20165	.97946	.20588	.8573	.0210	.9591	22
39	.15050	.98862	.15213	.5734	.0115	.6490	.16763	.98585	.17003	.8811	.0143	.9655	.18481	.98277	.18805	.3178	.0175	.4110	.20193	.97940	.20618	.8501	.0210	.9521	21
40	.15068	.98858	.15243	6.5605	.0115	6.6363	.16791	.98580	.17033	5.8708	.0144	5.9554	.18509	.98272	.18835	5.3093	.0176	5.4026	.20222	.97934	.20648	4.8430	.0211	4.9452	20
41	.15097	.98854	.15272	.5478	.0116	.6237	.16820	.98575	.17063	.8605	.0144	.9452	.18538	.98267	.18865	.3008	.0176	.3943	.20250	.97928	.20679	.8359	.0211	.9382	19
42	.15126	.98849	.15302	.5350	.0116	.6111	.16849	.98570	.17093	.8502	.0145	.9351	.18567	.98261	.18895	.2923	.0177	.3860	.20279	.97922	.20709	.8288	.0212	.9313	18
43	.15155	.98845	.15332	.5223	.0117	.5985	.16878	.98565	.17123	.8400	.0145	.9250	.18595	.98256	.18925	.2839	.0177	.3777	.20307	.97916	.20739	.8217	.0213	.9243	17
44	.15183	.98840	.15362	.5097	.0117	.5860	.16906	.98560	.17153	.8298	.0146	.9150	.18624	.98250	.18955	.2755	.0178	.3695	.20336	.97910	.20770	.8147	.0213	.9175	16
45	.15212	.98836	.15391	6.4971	.0118	6.5736	.16935	.98556	.17183	5.8196	.0146	5.9049	.18652	.98245	.18985	5.2671	.0179	5.3612	.20364	.97904	.20800	4.8077	.0214	4.9106	15
46	.15241	.98832	.15421	.4845	.0118	.5612	.16964	.98551	.17213	.8095	.0147	.8950	.18681	.98240	.19016	.2588	.0179	.3530	.20393	.97899	.20830	.8007	.0215	.9037	14
47	.15270	.98827	.15451	.4720	.0119	.5488	.16992	.98546	.17243	.7994	.0147	.8850	.18709	.98234	.19046	.2505	.0180	.3449	.20421	.97893	.20861	.7937	.0215	.8969	13
48	.15298	.98823	.15481	.4596	.0119	.5365	.17021	.98541	.17273	.7894	.0148	.8751	.18738	.98229	.19076	.2422	.0180	.3367	.20450	.97887	.20891	.7867	.0216	.8901	12
49	.15328	.98818	.15511	.4472	.0119	.5243	.17050	.98536	.17303	.7794	.0148	.8652	.18767	.98223	.19106	.2339	.0181	.3286	.20478	.97881	.20921	.7798	.0216	.8833	11
50	.15356	.98814	.15540	6.4348	.0120	6.5121	.17078	.98531	.17333	5.7694	.0149	5.8554	.18795	.98218	.19136	5.2257	.0181	5.3205	.20506	.97875	.20952	4.7728	.0217	4.8765	10
51	.15385	.98809	.15570	.4225	.0120	.4999	.17107	.98526	.17363	.7594	.0150	.8456	.18824	.98212	.19166	.2174	.0182	.3124	.20535	.97869	.20982	.7659	.0218	.8967	9
52	.15413	.98805	.15600	.4103	.0121	.4878	.17136	.98521	.17393	.7495	.0150	.8358	.18852	.98207	.19197	.2092	.0182	.3044	.20563	.97863	.21012	.7591	.0218	.8630	8
53	.15442	.98800	.15630	.3980	.0121	.4757	.17164	.98516	.17423	.7396	.0151	.8261	.18881	.98201	.19227	.2011	.0183	.2963	.20592	.97857	.21043	.7522	.0219	.8563	7
54	.15471	.98796	.15659	.3859	.0122	.4637	.17193	.98511	.17453	.7297	.0151	.8163	.18909	.98196	.19257	.1929	.0184	.2883	.20620	.97851	.21073	.7453	.0220	.8496	6
55	.15500	.98791	.15689	6.3737	.0122	6.4517	.17221	.98506	.17483	5.7199	.0152	5.8067	.18938	.98190	.19287	5.1848	.0184	5.2803	.20649	.97845	.21104	4.7385	.0220	4.8429	5
56	.15528	.98787	.15719	.3616	.0123	.4398	.17250	.98501	.17513	.7101	.0152	.7970	.18967	.98186	.19317	.1767	.0185	.2724	.20677	.97839	.21134	.7317	.0221	.8362	4
57	.15557	.98782	.15749	.3496	.0123	.4279	.17279	.98496	.17543	.7004	.0153	.7874	.18995	.98179	.19347	.1686	.0185	.2645	.20706	.97833	.21164	.7249	.0221	.8296	3
58	.15586	.98778	.15779	.3376	.0124	.4160	.17307	.98491	.17573	.6906	.0153	.7778	.19024	.98174	.19378	.1606	.0186	.2566	.20734	.97827	.21195	.7181	.0222	.8229	2
59	.15615	.98773	.15809	.3257	.0124	.4042	.17336	.98486	.17603	.6809	.0154	.7683	.19052	.98168	.19408	.1525	.0186	.2487	.20763	.97821	.21225	.7114	.0223	.8163	1
60	.15643	.98769	.15838	6.3137	.0125	6.3924	.17365	.98481	.17633	5.6713	.0154	5.7588	.19081	.98163	.19438	5.1445	.0187	5.2408	.20791	.97815	.21256	4.7046	.0223	4.8097	0
′	cos	sin	cot	tan	cosec	sec	cos	sin	cot	tan	cosec	sec	cos	sin	cot	tan	cosec	sec	cos	sin	cot	tan	cosec	sec	′
	81°						80°						79°						78°						

TABLE 9 Continued

	12°						13°						14°						15°						
′	sin	cos	tan	cot	sec	cosec	sin	cos	tan	cot	sec	cosec	sin	cos	tan	cot	sec	cosec	sin	cos	tan	cot	sec	cosec	′
0	.20791	.97815	.21256	4.7046	1.0223	4.8097	.22495	.97437	.23087	4.3315	1.0263	4.4454	.24192	.97029	.24933	4.0108	1.0306	4.1336	.25882	.96592	.26795	3.7320	1.0353	3.8637	60
1	.20820	.97809	.21286	.6979	.0224	.8032	.22523	.97430	.23117	.3257	.0264	.4398	.24220	.97022	.24964	.0058	.0307	.1287	.25910	.96585	.26826	.7277	.0353	.8595	59
2	.20848	.97803	.21316	.6912	.0225	.7966	.22552	.97424	.23148	.3200	.0264	.4342	.24249	.97015	.24995	.0009	.0308	.1239	.25938	.96577	.26857	.7234	.0354	.8553	58
3	.20876	.97797	.21347	.6845	.0225	.7901	.22580	.97417	.23179	.3143	.0265	.4287	.24277	.97008	.25025	3.9959	.0308	.1191	.25966	.96570	.26888	.7191	.0355	.8512	57
4	.20905	.97790	.21377	.6778	.0226	.7835	.22608	.97411	.23209	.3086	.0266	.4231	.24305	.97001	.25056	.9910	.0309	.1144	.25994	.96562	.26920	.7147	.0356	.8470	56
5	.20933	.97784	.21408	4.6712	.0226	4.7770	.22637	.97404	.23240	4.3029	.0266	4.4176	.24333	.96994	.25087	3.9861	.0310	4.1096	.26022	.96555	.26951	3.7104	.0357	3.8428	55
6	.20962	.97778	.21438	.6646	.0227	.7706	.22665	.97398	.23270	.2972	.0267	.4121	.24361	.96987	.25118	.9812	.0311	.1048	.26050	.96547	.26982	.7062	.0358	.8387	54
7	.20990	.97772	.21468	.6580	.0228	.7641	.22693	.97391	.23301	.2916	.0268	.4065	.24390	.96980	.25149	.9763	.0311	.1001	.26078	.96540	.27013	.7019	.0358	.8346	53
8	.21019	.97766	.21499	.6514	.0228	.7576	.22722	.97384	.23332	.2859	.0268	.4011	.24418	.96973	.25180	.9714	.0312	.0953	.26107	.96532	.27044	.6976	.0359	.8304	52
9	.21047	.97760	.21529	.6448	.0229	.7512	.22750	.97378	.23363	.2803	.0269	.3956	.24446	.96966	.25211	.9665	.0313	.0906	.26135	.96524	.27076	.6933	.0360	.8263	51
10	.21076	.97754	.21560	4.6382	.0230	4.7448	.22778	.97371	.23393	4.2747	.0270	4.3901	.24474	.96959	.25242	3.9616	.0314	4.0859	.26163	.96517	.27107	3.6891	.0361	3.8222	50
11	.21104	.97748	.21590	.6317	.0230	.7384	.22807	.97364	.23424	.2691	.0271	.3847	.24502	.96952	.25273	.9568	.0314	.0812	.26191	.96509	.27138	.6848	.0362	.8181	49
12	.21132	.97741	.21621	.6252	.0231	.7320	.22835	.97358	.23455	.2635	.0271	.3792	.24531	.96944	.25304	.9520	.0315	.0765	.26219	.96502	.27169	.6806	.0362	.8140	48
13	.21161	.97735	.21651	.6187	.0232	.7257	.22863	.97351	.23485	.2579	.0272	.3738	.24559	.96937	.25335	.9471	.0316	.0718	.26247	.96494	.27201	.6764	.0363	.8100	47
14	.21189	.97729	.21682	.6122	.0232	.7193	.22892	.97344	.23516	.2524	.0273	.3684	.24587	.96930	.25366	.9423	.0317	.0672	.26275	.96486	.27232	.6722	.0364	.8059	46
15	.21218	.97723	.21712	4.6057	.0233	4.7130	.22920	.97338	.23547	4.2468	.0273	4.3630	.24615	.96923	.25397	3.9375	.0317	4.0625	.26303	.96479	.27263	3.6679	.0365	3.8018	45
16	.21246	.97717	.21742	.5993	.0234	.7067	.22948	.97331	.23577	.2413	.0274	.3576	.24643	.96916	.25428	.9327	.0318	.0579	.26331	.96471	.27294	.6637	.0366	.7978	44
17	.21275	.97711	.21773	.5928	.0234	.7004	.22977	.97324	.23608	.2358	.0275	.3522	.24672	.96909	.25459	.9279	.0319	.0532	.26359	.96463	.27326	.6596	.0367	.7937	43
18	.21303	.97704	.21803	.5864	.0235	.6942	.23005	.97318	.23639	.2303	.0276	.3469	.24700	.96901	.25490	.9231	.0320	.0486	.26387	.96456	.27357	.6554	.0367	.7897	42
19	.21331	.97698	.21834	.5800	.0235	.6879	.23033	.97311	.23670	.2248	.0276	.3415	.24728	.96894	.25521	.9184	.0320	.0440	.26415	.96448	.27388	.6512	.0368	.7857	41
20	.21360	.97692	.21864	4.5736	.0236	4.6817	.23061	.97304	.23700	4.2193	.0277	4.3362	.24756	.96887	.25552	3.9136	.0321	4.0394	.26443	.96440	.27419	3.6470	.0369	3.7816	40
21	.21388	.97686	.21895	.5673	.0237	.6754	.23090	.97298	.23731	.2139	.0278	.3309	.24784	.96880	.25583	.9089	.0322	.0348	.26471	.96433	.27451	.6429	.0370	.7776	39
22	.21417	.97680	.21925	.5609	.0237	.6692	.23118	.97291	.23762	.2084	.0278	.3256	.24813	.96873	.25614	.9042	.0323	.0302	.26499	.96425	.27482	.6387	.0371	.7736	38
23	.21445	.97673	.21956	.5546	.0238	.6631	.23146	.97284	.23793	.2030	.0279	.3203	.24841	.96865	.25645	.8994	.0323	.0256	.26527	.96417	.27513	.6346	.0371	.7697	37
24	.21473	.97667	.21986	.5483	.0239	.6569	.23175	.97277	.23823	.1976	.0280	.3150	.24869	.96858	.25676	.8947	.0324	.0211	.26556	.96409	.27544	.6305	.0372	.7657	36
25	.21502	.97661	.22017	4.5420	.0239	4.6507	.23202	.97271	.23854	4.1921	.0280	4.3098	.24897	.96851	.25707	3.8900	.0325	4.0165	.26584	.96402	.27576	3.6263	.0373	3.7617	35
26	.21530	.97655	.22047	.5357	.0240	.6446	.23231	.97264	.23885	.1867	.0281	.3045	.24925	.96844	.25738	.8853	.0326	.0120	.26612	.96394	.27607	.6222	.0374	.7577	34
27	.21559	.97648	.22078	.5294	.0241	.6385	.23260	.97257	.23916	.1814	.0282	.2993	.24953	.96836	.25769	.8807	.0327	.0074	.26640	.96386	.27638	.6181	.0375	.7538	33
28	.21587	.97642	.22108	.5232	.0241	.6324	.23288	.97250	.23946	.1760	.0283	.2941	.24982	.96829	.25800	.8760	.0327	.0029	.26668	.96378	.27670	.6140	.0376	.7498	32
29	.21615	.97636	.22139	.5169	.0242	.6263	.23316	.97244	.23977	.1706	.0283	.2838	.25010	.96822	.25831	.8713	.0328	3.9984	.26696	.96371	.27701	.6100	.0376	.7459	31
30	.21644	.97630	.22169	4.5107	.0243	4.6201	.23344	.97237	.24008	4.1653	.0284	4.2836	.25038	.96815	.25862	3.8667	.0329	3.9939	.26724	.96363	.27732	3.6059	.0377	3.7420	30
31	.21672	.97623	.22200	.5045	.0243	.6142	.23373	.97230	.24039	.1600	.0285	.2785	.25066	.96807	.25893	.8621	.0330	.9894	.26752	.96355	.27764	.6018	.0378	.7380	29
32	.21701	.97617	.22230	.4983	.0244	.6081	.23401	.97223	.24069	.1546	.0285	.2733	.25094	.96800	.25924	.8574	.0330	.9850	.26780	.96347	.27795	.5977	.0379	.7341	28
33	.21729	.97611	.22261	.4921	.0245	.6021	.23429	.97216	.24100	.1493	.0286	.2681	.25122	.96793	.25955	.8528	.0331	.9805	.26808	.96340	.27826	.5937	.0380	.7302	27
34	.21757	.97604	.22291	.4860	.0245	.5961	.23458	.97210	.24131	.1440	.0287	.2630	.25151	.96785	.25986	.8482	.0332	.97[illegible]	.26836	.96332	.27858	.5896	.0381	.7263	26
35	.21786	.97598	.22322	4.4799	.0246	4.5901	.23486	.97203	.24162	4.1388	.0288	4.2579	.25179	.96778	.26017	3.8436	.0333	3.9716	.26864	.96324	.27889	3.5856	.0382	3.7224	25
36	.21814	.97592	.22353	.4737	.0247	.5841	.23514	.97196	.24192	.1335	.0288	.2527	.25207	.96771	.26048	.8390	.0334	.9672	.26892	.96316	.27920	.5816	.0382	.7186	24
37	.21843	.97575	.22383	.4676	.0247	.5782	.23542	.97189	.24223	.1282	.0289	.2476	.25235	.96763	.26079	.8345	.0334	.9627	.26920	.96308	.27952	.5776	.0383	.7147	23
38	.21871	.97579	.22414	.4615	.0248	.5722	.23571	.97182	.24254	.1230	.0290	.2425	.25263	.96756	.26110	.8299	.0335	.9583	.26948	.96301	.27983	.5736	.0384	.7108	22
39	.21899	.97573	.22444	.4555	.0249	.5663	.23599	.97175	.24285	.1178	.0291	.2375	.25291	.96749	.26141	.8254	.0336	.9539	.26976	.96293	.28014	.5696	.0385	.7070	21
40	.21928	.97566	.22475	4.4494	.0249	4.5604	.23627	.97169	.24316	4.1126	.0291	4.2324	.25319	.96741	.26172	3.8208	.0337	3.9495	.27004	.96285	.28046	3.5656	.0386	3.7031	20
41	.21956	.97560	.22505	.4434	.0250	.5545	.23655	.97162	.24346	.1073	.0292	.2273	.25348	.96734	.26203	.8163	.0338	.9451	.27032	.96277	.28077	.5616	.0387	.6993	19
42	.21985	.97553	.22536	.4373	.0251	.5486	.23684	.97155	.24377	.1022	.0293	.2223	.25376	.96727	.26234	.8118	.0338	.9408	.27060	.96269	.28109	.5576	.0387	.6955	18
43	.22013	.97547	.22566	.4313	.0251	.5428	.23712	.97148	.24408	.0970	.0293	.2173	.25404	.96719	.26266	.8073	.0339	.9364	.27088	.96261	.28140	.5536	.0388	.6917	17
44	.22041	.97541	.22597	.4253	.0252	.5369	.23740	.97141	.24439	.0918	.0294	.2122	.25432	.96712	.26297	.8027	.0340	.9320	.27116	.96253	.28171	.5497	.0389	.6878	16
45	.22070	.97534	.22628	4.4194	.0253	4.5311	.23768	.97134	.24470	4.0867	.0295	4.2072	.25460	.96704	.26328	3.7983	.0341	3.9277	.27144	.96245	.28203	3.5457	.0390	3.6840	15
46	.22098	.97528	.22658	.4134	.0253	.5253	.23797	.97127	.24501	.0815	.0296	.2022	.25488	.96697	.26359	.7938	.0341	.9234	.27172	.96238	.28234	.5418	.0391	.6802	14
47	.22126	.97521	.22689	.4074	.0254	.5195	.23825	.97120	.24531	.0764	.0296	.1972	.25516	.96690	.26390	.7893	.0342	.9190	.27200	.96230	.28266	.5378	.0392	.6765	13
48	.22155	.97515	.22719	.4015	.0255	.5137	.23853	.97113	.24562	.0713	.0297	.1923	.25544	.96682	.26421	.7848	.0343	.9147	.27228	.96222	.28297	.5339	.0393	.6727	12
49	.22183	.97508	.22750	.3956	.0255	.5079	.23881	.97106	.24593	.0662	.0298	.1873	.25573	.96675	.26452	.7804	.0344	.9104	.27256	.96214	.28328	.5300	.0393	.6689	11
50	.22211	.97502	.22781	4.3897	.0256	4.5021	.23910	.97099	.24624	4.0611	.0299	4.1824	.25601	.96667	.26483	3.7759	.0345	3.9061	.27284	.96206	.28360	3.5261	.0394	3.6651	10
51	.22240	.97495	.22811	.3838	.0257	.4964	.23938	.97092	.24655	.0560	.0299	.1774	.25629	.96660	.26514	.7715	.0345	.9018	.27312	.96198	.28391	.5222	.0395	.6614	9
52	.22268	.97489	.22842	.3779	.0257	.4907	.23966	.97086	.24686	.0509	.0300	.1725	.25657	.96652	.26546	.7671	.0346	.8976	.27340	.96190	.28423	.5183	.0396	.6576	8
53	.22297	.97483	.22872	.3721	.0258	.4850	.23994	.97079	.24717	.0458	.0301	.1676	.25685	.96645	.26577	.7627	.0347	.8933	.27368	.96182	.28454	.5144	.0397	.6539	7
54	.22325	.97476	.22903	.3662	.0259	.4793	.24023	.97072	.24747	.0408	.0302	.1627	.25713	.96638	.26608	.7583	.0348	.8890	.27396	.96174	.28486	.5105	.0398	.6502	6
55	.22353	.97470	.22934	4.3604	.0260	4.4736	.24051	.97065	.24778	4.0358	.0302	4.1578	.25741	.96630	.26639	3.7539	.0349	3.8848	.27424	.96166	.28517	3.5066	.0399	3.6464	5
56	.22382	.97463	.22964	.3546	.0260	.4679	.24079	.97058	.24809	.0307	.0303	.1529	.25769	.96623	.26670	.7495	.0349	.8805	.27452	.96158	.28549	.5028	.0399	.6427	4
57	.22410	.97457	.22995	.3488	.0261	.4623	.24107	.97051	.24840	.0257	.0304	.1481	.25798	.96615	.26701	.7451	.0350	.8763	.27480	.96150	.28580	.4989	.0400	.6390	3
58	.22438	.97450	.23025	.3430	.0262	.4566	.24136	.97044	.24871	.0207	.0305	.1432	.25826	.96608	.26732	.7407	.0351	.8721	.27508	.96142	.28611	.4951	.0401	.6353	2
59	.22467	.97443	.23056	.3372	.0262	.4510	.24164	.97037	.24902	.0157	.0305	.1384	.25854	.96600	.26764	.7364	.0352	.8679	.27536	.96134	.28643	.4912	.0402	.6316	1
60	.22495	.97437	.23087	4.3315	.0263	4.4454	.24192	.97029	.24933	4.0108	.0306	4.1336	.25882	.96592	.26795	3.7320	.0353	3.8637	.27564	.96126	.28674	3.4874	.0403	3.6279	0
′	cos	sin	cot	tan	cosec	sec	cos	sin	cot	tan	cosec	sec	cos	sin	cot	tan	cosec	sec	cos	sin	cot	tan	cosec	sec	′
	77°						76°						75°						74°						

TABLE 9 Continued

	16°						17°						18°						19°						
′	sin	cos	tan	cot	sec	cosec	sin	cos	tan	cot	sec	cosec	sin	cos	tan	cot	sec	cosec	sin	cos	tan	cot	sec	cosec	′
0	.27564	.96126	.28674	3.4874	1.0403	3.6279	.29237	.95630	.30573	3.2708	1.0457	3.4203	.30902	.95106	.32492	3.0777	1.0515	3.2361	.32557	.94552	.34433	2.9042	1.0576	3.0715	60
1	.27592	.96118	.28706	.4836	.0404	.6243	.29265	.95622	.30605	.2674	.0458	.4170	.30929	.95097	.32524	.0746	.0516	.2332	.32584	.94542	.34465	.9015	.0577	.0690	59
2	.27620	.96110	.28737	.4798	.0405	.6206	.29293	.95613	.30637	.2640	.0459	.4138	.30957	.95088	.32556	.0716	.0517	.2303	.32612	.94533	.34498	.8987	.0578	.0664	58
3	.27648	.96102	.28769	.4760	.0406	.6169	.29321	.95605	.30668	.2607	.0460	.4106	.30985	.95079	.32588	.0686	.0518	.2274	.32639	.94523	.34530	.8960	.0579	.0638	57
4	.27675	.96094	.28800	.4722	.0406	.6133	.29348	.95596	.30700	.2573	.0461	.4073	.31012	.95070	.32621	.0655	.0519	.2245	.32667	.94514	.34563	.8933	.0580	.0612	56
5	.27703	.96086	.28832	3.4684	.0407	3.6096	.29376	.95588	.30732	3.2539	.0461	3.4041	.31040	.95061	.32653	3.0625	.0520	3.2216	.32694	.94504	.34595	2.8905	.0581	3.0586	55
6	.27731	.96078	.28863	.4646	.0408	.6060	.29404	.95579	.30764	.2505	.0462	.4009	.31068	.95051	.32685	.0595	.0521	.2188	.32722	.94495	.34628	.8878	.0582	.0561	54
7	.27759	.96070	.28895	.4608	.0409	.6024	.29432	.95571	.30796	.2472	.0463	.3977	.31095	.95042	.32717	.0565	.0522	.2159	.32749	.94485	.34661	.8851	.0584	.0535	53
8	.27787	.96062	.28926	.4570	.0410	.5987	.29460	.95562	.30828	.2438	.0464	.3945	.31123	.95033	.32749	.0535	.0523	.2131	.32777	.94476	.34693	.8824	.0585	.0509	52
9	.27815	.96054	.28958	.4533	.0411	.5951	.29487	.95554	.30859	.2405	.0465	.3913	.31150	.95024	.32782	.0505	.0524	.2102	.32804	.94466	.34726	.8797	.0586	.0484	51
10	.27843	.96045	.28990	3.4495	.0412	3.5915	.29515	.95545	.30891	3.2371	.0466	3.3881	.31178	.95015	.32814	3.0475	.0525	3.2074	.32832	.94457	.34758	2.8770	.0587	3.0458	50
11	.27871	.96037	.29021	.4458	.0413	.5879	.29543	.95536	.30923	.2338	.0467	.3849	.31206	.95006	.32846	.0445	.0526	.2045	.32859	.94447	.34791	.8743	.0588	.0433	49
12	.27899	.96029	.29053	.4420	.0413	.5843	.29571	.95528	.30955	.2305	.0468	.3817	.31233	.94997	.32878	.0415	.0527	.2017	.32887	.94438	.34824	.8716	.0589	.0407	48
13	.27927	.96021	.29084	.4383	.0414	.5807	.29598	.95519	.30987	.2271	.0469	.3785	.31261	.94988	.32910	.0385	.0528	.1989	.32914	.94428	.34856	.8689	.0590	.0382	47
14	.27955	.96013	.29116	.4346	.0415	.5772	.29626	.95511	.31019	.2238	.0470	.3754	.31289	.94979	.32943	.0356	.0529	.1960	.32942	.94418	.34889	.8662	.0591	.0357	46
15	.27983	.96005	.29147	3.4308	.0416	3.5736	.29654	.95502	.31051	3.2205	.0471	3.3722	.31316	.94970	.32975	3.0326	.0530	3.1932	.32969	.94409	.34921	2.8636	.0592	3.0331	45
16	.28011	.95997	.29179	.4271	.0417	.5700	.29682	.95493	.31083	.2172	.0472	.3690	.31344	.94961	.33007	.0296	.0531	.1904	.32996	.94399	.34954	.8609	.0593	.0306	44
17	.28039	.95989	.29210	.4234	.0418	.5665	.29710	.95485	.31115	.2139	.0473	.3659	.31372	.94952	.33039	.0267	.0532	.1876	.33024	.94390	.34987	.8582	.0594	.0281	43
18	.28067	.95980	.29242	.4197	.0419	.5629	.29737	.95476	.31146	.2106	.0474	.3627	.31399	.94942	.33072	.0237	.0533	.1848	.33051	.94380	.35019	.8555	.0595	.0256	42
19	.28094	.95972	.29274	.4160	.0420	.5594	.29765	.95467	.31178	.2073	.0475	.3596	.31427	.94933	.33104	.0208	.0534	.1820	.33079	.94370	.35052	.8529	.0596	.0231	41
20	.28122	.95964	.29305	3.4124	.0420	3.5559	.29793	.95459	.31210	3.2041	.0476	3.3565	.31454	.94924	.33136	3.0178	.0535	3.1792	.33106	.94361	.35085	2.8502	.0598	3.0206	40
21	.28150	.95956	.29337	.4087	.0421	.5523	.29821	.95450	.31242	.2008	.0477	.3534	.31482	.94915	.33169	.0149	.0536	.1764	.33134	.94351	.35117	.8476	.0599	.0181	39
22	.28178	.95948	.29368	.4050	.0422	.5488	.29848	.95441	.31274	.1975	.0478	.3502	.31510	.94906	.33201	.0120	.0537	.1736	.33161	.94341	.35150	.8449	.0600	.0156	38
23	.28206	.95940	.29400	.4014	.0423	.5453	.29876	.95433	.31306	.1942	.0478	.3471	.31537	.94897	.33233	.0090	.0528	.1708	.33189	.94332	.35183	.8423	.0601	.0131	37
24	.28234	.95931	.29432	.3977	.0424	.5418	.29904	.95424	.31338	.1910	.0479	.3440	.31565	.94888	.33265	.0061	.0539	.1681	.33216	.94322	.35215	.8396	.0602	.0106	36
25	.28262	.95923	.29463	3.3941	.0425	3.5383	.29932	.95415	.31370	3.1877	.0480	3.3409	.31592	.94878	.33298	3.0032	.0540	3.1653	.33243	.94313	.35248	2.8370	.0603	3.0081	35
26	.28290	.95915	.29495	.3904	.0426	.5348	.29959	.95407	.31402	.1845	.0481	.3378	.31620	.94869	.33330	.0003	.0541	.1625	.33271	.94303	.35281	.8344	.0604	.0056	34
27	.28318	.95907	.29526	.3868	.0427	.5313	.29987	.95398	.31434	.1813	.0482	.3347	.31648	.94860	.33362	2.9974	.0542	.1598	.33298	.94293	.35314	.8318	.0605	.0031	33
28	.28346	.95898	.29558	.3832	.0428	.5279	.30015	.95389	.31466	.1780	.0483	.3316	.31675	.94851	.33395	.9945	.0543	.1570	.33326	.94283	.35346	.8291	.0606	.0007	32
29	.28374	.95890	.29590	.3795	.0428	.5244	.30043	.95380	.31498	.1748	.0484	.3286	.31703	.94841	.33427	.9916	.0544	.1543	.33353	.94274	.35379	.8265	.0607	2.9982	31
30	.28401	.95882	.29621	3.3759	.0429	3.5209	.30070	.95372	.31530	3.1716	.0485	3.3255	.31730	.94832	.33459	2.9887	.0545	3.1515	.33381	.94264	.35412	2.8239	.0608	2.9957	30
31	.28429	.95874	.29653	.3723	.0430	.5175	.30098	.95363	.31562	.1684	.0486	.3224	.31758	.94823	.33492	.9858	.0546	.1488	.33408	.94254	.35445	.8213	.0609	.9933	29
32	.28457	.95865	.29685	.3687	.0431	.5140	.30126	.95354	.31594	.1652	.0487	.3194	.31786	.94814	.33524	.9829	.0547	.1461	.33435	.94245	.35477	.8187	.0611	.9908	28
33	.28485	.95857	.29716	.3651	.0432	.5106	.30154	.95345	.31626	.1620	.0488	.3163	.31813	.94805	.33557	.9800	.0548	.1433	.33463	.94235	.35510	.8161	.0612	.9884	27
34	.28513	.95849	.29748	.3616	.0433	.5072	.30181	.95337	.31658	.1588	.0489	.3133	.31841	.94795	.33589	.9772	.0549	.1406	.33490	.94225	.35543	.8135	.0613	.9859	26
35	.28541	.95840	.29780	3.3580	.0434	3.5037	.30209	.95328	.31690	3.1556	.0490	3.3102	.31868	.94786	.33621	2.9743	.0550	3.1379	.33518	.94215	.35576	2.8109	.0614	2.9835	25
36	.28569	.95832	.29811	.3544	.0435	.5003	.30237	.95319	.31722	.1524	.0491	.3072	.31896	.94777	.33654	.9714	.0551	.1352	.33545	.94206	.35608	.8083	.0615	.9810	24
37	.28597	.95824	.29843	.3509	.0436	.4969	.30265	.95310	.31754	.1492	.0492	.3042	.31923	.94767	.33686	.9686	.0552	.1325	.33572	.94196	.35641	.8057	.0616	.9786	23
38	.28624	.95816	.29875	.3473	.0437	.4935	.30292	.95301	.31786	.1460	.0493	.3011	.31951	.94758	.33718	.9657	.0553	.1298	.33600	.94186	.35674	.8032	.0617	.9762	22
39	.28652	.95807	.29906	.3438	.0438	.4901	.30320	.95293	.31818	.1429	.0494	.2981	.31978	.94749	.33751	.9629	.0554	.1271	.33627	.94176	.35707	.8006	.0618	.9738	21
40	.28680	.95799	.29938	3.3402	.0438	3.4867	.30348	.95284	.31850	3.1397	.0495	3.2951	.32006	.94740	.33783	2.9600	.0555	3.1244	.33655	.94167	.35739	2.7980	.0619	2.9713	20
41	.28708	.95791	.29970	.3367	.0439	.4833	.30375	.95275	.31882	.1366	.0496	.2921	.32034	.94730	.33816	.9572	.0556	.1217	.33682	.94157	.35772	.7954	.0620	.9689	19
42	.28736	.95782	.30001	.3332	.0440	.4799	.30403	.95266	.31914	.1334	.0497	.2891	.32061	.94721	.33848	.9544	.0557	.1190	.33709	.94147	.35805	.7929	.0622	.9665	18
43	.28764	.95774	.30033	.3296	.0441	.4766	.30431	.95257	.31946	.1303	.0498	.2861	.32089	.94712	.33880	.9515	.0558	.1163	.33737	.94137	.35838	.7903	.0623	.9641	17
44	.28792	.95765	.30065	.3261	.0442	.4732	.30459	.95248	.31978	.1271	.0499	.2831	.32116	.94702	.33913	.9487	.0559	.1137	.33764	.94127	.35871	.7878	.0624	.9617	16
45	.28820	.95757	.30096	3.3226	.0443	3.4698	.30486	.95239	.32010	3.1240	.0500	3.2801	.32144	.94693	.33945	2.9459	.0560	3.1110	.33792	.94118	.35904	2.7852	.0625	2.9593	15
46	.28847	.95749	.30128	.3191	.0444	.4665	.30514	.95231	.32042	.1209	.0501	.2772	.32171	.94684	.33978	.9431	.0561	.1083	.33819	.94108	.35936	.7827	.0626	.9569	14
47	.28875	.95740	.30160	.3156	.0445	.4632	.30542	.95222	.32074	.1177	.0502	.2742	.32199	.94674	.34010	.9403	.0562	.1057	.33846	.94098	.35969	.7801	.0627	.9545	13
48	.28903	.95732	.30192	.3121	.0446	.4598	.30569	.95213	.32106	.1146	.0503	.2712	.32226	.94665	.34043	.9375	.0563	.1030	.33874	.94088	.36002	.7776	.0628	.9521	12
49	.28931	.95723	.30223	.3087	.0447	.4565	.30597	.95204	.32138	.1115	.0504	.2683	.32254	.94655	.34075	.9347	.0565	.1004	.33901	.94078	.36035	.7751	.0629	.9497	11
50	.28959	.95715	.30255	3.3052	.0448	3.4532	.30625	.95195	.32171	3.1084	.0505	3.2653	.32282	.94646	.34108	2.9319	.0566	3.0977	.33928	.94068	.36068	2.7725	.0630	2.9474	10
51	.28987	.95707	.30287	.3017	.0448	.4498	.30653	.95186	.32203	.1053	.0506	.2624	.32309	.94637	.34140	.9291	.0567	.0951	.33956	.94058	.36101	.7700	.0632	.9450	9
52	.29014	.95698	.30319	.2983	.0449	.4465	.30680	.95177	.32235	.1022	.0507	.2594	.32337	.94627	.34173	.9263	.0568	.0925	.33983	.94049	.36134	.7675	.0633	.9426	8
53	.29042	.95690	.30350	.2948	.0450	.4432	.30708	.95168	.32267	.0991	.0508	.2565	.32364	.94618	.34205	.9235	.0569	.0898	.34011	.94039	.36167	.7650	.0634	.9402	7
54	.29070	.95681	.30382	.2914	.0451	.4399	.30736	.95159	.32299	.0960	.0509	.2535	.32392	.94608	.34238	.9208	.0570	.0872	.34038	.94029	.36199	.7625	.0635	.9379	6
55	.29098	.95673	.30414	3.2879	.0452	3.4366	.30763	.95150	.32331	3.0930	.0510	3.2506	.32419	.94599	.34270	2.9180	.0571	3.0846	.34065	.94019	.36232	2.7600	.0636	2.9355	5
56	.29126	.95664	.30446	.2845	.0453	.4334	.30791	.95141	.32363	.0899	.0511	.2477	.32447	.94590	.34303	.9152	.0572	.0820	.34093	.94009	.36265	.7575	.0637	.9332	4
57	.29154	.95656	.30478	.2811	.0454	.4301	.30819	.95132	.32395	.0868	.0512	.2448	.32474	.94580	.34335	.9125	.0573	.0793	.34120	.93999	.36298	.7550	.0638	.9308	3
58	.29181	.95647	.30509	.2777	.0455	.4268	.30846	.95124	.32428	.0838	.0513	.2419	.32502	.94571	.34368	.9097	.0574	.0767	.34147	.93989	.36331	.7525	.0639	.9285	2
59	.29209	.95639	.30541	.2742	.0456	.4236	.30874	.95115	.32460	.0807	.0514	.2390	.32529	.94561	.34400	.9069	.0575	.0741	.34175	.93979	.36364	.7500	.0641	.9261	1
60	.29237	.95630	.30573	3.2708	.0457	3.4203	.30902	.95106	.32492	3.0777	.0515	3.2361	.32557	.94552	.34433	2.9042	.0576	3.0715	.34202	.93969	.36397	2.7475	.0642	2.9238	0
′	cos	sin	cot	tan	cosec	sec	cos	sin	cot	tan	cosec	sec	cos	sin	cot	tan	cosec	sec	cos	sin	cot	tan	cosec	sec	′
	73°						72°						71°						70°						

TABLE 9 Continued

	20°						21°						22°						23°						
′	sin	cos	tan	cot	sec	cosec	sin	cos	tan	cot	sec	cosec	sin	cos	tan	cot	sec	cosec	sin	cos	tan	cot	sec	cosec	′
0	.34202	.93969	.36397	2.7475	1.0642	2.9238	.35837	.93358	.38386	2.6051	1.0711	2.7904	.37461	.92718	.40403	2.4751	1.0785	2.6695	.39073	.92050	.42447	2.3558	1.0864	2.5593	60
1	.34229	.93959	.36430	.7450	.0643	.9215	.35864	.93348	.38420	.6028	.0713	.7883	.37488	.92707	.40436	.4730	.0787	.6675	.39100	.92039	.42482	.3539	.0865	.5575	59
2	.34257	.93949	.36463	.7425	.0644	.9191	.35891	.93337	.38453	.6006	.0714	.7862	.37514	.92696	.40470	.4709	.0788	.6656	.39126	.92028	.42516	.3520	.0866	.5558	58
3	.34284	.93939	.36496	.7400	.0645	.9168	.35918	.93327	.38486	.5983	.0715	.7841	.37541	.92686	.40504	.4689	.0789	.6637	.39153	.92016	.42550	.3501	.0868	.5540	57
4	.34311	.93929	.36529	.7376	.0646	.9145	.35945	.93316	.38520	.5960	.0716	.7820	.37568	.92675	.40538	.4668	.0790	.6618	.39180	.92005	.42585	.3482	.0869	.5523	56
5	.34339	.93919	.36562	2.7351	.0647	2.9122	.35972	.93306	.38553	2.5938	.0717	2.7799	.37595	.92664	.40572	2.4647	.0792	2.6599	.39207	.91993	.42619	2.3463	.0870	2.5506	55
6	.34366	.93909	.36595	.7326	.0648	.9098	.36000	.93295	.38587	.5916	.0719	.7778	.37622	.92653	.40606	.4627	.0793	.6580	.39234	.91982	.42654	.3445	.0872	.5488	54
7	.34393	.93899	.36628	.7302	.0650	.9075	.36027	.93285	.38620	.5893	.0720	.7757	.37649	.92642	.40640	.4606	.0794	.6561	.39260	.91971	.42688	.3426	.0873	.5471	53
8	.34421	.93889	.36661	.7277	.0651	.9052	.36054	.93274	.38654	.5871	.0721	.7736	.37676	.92631	.40673	.4586	.0795	.6542	.39287	.91959	.42722	.3407	.0874	.5453	52
9	.34448	.93879	.36694	.7252	.0652	.9029	.36081	.93264	.38687	.5848	.0722	.7715	.37703	.92620	.40707	.4565	.0797	.6523	.39314	.91948	.42757	.3388	.0876	.5436	51
10	.34475	.93869	.36727	2.7228	.0653	2.9006	.36108	.93253	.38720	2.5826	.0723	2.7694	.37730	.92609	.40741	2.4545	.0798	2.6504	.39341	.91936	.42791	2.3369	.0877	2.5419	50
11	.34502	.93859	36760	.7204	.0654	.8983	.36135	.93243	.38754	.5804	.0725	.7674	.37757	.92598	.40775	.4525	.0799	.6485	.39367	.91925	.42826	.3350	.0878	.5402	49
12	.34530	.93849	.36793	.7179	.0655	.8960	.36162	.93232	.38787	.5781	.0726	.7653	.37784	.92587	.40809	.4504	.0801	.6466	.39394	.91913	.42860	.3332	.0880	.5384	48
13	.34557	.93839	.36826	.7155	.0656	.8937	.36189	.93222	.38821	.5759	.0727	.7632	.37811	.92576	.40843	.4484	.0802	.6447	.39421	.91902	.42894	.3313	.0881	.5367	47
14	.34584	.93829	.36859	.7130	.0658	.8915	.36217	.93211	.38854	.5737	.0728	.7611	.37838	.92565	.40877	.4463	.0803	.6428	.39448	.91891	.42929	.3294	.0882	.5350	46
15	.34612	.93819	.36892	2.7106	.0659	2.8892	.36244	.93201	.38888	2.5715	.0729	2.7591	.37865	.92554	.40911	2.4443	.0804	2.6410	.39474	.91879	.42963	2.3276	.0884	2.5333	45
16	.34639	.93809	.36925	.7082	.0660	.8869	.36271	.93190	.38921	.5693	.0731	.7570	.37892	.92543	.40945	.4423	.0806	.6391	.39501	.91868	.42998	.3257	.0885	.5316	44
17	.34666	.93799	.36958	.7058	.0661	.8846	.36298	.93180	.38955	.5671	.0732	.7550	.37919	.92532	.40979	.4403	.0807	.6372	.39528	.91856	.43032	.3238	.0886	.5299	43
18	.34693	.93789	.36991	.7033	.0662	.8824	.36325	.93169	.38988	.5649	.0733	.7529	.37946	.92521	.41013	.4382	.0808	.6353	.39554	.91845	.43067	.3220	.0888	.5281	42
19	.34721	.93779	.37024	.7009	.0663	.8801	.36352	.93158	.39022	.5627	.0734	.7509	.37972	.92510	.41047	.4362	.0810	.6335	.39581	.91833	.43101	.3201	.0889	.5264	41
20	.34748	.93769	.37057	2.6985	.0664	2.8778	.36379	.93148	.39055	2.5605	.0736	2.7488	.37999	.92499	.41081	2.4342	.0811	2.6316	.39608	.91822	.43136	2.3183	.0891	2.5247	40
21	.34775	.93758	.37090	.6961	.0666	.8756	.36406	.93137	.39089	.5583	.0737	.7468	.38026	.92488	.41115	.4322	.0812	.6297	.39635	.91810	.43170	.3164	.0892	.5230	39
22	.34803	.93748	.37123	.6937	.0667	.8733	.36433	.93127	.39122	.5561	.0738	.7447	.38053	.92477	.41149	.4302	.0813	.6279	.39661	.91798	.43205	.3145	.0893	.5213	38
23	.34830	.93738	.37156	.6913	.0668	.8711	.36460	.93116	.39156	.5539	.0739	.7427	.38080	.92466	.41183	.4282	.0815	.6260	.39688	.91787	.43239	.3127	.0895	.5196	37
24	.34857	.93728	.37190	.6889	.0669	.8688	.36488	.93105	.39189	.5517	.0740	.7406	.38107	.92455	.41217	.4262	.0816	.6242	.39715	.91775	.43274	.3109	.0896	.5179	36
25	.34884	.93718	.37223	2.6865	.0670	2.8666	.36515	.93095	.39223	2.5495	.0742	2.7386	.38134	.92443	.41251	2.4242	.0817	2.6223	.39741	.91764	.43308	2.3090	.0897	2.5163	35
26	.34912	.93708	.37256	.6841	.0671	.8644	.36542	.93084	.39257	.5473	.0743	.7366	.38161	.92432	.41285	.4222	.0819	.6205	.39768	.91752	.43343	.3072	.0899	.5146	34
27	.34939	.93698	.37289	.6817	.0673	.8621	.36569	.93074	.39290	.5451	.0744	.7346	.38188	.92421	.41319	.4202	.0820	.6186	.39795	.91741	.43377	.3053	.0900	.5129	33
28	.34966	.93687	.37322	.6794	.0674	.8599	.36596	.93063	.39324	.5430	.0745	.7325	.38214	.92410	.41353	.4182	.0821	.6168	.39821	.91729	.43412	.3035	.0902	.5112	32
29	.34993	.93677	.37355	.6770	.0675	.8577	.36623	.93052	.39357	.5408	.0747	.7305	.38241	.92399	.41387	.4162	.0823	.6150	.39848	.91718	.43447	.3017	.0903	.5095	31
30	.35021	.93667	.37388	2.6746	.0676	2.8554	.36650	.93042	.39391	2.5386	.0748	2.7285	.38268	.92388	.41421	2.4142	.0824	2.6131	.39875	.91706	.43481	2.2998	.0904	2.5078	30
31	.35048	.93657	.37422	.6722	.0677	.8532	.36677	.93031	.39425	.5365	.0749	.7265	.38295	.92377	.41455	.4122	.0825	.6113	.39901	.91694	.43516	.2980	.0906	.5062	29
32	.35075	.93647	.37455	.6699	.0678	.8510	.36704	.93020	39458	.5343	.0750	.7245	.38322	.92366	.41489	.4102	.0826	.6095	.39928	.91683	.43550	.2962	.0907	.5045	28
33	.35102	.93637	.37488	.6675	.0679	.8488	.36731	.93010	.39492	.5322	.0751	.7225	.38349	.92354	.41524	.4083	.0828	.6076	.39955	.91671	.43585	.2944	.0908	.5028	27
34	.35130	.93626	.37521	.6652	.0681	.8466	.36758	.92999	.39525	.5300	.0753	.7205	.38376	.92343	.41558	.4063	.0829	.6058	.39981	.91659	.43620	.2925	.0910	.5011	26
35	.35157	.93616	.37554	2.6628	.0682	2.8444	.36785	.92988	.39559	2.5278	.0754	2.7185	.38403	.92332	.41592	2.4043	.0830	2.6040	.40008	.91648	.43654	2.2907	.0911	2.4995	25
36	.35184	.93606	.37587	.6604	.0683	.8422	.36812	.92978	.39593	.5257	.0755	.7165	.38429	.92321	.41626	.4023	.0832	.6022	.40035	.91636	.43689	.2889	.0913	.4978	24
37	.35211	.93596	.37621	.6581	.0684	.8400	.36839	.92967	.39626	.5236	.0756	.7145	.38456	.92310	.41660	.4004	.0833	.6003	.40061	.91625	.43723	.2871	.0914	.4961	23
38	.35239	.93585	.37654	.6558	.0685	.8378	.36866	.92956	.39660	.5214	.0758	.7125	.38483	.92299	.41694	.3984	.0834	.5985	.40088	.91613	.43758	.2853	.0915	.4945	22
39	.35266	.93575	.37687	.6534	.0686	.8356	.36893	.92945	.39694	.5193	.0759	.7105	.38510	.92287	.41728	.3964	.0836	.5967	.40115	.91601	.43793	.2835	.0917	.4928	21
40	.35293	.93565	.37720	2.6511	.0688	2.8334	.36921	.92935	.39727	2.5171	.0760	2.7085	.38537	.92276	.41762	2.3945	.0837	2.5949	.40141	.91590	.43827	2.2817	.0918	2.4912	20
41	.35320	.93555	.37754	.6487	.0689	.8312	.36948	.92924	.39761	.5150	.0761	.7065	.38564	.92265	.41797	.3925	.0838	.5931	.40168	.91578	.43862	.2799	.0920	.4895	19
42	.35347	.93544	.37787	.6464	.0690	.8290	.36975	.92913	.39795	.5129	.0763	.7045	.38591	.92254	.41831	.3906	.0840	.5913	.40195	.91566	.43897	.2781	.0921	.4879	18
43	.35375	.93534	.37820	.6441	.0691	.8269	.37002	.92902	.39828	.5108	.0764	.7026	.38617	.92242	.41865	.3886	.0841	.5895	.40221	.91554	.43932	.2763	.0922	.4862	17
44	.35402	.93524	.37853	.6418	.0692	.8247	.37029	.92892	.39862	.5086	.0765	.7006	.38644	.92231	.41899	.3867	.0842	.5877	.40248	.91543	.43966	.2745	.0924	.4846	16
45	.35429	.93513	.37887	2.6394	.0694	2.8225	.37056	.92881	.39896	2.5065	.0766	2.6986	.38671	.92220	.41933	2.3847	.0844	2.5958	.40205	.91531	.44001	2.2727	.0925	2.4829	15
46	.35456	.93503	.37920	.6371	.0695	.8204	.37083	.92870	.39930	.5044	.0768	.6967	.38698	.92209	.41968	.3828	.0845	.5841	.40301	.91519	.44036	.2709	.0927	.4813	14
47	.35483	.93493	.37953	.6348	.0696	.8182	.37110	.92859	.39963	.5023	.0769	.6947	.38725	.92197	.42002	.3808	.0846	.5823	.40328	.91508	.44070	.2691	.0928	.4797	13
48	.35511	.93482	.37986	.6325	.0697	.8160	.37137	.92848	.39997	.5002	.0770	.6927	.38751	.92186	.42036	.3789	.0847	.5805	.40354	.91496	.44105	.2673	.0929	.4780	12
49	.35538	.93472	.38020	.6302	.0698	.8139	.37164	.92838	.40031	.4981	.0771	.6908	.38778	.92105	.42070	.3770	.0849	.5787	.40381	.91484	.44140	.2655	.0931	.4764	11
50	.35565	.93462	.38053	2.6279	.0699	2.8117	.37191	.92827	.40065	2.4960	.0773	2.6888	.38805	.92164	.42105	2.3750	.0850	2.5770	.40408	.91472	.44175	2.2637	.0932	2.4748	10
51	.35592	.93451	.38086	.6256	.0701	.8096	.37218	.92816	.40098	.4939	.0774	.6869	.38832	.92152	.42139	.3731	.0851	.5752	.40434	.91461	.44209	.2619	.0934	.4731	9
52	.35619	.93441	.38120	.6233	.0702	.8074	.37245	.92805	.40132	.4918	.0775	.6849	.38859	.92141	.42173	.3712	.0853	.5734	.40461	.91449	.44244	.2602	.0935	.4715	8
53	.35647	.93431	.38153	.6210	.0703	.8053	.37272	.92794	.40166	.4897	.0776	.6830	.38886	.92130	.42207	.3692	.0854	.5716	.40487	.91437	.44279	.2584	.0936	.4699	7
54	.35674	.93420	.38186	.6187	.0704	.8032	.37299	.92784	.40200	.4876	.0778	.6810	.38912	.92118	.42242	.3673	.0855	.5699	.40514	.91425	.44314	.2566	.0938	.4683	6
55	.35701	.93410	.38220	2.6164	.0705	2.8010	.37326	.92773	.40233	2.4855	.0779	2.6791	.38939	.92107	.42276	2.3654	.0857	2.5681	.40541	.91414	.44349	2.2548	.0939	2.4666	5
56	.35728	.93400	.38253	.6142	.0707	.7989	.37353	.92762	.40267	.4834	.0780	.6772	.38966	.92096	.42310	.3635	.0858	.5663	.40567	.91402	.44383	.2531	.0941	.4650	4
57	.35755	.93389	.38286	.6119	.0708	.7968	.37380	.92751	.40301	.4813	.0781	.6752	.38993	.92084	.42344	.3616	.0859	.5646	.40594	.91390	.44418	.2513	.0942	.4634	3
58	.35782	.93379	.38320	.6096	.0709	.7947	.37407	.92740	.40335	.4792	.0783	.6733	.39019	.92073	.42379	.3597	.0861	.5628	.40620	.91378	.44453	.2495	.0943	.4618	2
59	.35810	.93368	.38353	.6073	.0710	.7925	.37434	.92729	.40369	.4772	.0784	.6714	.39046	.92062	.42413	.3577	.0862	.5610	.40647	.91366	.44488	.2478	.0945	.4602	1
60	.35837	.93358	.38386	2.6051	.0711	2.7904	.37461	.92718	.40403	2.4751	.0785	2.6695	.39073	.92050	.42447	2.3558	.0864	2.5593	.40674	.91354	.44523	2.2460	.0946	2.4586	0
′	cos	sin	cot	tan	cosec	sec	cos	sin	cot	tan	cosec	sec	cos	sin	cot	tan	cosec	sec	cos	sin	cot	tan	cosec	sec	′
	69°						68°						67°						66°						

TABLE 9 Continued

	24°						25°						26°						27°						
′	sin	cos	tan	cot	sec	cosec	sin	cos	tan	cot	sec	cosec	sin	cos	tan	cot	sec	cosec	sin	cos	tan	cot	sec	cosec	′
0	.40674	.91354	.44523	2.2460	1.0946	2.4586	.42262	.90631	.46631	2.1445	1.1034	2.3662	.43837	.89879	.48773	2.0503	1.1126	2.2812	.45399	.89101	.50952	1.9626	1.1223	2.2027	60
1	.40700	.91343	.44558	.2443	.0948	.4570	.42288	.90618	.46666	.1429	.1035	.3647	.43863	.89867	.48809	.0488	.1127	.2798	.45425	.89087	.50989	.9612	.1225	.2014	59
2	.40727	.91331	.44593	.2425	.0949	.4554	.42314	.90606	.46702	.1412	.1037	.3632	.43889	.89854	.48845	.0473	.1129	.2784	.45451	.89074	.51026	.9598	.1226	.2002	58
3	.40753	.91319	.44627	.2408	.0951	.4538	.42341	.90594	.46737	.1396	.1038	.3618	.43915	.89841	.48881	.0458	.1131	.2771	.45477	.89061	.51062	.9584	.1228	.1989	57
4	.40780	.91307	.44662	.2390	.0952	.4522	.42367	.90581	.46772	.1380	.1040	.3603	.43942	.89828	.48917	.0443	.1132	.2757	.45503	.89048	.51099	.9570	.1230	.1977	56
5	.40806	.91295	.44697	2.2373	.0953	2.4506	.42394	.90569	.46808	2.1364	.1041	2.3588	.43968	.89815	.48953	2.0427	.1134	2.2744	.45528	.89034	.51136	1.9556	.1231	2.1964	55
6	.40833	.91283	.44732	.2355	.0955	.4490	.42420	.90557	.46843	.1348	.1043	.3574	.43994	.89803	.48989	.0412	.1135	.2730	.45554	.89021	.51172	.9542	.1233	.1952	54
7	.40860	.91271	.44767	.2338	.0956	.4474	.42446	.90544	.46879	.1331	.1044	.3559	.44020	.89790	.49025	.0397	.1137	.2717	.45580	.89008	.51209	.9528	.1235	.1939	53
8	.40886	.91260	.44802	.2320	.0958	.4458	.42473	.90532	.46914	.1315	.1046	.3544	.44046	.89777	.49062	.0382	.1139	.2703	.45606	.88995	.51246	.9514	.1237	.1927	52
9	.40913	.91248	.44837	.2303	.0959	.4442	.42499	.90520	.46950	.1299	.1047	.3530	.44072	.89764	.49098	.0367	.1140	.2690	.45632	.88981	.51283	.9500	.1238	.1914	51
10	.40939	.91236	.44872	2.2286	.0961	2.4426	.42525	.90507	.46985	2.1283	.1049	2.3515	.44098	.89751	.49134	2.0352	.1142	2.2676	.45658	.88968	.51319	1.9486	.1240	2.1902	50
11	.40966	.91224	.44907	.2268	.0962	.4418	.42552	.90495	.47021	.1267	.1050	.3501	.44124	.89739	.49170	.0338	.1143	.2663	.45684	.88955	.51356	.9472	.1242	.1889	49
12	.40992	.91212	.44942	.2251	.0963	.4395	.42578	.90483	.47056	.1251	.1052	.3486	.44150	.89726	.49206	.0323	.1145	.2650	.45710	.88942	.51393	.9458	.1243	.1877	48
13	.41019	.91200	.44977	.2234	.0965	.4379	.42604	.90470	.47092	.1235	.1053	.3472	.44177	.89713	.49242	.0308	.1147	.2636	.45736	.88928	.51430	.9444	.1245	.1865	47
14	.41045	.91188	.45012	.2216	.0966	.4363	.42630	.90458	.47127	.1219	.1055	.3457	.44203	.89700	.49278	.0293	.1148	.2623	.45761	.88915	.51466	.9430	.1247	.1852	46
15	.41072	.91176	.45047	2.2199	.0968	2.4347	.42657	.90445	.47163	2.1203	.1056	2.3443	.44229	.89687	.49314	2.0278	.1150	2.2610	.45787	.88902	.51503	1.9416	.1248	2.1840	45
16	.41098	.91164	.45082	.2182	.0969	.4332	.42683	.90433	.47199	.1187	.1058	.3428	.44255	.89674	.49351	.0263	.1151	.2596	.45813	.88888	.51540	.9402	.1250	.1828	44
17	.41125	.91152	.45117	.2165	.0971	.4316	.42709	.90421	.47234	.1171	.1059	.3414	.44281	.89661	.49387	.0248	.1153	.2583	.45839	.88875	.51577	.9388	.1252	.1815	43
18	.41151	.91140	.45152	.2147	.0972	.4300	.42736	.90408	.47270	.1155	.1061	.3399	.44307	.89649	.49423	.0233	.1155	.2570	.45865	.88862	.51614	.9375	.1253	.1803	42
19	.41178	.91128	.45187	.2130	.0973	.4285	.42762	.90396	.47305	.1139	.1062	.3385	.44333	.89636	.49459	.0219	.1156	.2556	.45891	.88848	.51651	.9361	.1255	.1791	41
20	.41204	.91116	.45222	2.2113	.0975	2.4269	.42788	.90383	.47341	2.1123	.1064	2.3371	.44359	.89623	.49495	2.0204	.1158	2.2543	.45917	.88835	.51687	1.9347	.1257	2.1778	40
21	.41231	.91104	.45257	.2096	.0976	.4254	.42815	.90371	.47376	.1107	.1065	.3356	.44385	.89610	.49532	.0189	.1159	.2530	.45942	.88822	.51724	.9333	.1258	.1766	39
22	.41257	.91092	.45292	.2079	.0978	.4238	.42841	.90358	.47412	.1092	.1067	.3342	.44411	.89597	.49568	.0174	.1161	.2517	.45968	.88808	.51761	.9319	.1260	.1754	38
23	.41284	.91080	.45327	.2062	.0979	.4222	.42867	.90346	.47448	.1076	.1068	.3328	.44437	.89584	.49604	.0159	.1163	.2503	.45994	.88795	.51798	.9306	.1262	.1742	37
24	.41310	.91068	.45362	.2045	.0981	.4207	.42893	.90333	.47483	.1060	.1070	.3313	.44463	.89571	.49640	.0145	.1164	.2490	.46020	.88781	.51835	.9292	.1264	.1730	36
25	.41337	.91056	.45397	2.2028	.0982	2.4191	42920	.90321	.47519	2.1044	.1072	2.3299	.44489	.89558	.49677	2.0130	.1166	2.2477	.46046	.88768	.51872	1.9278	.1265	2.1717	35
26	.41363	.91044	.45432	.2011	.0984	.4176	.42946	.90308	.47555	.1028	.1073	.3285	.44516	.89545	.49713	.0115	.1167	.2464	.46072	.88755	.51909	.9264	.1267	.1705	34
27	.41390	.91032	.45467	.1994	.0985	.4160	.42972	.90296	.47590	.1013	.1075	.3271	.44542	.89532	.49749	.0101	.1169	.2451	.46097	.88741	.51946	.9251	.1269	.1693	33
28	.41416	.91020	.45502	.1977	.0986	.4145	.42998	.90283	.47626	.0997	.1076	.3256	.44568	.89519	.49785	.0086	.1171	.2438	.46123	.88728	.51983	.9237	.1270	.1681	32
29	.41443	.91008	.45537	.1960	.0988	.4130	.43025	.90271	.47662	.0981	.1078	.3242	.44594	.89506	.49822	.0071	.1172	.2425	.46149	.88714	.52020	.9223	.1272	.1669	31
30	.41469	.90996	.45573	2.1943	.0989	2.4114	.43051	.90258	.47697	2.0965	.1079	2.3228	.44620	.89493	.49858	2.0057	.1174	2.2411	.46175	.88701	.52057	1.9210	.1274	2.1657	30
31	.41496	.90984	.45608	.1926	.0991	.4099	.43077	.90246	.47733	.0950	.1081	.3214	.44646	.89480	.49894	.0042	.1176	.2398	.46201	.88688	.52094	.9196	.1275	.1645	29
32	.41522	.90972	.45643	.1909	.0992	.4083	.43104	.90233	.47769	.0934	.1082	.3200	.44672	.89467	.49931	.0028	.1177	.2385	.46226	.88674	.52131	.9182	.1277	.1633	28
33	.41549	.90960	.45678	.1892	.0994	.4068	.43130	.90221	.47805	.0918	.1084	.3186	.44698	.89454	.49967	.0013	.1179	.2372	.46252	.88661	.52168	.9169	.1279	.1620	27
34	.41575	.90948	.45713	.1875	.0995	.4053	.43156	.90208	.47840	.0903	.1085	.3172	.44724	.89441	.50003	1.9998	.1180	.2359	.46278	.88647	.52205	.9155	.1281	.1608	26
35	.41602	.90936	.45748	2.1859	.0997	2.4037	.43182	.90196	.47876	2.0887	.1087	2.3158	.44750	.89428	.50040	1.9984	.1182	2.2346	.46304	.88634	.52242	1.9142	.1282	2.1596	25
36	.41628	.90924	.45783	.1842	.0998	.4022	.43208	.90183	.47912	.0872	.1088	.3143	.44776	.89415	.50076	.9969	.1184	.2333	.46330	.88620	.52279	.9128	.1284	.1584	24
37	.41654	.90911	.45819	.1825	.1000	.4007	.43235	.90171	.47948	.0856	.1090	.3129	.44802	.89402	.50113	.9955	.1185	.2320	.46355	.88607	.52316	.9115	.1286	.1572	23
38	.41681	.90899	.45854	.1808	.1001	.3992	.43261	.90158	.47983	.0840	.1092	.3115	.44828	.89389	.50149	.9940	.1187	.2307	.46381	.88593	.52353	.9101	.1287	.1560	22
39	.41707	.90887	.45889	.1792	.1003	.3976	.43287	.90145	.48019	.0825	.1093	.3101	.44854	.89376	.50185	.9926	.1189	.2294	.46407	.88580	.52390	.9088	.1289	.1548	21
40	.41734	.90875	.45924	2.1775	.1004	2.3961	.43313	.90133	.48055	2.0809	.1095	2.3087	.44880	.89363	.50222	1.9912	.1190	2.2282	.46433	.88566	.52427	1.9074	.1291	2.1536	20
41	.41760	.90863	.45960	.1758	.1005	.3946	.43340	.90120	.48091	.0794	.1096	.3073	.44906	.89350	.50258	.9897	.1192	.2269	.46458	.88553	.52464	.9061	.1293	.1525	19
42	.41787	.90851	.45995	.1741	.1007	.3931	.43366	.90108	.48127	.0778	.1098	.3059	.44932	.89337	.50295	.9883	.1193	.2256	.46484	.88539	.52501	.9047	.1294	.1513	18
43	.41813	.90839	.46030	.1725	.1008	.3916	.43392	.90095	.48162	.0763	.1099	.3046	.44958	.89324	.50331	.9868	.1195	.2243	.46510	.88526	.52538	.9034	.1296	.1501	17
44	.41839	.90826	.46065	.1708	.1010	.3901	.43418	.90082	.48198	.0747	.1101	.3032	.44984	.89311	.50368	.9854	.1197	.2230	.46536	.88512	.52575	.9020	.1298	.1489	16
45	.41866	.90814	.46101	2.1692	.1011	2.3886	.43444	.90070	.48234	2.0732	.1102	2.3018	.45010	.89298	.50404	1.9840	.1198	2.2217	.46561	.88499	.52612	1.9007	.1299	2.1477	15
46	.41892	.90812	.46136	.1675	.1013	.3871	.43471	.90057	.48270	.0717	.1104	.3004	.45036	.89285	.50441	.9825	.1200	.2204	.46587	.88485	.52650	.8993	.1301	.1465	14
47	.41919	.90790	.46171	.1658	.1014	.3856	.43497	.90044	.48306	.0701	.1106	.2990	.45062	.89272	.50477	.9811	.1202	.2192	.46613	.88472	.52687	.8980	.1303	.1453	13
48	.41945	.90778	.46206	.1642	.1016	.3841	.43523	.90032	.48342	.0686	.1107	.2976	.45088	.89258	.50514	.9797	.1203	.2179	.46639	.88458	.52724	.8967	.1305	.1441	12
49	.41972	.90765	.46242	.1625	.1017	.3826	.43549	.90019	.48378	.0671	.1109	.2962	.45114	.89245	.50550	.9782	.1205	.2166	.46664	.88444	.52761	.8953	.1306	.1430	11
50	.41998	.90753	.46277	2.1609	.1019	2.3811	.43575	.90006	.48414	2.0655	.1110	2.2949	.45140	.89232	.50587	1.9768	.1207	2.2153	.46690	.88431	.52798	1.8940	.1308	2.1418	10
51	.42024	.90741	.46312	.1592	.1020	.3796	.43602	.89994	.48449	.0640	.1112	.2935	.45166	.89219	.50623	.9754	.1208	.2141	.46716	.88417	.52836	.8927	.1310	.1406	9
52	.42051	.90729	.46348	.1576	.1022	.3781	.43628	.89981	.48485	.0625	.1113	.2921	.45191	.89206	.50660	.9739	.1210	.2128	.46741	.88404	.52873	.8913	.1312	.1394	8
53	.42077	.90717	.46383	.1559	.1023	.3766	.43654	.89968	.48521	.0609	.1115	.2907	.45217	.89193	.50696	.9725	.1212	.2115	.46767	.88390	.52910	.8900	.1313	.1382	7
54	.42103	.90704	.46418	.1543	.1025	.3751	.43680	.89956	.48557	.0594	.1116	.2894	.45243	.89180	.50733	.9711	.1213	.2103	.46793	.88376	.52947	.8887	.1315	.1371	6
55	.42130	.90692	.46454	2.1527	.1026	2.3736	.43706	.89943	.48593	2.0579	.1118	2.2880	.45269	.89166	.50769	1.9697	.1215	2.2090	.46819	.88363	.52984	1.8873	.1317	2.1359	5
56	.42156	.90680	.46489	.1510	.1028	.3721	.43732	.89930	.48629	.0564	.1120	.2866	.45295	.89153	.50806	.9683	.1217	.2077	.46844	.88349	.53022	.8860	.1319	.1347	4
57	.42183	.90668	.46524	.1494	.1029	.3706	.43759	.89918	.48665	.0548	.1121	.2853	.45321	.89140	.50843	.9668	.1218	.2065	.46870	.88336	.53059	.8847	.1320	.1335	3
58	.42209	.90655	.46560	.1478	.1031	.3691	.43785	.89905	.48701	.0533	.1123	.2839	.45347	.89127	.50879	.9654	.1220	.2052	.46896	.88322	.53096	.8834	.1322	.1324	2
59	.42235	.90643	.46595	.1461	.1032	.3677	.43811	.89892	.48737	.0518	.1124	.2825	.45373	.89114	.50916	.9640	.1222	.2039	.46921	.88308	.53134	.8820	.1324	.1312	1
60	.42262	.90631	.46631	2.1445	.1034	2.3662	.43837	.89879	.48773	2.0503	.1126	2.2812	.45399	.89101	.50952	1.9626	.1223	2.2027	.46947	.88295	.53171	1.8807	.1326	2.1300	0
′	cos	sin	cot	tan	cosec	sec	cos	sin	cot	tan	cosec	sec	cos	sin	cot	tan	cosec	sec	cos	sin	cot	tan	cosec	sec	′
	65°						64°						63°						62°						

TABLE 9 Continued

′	28°						29°						30°						31°						
′	sin	cos	tan	cot	sec	cosec	sin	cos	tan	cot	sec	cosec	sin	cos	tan	cot	sec	cosec	sin	cos	tan	cot	sec	cosec	′
0	.46947	.88295	.53171	1.8807	1.1326	2.1300	.48481	.87462	.55431	1.8040	1.1433	2.0627	.50000	.86603	.57735	1.7320	1.1547	2.0000	.51504	.85717	.60086	1.6643	1.1666	1.9416	60
1	.46973	.88281	.53208	.8794	.1327	.1289	.48506	.87448	.55469	.8028	.1435	.0616	.50025	.86588	.57774	.7309	.1549	1.9990	.51529	.85702	.60126	.6632	.1668	.9407	59
2	.46998	.88267	.53245	.8781	.1329	.1277	.48532	.87434	.55507	.8016	.1437	.0605	.50050	.86573	.57813	.7297	.1551	.9980	.51554	.85687	.60165	.6621	.1670	.9397	58
3	.47024	.88254	.53283	.8768	.1331	.1266	.48557	.87420	.55545	.8003	.1439	.0594	.50075	.86559	.57851	.7286	.1553	.9970	.51578	.85672	.60205	.6610	.1672	.9388	57
4	.47050	.88240	.53320	.8754	.1333	.1254	.48583	.87405	.55583	.7991	.1441	.0583	.50101	.86544	.57890	.7274	.1555	.9960	.51603	.85657	.60244	.6599	.1674	.9378	56
5	.47075	.88226	.53358	1.8741	.1334	2.1242	.48608	.87391	.55621	1.7979	.1443	2.0573	.50126	.86530	.57929	1.7262	.1557	1.9950	.51628	.85642	.60284	1.6588	.1676	1.9369	55
6	.47101	.88213	.53395	.8728	.1336	.1231	.48633	.87377	.55659	.7966	.1445	.0562	.50151	.86515	.57968	.7251	.1559	.9940	.51653	.85627	.60324	.6577	.1678	.9360	54
7	.47127	.88199	.53432	.8715	.1338	.1219	.48659	.87363	.55697	.7954	.1446	.0551	.50176	.86500	.58007	.7239	.1561	.9930	.51678	.85612	.60363	.6566	.1681	.9350	53
8	.47152	.88185	.53470	.8702	.1340	.1208	.48684	.87349	.55735	.7942	.1448	.0540	.50201	.86486	.58046	.7228	.1562	.9920	.51703	.85597	.60403	.6555	.1683	.9341	52
9	.47178	.88171	.53507	.8689	.1341	.1196	.48710	.87335	.55774	.7930	.1450	.0530	.50226	.86471	.58085	.7216	.1564	.9910	.51728	.85582	.60443	.6544	.1685	.9332	51
10	.47204	.88158	.53545	1.8676	.1343	2.1185	.48735	.87320	.55812	1.7917	.1452	2.0519	.50252	.86457	.58123	1.7205	.1566	1.9900	.51753	.85566	.60483	1.6534	.1687	1.9322	50
11	.47229	.88144	.53582	.8663	.1345	.1173	.48760	.87306	.55850	.7905	.1454	.0508	.50277	.86442	.58162	.7193	.1568	.9890	.51778	.85551	.60522	.6523	.1689	.9313	49
12	.47255	.88130	.53619	.8650	.1347	.1162	.48786	.87292	.55888	.7893	.1456	.0498	.50302	.86427	.58201	.7182	.1570	.9880	.51803	.85536	.60562	.6512	.1691	.9304	48
13	.47281	.88117	.53657	.8637	.1349	.1150	.48811	.87278	.55926	.7881	.1458	.0487	.50327	.86413	.58240	.7170	.1572	.9870	.51827	.85521	.60602	.6501	.1693	.9295	47
14	.47306	.88103	.53694	.8624	.1350	.1139	.48837	.87264	.55964	.7868	.1459	.0476	.50352	.86398	.58279	.7159	.1574	.9860	.51852	.85506	.60642	.6490	.1695	.9285	46
15	.47332	.88089	.53732	1.8611	.1352	2.1127	.48862	.87250	.56003	1.7856	.1461	2.0466	.50377	.86383	.58318	1.7147	.1576	1.9850	.51877	.85491	.60681	1.6479	.1697	1.9276	45
16	.47357	.88075	.53769	.8598	.1354	.1116	.48887	.87235	.56041	.7844	.1463	.0455	.50402	.86369	.58357	.7136	.1578	.9840	.51902	.85476	.60721	.6469	.1699	.9267	44
17	.47383	.88061	.53807	.8585	.1356	.1104	.48913	.87221	.56079	.7832	.1465	.0444	.50428	.86354	.58396	.7124	.1580	.9830	.51927	.85461	.60761	.6458	.1701	.9258	43
18	.47409	.88048	.53844	.8572	.1357	.1093	.48938	.87207	.56117	.7820	.1467	.0434	.50453	.86339	.58435	.7113	.1582	.9820	.51952	.85446	.60801	.6447	.1703	.9248	42
19	.47434	.88034	.53882	.8559	.1359	.1082	.48964	.87193	.56156	.7808	.1469	.0423	.50478	.86325	.58474	.7101	.1584	.9811	.51977	.85431	.60841	.6436	.1705	.9239	41
20	.47460	.88020	.53919	1.8546	.1361	2.1070	.48989	.87178	.56194	1.7795	.1471	2.0413	.50503	.86310	.58513	1.7090	.1586	1.9801	.52002	.85416	.60881	1.6425	.1707	1.9230	40
21	.47486	.88006	.53957	.8533	.1363	.1059	.49014	.87164	.56232	.7783	.1473	.0402	.50528	.86295	.58552	.7079	.1588	.9791	.52026	.85400	.60920	.6415	.1709	.9221	39
22	.47511	.87992	.53995	.8520	.1365	.1048	.49040	.87150	.56270	.7771	.1474	.0392	.50553	.86281	.58591	.7067	.1590	.9781	.52051	.85385	.60960	.6404	.1712	.9212	38
23	.47537	.87979	.54032	.8507	.1366	.1036	.49065	.87136	.56309	.7759	.1476	.0381	.50578	.86266	.58630	.7056	.1592	.9771	.52076	.85370	.61000	.6393	.1714	.9203	37
24	.47562	.87965	.54070	.8495	.1368	.1025	.49090	.87121	.56347	.7747	.1478	.0370	.50603	.86251	.58670	.7044	.1594	.9761	.52101	.85355	.61040	.6383	.1716	.9193	36
25	.47588	.87951	.54107	1.8482	.1370	2.1014	.49116	.87107	.56385	1.7735	.1480	2.0360	.50628	.86237	.58709	1.7033	.1596	1.9752	.52126	.85340	.61080	1.6372	.1718	1.9184	35
26	.47613	.87937	.54145	.8469	.1372	.1002	.49141	.87093	.56424	.7723	.1482	.0349	.50653	.86222	.58748	.7022	.1598	.9742	.52151	.85325	.61120	.6361	.1720	.9175	34
27	.47639	.87923	.54183	.8456	.1373	.0991	.49166	.87078	.56462	.7711	.1484	.0339	.50679	.86207	.58787	.7010	.1600	.9732	.52175	.85309	.61160	.6350	.1722	.9166	33
28	.47665	.87909	.54220	.8443	.1375	.0980	.49192	.87064	.56500	.7699	.1486	.0329	.50704	.86192	.58826	.6999	.1602	.9722	.52200	.85294	.61200	.6340	.1724	.9157	32
29	.47690	.87895	.54258	.8430	.1377	.0969	.49217	.87050	.56539	.7687	.1488	.0318	.50729	.86178	.58865	.6988	.1604	.9713	.52225	.85279	.61240	.6329	.1726	.9148	31
30	.47716	.87882	.54295	1.8418	.1379	2.0957	.49242	.87035	.56577	1.7675	.1489	2.0308	.50754	.86163	.58904	1.6977	.1606	1.9703	.52250	.85264	.61280	1.6318	.1728	1.9139	30
31	.47741	.87868	.54333	.8405	.1381	.0946	.49268	.87021	.56616	.7663	.1491	.0297	.50779	.86148	.58944	.6965	.1608	.9693	.52275	.85249	.61320	.6308	.1730	.9130	29
32	.47767	.87854	.54371	.8392	.1382	.0935	.49293	.87007	.56654	.7651	.1493	.0287	.50804	.86133	.58983	.6954	.1610	.9683	.52299	.85234	.61360	.6297	.1732	.9121	28
33	.47792	.87840	.54409	.8379	.1384	.0924	.49318	.86992	.56692	.7639	.1495	.0276	.50829	.86118	.59022	.6943	.1612	.9674	.52324	.85218	.61400	.6286	.1734	.9112	27
34	.47818	.87826	.54446	.8367	.1386	.0912	.49343	.86978	.56731	.7627	.1497	.0266	.50854	.86104	.59061	.6931	.1614	.9664	.52349	.85203	.61440	.6276	.1737	.9102	26
35	.47844	.87812	.54484	1.8354	.1388	2.0901	.49369	.86964	.56769	1.7615	.1499	2.0256	.50879	.86089	.59100	1.6920	.1616	1.9654	.52374	.85188	.61480	1.6265	.1739	1.9093	25
36	.47869	.87798	.54522	.8341	.1390	.0890	.49394	.86949	.56808	.7603	.1501	.0245	.50904	.86074	.59140	.6909	.1618	.9645	.52398	.85173	.61520	.6255	.1741	.9084	24
37	.47895	.87784	.54559	.8329	.1391	.0879	.49419	.86935	.56846	.7591	.1503	.0235	.50929	.86059	.59179	.6898	.1620	.9635	.52423	.85157	.61560	.6244	.1743	.9075	23
38	.47920	.87770	.54597	.8316	.1393	.0868	.49445	.86921	.56885	.7579	.1505	.0224	.50954	.86044	.59218	.6887	.1622	.9625	.52448	.85142	.61601	.6233	.1745	.9066	22
39	.47946	.87756	.54635	.8303	.1395	.0857	.49470	.86906	.56923	.7567	.1507	.0214	.50979	.86030	.59258	.6875	.1624	.9616	.52473	.85127	.61641	.6223	.1747	.9057	21
40	.47971	.87742	.54673	1.8291	.1397	2.0846	.49495	.86892	.56962	1.7555	.1508	2.0204	.51004	.86015	.59297	1.6864	.1626	1.9606	.52498	.85112	.61681	1.6212	.1749	1.9048	20
41	.47997	.87728	.54711	.8278	.1399	.0835	.49521	.86877	.57000	.7544	.1510	.0194	.51029	.86000	.59336	.6853	.1628	.9596	.52522	.85096	.61721	.6202	.1751	.9039	19
42	.48022	.87715	.54748	.8265	.1401	.0824	.49546	.86863	.57039	.7532	.1512	.0183	.51054	.85985	.59376	.6842	.1630	.9587	.52547	.85081	.61761	.6191	.1753	.9030	18
43	.48048	.87701	.54786	.8253	.1402	.0812	.49571	.86849	.57077	.7520	.1514	.0173	.51079	.85970	.59415	.6831	.1632	.9577	.52572	.85066	.61801	.6181	.1756	.9021	17
44	.48073	.87687	.54824	.8240	.1404	.0801	.49596	.86834	.57116	.7508	.1516	.0163	.51104	.85955	.59454	.6820	.1634	.9568	.52597	.85050	.61842	.6170	.1758	.9013	16
45	.48099	.87673	.54862	1.8227	.1406	2.0790	.49622	.86820	.57155	1.7496	.1518	2.0152	.51129	.85941	.59494	1.6808	.1636	1.9558	.52621	.85735	.61882	1.6160	.1760	1.9004	15
46	.48124	.87659	.54900	.8215	.1408	.0779	.49647	.86805	.57193	.7484	.1520	.0142	.51154	.85926	.59533	.6797	.1638	.9549	.52646	.85020	.61922	.6149	.1762	.8995	14
47	.48150	.87645	.54937	.8202	.1410	.0768	.49672	.86791	.57232	.7473	.1522	.0132	.51179	.85911	.59572	.6786	.1640	.9539	.52671	.85004	.61962	.6139	.1764	.8986	13
48	.48175	.87631	.54975	.8190	.1411	.0757	.49697	.86776	.57270	.7461	.1524	.0122	.51204	.85896	.59612	.6775	.1642	.9530	.52695	.84989	.62003	.6128	.1766	.8977	12
49	.48201	.87617	.55013	.8177	.1413	.0746	.49723	.86762	.57309	.7449	.1526	.0111	.51229	.85881	.59651	.6764	.1644	.9520	.52720	.84974	.62043	.6118	.1768	.8968	11
50	.48226	.87603	.55051	1.8165	.1415	2.0735	.49748	.86748	.57348	1.7437	.1528	2.0101	.51254	.85866	.59691	1.6753	.1646	1.9510	.52745	.84959	.62083	1.6107	.1770	1.8959	10
51	.48252	.87588	.55089	.8152	.1417	.0725	.49773	.86733	.57386	.7426	.1530	.0091	.51279	.85851	.59730	.6742	.1648	.9501	.52770	.84943	.62123	.6097	.1772	.8950	9
52	.48277	.87574	.55127	.8140	.1419	.0714	.49798	.86719	.57425	.7414	.1531	.0081	.51304	.85836	.59770	.6731	.1650	.9491	.52794	.84928	.62164	.6086	.1775	.8941	8
53	.48303	.87560	.55165	.8127	.1421	.0703	.49823	.86704	.57464	.7402	.1533	.0071	.51329	.85821	.59809	.6720	.1652	.9482	.52819	.84912	.62204	.6076	.1777	.8932	7
54	.48328	.87546	.55203	.8115	.1422	.0692	.49849	.86690	.57502	.7390	.1535	.0061	.51354	.85806	.59849	.6709	.1654	.9473	.52844	.84897	.62244	.6066	.1779	.8924	6
55	.48354	.87532	.55241	1.8102	.1424	2.0681	.49874	.86675	.57541	1.7379	.1537	2.0050	.51379	.85791	.59888	1.6698	.1656	1.9463	.52868	.84882	.62285	1.6055	.1781	1.8915	5
56	.48379	.87518	.55279	.8090	.1426	.0670	.49899	.86661	.57580	.7367	.1539	.0040	.51404	.85777	.59928	.6687	.1658	.9454	.52893	.84866	.62325	.6045	.1783	.8906	4
57	.48405	.87504	.55317	.8078	.1428	.0659	.49924	.86646	.57619	.7355	.1541	.0030	.51429	.85762	.59967	.6676	.1660	.9444	.52918	.84851	.62366	.6034	.1785	.8897	3
58	.48430	.87490	.55355	.8065	.1430	.0648	.49950	.86632	.57657	.7344	.1543	.0020	.51454	.85747	.60007	.6665	.1662	.9435	.52942	.84836	.62406	.6024	.1787	.8888	2
59	.48455	.87476	.55393	.8053	.1432	.0637	.49975	.86617	.57696	.7332	.1545	.0010	.51479	.85732	.60046	.6654	.1664	.9425	.52967	.84820	.62446	.6014	.1790	.8879	1
60	.48481	.87462	.55431	1.8040	.1433	2.0627	.50000	.86603	.57735	1.7320	.1547	2.0000	.51504	.85717	.60086	1.6643	.1660	1.9416	.52992	.84805	.62487	1.6003	.1792	1.8871	0
′	cos	sin	cot	tan	cosec	sec	cos	sin	cot	tan	cosec	sec	cos	sin	cot	tan	cosec	sec	cos	sin	cot	tan	cosec	sec	′
	61°						60°						59°						58°						

TABLE 9 Continued

	32°						33°						34°						35°						
′	sin	cos	tan	cot	sec	cosec	sin	cos	tan	cot	sec	cosec	sin	cos	tan	cot	sec	cosec	sin	cos	tan	cot	sec	cosec	′
0	.52992	.84805	.62487	1.6003	1.1792	1.8871	.54464	.83867	.64941	1.5399	1.1924	1.8361	.55919	.82904	.67451	1.4826	1.2062	1.7883	.57358	.81915	.70021	1.4281	1.2208	1.7434	60
1	.53016	.84789	.62527	.5993	.1794	.8862	.54488	.83851	.64982	.5389	.1926	.8352	.55943	.82887	.67493	.4816	.2064	.7875	.57381	.81898	.70064	.4273	.2210	.7427	59
2	.53041	.84774	.62568	.5983	.1796	.8853	.54513	.83835	.65023	.5379	.1928	.8344	.55967	.82871	.67535	.4807	.2067	.7867	.57405	.81882	.70107	.4264	.2213	.7420	58
3	.53066	.84758	.62608	.5972	.1798	.8844	.54537	.83819	.65065	.5369	.1930	.8336	.55992	.82855	.67578	.4798	.2069	.7860	.57429	.81865	.70151	.4255	.2215	.7413	57
4	.53090	.84743	.62649	.5962	.1800	.8836	.54561	.83804	.65106	.5359	.1933	.8328	.56016	.82839	.67620	.4788	.2072	.7852	.57453	.81848	.70194	.4246	.2218	.7405	56
5	.53115	.84728	.62689	1.5952	.1802	1.8827	.54586	.83788	.65148	1.5350	.1935	1.8320	.56040	.82822	.67663	1.4779	.2074	1.7844	.57477	.81832	.70238	1.4237	.2220	1.7398	55
6	.53140	.84712	.62730	.5941	.1805	.8818	.54610	.83772	.65189	.5340	.1937	.8311	.56064	.82806	.67705	.4770	.2076	.7837	.57500	.81815	.70281	.4228	.2223	.7391	54
7	.53164	.84697	.62770	.5931	.1807	.8809	.54634	.83756	.65231	.5330	.1939	.8303	.56088	.82790	.67747	.4761	.2079	.7829	.57524	.81798	.70325	.4220	.2225	.7384	53
8	.53189	.84681	.62811	.5921	.1809	.8801	.54659	.83740	.65272	.5320	.1942	.8295	.56112	.82773	.67790	.4751	.2081	.7821	.57548	.81781	.70368	.4211	.2228	.7377	52
9	.53214	.84666	.62851	.5910	.1811	.8792	.54683	.83724	.65314	.5311	.1944	.8287	.56136	.82757	.67832	.4742	.2083	.7814	.57572	.81765	.70412	.4202	.2230	.7369	51
10	.53238	.84650	.62892	1.5900	.1813	1.8783	.54708	.83708	.65355	1.5301	.1946	1.8279	.56160	.82741	.67875	1.4733	.2086	1.7806	.57596	.81748	.70455	1.4193	.2233	1.7362	50
11	.53263	.84635	.62933	.5890	.1815	.8775	.54732	.83692	.65397	.5291	.1948	.8271	.56184	.82724	.67917	.4724	.2088	.7798	.57619	.81731	.70499	.4185	.2235	.7355	49
12	.53288	.84619	.62973	.5880	.1818	.8766	.54756	.83676	.65438	.5282	.1951	.8263	.56208	.82708	.67960	.4714	.2091	.7791	.57643	.81714	.70542	.4176	.2238	.7348	48
13	.53312	.84604	.63014	.5869	.1820	.8757	.54781	.83660	.65480	.5272	.1953	.8255	.56232	.82692	.68002	.4705	.2093	.7783	.57667	.81698	.70586	.4167	.2240	.7341	47
14	.53337	.84588	.63055	.5859	.1822	.8749	.54805	.83644	.65521	.5262	.1955	.8246	.56256	.82675	.68045	.4696	.2095	.7776	.57691	.81681	.70629	.4158	.2243	.7334	46
15	.53361	.84573	.63095	1.5849	.1824	1.8740	.54829	.83629	.65563	1.5252	.1958	1.8238	.56280	.82659	.68087	1.4687	.2098	1.7768	.57714	.81664	.70673	1.4150	.2245	1.7327	45
16	.53386	.84557	.63136	.5839	.1826	.8731	.54854	.83613	.65604	.5234	.1960	.8230	.56304	.82643	.68130	.4678	.2100	.7760	.57738	.81647	.70717	.4141	.2248	.7319	44
17	.53411	.84542	.63177	.5829	.1828	.8723	.54878	.83597	.65646	.5233	.1962	.8222	.56328	.82626	.68173	.4669	.2103	.7753	.57762	.81630	.70760	.4132	.2250	.7312	43
18	.53435	.84526	.63217	.5818	.1831	.8714	.54902	.83581	.65688	.5223	.1964	.8214	.56353	.82610	.68215	.4659	.2105	.7745	.57786	.81614	.70804	.4123	.2253	.7305	42
19	.53460	.84511	.63258	.5808	.1833	.8706	.54926	.83565	.65729	.5214	.1967	.8206	.56377	.82593	.68258	.4650	.2107	.7738	.57809	.81597	.70848	.4115	.2255	.7298	41
20	.53484	.84495	.63299	1.5798	.1835	1.8697	.54951	.83549	.65771	1.5204	.1969	1.8198	.56401	.82577	.68301	1.4641	.2110	1.7730	.57833	.81580	.70891	1.4106	.2258	1.7291	40
21	.53509	.84479	.63339	.5788	.1837	.8688	.54975	.83533	.65813	.5195	.1971	.8190	.56425	.82561	.68343	.4632	.2112	.7723	.57857	.81563	.70935	.4097	.2260	.7284	39
22	.53533	.84464	.63380	.5778	.1839	.8680	.54999	.83517	.65854	.5185	.1974	.8182	.56449	.82544	.68386	.4623	.2115	.7715	.57881	.81546	.70979	.4089	.2263	.7277	38
23	.53558	.84448	.63421	.5768	.1841	.8671	.55024	.83501	.65896	.5175	.1976	.8174	.56473	.82528	.68429	.4614	.2117	.7708	.57904	.81530	.71022	.4080	.2265	.7270	37
24	.53583	.84433	.63462	.5757	.1844	.8663	.55048	.83485	.65938	.5166	.1978	.8166	.56497	.82511	.68471	.4605	.2119	.7700	.57928	.81513	.71066	.4071	.2268	.7263	36
25	.53607	.84417	.63503	1.5747	.1846	1.8654	.55002	.83469	.65980	1.5156	.1980	1.8158	.56521	.82495	.68514	1.4595	.2122	1.7693	.57952	.81496	.71110	1.4063	.2270	1.7256	35
26	.53632	.84402	.63543	.5737	.1848	.8646	.55090	.83453	.66021	.5147	.1983	.8150	.56545	.82478	.68557	.4586	.2124	.7685	.57975	.81479	.71154	.4054	.2273	.7249	34
27	.53656	.84386	.63584	.5727	.1850	.8637	.55121	.83437	.66063	.5137	.1985	.8142	.56569	.82462	.68600	.4577	.2127	.7678	.57999	.81462	.71198	.4045	.2276	.7242	33
28	.53681	.84370	.63625	.5717	.1852	.8629	.55145	.83421	.66105	.5127	.1987	.8134	.56593	.82445	.68642	.4568	.2129	.7670	.58023	.81445	.71241	.4037	.2278	.7234	32
29	.53705	.84355	.63666	.5707	.1855	.8620	.55169	.83405	.66147	.5118	.1990	.8126	.56617	.82429	.68685	.4559	.2132	.7663	.58047	.81428	.71285	.4028	.2281	.7227	31
30	.53730	.84339	.63707	1.5697	.1857	1.8611	.55194	.83388	.66188	1.5108	.1992	1.8118	.56641	.82413	.68728	1.4550	.2134	1.7655	.58070	.81411	.71329	1.4019	.2283	1.7220	30
31	.53754	.84323	.63748	.5687	.1859	.8603	.55218	.83372	.66230	.5099	.1994	.8110	.56664	.82396	.68771	.4541	.2136	.7648	.58094	.81395	.71373	.4011	.2286	.7213	29
32	.53779	.84308	.63789	.5677	.1861	.8595	.55242	.83356	.66272	.5089	.1997	.8102	.56688	.82380	.68814	.4532	.2139	.7640	.58118	.81378	.71417	.4002	.2288	.7206	28
33	.53803	.84292	.63830	.5667	.1863	.8586	.55266	.83340	.66314	.5080	.1999	.8094	.56712	.82363	.68857	.4523	.2141	.7633	.58141	.81361	.71461	.3994	.2291	.7199	27
34	.53828	.84276	.63871	.5657	.1866	.8578	.55291	.83324	.66356	.5070	.2001	.8086	.56736	.82347	.68899	.4514	.2144	.7625	.58165	.81344	.71505	.3985	.2293	.7192	26
35	.53852	.84261	.63912	1.5646	.1868	1.8569	.55315	.83308	.66398	1.5061	.2004	1.8078	.56760	.82330	.68942	1.4505	.2146	1.7618	.58189	.81327	.71549	1.3976	.2296	1.7185	25
36	.53877	.84245	.63953	.5636	.1870	.8561	.55339	.83292	.66440	.5051	.2006	.8070	.56784	.82314	.68985	.4496	.2149	.7610	.58212	.81310	.71593	.3968	.2298	.7178	24
37	.53901	.84229	.63994	.5626	.1872	.8552	.55363	.83276	.66482	.5042	.2008	.8062	.56808	.82297	.69028	.4487	.2151	.7603	.58236	.81293	.71637	.3959	.2301	.7171	23
38	.53926	.84214	.64035	.5616	.1874	.8544	.55388	.83260	.66524	.5032	.2010	.8054	.56832	.82280	.69071	.4478	.2153	.7596	.58259	.81276	.71681	.3951	.2304	.7164	22
39	.53950	.84198	.64076	.5606	.1877	.8535	.55412	.83244	.66566	.5023	.2013	.8047	.56856	.82264	.69114	.4469	.2156	.7588	.58283	.81259	.71725	.3942	.2306	.7157	21
40	.53975	.84182	.64117	1.5596	.1879	1.8527	.55436	.83228	.66608	1.5013	.2015	1.8039	.56880	.82247	.69157	1.4460	.2158	1.7581	.58307	.81242	.71769	1.3933	.2309	1.7151	20
41	.53999	.84167	.64158	.5586	.1881	.8519	.55460	.83211	.66650	.5004	.2017	.8031	.56904	.82231	.69200	.4451	.2161	.7573	.58330	.81225	.71813	.3925	.2311	.7144	19
42	.54024	.84151	.64199	.5577	.1883	.8510	.55484	.83195	.66692	.4994	.2020	.8023	.56928	.82214	.69243	.4442	.2163	.7566	.58354	.81208	.71857	.3916	.2314	.7137	18
43	.54048	.84135	.64240	.5567	.1886	.8502	.55509	.83179	.66734	.4985	.2022	.8015	.56952	.82198	.69286	.4433	.2166	.7559	.58378	.81191	.71901	.3908	.2316	.7130	17
44	.54073	.84120	.64281	.5557	.1888	.8493	.55533	.83163	.66776	.4975	.2024	.8007	.56976	.82181	.69329	.4424	.2168	.7551	.58401	.81174	.71945	.3899	.2319	.7123	16
45	.54097	.84104	.64322	1.5547	.1890	1.8485	.55557	.83147	.66818	1.4966	.2027	1.7999	.57000	.82165	.69372	1.4415	.2171	1.7544	.58425	.81157	.71990	1.3891	.2322	1.7116	15
46	.54122	.84088	.64363	.5537	.1892	.8477	.55581	.83131	.66860	.4957	.2029	.7992	.57023	.82148	.69415	.4406	.2173	.7537	.58448	.81140	.72034	.3882	.2324	.7109	14
47	.54146	.84072	.64404	.5527	.1894	.8468	.55605	.83115	.66902	.4947	.2031	.7984	.57047	.82131	.69459	.4397	.2175	.7529	.58472	.81123	.72078	.3874	.2327	.7102	13
48	.54171	.84057	.64446	.5517	.1897	.8460	.55629	.83098	.66944	.4938	.2034	.7976	.57071	.82115	.69502	.4388	.2178	.7522	.58496	.81106	.72122	.3865	.2329	.7095	12
49	.54195	.84041	.64487	.5507	.1899	.8452	.55654	.83082	.66986	.4928	.2036	.7968	.57095	.82098	.69545	.4379	.2180	.7514	.58519	.81089	.72166	.3857	.2332	.7088	11
50	.54220	.84025	.64528	1.5497	.1901	1.8443	.55678	.83066	.67028	1.4919	.2039	1.7960	.57119	.82082	.69588	1.4370	.2183	1.7507	.58543	.81072	.72211	1.3848	.2335	1.7081	10
51	.54244	.84009	.64569	.5487	.1903	.8435	.55702	.83050	.67071	.4910	.2041	.7953	.57143	.82065	.69631	.4361	.2185	.7500	.58566	.81055	.72255	.3840	.2337	.7075	9
52	.54268	.83993	.64610	.5477	.1906	.8427	.55726	.83034	.67113	.4900	.2043	.7945	.57167	.82048	.69674	.4352	.2188	.7493	.58590	.81038	.72299	.3831	.2340	.7068	8
53	.54293	.83978	.64652	.5467	.1908	.8418	.55750	.83017	.67155	.4891	.2046	.7937	.57191	.82032	.69718	.4343	.2190	.7485	.58614	.81021	.72344	.3823	.2342	.7061	7
54	.54317	.83962	.64693	.5458	.1910	.8410	.55774	.83001	.67197	.4881	.2048	.7929	.57214	.82015	.69761	.4335	.2193	.7478	.58637	.81004	.72388	.3814	.2345	.7054	6
55	.54342	.83946	.64734	1.5448	.1912	1.8402	.55799	.82985	.67239	1.4872	.2050	1.7921	.57238	.81998	.69804	1.4326	.2195	1.7471	.58661	.80987	.72432	1.3806	.2348	1.7047	5
56	.54366	.83930	.64775	.5438	.1915	.8394	.55823	.82969	.67282	.4863	.2053	.7914	.57262	.81982	.69847	.4317	.2198	.7463	.58684	.80970	.72477	.3797	.2350	.7040	4
57	.54391	.83914	.64817	.5428	.1917	.8385	.55847	.82952	.67324	.4853	.2055	.7906	.57286	.81965	.69891	.4308	.2200	.7456	.58708	.80953	.72521	.3789	.2353	.7033	3
58	.54415	.83899	.64858	.5418	.1919	.8377	.55871	.82936	.67366	.4844	.2057	.7898	.57310	.81948	.69934	.4299	.2203	.7449	.58731	.80936	.72565	.3781	.2355	.7027	2
59	.54439	.83883	.64899	.5408	.1921	.8369	.55895	.82920	.67408	.4835	.2060	.7891	.57334	.81932	.69977	.4290	.2205	.7442	.58755	.80919	.72610	.3772	.2358	.7020	1
60	.54464	.83867	.64941	1.5399	.1922	1.8361	.55919	.82904	.67451	1.4826	.2062	1.7883	.57358	.81915	.70021	1.4281	.2208	1.7434	.58778	.80902	.72654	1.3764	.2361	1.7013	0
′	cos	sin	cot	tan	cosec	sec	cos	sin	cot	tan	cosec	sec	cos	sin	cot	tan	cosec	sec	cos	sin	cot	tan	cosec	sec	′
	57°						56°						55°						54°						

TABLE 9 Continued

	36°						37°						38°						39°						
′	sin	cos	tan	cot	sec	cosec	sin	cos	tan	cot	sec	cosec	sin	cos	tan	cot	sec	cosec	sin	cos	tan	cot	sec	cosec	′
0	.58778	.80902	.72654	1.3764	1.2361	1.7013	.60181	.79863	.75355	1.3270	1.2521	1.6616	.61566	.78801	.78128	1.2799	1.2690	1.6243	.62932	.77715	.80978	1.2349	1.2867	1.5890	60
1	.58802	.80885	.72699	.3755	.2363	.7006	.60205	.79846	.75401	.3262	.2524	.6610	.61589	.78783	.78175	.2792	.2693	.6237	.62955	.77696	.81026	.2342	.2871	.5884	59
2	.58825	.80867	.72743	.3747	.2366	.6999	.60228	.79828	.75447	.3254	.2527	.6603	.61612	.78765	.78222	.2784	.2696	.6231	.62977	.77678	.81075	.2334	.2874	.5879	58
3	.58849	.80850	.72788	.3738	.2368	.6993	.60251	.79811	.75492	.3246	.2530	.6597	.61635	.78747	.78269	.2776	.2699	.6224	.63000	.77660	.81123	.2327	.2877	.5873	57
4	.58873	.80833	.72832	.3730	.2371	.6986	.60274	.79793	.75538	.3238	.2532	.6591	.61658	.78729	.78316	.2769	.2702	.6218	.63022	.77641	.81171	.2320	.2880	.5867	56
5	.58896	.80816	.72877	1.3722	.2374	1.6979	.60298	.79776	.75584	1.3230	.2535	1.6584	.61681	.78711	.78363	1.2761	.2705	1.6212	.63045	.77623	.81219	1.2312	.2883	1.5862	55
6	.58920	.80799	.72921	.3713	.2376	.6972	.60320	.79758	.75629	.3222	.2538	.6578	.61703	.78693	.78410	.2753	.2707	.6206	.63067	.77605	.81268	.2305	.2886	.5856	54
7	.58943	.80782	.72966	.3705	.2379	.6965	.60344	.79741	.75675	.3214	.2541	.6572	.61726	.78675	.78457	.2746	.2710	.6200	.63090	.77586	.81316	.2297	.2889	.5850	53
8	.58967	.80765	.73010	.3697	.2382	.6959	.60367	.79723	.75721	.3206	.2543	.6565	.61749	.78657	.78504	.2738	.2713	.6194	.63113	.77568	.81364	.2290	.2892	.5854	52
9	.58990	.80747	.73055	.3688	.2384	.6952	.60390	.79706	.75767	.3198	.2546	.6559	.61772	.78640	.78551	.2730	.2716	.6188	.63135	.77549	.81413	.2283	.2895	.5839	51
10	.59014	.80730	.73100	1.3680	.2387	1.6945	.60413	.79688	.75812	1.3190	.2549	1.6552	.61795	.78622	.78598	1.2723	.2719	1.6182	.63158	.77531	.81461	1.2276	.2898	1.5833	50
11	.59037	.80713	.73144	.3672	.2389	.6938	.60437	.79670	.75858	.3182	.2552	.6546	.61818	.78604	.78645	.2715	.2722	.6176	.63180	.77513	.81509	.2268	.2901	.5828	49
12	.59060	.80696	.73189	.3663	.2392	.6932	.60460	.79653	.75904	.3174	.2554	.6540	.61841	.78586	.78692	.2708	.2725	.6170	.63203	.77494	.81558	.2261	.2904	.5822	48
13	.59084	.80679	.73234	.3655	.2395	.6925	.60483	.79635	.75950	.3166	.2557	.6533	.61864	.78568	.78739	.2700	.2728	.6164	.63225	.77476	.81606	.2254	.2907	.5816	47
14	.59107	.80662	.73278	.3647	.2397	.6918	.60506	.79618	.75996	.3159	.2560	.6527	.61886	.78550	.78786	.2692	.2731	.6159	.63248	.77458	.81655	.2247	.2910	.5811	46
15	.59131	.80644	.73323	1.3638	.2400	1.6912	.60529	.79600	.76042	1.3151	.2563	1.6521	.61909	.78532	.78834	1.2685	.2734	1.6153	.63270	.77439	.81703	1.2239	.2913	1.5805	45
16	.59154	.80627	.73368	.3630	.2403	.6905	.60552	.79582	.76088	.3143	.2565	.6514	.61932	.78514	.78881	.2677	.2737	.6147	.63293	.77421	.81752	.2232	.2916	.5799	44
17	.59178	.80610	.73412	.3622	.2405	.6898	.60576	.79565	.76134	.3135	.2568	.6508	.61955	.78496	.78928	.2670	.2739	.6141	.63315	.77402	.81800	.2225	.2919	.5794	43
18	.59201	.80593	.73457	.3613	.2408	.6891	.60599	.79547	.76179	.3127	.2571	.6502	.61978	.78478	.78975	.2662	.2742	.6135	.63338	.77384	.81849	.2218	.2922	.5788	42
19	.59225	.80576	.73502	.3605	.2411	.6885	.60622	.79530	.76225	.3119	.2574	.6496	.62001	.78460	.79022	.2655	.2745	.6129	.63360	.77365	.81898	.2210	.2926	.5783	41
20	.59248	.80558	.73547	1.3597	.2413	1.6878	.60645	.79512	.76271	1.3111	.2577	1.6489	.62023	.78441	.79070	1.2647	.2748	1.6123	.63383	.77347	.81946	1.2203	.2929	1.5777	40
21	.59272	.80541	.73592	.3588	.2416	.6871	.60668	.79494	.76317	.3103	.2579	.6483	.62046	.78423	.79117	.2639	.2751	.6117	.63405	.77329	.81995	.2196	.2932	.5771	39
22	.59295	.80524	.73637	.3580	.2419	.6865	.60691	.79477	.76364	.3095	.2582	.6477	.62069	.78405	.79164	.2632	.2754	.6111	.63428	.77310	.82043	.2189	.2935	.5766	38
23	.59318	.80507	.73681	.3572	.2421	.6858	.60714	.79459	.76410	.3087	.2585	.6470	.62092	.78387	.79212	.2624	.2757	.6105	.63450	.77292	.82092	.2181	.2938	.5760	37
24	.59342	.80489	.73726	.3564	.2424	.6851	.60737	.79441	.76456	.3079	.2588	.6464	.62115	.78369	.79259	.2617	.2760	.6099	.63473	.77273	.82141	.2174	.2941	.5755	36
25	.59365	.80472	.73771	1.3555	.2427	1.6845	.60761	.79424	.76502	1.3071	.2591	1.6458	.62137	.78351	.79306	1.2609	.2763	1.6093	.63495	.77255	.82190	1.2167	.2944	1.5749	35
26	.59389	.80455	.73816	.3547	.2429	.6838	.60784	.79406	.76548	.3064	.2593	.6452	.62160	.78333	.79354	.2602	.2766	.6087	.63518	.77236	.82238	.2160	.2947	.5743	34
27	.59412	.80437	.73861	.3539	.2432	.6831	.60807	.79388	.76594	.3056	.2596	.6445	.62183	.78315	.79401	.2594	.2769	.6081	.63540	.77218	.82287	.2152	.2950	.5738	33
28	.59435	.80420	.73906	.3531	.2435	.6825	.60830	.79371	.76640	.3048	.2599	.6439	.62206	.78297	.79449	.2587	.2772	.6077	.63563	.77199	.82336	.2145	.2953	.5732	32
29	.59459	.80403	.73951	.3522	.2437	.6818	.60853	.79353	.76686	.3040	.2602	.6433	.62229	.78279	.79496	.2579	.2775	.6070	.63585	.77181	.82385	.2138	.2956	.5727	31
30	.59482	.80386	.73996	1.3514	.2440	1.6812	.60876	.79335	.76733	1.3032	.2605	1.6427	.62251	.78261	.79543	1.2572	.2778	1.6064	.63608	.77162	.82434	1.2131	.2960	1.5721	30
31	.59506	.80368	.74041	.3506	.2443	.6805	.60899	.79318	.76779	.3024	.2607	.6420	.62274	.78243	.79591	.2564	.2781	.6058	.63630	.77144	.82482	.2124	.2963	.5716	29
32	.59529	.80351	.74086	.3498	.2445	.6798	.60922	.79300	.76825	.3016	.2610	.6414	.62297	.78224	.79639	.2557	.2784	.6052	.63653	.77125	.82531	.2117	.2966	.5710	28
33	.59552	.80334	.74131	.3489	.2448	.6792	.60945	.79282	.76871	.3009	.2613	.6408	.62320	.78206	.79686	.2549	.2787	.6046	.63675	.77107	.82580	.2109	.2969	.5705	27
34	.59576	.80316	.74176	.3481	.2451	.6785	.60968	.79264	.76918	.3001	.2616	.6402	.62342	.78188	.79734	.2542	.2790	.6040	.63697	.77088	.82629	.2102	.2972	.5699	26
35	.59599	.80299	.74221	1.3473	.2453	1.6779	.60991	.79247	.76964	1.2993	.2619	1.6396	.62365	.78170	.79781	1.2534	.2793	1.6034	.63720	.77070	.82678	1.2095	.2975	1.5694	25
36	.59622	.80282	.74266	.3465	.2456	.6772	.61014	.79229	.77010	.2985	.2622	.6389	.62388	.78152	.79829	.2527	.2795	.6029	.63742	.77051	.82727	.2088	.2978	.5688	24
37	.59646	.80264	.74312	.3457	.2459	.6766	.61037	.79211	.77057	.2977	.2624	.6383	.62411	.78134	.79876	.2519	.2798	.6023	.63765	.77033	.82776	.2081	.2981	.5683	23
38	.59669	.80247	.74357	.3449	.2461	.6759	.61061	.79193	.77103	.2970	.2627	.6377	.62433	.78116	.79924	.2512	.2801	.6017	.63787	.77014	.82825	.2074	.2985	.5677	22
39	.59692	.80230	.74402	.3440	.2464	.6752	.61084	.79176	.77149	.2962	.2630	.6371	.62456	.78097	.79972	.2504	.2804	.6011	.63810	.76996	.82874	.2066	.2988	.5672	21
40	.59716	.80212	.74447	1.3432	.2467	1.6746	.61107	.79158	.77196	1.2954	.2633	1.6365	.62479	.78079	.80020	1.2497	.2807	1.6005	.63832	.76977	.82923	1.2059	.2991	1.5666	20
41	.59739	.80195	.74492	.3424	.2470	.6739	.61130	.79140	.77242	.2946	.2636	.6359	.62501	.78061	.80067	.2489	.2810	.6000	.63854	.76958	.82972	.2052	.2994	.5661	19
42	.59762	.80177	.74538	.3416	.2472	.6733	.61153	.79122	.77289	.2938	.2639	.6352	.62524	.78043	.80115	.2482	.2813	.5994	.63877	.76940	.83022	.2045	.2997	.5655	18
43	.59786	.80160	.74583	.3408	.2475	.6726	.61176	.79104	.77335	.2931	.2641	.6346	.62547	.78025	.80163	.2475	.2816	.5988	.63899	.76921	.83071	.2038	.3000	.5650	17
44	.59809	.80143	.74628	.3400	.2478	.6720	.61199	.79087	.77382	.2923	.2644	.6340	.62570	.78007	.80211	.2467	.2819	.5982	.63921	.76903	.83120	.2031	.3003	.5644	16
45	.59832	.80125	.74673	1.3392	.2480	1.6713	.61222	.79069	.77428	1.2915	.2647	1.6334	.62592	.77988	.80258	1.2460	.2822	1.5976	.63944	.76884	.83169	1.2024	.3006	1.5639	15
46	.59856	.80108	.74019	.3383	.2483	.6707	.61245	.79051	.77475	.2907	.2650	.6328	.62615	.77970	.80306	.2452	.2825	.5971	.63966	.76865	.83218	.2016	.3010	.5633	14
47	.59879	.80090	.74064	.3375	.2486	.6700	.61268	.79033	.77521	.2900	.2653	.6322	.62638	.77952	.80354	.2445	.2828	.5965	.63989	.76847	.83267	.2009	.3013	.5628	13
48	.59902	.80073	.74809	.3367	.2488	.6694	.61290	.79015	.77568	.2892	.2656	.6316	.62660	.77934	.80402	.2437	.2831	.5959	.64011	.76828	.83317	.2002	.3016	.5622	12
49	.59926	.80056	.74855	.3359	.2491	.6687	.61314	.78998	.77614	.2884	.2659	.6309	.62683	.77915	.80450	.2430	.2834	.5953	.64033	.76810	.83366	.1995	.3019	.5617	11
50	.59949	.80038	.74900	1.3351	.2494	1.6681	.61337	.78980	.77661	1.2876	.2661	1.6303	.62706	.77897	.80498	1.2423	.2837	1.5947	.64056	.76791	.83415	1.1988	.3022	1.5611	10
51	.59972	.80021	.74946	.3343	.2497	.6674	.61360	.78962	.77708	.2869	.2664	.6297	.62728	.77879	.80546	.2415	.2840	.5942	.64078	.76772	.83465	.1981	.3025	.5606	9
52	.59995	.80003	.74991	.3335	.2499	.6668	.61383	.78944	.77754	.2861	.2667	.6291	.62751	.77861	.80594	.2408	.2843	.5936	.64100	.76754	.83514	.1974	.3029	.5600	8
53	.60019	.79986	.75037	.3327	.2502	.6661	.61405	.78926	.77801	.2853	.2670	.6285	.62774	.77842	.80642	.2400	.2846	.5930	.64123	.76735	.83563	.1967	.3032	.5595	7
54	.60042	.79968	.75082	.3319	.2505	.6655	.61428	.78908	.77848	.2845	.2673	.6279	.62796	.77824	.80690	.2393	.2849	.5924	.64145	.76716	.83613	.1960	.3035	.5590	6
55	.60065	.79951	.75128	1.3311	.2508	1.6648	.61451	.78890	.77895	1.2838	.2676	1.6273	.62819	.77806	.80738	1.2386	.2852	1.5919	.64160	.76698	.83662	1.1953	.3038	1.5584	5
56	.60088	.79933	.75173	.3303	.2510	.6642	.61474	.78873	.77941	.2830	.2679	.6267	.62841	.77788	.80786	.2378	.2855	.5913	.64189	.76679	.83712	.1946	.3041	.5579	4
57	.60112	.79916	.75219	.3294	.2513	.6636	.61497	.78855	.77988	.2822	.2681	.6261	.62864	.77769	.80834	.2371	.2858	.5907	.64212	.76660	.83761	.1939	.3044	.5573	3
58	.60135	.79898	.75264	.3286	.2516	.6629	.61520	.78837	.78035	.2815	.2684	.6255	.62887	.77751	.80882	.2364	.2861	.5901	.64234	.76642	.83811	.1932	.3048	.5568	2
59	.60158	.79881	.75310	.3278	.2519	.6623	.61543	.78819	.78082	.2807	.2687	.6249	.62909	.77733	.80930	.2356	.2864	.5896	.64256	.76623	.83860	.1924	.3051	.5563	1
60	.60181	.79863	.75355	1.3270	.2521	1.6616	.61566	.78801	.78128	1.2799	.2690	1.6243	.62932	.77715	.80978	1.2349	.2867	1.5890	.64279	.76604	.83910	1.1917	.3054	1.5557	0
′	cos	sin	cot	tan	cosec	sec	cos	sin	cot	tan	cosec	sec	cos	sin	cot	tan	cosec	sec	cos	sin	cot	tan	cosec	sec	′
	53°						52°						51°						50°						

TABLE 9 Continued

							41°						42°						43°						
′	sin	cos	tan	cot	sec	cosec	sin	cos	tan	cot	sec	cosec	sin	cos	tan	cot	sec	cosec	sin	cos	tan	cot	sec	cosec	′
0	.64279	.76604	.83910	1.1917	1.3054	1.5557	.65606	.75471	.86929	1.1504	1.3250	1.5242	.66913	.74314	.90040	1.1106	1.3456	1.4945	.68200	.73135	.93251	1.0724	1.3673	1.4663	60
1	.64301	.76586	.83959	.1910	.3057	.5552	.65628	.75452	.86980	.1497	.3253	.5237	.66935	.74295	.90093	.1100	.3460	.4940	.68221	.73115	.93306	.0717	.3677	.4658	59
2	.64323	.76567	.84009	.1903	.3060	.5546	.65650	.75433	.87031	.1490	.3257	.5232	.66956	.74276	.90146	.1093	.3463	.4935	.68242	.73096	.93360	.0711	.3681	.4654	58
3	.64345	.76548	.84059	.1896	.3064	.5541	.65672	.75414	.87082	.1483	.3260	.5227	.66978	.74256	.90198	.1086	.3467	.4930	.68264	.73076	.93415	.0705	.3684	.4649	57
4	.64368	.76530	.84108	.1889	.3067	.5536	.65694	.75394	.87133	.1477	.3263	.5222	.66999	.74236	.90251	.1080	.3470	.4925	.68285	.73056	.93469	.0699	.3688	.4644	56
5	.64390	.76511	.84158	1.1882	.3070	1.5530	.65716	.75375	.87184	1.1470	.3267	1.5217	.67021	.74217	.90304	1.1074	.3474	1.4921	.68306	.73036	.93524	1.0692	.3692	1.4640	55
6	.64412	.76492	.84208	.1875	.3073	.5525	.65037	.75356	.87235	.1463	.3270	.5212	.67043	.74197	.90357	.1067	.3477	.4916	.68327	.73016	.93578	.0686	.3695	.4635	54
7	.64435	.76473	.84257	.1868	.3076	.5520	.65759	.75337	.87287	.1456	.3274	.5207	.67064	.74178	.90410	.1061	.3481	.4911	.68349	.72996	.93633	.0680	.3699	.4631	53
8	.64457	.76455	.84307	.1861	.3080	.5514	.65781	.75318	.87338	.1450	.3277	.5202	.67086	.74158	.90463	.1054	.3485	.4906	.68370	.72976	.93687	.0674	.3703	.4626	52
9	.64479	.76436	.84357	.1854	.3083	.5509	.65803	.75299	.87389	.1443	.3280	.5197	.67107	.74139	.90515	.1048	.3488	.4901	.68391	.72956	.93742	.0667	.3707	.4622	51
10	.64501	.76417	.84407	1.1847	.3086	1.5503	.65285	.75280	.87441	1.1436	.3284	1.5192	.67129	.74119	.90568	1.1041	.3492	1.4897	.68412	.72937	.93797	1.0661	.3710	1.4617	50
11	.64523	.76398	.84457	.1840	.3089	.5498	.65847	.75261	.87492	.1430	.3287	.5187	.67150	.74100	.90621	.1035	.3495	.4892	.68433	.72917	.93851	.0655	.3714	.4613	49
12	.64546	.76380	.84506	.1833	.3092	.5493	.65869	.75241	.87543	.1423	.3290	.5182	.67172	.74080	.90674	.1028	.3499	.4887	.68455	.72897	.93906	.0649	.3718	.4608	48
13	.64568	.76361	.84556	.1826	.3096	.5487	.65891	.75222	.87595	.1416	.3294	.5177	.67194	.74061	.90727	.1022	.3502	.4882	.68476	.72877	.93961	.0643	.3722	.4604	47
14	.64590	.76342	.84606	.1819	.3099	.5482	.65913	.75203	.87646	.1409	.3297	.5171	.67215	.74041	.90780	.1015	.3506	.4877	.68497	.72857	.94016	.0636	.3725	.4599	46
15	.64612	.76323	.84656	1.1812	.3102	1.5477	.65934	.75184	.87698	1.1403	.3301	1.5166	.67237	.74022	.90834	1.1009	.3509	1.4873	.68518	.72837	.94071	1.0630	.3729	1.4595	45
16	.64635	.76304	.84706	.1805	.3105	.5471	.65956	.75165	.87749	.1396	.3304	.5161	.67258	.74002	.90887	.1003	.3513	.4868	.68539	.72817	.94125	.0624	.3733	.4590	44
17	.64657	.76286	.84056	.1798	.3109	.5466	.65978	.75146	.87801	.1389	.3307	.5156	.67280	.73983	.90940	.0996	.3517	.4863	.68561	.72797	.94180	.0618	.3737	.4586	43
18	.64679	.76267	.84806	.1791	.3112	.5461	.66000	.75126	.87852	.1383	.3311	.5151	.67301	.73963	.90993	.0990	.3520	.4858	.68582	.72777	.94235	.0612	.3740	.4581	42
19	.64701	.76248	.84856	.1785	.3115	.5456	.66022	.75107	.87904	.1376	.3314	.5146	.67323	.73943	.91046	.0983	.3524	.4854	.68603	.72757	.94290	.0605	.3744	.4577	41
20	.64723	.76229	.84906	1.1778	.3118	1.5450	.66044	.75088	.87955	1.1369	.3318	1.5141	.67344	.73924	.91099	1.0977	.3527	1.4849	.68624	.72737	.94345	1.0599	.3748	1.4572	40
21	.64745	.76210	.84956	.1771	.3121	.5445	.66066	.75069	.88007	.1363	.3321	.5136	.67366	.73904	.91153	.0971	.3531	.4844	.68645	.72717	.94400	.0593	.3752	.4568	39
22	.64768	.76191	.85006	.1764	.3125	.5440	.66087	.75049	.88058	.1356	.3324	.5131	.67387	.73885	.91206	.0964	.3534	.4839	.68666	.72697	.94455	.0587	.3756	.4563	38
23	.64790	.76173	.85056	.1757	.3128	.5434	.66109	.75030	.88110	.1349	.3328	.5126	.67409	.73865	.91259	.0958	.3538	.4835	.68688	.72677	.94510	.0581	.3759	.4559	37
24	.64812	.76154	.85107	.1750	.3131	.5429	.66131	.75011	.88162	.1343	.3331	.5121	.67430	.73845	.91312	.0951	.3542	.4830	.68709	.72657	.94565	.0575	.3763	.4554	36
25	.64834	.76135	.85157	1.1743	.3134	1.5424	.66153	.74992	.88213	1.1336	.3335	1.5116	.67452	.73826	.91366	1.0945	.3545	1.4825	.68730	.72637	.94620	1.0568	.3767	1.4550	35
26	.64856	.76116	.85207	.1736	.3138	.5419	.66175	.74973	.88265	.1329	.3338	.5111	.67473	.73806	.91419	.0939	.3549	.4821	.68751	.72617	.94675	.0562	.3771	.4545	34
27	.64878	.76097	.85257	.1729	.3141	.5413	.66197	.74953	.88317	.1323	.3342	.5106	.67495	.73787	.91473	.0932	.3552	.4816	.68772	.72597	.94731	.0556	.3774	.4541	33
28	.64900	.76078	.85307	.1722	.3144	.5408	.66218	.74934	.88369	.1316	.3345	.5101	.67516	.73767	.91526	.0926	.3556	.4811	.68793	.72577	.94786	.0550	.3778	.4536	32
29	.64923	.76059	.85358	.1715	.3148	.5403	.66240	.74915	.88421	.1309	.3348	.5096	.67537	.73747	.91580	.0919	.3560	.4806	.68814	.72557	.94841	.0544	.3782	.4532	31
30	.64945	.76041	.85408	1.1708	.3151	1.5398	.66262	.74895	.88472	1.1303	.3352	1.5092	.67559	.73728	.91633	1.0913	.3563	1.4802	.68835	.72537	.94896	1.0538	.3786	1.4527	30
31	.64967	.76022	.85458	.1702	.3154	.5392	.66284	.74876	.88524	.1296	.3355	.5087	.67580	.73708	.91687	.0907	.3567	.4797	.68856	.72517	.94952	.0532	.3790	.4523	29
32	.64989	.76003	.85509	.1695	.3157	.5387	.66305	.74857	.88576	.1290	.3359	.5082	.67602	.73688	.91740	.0900	.3571	.4792	.68878	.72497	.95007	.0525	.3794	.4518	28
33	.65011	.75984	.85559	.1688	.3161	.5382	.66327	.74838	.88628	.1283	.3362	.5077	.67623	.73669	.91794	.0894	.3574	.4788	.68899	.72477	.95062	.0519	.3797	.4514	27
34	.65033	.75965	.85609	.1681	.3164	.5377	.66349	.74818	.88680	.1276	.3366	.5072	.67645	.73649	.91847	.0888	.3578	.4783	.68920	.72457	.95118	.0513	.3801	.4510	26
35	.65055	.75946	.85660	1.1674	.3167	1.5371	.66371	.74799	.88732	1.1270	.3369	1.5067	.67666	.73629	.91901	1.0881	.3581	1.4778	.68941	.72437	.95173	1.0507	.3805	1.4505	25
36	.65077	.75927	.85710	.1667	.3170	.5366	.66393	.74780	.88784	.1263	.3372	.5062	.67688	.73610	.91955	.0875	.3585	.4774	.68962	.72417	.95229	.0501	.3809	.4501	24
37	.65100	.75908	.85761	.1660	.3174	.5361	.66414	.74760	.88836	.1257	.3376	.5057	.67709	.73590	.92008	.0868	.3589	.4769	.68983	.72397	.95284	.0495	.3813	.4496	23
38	.65121	.75889	.85811	.1653	.3177	.5356	.66436	.74741	.88888	.1250	.3379	.5052	.67730	.73570	.92062	.0862	.3592	.4764	.69004	.72377	.95340	.0489	.3816	.4492	22
39	.65144	.75870	.85862	.1647	.3180	.5351	.66458	.74722	.88940	.1243	.3383	.5047	.67752	.73551	.92116	.0856	.3596	.4760	.69025	.72357	.95395	.0483	.3820	.4487	21
40	.65166	.75851	.85912	1.1640	.3184	1.5345	.66479	.74702	.88992	1.1237	.3386	1.5042	.67773	.73531	.92170	1.0849	.3600	1.4755	.69046	.72337	.95451	1.0476	.3824	1.4483	20
41	.65188	.75832	.85963	.1633	.3187	.5340	.66501	.74683	.89044	.1230	.3390	.5037	.67794	.73511	.92223	.0843	.3603	.4750	.69067	.72317	.95506	.0470	.3828	.4479	19
42	.65210	.75813	.86013	.1626	.3190	.5335	.66523	.74664	.89097	.1224	.3393	.5032	.67816	.73491	.92277	.0837	.3607	.4746	.69088	.72297	.95562	.0464	.3832	.4474	18
43	.65232	.75794	.86064	.1619	.3193	.5330	.66545	.74644	.89149	.1217	.3397	.5027	.67837	.73472	.92331	.0830	.3611	.4741	.69109	.72277	.95618	.0458	.3836	.4470	17
44	.65254	.75775	.86115	.1612	.3197	.5325	.66566	.74625	.89201	.1211	.3400	.5022	.67859	.73452	.92385	.0824	.3614	.4736	.69130	.72256	.95673	.0452	.3839	.4465	16
45	.65276	.75756	.86165	1.1605	.3200	1.5319	.66588	.74606	.89253	1.1204	1.3404	1.5018	.67880	.73432	.92439	1.0818	.3618	1.4732	.69151	.72236	.95729	1.0446	.3843	1.4461	15
46	.65298	.75737	.86216	.1599	.3203	.5314	.66610	.74586	.89306	.1197	.3407	.5013	.67901	.73412	.92493	.0812	.3622	.4727	.69172	.72216	.95785	.0440	.3847	.4457	14
47	.65320	.75718	.86267	.1592	.3207	.5309	.66631	.74567	.89358	.1191	.3411	.5008	.67923	.73393	.92547	.0805	.3625	.4723	.69193	.72196	.95841	.0434	.3851	.4452	13
48	.65342	.75700	.86318	.1585	.3210	.5304	.66653	.74548	.89410	.1184	.3414	.5003	.67944	.73373	.92601	.0799	.3629	.4718	.69214	.72176	.95896	.0428	.3855	.4448	12
49	.65364	.75680	.86368	.1578	.3213	.5299	.66675	.74528	.89463	.1178	.3418	.4998	.67965	.73353	.92655	.0793	.3633	.4713	.69235	.72156	.95952	.0422	.3859	.4443	11
50	.65386	.75661	.86419	1.1571	.3217	1.5294	.66697	.74509	.89515	1.1171	.3421	1.4993	.67987	.73333	.92709	1.0786	.3636	1.4709	.69256	.72136	.96008	1.0416	.3863	1.4439	10
51	.65408	.75642	.86470	.1565	.3220	.5289	.66718	.74489	.89567	.1165	.3425	.4988	.68008	.73314	.92763	.0780	.3640	.4704	.69277	.72115	.96064	.0410	.3867	.4435	9
52	.65430	.75623	.86521	.1558	.3223	.5283	.66740	.74470	.89620	.1158	.3428	.4983	.68029	.73294	.92817	.0774	.3644	.4699	.69298	.72095	.96120	.0404	.3870	.4430	8
53	.65452	.75604	.86572	.1551	.3227	.5278	.66762	.74450	.89672	.1152	.3432	.4979	.68051	.73274	.92871	.0767	.3647	.4695	.69319	.72075	.96176	.0397	.3874	.4426	7
54	.65474	.75585	.86623	.1544	.3230	.5273	.66783	.74431	.89725	.1145	.3435	.4974	.68072	.73254	.92926	.0761	.3651	.4690	.69340	.72055	.96232	.0391	.3878	.4422	6
55	.65496	.75566	.86674	1.1537	.3233	1.5268	.66805	.74412	.89777	1.1139	.3439	1.4969	.68093	.73234	.92980	1.0755	.3655	1.4686	.69361	.72035	.96288	1.0385	.3882	1.4417	5
56	.65518	.75547	.86725	.1531	.3237	.5263	.66826	.74392	.89830	.1132	.3442	.4964	.68115	.73215	.93034	.0749	.3658	.4681	.69382	.72015	.96344	.0379	.3886	.4413	4
57	.65540	.75528	.86775	.1524	.3240	.5258	.66848	.74373	.89882	.1126	.3446	.4959	.68136	.73195	.93088	.0742	.3662	.4676	.69403	.71994	.96400	.0373	.3890	.4408	3
58	.65562	.75509	.86826	.1517	.3243	.5253	.66870	.74353	.89935	.1119	.3449	.4954	.68157	.73175	.93143	.0736	.3666	.4672	.69424	.71974	.96456	.0367	.3894	.4404	2
59	.65584	.75490	.86878	.1510	.3247	.5248	.66891	.74334	.89988	.1113	.3453	.4949	.68178	.73155	.93197	.0730	.3669	.4667	.69445	.71954	.96513	.0361	.3898	.4400	1
60	.65606	.75401	.86929	1.1504	.3250	1.5242	.66913	.74314	.90040	1.1106	.3456	1.4945	.68200	.73135	.93251	1.0724	.3673	1.4663	.69466	.71934	.96569	1.0355	.3902	1.4395	0
′	cos	sin	cot	tan	cosec	sec	cos	sin	cot	tan	cosec	sec	cos	sin	cot	tan	cosec	sec	cos	sin	cot	tan	cosec	sec	′
			49°						48°						47°						46°				

TABLE 9 Continued

44°

′	sin	cos	tan	cot	sec	cosec	′
0	.69466	.71934	.96569	1.0355	1.3902	1.4395	60
1	.69487	.71914	.96625	.0349	.3905	.4391	59
2	.69508	.71893	.96681	.0343	.3909	.4387	58
3	.69528	.71873	.96738	.0337	.3913	.4382	57
4	.69549	.71853	.96794	.0331	.3917	.4378	56
5	.69570	.71833	.96850	1.0325	.3921	1.4374	55
6	.69591	.71813	.96907	.0319	.3925	.4370	54
7	.69612	.71792	.96963	.0313	.3929	.4365	53
8	.69633	.71772	.97020	.0307	.3933	.4361	52
9	.69654	.71752	.97076	.0301	.3937	.4357	51
10	.69675	.71732	.97133	1.0295	.3941	1.4352	50
11	.69696	.71711	.97189	.0289	.3945	.4348	49
12	.69716	.71691	.97246	.0283	.3949	.4344	48
13	.69737	.71671	.97302	.0277	.3953	.4339	47
14	.69758	.71650	.97359	.0271	.3957	.4335	46
15	.69779	.71630	.97416	1.0265	.3960	1.4331	45
16	.69800	.71610	.97472	.0259	.3964	.4327	44
17	.69821	.71589	.97529	.0253	.3968	.4322	43
18	.69841	.71569	.97586	.0247	.3972	.4318	42
19	.69862	.71549	.97643	.0241	.3976	.4314	41
20	.69883	.71529	.97700	1.0235	.3980	1.4310	40
21	.69904	.71508	.97756	.0229	.3984	.4305	39
22	.69925	.71488	.97813	.0223	.3988	.4301	38
23	.69945	.71468	.97870	.0218	.3992	.4297	37
24	.69966	.71447	.97927	.0212	.3996	.4292	36
25	.69987	.71427	.97984	1.0206	.4000	1.4288	35
26	.70008	.71406	.98041	.0200	.4004	.4284	34
27	.70029	.71386	.98098	.0194	.4008	.4280	33
28	.70049	.71366	.98155	.0188	.4012	.4276	32
29	.70070	.71345	.98212	.0182	.4016	.4271	31
30	.70091	.71325	.98270	1.0176	.4020	1.4267	30
31	.70112	.71305	.98327	.0170	.4024	.4263	29
32	.70132	.71284	.98384	.0164	.4028	.4259	28
33	.70153	.71264	.98441	.0158	.4032	.4254	27
34	.70174	.71243	.98499	.0152	.4036	.4250	26
35	.70194	.71223	.98556	1.0146	.4040	1.4246	25
36	.70215	.71203	.98613	.0141	.4044	.4242	24
37	.70236	.71182	.98671	.0135	.4048	.4238	23
38	.70257	.71162	.98728	.0129	.4052	.4233	22
39	.70277	.71141	.98786	.0123	.4056	.4229	21
40	.70298	.71121	.98843	1.0117	.4060	1.4225	20
41	.70319	.71100	.98901	.0111	.4065	.4221	19
42	.70339	.71080	.98958	.0105	.4069	.4217	18
43	.70360	.71059	.99016	.0099	.4073	.4212	17
44	.70381	.71039	.99073	.0093	.4077	.4208	16
45	.70401	.71018	.99131	1.0088	.4081	1.4204	15
46	.70422	.70998	.99189	.0082	.4085	.4200	14
47	.70443	.70977	.99246	.0076	.4089	.4196	13
48	.70463	.70957	.99304	.0070	.4093	.4192	12
49	.70484	.70936	.99362	.0064	.4097	.4188	11
50	.70505	.70916	.99420	1.0058	.4101	1.4183	10
51	.70525	.70895	.99478	.0052	.4105	.4179	9
52	.70546	.70875	.99536	.0047	.4109	.4175	8
53	.70566	.70854	.99593	.0041	.4113	.4171	7
54	.70587	.70834	.99651	.0035	.4117	.4167	6
55	.70608	.70813	.99709	1.0029	.4122	1.4163	5
56	.70628	.70793	.99767	.0023	.4126	.4159	4
57	.70649	.70772	.99826	.0017	.4130	.4154	3
58	.70669	.70752	.99884	.0012	.4134	.4150	2
59	.70690	.70731	.99942	.0006	.4138	.4146	1
60	.70711	.70711	1.00000	1.0000	.4142	1.4142	0
′	cos	sin	cot	tan	cosec	sec	′

45°

TABLE 10 Metric Tap Drill Size

Metric Tap Size	Recommended Metric Drill				Closest Recommended Inch Drill			
	Drill Size (mm)	Inch Equivalent	Probable Hole Size (in.)	Probable Percent of Thread	Drill Size	Inch Equivalent	Probable Hole Size (in.)	Probable Percent of Thread
M1.6 × .35	1.25	.0492	.0507	69	—	—	—	—
M1.8 × .35	1.45	.0571	.0586	69	—	—	—	—
M2 × .4	1.60	.0630	.0647	69	#52	.0635	.0652	66
M2.2 × .45	1.75	.0689	.0706	70	—	—	—	—
M2.5 × .45	2.05	.0807	.0826	69	#46	.0810	.0829	67
*M3 × .5	2.50	.0984	.1007	68	#40	.0980	.1003	70
M3.5 × .6	2.90	.1142	.1168	68	#33	.1130	.1156	72
*M4 × .7	3.30	.1299	.1328	69	#30	.1285	.1314	73
M4.5 × .75	3.70	.1457	.1489	74	#26	.1470	.1502	70
*M5 × .8	4.20	.1654	.1686	69	#19	.1660	.1692	68
*M6 × 1	5.00	.1968	.2006	70	#9	.1960	.1998	71
M7 × 1	6.00	.2362	.2400	70	$\frac{15}{64}$	.2344	.2382	73
*M8 × 1.25	6.70	.2638	.2679	74	$\frac{17}{64}$	.2656	.2697	71
M8 × 1	7.00	.2756	.2797	69	J	.2770	.2811	66
*M10 × 1.5	8.50	.3346	.3390	71	Q	.3320	.3364	75
M10 × 1.25	8.70	.3425	.3471	73	$\frac{11}{32}$	.3438	.3483	71
*M12 × 1.75	10.20	.4016	.4063	74	Y	.4040	.4087	71
M12 × 1.25	10.80	.4252	.4299	67	$\frac{27}{64}$	.4219	.4266	72
M14 × 2	12.00	.4724	.4772	72	$\frac{15}{32}$	.4688	.4736	76
M14 × 1.5	12.50	.4921	.4969	71	—	—	—	—
*M16 × 2	14.00	.5512	.5561	72	$\frac{35}{64}$	.5469	.5518	76
M16 × 1.5	14.50	.5709	.5758	71	—	—	—	—
M18 × 2.5	15.50	.6102	.6152	73	$\frac{39}{64}$	.6094	.6144	74
M18 × 1.5	16.50	.6496	.6546	70	—	—	—	—
*M20 × 2.5	17.50	.6890	.6942	73	$\frac{11}{16}$	.6875	.6925	74
M20 × 1.5	18.50	.7283	.7335	70	—	—	—	—
M22 × 2.5	19.50	.7677	.7729	73	$\frac{49}{64}$	.7656	.7708	75
M22 × 1.5	20.50	.8071	.8123	70	—	—	—	—
*M24 × 3	21.00	.8268	.8327	73	$\frac{53}{64}$	.8281	.8340	72
M24 × 2	22.00	.8661	.8720	71	—	—	—	—
M27 × 3	24.00	.9449	.9511	73	$\frac{15}{16}$	.9375	.9435	78
M27 × 2	25.00	.9843	.9913	70	$\frac{63}{64}$	.9844	.9914	70
*M30 × 3.5	26.50	1.0433						
M30 × 2	28.00	1.1024						
M33 × 3.5	29.50	1.1614						
M33 × 2	31.00	1.2205						
M36 × 4	32.00	1.2598						
M36 × 3	33.00	1.2992						
M39 × 4	35.00	1.3780						
M39 × 3	36.00	1.4173						

Reaming Recommended to the Drill Size Shown (M27 × 2 through M39 × 3)

Formula for metric tap drill size: $\underset{\text{(mm)}}{\text{Basic major diameter}} - \dfrac{\text{\% Thread} \times \text{Pitch (mm)}}{76.980} = \text{Drilled hole size (mm)}$

Formula for percent of thread: $\dfrac{76.980}{\text{Pitch (mm)}} \times \left[\underset{\text{(mm)}}{\text{Basic major diameter}} - \underset{\text{(mm)}}{\text{Drilled hole size}}\right] = \text{Percent of thread}$

SOURCE: Material courtesy of TRW Inc., *New Greenfield Geometric ISO Metric Screw Thread Manual, 1973.*

GLOSSARY

Abrasive A substance such as finely divided aluminum oxide or silicon carbide used for grinding (abrading), smoothing, or polishing.

Acicular Needlelike; resembling needles or straws dropped at random.

Acute angle An angle of less than 90 degrees.

Alignment The proper positioning or state of adjustment of parts in relation to each other, especially in line, as in axial alignment.

Allotropic Materials that can exist in several different crystalline forms are said to be allotropic.

Alloy A combination of two or more substances, specifically metals such as alloy steels or aluminum alloys.

Aluminum oxide Also alumina (Al_2O_3). Occurs in nature as corundum and is used extensively as an abrasive. Today most aluminum oxide abrasives are manufactured.

Ammonia A pungent colorless gaseous alkaline compound of nitrogen and hydrogen (NH_3). It is very soluble in water.

Amorphous Having no definite form or outline. Materials such as glass that have no definite crystalline structure.

Angular Having one or more angles; measured by an angle; forming an angle.

Angularity The quality or characteristic of being angular.

Angular measure The means by which an arc of a circle is divided and measured. This can be in degrees (360 degrees in a full circle), minutes (60 minutes in one degree), and seconds (60 seconds in one minute), or in radians. See *Radian*.

Anhydrous Free from water.

Anneal A heat treatment in which metals are heated and then cooled very slowly for the purpose of decreasing hardness. Annealing is used to improve machinability and to remove stress from weldments, forgings, and castings. It is also used to remove stresses resulting from cold work, and to refine and make uniform the microscopic internal structures of metals.

Arbor A rotating shaft upon which a cutting tool is fastened. Often used as a term for *mandrel*.

As rolled When metal bars are hot rolled and allowed to cool in air, they are said to be in the "as rolled" or natural condition.

Austenite A solid solution of iron and carbon or iron carbide in which gamma iron, characterized by a face-centered cubic crystal, is the solvent.

Axial Having the characteristics of an axis (that is, centerline or center of rotation); situated around and in relation to an axis as in axial alignment.

Axial rake An angular cutting surface that is rotated about the axial centerline of a cutting tool such as a drill or reamer.

Axis Centerline or center of rotation of an object or part; the rotational axis of a machine spindle which extends beyond the spindle and through the workpiece. Machining of the object imparts the machine axis to that area of metal cutting. The line along which a major machine tool component such as a mill table, saddle, or spindle travels.

Backlash A condition created due to clearance between a thread and nut. The amount of thread turn before a component begins to move.

Beam The scale on a vernier caliper or height gage that is graduated in true or full sized units.

Bellmouth A condition in a machined hole where the end is flared out in a bell shape to a dimension larger than the nominal size of the hole.

Bezel A rim that holds a transparent face of a dial indicator that can be rotated to bring the index mark to zero.

Bimetallic Made from two different metals.

Blind hole A hole that does not go completely through an object.

Blotter A paper disk placed between a grinding wheel and the retaining flange, often marked with wheel type and speed rating.

Bolster plate A structural part of a press designed to support or reinforce the platen (base surface) on which the workpiece is placed for press work.

Bore (1) A machined hole. (2) The process of enlarging a drilled hole to a larger size.

Boring The process of removing metal from a hole by using a single-point tool. The workpiece can rotate with a stationary bar, or the bar can rotate on a stationary workpiece to bore a hole.

Brinell hardness The hardness of a metal or alloy measured by hydraulically pressing a hard ball (usually 10 mm diameter) with a standard load into the specimen. A number is derived by measuring the indentation with a special microscope.

Brittleness That property of a material that causes it to suddenly break at a given stress without bending or distortion of the edges of the broken surface. Glass, ceramics, and cast iron are somewhat brittle materials.

Broaching The process of removing unwanted metal by pulling or pushing a tool on which cutting teeth project through or along the surface of a workpiece. The cutting teeth are each progressively longer by a few thousandths of an inch to give each tooth a chip load. One of the most frequent uses of broaching is for producing internal shapes such as keyseats and splines.

Buffing wheel A disk made up of layers of cloth sewed together. Fine abrasive is applied to the periphery of the cloth wheel to provide a polishing surface as the wheel is rotated at a high speed.

Burnish To make shiny by rubbing. No surface material is removed by this finishing process. External and internal surfaces are often smoothed with high pressure rolling. Hardened plugs are sometimes forced through bores to finish and size them by burnishing.

Burr (1) A small rotary file. (2) A thin edge of metal, usually very sharp, left from a machining operation. See *Deburr*.

Bushing A hollow cylinder that is used as a spacer, a reducer for a bore size, or for a bearing. Bushings can be made of metals or nonmetals such as plastics or formica.

Button die A thread-cutting die that is round and usually slightly adjustable. It is held in a diestock or holder by means of a cone point setscrew that fits into a detent on the periphery of the die.

Calibration The adjustment of a measuring instrument such as a micrometer or dial indicator so it will measure accurately.

Cam A rotating or sliding part with a projection or projecting geometry that imparts motion to another part as it slides or rotates past.

Carburizing compound A carbonaceous material that introduces carbon into a heated solid ferrous alloy by the process of diffusion.

Cavity A machine feature, such as a hole, groove, or slot, enclosed on all sides in two dimensions. The space in a casting mold where molten metal will flow to form a cast part.

Celsius A temperature scale used in the SI metric systems of measurement where the freezing point of water is 0° and the boiling point is 100°. Also called centigrade.

Cementite Iron carbide, a compound of iron and carbon (Fe_3C) found in steel and cast iron.

Centerline A reference line on a drawing or part layout from which all dimensions are located.

Chamfer A bevel cut on a sharp edge of a part to improve resistance to damage and as a safety measure to prevent cuts.

Chasing a thread In machining terminology, chasing a thread is making successive cuts in the same groove with a single point tool. Also when cleaning or repairing a damaged thread.

Chatter Vibration of workpiece, machine, tool, or a combination of all three due to looseness or weakness in one or more of these areas. Chatter may be found in either grinding or machine operations and is usually noted as vibratory sound and seen on the workpiece as wave marks.

Checked A term used mostly in grinding operations, indicating a surface having many small cracks (checks). The term *heat checked* or *crazed* is used in reference to friction clutch surfaces.

Chips The particles that are removed when materials are cut; also called filings.

Chip trap A deformed end of a lathe cutting tool that prevents the chip from flowing across and away from the tool.

Circularity The extent to which an object has the form of a circle; the measured accuracy or roundness of a circular or cylindrical object such as a shaft. A lack of circularity is referred to in shops as out of round, egg-shaped, or having a flat spot.

Circumference The periphery or outer edge of a circle. Its length is calculated by multiplying π (3.1416) times the diameter of the circle.

Clutch A component usually found in a mechanical drive that permits a driven component and driving component to be mechanically disconnected and reconnected at will.

Coarseness A definition of grit size in grinding or spacing of teeth on files and other cutting tools.

Coincident Two graduations on separate graduated scales being in line with each other, such as the coincident line of a vernier and true scale graduations.

Cold finish Refers to the surface finish obtained on metal by any of several means of cold working, such as rolling or drawing.

Cold working Any process such as rolling, forging, or forming a cold metal in which the metal is stressed beyond its yield point. Grains are deformed and elongated in the process, causing the metal to have a higher hardness and lower ductility.

Complementary angles Two angles whose sum is 90 degrees. Often referred to in machine shop work since most angular machining is done within one quadrant or 90 degrees.

Concave An internal arc or curve; a dent.

Concentricity The extent to which an object has a common center or axis. Specifically, in machine work, the extent to which two or more surfaces or a shaft rotate in relation to each other; the amount of runout on a rotating member.

Contour Machining an uneven but continuous path on a workpiece in two or three dimensions.

Convex An external arc or curve; a bulge.

Coolant A cutting fluid used to cool the tool and workpiece, especially in grinding operations; usually water based.

Coordinate A method of specifying point locations in a two dimensional plane system defined by two perpendicular axes.

Cosine error A condition where the axis of a measuring instrument is out of line with the axis of the measurement to be taken, resulting in an error equal to the measuring instrument reading multiplied by the cosine of the misalignment angle.

Crest of thread Outer edge (point or flat) of a thread form.

Critical temperatures The upper and lower transformation points of iron between which is the transformation range in which ferrite changes to austenite as the temperature rises.

Cutting fluid Any of several materials used in cutting metals: cutting oils, soluble or emulsified oils (water based), and sulfurized oils.

Cyanogen $(CN)_2$ A colorless, flammable, poisonous gas with characteristic odor. It forms cyanic and hydrocyanic acids when in contact with water. Cyanogen compounds are often used for case hardening.

Deburr To remove a sharp edge or corner caused by a machining process.

Decarburization The loss of surface carbon from ferrous metals when heated to high temperatures in an atmosphere containing oxygen.

Decibel A unit for expressing the relative intensity of sounds on a scale from zero (least perceptible sound) to about 130 (the average pain level).

Degrees The circle is divided into 360 degrees, four 90-degree quadrants. Each degree is divided into 60 minutes and each minute into 60 seconds. Degrees are measured with protractors, optical comparators, and sine bars, to name a few methods. Degrees are also divisions of temperature scales.

Dendrite A formation that resembles a pine tree in the microstructure of solidifying metals. Each dendrite usually forms a single grain or crystal.

Diagonal A straight line from corner to corner on a square, rectangle, or any parallelogram.

Diameter Twice the radius; the length of any straight line going through the center of a figure or body; specifically, a circle in the drafting and layout.

Diametral pitch The ratio of the number of teeth on gears to the number of inches of pitch diameter.

Die (1) Cutting tool for producing external threads. (2) A device that is mounted in a press for cutting and forming sheet metal.

Die cast metal Metal alloys, often called pot metals, that are forced into a die in a molten state by hydraulic pressure. Thousands of identical parts can be produced from a single die or mold by this process of die casting.

Dimension A measurement in one direction; one of three coordinates—length, width, and depth. Thickness, radius, and diameter are given as dimensions on drawings.

Discrimination The level of measurement to which an instrument is capable within a given measuring system. A .001-in. micrometer can be read to within one thousandth of an inch. With a vernier, it can discriminate to one ten-thousandth of an inch.

Distortion The alteration of the shape of an object that would normally affect its usefulness. Bending, twisting, and elongation are common forms of distortion in metals.

Dovetail An angular shape used on many types of interlocking slide components, especially on machine tools.

Ductility The property of a metal to be deformed permanently without rupture while under tension. A metal that can be drawn into a wire is ductile.

Ebonized Certain cold-drawn or cold-rolled bars that have black stained surfaces are said to be ebonized. This is not the same as the black scaly surface of hot rolled steel products.

Eccentricity A rotating member whose axis of rotation is different or offset from the primary axis of the part or mechanism. Thus, when one turned section of a shaft centers on a different axis than the shaft, it is said to be eccentric or to have "runout." For example, the throws or cranks on an engine crankshaft are eccentric to the main bearing axis.

Edgefinder A tool fastened in a machine spindle that locates the position of the workpiece edge in relation to the spindle axis.

EDM Electro-discharge machining. With this process, a graphite or metal electrode is slowly fed into the workpiece that is immersed in oil. A pulsed electrical charge causes sparks to jump to the workpiece, each tearing out a small particle. In this way, the electrode gradually erodes its way through the workpiece that can be a soft or an extremely hard material such as tungsten carbide.

Elasticity The property of a material to return to its original shape when stretched or compressed.

Emulsifying oils An oil containing an emulsifying agent such as detergent so it will mix with water. Oil emulsions are used extensively for coolants in machining operations.

Expansion The enlargement of an object, usually caused by an increase in temperature. Metals expand when heated and contract when cooled in varying amounts, depending on the coefficient of expansion of the particular metal.

Extruding A form of metal working in which a metal bar, either cold or heated, is forced through a die that forms a special cross-sectional shape such as an angle or channel. Extrusions of soft metals such as aluminum and copper are very common.

Face (1) The side of a metal disc or end of a shaft when turning in a lathe. A facing operation is usually at 90 degrees to the spindle axis of the lathe. (2) The periphery or outer cylindrical surface of a straight grinding wheel.

Fahrenheit A temperature scale that is calibrated with the freezing point of water at 32° and the boiling point at 212°. The Fahrenheit scale is gradually being replaced with the Celsius scale used with the metric system of measurement.

Ferrite The microstructure of iron or steel that is mostly pure iron and appears light gray or white when etched and viewed with a microscope.

Ferromagnetic Metals or other substances that have unusually high magnetic permeability, a saturation point with some residual magnetism, and high hysteresis. Iron and nickel are both ferromagnetic.

Ferrous From the Latin word *ferrum*, meaning iron. An alloy containing a significant amount of iron.

Fillet (1) A concave junction of two-surfaces. (2) An inside corner radius of a shoulder on a shaft. (3) An inside corner weld.

Finishing (surface) The control of roughness by turning, grinding, milling, lapping, superfinishing, or a combination of any of these processes. Surface texture is designated in terms of roughness profile in microinches, waviness, and lay (direction of roughness).

Fixture A device that holds workpieces and aligns them with the tool or machine axis with repeatable accuracy.

Flammable Any material that will readily burn or explode when brought into contact with a spark or flame.

Flash Excess material that is extruded between die halves in die castings or forging dies; also, the upset material formed when welding bandsaws.

Floating Free to move about over a given area; for example, a floating edge finder tip, floating die holder, or floating reamer holder.

Flute The groove in a drill, tap, reamer, or milling cutter.

Forging A method of metal working in which the metal is hammered into the desired shape or is forced into a mold by pressure or hammering, usually after being heated to a more plastic state. Hot forging requires less force to form a given point than does cold forging, which is usually done at room temperature.

Formica A trademark used to designate several plastic laminated products; especially, a laminate used to make gears.

Forming A method of working sheet metal into useful shapes by pressing or bending.

FPM or SFM Surface feet per minute on a moving workpiece or tool.

Friction Rubbing of one part against another; resistance to relative motion between two parts in contact, usually generating heat.

Galling Cold welding of two metal surfaces in intimate contact under pressure. Also called seizing, it is more severe and more likely to happen between two similar soft metals, especially when they are clean and dry.

Gib A part of a slide mechanism used to adjust the clearance between two sliding parts.

Glazing (1) A work-hardened surface on metals resulting from using a dull tool or a too rapid cutting speed. (2) A dull grinding wheel whose surface grains have worn flat causing the workpiece to be overheated and "burned" (discolored).

Graduations Division marks on a rule, measuring instrument, or machine dial.

Grain In metals, a single crystal consisting of parallel rows of atoms called a space lattice.

Grain boundary The outer perimeter of a single grain where it contacts adjacent grains.

Grain growth Called recrystallization. Metal grains begin to reform to larger and more regular size and shape at certain temperatures, depending to some extent on the amount of prior cold working.

Grit (1) Any small, hard particles such as sand or grinding compound. Dust from grinding operations settles on machine surfaces as grit, which can damage sliding surfaces. (2) Diamond dust, aluminum oxide, or silicon carbide particles used for grinding wheels.

Ground and polished (G & P) A finishing process for some steel alloy shafts during their manufacture. The rolled, drawn, or turned shafting is placed on a centerless grinder and precision ground, after which a polishing operation produces a fine finish.

Gullet The bottom of the space between teeth on saws and circular milling cutters.

Hardenability The property that determines the depth and distribution of hardness in a ferrous alloy induced by heating and quenching.

Hardening Metals are hardened by cold working or heat treating. Hardening causes metals to have a higher resistance to penetration and abrasion.

Harmonic chatter A harmonic frequency is a multiple of the fundamental frequency of sound. Any machine part, such as a boring bar, has a fundamental frequency and will vibrate at that frequency and also at several harmonic or multiple frequencies. Thus, chatter or vibration of a tool may be noted at several different spindle speeds.

Hazard A situation that is dangerous to any person in the vicinity. Also, a danger to property, such as a fire hazard.

Heat treated Metal whose structure has been altered or modified by the application of heat.

Helical The geometry of a helix where a point both rotates and moves parallel to the axis of a cylinder. Examples include threads, springs, and drill flutes.

Helix The path described by a point rotating about a cylinder while at the same time being moved along the cylinder. The distance of movement compared to each revolution is the lead of the helix.

High-pressure lube A petroleum grease or oil containing graphite or molybdenum disulfide that continues to lubricate even after the grease has been wiped off.

Hog To remove large amounts of material from a workpiece with deep heavy cuts.

Horizontal Parallel to the horizon or base line; level.

Hot rolled Metal flattened and shaped by rolls while at a red heat.

Hub A thickening near the axis of a wheel, gear, pulley, sprocket, and other shaft-driven members that provides a bore in its center to receive a shaft. The hub also provides extra strength to transfer power to or from the shaft by means of a key and keyseat.

Increment A single step of a number of steps; a succession of regular additions; a minute increase.

Inert gas A gas, such as argon or helium, that will not readily combine with other elements.

Infeed The depth a tool is moved into the workpiece.

Interface The point or area of contact between tool and workpiece; also the contact point or area of two mating parts in an assembly.

Interference fit Force fit of a shaft and bore, bearings, and housings or shafts. Negative clearance in which the fitted part is very slightly larger than the bore.

Internal stress Also called residual stress. Stress in

metals that is built in by heat treatment or by cold working.

Involute Geometry found in modern gears that permits mating gear teeth to engage each other with rolling rather than sliding friction.

Iron carbide Also called cementite (Fe_3C), a compound of iron and carbon, which is quite hard.

Jig A device that guides a cutting tool and aligns it to the workpiece.

Journal The part of a rotating shaft or axle that turns in a bearing.

Kerf The width of a cut produced by a saw.

Key A removable metal part that, when assembled into keyseats, provides a positive drive for transmitting torque between shaft and hub.

Keyseat An axially located rectangular groove in a shaft or hub.

Keystock Square or rectangular cold-rolled steel bars used for making and fitting keys in keyseats.

Keyway Same as keyseat (British terminology).

Knurl Diamond or straight impressions on a metal surface produced by rolling with pressure. The rolls used are called knurls.

Laminated Composed of multiple layers of the same or different materials.

Lattice The regular rows of atoms in a metal crystal.

Lead The distance a thread or nut advances along a threaded rod in one revolution.

Loading A grinding wheel whose voids are being filled with metals, causing the cutting action of the wheel to be diminished.

Lobe The offset or projection on a cam that contacts the part to which motion is to be imparted.

Longitudinal Lengthwise, as the longitudinal axis of the spindle or machine.

Machinability The relative ease of machining, which is related to the hardness of the material to be cut.

Magnetic Having the property of magnetic attraction and permeability.

Malleability The ability of a metal to deform permanently without rupture when loaded in compression.

Mandrel A cylindrical bar upon which the workpiece is affixed and subsequently machined between centers. Mandrels, often erroneously called arbors, are used in metal turning and cylindrical grinding operations.

Mar To scratch or otherwise damage a machined surface.

Martensite The hardest constituent of steel formed by quenching carbon steel from the austenitized state to room temperature. The microstructure can be seen as acicular or needlelike.

Mechanical properties (of metals) Some mechanical properties of metals are tensile strength, ductility, malleability, elasticity, and plasticity. Mechanical properties can be measured by mechanical testing.

Metal cementation Introducing a metal or material into the surface of another by heat treatment. Carburizing is one example of metal cementation.

Metallizing Applying a coating of metal on a surface by spraying molten metal on it. Also called spray weld and metal spray.

Metal spinning A process in which a thin disc of metal is rapidly turned in a lathe and forced over a wooden form or mandrel to form various conical or cylindrical shapes.

Metrology The science of weights and measures or measurement.

Microstructure Structure that is only visible at high magnification.

Mode A particular way in which something is done or a machine is operated, such as manual or automatic mode on machines.

Mushroom head (1) An oversized head on a fastener or tool that allows it to be easily pushed with the hand. (2) A deformed striking end of a chisel or punch that should be removed by grinding.

Neutral In machine work, neither positive or negative rake is a neutral or zero rake; a neutral fit is neither a clearance nor interference fit.

Nitriding A surface hardening treatment for ferrous alloys that is obtained by heating an alloy in the presence of disassociated ammonia gas, which releases nitrogen to the steel. The formation of iron nitride causes the hardened surface.

Nitrogenous gas Ammonia (NH_3) used in nitriding.

Nomenclature Pertaining to the names of individual parts of machines or tools; a list of machine parts indicating their names.

Nominal Usually refers to a standard size or quantity as named in standard references.

Nonferrous Metals other than iron or iron alloys; for example, aluminum, copper, and nickel.

Normalizing A treatment consisting of heating to a temperature above the critical range of steel followed by cooling in air. Normalizing produces in steel a "normal structure" consisting of free ferrite and cementite or free pearlite and cementite, depending on the carbon content.

Nose radius Refers to the rounding of the point of a lathe cutting tool. A large radius produces a better finish and is stronger than a small one.

Obtuse angle An angle greater than 90 degrees.

Orthographic drawing Projections of a single view of an object in which the view is projected along lines perpendicular to both the view and the drawing surface.

Oxide scale At a red heat, oxygen readily combines with iron to form a black oxide scale (Fe_3O_4), also called mill scale. At lower temperatures, 400 to 650°F (204 to 343°C), various oxide scale colors (straw, yellow, gold, violet, blue, and gray) are produced, each color within a narrow temperature range. These colors are used by some heat treaters to determine temperatures for tempering.

Oxidize To combine with oxygen; to burn or corrode by oxidation.

Oxyacetylene Mixture of oxygen and acetylene gases to produce an extremely hot flame used for heating and welding.

Parallax error An error in measurement caused by reading a measuring device, such as a rule, at an improper angle.

Parallel The condition in which lines or planes are equidistant from each other.

Parting Also called cutting off; a lathe operation in which a thin blade tool is fed into a turning workpiece to make a groove that is continued to the center to sever the material.

Pearlite Alternating layers of cementite and ferrite in carbon steel. Under a microscope, the microstructure of pearlite sometimes appears like mother-of-pearl, hence the name. It is found in carbon steels that have been slowly cooled.

Pecking A process used in drilling deep holes to remove chips before they can seize and jam the drill. The drill is fed into the hole a short distance to accumulate some chips in the flutes and then drawn out of the hole, allowing the chips to fly off. This process is repeated until the correct depth of the hole is reached.

Pedestal A base or floor stand under a machine tool.

Penetrant A thin liquid that is able to enter small cracks and crevices. Penetrant oils are used to loosen rusted threads; dye penetrants are used to find hidden cracks.

Periphery The perimeter or external boundary of a surface or body.

Perpendicular At 90 degrees to the horizontal or base line.

Pin Straight, tapered, or cotter pins are used as fasteners of machine parts or for light drives.

Pinion The smaller gear of a gear set, especially in bevel gears.

Pinning A condition where chips of workpiece material jam in the teeth of a file.

Pitch In saw teeth, the number per inch; in threads, one divided by the number per inch; the diameter or radius to the centerline of a number of features or feature located on the circumference of a circle, such as pitch circle or pitch diameter.

Pitch diameter For threads, the pitch diameter is an imaginary circle, which on a perfect thread occurs at the point where the widths of the thread and groove are equal. On gears, it is the diameter of the pitch circle.

Potassium cyanide A very poisonous crystalline salt (KCN) used in electroplating and for case hardening steel.

Pot metals Die-casting alloys, which can be zinc, lead, or aluminum-based, among others.

Precipitation hardening A process of hardening an alloy by heat treatment in which a constituent or phase precipitates from a solid solution at room temperature or at a slightly elevated temperature.

Precision A relative but higher level of accuracy within certain tolerance limits. Precision gage blocks are accurate within a few millionths of an inch, yet precision lathe work in some shops may be within a few thousandths of an inch tolerance.

Pressure Generally expressed in units as pounds per square inch (psi), and is called unit pressure, while force is the total load.

Profile An outline view, also a side or elevation view.

Proportion An equality of two ratios.

Prototype A full-scale original model on which something is patterned.

Pulley A flat-faced wheel used to transmit power by means of a flat belt. Grooved pulleys are called sheaves.

Quench A rapid cooling of heated metal for the purpose of imparting certain properties, especially hardness. Quenchants are water, oil, fused salts, air, and molten lead.

Quench cracking Cracking of heat metal during the quenching operation caused by internal stresses.

Quick-change gearbox A set of gears and selector levers by which the ratio of spindle rotation to lead screw rotation on a lathe can be quickly set. Many ratios in terms of feeds or threads per inch can be selected without the use of change gears.

Quick-change tool post A lathe toolholding device in which preset cutting tools are clamped in toolholders that can be placed on the tool post or interchanged with others to an accurately repeatable location.

Quill The nonrotating but retracting and extending portion of a drill press or milling machine containing the bearings and machine spindle.

Radial Radiating outward from the center.

Radial rake On cylindrical or circular cutting tools, such as milling cutters or taps, the rake angle that is off the radius is called the radial rake.

Radian A unit of angular measurement that is equal to the angle at the center of a circle subtended by an arc equal in length to the radius.

Radius On a circle, a distance equal to one-half the diameter.

Rake A tool angle that provides a keenness to the cutting edge.

Rapid traverse A rapid-travel arrangement on a machine tool used to quickly bring the workpiece or cutting tool into close proximity before the cut is started.

Recessing Grooving.

Reciprocating A movement back and forth along a given axis.

Recrystallize Metal crystals become flattened and distorted as the metal hardens when it is being cold worked. At a particular temperature range (about 950°F or 510°C for low-carbon steel) for each metal, called its recrystallization temperature, the distorted grains begin to reform into regular, larger, softer grains or crystals.

Reference point On a layout or drawing, there must be a point of reference from which all dimensions originate for a part to avoid an accumulating error. This could be a machined edge, datum, or centerline.

Relief angle An angle that provides cutting edge clearance for the cutting action.

Resistance In an electrical unit, the opposition to current flow.

Resulfurized Sulfur in steel is normally a contaminant that causes steel to be "hot short" (separates while being hot forged) because of iron sulfide inclusions. As much sulfur as possible is therefore removed at the steel mill. Resulfurized steel is free machining because sulfur and manganese (to control the sulfur) are deliberately added to make manganese sulfide (instead of iron sulfide) inclusions that make a sort of lubricant for the chip and do not cause hot shortness.

Right angle A 90-degree angle.

Ring test A means of detecting cracks in grinding wheels. The wheel is lightly struck and, if a clear tone is heard, the wheel is not cracked.

Rockwell A hardness test that uses a penetrator and known weights. Several scales are used to cover the very soft to the very hard materials. The Rockwell C scale is mostly used for steel.

Root The bottom of a thread or gear tooth.

Root truncation The flat at the bottom of a thread groove.

Rotameter A device used to indicate the flow of gases in terms of cubic feet or inches in a given period of time.

Roughing In machining operations, the rapid removal of unwanted material on a workpiece, leaving a small amount for finishing. Since coarse feeds are used, the surface is often rough.

RPM Revolutions per minute.

Runout An ecentricity of rotation as that of a cylindrical part held in a lathe chuck being offcenter as it rotates. The amount of runout of a rotating member is often checked with a dial indicator.

Scaling The tendency of metals to form oxides on their surfaces when held at a high temperature in air. The oxides usually form as a loose scale.

Scriber A sharp-pointed tool used for making scratch marks on metal for the purpose of layout.

Sector A portion of a circle between two rays defining a specific angle.

Seize A condition where two metal parts are pressed together without the aid of lubrication, resulting in frictional forces tearing metal from each part and causing a mechanical welding (seizing) of the two.

Semiprecision Using a method of layout, measurement, or machining in which the tolerances are greater than that capable by the industry for convenience or economy.

Serrated Small grooves, often in a diamond pattern, used mostly for a gripping surface.

Set The width of saw tooth. The set of saw teeth is wider than the blade width.

Setup The arrangement by which the machinist fastens the workpiece to a machine table or work-holding device and aligns the cutting tool for metal removal. A poor setup is said to be when the workpiece could move from the pressure of the cutting tool, thus damaging the workpiece or tool, or when chatter results from lack of rigidity.

sfpm Surface feet per minute.

Shallow hardening Some steels such as plain carbon steel (depending on their mass), when heated and quenched, harden to a depth of less than $\frac{1}{8}$ in. These are shallow hardening steels.

Shank The part of a tool that is held in a work-holding device or in the hand.

Shearing action A concentration of forces in which the bending moment is virtually zero and the metal tends to tear or be cut along a transverse axis at the point of applied pressure.

Sheaves Grooved pulleys such as those used for V-belts or cables.

Sherardized Zinc inoculated steel, a process by which the surface of steel is given a protective coating of zinc. It is not the same as galvanized or zinc-dipped steel. Zinc powder is packed around the steel while it is heated to a relatively low temperature in the sherardizing process.

Shim A thin piece of material used to take up space between workpiece and work-holding device; a piece used to fill space between machinery and foundations in assemblies.

SI Système International. The metric system of weights and measures.

Silicon carbide A manufactured abrasive. Silicon carbide wheels are used for grinding nonferrous metals, cast iron, and tungsten carbide, but are not normally used for grinding steel.

Sine bar A small precision bar with a given length (5 or 10 in.) that remains constant at any angle. It is used with precision gage blocks to set up or determine angles within a few seconds of a degree.

Sintering Holding a compressed metal powder briquette at a temperature just below its melting point until it fuses into a solid mass of metal.

Slot Groove or depression as in a keyseat slot.

Snagging Rough grinding to remove unwanted metal from castings and other products.

Soluble oils Oils that have been emulsified and will combine with water.

Solution heating treating See *Precipitation hardening*.

Solvent A material, usually liquid, that dissolves another. Dissolved material is the solute.

Spark testing A means of determining the relative carbon content of plain carbon steels and identifying some other metals by observing the sparks given off while grinding the metal.

Specifications Requirements and limits for a particular job.

Speeds Machine speeds are expressed in revolutions per minute; cutting speeds are expressed in surface feet per minute.

Sphericity (1) A condition of circularity in all possible axes. (2) The quality of being in the shape of a ball. (3) The extent to which a true sphere can be produced with a given process.

Spheroidize anneal A heat treatment for carbon steels that forms the cementite into spheres, making it softer and usually more machinable than by other forms of annealing.

Spiral A path of a point in a rotating plane that is continuously receding from the center is called a flat spiral. The term *spiral* is often used, though incorrectly, to describe a helix.

Spline A shaft on which teeth have been machined parallel to the shaft axis that will engage similar internal teeth in a mating part to prevent turning.

Sprockets Toothed wheels used with chain for drive or conveyor systems.

Squareness The extent of accuracy that can be maintained when making a workpiece with a right angle.

Stepped shaft A shaft having more than one diameter.

Stick-slip A tendency of some machine parts that slide on ways to bind slightly when pressure to move them is applied, followed by sudden release that often causes the movement to be greater than desired.

Straightedge A comparison measuring device used to determine flatness. A precision straightedge usually has an accuracy of about plus or minus .0002 in. in a 24-in. length.

Strength The ability of a metal to resist external forces. This can be tensile, compressive, or shear strength.

Stress An external force applied to an object.

Stress relief anneal A heat treatment, usually under the critical range, for the purpose of relieving stresses caused by welding or cold working.

Stroke A single movement of many movements, as in a forward stroke with a hacksaw.

Surface plate A cast iron or granite surface having a precision flatness for precision layout, measurement, and setup.

Symmetrical Usually bilateral in machinery where two sides of an object are alike but usually as a mirror image.

Synthetic oils Artificially produced oils that have been given special properties such as resistance to high temperatures. Synthetic water soluble oils or emulsions are replacing water soluble petroleum oils for cutting fluids and coolants.

Tang The part of a file on which a handle is affixed.

Tapered thread A thread made on a taper such as a pipe thread.

Tap extractor A tool that is sometimes effective in removing broken taps.

Tapping A method of cutting internal threads by means of rotating a tap into a hole that is sufficiently under the nominal tap size to make a full thread.

Telescoping gage A transfer type tool that assumes the size of the part to be measured by expanding or telescoping. It is then measured with a micrometer.

Temper (1) The cold worked condition of some nonferrous metals. (2) Also called draw, a method of toughening hardened carbon steel by reheating it.

Temperature The level of heat energy in a material as measured by a thermometer or thermostat and recorder with any of several temperature scales: Celsius, Fahrenheit, or Kelvin.

Template A metal, cardboard, or wooden form used to transfer a shape or layout when it must be repeated many times.

Tensile strength The maximum unit load that can be applied to a material before ultimate failure occurs.

Tension A stretching or pulling force.

Terminating threads Methods of ending the thread, such as undercutting, drilled holes, or tool removal.

Test bar A precision ground bar that is placed between centers on a lathe to test for center alignment using a dial indicator.

Thermal cracking Checking or cracking caused by heat.

Thread axis The centerline of the cylinder on which the thread is made.

Thread chaser A tool used to restore damaged threads.

Thread crest The top of the thread.

Thread die A device used to cut external threads.

Thread engagement The distance a nut or mating part is turned onto the thread.

Thread fit Systems of thread fits for various thread forms range from interference fits to very loose fits; extensive references on thread fits may be found in machinist's handbooks.

Thread lead The distance a nut travels in one revolution. The pitch and lead are the same on single lead threads but not on multiple lead threads.

Thread pitch The distance from a point on one thread to a corresponding point on the next thread.

Thread relief Usually an internal groove that provides a terminating point for the threading tool.

T-nut A threaded nut in a T shape that is designed to fit into the T-slot on a machine tool table.

Tolerance The allowance of acceptable error within which the mechanism will still fit together and be totally functional.

Tool geometry The proper shape of a cutting tool that makes it work effectively for a particular application.

Tooling Generally any machine tool accessory separate from the machine itself. Tooling includes cutting tools, holders, work-holding accessories, jigs, and fixtures.

Toolmaker An experienced general machinist often involved with high precision work making other tools, die, jigs, and fixtures used to support regular machining and manufacturing.

Torque A force that tends to produce rotation or torsion. Torque is measured by multiplying the applied force by the distance at which it is acting to the axis of the rotating part.

Toxic fumes Gases resulting from heating certain materials are toxic, sometimes causing illness (as metal fume fever from zinc fumes) or permanent damage (as from lead or mercury fumes).

Transfer measurement A step in measurement in which a transfer measuring tool such as telescoping gage is set to the unknown dimension and subsequently measured with a direct measuring tool such as a micrometer.

Transformation temperature Same as critical temperature; the point at which ferrite begins to transform to austenite.

Traverse To move a machine table or part from one point to another.

Trueing (1) In machine work, the use of a dial indicator to set up work accurately. (2) In grinding operations, to dress a wheel with a diamond.

Truncation To remove the point of a triangle (as of a thread), cone, or pyramid.

T-slot The slot in a machine tool table, shaped like a T, and used to hold T-nuts and studs for various clamping setups or their hold-down requirements.

Tungsten carbide An extremely hard compound that is formed with cobalt and tungsten carbide powders by briquetting and sintering into tool shapes.

Turning Machine operations in which the work is rotated against a single point tool.

Vernier A means of dividing a unit measurement on a graduated scale by means of a short scale made to slide along the divisions of a graduated instrument.

Vibration An oscillating movement caused by loose bearings or machine supports, off-center weighting on rotating elements, bent shafts, or nonrigid machining setups.

Vise A workholding device. Some types are bench, drill press, and machine vises.

Wedge angle Angle of keenness; cutting edge.

Wheel dressing Trueing and sharpening the grinding surface of an abrasive wheel by means of a dressing tool such as a diamond or Desmond dresser.

A device used to align a machine spindle to a punch mark.

Wrought Hold or cold worked; forged.

Zero back rake Also neutral rake; neither positive nor negative; level.

Zero index Also zero point. The point at which micrometer dials on a machine are set to zero and the cutting tool is located to a given reference, such as workpiece edge.

INDEX